## List of Basic Procedures

Procedure for solving word or "story" problems, p. 45

Procedure to solve a system of $n$ linear equations in $n$ variables by row reduction, p. 217

Procedure for solving a linear system by row reduction, p. 231

Procedure for finding the inverse of an $n$ x $n$ matrix, p. 241

Procedure for solving a system of $n$ linear equations in $n$ variables by Cramer's rule, p.258

Procedure to find the inverse of an $n$ x $n$ matrix using the adjoint, p. 267

Procedure for computing the steady-state vector of a regular transition matrix. (See Example 3b, p. 288)

Procedure to solve geometrically a linear programming problem, p. 307

Procedure for finding the pivot, p. 332

Procedure for moving from corner to corner, p. 337

Procedure of the simplex method, p. 342

Procedure to solve a minimization problem in standard form, p. 372

Procedure: the big $M$ method, p. 381

Procedure to compute a conditional probability using Bayes' formula, p. 473

Procedure to evaluate a derivative using the definition, p. 537

Procedure to evaluate a derivative using implicit differentiation, p. 565

Procedure for performing logarithmic differentiation, p. 577

Procedure for finding the extrema of a continuous function $f$, p. 624

Procedure to find the extrema of a continuous function using the second derivative test, p. 630

Procedure for solving applied optimization problems, p. 631

Procedure for sketching the graph of a function, p. 662

Procedure to approximate functional values using differentials, p.666

Procedure for solving related rates problems, p.670

Procedure to evaluate the definite integral $\int_a^b f(x)\, dx$, p. 745

Procedure for solving a separable differential equation, p. 775

Procedure to find extrema, p. 874

# Introduction to
# Mathematical Analysis
## for Business and Economics

# Introduction to Mathematical Analysis
## for Business and Economics

**André L. Yandl**
Seattle University

**Brooks/Cole Publishing Company**
Pacific Grove, California

# To my wife, Shirley

Consulting Editor: Robert Wisner

**Brooks/Cole Publishing Company**

A Division of Wadsworth, Inc.

Printed in the United States of America

10 9 8 7 6 5 4 3 2 1

**Library of Congress Cataloging-in-Publication Data**

Yandl, André L.
    Introduction to mathematical analysis for business and economics / André L. Yandl.
       p.      cm.
      ISBN 0-534-13494-7
      1. Mathematical analysis.    I. Title.
   QA300.Y36    1990                                     90-31734
   515—dc20                                          CIP

Sponsoring Editor: *Jeremy Hayhurst*
Project Development Editor: *Patricia P. Gadban*
Signing Representative: *Karen Buttles*
Editorial Assistants: *Nancy Champlin, Gay Bond*
Production Editor: *Timothy A. Phillips*
Manuscript Editor: *Margaret Chang*
Permissions Editor: *Carline Haga*
Interior and Cover Design: *Katherine Minerva*
Cover Photo: *Lee Hocker*
Art Coordinator: *Lisa Torri*
Interior Illustration: *Lori Heckelman*
Typesetting: *Jonathan Peck Typographers, Ltd.*
Cover Printing: *Lehigh Press*
Printing and Binding: *Arcata Graphics/Fairfield*

# Preface

This book is the tangible result of my thirty-four years of teaching experience at Seattle University. For many of those years I have taught a two-quarter sequence in algebra and calculus to business students. That experience is the foundation and, I hope, the strength of this text.

I have seen a lot of changes in courses and curricula in those years of teaching. Indeed, from 1968 to 1970 I was a member of the MAA's CUPM committee chaired by Ralph Boas to examine the national undergraduate curricula and make recommendations for improving those curricula at all levels. I had the opportunity to serve on two interesting panels, that for mathematics in the two-year colleges, and the parallel committee on basic mathematics. I am pleased to recall that many of the needs of today's students and their instructors were anticipated then. My hope is that the best of those ideas have been implemented.

In the twenty intervening years, two new courses have shaped the curricula and syllabi for the business mathematics courses taught in most schools today. Calculus for business students, or applied calculus, has become a bona fide course because business students simply have different needs from those of math, engineering, and science majors; and "finite" mathematics topics have become consolidated into a coherent program which has obvious utility for students in business and managerial science. Finite mathematics—the application of sets, matrices, some linear algebra, probability, and statistics to practical problems—is now standard in most business, economics, and sometimes social science curricula. Alas, changes in curricula have not provided any more time for classroom instruction, and courses necessarily emphasize some of these topics to the exclusion of others. This text is no exception, and it will be necessary to select the topics that best fit your syllabus. A chart showing the interdependence of chapters is given on page x. I am hopeful that even the material not explicitly covered in class will be useful to some students in some way, perhaps as motivation for exploring more mathematics that is useful in the business world, or as reference material for future courses.

## Philosophy and Rationale

I am well aware that most students resist actually reading their texts, preferring to start by working the exercises assigned in class, resorting to examples in the section where necessary, and, as a last resort, reading from the beginning of the section to the end. I hope that the efforts spent to make this material readable and comprehensible will convince many students that a better strategy is to read before and after class, to read with pencil and paper at hand, and to study theorems to see precisely what they say (and do not say) before attempting the exercises.

I strongly believe that a textbook and an instructor should complement one another. Either the text should provide necessary details with the instructor providing a clear outline and reinforcement for the important points, or the text should be a brief outline in notelike form to which the instructor adds detail and explanation. Textbooks that take the latter point of view are preferred by some instructors who believe that a minimum of verbiage is desirable in order to avoid confusing students. These texts are also short and preferred by publishers because the page length, and therefore cost of producing the book, is minimized. They look accessible, are easy to carry, and seem to minimize the effort required by the course. Many instructors are seduced.

I believe that a far better approach is to let the instructor guide the student toward the mathematical priorities while letting the text provide details, reinforcement, and rationale as well as be a repository of good examples and exercises. My teaching experience tells me that short, overly concise, "notelike" texts do most students a disservice. When students use books with little or no detail they appear to be successful because they are able to memorize how to do the problems from the templates provided; however they do not retain what they've learned because they never understood the reasoning behind the procedures. I was fortunate to find a publisher who agreed with my views.

I have tried to explain each topic in enough detail so that students are able to learn to follow mathematical arguments without fear. Whenever appropriate, I have attempted to appeal to their own experiences with business transactions (see Example 1, Section 10.3) and to their visual intuition (see, for example, page 316, the use of the "squirrel on the roof" in motivating the simplex method).

## Structure and Coverage

The basic structure and content of this book is comparable to other books designed to support algebra, finite mathematics, and applied calculus courses, but there are enough differences to justify some comments. (1) This text stresses algebra concepts earlier and more thoroughly than many texts for a simple reason. Many students, especially mature returning students, have forgotten their previous mathematics and need to be reminded of what they once knew. They are unable to succeed without mastery of basic algebra, whether it is formally covered in the course, given as assigned reading, or used just for reference when needed. (2) Students typically use texts as a source of worked examples to help solve the problems they are assigned, and it is here that I have spent much effort in writing. Many of my examples were created as a result of explaining concepts for which students requested extra help. It seems that most instructors do not have the opportunity to work enough examples in class, and it is here that a text can and should provide support. I have provided an abundance of those that I feel are most illustrative of the concepts and most helpful to students (see, for example, Section 8.7). Most examples end with a reference to an exercise in the text at the end of the section. These references serve two purposes. Students are able to immediately apply the technique and test whether or not they understand the example given, and they are able to match certain types of exercises with corresponding examples so that the student may refer to that example when doing assigned exercises. (3) Some students, even those successful in other courses, are not successful in mathematics despite spending a lot of time studying. I hope to make my students more efficient and proficient in approaching mathematics by providing some strategies to help them succeed. In this text, this objective is addressed by the introductory section titled "Getting Started." Students should be urged, even required, to read it. The main emphasis is on helping students to understand why mathematical methods work, and provide some insight to a sensible *method* of studying mathematics.

## Optional Sections

With all the material inherent in this course, it is necessary to explain why the sections designated optional are present. There are three reasons for the optional sections. Many of them deal with material that students have almost certainly encountered but have forgotten. Section 0.5, for example, on synthetic division, is present as a refresher for students who have long since sold their intermediate algebra texts. Other optional sections are present because, given my teaching experience, I believe they are valuable. Section 9.6 on tabular integration is such a section. Students learn this method very easily and succeed in doing problems they were unable to do using integration by parts. For my classes, this positive reinforcement makes the time spent teaching the topic well worthwhile. I invite, urge, and challenge you to try this section in your classes. The results may surprise you. Still other optional sections are designed for students with more interest and aptitude, because these students tend to want to know *why* techniques work the way they do. Sections 8.4 and 9.9, for example, expand on the theory and methods presented in earlier sections, and I hope they will be satisfying to the more mathematically inclined. All of the optional topics may be omitted without loss of continuity.

## Pedagogy and Design

Most students, particularly those in nonmathematical majors, require a great deal of assistance in learning mathematics. I believe that the text optimizes students' chances of mastering the material in the following ways: (1) Each chapter concludes with a careful chapter summary that can refresh students' memory and boost their confidence. These summaries include lists of important symbols and terms, a summary of the chapter discussion, and sample examination exercises. (2) Step-by-step procedures for solving important problems are listed on the endpapers of the text for easy reference. (3) Icons such as the "slippery road" ⟨§⟩ and "3 × 5 cards" ▭, alert students to possible pitfalls and the material which should be committed to memory. (4) Definitions, theorems, and basic formulas are boxed, labeled, and identified by the 3 × 5 card icon, ▭. These icon-flagged statements are reproduced with examples on real 3 × 5 cards and available in the accompanying student study guide for the text. I always encourage my students to make and use these flash cards for this and future courses. (5) Some discussion in the text is labeled "A Closer Look" and delineated with a grey screen. These portions of the text give a deeper understanding of the material (see, for example, pages 155, 369–372, 458–459) but may be omitted without loss of continuity. (6) Examples reference exercises that can and should be worked immediately after the student has understood the example. (7) Exercises are numerous and every effort has been made to be as realistic as possible without making them overly difficult. Starred exercises are indicated for one of the following reasons: (a) they require material from one of the optional sections, (b) they are more theoretical or difficult, or (c) they are most appropriate for better-prepared students. (8) For most new concepts the algebraic manipulations are as simple as possible because the emphasis should be on understanding, not applying, the concept. However, many students at this level need much more exposure to and experience with algebra, and therefore some of the examples are explicitly designed to challenge and help sharpen their algebraic skills.

## Calculator Usage

One of the obvious changes in thirty-four years of teaching is the relatively recent emergence of calculators in the classroom. I embrace them. Many of the text's exercises deliberately involve large numbers to encourage the use of calculators. Similarly, simple reference tables have been minimized. Before the widespread use of calculators students had to spend considerable time doing simple arithmetic and hand calculation. Calculators have saved students much of this time and I believe that it can be profitably invested in actually reading their texts and mastering concepts. Of course, students should refer to their owner's manuals before using their calculators, but many have lost their manuals or acquired their calculators secondhand. Accordingly, a short prescription for using calculators is given in "Getting Started."

## Chapters in Brief

Chapter "0" is so designated because this material is really prerequisite for the course, but important and necessary review for most students. The amount of time spent on this material depends on how well the majority of students remember their intermediate algebra. How a class approaches reviewing this material is flexible. Section 0.3, for example, on Venn diagrams, may be covered just before Chapter 6 on probability. However, I find it useful to include this material early because it nicely illustrates how a complex idea can be made simple with a clear geometric argument.

Chapter 1 introduces the function and inverse function concepts and also the elementary functions that are used later on in the course. This material is mainly used in the calculus portion of the text and sparingly in Chapters 5 and 6. It can be taught immediately before Chapter 5, but my feeling is that the concept of function is so central to mathematics that it should be taught earlier. A section on sigma notation is included here because many business faculty like students to be proficient in using it as soon as possible. It may be covered immediately before the definition of the definite integral and some instructors may wish to postpone it until after Section 9.7.

Chapter 2 is devoted to systems of linear equations and inequalities. The treatment is standard but the presentation is aimed at preparing the student for matrices and solving systems of linear equations by row reduction. The last section includes some important business applications.

Chapter 3 begins the study of matrices. I discuss cryptography in addition to the more usual business applications for two reasons. It provides an opportunity to demonstrate how a simple mathematical tool can be used effectively in a practical situation, and students usually enjoy the intrigue of coding and uncoding messages. Markov chains are included here because not all instructors have the luxury of including probability in their syllabi.

Chapter 4 introduces linear programming via the geometric approach. The simplex method is discussed in more detail than most books attempt at this level. I introduce the three-dimensional coordinate system because I want to explain geometrically *why* the simplex method works. I also feel that it is useful not to restrict the discussion to the plane, especially when the big $M$ method is discussed.

Sequences and mathematics of finance are discussed in Chapter 5. Although I want to show the students how to use standard formulas to solve annuity problems, I strongly encourage them to solve the problems using only two basic formulas. I believe that they better understand the unifying ideas this way. I encourage them to use their calculators by restricting the use of annuity tables to two pages.

Chapter 6 contains basic probability material, including Bayes' formula, the binomial distribution, and more on the Markov chains.

Calculus begins in Chapter 7, which is devoted to the derivative, continuity, and calculating derivatives of the elementary functions.

Chapter 8 presents the diverse applications of differentiation, including marginal analysis, elasticity, then optimization. Curve sketching, differentials, and related rates follow. After much consideration, I chose to place curve sketching after optimization because I feel that it is helpful for students to understand how to find the extrema of a function before sketching its graph. I know some instructors will prefer the inverse of this order of the material. If so, perhaps it will help to refer to the example of sketching the graph of a polynomial (Example 2, page 607). This example should give the student an idea of how to visualize the relative extrema of the function. For the benefit of skeptical students some optional proofs are given, which I hope instructors will use when appropriate.

Chapter 9 introduces integration, the coverage of which is more extensive than many books for this course. The definite integral is defined as the limit of Riemann sums. I strongly believe that this approach is necessary if we are to expect the students to understand the applications. Why would consumers' and producers' surplus be represented by areas otherwise? I have discussed the definite integral as an area (Section 9.8) and also as a distance (Section 9.9) because both of these interpretations contribute to understanding the fundamental theorem of calculus. Some instructors will choose to simplify the presentation by omitting Sections 9.5, 9.6, and 9.9.

I purposely begin with examples that the student cannot yet solve using the basic integration formulas because it helps if the student first learns to *guess* the answer and then check by differentiation. I stress the use of differentiation to obtain antiderivatives by guessing because may students will ultimately use computer software to solve integration problems. Regardless, most simple integration problems will continue to be solved by hand just as simple arithmetic is more conveniently done without a calculator.

Chapter 10 introduces some applications of integration, including an optional section on differential equations. I have tried to present the examples in such a way that most students will understand why the integral is used in many applied situations. (See Example 1, Section 10.3 and Examples 1 and 2, Section 10.4.)

Chapter 11 treats functions of several variables in a manner that is accessible to students at this level. In Section 11.7, the formulas for the slope and the $y$-intercept of the line of best fit are developed. Some instructors may wish to omit this and simply have the students set $f_b(m, b) = 0$ and $f_m(m, b) = 0$ and solve the resulting simultaneous system (see Example 1, Section 11.7).

## Topic Dependencies

*Note:* A blue arrow indicates that Topic A is prerequisite to Topic B. A black arrow indicates that Topic A *should* be covered before Topic B but may be omitted when students are well prepared.

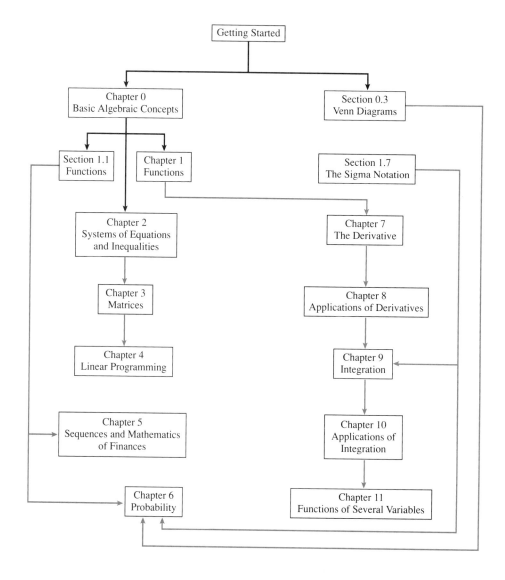

### Accompanying Texts

The material in this text is available in two separate volumes, *Finite Mathematics*, Brooks/Cole, 1991, and *Applied Calculus*, Brooks/Cole, 1991.

A Student Solutions Manual is available for this and the separate volumes giving complete worked solutions to all the odd-numbered exercises in the text(s). Another solutions manual for the instructor gives complete solutions to all even-numbered exercises as well as sample examination questions.

A separate Study Guide includes flash $3 \times 5$ cards, objectives, and other aids for learning. Computerized testing for both IBM family and Apple Macintosh computers is also available to adopting schools. [Contact your Wadsworth-Brooks/Cole sales representative, or call 1-800-876-2350, for more details.]

# Preface to Students

Many of the teaching ideas used in this book originated with conversations I have had with my students. I usually leave home early to miss the traffic and arrive on campus at 7:00 A.M. I often go to the school cafeteria for a cup of tea and almost always some of my students join me, sometimes when they're *cramming* for an exam and *anxious* and sometimes when they're just curious about material from a lecture. These numerous hours of one-on-one or one-on-few conversations have helped me to better understand the many difficulties some students have. Often, the real problem is simple: they haven't done enough work. But I've also found that many students are not afraid of doing a little work if it helps them understand *why* they are being taught this material and *why* things work as they do. I have also found that students (like you) *can be genuinely enthusiastic* about mathematics *when they experience success*. My goal is to aid your teacher in conveying that experience of success to you. To aid in this process, this text includes some topics that often do not appear in other books for this course. Topics such as tabular integration are included because my students experience a great amount of satisfaction from being able to solve problems which previously seemed too difficult with other commonly taught methods. In general, I encourage you to use whichever method gives a correct answer and is understandable for you. This does not mean that we will stress "monkey-see, monkey-do." You will have to do some work in this course, but the result should be more satisfying to you than if you had blindly followed a method without understanding it and why it works.

Many of the examples you'll encounter were originally made up during my conversations with students. I selected those that seemed to stir their interest the most, such as the problem of increasing tuition (see Example 2, page 634). For another example, I made a real effort to explain the simplex method, geometrically and visually, because I know that my students learn it best that way. Many of my students are surprised to learn that the best strategy for studying mathematics may be quite different from that which has helped them succeed in other disciplines. I have included some hints on how to study mathematics in the preliminary section titled "Getting Started." I urge you to carefully read that section before you begin with the core material.

Very often, students come to my office with a question about an example in a textbook. Almost always the question is: "How did the author get from this step to that step?" Usually I suggest that they take a pencil and paper and try to work out the details omitted by the author. Most of the examples given here have very detailed solutions so you should not find it difficult to see how to arrive at each step. It is always good practice, however, to study with paper and pencil handy. For example, in a "story" or word problem, I may show you how to get the equation, and give you the answer without giving all the steps to the solution using the equation. In such a case, it is best for you to try to solve the equation anyway to convince yourself that you are able to do so. If you cannot solve it, go back to the discussion on how to solve this type of equation, then, read the corresponding examples, and try again.

I feel very strongly that you should understand *why* procedures that you are asked to follow work the way they do. Certainly you will enjoy studying more if you actually *understand* what is being done. With this approach, you will better remember what you learned and you will be able to use your acquired knowledge when needed, both in this course and in the real world.

I have tried to help you differentiate between useful explanations and essential facts by putting all formulas, definitions, theorems, and procedures for solving each type of problem in colored boxes. On the inside cover of the book, you will find a list of the procedures used to solve certain types of problems, for easy reference. You will find the $3 \times 5$ icon beside a statement important enough for you to memorize. Some parts of the text have been labeled "A Closer Look" and appear in grey color screens. Your instructor will advise you about these sections.

It is likely that your instructor will not have time to cover all of the material, especially the examples, in class. Therefore, it is *essential that you read* each section you've been assigned *carefully*. It is also essential that you test your understanding of the material as you proceed. To help you do this, I refer you to an exercise at the end of the section at the end of a worked example. Immediately after studying the material and reading an example, try to *do the exercise without looking at the corresponding example*. I have referred you only to odd-numbered exercises so that you may check your answers in the back of the book. If you are not able to solve the exercise, go back and study the corresponding discussion, read the example, and try again. Remember that if your answer doesn't match the answer in the book exactly, you may still be right and further simplification of the algebra will yield my answer. Do not proceed to the next topic until you succeed in doing the exercise correctly. The symbol ■ signifies the end of the example if no exercise is indicated. The same symbol is used to indicate the end of a proof.

I truly believe that in my thirty-four years of teaching, I have seen *all* the types of errors that students can possibly make when doing elementary mathematics. I have tried to warn you *not* to make these mistakes whenever possible. These warnings can be found in the text, or in the footnotes, and are indicated in the margin by the icon

⟨§⟩, which you can think of as a "slippery road ahead" sign.

In addition, I have purposely included throughout the text some exercises that involve large numbers to encourage you to use your calculator. If you are not proficient in using a calculator, read the appropriate material in the "Getting Started" section, read your owner's manual, and practice as often as possible. The time you will save can be used to think about the concepts you are learning.

At the beginning of each of the chapter review sections, I have listed the new terms and symbols that appear in the chapter. Make sure that you are familiar with the meaning of each of these terms. Following the list is an extensive summary of the chapter. These summaries are detailed because I believe that a summary should contain almost all the really important information. To really test your knowledge, mentally construct your own summary before reading those given in the text. Sample exam questions are given after the summaries. Some of these are straightforward; others are quite challenging. The types of problems that you will have when you take an exam depend, of course, on your instructor. However, if you can do all of the sample problems that appear in the review sections, you can be confident you have mastered the material.

Finally, I wish you the very best success in this course and in your future endeavors.

## Acknowledgments

I thank the following people whose criticism at the different stages of the manuscript was invaluable to me in developing and improving the book:

Patricia M. Bannatine, *Marquette University*; Theresa J. Barz, *St. John's University*; Joseph Brody, *Concordia University*; Lynn Cleary, *University of Maryland*; John E. Gilbert, *University of Texas, Austin*; Madelyn T. Gould, *DeKalb University*; Bennette R. Harris, *University of Wisconsin, Whitewater*; Judith H. Hector, *Walters State Community College*; Alec Ingraham, *New Hampshire College*; Gerald Leibowitz, *University of Connecticut*; Lawrence P. Maher, *North Texas State University*; H. T. Mathews, *University of Tennessee, Knoxville*; John Morrison, *University of Delaware*; William C. Ramaley, *Ft. Lewis College*; Gordon Shilling, *University of Texas, Arlington*; S. J. Stack, *Wilfrid Laurier University*; Louis M. Weiner, *Northeastern Illinois University*; and Carroll G. Wells, *Western Kentucky University*.

During the writing of this book, I have received help and encouragement from many colleagues and friends. I wish to acknowledge a debt to my former dean, Terry van der Werff, who first suggested that I write another book. His interest and support greatly influenced my writing. I also express my appreciation to the Seattle University sabbatical and summer faculty fellowship committees for the support they gave me during the writing of this book.

I am indebted to my colleagues: Mary B. Ehlers, who worked all of the exercises; Sr. Kathleen Sullivan, who helped prepare the study guide; C. C. Chang, Wynne A. Guy, Janet Mills, Ahmad Mirbagheri, Carl Swenson, Burnett Toskey, and Alan Troy; who gave me much support during the many years of writing. I owe many thanks to Kristi Weir of Bellevue Community College's Department of Economics for the many informative conversations we had on the teaching of mathematics to economics students while she was on the faculty of Seattle University. I am also indebted to my students, who in the past thirty-four years have helped me develop the ideas incorporated into this book, and Lisa Dougherty, my niece, for her excellent typing of the entire calculus manuscript.

I also wish to express my deepest appreciation to the staff of Brooks/Cole Publishing Company: Acquisitions Editor, Jeremy Hayhurst, for his sponsorship, and advice and support throughout the production of the book; Pat Gadban, Senior Project Development Editor, for the careful editing of the early draft of the manuscript and many helpful suggestions for improvement; Professor Robert Wisner of New Mexico State University, Mathematics Series Editor, for his careful reading of the manuscript and constructive criticism; Tim Phillips, Production Editor, for coordinating all the tasks and providing me with needed assistance and enthusiastic support; Faith Stoddard, Assistant Mathematics Editor, for coordinating the production of the accompanying solutions manuals and the study guide; Katherine Minerva, Senior Designer, for a beautiful and effective interior and cover design; Nancy Champlin, Editorial Assistant, for providing editorial support for the text, solutions manuals, and study guide; and last but not least, William M. Roberts, President, and Michael V. Needham, past President, Craig F. Barth, Vice President, and the entire Brooks/Cole staff, for the support that they gave me and for making me feel so welcome during my visits to Pacific Grove.

Finally, my thanks go to my wife, Shirley, and to my three sons Steven, Michael, and Kris, and their families, for their patience during the last few years, while much of my time was spent writing instead of with them.

*André L. Yandl*

# Contents

Chapter 11                          Functions of Several Variables      831

# Getting Started

## Philosophy and Rationale

The purpose of having this course as an integral part of your program of study is to help you understand the use of mathematics in subsequent courses such as statistics, probability, and economics. It is important that you understand the basic concepts rather than simply memorize some examples in order to pass the exams. Many of you may find mathematics difficult. Nevertheless, if you study correctly, you can improve your understanding of the subject significantly. In this introduction you will learn how to study mathematics, how to use this book effectively, and how to use a calculator. What follows is an alternative to memorization through the understanding of mathematical processes.

## Getting Started

It is true that to work successfully with mathematics you must know many definitions, theorems, and formulas. However, it is not necessary or desirable to memorize every detail. In fact, it is possible to recall a great amount of information from relatively few memorized facts. *The trick is to look for patterns that occur frequently in mathematics.*

### Patterns

In basic algebra courses, most of you encountered the following formulas:

$$(a + b)^2 = a^2 + 2ab + b^2 \tag{1}$$
$$(a - b)^2 = a^2 - 2ab + b^2 \tag{2}$$
$$(a + b)^3 = a^3 + 3a^2b + 3ab^2 + b^3 \tag{3}$$
$$(a - b)^3 = a^3 - 3a^2b + 3ab^2 - b^3 \tag{4}$$
$$a^2 - b^2 = (a - b)(a + b) \tag{5}$$
$$a^3 - b^3 = (a - b)(a^2 + ab + b^2) \tag{6}$$

Many of you learned all of these, remembered two or three of them a few weeks after completion of the course, and then later promptly forgot all of them!

There is a way to remember formulas more effectively. Consider the last two formulas. The left sides are in the form $a^n - b^n$, where $n$ is a positive integer. In both cases, $a - b$ is a factor on the right side. This is the case not only for $n = 2$ and $n = 3$, but for all positive integers $n$. Therefore, if you are asked to factor $a^5 - b^5$, you would know that $a - b$ is one of the factors. However, how do you get the other factor? You could divide $a^5 - b^5$ (the left side) by $a - b$ (the factor you know) to obtain $a^4 + a^3b + a^2b^2 + ab^3 + b^4$. Thus,

$$a^5 - b^5 = (a - b)(a^4 + a^3b + a^2b^2 + ab^3 + b^4)$$

Recall that the degree of a term is the sum of all exponents assigned to the variables in that term. For example, the degree of $3x^2y^3$ is 5 since $2 + 3 = 5$. If $n$ is a positive integer, one factor of $a^n - b^n$ is $a - b$. The other factor consists of the sum of all terms of degree $n - 1$ in the variables $a$ and $b$, with nonnegative integer exponents and with coefficient 1.

**EXAMPLE 1**   Factor $a^7 - b^7$.

**Solution**   The first factor is $a - b$. Each term of the second factor is of degree 6. Hence, the terms are $a^6$, $a^5b$, $a^4b^2$, $a^3b^3$, $a^2b^4$, $ab^5$, and $b^6$. The factorization is

$$a^7 - b^7 = (a - b)(a^6 + a^5b + a^4b^2 + a^3b^3 + a^2b^4 + ab^5 + b^6). \qquad \blacksquare$$

We have been looking at the formulas (5) and (6). Now consider (1) and (3). In both, the left side is in the form $(a + b)^n$, where $n$ is a positive integer. In the expansion of $(a + b)^2$, there are three terms of degree 2; while in the expansion of $(a + b)^3$, there are four terms of degree 3. The pattern we see is that in the expansion of $(a + b)^n$, where $n$ is a positive integer, there are $n + 1$ terms of degree $n$. Thus, in the expansion of $(a + b)^4$, there are five terms of degree 4. Without their coefficients, the terms are $a^4$, $a^3b$, $a^2b^2$, $ab^3$, and $b^4$.

There are several ways to find the appropriate coefficients, known as the *binomial coefficients*. One way is to represent the coefficients as an array known as *Pascal's triangle*. (See Figure 1.)

```
              1
            1   1
          1   2   1
        1   3   3   1
      1   4   6   4   1
    1   5  10  10   5   1
    .   .   .   .   .   .
```

**FIGURE 1**

In each row of this triangular array the first and last entries are 1s. Each of the other entries is the sum of the two nearest entries in the row just above. Thus, the first three elements of the next row in Figure 1 would be 1 (the border element), 6 (since $1 + 5 = 6$), and 15 (since $5 + 10 = 15$). To write the expansion of $(a + b)^4$, find the row of Pascal's triangle whose second entry is 4. The entries for this row are 1  4  6  4  1. Thus,

$$(a + b)^4 = a^4 + 4a^3b + 6a^2b^2 + 4ab^3 + b^4$$

Although Pascal's triangle is an interesting way of obtaining the coefficients, it is not very efficient. Trying to find the 51 coefficients in the expansion of $(a + b)^{50}$ by writing 51 rows of Pascal's triangle would be both time- and space-consuming. However, there is an easier way to find the binomial coefficients using the following method:

1. In the expansion of $(a + b)^n$, the first term is $a^n$ and may be written $1a^nb^0$.

2. Having obtained one term of the expansion, the coefficient of the next term can be determined as follows. Multiply the coefficient of the term just obtained by the exponent assigned to $a$ in that term and divide the result by one more than the exponent assigned to $b$ in that term. In other words, if $ka^ib^j$ is a term of the expansion, the coefficient of the next term is $\frac{ki}{j+1}$. In fact, the next term will be $\frac{ki}{j+1}a^{i-1}b^{j+1}$ and the last term will be $b^n$.

**EXAMPLE 2**   Write the expansion of $(a + b)^5$.

**Solution**   Without coefficients, the terms are $a^5$, $a^4b$, $a^3b^2$, $a^2b^3$, $ab^4$, and $b^5$. Note that the first term is $a^5$ and may be written $1a^5b^0$. Thus, the coefficient of the second term is 5, since $\frac{(1)(5)}{0+1} = 5$, and the second term is $5a^4b^1$. The next coefficient is 10, since $\frac{(5)(4)}{1+1} = 10$, and the third term is $10a^3b^2$. The other three coefficients are similarly obtained to complete the expansion:

$$(a + b)^5 = a^5 + 5a^4b + 10a^3b^2 + 10a^2b^3 + 5ab^4 + b^5 \qquad \blacksquare$$

Now consider formulas (2) and (4). The left sides are in the form $(a - b)^n$. Here, we note that $(a - b) = (a + (-b))$ and, starting with the expansion of $(a + b)^n$, we replace every $b$ by $-b$.

**EXAMPLE 3**   Write the expansion of $(a - b)^5$.

**Solution**   Start with the expansion of $(a + b)^5$ obtained in Example 2 and replace every $b$ by $-b$. Thus,

$$(a - b)^5 = a^5 + 5a^4(-b) + 10a^3(-b)^2 + 10a^2(-b)^3 + 5a(-b)^4 + (-b)^5$$
$$= a^5 - 5a^4b + 10a^3b^2 - 10a^2b^3 + 5ab^4 - b^5 \qquad \blacksquare$$

Notice the symmetry in the row of coefficients for the expansion of $(a + b)^n$. The first and last coefficients are both 1. The second and next-to-last coefficients are equal, and so forth.

Here is an example of another pattern. The squares of the first seven nonnegative integers are 0, 1, 4, 9, 16, 25, and 36. The differences of successive terms are $1 - 0 = 1$, $4 - 1 = 3$, $9 - 4 = 5$, $16 - 9 = 7$, $25 - 16 = 9$, and $36 - 25 = 11$. Observe that 1, 3, 5, 7, 9, and 11 are the first six odd integers. It is easy to show that this pattern will continue. To this end, note that if $n$ is any nonnegative integer, then

$$(n + 1)^2 - n^2 = n^2 + 2n + 1 - n^2 = 2n + 1$$

This pattern can be used to sketch the graph of $y = x^2$. Start at the point (0, 0). Move to the right 1, and up 1, to obtain (1, 1). Move to the right 1 and up 3, to obtain (2, 4). Move to the right 1 and up 5, to obtain (3, 9). Move to the right 1 and up 7, to obtain (4, 16). Move to the right 1 and up 9, to obtain (5, 25), and so on. (See Figure 2.)

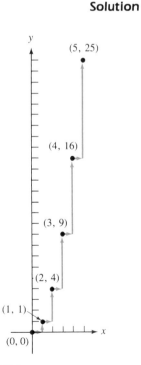

**FIGURE 2**

To introduce the next illustration, we need the following definition.

> **DEFINITION 1:**    An *Arithmetic Progression* is a sequence of numbers where the difference between any two consecutive terms of the sequence is constant. For example, 2, 5, 8, 11, 14, 17, 20, . . . is an arithmetic progression since $5 - 2 = 8 - 5 = 11 - 8 = 14 - 11 = 17 - 14 = 20 - 17 = \ldots$ .

Suppose that you must find the sum of the first ten terms of this arithmetic progression. Of course, you could simply add them. However, for cases in which there are many terms it is best to use formulas. One of these is

$$a + (a + d) + (a + 2d) + \cdots + (a + (n - 1)d) = \frac{[2a + (n - 1)d]n}{2}$$

This formula need not be memorized; there is a better way to remember how to do this. Let us go back to finding the sum of the first ten terms of the progression 2, 5, 8, 11, 14, . . . .

We first write

$$S = 2 + 5 + 8 + 11 + 14 + 17 + 20 + 23 + 26 + 29$$

We can also perform the addition in reverse order,

$$S = 29 + 26 + 23 + 20 + 17 + 14 + 11 + 8 + 5 + 2$$

Adding these two equalities, we obtain

$$2S = (2 + 29) + (5 + 26) + (8 + 23) + (11 + 20) + (14 + 17) + (17 + 14) \\ + (20 + 11) + (23 + 8) + (26 + 5) + (29 + 2) = 31 + 31 + 31 + 31 + 31 \\ + 31 + 31 + 31 + 31 = (31)(10)$$

Thus,

$$S = \left(\frac{31}{2}\right)(10) = 155$$

But, $\frac{31}{2} = \frac{2 + 29}{2}$. So, $\frac{31}{2}$ is the average of the first and last terms of the sum we evaluated; also, 10 is the number of terms. Therefore, *to find the sum of the first n terms of an arithmetic progression, we calculate the average of the first and nth terms and multiply this average by the number of terms.* Most of you will find this much easier to remember than the formula.

## Mnemonic Devices

In addition to noticing patterns, it is often useful to think up a "trick" or mnemonic device* to recall a fact we need to know. For example, the graph of the equation $y = ax^2 + bx + c$, where $a$, $b$, and $c$ are constants and $a \neq 0$, is a parabola. It is concave up if $a > 0$ and concave down if $a < 0$. (See Figure 3.)

---

*A mnemonic device is any device or "trick" that helps you remember—such as a string around your finger or a rhyme ("30 days has September . . .").

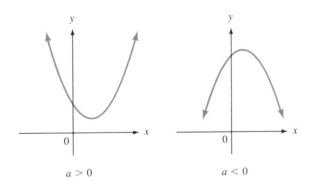

**FIGURE 3**

Some students remember the shapes of these graphs by associating the graph on the left with a smiling face and the graph on the right with a sad face. (See Figure 4.) (Of course, it is usually best to think up your own mnemonic device, since otherwise it may be just as difficult to remember the trick as it is to remember the fact.)

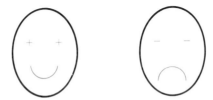

**FIGURE 4**

### Geometric Interpretations

Another idea that many students find effective in studying mathematics is to draw, whenever possible, a geometric interpretation of the problem at hand. As an illustration, we use the formula $(a + b)^2 = a^2 + 2ab + b^2$. In Figure 5, we have drawn two squares: one with side $a$ units, the other with side $(a + b)$ units. We see from the figure that the area of the larger square is equal to the area of a square with side $a$ units, plus the area of a square with side $b$ units, plus the areas of two rectangles with sides $a$ units by $b$ units. The areas are $a^2$, $b^2$, and $2ab$ square units, respectively. Thus, the formula $(a + b)^2 = a^2 + 2ab + b^2$ is illustrated geometrically.

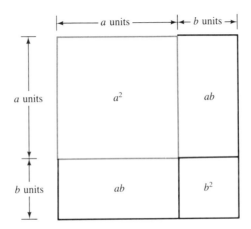

**FIGURE 5**

### Individual Relationships

You may also relate a concept to a familiar situation. For example, when we multiply a sum of several terms by a number, it is necessary to multiply each term by that number. On the other hand, when we multiply a product of several factors by a number, it is sufficient to multiply only one factor by that number. You could relate this to the following situation. Suppose that you had two part-time jobs. Your total income is the sum of the incomes you get from each job. If you need to double your total income, you can accomplish this by doubling *each* of the two incomes. On the other hand, if you have only one part-time job, your weekly income is found by multiplying your hourly wage by the number of hours you work in a week. If you must double your income, you may accomplish this by either doubling the number of hours you work each week, or by talking your employer into doubling your hourly wage. If you relate mathematical concepts to real-life situations as we have just illustrated, you will find that you can understand more mathematical concepts than you expected.

In addition to looking for patterns, thinking up mnemonic devices, trying to give geometric interpretations whenever feasible, and relating mathematical concepts to your own experiences, you must also acquire good study habits.

### Suggestions for Good Study Habits

1. Study daily. Two hours each day is better than 14 hours on the weekend.
2. Have regular review periods.
3. Schedule study sessions so that they either immediately precede or follow the lecture.
4. Go over each set of lecture notes as soon as possible after class and outline the main ideas. It is useful to copy key facts, definitions, theorems, and formulas on $3 \times 5$ cards. These cards may be used not only for study, but also for review in subsequent courses for which the present course is a prerequisite.
5. Concentrate and think clearly during the lectures.
6. Do your studying *before* you start your written assignment.
7. Read slowly and do not go to a new concept until you clearly understand the previous one.
8. Many students spend too much time "studying" and not enough time "thinking." Keep asking yourself why a certain step was taken by the professor in the lecture or by the author in the text. Once you understand why a step is taken, you will be more likely to take it yourself when it is needed.
9. Learn what can be done and what cannot be done in mathematics. There are rules, just like in a game. No one would try to play poker, or baseball, or chess without learning the rules. Do the same with mathematics!
10. Attend class regularly. This is especially important in mathematics.
11. Have occasional discussions with the professor.
12. Each weekend, write a letter to someone explaining, in your own terms, the basic concepts covered during the week. You need not send the letter. The purpose of this exercise is to reflect on what was done during that week. A good way to understand a concept is to try to explain it to someone else.

## How to Use a Calculator

Most arithmetic calculations that you will encounter in this book should be done using a calculator. The time saved is better spent on studying the concepts and trying to understand the applications. Therefore, this section includes a brief introduction to calculators. Since there are many different brands, you are urged to consult the owner's manual because there may be some variation in the way each brand handles a specific problem.

Calculators are manufactured for many different types of users. For the person who carries a calculator to the local store simply to compare prices, a four-function calculator, perhaps one with memory, may be sufficient. On the other hand, a student majoring in a scientific field may need a scientific calculator or a programmable calculator. There are also special-purpose calculators available to handle problems peculiar to statistics, business, and so on. You will have to decide which type you are more likely to need.

Calculators are classified according to the type of logic they use in doing calculations. Basically, there are three types of logic used by calculators.

A calculator using arithmetic logic works from left to right and performs the operations in the order in which they appear. For example, an arithmetic calculator will give the answer 20 for $3 + 7 \times 2$ because it will add the 3 and 7 first to obtain 10, then multiply the 10 by 2 to get 20.

In algebra, however, the convention is to perform multiplication before addition, which is the way a calculator using algebraic logic works. Thus, an algebraic calculator will yield the answer 17 for the problem above. It will multiply the 7 by 2 first to get 14, and then it will add 3 to this result to get 17.

Finally, there are calculators that use reverse Polish notation (RPN) logic. These can be identified by the ENTER key. When we use this type of calculator the numbers are entered first, then the operation symbols are entered.

In algebra, we have

$$5 + 7 \times 3 = 26$$

because we perform the multiplication first to get 21 and then we add this result to 5 to get 26. We have illustrated the sequence of key strokes that must be used with each type of calculator to obtain the answer 26.

Arithmetic logic:

| 7 | × | 3 | = | + | 5 | = |

Algebraic logic:

| 5 | + | 3 | × | 7 | = |

RPN logic:

| 3 | ENTER | 7 | × | 5 | + |

Several examples are given to illustrate the use of calculators. In these examples, a scientific calculator that uses algebraic logic is used. Those of you who own calculators that use RPN logic should work the same examples consulting your owner's manuals, if necessary.

**EXAMPLE 4**     Use a calculator to perform the following addition problem: $43 + 29$.

**Solution**     First make sure that the calculator is on, or if it is already on, that it is cleared by pushing the key labeled ON/C . The display should show 0. Proceed as follows:

| ENTER | PRESS | DISPLAY |
|-------|-------|---------|
| 43    |       | 43      |
|       | +     | 43      |
| 29    |       | 29      |
|       | =     | 72      |

The sum is 72.     ■

If a wrong number is entered, it is not necessary to start over. The calculator should have a key labeled CE , which means clear entry. For example, in the foregoing problem if 39 is entered instead of the desired 29, we can clear the 39 by using the clear entry key. We should note that a subtotal is displayed each time the + or − key is pushed. The sequence of steps is the following:

| ENTER | PRESS | DISPLAY |
|-------|-------|---------|
| 43    |       | 43      |
|       | +     | 43      |
| 39    |       | 39      |
|       | CE    | 0       |
| 29    |       | 29      |
|       | =     | 72      |

**EXAMPLE 5**     Perform the following:

$$2.45 + 3.4 + 7 - 5.32$$

**Solution**

| ENTER | PRESS | DISPLAY | ENTER | PRESS | DISPLAY |
|-------|-------|---------|-------|-------|---------|
| 2.45  |       | 2.45    | 7     |       | 7       |
|       | +     | 2.45    |       | −     | 12.85   |
| 3.4   |       | 3.4     | 5.32  |       | 5.32    |
|       | +     | 5.85    |       | =     | 7.53    |

The answer is 7.53.   ■

Some calculators provide parentheses so that certain operations may be grouped.

**EXAMPLE 6**   Calculate the value of $\dfrac{9^2 + 3}{50 - 6^2}$.

**Solution**   Since the values of the numerator and the denominator have to be determined and the results divided, we use parentheses as shown in the following expression:

$$\frac{(9^2 + 3)}{(50 - 6^2)}$$

We now display the sequence of steps using the calculator.

| ENTER | PRESS | DISPLAY | ENTER | PRESS | DISPLAY |
|-------|-------|---------|-------|-------|---------|
|       | ON/C  | 0       |       | (     | 0       |
|       | (     | 0       | 50    |       | 50      |
| 9     |       | 9       |       | −     | 50      |
|       | $x^2$ | 81      | 6     |       | 6       |
|       | +     | 81      |       | $x^2$ | 36      |
| 3     |       | 3       |       | )     | 14      |
|       | )     | 84      |       | =     | 6       |
|       | ÷     | 84      |       |       |         |

The answer is 6.   ■

Annuities will be discussed in detail in Chapter 5. The amount $A$ dollars of an annuity is the future value of all the payments, plus the compound interest on each of these payments. It is given by the formula

$$A = R\left[\frac{(1 + i)^n - 1}{i}\right]$$

where $R$ dollars is the payment per period, $i$ is the rate of interest per period, and $n$ is the number of payments. The next example illustrates how to use this formula.

**EXAMPLE 7**    I.M. Thrift deposits $200 at the end of every three months in a bank paying 8% interest compounded quarterly. What will the balance in the account be after 15 years?

**Solution**    Using the formula

$$A = R\left[\frac{(1 + i)^n - 1}{i}\right]$$

with $R = 200$, $i = 0.02$ (since $\frac{.08}{4} = .02$), and $n = 60$ (since $15(4) = 60$), we get $A = 200[(1 + .02)^{60} - 1]/.02$. We now use the calculator to find the value of $A$.

| ENTER | PRESS | DISPLAY |
|---|---|---|
| | ON/C | 0 |
| 200 | | 200 |
| | × | 200 |
| | ( | 0 |
| 1.02 | | 1.02 |
| | $y^x$ | 1.02 |
| 60 | | 60 |
| | − | 3.28103079 |
| 1 | | 1 |
| | ) | 2.28103079 |
| | ÷ | 456.206158 |
| .02 | | .02 |
| | = | 22810.3079 |

The amount in the account will be $22,810.31, rounded to two decimal places.  ■

## Calculators and Significant Digits*

One simple method for specifying the accuracy of a number is to state how many **significant digits** it has. The significant digits in a number are the ones from the first nonzero digit to the last nonzero digit (reading from left to right). Thus 1074 has four significant digits, 1070 has three, 1100 has two, and 1000 has one significant digit.

This rule may sometimes lead to ambiguities. For example, if a distance is 200 km to the nearest kilometer, then the number 200 really has three significant digits, not just one. This ambiguity is avoided if we use *scientific notation*—that is, if we express the number as a multiple of a power of 10:

$$2.00 \times 10^2$$

| Number | Significant digits |
|--------|:------------------:|
| 12,300 | 3 |
| 3000 | 1 |
| 3000.58 | 6 |
| 0.03416 | 4 |
| 4200 | 2 |
| $4.2 \times 10^3$ | 2 |
| $4.20 \times 10^3$ | 3 |

When working with approximate values, students often make the mistake of giving a final answer with **more** significant digits than the original data. This is incorrect because you cannot "create" precision by using a calculator. The final result can be no more accurate than the measurements given in the problem. For example, suppose we are told that the two shorter sides of a right triangle are measured to be 1.25 and 2.33 inches long. By the Pythagorean Theorem, we find, using a calculator, that the hypotenuse has length

$$\sqrt{1.23^2 + 2.33^2} \approx 2.644125564 \text{ in.}$$

But since the given lengths were expressed to three significant digits, the answer cannot be any more accurate. We can therefore say only that the hypotenuse is 2.64 in. long, rounding to the nearest hundredth.

In general, the final answer should be expressed with the same accuracy as the *least-accurate* measurement given in the statement of the problem. The following rules make this principle more precise.

### Rules for Working with Approximate Data

1. When multiplying or dividing, round off the final result so that it has as many *significant digits* as the given value with the fewest number of significant digits.

---

*Reprinted with permission from Stewart/Redlin/Watson–Mathematics for Calculus, 1989

2.  When adding or subtracting, round off the final result so that it has its last significant digit in the *decimal place* in which the least-accurate given value has its last significant digit.
3.  When taking powers or roots, round off the final result so that it has the same number of *significant digits* as the given value.

As an example, suppose that a rectangular table top is measured to be 122.64 in. by 37.3 in. We express its area and perimeter as follows:

$$\text{Area} = \text{length} \times \text{width} = 122.64 \times 37.3 \approx 4570 \text{ in.}^2 \quad \text{(three significant digits)}$$

$$\text{Perimeter} = 2(\text{length} + \text{width}) = 2(122.64 + 37.3) \approx 319.9 \text{ in.} \quad \text{(tenths digit)}$$

Note that in the formula for the perimeter, the value 2 is an exact value, not an approximate measurement. It therefore does not affect the accuracy of the final result. In general, if a problem involves only exact values, we may express the final answer with as many significant digits as we wish.

Note also that to make the final result as accurate as possible, you should wait until the last step to round off your answer. If necessary, use the memory feature of your calculator to retain the results of intermediate calculations.

## Exercises

*In Exercises 1–10, write the given expression as a product. Each expression is in the form $a^n - b^n$, where n is a positive integer.*

**1.** $x^7 - y^7$

**2.** $16x^4 - 81y^4$

**3.** $8x^3 - 125y^3$

**4.** $\dfrac{x^2}{4} - 9y^2$

**5.** $32a^5 - 243c^5$

**6.** $x^6 - 64$

**7.** $x^3 + y^3$ (*Hint:* $x^3 + y^3 = x^3 - (-y^3).$)

**8.** $243x^5 + 32y^5$

**9.** $\dfrac{x^3}{8} + 27y^3$

**10.** $32 - a^5$

**11.** Write the row of Pascal's triangle that corresponds to $n = 6$.

**12.** Same as Exercise 11 for $n = 7$.

*For Exercises 13–22, write the expansion of the given expressions.*

**13.** $(x + y)^6$

**14.** $(x - y)^6$

**15.** $(x + y)^7$

**16.** $(x - y)^7$

**17.** $(2x + 3y)^3$

**18.** $(2x - 3y)^3$

**19.** $(3a - 2b)^4$

**20.** $\left(\dfrac{x}{2} + \dfrac{y}{3}\right)^5$

**21.** $\left(2x + \dfrac{y}{2}\right)^8$

**22.** $\left(2x - \dfrac{y}{2}\right)^8$

*In Exercises 23–32, a term of the expansion of $(a + b)^n$ for some positive integer n is given.*

*a.* Find the value of n.
*b.* Find the next term.
*c.* Find the preceding term.

**23.** $495a^8b^4$

**24.** $286a^3b^{10}$

**25.** $1001a^4b^{10}$

**26.** $5005a^6b^9$

**27.** $6188a^5b^{12}$

**28.** $3876a^4b^{15}$

**29.** $4845a^{16}b^4$

**30.** $12{,}870a^8b^8$

**31.** $6435a^7b^8$

**32.** $715a^9b^4$

**33.** The first term of an arithmetic progression is 3 and the 50th term is 101. What is the sum of the first 50 terms of the progression?

**34.** The first term of an arithmetic progression is $-4$ and the 100th term is 293. What is the sum of the first 100 terms of the progression?

**35.** What is the sum of the first 50 positive integers? (*Hint:* Note that 1, 2, 3, 4, 5, . . . is an arithmetic progression.)

**36.** What is the sum of the first 850 positive integers?

**37.** Compare the sums
$1^3 + 2^3 + 3^3 + \cdots + n^3$ and $1 + 2 + 3 + \cdots + n$
for the following values of $n$:

    **a.** $n = 2$                **b.** $n = 3$

    **c.** $n = 4$                **d.** $n = 5$

    **e.** $n = 6$                **f.** $n = 7$

Do you notice a pattern? Recalling that
$$1 + 2 + 3 + \cdots + n = \frac{(n + 1)}{2}n,$$
guess a formula for
$$1^3 + 2^3 + 3^3 + \cdots + n^3.$$

**38.** Given a sequence of numbers $a_1, a_2, a_3, \ldots$, the sum $a_1 + \cdots + a_n$ is called the *nth partial sum of the sequence.*
    **a.** Write the first 20 odd positive integers.
    **b.** Find the 20 partial sums. Do you notice a pattern?

*Using a calculator, find the value of each of the following expressions.*

**39.** $53(47) + 36(72)$

**40.** $23.4(64.3) + 17.2(41.7)$

**41.** $72(23.5) + 64(90.2) - 13.3(14.7)$

**42.** $[17.3(82.5) + 36.4(17.9)]/82.3$

**43.** $[23.5(45.3) - 17.6(34.8)]/17.5$

**44.** $[47.6(19.8) + 18.3(23.4)]/(17.4 + 35.6)$

**45.** $[59.5(23.6) + 27.5(41.7)]/(19.8 + 64.2)$

**46.** $[58.3(39.6) + 32.4(37.5)]/(23.6 + 17.4)$

**47.** $45(13.2 + 17.8) - 16(14.9 + 13.1)$

**48.** $27(24.6 + 42.9) + 25(23.4 + 19.1)$

**49.** $32^2 + 53^2$            **50.** $41^2 + 67^2$

**51.** $45^2 - 78^2$          **52.** $3^4 + 5^3$

**53.** $5.2^5 - 7.4^4$        **54.** $21.1^4 + 35.3^3$

**55.** $(31.2^2 + 14.7^3)6.2^3$

**56.** $(3.5^2 - 3.4^4)(3.4^2 - 5.2^3)$

**57.** $(8.3^2 + 63.2^3)(51.3^2 + 63.2^3 - 17.4^4)$

**58.** $(1.05^{21} + 1.02^{17})(2.1 + 3.2^5)$

**59.** $(1.04^{40} - 1.03^{38})(1.02^{12} - 23^3)$

**60.** $\dfrac{2^5 + 3^6}{4^3 + 6^2}$

**61.** $\dfrac{5.1^5 + 6.2^4}{3.2^2 + 4.1^3}$

**62.** $\dfrac{3.2^6 + 2.7^4}{5.2^4 - 2.6^3}$

**63.** $\dfrac{32^3 + .43^4}{5.2^2 + 4.5^4}$

**64.** $\dfrac{3.2^6 + 4.3^6}{3.14^2 + .6^4}$

**65.** $\dfrac{4^{12} + 1.02^{42}}{1.04^{25} - 13.2}$

**66.** $\dfrac{3.4^{10} + 1.05^{35}}{1.03^{15} - 1.01^{16} + 23.72}$

**67.** $\dfrac{1.6^{20} + 1.05^{30}}{1.015^{20} + 1.025^{10} - 2313.2}$

*If P dollars earns interest at a rate of i (written as a decimal) per time period and if interest is compounded at the end of each of n periods, then the balance in the account will be $P(1 + i)^n$. Use this formula in working Exercises 68–71.*

**68.** If \$500 is invested at 6%, what is the amount after 15 years if interest is compounded **a.** annually, **b.** semiannually, **c.** quarterly, and **d.** monthly? Write your answer to the nearest cent.

**69.** If \$3500 is invested at 8%, what is the amount after 15 years if interest is compounded **a.** annually, **b.** semiannually, **c.** quarterly, and **d.** monthly? Write your answer to the nearest cent.

**70.** Beth wishes to invest \$750 for a period of 5 years. Bank A pays 5.85% interest compounded monthly and bank B pays 6% interest compounded semiannually. Which bank should she select? Assuming she selects the bank that will pay the highest return, what will her balance be at the end of 5 years?

**71.** Gregory wishes to invest \$500 for a period of 8 years. Bank A pays 6.95% interest compounded monthly and bank B pays 7% interest compounded quarterly. Which bank should he select? Assuming he selects the bank that will pay the highest return, what will his balance be at the end of 8 years?

*Use the formula presented in Example 7 in working Exercises 72–75.*

**72.** Mr. Planner deposits \$500 at the end of each month in a bank paying 6% interest compounded monthly. What will the balance in the account be after 12 years?

**73.** Ms. Wise deposits \$350 every 6 months in a bank paying 9% interest compounded semiannually. What will the balance in the account be after 20 years?

**74.** Miss Prudente deposits the same amount at the end of every 3 months in a bank paying 8% interest compounded quarterly. She has $30,200.99 in her account after 10 years. How much did she deposit each time?

**75.** J. De Largent deposits the same sum at the end of each month in a Montreal bank paying 9% interest compounded monthly. He has $77,313.47 in his account after 12 years. How much did he deposit at the end of each month?

**76.** Use a calculator to find $231(9) + 507(5)$. When the answer is displayed, turn your calculator upside down to find out how you should not feel while driving.

**77.** If you wish to know who one of the largest customers of the Arab countries is, perform the following function on your calculator and turn the calculator upside down before you read it: $78.25469(5) + 15(21.3)$.

# Basic
# Algebraic Concepts

0

This text cannot be studied successfully without a good grasp of basic algebraic concepts. Therefore, we will briefly review some of the topics essential to understanding the succeeding chapters. The sections in this chapter cover some of the algebraic terms, notations, and manipulative skills that you are expected to have mastered in order to proceed with finite mathematics and calculus.

# 0.1 The Real Numbers

The number system that is used in elementary mathematics is the *system of real numbers*. Except for occasional mention of complex numbers when we discuss polynomial equations, we will restrict our attention almost exclusively to real numbers.

## Some Terms Used When Working with Real Numbers

Certain real numbers have special names. There are the *positive integers*

$$1, 2, 3, 4, \ldots$$

the *negative integers*

$$-1, -2, -3, -4, \ldots$$

and the *integers*

$$\ldots, -4, -3, -2, -1, 0, 1, 2, 3, 4, \ldots$$

Numbers which can be expressed as ratios $\frac{p}{q}$, where $p$ and $q$ are integers with $q \neq 0$, are called *rational numbers*. $\left(\text{Every integer } n \text{ is a rational number since it can be expressed as } \frac{n}{1}.\right)$ There are real numbers which are not rational; they are called *irrational numbers*. For example, the number $\pi$ is irrational.

Each real number has a decimal representation. For example, the decimal representation of $\frac{3}{4}$ is .75, and that of $\frac{22}{3}$ is 7.3333 . . . . A decimal in which all the digits, after a certain one, consist of a set of one or more digits repeated indefinitely, and where no other digits appear between two consecutive repetitions of that set, is called a *periodic decimal*. A convenient way to express a periodic decimal is to use a bar over the repeating group of digits. For example, we write $.\overline{3}$ instead of 0.3333 . . . and $12.1\overline{35}$ instead of 12.1353535 . . . . It can be shown that a real number is rational if, and only if, its decimal representation is periodic.

DEFINITION 0.1:　If $a$ and $b$ are real numbers and $b^2 = a$, then $b$ is called a *square root of a*.

Since $b^2$ is nonnegative for any real number $b$, $a$ must be nonnegative. If $a$ is positive, it has two square roots. For example, $-2$ and $2$ are both square roots of 4 since $(-2)^2 = 2^2 = 4$.

> **DEFINITION 0.2:** The positive square root of a positive number $c$ is called its *principal square root* and is denoted $\sqrt{c}$.

For example, the principal square root of 9 is 3 and we write $\sqrt{9} = 3$. Be careful not to write $\sqrt{9} = \pm 3$ since $\sqrt{9}$ denotes the principal, hence the positive, square root of 9. Thus, $\sqrt{9} \neq -3$.

## Properties of Real Numbers

The system of real numbers has the following properties:

1. **Closure properties:** For any pair of real numbers $a$ and $b$, the sum $a + b$ and the product $ab$ are real numbers.

2. **Associative properties:** We have

$$(a + b) + c = a + (b + c)$$

and

$$(ab)c = a(bc)$$

for every three real numbers $a$, $b$, and $c$.

3. **Commutative properties:** For every two real numbers $a$ and $b$, we have

$$a + b = b + a$$

and

$$ab = ba$$

4. **Identity elements:** There exist real numbers 0 and 1 with the property that for every real number $a$,

$$a + 0 = a$$

and

$$(a)(1) = a$$

5. **Inverse elements:** For each real number $a$, there exists a real number $-a$, such that $a + (-a) = 0$; and for each nonzero real number $b$, there exists a real number, denoted $\frac{1}{b}$, or $b^{-1}$, such that $(b)\left(\frac{1}{b}\right) = 1$. This number is often called the *reciprocal of b*.

6. **Distributive property:** For every three real numbers $a$, $b$, and $c$, we have

$$a(b + c) = ab + ac$$

Using the properties listed above, it is possible to derive other properties of the real numbers. An important property of the system of real numbers which is frequently used is the following:

If $a$ and $b$ are real numbers whose product $ab$ is equal to 0, then either $a = 0$ or $b = 0$.

(See Exercises *46 and *47.)

## Operations with Real Numbers

**DEFINITION 0.3:**   If $a$ and $b$ are real numbers, the *difference* between $a$ and $b$ is defined to be the number $a + (-b)$. It is denoted $a - b$. The operation used to find the difference between two real numbers is called *subtraction*. If $a$ and $b$ are real numbers and $b \neq 0$, their *quotient* is defined as $a\left(\frac{1}{b}\right)$. It is denoted $a/b$ $\left(\text{or } \frac{a}{b}\right)$. The operation for finding a quotient is called *division*.

Exponents will be discussed in greater detail in Section 1.4. For now, we simply remind you that the *power of a number* is the product obtained by multiplying the number by itself $n$ times. Thus,

$$a \cdot a \cdot a \cdot a \cdot \cdots \cdot a \text{ (} n \text{ factors)}$$

is the *nth power of a* and is writen $a^n$. The superscript $n$ is called the *exponent* and $a$ is called the *base*. Observe that $a^1$ means $a$. Therefore, if the exponent of a quantity is not written, it is understood to be 1. It is important to be able to manipulate expressions involving exponents. Recall the following facts which are usually learned in an intermediate algebra course. If $a \neq 0$ and $n$ is a positive integer, we define

$$a^0 = 1, \text{ and } a^{-n} = \frac{1}{a^n}$$

The following laws of exponents for integers $m$ and $n$ are stated without proof. Later, we will discuss the validity of these laws for other types of exponents.

**Laws of Exponents for Integers $m$ and $n$**

**1.** $a^m \cdot a^n = a^{m+n}$

**2.** If $a \neq 0$, then $\dfrac{a^m}{a^n} = a^{m-n}$

**3.** $(a^m)^n = a^{mn}$

**4.** $(ab)^n = a^n b^n$

**5.** If $b \neq 0$, then $\left(\dfrac{a}{b}\right)^n = \dfrac{a^n}{b^n}$

In working with exponents, it is essential to remember the following convention. Unless otherwise specified by parentheses, in a sequence of operations multiplication is performed before addition or subtraction. Thus, $(-3)^2 = 9$ since $(-3)^2 = (-3)(-3) = 9$. However, $-3^2 = -9$ since $-3^2 = -3 \cdot 3 = -(3 \cdot 3) = -9$. In other words, $-3^2 \neq 9$.

**EXAMPLE 1** Simplify $\dfrac{(4x^2y^3)^4}{(2x^3y^2)^3}$.

**Solution** Using the laws of exponents stated above, we obtain

$$\frac{(4x^2y^3)^4}{(2x^3y^2)^3} = \frac{(2^2)^4(x^2)^4(y^3)^4}{2^3(x^3)^3(y^2)^3}$$

$$= \frac{2^8 x^8 y^{12}}{2^3 x^9 y^6} = 2^{8-3}x^{8-9}y^{12-6} = 2^5 x^{-1}y^6$$

$$= 2^5\left(\frac{1}{x^1}\right)y^6 = \frac{32y^6}{x}$$

**Do Exercise 27.** ■

## Complex Numbers

Although real numbers are used almost exclusively in this book, you may encounter some complex numbers in solving equations such as $x^2 - 2x + 5 = 0$. While it is expected that you will have worked with the quadratic formula in previous courses, the formula will be reviewed briefly in Section 0.7. For our present purpose, however, we simply state that the solutions of the equation $ax^2 + bx + c = 0$, where $a \neq 0$, are given by the formula

$$x = \frac{-b \pm \sqrt{b^2 - 4ac}}{2a}$$

Thus, the solutions of $x^2 - 2x + 5 = 0$ are found by replacing $a$ by 1, $b$ by $-2$, and $c$ by 5 in the quadratic formula. We obtain

$$x = \frac{-(-2) \pm \sqrt{(-2)^2 - 4(1)(5)}}{2(1)}$$

$$= \frac{2 \pm \sqrt{-16}}{2} = \frac{2 \pm 4\sqrt{-1}}{2} = 1 \pm 2\sqrt{-1}$$

Because the square of any real number is nonnegative, $\sqrt{-1}$ cannot represent a real number. We therefore have a need for numbers in the form $a + b\sqrt{-1}$, where $a$ and $b$ are real numbers. We replace $\sqrt{-1}$ by the symbol $i$ and define a *complex number* to be a number in the form $a + bi$, where $a$ and $b$ are real numbers and $i$ is the (nonreal) number having the property $i^2 = -1$. Each real number $a$ is a complex number since it can be written as $a + (0)i$. If $b \neq 0$, the number $a + bi$ is called an *imaginary number*. In this text, unless otherwise specified, whenever we say *number* we shall mean a real number.

## Review of Arithmetic of Signed Numbers

We assume that the students using this book have successfully completed a course in intermediate algebra or the equivalent. Consequently, the basic rules of arithmetic of signed numbers and other elementary rules of algebra are used throughout this text. As a quick review, however, we will give some of these rules, repeating a few that have been previously noted. All of them depend on the properties of real numbers stated at the beginning of this section. Each property is followed by numerical examples.

1.  $-a = (-1)a$

*Examples:*
a. $(-1)8 = -8$
b. $-23 = (-1)23$

2.  $a - (-b) = a + b$

*Examples:*
a. $7 - (-9) = 7 + 9 = 16$
b. $-4 - (-7) = -4 + 7 = 3$

3.  $-(-a) = a$

*Example:* $-(-5) = 5$

4.  $a - b = a + (-b)$

*Examples:*
a. $8 - (-3) = 8 + [-(-3)] = 8 + 3 = 11$
b. $-2 - 6 = -2 + (-6) = -8$

5.  $a(b - c) = ab - ac$

This equation shows that multiplication is distributive over subtraction as well as addition.

*Examples:*
a. $6(7 - 2) = 6 \cdot 7 - 6 \cdot 2 = 42 - 12 = 30$
b. $-3(-2 - 5) = -3 \cdot (-2) - (-3) \cdot 5 = 6 - (-15) = 21$

6.  $-(a + b) = -a - b$

*Examples:*
a. $-(5 + 9) = -5 - 9 = -14$
b. $-(6 + (-1)) = -6 - (-1) = -6 + 1 = -5$

7.  $-(a - b) = -a + b$

*Examples:*
a. $-(7 - 4) = -7 + 4 = -3$
b. $-(5 - 9) = -5 + 9 = 4$

Rules 6 and 7 demonstrate that when we remove parentheses preceded by a negative sign, we must change the sign of each term within the parentheses.

8.  $\dfrac{a}{b} = a\left(\dfrac{1}{b}\right)$, provided $b \neq 0$

   *Examples:*

   **a.** $\dfrac{5}{3} = 5\left(\dfrac{1}{3}\right)$

   **b.** $7\left(\dfrac{1}{4}\right) = \dfrac{7}{4}$

9.  $\dfrac{a}{b} = \dfrac{ac}{bc}$, provided $bc \neq 0$

   *Examples:*

   **a.** $\dfrac{6}{21} = \dfrac{6(1/3)}{21(1/3)} = \dfrac{2}{7}$

   **b.** $\dfrac{7}{25} = \dfrac{7(4)}{25(4)} = \dfrac{28}{100} = 0.28$

10. $\dfrac{0}{a} = 0$, provided $a \neq 0$

   *Example:* $\dfrac{0}{8} = 0$

11. If $a \neq 0$, $\dfrac{a}{0}$ is not defined

   *Example:* $\dfrac{3}{0}$ is not defined

12. $\dfrac{0}{0}$ is indeterminate

   Concerning Rules 10, 11, and 12, in general $\dfrac{a}{b} = x$ if, and only if, $a = bx$ and there is only one such number $x$. Thus, if $a \neq 0$, 0 is the unique number with the property $0 = a \cdot 0$ and Rule 10 is justified. On the other hand, if $a \neq 0$, there is no real number $x$ such that $a = 0 \cdot x$, since $0 \cdot x = 0$ for any real number $x$. Thus, $\dfrac{a}{0}$ is not defined. However, there are infinitely many values of $x$ for which the statement $0 = 0 \cdot x$ is true. Thus, we say that $\dfrac{0}{0}$ is indeterminate.*

*Note that saying a symbol is undefined or indeterminate is quite different from saying that the value of that symbol is zero. The number 0 is well defined, as every student who has received a "0" grade on an exam knows!

13.   If $a \neq 0$, $\left(\dfrac{a}{a}\right) = 1$

      *Example:* $\dfrac{-4}{-4} = 1$

14.   If $a \neq 0$, $a\left(\dfrac{b}{a}\right) = b$

      *Example:* $7\left(\dfrac{-2}{7}\right) = -2$

15.   $\dfrac{a}{b} \cdot \dfrac{c}{d} = \dfrac{ac}{bd}$, provided $bd \neq 0$

That is, we multiply quotients by multiplying numerators and multiplying denominators.

      *Examples:*

      **a.** $\dfrac{3}{4} \cdot \dfrac{7}{5} = \dfrac{3(7)}{4(5)} = \dfrac{21}{20}$

      **b.** $\dfrac{5}{9} \cdot \dfrac{3}{10} = \dfrac{5(3)}{9(10)} = \dfrac{5(3)(1/3)(1/5)}{9(10))(1/3)(1/5)} = \dfrac{1}{3(2)} = \dfrac{1}{6}$

16.   $\dfrac{a}{b} + \dfrac{c}{b} = \dfrac{a + c}{b}$, provided $b \neq 0$

      *Examples:*

      **a.** $\dfrac{3}{7} + \dfrac{5}{7} = \dfrac{3 + 5}{7} = \dfrac{8}{7}$

      **b.** $\dfrac{-4}{3} + \dfrac{-8}{3} = \dfrac{-4 + (-8)}{3} = \dfrac{-12}{3} = -4$

17.   $\dfrac{a}{b} - \dfrac{c}{b} = \dfrac{a - c}{b}$, provided $b \neq 0$

      *Examples:*

      **a.** $\dfrac{5}{4} - \dfrac{8}{4} = \dfrac{5 - 8}{4} = \dfrac{-3}{4}$

      **b.** $\dfrac{-3}{5} - \dfrac{-8}{5} = \dfrac{-3 - (-8)}{5} = \dfrac{-3 + 8}{5} = \dfrac{5}{5} = 1$

18.   $\dfrac{a}{b} + \dfrac{c}{d} = \dfrac{ad + bc}{bd}$, provided $bd \neq 0$

*Examples:*

**a.** $\dfrac{3}{5} + \dfrac{4}{7} = \dfrac{3(7) + 4(5)}{5(7)} = \dfrac{21 + 20}{35} = \dfrac{41}{35}$

It is often advantageous to use the least common multiple of the denominators, as in the following:

**b.** $\dfrac{5}{21} + \dfrac{8}{35} = \dfrac{5}{3(7)} + \dfrac{8}{5(7)} = \dfrac{5(5)}{3(7)(5)} + \dfrac{8(3)}{5(7)(3)} = \dfrac{25}{105} + \dfrac{24}{105}$

$$= \dfrac{25 + 24}{105} = \dfrac{49}{105} = \dfrac{49(1/7)}{105(1/7)} = \dfrac{7}{15}$$

19. $\dfrac{a}{b} - \dfrac{c}{d} = \dfrac{ad - bc}{bd}$, provided $bd \neq 0$

*Example:*

$$\dfrac{4}{7} - \dfrac{5}{3} = \dfrac{4(3) - 5(7)}{7(3)} = \dfrac{12 - 35}{21} = \dfrac{-23}{21} = -\dfrac{23}{21}$$

20. $\dfrac{\frac{a}{b}}{\frac{c}{d}} = \dfrac{a}{b} \cdot \dfrac{d}{c} = \dfrac{ad}{bc}$, provided $bcd \neq 0$

To divide by a quotient, invert the denominator, then multiply.

*Examples:*

**a.** $\dfrac{\frac{6}{5}}{\frac{3}{35}} = \dfrac{6}{5} \cdot \dfrac{35}{3} = \dfrac{6(35)}{5(3)} = 2(7) = 14$

**b.** $\dfrac{\frac{3}{1}}{\frac{5}{7}} = \dfrac{\frac{3}{1}}{\frac{5}{7}} = \dfrac{3}{1} \cdot \dfrac{7}{5} = \dfrac{3(7)}{1(5)} = \dfrac{21}{5}$

**c.** $\dfrac{\frac{3}{5}}{\frac{4}{1}} = \dfrac{\frac{3}{5}}{\frac{4}{1}} = \dfrac{3}{5} \cdot \dfrac{1}{4} = \dfrac{3(1)}{5(4)} = \dfrac{3}{20}$

   You are urged to master the basic rules stated in this section and to know which of the rules you are using when performing algebraic simplifications. Phrases such as "cross multiplying," "canceling," and so on, should be used with care. For example, some students who routinely simplify the fraction $\frac{ab}{ac}$ by "canceling" the $a$'s—$\frac{a\!\!\!/b}{a\!\!\!/c} = \frac{b}{c}$—will often do this,

$$\dfrac{a + b}{a + c} = \dfrac{a\!\!\!/ + b}{a\!\!\!/ + c} = \dfrac{b}{c}$$

or this,

$$\frac{15}{54} = \frac{1\cancel{5}}{\cancel{5}4} = \frac{1}{4}.$$

Of course these two simplifications are incorrect! (See Exercise 48.)

## Exercise Set 0.1

*In Exercises 1–13, simplify the given expressions.*

**1.** $5(7 - 9)$

**2.** $-3(-7 + 10)$

**3.** $-[3 - (4 - 8)]$

**4.** $\dfrac{3}{4} + \dfrac{-8}{4}$

**5.** $\dfrac{-5}{9} - \dfrac{-7}{9}$

**6.** $\dfrac{7}{9} + \dfrac{-13}{25}$

**7.** $\dfrac{-15}{19} - \dfrac{-17}{23}$

**8.** $\dfrac{4}{21} + \dfrac{-3}{28}$ (Use the least common denominator.)

**9.** $\dfrac{-7}{65} - \dfrac{-8}{169}$ (Use the least common denominator.)

**10.** $\dfrac{-5}{7} \cdot \dfrac{-4}{15}$

**11.** $\dfrac{\dfrac{-21}{55}}{\dfrac{7}{25}}$

**12.** $\dfrac{\dfrac{14}{-75}}{147}$

**13.** $\dfrac{\dfrac{18}{15}}{7}$

*In Exercises 14–23, use long division to obtain the decimal representation of the given rational number.*

**14.** $\dfrac{5}{3}$

**15.** $\dfrac{11}{6}$

**16.** $\dfrac{15}{7}$

**17.** $\dfrac{13}{8}$

**18.** $\dfrac{215}{9}$

**19.** $\dfrac{17}{11}$

**20.** $\dfrac{19}{12}$

**21.** $\dfrac{215}{13}$

**22.** $\dfrac{334}{15}$

**23.** $\dfrac{219}{17}$

*In Exercises 24–31, simplify the given expressions.*

**24.** $(4x^2y^4)^2$

**25.** $-x^2(-3x)^2$

**26.** $\dfrac{(9x^2y^4)^3}{(27x^4y^2)^2}$

**27.** $\dfrac{(8a^3b^2)^5}{(4a^3b^3)^4}$

**28.** $\dfrac{(5x^3y^6)^2}{(25x^2y^3)^4}$

**29.** $\dfrac{(-x^3y^2)^4}{[2x^3(-y)^3]^3}$

**30.** $\dfrac{(-9a^2b^3)^4}{(81a^3b^4)^3}$

**31.** $\dfrac{(-7x^4y^3)^5}{(49x^3y^4)^3}$

*In Exercises 32–36, use the quadratic formula to solve the given equations. State whether the solutions are real or imaginary.*

**32.** $2x^2 + 5x - 18 = 0$

**33.** $2x^2 - 2x + 1 = 0$

**34.** $4x^2 - 4x + 5 = 0$

**35.** $-3x^2 + 5x + 12 = 0$

**36.** $25x^2 + 5x + 1 = 0$

**37.** Prove that the sum of two rational numbers is a rational number.

**38.** Prove that the difference of two rational numbers is a rational number.

**39.** Prove that the product of two rational numbers is a rational number.

**40.** Prove that the quotient of two rational numbers is a rational number, provided the divisor is not 0.

**41.** Prove that the sum of a rational number and an irrational number is an irrational number.

**42.** Prove that the product of a nonzero rational number and an irrational number is an irrational number.

**43.** Show that subtraction is not a commutative operation.

**44.** Is subtraction associative? Justify your answer.

**45.** Is addition closed on the set of odd integers? Justify your answer.

**\*46.** Using the properties listed in this section, prove that if $a$ is any real number, then $a(0) = 0$. (*Hint:* Start with $0 + 0 = 0$. Then $a(0 + 0) = a(0)$. Let $a(0) = c$, so that we have $c + c = c$. Now add $-c$ to both sides.)

**\*47.** Prove that if $a$ and $b$ are real numbers and $ab = 0$, then $a = 0$ or $b = 0$. (*Hint:* If $a = 0$, we are done. If $a \neq 0$, multiply both sides by $\frac{1}{a}$ and use the result of Exercise *46.)

**48.** One should always aspire to finding the correct answer to a problem. However *obtaining the correct answer does not necessarily mean that the method used was correct*, as illustrated by the following:

A student who was asked to simplify the fraction $\frac{16}{64}$ canceled the 6s to get

$$\frac{16}{64} = \frac{1\!\!\!/6}{6\!\!\!/4} = \frac{1}{4}$$

**a.** Show that the answer is correct.

**b.** Show that canceling the 6s in the fraction $\frac{166}{664}$ also yields the correct answer.

**c.** Generalize this using more 6s.

**d.** Give three examples to illustrate that the student's method, in spite of producing correct answers in the above examples, will almost always yield an incorrect answer.

## 0.2  Elementary Introduction to Sets

When we use mathematics to solve a problem in any field, the first step is to construct a mathematical model for the problem at hand. For example, when solving a story problem in basic algebra, it is necessary to write an equation (the mathematical model) which represents the conditions stated in the problem. Many applications of mathematics involve collections of objects; these collections are called *sets*. Mathematical models in such cases will, therefore, involve sets. It is important that you have some knowledge of elementary facts about sets and about operations involving sets. This section provides some basic terminology and notation which will be used frequently in later sections.

### Set Description

> DEFINITION 0.4:   A *set* is a collection of objects.

To describe a set, we usually state a necessary and sufficient property* that an object must have to be a member of that set. For example, the set of all smart people is not well defined; one person may be declared smart by one teacher and slow by another. On the other hand, the set of the first five counting numbers is well defined.

In general, we use capital letters to denote sets and small letters to denote members of sets. If $A$ is the name of a set and $b$ is the name of an object, the sentence "$b$ is an element of $A$" is abbreviated $b \in A$. Similarly, $b \notin A$ is an abbreviation for "$b$ is not an element of $A$." The terms *member* and *element* are synonymous.

When a set has a small number of elements, we can describe it by displaying the names of its members, listing each only once, separating consecutive ones by commas and putting the list within braces. This technique is often called the *roster method*.

---

*If $p$ and $q$ are properties, and $q$ is satisfied whenever $p$ is satisfied, we say that $p$ is sufficient for $q$. On the other hand, if $q$ must be true for $p$ to be true, we say that $q$ is necessary for $p$. For example, "having wings" is a necessary property to "be a normal bird," but it is not sufficient (bats and bees have wings also). However, "having feathers" is a property that is both necessary and sufficient to "be a normal bird" in the animal world. As another example, to "be a woman" is a sufficient condition to "be a human," but it is not necessary (men are human beings also!).

For example, the set whose members are the first three counting numbers is described as follows:

$$\{1, 2, 3\}$$

A collection with no members at all is also a set. It is called *the empty set* (it can be shown that there is only one such set). The symbol Ø (or { }) is used to denote the empty set.

In algebra we often use letters to represent numbers. For example, you have seen statements such as

$$(a + b)^2 = a^2 + 2ab + b^2$$
$$3x + 6 = 7 - 10x$$

and

$$x + y = y + x$$

At first, statements such as these may seem strange. However, a moment of reflection should reveal that even in the statement $2 + 8 = 4 + 6$, the 2, 8, 4, and 6 are symbols which represent some specific numbers. On the other hand, in the statement

$$x + 5 = 2x + 1$$

$x$ may be regarded as a symbol that does not denote a specific number; it may be replaced by any one member from a certain set of numbers. In general, such a symbol is called a *variable*. In any discussion in which a variable is involved, we must have a set so that we may replace the variable by the name of some member of that set. This set is called the *replacement set* for the variable. For example, if we consider the statement $x + 5 = 2x + 1$ with replacement set $\{3, 4, 5\}$, we may replace $x$ by any of the numbers 3, 4, or 5. If we replace $x$ by 3, $2x$ must be replaced by 6 and we get

$$3 + 5 = 6 + 1$$

which is false. If we replace $x$ by 4, $2x$ must be replaced by 8 and we get

$$4 + 5 = 8 + 1$$

which is true. Hence, 4 is a *solution* of the sentence. If $x$ is replaced by 5, again we get a false statement.

Be careful not to replace $x$ by one number on one side of the equal sign and by a different number on the other side.

DEFINITION 0.5:    If a sentence involves a variable, it is called an *open sentence*.

Often, the replacement set which should accompany the open sentence is not given, but its membership is understood from context. For example, the replacement set for the open sentence "$x$ is the president of General Motors Co." is understood to be a set of names of people.

When a set has many members, the roster method is not a very practical way to describe it. It is common practice to use an open sentence to describe the set whose members are exactly those solutions of the open sentence. We use the following procedure:

If $p(x)$ (read "$p$ of $x$") is an open sentence in the variable $x$, the symbol

$$\{x \mid p(x)\}$$

(read "the set of all $x$'s such that $p(x)$") describes the set of solutions of the open sentence. Of course, the replacement set should be described if it is not clear from context.

**EXAMPLE 1**  List five members of the set $\{x \mid x$ is, or has been, a professional basketball player$\}$.

**Solution**  Jerry West, Kareem Abdul Jabbar, Elgin Baylor, Jack Sikma, and Ralph Sampson are five members of the set. There are, of course, many others.
**Do Exercise 21.**

**EXAMPLE 2**  Describe the set $\{3, 4\}$ using an open sentence.

**Solution**  $\{3, 4\} = \{y \mid y$ is a whole number and $2 < y < 5\}$, or

$$\{3, 4\} = \{t \mid (t - 3)(t - 4) = 0\}$$

**EXAMPLE 3**  Let $A = \{u \mid u$ is a real number and $2 < u < 5\}$. Which of the following statements is true?

**a.** $\pi \in A$    **b.** $\dfrac{10}{3} \in A$    **c.** $6 \in A$

**Solution**  **a.** $\pi \in A$ is true since $\pi$ is a real number larger than 2 and less than 5.

**b.** $\dfrac{10}{3} \in A$ is true since $\dfrac{10}{3}$ is a rational (hence real) number which is greater than 2 and less than 5.

**c.** $6 \in A$ is false since 6 is not less than 5.

**Do Exercise 31.**

## Elementary Set Algebra

> **DEFINITION 0.6:**  Suppose that $A$ and $B$ are sets and that each element of $A$ is also an element of $B$, we then say that $A$ is a *subset* of $B$. We use the symbol $A \subseteq B$ to denote this.

For example, if $A = \{1, 2, 3\}$ and $B = \{0, 1, 2, 3, 4\}$, then $A \subseteq B$ (read: "$A$ is a subset of $B$") since each element of $A$ is also in $B$. According to the definition, it is clear that if $A$ is any set, then $A \subseteq A$. Also if $A$, $B$, and $C$ are sets, such that $A \subseteq B$ and $B \subseteq C$, then $A \subseteq C$.

> **DEFINITION 0.7:**   If the set $A$ is a subset of the set $B$, but $B$ has at least one element which is not in $A$, then we say that $A$ is a *proper subset* of $B$ and we write $A \subset B$.

It can be shown that if $A$ is an arbitrary set, the empty set is a subset of $A$. Symbolically, if $A$ is any set, $\emptyset \subseteq A$.

**EXAMPLE 4**   Let $A = \{a, b, c\}$. List all the subsets of $A$.

**Solution**   The subsets of $A$ are $\emptyset$, $\{a\}$, $\{b\}$, $\{c\}$, $\{a, b\}$, $\{a, c\}$, $\{b, c\}$, and $A$.
**Do Exercise 37.**                                                                 ∎

In general, it is true that if a set has $n$ elements, then it has $2^n$ subsets. Set $A$ has three elements and eight subsets: $2^3 = 8$. There is an interesting relationship between the number of subsets of a set with $n$ members and the coefficients in the expansion of $(a + b)^n$. In the preceding example, we have shown that the set $\{a, b, c\}$ has *one* subset with no element (the empty set), *three* subsets each with one member, *three* subsets each with two members, and *one* subset with three members (the set itself). In the "Getting Started" section we saw that the coefficients of the expansion of $(a + b)^3$ are 1, 3, 3, and 1. We also saw that the coefficients of the expansion of $(a + b)^5$ are 1, 5, 10, 10, 5, and 1. It is also true that a set with five members has *one* subset with no member, *five* subsets each with one member, *ten* subsets each with two members, *ten* subsets each with three members, *five* subsets each with four members, and *one* subset with five members. As expected, $1 + 5 + 10 + 10 + 5 + 1 = 32 = 2^5$, the total number of subsets. (See Exercise 39.)

We will now discuss several ways to derive new sets from given sets.

> **DEFINITION 0.8:**   The set whose members are the elements under consideration in a given discussion is called the *universe*.

For example, in an algebra course, the universe may be the set of complex numbers; while in a sociology course, the universe may be the set of all people.

> **DEFINITION 0.9:**   If $A$ is a set and $U$ is the universe, the set which consists of all members of $U$ which are not in $A$ is called the *complement of A* (with respect to $U$) and is denoted $A'$.

For example, if $U$ is the set of all counting numbers and $A$ is the set of even numbers, then $A'$ is the set of odd numbers.

> **DEFINITION 0.10:**   The *union* of the sets $A$ and $B$ is the set of all elements that are in $A$, or in $B$, or in both. We use the symbol $A \cup B$ to denote this set.

That is,

$$A \cup B = \{x \mid x \in A, \text{ or } x \in B, \text{ or both}\}$$

> **DEFINITION 0.11:**   The *intersection* of the sets $A$ and $B$ is the set of all elements that belong to both $A$ and $B$. We use the symbol $A \cap B$ to denote this set.

In other words,

$$A \cap B = \{x \mid x \in A \text{ and } x \in B\}$$

**EXAMPLE 5**   Let $A = \{1, 2, 3\}$ and $B = \{2, 3, 4, 5\}$. Find $A \cup B$ and $A \cap B$.

**Solution**   $A \cup B = \{1, 2, 3, 4, 5\}$ and $A \cap B = \{2, 3\}$.
**Do Exercise 53.** ■

Verify that if $A$ and $B$ are any sets, then $A \cap B \subseteq A \cup B$.

**EXAMPLE 6**   Let the universe be the set $\{i, j, k, l, m, n, o, p, q, r\}$, $R = \{j, l, n, o, q, r\}$, $S = \{i, j, q, r\}$, $T = \{j, m, o\}$, and $V = \emptyset$. Find:
a. $R'$    b. $V'$    c. $(R \cup T)'$
d. $R' \cap T'$    e. $R \cap S$    f. $R \cap T$
g. $R \cap (S \cup T)$    h. $(R \cap S) \cup (R \cap T)$

**Solution**   a. $R' = \{j, l, n, o, q, r\}' = \{i, k, m, p\}$
b. $V' = \emptyset' = \{i, j, k, l, m, n, o, p, q, r\}$
c. $(R \cup T)' = (\{j, l, n, o, q, r\} \cup \{j, m, o\})' = \{j, l, m, n, o, q, r\}' = \{i, k, p\}$
d. $R' \cap T' = \{i, k, m, p\} \cap \{i, k, l, n, p, q, r\} = \{i, k, p\}$
e. $R \cap S = \{j, l, n, o, q, r\} \cap \{i, j, q, r\} = \{j, q, r\}$
f. $R \cap T = \{j, l, n, o, q, r\} \cap \{j, m, o\} = \{j, o\}$
g. $R \cap (S \cup T) = \{j, l, n, o, q, r\} \cap (\{i, j, q, r\} \cup \{j, m, o\}) = \{j, l, n, o, q, r\} \cap \{i, j, m, o, q, r\} = \{j, o, q, r\}$
h. $(R \cap S) \cup (R \cap T) = \{j, q, r\} \cup \{j, o\} = \{j, o, q, r\}$

**Do Exercises 49, 57, and 60.** ■

You should compare the answers in **c** and **d**, and in **g** and **h** of the preceding example.

> **DEFINITION 0.12:**   If two sets have no common elements, their intersection is the empty set. In that case, we say that the two sets are *disjoint*.

## Exercise Set 0.2

*In Exercises 1–10, use the roster method to describe the given collections.*

1. The first six counting numbers.
2. The first ten counting numbers.
3. The first counting number.
4. The last five letters of the English alphabet.
5. The states in the United States with names beginning with the letter "W."
6. All counting numbers between 15 and 20.
7. All letters of the English alphabet between b and f.
8. All counting numbers between 3 and 5.
9. The presidents of the United States who succeeded John F. Kennedy.
10. The vowels of the English alphabet.

*In Exercises 11–20, an open sentence and a replacement set for the variable are given. In each case, determine the solution set. Recall that each solution must be a member of the replacement set.*

11. $2x + 3 = x + 10$; $\{5, 6, 7, 8\}$
12. $x + 5 = 2x - 1$; $\{4, 5, 6, 7, 8\}$
13. $\dfrac{x}{2} = x - 3$; $\{4, 5, 6\}$
14. $x^2 + 5x - 6 = 0$; $\{0, 1, 2\}$
15. $3x + 10 = 5x + 6$; $\{1, 2, 3\}$
16. $5x - 6 = 2x + 3$; $\{1, 2, 3, 4\}$
17. $4x + 1 = 5x - 4$; $\{2, 6, 11\}$
18. $3x + 7 = 5x + 1$; $\{1, 2, 3, 4\}$
19. $6 - 3x = 5x - 2$; $\{0, 1, 2\}$
20. $x^3 - 8 = 0$; $\{0, 2, 5\}$

*In Exercises 21–30, a set is described. In each case, list three members of the given set.*

21. $\{x \mid x$ has been a president of the United States$\}$
22. $\{x \mid x$ is a professional football player$\}$
23. $\{y \mid y$ is a whole number larger than 2 and less than 5$\}$

24. $\{t \mid t$ is a real number larger than 5$\}$
25. $\{z \mid z$ is a counting number$\}$
26. $\{u \mid (u - 1)(u - 3)(u - 5)(u - 7) = 0\}$. (*Hint:* Recall that a product of real numbers is 0 if, and only if, at least one of the factors is 0.)
27. $\{x \mid x$ is a whole number less than 5$\}$
28. $\{x \mid x$ is an American movie actor$\}$
29. $\{y \mid y$ is a television personality$\}$
30. $\{s \mid s$ is a professional baseball player$\}$

*In Exercises 31–35, indicate whether the given statement is true or false.*

31. $2 \in \{x \mid x$ is a whole number greater than 3$\}$
32. $5 \in \{x \mid (x - 1)(x - 5)(x - 7) = 0\}$
33. $\pi \in \{y \mid y$ is a real number greater than 2$\}$
34. $\dfrac{9}{2} \in \{t \mid t$ is a positive real number$\}$
35. $m \in \{x \mid x$ is a vowel in the English alphabet$\}$
36. Let $A = \{1\}$. How many subsets does this set have? List all the subsets.
37. Same as Exercise 36 for $A = \{1, 2\}$.
38. Same as Exercise 36 for $A = \{1, 2, 3, 4\}$.
39. Same as Exercise 36 for $A = \{1, 2, 3, 4, 5\}$.
40. How many proper subsets does the set $\{1\}$ have?
41. Same as Exercise 40 for the set $\{1, 2\}$.
42. Same as Exercise 40 for the set $\{1, 2, 3\}$.
43. Same as Exercise 40 for the set $\{1, 2, 3, 4\}$.
44. Same as Exercise 40 for the set $\{1, 2, 3, 4, 5\}$.
45. If a set has $n$ elements, how many proper subsets does it have?
46. If a set has four elements, how many subsets does it have
    a. each with no element?
    b. each with one element?
    c. each with two elements?
    d. each with three elements?
    e. each with four elements?

**47.** If a set has six elements, how many subsets does it have

    **a.** each with no element?

    **b.** each with one element?

    **c.** each with two elements?

    **d.** each with three elements?

    **e.** each with four elements?

    **f.** each with five elements?

    **g.** each with six elements?

**48.** Let $S$ be a set with $n$ members and let $k$ be an integer between 0 and $n$. Explain why the number of subsets, each with $k$ elements, is the same as the number of subsets each with $n - k$ elements.

*In Exercises 49–60 the universe is the set* $\{0, 1, 2, 3, 4, 5, 6, 7, 8, 9\}$, $A = \{1, 2, 3, 4, 5, 6\}$, $B = \{2, 5\}$, $C = \{3, 4, 8\}$, *and* $D = \emptyset$.

**49.** Find $A'$.

**50.** Find $B'$.

**51.** Find $C'$.

**52.** Find $D'$.

**53.** Find $A \cup B$ and $A \cap B$.

**54.** Find $A \cup C$ and $A \cap C$.

**55.** Find $B \cup C$ and $B \cap C$.

**56.** Find $A \cap (B \cup C)$.

**57.** Find $(A \cap B)'$ and $A' \cup B'$.

**58.** Find $(B \cup C)'$ and $B' \cap C'$.

**59.** Find $(A \cap C)'$ and $A' \cup C'$.

**60.** Find $A \cup (B \cap C)$ and $(A \cup B) \cap (A \cup C)$.

# 0.3 Venn Diagrams

In this section we illustrate the concepts of the preceding section geometrically. We also show how sets can be used to solve certain survey problems.

### Description of Venn Diagrams

The geometric illustration is made using a *Venn diagram*. If the universe for a certain discussion is set $U$, we represent that set by a rectangle. In general, each set in that discussion is represented by points on a plane bounded by one or more simple closed curves. Circles, ellipses, squares, and triangles are examples of simple closed curves. (See Figure 0.1a.) On occasion, it is useful to represent a single set using several simple closed curves.

C is represented by two
disjoint circles

**FIGURE 0.1**

(a)

(b)

For example, we have represented two disjoint sets $A$ and $B$ with two separate circles. We want to represent a third set $C$ and to emphasize that $C$ has points in common with both $A$ and $B$, but it has no points outside of either $A$ or $B$. In that case, we represent $C$ by the union of two circles. (See Figure 0.1b.) The only stipulation is that all configurations representing sets must lie within the rectangle which represents the universe.

We should understand that there are infinitely many points inside a simple closed curve, while there may be only a finite number of elements in the set represented by that simple closed curve. This creates no difficulty, because a Venn diagram is often used only to illustrate the relationship between sets, and not their sizes. For example, in Figure 0.2a we illustrate the fact that the set $A$ is a subset of the set $B$, in Figure 0.2b we show that sets $A$ and $B$ are disjoint, while Figure 0.2c displays the set which is the intersection of the sets $A$ and $B$.

  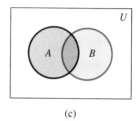

**FIGURE 0.2**          (a)                    (b)                    (c)

If it is important to know how many elements are in a set, we place a number within the simple closed curve which represents that set. Since we usually have several sets represented in a Venn diagram, we have several simple closed curves and there is often a number within more than one of these. Therefore, the following convention is used: Any number placed in a Venn diagram gives the number of elements in the smallest of all the sets which are represented by the simple closed curves which enclose that number. As an illustration, three sets $A$, $B$, and $C$ are represented by circles in Figure 0.3. The number 5 is within several simple closed curves, each of which represents a set. The smallest of these sets is the set of all points which belong to both $A$ and $B$, but which do not belong to $C$. Hence, we know that there are five elements in the intersection of the sets $A$, $B$, and $C'$. The number 10 in the upper right corner is in the interior of the rectangle representing the universe and is in the exterior of each of the three circles. So the number 10 indicates that there are ten elements in the universe that are not in any of the sets $A$, $B$, or $C$.

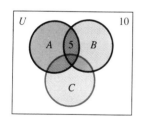

**FIGURE 0.3**

**EXAMPLE 1**   Use a Venn diagram to illustrate the fact that if $A$ and $B$ are arbitrary sets, then

$$(A \cup B)' = A' \cap B'$$

(See also Example 6(c, d) and Exercise 58 in the preceding section.)

**Solution**   In Figure 0.4a the shaded region illustrates $A \cup B$, in Figure 0.4b it illustrates $(A \cup B)'$, in Figures 0.4c, 0.4d, and 0.4e, the shaded regions represent $A'$, $B'$, and $A' \cap B'$, respectively. Comparing the shaded regions in Figures 0.4b and 0.4e, we conclude that these regions are the same. Hence, we have

$$(A \cup B)' = A' \cap B'$$

**Do Exercise 7.**

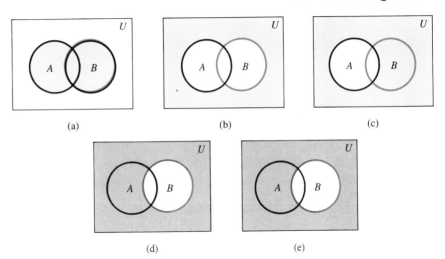

(a)                    (b)                    (c)

(d)                    (e)

**FIGURE 0.4**

The foregoing equality is one of *De Morgan's Laws.* It can be proved rigorously, but we shall restrict our argument to its geometric illustration given in Example 1.

## Solving Survey Problems

Venn diagrams can be used in solving survey problems as illustrated in the following two examples.

**EXAMPLE 2**  A large university is planning to build a physical education complex and wants to know how to divide space for racquetball, squash, and tennis courts. To that end, 250 students were selected randomly and it was found that among these,

> 55 play racquetball;
> 25 play squash;
> 65 play tennis;
> 15 play racquetball and squash;
> 10 play tennis and squash;
> 25 play tennis and racquetball; and
> 5 play all three games.

Use this data and a Venn diagram to answer the following questions.

**a.** How many of the students surveyed play none of the three games?
**b.** How many of them play only racquetball?
**c.** How many play racquetball and tennis but not squash?

**Solution**  Let $R$, $S$, and $T$ be the sets of students surveyed who play racquetball, squash, and tennis, respectively. Let $U$ (the universe) be the set of all students surveyed. We know that $U$ has 250 members; $R$, $S$, and $T$ have 55, 25, and 65 members, respectively. Further, $R \cap S$ has 15 members, $T \cap S$ has 10 members, $T \cap R$ has 25 members, and $R \cap S \cap T$ has 5 members. Since this latter set is the smallest of the sets we are considering, we begin with it and place a 5 in the region that is within all three circles. (See Figure 0.5a.) We know that there are 25 members in $T \cap R$. Hence, there are

25 members within the circles representing $T$ and $R$, respectively. But five of them are already accounted for because they are in $R \cap T \cap S$. Thus, we place a 20 in the region that is within the circles representing $T$ and $R$, but outside the circle representing $S$. (See Figure 0.5b.) The reason for each of the other entries should now be clear. (See Figure 0.5c.)

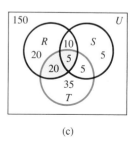

**FIGURE 0.5**

(a)                              (b)                              (c)

**a.** Adding all the numbers which have been placed within at least one circle, we get 100. Thus, 100 of the students play at least one of the three games. Therefore, 150 of the students surveyed play none of the three games since $250 - 100 = 150$.

**b.** The number placed within the circle representing $R$ but outside the other two circles is 20. We conclude that 20 of the students surveyed play only racquetball.

**c.** The number placed in the region within the circles representing $R$ and $T$ but outside the other circle is also 20. Thus, 20 of the students surveyed play racquetball and tennis but not squash.

**Do Exercise 13.**                                                                          ■

**EXAMPLE 3**   A company hired a college student to survey a sample of the student population at a certain eastern university. The student claims that he selected 200 students at random and obtained the following results. Among the 200 students surveyed,

120 drink Coke;
100 drink Seven-Up;
115 drink Pepsi;
 55 drink Coke and Seven-Up;
 60 drink Coke and Pepsi;
 50 drink Seven-Up and Pepsi; and
 35 drink all three.

The manager of the project received a report that the hired student was seen spending a great deal of time in the cafeteria socializing and that perhaps he had made up the data. The manager, who had taken a mathematics for business students course while she was in college, used a Venn diagram to check the consistency of the data. Shortly after that, she fired the student. What was her reason?

**Solution**   Let $C$, $S$, and $P$ denote the sets of students who reportedly drink Coke, Seven-Up, and Pepsi, respectively. Then the sets $C$, $S$, $P$, $C \cap S$, $C \cap P$, $S \cap P$, and $C \cap S \cap P$ should have 120, 100, 115, 55, 60, 50, and 35 members, respectively. We begin by placing a 35 in the region that is within the three circles representing the three sets, because that region represents the intersection of the three sets. (See Figure 0.6.) We then place a 20 in the region that is within the circles representing $C$ and $S$, but outside the circle representing $P$. The justification for this entry is that we must have 55 elements in the region within the circles representing $C$ and $S$, but 35 are already accounted for in the region within all three circles and $55 - 35 = 20$. We continue in this manner and place numbers in all the appropriate regions. Adding all the entries in Figure 0.6, we get

$$40 + 20 + 30 + 25 + 35 + 15 + 40 = 205$$

Since the hired student claims that he surveyed only 200 students, either he has made up the data or his work was performed carelessly. Hence, he deserves to be fired!

**Do Exercise 19.**

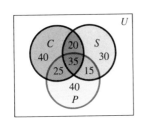

**FIGURE 0.6**

# Exercise Set 0.3

*In Exercises 1–6, A, B, and C are sets. Use a Venn diagram to illustrate the given set.*

**1.** $A \cap (B \cup C)$

**2.** $A' \cap (B \cup C)$

**3.** $A \cup (B' \cap C)$

**4.** $(A \cap B) \cup C$

**5.** $(A \cap B)' \cup C'$

**6.** $B' \cup (A \cap B)$

*In Exercises 7–12, use a Venn diagram to illustrate geometrically the truth of the given statement.*

**7.** $(A \cap B)' = A' \cup B'$

**8.** $(A \cap B) \subseteq A$

**9.** $A \subseteq (A \cup B)$

**10.** $A \cap (B \cup C) = (A \cap B) \cup (A \cap C)$

**11.** $A \cup (B \cap C) = (A \cup B) \cap (A \cup C)$

**12.** $(A \cap B) \subseteq (A \cup B)$

**13.** Suppose that 350 students at a large western university were surveyed with the following results. Among these students,

    33 read Scientific American;
    40 read National Geographic;
    43 read Playboy;
    20 read National Geographic and Playboy;
    15 read Scientific American and Playboy;
    19 read Scientific American and National Geographic; and
    7 read all three magazines.

Use a Venn diagram to find how many of the students surveyed
**a.** read only Scientific American.
**b.** read Scientific American and Playboy but do not read National Geographic.
**c.** read none of the three magazines.

**14.** A survey of 400 people in a small town in eastern Washington provided the following results. Among these 400 people,

    38 drink coffee;
    48 drink tea;
    39 drink mineral water;
    15 drink both coffee and tea;
    13 drink both coffee and mineral water;
    18 drink both tea and mineral water; and
    5 drink all three.

Using a Venn diagram, find how many of the people surveyed drink

**a.** mineral water but neither coffee nor tea.
**b.** tea and coffee but not mineral water.
**c.** none of the three.

**15.** At the end of their freshman year, 200 randomly selected students at a small western university were surveyed. They were asked whether they had taken English, mathematics, or philosophy during the year. The following results were recorded:

    20 had taken none of the three subjects;
    50 had taken only English;
    30 had taken only mathematics;
    20 had taken only philosophy;
    10 had taken English and philosophy but not mathematics;
    40 had taken English and mathematics but not philosophy; and
    15 had taken mathematics and philosophy but not English.

Using a Venn diagram, determine how many among the students surveyed took

a. all three subjects.    b. English and mathematics.
c. English.               d. mathematics.
e. philosophy.

**16.** A Safety Inspection Center selected 350 cars randomly and checked their tires, brakes, and headlights. It was found that

43 had defective tires;
45 had defective brakes;
24 had defective lights;
23 had defective tires and brakes;
17 had defective brakes and lights;
12 had defective tires and lights; and
10 had defective tires, lights, and brakes.

Using a Venn diagram, determine how many of the cars inspected had

a. only defective tires.
b. defective tires and brakes but safe lights.
c. safe brakes, lights, and tires.

**17.** The director of the Sports program at a small private university randomly selected 250 students and collected the following data. Among these 250 students,

53 smoked;
78 drank alcoholic beverages;
58 played some sport;
25 smoked and played some sport;
30 drank alcoholic beverages and played some sport;
28 smoked and drank alcoholic beverages; and
10 did all three.

Using a Venn diagram, determine how many of the 250 students surveyed

a. play some sport but do not smoke or drink alcoholic beverages.
b. play some sport and drink some alcoholic beverages but do not smoke.
c. do none of the three.

**18.** A clinic hired a college student to survey a sample of the student population at a certain western university. The student claims that she selected 300 of her peers at random and obtained the following results. Among the 300 students surveyed,

134 eat red meat;
183 eat fish;
151 eat poultry;
 63 eat red meat and fish;
 50 eat red meat and poultry;
 63 eat fish and poultry; and
 25 eat all three.

The manager of the project saw that the hired student was spending a lot of time in the cafeteria socializing and he worried that the student had made up the data. The manager used a Venn diagram to check the consistency of the data. Shortly after that, he fired the student. What was his reason?

**19.** The manager of a large store noticed that one of the young clerks was working hard and thought to promote him to assistant manager. As a test, she asked him to survey 230 customers and he came up with the following data. Among the 230 people who entered the store,

173 were males;
193 made a purchase;
113 used a credit card;
123 were males who made a purchase; and
 63 males used a credit card.

The manager used the data and a Venn diagram to determine how many, among the people who entered the store,

a. were males who did not make a purchase.
b. were males who paid cash for their purchases.
c. were females who used credit cards.
d. were females who paid cash for their purchases.

She then decided that the young clerk was not ready for the promotion. What were her answers to the four calculations and why did she decide not to promote the young clerk?

# 0.4  Polynomials

A real polynomial in $x$ is any expression that can be obtained from the symbol $x$ and real numbers using only the operations of addition, subtraction, and multiplication a finite number of times. For example $5x^3$ and $3x$ are polynomials since $5x^3 = 5 \cdot x \cdot$

$x \cdot x$ and $3x = 3 \cdot x$. It follows that $5x^3 + 3x - 3$ is also a polynomial. The formal definition follows.

> **DEFINITION 0.13:** A *real polynomial in x* is an expression that can be written in the form
>
> $$a_0x^n + a_1x^{n-1} + a_2x^{n-2} + \cdots + a_{n-1}x + a_n$$
>
> where $n$ is a nonnegative integer and $a_0, a_1, a_2, \cdots, a_{n-1}, a_n$ are real numbers.

The symbol $x$ is called an *indeterminate* (*variable*) and may be replaced by real numbers. If $a_0 \neq 0$, the polynomial is of *degree n* and $a_0$ is the *leading coefficient*. Nonzero constants are considered polynomials of degree 0. The constant 0 has no degree.

The degree of a term is the exponent that is attached to the variable in that term. For example, the third term of the polynomial $-2y^6 + 3y^5 - 5y^4 + 10y^2 + 7y + 13$ has degree 4. A *complex polynomial* is defined in exactly the same way simply by replacing the word *real* by the word *complex* in the foregoing definition. When we discuss division of arbitrary polynomials, it is convenient to have a compact notation for such expressions. We use $f(x)$ (read "$f$ of $x$"), $g(x)$, $h(x)$, and so on, to denote arbitrary polynomials in $x$.

A polynomial is said to be *linear* if it is of degree 1, and *quadratic* if it is of degree 2. After collecting like terms, it is called a *monomial* if it has only one term, a *binomial* if it has two terms, and a *trinomial* if it has three terms. For example, $2x + 3$ is a linear polynomial (binomial) in $x$, $-y^2 + 3y - 7$ is a quadratic polynomial (trinomial) in $y$ with a leading coefficient of $-1$, and $9z^5 + 3z - 2$ is a trinomial of degree 5 with a leading coefficient of 9.*

## Addition, Subtraction, Multiplication, and Division of Polynomials

*Addition:* When we add two polynomials in $x$, we make use of the basic properties of real numbers. We illustrate with the following simple example.

**EXAMPLE 1**  Add $2x^3 + 5x^2 - 6x + 1$ and $3x^3 + 8x + 7$.

**Solution**  Using the associative and commutative properties of real numbers we get

$$(2x^3 + 5x^2 - 6x + 1) + (3x^3 + 8x + 7)$$
$$(2x^3 + 3x^3) + 5x^2 + (-6x + 8x) + (1 + 7)$$

---

*$5x + 2 + 3x$ is not a trinomial because, after collecting like terms, it is written $8x + 2$. In addition, expressions like $10x^3 + 5x^{3/2} + 2x^{1/2} - 6$ and $4x^{-3} + 3x^{-2} + 7x$ are not polynomials because not all of the exponents are nonnegative integers.

Using the distributive property, this may be written

$$(2 + 3)x^3 + 5x^2 + (-6 + 8)x + 8$$

Thus, the sum of the two polynomials is

$$5x^3 + 5x^2 + 2x + 8$$

In general, once we understand that the basic properties of real numbers are being used, it is easier to take a shortcut by using the same format as for addition. We place the two polynomials to be added underneath each other, making sure to have the like degree terms match, and then we add.

**EXAMPLE 2** Add $3x^4 + 5x^2 - 16x + 7$ and $6x^5 + 3x^3 + 10x - 3$.

**Solution** We use the following format:

$$
\begin{array}{r}
3x^4 + \phantom{6x^5} \quad 5x^2 - 16x + 7 \\
6x^5 + \phantom{3x^4} \quad 3x^3 \phantom{5x^2} + 10x - 3 \\
\hline
6x^5 + 3x^4 + 3x^3 + 5x^2 - \phantom{1}6x + 4
\end{array}
$$

In displaying each polynomial, write the terms so that the degrees of the terms are decreasing. Further, if a term of a certain degree appears in one polynomial but is missing in the other, leave a space for the missing term so that terms of like degrees can be written underneath each other.

*Subtraction:* If we recall that by definition the difference between $a$ and $b$ is the sum of $a$ and the additive inverse of $b$, we can easily see that to subtract two polynomials we simply need to change the sign of every term of the polynomial to be subtracted and proceed as in addition.

**EXAMPLE 3** Subtract the polynomial $3x^4 + 2x^3 - 5x^2 + 6x - 13$ from $x^4 + 3x^3 + 6x^2 - 5$.

**Solution** Change the sign of every term of the polynomial to be subtracted and add the resulting polynomial to the other polynomial as follows:

$$
\begin{array}{r}
x^4 + 3x^3 + \phantom{1}6x^2 \phantom{+ 5x^2} - 5 \\
-3x^4 - 2x^3 + \phantom{1}5x^2 - 6x + 13 \\
\hline
-2x^4 + \phantom{1}x^3 + 11x^2 - 6x + \phantom{1}8
\end{array}
$$

Thus, the difference of the two polynomials is

$$-2x^4 + x^3 + 11x^2 - 6x + 8$$

**Do Exercise 5.**

*Multiplication:* Again, the basic properties of real numbers are used in the following example.

EXAMPLE 4    Multiply the polynomial $2x + 3$ by the polynomial $5x^2 + 3x + 2$.

Solution    Using the distributive property, we get

$$(2x + 3)(5x^2 + 3x + 2) = (2x + 3)5x^2 + (2x + 3)3x + (2x + 3)2$$
$$= (2x)(5x^2) + 3(5x^2) + (2x)(3x) + (3)(3x) + (2x)(2) + (3)(2)$$

Now using the associative and commutative properties and the laws of exponents, we get

$$10x^3 + 15x^2 + 6x^2 + 9x + 4x + 6$$

Therefore, the product is

$$10x^3 + 21x^2 + 13x + 6$$

As with multiplication, the work can be simplified using the following format:

$$
\begin{array}{r}
5x^2 + 3x + 2 \\
2x + 3 \\
\hline
15x^2 + 9x + 6 \\
10x^3 + 6x^2 + 4x \\
\hline
10x^3 + 21x^2 + 13x + 6
\end{array}
$$

Here is another example.

EXAMPLE 5    Multiply $2x^2 + 5x - 6$ by $-3x^3 + 6x - 4$.

Solution    We use the compact format as we did in the preceding example.

$$
\begin{array}{r}
2x^2 + 5x - 6 \\
-3x^3 + 6x - 4 \\
\hline
-8x^2 - 20x + 24 \\
12x^3 + 30x^2 - 36x \\
-6x^5 - 15x^4 + 18x^3 \\
\hline
-6x^5 - 15x^4 + 30x^3 + 22x^2 - 56x + 24
\end{array}
$$

Hence, the product of the two polynomials is

$$-6x^5 - 15x^4 + 30x^3 + 22x^2 - 56x + 24$$

**Do Exercise 19.**

*Division:* Recall that when we divide a positive integer $a$ (*the dividend*) by a positive integer $b$ (*the divisor*), we obtain a positive integer $q$ (*the quotient*) and a positive integer $r$ (*the remainder*) such that

$$0 \le r < b \quad \text{and} \quad a = bq + r$$

The last equality is often written

$$\frac{a}{b} = q + \frac{r}{b}$$

The fact that the numbers $q$ and $r$ exist and are unique is guaranteed by a theorem proved in more advanced books. Similarly, when we divide a polynomial $f(x)$ (the dividend) by a nonconstant polynomial $g(x)$ (the divisor), we get a polynomial $q(x)$ (the quotient) and a polynomial $r(x)$ (the remainder) such that $f(x) = g(x)q(x) + r(x)$ and the degree of $r(x)$ is less than the degree of $g(x)$, or $r(x) = 0$. We can also write

$$\frac{f(x)}{g(x)} = q(x) + \frac{r(x)}{g(x)}$$

It can be proved that as long as the divisor is not 0, the quotient polynomial $q(x)$ and the remainder polynomial $r(x)$ always exist and are uniquely determined. The division process is called the *division algorithm*. In spite of the heavy use of calculators, it is wise to remember how to perform long division of integers because the steps involved in the division algorithm of polynomials are the same as those for the division of integers.

**EXAMPLE 6**    Divide the polynomial $3x^3 + 5x^2 + 4x + 3$ by the polynomial $x^2 + 2x - 1$.

**Solution**    We write

$$
\begin{array}{r}
3x \text{ (first term of } q(x) \text{ since } \dfrac{3x^3}{x^2} = 3x) \\
\hline
g(x) = x^2 + 2x - 1 \overline{)3x^3 + 5x^2 + 4x + 3} \quad = f(x) \\
\underline{3x^3 + 6x^2 - 3x} \quad\quad = 3xg(x) \\
-x^2 + 7x + 3 \quad = f(x) - 3xg(x)
\end{array}
$$

The degree of this "first remainder" is not less than the degree of the divisor $g(x)$, so we continue. The second term of the quotient will be $-1$ since $-x^2/x^2 = -1$.

$$
\begin{array}{r}
3x - 1 \quad\quad = q(x) \\
\hline
g(x) = x^2 + 2x - 1 \overline{)3x^3 + 5x^2 + 4x + 3} = f(x) \\
\underline{3x^3 + 6x^2 - 3x} \quad = 3xg(x) \\
-x^2 + 7x + 3 = f(x) - 3xg(x) \\
\underline{-x^2 - 2x + 1} = (-1)g(x) \\
9x + 2 = f(x) - (3x - 1)g(x)
\end{array}
$$

The degree of the "second remainder" $9x + 2$ is 1, which is less than the degree of the divisor $g(x)$. Thus, it is the final remainder and we stop. The quotient is $3x - 1$ and the remainder is $9x + 2$. We may write

$$\frac{3x^3 + 5x^2 + 4x + 3}{x^2 + 2x - 1} = 3x - 1 + \frac{9x + 2}{x^2 + 2x - 1}$$

To check our work, we proceed as follows:

$$(x^2 + 2x - 1)(3x - 1) + (9x + 2) = (3x^3 + 5x^2 - 5x + 1) + (9x + 2)$$
$$= 3x^3 + 5x^2 + 4x + 3 = f(x)$$

Thus, our result is correct.
**Do Exercise 33.**    ■

The result is an *identity*. That is, it is true for all feasible replacements of the variable *x* (those replacements which do not yield a divisor equal to 0). Another way to check the correctness for our result is to replace *x* by an arbitrary number, say 2. Since $f(x) = 3x^3 + 5x^2 + 4x + 3$, $f(2) = 55$. Similarly, $g(x) = x^2 + 2x - 1$, thus $g(2) = 7$. Further, we have

$$q(x) = 3x - 1 \text{ and } r(x) = 9x + 2$$

which yield

$$q(2) = 5 \text{ and } r(2) = 20$$

These results yield $55 = (7)(5) + 20$, which is correct. Nevertheless, 20 is not the remainder that we would obtain if we divided 55 by 7, because 20 is larger than 7. This is because in working the division algorithm for polynomials, the *degree* of the remainder must be less than the degree of the divisor. This tells us nothing, however, about the values of these quantities when we replace the variable by a number. Also, checking the division process by substituting a number for the variable is not foolproof because two different polynomials may have the same value for some value of the variable. For example, if $f(x) = 2x + 3$ and $g(x) = 3x - 1$, then $f(x) \neq g(x)$ although $f(4) = g(4)$ since both $f(4)$ and $g(4)$ are equal to 11. With the help of a programmable calculator, the following procedure may easily be used as a foolproof way of checking the division. To verify that

$$f(x) = g(x)q(x) + r(x)$$

is an identity, where both sides of the equality are polynomials of degree *n*, replace *x* successively by $n + 1$ distinct real numbers. If a true statement results each time, then the equality is an identity.

**EXAMPLE 7** In "Getting Started," we found that if *n* is any positive integer, $x - a$ is a factor of $x^n - a^n$. Therefore, $x - 2$ is a factor of $x^4 - 16$. Find the other factor using division.

**Solution** The dividend is of degree 4, but has no terms of degrees 3, 2, or 1. Therefore, in arranging our work, we leave spaces for those missing terms for a reason which will be evident as we carry out the steps.

$$
\begin{array}{l}
\phantom{g(x) = x - 2\,)}\,x^3 + 2x^2 + 4x + \phantom{0}8 \qquad\qquad = q(x) \\
g(x) = x - 2\,\overline{)x^4 \phantom{+ 2x^2 + 4x + 0}\quad - 16} \quad = f(x) \\
\phantom{g(x) = x - 2\,)}\,\underline{x^4 - 2x^3} \qquad\qquad\qquad\quad = x^3 g(x) \\
\phantom{g(x) = x - 2\,)xxx}\,2x^3 \qquad\qquad\quad -16 = f(x) - x^3 g(x) \\
\phantom{g(x) = x - 2\,)xxx}\,\underline{2x^3 - 4x^2} \qquad\qquad\quad = 2x^2 g(x) \\
\phantom{g(x) = x - 2\,)xxxxxxx}\,4x^2 \qquad -16 = f(x) - (x^3 + 2x^2)g(x) \\
\phantom{g(x) = x - 2\,)xxxxxxx}\,\underline{4x^2 - 8x} \qquad\qquad = 4xg(x) \\
\phantom{g(x) = x - 2\,)xxxxxxxxxxx}\,8x - 16 = f(x) - (x^3 + 2x^2 + 4x)g(x) \\
\phantom{g(x) = x - 2\,)xxxxxxxxxxx}\,\underline{8x - 16 = 8g(x)} \\
\phantom{g(x) = x - 2\,)xxxxxxxxxxxxxx}\,0 = f(x) - (x^3 + 2x^2 + 4x + 8)g(x) \\
\phantom{g(x) = x - 2\,)xxxxxxxxxxxxxx}\,= f(x) - q(x)g(x) = r(x)
\end{array}
$$

As expected, the remainder is 0 since $x - 2$ is a factor of $x^4 - 16$. We can now write

$$x^4 - 16 = (x - 2)(x^3 + 2x^2 + 4x + 8)$$

You should leave out the documentation and write this division as follows.

$$
\begin{array}{r}
x^3 + 2x^2 + 4x + 8 \phantom{xxx} \\
x - 2 \overline{)x^4 \phantom{xxxxxxxxxxxxxx} - 16} \\
\underline{x^4 - 2x^3 \phantom{xxxxxxxxxx}} \\
2x^3 \phantom{xxxxxx} - 16 \\
\underline{2x^3 - 4x^2 \phantom{xxxx}} \\
4x^2 \phantom{xx} - 16 \\
\underline{4x^2 - 8x} \\
8x - 16 \\
\underline{8x - 16} \\
0
\end{array}
$$

∎

## Remainder Theorem

When we divide an arbitrary polynomial by a polynomial of degree 1, the remainder must be 0 or the degree of the remainder must be 0. That is, the remainder must be a constant. Thus, if $f(x)$ is an arbitrary polynomial and $a$ is an arbitrary real number, then there exists a unique polynomial $q(x)$ (the quotient) and a unique real number $r$ (the remainder) such that

$$f(x) = (x - a)q(x) + r$$

Since the foregoing equality is an identity, it must be true for all replacements of $x$ by real numbers. In particular, it must be true if we replace $x$ by $a$.

$$
\begin{aligned}
f(a) &= (a - a)q(a) + r \\
&= 0 + r = r
\end{aligned}
$$

Thus, the remainder is equal to the value of the polynomial $f(x)$ when $x = a$. This fact is usually stated as the *Remainder Theorem* in basic algebra books. At times, we are required to evaluate a certain polynomial $p(x)$ at some number $a$. If the degree of this polynomial is high, the calculations may be lengthy. It is more efficient to divide the polynomial by $x - a$ to get the remainder which is equal to $p(a)$. This method is especially effective using *synthetic division*. (See Section 0.5.)

**EXAMPLE 8**    Let $p(x) = 2x^3 + 5x^2 - 6x + 4$.

**a.** Evaluate $p(6)$.
**b.** Divide $p(x)$ by $x - 6$ and verify that the remainder is equal to $p(6)$.

**Solution**    **a.** $p(6) = 2(6^3) + 5(6^2) - 6(6) + 4$
$= 432 + 180 - 36 + 4 = 580$.
**b.** Using the long division format, we get

$$
\begin{array}{r}
2x^2 + 17x + \phantom{0}96 \\
x - 6\overline{)2x^3 + \phantom{0}5x^2 - 6x + \phantom{00}4} \\
\underline{2x^3 - 12x^2} \phantom{00000000000} \\
17x^2 - \phantom{0}6x + \phantom{00}4 \\
\underline{17x^2 - 102x} \phantom{0000} \\
96x + \phantom{00}4 \\
\underline{96x - 576} \\
580
\end{array}
$$

The remainder is 580, as is the value of $p(6)$ found in part (a).
**Do Exercise 37.**                                                                   ■

In the previous example, dividing $p(x)$ by $x - 6$ was not the most efficient way to evaluate $p(6)$. However, we shall see in the next section that there is a more effective method for dividing a polynomial $p(x)$ by $x - a$, and we shall rework Example 8 using that method.

# Exercise Set 0.4

*In Exercises 1–9, polynomials $f(x)$ and $g(x)$ are given. In each case, find their sum $f(x) + g(x)$ and their difference $f(x) - g(x)$.*

**1.** $f(x) = 2x^3 + 5x^2 - 10$; $g(x) = -3x^4 + 4x^3 + 10x - 7$

**2.** $f(x) = -4x^4 + 2x^2 + 7$;
$g(x) = 5x^4 + 6x^3 - 2x^2 + 7x + 1$

**3.** $f(x) = 7x^6 + 5x - 8$; $g(x) = 5x^3 + 5x - 13$

**4.** $f(x) = 6x^4 + 5x^3 - 6x^2 + 3x - 2$;
$g(x) = -6x^4 - 5x^3 + 7$

**5.** $f(x) = 4x^5 + 6x^2 + 2x - 5$;
$g(x) = -3x^4 + x^3 + 5x^2 + x + 5$

**6.** $f(x) = x^5 + x^4 - 3x^2 + 2$;
$g(x) = -x^5 + 3x^4 + x^3 + x^2 - 3$

**7.** $f(x) = x^4 + 3x - 6$; $g(x) = 2x^6 + x^4 + 7x^2 - 6x + 7$

**8.** $f(x) = \frac{1}{2}x^3 - \frac{1}{3}x^2 + \frac{1}{4}x + \frac{1}{5}$;

$g(x) = \frac{1}{3}x^3 + \frac{1}{4}x^2 - \frac{1}{5}x + \frac{1}{6}$

**9.** $f(x) = \frac{2}{3}x^3 + \frac{5}{6}x^2 - \frac{7}{8}x + \frac{12}{13}$;

$g(x) = \frac{3}{4}x^3 - \frac{5}{7}x^2 + \frac{8}{7}x - \frac{5}{4}$

**10.** Multiply $2x + 3$ by $5x - 2$.

**11.** Multiply $3x - 4$ by $4x + 5$.

**12.** Multiply $x + 3$ by $x + 3$.

**13.** Multiply $x - 2$ by $x + 2$.

**14.** Multiply $x - 1$ by $x^2 + x + 1$.

**15.** Multiply $x - 3$ by $x^2 + 6x - 4$.

**16.** Multiply $x - 2$ by $x^2 + 2x + 4$.

**17.** Multiply $2x + 3$ by $3x^2 - 5x + 7$.

**18.** Multiply $3x - 7$ by $2x^3 + 5x^2 - 7x + 3$.

**19.** Find the product $f(x)g(x)$ where $f(x)$ and $g(x)$ are the polynomials of Exercise 1.

**20.** Find the product $f(x)g(x)$ where $f(x)$ and $g(x)$ are the polynomials of Exercise 2.

**21.** Find the product $f(x)g(x)$ where $f(x)$ and $g(x)$ are the polynomials of Exercise 3.

**22.** Find the product $f(x)g(x)$ where $f(x)$ and $g(x)$ are the polynomials of Exercise 4.

**23.** Find the product $f(x)g(x)$ where $f(x)$ and $g(x)$ are the polynomials of Exercise 5.

**24.** Find the product $f(x)g(x)$ where $f(x)$ and $g(x)$ are the polynomials of Exercise 6.

**25.** Find the product $f(x)g(x)$ where $f(x)$ and $g(x)$ are the polynomials of Exercise 7.

**26.** Find the product $f(x)g(x)$ where $f(x)$ and $g(x)$ are the polynomials of Exercise 8.

**27.** Divide $3x^2 + 5x - 6$ by $x - 3$.

**28.** Divide $5x^2 + 7x - 13$ by $x + 2$.

**29.** Divide $x^2 - 4$ by $x - 2$.

**30.** Divide $x^3 + 3x^2 + 3x + 1$ by $x + 1$.

**31.** Divide $5x^4 - 3x^3 + 2x^2 - 5x + 11$ by $x - 5$.

**32.** Divide $3x^5 + 2x^3 - 5x + 6$ by $x + 3$.

**33.** Divide $x^4 - 3x^2 + 6$ by $x^2 + 3x - 2$.

**34.** Divide $2x^5 + 3x^4 - 2x^2 + 7$ by $x^2 + 5x + 2$.

**35.** Divide $3x^6 + 4x^4 - 5x^2 + 6x - 7$ by $x^2 - 4x + 2$.

*In Exercises 36–40, a polynomial p(x) and a number a are given.
In each case,*

*a. Evaluate p(a).*

*b. Divide p(x) by x − a and verify that the remainder is equal
to p(a).*

**36.** $p(x) = 3x^4 + 2x^3 + 3x^2 - 6x + 2;\ a = 3$

**37.** $p(x) = x^5 - 4x^3 + 3x^2 - 7x + 13;\ a = 5$

**38.** $p(x) = 5x^6 - 7x^3 + 8x - 6;\ a = -2$

**39.** $p(x) = 3x^7 + 5x^6 - 4x^5 + 4x^3 - 6x^2 + 5x - 7;\ a = -3$

**40.** $p(x) = x^{10} - 4;\ a = 3$

## 0.5  Synthetic Division (Optional)

In this section we present a more efficient way to find the quotient $q(x)$ and the remainder $r$ when an arbitrary polynomial $f(x)$ is divided by a linear binomial of the form $x - a$.

Let

$$f(x) = a_0x^n + a_1x^{n-1} + a_2x^{n-2} + \cdots + a_{n-1}x + a_n$$

The quotient will be of degree $n - 1$, since the divisor is of degree 1. Suppose that

$$q(x) = b_0x^{n-1} + b_1x^{n-2} + b_2x^{n-3} + \cdots + b_{n-2}x + b_{n-1}$$

By the remainder theorem, we know that the remainder $r$ is equal to $f(a)$. Since we have

$$f(x) = (x - a)q(x) + r$$

we may write

$$a_0x^n + a_1x^{n-1} + a_2x^{n-2} + \cdots + a_{n-1}x + a_n$$
$$= (x - a)(b_0x^{n-1} + b_1x^{n-2} + b_2x^{n-3} + \cdots + b_{n-2}x + b_{n-1}) + r$$
$$= b_0x^n + (b_1 - ab_0)x^{n-1} + (b_2 - ab_1)x^{n-2} + \cdots +$$
$$(b_{n-1} - ab_{n-2})x - ab_{n-1} + r$$

The preceding equality is an identity, so corresponding coefficients must be equal and we have

$$a_0 = b_0,\ a_1 = b_1 - ab_0,\ a_2 = b_2 - ab_1,\ \ldots,\ a_{n-1} = b_{n-1} - ab_{n-2},$$

and

$$a_n = r - ab_{n-1}$$

Thus,

$$b_0 = a_0,\ b_1 = a_1 + ab_0,\ b_2 = a_2 + ab_1,\ \ldots,\ b_{n-1} = a_{n-1} + ab_{n-2},$$

and

$$r = a_n + ab_{n-1}$$

The successive coefficient $b_i$'s and the remainder $r$ may be computed rapidly, using the following format:

$$
\begin{array}{c|ccccccc}
a & a_0 & a_1 & a_2 & a_3 & \cdots & a_{n-1} & a_n \\
  &     & ab_0 & ab_1 & ab_2 & \cdots & ab_{n-2} & ab_{n-1} \\
\hline
  & b_0 & b_1 & b_2 & b_3 & \cdots & b_{n-1} & r
\end{array}
$$

We have written the number $a$ on the first line, then the coefficients of $f(x)$. In the first space of the third line, below $a_0$, we write $b_0$, which is equal to $a_0$. Multiply $a$ by $b_0$ and enter the product in the first space of the second line, below $a_1$. Add $a_1$ and $ab_0$ to get $b_1$, which is written in the second space of the third line. Multiply $b_1$ by $a$ and write the product on the second line, below $a_2$. Add $a_2$ and $ab_1$ to get $b_2$, which is written in the third space of the third line. Continue in the same manner until the last number of the third line, below $a_n$, is obtained. This last number is $r$, the remainder. The other numbers are the coefficients of $q(x)$ written in order. The method which we just described is called *synthetic division*.

**EXAMPLE 1**   Divide the polynomial $2x^3 + 4x^5 + 4x - 3$ by $x - 2$.

**Solution**   The polynomial should be written in the form $4x^5 + 2x^3 + 4x - 3$. Since the terms of degree 4 and 2 are missing, the coefficients in order are 4, 0, 2, 0, 4, and $-3$. Using the method described above, we have

$$
\begin{array}{c|cccccc}
2 & 4 & 0 & 2 & 0 & 4 & -3 \\
  &   & 8 & 16 & 36 & 72 & 152 \\
\hline
  & 4 & 8 & 18 & 36 & 76 & 149
\end{array}
$$

Thus, the remainder is 149 and the coefficients of the quotient, written in order, are 4, 8, 18, 36, and 76. Therefore, the quotient is $4x^4 + 8x^3 + 18x^2 + 36x + 76$. Note that the first number in the second line is 8, since $2 \cdot 4 = 8$; the second number is 16, since $2 \cdot 8 = 16$; and so on.
**Do Exercise 7.**   ■

**EXAMPLE 2**   Divide $3y^4 + 5y^3 - 3y + 7$ by $y + 3$.

**Solution**   We first write $y + 3 = y - (-3)$. Thus, in using synthetic division, $a = -3$. We now proceed as in the previous example.

$$
\begin{array}{c|ccccc}
-3 & 3 & 5 & 0 & -3 & 7 \\
   &   & -9 & 12 & -36 & 117 \\
\hline
   & 3 & -4 & 12 & -39 & 124
\end{array}
$$

Thus, the quotient is $3y^3 - 4y^2 + 12y - 39$ and the remainder is 124.
**Do Exercise 5.**   ■

**EXAMPLE 3**   In Example 8 of the preceding section, you were given $p(x) = 2x^3 + 5x^2 - 6x + 4$ and you were asked to evaluate $p(6)$. Then, you were asked to divide $p(x)$ by $x - 6$ and to compare the remainder to the value of $p(6)$. Now, find the value of $p(6)$ using synthetic division.

**Solution**   Use the same process as in Example 2.

$$
\begin{array}{r|rrrr}
6 & 2 & 5 & -6 & 4 \\
  &   & 12 & 102 & 576 \\
\hline
  & 2 & 17 & 96 & 580
\end{array}
$$

Since the last number of the third line is 580, the value of $p(6)$ is 580.
**Do Exercise 15.**

**EXAMPLE 4**   If $p(x) = 3x^7 - 4x^5 + 6x^4 - 7x^2 + 5x + 17$, evaluate $p(-3)$.

**Solution**   You simply need to find the remainder of the division of $p(x)$ by $x + 3$, since $x - (-3) = x + 3$. Using synthetic division, write

$$
\begin{array}{r|rrrrrrrr}
-3 & 3 & 0 & -4 & 6 & 0 & -7 & 5 & 17 \\
   &   & -9 & 27 & -69 & 189 & -567 & 1722 & -5181 \\
\hline
   & 3 & -9 & 23 & -63 & 189 & -574 & 1727 & -5164
\end{array}
$$

Therefore, $p(-3) = -5164$.
**Do Exercise 19.**

If $f(x)$, $g(x)$, and $h(x)$ are polynomials and $f(x) = g(x)h(x)$, then we say that $g(x)$ and $h(x)$ are *factors* of $f(x)$.

**EXAMPLE 5**   Using synthetic division, verify that $x - 2$ is a factor of $5x^4 + 2x^2 - 45x + 2$.

**Solution**   You need only verify that when $5x^4 + 2x^2 - 45x + 2$ is divided by $x - 2$, the remainder is 0. Using synthetic division, you will obtain

$$
\begin{array}{r|rrrrr}
2 & 5 & 0 & 2 & -45 & 2 \\
  &   & 10 & 20 & 44 & -2 \\
\hline
  & 5 & 10 & 22 & -1 & 0
\end{array}
$$

The last number of the third row is 0. Hence, when the polynomial $5x^4 + 2x^2 - 45x + 2$ is divided by $x - 2$ the remainder is 0. We conclude that $x - 2$ is a factor of $5x^4 + 2x^2 - 45x + 2$.
**Do Exercise 25.**

# Exercise Set 0.5

*Use synthetic division to do the following:*

**1.** Divide $5x^2 + 5x - 7$ by $x - 3$.

**2.** Divide $3x^2 + 7x + 13$ by $x - 5$.

**3.** Divide $4x^3 + 5x^2 - 6x + 7$ by $x - 2$.

**4.** Divide $5x^3 - 4x^2 - 8x - 10$ by $x - 4$.

**5.** Divide $3x^3 + 7x^2 - 4x + 8$ by $x + 2$.

**6.** Divide $-2x^3 + 4x^2 - 5x + 14$ by $x + 3$.

**7.** Divide $4x^5 + 5x^3 - 5x^2 + 3x - 2$ by $x - 6$.

**8.** Divide $3x^5 + 4x^4 - 6x^2 + 2x - 6$ by $x + 5$.

**9.** Divide $6x^7 + 5x^6 - 3x^5 + 6x^4 - 8x^3 + 5x^2 - 6x + 1$ by $x - 4$.

**10.** Divide $5x^8 - 6x^4 + 5x^2 + 10x - 17$ by $x + 3$.

**11.** Divide $6x^9 - 8x^7 + 4x^5 - 5x^3 + 6x - 5$ by $x + 5$.

**12.** Divide $3x^{10} + 4x^6 + 4x^2 - 4x + 10$ by $x - 7$.

13. Divide $7x^{11} - 7x^7 + 6x^3 - 5x^2 - 4x + 3$ by $x + 4$.

14. If $p(x) = 4x^3 + 5x^2 - 6x + 7$, evaluate $p(4)$.

15. If $p(x) = 5x^3 - 4x^2 - 8x - 10$, evaluate $p(3)$.

16. If $p(x) = 3x^3 + 7x^2 - 4x + 8$, evaluate $p(5)$.

17. If $p(x) = -2x^3 + 4x^2 - 5x + 14$, evaluate $p(2)$.

18. If $p(x) = 4x^5 + 5x^3 - 5x^2 + 3x - 2$, evaluate $p(6)$.

19. If $p(x) = x^5 + 4x^4 - 6x^2 + 2x - 6$, evaluate $p(-3)$.

20. If $p(x) = 6x^7 + 5x^6 - 3x^5 + 6x^4 - 8x^3 + 5x^2 - 6x + 1$, evaluate $p(-1)$.

21. If $p(x) = 5x^8 - 6x^4 + 5x^2 + 10x - 17$, evaluate $p(-2)$.

22. If $p(x) = 6x^9 - 8x^7 + 4x^5 - 5x^3 + 6x - 5$, evaluate $p(-3)$.

23. If $p(x) = x^{10} + 4x^6 + 4x^2 - 4x + 10$, evaluate $p(4)$.

24. If $p(x) = 7x^{11} - 7x^7 + 6x^3 - 5x^2 - 4x + 3$, evaluate $p(-2)$.

25. Verify that $x - 1$ is a factor of $5x^3 + 3x^2 + 7x - 15$.

26. Verify that $x - 2$ is a factor of $x^4 + 5x^2 - 10x - 16$.

27. Verify that $x + 1$ is a factor of $3x^4 + 5x^2 - 6x - 14$.

28. Verify that $x + 2$ is a factor of $5x^4 + 3x^3 - 5x^2 + 6x - 24$.

## 0.6 Factoring

The *prime* factorization of a positive integer is a factorization of that integer where all of its factors are *prime numbers*. In this case, the factorization is said to be complete. For example, if we write $105 = 3 \cdot 5 \cdot 7$, the factorization of 105 is complete since 3, 5, and 7 are prime numbers. Note that 105 can be written as a product of smaller factors if the factors are not required to be integers. For example, $105 = 2 \cdot \left(\frac{3}{2}\right) \cdot 5 \cdot 7$.

Similarly, to *factor a polynomial* means to write it as a product of polynomials and the factorization is *complete* if each factor cannot be factored further. Recall that $3 \cdot 5 \cdot 7$ can be factored further if the factors are nonintegers. Similarly with polynomials, we must specify the types of coefficients that may be used for the factor polynomials. For example, we can write

$$x^4 - 25 = (x^2 - 5)(x^2 + 5) \tag{1}$$

This factorization is complete if the coefficients are integers and rational numbers. However, if the coefficients are allowed to be irrational numbers, then we can write

$$x^4 - 25 = (x - \sqrt{5})(x + \sqrt{5})(x^2 + 5) \tag{2}$$

This factorization is complete if the coefficients are real numbers. But, if we use complex numbers as coefficients, it can be shown that

$$x^4 - 25 = (x - \sqrt{5})(x + \sqrt{5})(x - \sqrt{5}i)(x + \sqrt{5}i) \tag{3}$$

In general, in any factorization problem, we must be told the type of numbers that may be used as coefficients for the polynomial factors. If $f(x)$ is a polynomial, *to factor $f(x)$ over the integers (real, complex numbers)*, means to write $f(x) = f_1(x) \cdot f_2(x) \cdot \cdots \cdot f_n(x)$ where all factors $f_i(x)$ are polynomials with integer (real, complex number) coefficients. For example, (1) is a complete factorization of $x^4 - 25$ over the integers, but it is not complete over the real numbers. On the other hand, (2) is the complete factorization of $x^4 - 25$ over the real numbers, while it is not complete over the complex numbers.

## Factoring Using Formulas

To be able to factor polynomials effectively, you need some expertise in multiplying polynomials. In fact, the first step in learning factorization should be to review the basic multiplication formulas learned in previous algebra courses. We need only note that a multiplication formula read from right to left is a factorization formula.

$$x(y + z) = xy + xz \tag{1}$$
$$(x + a)(x + b) = x^2 + (a + b)x + ab \tag{2}$$
$$(x - a)(x + a) = x^2 - a^2 \tag{3}$$
$$(x + a)^2 = x^2 + 2ax + a^2 \tag{4}$$
$$(x - a)^2 = x^2 - 2ax + a^2 \tag{5}$$
$$(x + a)^3 = x^3 + 3ax^2 + 3a^2x + a^3 \tag{6}$$
$$(x - a)^3 = x^3 - 3ax^2 + 3a^2x - a^3 \tag{7}$$
$$(x - a)(x^2 + ax + a^2) = x^3 - a^3 \tag{8}$$
$$(x + a)(x^2 - ax + a^2) = x^3 + a^3 \tag{9}$$

"Getting Started" contains ways to help you memorize and generalize these formulas.

**EXAMPLE 1**   Factor $2x^3 - 50x$ completely, over the integers.

**Solution**   In any factorization problem, if a common factor appears in each term of the given polynomial, that factor should be taken out first, using Formula (1). We write

$$2x^3 - 50x = 2x(x^2 - 25)$$

The second factor on the right is a difference of two squares. Thus, we use Formula (3) and write

$$2x^3 - 50x = 2x(x - 5)(x + 5)$$

**Do Exercise 13.**                                                                          ■

**EXAMPLE 2**   Factor $x^2 - 6x + 5$ over the integers.

**Solution**   Formula (2) may be written

$$x^2 + (a + b)x + ab = (x + a)(x + b)$$

Comparing the given polynomial to the left side of this equality, we conclude that $ab = 5$ and $a + b = -6$. Now, the only possible pairs of integers whose product is equal to 5 are $a = 1$, $b = 5$ (which yields $a + b = 6$), and $a = -1$, $b = -5$ (which yields $a + b = -6$).
   Since we must have $a + b = -6$, the correct choice is

$$x^2 - 6x + 5 = (x - 1)(x - 5)$$                                                    ■

**EXAMPLE 3**   Factor $4x^2 - 20xy + 25y^2$ completely.

**Solution**   The first and third terms are perfect squares and the middle term is negative. Hence, we attempt to use Formula (5). Since $4x^2 = (2x)^2$ and $25y^2 = (5y)^2$, we need only check that the middle term of the given expression is twice the product $2x \cdot 5y$. In fact, we have $20xy = 2(2x)(5y)$. Using Formula (5), we obtain

$$4x^2 - 20xy + 25y^2 = (2x - 5y)^2$$

∎

**Do Exercise 7.**

## Factoring Using the Factor Theorem

The factorization in the first three examples was straightforward because the given polynomials were simple and we could use the basic formulas. In general, factorization of polynomials is done much like factorization of integers, by trial and error. For example, suppose we were asked to write 3960 as a product of prime factors. Since the unit digit is 0, 10 is a factor and we first write $3960 = 396 \cdot 10$. Since 396 and 10 are both divisible by 2, we get $3960 = 198 \cdot 2 \cdot 5 \cdot 2$. However, noting that 198 is also divisible by 2, we can also write $3960 = 2 \cdot 99 \cdot 2 \cdot 5 \cdot 2$. Since $99 = 11 \cdot 9 = 11 \cdot 3 \cdot 3$, we finally write $3960 = 2^3 \cdot 3^2 \cdot 5 \cdot 11$, which is the prime factorization of 3960.

**A CLOSER LOOK**   In trying to find factors of positive integers, we often make use of the following rules.

1. If the unit digit of a positive integer is divisible by 2, the integer itself is divisible by 2.
2. If the sum of the digits of a positive integer is divisible by 3, the integer itself is divisible by 3.
3. If the unit digit of a positive integer is 0 or 5, the integer is divisible by 5.
4. If the sum of the digits of a positive integer is divisible by 9, the integer itself is divisible by 9.
5. If the sum of every other digit of a positive integer minus the sum of the remaining digits is divisible by 11, then the initial positive integer is divisible by 11.

For example, consider 318,321,938. Since $(3 + 8 + 2 + 9 + 8) - (1 + 3 + 1 + 3) = 30 - 8 = 22 = 11 \cdot 2$, the given integer is divisible by 11. With the help of a calculator, we find that $318,321,938 = 11 \cdot 28,938,358$.

Because of calculators, the rules to check the divisibility of positive integers are seldom discussed in modern mathematics books. We stated them here so that you can appreciate the fact that it is convenient to be able to tell the existence of a factor *before* attempting the factorization. The case for polynomials is similar.

Suppose we are asked to factor a certain polynomial. It would be helpful to be able to determine whether another polynomial is a factor. It is easy to check whether a polynomial of the form $x - a$ is a factor. In fact, if $x - a$ is a factor of the polynomial $f(x)$, then there exists a polynomial $g(x)$ such that $f(x) = (x - a)g(x)$ is an identity. Thus, if we replace $x$ by $a$ in this equality, we get the following true statement: $f(a) = (a - a)g(a) = 0 \cdot g(a) = 0$. This proves one half of the Factor Theorem. For the proof of the other half of the theorem, see Exercise *46.

> **FACTOR THEOREM 0.1:** The linear polynomial $x - a$ is a factor of the polynomial $f(x)$ if, and only if, $f(a) = 0$.

**EXAMPLE 4**    Factor the polynomial $2x^3 + 5x^2 - x - 6$ completely, over the integers.

**Solution**    We need to determine whether this polynomial has some factor in the form $x - a$. Because the constant term in the given polynomial is $-6$, we suspect that the possible values of $a$ are $-1$, $1$, $-2$, $2$, $-3$, $3$, $-6$, and $6$. If we let $f(x)$ denote the given polynomial, we get $f(-1) = 2(-1)^3 + 5(-1)^2 - (-1) - 6 = -2$. Since $f(-1) \neq 0$, we conclude that $(x - (-1))$—that is, $(x + 1)$—is not a factor. Now we calculate the value of $f(1)$. We easily find that $f(1) = 0$. Thus, $x - 1$ is a factor. To find the other factor, we divide $f(x)$ by $x - 1$, which can be done using synthetic division or long division.

$$
\begin{array}{r|rrrr}
1 & 2 & 5 & -1 & -6 \\
  &   & 2 & 7 & 6 \\
\hline
  & 2 & 7 & 6 & 0 \\
\end{array}
$$

We have

$$2x^3 + 5x^2 - x - 6 = (x - 1)(2x^2 + 7x + 6)$$

Let $g(x) = 2x^2 + 7x + 6$ and suppose that $x - b$ is a factor of $g(x)$ with $b$ as an integer. The fact that the constant term of $g(x)$ is 6 indicates that the possible values of $b$ are the divisors of 6. We quickly find that $g(1)$ and $g(2)$ are not equal to 0. We do not need to check $g(-1)$ since we already know that $(x + 1)$ is not a factor. However, $g(-2) = 2(-2)^2 + 7(-2) + 6 = 0$. We conclude that $x - (-2)$—that is, $x + 2$—is a factor of $g(x)$. Thus, we write

$$2x^2 + 7x + 6 = (x + 2)(? + ?)$$

Since the first term of $g(x)$ is $2x^2$ and the first term of $x + 2$ is $x$, we replace the first question mark with $2x$. Similarly, we replace the second question mark with 3 since $\frac{6}{2} = 3$. We can now write

$$2x^3 + 5x^2 - x - 6 = (x - 1)(x + 2)(2x + 3)$$

This identity can easily be verified by performing the multiplications on the right side. **Do Exercise 33.**    ■

## Factoring Using Other Methods

**EXAMPLE 5**    Factor $x^4 + x^2y^2 + y^4$ completely, over the integers.

**Solution**    Clearly, $x^4 = (x^2)^2$ and $y^4 = (y^2)^2$. However, we cannot use Formula (4) because the middle term $x^2y^2$ is not twice the product of $x^2$ and $y^2$. Here, we can use a method that is very common in life: when we lack something which we need, we simply borrow it, then give it back. Similarly, in this factorization problem we add $x^2y^2$ in order to get $2x^2y^2$ and then subtract $x^2y^2$. We write

$$x^4 + x^2y^2 + y^4 = (x^4 + 2x^2y^2 + y^4) - x^2y^2$$
$$= (x^2 + y^2)^2 - (xy)^2 \qquad \text{[by Formula (4)]}$$
$$= [(x^2 + y^2) - xy][(x^2 + y^2) + xy] \qquad \text{[by Formula (3)]}$$
$$= (x^2 - xy + y^2)(x^2 + xy + y^2)$$

This factorization over the integers is complete.
**Do Exercise 37.**                                   ■

In attempting to factor an algebraic expression, it is often useful to group together terms which have common monomial factors. This method is often called *factoring by grouping*.

**EXAMPLE 6**    Factor the expression $2x^2 - 6yz + 3xz - 4xy$.

**Solution**    We rearrange the terms as follows:

$$2x^2 - 6yz + 3xz - 4xy = (2x^2 - 4xy) + (3xz - 6yz)$$
$$= 2x(x - 2y) + 3z(x - 2y)$$

Since $(x - 2y)$ is a common factor, we get

$$2x^2 - 6yz + 3xz - 4xy = (x - 2y)(2x + 3z)$$

**Do Exercise 39.**                                   ■

**EXAMPLE 7**    Factor $x^3 - 3x^2 + 2$ completely, over the real numbers.

**Solution**    Let $f(x) = x^3 - 3x^2 + 2$. Then, $f(1) = 0$. So, $x - 1$ is a factor of $f(x)$. To find the other factor, we divide $f(x)$ by $(x - 1)$ and obtain $x^2 - 2x - 2$. It follows that

$$x^3 - 3x^2 + 2 = (x - 1)(x^2 - 2x - 2)$$

If we let $g(x) = x^2 - 2x - 2$, $x - b$ will be a factor of $g(x)$ if and only if $g(b) = 0$. Hence, the values of $b$ may be found by solving the equation

$$x^2 - 2x - 2 = 0$$

Recall that if $ax^2 + bx + c = 0$, where $a$, $b$, and $c$ are constants and $a \neq 0$, then

$$x = \frac{-b \pm \sqrt{b^2 - 4ac}}{2a}$$

This formula is known as the *quadratic formula* and can be used to solve the equation $x^2 - 2x - 2 = 0$ to get

$$x = \frac{-(-2) \pm \sqrt{(-2)^2 - 4(1)(-2)}}{2(1)} = \frac{2 \pm \sqrt{12}}{2} = \frac{2 \pm 2\sqrt{3}}{2} = 1 \pm \sqrt{3}$$

Thus, $(x - (1 - \sqrt{3}))$ and $(x - (1 + \sqrt{3}))$ are factors of $g(x)$. We now may write

$$x^3 - 3x^2 + 2 = (x - 1)(x - 1 + \sqrt{3})(x - 1 - \sqrt{3})$$

**Do Exercise 33.**                                   ■

## Exercise Set 0.6

*In Exercises 1–5, check whether the given integer is divisible by (a) 2, (b) 3, (c) 5, (d) 9, and (e) 11. Then, write the prime factorization of that integer.*

**1.** 540

**2.** 43,560

**3.** 178,200

**4.** 16,335

**5.** 303,750

*In Exercises 6–25, factor the given polynomials completely, over the integers.*

**6.** $x^2 + 12x + 36$

**7.** $x^2 - 14x + 49$

**8.** $x^2 + 2x - 15$

**9.** $x^2 - x - 6$

**10.** $x^2 + 3x - 28$

**11.** $x^2 - 2x - 35$

**12.** $x^2 + 6x - 55$

**13.** $x^2 - 4$

**14.** $2x^2 - 5x - 3$

**15.** $2x^2 - 5x - 12$

**16.** $3x^2 - 13x - 10$

**17.** $3x^2 + 14x + 15$

**18.** $x^3 - 2x^2 - 5x + 6$

**19.** $x^3 + 4x^2 - 7x - 10$

**20.** $x^3 - 6x^2 - x + 30$

**21.** $x^3 - 2x^2 - 29x - 42$

**22.** $2x^3 + 9x^2 - 6x - 5$

**23.** $3x^3 + 13x^2 + 13x + 3$

**24.** $x^4 + x^3 - 7x^2 - x + 6$

**25.** $x^4 - 5x^3 - 7x^2 + 29x + 30$

*In Exercises 26–38, factor the given polynomials completely, over the real numbers.*

**26.** $x^2 - 7$

**27.** $x^2 - 3$

**28.** $x^2 - 2x - 6$

**29.** $x^2 - 4x + 1$

**30.** $x^2 + 2x - 1$

**31.** $x^4 - 9$

**32.** $x^3 - 4x^2 + 2x + 4$

**33.** $x^3 - 3x^2 - 5x - 1$

**34.** $x^3 - x^2 - 7x + 3$

**35.** $x^3 - 7x^2 + 6x + 20$

**36.** $x^3 - x^2 - 11x + 3$

**37.** $x^4 + 4x^2 + 16$

**38.** $x^4 + 5x^2 + 25$

*In Exercises 39–45, factor the given expressions by grouping.*

**39.** $xz + 6wy + 3wx + 2yz$

**40.** $xy - 2ab - ay + 2bx$

**41.** $2xz - 2wy + yz - 4wx$

**42.** $6xy + 6ab + 4ay + 9bx$

**43.** $6wx - 6yz - 9wz + 4xy$

**44.** $x^2 - 4xy + 4y^2 - 9z^2$

**45.** $x^3 + x^2 - y^3 - y^2$

**\*46.** Prove that if $f(x)$ is a polynomial and $f(a) = 0$, then $x - a$ is a factor of $f(x)$. (*Hint:* Use the Remainder Theorem stated in Section 0.4.)

# 0.7  Equations in One Variable

## Equivalent Equations

**DEFINITION 0.14:**    An open sentence in a variable that is a statement of equality is called an *equation in one variable*.

For example,

$$2x + 9 = 13x - 24$$

and

$$y^2 - 5y + 6 = 0$$

are equations in one variable. In general, with each equation in one variable we must have a replacement set. That is, we must have a set of numbers to use as replacements for the variable. The replacement set for a variable is usually not described. In this book, however, unless otherwise specified, it is the *largest* set of real numbers that may be used as replacements for the variable so that each replacement yields quantities that are both defined and real. For example, the replacement set for the equation

$$\frac{3}{x - 2} + \frac{x + 4}{2x + 1} = 4$$

is the set of all real numbers except 2 and $\frac{-1}{2}$, since replacing $x$ by either of these two numbers would yield a division by 0, which is not defined. As another example, the replacement set for the equation

$$\sqrt{x - 3} = 2x - 21$$

is the set of all real numbers greater than or equal to 3, because replacing $x$ by a number less than 3 would yield a negative quantity under the radical sign and square roots of negative numbers are not real numbers.

> **DEFINITION 0.15:** A *solution* of an equation is a member of the replacement set which yields a true statement when it is used as a replacement for the variable.

For example, the number 3 is a solution of the equation

$$2x + 9 = 13x - 24$$

since

$$2(3) + 9 = 13(3) - 24$$

is true. Also, $-2$ and 5 are solutions of the equation

$$y^2 - 3y - 10 = 0$$

since

$$(-2)^2 - 3(-2) - 10 = 0 \text{ and } 5^2 - 3(5) - 10 = 0$$

are both true.

The set of all solutions of an equation is called the *solution set* of the equation. It is important to note that the solution set of an equation depends on the replacement set. For example, the solution set of the equation

$$x^2 + 1 = 0$$

is empty if the replacement set is the set of real numbers, but has two members if the replacement set is the set of complex numbers.

> **DEFINITION 0.16:** If the solution set of an equation is equal to its replacement set, the equation is called an *identity*; otherwise, it is called a *conditional equation*.

For example,

$$(x + 1)^2 = x^2 + 2x + 1$$

is an identity, but

$$3x + 2 = 5x - 12$$

is a conditional equation.

> **DEFINITION 0.17:**   Two equations whose replacement sets are the same are said to be *equivalent* if, and only if, their solution sets are equal. The process of finding the solution set of an equation is 'called *solving the equation*.

In general, when given an equation, we attempt to form a sequence of equivalent equations until we get an equation which is simple enough so that its solution set is found by inspection. Since the last equation is equivalent to the original one, its solution set is also the solution set of the original equation. For example, the following is a sequence of equivalent equations,

$$5x - 11 = 3x - 3$$
$$2x - 11 = -3$$
$$2x = 8$$
$$x = 4$$

Obviously, {4} is the solution set of the last equation. Therefore, {4} is also the solution set of the equation

$$5x - 11 = 3x - 3$$

> If an equation is given, an equivalent equation will be found by doing any one of the following:
>
> 1. Add the same polynomial to both sides.
> 2. Subtract the same polynomial from both sides.
> 3. Multiply both sides by the same nonzero number.
> 4. Divide both sides by the same nonzero number.

Be aware that although any polynomial may be added to, or subtracted from, both sides of an equation, you should not multiply or divide both sides of an equation by a quantity involving the variable. Consider, for example, the following equation:

$$(x - 1)(x - 3) = 0$$

It is obvious that its solution set is {1, 3}. If we divide both sides by $(x - 1)$, we obtain the equation

$$x - 3 = 0$$

with solution set {3}. We have lost one solution. The new equation is *defective* relative to the previous one. Now, if we multiply both sides of the equation

$$(x - 1)(x - 3) = 0$$

by $(x - 7)$, we obtain the equation

$$(x - 1)(x - 3)(x - 7) = 0$$

whose solution set is {1, 3, 7}. We have gained one solution. The new equation is *redundant* with respect to the previous one.

## Linear Equations in One Variable

An equation that can be written in the form

$$ax + b = cx + d$$

where $a$, $b$, $c$, and $d$ are real numbers and either $a$ or $c$ does not equal 0, is called a *linear equation in x*. This type of equation is easy to solve as is shown in the next two examples.

**EXAMPLE 1**   Solve the equation $3x + 5 = 7x - 11$.

**Solution**   Subtract 5 from both sides of the equation.

$$3x = 7x - 16$$

Now subtract $7x$ from both sides of the new equation.

$$-4x = -16$$

Dividing both sides of this last equation by $-4$ yields

$$x = 4$$

The solution set of the last equation is obviously {4}. Thus, the solution set of the equation $3x + 5 = 7x - 11$ is also {4}.
**Do Exercise 7.**                                    ■

**EXAMPLE 2**   Solve the equation

$$\frac{x + 12}{2} = \frac{2x + 10}{3} + 3$$

**Solution**   First multiply both sides of the equation by 6 in order to eliminate fractions, obtaining

$$3x + 36 = (4x + 20) + 18$$

Adding $-4x - 36$ to both sides we get

$$-x = 2$$

Multiplying both sides of the last equation by $-1$ yields

$$x = -2$$

The solution set of the last equation, and of the original equation, is {−2}.
**Do Exercise 13.**                                    ■

## Quadratic Equations in One Variable

If $a$, $b$, and $c$ are real numbers with $a \neq 0$, an equation that can be written in the form

$$ax^2 + bx + c = 0$$

is called a *quadratic equation*. As we pointed out in the preceding section, you have undoubtedly learned to solve such equations, as well as linear equations, in previous courses. We shall, however, give several examples to serve as a quick review of the available techniques.

**EXAMPLE 3**    Solve the equation

$$2x^2 + 5x - 18 = 0$$

by factoring.

**Solution**    The left side of the equation may be factored as follows:

$$(2x + 9)(x - 2) = 0$$

Using the fact that a product of two real numbers is equal to 0 if, and only if, at least one of the two numbers is equal to 0, we get

$$2x + 9 = 0 \quad \text{or} \quad x - 2 = 0$$

The solution set of the foregoing open sentence is easily seen to be $\left\{\frac{-9}{2}, 2\right\}$. Hence, the solution set of the original quadratic equation is also $\left\{\frac{-9}{2}, 2\right\}$.
**Do Exercise 15.**                                                                                          ■

**EXAMPLE 4**    Solve the quadratic equation

$$3x^2 - 12x + 9 = 0$$

by completing squares.

**Solution**    Dividing both sides by 3 to change the leading coefficient to 1, we get

$$x^2 - 4x + 3 = 0$$

We then subtract 3 from both sides

$$x^2 - 4x = -3$$

The square of half the coefficient of $x$ is 4, since $4 = \left(\frac{-4}{2}\right)^2$. Thus, we add 4 to both sides, so that the left side will be a perfect square. (See Exercise 56.) We obtain

$$x^2 - 4x + 4 = -3 + 4$$

which may be written

$$(x - 2)^2 = 1$$

This equation is equivalent to the open sentence

$$x - 2 = 1 \quad \text{or} \quad x - 2 = -1$$

Thus,

$$x = 3 \quad \text{or} \quad x = 1$$

It follows that the solution set of the given equation is $\{1, 3\}$.
**Do Exercise 19.**                                                             ■

Using exactly the same steps as we did in the preceding example, we can show that if $a$, $b$, and $c$ are real numbers with $a \neq 0$, the equations

$$ax^2 + bx + c = 0$$

and

$$x = \frac{-b \pm \sqrt{b^2 - 4ac}}{2a}$$

are equivalent. This last equation is called the *quadratic formula*.

**EXAMPLE 5**   Solve the equation of Example 4 using the quadratic formula.

**Solution**   For the equation of Example 4, $a = 3$, $b = -12$, and $c = 9$. Thus, the equation is equivalent to

$$x = \frac{-(-12) \pm \sqrt{(-12)^2 - 4(3)(9)}}{2(3)}$$

$$= \frac{12 \pm \sqrt{36}}{6}$$

$$= \frac{12 \pm 6}{6}$$

That is,

$$x = \frac{12 + 6}{6} = 3 \quad \text{or} \quad x = \frac{12 - 6}{6} = 1$$

Therefore, the solution set of the equation is $\{1, 3\}$.
**Do Exercise 25.**                                                            ■

## Equations Involving Radicals

We have said earlier that multiplying both sides of an equation by a quantity which involves the variable may introduce extraneous solutions. However, it is sometimes necessary to do so as illustrated in the following example.

**EXAMPLE 6**   Solve the equation

$$\sqrt{x - 1} = x - 3$$

**Solution**   To eliminate the radical on the left side of the equation, we must square both sides of the equation. In doing so, we multiply by a quantity involving the variable, which may introduce extraneous solutions. We obtain

$$x - 1 = x^2 - 6x + 9$$

This equation is equivalent to

$$x^2 - 7x + 10 = 0$$

which may be written

$$(x - 5)(x - 2) = 0$$

It is easy to see that the solution set of the last equation is $\{2, 5\}$. However, this last equation may be redundant with respect to the original one. Hence, we *must* check the validity of the solutions in the original equation. Replacing $x$ by 2 in the given equation, we get

$$\sqrt{2 - 1} = 2 - 3, \quad \text{or} \quad 1 = -1$$

which is false. Hence, 2 is an extraneous solution and we reject it. Replacing $x$ by 5 in the original equation, we get

$$\sqrt{5 - 1} = 5 - 3, \quad \text{or} \quad 2 = 2$$

which is true. Hence 5 is a solution and the solution set of the given equation is $\{5\}$. We can be sure that there are no other solutions since the solution set of the equation $\sqrt{x - 1} = x - 3$ must be a subset of the solution set of the equation $x - 1 = x^2 - 6x + 9$.
**Do Exercise 31.**                                                                 ■

## Story Problems

EXAMPLE 7    We wish to enclose a rectangular region measuring 15,000 square feet with 400 feet of fence. One side of the area is along an existing wall and does not require fencing. What should the dimensions of the rectangle be?

Solution    Let one side perpendicular to the wall be $x$ feet long. Since there are two sides perpendicular to the wall, $2x$ feet of fence will be used for these two sides. Hence, there will be $(400 - 2x)$ feet of fence available for the third side. Therefore, the length of the side parallel to the wall should be $(400 - 2x)$ feet. The area of the rectangle will be $x(400 - 2x)$ square feet. Since the area is 15,000 square feet, we write

$$x(400 - 2x) = 15,000$$

This equation may be written

$$2x^2 - 400x + 15,000 = 0$$

and

$$2(x - 50)(x - 150) = 0$$

The solution set of the equation is $\{50, 150\}$. Thus, we could have the two sides perpendicular to the wall be 50 feet each and the side parallel to the wall be 300 feet long. Or, we could have the two sides perpendicular to the wall be 150 feet each and the side parallel to the wall be 100 feet long.

Clearly, in the first case the area is 15,000 square feet (since $300 \cdot 50 = 15,000$) and the length of the fence is 400 feet (since $50 + 300 + 50 = 400$). These are the conditions stated in the original problem. In the other case, the area is also 15,000

square feet (since $150 \cdot 100 = 15{,}000$) and the length of the fence is 400 feet since $(150 + 100 + 150 = 400)$. Thus, both solutions are correct.
**Do Exercise 37.**

     ■

The preceding is a *word problem*. In general, a written or verbal statement expressing some condition or conditions of equality that exist among some quantities, of which at least one quantity is unknown, is called a word (or story) problem.

### General Procedure for Solving Word or "Story" Problems

**Step 1.** Read the problem carefully, and write a short summary of the problem on scratch paper if necessary.

**Step 2.** Write down the particular idea or formula to be used in the problem. For example, total cost of production = fixed cost + cost per unit · number of units produced.

**Step 3.** Ask yourself "What are the quantities involved that are unknown?"

**Step 4.** Represent one of these quantities by some symbol, say $x$, and be specific in writing down in what units the quantities are to be expressed. For example, to say "let $x$ be the weight of . . ." is not enough. Is it $x$ pounds? $x$ ounces? $x$ kilograms?

Often, there is more than one way to set up a problem. A different choice of unknown quantity for the variable will lead to a different, possibly simpler, equation. You will learn to make the better choice with practice. Right now, the solution of the problem does not depend on what you choose to call $x$. Therefore, it seems natural at the beginning to call $x$ the unknown that is to be found.

**Step 5.** Write all unknown quantities in terms of $x$ (or whatever symbol is used in Step 4).

**Step 6.** Form an equation that, according to the statement of the problem, expresses the relation between the unknown quantities introduced in Step 5 and the known quantities of the problem.

**Step 7.** Solve the equation. Go back and find the corresponding values of the other relevant quantities. (See Step 5.)

**Step 8.** Check your results.

To check the correctness of the solution you must verify that the results satisfy the conditions stated in the original problem. If you check only that the results are solutions to the equation obtained in Step 6, then you have only checked the correctness of the work in solving the equation—you have not checked the accuracy of the analysis of the original problem to see if the solution fits.

We conclude this section with several applications.

**EXAMPLE 8** A waiter earns a basic salary of $500 per month plus tips. On the average, he makes $20 per hour in tips. How many hours must he work each month to earn $2900?

**Solution** Let $x$ be the number of hours he works per month. His average monthly take in tips is $20x$. Adding this to his basic salary, we get $(20x + 500)$. But this must equal $2900. Thus,

$$20x + 500 = 2900$$

It is easy to see that the solution of this equation is 120. Therefore, the waiter must work 120 hours to earn $2900 per month. You should check the correctness of this result to the original problem. ∎

**EXAMPLE 9** Mr. J. De Largent has $45,000 to invest. He wants to earn at least $3450 in interest per year. He has a choice of investing in industrial bonds paying 9% per year, investing in safer government bonds at 6% per year, or investing in a combination of the two. How should he invest his money to minimize his risk and yet accomplish his goal?

**Solution** If he wanted to be absolutely safe, he could invest all of his money in government bonds. However, he would earn only $2700, since $45,000\left(\frac{6}{100}\right) = 2700$. Let $x$ be the maximum amount he can invest in government bonds and still accomplish his goal. Then, he can invest $(45,000 - x)$ in the more risky industrial bonds. He will earn $x\left(\frac{6}{100}\right)$ dollars from the government bonds and $(45,000 - x)\left(\frac{6}{100}\right)$ dollars from the industrial bonds. Thus, his yearly earnings will be $\left[\frac{6x}{100} + \frac{(45,000 - x)9}{100}\right]$. Since he wishes to earn at least $3450, we have the equation:

$$\frac{6x}{100} + \frac{(45,000 - x)9}{100} = 3450$$

Multiplying both sides of this equation by 100 and solving for $x$, we obtain

$$6x + (45,000 - x)9 = 345,000$$

$$-3x = -60,000, \text{ and } x = \frac{-60,000}{-3} = 20,000$$

Hence, he should invest $20,000 in government bonds and $25,000 in industrial bonds (since $45,000 - 20,000 = 25,000$). You should verify that this strategy will enable him to reach his goal and yet minimize his risk. ∎

**EXAMPLE 10** A motel in Reno, Nevada has 80 rooms. The manager knows that all the rooms will be rented if she charges $40 a room per day. She also knows that for each $7 increase in rent, one less room will be rented and, because of the competition, no room should be rented for more than $120. It costs $5 per day per occupied room for cleaning. On a given day, how much was the rent per room if the profit on that day was $5250? This is a larger profit than if all 80 rooms are rented at $40 per room.

**Solution** Suppose each room is rented for $x$. Then, the increase per room is $(x - 40)$ and the number of $7 increases is $\frac{x - 40}{7}$. Therefore, the number of rented rooms is $\left(80 - \frac{x - 40}{7}\right)$. Since it costs $5 to clean each rented room, the profit per rented room is $(x - 5)$. The total profit is

(number of rooms rented) · (profit per room)

or,

$$\$\left[80 - \frac{x - 40}{7}\right](x - 5)$$

But on that day the total profit is $5250. So, we have

$$\left[80 - \frac{x - 40}{7}\right](x - 5) = 5250$$

Multiplying both sides of this equation by 7 and solving for $x$ we obtain

$$[560 - (x - 40)](x - 5) = 36{,}750$$
$$(600 - x)(x - 5) = 36{,}750$$
$$x^2 - 605x + 39{,}750 = 0$$
$$(x - 75)(x - 530) = 0$$

Thus, $x = 75$ or $x = 530$. We reject the larger of the two solutions since the rent should not exceed $120. We conclude that on the day the profit was $5250, the rent per room was $75. You are urged to check the correctness of this result. ■

EXAMPLE 11    The manager of a barbershop knows that if he charges $p$ per haircut, the number of customers he will get per day is $x$, where $x$ and $p$ are related as follows:

$$p = 24 - .25x$$

The average daily cost of running the barbershop when $x$ customers get haircuts is $(100 + 5x)$. How many customers had a haircut in the shop on a day when the profit was $260?

Solution    In general, the profit is equal to the revenue minus the cost. Also, the revenue is equal to the price per unit multiplied by the number of units. In this case, the revenue per day is $px$. But,

$$px = (24 - .25x)x = 24x - .25x^2$$

The cost is $(100 + 5x)$. Thus, the profit is $[(24x - .25x^2) - (100 + 5x)]$. Since the profit made on the day in question was $260,

$$[(24x - .25x^2) - (100 + 5x)] = 260$$

This equation may be solved as follows:

$$-.25x^2 + 19x - 360 = 0$$

Multiplying both sides by $-4$ we obtain

$$x^2 - 76x + 1440 = 0$$
$$(x - 40)(x - 36) = 0$$
$$x = 40 \quad \text{or} \quad x = 36$$

Thus, the shop will realize a $260 profit on a day when the number of customers getting haircuts is either 36 or 40. However, when the number of customers is 36, the price per haircut will be $15 since $24 - .25(36) = 15$ and when the number of customers is 40, the price per haircut will be $14 since $24 - .25(40) = 14$. Verify that both answers are correct by using them in the original problem.   ∎

**EXAMPLE 12**  When the price of a certain brand of dog food is $p$ per case, the quantity (in thousands) the consumers are willing to buy is $D$ and the quantity (also in thousands) the producers are willing to supply is $S$, where $S$, $D$, and $p$ are related as follows:

$$D = 200 - \frac{1}{2}p^2 \quad \text{and} \quad S = \frac{23}{3}p + \frac{1}{4}p^2$$

Find the price per case for which the demand $D$ and supply $S$ are equal.

**Solution**  We start with the equation $D = S$. Thus,

$$200 - \frac{1}{2}p^2 = \frac{23}{3}p + \frac{1}{4}p^2$$

We first multiply both sides of the equation by 12 and then solve for $p$ as follows:

$$2400 - 6p^2 = 92p + 3p^2$$
$$9p^2 + 92p - 2400 = 0$$
$$(p - 12)(9p + 200) = 0$$

Thus, $p = 12$ or $p = \frac{-200}{9}$. Obviously, the negative solution is not applicable. Therefore, the price per case for which supply and demand are equal is $12. This is called the *equilibrium price*. This concept will be studied more extensively later.

**Do Exercise 41.**   ∎

# Exercise Set 0.7

*In Exercises 1–5, find the replacement set of the equation but do not solve the equation.*

**1.** $\frac{x-3}{x+2} + \frac{5}{x-3} = \frac{3}{x+5}$

**2.** $\frac{4}{y^2-4} - \frac{y}{y+7} = \frac{y-4}{y-3}$

**3.** $\frac{z+3}{z+1} + \frac{z-3}{z-1} = \frac{z+5}{z+6} - \frac{z-7}{z+1}$

**4.** $\frac{x-1}{x^2+5x-6} = \frac{x+1}{x^2+3x-10}$

**5.** $\sqrt{x-4} + x + 1 = \sqrt{x+3}$

*In Exercises 6–35, solve the given equations.*

**6.** $2x + 10 = 5x - 11$        **7.** $5 - 3y = 2y + 20$

**8.** $5x - 6 = 3x + 10$

**9.** $2x - (3 + 5x) = 4x - 17$

**10.** $2 - 3(x + 4) = 5(2 - x) - 14$

**11.** $3y - (2 - 4y) = 2(y + 3) - 18$

**12.** $\frac{z+4}{2} + 2 = \frac{2z+11}{3}$

**13.** $\frac{x+3}{5} - \frac{x-8}{3} = \frac{x+4}{2}$

**14.** $y - \frac{3}{2} = \frac{y+3}{3} + \frac{3}{2}$

**15.** $x^2 - 6x + 5 = 0$        **16.** $y^2 + 3y - 18 = 0$

**17.** $z^2 + 7z - 18 = 0$        **18.** $x^2 - 3x - 10 = 0$

**19.** $y^2 + 10y - 75 = 0$        **20.** $-z^2 + 3z + 18 = 0$

**21.** $(x - 1)(x + 2) = (x + 3)(x - 3) + 13$

**22.** $(y - 2)(y + 3) = (y + 3)(y + 1)$

**23.** $2z^2 + 3z - (z + 1)(z - 1) = 11$

**24.** $3x^2 - 5x - 28 = 0$        **25.** $-5x^2 + 7x + 196 = 0$

**26.** $4y^2 + 3y - 76 = 0$        **27.** $-3z^2 - 11z + 60 = 0$

**28.** $6 + \frac{2}{x-3} = x + 4$        **29.** $\frac{6}{x+1} + \frac{5}{x+3} = 3$

**30.** $\sqrt{x-3} + x = 5$          **31.** $\sqrt{x+1} + 1 = x - 4$

**32.** $\sqrt{2x-1} + 2 = 3x - 10$

**33.** $\sqrt{x+6} + 3 = \sqrt{27+x}$

**34.** $\sqrt{3x+7} + x = 7$

**35.** $2x - \sqrt{5x-1} = 3x - 5$

**36.** A man invested part of $25,000 at 11% and the remainder at 6%. His income in one year from these two investments was the same as if he had invested the whole sum at 9%. How much did he invest at each rate?

**37.** A woman invested part of $40,000 at 12% and the remainder at 8%. Her income in one year from these two investments was the same as if she had invested the whole sum at 9%. How much did she invest at each rate?

**38.** If the price of a brand of mineral water is $p per case, the quantity (in thousands) the consumers are willing to buy is $D$ and the quantity (also in thousands) the producers are willing to supply is $S$, where $S$, $D$, and $p$ are related as follows:

$$D = 675 - \frac{1}{3}p^2 \quad \text{and} \quad S = 19.5p + \frac{2}{3}p^2$$

Find the price per case for which the demand $D$ and supply $S$ are equal.

**39.** When the price of a certain racquetball racquet is $p, the quantity (in hundreds) the consumers are willing to buy is $D$ and the quantity (also in hundreds) the producers are willing to supply is $S$, where $S$, $D$, and $p$ are related as follows:

$$D = 1600 - \frac{1}{4}p^2 \quad \text{and} \quad S = \frac{37}{11}p + \frac{1}{2}p^2$$

Find the price per racquet for which the demand $D$ and supply $S$ are equal.

**40.** The Board of Trustees of a small private university met to discuss an increase in tuition for the following year. The tuition for the past year was $150 per credit hour and the total number of credits taught were 180,000. It is known that for each $1 increase per credit hour, the total number of credits taught will decrease by 800. The trustees established that the total revenue from tuition must increase by $604,800 and that the tuition should not increase by more than 10%. How much will the university charge per credit hour the following year?

**41.** The manager of an apple orchard in Wenatchee, Washington, knows that if 30 trees are planted per acre, the average yield per tree will be 475 apples. He also knows that for each additional tree per acre, the average yield per tree is expected to decrease by 7. Because of local regulations, the number of trees per acre cannot exceed 50. If the total yield per acre is 16,318 apples, how many trees have been planted on each acre?

**42.** The manager of a wine shop knows that if she charges $p per bottle of a certain wine and if she keeps the price below $20 per bottle, the number of bottles she will sell per month is $x$, where $x$ and $p$ are related as follows:

$$p = 40 - .2x$$

The cost of selling $x$ bottles of this particular wine in a month is $C, where $C$ and $x$ are related as follows:

$$C = 20 + 6x$$

How many bottles of this wine were sold during a certain month if the profit during that month was $580? What was the price per bottle?

**43.** The racquetball pro at an Athletic Club knows that if he charges $p per hour for private lessons he will teach on the average $x$ hours per month as long as he does not charge more than $75 per hour, where $x$ and $p$ are related as follows:

$$p = 120 - 1.5x$$

Whenever he teaches $x$ hours in a month, the club charges him $(100 + 5x)$. How many hours did he teach in a month if his profit was $1270? How much did he charge per hour?

**44.** A grocer has two kinds of chocolate. The first kind sells at $4.20 a pound, the second kind at $3.70 a pound. How much should he use of each kind to get 50 pounds of a mixture that he could sell at $4 a pound?

**45.** A grocer has two kinds of nuts. The first kind sells at $5.10 a pound, the second kind at $4.50 a pound. How much should she use of each kind to get 120 pounds of a mixture that she could sell at $4.90 a pound?

**46.** A van radiator contains 21 quarts of a solution that is 30% antifreeze. How much of the solution should be drained out and replaced by pure antifreeze in order to get 21 quarts of a solution that is 40% antifreeze?

**47.** A truck radiator with a 30-quart capacity has been prepared for fall driving with 3 quarts of antifreeze, but winter driving requires a solution that is 20% antifreeze. How many quarts of the fall solution should be withdrawn and replaced by pure antifreeze?

**48.** A man has $20,000 invested in three bonds that pay different rates of interest. He has three times as much invested in a 6% bond as he has in a 8% bond, and the remainder is invested in a 9% bond. His annual interest from these three investments is $1380. How much did he invest in each bond?

**49.** A garage owner buys a certain number of spark plugs at a cost of three for $2. He sells half of them at $.80 a piece and, to attract business, he puts the other half on sale at $.60 a piece. When they are sold, he has realized a $10 profit on the spark plugs. How many of them did he sell?

**50.** We wish to enclose 7800 square feet in a rectangular region against an existing wall. The side against the wall does not require fencing. If we use 250 feet of fence, what are the dimensions of the rectangle?

**51.** Betty has just been promoted to the position of vice president and her office staff decided to have a party to celebrate. John was given $84 to buy ginger ale and cranberry juice to prepare 18 gallons of punch. If the ginger ale and cranberry juice cost $4 and $6 per gallon, respectively, and John spent all the money given to him, how much ginger ale was in 1 gallon of punch?

**52.** The distance from the ground to an object thrown vertically upward with an initial velocity of 256 ft/sec is given by the formula $h = 256t - 16t^2$, where $t$ is the time in seconds and $h$ is the distance in feet above the ground $t$ seconds after the throw. In how many seconds will the object return to the ground?

**53.** A square sheet of metal is used to construct a tray by cutting four corners and turning up the four flaps. If the tray has a volume of 18,000 square centimeters and is 5 centimeters high, what were the dimensions of the original sheet of metal?

**54.** The distance between Seattle and Spokane is 315 miles. Ralph and Norma leave Seattle at the same time. However, Norma arrives in Spokane 1 hour and 10 minutes before Ralph with an average speed that was 9 miles per hour faster than his. How fast was each traveling?

**55.** A homeowner wants to add a 500-square-foot rectangular room to his house against a wall 35 feet long. The cost of building the exterior walls is $120 per linear foot, and the cost of removing part of the existing wall is $40 per linear foot. If the total cost is $9200, what are the dimensions of the room?

**56.** Suppose that $h$ is a nonzero real number. What real number should be added to $x^2 + hx$ to get a quantity that may be written $(x + k)^2$? (*Hint:* Expand $(x + k)^2$ and compare the first two terms of the expansion to $x^2 + hx$.)

# 0.8  Inequalities

In this section we discuss linear inequalities in one variable. First, we must recall the meaning of $<$ (less than) and $>$ (greater than) and their fundamental properties.

## Properties of Inequalities

The following properties of real numbers are needed.

1. If $c$ is an arbitrary real number, exactly one of the following is true: **a.** $c$ is positive; **b.** $-c$ is positive; **c.** $c = 0$.
2. The sum and the product of any two positive real numbers are positive real numbers.

> **DEFINITION 0.18:**   If $a$ and $b$ are real numbers and $b - a$ is positive, we say that $a$ *is less than* $b$ and we write $a < b$.
> In that case, we also say that $b$ *is greater than* $a$ and write $b > a$.
> If $a < b$, or $a = b$, we write $a \leq b$ (read: "$a$ is less than or equal to $b$"). *Greater than or equal to* is defined similarly.

### The Basic Properties of Inequalities

In the following statements, $a$, $b$, $c$, and $d$ denote real numbers.

1. Precisely one of the following statements is true at one time: $a < b$, $b < a$, $a = b$.
2. $a < b$ $(a \leq b)$ if, and only if, $a + c < b + c$ $(a + c \leq b + c)$. (Adding the same number to both sides of an inequality preserves the inequality.)
3. If $0 < c$, then $a < b$ $(a \leq b)$ if, and only if, $ac < bc$ $(ac \leq bc)$. (Multiplying both sides of an inequality by the same positive number preserves the inequality.)
4. If $c < 0$, then $a < b$ $(a \leq b)$ if, and only if, $ac > bc$ $(ac \geq bc)$. (Multiplying both sides of an inequality by a negative number reverses the inequality—that is, reverses the inequality symbol.)
5. If $a < b$ and $b < c$, then $a < c$. Also if $a \leq b$ and $b \leq c$ then $a \leq c$.
6. If $0 < a < b$, then $\frac{1}{b} < \frac{1}{a}$.
7. If $a < b$ and $c < d$, then $a + c < b + d$. Also, if $a \leq b$ and $c \leq d$, then $a + c \leq b + d$.*

In the following example, we prove two of the properties listed above. The proofs of the remaining properties are left as exercises.

**EXAMPLE 1**

1. Prove that if $a$, $b$, and $c$ are real numbers, then $a < b$ and $b < c$ implies $a < c$.
2. Prove that if $a$, $b$, and $c$ are real numbers, then $a < b$ and $0 < c$ implies $ac < bc$.

**Solution**

1. By definition, $a < b$ if, and only if, $b - a$ is positive. Similarly, $b < c$ if, and only if, $c - b$ is positive. Thus, $(b - a) + (c - b)$ is positive. But, $(b - a) + (c - b) = c - a$. Therefore, $c - a$ is positive and we conclude that $a < c$.
2. Since $a < b$, $b - a$ is positive, and since $0 < c$, $c$ is positive. Thus, $(b - a)c$ is positive. However, $(b - a)c = bc - ac$. Therefore, $bc - ac$ is positive, and so $ac < bc$.

**Do Exercise 41.** ■

### Linear Inequalities

**DEFINITION 0.19:**   We say that two open sentences with same replacement set are *equivalent* whenever they have the same solution set.

*Note that Property 7 allows us to add corresponding sides of inequalities. However, *it is wrong to subtract corresponding sides of inequalities*, as is shown by the following: $10 < 11$ is true, $2 < 5$ is also true; however, $10 - 2 < 11 - 5$ is false since $8 > 6$.

As it is with equations, solving a linear inequality means finding its solution set. The procedures used to solve a linear inequality are the same as those used to solve equations with one important exception. When we multiply (or divide) both sides of an inequality by a negative number, we must reverse the inequality symbol.

EXAMPLE 2    Solve the inequality $5x + 2 < 2x + 11$.

Solution    Use Property 2 and subtract 2 from each side of

$$5x + 2 < 2x + 11$$

to obtain

$$5x < 2x + 9$$

Now subtract $2x$ from each side to get

$$3x < 9$$

Multiplying both sides by $\frac{1}{3}$ (by Property 3), we get the equivalent inequality

$$x < 3$$

The solution set is the set of all real numbers less than 3.
**Do Exercise 1.**

Sometimes we encounter statements such as $a < b$ and $b < c$. This particular statement is abbreviated as $a < b < c$. We may also write $a \leq b \leq c$ instead of $a \leq b$ and $b \leq c$. The meaning of statements such as $a < b \leq c$ should now be clear. The solution sets of inequalities often are special sets, as described in the following.

**DEFINITION 0.20:**   Let $a$ and $b$ be two real numbers such that $a < b$.

a. The set $\{x \mid x$ is a real number and $a \leq x \leq b\}$ is called a *closed interval* and is denoted $[a, b]$, where we use two square brackets to indicate that the end points $a$ and $b$ belong to that interval.

b. The set $\{x \mid x$ is a real number and $a < x < b\}$ is called an *open interval* and is denoted $(a, b)$, where we use two parentheses to indicate that the end points $a$ and $b$ do not belong to that interval.

c. The sets $\{x \mid x$ is a real number and $a \leq x < b\}$ and $\{x \mid x$ is a real number and $a < x \leq b\}$ are called *half-open intervals* and are denoted $[a, b)$ and $(a, b]$, respectively. The intervals $[a, b]$, $(a, b)$, $[a, b)$, and $(a, b]$ are *bounded intervals*.

d. The sets $\{x \mid x$ is a real number and $x < a\}$, $\{x \mid x$ is a real number and $x \leq a\}$, $\{x \mid x$ is a real number and $x > a\}$, and $\{x \mid x$ is a real number and $x \geq a\}$ are denoted $(-\infty, a)$, $(-\infty, a]$, $(a, \infty)$, and $[a, \infty)$, respectively, and are called *unbounded intervals*—closed or open depending on whether the end point $a$ is or is not in the set.

The symbols $-\infty$ and $\infty$ are called minus infinity and infinity, respectively. You should not think of them as numbers. The solution set of the inequality of Example 2 is the unbounded open interval $(-\infty, 3)$.

## Double Inequalities

**EXAMPLE 3**   Solve the double inequality $2x + 1 < 3x + 2 < 38 - 6x$.

**Solution**   We write the double inequality as follows:

$$2x + 1 < 3x + 2 \quad \text{and} \quad 3x + 2 < 38 - 6x$$

Using Properties of Inequalities 1–7, we obtain the following sequence of equivalent open sentences:

$$2x + 1 + (-1) < 3x + 2 + (-1) \quad \text{and} \quad 3x + 2 + (-2) < 38 - 6x + (-2)$$
$$2x < 3x + 1 \quad \text{and} \quad 3x < 36 - 6x$$
$$2x + (-3x) < 3x + 1 + (-3x) \quad \text{and} \quad 3x + 6x < 36 - 6x + 6x$$
$$-x < 1 \quad \text{and} \quad 9x < 36$$
$$(-1)(-x) > (-1)(1) \quad \text{and} \quad \frac{1}{9}(9x) < \frac{1}{9}(36)$$
$$x > -1 \quad \text{and} \quad x < 4$$

That is, $-1 < x < 4$. Since the last and first inequalities are equivalent, they have identical solution sets. Thus, the solution set of the original inequality is the open interval $(-1, 4)$. ∎

It is often helpful to have a graphic representation of the solution set of an inequality. If $S$ is a set of real numbers, its graph is the set of all points on a number line whose coordinates are members of the set $S$. For example, Figure 0.7 displays the solution set of Example 3.

**FIGURE 0.7**

The heavy line shows the points on the graph. We have placed a small open circle around the end points $-1$ and $4$ to indicate that they are *not* on the graph.
**Do Exercise 7.**

## Nonlinear Inequalities

**EXAMPLE 4**   Solve the inequality $x^2 + 5x - 6 < 0$.

**Solution**   The inequality $x^2 + 5x - 6 < 0$ may be written

$$(x + 6)(x - 1) < 0$$

Using the fact that the product of two real numbers is negative if, and only if, one of the factors is positive and the other is negative, we obtain the following sequence of equivalent inequalities:

$$(x + 6)(x - 1) < 0$$
$$(x + 6 > 0 \text{ and } x - 1 < 0) \quad \text{or} \quad (x + 6 < 0 \text{ and } x - 1 > 0)$$
$$(x > -6 \text{ and } x < 1) \quad \text{or} \quad (x < -6 \text{ and } x > 1)$$

Since no real number is less than $-6$ and greater than 1, the solution set of the given inequality is

$$\{x \mid x \text{ is a real number and } -6 < x < 1\}$$

This set is the open interval $(-6, 1)$. Its graph is displayed in Figure 0.8.
**Do Exercise 11.**

$-7\ -6\ -5\ -4\ -3\ -2\ -1\ \ 0\ \ 1\ \ 2$

**FIGURE 0.8**

Inequalities similar to that of Example 4 often involve more than one factor. It is much simpler to obtain the solution sets of such inequalities graphically, by displaying the sign of each factor (positive or negative) under a "number line" for values of $x$. We can then use the fact that a product of nonzero real numbers is negative if, and only if, the number of negative factors is odd. We could have solved Example 4 using the following display:

| $x$ | | $-6$ | | $1$ | |
|---|---|---|---|---|---|
| $x + 6$ | $-$ | $0$ | $+$ | $+$ | $+$ |
| $x - 1$ | $-$ | | $-$ | $0$ | $+$ |
| $(x + 6)(x - 1)$ | $+$ | $0$ | $-$ | $0$ | $+$ |

In this table, the first line represents values of $x$. We entered the number $-6$ and 1 because $x + 6 = 0$ when $x = -6$, and $x - 1 = 0$ when $x = 1$. The second line represents the values of $x + 6$. We entered a 0 under the $-6$ since $x + 6 = 0$ when $x = -6$. We displayed a minus sign to the left to indicate that $x + 6$ is negative when $x$ is less than $-6$. We also displayed some plus signs to the right to indicate that $x + 6$ is positive when $x$ is greater than $-6$. We proceeded similarly for $x - 1$ on the third line. The fourth line represents values of the product $(x + 6)(x - 1)$ which are positive when $x + 6$ and $x - 1$ are both positive or both negative, and negative when these factors have opposite signs. We also entered 0 below $-6$ and 1 because the product is 0 when $x$ is replaced by either of these two numbers. Since we are solving the inequality $x^2 + 5x - 6 < 0$, we wish to find values of $x$ for which this product is negative. From the table we see that this will occur whenever $-6 < x < 1$. Therefore, the solution set is the open interval $(-6, 1)$.

To use the method of the last example, one side of the inequality *must* be 0. We illustrate with the following:

**EXAMPLE 5**    Solve the inequality $\dfrac{2x^2 + 4x - 10}{x + 3} \leq x - 2$.

**Solution**    We subtract $x - 2$ from both sides to get

$$\frac{2x^2 + 4x - 10}{x + 3} - (x - 2) \leq 0$$

From this, we get

$$\frac{2x^2 + 4x - 10 - (x - 2)(x + 3)}{x + 3} \leq 0$$

which simplifies to

$$\frac{x^2 + 3x - 4}{x + 3} \leq 0$$

or to

$$\frac{(x + 4)(x - 1)}{x + 3} \leq 0$$

The solution set can now be obtained.

| $x$ | | −4 | | −3 | | 1 | |
|---|---|---|---|---|---|---|---|
| $x + 4$ | − | 0 | + | + | + | + | + |
| $x - 1$ | − | − | − | − | − | 0 | + |
| $x + 3$ | − | − | − | 0 | + | + | + |
| $\dfrac{(x + 4)(x - 1)}{x + 3}$ | − | 0 | + | U | − | 0 | + |

We entered a $U$ in the bottom row under the $-3$ because $(x + 4)(x - 1)/(x + 3)$ is undefined when $x = -3$. The solution set is $(-\infty, -4] \cup (-3, 1]$. The graph of the solution set is shown in Figure 0.9. The small filled-in circles around the end points $-4$ and $1$ indicate that these two numbers *are* in the solution set.
**Do Exercise 35.** ■

**FIGURE 0.9**

In solving inequalities in the form $\frac{f(x)}{g(x)} < 0$, where $f(x)$ and $g(x)$ are polynomials, it is necessary to factor $f(x)$ and $g(x)$ which sometimes may be difficult. In this regard, it is useful to recall the following.

1. If $f(x)$ is a polynomial, then $x - a$ is a factor of $f(x)$ if, and only if, $f(a) = 0$. In particular, if $r_1$ and $r_2$ are solutions of the equation $ax^2 + bx + c = 0$, with $a \neq 0$, then $ax^2 + bx + c = a(x - r_1)(x - r_2)$.
2. The solutions of a quadratic equation can always be found using the quadratic formula.
3. If the solutions of the equation $ax^2 + bx + c = 0$ are not real (that is, $b^2 - 4ac < 0$), then $ax^2 + bx + c$ is either positive for all values of $x$, or negative for all values of $x$.

**EXAMPLE 6**    Solve the inequality $\dfrac{8 - x^3}{3x^2 - 6x - 6} \geq 0$.

**Solution**    First write

$$8 - x^3 = 2^3 - x^3 = (2 - x)(4 + 2x + x^2)$$

and observe that $x^2 + 2x + 4 = 0$ has no real solution (since $2^2 - 4 \cdot 1 \cdot 4 < 0$) and $x^2 + 2x + 4$ is positive when $x = 0$. Thus, $x^2 + 2x + 4 > 0$ for all values of $x$.

The equation $3x^2 - 6x - 6 = 0$ may be solved using the quadratic formula. We obtain

$$x = \frac{-(-6) \pm \sqrt{(-6)^2 - 4(3)(-6)}}{2(3)}$$

$$= \frac{6 \pm \sqrt{108}}{6} = \frac{6 \pm 6\sqrt{3}}{6} = 1 \pm \sqrt{3}$$

Hence,

$$3x^2 - 6x - 6 = 3(x - (1 + \sqrt{3}))(x - (1 - \sqrt{3}))$$

The original inequality may now be written,

$$\frac{(2 - x)(x^2 + 2x + 4)}{3(x - (1 + \sqrt{3}))(x - (1 - \sqrt{3}))} \geq 0$$

We will use the following table to obtain the solution set.

| $x$ | | | | $1-\sqrt{3}$ | | $2$ | | $1+\sqrt{3}$ | |
|---|---|---|---|---|---|---|---|---|---|
| $2 - x$ | + | + | + | 0 | − | − | − | | |
| $x^2 + 2x + 4$ | + | + | + | + | + | + | + | | |
| $x - (1 + \sqrt{3})$ | − | − | − | − | − | 0 | + | | |
| $x - (1 - \sqrt{3})$ | − | 0 | + | + | + | 0 | + | | |
| $\dfrac{(x - 2)(x^2 + 2x + 4)}{3(x - (1 + \sqrt{3}))(x - (1 - \sqrt{3}))}$ | + | U | − | 0 | + | U | − | | |

We locate the plus signs and the number 0 on the last line of the table and find the solutions directly above on the first line. The solution set is $(-\infty, 1-\sqrt{3}) \cup [2, 1+\sqrt{3})$. We could have omitted the factor $x^2 + 2x + 4$ in the table, as was the factor 3, because its values are always positive and therefore do not affect the sign of the expression in the bottom line. However, when a factor does not change sign but is negative for all values of $x$, it *must* be included in the table.
**Do Exercise 39.**  ■

## Compact Solution of Nonlinear Inequalities

It is possible to use the methods of the preceding examples by displaying only the first and last lines of the table. Note the following properties.

1. If $f(x)$ is a polynomial and the real solutions of the equation $f(x) = 0$ are $r_1$, $r_2, \cdots, r_k$, then $f(x)$ can be factored as

$$f(x) = (x - r_1)^{n_1}(x - r_2)^{n_2} \cdots (x - r_k)^{n_k}g(x)$$

where $n_1, n_2, \cdots, n_k$ are positive integers, and $g(x)$ is a polynomial which has no real roots.

2. As the values of $x$ increase and pass through $r_1$, the values of $f(x)$ will change sign only if $n_i$ is odd. It is easy to see this because when $n_i$ is even, $(x - r_i)^{n_i} > 0$ when $x \neq r_i$, and therefore that factor does not affect the sign of $f(x)$.

We illustrate how these may be used to simplify the solution in the next example.

**EXAMPLE 7**  Solve the inequality $\dfrac{(x-2)^4(x+1)^5(x^2+1)}{(x-3)^7(-x^2+3x-9)} > 0$.

**Solution**  Let

$$f(x) = \frac{(x-2)^4(x+1)^5(x^2+1)}{(x-3)^7(-x^2+3x-9)}$$

Obviously, the only real solutions of the equation $(x-2)^4(x+1)^5(x^2+1) = 0$ are $-1$ and $2$, and $3$ is the only real solution of $(x-3)^7(-x^2+3x-9) = 0$. Thus, $f(-1) = f(2) = 0$ and $f(3)$ is not defined. The exponent in $(x-2)^4$ is even, thus the sign of the values of the left side of the inequality will not change as $x$ increases through $2$. However, in $(x+1)^5$ and $(x-3)^7$ the exponents are odd. Thus, the sign of the values of the left side of the inequality will change as $x$ increases through $-1$ and will change again as $x$ increases through $3$.

In the first line of the table, we will display only the values $-1$, $2$, and $3$. However, we will place $-1$ and $3$ within small squares to remind us that the sign of the values of $f(x)$ must change as $x$ increases through these numbers. The table will have only two lines. Under the $-1$ and under the $2$ we will enter $0$ in the second line. Under the $3$ we will enter a $U$. We must determine the sign of the value of $f(x)$ for some value of $x$. The simplest way is to find $f(0)$. We have

$$f(0) = \frac{(0-2)^4(0+1)^5(0^2+1)}{(0-3)^7(-0^2+3(0)-9)} = \frac{(-2)^4(1)^5(1)}{(-3)^7(-9)}$$

We need not calculate the exact value of $f(0)$ because we are interested only in knowing whether it is positive or negative. The value of $(-2)^4(1)^5(1)/(-3)^7(-9)$ is positive since the number of negative factors is even ($4 + 7 + 1 = 12$). Since $0$ is between $-1$ and $2$ on the first line, we enter some plus signs between these numbers on the second line. Recalling that the signs of the values of $f(x)$ change at $-1$ and at $3$, we enter minus signs on the second line, to the left of $-1$ and to the right of $3$ and plus signs between $2$ and $3$. The solution set can now be found.

| $x$ | | $\boxed{-1}$ | | $2$ | | $\boxed{3}$ | |
|---|---|---|---|---|---|---|---|
| $\dfrac{(x-2)^4(x+1)^5(x^2+1)}{(x-3)^7(-x^2+3x-9)}$ | $-$ $-$ | $0$ | $+$ $+$ | $0$ | $+$ $+$ | $U$ | $-$ $-$ |

The solution set is $(-1, 2) \cup (2, 3)$. Its graph is displayed in Figure 0.10.
**Do Exercise 55.**

**FIGURE 0.10**

## Caution Against Common Errors

Before giving you story problems involving inequalities, we caution you against making some common errors.

1. In solving inequalities such as

$$\frac{(x-3)(x+2)}{(x-5)} \leq 0$$

students often multiply both sides of the inequality by $x - 5$ to obtain the inequality $(x - 3)(x + 2) \leq 0$, which is free of fractions. However, $\frac{(x - 3)(x + 2)}{(x - 5)} \leq 0$ and $(x - 3)(x + 2) \leq 0$ are not equivalent. This is because, when we multiply both sides of an inequality by a negative number, we must reverse the inequality symbol. However, when both sides of the inequality are multiplied by $x - 5$, that quantity may be positive or negative, depending on the value of $x$. It is important to remember in solving inequalities not to multiply or divide both sides by a quantity involving a variable; unless, of course, we know that the quantity has the same sign for all values of $x$. For example, we know that $x^2 + 1$ is positive for all values of $x$. Consequently, multiplying both sides of an inequality by $x^2 + 1$ would yield an equivalent inequality.

2. You must also be careful in combining double inequalities. If $a$ and $b$ are real numbers with $a < b$, it is correct to write the statement $x > a$ and $x < b$ in the more compact form $a < x < b$. Note that this open sentence uses the word "and." Another example, the set $(-\infty, -1) \cup (2, \infty)$, whose graph is displayed in Figure 0.11, is defined by the open sentence $x < -1$ or $x > 2$. (Notice the word "or.") You cannot write that sentence in the form $-1 > x > 2$ because it would mean $-1 > x$ and $x > 2$ and would imply the false statement $-1 > 2$.

**FIGURE 0.11**

3. If you use the methods of Examples 4, 5, and 6, you must be careful where you display the plus and minus signs on each line. If the factor is in the form $x - a$, 0 should be written under the $a$, plus signs should be displayed to the right of 0, and minus signs to the left. However, if the factor is in the form $a - x$, again 0 should be written under the $a$, but plus signs should be displayed to the *left* of 0, and minus signs to the *right*. (See the second line in the table of Example 6.) We suggest that you actually substitute a value of $x$ in $x - a$, or in $a - x$, to determine the correct entries on the line.

4. Finally, when you use the method of Example 7, you must be very careful to determine the sign of the value of the left side of the inequality for some value of $x$ accurately. Every entry on the bottom line depends on that result. We suggest that as a check you use at least two distinct values of $x$.

## Applications

We conclude this section with some story problems involving inequalities.

**EXAMPLE 8**  Pierre and Suzie have jobs paying \$12 and \$8 per hour, respectively. They have agreed that Pierre would work 20 more hours than Suzie does each week until she finishes her work for an MBA. They must earn at least \$540 to meet their obligations. What is the minimum number of hours that Suzie should work each week?

**Solution**  Let $x$ be the number of hours that Suzie works each week. Then Pierre is working $(x + 20)$ hours. Thus, each week Suzie earns \$8x while Pierre earns \$12(x + 20). Their total income per week is $[8x + 12(x + 2)]$. Since they need at least \$540 per week, their income should be greater than or equal to \$540. Consequently,

$$8x + 12(x + 20) \geq 540$$
$$20x \geq 300$$
$$x \geq 15$$

Thus, Suzie should work at least 15 hours per week.
**Do Exercise 45.**

**EXAMPLE 9**  The market research department of some company has established that if the company's new product is to be put on the market at a price of $\$p$ per unit, consumers would be likely to buy $q$ units per day where $q = 315 - 9p$. The production department established that the cost of producing $q$ units per day is $\$C(q)$ where $C(q) = 1125 + 5q$. When the product is put on the market, what should the price per unit be in order for the company to experience a daily profit?

**Solution**  The company will experience a daily profit if each day the revenue is greater than the cost. That is, if $R > C$. But the revenue is obtained by multiplying the number of units sold by the price per unit. So

$$R > C$$

But the revenue is obtained by multiplying the number of units sold by the price per unit. So

$$R = qp = (315 - 9p)p = -9p^2 + 315p$$

Since we want to compare $R$ and $C$, we must express the cost $C$ in terms of $p$. Thus,

$$C = 1125 + 5q = 1125 + 5(315 - 9p) = 2700 - 45p$$

We must have

$$-9p^2 + 315p > 2700 - 45p$$

This inequality is equivalent to

$$p^2 - 40p + 300 < 0$$

which can be written

$$(p - 10)(p - 30) < 0$$

We now use the following table to find the solution set of the inequality.

| $p$ | 0 | 10 | | 30 | |
|---|---|---|---|---|---|
| $(p - 10)(p - 30)$ | + | 0 | − | 0 | + |

From the table, we see that $(p - 10)(p - 30) < 0$, whenever $10 < p < 30$. Thus, the company will make a profit if the price per unit is somewhere between $\$10$ and $\$30$. If the price per unit is exactly $\$10$, or exactly $\$30$, the profit will be 0. At these prices, the company breaks even. Later, we shall be able to determine the price $\$p$ which maximizes the daily profit.
**Do Exercise 47.**

# Exercise Set 0.8

*Solve the following inequalities and sketch the graphs of their solution sets.*

**1.** $2x + 13 < 5x + 28$

**2.** $3x - 4 \geq 7x + 12$

**3.** $6x + 5 \leq 2x + 33$

**4.** $2 - 3x > 5x - 22$

**5.** $\dfrac{x}{2} + \dfrac{1}{3} < \dfrac{3x}{5} - \dfrac{2}{7}$

**6.** $5 + 4x \leq 5x + 3 \leq 2x + 18$

**7.** $3x + 1 < 6x + 4 < 4x + 10$

**8.** $9x + 10 \leq 12x + 16 < 8x + 16$

**9.** $15 + 4x < 8x + 3 < 6x + 5$

**10.** $x^2 + 6x - 7 < 0$

**11.** $x^2 + 8x - 20 \leq 0$

**12.** $2x^2 + 2x - 24 > 0$

**13.** $x^2 - 3x - 10 \leq 0$

**14.** $x^3 - 27 > 0$

**15.** $8x^3 - 27 < 0$

**16.** $x^4 - 16 > 0$

**17.** $16x^4 - 81 \leq 0$

**18.** $x^2 + 4x - 6 < 0$

**19.** $3x^2 + 24x - 9 > 0$

**20.** $x^2 + 5x + 7 < 0$

**21.** $\dfrac{x^2 + 3x - 4}{x + 5} > 0$

**22.** $\dfrac{x^2 + x - 6}{x - 7} < 0$

**23.** $\dfrac{x^2 - 5x + 6}{5 - x} > 0$

**24.** $\dfrac{x^2 + x - 3}{x^2 - 25} < 0$

**25.** $\dfrac{x^2 - 16}{x^3 - 8} > 0$

**26.** $\dfrac{x^2 + 2x + 4}{x^2 - 2x + 5} < 0$

**27.** $\dfrac{x^2 + 3x + 8}{x + 5} < 0$

**28.** $\dfrac{x^2 - 4x + 10}{4 - x} > 0$

**29.** $\dfrac{2x^2 + 4x - 4}{x + 1} < 0$

**30.** $\dfrac{5x + 1}{x + 5} < 5 - x$

**31.** $\dfrac{x^2 + 3x - 6}{x + 3} > 0$

**32.** $\dfrac{3x^2 + 5x + 1}{5 - 2x} < 0$

**33.** $\dfrac{3x^2 - 6x + 2}{x^2 - 5} \leq 0$

**34.** $\dfrac{x^2 + 3x + 9}{4x^2 - 8x + 3} \leq 0$

**35.** $\dfrac{x^2 + 5x + 2}{x + 5} > 2x - 3$

**36.** $\dfrac{2x^2 - 3x + 7}{3 - x} < 5x + 2$

**37.** $\dfrac{3x + 2}{x - 3} > \dfrac{6x + 4}{x - 2}$

**38.** $\dfrac{2x + 7}{x - 3} > \dfrac{3x - 4}{x + 3}$

**39.** $\dfrac{-x^2 + 3x - 9}{x^2 + 3x + 2} < 0$

**40.** $\dfrac{-x^2 + 2x - 6}{2x^2 + 3x - 7} > 0$

**\*41.** Prove that if $a$, $b$, and $c$ are real numbers, $a < b$ and $c < 0$, then $bc < ac$.

**\*42.** Prove that if $0 < a < b$, then $\dfrac{1}{b} < \dfrac{1}{a}$.

**43.** A student tried to solve the inequality $x + 2 < 5x + 10$ as follows:

$$x + 2 < 5x + 10$$
$$x + 2 < 5(x + 2)$$
$$1 < 5$$

He concluded that the solution set was the set of all real numbers. Was he correct? Explain.

**44.** A student wishes to earn an "A" grade in her business calculus course. The professor has stated that an average of at least 92 is required for an "A" grade. The student received grades of 89, 93, 90, and 94 on the first four exams. What minimum score must she receive on the fifth and final exam in order to earn an "A"?

**45.** Roy and Betty have agreed that he will work 10 more hours per week than she does so that she can complete her college education. She earns $7 per hour on her part-time job, while he earns $6 per hour on his. They have calculated that they need at least $385 per week to meet their obligations. What is the minimum number of hours that Betty should work per week?

**46.** Company X rents a 1987 Chevrolet Citation for $20 per day plus $.12 per mile, while company Y rents the same model car for $14 per day plus $.15 per mile. How many miles per day should a renter drive so that renting from company X would be to his advantage?

**47.** Some company has established that if it prices its new calculator at $p$ per unit, consumers would be likely to buy $q$ units per day where

$$q = 252 - 7p$$

The production department established that the cost of producing $q$ calculators per day is $C(q)$ where

$$C(q) = 1092 + 4q$$

What should the price be per calculator for the company to experience a daily profit?

**48.** The market research department of the AZ Company has established that if the company's new talking doll is priced at $p per doll, consumers would be likely to buy $q$ dolls per month where

$$q = 1541 - 23p$$

The production department established that the cost of producing $q$ dolls per month is $$C(q)$ where

$$C(q) = 14{,}467 + 13q$$

When the doll is put on the market, what should the price be per doll for the company to experience a monthly profit?

**49.** The Baldhead Company has established that if the company's new hair dryer is priced at $p per unit, consumers would be likely to buy $q$ units per month where

$$q = 935 - 17p$$

The production department established that the cost of producing $q$ dryers per month is $$C(q)$ where

$$C(q) = 5100 + 15q$$

When the hair dryer is put on the market, what should the price be per unit for the company to experience a monthly profit?

**50.** A manufacturer can manufacture and sell $q$ thousand units of a commodity at a price of $p per unit and a cost of $C$ thousand dollars, where $p$ and $C$ are given by

$$p = 20 - .5q$$

and

$$C = 36 + 6q$$

How many units of the commodity should be produced and sold in order for the profit to be at least $60,000?

**51.** Same as Exercise 50 where

$$p = 30 - .25q$$

and

$$C = 200 + 5q$$

and the profit should be at least $400,000.

**52.** A van radiator contains 21 quarts of a solution which is 30% antifreeze. What amount of solution should be drained out and replaced by pure antifreeze in order to get 21 quarts of a solution that is at least 40% but not more than 55% antifreeze?

**53.** A car radiator contains 15 quarts of a solution which.is 25% antifreeze. What amount of solution should be drained out and replaced by pure antifreeze in order to get 15 quarts of a solution that is at least 35% but not more than 45% antifreeze?

**54.** Solve

$$\frac{(x - 2)^6(x + 3)^7(x^2 + x + 1)}{(x + 1)^3(x^2 + 6x + 10)^5} < 0$$

using the method of Example 7.

**55.** Solve

$$\frac{(x + 3)^3(x - 5)^7(x^2 + 2x + 6)}{(x + 5)^3(x^2 + 8x + 20)^5} < 0$$

using the method of Example 7.

**56.** Solve

$$\frac{(x - 1)^9(x + 8)^6(x^2 + 3x + 4)}{(x + 3)^5(x^2 + 5x + 10)^5} > 0$$

using the method of Example 7.

# 0.9  Absolute Value

In mathematical applications, real numbers represent quantities. Positive and negative quantities have opposite meaning. For example, if traveling 5 miles means traveling 5 miles north, then traveling $-5$ miles means traveling 5 miles south. In some situations, we are interested only in the magnitude of the quantity. We then use the absolute value of the real number.

## Properties of Absolute Value

The absolute value of a real number is its distance from 0. Thus, $|x| = x$ if $x \geq 0$, and $|x| = -x$ otherwise.

**DEFINITION 0.21:**    If $a$ is any real number

$$|a| = \max\{-a, a\}$$

where $\max\{-a, a\}$ denotes the larger of the numbers $a$ and $-a$.

We use two vertical bars to denote the absolute value of the number $a$. For example, $|5| = 5$ since $\max\{-5, 5\} = 5$ and $|-3| = 3$ since $\max\{-(-3), -3\} = \max\{3, -3\} = 3$. Note also that $|0| = 0$.

Using the definition, the following fundamental facts can be proved:

1. If $a$ and $b$ are real numbers and $b > 0$, then $|a| < b$ if, and only if, $-b < a < b$. Also, $|a| \leq b$ if, and only if, $-b \leq a \leq b$.
2. If $a$ and $b$ are any two real numbers, then

$$|a + b| \leq |a| + |b|$$

3. If $a$ and $b$ are any two real numbers, then

$$|a - b| \leq |a| + |b|$$

This inequality is often called the *triangle inequality* for reasons which we shall soon discuss. Since the square of a real number is always nonnegative, it is easy to see that for any real number $c$, we have

$$|c|^2 = c^2 = |c^2|$$

**DEFINITION 0.22:**    If the real numbers $a$ and $b$ are the coordinates of two points $A$ and $B$ on a number line, respectively, then the number $b - a$ is the *directed distance* from $A$ to $B$ and $|b - a|$ is the *distance* between $A$ and $B$. Note that

$$|b - a| = |-(b - a)| = |a - b|$$

**EXAMPLE 1**    Suppose that the points $A$ and $B$ have coordinates 3 and $-2$, respectively.

  **a.** Find the directed distance from $A$ to $B$.
  **b.** Find the directed distance from $B$ to $A$.
  **c.** Find the distance between $A$ and $B$.

**Solution**    **a.** The directed distance from $A$ to $B$ is $-5$ units since $-2 - 3 = -5$.
  **b.** The directed distance from $B$ to $A$ is 5 units since $3 - (-2) = 5$.
  **c.** The distance between $A$ and $B$ is 5 units since $|-2 - 3| = |-5| = 5$. (See Figure 0.12.)

**Do Exercise 25.**

**FIGURE 0.12**

Suppose that the points $A$ and $B$ on a number line have coordinates $a$ and $b$, respectively. The origin $O$ has coordinate 0. We know from geometry that the length of one side of a triangle is less than the sum of the lengths of the other two sides. Similarly, if three points are on a straight line, the distance between two of these points is less than or equal to the sum of the distances between each of these and the third point. Consequently, the distance between $A$ and $B$ is less than or equal to the sum of the distances between the origin and $A$ and $B$, respectively. But the distance between $A$ and $B$ is $|a - b|$, the distance between $A$ and the origin is $|a - 0| = |a|$, and the distance between $B$ and the origin is $|b - 0| = |b|$. Thus, we have the triangle inequality which we stated earlier

$$|a - b| \leq |a| + |b|$$

## Open Sentences Involving Absolute Value

We will now illustrate how to solve inequalities involving absolute values.

**EXAMPLE 2** Solve the inequality $|x + 2| < 3$.

**Solution** We know that $|x + 2| < 3$ if, and only if,

$$-3 < x + 2 < 3$$

Thus,

$$-3 + (-2) < x + 2 + (-2) < 3 + (-2)$$

That is,

$$-5 < x < 1$$

Therefore, the solution set of the given inequality is the open interval $(-5, 1)$.
**Do Exercise 1.** ∎

You might find the geometric interpretation of the preceding example useful. That is, we write $|x + 2| < 3$ in the equivalent form $|x - (-2)| < 3$. Then we argue that the distance between $x$ and $-2$ on the number line is less than 3. Consequently, $x$ must be between $-5$ and 1 since $-5$ is 3 units to the left of $-2$ and 1 is 3 units to the right of $-2$. (See Figure 0.13.)

**FIGURE 0.13**

**EXAMPLE 3** Solve the inequality $|2x + 3| > 5$.

**Solution** If we think of $2x + 3$ as a number on the number line, the distance between that number and the origin must be greater than 5. Thus, $2x + 3$ must be either to the left of $-5$ or to the right of 5. (See Figure 0.14.) That is, we must have

$$2x + 3 < -5 \quad \text{or} \quad 5 < 2x + 3$$

**FIGURE 0.14**

Consequently,

$$2x < -8 \quad \text{or} \quad 2 < 2x$$

That is,

$$x < -4 \quad \text{or} \quad 1 < x$$

The solution set is

$$(-\infty, -4) \cup (1, \infty)$$

**Do Exercise 3.**                                    ■

**EXAMPLE 4**   Solve the inequality $|x + 2| < |2x - 1|$.

**Solution**   Since the inequality involves absolute values on both sides, we use the fact that it is equivalent to the following:

$$(x + 2)^2 < (2x - 1)^2 \qquad \text{(See Exercise *33.)}$$

That is,

$$x^2 + 4x + 4 < 4x^2 - 4x + 1$$

and

$$0 < 3x^2 - 8x - 3$$

This inequality yields

$$0 < (x - 3)(3x + 1)$$

We now solve this inequality using the technique which we discussed in the preceding section.

| $x$ | | $\dfrac{-1}{3}$ | $0$ | $3$ | |
|---|---|---|---|---|---|
| $(x - 3)(3x + 1)$ | $+$ | $0$ | $-$ | $0$ | $+$ |

Since we wanted the product $(x - 3)(3x + 1)$ to be positive, we must have $x < \frac{-1}{3}$ or $x > 3$. Therefore, the solution set is $\left(-\infty, \frac{-1}{3}\right) \cup (3, \infty)$.

**Do Exercise 15.**                                 ■

**EXAMPLE 5**   Solve the equation $|x + 3| + |x - 5| = 12$.

**Solution**   Since $x + 3 = 0$ when $x = -3$, and $x - 5 = 0$ when $x = 5$, we consider the cases where $x \le -3$, $-3 < x < 5$, and $x \ge 5$.

**Case 1:** $x \le -3$. Then $x < 5$ and both $x + 3$ and $x - 5$ are nonpositive. Consequently,

$$|x + 3| = -(x + 3) = -x - 3 \quad \text{and} \quad |x - 5| = -(x - 5) = 5 - x$$

Thus, the given equation can be written

$$(-x - 3) + (5 - x) = 12$$

That is,

$$x = -5$$

Since $-5 < -3$, this solution is valid.

**Case 2:** $-3 < x < 5$. Then $x + 3$ is positive and $x - 5$ is negative. Consequently,

$$|x + 3| = x + 3 \quad \text{and} \quad |x - 5| = -(x - 5) = 5 - x$$

Thus, the given equation may be written

$$(x + 3) + (5 - x) = 12$$

This equation has no solution.

**Case 3:** $x \ge 5$. Then, both $x + 3$ and $x - 5$ are nonnegative. So

$$|x + 3| = x + 3 \quad \text{and} \quad |x - 5| = x - 5$$

It follows that the given equation may be written

$$(x + 3) + (x - 5) = 12$$

Thus,

$$x = 7$$

Since $7 > 5$, this solution is also valid. Hence, the given equation has two solutions, $-5$ and $7$.

**Do Exercise 19.** ■

## Exercise Set 0.9

*In Exercises 1–18, solve the given inequality and sketch the graph of the solution set.*

**1.** $|x - 3| < 4$

**2.** $|x - 6| \le 2$

**3.** $|x + 7| > 3$

**4.** $|x + 5| \ge 2$

**5.** $|3x - 2| < 4$

**6.** $|5x - 1| < 9$

**7.** $|2 + 4x| \ge 14$

**8.** $|5 - 3x| > 8$

**9.** $|2 - 3x| < 0$

**10.** $|3 - 2x| \ge 0$

**11.** $|2 - 5x| \le 4x + 2$

**12.** $|3 + 2x| < 5x - 3$

**13.** $|x + 13| < 5x + 5$

**14.** $|6x + 1| \le |3x + 16|$

**15.** $|1 - 4x| > |3x + 1|$

**16.** $|2 + 3x| \ge |5x + 6|$

**17.** $|x + 2| + |x - 3| > 0$

**18.** $|4x - 12| + |3x - 9| > 0$

*In Exercises 19–23, solve the given equations.*

**19.** $|x + 2| + |x - 8| = 16$

**20.** $|x + 5| + |x + 3| = 10$

**21.** $|x + 3| + |x - 5| = 8$

**22.** $|2x + 1| + |x - 5| = 14$

**23.** $|3x + 2| + |5x - 7| = 19$

*In Exercises 24–28, two points A and B are given with their coordinates in parentheses.*

a. *Find the directed distance from A to B.*

b. *Find the directed distance from B to A.*

c. *Find the distance between A and B.*

**24.** $A(3); B(-4)$

**25.** $A(-6); B(7)$

**26.** $A(9); B(12)$

**27.** $A\left(\frac{-1}{2}\right); B\left(\frac{1}{3}\right)$

**28.** $A\left(\frac{-2}{5}\right); B\left(\frac{-3}{7}\right)$

*29. Prove that $|a| = 0$ if, and only if, $a = 0$.

*30. Prove that for any two real numbers $a$ and $b$,

$$|ab| = |a|\,|b|$$

*31. Prove that if $a$ and $b$ are real numbers with $b \neq 0$, then

$$\left|\frac{a}{b}\right| = \frac{|a|}{|b|}$$

*32. Prove that for any real number $c$, $|-c| = |c|$.

*33. Prove that for any real numbers $a$ and $b$, $|a| \leq |b|$ if, and only if, $a^2 \leq b^2$.

*34. Prove the inequality

$$\bigl||a| - |b|\bigr| \leq |a - b|$$

where $a$ and $b$ are arbitrary real numbers. (*Hint:* Start with $|x + y| \leq |x| + |y|$ and replace $x$ by $a - b$ and $y$ by $b$. Then replace $x$ by $a$ and $y$ by $b - a$.)

# 0.10 Chapter Review

IMPORTANT SYMBOLS AND TERMS

$\sqrt{\phantom{x}}$ [0.1]
$\subset$ [0.2]
$\cup$ [0.2]
$\cap$ [0.2]
$\in$ [0.2]
$\subseteq$ [0.2]
$\emptyset$ (or { }) [0.2]
$<$ [0.8]
$>$ [0.8]
$\leq$ [0.8]
$\geq$ [0.8]
$(-\infty, a)$ [0.8]
$(-\infty, a]$ [0.8]
$(a, \infty)$ [0.8]
$(a, \infty]$ [0.9]
$[a, b]$ [0.8]
$(a, b)$ [0.8]
$(a, b]$ [0.8]
$[a, b)$ [0.8]
$A'$ [0.2]
Absolute value [0.9]
Addition [0.4]
Additive inverse [0.1]
Associative properties [0.1]
Base [0.1]
Binomial [0.4]
Bounded interval [0.8]
Closed interval [0.8]
Closure properties [0.1]
Commutative properties [0.1]
Complement [0.2]
Complete factorization [0.6]
Completing squares [0.7]
Complex numbers [0.1]
Conditional equation [0.7]
Defective [0.7]
Degree [0.4]

Distributive property [0.1]
Dividend [0.4]
Division [0.1], [0.4]
Division algorithm [0.4]
Division of polynomials [0.4]
Divisor [0.4]
Element [0.2]
Empty set [0.2]
Equation in one variable [0.7]
Equilibrium price [0.7]
Equivalent [0.7], [0.8]
Equivalent inequalities [0.8]
Exponent [0.1]
Factor [0.6]
Factor over the integers (rational, real, and complex numbers) [0.6]
Factor Theorem [0.6]
Factorization by grouping [0.6]
Greater than [0.8]
Greater than or equal to [0.8]
Half-open intervals [0.8]
Identity [0.4], [0.7]
Identity elements [0.1]
Inverse element [0.1]
Inequalities [0.8]
Integer [0.1]
Intersection [0.2]
Irrational number [0.1]
Leading coefficient [0.4]
Less than [0.8]
Less than or equal to [0.8]
Linear [0.4]
Linear equation [0.7]

Open interval [0.8]
Open sentence [0.2]
Periodic decimal [0.1]
Polynomial [0.4]
Positive integer [0.1]
Power of a number [0.1]
Prime factorization [0.1]
Prime number [0.6]
Principal square root [0.1]
Proper subset [0.2]
Properties of inequalities [0.8]
Quadratic [0.4]
Quadratic equation [0.7]
Quadratic formula [0.1], [0.6], [0.7]
Quotient [0.1], [0.4]
Rational number [0.1]
Reciprocal [0.1]
Redundant [0.7]
Remainder [0.4]
Remainder Theorem [0.4]
Replacement set [0.2]
Roster method [0.2]
Set [0.2]
Solution [0.2], [0.7]
Solution set [0.7]
Solving an equation [0.7]
Square root [0.1]
Subset [0.2]
Subtraction [0.1], [0.4]
Synthetic division [0.5]
System of complex numbers [0.1]
System of real numbers [0.1]
Triangle inequality [0.9]
Trinomial [0.4]

**SUMMARY**

The real (and complex) numbers satisfy the closure properties, the associative properties, the commutative properties, the distributive property, and the zero product property (if $ab = 0$, then $a = 0$ or $b = 0$, or both). There are identity elements and inverse elements. The product $a \cdot a \cdot a \cdot a \cdot \cdots \cdot a$ ($n$ factors) is called the $n$th power of $a$ and is denoted $a^n$. The positive integer $n$ is called the exponent and $a$ is called the base. If $a \neq 0$ and $n$ is a positive integer, we define $a^0 = 1$ and $a^{-n} = 1/a^n$.

The following laws of exponents hold:

**a.** $a^m \cdot a^n = a^{m+n}$, **b.** if $a \neq 0$, $\dfrac{a^m}{a^n} = a^{m-n}$, **c.** $(a^m)^n = a^{mn}$,

**d.** $(ab)^n = a^n b^n$, and **e.** if $b \neq 0$, $\left(\dfrac{a}{b}\right)^n = \dfrac{a^n}{b^n}$.

An expression of the form $a_0 x^n + a_1 x^{n-1} + a_2 x^{n-2} + \cdots + a_{n-1} x + a_n$, where $n$ is a positive integer and $a_0, a_1, a_2, \ldots, a_{n-1}, a_n$ are numbers with $a_0 \neq 0$ is called a polynomial in $x$ of degree $n$, and $a_0$ is the leading coefficient. If $f(x)$ and $g(x)$ are two polynomials in $x$, we can find their sum, difference, and product. Also, there exists unique polynomials $q(x)$ (the quotient) and $r(x)$ (the remainder) such that the degree of $r(x)$ is less than the degree of $g(x)$, (or $r(x) = 0$), and

$$f(x) = g(x)q(x) + r(x)$$

which we can write,

$$\frac{f(x)}{g(x)} = q(x) + \frac{r(x)}{g(x)}$$

A polynomial $p(x)$ may be factored as a product of polynomials where each factor is linear or quadratic. It is useful to remember that if $p(c) = 0$, then $x - c$ is a factor.

A set is a collection of objects. If $A$ and $B$ are sets and $x \in A$ implies $x \in B$, we say that $A$ is a subset of $B$ and write $A \subseteq B$. In any discussion we have a universe $U$ and each set under consideration is a subset of $U$. If $A$ is a set, its complement is denoted $A'$ and is the set of all elements in $U$ but not in $A$. If $A$ and $B$ are sets, their union $A \cup B$ and intersection $A \cap B$ are defined as follows:

$$A \cup B = \{x \mid x \in A \text{ or } x \in B, \text{ or both}\}$$

and

$$A \cap B = \{x \mid x \in A \text{ and } x \in B\}$$

Venn diagrams may be used to represent sets pictorially.

A statement of equality containing one variable is called an equation. Two equations with the same replacement set are said to be equivalent if they have the same solution set. An equivalent equation is obtained if we do any of the following.

**a.** Add the same quantity to both sides of the equation.
**b.** Subtract the same quantity from both sides of the equation.
**c.** Multiply both sides by the same nonzero constant.
**d.** Divide both sides by the same nonzero constant.

If an equation involves a radical and we square both sides to eliminate the radical, we may obtain extra solutions. Hence, we must check which of the solutions of the new equation are solutions of the original one.

The solutions of $ax^2 + bx + c = 0$, where $a \neq 0$, may be found using the quadratic formula

$$x = \frac{-b \pm \sqrt{b^2 - 4ac}}{2a}$$

by factoring, or by completing squares. Sometimes, a quadratic equation has no real solution. In such cases, the solutions will be in the form $r + si$, where $r$ and $s$ are real numbers and $i^2 = -1$. The symbol $i$ is called the imaginary unit. If $a$ and $b$ are real numbers and $b - a$ is positive, we say that $a$ is less than $b$ and write $a < b$. We also say that $b$ is greater than $a$ and write $b > a$. If $a < b$ or $a = b$, we write $a \leq b$ (read: "$a$ is less than or equal to $b$"). Greater than or equal to is defined similarly. For real numbers $a$, $b$, $c$, and $d$,

1. Exactly one of the following is true: $a < b$, $a = b$, $a > b$.
2. $a < b$ if, and only if, $a + c < b + c$ (similarly for $\leq$).
3. If $0 < c$, then $a < b$ if, and only if, $ac < bc$ (similarly for $\leq$).
4. If $c < 0$, then $a < b$ if, and only if, $ac > bc$ (similarly for $\leq$).
5. If $a < b$ and $b < c$, then $a < c$ (similarly for $\leq$).
6. If $0 < a < b$, then $\frac{1}{b} < \frac{1}{a}$.
7. If $a < b$ and $c < d$, then $a + c < b + d$ (similarly for $\leq$). However, $a < b$ and $c < d$ does not imply $a - c < b - d$.

The process of solving linear inequalities in one variable is similar to that of solving linear equations. However, if we multiply or divide both sides of an inequality by a negative number, we must reverse the sense of the inequality. To solve an inequality which is not linear, we try to get an equivalent inequality in the form

$$\frac{f(x)}{g(x)} < 0 \quad \text{or} \quad \frac{f(x)}{g(x)} > 0$$

where $f(x)$ and $g(x)$ are polynomials ($g(x)$ may be a constant). We then factor $f(x)$ and $g(x)$ completely and tabulate the signs of the factors for different values of $x$. We then use the rule of signs for multiplication and division to determine the values of $x$ for which $\frac{f(x)}{g(x)}$ is positive, negative, 0, and undefined. We can determine the solution set from the table. Certain sets of real numbers are called intervals. If $a$ is a real number, its absolute value is denoted $|a|$ and is the max$\{-a, a\}$. Geometrically, $|a - b|$ represents the distance on a number line between the points whose coordinates are $a$ and $b$, respectively. The following are true for real numbers $a$ and $b$.

1. $|a| = |-a|$
2. $|a| < b$ if, and only if, $-b < a < b$ (similarly for $\leq$)
3. $|a + b| \leq |a| + |b|$
4. $|a - b| \leq |a| + |b|$
5. $|a|^2 = a^2$

**SAMPLE EXAM QUESTIONS** *The starred questions require material from optional sections.*

1. Simplify.

a. $-4(-9 + 17)$  b. $-3[2 - (-5 + 4)]$  c. $\frac{4}{5} + 6$

d. $\frac{4}{7} + \frac{5}{3}$  e. $-7 + \frac{4}{43}$  f. $\frac{9}{8} + \frac{7}{28}$

$$\textbf{g.}\ \ \frac{\dfrac{-3}{5}}{\dfrac{12}{}}\qquad\qquad\qquad \textbf{h.}\ \ \frac{\dfrac{-5}{7}}{\dfrac{15}{28}}$$

**2.** Simplify.

    **a.** $(9x^2y^5)^5$      **b.** $-y^3(-25y^2)^4$      **c.** $\dfrac{(5a^2b^3)^6}{15a^9b^8}$      **d.** $\dfrac{(-4x^3y^4)^5}{(8x^2y^5)^3}$

**3.** Solve each equation.

    **a.** $2x + 3 = -5x + 24$      **b.** $\dfrac{x + 3}{4} = \dfrac{x + 1}{5} + 1$

    **c.** $\dfrac{2}{y} + 3 = \dfrac{5}{y} - 6$ (*Hint:* Let $u = \dfrac{1}{y}$ and solve for $u$ first.)

**4.** Solve by using the indicated method.

    **a.** $x^2 + 8x - 9 = 0$, (factoring)

    **b.** $x^2 - 11x + 18 = 0$, (factoring)

    **c.** $2x^2 + 16x - 66 = 0$, (completing squares)

    **d.** $3x^2 - 3x + 90 = 0$, (completing squares)

    **e.** $-2x^2 + 5x - 2 = 0$, (quadratic formula)

    **f.** $\dfrac{1}{3}x^2 + \dfrac{1}{6}x - 50 = 0$, (quadratic formula)

    **g.** $2x^2 + 2x + 5 = 0$, (quadratic formula)

**5.** Solve each equation.

    **a.** $\sqrt{x + 4} = 2x - 7$

    **b.** $\sqrt{7x + 22} = x + 4$

**6.** Solve each equation by factoring.

    **a.** $x^3 - 3x^2 - x + 3 = 0$

    **b.** $x^3 - 2x^2 - 5x + 6 = 0$

**\*7.** Use synthetic division to do the following:

    **a.** Show that 2 is a solution of the equation

$$2x^5 + 5x^4 - 6x^3 - 5x^2 - 35x - 6 = 0$$

    **b.** Evaluate $p(-3)$ if $p(x) = 5x^7 - 4x^5 + 3x^3 + 6x^2 - 4x + 7$.

**8.** Let $f(x) = 3x^4 - 5x^2 + 6x - 10$, $g(x) = x^2 + 5x - 2$, $h(x) = 5x^5 - 6x^3 + 10x^2 - 3x + 4$. Find:

    **a.** The sum $f(x) + h(x)$.

    **b.** The difference $f(x) - h(x)$.

    **c.** The product $f(x)g(x)$.

    **d.** The product $f(x)h(x)$.

    **e.** The quotient and remainder of the division $\dfrac{f(x)}{g(x)}$.

    **f.** The quotient and remainder of the division $\dfrac{h(x)}{g(x)}$.

**9.** Suppose that $p(x)$ is a polynomial and that $p(2) = 7$. What can you say about the remainder of the division $\dfrac{p(x)}{x - 2}$?

**10.** Factor each completely, over the integers.

    **a.** $x^2 + 10x - 11$      **b.** $2x^2 + 5x - 18$      **c.** $4x^2 - 20x + 25$

**d.** $8x^3 - 125$               **e.** $x^4 + 4x^2 + 16$               **f.** $x^3 - 7x + 6$
**g.** $x^3 + 5x^2 - 4x - 20$

**11.** Factor each completely, over the real numbers.

   **a.** $x^3 - x^2 - 3x + 3$          **b.** $x^3 - 3x^2 + 5x - 15$          **c.** $x^3 - x^2 - 3x + 2$

**12.** What is the difference between the terms "identity" and "conditional equation"?

**13.** Solve each of the following inequalities and sketch the graphs of their solutions sets.

   **a.** $3x - 2 < 5x - 10$          **b.** $\dfrac{x+1}{3} + 1 \geq \dfrac{2x-3}{7} + 2$

   **c.** $|x - 2| < 5$               **d.** $|x + 3| \leq 2$
   **e.** $|2 - x| \geq 5$            **f.** $|3 + x| + |5 - 2x| < 0$

   **g.** $|2x - 3| < \dfrac{5}{4}$          **h.** $x^2 + 13x - 14 < 0$

   **i.** $\dfrac{x^2 - 4}{x + 3} \geq 0$          **j.** $\dfrac{x^2 + x - 2}{15 + 2x - x^2} < 0$

   **k.** $\dfrac{2x + 3}{x - 2} < 9$          **l.** $|2 - 3x| < 5x - 6$

**14.** Give an example to illustrate that $a < b$ and $c < d$ does not imply $a - c < b - d$.

**15.** Which of the following are true? Justify your answers.

   **a.** $2 \in (2, \infty)$          **b.** $\pi \in (3, 4)$          **c.** $\{2, 3, 5\} \in [2, 5]$
   **d.** $5 \in (-1, 5) \cap [5, 7]$          **e.** $(2, 3) \subseteq (-1, 4) \cap (1, 5)$
   **f.** $(4, 6) \cap [6, 9) \subseteq \{1, 3, 5\}$          **g.** $\{3, 6, 7\} \subseteq (-\infty, 4) \cup (5, 9)$
   **h.** $6 \subseteq \{5, 6, 7, 8\}$          **i.** $\emptyset \in [5, \infty)$

**16.** List all the subsets of the set $\{a, b, c\}$.

**17.** If the set $A$ has five members, how many proper subsets does it have?

**18.** Use a Venn diagram to illustrate pictorially that if $A$ and $B$ are sets, then $A \cup (B \cap C) = (A \cup B) \cap (A \cup C)$.

**19.** A television station hires a man to poll 5000 viewers to find out how many people like the programs "Golden Girls," "Facts of Life," and "Family Ties" listing only those viewers who indicate a liking for at least one of the three programs. The director is troubled by the fact that the man gathered the results very quickly and doubts that he conducted the survey carefully. However, she is willing to pay him his fee provided the figures are consistent. These are the results:

| | |
|---|---|
| Golden Girls | 2093 |
| Facts of Life | 3534 |
| Family Ties | 3033 |
| Facts of Life and Family Ties | 2103 |
| Facts of Life and Golden Girls | 1523 |
| Golden Girls and Family Ties | 1323 |
| All three | 1013 |

If the director asked you whether or not she should pay the man, what would you tell her? Justify your answer.

**20.** We want to enclose 4200 square meters in a rectangular region against an existing wall. The side against the wall does not require fencing. If we use 190 feet of fence, what are the dimensions of the rectangle?

# Functions

1

When we say that the distance traveled by an object in a fixed amount of time is a *function* of its speed, we mean that the distance traveled is dependent on the speed of the object and, therefore, if the speed is known, the distance traveled can be determined. Speed and distance are both variable quantities. In a formula such as $d = 5s$, $d$ and $s$ are both variables and for each value of $s$ (the speed), the corresponding value of $d$ can be found (the distance traveled in 5 units of time).

## 1.1  Function

> **DEFINITION 1.1:**  In general, a *function* from a set $S$ to a set $T$ is a rule that assigns to each element $x$ of set $S$ a unique element of set $T$. The set $S$ is called the *domain* of the function. If $f$ is the name of a function, then the unique element in $T$ corresponding to an element $x$ in $S$ is denoted $f(x)$ (read: "$f$ of $x$") and is called the *image* of $x$. The set $\{f(x) \mid x \in S\}$ is called the *range* of the function. It is the set of all possible values of $f(x)$ in $T$.

A function can be described by a table (Example 1), a formula (Example 2), words (Example 3), or a set of ordered pairs (Example 4).

**EXAMPLE 1**   A function $g$ from the set $\{1, 2, 3\}$ to the set $\{a, b, c\}$ is defined by the following table:

| $x$ | 1 | 2 | 3 |
|------|---|---|---|
| $g(x)$ | $a$ | $c$ | $a$ |

Each letter on the second line corresponds to the number directly above it.

**a.** What is the domain of $g$?
**b.** Find $g(1)$, $g(2)$, and $g(3)$.
**c.** What is the range of $g$?

**Solution**   **a.** The domain of $g$ is the set $\{1, 2, 3\}$.
**b.** $g(1) = a$ since $a$ is below 1 in the table. Similarly, $g(2) = c$ and $g(3) = a$.
**c.** The range of $g$ is the set $\{a, c\}$ (see part (b)).
**Do Exercise 1.**   ■

**EXAMPLE 2**   The function $h$ from the set of integers to the set of nonnegative integers is defined by the equation

$$h(x) = x^2$$

**a.** Find $h(-2)$, $h(0)$, $h(3)$, and $h(2.4)$.
**b.** What is the range of the function $h$?

**Solution**   **a.** $h(-2) = (-2)^2 = 4$, $h(0) = 0^2 = 0$, $h(3) = 3^2 = 9$, and $h(2.4)$ is not defined since 2.4 is not an integer and hence is not in the domain of $h$.

**b.** The range of the function $h$ is the set of all square integers $\{0, 1, 4, 9, 16, 25, 36, \ldots\}$.
**Do Exercise 7.**   ■

EXAMPLE 3   Let $N$ be the set of all nations in the world and let $C$ be the set of all cities. For each $x$ in $N$, let $f(x)$ be the capital of nation $x$.

**a.** Find $f(\text{U.S.})$, $f(\text{France})$, and $f(\text{Canada})$.
**b.** What is the range of $f$?

Solution   **a.** $f(\text{U.S.}) = \text{Washington, D.C.}$; $f(\text{France}) = \text{Paris}$; and $f(\text{Canada}) = \text{Ottawa}$.
**b.** The range of $f$ is the set of all capitals of nations of the world.
**Do Exercise 9.**   ■

## Ordered Pairs

Often, names of pairs of objects are listed in order. For example, in a telephone directory, one can find the name of a resident listed first and to the right of the name a corresponding phone number. In general, let $(a, b)$ denote an *ordered pair* where $a$ is the first member of the pair and $b$ is the second member; $a$ and $b$ need not be distinct, for example, $(2, 2)$ is an ordered pair. We can use ordered pairs to describe functions as in this example.

EXAMPLE 4   Let $f$ be the function from the set $\{a, b, c, d\}$ to the set $\{-1, 2, 5, 6, 13\}$ defined by the set of ordered pairs $\{(a, 5), (b, -1), (c, 13), (d, 5)\}$, where it is understood that each member of the domain is the first element of an ordered pair and that the image, or value, of that member is the second element of the same ordered pair.

**a.** Find $f(a)$, $f(b)$, $f(c)$, and $f(d)$.
**b.** Find the range of $f$.

Solution   **a.** Since the ordered pair $(a, 5)$ is in the set, we see that $f(a) = 5$. Similarly, $f(b) = -1$, $f(c) = 13$, and $f(d) = 5$.
**b.** The results of part **a** indicate that the range of the function is $\{-1, 5, 13\}$.   ■

EXAMPLE 5   Does the set of ordered pairs $\{(1, 4), (2, 5), (1, 3)\}$ describe a function?

Solution   The first elements of the ordered pairs constitute the domain of the function. Note that both $(1, 4)$ and $(1, 3)$ are in the given set. This set of ordered pairs does not describe a function because more than one element corresponds to 1 and by definition, each element of the domain of a function can correspond only to one element of the range. Here, both 3 and 4 correspond to 1.
**Do Exercise 11.**   ■

## Independent and Dependent Variables

DEFINITION 1.2:   If $f$ is the name of a function and we write $y = f(x)$, $x$ is called the *independent variable* and $y$ is called the *dependent variable*.

The replacement set for the independent variable is the domain of the function. Note that in the first four examples we described the domain and the rule defining the function. In most of our work, we shall give only the rule defining the function. The domain and range will be sets of real numbers. In particular, unless otherwise specified, the domain will be the largest set of real numbers that can be used as a replacement set for the independent variable so that the corresponding values of the dependent variable are defined and are real numbers.

**EXAMPLE 6**    Let the function $g$ be defined by

$$y = g(x) = \frac{2}{x - 3}$$

Find the domain of $g$.

**Solution**    Replacing $x$ by 3 would yield a division by 0, which is not defined. Hence, 3 is not in the domain of $g$. If $x$ is replaced by any other real number, the corresponding value of $y$ will be a real number. Thus, the domain of $g$ is the set of all real numbers except the number 3.
**Do Exercise 13.**                                                                                           ■

**EXAMPLE 7**    Let the function $f$ be defined by

$$y = f(x) = \frac{x - 3}{\sqrt{x^2 + 5x - 6}}$$

Find the domain of $f$.

**Solution**    To find the value of $y$, we must divide by $\sqrt{x^2 + 5x - 6}$. Thus, this expression must be real and nonzero. Since the square root of a negative number is not a real number, $x^2 + 5x - 6 > 0$. This inequality can be solved using the method described in Section 0.8. The solution set is $(-\infty, -6) \cup (1, \infty)$. Therefore, the domain of $f$ is that set.
**Do Exercise 17.**                                                                                           ■

## Graphing Functions

Recall from basic algebra that there is a one-to-one relationship between the set of all ordered pairs of real numbers and a Cartesian (coordinate) plane. If the domain and range of a function $f$ are sets of real numbers, then the *graph* of $f$ is the set of all points in the Cartesian plane whose coordinates are the ordered pairs $(x, f(x))$, where $x$ is in the domain of $f$.

**EXAMPLE 8**    Let $f$ be the function from the set $\{-2, 1, 3, 4\}$ to the set of real numbers defined by $f(x) = 2x - 3$. Sketch the graph of $f$.

**Solution**    Since $f(-2) = 2(-2) - 3 = -7$, the point whose coordinates are $(-2, -7)$ is on the graph of $f$. Similarly, $f(1) = 2(1) - 3 = -1$. Thus, the point with coordinates $(1, -1)$ is also on the graph of $f$. It should now be clear that the points with coordinates

(3, 3) and (4, 5) are also points of the graph of $f$. The graph of $f$ is shown in Figure 1.1. Note that the domain and the graph of a function have the same number of elements.

**Do Exercise 39.**

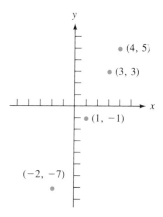

**FIGURE 1.1**

## Adding, Subtracting, Multiplying, and Dividing Functions

If the ranges of two functions $f$ and $g$ are sets of real numbers, it is possible to add, subtract, multiply, and divide these functions. If the function $S$ is the sum of $f$ and $g$, it is required that $S(x) = f(x) + g(x)$. However, this sum will be defined only if both of the terms $f(x)$ and $g(x)$ are defined. Thus, $x$ must belong to both the domain of $f$ and the domain of $g$. That is, $x$ must be in the intersection of the two domains.

> **DEFINITION 1.3:**   If $f$ and $g$ are functions with domains $A$ and $B$, respectively, we define their *sum S*, *difference D*, *product P*, and *quotient Q* as follows:
>
> $S(x) = f(x) + g(x)$ for each $x \in A \cap B$,
>
> $D(x) = f(x) - g(x)$ for each $x \in A \cap B$,
>
> $P(x) = f(x)g(x)$ for each $x \in A \cap B$, and
>
> $Q(x) = \dfrac{f(x)}{g(x)}$ for each $x \in A \cap B$, such that $g(x) \neq 0$.

**EXAMPLE 9**   Let the functions $f$ and $g$ be defined by

$$y = f(x) = \frac{x^2 - 1}{x^2 - 9} \quad \text{and} \quad y = g(x) = \frac{x + 3}{x - 1}$$

Find their sum $S$, difference $D$, product $P$, and quotient $Q$.

**Solution**   The domain of $f$ is the set of all real numbers except $-3$ and $3$, and that of $g$ is the set of all real numbers except $1$. Thus, the domain of the sum $S$ is the set of all real numbers except $-3$, $1$, and $3$.

$$S(x) = f(x) + g(x) = \frac{x^2 - 1}{x^2 - 9} + \frac{x + 3}{x - 1}$$

$$= \frac{2x^3 + 2x^2 - 10x - 26}{(x^2 - 9)(x - 1)}$$

The difference $D$ has the same domain and is defined by

$$D(x) = f(x) - g(x) = \frac{x^2 - 1}{x^2 - 9} - \frac{x + 3}{x - 1}$$

$$= \frac{-4x^2 + 8x + 28}{(x^2 - 9)(x - 1)}$$

The domain of the product $P$ is also the set of all real numbers except $-3$, $1$, and $3$. The product is defined by

$$P(x) = f(x)g(x) = \frac{(x^2 - 1)(x + 3)}{(x^2 - 9)(x - 1)}$$

$$= \frac{(x - 1)(x + 1)(x + 3)}{(x - 3)(x + 3)(x - 1)} = \frac{x + 1}{x - 3}$$

Note that we divided the numerator and denominator of the fraction defining $P(x)$ by the quantity $(x - 1)(x + 3)$. We can do this because the numbers $1$ and $-3$ are not in the domain of $P$ and, therefore, we are not dividing by zero. We must in this case indicate that the domain of $P$ is the set of all real numbers except $-3$, $1$, and $3$ since this fact is not apparent from the final description of $P$,

$$P(x) = \frac{x + 1}{x - 3}$$

Now, the quotient $Q$ is obtained as follows:

$$Q(x) = \frac{f(x)}{g(x)} = \frac{\dfrac{x^2 - 1}{x^2 - 9}}{\dfrac{x + 3}{x - 1}}$$

$$= \frac{(x^2 - 1)(x - 1)}{(x^2 - 9)(x + 3)} = \frac{x^3 - x^2 - x + 1}{x^3 + 3x^2 - 9x - 27}$$

Again, we should specify that the domain of $Q$ is the set of all real numbers except $-3$, $1$, and $3$, since this fact is not apparent from the final description of the function $Q$. Note also that in this case we are dividing $f(x)$ by $g(x)$ and $g(x) = 0$ only when $x = -3$. However, the number $-3$ was already excluded because it was not in the domain of the function $f$.

**Do Exercise 29.**

EXAMPLE 10   Let the functions $f$ and $g$ be defined by

| $x$ | $a$ | $b$ | $c$ | $d$ | $e$ |
|-----|-----|-----|-----|-----|-----|
| $f(x)$ | $-3$ | 2 | 0 | 5 | 3 |

and

| $x$ | $a$ | $b$ | $c$ | $e$ | $i$ |
|-----|-----|-----|-----|-----|-----|
| $g(x)$ | 4 | 2 | $-1$ | 0 | 4 |

Find their sum $S$, difference $D$, product $P$, and quotient $Q$.

Solution   The domain of $f$ is $\{a, b, c, d, e\}$ and that of $g$ is $\{a, b, c, e, i\}$. The intersection of the two domains is $\{a, b, c, e\}$. Hence, this set is also the domain of the functions $S$, $D$, and $P$.

We have

$$S(a) = f(a) + g(a) = -3 + 4 = 1$$
$$S(b) = f(b) + g(b) = 2 + 2 = 4$$
$$S(c) = f(c) + g(c) = 0 + (-1) = -1$$
$$S(e) = f(e) + g(e) = 3 + 0 = 3$$

Thus, $S$ is defined by the table

| $x$ | $a$ | $b$ | $c$ | $e$ |
|-----|-----|-----|-----|-----|
| $S(x)$ | 1 | 4 | $-1$ | 3 |

Similarly, $D$ is defined by

| $x$ | $a$ | $b$ | $c$ | $e$ |
|-----|-----|-----|-----|-----|
| $D(x)$ | $-7$ | 0 | 1 | 3 |

since $D(a) = f(a) - g(a) = -3 - 4 = -7$, and so on. We also find that

$$P(a) = f(a)g(a) = (-3)(4) = -12$$
$$P(b) = f(b)g(b) = (2)(2) = 4$$
$$P(c) = f(c)g(c) = (0)(-1) = 0$$
$$P(e) = f(e)g(e) = (3)(0) = 0$$

Thus, $P$ is defined by

| $x$ | $a$ | $b$ | $c$ | $e$ |
|-----|-----|-----|-----|-----|
| $P(x)$ | $-12$ | 4 | 0 | 0 |

To find the quotient $Q$, we proceed as follows.

$$Q(a) = \frac{f(a)}{g(a)} = \frac{-3}{4} = -.75$$

$$Q(b) = \frac{f(b)}{g(b)} = \frac{2}{2} = 1$$

$$Q(c) = \frac{f(c)}{g(c)} = \frac{0}{-1} = 0$$

$$Q(e) = \frac{f(e)}{g(e)}.$$

But $g(e) = 0$ and division by 0 is not defined, so $e$ is not in the domain of the function $Q$. Thus, the domain of $Q$ is the set $\{a, b, c\}$ and $Q$ is defined by

| $x$ | $a$ | $b$ | $c$ |
|---|---|---|---|
| $Q(x)$ | $-.75$ | 1 | 0 |

**Do Exercise 33.**

## Composition of Functions

Suppose now that we have two functions $f$ and $g$, and that the domain of $g$ and the range of $f$ have some common elements. Then for each $x$ in the domain of $f$ for which $f(x)$ is in the domain of $g$, $g(f(x))$ is defined. We obtain a new function called the *composition of g with f*.

> DEFINITION 1.4:   Let $f$ and $g$ be functions with domains $A$ and $B$, respectively. Let $C = \{x \mid x \in A \text{ and } f(x) \in B\}$. The *composition of g with f* is denoted $g \circ f$ and is the function with domain $C$ defined by
>
> $$(g \circ f)(x) = g(f(x))$$
>
> for each $x \in C$.

EXAMPLE 11    Let the functions $f$ and $g$ be defined by

$$f(x) = x - 5 \quad \text{and} \quad g(x) = \sqrt{x}$$

**a.** Describe the composition of $g$ with $f$.
**b.** Describe the composition of $f$ with $g$.

Solution    **a.** Clearly, the domain of $f$ is the set of all real numbers and the domain of $g$ is the set of all nonnegative real numbers. Thus the domain of the composition $g \circ f$ is the set $\{(x) \mid f(x) \text{ is nonnegative}\}$. We have

$$\{x \mid f(x) \text{ is nonnegative}\} = \{x \mid x - 5 \geq 0\}$$
$$= \{x \mid x \geq 5\} = [5, \infty)$$

Therefore, the domain of $g \circ f$ is the unbounded closed interval $[5, \infty)$. Also

$$(g \circ f)(x) = g(f(x)) = \sqrt{f(x)} = \sqrt{x - 5}$$

**b.** The domain of $g$ is the set of all nonnegative real numbers. Each element of the range of $g$ is in the domain of $f$ since the domain of $f$ is the set of all real numbers. Hence, the domain of the composition $f \circ g$ is the set of all nonnegative real numbers. We have

$$(f \circ g)(x) = f(g(x)) = g(x) - 5 = \sqrt{x} - 5.$$

Note that $\sqrt{x - 5} \neq \sqrt{x} - 5$. Thus, in general, $(g \circ f)(x) \neq (f \circ g)(x)$.

**Do Exercise 43.**                                                                                                 ■

**EXAMPLE 12**   Let the functions $h$ and $s$ be defined by

$$h(t) = 2t + 3 \quad \text{and} \quad s(u) = u^2 + 2u - 3$$

Describe the composition of $s$ with $h$.*

**Solution**   The domain of the function $s$ is the set of real numbers. Thus, $h(x)$ is in the domain of $s$ for each $x$ in the domain of $h$. We have

$$
\begin{aligned}
(s \circ h)(x) = s(h(x)) &= s(2x + 3) \\
&= (2x + 3)^2 + 2(2x + 3) - 3 \\
&= 4x^2 + 12x + 9 + 4x + 6 - 3 \\
&= 4x^2 + 16x + 12
\end{aligned}
$$

The domain of $s \circ h$ is the set of all real numbers.

**Do Exercise 47.**                                                                                                 ■

**EXAMPLE 13**   Let the function $f$ by defined by

$$f(x) = (2x + 3)^3$$

Find two functions $g$ and $h$ such that $h \circ g = f$.

**Solution**   In calculating the value of $f(x)$, we first find $2x$, then $2x + 3$, then we raise the last quantity obtained to the third power. We illustrate this sequence of steps with the following diagram

$$x \rightarrow 2x \rightarrow 2x + 3 \rightarrow (2x + 3)^3$$

The function $f$ is, in fact, the composition of three functions. However, we consider the first function to be $g$ defined by $g(x) = 2x + 3$, and the second function to be $h$

---

*We are using $t$ and $u$ for the names of the independent variables of the functions $h$ and $s$, respectively. You should keep in mind that the names of the variables in the description of a function are arbitrary. The relationship between the independent variable and the dependent variable is the description of the function, regardless of which names are used for the variables.

defined by $h(x) = x^3$, because the last step illustrated in the diagram required cubing $2x + 3$. Thus,

$$f(x) = (2x + 3)^3 = (g(x))^3 = h(g(x)) = (h \circ g)(x)$$

**Do Exercise 49.**

■

# Exercise Set 1.1

*In Exercises 1–10, two sets S and T are given and a function f from S to T is described. In each case, find **a.** the domain of f and **b.** f(x) for the given values of x.*

**1.** $S = \{2, 4, 6, 8\}$, $T = \{w, x, y, z\}$, and $f$ is defined by the table:

| $x$ | 2 | 4 | 6 | 8 |
|------|---|---|---|---|
| $f(x)$ | $y$ | $x$ | $z$ | $w$ |

$x = 2$, $x = 6$.

**2.** $S = \{a, b, c, d, e\}$, $T = \{-1, 3, 6, 9, 13\}$ and $f$ is defined by the table:

| $x$ | $a$ | $b$ | $c$ | $d$ | $e$ |
|------|-----|-----|-----|-----|-----|
| $f(x)$ | $-1$ | 13 | 9 | $-1$ | 9 |

$x = a$, $x = c$, $x = e$.

**3.** $S = \{0, 1, 2, 3, 4\}$, $T = \{-2, -1, 0, 2, 3, 4, 5, 6, 7, 8, 9\}$, and $f$ is defined by $f(x) = 2x$; $x = 1$, $x = 3$, $x = 4$.

**4.** $S = \{-2, -1, 1, 2, 3, 4\}$, $T = \{-7, -4, 2, 5, 8, 11\}$, and $f$ is defined by $f(x) = 3x - 1$; $x = -1$, $x = 2$, $x = 4$.

**5.** $S = \{0, 1, 2, 3, 4\}$, $T = \{0, 1, 7, 8, 10, 27, 30, 64, 79\}$, and $f$ is defined by $f(x) = x^3$; $x = 0$, $x = 2$, $x = 4$.

**6.** $S$ is the set of positive integers, $T$ is the set of rational numbers, and $f$ is defined by $f(x) = \frac{x}{(x + 1)}$; $x = 2$, $x = 6$, $x = 9$.

**7.** $S$ is the set of real numbers, $T$ is the set of real numbers, and $f$ is defined by $f(x) = x^2 + 1$; $x = -2$, $x = 2$, $x = 6$.

**8.** $S$ is the set of positive integers, $T$ is the set of prime numbers, and $f(x)$ is the smallest prime number larger than $x$; $x = 4$, $x = 8$, $x = 14$.

**9.** $S$ is the set of real numbers, $T$ is the set of integers and, $f(x)$ is the largest integer less than or equal to $x$. This function is often denoted $[x]$; $x = -2.3$, $x = \pi$, $x = 4.2$, $x = 9$.

**10.** $S = \{1, 2, 3, 4\}$, $T = \{a, b, c, d, e, k\}$, and $f$ is defined as in Example 4 by $\{(1, c), (2, a), (3, k), (4, d)\}$; $x = 2$, $x = 4$.

**11.** Does the set $\{(a, 1), (b, 3), (a, 4), (d, 5), (e, 7), (g, 8), (h, 2)\}$ describe a function? Justify your answer.

**12.** Does the set $\{(x, y) \mid x = y^2, y$ is a real number$\}$ define a function? Justify your answer.

*In Exercises 13–18, an equation defines a function. Find the domain of each.*

**13.** $f(x) = \dfrac{2}{x - 4}$

**14.** $g(x) = \dfrac{x + 1}{x + 3}$

**15.** $h(x) = \dfrac{x - 2}{x^2 - 16}$

**16.** $f(x) = \sqrt{x^2 + 3x - 10}$

**17.** $g(x) = \sqrt{x^3 - 8}$

**18.** $h(x) = \dfrac{5 + 2x}{\sqrt{x^2 - 4x + 3}}$

**19.** Find $f(5)$, $f(0)$, $f(-3)$ where $f$ is the function of Exercise 13.

**20.** Find $g(-2)$, $g(0)$, $g(\frac{1}{2})$ where $g$ is the function of Exercise 14.

**21.** Find $h(-2)$, $h(1)$, $h(5)$ where $h$ is the function of Exercise 15.

**22.** Find $f(3)$, $f(5)$, $f(-8)$ where $f$ is the function of Exercise 16.

**23.** Find $g(2)$, $g(3)$, $g(5)$ where $g$ is the function of Exercise 17.

**24.** Find $h(-1)$, $h(0)$, $h(1)$ where $h$ is the function of Exercise 18.

*In Exercises 25–35, describe the sum S, the difference D, the product P, and the quotient Q of the functions f and g. In each case specify the domain of the resulting function.*

**25.** $f(x) = 2x + 3$ and $g(x) = x^2 + 3x - 4$

**26.** $f(x) = x^2 - 9$ and $g(x) = -x^2 + 3x - 4$

**27.** $f(x) = \dfrac{2}{x-3}$ and $g(x) = \dfrac{3}{x+2}$

**28.** $f(x) = \dfrac{x}{x^2-16}$ and $g(x) = \dfrac{3x}{x+4}$

**29.** $f(x) = \dfrac{x-2}{x+5}$ and $g(x) = \dfrac{x+5}{x^2-4}$

**30.** $f(x) = \dfrac{2x-3}{x+6}$ and $g(x) = \dfrac{3x-7}{x-3}$

**31.** $f(x) = \dfrac{x^2-25}{x+7}$ and $g(x) = \dfrac{x-3}{x-5}$

**32.** $f(x) = 2x + 3$ and $g(x) = \dfrac{x}{2x+3}$

**33.** $f$ and $g$ are defined by the tables

| $x$ | 1 | 3 | 6 | 8 |
|-----|---|---|----|---|
| $f(x)$ | 2 | 5 | -1 | 4 |

| $x$ | 0 | 1 | 4 | 6 | 8 |
|-----|----|---|---|---|---|
| $g(x)$ | -1 | 3 | 2 | 0 | 5 |

**34.** $f$ and $g$ are defined by the tables

| $x$ | $a$ | $b$ | $c$ | $d$ |
|-----|---|----|---|---|
| $f(x)$ | 2 | .5 | 4 | 0 |

| $x$ | $b$ | $c$ | $d$ | $e$ | $n$ |
|-----|---|---|---|---|---|
| $g(x)$ | 4 | 0 | 3 | 2 | 9 |

**35.** The domain of $f$ is $\{1, 2, 3, 4, 5, 6\}$, that of $B$ is $\{1, 3, 5, 7, 9\}$. The function $f$ is defined by

$$\{(1, 4), (2, 3), (3, 1), (4, 0), (5, 2), (6, -1)\}$$

and $g$ is defined by

$$\{(1, -2), (3, 5), (5, 0), (7, -4), (9, 3)\}.$$

**36.** Plot the graph of the function $f$ of Exercise 3.

**37.** Plot the graph of the function $f$ of Exercise 4.

**38.** Plot the graph of the function $f$ of Exercise 5.

**39.** Plot the graph of the function $f$ of Exercise 33.

**40.** Plot the graph of the function $g$ of Exercise 33.

**41.** Plot the graph of the function $f$ of Exercise 35.

**42.** Plot the graph of the function $g$ of Exercise 35.

*In Exercises 43–48, find formulas for the compositions $f \circ g$, $g \circ f$, $f \circ f$, and $g \circ g$.*

**43.** $f(x) = 2x - 3$ and $g(x) = x^2 + 1$

**44.** $f(x) = 3x - 2$ and $g(x) = -x^2 + 3x - 5$

**45.** $f(x) = x^2 + 2x - 3$ and $g(x) = -x^2 + 3x - 2$

**46.** $f(u) = 3u + 5$ and $g(v) = \dfrac{v-5}{3}$

**47.** $f(t) = |t - 3|$ and $g(z) = |z + 2|$

**48.** $f(x) = x^2 + 1$ and $g(x) = \sqrt{x - 1}$

**49.** Let $f(x) = \sqrt{5x + 2}$. Find functions $g$ and $h$ such that $f = g \circ h$.

**50.** Same as Exercise 49 for $f(x) = (2x - 3)^4$.

**51.** Same as Exercise 49 for $f(x) = (3x + 2)^9$.

**52.** Same as Exercise 49 for $f(x) = \dfrac{5}{x-3}$.

**53.** Does $\{(2, 5), (9, 5), (5, 3), (0, 4)\}$ define a function? Justify your answer.

**54.** Same as Exercise 53 for $\{(1, 4), (3, 7), (9, 0), (1, 3), (8, 2)\}$.

**55.** Same as Exercise 53 for $\{(2, 5), (3, 5), (4, 5), (5, 5), (6, 5)\}$.

## 1.2  Inverse of a Function

We have seen that a function is a rule that assigns to each element of a set $S$ a unique element of a set $T$. Although for each member of the domain there is one, and only one, corresponding member of the range, it is often the case that some member of the range is the image of several members of the domain. When this does *not* occur—that is, when each element of the range is the image of one, and only one, member of the domain—the function is said to be *one-to-one*.

One-to-One Functions

> **DEFINITION 1.5:**   A function $f$ is said to be a *one-to-one* function provided that $f(x_1) = f(x_2)$ can occur only if $x_1 = x_2$.

**EXAMPLE 1**   Let the function $f$ be defined by $f(x) = x^2$. Is $f$ one-to-one?

**Solution**   Note that $f(3) = 3^2 = 9$ and $f(-3) = (-3)^2 = 9$. Thus, $f(3) = f(-3)$. Since $3 \neq -3$ and $f(3) = f(-3)$, the function $f$ is not one-to-one.
**Do Exercise 1.**                                                                    ■

Inverse of a One-to-One Function

If a function $f$ is one-to-one, the rule defining the function not only assigns to each member of the domain a unique member of the range, but also assigns to each element of the range one, and only one, element of the domain. In fact, the rule defines two functions.

> **DEFINITION 1.6:**   Suppose that $f$ is a one-to-one function with domain $A$ and range $B$. The *inverse of $f$* is the function denoted $f^{-1}$ whose domain is $B$ and whose range is $A$ and having the property that $y = f^{-1}(x)$ if, and only if, $f(y) = x$.

Note that the domain of $f$ is the range of the inverse function $f^{-1}$ and the range of $f$ is the domain of $f^{-1}$. It should be clear that $f^{-1}$ is also one-to-one, that $f^{-1}(f(x)) = x$ for each $x$ in the domain of $f$, and that $f(f^{-1}(x)) = x$ for each $x$ in the domain of $f^{-1}$ (the range of $f$).

**EXAMPLE 2**   Let the function $f$ be defined by the table:

| $x$ | $a$ | $b$ | $c$ | $d$ |
|---|---|---|---|---|
| $f(x)$ | 2 | 5 | 1 | 7 |

**a.** Show that $f$ is one-to-one.
**b.** Describe $f^{-1}$.

**Solution**   **a.** The domain of $f$ is $\{a, b, c, d\}$ and the range of $f$ is $\{1, 2, 5, 7\}$. From the table, we see that no element of the range is the image of two distinct elements of the domain. Thus, $f$ is one-to-one.
**b.** The domain of $f^{-1}$ is $\{1, 2, 5, 7\}$ and the range of $f^{-1}$ is $\{a, b, c, d\}$. The inverse function is, therefore, defined by the following table:

| $x$ | 1 | 2 | 5 | 7 |
|---|---|---|---|---|
| $f^{-1}(x)$ | $c$ | $a$ | $b$ | $d$ |

We placed $c$ below 1 because 1 was below $c$ in the table defining $f$. The reason for the other entries in the table should be clear.

**Do Exercise 7.**   ■

**EXAMPLE 3**   Let the function $f$ be defined by $f(x) = 5x + 2$.

**a.** Show that $f$ is one-to-one.
**b.** Find $f^{-1}(x)$.

**Solution**   **a.** Suppose $f(x_1) = f(x_2)$. Then

$$5x_1 + 2 = 5x_2 + 2$$

That is,

$$5x_1 = 5x_2$$

and so

$$x_1 = x_2$$

We have shown that if $f(x_1) = f(x_2)$, then $x_1 = x_2$. Thus, $f$ is one-to-one.

**b.** We know that $f(f^{-1}(x)) = x$. We also know from the definition of $f$ that $f(f^{-1}(x)) = 5f^{-1}(x) + 2$. Thus, $5f^{-1}(x) + 2 = x$. Solving for $f^{-1}(x)$, we get

$$f^{-1}(x) = \frac{x - 2}{5}$$

**Do Exercise 15.**   ■

In general, if $f$ is a one-to-one function and we want to find $f^{-1}(x)$, we see that $y = f^{-1}(x)$ if, and only if, $f(y) = x$. Thus, in the formula $y = f(x)$ defining $f$, we replace $y$ by $x$ and $x$ by $y$ and solve for $y$.

**EXAMPLE 4**   Assume that the function $f$ defined by

$$y = f(x) = \frac{x - 2}{x + 3}$$

is one-to-one and find $f^{-1}(x)$.

**Solution**   In the formula defining $f$, replace $y$ by $x$ and $x$ by $y$ to obtain

$$x = \frac{y - 2}{y + 3}$$

Thus,

$$x(y + 3) = y - 2$$

and

$$xy + 3x = y - 2$$

Therefore,

$$3x + 2 = y - xy$$
$$= y(1 - x)$$

If $1 - x \neq 0$, then

$$y = \frac{3x + 2}{1 - x}$$

We conclude that

$$f^{-1}(x) = \frac{3x + 2}{1 - x}$$

**Do Exercise 17.**                                                                                                   ■

## Restriction of a Function

Recall from basic algebra that a positive real number has two distinct square roots. The rule that assigns to each positive number one of its square roots is *not* a function. However, it is customary to call the positive square root of a number its principal square root. Thus, if we assign to each positive real number its principal square root, we have a function. We would like this function to be the inverse of the squaring function. However, as was shown in Example 1, the squaring function is not one-to-one and, therefore, does not have an inverse. Let us, then, restrict the squaring function to the set of nonnegative real numbers. That is, suppose $P$ is the set of nonnegative real numbers and define a function $f$ from $P$ to $P$ by $f(x) = x^2$. This function is clearly one-to-one and, therefore, has an inverse—the "square root" function. Restricting functions to subsets of their domains is done frequently in mathematics.

> **DEFINITION 1.7:**    Let $f$ be a function with domain $S$ and let $A$ be a nonempty subset of $S$. The *restriction* of $f$ to $A$ is the function $g$ with domain $A$, defined by $g(x) = f(x)$ for all $x$ in $A$.

**EXAMPLE 5**    Let $f$ be defined by the table:

| $x$ | $a$ | $b$ | $c$ | $d$ | $e$ |
|---|---|---|---|---|---|
| $f(x)$ | 2 | 5 | 2 | 7 | 1 |

**a.** Show that $f$ is not one-to-one.
**b.** Choose a subset $B$ of the domain of $f$ so that the restriction of $f$ to $B$ is one-to-one.

**Solution**    **a.** From the table, we see that $f(a) = f(c)$ since both are equal to 2. Thus, the function $f$ is not one-to-one.

**b.** Let $B = \{a, b, d, e\}$. Then, the restriction of $f$ to $B$, say $g$, is defined by the following table:

| $x$ | $a$ | $b$ | $d$ | $e$ |
|---|---|---|---|---|
| $g(x)$ | 2 | 5 | 7 | 1 |

Clearly, $g$ is one-to-one since no element in the set $\{2, 5, 7, 1\}$ is the image of more than one element from the set $\{a, b, d, e\}$.

Note that in this example, we could have chosen the set $\{b, c, d, e\}$ instead of $\{a, b, d, e\}$.

**Do Exercise 23.**                                                                            ∎

**EXAMPLE 6**   Let the function $g$ be defined by $g(x) = x^2 - 4x + 9$.

**a.** Show that $g$ is not a one-to-one function.
**b.** Find a subset $B$ of the domain of $g$ so that $g$ restricted to $B$ is one-to-one.
**c.** Let the restriction of $g$ to $B$ be denoted $h$ and find a formula for $h^{-1}$.

**Solution**   **a.** It is useful to use the technique of completing squares and write

$$g(x) = x^2 - 4x + 9 = (x^2 - 4x + 4) + 5 = (x - 2)^2 + 5$$

It is now clear that the domain of $g$ is the set of all real numbers, and that the range of $g$ is the unbounded closed interval $[5, \infty)$. To see this, note that $(x - 2)^2 \geq 0$ and therefore $(x - 2)^2 + 5 \geq 5$ for all values of $x$. Note now that if $c$ is any positive number,

$$g(2 + c) = (2 + c - 2)^2 + 5 = c^2 + 5$$

and

$$g(2 - c) = (2 - c - 2)^2 + 5 = (-c)^2 + 5 = c^2 + 5$$

Since $c$ is positive, $2 + c$ and $2 - c$ are distinct. The fact that $g(2 + c) = g(2 - c)$ indicates that $g$ is not one-to-one.

**b.** From part **a**, it is clear that if the restriction of $g$ is to be one-to-one, we cannot have both $2 + c$ and $2 - c$ in the domain of the restriction for any positive number $c$. Thus, we let $B = \{2 + c \mid c \geq 0\} = [2, \infty)$.

It should now be clear that the restriction of $g$ to $B$ is one-to-one. To verify this, proceed as follows. Suppose $g(x_1) = g(x_2)$ with $x_1$ and $x_2$ in $B$. Then

$$x_1^2 - 4x_1 + 9 = x_2^2 - 4x_2 + 9$$

which yields

$$x_1^2 - x_2^2 - 4x_1 + 4x_2 = 0$$

This equation can be written

$$(x_1 - x_2)(x_1 + x_2) - 4(x_1 - x_2) = 0$$

or

$$(x_1 - x_2)(x_1 + x_2 - 4) = 0$$

Therefore,

$$x_1 = x_2 \quad \text{or} \quad x_1 + x_2 - 4 = 0$$

Since both $x_1$ and $x_2$ are in $B$, both are greater than or equal to 2 and their sum is greater than or equal to 4, with equality only if $x_1 = x_2 = 2$. Hence, $x_1 + x_2 - 4 = 0$ where $x_1$ and $x_2$ are in $B$ implies that $x_1 = x_2$. Thus, the restriction of $g$ to $B$ is indeed one-to-one.

**c.** If $h$ is the restriction of $g$ to $B$, then the domain of $h$ is $B$, the closed unbounded interval $[2, \infty)$. The range of $h$ is the closed unbounded interval $[5, \infty)$. Thus, the domain of $h^{-1}$ is $[5, \infty)$ and its range is $[2, \infty)$. Also, $y = h(x) = (x - 2)^2 + 5$. To find the formula for $h^{-1}$, replace $y$ by $x$ and $x$ by $y$ in the foregoing equation to obtain

$$x = (y - 2)^2 + 5$$

Thus,

$$(y - 2)^2 = x - 5$$

Taking the principal square root of each side, we get

$$y - 2 = \sqrt{x - 5}$$

and

$$y = \sqrt{x - 5} + 2$$

Hence,

$$h^{-1}(x) = \sqrt{x - 5} + 2$$

**Do Exercise 25.**

We conclude this section with a geometric discussion to illustrate pictorially some of the concepts.

## The Vertical Line Test

Recall that if both the domain and range of a function $f$ are sets of real numbers, the graph of $f$ is the set of all points in the Cartesian (or rectangular) coordinate system whose coordinates are $(x, f(x))$ with $x$ in the domain of $f$. Suppose now, that $G$ is a set of points in the coordinate plane and we ask whether $G$ is the graph of some function. Elements of the domain of a function are represented by points on the $x$-axis, and elements of the range are represented by points on the $y$-axis. If $G$ is the graph of a function $f$, then no perpendicular to the $x$-axis—that is, no vertical line—can intersect $G$ at more than one point. If there were a vertical line intersecting $G$ at two distinct points, say $(a, b)$ and $(a, c)$, then there would be some element $a$ in the domain to which two distinct elements are assigned. Therefore, the definition of a function would not be satisfied. (See Figure 1.2.)

 **Vertical Line Test**: A set of points in the $xy$-plane is the graph of a function with independent variable $x$ and dependent variable $y$ if, and only if, each vertical line intersects that set at no more than one point.

## The Horizontal Line Test

Suppose $G$ is the graph of some function $f$ and we ask whether this function is one-to-one. If $f$ is one-to-one, we cannot have $f(a) = f(b)$ with $a \neq b$. Therefore, we cannot have a perpendicular to the $y$-axis—that is, a horizontal line—intersecting $G$

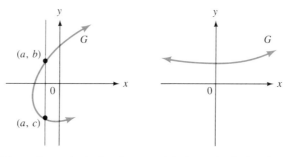

G is not the graph of a function

(a)

G is the graph of a function

(b)

**FIGURE 1.2**

at more than one point because if some horizontal line were to intersect $G$ at two distinct points, say $(a, c)$ and $(b, c)$, then we would have $c = f(a) = f(b)$ with $a \neq b$, contradicting the definition of a one-to-one function. (See Figure 1.3.)

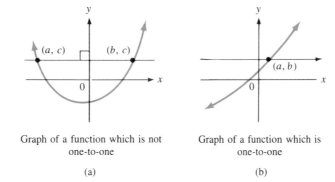

Graph of a function which is not
one-to-one

(a)

Graph of a function which is
one-to-one

(b)

**FIGURE 1.3**

**Horizontal Line Test**: A function is one-to-one, hence has an inverse, whenever each horizontal line intersects its graph at no more than one point.

It is easy to see that if two points have coordinates $(a, b)$ and $(b, a)$, respectively, then these two points are symmetric with respect to the line passing through the origin and forming a 45° angle with the positive $x$-axis. (See Figure 1.4.)

## Graphing Inverse Functions

Using the definition of the inverse of a one-to-one function, we can see that if the point with coordinates $(a, b)$ is on the graph of a one-to-one function $f$, then the point with coordinates $(b, a)$ is on the graph of $f^{-1}$, and conversely. Consequently, the graph of a one-to-one function $f$ and the graph of its inverse $f^{-1}$ are symmetric with respect to the line passing through the origin and making a 45° angle with the positive $x$-axis. (See Figure 1.5.)

**FIGURE 1.4**

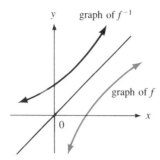

**FIGURE 1.5**

**EXAMPLE 7**   We have shown in Example 3 that if the function $f$ is defined by $f(x) = 5x + 2$, then its inverse is defined by $f^{-1}(x) = \frac{x-2}{5}$. Sketch the graphs of both $f$ and $f^{-1}$, and note the symmetry.

**Solution**   Recall from basic algebra that if a function $g$ is defined by an equation of the form $y = g(x) = mx + b$, where $m$ and $b$ are real numbers, then it is called a linear function and its graph is a straight line. Note that both $f$ and $f^{-1}$ are linear functions. Therefore, we need only find two points to draw each graph. Note also that $f(0) = 5(0) + 2 = 2$ and $f(2) = 5(2) + 2 = 12$. Thus, the points $(0, 2)$ and $(2, 12)$ are on the graph of $f$. Similarly, $f^{-1}(2) = \frac{2-2}{5} = 0$ and $f^{-1}(12) = \frac{12-2}{5} = 2$. Hence, the points $(2, 0)$ and $(12, 2)$ are on the graph of $f^{-1}$. The two graphs are sketched in Figure 1.6. Notice the symmetry with respect to the line passing through the origin and making a 45° angle with the positive $x$-axis.
**Do Exercise 39.**

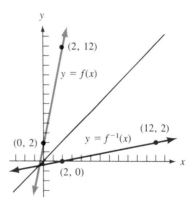

**FIGURE 1.6**

**EXAMPLE 8**   **a.** Sketch the graph of the function $g$ of Example 6, and show graphically that it is not one-to-one.
  **b.** Sketch the graph of the restriction of the function $g$ of Example 6 to the interval $[2, \infty)$.
  **c.** Sketch the graph of the inverse function $h^{-1}$ found in Example 6.

**Solution**    **a.** Recall from basic algebra that to sketch a graph with infinitely many points we usually find a few points of the graph and then sketch a smooth curve passing through them. Later in the text, we discuss more advanced techniques, but at the present time, we proceed as in elementary algebra. Consider a few values of $x$ and calculate the corresponding values of $g(x)$. Some results are given in the following table:

| $x$ | $-1$ | 0 | 1 | 2 | 3 | 4 | 5 |
|-----|------|---|---|---|---|---|---|
| $g(x)$ | 14 | 9 | 6 | 5 | 6 | 9 | 14 |

Now plot the points $(-1, 14)$, $(0, 9)$, $(1, 6)$, $(2, 5)$, $(3, 6)$, $(4, 9)$, and $(5, 14)$. Sketch a smooth curve through these points to obtain the graph of $g$. Observe that some horizontal line intersects the graph at two distinct points. Thus, the function $g$ is not one-to-one. (See Figure 1.7a.)

**b.** The graph of the restriction of $g$ to the interval $[2, \infty)$ is the half of the graph of $g$ which is directly above that interval on the $x$-axis. (See Figure 1.7b.)

**c.** To sketch the graph of $h^{-1}$, we consider several values of $x$ and calculate the corresponding values of $h^{-1}(x)$. Since $h^{-1}(x) = \sqrt{x - 5} + 2$, we choose values of $x$ to yield perfect squares for $x - 5$. The results are

| $x$ | 5 | 6 | 9 | 14 |
|-----|---|---|---|-----|
| $h^{-1}(x)$ | 2 | 3 | 4 | 5 |

We now plot the points $(5, 2)$, $(6, 3)$, $(9, 4)$, and $(14, 5)$ and sketch a smooth curve through these points to obtain the graph of $h^{-1}$. (See Figure 1.7b.) Observe the symmetry of the two graphs with respect to the line passing through the origin and making a $45°$ angle with the positive $x$-axis.

**Do Exercise 43.**

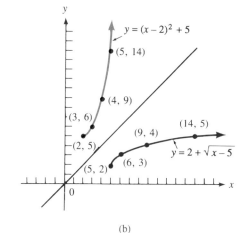

**FIGURE 1.7**                                         (a)                                                                  (b)

# Exercise Set 1.2

*In Exercises 1–6, show that the function is not one-to-one.*

**1.** The function $f$ defined by $f(x) = x^4$.

**2.** The function $g$ defined by $g(x) = (x - 3)^2$.

**3.** The function $h$ defined by $h(x) = (x + 3)^2$.

**4.** The function $f$ defined by

| $x$ | 1 | 3 | 5 | 7 | 9 | 11 |
|---|---|---|---|---|---|---|
| $f(x)$ | $b$ | $c$ | $a$ | $b$ | $c$ | $d$ |

**5.** The function $g$ defined by

| $x$ | $a$ | $c$ | $e$ | $i$ | $k$ | $m$ | $n$ |
|---|---|---|---|---|---|---|---|
| $g(x)$ | 2 | 5 | 0 | 4 | 5 | 0 | 3 |

**6.** Let $P$ be the set of all human beings and let $W$ be the set of all women. Let $M$ be the function from $P$ to $W$ defined as follows. For each $x$ in $P$, $M(x)$ is $x$'s mother.

*In Exercises 7–16, a function f is defined. In each case:*
*a. Show that f is one-to-one.*
*b. Describe $f^{-1}$.*

**7.** The function $f$ defined by

| $x$ | $a$ | $b$ | $c$ | $d$ | $e$ |
|---|---|---|---|---|---|
| $f(x)$ | 2 | 5 | 3 | 1 | $-7$ |

**8.** The function $f$ defined by

| $x$ | 2 | 3 | 5 | 7 | 11 | 13 | 17 |
|---|---|---|---|---|---|---|---|
| $f(x)$ | $a$ | $c$ | $n$ | $r$ | $d$ | $e$ | $b$ |

**9.** The function $f$ defined by

| $x$ | 1 | 2 | 3 | 4 | 5 | 6 | 7 |
|---|---|---|---|---|---|---|---|
| $f(x)$ | $s$ | $t$ | $u$ | $v$ | $w$ | $y$ | $z$ |

**10.** The domain of $f$ is $\{-8, -7, -6, -5, -4, -3, -2\}$ and $f(x) = x^2$.

**11.** The domain of $f$ is $\{-3, -1, 0, 2, 4, 5\}$ and $f(x) = x^2$.

**12.** The domain of $f$ is the interval $[4, \infty)$ and $f(x) = \sqrt{x}$.

**13.** The domain of $f$ is the interval $[3, \infty)$ and $f(x) = \sqrt{x - 3}$.

**14.** The function $f$ defined by $f(x) = 3x - 4$.

**15.** The function $f$ defined by $f(x) = -2x + 7$.

**16.** The function $f$ defined by $f(x) = x^3$.

*In Exercises 17–21, assume that the given function is one-to-one and find a formula for its inverse.*

**17.** The function $f$ defined by $y = f(x) = \dfrac{x + 5}{x + 3}$.

**18.** The function $g$ defined by $y = g(x) = \dfrac{2x + 5}{3 - x}$.

**19.** The function $h$ defined by $y = h(x) = \dfrac{3x + 2}{1 - x}$.

**20.** The function $f$ defined by $y = f(x) = \dfrac{5x}{x + 7}$.

**21.** The function $g$ defined by $y = g(x) = \dfrac{x}{x - 1}$. What do you notice comparing the formulas for $g$ and $g^{-1}$?

*In Exercises 22–32, a function that is not one-to-one is defined. In each case, do the following:*
*a. Find a subset B of the domain of the function so that B has as many elements as possible and the restriction of the function to B is one-to-one.*
*b. Describe the inverse of the restriction of the function to B.*

**22.** The function $f$ defined by

| $x$ | $a$ | $b$ | $c$ | $d$ | $e$ | $i$ | $k$ | $l$ | $m$ |
|---|---|---|---|---|---|---|---|---|---|
| $f(x)$ | 2 | 5 | 3 | 1 | 3 | 0 | 5 | 7 | $-3$ |

**23.** The function $g$ defined by

| $x$ | $-1$ | 3 | 5 | 7 | 8 | 9 | 11 | 15 |
|---|---|---|---|---|---|---|---|---|
| $g(x)$ | $c$ | $a$ | $a$ | $b$ | $d$ | $a$ | $c$ | $e$ |

**24.** The function $h$ defined by $h(x) = x^2 + 6x + 16$.

**25.** The function $f$ defined by $f(x) = x^2 - 4x + 5$.

**26.** The function $g$ defined by $g(x) = x^2 - 8x - 3$.

**27.** The function $h$ defined by $h(x) = x^2 - 10x + 27$.

**28.** The function $f$ defined by $f(x) = 2x^2 + 8x - 1$.

**29.** The function $g$ defined by $g(x) = -x^2 + 4x - 3$.

**30.** The function $h$ defined by $h(x) = -x^2 + 6x + 3$.

**31.** The function $f$ defined by $f(x) = (x - 3)^4 + 2$.

**32.** The function $g$ defined by $g(x) = (5 - x)^4 - 3$.

**33.** Sketch the graph of $g$ of Exercise 2 and show geometrically that it is not one-to-one.

**34.** Sketch the graph of $h$ of Exercise 3 and show geometrically that it is not one-to-one.

**35.** Sketch the graph of both $f$ and $f^{-1}$ where $f$ is the function of Exercise 10.

**36.** Sketch the graph of both $f$ and $f^{-1}$ where $f$ is the function of Exercise 11.

**37.** Sketch the graph of both $f$ and $f^{-1}$ where $f$ is the function of Exercise 12.

**38.** Sketch the graph of both $f$ and $f^{-1}$ where $f$ is the function of Exercise 13.

**39.** Sketch the graph of both $f$ and $f^{-1}$ where $f$ is the function of Exercise 14.

**40.** Sketch the graph of both $f$ and $f^{-1}$ where $f$ is the function of Exercise 15.

**41.** Sketch the graph of both $f$ and $f^{-1}$ where $f$ is the function of Exercise 16.

*In Exercises 42–48, sketch the graphs of **a.** the given function, **b.** an appropriate one-to-one restriction of the given function, and **c.** the inverse of the restriction graphed in part **b**.*

**42.** The function $h$ of Exercise 24.

**43.** The function $f$ of Exercise 25.

**44.** The function $g$ of Exercise 26.

**45.** The function $h$ of Exercise 27.

**46.** The function $f$ of Exercise 28.

**47.** The function $g$ of Exercise 29.

**48.** The function $h$ of Exercise 30.

# 1.3   Some Special Functions

To use mathematics effectively in applications we need a certain number of functions to serve as mathematical models. This section is devoted almost exclusively to rational functions.

## Linear Functions

Recall that if $m$ and $b$ are any two real constants, the function defined by

$$y = f(x) = mx + b$$

is called a *linear function*. The graph of a linear function is a straight line and the real number $m$ is called the *slope* of that line. The number $b$ is often called the *y-intercept* of the line because the point with coordinates $(0, b)$ is the intersection of the line with the y-axis. To find the *x-intercept*, simply replace $y$ by 0 in the equation $y = mx + b$ and solve for $x$. The following facts about linear functions should be familiar.

### Facts About Linear Functions

1. If the points with coordinates $(x_1, y_1)$ and $(x_2, y_2)$ are on a line, the slope $m$ of the line is

$$m = \frac{y_2 - y_1}{x_2 - x_1}$$

provided that $x_2 - x_1 \neq 0$. In the case of $x_2 - x_1 = 0$, the line is vertical. Its slope is not defined and the equation of such a line is $x = c$, where $c$ is the abscissa of the point of intersection of the line with the x-axis.

2. If two lines have slopes $m_1$ and $m_2$, respectively, then the two lines are parallel if, and only if, $m_1 = m_2$. The lines are perpendicular if, and only if, $m_1 m_2 = -1$.

3. If a line is parallel to the $x$-axis, its slope is 0. Replacing $m$ by 0 in the equation of a straight line, we get $y = b$. Thus, if a line is parallel to the $x$-axis and passes through the point $(0, b)$, its equation is $y = b$.

A typical problem you will encounter is having to find the equation of a line given some information about that line. Though there are several ways to do this, we suggest the following method.

To find the equation of a straight line, select a point arbitrarily on the line and call its coordinates $(x, y)$. Using given information, find the slope of the line in two ways, at least one of which makes use of the point $(x, y)$. Get the equation of the line by setting the two answers obtained for the slope equal to each other. Of course, this equation can often be simplified.

**EXAMPLE 1**   Find the equation of each of these lines:

**a.** The line passing through the points $(-1, 3)$ and $(2, 9)$.
**b.** The line passing through the point $(-2, 5)$ and parallel to the line of part **a**.
**c.** The line having $x$-intercept 3 and perpendicular to the graph of $3x + 2y = 12$. (See Figures 1.8 and 1.9.)

**Solution**   **a.** Select a point arbitrarily on the line and let its coordinates be $(x, y)$. Since the points with coordinates $(x, y)$ and $(-1, 3)$ are on the line, the slope is given by

$$m = \frac{y - 3}{x - (-1)} = \frac{y - 3}{x + 1}$$

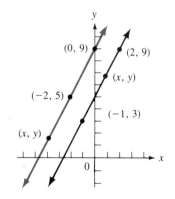

Using the fact that the points $(-1, 3)$ and $(2, 9)$ are also on the line, we find that the slope is 2 since $\frac{9 - 3}{2 - (-1)} = \frac{6}{3} = 2$. Setting the two answers equal, we obtain

$$\frac{y - 3}{x + 1} = 2$$

This may be written

$$y - 3 = 2(x + 1)$$

This is called the *point-slope* form of the line. The equation can also be written

$$y = 2x + 5$$

**FIGURE 1.8**

which is called the *slope-intercept* form of the line.

If we replace $y$ by 0 in the equation of the line, then

$$0 = 2x + 5$$

which yields $x = \frac{-5}{2}$. Hence, the point $\left(\frac{-5}{2}, 0\right)$ is the intersection of the line with the $x$-axis. Therefore, the $x$-intercept is $\frac{-5}{2}$.

**b.** Since the line of part **a** has slope 2, any line parallel to it will also have slope 2. Let $(x, y)$ be an arbitrary point on the line. Then the points $(-2, 5)$ and $(x, y)$ are on the line and the slope is $\frac{y - 5}{x - (-2)}$. Since the slope is 2, we write

$$\frac{y - 5}{x + 2} = 2$$

This can be written

$$y = 2x + 9$$

**c.** Since the line has $x$-intercept 3, the point $(3, 0)$ is on the line. Let $(x, y)$ be an arbitrary point on the line. Then the slope of the line is

$$m = \frac{y - 0}{x - 3} = \frac{y}{x - 3}$$

The equation $3x + 2y = 12$, may be written

$$y = \left(\frac{-3}{2}\right)x + 6$$

Therefore, the graph of that equation has slope $\frac{-3}{2}$. If a line perpendicular to that graph has slope $m$, we must have $\left(\frac{-3}{2}\right)m = -1$. Thus, $m = \frac{2}{3}$. Setting the two answers we obtained for $m$ equal to each other, we get

$$\frac{y}{x - 3} = \frac{2}{3}$$

This equation can be written in the point-slope form as

$$y - 0 = \frac{2}{3}(x - 3)$$

and in slope-intercept form as

$$y = \left(\frac{2}{3}\right)x - 2$$

**Do Exercise 1.**

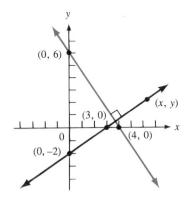

**FIGURE 1.9**

**EXAMPLE 2**   The owner of a business knows from experience that if the price per unit of a certain commodity is $50, consumers will purchase 240 units; while if the price is increased to $75 per unit, consumers will purchase only 185 units. Assuming that the demand equation is linear, find a formula for the revenue $R$ in terms of the quantity $q$ units purchased by the consumers.

**Solution**   The graph of the demand equation is a line. The points with coordinates $(240, 50)$ and $(185, 75)$ are on that line. Let $(q, p)$ be the coordinates of an arbitrary point on the line. Then the slope of the line is

$$m = \frac{p - 50}{q - 240}$$

and

$$m = \frac{50 - 75}{240 - 185} = \frac{-25}{55} = \frac{-5}{11}$$

Thus,

$$\frac{p - 50}{q - 240} = \frac{-5}{11}$$

This can be written

$$p - 50 = \frac{-5}{11}(q - 240)$$

Thus,

$$p = \frac{-5}{11}(q - 240) + 50$$

and

$$p = \frac{-5}{11}q + \frac{1750}{11}$$

But the revenue is obtained by multiplying the price $p$ per unit by the number $q$ of units. That is,

$$R = pq = \left(\frac{-5}{11}q + \frac{1750}{11}\right)q$$

$$= \frac{-5}{11}q^2 + \frac{1750}{11}q$$

**Do Exercise 39a and b.**    ■

## Quadratic Functions

Recall from basic algebra that if $a$, $b$, and $c$ are real numbers with $a \neq 0$, the function defined by

$$y = f(x) = ax^2 + bx + c$$

is a *quadratic function*. The graph of a quadratic function is a *parabola*, concave up if $a > 0$ (see Figure 1.10a) and concave down if $a < 0$ (see Figure 1.10b). The *vertex*

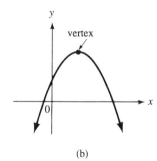

**FIGURE 1.10**                    (a)                         (b)

of the parabola is its lowest point when $a > 0$ and its highest point when $a < 0$. The abscissa of the vertex is the number $\frac{-b}{2a}$ and the parabola is symmetric with respect to a vertical line passing through the vertex.

**EXAMPLE 3**    The function $f$ is defined by

$$y = f(x) = -2x^2 + 8x + 1$$

Sketch its graph.

**Solution**    The graph is a parabola which is concave down since $-2 < 0$. Therefore, the vertex will be the highest point of the graph. Its abscissa is 2 since $\frac{-8}{2(-2)} = 2$. To find the ordinate of the vertex, we calculate the value of $y$ when $x = 2$. We find $y = -2(2^2) + 8(2) + 1 = 9$. We conclude that the vertex has coordinates $(2, 9)$. Hence the parabola is symmetric with respect to the vertical line whose equation is $x = 2$; therefore, we need sketch only the half of the parabola to the left of that line. The other half is obtained by symmetry. We calculate the values of $y$ when $x$ is replaced successively by $-1$, $0$, and $1$. The results are tabulated as

| $x$ | $-1$ | $0$ | $1$ |
|-----|------|-----|-----|
| $y$ | $-9$ | $1$ | $7$ |

Therefore, the points $(-1, -9)$, $(0, 1)$, $(1, 7)$, and $(2, 9)$ are on the graph of $f$. We plot these points and draw a smooth curve through them. (See Figure 1.11a.) The other half of the graph results from symmetry. (See Figure 1.11b.)
**Do Exercise 13.**

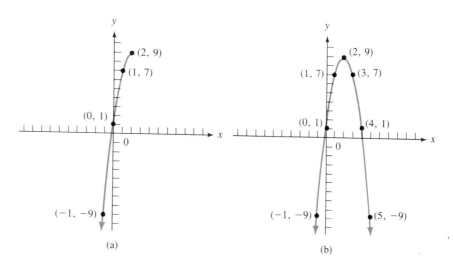

**FIGURE 1.11**

    (a)          (b)

**EXAMPLE 4**    We wish to enclose a rectangular region against an existing wall. No fence is required against the wall. Fencing materials cost $20 per foot for the side parallel to the wall and $15 per foot for the other two sides. Find the dimensions of the largest area that can be enclosed for $840.

**Solution**  Let $x$ feet be the length of each side perpendicular to the wall. The total length of these two sides is $2x$ feet, and the cost of materials for these two sides is $30x$ since $15(2x) = 30x$. Thus, $840 - 30x$ dollars is available for the side parallel to the wall. Since $20 a foot is to be spent on that side, its length is $\frac{840 - 30x}{20}$ feet. Thus, the area $A$ square feet of the enclosed region is

$$A = f(x) = \left(\frac{840 - 30x}{20}\right)x = \frac{-3}{2}x^2 + 42x$$

The graph of this equation is a parabola which is concave down since $\frac{-3}{2}$, the coefficient of $x^2$, is negative. The vertex is the highest point of the graph and the $y$-coordinate of the vertex gives the maximum value of $A$. The $x$-coordinate of the vertex is 14 since in the foregoing equation

$$b = 42, \ a = \frac{-3}{2}, \ \text{and} \ \frac{-b}{2a} = \frac{-42}{2\left(\frac{-3}{2}\right)} = \frac{-42}{-3} = 14$$

Thus, the sides perpendicular to the wall must be 14 feet. The other side must be 21 feet since $\frac{840 - 30(14)}{20} = 21$.

**Do Exercise 41.**  ■

Here is another example.

**EXAMPLE 5**  The racquetball pro at an athletic club knows that if she charges $40 an hour for private lessons she will teach 45 hours a month, but if she charges only $20 an hour she will teach 110 hours. The club charges the pro $50 + 12x$ dollars for the use of $x$ hours of court time during a month.

  **a.** Find the price $\$p$ per hour of private lesson in terms of the number of $x$ hours the pro is teaching per month, assuming a linear relationship between $p$ and $x$.
  **b.** Find an equation giving the monthly revenue $\$R$ in terms of $x$.
  **c.** How many hours should the pro teach each month to maximize her profit?

**Solution**  **a.** The graph of the equation will be a line. Let $(x, p)$ be an arbitrary point on the line. The points $(45, 40)$ and $(110, 20)$ are on the line. Hence, the slope of the line can be obtained in the following ways:

$$m = \frac{p - 40}{x - 45} \quad \text{and} \quad m = \frac{40 - 20}{45 - 110} = \frac{-4}{13}$$

Thus,

$$\frac{p - 40}{x - 45} = \frac{-4}{13}$$

This equation may be written

$$p = \frac{-4}{13}x + \frac{700}{13}$$

**b.** The revenue $R$ is obtained by multiplying the number of $x$ hours taught by the price $p$ per hour. Therefore,

$$R = xp = x\left(\frac{-4}{13}x + \frac{700}{13}\right), \text{ so } R = \frac{-4}{13}x^2 + \frac{700}{13}x$$

**c.** Since the club charges the pro $50 + 12x$ dollars to teach $x$ hours in a month, the monthly profit $P$ is given by

$$P = \frac{-4}{13}x^2 + \frac{700}{13}x - (50 + 12x) = \frac{-4}{13}x^2 + \frac{544}{13}x - 50$$

The graph of this equation is a parabola that is concave down since $\frac{-4}{13}$, the coefficient of $x^2$, is negative. The vertex is the highest point of the graph and its $y$-coordinate gives the maximum value of the monthly profit. To find the $x$-coordinate of the vertex, note that $a = \frac{-4}{13}$ and $b = \frac{544}{13}$. The $x$-coordinate of the vertex is 68 since

$$\frac{-\dfrac{544}{13}}{2\left(\dfrac{-4}{13}\right)} = 68$$

Thus, the pro should teach 68 hours per month to maximize her profit.
**Do Exercise 45.**                                                                  ■

## Polynomial Functions

Linear and quadratic functions belong to a larger class of functions called *polynomial functions*.

> **DEFINITION 1.8:**    If $n$ is a nonnegative integer and $a_0, a_1, a_2, \ldots, a_n$ are real numbers with $a_0 \neq 0$, then the function $p$ defined by
>
> $$y = p(x) = a_0x^n + a_1x^{n-1} + a_2x^{n-2} + \cdots + a_{n-1}x + a_n$$
>
> is a *polynomial function of degree n*. The constant $a_0$ is called the *leading coefficient*.

For example, any quadratic function is a polynomial function of degree 2, and the function defined by $f(x) = -3x^5 + 6x^3 - .42x^2 + 7x - 15$ is a polynomial function of degree 5 with leading coefficient $-3$. Polynomial functions are *continuous* functions. When a function is continuous, its graph can be drawn without lifting the pencil. That is, the graph has no breaks. (See Figure 1.12.)

Points such as $A$ and $B$ of Figure 1.12b are called *critical points*. The graph of a polynomial function $p$ of degree $n$, has at most $n - 1$ critical points.

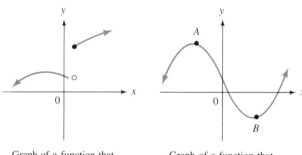

Graph of a function that
is not continuous

Graph of a function that
is continuous

**FIGURE 1.12**

(a)

(b)

**EXAMPLE 6**   The function $g$ is defined by $y = g(x) = \dfrac{x^3}{3} - \dfrac{x^2}{2} - 2x + \dfrac{7}{3}$. Sketch its graph.

**Solution**   We replace $x$ successively by $-2, -1, 0, 1, 2, 3$, and $4$ and calculate the corresponding values of $y$. The results are

| $x$ | $-2$ | $-1$ | $0$ | $1$ | $2$ | $3$ | $4$ |
|---|---|---|---|---|---|---|---|
| $g(x)$ | $\dfrac{5}{3}$ | $\dfrac{7}{2}$ | $\dfrac{7}{3}$ | $\dfrac{1}{6}$ | $-1$ | $\dfrac{5}{6}$ | $\dfrac{23}{3}$ |

Therefore, the points with coordinates $\left(-2, \frac{5}{3}\right)$, $\left(-1, \frac{7}{2}\right)$, $\left(0, \frac{7}{3}\right)$, and so on, are on the graph of the polynomial function $g$. We plot these points and draw a smooth curve passing through them.* (See Figure 1.13.)
**Do Exercise 19.**    ■

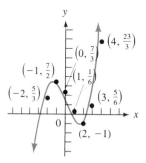

**FIGURE 1.13**

---

*The graph of $g$ sketched in Figure 1.13 has two critical points. Since the degree of the polynomial function $g$ is 3, its graph can have at most two critical points. Therefore, we know that all critical points of the graph of $g$ have been plotted and we need not wonder whether there are critical points for values of $x$ less than $-2$, or for values of $x$ larger than 6; that is, the graph plotted in Figure 1.13 is an accurate representation of the graph of $g$.

If $p$ and $q$ are polynomial functions, their quotient is called a *rational function*. For example, if the functions $f$ and $g$ are defined by

$$f(x) = \frac{x^2 + 6x - 7}{3x + 5} \quad \text{and} \quad g(x) = \frac{-2x^3 + 5x^2 - 3x + 2}{2x^2 - 13x + 17}$$

then $f$ and $g$ are rational functions.

## Multiple Formulas and the Greatest Integer Function

We conclude this section with a few remarks on functions. In our work, most of the functions will be defined by formulas. Sometimes, however, we need more than one formula to define a function, as illustrated in the following example.

**EXAMPLE 7**   Betty has a job paying $10 per hour. She gets paid time and a half for overtime (time over 40 hours per week). If she works $x$ hours during a certain week, express her earnings $\$I(x)$ in terms of $x$.

**Solution**   If $0 \le x \le 40$, then clearly $I(x) = 10x$. However, if $40 \le x \le 168$, Betty will get paid $400 for the first 40 hours, plus the overtime pay. Note that she has worked $x - 40$ hours of overtime for which she receives $\$15(x - 40)$ since her overtime pay is $15 per hour. In that case, $I(x) = 400 + 15(x - 40) = 15x - 200$. We express our result as

$$I(x) = \begin{cases} 10x, & \text{if } 0 \le x \le 40 \\ 15x - 200, & \text{if } 40 < x \le 168 \end{cases}$$

**Do Exercise 37.**   ■

You have undoubtedly noticed the way merchants tend to price their commodities. The price of a steak dinner at Pat's House of Steaks is $7.99, not $8. Obviously, the manager hopes that customers will read the 7 and overlook the .99. In the same spirit, we define the greatest integer function as follows.

> **DEFINITION 1.9:**   For each real number $x$, let $[x]$ denote the greatest integer that is less than or equal to $x$. Then the function defined by $f(x) = [x]$ for all real numbers $x$ is called the *greatest integer function*.

**EXAMPLE 8**   Let $f$ be the greatest integer function.

    **a.** Find $f(-2.3)$, $f(-2)$, $f(3)$, and $f(3.2)$.
    **b.** Sketch the graph of $f$.

**Solution**   **a.** $f(-2.3) = [-2.3] = -3$, since $-3$ is the largest integer less than or equal to $-2.3$. Similarly, $f(-2) = -2$, $f(3) = 3$, and $f(3.2) = 3$.

**b.** To sketch the graph of $f$, first observe that if $n$ is any integer, then $f(n) = [n] = n$. Thus, the point $(n, n)$ is on the graph of $f$. Further, if $x$ is a real number such that $n < x < n + 1$, then $f(x) = n$. Thus, the point $(x, n)$ is on the graph of $f$. Since all points of the graph of $f$ with $x$-coordinates between two consecutive integers have the same $y$-coordinate, these points will form a line segment parallel to the $x$-axis. The graph is sketched in Figure 1.14.

**Do Exercise 31.**

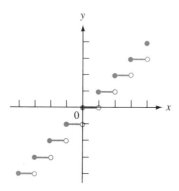

**FIGURE 1.14**

## Exercise Set 1.3

*In Exercises 1–12, a straight line is described. Find its equation.*

**1.** The line passing through the points $(-1, -3)$ and $(3, 5)$.

**2.** The line passing through the points $(3, 4)$ and $(5, 2)$.

**3.** The line passing through the points $(-5, -2)$ and $(-3, 6)$.

**4.** The line passing through the points $(-4, -6)$ and $(-4, 8)$.

**5.** The line passing through the points $(3, 2)$ and $(-13, 2)$.

**6.** The line passing through the point $(2, 5)$ and parallel to the line of Exercise 1.

**7.** The line passing through the point $(3, 8)$ and parallel to the line of Exercise 2.

**8.** The line passing through the point $(2, 7)$ and parallel to the line of Exercise 3.

**9.** The line passing through the point $(3, 9)$ and perpendicular to the line of Exercise 4.

**10.** The line passing through the point $(12, -15)$ and perpendicular to the line whose equation is $2x + 5y = 25$.

**11.** The line with $x$-intercept $-7$ and perpendicular to the graph of the equation $12x - 6y = 36$.

**12.** The line with $x$-intercept 5 and $y$-intercept $-4$.

*In Exercises 13–18, sketch the graph of the given function showing the vertex and symmetry.*

**13.** The function $f$ defined by $f(x) = x^2 + 10x - 4$.

**14.** The function $g$ defined by $g(x) = 2x^2 - 12x + 7$.

**15.** The function $h$ defined by $h(x) = 3x^2 - 18x + 5$.

**16.** The function $f$ defined by $f(x) = -x^2 + 4x - 3$.

**17.** The function $g$ defined by $g(x) = -4x^2 + 24x - 9$.

**18.** The function $h$ defined by $h(x) = -.5x^2 - 3x + 4$.

*In Exercises 19–25, a polynomial function is defined.*
*a. Find the maximum number of critical points of the graph.*
*b. Sketch the graph.*

**19.** The function $f$ defined by $f(x) = x^3 + 3x^2 - x - 3$.

**20.** The function $g$ defined by $g(x) = x^3 + x^2 - 9x - 9$.

**21.** The function $h$ defined by $h(x) = x^3 + 5x^2 + 2x - 8$.

**22.** The function $f$ defined by $f(x) = x^3 - 19x - 30$.

**23.** The function $f$ defined by $f(x) = x^3$.

**24.** The function $g$ defined by $g(x) = x^3 - 3x^2 + 3x - 1$.
(*Hint:* Recall the expansion of $(a - b)^3$.)

**25.** The function $h$ defined by $h(x) = x^4$.

*In Exercises 26–29, a rational function is defined.*
*a. Find the domain of the function.*
*b. Find the value of the function at the given values of x.*

**26.** $f(x) = \dfrac{x^2 - 5x + 6}{x - 3}$; $x = -1, x = 2, x = 4$

**27.** $f(x) = \dfrac{x^2 - 3x + 4}{x^2 + 2x - 3}$; $x = 2, x = 4, x = 6$

**28.** $f(x) = \dfrac{x + 2}{x^2 + 2x - 15}$; $x = -4, x = -2, x = 4$

**29.** $f(x) = \dfrac{x^2 - 5x + 3}{x^3 - x^2 - x + 1}$; $x = -2, x = 0, x = 2$

**30.** Let $f$ be the greatest integer function, and let $g$ and $h$ be defined by $g(x) = 2x$ and $h(x) = \frac{x}{2}$.
   **a.** Sketch the graph of the composition $f \circ g$.
   **b.** Sketch the graph of the composition $g \circ f$.
   **c.** Sketch the graph of the composition $f \circ h$.
   **d.** Sketch the graph of the composition $h \circ f$.

**31.** Let $f$ be the greatest integer function, and let $g$ and $h$ be defined by $g(x) = |x|$ and $h(x) = x - 3$.
   **a.** Sketch the graph of the composition $f \circ g$.
   **b.** Sketch the graph of the composition $g \circ f$.
   **c.** Sketch the graph of the composition $f \circ h$.
   **d.** Sketch the graph of the composition $h \circ f$.

**32.** Let $P(x)$ cents be the amount of postage necessary to mail a letter first class within the United States, where the letter weighs $x$ ounces. Call up your local post office to obtain the value of $P(x)$ for different values of $x$, then sketch the graph of the function $P$.

**33.** Let $f(x) = (-1)^{[x/3]}$. Calculate the values of $f(0), f(1),$ $f(2), f(3), f(4), f(5), f(6), f(7), f(8), f(9), f(10), f(11),$ and $f(12)$. Do you notice a pattern?

**34.** Let $g$ be the absolute value function defined by $g(x) = |x|$. Sketch its graph.

**35.** Let $h(x) = |x - 2|$. Sketch the graph of the function $h$.

**36.** Roy has a job paying $12 per hour. He gets paid time and a half for overtime (time over 40 hours per week). If he works $x$ hours during a week, express his earnings $I(x)$ in terms of $x$. Sketch the graph of the function $I$.

**37.** Ralph has a job paying $14 per hour for the first 40 hours of the week. He gets time and a half for the next 8 hours and double time after that. If he works $x$ hours during a week, express his earnings $I(x)$ in terms of $x$. Sketch the graph of the function $I$.

**38.** The owner of a dude ranch knows that if he charges $40 per day to rent a horse, 30 horses will be rented; and if he charges $25 a day per horse, 60 horses will be rented.
   **a.** Find an equation relating the price $p$ per horse per day and the number $x$ of horses rented per day, assuming this equation is linear.
   **b.** Find an equation giving the revenue $R$ in terms of the number $x$ of horses rented.
   **c.** If the daily cost to have $x$ horses is $(90 + 10x)$, how many horses should the owner have to maximize his daily profit? What is the maximum daily profit?

**39.** A store's market research department found that if the store priced a new electric razor at $45, 90 razors would be sold per month. On the other hand, 135 would be sold per month if the price were $30. The cost to the store to sell $x$ razors per month is $(200 + 21x)$.
   **a.** Find an equation relating the price $p$ per razor and the number $x$ of razors sold per month, assuming that this equation is linear.
   **b.** Find an equation giving the revenue $R$ in terms of the number $x$ of razors sold in a month.
   **c.** How many razors should be sold to maximize the monthly profit? What is the maximum monthly profit?

**40.** We wish to enclose a rectangular region against an existing wall. No fence is required along the wall. The cost of fencing materials for one of the sides perpendicular to the wall is $12 per linear foot, while the cost for the other two sides to be fenced is $9 per linear foot. What are the dimensions of the maximum area that can be enclosed if $966 is spent on fencing materials?

**41.** A farmer wishes to enclose a rectangular region. Fencing materials for one side will cost $13 per linear foot, while the other three sides will cost $8 per linear foot. Give the dimensions of the largest area that can be enclosed if $840 is available for fencing.

**42.** The members of the Oregon Ducklings, a Little League baseball team, have been collecting aluminum cans to help pay for the cost of equipment. They have 180 pounds of aluminum, for which they can get $.80 a pound now. If they continue to collect, they can get 6 more pounds per day, but the price per pound will decrease $.02 per day. Knowing that they can make only one transaction, how many days should they wait to maximize their income? What is the maximum income?

**43.** The Lebanese University of New York charged $160 per credit hour in the 1988–1989 academic year and students registered for a total of 225,000 credit hours. The president of the university wanted to raise tuition. Studies showed him that an increase of $x$ dollars per credit hour would cause a decrease of 1250x in the number of credit hours taught. He correctly calculated the increase that would create the maximum revenue from tuition in the next academic year. What was his answer?

44. The owner of an apple orchard knows that if 26 apple trees are planted on 1 acre of land, each tree will yield an average of 400 apples. For each additional tree per acre, the average yield per tree will decrease by 8. How many trees should be planted per acre to maximize the yield?

45. The manager of a sports equipment shop knows that 10 table tennis paddles will be sold per week if the price is $15 per paddle, while 30 paddles will be sold a week if the price is $5 per paddle.

   a. Find an equation to give the price $p per paddle in terms of the number of x paddles sold each week, assuming that the equation is linear.
   b. Find an equation giving the revenue $R in terms of x.
   c. If the cost of selling x paddles a week is $(25 + 4x), how many paddles should be sold each week to maximize the profit?

# 1.4  The Exponential Function

## The Laws of Exponents

For each positive integer $n$ and each real number $a$, $a^n$ denotes the product $a \cdot a \cdot a \cdot \cdots \cdot a$ ($n$ factors), and is called the *nth power of a*. The number $a$ is called the *base* and the positive integer $n$ is called the *exponent*. We will extend this definition and give a meaning to $a^x$ where $x$ is any real number.

   If $a$ and $b$ are real numbers, and $m$ and $n$ are positive integers, then the following are easily verified:

1. $a^m \cdot a^n = a^{m+n}$
2. $\dfrac{a^m}{a^n} = a^{m-n}$ (provided $a \neq 0$ and $m > n$)
3. $(a^m)^n = a^{mn}$
4. $(ab)^m = a^m b^m$
5. $\left(\dfrac{a}{b}\right)^m = \dfrac{a^m}{b^m}$ (provided $b \neq 0$)

These five equalities are often called the *laws of exponents*.
   We need to define $a^x$ in such a way that the laws of exponents always hold. Begin with

$$\frac{a^m}{a^n} = a^{m-n} \text{ (provided } a \neq 0 \text{ and } m > n)$$

and remove the restriction $m > n$. If $m = n$, we have

$$\frac{a^m}{a^m} = a^{m-m} \text{ (provided } a \neq 0)$$

which simplifies to

$$1 = a^0$$

Thus, we define $a^0 = 1$ whenever $a \neq 0$.

Let $n$ be a positive integer and $a \neq 0$. Replace $m$ by $-n$ in

$$a^m \cdot a^n = a^{m+n}$$

to get

$$a^{-n} \cdot a^n = a^{-n+n} = a^0 = 1$$

Thus,

$$a^{-n} = \frac{1}{a^n}$$

Thus, we define $a^{-n} = 1/a^n$ where $n$ is any positive integer and $a$ is any nonzero real number.

If $n$ is a positive integer, and $a$ and $b$ are real numbers such that $a^n = b$, we say that $a$ is a *square root* of $b$ when $n = 2$, a *cube root* of $b$ when $n = 3$, and an *nth root* of $b$ when $n > 3$.

In more advanced courses it can be shown that any nonzero complex number has exactly $n$ nth roots. For example, $-2$ and $2$ are the two square roots of 4 since $(-2)^2 = 2^2 = 4$. However, in general, at least $n - 2$ of the $n$ nth roots of a number are not real. In this book, unless otherwise specified, we will consider only nth roots of nonnegative real numbers and call the nonnegative nth root of a nonnegative real number its *principal nth root*. For example, the principal fourth root of 16 is 2 since $2^4 = 16$. Note that $-2$ is also a fourth root of 16 and the other two fourth roots of 16 are imaginary numbers. If $a \geq 0$ and $n$ is an integer greater than 1, the principal nth root of $a$ is denoted $\sqrt[n]{a}$. If $n = 2$, we write $\sqrt{a}$ instead of $\sqrt[2]{a}$. Note also that if $x$ is any real number, $x^2$ is nonnegative so that $\sqrt{x^2}$ is a real number. However, if $x$ is a negative number, then $\sqrt{x^2} \neq x$ since the principal square root of $x^2$ is nonnegative. Thus, we write

$$\sqrt{x^2} = |x|$$

for all real numbers $x$.

Suppose now that $n$ is an integer greater than 1 and $a$ is a nonnegative real number. Replace $m$ by $\frac{1}{n}$ in

$$(a^m)^n = a^{mn}$$

to get

$$(a^{1/n})^n = a^{1/n \cdot n} = a^1 = a$$

Therefore, $a^{1/n}$ must be the principal nth root of $a$. We write

$$a^{1/n} = \sqrt[n]{a}$$

Thus,

$$(a^{1/n})^m = (\sqrt[n]{a})^m$$

and if the third law of exponents is to hold,

$$a^{m/n} = (\sqrt[n]{a})^m$$

**DEFINITION 1.10:**    If $a$ is a nonnegative real number, and $m$ and $n$ are positive integers with $n > 1$, then

$$a^{m/n} = \sqrt[n]{a^m} = (\sqrt[n]{a})^{m*}$$

In more advanced calculus texts, a formal definition is given for $a^x$ where $a$ is a nonnegative real number and $x$ is any real number, rational or irrational. This definition is beyond the scope of this text. However, the important fact is that the laws of exponents hold in all cases.

If $a$ and $b$ are any two nonnegative real numbers and $x$ and $y$ are any two real numbers, then

$$a^x \cdot a^y = a^{x+y}$$

$$\frac{a^x}{a^y} = a^{x-y}$$

$$(a^x)^y = a^{xy}$$

$$(ab)^x = a^x \cdot b^x$$

$$\left(\frac{a}{b}\right)^x = \frac{a^x}{b^x}$$

Of course, we need to make the necessary restrictions to avoid division by 0 and to avoid $0^0$, which is not defined.

## The Exponential Function with Base $b$

We are now ready to define a new type of function.

**DEFINITION 1.11:**    If $b$ is a positive real number and $b \neq 1$, the *exponential function* with base $b$ is defined by

$$f(x) = b^x$$

for all real numbers $x$.

As stated earlier, we require the base $b$ to be nonnegative to avoid imaginary numbers (for example $(-3)^{1/2}$ is imaginary). We also require that base $b$ be different from 0 and from 1, since either of these two numbers used as a base would yield a constant function.

---

*The definition of $a^{m/n}$ is also valid when $a$ is negative and $n$ is odd.

**EXAMPLE 1**    Sketch the graph of the exponential function with base 2.

**Solution**    We have $y = 2^x$ for all real numbers $x$. Replace $x$ successively by $-3, -2, -1, 0,$ 1, 2, and 3 and calculate $y$. The results are

| $x$ | $-3$ | $-2$ | $-1$ | 0 | 1 | 2 | 3 |
|-----|------|------|------|---|---|---|---|
| $y$ | $\dfrac{1}{8}$ | $\dfrac{1}{4}$ | $\dfrac{1}{2}$ | 1 | 2 | 4 | 8 |

Thus, the points $\left(-3, \frac{1}{8}\right), \left(-2, \frac{1}{4}\right), \left(-1, \frac{1}{2}\right),$ (0, 1), (1, 2), (2, 4), and (3, 8) are on the graph of the function. We plot these seven points and draw a smooth curve through them to get a sketch of the graph of $y = 2^x$. (See Figure 1.15.)
**Do Exercise 11.**

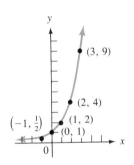

**FIGURE 1.15**

Properties of the Exponential Function with Base $b$, Where $b$ is an Arbitrary Positive Number and $b \neq 1$

1. The domain is the set of all real numbers.
2. The range is the set of all positive real numbers.
3. The graph of $y = b^x$ passes through the point (0, 1) since $b^0 = 1$.
4. If $b > 1$, the exponential function with base $b$ is an increasing function. That is, $b^{x_1} < b^{x_2}$ whenever $x_1 < x_2$. (See Figure 1.16a.)
5. If $b < 1$, the exponential function with base $b$ is a decreasing function. That is, $b^{x_1} > b^{x_2}$ whenever $x_1 < x_2$. (See Figure 1.16b.)
6. The graph of $y = b^x$ can be drawn without lifting the pencil. That is, the exponential function with base $b$ is a continuous function.
7. $b^{x+y} = b^x \cdot b^y$.
8. $b^{x-y} = \dfrac{b^x}{b^y}$.
9. $b^{xy} = (b^x)^y$.

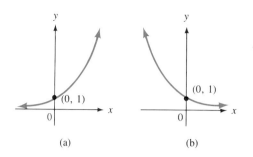

**FIGURE 1.16**                    (a)                    (b)

The expression $\left(1 + \frac{1}{n}\right)^n$, where $n$ is a positive integer, occurs frequently in mathematics. As $n$ increases without bound, this expression approaches a useful irrational number. This number is denoted $e$ and its value is approximately 2.71828. The proof that the expression $\left(1 + \frac{1}{n}\right)^n$ approaches a specific number as the value of $n$ increases without bound requires concepts not yet covered at this level. However, the value of the expression $\left(1 + \frac{1}{n}\right)^n$ can be obtained for any value of $n$. For $n = 10$, using a calculator we get

| ENTER | PRESS | DISPLAY |
|---|---|---|
|  | ON/C | 0 |
|  | ( | 0 |
| 1 |  | 1 |
|  | + | 1 |
| 1 |  | 1 |
|  | ÷ | 1 |
| 10 |  | 10 |
|  | ) | 1.1 |
|  | $y^x$ | 1.1 |
| 10 |  | 10 |
|  | = | 2.59374246 |

You may gain some insight for what happens by completing this table.

| $n$ | 1 | 10 | 100 | 1000 | 10,000 | 100,000 |
|---|---|---|---|---|---|---|
| $\left(1 + \frac{1}{n}\right)^n$ | 2.0000 | 2.5937 | 2.7048 | 2.7169 |  |  |

> **DEFINITION 1.12:**   The exponential function with base $e$ is often called the *natural exponential function*.

This function is so useful that most calculators have a function key $e^x$.

## Applications

Exponential functions are important in mathematics. In this section, we give two illustrations of how they occur in a natural way. The first includes a brief discussion of compound interest and the second is an elementary treatment of population growth and decay, and exponential depreciation.

### Compound Interest

Suppose $P$ is deposited in a bank that pays interest at a stated annual rate $r$, written as a decimal. For example, if the annual rate of interest is 7%, then $r = 0.07$. Suppose further that there are $k$ interest periods of equal lengths per year. Then, the interest for each period is added to the principal at the end of that period and this interest, as well as the principal, earns interest during the following periods. In such cases, we say that the interest is *compounded*. The rate of interest per conversion period is $i$, where $i = \frac{r}{k}$. Let $A_n$ be the amount in the bank account at the end of the $n$th conversion period. We shall derive a formula for $A_n$ in terms of the principal $P$, the rate of interest per period $i$, and the number of conversion periods $n$.

During the first period, the interest earned is $P \cdot i$. Thus,

$$A_1 = P + P \cdot i = P(1 + i)$$

The interest earned during the second period is $A_1 \cdot i$. Therefore,

$$A_2 = A_1 + A_1 i = A_1(1 + i)$$
$$= P(1 + i)(1 + 1) = P(1 + i)^2$$

The interest earned during the third period is $A_2 \cdot i$. It follows that

$$A_3 = A_2 + A_2 i = A_2(1 + i)$$
$$= P(1 + i)^2(1 + i) = P(1 + i)^3$$

Continuing in this way,

$$A_n = P(1 + i)^n$$

**EXAMPLE 2**   Suppose that $500 is deposited at 8% annual interest, compounded quarterly. Find the amount in the account 5 years later. What is the total interest earned?

**Solution**   We use the formula

$$A_n = P(1 + i)^n$$

with $P = 500$, $i = 0.02$ (since $i = \frac{r}{k} = \frac{0.08}{4} = 0.02$), and $n = 20$ (since $n = kt = 4 \cdot 5 = 20$). We get

$$A_{20} = 500(1 + 0.02)^{20}$$
$$= 500(1.02)^{20} \text{*}$$
$$\doteq 500(1.4859474) = 742.9737\text{**}$$

To the nearest cent, the amount after 5 years is $742.97. The interest is $242.97 since $742.97 - 500 = 242.97$.

**Do Exercise 19.**                                                                    ■

## Nominal and Effective Rates

> **DEFINITION 1.13:**   The advertised yearly rate of interest is often called the *nominal interest rate*, while the actual percentage by which an account grows during each year is called the *effective interest rate*.

**EXAMPLE 3**   A bank advertises a nominal interest rate of 6% compounded monthly. What is the effective interest rate?

**Solution**   For convenience, assume that $100 is deposited in the bank and use the formula

$$A_n = P(1 + i)^n$$

---

*The value of $(1.02)^{20}$ may be found using tables or the $y^x$ key on a scientific calculator. We refer you to any book of mathematical tables. Find the page giving values of $(1 + i)^n$. To find $(1.02)^{20}$ look for 2% across the top and 20 (for 20 conversion periods) down the side. Where the column and row intersect you should find 1.485947, which is the approximate value of $(1.02)^{20}$. In the example, we used a calculator as follows:

| ENTER | PRESS | DISPLAY |
|-------|-------|---------|
|       | ON/C  | 0.      |
| 1.02  |       | 1.02    |
|       | $y^x$ | 1.02    |
| 20    |       | 20.     |
|       | ×     | 1.4859474 |
| 500   |       | 500     |
|       | =     | 742.9737 |

**The symbol $\doteq$ means "approximately equal to."

with $P = 100$, $i = 0.005$ $\left(\text{since } \frac{0.06}{12} = 0.005\right)$, and $n = 12$. We obtain

$$A_{12} = 100(1 + 0.005)^{12}$$
$$= 100(1.005)^{12} \doteq 100(1.06167781) = 106.167781$$

Therefore, the interest earned during 1 year when $100 is deposited is, to the nearest cent, $6.17. The effective interest rate is approximately 6.17%.
**Do Exercise 25.** ■

Let us write the formula for $A_n$ in terms of the nominal interest rate $r$, the number of conversion periods per year $k$, and the number of years $t$. We get

$$A_n = P(1 + i)^n = P\left(1 + \frac{r}{k}\right)^{kt}$$

Let $m = \frac{k}{r}$ so that $\frac{1}{m} = \frac{r}{k}$ and $k = mr$. Then, the formula may be written

$$A_n = P\left(1 + \frac{1}{m}\right)^{mrt}$$

and, using one of the laws of exponents,

$$A_n = P\left[\left(1 + \frac{1}{m}\right)^{m}\right]^{rt}$$

Since $m = \frac{k}{r}$, for a fixed nominal interest rate $r$, the value of $m$ increases as the number $k$ of conversion periods per year increases. Note that the value of $\left(1 + \frac{1}{m}\right)^{m}$ increases as the value of $m$ increases. (See Exercise 47.) Therefore, a bank that compounds interest frequently will attract more customers than a bank offering the same nominal interest rate compounded less frequently.

What happens to $A_n$ as the number of conversion periods increases without bound? That is, what will the formula be if interest is *compounded continuously*? We stated earlier that the value of $\left(1 + \frac{1}{m}\right)^{m}$ approaches the number $e$ as the value of $m$ increases without bound. It follows that the value of $P\left[\left(1 + \frac{1}{m}\right)^{m}\right]^{rt}$ approaches $Pe^{rt}$ as the value of $m$ increases without bound. From this we can derive the formula for interest compounded continuously. If $P$ is invested at a nominal interest rate $r$ and interest is compounded continuously, then the balance after $t$ years is

$$A(t) = Pe^{rt}$$

**EXAMPLE 4**   Suppose $5000 is invested at a nominal interest rate of 6% compounded continuously. Find the value of that investment after 10 years.

**Solution**   Use the formula

$$A(t) = Pe^{rt}$$

with $P = 5000$, $r = 0.06$, and $t = 10$. Using the $e^x$ function key on a calculator we find that

$$A(10) = 5000e^{(0.06)10} = 5000e^{0.6}$$

$$\doteq 5000(1.8221188) = 9110.594002$$

To the nearest cent, the answer is $9110.59.
**Do Exercise 29.**                                                                            ∎

### Population Growth and Exponential Depreciation

Suppose that the initial size of a population is $P_0$ and that the increase in its size in a unit of time is a certain percentage of its initial size. Then, as in the formula for compound interest, we find that after $t$ units of time, the size $P(t)$ of the population is given by

$$P(t) = P_0(1 + r)^t$$

where $r$, expressed as a decimal, is the rate of growth of the population per unit of time.

**EXAMPLE 5**   At the beginning of 1975 the population of the earth was about 4 billion. Assuming population growth at 2% per year, predict the approximate size of the earth's population at the end of the year 2000.

**Solution**   The number of years from the beginning of 1975 to the end of 2000 is 26. Thus, we use

$$P(t) = P_0(1 + r)^t$$

with $P_0 = 4$, $r = 0.02$, and $t = 26$. We get

$$P(26) = 4(1 + 0.02)^{26}$$

$$= 4(1.02)^{26}$$

$$\doteq 4(1.67341811)$$

$$= 6.69367244$$

Thus, at the end of the year 2000, the population should be approximately 6.7 billion.
**Do Exercise 31.**                                                                            ∎

Using techniques of calculus, it can be shown that if the rate of change of the size of a population at instant $t$ is proportional to the size of the population at that instant, then the size $P(t)$ is given by

$$P(t) = P_0 e^{kt}$$

where $P_0$ is the size of the population at instant 0, $k$ is a constant that depends on the population, and $t$ is the number of time units. Since $e$ and $k$ are both constants, we can write $b$ for $e^k$ and obtain the formula

$$P(t) = P_0 b^t$$

since $e^{kt} = (e^k)^t = b^t$.

In the formula $P(t) = P_0(1 + r)^t$ written earlier, we can also replace $1 + r$ with $b$ and get

$$P(t) = P_0 b^t$$

Note that both formulas are identical, although the base $b$ represents different quantities in each case.

Since the size of the population at instant $t$ is a constant multiple of the value of the exponential function with base $b$ at $t$, we say that the population is *growing exponentially*. Recalling that the exponential function with base $b$ is increasing when $b > 1$ and decreasing when $b < 1$, we see that the population experiences an exponential growth when $b > 1$ and an exponential decay whenever $b < 1$.

**EXAMPLE 6**   The number of bacteria in a certain culture is 160,000 at 1:00 P.M. At 3:00 P.M. the same day, the count is 320,000. Assuming that the population grows exponentially, what will the count be at 7:00 P.M. that day?

**Solution**   We use $P(t) = P_0 b^t$, where $P_0$ is 160,000 (since this was the initial count) and $t$ is in hours. It is given that $P(2) = 320,000$ (2 hours elapsed from 1:00 to 3:00 P.M.), and we are asked to find $P(6)$. To find the value of $b$, we proceed as follows:

$$P(2) = 160,000 b^2$$

Thus,

$$320,000 = 160,000 b^2$$

It follows that

$$2 = b^2$$

and

$$b = 2^{1/2}$$

Hence,

$$P(t) = 160,000(2^{1/2})^t$$

Therefore,

$$\begin{aligned} P(6) &= 160,000(2^{1/2})^6 \\ &= 160,000(2^3) \\ &= 1,280,000 \end{aligned}$$

At 7:00 P.M. the number of bacteria in the culture will be 1,280,000.
**Do Exercise 35.**

**DEFINITION 1.14:**   If a substance decays exponentially, its *half-life* is the amount of time it takes for its size to be half of the initial amount.

**EXAMPLE 7**   The half-life of a certain radioactive substance is 25 years. If 20 grams of that substance was present in 1985, how much will remain in the year 2025?

**Solution**    We use $P(t) = P_0 b^t$, where $P_0 = 20$ since 20 grams was the initial amount in 1985. We know that $P(25) = 10$ because the half-life of the substance is 25 years. Using this data, we get

$$P(25) = 20b^{25} = 10$$

It follows that

$$b^{25} = 0.5$$

and

$$b = (0.5)^{1/25}$$

Hence,

$$P(t) = 20[(0.5)^{1/25}]^t$$

The time elapsed from 1985 to 2025 is 40 years. Thus,

$$\begin{aligned} P(40) &= 20[(0.5)^{1/25}]^{40} \\ &= 20(0.5)^{1.6} \\ &\doteq 20(0.329876978) \\ &= 6.59753956 \end{aligned}$$

Approximately 6.6 grams of the substance will be left in the year 2025.
**Do Exercise 37.**                                                                     ■

The following is an example of *exponential depreciation*.

**EXAMPLE 8**    The value of a building that cost its owner $90,000 is declining 20% per year. What will the building be worth at the end of **a.** 1 year? **b.** 2 years? **c.** 3 years? **d.** $t$ years?

**Solution**    Let $V_t$ be the value of the building after $t$ years.

**a.** The value of the building declined 20% the first year. Thus, at the end of the first year, the value of the building is 80% of the original value. Thus,

$$V_1 = 90{,}000(.8) = 72{,}000$$

The value of the building is $72,000 after 1 year.
**b.** At the end of the second year, the value of the building is 80% of $V_1$. That is,

$$V_2 = V_1(.8) = [90{,}000(.8)](.8) = 90{,}000(.8)^2 = 57{,}600$$

At the end of the second year the building is worth $57,600.
**c.** At the end of the third year, the value of the building is 80% of $V_2$. That is,

$$\begin{aligned} V_3 &= V_2(.8) = [90{,}000(.8)^2](.8) = 90{,}000(.8)^3 \\ &= 46{,}080 \end{aligned}$$

At the end of the third year, the building is worth $46,080.

**d.** Continuing in the same manner, it is easy to see that at the end of $t$ years, the value of the building is given by

$$V_t = 90,000(.8)^t$$

**Do Exercise 43.**

---

# Exercise Set 1.4

*In Exercises 1–10, simplify the given expression.*

**1.** $5(8^{2/3})(9^{3/2})$

**2.** $-2(16^{3/4})(25^{5/2})$

**3.** $7(27^{4/3})(49^{3/2})$

**4.** $\dfrac{-9(8^{5/3})(81^{5/4})}{4^{5/2}}$

**5.** $\dfrac{(16^{-5/4})(8^{7/3})}{27^{-5/3}}$

**6.** $\sqrt[3]{27} \cdot \sqrt[4]{256}$

**7.** $\sqrt[5]{59049} \cdot (\sqrt[3]{4096})^2$

**8.** $\dfrac{\sqrt[4]{512} \cdot \sqrt[3]{32}}{\sqrt[6]{128}}$

**9.** $\dfrac{\sqrt[3]{81} \cdot \sqrt[5]{729}}{(\sqrt[5]{2187})^6}$

**10.** $\sqrt[4]{32} \cdot 7^0 \cdot 4^{5/2}$

**11.** Sketch the graph of the exponential function with base 3.

**12.** Sketch the graph of the exponential function with base 5.

**13.** Sketch the graph of the exponential function with base $\frac{1}{2}$.

**14.** Sketch the graph of the exponential function with base $\frac{1}{3}$.

**15.** Sketch the graph of the exponential function with base $\frac{1}{4}$.

**16.** Let the function $f$ be defined by $f(x) = 3x$ and let $g$ be the exponential function with base 2. Sketch the graph of the composition $g \circ f$ and that of the composition $f \circ g$.

**17.** Let the function $g$ be defined by $g(x) = 2x - 3$ and let $h$ be the exponential function with base 3. Sketch the graph of the composition $h \circ g$ and that of the composition $g \circ h$.

**18.** Suppose that $1000 is deposited in a bank that pays 6% compounded monthly. Find the amount in that account (to the nearest cent) 7 years later. What is the compound interest earned during the 7 years?

**19.** Suppose that $5000 is deposited in a bank that pays 8% compounded quarterly. Find the amount in that account (to the nearest cent) 9 years later. What is the compound interest earned during the 9 years?

**20.** Suppose that $3000 is deposited at 6% compounded quarterly. Find the amount in that account 4 years later. What is the interest earned?

**21.** Suppose that $8000 is deposited at 9% compounded monthly. Find the amount in that account 28 months later. What is the interest earned?

**22.** You have $5000 to deposit in a bank. Bank A pays 7.9% interest compounded monthly, while bank B pays 8% interest compounded quarterly. Which bank would you choose?

**23.** You have $3000 to invest. Banks A, B, and C pay 7.95% interest compounded daily, 8% interest compounded monthly, and 8.1% interest compounded quarterly, respectively. Which bank would you choose?

**24.** You have $8000 to invest. Banks A, B, and C pay 9% interest compounded daily, 9.05% interest compounded monthly, and 9.1% interest compounded quarterly, respectively. Which bank would you choose?

**25.** A bank advertises a nominal interest rate of 8% compounded quarterly. What is the effective interest rate?

**26.** A bank advertises a nominal interest rate of 9% compounded monthly. What is the effective interest rate?

**27.** A bank advertises a nominal interest rate of 7% compounded daily. What is the effective interest rate?

**28.** Suppose that $8000 is invested at a nominal interest rate of 8% compounded continuously. Find the value of that investment after 12 years.

**29.** Suppose that $4000 is invested at a nominal interest rate of 9% compounded continuously. Find the value of that investment after 15 years.

**30.** Suppose that $7500 is invested at a nominal interest rate of 7% compounded continuously. Find the value of that investment after 25 years.

**31.** The population on an island was estimated to be 75,000 at the beginning of 1986. Assuming that the population is growing at a rate of 2.5% per year, predict the approximate size of the population at the end of 1995.

**32.** The population on an island was estimated to be 90,000 at the beginning of 1987. Assuming that the population is growing at the rate of 1.5% per year, predict the approximate size of the population at the end of 1998.

**33.** The population on an island was estimated to be 250,000 at the beginning of 1983. Assuming that the population is growing at the rate of 3.5% per year, predict the approximate size of the population at the end of 2000.

*In Exercises 34–39, assume that the population grows, or decays, exponentially.*

**34.** The number of bacteria in a culture is 250,000 at 2:00 A.M. At 5:00 A.M. the same day, the count is 325,000. What will the count be at 2:00 P.M. that day?

**35.** The number of bacteria in a culture is 150,000 at 3:00 A.M. on Monday. At 2:00 A.M. Tuesday, the count is 250,000. What will the count be at 5:00 P.M. Thursday?

**36.** The number of wild rabbits on an island was estimated to be 150,000 at the beginning of 1985. The estimate at the beginning of 1987 was 225,000. Predict the approximate size of the rabbit population at the beginning of 1995.

**37.** Carbon 14 has a half-life of 5730 years. If there were 5 grams of carbon 14 initially, how much will there be after 17,190 years?

**38.** Cobolt 57 has a half-life of 270 days. If there were 8 grams of cobolt 57 initially, how much will there be after 1080 days?

**39.** Hydrogen 3 (or Tritium) has a half-life of 12.3 years. Suppose there were 3 grams of hydrogen 3 at the beginning of 1975. How much will there be at the end of the year 2000?

**40.** A secretarial school has established that on the average, if $t$ weeks is the amount of time one student has attended the school and $W(t)$ is the number of words per minute that a student is able to type, then $W(t) = 90(1 - e^{-0.28t})$. Find the number of words per minute a student should type, on the average, after attending that school for a period of **a.** 3 weeks, **b.** 6 weeks, and **c.** 12 weeks.

**41.** Same as Exercise 40 if $W(t) = 110(1 - e^{-0.2t})$.

**42.** On the average, should a student attend the secretarial school of Exercise 40, or that of Exercise 41, if that student can go to school only for a period of **a.** 2 weeks, **b.** 5 weeks, and **c.** 8 weeks?

**43.** The original value of a truck is $30,000. If the value of the truck is declining 15% per year, what will the truck be worth at the end of **a.** 2 years? **b.** 4 years? **c.** 8 years?

**44.** The value of a machine that cost a roofing company $120,000 is declining 18% per year. What will the machine be worth at the end of **a.** 3 years? **b.** 6 years? **c.** 12 years?

**45.** The original value of a car is $15,000. If the value of the car is declining 18% per year, what will the car be worth at the end of **a.** 4 years? **b.** 6 years? **c.** 8 years?

**46.** The value of a boat that cost its owner $60,000 is declining 9% per year. What will the boat be worth at the end of **a.** 2 years? **b.** 5 years? **c.** 15 years?

**47.** Using a calculator, fill in the blanks in the following table and verify that your entries on the bottom line are increasing as the values of $m$ increase.

| $m$ | 2 | 5 | 15 | 50 | 150 | 500 | 1500 |
|---|---|---|---|---|---|---|---|
| $\left(1 + \dfrac{1}{m}\right)^m$ | | | | | | | |

# 1.5  The Logarithmic Function

## The Logarithmic Function with Base $b$

The exponential function with base $b$ was defined in the preceding section where we stated that this function is decreasing in the case $0 < b < 1$ and increasing in the case $b > 1$. It follows that the exponential function with base $b$ is a one-to-one function and therefore has an inverse.

> **DEFINITION 1.15:**   The inverse of the exponential function with base $b$ is called the *logarithmic function with base b* and is denoted $\log_b$.

(It is customary to write $\log_b x$ instead of the usual functional notation $\log_b(x)$.)

Using the relationship between a function and its inverse as described earlier, we see that

1. The domain of $\log_b$ is the set of all positive real numbers since it is equal to the range of the exponential function with base $b$.
2. The range of $\log_b$ is the set of all real numbers since it is equal to the domain of the exponential function with base $b$.
3. $b^{\log_b x} = x$ for every positive real number $x$. To see this, let $f(x) = b^x$, so that $f^{-1}(x) = \log_b x$ and $b^{\log_b x} = f(\log_b x) = f(f^{-1}(x)) = x$.
4. $\log_b b^x = x$ for every real number $x$. To see this, let $f(x) = b^x$, so that $f^{-1}(x) = \log_b x$ and $\log_b b^x = f^{-1}(b^x) = f^{-1}(f(x)) = x$.
5. The graphs of $y = \log_b x$ and $y = b^x$ are symmetric with respect to the graph of $y = x$. (See Figure 1.17.)
6. The graph of $y = \log_b x$ passes through the point $(1, 0)$ since the graph of $y = b^x$ passes through the point $(0, 1)$.
7. $\log_b b = 1$ since $b^1 = b$ and therefore, $\log_b b = \log_b b^1 = 1$.

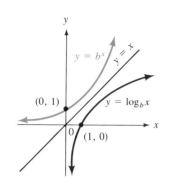

**FIGURE 1.17**

## Laws of Logarithms

There are, in addition to the preceding properties, the *laws of logarithms*.

> ### The Laws of Logarithms
>
> 1. $\log_b xy = \log_b|x| + \log_b|y|$, whenever $xy$ is a positive real number.
> 2. $\log_b \dfrac{x}{y} = \log_b|x| - \log_b|y|$, whenever $\dfrac{x}{y}$ is a positive real number.
> 3. $\log_b x^y = y\log_b x$ for all positive real numbers $x$ and arbitrary real numbers $y$.
> 4. $\log_b a = \dfrac{\log_c a}{\log_c b}$ for all positive real numbers $a$, $b$, and $c$ with $b$ and $c$ not equal to 1.

We shall prove only the first of the four laws; the proofs of the other three are left as exercises.

***Proof:*** If $x$ and $y$ are real numbers and $xy > 0$, using Property 7 of the exponential function, we get

$$b^{(\log_b|x| + \log_b|y|)} = b^{\log_b|x|} \cdot b^{\log_b|y|}$$
$$= |x|\,|y| = |xy| = xy, \text{ since } xy > 0$$

Thus,

$$\log_b xy = \log_b b^{(\log_b|x| + \log_b|y|)}$$
$$= \log_b|x| + \log_b|y|$$

The proof of the first law is now complete. Note that we used the fact that $b^{\log_b z} = z$ for any positive real number $z$ and that $\log_b b^z = z$ for any real number $z$.

## Common and Natural Logarithmic Functions

Although we may use any positive real number not equal to 1 as a base of a logarithmic function, the two numbers used most frequently as bases are 10 and $e$. The logarithmic function with base 10 is called the *common logarithmic function*, and we write $\log x$ instead of $\log_{10} x$.* The logarithmic function with base $e$ is called the *natural logarithmic function* and we write $\ln x$ instead of $\log_e x$. It follows that

**a.** $\log 10 = 1$, since $\log 10 = \log_{10} 10 = 1$ (by Property 7 of the function $\log_b$), and

**b.** $\ln e = 1$, since $\ln e = \log_e e = 1$ (also by Property 7 of the function $\log_b$).

We will use the natural logarithmic function almost exclusively. However, we will give some examples using both logarithmic functions so that you can become familiar with the basic laws of logarithms.

**EXAMPLE 1**    Given that $\log 2 \doteq .3010$ and $\log 3 \doteq .4771$, and that $\ln 2 \doteq .6931$ and $\ln 3 \doteq 1.0986$, find the following:

**a.** $\log 8$          **b.** $\log 12$          **c.** $\ln \sqrt[3]{2}$          **d.** $\ln\left(\dfrac{9}{8}\right)$

**e.** $\log 5$          **f.** $\ln(3e^4)$          **g.** $\log_2 3$          **h.** $\log_5 12$

**Solution**    **a.** Since $8 = 2^3$, $\log 8 = \log 2^3 = 3\log 2 \doteq 3(.3010) = .9030$.

**b.** Observe first that $12 = 3 \cdot 4 = 3 \cdot 2^2$. Therefore, $\log 12 = \log(3 \cdot 2^2) = \log 3 + \log 2^2 = \log 3 + 2\log 2 \doteq .4771 + 2(.3010) = .4771 + .6020 = 1.0791$.

**c.** Recall that $\sqrt[3]{2} = 2^{1/3}$. Hence, $\ln \sqrt[3]{2} = \ln 2^{1/3} = \left(\frac{1}{3}\right)\ln 2 \doteq \left(\frac{1}{3}\right)(.6931) = .2310$.

**d.** We write $\frac{9}{8} = 3^2/2^3$. Thus, $\ln \frac{9}{8} = \ln (3^2/2^3) = \ln 3^2 - \ln 2^3 = 2\ln 3 - 3\ln 2 \doteq 2(1.0986) - 3(.6931) = .1179$.

**e.** Confronted with the task of finding $\log 5$ when only $\log 2$ and $\log 3$ are given, students are often tempted to start with $2 + 3 = 5$. However, this does not lead anywhere because there is no law to give the value of $\log(x + y)$ in terms of $\log x$ and of $\log y$. Instead, start with $5 = \frac{10}{2}$. Thus, $\log 5 = \log\left(\frac{10}{2}\right) = \log 10 - \log 2 \doteq 1 - .3010 = .6990$.

**f.** We have $\ln(3e^4) = \ln 3 + \ln e^4 = \ln 3 + 4 \ln e \doteq 1.0986 + 4(1) = 1.0986 + 4 = 5.0986$.

**g.** Using Law 4, we write $\log_2 3 = \frac{\log 3}{\log 2} \doteq \frac{.4771}{.3010} \doteq 1.5850$. Note that we also could have used the natural logarithmic function to write $\log_2 3 = \frac{\ln 3}{\ln 2} \doteq \frac{1.0986}{.6931} \doteq 1.5850$. As expected, we obtained the same value for $\log_2 3$.

---

*Prior to the electronic calculator, common logarithms were used for computational purposes. The laws of logarithms illustrate the fact that a logarithmic function changes a product to a sum, a quotient to a difference, and a power to a product. Values of $\ln x$ and $\log x$ may be obtained using a calculator. Tables of values of common and natural logarithms are given in appendices of most standard mathematics books. Since it is no longer fashionable to use logarithms to perform calculations, we will not illustrate how this is done. If you are interested, you may find examples of such calculations in many mathematics books published prior to 1970.

**h.** We again use Law 4 to write $\log_5 12 = \frac{\log 12}{\log 5}$. We found in part **b** that $\log 12 = 1.0791$ and in part **e** that $\log 5 = .6990$. Thus, $\log_5 12 \doteq \frac{1.0791}{.6990} \doteq 1.54377$.

**Do Exercises 1 and 9.\***

EXAMPLE 2    Solve the equation $\log_4(3x - 2) = 3$.

Solution    Since the domain of the function $\log_4$ is the set of positive real numbers, we must have $3x - 2 > 0$. That is, $x > \frac{2}{3}$. Recall that the function $\log_4$ is the inverse of the exponential function to the base 4. Consequently, $4^{\log_4 z} = z$ for each positive real number $z$. Using this fact, and the fact that the exponential function is one-to-one, we conclude that the following equations are equivalent:

$$\log_4(3x - 2) = 3$$
$$4^{\log_4(3x-2)} = 4^3$$
$$3x - 2 = 4^3$$
$$3x = 66$$
$$x = 22$$

Since 22 is larger than $\frac{2}{3}$, it is the solution of the equation $\log_4(3x - 2) = 3$.
**Do Exercise 11.**

When solving equations involving one or more functions of the unknown, check that the answers belong to the domain(s) of the function(s).

EXAMPLE 3    Solve the equation $\log_2(x - 3) = 1 - \log_2(x - 4)$.

Solution    Note that both $x - 3$ and $x - 4$ must be positive. Thus $x$ must be greater than 3 and 4. A solution will be feasible only if it is greater than 4. The equation can be written

$$\log_2(x - 3) + \log_2(x - 4) = 1$$

and so

$$\log_2(x - 3)(x - 4) = 1$$

This equation is equivalent to

$$2^{\log_2(x-3)(x-4)} = 2^1$$

and to

$$(x - 3)(x - 4) = 2$$

That is,

$$x^2 - 7x + 10 = 0$$

---

\*Most of the values of the logarithmic functions in tables, or with a calculator, are approximate values. You should, therefore, be aware that the answers in the examples and the exercises are approximations, usually correct to four decimal places.

The solutions are easily found by factoring. They are 2 and 5. However, only 5 is greater than 4. Thus, the only solution of the initial equation is 5.
**Do Exercise 17.**                                                                            ■

## Solving for Unknown Exponents

Logarithms are often used to solve equations where the unknown appears as an exponent, as illustrated in the following example.

**EXAMPLE 4**    Solve the equation $3^{2x} = 2^{x+1}$.

**Solution**    We may use either of the two logarithmic functions since we were given their values at 2 and at 3 in Example 1. Since a logarithmic function is one-to-one, we have

$$3^{2x} = 2^{x+1}$$

if, and only if,

$$\ln 3^{2x} = \ln 2^{x+1}$$

This equation is equivalent to

$$2x \ln 3 = (x + 1)\ln 2$$

Thus,

$$2x(1.0986) \doteq (x + 1)(.6931)$$

We have replaced the exponential equation with a linear equation that can easily be solved. We have

$$(2.1972)x \doteq (.6931)x + .6931$$

From this, we get

$$(1.5041)x \doteq .6931$$

Thus,

$$x \doteq \frac{.6931}{1.5041} \doteq .4608$$

The solution of the exponential equation is approximately .4608.
**Do Exercise 27.**                                                                            ■

We are now ready to give other examples of compound interest and population growth problems.

**EXAMPLE 5**    A bank pays 8% interest compounded quarterly. How long will it take for a deposit to double in value?

**Solution**  Use the formula $A = P(1 + i)^n$. Replace $i$ by .02 $\left(\text{since } \frac{.08}{4} = .02\right)$, $A$ by $2P$ (since the deposit is to double in value), and then solve for $n$. We get

$$2P = P(1 + .02)^n$$

That is,

$$2 = 1.02^n$$

This is equivalent to

$$\ln 2 = \ln 1.02^n$$

Thus,

$$\ln 2 = n \ln 1.02$$

and

$$n = \frac{\ln 2}{\ln 1.02} \doteq \frac{.6931}{.0198} \doteq 35$$

It will take approximately 35 quarters, or $8\frac{3}{4}$ years, for a deposit to double.
**Do Exercise 29.**  ■

**EXAMPLE 6**  We initially have 3 grams of sodium-24; 7.5 hours later, 2.12 grams remain. Calculate the half-life of sodium-24.

**Solution**  Use the formula $P(t) = P_0 b^t$. It is given that $P_0 = 3$ and $P(7.5) = 2.12$. We are asked to find the value of $t$ for which $P(t) = P_0/2 = 1.5$. We have

$$P(7.5) = 3b^{7.5}$$

Thus,

$$2.12 = 3b^{7.5}$$

and

$$b^{7.5} = \frac{2.12}{3} \doteq .70667$$

Using the laws of exponents, we obtain

$$(b^{7.5})^{1/7.5} \doteq .70667^{1/7.5}$$

and

$$b^1 = b \doteq .70667^{1/7.5}$$

It follows that

$$P(t) \doteq 3(.70667^{t/7.5})$$

Replacing $P(t)$ by 1.5, we get

$$1.5 \doteq 3(.70667^{t/7.5})$$

Thus,

$$.5 \doteq .70667^{t/7.5}$$

and

$$\ln .5 \doteq \ln .70667^{t/7.5} \doteq \frac{t}{7.5} \ln .70667$$

It follows that

$$t \doteq \frac{(7.5) \ln .5}{\ln .70667}$$

$$\doteq \frac{(7.5)(-.6931)}{-.3472} \doteq 14.97$$

The half-life of sodium-24 is approximately 15 hours.
**Do Exercise 41.**

# Exercise Set 1.5

*In Exercises 1–10, use the following: log 2 $\doteq$ .3010, log 3 $\doteq$
.4771, ln 2 $\doteq$ .6931, and ln 3 $\doteq$ 1.0986 to find:*

**1. a.** $\log 64$    **b.** $\log 36$    **c.** $\log 54$

**2. a.** $\log 1.5$    **b.** $\log .75$    **c.** $\log \dfrac{27}{16}$

**3. a.** $\log \sqrt[5]{4}$    **b.** $\log \sqrt[4]{27}$    **c.** $\log \sqrt{.75}$

**4. a.** $\log 25$    **b.** $\log 200$    **c.** $\log .03$

**5. a.** $\log_3 2$    **b.** $\log_4 27$    **c.** $\log_9 64$

**6. a.** $\ln 64$    **b.** $\ln 36$    **c.** $\ln 54$

**7. a.** $\ln 1.5$    **b.** $\ln .75$    **c.** $\ln \dfrac{27}{16}$

**8. a.** $\ln \sqrt[5]{4}$    **b.** $\ln \sqrt[4]{27}$    **c.** $\ln \sqrt[7]{.75}$

**9. a.** $\ln 5e^3$    **b.** $\ln \sqrt[3]{e^2}$    **c.** $\ln \dfrac{27}{e}$

**10. a.** $\log_9 8$    **b.** $\log_8 243$    **c.** $\log_{.5} 27$

*In Exercises 11–28, solve the given equation.*

**11.** $\log_3(2x + 3) = 4$    **12.** $\log_2(5x - 6) = 3$

**13.** $\log_5(5 - 3x) = 2$

**14.** $\log_3(2x - 5) = 2 + \log_3(3x - 7)$

**15.** $\log_2(5 - 6x) = 3 + \log_2(3x - 7)$

**16.** $\log(x + 4) = 1 - \log(x + 1)$

**17.** $\log_{12}(x - 2) = 1 - \log_{12}(x - 3)$

**18.** $\log_2(x + 7) = 4 - \log_2(x + 1)$

**19.** $\log_2(x + 7) = 2 - \log_2(x + 4)$

**20.** $\ln(x - e) - 2 = \ln 6 - \ln(x - 2e)$

**21.** $\log|x + 4| = 1 - \log|x + 1|$

**22.** $\log_2|x + 7| = 2 - \log_2|x + 4|$

**23.** $\log_x(x + 12) = 2$    **24.** $\log_x(2x + 15) = 2$

**25.** $\log_x(3 + x - 3x^2) = 3$    **26.** $e^{3x-4} = 2^{2x+1}$

**27.** $3^{2x-1} = 2e^{3x+5}$    **28.** $3^{x/2} = 2(10^{3x})$

**29.** A bank pays 8% interest compounded quarterly. How long
will it take for a deposit to triple in value?

**30.** A bank pays 6% interest compounded monthly. How long
will it take for a deposit to double in value?

**31.** A bank pays 6% interest compounded monthly. How long
will it take for a deposit to triple in value?

**32.** A bank pays 7% interest compounded continuously. How
long will it take for a deposit to double in value?

**33.** A bank pays 8% interest compounded continuously. How
long will it take for a deposit to triple in value?

**34.** The population of the earth was estimated to be 4 billion at the beginning of 1975. Assuming that the population grows 2% per year, when can we expect the population to be approximately **a.** 6 billion? **b.** 8 billion?

**35.** On an island, it was estimated that there were 20,000 wild rabbits at the beginning of 1987. Assuming that the rabbit population grows 30% per year, when can we expect the population to be 60,000?

**36.** A secretarial school has established that on the average, if $t$ weeks is the amount of time one student has attended the school and $W(t)$ is the number of words per minute that student is able to type, then $W(t) = 90(1 - e^{-0.15t})$. Approximately how many weeks must a student attend the school to be able to type 70 words per minute?

**37.** Same as Exercise 36 with $W(t) = 110(1 - e^{-0.2t})$ and the student being able to type 80 words per minute.

*In Exercises 38–41, assume that the population grows, or decays, exponentially.*

**38.** In 1980, the population on an island was estimated to be 70,000; in 1985, the estimate was 80,000. When can we expect the population to be approximately 95,000?

**39.** The population of a city was 150,000 in 1970 and 195,000 in 1985. When can we expect the population to be approximately 225,000?

**40.** If the half-life of mercury 197 is 65 hours, how long does it take mercury 197 to lose three fourths of its mass?

**41.** The half-life of radium is 1690 years. How long does it take radium to lose one fourth of its mass?

**\*42.** Prove that if $x$ and $y$ are real numbers, and $\frac{x}{y} > 0$, then

$$\log_b \frac{x}{y} = \log_b|x| - \log_b|y|.$$

**\*43.** Prove that for all positive real numbers $x$ and arbitrary real numbers $y$,

$$\log_b x^y = y \log_b x.$$

**\*44.** Prove that for all positive real numbers $a$, $b$, and $c$ with $b$ and $c \neq 1$: $\log_b a = \log_c a / \log_c b$.

**\*45.** Note that the equality of Exercise *44 may be written $(\log_c b)(\log_b a) = \log_c a$. Prove the following generalization. For any positive real numbers $a$, $b$, $c$, and $d$ with $a$, $b$, and $c \neq 1$: $(\log_a b)(\log_b c)(\log_c d) = \log_a d$. (*Hint:* Let $\log_a b = x$, $\log_b c = y$, $\log_c d = z$, $\log_a d = w$, and note that $\log_a b = x$ is equivalent to $a^x = b$, and so on.) State a further generalization.

**46.** Suppose that money is invested at $n\%$ interest. Let $x$ be the number of years it will take for the investment to double.
   **a.** Find a formula giving $x$ in terms of $n$.
   **b.** Using a calculator, find the values of the product $xn$ for $n = 5$, $n = 6$, $n = 7$, $n = 8$, $n = 9$, and $n = 10$.
   **c.** Based on your results in part **b**, find a positive integer $K$ such that the following rule may be stated: "money invested at $n\%$ interest will double in approximately $\frac{K}{n}$ years."

**47.** Use the rule obtained in Exercise 46c to approximate how long it will take for money invested at 5.5% to double.

**48.** Use the rule obtained in Exercise 46c to approximate how long it will take for money invested at 6.8% to double.

---

## 1.6  More on Graphs

Graphs are frequently used to show relationships between variable quantities. A business publication may use a graph to illustrate the increase in value of some stock or a political party may circulate a brochure during an election year on which there is a graph showing the increase in cost of living while the opposite party was in control. These are effective strategies because a graph displays information visually. We have seen that the graph of a linear function is a line and that of a quadratic function is a parabola. In this section we briefly discuss how to use graphs to illustrate relationships between functions and to display information about functions.

### Comparing Graphs

EXAMPLE 1    The management of a business firm knows that the cost of producing $q$ thousand gadgets in a month is $C = 7q + 50$, where $C$ is in thousands of dollars. It is also known that the revenue from these is $R = -0.2q^2 + 14q$, where $R$ is also in thousands of dollars. The firm is limited to a maximum production of 35,000 gadgets per month. Illustrate graphically how the firm can operate at a profit.

Solution    Graph both functions on the same coordinate system, using the horizontal axis for the independent variable $q$, and the vertical axis for the dependent variables $C$ and $R$. Since the maximum monthly production is 35,000, consider only values of $q$ such that $0 \le q \le 35$. The points of intersection are easily found to be (10, 120) and (25, 225). The two graphs are sketched in Figure 1.18. The parabola (graph of the revenue function) is above the line (graph of the cost function) when $10 < q < 25$. So to operate at a profit, the firm must produce more than 10,000—but less than 25,000—gadgets per month.
**Do Exercise 1.**

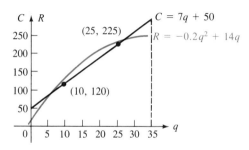

**FIGURE 1.18**

### Shifting Graphs

In Example 1, $C = 7q + 50$ is the cost equation. When the production is 0, the fixed cost is $50,000. It is possible that the fixed cost might increase, while the variable cost remains the same.

EXAMPLE 2    The fixed cost of producing $q$ thousand gadgets in Example 1 has increased by $15,000 per month. Graph the original and the new cost functions on the same coordinate plane.

Solution    The new cost equation is

$$C = (7q + 50) + 15 = 7q + 65$$

The graph of this equation and that of the original equation are shown in Figure 1.19. Observe that the new graph may be obtained by shifting the original graph vertically 15 units, because if the point $(q_1, C_1)$ is on the original graph, then $(q_1, C_1 + 15)$ is on the new graph.
**Do Exercise 5.**

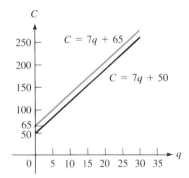

**FIGURE 1.19**

**EXAMPLE 3**   Due to government subsidies, the firm of Example 1 found that the revenue from $q$ thousand gadgets now is the same as what it was from $q + 2$ thousand gadgets previously. Sketch both revenue functions on the same coordinate plane.

**Solution**   Let $R = f(q) = -0.2q^2 + 14q$ and $R = g(q)$ be the original and new revenue equations, respectively. Then, $g(q) = f(q + 2)$. The graphs are shown in Figure 1.20. Observe that if a point $(q_1, R_1)$ is on the graph of the new revenue function, then the point $(q_1 + 2, R_1)$ is·on the graph of the original function, and conversely. Also, the point $(q_1, R_1)$ is found by shifting $(q_1 + 2, R_1)$ 2 units to the left. Thus, the new graph is 2 horizontal units to the left of the original graph. Because the values of $q$ for the new function must be nonnegative, we shift only the part of the original graph over the interval $[2, 37]$. (See Figure 1.20.)
**Do Exercise 7.**

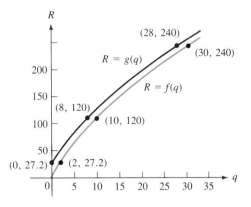

**FIGURE 1.20**

In the above example, we obtained the graph of $R = g(q) = f(q + 2)$ by shifting the graph of $R = f(q)$ 2 units horizontally to the left. The last two examples illustrate the following general procedure.

1. The graph of $y = f(x) + b$ may be obtained by shifting the graph of $y = f(x)$ vertically, $b$ units up if $b$ is positive and $-b$ units down if $b$ is negative.
2. The graph of $y = g(x + c)$ may be obtained by shifting the graph of $y = g(x)$ horizontally, $c$ units to the left if $c$ is positive and $-c$ units to the right if $c$ is negative.

**EXAMPLE 4**  Obtain the graph of $y = |x + 4| - 3$ by shifting the graph of $y = |x|$ first horizontally, then vertically.

**Solution**  The graph of $y = |x|$ is shown in Figure 1.21a. Obtain the graph of $y = |x + 4|$ by a horizontal shift 4 units to the left. (See Figure 1.21b.) Finally, the graph of $y = |x + 4| - 3$ is obtained by shifting the graph of $y = |x + 4|$ 3 units down. (See Figure 1.21c.) When shifting graphs, it is advisable to check your work as is done here. The point $(-4, -3)$ is the lowest point of the new graph. The point $(0, 0)$ was the lowest point of the graph of $y = |x|$. Clearly, $(-4, -3)$ is obtained by shifting $(0, 0)$ first 4 units to the left, then 3 units down.
**Do Exercise 9.**

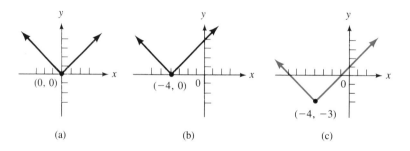

**FIGURE 1.21**          (a)                    (b)                    (c)

**EXAMPLE 5**  In Figure 1.22, the graph of $y = g(x)$ was obtained from the graph of $y = f(x) = x^2$ by making a horizontal shift followed by a vertical shift. What is the equation defining $g(x)$?

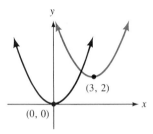

**FIGURE 1.22**

**Solution**  The vertex of the graph of $y = g(x)$ is the point $(3, 2)$, which is obtained from the vertex of the graph of $y = x^2$ by a horizontal shift 3 units to the right followed by a vertical shift 2 units up. So, the equation defining the function $g$ is $g(x) = (x - 3)^2 + 2$.
**Do Exercise 19.**

The procedures illustrated in these examples are useful in sketching the graph of a function if we can do it by shifting a known or already obtained graph. Here is another example.

**EXAMPLE 6**  Find the graph of $y = f(x) = x^3 + 6x^2 + 12x + 4$ by shifting the graph of $y = x^3$.

**Solution**  Start with

$$(x + 2)^3 = x^3 + 3x^2 \cdot 2 + 3x \cdot 2^2 + 2^3$$
$$= x^3 + 6x^2 + 12x + 8$$

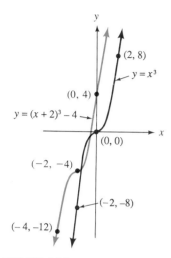

So, $y = f(x) = (x + 2)^3 - 4$. Thus the graph of the given equation may be found by shifting the graph of $y = x^3$ first horizontally 2 units to the left, then vertically 4 units down. (See Figure 1.23.)
**Do Exercise 23.**

**FIGURE 1.23**

## Rising and Falling Graphs

A function $f$ is said to be *increasing* (*decreasing*) if the value of $f(x)$ increases (decreases) as the value of $x$ increases. It is clear that, when a function is increasing, its graph rises going from left to right and if a function is decreasing, its graph falls.

**EXAMPLE 7**  The graph of the equation $v = g(t)$ in Figure 1.24 represents the daily value $\$v$ of a stock over the period March 1–March 31, 1990, where $t$ is the number of days elapsed since March 1, 1990. Over what periods was the value increasing? decreasing?

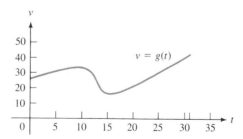

**FIGURE 1.24**

**Solution**  The graph is rising over the intervals [0, 10] and [15, 31]; it is falling over the interval [10, 15]. So, the value of the stock was increasing from $t = 0$ to $t = 10$, and again from $t = 15$ to $t = 31$. Since $t = 0$ represents March 1, the value of the stock was increasing from March 1 to March 10, and then from March 15 to March 31. It was decreasing from March 10 to March 15.
**Do Exercise 27.**

A deeper discussion of graph sketching will be given in Chapter 8.

# Exercise Set 1.6

1. A company manufactures portable radios. Its monthly production cost $C thousand is

    $$C = 4q + 10$$

    where $q$ thousand is the number of radios manufactured in a month. It is also known that the revenue from these is

    $$R = -0.125q^2 + 7q$$

    where $R$ is also in thousands of dollars. The company is limited to a maximum production of 28,000 radios per month. Illustrate graphically how it can operate at a profit.

2. The management of a sporting equipment company knows that the cost of producing and selling $q$ thousand golf clubs in any year is

    $$C = 0.6q^2 + 6q + 360$$

    where $C$ is in thousands of dollars. It is also known that the revenue from these is

    $$R = -1.2q^2 + 60q$$

    where $R$ is also in thousands of dollars. The company is limited to a maximum yearly production of 25,000 golf clubs. Illustrate graphically how it can make a profit.

3. A traveler needs to rent a car for the day. The car rental agency offers the following two options on the car he wants to rent: $30 plus $.40 per mile and $40 plus $.35 per mile. Determine graphically which option he should choose if he plans to drive a. less than 200 miles and b. more than 200 miles.

4. Another traveler needs to rent a car for the day. The car rental agency offers the following three options on the car she wants to rent: $20 plus $.40 per mile, $34 plus $.36 per mile, and $43 plus $.30 per mile. Determine graphically the conditions under which she should choose a. the first option, b. the second option, and c. the third option.

5. The fixed cost of producing $q$ thousand radios in Exercise 1 has increased by $12,000 per month. Graph the original and the new cost functions on the same coordinate plane.

6. Due to government subsidies, the company of Exercise 1 found that the revenue from $q$ thousand radios now is the same as what it was from $q + 3$ thousand radios previously. Sketch both revenue functions on the same coordinate plane.

7. The sporting equipment company of Exercise 2 found that the revenue from $q$ thousand golf clubs now is the same as what it was from $q + 1$ thousand golf clubs previously. Sketch both revenue functions on the same coordinate plane.

*In Exercises 8–15, two equations are given. Find the graph of $y = g(x)$ by shifting the graph of $y = f(x)$ first horizontally, then vertically.*

8. $y = f(x) = |x|$ and $y = g(x) = |x - 5| + 2$

9. $y = f(x) = |x|$ and $y = g(x) = |x - 2| - 5$

10. $y = f(x) = x^2$ and $y = g(x) = (x + 3)^2 + 2$

11. $y = f(x) = x^2$ and $y = g(x) = (x - 5)^2 + 3$

12. $y = f(x) = x^3$ and $y = g(x) = (x - 1)^3 + 4$

13. $y = f(x) = x^3$ and $y = g(x) = (x + 2)^3 - 3$

14. $y = f(x) = [x]$ and $y = g(x) = [x + 2] - 3$ (*Hint:* Recall that $[x]$ denotes the greatest integer less than or equal to $x$.)

15. $y = f(x) = [x]$ and $y = g(x) = [x - 5] + 2$

*In Exercises 16–19, the graph of $y = g(x)$ was derived from the graph of $y = f(x)$ by making a horizontal shift followed by a vertical shift. What is the equation defining $g(x)$?*

16. $y = f(x) = |x|$

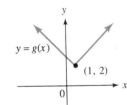

17. $y = f(x) = |x|$

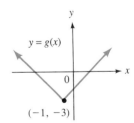

18. $y = f(x) = x^2$

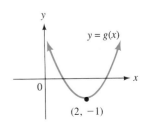

**19.** $y = f(x) = x^2$

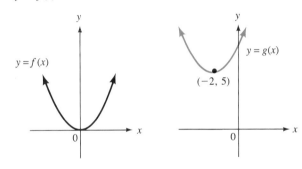

*In Exercises 20–25, two equations are given. First sketch the graph of the equation $y = f(x)$. Then, find the graph of $y = g(x)$ by shifting the graph of $y = f(x)$.*

**20.** $y = f(x) = x^2;\ y = g(x) = x^2 + 6x + 4$

**21.** $y = f(x) = x^2;\ y = g(x) = x^2 - 4x + 5$

**22.** $y = f(x) = x^3;\ y = g(x) = x^3 - 6x^2 + 12x - 5$

**23.** $y = f(x) = x^3;\ y = g(x) = x^3 + 3x^2 + 3x + 6$

**24.** $y = f(x) = x^4;\ y = g(x) = x^4 + 4x^3 + 6x^2 + 4x + 7$

**25.** $y = f(x) = x^4;\ y = g(x) = x^4 + 8x^3 + 24x^2 + 32x + 19$

**26.** The graph of the equation $v = g(t)$ below represents the daily value $v of a stock over the period February 1– February 28, 1990, where $t$ is the number of days elapsed since February 1, 1990. Over what periods was the value increasing? decreasing?

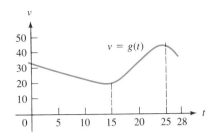

**27.** The graph of the equation $y = g(t)$ below represents the average monthly rate of unemployment in a country from January 1, 1989 to January 1, 1990, where $t$ is the number of months elapsed since January 1, 1989. Over what periods was unemployment increasing? decreasing?

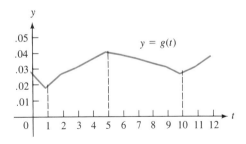

**28.** The graph of the equation $y = g(t)$ represents the average yearly rate of unemployment in a country from January 1, 1979 to January 1, 1990, where $t$ is the number of years elapsed since January 1, 1979. Over what periods was employment increasing? decreasing?

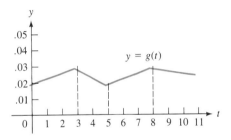

**29.** Sketch the graphs of $y = x^2$ and $y = -x^2$ on the same coordinate plane. In general, how are the graphs of $y = f(x)$ and $y = -f(x)$ related?

**30.** Let $f(x) = |x + 2| - 4$. Start with the graph of $y = |x|$, shift it to obtain the graph of $y = f(x)$, then obtain the graph of $y = -f(x)$. (*Hint:* See Exercise 29.)

**31.** Let $f(x) = [x - 3] + 5$. Start with the graph of $y = [x]$, shift it to obtain the graph of $y = f(x)$, then obtain the graph of $y = -f(x)$. (*Hint:* See Exercise 29.)

# 1.7  The Sigma Notation

In mathematics, it is often desirable to compactly write a sum with many terms. For example, the expression

$$1 + 2^2 + 3^2 + \cdots + 50^2$$

may be used to write the sum of the squares of the first 50 positive integers. However, this type of notation is often not practical. For example, the individual terms may not be clearly defined from the first few terms. There is another way of expressing such sums. The symbol commonly used for this purpose is the capital sigma ($\Sigma$) of the Greek alphabet. Once this new notation becomes familiar, you will find not only that much labor is saved in writing sums, but also that the sums are clearer.

We now explain how this symbol is used. Suppose that $A$ is a finite set and $f$ is a function with domain $A$ and range $B$ where $B$ is a set of real numbers. Then for each $x$ in $A$, there is one, and only one, real number $f(x)$. The symbol

$$\sum_{x \in A} f(x)$$

denotes the sum whose terms are these $f(x)$'s. We stress that the terms of that sum need not be distinct and that *there are exactly as many terms in the sum as there are elements in A*. Also, if $A$ is a set consisting of $n$ consecutive integers, with $a$ and $b$ its smallest and largest members, respectively, then we use the symbol $\sum_{i=a}^{b} f(i)$ instead of $\sum_{x \in A} f(x)$.

**EXAMPLE 1**   Let $A = \{2, 4, 6, 8\}$ and let $f$ be the function with domain $A$ and defined by $f(x) = x^2$. Evaluate $\sum_{x \in A} f(x)$.

**Solution**   Note that $f(2) = 2^2 = 4$, $f(4) = 4^2 = 16$, $f(6) = 6^2 = 36$, and $f(8) = 8^2 = 64$. Thus,

$$\sum_{x \in A} f(x) = \sum_{x \in A} x^2 = 4 + 16 + 36 + 64 = 120$$

**Do Exercise 1.**    ◼

**EXAMPLE 2**   Let $B = \{2, 3, 4, 5, 6\}$ and let $g$ be the function with domain $B$ and defined by $g(i) = 3i - 1$. Find $\sum_{i=2}^{6} g(i)$.

**Solution**   Note that $g(2) = 3(2) - 1 = 5$, $g(3) = 3(3) - 1 = 8$, $g(4) = 3(4) - 1 = 11$, $g(5) = 3(5) - 1 = 14$, and $g(6) = 3(6) - 1 = 17$. It follows that

$$\sum_{i=2}^{6} g(i) = \sum_{i=2}^{6} (3i - 1) = 5 + 8 + 11 + 14 + 17 = 55$$

**Do Exercise 9.**    ◼

**EXAMPLE 3**   Evaluate $\sum_{i=1}^{7} 3$.

**Solution**   In this case we have $A = \{1, 2, 3, 4, 5, 6, 7\}$ and the function with domain $A$ is the constant function defined by $f(i) = 3$. Thus, $f(1) = f(2) = f(3) = f(4) = f(5) = f(6) = f(7) = 3$, and

$$\sum_{i=1}^{7} 3 = \sum_{i=1}^{7} f(i) = 3 + 3 + 3 + 3 + 3 + 3 + 3 = 21$$

**Do Exercise 23.**    ◼

In general, if $n$ is a positive integer and $k$ is a constant, using the same steps as in the preceding example, we can show that

$$\sum_{i=1}^{n} k = nk$$

**EXAMPLE 4**   Express the sum $1 + 3 + 5 + 7 + 9 + 11$ using the sigma notation.

**Solution**   Noting that the given sum is the sum of the first six odd positive integers and that an odd integer can be expressed as $2i + 1$ where $i$ is an integer, we write

$$1 + 3 + 5 + 7 + 9 + 11 = \sum_{i=0}^{5} (2i + 1)$$

which can also be written

$$1 + 3 + 5 + 7 + 9 + 11 = \sum_{i=1}^{6} (2i - 1)$$

**Do Exercise 25.**

Mathematical induction* may be used to prove the following formulas:

$$\sum_{i=1}^{n} i = \frac{n(n + 1)}{2} \tag{1}$$

$$\sum_{i=1}^{n} i^2 = \frac{n(n + 1)(2n + 1)}{6} \tag{2}$$

$$\sum_{i=1}^{n} i^3 = \frac{n^2(n + 1)^2}{4} \tag{3}$$

We can use these formulas to evaluate certain sums, as illustrated in the following example.

**EXAMPLE 5**   Evaluate **a.** $\displaystyle\sum_{i=1}^{20} i$, **b.** $\displaystyle\sum_{i=5}^{25} i^2$, and **c.** $\displaystyle\sum_{i=1}^{7} (i^3 + 2i^2 + 3i - 5)$.

**Solution**   **a.** Using Formula (1),

$$\sum_{i=1}^{20} i = \frac{20(20 + 1)}{2} = 10(21) = 210$$

**b.** To use Formula (2), first note that

$$\sum_{i=5}^{25} i^2 = \sum_{i=1}^{25} i^2 - \sum_{i=1}^{4} i^2$$

*You may find a discussion of Mathematical Induction in Appendix A.

Thus,

$$\sum_{i=5}^{25} i^2 = \frac{25(25 + 1)[2(25) + 1]}{6} - \frac{4(4 + 1)[2(4) + 1]}{6}$$

$$= \frac{25(26)(51)}{6} - \frac{4(5)(9)}{6}$$

$$= 5525 - 30 = 5495$$

c. Using the basic properties of the real numbers and Formulas (1), (2), and (3) we obtain

$$\sum_{i=1}^{7} (i^3 + 2i^2 + 3i - 5) = \sum_{i=1}^{7} i^3 + 2 \sum_{i=1}^{7} i^2 + 3 \sum_{i=1}^{7} i - \sum_{i=1}^{7} 5$$

$$= \frac{7^2(7 + 1)^2}{4} + 2\frac{7(7 + 1)[2(7) + 1]}{6} + 3\frac{7(7 + 1)}{2} - 7(5)$$

$$= 784 + 280 + 84 - 35 = 1113$$

**Do Exercises 17 and 21.**    ■

# Exercise Set 1.7

*In Exercises 1–8, a set A is given and a function f is defined on A. In each case, evaluate $\sum_{x \in A} f(x)$.*

**1.** $A = \{1, 3, 7, 10\}; f(x) = 3x + 1$

**2.** $A = \{0, 2, 5, 8, 11\}; f(x) = 2x - 4$

**3.** $A = \{-2, -1, 0, 3, 6\}; f(x) = x^2 + 1$

**4.** $A = \{-3, 1, 4, 5, 8\}; f(x) = -x^2 + 2$

**5.** $A = \{-4, 0, 2, 4, 5, 9\}; f(x) = x^3$

**6.** $A = \{-5, -1, 0, 3, 6, 10\}; f(x) = -x^3 + 3$

**7.** $A = \{-4, -3, -2, -1, 0, 1, 2, 3, 4\}; f(x) = x^3$

**8.** $A = \{-5, -2, 0, 1, 3, 6\}; f(x) = 5$

*In Exercises 9–24, evaluate the given sum.*

**9.** $\sum_{i=2}^{6} (3i + 5)$

**10.** $\sum_{i=3}^{10} (2i - 7)$

**11.** $\sum_{i=5}^{8} (i^2 + 3i - 2)$

**12.** $\sum_{i=6}^{10} (-i^2 + 3)$

**13.** $\sum_{i=-2}^{2} (i^2 + 1)$

**14.** $\sum_{i=1}^{150} i$

**15.** $\sum_{i=1}^{150} i^2$

**16.** $\sum_{i=1}^{150} i^3$

**17.** $\sum_{i=41}^{160} i$

**18.** $\sum_{i=41}^{120} i^2$

**19.** $\sum_{i=61}^{170} i^3$

**20.** $\sum_{i=1}^{100} (i^3 + 5i^2 - 7i + 6)$

**21.** $\sum_{i=1}^{120} (2i^3 - 5i^2 + 6i - 7)$

**22.** $\sum_{i=1}^{150} 6$

**23.** $\sum_{i=0}^{185} -9$

**24.** $\sum_{i=1}^{120} (-1)^i$

*In Exercises 25–35, express the given sums using the $\sum$ notation.*

**25.** $1 + 3 + 5 + 7 + 9 + 11 + 13 + 15 + 17$

**26.** $3 + 6 + 9 + 12 + 15 + 18$

**27.** $5 + 8 + 11 + 14 + 17 + 20 + 23$

**28.** $5 + 10 + 15 + 20 + 25 + 30 + 35 + 40$

**29.** $1 + 4 + 9 + 16 + 25 + 36 + 49 + 64$

**30.** $16 + 25 + 36 + 49 + 64 + 81$

**31.** $8 + 27 + 64 + 125 + 216$

**32.** $2 + 2 + 2 + 2 + 2 + 2 + 2 + 2 + 2$

**33.** $1 - 1 + 1 - 1 + 1 - 1 + 1 - 1 + 1 - 1$

**34.** $-8 - 1 + 1 + 8 + 27 + 64 + 125$

**35.** $1 - 4 + 9 - 16 + 25 - 36 + 49 - 64 + 81$

# 1.8  Chapter Review

SUMMARY    A function is a rule that assigns to each element of a first set (the domain) a unique element of a second set. If $f$ is the name of a function, and $y$ corresponds to $x$, we write $y = f(x)$. The set $\{f(x) \mid x \in \text{domain of } f\}$ is the range of $f$. If the domain $D_f$ and range of a function $f$ are sets of real numbers, the set of points $(x, f(x))$ where $x \in D_f$ is called the graph of the function $f$. If $f(x_1) = f(x_2)$ can occur only when $x_1 = x_2$, $f$ is said to be one-to-one. In this case, $f$ has an inverse, denoted $f^{-1}$. We have

$$f(f^{-1}(x)) = x \text{ for each } x \text{ in the domain of } f^{-1}$$

and

$$f^{-1}(f(x)) = x \text{ for each } x \text{ in the domain of } f.$$

A curve in the $xy$-plane is the graph of some function if, and only if, each vertical line intersects the curve at no more than one point. A function is one-to-one if, and only if, each horizontal line intersects its graph at no more than one point. The graph of a one-to-one function and that of its inverse are symmetric with respect to the graph of $y = x$. If $f$ and $g$ are functions with domains $D_f$ and $D_g$, respectively, and with ranges the sets of real numbers, their sum $S$, difference $D$, product $P$, and quotient $Q$ are defined as follows:

$$S(x) = f(x) + g(x) \text{ for each } x \in D_f \cap D_g$$

$$D(x) = f(x) - g(x) \text{ for each } x \in D_f \cap D_g$$

$$P(x) = f(x) \cdot g(x) \text{ for each } x \in D_f \cap D_g$$

$$Q(x) = \frac{f(x)}{g(x)} \text{ for each } x \in D_f \cap D_g \text{ with } g(x) \neq 0$$

If $f$ and $g$ are functions with domains $A$ and $B$, respectively, and $C = \{x \mid x \in A \text{ and } f(x) \in B\}$, then the function defined by the equation $y = (g \circ f)(x) = g(f(x))$ for each $x \in C$ is the composition of $g$ with $f$. We have introduced the following functions.

1. The linear functions $y = f(x) = mx + b$, where $m$ and $b$ are real numbers. The graph of a linear function is a line. If $(x_1, y_1)$ and $(x_2, y_2)$ are two points on a line, the slope $m$ of the line is

$$m = \frac{y_2 - y_1}{x_2 - x_1}, \text{ provided } x_2 - x_1 \neq 0$$

If two lines have slopes $m_1$ and $m_2$, respectively, the lines are parallel if, and only if, $m_1 = m_2$; they are perpendicular if, and only if, $m_1 \cdot m_2 = -1$.

2. The quadratic function is $y = f(x) = ax^2 + bx + c$, where $a$, $b$, and $c$ are constants and $a \neq 0$. Its graph is a parabola which is concave up if $a > 0$, and concave down if $a < 0$. The abscissa of the vertex is $\frac{-b}{2a}$

3. If $a_0, a_1, a_2, \ldots, a_n$ are real numbers with $a_0 \neq 0$, and $n$ is a positive integer, then

$$p(x) = a_0 x^n + a_1 x^{n-1} + a_2 x^{n-2} + \cdots + a_{n-1} x + a_n$$

defines a polynomial function of degree $n$. Its graph has at most $n - 1$ critical points.

4. A rational function is the quotient of two polynomial functions.

5. If $b$ is a positive number and $b \neq 1$, the exponential function with base $b$ is defined by $f(x) = b^x$ for all real numbers $x$. The exponential function serves as a mathematical model in many applications. The formula

$$A = P(1 + i)^n$$

is used to calculate the compound amount when money has been invested. It is also used to approximate the size of a population growing at a given rate.

6. The inverse of the exponential function with base $b$ is the logarithmic function to the base $b$ and is denoted $\log_b$. Thus, if $b > 0$ and $b \neq 1$,

$$y = \log_b x \text{ if, and only if, } x = b^y$$

The domain of the logarithmic function to the base $b$ is the set of all positive real numbers and its range is the set of all real numbers. The graph passes through $(1, 0)$. Also,

$$\log_b(xy) = \log_b |x| + \log_b |y|, \text{ whenever } xy > 0$$

$$\log_b\left(\frac{x}{y}\right) = \log_b |x| - \log_b |y|, \text{ whenever } \frac{x}{y} > 0$$

$$\log_b(x^y) = y \log_b x, \text{ whenever } x > 0$$

$$\log_b b = 1$$

$\log_b 1 = 0$, and

$$\log_b c = \frac{\log_a c}{\log_a b}, \text{ where } a, b \text{ are positive and not equal to } 1.$$

The bases most often used are 10 and $e$, where $e$ is an irrational number approximately equal to 2.71828. An equation of the form $a^{f(x)} = b^{g(x)}$, where $a$ and $b$ are positive and not equal to 1, may be solved by first taking the logarithm of both sides.

The graph of the equation $y = f(x + b)$ may be obtained by shifting the graph of $y = f(x)$ horizontally, while the graph of $y = g(x) + c$ may be obtained by shifting the graph of $y = g(x)$ vertically.

If $A$ is a finite set and $f$ is a function with domain $A$ and range a set of numbers, the symbol $\sum_{x \in A} f(x)$ denotes the sum obtained by finding the value of $f(x)$ for each $x \in A$ and adding all these values. If $A$ is a set of consecutive integers whose smallest member is $a$ and largest member is $b$, then $\sum_{i=a}^{b} f(i)$ denotes the sum $f(a) + f(a + 1) + f(a + 2) + \cdots + f(b)$.

**SAMPLE EXAM QUESTIONS**

*In Exercises 1–5, two sets S and T are given and a function f from S to T is described. In each case:*

*a.  Find the domain of f.*

*b.  Find the range of f.*

*c.  Determine whether the function f is one-to-one. If it is one-to-one, find its inverse. If it is not one-to-one, find a restriction that is one-to-one and find the inverse of the restriction.*

**1.** $S = \{a, b, c, d\}$, $T = \{1, 3, 5, 7, 9\}$, and the function $f$ is defined by

| $x$ | $a$ | $b$ | $c$ | $d$ |
|-----|-----|-----|-----|-----|
| $f(x)$ | 3 | 9 | 1 | 3 |

**2.** $S = [1, 3]$, $T = [2, 6]$, $f(x) = 2x$ for each $x \in S$.

**3.** $S$ and $T$ are the set of real numbers and $f(x) = x^3 - 2$.

**4.** $S = [-1, 2]$, $T = [-2, 5]$, and $f(x) = x^2$ for each $x \in S$. (*Hint:* Sketch the graph of $f$.)

**5.** Let $S$ be the set of all living human beings at a certain instant. Let $T$ be the set of all mothers of these human beings. (Some members of $T$ may be deceased.) For each $x \in S$, $f(x)$ is $x$'s mother.

*In Exercises 6–11, an equation defines a function. In each case, find the domain of that function.*

**6.** $f(x) = \dfrac{x - 1}{x + 2}$

**7.** $g(x) = \sqrt{x - 2}$

**8.** $h(x) = \sqrt{x^2 + 9x - 10}$

**9.** $f(x) = \dfrac{x + 1}{\sqrt{x^2 + 7x - 18}}$

**10.** $h(x) = \dfrac{x}{x^2 + x - 2}$

**11.** $g(x) = \dfrac{x^2 - 9}{x - 3}$

*In Exercises 12–14, find the sum, difference, product, and quotient of the given functions. In each case, specify the domain of the function you are describing.*

**12.** $f(x) = \dfrac{x^2 + 5x + 6}{x - 2}$; $g(x) = \dfrac{x^2 - 4}{x + 3}$

**13.** $f(x) = x^2 + x - 2$; $g(x) = \dfrac{x + 3}{x - 1}$    **14.** $f(x) = \dfrac{x^2 + x - 2}{x^2 + 2x - 15}$; $g(x) = \dfrac{x^2 + 7x + 10}{x^2 - 4x + 3}$

*In Exercises 15 and 16, find formulas for the compositions $f \circ g$ and $g \circ f$.*

**15.** $f(x) = x^2 + 2x - 3$; $g(x) = 2x + 3$    **16.** $f(x) = \dfrac{2x - 5}{x + 7}$; $g(x) = \dfrac{7x + 5}{2 - x}$

*In Exercises 17–22, sketch the graph of the given function.*

**17.** $f(x) = 2x + 3$

**18.** $g(x) = 2x^2 - 12x + 11$

**19.** $h(x) = -3x^2 + 24x - 33$

**20.** $f(x) = [x]^2$ (greatest integer function squared).

**21.** $g$ is defined by the set of ordered pairs $\{(-1, 2), (0, 3), (1, 4), (2, -3), (3, 0)\}$.

**22.** $h$ is defined by the set of ordered pairs $\{(x, y) \mid x \in \{-1, 0, 1, 2, 3, 4\}$ and $y = x^2 + 1\}$.

**23.** Let $f(x) = (2x + 3)^4$. Find functions $g$ and $h$ such that $f = g \circ h$.

**24.** Same as Exercise 23 for $f(x) = \sqrt{x^2 + 1}$.

**25.** Let $g$ be defined by $g(x) = 3x + 2$.
   **a.** Show that $g$ is a one-to-one function.
   **b.** Find a formula for $g^{-1}(x)$.
   **c.** Draw the graphs of $g$ and $g^{-1}$ on the same coordinate plane.

**26.** Determine which of the following curves are graphs of **a.** a function and **b.** a one-to-one function.

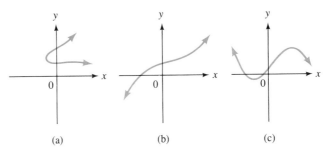

(a)          (b)          (c)

**27.** Find the equation of each of the following lines in the requested form.
   **a.** Passing through $(-1, 2)$ and $(3, 4)$. Slope-intercept form.
   **b.** Passing through $(2, 5)$ and parallel to the graph of $2x + 3y = 5$. Point-slope form.
   **c.** Passing through $(-1, 3)$ and perpendicular to the graph of $x + 2y = 5$. Slope-intercept form.
   **d.** Passing through $(1, 3)$ and perpendicular to the $x$-axis.
   **e.** Passing through $(-1, 4)$ and perpendicular to the $y$-axis.

**28.** Julie has a job paying \$12 per hour. She is paid time and a half for the first 20 hours of overtime (between 40 and 60 hours in a given week), double time after that.
   **a.** If she works $x$ hours during a given week, determine her income $\$I(x)$ in terms of $x$.
   **b.** Find $I(36)$.          **c.** Find $I(52)$.          **d.** Find $I(65)$.

**29.** Sketch the graph of the function $f$ defined by $f(x) = \left(\frac{1}{3}\right)^x$.

**30.** Sketch the graph of each function. (Use a calculator.)
   **a.** $y = f(x) = e^{-2x}$          **b.** $y = g(x) = e^{x/2}$          **c.** $y = h(x) = (2.5)^x$

**31.** If $700 is deposited at 6% interest compounded monthly, find the amount in the account (to the nearest cent) 3 years later. What is the interest earned?

**32.** The rabbit population on an island was estimated to be 15,000 at the beginning of 1987. Assuming that the population grows 12% per year, predict the approximate size of the rabbit population at the end of 1998.

**33.** Given $\log 2 \doteq .3010$ and $\log 3 \doteq .4771$, find:
  **a.** $\log 6$                **b.** $\log 8$                **c.** $\log \sqrt{3}$
  **d.** $\log \dfrac{4}{27}$                **e.** $\log 250$                **f.** $\log_2 10$

**34.** Solve the given equations.
  **a.** $\log_2(5x - 8) = 5$                **b.** $\ln(3x - 5) = 2$
  **c.** $\log(x - 7) + \log(x + 8) = 2$                **d.** $2^{3x+1} = 3^{2x-3}$

**35.** If a bank pays 6% compounded monthly, how long will it take for a deposit to quadruple?

**36.** If $A = \{-1, 2, 3, 5, 9\}$ and $f(x) = x^2 + 1$, find $\displaystyle\sum_{x \in A} f(x)$.

**37.** Evaluate each of the following:

  **a.** $\displaystyle\sum_{i=1}^{20} i^2$                **b.** $\displaystyle\sum_{i=1}^{30} 5$                **c.** $\displaystyle\sum_{i=1}^{100} \left( \dfrac{1}{i} - \dfrac{1}{i+1} \right)$

**38.** The Mason Company produces a gadget at a cost of $3.50 each. The fixed cost per month is $7000. The gadget sells for $5 each. Assuming that all gadgets produced can be sold, how many must be produced to realize a $9500 profit per month?

**39.** Mr. Greenthumb has a rectangular vegetable garden 60 feet by 40 feet. He will double the area of his garden by adding strips of equal width to one side and one end to get a new rectangular garden. Find the width of the strips.

**40.** The manager of a store wishes to mix two kinds of nuts to get 150 pounds of a mixture worth $4.10 a pound. Kind A is worth $3.50 per pound and kind B is worth $5 a pound. How many pounds of each kind must he use?

**41.** The manager of a store knows that 40 tennis racquets will be sold each week if the price is $90 per racquet, while 80 will be sold during the same period if the price is $70 per racquet.
  **a.** Derive an equation giving the price $p$ per racquet in terms of the number $q$ of racquets sold each week assuming this equation is linear.
  **b.** Derive an equation giving the revenue $R$ per week in terms of $q$.
  **c.** If the cost to sell $q$ racquets per week is $(60q + 250)$, how many racquets should be sold to maximize the profit? What is the price per racquet that will yield this maximum profit? What is the maximum profit?

**42.** A company manufactures hair dryers. The cost of producing and selling $q$ thousand dryers in any month is $C$ thousand, where $C = 15q + 75$. The revenue from these is $R$ thousand, where $R = -.25q^2 + 25q$. The company is limited to a maximum production of 50,000 dryers per month. Illustrate graphically how the company should operate to realize a profit.

**43.** Sketch the graph of $y = |x|$. Then shift this graph to obtain the graph of $y = |x - 3| + 5$.

**44.** The graph of a parabola may be obtained by shifting the graph of $y = x^2$ horizontally, then vertically. The vertex of this parabola has coordinates $(-3, 5)$. What is the equation of the parabola?

**45.** The graph of $y = f(x)$ may be obtained by shifting the graph of $y = |x|$ horizontally, then vertically. The lowest point of this graph has coordinates $(2, -8)$. What is the equation defining $f(x)$?

# 2

# Systems of Equations and Inequalities

Many mathematical models that we encounter involve more than one variable. For example, a farmer may grow both apples and pears. In a mathematical model, $x$ and $y$ may represent the numbers of pounds of apples and pears, respectively, that he will produce. The manager of a grocery store may mix three kinds of nuts where $x$, $y$, and $z$, represent the number of pounds of each type $A$, $B$, and $C$, respectively. In this and subsequent chapters, we will consider open sentences with more than one variable.

# 2.1  Systems of Two Linear Equations in Two Variables

> **DEFINITION 2.1:**  A *linear equation in two variables x and y* is an equation that can be written in the form $ax + by = c$, where $a$, $b$, and $c$ are real numbers and $ab \neq 0$.

The replacement set for $x$ and $y$ is the set of all real numbers. An ordered pair of real numbers $(x_0, y_0)$ is said to be a *solution of the equation* if, and only if, $ax_0 + by_0 = c$ is a true statement. For example, the ordered pair $(-2, 3)$ is a solution of the equation $2x + 3y = 5$ since $2(-2) + 3(3) = 5$ is true. On the other hand, the ordered pair $(3, 1)$ is not a solution of the same equation since $2(3) + 3(1) = 5$ is not true. When using other symbols for the two variables, consider one of them as the first variable and the other as the second, since a solution is an *ordered* pair of real numbers.

> **DEFINITION 2.2:**  The set of all ordered pairs of real numbers that are solutions of the equation $ax + by = c$ is called the *solution set of the equation*. A linear equation in two variables has infinitely many solutions.

To see this, we need only replace one of the two variables by any real number and solve for the other.

**EXAMPLE 1**  Find three solutions of the equation $2x + 3y = 18$.

**Solution**  First, we replace $x$ by 0 and get $0 + 3y = 18$ which yields $y = 6$. Thus, the ordered pair $(0, 6)$ is a solution. Next, we replace $y$ by 0 and get $2x + 0 = 18$ which yields $x = 9$. Therefore, the ordered pair $(9, 0)$ is also a solution. We chose to replace $x$ and $y$ by 0 because this required only simple calculations. However, we may replace either variable by any real number we wish. If we replace $x$ by 4.5, we get $2(4.5) + 3y = 18$ which yields $9 + 3y = 18$ and $y = 3$. Thus, the ordered pair $(4.5, 3)$ is a third solution. ■

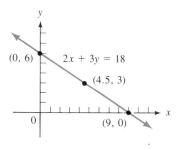

**FIGURE 2.1**

## Graphing Systems of Two Linear Equations

We have seen that the graph of an equation of the form $ax + by = c$ is a straight line. For example, the graph of the equation of the previous example is the line passing through the points $(0, 6)$ and $(9, 0)$ (see Figure 2.1). Note that the point $(4.5, 3)$ is also on the graph. In fact, the coordinates of each point on the graph provide a solution of the equation.

If $a_1$, $b_1$, $c_1$, $a_2$, $b_2$, and $c_2$ are real numbers, the open sentence $a_1x + b_1y = c_1$ and $a_2x + b_2y = c_2$ is called a *system of two linear equations in two variables*. It is usually written as

$$\begin{cases} a_1x + b_1y = c_1 \\ a_2x + b_2y = c_2 \end{cases}$$

An ordered pair of real numbers $(x_0, y_0)$ is a solution of the foregoing system if, and only if, it is a solution of both equations.

> **DEFINITION 2.3:**   The *solution set* of the system is the intersection of the solution sets of both equations.

**EXAMPLE 2**   Solve the system

$$\begin{cases} 2x + 3y = \phantom{-}8 \\ 3x - 2y = -1 \end{cases}$$

graphically.

**Solution**   We know that the graph of each equation is a straight line. We sketch both graphs on the same Cartesian coodinate plane (see Figure 2.2). These two lines are graphical representations of the solution sets of the two equations of the system. Thus, the intersection of the two lines is a graphical representation of the solution of the system. It appears that this intersection is the single point $(1, 2)$. (See Figure 2.2.) We verify that $(1, 2)$ is indeed a solution of our system. We replace $x$ by 1 and $y$ by 2 in both equations. We get

$$2(1) + 3(2) = 8$$
$$3(1) - 2(2) = -1$$

Note that both of the equalities are true. Hence, $(1, 2)$ is a solution of the system. In fact, it is the only solution.    ∎

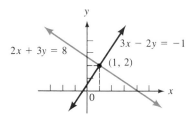

**FIGURE 2.2**

In general, it is not practical to solve a system of linear equations graphically. You would have to estimate the coordinates of the point of intersection of the two graphs, which may not be accurate. We shall, therefore, describe several algebraic methods to solve such systems. However, it is useful to consider the geometric approach because it sheds light on the nature of the solution set. Suppose, for example, that we have a system of two linear equations in the variables $x$ and $y$. Let $L_1$ and $L_2$ be the graphs of the equations. Then, only one of three things can happen.

Characteristics of Graphs of Two Linear Equations

1. $L_1$ and $L_2$ intersect at exactly one point (see Figure 2.3a). In that case, the solution set of the system has exactly one member. We say that the system is *consistent and independent*.
2. $L_1$ and $L_2$ are parallel (see Figure 2.3b). In that case, we have no point of intersection, hence we have no solution. The solution set is the empty set and we say that the system is *inconsistent*.
3. $L_1$ and $L_2$ coincide. Then the coordinates of each point of $L_1$ are solutions of the given system, and the solution set has infinitely many members (see Figure 2.3c). In that case, we say that the system is *consistent and dependent*.

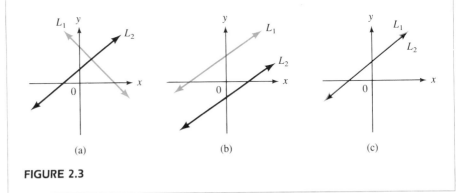

(a)    (b)    (c)

**FIGURE 2.3**

## Solving by Elimination of Variables

You can solve a system of two linear equations in two variables using a number of algebraic methods. You can *eliminate a variable by addition or subtraction*. Given a system of two linear equations in two variables, an *equivalent system* is obtained if you multiply both sides of either equation by a nonzero constant or replace either equation by the sum, or the difference, of the two equations. Performing these two operations on a system results in an equivalent system where one of the two equations will have only one variable (the other variable having been eliminated). We solve the equation that has only one variable and replace this variable by its value in the other equation. We then solve this equation.

**EXAMPLE 3**    Solve the system

$$\begin{cases} 2x + 3y = -5 \\ x + 5y = -13 \end{cases}$$

**Solution**    We first decide which variable to eliminate. Since the coefficient of $x$ in the second equation is 1, we choose to eliminate $x$ and multiply both sides of the second equation by $-2$ for a reason which will soon be clear. We obtain the equivalent system

$$\begin{cases} 2x + 3y = -5 \\ -2x - 10y = 26 \end{cases}$$

We now replace the second equation by the sum of the two equations to get the new equivalent system

$$\begin{cases} 2x + 3y = -5 \\ -7y = 21 \end{cases}$$

Note that $x$ has been eliminated from the second equation. We solve for $y$ in the second equation to get

$$\begin{cases} 2x + 3y = -5 \\ y = -3 \end{cases}$$

Replacing $y$ by $-3$ in the first equation, we obtain

$$\begin{cases} 2x - 9 = -5 \\ y = -3 \end{cases}$$

Adding 9 to both sides of the first equation then dividing both sides by 2 yields

$$\begin{cases} x = 2 \\ y = -3 \end{cases}$$

The solution of this last system is $(2, -3)$. Since each system is equivalent to the preceding one, the solution of the original system is also $(2, -3)$.
**Do Exercise 3.**                                                                 ∎

**EXAMPLE 4**    Solve the system

$$\begin{cases} 7x + 2y = 31 \\ 3x - 5y = 25 \end{cases}$$

**Solution**    Decide which variable to eliminate. Since none of the coefficients is 1 and there is no advantage in choosing one variable over the other, eliminate $y$ this time. We want an equivalent system where the $y$ coefficients will have opposite signs but will have equal absolute values. Thus, we multiply both sides of the first equation by 5 and both sides of the second equation by 2. We obtain

$$\begin{cases} 35x + 10y = 155 \\ 6x - 10y = 50 \end{cases}$$

Replacing the second equation by the sum of the two equations, we get the equivalent system

$$\begin{cases} 35x + 10y = 155 \\ 41x = 205 \end{cases}$$

Solving the second equation for $x$ we get

$$\begin{cases} 35x + 10y = 155 \\ x = 5 \end{cases}$$

Replacing $x$ by 5 in the first equation, we obtain

$$\begin{cases} 35(5) + 10y = 155 \\ x = 5 \end{cases}$$

Subtracting 175 from both sides of the first equation, then dividing both sides by 10 yields

$$\begin{cases} y = -2 \\ x = \phantom{-}5 \end{cases}$$

The solution of the last system is $(5, -2)$. Since each system is equivalent to the preceding one, the solution of the given system is also $(5, -2)$.
**Do Exercise 9.**

You can also solve a system of two linear equations in two variables by *eliminating a variable by substitution*. Solve for one of the variables in one of the two equations and substitute the expression obtained for that variable in the *other* equation.

**EXAMPLE 5**   Solve the system

$$\begin{cases} 3x + 2y = \phantom{-}6 \\ 5x + \phantom{2}y = -4 \end{cases}$$

**Solution**   Since the coefficient of $y$ in the second equation is 1, it is easiest to solve for $y$ in that equation. We get

$$y = -5x - 4$$

We now substitute $-5x - 4$ for $y$ in the first equation and obtain the equivalent system

$$\begin{cases} 3x + 2(-5x - 4) = 6 \\ \phantom{3x + 2(-5x - 4) = }y = -5x - 4 \end{cases}$$

This system may be written

$$\begin{cases} -7x - 8 = 6 \\ \phantom{-7x - 8 =}y = -5x - 4 \end{cases}$$

Note that $y$ has been eliminated from the first equation. We solve for $x$ in that equation to get

$$\begin{cases} x = -2 \\ y = -5x - 4 \end{cases}$$

Replacing $x$ by $-2$ in the second equation, we get

$$\begin{cases} x = -2 \\ y = -5(-2) - 4 \end{cases}$$

That is,

$$\begin{cases} x = -2 \\ y = \phantom{-}6 \end{cases}$$

Thus, the solution of the last and of the original system is $(-2, 6)$.
**Do Exercise 19.**

**EXAMPLE 6**   Solve the system

$$\begin{cases} 5x + 3y = -3 \\ 3x - 7y = 29 \end{cases}$$

**Solution**   Since none of the coefficients is 1, there is no advantage in choosing one variable to eliminate over the other. We solve for $x$ in the second equation and get

$$x = \frac{7y + 29}{3}$$

Substituting the foregoing expression for $x$ in the first equation, we obtain the equivalent system

$$\begin{cases} 5 \cdot \dfrac{7y + 29}{3} + 3y = -3 \\ \qquad\qquad 3x - 7y = 29 \end{cases}$$

We simplify the first equation by multiplying both sides by 3 to get

$$\begin{cases} 5(7y + 29) + 9y = -9 \\ \qquad\qquad 3x - 7y = 29 \end{cases}$$

This system may be written

$$\begin{cases} \qquad 44y = -154 \\ 3x - 7y = 29 \end{cases}$$

Dividing both sides of the first equation by 44 we obtain

$$\begin{cases} \qquad y = \dfrac{-7}{2} \\ 3x - 7y = 29 \end{cases}$$

Replacing $y$ by $\frac{-7}{2}$ in the second equation and solving for $x$ we get

$$\begin{cases} y = \dfrac{-7}{2} \\ x = \dfrac{3}{2} \end{cases}$$

The solution of this system and consequently of the original system is $\left(\frac{3}{2}, \frac{-7}{2}\right)$.
**Do Exercise 21.**                                                    ■

## Solving by Cramer's Rule

With the help of determinants, which we define below, we can solve a system of two linear equations in two variables efficiently using a method credited to the Swiss mathematician Gabriel Cramer (1704–1752) and called, appropriately, *Cramer's rule*. First, we define matrix and determinant.

**DEFINITION 2.4:** A square array of numbers of the form

$$\begin{bmatrix} a & b \\ c & d \end{bmatrix}$$

is called a $2 \times 2$ (read: "two-by-two") *matrix*. The *determinant* of this $2 \times 2$ matrix is the number $ad - bc$ and is usually denoted

$$\begin{vmatrix} a & b \\ c & d \end{vmatrix}$$

For example,

$$\begin{vmatrix} 3 & -5 \\ 2 & 4 \end{vmatrix} = 3(4) - 2(-5) = 12 + 10 = 22$$

**Cramer's Rule**

It can be shown (see Exercise 65) that if we let

$$\begin{vmatrix} a_1 & b_1 \\ a_2 & b_2 \end{vmatrix} = C, \quad \begin{vmatrix} c_1 & b_1 \\ c_2 & b_2 \end{vmatrix} = A_x, \quad \begin{vmatrix} a_1 & c_1 \\ a_2 & c_2 \end{vmatrix} = A_y,$$

then the systems

$$\begin{cases} a_1 x + b_1 y = c_1 \\ a_2 x + b_2 y = c_2 \end{cases}$$

and

$$\begin{cases} x = \dfrac{A_x}{C} \\ y = \dfrac{A_y}{C} \end{cases}$$

are equivalent, provided that $C \neq 0$.

Cramer's rule is easy to remember if we note the following:

1. $C$ is the determinant of the matrix of coefficients of the variables in the original system.
2. $A_x$, which appears in the numerator of the fraction giving the value of $x$, is the determinant of the matrix obtained from the matrix of coefficients by replacing the column of coefficients of $x$ by the column of constants on the right side of the original system.
3. $A_y$, which appears in the numerator of the fraction giving the value of $y$, is the determinant of the matrix obtained from the matrix of coefficients by replacing the column of coefficients of $y$ by the column of constants on the right side of the original system.

**EXAMPLE 7**   Solve the system

$$\begin{cases} 2x + 3y = 6 \\ 5x + 4y = 1 \end{cases}$$

using Cramer's rule.

**Solution**   We find

$$C = \begin{vmatrix} 2 & 3 \\ 5 & 4 \end{vmatrix} = 2(4) - 5(3) = 8 - 15 = -7$$

$$A_x = \begin{vmatrix} 6 & 3 \\ 1 & 4 \end{vmatrix} = 6(4) - 1(3) = 24 - 3 = 21$$

$$A_y = \begin{vmatrix} 2 & 6 \\ 5 & 1 \end{vmatrix} = 2(1) - 5(6) = 2 - 30 = -28$$

Thus,

$$\begin{cases} x = \dfrac{A_x}{C} = \dfrac{21}{-7} = -3 \\ y = \dfrac{A_y}{C} = \dfrac{-28}{-7} = 4 \end{cases}$$

We see that the solution of the original system is $(-3, 4)$.
**Do Exercise 35.**

## Inconsistent Systems and Consistent and Dependent Systems

The systems we have considered in the preceding examples were consistent and independent—that is, each system had exactly one solution. We now give an example of a system that is inconsistent (has no solution) and one that is consistent and dependent (its solution set is infinite).

**EXAMPLE 8**   Solve the system

$$\begin{cases} 2x + 3y = 13 \\ 6x + 9y = -2 \end{cases}$$

**Solution**   Multiplying the first equation by $-3$ and adding the result to the second equation to get a new second equation we obtain the equivalent system

$$\begin{cases} 2x + 3y = 13 \\ 0 = -41 \end{cases}$$

The second equality in this system is a compact form of the equation $0 \cdot x + 0 \cdot y = -41$. Clearly, this equation has no solution since any replacement of $x$ and $y$ by real numbers will lead to the false statement $0 = -41$. Thus, the solution set of this system and, therefore, of the original system, is the empty set. The system is inconsistent. Graph both linear equations of the system and note that the graphs are parallel lines.
**Do Exercise 5.**

**EXAMPLE 9** Solve the system

$$\begin{cases} 3x - 5y = 7 \\ -6x + 10y = -14 \end{cases}$$

**Solution** Multiplying the first equation by 2 and adding the result to the second equation to get a new second equation we have the equivalent system

$$\begin{cases} 3x - 5y = 7 \\ 0 = 0 \end{cases}$$

The second equality of this system is a compact form of the equation $0 \cdot x + 0 \cdot y = 0$. Any ordered pair of real numbers is a solution of this equation since replacing $x$ and $y$ by arbitrary real numbers always yields the true statement $0 = 0$. Thus, the solution set of the system is the solution set of the first equation $3x - 5y = 7$. If we let $x = c$, where $c$ is an arbitrary real number, we obtain

$$3c - 5y = 7$$

Solving for $y$,

$$y = \frac{3c - 7}{5}$$

Hence, the solution set of the original system is

$$\left\{ \left( c, \frac{3c - 7}{5} \right) \;\middle|\; c \text{ is an arbitrary real number} \right\}$$

For example, if we replace $c$ successively by $-6, 0, 4, 9$, and $14$ we get the solutions $(-6, -5)$, $\left( 0, \frac{-7}{5} \right)$, $(4, 1)$, $(9, 4)$, and $(14, 7)$. Of course, we could get as many solutions as we wish. Graph the two linear equations of the original system and note that the two graphs coincide.
**Do Exercise 11.**

## Nonlinear Systems

Sometimes a system of two nonlinear equations in two variables may be reduced to a system of linear equations by making an appropriate substitution. The next two examples involve such systems.

**EXAMPLE 10** Solve the system

$$\begin{cases} 2x^2 - 3y^2 = -10 \\ 3x^2 + 2y^2 = 11 \end{cases}$$

**Solution** The equations are nonlinear. However, if we let $u = x^2$ and $v = y^2$, we obtain the following system of two linear equations in $u$ and $v$:

$$\begin{cases} 2u - 3v = -10 \\ 3u + 2v = 11 \end{cases}$$

Using any of the methods discussed, we easily find that the unique solution of this system is $(1, 4)$. However, this ordered pair is not a solution of the original system.

Recalling that $u = x^2$ and $v = y^2$, we write $1 = x^2$ and $4 = y^2$. Thus, the possible values of $x$ are $-1$ and $1$, while the possible values of $y$ are $-2$ and $2$. Consequently, the solution set of the original system is $\{(-1, -2), (-1, 2), (1, -2), (1, 2)\}$. ∎

**Do Exercise 53.**

EXAMPLE 11   Solve the system

$$\begin{cases} \dfrac{2}{x} - \dfrac{6}{y} = 3 \\ \dfrac{4}{x} - \dfrac{3}{y} = 3 \end{cases}$$

**Solution**   We let $\frac{1}{x} = u$ and $\frac{1}{y} = v$, thus obtaining the system

$$\begin{cases} 2u - 6v = 3 \\ 4u - 3v = 3 \end{cases}$$

This system may be solved using any of the three methods we introduced earlier. Its solution is $\left(\frac{1}{2}, -\frac{1}{3}\right)$. Thus $\frac{1}{x} = \frac{1}{2}$ and $\frac{1}{y} = \frac{-1}{3}$. It follows that $x = 2$ and $y = -3$. The solution of the original system is $(2, -3)$. ∎

**Do Exercise 55.**

## Applications of Systems of Linear Equations

The next two examples will illustrate some applications whose mathematical models are systems of linear equations.

EXAMPLE 12   Prior to a crucial NFL game between Denver and Seattle, a clever gambler met two enthusiastic football fans. The first fan was from Denver, and he was giving 2 to 1 odds on the Broncos; the second fan was from Seattle and he was giving 3 to 1 odds on the Seahawks. The clever gambler placed a bet with each fan so that she would win $100 as long as the game did not end up in a tie. How much did she bet with each fan?

**Solution**   Suppose that the clever gambler bet $x$ with the Denver fan and $y$ with the Seattle fan. If Denver won the game, then the gambler would have to pay $x$ to the Denver fan but would receive $3y$ from the Seattle fan. Hence, her profit would be $\$(-x + 3y)$. Since she wanted to win $100, we obtain the first equation $-x + 3y = 100$. In exactly the same way, we see that if Seattle won the game, the gambler would win $2x$ from the Denver fan but would have to pay $y$ to the Seattle fan. Thus, $2x - y = 100$. We now have the system

$$\begin{cases} -x + 3y = 100 \\ 2x - y = 100 \end{cases}$$

This system can easily be solved. Its solution is $(80, 60)$. We conclude that the clever gambler bet $80 with the Denver fan and $60 with the Seattle fan. This example shows that sports fans should never bet on games, especially with strangers! ∎

**Do Exercise 59.**

EXAMPLE 13   A grocer has walnuts worth $4 a pound and Brazil nuts worth $5 a pound. How many pounds of each type must he add to get 150 pounds of a mixture worth $4.30 a pound?

Solution   Let $x$ be the number of pounds of walnuts and $y$ be the number of pounds of Brazil nuts the grocer must mix. Since the mixture must weigh 150 pounds, the first equation is $x + y = 150$. The value of the walnuts is $\$4x$ and the value of the Brazil nuts is $\$5y$. Since the mixture must be worth $4.30 per pound, its value must be $645 since $(4.3)(150) = 645$. Clearly, the total value of the mixture is equal to the sum of the values of the components. Hence, the second equation is $4x + 5y = 645$. We obtain the system

$$\begin{cases} x + y = 150 \\ 4x + 5y = 645 \end{cases}$$

This system may be solved using any of the discussed methods. Its solution is $(105, 45)$. Thus, the grocer should mix 105 pounds of walnuts with 45 pounds of Brazil nuts to get 150 pounds of a mixture worth $4.30 a pound.
**Do Exercise 61.**

# Exercise Set 2.1

*Solve each of the systems of linear equations in Exercises 1–17 using the method of elimination of a variable by addition or subtraction.*

1. $\begin{cases} 2x - 3y = 8 \\ 3x + 5y = -7 \end{cases}$

2. $\begin{cases} 3x + y = -1 \\ 2x - 5y = -12 \end{cases}$

3. $\begin{cases} -2x + 3y = -9 \\ x - 4y = 7 \end{cases}$

4. $\begin{cases} 5x + 2y = -5 \\ 3x - 7y = -44 \end{cases}$

5. $\begin{cases} 2x - 7y = 17 \\ 6x - 21y = -2 \end{cases}$

6. $\begin{cases} 3x - 5y = 25 \\ 5x + 6y = 13 \end{cases}$

7. $\begin{cases} 7x - 3y = 37 \\ 2x + 6y = -10 \end{cases}$

8. $\begin{cases} 8x - 6y = -4 \\ 4x - 3y = -2 \end{cases}$

9. $\begin{cases} 5x + 8y = -34 \\ 3x - 5y = 9 \end{cases}$

10. $\begin{cases} 3x + 2y = 2 \\ 9x - 8y = -1 \end{cases}$

11. $\begin{cases} x + 3y = 5 \\ 4x + 12y = 20 \end{cases}$

12. $\begin{cases} .5x + .3y = 2.5 \\ 1.2x - 1.4y = -4.6 \end{cases}$

13. $\begin{cases} 5x - 3y = 38 \\ -4x + 7y = -35 \end{cases}$

14. $\begin{cases} 1.3x - 2.5y = 6 \\ 2.6x - 5y = -3 \end{cases}$

15. $\begin{cases} 5r + 8s = 46 \\ 2r - 5s = -37 \end{cases}$

16. $\begin{cases} 7u + 2v = 3 \\ -3u + 5v = 28 \end{cases}$

17. $\begin{cases} \dfrac{1}{4}t - \dfrac{1}{3}u = -\dfrac{1}{12} \\ \dfrac{1}{5}t + \dfrac{1}{6}u = \dfrac{3}{20} \end{cases}$

*For Exercises 18–34, solve the systems of Exercises 1–17 using the method of elimination of a variable by substitution. For Exercise n solve the system of Exercise n − 17.*

*For Exercises 35–51, solve the systems of Exercises 1–17 using Cramer's rule. Whenever the determinant of the coefficient matrix is 0, use a different method. For Exercise n, solve the system of Exercise n − 34.*

*In Exercises 52–57 solve the given system of nonlinear equations.*

52. $\begin{cases} x^2 + y^2 = 13 \\ 2x^2 - 5y^2 = -37 \end{cases}$

53. $\begin{cases} 2x^2 - y^2 = -17 \\ 3x^2 + 2y^2 = 62 \end{cases}$

54. $\begin{cases} 2x^2 - 13y^2 = 59 \\ 3x^2 + 5y^2 = 113 \end{cases}$

55. $\begin{cases} \dfrac{5}{x} + \dfrac{8}{y} = -34 \\ \dfrac{3}{x} - \dfrac{5}{y} = 9 \end{cases}$

56. $\begin{cases} \dfrac{3}{x} + \dfrac{2}{y} = 2 \\ \dfrac{9}{x} - \dfrac{8}{y} = -1 \end{cases}$

57. $\begin{cases} -\dfrac{2}{x} + \dfrac{3}{y} = -9 \\ \dfrac{1}{x} - \dfrac{4}{y} = 7 \end{cases}$

58. Prior to a crucial game between the Minnesota Vikings and the Pittsburgh Steelers a clever gambler met two football fans. The first fan was giving 9 to 6 odds on the Vikings. The second fan was giving 7 to 5 odds on the Steelers. The clever gambler placed a bet with each fan so that he would win $110 as long as the game did not end up in a tie. How much did he bet with each fan?

59. Prior to an important NFL game between the Dallas Cowboys and the Los Angeles Rams an admirer of Jimmy the

Greek met two football fans. The first fan was giving 8 to 5 odds on the Cowboys, while the second fan was giving 7 to 4 odds on the Rams. She placed a bet with each fan so that she would win $360 as long as the game did not end up in a tie. How much did she bet with each fan?

**60.** A grocer has walnuts worth $3.50 a pound and Brazil nuts worth $4.50 a pound. How many pounds of each type must she mix in order to get 200 pounds of a mixture worth $4.15 a pound?

**61.** A grocer has two brands of coffee. Brand A is worth $4.25 a pound and brand B is worth $3.75 a pound. How many pounds of each brand must he mix in order to get 100 pounds of a mixture worth $3.90 a pound?

**62.** Julie invested part of the $10,000 she won at the lottery at a 6% annual rate of interest and the rest in a higher-risk bond paying an annual rate of interest of 9%. Her total annual income on the two investments is $690. How much did she invest at each rate?

**63.** John invested part of the $50,000 he inherited at a 6.5% annual rate of interest and the rest in a higher risk bond paying an annual rate of interest of 9.5%. His total annual income on the two investments is $3790. How much did he invest at each rate?

**64.** During last season, 6250 tickets were sold for a baseball game in the Seattle Kingdome. An adult pays $9 for a ticket, while a child under 16 pays only $3. If the total revenue from tickets for that game was $51,900, how many children and how many adults bought tickets?

**\*65.** Prove Cramer's rule. (*Hint:* Use the method of elimination of a variable by addition to solve the system

$$\begin{cases} a_1x + b_1y = c_1 \\ a_2x + b_2y = c_2 \end{cases}$$

## 2.2 Systems of Three Linear Equations in Three Variables

In the present section we will consider systems of three linear equations in three variables. An example of such a system is

$$\begin{cases} 2x + 3y - 2z = 15 \\ 3x - y + 4z = -8 \\ -x + 4y + 2z = 7 \end{cases}$$

The ordered triple $(1, 3, -2)$ is a *solution* of this system since we get three true statements when we replace $x$, $y$, and $z$ by 1, 3, and $-2$, respectively, in the three equations. The set of all solutions is called the *solution set* of the system.

### Solving Linear Systems of Equations in Three Variables

The methods used to solve such systems are the same as those used in solving systems of two linear equations in two variables.\*

---

\*Although it is possible to interpret the solution set of a system of three linear equations in three variables as the intersection of three planes in a three-dimensional space, we shall not pursue this approach in this book. However, the student may find this geometric interpretation useful when considering the nature of the solution set of such systems. Clearly, the intersection of three planes will be either a single point, a straight line, a plane (if all three planes coincide), or the empty set. Thus, a system of three linear equations in three variables may have a single solution, an infinite number of solutions, or no solution.

As in the case of two systems of two equations, given a system of three linear equations in three variables, we obtain an *equivalent system* if we either multiply both sides of any equation by a nonzero constant or add to any equation one of the other two equations.

To shorten the explanations in the examples, we abbreviate the statement "both sides of the $i$th equation have been multiplied by the nonzero number $k$ to yield the new $i$th equation" using the notation "$kE_i \rightarrow E_i$." We shall also abbreviate the statement "the members of the $i$th equation have been multiplied by the number $k$ and the results have been added to the corresponding members of the $j$th equation to yield the new $j$th equation" using the notation "$kE_i + E_j \rightarrow E_j$." The meaning of "$k_1 E_i + k_2 E_j \rightarrow E_j$" should be easy to determine.

To find the solution to a given system of three linear equations in three variables, we eliminate one of the variables from two of the equations. We then eliminate one of the remaining variables from one of these two equations, and so on, until we obtain an equivalent system in the form

$$\begin{cases} x = x_0 \\ y = y_0 \\ z = z_0 \end{cases}$$

which gives the solution of the original system.

**EXAMPLE 1**   Solve the system

$$\begin{cases} 2x + 3y - 5z = -14 \\ 3x - 2y + 3z = \phantom{-}17 \\ 4x + 3y - 2z = \phantom{-}-1 \end{cases}$$

**Solution**   We wish to eliminate $x$ from equations 2 and 3. We can eliminate $x$ from equation 2 by multiplying equation 1 by 3, equation 2 by $-2$, and adding the results to yield a new equation 2. In the explanation following, we abbreviate this as $3E_1 + (-2)E_2 \rightarrow E_2$. We can also eliminate $x$ from the third equation by multiplying the first equation by $-2$ and adding the result to equation 3 to yield the new equation 3. This is abbreviated as $(-2)E_1 + E_3 \rightarrow E_3$.

$$\begin{cases} 2x + 3y - 5z = -14 \\ 3x - 2y + 3z = \phantom{-}17 \\ 4x + 3y - 2z = \phantom{-}-1 \end{cases} \qquad \begin{array}{l} \textit{Explanation} \\ 3E_1 + (-2)E_2 \rightarrow E_2 \\ (-2)E_1 + E_3 \rightarrow E_3 \end{array}$$

$$\begin{cases} 2x + \phantom{1}3y - \phantom{2}5z = -14 \\ \phantom{2x +} 13y - 21z = -76 \\ \phantom{2x +} -3y + \phantom{2}8z = \phantom{-}27 \end{cases}$$

Thus, in the new equivalent system $x$ has been eliminated from equations 2 and 3. We now wish to eliminate $y$ from equation 3. This can be accomplished by multiplying equation 2 by 3, equation 3 by 13, and adding the results to get the new equation 3.

We abbreviate this as $3E_2 + 13E_3 \rightarrow E_3$. The other necessary steps are described in abbreviated form in the explanation column.

*Explanation*

$$\begin{cases} 2x + 3y - 5z = -14 \\ \phantom{2x +} 13y - 21z = -76 \\ \phantom{2x +} -3y + 8z = 27 \end{cases} \qquad 3E_2 + 13E_3 \rightarrow E_3$$

$$\begin{cases} 2x + 3y - 5z = -14 \\ \phantom{2x +} 13y - 21z = -76 \\ \phantom{2x + 13y -} 41z = 123 \end{cases} \qquad \frac{1}{41}E_3 \rightarrow E_3$$

$$\begin{cases} 2x + 3y - 5z = -14 \\ \phantom{2x +} 13y - 21z = -76 \\ \phantom{2x + 13y -} z = 3 \end{cases} \qquad 21E_3 + E_2 \rightarrow E_2$$

$$\begin{cases} 2x + 3y - 5z = -14 \\ \phantom{2x +} 13y = -13 \\ \phantom{2x + 13y -} z = 3 \end{cases} \qquad \frac{1}{13}E_2 \rightarrow E_2$$

$$\begin{cases} 2x + 3y - 5z = -14 \\ \phantom{2x +} y = -1 \\ \phantom{2x + 3y -} z = 3 \end{cases} \qquad (-3)E_2 + 5E_3 + E_1 \rightarrow E_1$$

$$\begin{cases} 2x = 4 \\ y = -1 \\ z = 3 \end{cases} \qquad \frac{1}{2}E_1 \rightarrow E_1$$

$$\begin{cases} x = 2 \\ y = -1 \\ z = 3 \end{cases}$$

The solution of this last system is $(2, -1, 3)$. Since each system is equivalent to the preceding one, the solution of the original system is also $(2, -1, 3)$.
**Do Exercise 1.**                                                       ■

**EXAMPLE 2**   Solve the system

$$\begin{cases} x + 2y - 5z = -6 \\ -2x + 3y - z = -13 \\ 5x + 7y - 13z = -12 \end{cases}$$

**Solution**
*Explanation*

$$\begin{cases} x + 2y - 5z = -6 \\ -2x + 3y - z = -13 \\ 5x + 7y - 13z = -12 \end{cases} \qquad \begin{array}{l} 2E_1 + E_2 \rightarrow E_2 \\ (-5)E_1 + E_3 \rightarrow E_3 \end{array}$$

$$\begin{cases} x + 2y - 5z = -6 \\ \phantom{x + 2}7y - 11z = -25 \\ \phantom{x + 2}-3y + 12z = \phantom{-}18 \end{cases} \qquad 3E_2 + 7E_3 \rightarrow E_3$$

$$\begin{cases} x + 2y - 5z = -6 \\ \phantom{x + 2}7y - 11z = -25 \\ \phantom{x + 2y - 5}51z = \phantom{-}51 \end{cases} \qquad \frac{1}{51}E_3 \rightarrow E_3$$

$$\begin{cases} x + 2y - 5z = -6 \\ \phantom{x + 2}7y - 11z = -25 \\ \phantom{x + 2y - 115}z = \phantom{-}1 \end{cases} \qquad 11E_3 + E_2 \rightarrow E_2$$

$$\begin{cases} x + 2y - 5z = -6 \\ \phantom{x + 2}7y \phantom{- 11z} = -14 \\ \phantom{x + 2y - 115}z = \phantom{-}1 \end{cases} \qquad \frac{1}{7}E_2 \rightarrow E_2$$

$$\begin{cases} x + 2y - 5z = -6 \\ \phantom{x + 2}y \phantom{- 11z} = -2 \\ \phantom{x + 2y - 115}z = \phantom{-}1 \end{cases} \qquad (-2)E_2 + 5E_3 + E_1 \rightarrow E_1$$

$$\begin{cases} x = \phantom{-}3 \\ y = -2 \\ z = \phantom{-}1 \end{cases}$$

The solution is $(3, -2, 1)$.
**Do Exercise 3.**                                                                        ■

It is also possible to eliminate a variable by substitution. We solve for one of the variables in terms of the other two variables using one of the three equations, then we substitute the expression obtained for that variable in the other two equations. We then have a system of two linear equations in two variables that we can solve by any of the methods discussed earlier.

**EXAMPLE 3**   Solve the system

$$\begin{cases} 2x + 4y - 5z = -8 \\ 3x - \phantom{4}y + 2z = \phantom{-}14 \\ 5x + 3y - 7z = -2 \end{cases}$$

**Solution**   Since the coefficient of $y$ in the second equation is $-1$, it is best to solve for $y$ in terms of $x$ and $z$ in that equation. We have

$$y = 3x + 2z - 14$$

Substituting this expression for $y$ in equations 1 and 3, we get

$$\begin{cases} 2x + 4(3x + 2z - 14) - 5z = -8 \\ 5x + 3(3x + 2z - 14) - 7z = -2 \end{cases}$$

This system can be simplified to

$$\begin{cases} 14x + 3z = 48 \\ 14x - \phantom{3}z = 40 \end{cases}$$

The solution of this system is easily found to be $(3, 2)$. Since $y = 3x + 2z - 14$, we get $y = 3(3) + 2(2) - 14 = -1$. Hence, the solution of the original system is $(3, -1, 2)$.

**Do Exercise 17.**                                                                  ∎

## Consistent and Dependent Systems and Inconsistent Systems

The next two examples will involve a system whose solution set is infinite and another whose solution set is empty.

EXAMPLE 4   Solve the system

$$\begin{cases} x + y - 2z = 1 \\ x - y + 3z = 3 \\ 3x + y - \phantom{3}z = 5 \end{cases}$$

**Solution**   We first eliminate $x$ from equations 2 and 3.

*Explanation*

$$\begin{cases} x + y - 2z = 1 \\ x - y + 3z = 3 \\ 3x + y - \phantom{3}z = 5 \end{cases}$$   
$(-1)E_1 + E_2 \rightarrow E_2$
$(-3)E_1 + E_3 \rightarrow E_3$

$$\begin{cases} x + y - 2z = 1 \\ \phantom{x} -2y + 5z = 2 \\ \phantom{x} -2y + 5z = 2 \end{cases}$$

Notice that equations 2 and 3 of the system become identical. Thus, one may be left out, and we have a system of two linear equations in three variables.

$$\begin{cases} x + y - 2z = 1 \\ \phantom{x} -2y + 5z = 2 \end{cases}$$

We may give $z$ any value we wish and solve for the other two variables. Thus, let $z = c$ where $c$ is an arbitrary real number. Then the system may be written

$$\begin{cases} x + y = 1 + 2c \\ \phantom{x} -2y = 2 - 5c \end{cases}$$

This may be solved as follows:

*Explanation*

$$\begin{cases} x + y = 1 + 2c \\ \phantom{x} -2y = 2 - 5c \end{cases}$$   
$-\dfrac{1}{2}E_2 \rightarrow E_2$

$$\begin{cases} x + y = 1 + 2c \\ \quad\quad y = -1 + \dfrac{5}{2}c \quad (-1)E_2 + E_1 \to E_1 \end{cases}$$

$$\begin{cases} x = 2 - \dfrac{1}{2}c \\ y = -1 + \dfrac{5}{2}c \end{cases}$$

Thus, the solution set of the original system is $\left\{ \left( 2 - \frac{c}{2}, -1 + \frac{5c}{2}, c \right) \mid \right.$ $c$ is any real number$\left.\right\}$. We can produce as many solutions as we wish by giving $c$ different values. This system is consistent and dependent. For example, if $c = 0$, we get the solution $(2, -1, 0)$; if $c = 2$, we get $(1, 4, 2)$; and if $c = -2$, we obtain $(3, -6, -2)$. You should verify that these three ordered triples are indeed solutions of the original system.
**Do Exercise 21.**

**EXAMPLE 5**   Solve the system

$$\begin{cases} x + \quad y - \quad 2z = 3 \\ 2x - \quad 3y + \quad 3z = 2 \\ 5x - 10y + 11z = 7 \end{cases}$$

**Solution**   Eliminate $x$ from equations 1 and 2, then $y$ from equation 3 as follows:

*Explanation*

$$\begin{cases} x + \quad y - \quad 2z = 3 \\ 2x - \quad 3y + \quad 3z = 2 \quad\quad (-2)E_1 + E_2 \to E_2 \\ 5x - 10y + 11z = 7 \quad\quad (-5)E_1 + E_3 \to E_3 \end{cases}$$

$$\begin{cases} x + \quad y - \quad 2z = \quad 3 \\ \quad\quad -5y + \quad 7z = -4 \\ \quad\quad -15y + 21z = -8 \quad\quad (-3)E_2 + E_3 \to E_3 \end{cases}$$

$$\begin{cases} x + \quad y - 2z = \quad 3 \\ \quad\quad -5y + 7z = -4 \\ \quad\quad\quad\quad 0 = \quad 4 \end{cases}$$

The last system has no solution since the third equality can be written $0 \cdot x + 0 \cdot y + 0 \cdot z = 4$, which yields a false statement each time $x$, $y$, and $z$ are replaced by real numbers. Hence the solution set of our original system is empty. The system is inconsistent.
**Do Exercise 15.**

## Applications

We now give examples of word problems that make use of systems of three linear equations in three variables.

**EXAMPLE 6**   John, Mary, and Julie went to the same grocery store to buy sugar, coffee, and butter. John bought 5 pounds of sugar, 6 pounds of coffee, and 4 pounds of butter for $33; Mary bought 10 pounds of sugar, 2 pounds of coffee, and 12 pounds of butter for $40; while Julie paid $33 for 15 pounds of sugar, 3 pounds of coffee, and 6 pounds of butter. Find the cost per pound of each item.

**Solution**   Let $x$, $y$, and $z$ be the prices in dollars per pound of sugar, coffee, and butter, respectively. Thus, John paid $\$5x$ for the sugar, $\$6y$ for the coffee, and $\$4z$ for the butter. Therefore, he spent $\$(5x + 6y + 4z)$. Since it was given that he spent $33, we have the first equation $5x + 6y + 4z = 33$. The other two equations are obtained in exactly the same way, giving us the system

$$\begin{cases} 5x + 6y + 4z = 33 \\ 10x + 2y + 12z = 40 \\ 15x + 3y + 6z = 33 \end{cases}$$

We now solve the system

$$\begin{cases} 5x + 6y + 4z = 33 \\ 10x + 2y + 12z = 40 \\ 15x + 3y + 6z = 33 \end{cases}$$

*Explanation*

$(-2)E_1 + E_2 \rightarrow E_2$
$(-3)E_1 + E_3 \rightarrow E_3$

$$\begin{cases} 5x + 6y + 4z = 33 \\ -10y + 4z = -26 \\ -15y - 6z = -66 \end{cases}$$

$3E_2 + (-2)E_3 \rightarrow E_3$

$$\begin{cases} 5x + 6y + 4z = 33 \\ -10y + 4z = -26 \\ 24z = 54 \end{cases}$$

$\dfrac{1}{2}E_2 \rightarrow E_2$

$\dfrac{1}{24}E_3 \rightarrow E_3$

$$\begin{cases} 5x + 6y + 4z = 33 \\ -5y + 2z = -13 \\ z = 2.25 \end{cases}$$

$(-2)E_3 + E_2 \rightarrow E_2$

$$\begin{cases} 5x + 6y + 4z = 33 \\ -5y = -17.5 \\ z = 2.25 \end{cases}$$

$\dfrac{-1}{5}E_2 \rightarrow E_2$

$$\begin{cases} 5x + 6y + 4z = 33 \\ y = 3.5 \\ z = 2.25 \end{cases}$$

$(-6)E_2 + (-4)E_3 + E_1 \rightarrow E_1$

$$\begin{cases} 5x = 3 \\ y = 3.5 \\ z = 2.25 \end{cases}$$

$\dfrac{1}{5}E_1 \rightarrow E_1$

$$\begin{cases} x = .6 \\ y = 3.5 \\ z = 2.25 \end{cases}$$

We conclude that sugar, coffee, and butter cost $.60, $3.50, and $2.25 a pound, respectively.

**Do Exercise 31.**

A CLOSER LOOK

**EXAMPLE 7**   A refinery must provide one of its customers with 10,000 gallons of 88 octane gasoline. City regulations require that the vapor pressure index of the gasoline be 24. The refinery must mix three kinds of gasoline to obtain the appropriate mixture. (See the following table.)

|                        | Octane rating | Vapor pressure index | Price per gallon |
|------------------------|---------------|----------------------|------------------|
| Regular unleaded       | 79            | 33                   | $.80             |
| Premium unleaded       | 87            | 25                   | $.88             |
| Super premium unleaded | 98            | 14                   | $1.10            |

How many gallons of each gasoline should be mixed to fill the customer's order at a minimum cost?

**Solution**   Let $x$, $y$, and $z$ gallons be the quantities of regular unleaded, premium unleaded, and super premium unleaded, respectively, that must be used. Since 10,000 gallons must be provided, we have

$$x + y + z = 10,000$$

Since the octane rating of 10,000 gallons of mixture must be 88, and the octane rating of $x$ gallons is 79, that of $y$ gallons is 87, and that of $z$ gallons is 98, we have

$$79x + 87y + 98z = 88(10,000)$$

Using the fact that the vapor pressure index of the 10,000 gallons of mixture must be 24, and that the vapor pressure indices of the $x$, $y$, and $z$ gallons are 33, 25, and 14, respectively, we get

$$33x + 25y + 14z = 24(10,000)$$

To solve the system

$$\begin{cases} x + y + z = 10,000 \\ 79x + 87y + 98z = 88(10,000) \\ 33x + 25y + 14z = 24(10,000) \end{cases}$$

we proceed as follows:

*Explanation*

$$\begin{cases} x + y + z = 10{,}000 \\ 79x + 87y + 98z = 880{,}000 \\ 33x + 25y + 14z = 240{,}000 \end{cases} \quad \begin{array}{l} (-79)E_1 + E_2 \rightarrow E_2 \\ (-33)E_1 + E_3 \rightarrow E_3 \end{array}$$

$$\begin{cases} x + y + z = 10{,}000 \\ \phantom{x +} 8y + 19z = 90{,}000 \\ \phantom{x +} -8y - 19z = -90{,}000 \end{cases}$$

The second and third equations are obviously equivalent. Thus, we delete the third equation to obtain

$$\begin{cases} x + y + z = 10{,}000 \\ \phantom{x +} 8y + 19z = 90{,}000 \end{cases}$$

Since $z$ represents the number of gallons of super premium unleaded gasoline, it cannot be negative. Thus, we let $z = a$, where $a$ is an arbitrary nonnegative number. From the second equation

$$y = 11{,}250 - \frac{19a}{8}$$

Replacing $y$ by $11{,}250 - \frac{19a}{8}$ and $z$ by $a$ in the first equation, we get

$$x = 10{,}000 - \left(11{,}250 - \frac{19a}{8}\right) - a = \frac{11a}{8} - 1250$$

We must be sure that the values of $x$ and $y$ are nonnegative also. Thus, we require that

$$11{,}250 - \frac{19a}{8} \geq 0 \quad \text{and} \quad \frac{11a}{8} - 1250 \geq 0$$

This double inequality may be written

$$\frac{10{,}000}{11} \leq a \leq \frac{90{,}000}{19}$$

The solution set is

$$\left\{ \left( \frac{11a}{8} - 1250, \; 11{,}250 - \frac{19a}{8}, \; a \right) \; \middle| \; \frac{10{,}000}{11} \leq a \leq \frac{90{,}000}{19} \right\}$$

Since the prices per gallon of the regular unleaded, premium unleaded, and super premium unleaded are \$.80, \$.88, and \$1.10, respectively, the total cost in terms of $x$, $y$, and $z$ is \$$(.8x + .88y + 1.10z)$. We may express this cost in terms of $a$ as follows:

$$C = .8\left(\frac{11a}{8} - 1250\right) + .88\left(11{,}250 - \frac{19a}{8}\right) + 1.10a = 8900 + .11a$$

Thus, the cost will be at a minimum when the value of $a$ is minimized. That is, when $a = \frac{10,000}{11}$. The cost will be \$9000 since $8900 + .11\left(\frac{10,000}{11}\right) = 9000$. Also, $x = \frac{11}{8}\left(\frac{10,000}{11}\right) - 1250 = 0$, and $y = 11,250 - \frac{19}{8}\left(\frac{10,000}{11}\right) = \frac{100,000}{11}$. Thus, the refinery should use $\frac{100,000}{11}$ gallons of premium unleaded gasoline and $\frac{10,000}{11}$ gallons of super premium unleaded gasoline.

**Do Exercise 39.**

# Exercise Set 2.2

*In Exercises 1–15, use the method of elimination of variables by addition or subtraction to solve the given systems.*

1. $\begin{cases} 2x + 3y - 5z = -1 \\ 3x - 2y + 4z = -3 \\ 6x + 4y - 3z = -1 \end{cases}$

2. $\begin{cases} x + 3y + 7z = 5 \\ 2x - 4y + 2z = 8 \\ 5x - 2y + 3z = 10 \end{cases}$

3. $\begin{cases} 2x + 3y + 2z = 8 \\ -x + 4y - 3z = 1 \\ 3x - 2y + 7z = 11 \end{cases}$

4. $\begin{cases} 5x + 2y - 3z = 21 \\ 2x - 7y + 5z = -4 \\ 3x \qquad - 4z = 17 \end{cases}$

5. $\begin{cases} x + y + 3z = 2 \\ 3x + 2y - z = -3 \\ 11x + 8y + 3z = 6 \end{cases}$

6. $\begin{cases} 3x \qquad + 2z = -9 \\ 2y + 5z = 4 \\ 3x + y + 3z = -7 \end{cases}$

7. $\begin{cases} -2x + 3y + z = 4 \\ 3x - 2y + 2z = 14 \\ 4x + 3y - z = 6 \end{cases}$

8. $\begin{cases} x + 2y - 5z = 3 \\ 2x - 3y + 3z = -2 \\ 5x + 3y - 12z = 7 \end{cases}$

9. $\begin{cases} 3x + 2y - 5z = -2 \\ 2x + 3y - 8z = 5 \\ -x + 2y - 3z = 8 \end{cases}$

10. $\begin{cases} 2x - 3y + 4z = 1 \\ 4x + 5y - 3z = 2 \\ 2x - 14y + 15z = 7 \end{cases}$

11. $\begin{cases} x + y + z = 6 \\ 2x + y + 3z = 12 \\ x - y + 5z = 10 \end{cases}$

12. $\begin{cases} 2x - 2y + 3z = 1 \\ x - 3y - 2z = -9 \\ x + y + z = 6 \end{cases}$

13. $\begin{cases} x + y = -4 \\ x + z = 1 \\ 3x - y + 2z = 4 \end{cases}$

14. $\begin{cases} 2x + 3y - z = -1 \\ 3x + y - 2z = -13 \\ x - 4y + 2z = 11 \end{cases}$

15. $\begin{cases} 3x + y + z = 0 \\ x + y = 1 \\ 7x + 3y + 3z = 2 \end{cases}$

*Solve each of the systems of linear equations in Exercises 16–27 by elimination of a variable by substitution.*

16. The system of Exercise 2.

17. The system of Exercise 3.

18. The system of Exercise 5.

19. The system of Exercise 6.

20. The system of Exercise 7.

21. The system of Exercise 8.

22. The system of Exercise 9.

23. The system of Exercise 11.

24. The system of Exercise 12.

25. The system of Exercise 13.

26. The system of Exercise 14.

27. The system of Exercise 15.

28. A confectioner wishes to mix three kinds of candy to get 100 pounds of a mixture worth \$3.95 a pound. Kind A is worth \$3.50 a pound, while kinds B and C are worth \$4 and \$5 a pound, respectively. He wants to have the weight of kind A equal to the sum of the weights of kinds B and C. How many pounds of each kind of candy must he use?

29. A confectioner wishes to mix three kinds of candy to get 400 pounds of a mixture worth \$4.50 a pound. Kind A is worth \$4.20 a pound, while kinds B and C are worth \$5.25 and \$6 a pound, respectively. She wants the weight of kind A to be three times the sum of the weights of kinds B and C. How many pounds of each kind of candy must she use?

30. Mark won \$10,000 at the Washington State lottery. He invested his winnings in three bonds. Bond A was the safest investment paying 6.5% interest per year, while bonds B

and C were not as safe but paid 8% and 12% interest per year, respectively. Mark wanted to be conservative so he invested four times as much in bond A than in the other two bonds combined. His yearly income from these investments is $700. How much did he invest in each bond?

31. Janet inherited $50,000. She invested that sum in three bonds. Bond A paid 6.2% interest per year, while bonds B and C paid 9% and 11% interest per year, respectively. She decided to invest as much in bond A as in the other two bonds combined. Her yearly income from these investments is $4000. How much did she invest in each bond?

32. Shirley, Norma, and Roy went to the same grocery store to buy flour, sugar, and butter. Shirley bought 20 pounds of flour, 40 pounds of sugar, and 5 pounds of butter for $39.25; Norma bought 10 pounds of flour, 20 pounds of sugar, and 10 pounds of butter for $36.50; while Roy paid $40 for 50 pounds of flour, 20 pounds of sugar, and 8 pounds of butter. Find the cost per pound of each item.

33. Ann, Greg, and George went to the same store to purchase items $X$, $Y$, and $Z$. Ann bought 2 units of $X$, 3 units of $Y$, and 1 unit of $Z$ for a total cost of $57.75; while Greg bought 1 unit of $X$, 2 units of $Y$, and 3 units of $Z$ for $67.50. George paid $85.25 for 2 units of $X$, 3 units of $Y$, and 3 units of $Z$. What is the price per unit of each item?

34. Three kinds of cargo were loaded on a ferry boat. Each unit of each kind requires a certain amount of space, weighs a certain number of pounds, and costs a certain amount, as given in the following table.

| | Space in cubic feet | Weight in pounds | Cost in dollars |
|---|---|---|---|
| Kind A | 5 | 50 | 32 |
| Kind B | 12 | 180 | 85 |
| Kind C | 7 | 85 | 96 |

If the total value of the cargo was $31,720, while the total weight and space were 51,200 pounds and 3740 cubic feets respectively, how many units of each kind were loaded on the ferryboat?

35. A construction company builds three types of houses. The number of units of roofing, concrete, and lumber required for each type of house are given in the following table.

| | Roofing | Concrete | Lumber |
|---|---|---|---|
| Type A | 3 | 5 | 4 |
| Type B | 2 | 6 | 7 |
| Type C | 5 | 4 | 6 |

If in a certain month the company used 51 units of roofing, 67 units of concrete, and 77 units of lumber, how many houses of each type were built that month?

36. An absentminded professor forgot how many students he has in his class. However, he remembers clearly that he has as many freshmen as he has sophomores and juniors together, that the number of freshmen exceeds the number of sophomores by 5 and of juniors by 15. How many students does he have?

37. The absentminded professor of Exercise 36 has a wife who does not like to reveal her age. One of her inconsiderate friends asked her what her age was and she replied: "My husband is 4 years older than I am. I was 24 when my son was born and in 20 years my age combined with my son's age will be only three halves of my husband's age." Find her age and those of her husband and son.

*38. A refinery must provide one of its customers with 25,200 gallons of 89 octane gasoline. State regulations require that the vapor pressure index of the gasoline be 21. The refinery must mix three kinds of gasoline to obtain the appropriate mixture. Information on the three kinds of gasoline is given in the table.

| | Octane rating | Vapor pressure index | Price per gallon |
|---|---|---|---|
| Regular unleaded | 76 | 34 | $.85 |
| Premium unleaded | 88 | 22 | $1.04 |
| Super premium unleaded | 97 | 13 | $1.15 |

How many gallons of each gasoline should be mixed to fill the customer's order at a minimum cost?

*39. A hospital dietician wishes to mix three food items to get 200 pounds of a mixture containing 290 units of vitamins and 460 calories per pound. The relevant information on the three food items is shown in the table.

| | Units of vitamins per pound | Number of calories per pound | Price per pound |
|---|---|---|---|
| Item A | 150 | 600 | $1.25 |
| Item B | 250 | 500 | $1.35 |
| Item C | 420 | 330 | $1.60 |

How many pounds of each item must she mix to minimize the cost?

*40. A company manufactures three products $X$, $Y$, and $Z$. The manufacturing of each unit of each product requires a certain number of hours from each of three departments A, B, and C, as is shown in the table.

| Department | Hours required per unit of product | | |
|---|---|---|---|
| | X | Y | Z |
| A | 7 | 2 | 1 |
| B | 2.5 | .7 | .3 |
| C | 2.9 | .85 | .5 |

Each month the company has 230 hours available in department A, 81 hours in department B, and 97 hours in department C. Because of contractual obligations, all available hours must be used. If the profits per unit of products X, Y, and Z are, respectively, $239, $100, and $120, how many units of each product should be manufactured and sold per month to maximize the profit? What is the maximum profit?

## 2.3  Graphical Solutions of Systems of Equations

We have seen that an ordered pair (or triple) of real numbers is a solution of a system of equations in two (or three) variables if, and only if, it is a solution of each equation of the system. Consequently, the solution set of the system is the intersection of the solution sets of these equations. If the graphs of the equations of the system are sketched on the same set of coordinate axes, then we can determine geometrically the nature of the solution set of the system by considering the intersection of these graphs. In an earlier section (Example 2, Section 2.1) we gave a graphical solution of a system of two linear equations in two variables. In this section, we consider examples of systems of two equations in two variables where at least one of the two equations is nonlinear. The solutions of these systems will be found by sketching the graphs of the two equations and reading the coordinates of the points of intersection of these graphs. Since small errors in reading the coordinates of a point from a graph as well as small errors in drawing the graphs are unavoidable, solutions found graphically should always be checked by substitution in both equations.

**EXAMPLE 1**   Solve this system of equations graphically.

$$\begin{cases} y = x^2 - 2x - 2 \\ y = x + 2 \end{cases}$$

**Solution**   The graph of the first equation is a parabola that is concave up since the leading coefficient is positive. The abscissa of the vertex is 1 since in this case $\frac{-b}{2a} = \frac{-(-2)}{2(1)} = \frac{2}{2} = 1$. The ordinate of the vertex is found by replacing $x$ by 1 in the first equation. It is $-3$. Hence, the vertex is the point $(1, -3)$. The parabola is sketched in Figure 2.4. The graph of the second equation is a straight line, which is also sketched in Figure 2.4.

From the graphs, it appears that there are two points of intersection with coordinates $(-1, 1)$ and $(4, 6)$. We substitute $-1$ for $x$ and 1 for $y$ in both equations of the system and obtain

$$\begin{cases} 1 = (-1)^2 - 2(-1) - 2 \\ 1 = -1 + 2 \end{cases}$$

which are both true. Similarly, we substitute 4 for $x$ and 6 for $y$ in both equations of the system and get

$$\begin{cases} 6 = 4^2 - 2(4) - 2 \\ 6 = 4 + 2 \end{cases}$$

which are also both true. Hence, the solutions of the system are $(-1, 1)$ and $(4, 6)$.
**Do Exercise 1.**

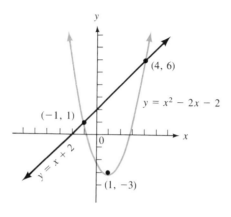

**FIGURE 2.4**

The system of the previous example could have easily been solved algebraically. For example, we could have substituted $x + 2$ for $y$ in the first equation to obtain

$$x + 2 = x^2 - 2x - 2$$

which is equivalent to

$$x^2 - 3x - 4 = 0$$

and to

$$(x + 1)(x - 4) = 0$$

The solutions of these equations are $-1$ and 4. Replacing $x$ successively by $-1$ and 4 in either of the two equations of the original system yields $y = 1$ and $y = 6$. Thus, the solutions are $(-1, 1)$ and $(4, 6)$.

In general, it is not advisable to solve a system of equations graphically if a simple algebraic solution is available. However, graphical solutions are useful in shedding

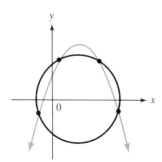

**FIGURE 2.5**

light on the nature of the solution set. For example, if the graph of one equation is a parabola and that of the other equation is a circle, we know that there will be at most four solutions. (See Figure 2.5.) Also, obtaining an approximate solution of a system graphically may be the best we can do when an algebraic solution is very difficult or not available as in the next two examples.

**EXAMPLE 2**    Solve this system of equations graphically.

$$\begin{cases} 15x - 8y + 19 = 0 \\ \qquad\qquad y = 2^x \end{cases}$$

**Solution**    The graph of the first equation is a straight line. We easily find that the points $\left(0, \frac{19}{8}\right)$ and $(3, 8)$ are on the graph. Using this information we sketch the graph of the first equation in Figure 2.6. The graph of the second equation is that of an exponential function. We replace $x$ successively by $-2, -1, 0, 1, 2, 3,$ and $4$ to find the corresponding values of $y$. The results are tabulated below.

| $x$ | $-2$ | $-1$ | 0 | 1 | 2 | 3 | 4 |
|---|---|---|---|---|---|---|---|
| $y$ | $\dfrac{1}{4}$ | $\dfrac{1}{2}$ | 1 | 2 | 4 | 8 | 16 |

Hence, the points $\left(-2, \frac{1}{4}\right), \left(-1, \frac{1}{2}\right), (0, 1), (1, 2), (2, 4), (3, 8),$ and $(4, 16)$ are on the graph. We plot these points and sketch a smooth curve through them to get the graph of $y = 2^x$. Both graphs are shown in Figure 2.6. From the figure it appears that the graphs intersect at the points $\left(-1, \frac{1}{2}\right)$ and $(3, 8)$.

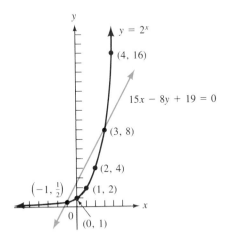

**FIGURE 2.6**

Replacing $x$ by $-1$ and $y$ by $\frac{1}{2}$ in both equations of the system, we obtain

$$\begin{cases} 15(-1) - 8\left(\dfrac{1}{2}\right) + 19 = 0 \\ \qquad\qquad\qquad \dfrac{1}{2} = 2^{-1} \end{cases}$$

which are both true. Thus, $\left(-1, \frac{1}{2}\right)$ is indeed a solution of the system. Replacing $x$ by 3 and $y$ by 8 in both equations, we get

$$\begin{cases} 15(3) - 8(8) + 19 = 0 \\ \qquad\qquad\quad 8 = 2^3 \end{cases}$$

which are also both true. Hence $(3, 8)$ is the other solution of the system. The solutions are $\left(-1, \frac{1}{2}\right)$ and $(3, 8)$.

**Do Exercise 11.**                                                                       ■

**EXAMPLE 3**   Solve this system graphically.

$$\begin{cases} y = \ln x \\ y = 2^{-x} \end{cases}$$

**Solution**   We replace $x$ successively by .5, 1, 2, and 3 in the first equation and with the help of a calculator we find the approximate corresponding values of $y$. The results are tabulated below.

| $x$ | .5 | 1 | 2 | 3 |
|-----|-----|-----|-----|-----|
| $y$ | $-0.693147$ | 0 | 0.693147 | 1.098612 |

Using the entries in this table, we sketch the graph of the first equation in Figure 2.7. Now we replace $x$ successively by $-1$, 0, .5, 1, 2, and 3 in the second equation and find the corresponding values of $y$. The results are tabulated below.

| $x$ | $-1$ | 0 | .5 | 1 | 2 | 3 |
|-----|-----|-----|-----|-----|-----|-----|
| $y$ | 2 | 1 | 0.7071 | 0.5 | 0.25 | 0.125 |

Using these results we sketch the graph of the second equation. Both graphs are shown in Figure 2.7.

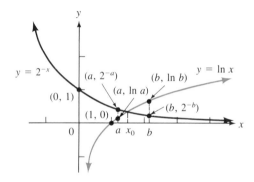

**FIGURE 2.7**

Notice that the graph of $y = 2^{-x}$ is falling while the graph of $y = \ln x$ is rising. If the graphs intersect at $(x_0, y_0)$, then $a < x_0$ whenever $\ln a < 2^{-a}$ and $x_0 < b$ whenever $2^{-b} < \ln b$. Using a calculator*, find $2^{-x} - \ln x$ for some value of $x$. If the answer is positive, then $2^{-x} > \ln x$ and $x_0 > x$. If the answer is negative, then $2^{-x} < \ln x$ and $x_0 < x$. From the graphs, it appears that $x_0$ is near 1.5. So we start with $x = 1.5$. We find

$$2^{-1.5} - \ln 1.5 \doteq 0.3535 - 0.4054 = -0.0519 < 0$$

So $x_0 < 1.5$ and we try 1.4. We find

$$2^{-1.4} - \ln 1.4 \doteq 0.3789 - 0.3364 = 0.0425 > 0$$

Therefore $1.4 < x_0$. We now know that $1.4 < x_0 < 1.5$. We try 1.45 and find

$$2^{-1.45} - \ln 1.45 \doteq 0.3660 - 0.37156 = -0.00556 < 0$$

Thus, $x_0 < 1.45$. We try 1.44 and find

$$2^{-1.44} - \ln 1.44 \doteq 0.36856 - 0.36464 = 0.00392 > 0 \qquad \text{(See Figure 2.8.)}$$

Therefore, $1.44 < x_0$. We try 1.445 and find

$$2^{-1.445} - \ln 1.445 \doteq 0.36729 - 0.36810 = -0.00081 < 0$$

Thus, $1.44 < x_0 < 1.445$. We conclude that $x_0 = 1.44$ is correct up to two decimal places. An approximate solution of the system is $(1.44, 0.37)$, rounded off to two decimal places.

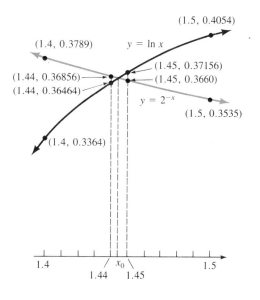

**FIGURE 2.8**

---

*Calculations such as those of Example 3 may be carried out most efficiently using a programmable calculator.

## Exercise Set 2.3

*In Exercises 1–15, solve the given system of equations graphically.*

**1.** $\begin{cases} y = x^2 + 6x - 3 \\ y = 2x + 2 \end{cases}$

**2.** $\begin{cases} y = x^2 - 4x + 2 \\ y = -2x + 5 \end{cases}$

**3.** $\begin{cases} y = -x^2 + 2x + 5 \\ y = 2x + 1 \end{cases}$

**4.** $\begin{cases} y = -2x^2 + 8x + 7 \\ y = -2x + 15 \end{cases}$

**5.** $\begin{cases} y = x^2 + 3x + 1 \\ y = 2x^2 + 4x - 1 \end{cases}$

**6.** $\begin{cases} y = x^2 - 4x + 5 \\ y = 2x^2 - 5x + 3 \end{cases}$

**7.** $\begin{cases} y = -x^2 + 4x + 6 \\ y = x^2 + 6x + 2 \end{cases}$

**8.** $\begin{cases} y = \dfrac{7}{6}x^2 + \dfrac{1}{3} \\ y = 2^x + 1 \end{cases}$

**9.** $\begin{cases} y = 3^{-x} + 2 \\ y = -x^2 + \dfrac{1}{27}x + \dfrac{163}{27} \end{cases}$

**10.** $\begin{cases} y = 3^x \\ y = \dfrac{35}{4} + 2^{-x} \end{cases}$

**11.** $\begin{cases} y = 2^x \\ y = -x^2 + 7x - 4 \end{cases}$

**12.** $\begin{cases} y = -x^2 + 9x - 11 \\ y = 3^{x/2} \end{cases}$

**13.** $\begin{cases} y = 4^{-x} \\ y = \dfrac{1}{6}(3^{2x}) \end{cases}$

**14.** $\begin{cases} y = \ln(x + 1) \\ y = 3^{-2x} \end{cases}$

**15.** $\begin{cases} y = 2\ln(x + 3) \\ y = 2^{-x/2} \end{cases}$

# 2.4 Systems of Linear Inequalities in Two Variables

## Graphing Linear Inequalities in Two Variables

In this section, we introduce a method for finding the solution set of a system of linear inequalities in two variables graphically. Suppose we are asked to sketch the graph of the inequality $2x + 3y < 18$. By definition, a point $P$ with coordinates $(x_0, y_0)$ is on the graph of that inequality if, and only if, $2x_0 + 3y_0 < 18$ is a true statement. That is, if, and only if, $(x_0, y_0)$ is a solution of the inequality $2x + 3y < 18$. Let $L$ be the line with equation $2x + 3y = 18$. Note that this equation may be written $y = \left(-\dfrac{2}{3}\right)x + 6$. We have the following equivalent statements:

1. The point $(x_0, y_0)$ is on the graph of the inequality $2x + 3y < 18$,
2. $2x_0 + 3y_0 < 18$,
3. $y_0 < \dfrac{-2x_0}{3} + 6$.

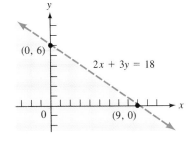

**FIGURE 2.9**

Since the point $\left(x_0, \left(\dfrac{-2}{3}\right)x_0 + 6\right)$ is on $L$, we conclude that $P$ is on the graph of the given inequality if, and only if, $P$ is below the line $L$. Thus, the graph of $2x + 3y < 18$ consists of all those points that are below the line $L$. (See Figure 2.9.) The graph is pictured as a shaded region in the plane. The line $L$ is shown as a dashed line to indicate that it is not part of the graph of the inequality. If we were graphing $2x + 3y \leq 18$, we would have drawn $L$ as a solid line to indicate that it is included in the graph.

Notice that the line $L$ divides the plane into three disjoint sets, namely:

1. The line $L$, which is the graph of $2x + 3y = 18$;
2. The region below $L$, which is the graph of $2x + 3y < 18$; and
3. The region above $L$, which is the graph of $2x + 3y > 18$.

The next two examples depict more complicated situations.

**EXAMPLE 1**   Sketch the graph of the system of inequalities.

$$\begin{cases} x < 6 \\ y > -4 \\ 3x + 2y \le 12 \end{cases}$$

**Solution**   A point $(x_0, y_0)$ is on the graph of the system if, and only if, it is on the graph of each inequality. Hence, the graph of the system of inequalities is the intersection of the three graphs. Clearly, the graph of the first inequality is the *half-plane* to the left of the line with equation $x = 6$. This is indicated in Figure 2.10 by drawing an arrow pointing to the left of that line. Similarly, the graph of the second inequality is the half-plane above the line with the equation $y = -4$. We indicate this in Figure 2.10 by drawing an arrow pointing upward.

Finally, we draw the line $L$ with equation $3x + 2y = 12$. This line divides the plane into three disjoint subsets, one of which is a half-plane which is the graph of the third inequality of the system. To find which of the two half-planes is the graph of that inequality, we test an arbitrary point not on $L$. Although the choice of which point to use as a check is arbitrary, the origin is the most convenient because the calculations are easiest. The coordinates of the origin are $(0, 0)$. We replace $x$ and $y$ by 0 in the third inequality of the system and get $3 \cdot 0 + 2 \cdot 0 < 12$, which is certainly true. Therefore, the origin is on the graph of the third inequality. Consequently, the half-plane that contains the origin is the graph of the third inequality. We indicate this by placing an arrow by the line $L$ pointing toward the origin. The graph of the given system is the intersection of the three half-planes, which is the region shaded in Figure 2.10. Note that the graphs of $x = 6$ and $y = -4$ have been drawn as dashed lines because they are not included in the graphs of $x < 6$ and $y > -4$, respectively. On the other hand, the graph of $3x + 2y = 12$ has been drawn as a solid line to show that it is part of the graph of $3x + 2y \le 12$.
**Do Exercise 5.**

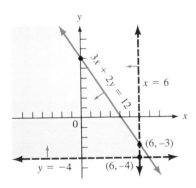

**FIGURE 2.10**

**EXAMPLE 2**   Sketch the graph of the system of inequalities

$$\begin{cases} 7x + y \ge 15 \\ x + y < 9 \\ x - 2y < 0 \end{cases}$$

**Solution**   Let $L_1$, $L_2$, and $L_3$ be the graphs of the equations $7x + y = 15$, $x + y = 9$, and $x - 2y = 0$, respectively. To find the intersection of $L_1$ and $L_2$, we solve the system

$$\begin{cases} 7x + y = 15 \\ x + y = 9 \end{cases}$$

The solution of the system is $(1, 8)$. Thus, the point $(1, 8)$ is the intersection of $L_1$ and $L_2$. Similarly, to find the intersection of $L_1$ and $L_3$ we solve the system

$$\begin{cases} 7x + y = 15 \\ x - 2y = 0 \end{cases}$$

We find that the solution is $(2, 1)$. Thus, the point $(2, 1)$ is the intersection of $L_1$ and $L_3$. Finally, to find the intersection of $L_2$ and $L_3$ we solve the system

$$\begin{cases} x + y = 9 \\ x - 2y = 0 \end{cases}$$

The solution is $(6, 3)$. Thus, the point $(6, 3)$ is the intersection of $L_2$ and $L_3$. The usefulness of finding the three points of intersection is now evident when we sketch the three lines since we have obtained two points on each of the three lines. (See Figure 2.11.)

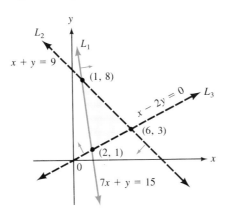

**FIGURE 2.11**

The line $L_1$ divides the plane into three disjoint regions, one of which is the graph of the inequality $7x + y > 15$. Using the origin, we replace $x$ and $y$ by 0 in the given inequality and obtain the false statement $7 \cdot 0 + 0 > 15$. The origin is not, therefore, on the graph of the first inequality and the half-plane that does not contain the origin is the graph of the first inequality. We indicate this by placing an arrow next to the line $L_1$ pointing away from the origin.

Similarly, we find that the origin does belong to the graph of the second inequality. Thus, we place an arrow next to the line $L_2$ pointing toward the origin. Finally, we see that the line $L_3$ divides the plane into three disjoint regions, one of which is the graph of the third inequality. We may not use the origin to determine which half-plane is that graph since the origin is on the line $L_3$. Since the third inequality is equivalent to $y > \left(\frac{1}{2}\right)x$, we conclude that the half-plane above the line $L_3$ is its graph. We indicate this by drawing an arrow pointing upward by the line $L_3$.

The graph of the given system is the intersection of the three half-planes we have obtained. It is the triangle shaded in Figure 2.11. Note that the side of the triangle that is part of the line $L_1$ is included in the graph and consequently has been drawn as a solid line, while the other two sides are not included in the graph and have been drawn as dashed lines.

**Do Exercise 11.**

## An Application

EXAMPLE 3   A company manufactures two types of gadgets, A and B. A requires 5 hours to assemble and 3 hours to paint, while B requires 2 hours to assemble and 1 hour to paint. The company employs 63 assemblers and 33 painters each working 8 hours a day. Let $x$ and $y$ denote the respective daily production of gadgets A and B. Graph the set of possible ordered pairs $(x, y)$.

**Solution** Obviously,

$$x \geq 0 \quad \text{and} \quad y \geq 0$$

since we cannot have negative production. Since each employee works 8 hours per day, the daily number of hours available for assembly is 504 because $63 \cdot 8 = 504$, and the daily number of hours available for painting is 264 because $33 \cdot 8 = 264$. The production of $x$ gadgets A requires $5x$ assembly hours and $3x$ painting hours, while the production of $y$ gadgets B requires $2y$ assembly hours and $y$ painting hours. Thus, the total number of assembly hours needed is $5x + 2y$, while the total number of painting hours required is $3x + y$. Since these totals cannot exceed 504 and 264, respectively, we have the following system of inequalities:

$$\begin{cases} 5x + 2y \leq 504 \\ 3x + y \leq 264 \\ x \geq 0 \\ y \geq 0 \end{cases}$$

Using the same procedure as we did in Examples 1 and 2, we represent the solution set of this system as the shaded region in Figure 2.12. A question that naturally arises is this: Subject to the given restrictions and knowing the cost of production and the revenue for each of the gadgets, how should $x$ and $y$ be chosen to maximize the daily profit? We shall answer this question in Chapter 4.
**Do Exercise 21.**

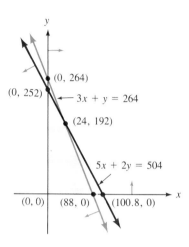

**FIGURE 2.12**

---

# Exercise Set 2.4

*In Exercises 1–17, sketch the graph of the given system of linear inequalities.*

**1.** $\begin{cases} x + y > 1 \\ x < 2 \end{cases}$

**2.** $\begin{cases} x + 2y < 4 \\ y > -2 \end{cases}$

**3.** $\begin{cases} 2x + y < 6 \\ x - y > -3 \end{cases}$

**4.** $\begin{cases} 3x - 5y \geq 30 \\ x + y < 2 \end{cases}$

**5.** $\begin{cases} x + 2y \leq 4 \\ x > -1 \\ y > -2 \end{cases}$

**6.** $\begin{cases} 2x + y > 4 \\ y < 5 \\ x < 3 \end{cases}$

**7.** $\begin{cases} x - 3y < -3 \\ x < 1 \\ y > -1 \end{cases}$

**8.** $\begin{cases} 2x + 3y \geq 12 \\ x > 1 \\ y > 2 \end{cases}$

**9.** $\begin{cases} x + y \leq 1 \\ 2x - y \geq -4 \\ y \geq -2 \end{cases}$

**10.** $\begin{cases} x + 2y \geq 2 \\ x - 2y \geq 0 \\ x \leq 3 \end{cases}$

**11.** $\begin{cases} 2x + y \leq 6 \\ x - y > -3 \\ 2x + 3y > -6 \end{cases}$

**12.** $\begin{cases} 3x - 5y \leq 30 \\ x + y \geq 2 \\ 3x + 7y \leq 30 \end{cases}$

**13.** $\begin{cases} x + y \leq 1 \\ 2x - y \geq -4 \\ x - 2y \leq 4 \end{cases}$

**14.** $\begin{cases} x - y \geq -1 \\ x + y < 5 \\ 3x - 5y \leq -1 \end{cases}$

**15.** $\begin{cases} x + y \geq 1 \\ 2x - y \geq -4 \\ 7x - 8y \leq 22 \end{cases}$

**16.** $\begin{cases} 2x - y \geq -4 \\ x - 2y \leq 4 \\ x + y \geq 1 \\ x + y \leq 7 \end{cases}$

**17.** $\begin{cases} 3x - 4y > -19 \\ 5x + 3y > 7 \\ 2x - y < 5 \\ 4x + y < 19 \end{cases}$

18. A farmer has 200 acres on which he raises peanuts and corn. He has accepted orders requiring 10 acres of peanuts and 25 acres of corn. Moreover, he must follow regulations that the acreage for corn must be at least twice the acreage of peanuts. Let $x$ and $y$ denote the number of acres allocated to peanuts and corn, respectively. Graph the set of possible ordered pairs $(x, y)$.

19. A farmer has 800 acres on which she will raise wheat and barley. She has accepted orders requiring 150 acres of wheat and 100 acres of barley. However, she must follow regulations that the acreage for wheat cannot exceed three times the acreage of barley. Let $x$ and $y$ denote the number of acres allocated to wheat and barley, respectively. Graph the set of possible ordered pairs $(x, y)$.

20. An electronics company makes two models of VCRs. Model A takes 2 hours to assemble and 15 minutes to test, while model B takes 3 hours to assemble and 30 minutes to test. The plant is designed in such a way that a maximum of 60,000 hours per month are available for assembly, while a maximum of 8000 hours a month are available for testing. The company has accepted orders totaling 10,000 units of the A model and 2000 units of the B model. Let $x$ and $y$ denote the numbers of A and B models, respectively, the company can make per month. Graph the set of possible ordered pairs $(x, y)$.

21. A camera company makes two models of movie cameras. The deluxe model takes 3 hours to assemble and 10 minutes to test, while the standard model takes 2 hours to assemble and 6 minutes to test. The plant is designed in such a way that a maximum of 38,000 hours per month are available for assembly, while a maximum of 2000 hours a month are available for testing. The company has accepted orders totaling 2500 of the deluxe model and 3600 of the standard model. Let $x$ and $y$ denote the numbers of deluxe and standard models, respectively, the company can make per month. Graph the set of possible ordered pairs $(x, y)$.

22. An owner of a gasoline station can have at most 16,000 gallons of gasoline per day in his tanks. On a certain day, he estimates that his regular customers will purchase at least 5000 gallons of regular gasoline and at least 2000 gallons of premium gasoline. He also knows from experience that he cannot sell more than three times as much regular gasoline as premium gasoline. Let $x$ and $y$ denote, respectively, the number of gallons of regular and premium gasoline he can sell per day. Graph the set of possible ordered pairs $(x, y)$.

23. An investor wishes to buy a maximum of 300 shares of common stock. Her broker recommends stocks A and B and suggests that to reduce the risk of a fluctuating market at least 50 shares of stock A and not more than 200 shares of stock B should be bought. He also recommends that the number of shares of stock A should not exceed twice that of stock B. Let $x$ and $y$ represent the numbers of shares of stocks A and B, respectively, that the investor will buy. Graph the set of all possible ordered pairs $(x, y)$.

24. An investor wishes to buy a maximum of 500 shares of common stock. His broker recommends stocks A and B and suggests that to reduce the risk of a fluctuating market at least 50 shares of stock A and 100 shares of stock B should be bought. She also recommends that the number of shares of stock B should not exceed four times that of stock A. Let $x$ and $y$ represent the numbers of shares of stocks A and B, respectively, that the investor will buy. Graph the set of all possible ordered pairs $(x, y)$.

25. Shirley makes dolls and blankets to be sold at her church bazaar. She can make a doll in 3 hours, while it takes her 5 hours to make a blanket. Her costs to make each doll and each blanket are $4 and $7, respectively. If she can spend at most 120 hours a month on this church work and if the church provides at most $166 a month to cover the costs of making the dolls and blankets, sketch the graph of the set of ordered pairs $(x, y)$ where $x$ and $y$ are the numbers of dolls and blankets that Shirley can make each month.

26. A handyman makes baskets and vases to be sold at the local flea market. He can make a basket in 4 hours, while it takes him 3 hours to make a vase. His costs to make each basket and each vase are $2 and $5, respectively. If he can spend at most 36 hours a week on this project and if his cost on this work is not to exceed $32 per week, sketch the graph of the set of ordered pairs $(x, y)$ where $x$ and $y$ are the numbers of baskets and vases that the handyman can make each week.

27. Same as Exercise 26 with the additional requirement that the handyman has accepted orders of three baskets and two vases per week.

28. Andre's motel has 120 rooms and a restaurant that can seat 45 people. Some of the rooms may be furnished to accommodate travelers who like to go first class, while the remaining rooms will be furnished to satisfy the average guest. The manager of the motel knows from experience that 60% of the first-class guests will eat in the motel restaurant, while only 30% of the other guests will eat there. If the manager furnishes $x$ rooms for first-class guests and $y$ rooms for ordinary guests, sketch the graph of the set of possible ordered pairs $(x, y)$.

29. Same as Exercise 28 with the additional requirements that the number of rooms for first-class guests must be at least 10 and the number of rooms for ordinary guests must not exceed 100.

**30.** A nursing home dietician is planning the menus for the patients. She wants to provide a diet that has a minimum of 120 units of fat, 60 units of carbohydrates, and 90 units of proteins per patient per week. These goals can be met using two types of food. Type I contains 3 units of fats, 2 units of carbohydrates, and 6 units of proteins per pound; each pound of type II contains 12 units of fats, 4 units of carbohydrates, and 3 units of proteins. If the dietician plans to use $x$ pounds of type I and $y$ pounds of type II food per patient per week, graph the set of possible ordered pairs $(x, y)$.

---

# 2.5  Applications to Business and Economics

## Break-Even Analysis

If the cost to a company of producing $q$ units of a commodity is $C(q)$ dollars and the revenue obtained from the sales of $q$ units of this commodity is $R(q)$ dollars, then the company makes a profit whenever $R(q) > C(q)$. However, the company operates at a loss whenever $R(q) < C(q)$ and breaks even when $R(q) = C(q)$. The number of units produced and sold in the latter case is called the *break-even point*. In general, at the start of production of a commodity, a company operates at a loss because the revenue is not large enough to cover the *fixed costs* (the costs incurred by the company at a zero production level, such as rent and depreciation). To find the break-even point, we simply solve the system

$$\begin{cases} y = C(q) \\ y = R(q) \end{cases}$$

**EXAMPLE 1**   The cost $C(q) of producing $q$ units of a certain commodity in a month is given by the equation $y = C(q) = 5q + 12,000$. If each unit of that commodity sells for $8, find the following:

**a.** The fixed costs per month.
**b.** The number of units of the commodity that should be produced and sold per month to make sure that the business breaks even.

**Solution**   **a.** By definition, the fixed costs may be obtained by replacing $q$ by 0 in the formula giving the cost function. Thus, the fixed costs are $12,000 per month, since

$$C(0) = 5(0) + 12,000 = 12,000$$

**b.** Since each unit of the commodity sells for $8, the revenue from the sales of $q$ units is given by

$$y = R(q) = 8q$$

To find the break-even point, we solve the system

$$\begin{cases} y = 5q + 12,000 \\ y = 8q \end{cases}$$

Substituting $8q$ for $y$ in the first equation we obtain

$$8q = 5q + 12,000$$

Thus,

$$3q = 12,000$$

and

$$q = 4000$$

Therefore, 4000 units of the commodity should be manufactured and sold each month to guarantee that the company breaks even. In Figure 2.13 we show that the company operates at a loss when production is between 0 and 4000 units and makes a profit when production is greater than 4000 units.

**Do Exercise 1.**

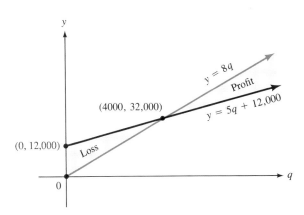

**FIGURE 2.13**

**EXAMPLE 2**   A company making personal computers found that the cost $\$C(q)$ of producing $q$ PCs per month is given by

$$y = C(q) = .5q^2 + 1000q + 1,125,000$$

provided that $q$ is between 0 and 2000. It also found that the revenue $\$R(q)$ from the sales of $q$ PCs per month is given by $y = R(q) = -q^2 + 4000q$ provided that $q$ is between 0 and 2000.

**a.** Find the level of production at which the company will break even.
**b.** At what level of production will the company make a profit?

**Solution**   **a.** To find the break-even point, it is sufficient to solve the system

$$\begin{cases} y = .5q^2 + 1000q + 1,125,000 \\ y = -q^2 + 4000q \end{cases}$$

Substituting $-q^2 + 4000q$ for $y$ in the first equation, we obtain

$$-q^2 + 4000q = .5q^2 + 1000q + 1,125,000$$

This equation may be written

$$1.5q^2 - 3000q + 1,125,000 = 0$$

Factoring the left side of this equation, we get

$$1.5(q - 500)(q - 1500) = 0$$

Thus, $q = 500$ or $q = 1500$. We conclude that there are two break-even points. The company may manufacture 500 PCs per month with costs and revenues equal to \$1,750,000, or it may produce 1500 PCs per month with costs and revenues equal to \$3,750,000.

**b.** The company is making a profit whenever the revenues are larger than the costs. That is, whenever

$$R(q) > C(q)$$

Thus, we have

$$-q^2 + 4000q > .5q^2 + 1000q + 1,125,000$$

The previous inequality is equivalent to

$$1.5(q - 500)(q - 1500) < 0$$

Using the method which we presented in Section 0.8, we find that the solution set of the inequality is

$$\{q \mid 500 < q < 1500\}$$

Thus, the company will operate at a profit if it produces more than 500 but fewer than 1500 PCs per month. (See Figure 2.14.)

**Do Exercise 3.**

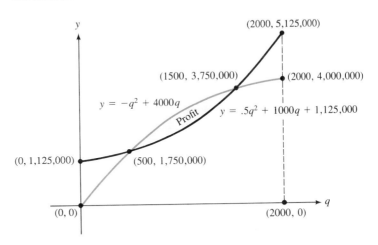

**FIGURE 2.14**

## Commodity Demand and Supply

One of the most basic principles of economic theory is that the market price of a commodity is determined by the demand and supply. In general, the quantity $q$ units of a product that consumers are willing to purchase depends on the price $p$ dollars per unit at which the product is made available. The relationship between $q$ and $p$ is called the *demand function D* and the graph of the equation $q = D(p)$ is called the *demand curve.* Similarly, the quantity $q$ units of a product that its supplier is willing to make available depends on the price $p$ dollars per unit at which they are able to sell it. The relationship between $q$ and $p$ is called the *supply function S* and the graph of the equation $q = S(p)$ is called the *supply curve.* Increasing the price per unit tends to decrease the number of units consumers are willing to buy, so in general the demand function is decreasing while the supply function is increasing.

In the usual economy, the price will fluctuate and tend to adjust so that the quantity suppliers are willing to provide is equal to the quantity consumers are willing to purchase. When this occurs, we say that the *market is in equilibrium.* The corresponding price and quantity are called the *equilibrium price* and the *equilibrium quantity*, respectively. In the simplest case where the demand and supply functions are linear, the equilibrium price and quantity may be found by solving a system of linear equations as in the following example.

**EXAMPLE 3**   The demand for a commodity is described by the equation $q = D(p) = 15{,}000 - 150p$, where $q$ is the quantity demanded (measured in thousands of units) and $p$ is the price (measured in dollars per unit). The supply function for this commodity is described by the equation $q = S(p) = -200 + 100p$, where $q$ is the quantity supplied (also measured in thousands of units) and $p$ is the price (measured in dollars per unit). Find the equilibrium price and quantity.

**Solution**   The supply function indicates that the price cannot be less than $2 per unit and the demand function shows that the price cannot be greater than $100 per unit since quantities cannot be negative; therefore we have $-200 + 100p \geq 0$ and $15{,}000 - 150p \geq 0$. Also, both functions are linear and their graphs are straight lines in the $pq$-plane. The demand curve is a line with slope $-150$. Thus, as expected, the demand function is decreasing. The supply curve is a line with slope $100$. Hence, the supply function is increasing. Since the two lines have different slopes, they are not parallel and will have a unique point of intersection. To find the coordinates of this point, we solve the system

$$\begin{cases} q = 15{,}000 - 150p \\ q = -200 + 100p \end{cases}$$

Substituting $-200 + 100p$ for $q$ in the first equation we get

$$-200 + 100p = 15{,}000 - 150p$$

That is,

$$250p = 15{,}200$$

and

$$p = \frac{15{,}200}{250} = 60.80$$

Replacing $p$ by $60.80$ in the first equation of the system we get

$$q = 15{,}000 - 150(60.80) = 5880$$

Thus, we conclude that the equilibrium quantity of 5880 thousand units is demanded and supplied at the equilibrium price of $60.80 per unit. (See Figure 2.15.)
**Do Exercise 7.**

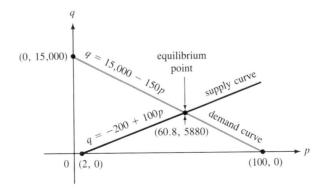

**FIGURE 2.15**

Sometimes the quantity demanded or supplied depends not only on the price of the commodity, but also on some other variable, such as the consumer's average income. In such cases, the equilibrium price and quantity will vary according to the value of the variable as in the following example.

EXAMPLE 4    A demand function for a commodity is described by $q = D(p) = -100p + 1500 + 10I$, where $q$ is the quantity demanded (measured in thousands of units), $p$ is the price (measured in dollars per unit), and $I$ is the consumer's average income. The supply function for the same commodity is defined by $q = S(p) = 300\sqrt{8p - 60}$, where $q$ is the quantity supplied (measured in thousands of units) and $p$ is the price (measured in dollars per unit). Note that $p \geq 7.5$ since $8p - 60$ must be nonnegative.

    **a.** Find the equilibrium price and quantity when the consumer's average income is given by $I = 350$.
    **b.** Same as part **a** with $I = 590$.
    **c.** Sketch the demand and supply curves on the same coordinate plane.

Solution    **a.** Since $q = D(p) = -100p + 1500 + 10I$, when $I = 350$, we have

$$q = D(p) = -100p + 5000$$

To find the equilibrium quantity and price, we solve the system

$$\begin{cases} q = -100p + 5000 \\ q = 300\sqrt{8p - 60} \end{cases}$$

Substituting $300\sqrt{8p - 60}$ for $q$ in the first equation we get

$$300\sqrt{8p - 60} = -100p + 5000 \qquad (*)$$

Note that the foregoing equation shows that supply equals demand when $p$ is the equilibrium price. Dividing both sides of equation (*) by 100 and then squaring both sides yields

$$9(8p - 60) = (-p + 50)^2$$

and

$$72p - 540 = p^2 - 100p + 2500$$

That is,

$$p^2 - 172p + 3040 = 0$$

Factoring the left side of this equation yields

$$(p - 20)(p - 152) = 0$$

Thus, $p = 20$ or $p = 152$.

Since we squared both sides of an equation, we may have introduced some extraneous solutions. We must check the validity of the solutions we obtained by replacing $p$ by 20 in equation (*) to get

$$300\sqrt{8(20) - 60} = -100(20) + 5000$$

which yields the true statement

$$3000 = 3000$$

Thus, 20 is a solution.

Replacing $p$ by 152 in equation (*) we get

$$300\sqrt{8(152) - 60} = -100(152) + 5000$$

which yields the false statement

$$10,200 = -10,200$$

Thus, we reject 152 and conclude that 20 is the only solution. Therefore, the equilibrium price is $20 per unit. We find the equilibrium quantity by replacing $p$ by 20 in either the demand or supply equation. We get

$$q = D(20) = -100(20) + 5000$$
$$= 3000$$

Hence, the equilibrium quantity is 3,000,000 units (since $q$ is in thousands).

**b.** When $I = 590$, we get

$$q = D(p) = -100p + 1500 + 10(590)$$
$$= -100p + 7400$$

To find the new equilibrium price and quantity, we solve the system

$$\begin{cases} q = -100p + 7400 \\ q = 300\sqrt{8p - 60} \end{cases}$$

We solve it in exactly the same way as we solved the system of part **a** and find that the only solution is $p = 32$. Replacing $p$ by 32 in the first equation we get

$$q = -100(32) + 7400 = 4200$$

The equilibrium price is $32 per unit and the equilibrium quantity is 4200 thousand units.

**c.** The supply curve and the two demand curves are sketched in Figure 2.16. Note that the second demand curve was obtained by shifting the first demand curve to the right.

**Do Exercise 9.**

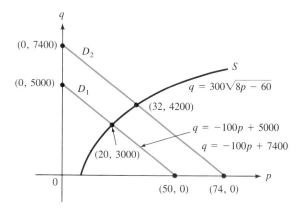

**FIGURE 2.16**

## Taxes and Subsidies

Often, the government imposes taxes on certain commodities and in other cases the government pays subsidies to producers to encourage them to supply essential commodities to consumers at reasonable prices. Taxes and subsidies are variables which affect the equilibrium price and equilibrium quantity of a commodity. The next two examples illustrate this.

**EXAMPLE 5**   The demand and supply functions for a commodity are defined by

$$q = D(p) = -3p + 139$$
$$q = S(p) = 2p - 11$$

where $q$ is the number of units (measured in thousands) and $p$ is the price (measured in dollars per unit).

**a.** Find the equilibrium price and quantity.
**b.** Suppose that the government imposes a tax of $7.50 per unit on the commodity. Find the new equilibrium price and quantity.

**Solution**   **a.** To find the equilibrium point, we solve the system

$$\begin{cases} q = -3p + 139 \\ q = 2p - 11 \end{cases}$$

Substituting $2p - 11$ for $q$ in the first equation we obtain

$$2p - 11 = -3p + 139$$

from which we get

$$p = 30$$

Replacing $p$ by 30 in the first equation gives

$$q = -3(30) + 139$$
$$= 49$$

The equilibrium price is \$30 per unit and the equilibrium quantity is 49,000 units.

**b.** We assume that the tax imposed is included in the price $p$ paid by the consumer.* Thus, the demand equation does not change. However, when the consumers pay \$$p$ per unit, only \$$(p - 7.5)$ goes to the suppliers and \$7.50 goes to the government. The supply equation becomes

$$q = 2(p - 7.5) - 11$$

or

$$q = 2p - 26$$

To find the new equilibrium point, we solve the system

$$\begin{cases} q = -3p + 139 \\ q = 2p - 26 \end{cases}$$

We find that $p = 33$ and $q = 40$. The new equilibrium price is \$33 per unit and the new equilibrium quantity is 40,000 units. Note that the price per unit increased by \$3 and the quantity demanded and supplied decreased by 9000 units. (See Figure 2.17.)

**Do Exercise 13.**

---

*If the tax is not included in the price, but is added after the purchase, it will not affect the supply equation. However, $p$ must be replaced by $p + T$ in the demand equation. The new equilibrium price is obtained by solving

$$\begin{cases} q = D(p + T) \\ q = S(p) \end{cases}$$

See Exercise 15.

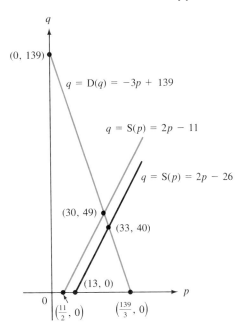

**FIGURE 2.17**

Observe that the graph of the supply equation has been shifted 7.5 units to the right when the tax was added.

EXAMPLE 6    The demand and supply functions for a commodity are defined by $q = D(p) = -2p + 89$ and $q = S(p) = 6p - 7$, where $p$ is the price (measured in dollars per unit) and $q$ is the quantity (measured in thousands of units).

**a.** Find the equilibrium price and quantity.

**b.** Suppose that the government pays the suppliers a subsidy of $4 per unit. Find the new equilibrium price and quantity.

Solution    **a.** We must solve the system

$$\begin{cases} q = -2p + 89 \\ q = 6p - 7 \end{cases}$$

We find that $p = 12$ and $q = 65$ and conclude that the equilibrium price is $12 per unit and the equilibrium quantity is 65,000.

**b.** The subsidy paid to the supplier does not affect the demand equation. However, when the price is $p per unit, the producers receive $(p + 4)$ since the consumers pay $p and the government pays $4 for each unit produced and sold. The supply equation becomes

$$q = 6(p + 4) - 7$$

which may be written

$$q = 6p + 17$$

To find the new equilibrium point, we solve the following system

$$\begin{cases} q = -2p + 89 \\ q = 6p + 17 \end{cases}$$

We find that $p = 9$ and $q = 71$. We conclude that the new equilibrium price and quantity are \$9 per unit and 71,000 units, respectively. Note that the equilibrium price has decreased by \$3 per unit and the quantity demanded and supplied has increased by 6000 units.

**Do Exercise 17.**                                                        ∎

**EXAMPLE 7**    The demand equation for a commodity is $q = -p^2 - p + 210$, where $q$ is the number of units demanded (in millions) when \$p$ is the price per unit. The suppliers will produce and sell 48,000,000 units when the price is \$8 per unit, but only 18,000,000 units when the price is \$5 per unit.

    **a.** Find the supply equation assuming that it is linear.
    **b.** Find the equilibrium price and quantity.
    **c.** What should the government subsidy be per unit to increase the demand by 42,000,000?

**Solution**    **a.** It is given that $q = 48$ when $p = 8$ and that $q = 18$ when $p = 5$. Therefore, the points (8, 48) and (5, 18) are on the supply curve. Noting that the supply curve is a straight line, we find that its slope is 10 since $(48 - 18)/(8 - 5) = 10$. Therefore, the point-slope equation of the line is

$$q - 18 = 10(p - 5)$$

This equation may be written

$$q = 10p - 32$$

    **b.** We find the equilibrium price and quantity solving the system

$$\begin{cases} q = -p^2 - p + 210 \\ q = 10p - 32 \end{cases}$$

Subtracting the first equation from the second, we obtain

$$0 = p^2 + 11p - 242$$

This equation may be factored as

$$0 = (p - 11)(p + 22)$$

Thus, $p = 11$ or $p = -22$. We consider only the positive value of $p$ and conclude that the equilibrium price is \$11. Replacing $p$ by 11 in the supply equation, we obtain $q = 10 \cdot 11 - 32 = 78$. Hence, the equilibrium quantity is 78,000,000.

    **c.** The demand is increased by 42,000,000 if the value of $q$ is increased by 42. Therefore, the new value of $q$ will be 120 since $78 + 42 = 120$. The subsidy paid by the government does not affect the demand equation so we can find the price paid by the consumers when $q = 120$. We find that

$$120 = -p^2 - p + 210$$

The foregoing equation may be written

$$p^2 + p - 90 = 0$$

The only positive solution of this equation is 9. Thus, the new equilibrium price (paid by the consumers) is \$9 per unit. Suppose now that the government pays a subsidy of \$$S$ per unit, so that the suppliers receive \$$(9 + S)$ per unit. Recalling that the value of $q$ is now 120, we replace $p$ and $q$ by $9 + S$ and 120, respectively, in the supply equation and obtain

$$120 = 10(9 + S) - 32$$

Thus,

$$120 = 10S + 58$$

and

$$S = 6.2$$

We conclude that the government should pay a subsidy of \$6.20 per unit to increase the demand by 42,000,000 units.

**Do Exercise 19.**   ■

## Multiple Commodities

In the previous examples, we have assumed that the quantities of a commodity demanded and supplied are affected only by the price of that commodity, and in some cases by other variables such as consumer's average income, taxes, and subsidies. In many instances, the quantities of a commodity demanded and supplied depend not only on the price of that commodity but also on the price of related commodities. For example, an increase in the price of beef may create an increase in the demand for poultry. We shall conclude this section with an example involving more than one commodity. In all cases, the underlying principle is that the market equilibrium is reached when the supply of each commodity equals the demand for that commodity.*

EXAMPLE 8     Let $D_x$ and $S_x$ denote the quantities (measured in thousands of units) demanded and supplied, respectively, of commodity X. Let $D_y$ and $S_y$ denote the quantities (also measure in thousnds of units) demanded and supplied, respectively, of commodity Y. Suppose that $p_x$ and $p_y$ represent the prices (in dollars per unit) of commodities X and

---

*In principle, if we knew the demand and supply functions for every commodity in the economy, we could then compute the so-called *Walrasian general equilibrium*. In practice, if we consider $n$ commodities and have $n$ demand functions and $n$ supply functions, we can get a system of $n$ equations in $n$ unknowns by setting $D_i = S_i$ for each $i = 1, 2, 3, \ldots, n$. We can then solve this system to get the equilibrium price of each of the $n$ commodities. Finally, we obtain the equilibrium quantities traded for each of the $n$ commodities. Cases where $n = 3$ are considered in the exercises.

Y, respectively. It is known that the demand and supply functions for these two commodities are interrelated as is shown by

$$D_x = 168 - 8p_x - 15p_y$$

$$S_x = -13 + 11p_x + 20p_y$$

$$D_y = 119 - 13p_x - 14p_y$$

$$S_y = -5 + 10p_y.$$

**a.** Find the equilibrium prices of these commodities.

**b.** Find the equilibrium quantities of these commodities.

**Solution** **a.** The market will be in equilibrium if $D_x = S_x$ and $D_y = S_y$. Thus, we set

$$168 - 8p_x - 15p_y = -13 + 11p_x + 20p_y$$

and

$$119 - 13p_x - 14p_y = -5 + 10p_y$$

These two equations may be simplified to obtain

$$\begin{cases} 19p_x + 35p_y = 181 \\ 13p_x + 24p_y = 124 \end{cases}$$

The solution is

$$\begin{cases} p_x = 4 \\ p_y = 3 \end{cases}$$

Thus, the equilibrium price of commodity X is $4 per unit and that of commodity Y is $3 per unit.

**b.** Replacing each $p_x$ by 4 and each $p_y$ by 3 in the original equations giving the demand for each commodity, we obtain

$$D_x = 168 - 8 \cdot 4 - 15 \cdot 3 = 91$$

and

$$D_y = 119 - 13 \cdot 4 - 14 \cdot 3 = 25$$

Note that we could have just as well used the supply equations to get these results since at equilibrium the demand and supply are equal. We conclude that the equilibrium quantity of commodity X is 91,000 and that of commodity Y is 25,000.

**Do Exercise 23.** ∎

---

# Exercise Set 2.5

**1.** The cost $\$C(q)$ of producing $q$ units of a commodity in a week is given by

$$y = C(q) = 3q + 25,000.$$

If each unit of that commodity sells for $5, find the following:

**a.** The fixed cost per week.

**b.** The number of units of the commodity that should be produced and sold per week to make sure that the business breaks even.

**2.** The Triplenight Company found that the cost $\$C(q)$ of making $q$ racquetball racquets per month is given by

$$y = 45q + 30,000.$$

Each racquet may be sold for \$60. Find the following:

**a.** The fixed cost per month.

**b.** The number of racquets that should be made and sold each month so that the company breaks even.

**3.** A company manufacturing motorcycles found that the cost \$$C(q)$ of making $q$ motorcycles per month is given by

$$y = C(q) = .25q^2 + 2000q + 525,000$$

provided that $q$ is between 0 and 1500. It also found that the revenue \$$R(q)$ from the sales of $q$ motorcycles per month is given by

$$y = R(q) = -2.25q^2 + 4500q$$

provided that $q$ is between 0 and 1500.

**a.** Find the level of production at which the company will break even.

**b.** At what level of production will the company make a profit?

**4.** A company manufacturing television sets found that the cost \$$C(q)$ of making $q$ deluxe models per week is given by

$$y = C(q) = .2q^2 + 560q + 120,000$$

provided that $q$ is between 0 and 700. It also found that the revenue \$$R(q)$ from the sales of $q$ deluxe sets per week is given by

$$y = R(q) = -q^2 + 1400q$$

provided that $q$ is between 0 and 700.

**a.** Find the level of production at which the company will break even.

**b.** At what level of production will the company make a profit?

**5.** A company manufacturing VCRs found that the cost \$$C(q)$ of making $q$ recorders per day is given by

$$y = C(q) = .5q^2 + 215q + 18,000$$

provided that $q$ is between 0 and 100. It also found that the revenue \$$R(q)$ from the sales of $q$ VCRs per day is given by

$$y = R(q) = -4q^2 + 800q$$

provided that $q$ is between 0 and 100.

**a.** Find the level of production at which the company will break even.

**b.** At what level of production will the company make a profit?

**6.** The demand for a commodity can be described by

$$q = D(p) = 35,000 - 175p$$

where $q$ is the quantity demanded (measured in thousands of units) and $p$ is the price (measured in dollars per unit). The supply function for this commodity is described by

$$q = S(p) = -750 + 150p$$

where $q$ is the quantity supplied (also measured in thousands of units) and $p$ is the price (measured in dollars per unit). Find the equilibrium price and quantity.

**7.** The demand for a product is described by

$$q = D(p) = 75,000 - 120p$$

where $q$ is the number of units demanded and $p$ is the price (measured in dollars per unit). The supply function for this product has been found to be

$$q = S(p) = -3750 + 105p$$

where $q$ is the number of units supplied and $p$ is the price (measured in dollars per unit). Find the equilibrium price and quantity.

**8.** The demand for a commodity can be described by

$$q = D(p) = -p^2 - 10p + 11,000$$

where $q$ is the quantity demanded (measured in thousands of units) and $p$ is the price (measured in dollars per unit). The supply function for this commodity is

$$q = S(p) = -100 + 115p$$

where $q$ is the quantity supplied (also measured in thousands of units) and $p$ is the price (measured in dollars per unit). Find the equilibrium price and quantity.

**9.** The demand function for a commodity is

$$q = D(p) = -60p + 2640 + 12I$$

where $q$ is the quantity demanded (measured in thousands of units), $p$ is the price (measured in dollars per unit), and $I$ is the consumer's average income measured in appropriately defined units. The supply function for the commodity is

$$q = S(p) = 42p - 60$$

where $q$ is the quantity supplied measured in thousands of units and $p$ is the price (measured in dollars per unit).

**a.** Find the equilibrium price and quantity when the consumer's average income is given by $I = 30$.

**b.** Same as part **a** with $I = 81$.

**c.** Sketch the demand and supply curves on the same coordinate plane.

**10.** The demand function for a product is

$$q = D(p) = -p^2 - 50p + 40{,}000 + 20I$$

where $q$ is the number of units demanded when the price is $\$p$ per unit and $I$ is the consumer's average income. The supply function for the product is

$$q = S(p) = 250p - 400$$

where $q$ is the number of units supplied when the price is $\$p$ per unit.

**a.** Find the equilibrium price and quantity when the consumer's average income is given by $I = 500$.

**b.** Same as part **a** with $I = 775$.

**c.** Sketch the demand and supply curves on the same coordinate plane.

**11.** The demand function for a commodity is

$$q = D(p) = -p^2 - 20p + 4000 + 2I$$

where $q$ is the quantity demanded (measured in thousands of units), $p$ is the price (measured in dollars per unit), and $I$ is the consumer's average income. The supply function for the commodity is

$$q = S(p) = p^2 + 28p - 510,$$

where $q$ is the quantity supplied measured in thousands of units and $p$ is the price (measured in dollars per unit).

**a.** Find the equilibrium price and quantity when the consumer's average income is given by $I = 850$.

**b.** Same as part **a** with $I = 1201$.

**c.** Sketch the demand and supply curves on the same coordinate plane.

**12.** The demand function for a commodity is

$$q = D(p) = -60p + 20{,}635 + 127I$$

where $q$ is the number of units of the commodity demanded when the price is $\$p$ per unit and $I$ is the consumer's average income. The supply function for the commodity is

$$q = S(p) = 250\sqrt{5p - 80}$$

where $q$ is the number of units supplied when the price of the commodity is $\$p$ per unit. Note that $p \geq 16$ since $5p - 80 \geq 0$.

**a.** Find the equilibrium price and quantity when the consumer's average income is given by $I = 75$.

**b.** Same as part **a** with $I = 125$.

**c.** Sketch the demand and supply curves on the same coordinate plane.

**13.** The demand and supply functions for a commodity are

$$q = D(p) = -5p + 159$$
$$q = S(p) = 3p - 9$$

where $q$ is the number of units (measured in thousands) and $p$ is the price (measured in dollars per unit).

**a.** Find the equilibrium price and quantity.

**b.** Suppose that the government imposes a tax of $\$8$ per unit on the commodity. Find the new equilibrium price and quantity.

**14.** A dealer can sell 650 units of a certain product when the price is $\$10$ per unit but can sell only 130 units when the price is $\$50$ per unit. The supply function for this product is

$$q = S(p) = 8p - 60$$

where $q$ is the number of units supplied when the price is $\$p$ per unit.

**a.** Find the demand equation assuming that it is linear.

**b.** Find the equilibrium price and quantity.

**c.** Suppose that the government imposes a tax of $\$5.25$ per unit on the commodity. Find the new equilibrium price and quantity.

**15.** When the price of a certain commodity is $\$12$ per unit, the demand and supply are 1478 and 190 units, respectively; a price of $\$40$ per unit changes the demand and supply to 582 and 862 units, respectively. Assuming that the demand and supply functions are linear, find the following:

**a.** The demand equation.

**b.** The supply equation.

**c.** The equilibrium price and quantity.

**d.** Suppose that the government imposes a tax of $\$7$ per unit on the commodity. Find the new equilibrium price and quantity. (Assume the tax is not included in the price but is added after the purchase.)

**16.** The demand and supply functions for a commodity are

$$q = D(p) = -12p + 7000 \text{ and } q = S(p) = 18p - 2000$$

where $p$ is the price per unit in dollars and $q$ is the quantity demanded (in the first equation) and supplied (in the second equation).

**a.** Find the equilibrium price and quantity.

**b.** Suppose that the government pays the suppliers a subsidy of $\$50$ per unit. Find the new equilibrium price and quantity.

17. The demand and supply functions for a commodity are known to be linear. If the price of the commodity is $5 per unit, the quantities demanded and supplied are 510 and 35, respectively. If the price is $20 per unit, the quantities demanded and supplied are 360 and 260, respectively.

   **a.** Find the demand equation.
   **b.** Find the supply equation.
   **c.** Find the equilibrium price and quantity.
   **d.** Suppose that the government pays the suppliers a subsidy of $5 per unit. Find the new equilibrium price and quantity.

18. The demand equation for a commodity is

$$q = D(p) = -p^2 - 10p + 11,000$$

   where $q$ is the number of units demanded when the price is $p$ per unit. The supply equation is linear, and it is known that the suppliers will produce and sell 800 units when the price is $10 per unit, but when the price is $8 per unit they will produce and sell only 560 units.

   **a.** Find the supply equation.
   **b.** Find the equilibrium price and quantity.
   **c.** If the government provides a subsidy of $20 per unit, find the new equilibrium price and quantity.

19. The demand equation for milk in a French province is given by

$$q = D(p) = -p^2 - p + 156$$

   where $q$ is the number of liters of milk demanded (in millions) when the price is $p$ francs per liter. Milk producers are willing to produce and sell 34 million liters of milk when the price is 4 francs per liter and 59 million liters when the price is 6 francs per liter.

   **a.** Find the supply equation assuming it is linear.
   **b.** Find the equilibrium price and quantity.
   **c.** The French government wishes to increase the demand for milk in that province to 114 million liters. What subsidy per liter must be provided to accomplish that goal?

20. The demand equation for potatoes in a state is given by

$$q = D(p) = -p^2 - 3p + 270$$

   where $q$ is the number of pounds (in millions) demanded when the price is $p$ cents per pound. The farmers are willing to grow and sell 65,000,000 and 95,000,000 pounds of potatoes when the prices are $0.05 and $0.07 per pound, respectively.

   **a.** Find the supply equation assuming it is linear.
   **b.** Find the equilibrium price and quantity.

   **c.** The state government wishes to increase the demand by 42,000,000 pounds. What subsidy per pound must be provided to accomplish that goal?

21. The demand equation for a product is

$$q = D(p) = -p^2 - 2p + 575$$

   where $q$ is the number of units (in thousands) demanded when the price is $p$ per unit. The supply equation is linear and it is known that the suppliers will produce and sell 231,000 units when the price is $10 per unit, but when the price is $5 per unit they will produce and sell only 81,000 units.

   **a.** Find the supply equation.
   **b.** Find the equilibrium price and quantity.
   **c.** If the government wants to increase the demand by 81,000, what subsidy per unit must it provide?

22. The demand equation for a commodity is

$$q = D(p) = -p^2 - 2p + 323$$

   where $q$ is the number of units demanded (in millions) when the price is $p$ per unit. The supply equation is linear and it is known that the suppliers will produce and sell 140,000,000 units when the price is $7 per unit, but when the price is $4 per unit they will produce and sell only 77,000,000 units.

   **a.** Find the supply equation.
   **b.** Find the equilibrium price and quantity.
   **c.** If the government wants to increase the quantity demanded by 21,000,000 units, what subsidy per unit must it provide?

23. Let $D_x$ and $S_x$ denote the number of gallons of ice cream (measured in thousands) demanded and supplied, respectively. Let $D_y$ and $S_y$ denote the number of containers of ice cream topping (measured in hundreds) demanded and supplied, respectively. Suppose that $p_x$ and $p_y$ represent the prices (in dollars per unit) of ice cream and topping, respectively. It is known that the demand and supply functions for ice cream and topping are interrelated as shown by

$$D_x = 16 - 2p_x - p_y$$
$$S_x = -10 + 3p_x + 3p_y$$
$$D_y = 30 - 2p_x - 3p_y$$
$$S_y = -17 + 8p_x + 3p_y$$

   **a.** Find the equilibrium prices of ice cream and topping.
   **b.** Find the equilibrium quantities of ice cream and topping.

24. Let $D_x$ and $S_x$ denote the numbers of half-gallons of gourmet ice cream (measured in thousands) demanded and supplied,

respectively. Let $D_y$ and $S_y$ denote the number of half-gallons of regular ice cream (also measured in thousands) demanded and supplied, respectively. Suppose that $p_x$ and $p_y$ represent the prices (in dollars per half-gallon) of gourmet and regular ice cream, respectively. It is known that the demand and supply functions for gourmet and regular ice cream are interrelated as is shown by

$$D_x = 310 - 40p_x + 20p_y$$
$$S_x = -110 + 60p_x - 10p_y$$
$$D_y = 445 + 16p_x - 50p_y$$
$$S_y = -206 - 4p_x + 250p_y$$

**a.** Find the equilibrium prices of gourmet and regular ice cream.
**b.** Find the equilibrium quantities of gourmet and regular ice cream.

25. Suppose that when three commodities X, Y, and Z are priced at $x$, $y$, and $z$ dollars per unit, respectively, the quantities demanded are $D_x$, $D_y$, and $D_z$, respectively (measured in thousands) and the quantities supplied are $S_x$, $S_y$, and $S_z$, respectively (also measured in thousands). The demand and supply functions for these commodities are interrelated as expressed by

$$D_x = 520 - 30x + 92y - 25z$$
$$S_x = -120 + 70x - 8y + 75z$$

$$D_y = 140 + 32x - 52y + 18z$$
$$S_y = -7 - 8x + 98y - 12z$$
$$D_z = 478 - 20x + 28y - 5z$$
$$S_z = -21 + 40x - 12y + 65z$$

Find the equilibrium prices and quantities for all three commodities.

26. Same as Exercise 25 with the following equations:

$$D_x = 150 - 10x + 85y - 17z$$
$$S_x = -8 + 90x - 15y + 83z$$
$$D_y = 280 + 42x - 20y + 81z$$
$$S_y = -17 - 8x + 120y - 19z$$
$$D_z = 310 - 17x + 51y - 12z$$
$$S_z = -44 + 63x - 9y + 88z$$

27. Same as Exercise 25 with the following equations:

$$D_x = 800 - 23x + 95y - 18z$$
$$S_x = -35 + 77x - 5y + 82z$$
$$D_y = 320 + 50x - 12y + 35z$$
$$S_y = -68 - 10x + 118y - 5z$$
$$D_z = 1310 - 10x + 42y - 20z$$
$$S_z = -112 + 80x - 8y + 100z$$

# 2.6 Chapter Review

$kE_i \rightarrow E_i$ [2.2]
$kE_i + E_j \rightarrow E_j$ [2.2]
$k_1E_i + k_2E_j \rightarrow E_j$ [2.2]
Break-even point [2.5]
Consistent and dependent [2.1], [2.2]
Consistent and independent [2.1], [2.2]
Cramer's rule [2.1]
Demand curve [2.5]
Demand function [2.5]
Determinant [2.1]
Elimination of variables by addition [2.1]
Elimination of variables by substitution [2.1]

Equilibrium price [2.5]
Equilibrium quantity [2.5]
Equivalent systems [2.1], [2.2]
Fixed cost [2.5]
Graphical solution [2.1], [2.3]
Half-plane [2.4]
Inconsistent [2.1]
Linear equation in two variables [2.1]
Market equilibrium [2.5]
Matrix [2.1]
Nonlinear systems [2.1]
Ordered pair [2.1]
Solution [2.1]
Solution set [2.1], [2.2]

Subsidy [2.5]
Supply curve [2.5]
Supply function [2.5]
System of linear inequalities in two variables [2.4]
System of three linear equations in three variables [2.2]
System of two linear equations in two variables [2.1]
Tax [2.5]
Walrasian general equilibrium [2.5]

SUMMARY An equation that can be written in the form $ax + by = c$, where $a$, $b$, and $c$ are real numbers and $ab \neq 0$ is called a linear equation in two variables $x$ and $y$. An ordered pair of real numbers $(x_0, y_0)$ is a solution of the equation if the statement $ax_0 + by_0 = c$ is true. The set of all solutions is called the solution set of the equation. If $a_1$, $b_1$, $c_1$, $a_2$, $b_2$, and $c_2$ are real numbers

$$\begin{cases} a_1x + b_1y = c_1 \\ a_2x + b_2y = c_2 \end{cases}$$

is called a system of two linear equations in two variables. The intersection of the solution sets of the two equations is the solution set of the system. We may solve the system by

**a.** finding the intersection of the graphs of the two equations,
**b.** eliminating one variable by addition or subtraction,
**c.** eliminating one variable by substitution, or
**d.** using Cramer's rule.

A square array of numbers of the form

$$\begin{bmatrix} a & b \\ c & d \end{bmatrix}$$

is called a $2 \times 2$ matrix. The determinant of this $2 \times 2$ matrix is the number $ad - bc$ and is denoted

$$\begin{vmatrix} a & b \\ c & d \end{vmatrix}$$

First make sure that the system is written as

$$\begin{cases} a_1x + b_1y = c_1 \\ a_2x + b_2y = c_2 \end{cases}$$

Then let

$$\begin{vmatrix} a_1 & b_1 \\ a_2 & b_2 \end{vmatrix} = C, \quad \begin{vmatrix} c_1 & b_1 \\ c_2 & b_2 \end{vmatrix} = A_x, \text{ and } \begin{vmatrix} a_1 & c_1 \\ a_2 & c_2 \end{vmatrix} = A_y$$

If $C \neq 0$, the solution of the system is $\left(\dfrac{A_x}{C}, \dfrac{A_y}{C}\right)$.

Sometimes the right substitutions in a nonlinear system of two equations can yield a linear system that can be solved. Then the values of the original variables can be obtained from the solutions of the linear system. We also consider systems of three linear equations in three variables.

$$\begin{cases} a_1x + b_1y + c_1z = d_1 \\ a_2x + b_2y + c_2z = d_2 \\ a_3x + b_3y + c_3z = d_3 \end{cases}$$

We can get an equivalent system by multiplying both sides of an equation by a nonzero constant, by adding to any equation one of the other two equations, or by combining these two operations. Starting with a given system, we try to obtain an equivalent system in the form

$$\begin{cases} x = x_0 \\ y = y_0 \\ z = z_0 \end{cases}$$

whose solution is obviously $(x_0, y_0, z_0)$. The necessary steps should be documented on the right side of each system using the following abbreviations: $kE_i \rightarrow E_i$ means "the $i$th equation is multiplied by the (nonzero) constant $k$ to get the new $i$th equation," and $kE_i + E_j \rightarrow E_j$ means "the $i$th equation is multiplied by the constant $k$ and the result is added to the $j$th equation to get the new $j$th equation." The meaning of $k_1E_i + k_2E_j \rightarrow E_j$ should be clear.

We can also eliminate a variable by substitution to obtain a linear system of two equations in two variables which then may be solved using any of the methods outlined earlier. A system of two equations in two variables, where at least one of the two equations is nonlinear, may be solved by graphing both equations and finding the coordinates of the points of intersection of the graphs.

An expression that can be written $ax + by < c$, $ax + by > c$, $ax + by \leq c$, or $ax + by \geq c$, where $a, b, c$ are real numbers and $ab \neq 0$ is called a linear inequality in two variables. An ordered pair of real numbers $(x_0, y_0)$ is a solution of one of these if a true statement results when $x$ and $y$ are replaced by $x_0$ and $y_0$, respectively. The graph of a linear inequality is a half-plane. If we have a system of $n$ inequalities in the two variables $x$ and $y$, we sketch the graph of each inequality on the same Cartesian coordinate plane. The intersection of the $n$ half-planes is the graph of the system. We can get the coordinates of a vertex of the graph by choosing the appropriate two inequalities, changing the inequality signs to equal signs, and solving the resulting linear system of two equations in two variables.

In applications, if we know the cost $(C)$ and revenue $(R)$ functions, the break-even point is found by solving

$$\begin{cases} y = C(x) \\ y = R(x) \end{cases}$$

If $q = D(p)$ and $q = S(p)$ define the demand and supply functions, respectively, the market is in equilibrium when $D(p) = S(p)$. The equilibrium price and equilibrium quantity are found by solving

$$\begin{cases} q = D(p) \\ q = S(p) \end{cases}$$

If a tax of $T$ dollars per unit is imposed by the government, the equilibrium price and quantity will be affected. If the tax is already included in the price the consumer pays, then the demand equation does not change, but we must replace $p$ by $p - T$ in the supply equation since the supplier receives only $p - T$ dollars per unit, the other $T$ dollars goes to the government. We get the new equilibrium point by solving the system

$$\begin{cases} q = D(p) \\ q = S(p - T) \end{cases}$$

If the tax is added after the purchase, the demand equation is affected since the consumer pays $p + T$ dollars per unit but the supply equation is unchanged. We get the new equilibrium point by solving the system

$$\begin{cases} q = D(p + T) \\ q = S(p) \end{cases}$$

If the government pays a subsidy of $S$ dollars per unit, the demand equation is not affected. However, we replace $p$ by $p + S$ in the supply equation since the suppliers receive $p$ dollars (paid by the consumer) plus $S$ dollars (paid by the government) for each unit sold. Suppose we have two commodities X and Y and when the prices of these commodities are $p_x$ and $p_y$ dollars per unit, respectively, the quantities demanded are $D_x$ and $D_y$, respectively, and the quantities supplied are $S_x$ and $S_y$, respectively. In general $D_x$, $D_y$, $S_x$, and $S_y$ are given in terms of $p_x$ and $p_y$. By letting $D_x = S_x$ and $D_y = S_y$, we obtain a system of two equations in the two variables $p_x$ and $p_y$. The solution of this system gives us the equilibrium prices.

**SAMPLE EXAM QUESTIONS**    *In Exercises 1–8, solve the given system using the indicated method.*

1. $\begin{cases} 2x + 3y = -5 \\ 3x - 5y = 21 \end{cases}$ (graphically)

2. $\begin{cases} 3x - 2y = -9 \\ 2x + 4y = 10 \end{cases}$ (graphically)

3. $\begin{cases} 5x + 2y = 0 \\ 4x - 3y = 23 \end{cases}$ (substitution)

4. $\begin{cases} 3x + 7y = 10 \\ 5x - 2y = 3 \end{cases}$ (substitution)

5. $\begin{cases} 2x - 3y = 7 \\ 3x + 9y = -3 \end{cases}$ (addition)

6. $\begin{cases} 3x - 2y = -5 \\ 6x - 4y = 4 \end{cases}$ (addition)

7. $\begin{cases} 5x + y = -13 \\ 3x - 2y = -13 \end{cases}$ (Cramer's rule)

8. $\begin{cases} 2x + 3y = 11 \\ 3x - 7y = 5 \end{cases}$ (Cramer's rule)

*In Exercises 9–12, solve the given system by first making an appropriate substitution.*

9. $\begin{cases} 2x^2 + y^2 = 17 \\ 5x^2 - y^2 = 11 \end{cases}$

10. $\begin{cases} \frac{1}{3}x^2 + y^2 = 4 \\ \frac{1}{9}x^2 + 5y^2 = 6 \end{cases}$

11. $\begin{cases} \frac{5}{x} + \frac{1}{y} = -13 \\ \frac{3}{x} - \frac{2}{y} = -13 \end{cases}$

12. $\begin{cases} \frac{5}{x} + \frac{2}{y} = 8 \\ \frac{3}{x} - \frac{4}{y} = 10 \end{cases}$

*Solve the systems of Exercises 13, 14, and 15 graphically.*

13. $\begin{cases} y = -2x^2 + 8x + 13 \\ 6x - y = -9 \end{cases}$

14. $\begin{cases} 31x + 40y = 98 \\ y = 2^{-x} \end{cases}$

15. $\begin{cases} y = x^2 + 5x + 6 \\ y = -x^2 + 3x + 10 \end{cases}$

*Solve the linear systems of Exercises 16–22 using the method of elimination of variables by addition.*

16. $\begin{cases} 2x - 2y + 3z = 1 \\ x - 3y - 2z = -9 \\ x + y + z = 6 \end{cases}$

17. $\begin{cases} 12x - 12y - z = 0 \\ 6x + 6y + 2z = 90 \\ 2x + 3y + z = 36 \end{cases}$

18. $\begin{cases} 6x - 2y - z = 6 \\ x - 3y - 4z = 5 \\ 3x + y + 4z = 0 \end{cases}$

19. $\begin{cases} x + 2y - 7z = 75 \\ 6x - 2y + 3z = 75 \\ 2x + y + z = 0 \end{cases}$

20. $\begin{cases} 4x - y + z = 11 \\ 3x - 2y - z = 6 \\ 7x - 3y = 17 \end{cases}$

21. $\begin{cases} 2x + 3y - z = 2 \\ 5x - 2y + z = 3 \\ 9x + 4y - z = 9 \end{cases}$

**22.** $\begin{cases} 5x \quad\;\; - \;\; z = 16 \\ 3x - \;y - 2z = \;\;9 \\ \;x + 2y + 3z = \;\;4 \end{cases}$

*Solve the linear systems of Exercises 23–26 by elimination of a variable by substitution.*

**23.** The system of Exercise 16.

**24.** The system of Exercise 17.

**25.** The system of Exercise 18.

**26.** The system of Exercise 19.

**27.** The system of Exercise 20.

**28.** The system of Exercise 21.

**29.** The system of Exercise 22.

*In Exercises 30–34, sketch the graph of the system of linear inequalities.*

**30.** $\begin{cases} \quad\;\; x > \;\;5 \\ \quad\;\; y > \;\;2 \\ 5x + 3y \le 46 \end{cases}$

**31.** $\begin{cases} \quad\;\; x > \;\;2 \\ 2x - \;y \le \;\;9 \\ \;x + y \le 12 \end{cases}$

**32.** $\begin{cases} 5x + 3y < \;28 \\ \;x - \;y > -4 \\ \;x - 5y < \;\;0 \end{cases}$

**33.** $\begin{cases} \quad\;\; x \le 7 \\ \quad\;\; y \le 8 \\ \quad\;\; x \ge 0 \\ \quad\;\; y \ge 0 \\ \;x + y \ge 5 \end{cases}$

**34.** $\begin{cases} \;x - \;\;y \ge -7 \\ \;x - 2y \le \;\;4 \\ 3x + \;\;y \ge 19 \end{cases}$

**35.** Roy owns two stores. In 1987 the first store earned 10% of the investment in it, while the second store lost 5%, for a net gain of $47,500. In 1988 the stores earned 8% and 6%, respectively, for a net gain of $93,000. What is each store worth?

**36.** A mixture of 8 pounds of grade A coffee mixed with 10 pounds of grade B coffee is worth $3.50 a pound, and 14 pounds of grade A mixed with 4 pounds of grade B is worth $3.92 a pound. What is the price of each grade of coffee?

**37.** The wages per day of fifteen adults and eight teenagers amount to $1336. Two adults together earn $48 more per day than three teenagers. What are the wages of each adult and each teenager?

**38.** Six pounds of tea and fourteen pounds of coffee cost $77.40. If the price of coffee should decrease by 10% and that of tea should increase by 10%, the cost of the quantities would be $75.06. What is the price of each per pound?

**39.** If Kris works 2 days and Pete works 5 days on a certain job they can complete three eighths of the job. If Kris works 6 more days and Pete works 5 more days, they can finish the job. How long would it take either one to do the job alone?

**40.** A seamstress paid $99 for 3 yards of white, 5 yards of blue, and 2 yards of red material. Another seamstress paid $107 for 2 yards of white, 3 yards of blue, and 4 yards of red material. The first seamstress needed more material later and paid $13 for $\frac{1}{4}$ yard of white, $\frac{1}{3}$ yard of blue, and $\frac{1}{2}$ yard of red. Find the price per yard of each kind of material.

**41.** A company manufactures two types of rowing machines. Each deluxe model takes 4 hours to assemble and 2 hours to paint, while each standard model takes 3 hours to assemble and 1 hour and 12 minutes to paint. The company employs 78 assemblers and 36 painters each working 8 hours a day. Let $x$ and $y$ denote the respective daily production of deluxe and standard models, respectively. Sketch the graph of the set of possible ordered pairs $(x, y)$.

**42.** A company makes two models of television sets. The deluxe model takes 6 hours to assemble and 20 minutes to test while the standard model takes 4 hours to assemble and 12 minutes to test. The plant is designed in such a way that a maximum of 19,000 hours are available each month for assembly and a maximum of 1000 hours are available per month for testing. The company has accepted orders totaling 1250 deluxe models and 1800 standard models per month. Let $x$ and $y$ denote the numbers of deluxe and standard models, respectively, the company manufactures per month. Sketch the graph of the set of possible ordered pairs $(x, y)$.

**43.** Let $D_x$ and $S_x$ denote the numbers of pounds of gourmet chocolate demanded and supplied, respectively, when the price of gourmet chocolate is $p_x$ dollars. Let $D_y$ and $S_y$ denote the numbers of pounds of regular chocolate demanded and supplied, respectively, when the price of regular chocolate is $p_y$ dollars per pound. It is known that these quantities are interrelated as follows:

$$D_x = 387,500 - 50,000p_x + 25,000p_y$$
$$S_x = -137,500 + 75,000p_x - 12,500p_y$$
$$D_y = 556,250 + 20,000p_x - 62,500p_y$$
$$S_y = -257,500 - 5,000p_x + 312,500p_y$$

  **a.** Find the equilibrium prices per pound of gourmet and regular chocolate.
  **b.** Find the equilibrium quantities of gourmet and regular chocolate.

# Matrices

3

It is frequently useful to arrange a collection of objects, or numbers, in a rectangular array. For example, an instructor may have a seating chart on which the rows and columns of the classroom are represented. If the student occupies a seat located in the fifth row from the front and the seventh column from the left wall, then the student's name would appear in row 5, column 7 of the chart. A used-car dealer may keep a record of his inventory in a similarly compact way. He may draw a chart on which the first column is for Chevrolets, the second column for Fords, the third column for Dodges, and so on. He reserves the first row for 1978 models, the second row for 1979 models, and the third row for 1980 models. If he enters a 5 in row 2, column 3 and a 0 in row 1, column 2, we know that he has five 1979 Dodges and no 1978 Fords.

> **DEFINITION 3.1:**   In general, we define a *matrix* (plural *matrices*) to be a rectangular array of numbers, enclosed within brackets.

## 3.1  Matrix Notation and Matrix Arithmetic

Often matrices are denoted by boldface capital letters. For example we may use $\mathbf{A}$, $\mathbf{B}$, $\mathbf{S}$, $\mathbf{T}$, . . . to denote matrices. Since a matrix is a rectangular array of numbers, each of these numbers has a certain position in the array. If a number is in row $i$ and column $j$ of matrix $\mathbf{A}$, it is called the *i-j entry* (element) of the matrix and it is denoted $a_{ij}$ (read: "$a$ sub $i,j$"). The first subscript always indicates the number of the (horizontal) row, while the second subscript always gives the (vertical) column number. We use a comma if either $i$ or $j$ has more than one digit. For example, if an entry of matrix $\mathbf{B}$ is in row 25, column 7, we denote it by $b_{25,7}$ (without the comma $b_{257}$ would be ambiguous).

EXAMPLE 1   Let

$$\mathbf{B} = \begin{bmatrix} -1 & 5 & 8 & -7 \\ 2 & 3 & 5 & -4 \\ 6 & 7 & -3 & 13 \end{bmatrix}$$

Find $b_{13}$ and $b_{34}$.

Solution   Since $b_{13}$ is the element in row 1, column 3, we have $b_{13} = 8$. Similarly, $b_{34}$ is the entry in row 3, column 4. Thus, $b_{34} = 13$.   ∎

### Special Matrices

If a matrix has $m$ rows and $n$ columns, it is said to be of *dimension* $m \times n$ (read: "$m$ by $n$").

**DEFINITION 3.2:**    A $1 \times n$ matrix is called an *n-dimensional row vector*, while an $n \times 1$ matrix is called an *n-dimensional column vector*. A matrix of dimension $m \times m$ is called a *square matrix*. We often use the notation $[a_{ij}]_{m \times n}$ to denote an $m \times n$ matrix.

**EXAMPLE 2**    Find $\mathbf{B} = [b_{ij}]_{2 \times 3}$ if $b_{ij} = 3i - 2j + 1$.

**Solution**    The matrix $\mathbf{B}$ has two rows and three columns. Hence, there will be six entries that can be found using the formula $b_{ij} = 3i - 2j + 1$. To find $b_{11}$ (the element in row 1, column 1), we replace both $i$ and $j$ by 1 and obtain $b_{11} = 2$ since $3(1) - 2(1) + 1 = 2$. Similarly, we find

$$b_{12} = 3(1) - 2(2) + 1 = 0$$
$$b_{13} = 3(1) - 2(3) + 1 = -2$$
$$b_{21} = 3(2) - 2(1) + 1 = 5$$
$$b_{22} = 3(2) - 2(2) + 1 = 3$$
$$b_{23} = 3(2) - 2(3) + 1 = 1$$

Thus,

$$\mathbf{B} = \begin{bmatrix} b_{11} & b_{12} & b_{13} \\ b_{21} & b_{22} & b_{23} \end{bmatrix} = \begin{bmatrix} 2 & 0 & -2 \\ 5 & 3 & 1 \end{bmatrix}$$

**Do Exercise 3.**

**DEFINITION 3.3:**    The *diagonal elements* of a square matrix are those elements $a_{ij}$ for which $i = j$. A square matrix for which all diagonal elements are equal and all other elements are 0 is called a *scalar matrix*.

For example,

$$\begin{bmatrix} 3 & 0 & 0 & 0 \\ 0 & 3 & 0 & 0 \\ 0 & 0 & 3 & 0 \\ 0 & 0 & 0 & 3 \end{bmatrix}$$

is a scalar matrix.

**DEFINITION 3.4:**    If all diagonal elements of a scalar matrix are 1, the matrix is called a *multiplicative identity*.

For example,

$$\begin{bmatrix} 1 & 0 & 0 \\ 0 & 1 & 0 \\ 0 & 0 & 1 \end{bmatrix}$$

is the multiplicative identity matrix of order 3 and is often denoted $I_3$.

> **DEFINITION 3.5:**   If all entries of a matrix are 0, we say that the matrix is a *zero matrix* and denote it **0**. A zero matrix is frequently called an *additive identity*.

Finally, two matrices are said to be *equal* if, and only if, they have the same dimensions and the corresponding elements are equal. That is, $[a_{ij}]_{m \times n} = [b_{ij}]_{p \times q}$ if, and only if, $m = p$, $n = q$, and $a_{ij} = b_{ij}$ for all $i$ and $j$ where $1 \le i \le m$ and $1 \le j \le n$. When we make a statement about matrix elements $a_{ij}$ and we add "for all $i,j$," we specifically mean for all integers $i$ and $j$ where $1 \le i \le m$ and $1 \le j \le n$, where $m$ is the number of rows and $n$ is the number of columns.

**EXAMPLE 3**   Find $x$, $y$, $z$, and $w$ if

$$\begin{bmatrix} 2 & 3y \\ 3z & 5 \\ 6 & 4w \end{bmatrix} = \begin{bmatrix} 4x & 9 \\ 15 & 5 \\ 6 & 16 \end{bmatrix}$$

**Solution**   Both matrices have dimensions $3 \times 2$. We need only have equality between corresponding elements. Thus, we must have $2 = 4x$, $3y = 9$, $3z = 15$, $5 = 5$, $6 = 6$, and $4w = 16$. The fourth and fifth equalities are obviously true. The others will be true if $x = \frac{1}{2}$, $y = 3$, $z = 5$, and $w = 4$.
**Do Exercise 23.**                                                                          ■

## Matrix Arithmetic

It is often useful to multiply each entry of a matrix by a fixed number as illustrated in the following example.

**EXAMPLE 4**   A gas station owner uses matrices to keep a record of the number of gallons of gasoline he sells each month. He owns two stations and records the number of gallons sold at station A in the first row and those sold at station B in the second row. He records the number of gallons of regular, unleaded, and premium unleaded each month in columns 1, 2, and 3, respectively. The following matrix is the record for June 1990.

| Station | Regular | Unleaded | Premium unleaded |
|---------|---------|----------|------------------|
| A | 12,000 | 25,500 | 10,700 |
| B | 15,000 | 32,000 | 11,250 |

**a.** How many gallons of unleaded gasoline were sold at station B?

**b.** If his goal is to double the sales of all three kinds of gasoline at both stations, what will the goal matrix be for the next month?

**Solution**   **a.** Since we want to know how many gallons were sold at station B, we must look at the second row. The second column is used to record the sales of unleaded gasoline. The entry in row 2, column 2 gives us the number of gallons of unleaded gasoline sold at station B. That number is 32,000. Thus, 32,000 gallons of unleaded gasoline were sold at station B during the month of June 1990.

**b.** If the goal for the following month is to double the sales of all three kinds of gasoline at both stations, we must double every entry of the given matrix.

| Station | Regular | Unleaded | Premium unleaded |
|---------|---------|----------|------------------|
| A | 24,000 | 51,000 | 21,400 |
| B | 30,000 | 64,000 | 22,500 |

> **DEFINITION 3.6:**   Let $[a_{ij}]$ be an $m \times n$ matrix and let $c$ be any scalar (number). Then the *scalar multiple* of $[a_{ij}]$ by $c$, $c[a_{ij}]$ is the $m \times n$ matrix $[b_{ij}]$ where $b_{ij} = ca_{ij}$ for all $i,j$.

**EXAMPLE 5**   Find 3**A** if

$$\mathbf{A} = \begin{bmatrix} 2 & 5 & -3 \\ -1 & 0 & 7 \end{bmatrix}$$

**Solution**   $3\mathbf{A} = \begin{bmatrix} 3(2) & 3(5) & 3(-3) \\ 3(-1) & 3(0) & 3(7) \end{bmatrix} = \begin{bmatrix} 6 & 15 & -9 \\ -3 & 0 & 21 \end{bmatrix}$

**Do Exercise 13.**

We introduce the idea for adding and subtracting matrices in the following example.

**EXAMPLE 6**   A used-car dealer sells only Chevrolets, Fords, and Renaults. She owns a lot in Seattle and another in Portland. She has two matrices, one for the inventory at each lot. Columns 1, 2, and 3 are used for the Chevrolets, Fords, and Renaults, respectively. Rows 1, 2, 3, and 4 are used to record the 1984, 1985, 1986, and 1987 models, respectively. The matrix **S** is the record for the Seattle lot, while matrix **P** is the record for the Portland lot.

$$\mathbf{S} = \begin{bmatrix} 2 & 6 & 4 \\ 0 & 3 & 2 \\ 3 & 2 & 0 \\ 6 & 0 & 2 \end{bmatrix} \quad \text{and} \quad \mathbf{P} = \begin{bmatrix} 5 & 7 & 3 \\ 1 & 4 & 6 \\ 5 & 1 & 0 \\ 3 & 4 & 2 \end{bmatrix}$$

Find a single matrix **C** that will show the total number of cars of each kind.

**Solution** The dealer has two 1984 Chevrolets in Seattle and five in Portland. Thus, the total number of 1984 Chevrolets is seven. We shall enter a 7 as the 1-1 entry of the matrix **C**. We obtain the other entries similarly. We have

$$\mathbf{S} + \mathbf{P} = \begin{bmatrix} 2+5 & 6+7 & 4+3 \\ 0+1 & 3+4 & 2+6 \\ 3+5 & 2+1 & 0+0 \\ 6+3 & 0+4 & 2+2 \end{bmatrix} = \begin{bmatrix} 7 & 13 & 7 \\ 1 & 7 & 8 \\ 8 & 3 & 0 \\ 9 & 4 & 4 \end{bmatrix} \qquad \blacksquare$$

In general, matrices can be added and subtracted only if they have the same dimensions.

**DEFINITION 3.7:**   If $\mathbf{A} = [a_{ij}]$ and $\mathbf{B} = [b_{ij}]$, where $\mathbf{A}$ and $\mathbf{B}$ are two $m \times n$ matrices, their *sum* and *difference* are the matrices denoted $\mathbf{A} + \mathbf{B}$ and $\mathbf{A} - \mathbf{B}$, respectively. If

$$\mathbf{A} + \mathbf{B} = [s_{ij}] \text{ and } \mathbf{A} - \mathbf{B} = [d_{ij}]$$

then

$$s_{ij} = a_{ij} + b_{ij} \text{ and } d_{ij} = a_{ij} - b_{ij} \text{ for all } i,j$$

It is easy to verify that addition of matrices is commutative and associative. That is, for any three $m \times n$ matrices $\mathbf{A}$, $\mathbf{B}$, and $\mathbf{C}$, it is true that

$$\mathbf{A} + \mathbf{B} = \mathbf{B} + \mathbf{A}$$

and

$$\mathbf{A} + (\mathbf{B} + \mathbf{C}) = (\mathbf{A} + \mathbf{B}) + \mathbf{C}$$

This follows from the commutative and associative properties of addition of numbers. These properties carry over to addition of matrices because to add matrices, we simply add corresponding elements.

EXAMPLE 7   Let the matrices $\mathbf{A}$, $\mathbf{B}$, $\mathbf{C}$, and $\mathbf{D}$ be defined as follows:

$$\mathbf{A} = \begin{bmatrix} 2 & 3 & -1 \\ 5 & -2 & 6 \end{bmatrix} \quad \mathbf{B} = \begin{bmatrix} 4 & -2 & 6 \\ -7 & 0 & 3 \end{bmatrix}$$

$$\mathbf{C} = \begin{bmatrix} 3 & -4 \\ 1 & 0 \\ 3 & 4 \end{bmatrix} \quad \mathbf{D} = \begin{bmatrix} 5 & 8 \\ 9 & 6 \\ -2 & 3 \end{bmatrix}$$

Find the following:

**a.** $\mathbf{A} + \mathbf{B}$   **b.** $\mathbf{C} - \mathbf{D}$

**c.** $2\mathbf{A} + (-3)\mathbf{B}$   **d.** $\mathbf{A} + \mathbf{C}$

**Solution**    **a.** $\mathbf{A} + \mathbf{B} = \begin{bmatrix} 2 & 3 & -1 \\ 5 & -2 & 6 \end{bmatrix} + \begin{bmatrix} 4 & -2 & 6 \\ -7 & 0 & 3 \end{bmatrix}$

$= \begin{bmatrix} 2+4 & 3+(-2) & -1+6 \\ 5+(-7) & -2+0 & 6+3 \end{bmatrix}$

$= \begin{bmatrix} 6 & 1 & 5 \\ -2 & -2 & 9 \end{bmatrix}$

**b.** $\mathbf{C} - \mathbf{D} = \begin{bmatrix} 3 & -4 \\ 1 & 0 \\ 3 & 4 \end{bmatrix} - \begin{bmatrix} 5 & 8 \\ 9 & 6 \\ -2 & 3 \end{bmatrix}$

$= \begin{bmatrix} 3-5 & -4-8 \\ 1-9 & 0-6 \\ 3-(-2) & 4-3 \end{bmatrix} = \begin{bmatrix} -2 & -12 \\ -8 & -6 \\ 5 & 1 \end{bmatrix}$

**c.** $2\mathbf{A} + (-3)\mathbf{B} = 2\begin{bmatrix} 2 & 3 & -1 \\ 5 & -2 & 6 \end{bmatrix} + (-3)\begin{bmatrix} 4 & -2 & 6 \\ -7 & 0 & 3 \end{bmatrix}$

$= \begin{bmatrix} 2(2) & 2(3) & 2(-1) \\ 2(5) & 2(-2) & 2(6) \end{bmatrix} + \begin{bmatrix} (-3)4 & (-3)(-2) & (-3)6 \\ (-3)(-7) & (-3)0 & (-3)3 \end{bmatrix}$

$= \begin{bmatrix} 4 & 6 & -2 \\ 10 & -4 & 12 \end{bmatrix} + \begin{bmatrix} -12 & 6 & -18 \\ 21 & 0 & -9 \end{bmatrix}$

$= \begin{bmatrix} 4+(-12) & 6+6 & -2+(-18) \\ 10+21 & -4+0 & 12+(-9) \end{bmatrix}$

$= \begin{bmatrix} -8 & 12 & -20 \\ 31 & -4 & 3 \end{bmatrix}$

**d.** $\mathbf{A} + \mathbf{C} = \begin{bmatrix} 2 & 3 & -1 \\ 5 & -2 & 6 \end{bmatrix} + \begin{bmatrix} 3 & -4 \\ 1 & 0 \\ 3 & 4 \end{bmatrix}$

The addition in part **d** cannot be performed since **A** is a $2 \times 3$ matrix, while **C** is a $3 \times 2$ matrix and matrices can be added, or subtracted, only if they have the same dimensions.

**Do Exercise 19.**

---

# Exercise Set 3.1

**1.** Give the dimensions of each of the following matrices.

$\mathbf{A} = \begin{bmatrix} 1 & 2 & -3 \\ 3 & 0 & 9 \end{bmatrix}$, $\mathbf{B} = \begin{bmatrix} 2 & 7 \\ 4 & -6 \\ 3 & 5 \end{bmatrix}$,

$\mathbf{C} = \begin{bmatrix} 5 & 2 & 4 & 0 \\ 4 & 0 & -3 & 7 \\ 2 & 5 & 9 & 0 \end{bmatrix}$

**2.** For the matrices of Exercise 1, find:

   **a.** $a_{11}, a_{12}, a_{22}$, and $a_{23}$ if $\mathbf{A} = [a_{ij}]$
   **b.** $b_{12}, b_{21}, b_{22}$, and $b_{32}$ if $\mathbf{B} = [b_{ij}]$
   **c.** $c_{23}, c_{24}, c_{31}$, and $c_{34}$ if $\mathbf{C} = [c_{ij}]$

**3.** Write the $2 \times 3$ matrix $[a_{ij}]$ if $a_{ij} = 3i + 2j$ for all $i,j$.

**4.** Write the $3 \times 2$ matrix $[b_{ij}]$ if $b_{ij} = 5i - 3j$ for all $i,j$.

**5.** Write the $3 \times 3$ matrix $[c_{ij}]$ if $c_{ij} = i^2 + 2j$ for all $i,j$.

**6.** Write the $4 \times 3$ matrix $[d_{ij}]$ if $d_{ij} = 3i + j^2$ for all $i,j$.

**7.** Write the $4 \times 4$ matrix $[a_{ij}]$ if

$$a_{ij} = \begin{cases} 0 & \text{whenever } i = j \\ 2i + j & \text{whenever } i \neq j. \end{cases}$$

**8.** Write the $4 \times 4$ matrix $[b_{ij}]$ if

$$b_{ij} = \begin{cases} 0 & \text{whenever } i = j \\ i + 2j & \text{whenever } i \neq j. \end{cases}$$

**9.** Write the $3 \times 4$ matrix $[c_{ij}]$ if

$$c_{ij} = \begin{cases} i^2 & \text{whenever } i = j \\ 3i + j & \text{whenever } i \neq j. \end{cases}$$

**10.** Write the $4 \times 3$ matrix $[d_{ij}]$ if

$$d_{ij} = \begin{cases} j^3 & \text{whenever } i = j \\ i + 5j & \text{whenever } i \neq j. \end{cases}$$

**11.** Write the $4 \times 4$ matrix $[a_{ij}]$ if

$$a_{ij} = \begin{cases} (-1)^j & \text{whenever } i = j \\ 2i & \text{whenever } i \neq j. \end{cases}$$

*In Exercises 12–20, perform the indicated operations. If an operation is not possible, state why.*

**12.** $4 \begin{bmatrix} 2 & 5 \\ 7 & 0 \end{bmatrix}$

**13.** $-3 \begin{bmatrix} 3 & 4 & -2 \\ 2 & -4 & 0 \\ 7 & 6 & 12 \end{bmatrix}$

**14.** $6 \begin{bmatrix} 3 & 7 & -2 & 8 \\ 5 & -3 & 9 & 0 \end{bmatrix}$

**15.** $\begin{bmatrix} 2 & -4 & 5 & -7 \\ 3 & 6 & -2 & 9 \end{bmatrix} + \begin{bmatrix} 0 & 0 & -1 \\ -3 & 5 & 9 \end{bmatrix}$

**16.** $\begin{bmatrix} 4 & 2 \\ 2 & -1 \end{bmatrix} - \begin{bmatrix} 3 & -5 \\ 1 & -3 \end{bmatrix}$

**17.** $2 \begin{bmatrix} 3 & 2 \\ 9 & 0 \\ 4 & 2 \\ 6 & 5 \end{bmatrix} + 3 \begin{bmatrix} 8 & 5 \\ -1 & 3 \\ 3 & -2 \\ -3 & 0 \end{bmatrix}$

**18.** $(-2) \begin{bmatrix} 5 & 7 & 3 \\ 4 & 1 & 6 \end{bmatrix} + 4 \begin{bmatrix} 4 & -9 \\ 3 & 7 \end{bmatrix}$

**19.** $5 \begin{bmatrix} 3 & 8 & 1 \\ 5 & 1 & -2 \\ 9 & 2 & 3 \end{bmatrix} - 2 \begin{bmatrix} 2 & -9 & 4 \\ 0 & 3 & 7 \\ 5 & 0 & 2 \end{bmatrix}$

**20.** $4 \begin{bmatrix} 5 & 4 \\ 2 & 9 \\ 1 & 0 \\ 5 & 3 \end{bmatrix} + 2 \begin{bmatrix} 2 & -3 \\ 3 & 6 \\ 4 & 2 \\ 4 & 2 \end{bmatrix} - 3 \begin{bmatrix} 6 & 1 \\ 5 & -2 \\ 0 & 3 \\ 6 & 1 \end{bmatrix}$

*In Exercises 21–28, find the values of the variables that make the given matrix equations true statements.*

**21.** $\begin{bmatrix} 3 & x \\ z & 6 \end{bmatrix} = \begin{bmatrix} y & 5 \\ 7 & 6 \end{bmatrix}$

**22.** $\begin{bmatrix} 5 + x & 6 + 2y \\ z - 3 & w + 10 \end{bmatrix} = \begin{bmatrix} 2x + 2 & 3y + 4 \\ 3z - 13 & 3w + 12 \end{bmatrix}$

**23.** $\begin{bmatrix} 2 - x & 3 + y \\ 5 & 2 + z \\ 4w & 7 \end{bmatrix} = \begin{bmatrix} 3x - 2 & 2y + 1 \\ 5 & 3z - 6 \\ 2w + 6 & 7 \end{bmatrix}$

**24.** $\begin{bmatrix} 5 & 2x \\ 2z & 5 \end{bmatrix} + \begin{bmatrix} 3y & 2 \\ 3 & -w \end{bmatrix} = \begin{bmatrix} 11 & 0 \\ 3z + 6 & 3 \end{bmatrix}$

**25.** $\begin{bmatrix} x + y & 2 \\ 5 & 2x + y \end{bmatrix} + \begin{bmatrix} x + 2y & 1 \\ 2 & x - 3y \end{bmatrix}$
$= \begin{bmatrix} 5 & 3 \\ 7 & -12 \end{bmatrix}$

**26.** $\begin{bmatrix} 4x + y & 3x \\ 4 & 3y + 4 \end{bmatrix} + \begin{bmatrix} 3x - 1 & 3x \\ 2 & 2y + 15 \end{bmatrix} =$
$\begin{bmatrix} 2x - 2y & 6x \\ 6 & 2x \end{bmatrix}$

**27.** $\begin{bmatrix} x + y & 2 & 1 \\ 3 & 2x - y & 2 \\ 5 & 3y & 2y - 2z \end{bmatrix}$
$+ \begin{bmatrix} x + z & 3 & -1 \\ 4 & x - y & 3 \\ 1 & y & y \end{bmatrix}$
$= \begin{bmatrix} 7 - 2z & 5 & 0 \\ 7 & -z & 5 \\ 6 & 4y & 3 + x \end{bmatrix}$

**28.** $\begin{bmatrix} 2x + y & 1 \\ 3y & x + y \\ x & 7 \end{bmatrix} + \begin{bmatrix} 3x + y & 2 \\ 2y & 2 + x \\ z - 6 & 1 \end{bmatrix}$
$= \begin{bmatrix} 4x - 3z + 5 & 3 \\ 5y & 1 + z \\ y - z & 8 \end{bmatrix}$

**29.** A used-car dealer uses matrices to keep track of his inventory. He uses columns 1, 2, 3, and 4 for Fords, Chevrolets, Dodges, and Buicks, respectively; and rows 1, 2, 3, 4, and 5 for 1983, 1984, 1985, 1986, and 1987 models, respectively. He owns two lots. The following matrices show his inventory on these lots.

$$\begin{bmatrix} 1 & 0 & 4 & 6 \\ 3 & 2 & 1 & 4 \\ 4 & 0 & 2 & 4 \\ 0 & 3 & 1 & 5 \\ 2 & 4 & 1 & 0 \end{bmatrix}$$

$$\begin{bmatrix} 0 & 2 & 5 & 3 \\ 5 & 0 & 4 & 2 \\ 2 & 5 & 0 & 7 \\ 0 & 3 & 5 & 2 \\ 5 & 3 & 1 & 0 \end{bmatrix}$$

Write a single matrix giving his total inventory.

**30.** Four racquetball players decide to hold a round-robin tournament during the summer. They each play the other three players three games in each of the two halves of the tournament and use matrices to keep track of the results. In the following two matrices, $a_{ij}$ is the number of games player $i$ won against player $j$ in each half of the tournament, respectively.

$$\begin{bmatrix} 0 & 2 & 1 & 3 \\ 1 & 0 & 2 & 1 \\ 2 & 1 & 0 & 1 \\ 0 & 2 & 2 & 0 \end{bmatrix}$$

$$\begin{bmatrix} 0 & 1 & 3 & 1 \\ 2 & 0 & 3 & 1 \\ 0 & 0 & 0 & 2 \\ 2 & 2 & 1 & 0 \end{bmatrix}$$

**a.** Why is it that in both matrices, $a_{ij} + a_{ji} = 3$ for all $i,j$ with $i \neq j$?
**b.** Why is it that in both matrices $a_{ii} = 0$ for each $i$?
**c.** Write a single matrix showing the record for the tournament.
**d.** Which player won the most games for the whole tournament?

**31.** Same as Exercise 30, this time with five players and the following two matrices.

$$\begin{bmatrix} 0 & 2 & 2 & 1 & 3 \\ 1 & 0 & 3 & 1 & 2 \\ 1 & 0 & 0 & 2 & 0 \\ 2 & 2 & 1 & 0 & 2 \\ 0 & 1 & 3 & 1 & 0 \end{bmatrix}$$

$$\begin{bmatrix} 0 & 3 & 1 & 2 & 1 \\ 0 & 0 & 2 & 1 & 3 \\ 2 & 1 & 0 & 3 & 0 \\ 1 & 2 & 0 & 0 & 1 \\ 2 & 0 & 3 & 2 & 0 \end{bmatrix}$$

**32.** A manufacturer has plants at four different locations and warehouses at three different locations. In the following matrix, $a_{ij}$ gives the cost (in dollars) of transporting a unit of the product from plant $i$ to warehouse $j$.

$$\begin{bmatrix} 5 & 4 & 6 \\ 2 & 6 & 4 \\ 4 & 2 & 8 \\ 6 & 5 & 2 \end{bmatrix}$$

If the cost is increased uniformly by 20%, what is the corresponding matrix?

**33.** A sporting goods company has a plant in Montreal and another in Chicago. At each plant it manufactures rowing machines and exercise cycles. It has three models A, B, and C of each. The following two matrices give the numbers (in thousands) of each machine of each type manufactured in 1990, where in each matrix rows 1 and 2 are used for the rowing machines and exercise cycles, respectively, and columns 1, 2, and 3 are used for models A, B, and C, respectively.

$$\mathbf{M} = \begin{bmatrix} 40 & 24 & 32 \\ 36 & 16 & 20 \end{bmatrix}$$

$$\mathbf{C} = \begin{bmatrix} 20 & 15 & 25 \\ 35 & 30 & 20 \end{bmatrix}$$

**a.** Write a single matrix showing the total production for both plants in 1990.
**b.** If production is increased by 25% in the Montreal plant (matrix **M**) and by 20% in the Chicago plant (matrix **C**),

what is the matrix for total production for the following year?

34. The following matrices give the enrollment figures at some university in the fall, winter, and spring of 1990. Columns 1, 2, 3, and 4 are used for the number (in hundreds) of freshmen, sophomores, juniors, and seniors, respectively. Rows 1, 2, 3, 4, and 5 are used to record the number of business, engineering, arts and sciences, education, and nursing majors, respectively.

$$\begin{bmatrix} 2 & 2.1 & 2.5 & 2.3 \\ 2 & 2.2 & 2.3 & 2.2 \\ 3 & 2.9 & 2.8 & 2.7 \\ 1 & 1.1 & 1 & 1.2 \\ 2 & 1.9 & 2.1 & 2.1 \end{bmatrix}$$

$$\begin{bmatrix} 2.3 & 2 & 2.2 & 2.4 \\ 1.9 & 2.1 & 2.1 & 1.9 \\ 2.9 & 2.7 & 2.9 & 2.8 \\ 1.1 & 1 & 1.2 & 1.1 \\ 2.1 & 2 & 2.2 & 2.1 \end{bmatrix}$$

$$\begin{bmatrix} 2 & 1.9 & 2.1 & 2.2 \\ 2.1 & 2.3 & 1.9 & 2.2 \\ 2.8 & 2.6 & 3 & 2.6 \\ 1.2 & 1.2 & 1.1 & 1 \\ 2.2 & 2.1 & 2.3 & 2.1 \end{bmatrix}$$

Write a matrix that gives the average enrollment per class and per major for the year 1990.

# 3.2  Matrix Multiplication

In the preceding section, you learned to add and subtract matrices that have the same dimensions. You also learned to multiply a matrix by a scalar. In this section, you will learn about the multiplication of matrices. We begin with the simpler case where both factors are vectors. (Recall from Definition 3.2 that a $1 \times n$ matrix is called an $n$-dimensional row vector and an $n \times 1$ matrix is called an $n$-dimensional column vector.) To help you understand the definition, we begin with the following example.

EXAMPLE 1   The Essers are considering having their daughter attend Seattle University in Fall 1992. They obtain some information in the following tabular form.

| Cost (in dollars) | | |
|---|---|---|
| Tuition per credit | Living expenses per month | Average price of a book |
| 150 | 460 | 35 |

The number of items they will have to pay for are given in the following table.

| Number of credits | Number of months | Number of books |
|---|---|---|
| 45 | 9 | 15 |

a. Represent the item costs by a matrix.
b. Represent the numbers of each item needed by a matrix.
c. Find the estimated total cost for the academic year.

**Solution**  **a.** As in the table, the first, second, and third columns are used to represent the cost per credit hour, the average monthly living expenses, and average cost per book, respectively. We obtain the matrix

$$[\ 150 \quad 460 \quad 35\ ]$$

The foregoing is a 3-dimensional row vector.

**b.** Again, if we use the first, second, and third columns to represent the number of credits needed, the number of months, and the number of books needed, we get the matrix

$$[\ 45 \quad 9 \quad 15\ ]$$

This matrix is a 3-dimensional row vector also.

**c.** The total cost of tuition is \$6750 since $(150)(45) = 6750$. The estimated cost of living expenses is \$4140 since $(460)(9) = 4140$. The estimated cost of books is \$525 since $(35)(15) = 525$. Hence, the estimated total cost is \$11,415 because $6750 + 4140 + 525 = 11,415$.

$$(150)(45) + (460)(9) + (35)(15) = 11,415 \qquad ■$$

We see that the total cost is determined by multiplying the first entry of one vector by the first entry of the other vector, the second entry of one vector by the second entry of the other, and the third entry of one vector by the third entry of the other and then adding the three products. This is an example of the useful dot product of vectors, which we now define.

## Dot Product of Vectors

> **DEFINITION 3.8:**   Suppose that **A** and **B** are two $n$-dimensional vectors. If we multiply the first entry of **A** by the first entry of **B**, then we multiply the second entry of **A** by the second entry of **B**, and if we continue this process until we have multiplied the $n$th entry of **A** by the $n$th entry of **B**, and if we finally add the $n$ products, the final result is called the *dot product* (*inner product*) of the two vectors and is denoted **A** · **B**.

The two vectors *must* have the same dimensions. However, both can be row vectors ($1 \times n$ matrices), column vectors ($n \times 1$ matrices), or one of each.

**EXAMPLE 2**

Let $\mathbf{A} = [\ 2 \quad 3 \quad -4 \quad 0\ ]$, $\mathbf{B} = [\ -2 \quad 3 \quad 2 \quad 4\ ]$, and $\mathbf{C} = \begin{bmatrix} 3 \\ 5 \\ 2 \\ -1 \end{bmatrix}$

**a.** Find **A** · **B**.
**b.** Find **A** · **C**.

**Solution**   **a.** $\mathbf{A} \cdot \mathbf{B} = [\; 2 \quad 3 \quad -4 \quad 0\;] \cdot [\; -2 \quad 3 \quad 2 \quad 4\;]$

$$= 2(-2) + 3(3) + (-4)2 + 0(4) = -3$$

The dot product $\mathbf{A} \cdot \mathbf{B}$ is $-3$.

**b.** $\mathbf{A} \cdot \mathbf{C} = [2 \quad 3 \quad -4 \quad 0] \cdot \begin{bmatrix} 3 \\ 5 \\ 2 \\ -1 \end{bmatrix}$

$$= 2(3) + 3(5) + (-4)2 + 0(-1) = 13$$

The inner product $\mathbf{A} \cdot \mathbf{C}$ is 13. ■

## Matrix Multiplication

In the preceding section, we saw that if we know the dimensions of a matrix and if we can find each of its entries, then we can write the matrix. We begin our description of matrix multiplication by simply stating that if $\mathbf{A}$ and $\mathbf{B}$ are two matrices whose product is a matrix $\mathbf{C}$, then $c_{ij}$ (the $i$-$j$ entry of the product) is the dot product of the $i$th row of $\mathbf{A}$ by the $j$th column of $\mathbf{B}$. Therefore, we can find $c_{ij}$ only if the $i$th row of $\mathbf{A}$ and the $j$th column of $\mathbf{B}$ are vectors with the same dimensions. That is, $c_{ij}$ can be found only if the number of columns of $\mathbf{A}$ is the same as the number of rows of $\mathbf{B}$. In that case, $\mathbf{A}$ is said to be *conformable to* $\mathbf{B}$ for multiplication. Before stating the definition, we give the following example.

**EXAMPLE 3**   Let $\mathbf{A} = \begin{bmatrix} 2 & -1 & 3 \\ 4 & 0 & -5 \end{bmatrix}$ and $\mathbf{B} = \begin{bmatrix} 5 & -2 \\ 9 & 3 \\ 1 & 0 \end{bmatrix}$.

**a.** Verify that $\mathbf{A}$ is conformable to $\mathbf{B}$ for multiplication.
**b.** Find all possible $c_{ij}$'s where each $c_{ij}$ is the dot product of the $i$th row of $\mathbf{A}$ by the $j$th column of $\mathbf{B}$.
**c.** Arrange the results in matrix form.

**Solution**   **a.** $\mathbf{A}$ has three columns and $\mathbf{B}$ has three rows. Therefore, $\mathbf{A}$ is conformable to $\mathbf{B}$ for multiplication.
**b.** To find $c_{11}$, we calculate the dot product of row 1 of $\mathbf{A}$ by column 1 of $\mathbf{B}$. We get

$$[\; 2 \quad -1 \quad 3\;] \cdot \begin{bmatrix} 5 \\ 9 \\ 1 \end{bmatrix} = 2(5) + (-1)9 + 3(1) = 4$$

Similarly, we find $c_{12}$ by calculating the dot product of row 1 of $\mathbf{A}$ by column 2 of $\mathbf{B}$. We obtain

$$[\; 2 \quad -1 \quad 3\;] \cdot \begin{bmatrix} -2 \\ 3 \\ 0 \end{bmatrix} = 2(-2) + (-1)3 + 3(0) = -7$$

The entry $c_{21}$ is found by calculating the dot product of row 2 of $\mathbf{A}$ by column 1 of $\mathbf{B}$. We get

$$[\ 4 \quad 0 \quad -5\ ] \cdot \begin{bmatrix} 5 \\ 9 \\ 1 \end{bmatrix} = 4(5) + 0(9) + (-5)1 = 15$$

The dot product of row 2 of $\mathbf{A}$ by column 2 of $\mathbf{B}$ gives $c_{22}$.

$$[\ 4 \quad 0 \quad -5\ ] \cdot \begin{bmatrix} -2 \\ 3 \\ 0 \end{bmatrix} = 4(-2) + 0(3) + (-5)0 = -8$$

c. Arranging the $c_{ij}$'s we obtained in part **b** in matrix form we get

$$\mathbf{C} = \begin{bmatrix} c_{11} & c_{12} \\ c_{21} & c_{22} \end{bmatrix} = \begin{bmatrix} 4 & -7 \\ 15 & -8 \end{bmatrix}$$ ∎

As you will soon see, this matrix is the product $\mathbf{AB}$. The matrix $\mathbf{C}$ found in Example 3 has 2 rows because $\mathbf{A}$ had 2 rows and we could find $c_{ij}$ for $i = 1$ and for $i = 2$. Since $\mathbf{B}$ had 2 columns, we could find $c_{ij}$ for $j = 1$ and for $j = 2$.

In general, if $\mathbf{A}$ is $m \times n$ and $\mathbf{B}$ is $n \times p$, $\mathbf{A}$ is conformable to $\mathbf{B}$ for multiplication (since the number of columns of $\mathbf{A}$ is the same as the number of rows of $\mathbf{B}$) and the product $\mathbf{AB}$ can be found. Suppose that the product is $[c_{ij}]$. For each value of $j$, $c_{ij}$ can be found for $i = 1, 2, \ldots, m$. (See Example 3.) Thus, the product has $m$ rows. Also for each value of $i$, $c_{ij}$ can be found for $j = 1, 2, \ldots, p$. Therefore, the product has $p$ columns. Formally, we have the following definition.

> **DEFINITION 3.9:**   Suppose that $\mathbf{A}$ is an $m \times n$ matrix and $\mathbf{B}$ is an $n \times p$ matrix. Then the *product* $\mathbf{AB}$ is the $m \times p$ matrix $[c_{ij}]$ where for all $i$ ($i = 1$, $2, \ldots, m$), and all $j$ ($j = 1, 2, \ldots, p$), $c_{ij}$ is the dot product of the $i$th row of $\mathbf{A}$ by the $j$th column of $\mathbf{B}$.

To test whether a matrix $\mathbf{A}$ is conformable to a matrix $\mathbf{B}$ for multiplication and to determine the dimensions of the product $\mathbf{AB}$ in the case that it is, you may find the following procedure helpful. Let $\mathbf{A}$ be an $m \times n$ matrix and $\mathbf{B}$ be a $p \times q$ matrix. Write the dimensions in a staggered form as shown.

$$m \times n$$
$$p \times q$$

These must match ($n$ and $p$).

When they do, **A** is conformable to **B** for multiplication. In that case, cross them out and get the dimensions of the product **AB**. For example, if **A** is 5 × 3 and **B** is 3 × 4, we write

$$5 \times 3$$
$$3 \times 4$$
$$\uparrow$$
These match (3 and 3).

Therefore **A** is conformable to **B** for multiplication. Cross out the 3s,

$$5 \times \cancel{3}$$
$$\cancel{3} \times 4$$

The product **AB** is 5 × 4.

EXAMPLE 4    Let **A** and **B** be the matrices of Example 3.

a. Verify that **B** is conformable to **A** for multiplication.
b. Find the product **BA**.

**Solution**    a. We write

$$3 \times 2$$
$$2 \times 3$$
$$\uparrow$$
These match.

Thus **B** is conformable to **A** for multiplication. Cross out the 2s,

$$3 \times \cancel{2}$$
$$\cancel{2} \times 3$$

The product **BA** is 3 × 3.

b. Let **BA** = $[d_{ij}]$. Then

$$d_{11} = \begin{bmatrix} 5 & -2 \end{bmatrix} \cdot \begin{bmatrix} 2 \\ 4 \end{bmatrix} = 5(2) + (-2)4 = 2$$

$$d_{12} = \begin{bmatrix} 5 & -2 \end{bmatrix} \cdot \begin{bmatrix} -1 \\ 0 \end{bmatrix} = 5(-1) + (-2)0 = -5$$

$$d_{13} = \begin{bmatrix} 5 & -2 \end{bmatrix} \cdot \begin{bmatrix} 3 \\ -5 \end{bmatrix} = 5(3) + (-2)(-5) = 25$$

$$d_{21} = \begin{bmatrix} 9 & 3 \end{bmatrix} \cdot \begin{bmatrix} 2 \\ 4 \end{bmatrix} = 9(2) + 3(4) = 30$$

$$d_{22} = \begin{bmatrix} 9 & 3 \end{bmatrix} \cdot \begin{bmatrix} -1 \\ 0 \end{bmatrix} = 9(-1) + 3(0) = -9$$

$$d_{23} = \begin{bmatrix} 9 & 3 \end{bmatrix} \cdot \begin{bmatrix} 3 \\ -5 \end{bmatrix} = 9(3) + 3(-5) = 12$$

$$d_{31} = [\ 1 \quad 0\ ] \cdot \begin{bmatrix} 2 \\ 4 \end{bmatrix} = 1(2) + 0(4) = 2$$

$$d_{32} = [\ 1 \quad 0\ ] \cdot \begin{bmatrix} -1 \\ 0 \end{bmatrix} = 1(-1) + 0(0) = -1$$

$$d_{33} = [\ 1 \quad 0\ ] \cdot \begin{bmatrix} 3 \\ -5 \end{bmatrix} = 1(3) + 0(-5) = 3$$

Thus,

$$\mathbf{BA} = \begin{bmatrix} d_{11} & d_{12} & d_{13} \\ d_{21} & d_{22} & d_{23} \\ d_{31} & d_{32} & d_{33} \end{bmatrix} = \begin{bmatrix} 2 & -5 & 25 \\ 30 & -9 & 12 \\ 2 & -1 & 3 \end{bmatrix}$$

## Matrix Multiplication Is Not Commutative

In Examples 3 and 4 we found the products $\mathbf{AB}$ and $\mathbf{BA}$, respectively, for the same two matrices $\mathbf{A}$ and $\mathbf{B}$. The products were not equal since $\mathbf{AB}$ was a $2 \times 2$ matrix, while $\mathbf{BA}$ was $3 \times 3$.

In general, if $\mathbf{A}$ is conformable to $\mathbf{B}$ for multiplication, $\mathbf{B}$ need not be conformable to $\mathbf{A}$ for multiplication. For example, if $\mathbf{A}$ is $3 \times 5$ and $\mathbf{B}$ is $5 \times 2$, then $\mathbf{A}$ is conformable to $\mathbf{B}$ but $\mathbf{B}$ is not conformable to $\mathbf{A}$. Even in cases where both $\mathbf{AB}$ and $\mathbf{BA}$ can be found and have the same dimensions, the products are not necessarily equal. That is, multiplication of matrices is not commutative. We illustrate this with another example.

**EXAMPLE 5** Find $\mathbf{AB}$ and $\mathbf{BA}$ if $\mathbf{A} = \begin{bmatrix} 2 & 5 \\ 3 & 1 \end{bmatrix}$ and $\mathbf{B} = \begin{bmatrix} 3 & 2 \\ 4 & 7 \end{bmatrix}$.

**Solution** In exactly the same way as we performed multiplication in the preceding examples, we get

$$\mathbf{AB} = \begin{bmatrix} 2 & 5 \\ 3 & 1 \end{bmatrix} \cdot \begin{bmatrix} 3 & 2 \\ 4 & 7 \end{bmatrix} = \begin{bmatrix} 26 & 39 \\ 13 & 13 \end{bmatrix}$$

and

$$\mathbf{BA} = \begin{bmatrix} 3 & 2 \\ 4 & 7 \end{bmatrix} \cdot \begin{bmatrix} 2 & 5 \\ 3 & 1 \end{bmatrix} = \begin{bmatrix} 12 & 17 \\ 29 & 27 \end{bmatrix}$$

For practice, you will find it valuable to carry out the details of the multiplication in this example. Note that $\mathbf{AB} \neq \mathbf{BA}$, even though both are $2 \times 2$ matrices. However, the statement $\mathbf{AB} \neq \mathbf{BA}$ does not mean that we can never have equality. Consider the following example.

**EXAMPLE 6** A student and his girlfriend are discussing the fact that matrix multiplication is not commutative. The student says that he randomly wrote many $2 \times 2$ matrices and multiplied them in different order, but he never got the same answer. He concluded that the equality $\mathbf{AB} = \mathbf{BA}$ could never be obtained. His clever girlfriend bet him a lunch that she could get this equality as often as she pleased. She won the bet. Explain how she did it.

**Solution**  She probably started with an arbitrary square matrix, say $\mathbf{A} = \begin{bmatrix} 2 & 5 \\ 4 & 1 \end{bmatrix}$. Then, let

$$\mathbf{B} = \mathbf{A}^2 = \begin{bmatrix} 2 & 5 \\ 4 & 1 \end{bmatrix} \cdot \begin{bmatrix} 2 & 5 \\ 4 & 1 \end{bmatrix} = \begin{bmatrix} 4+20 & 10+5 \\ 8+4 & 20+1 \end{bmatrix} = \begin{bmatrix} 24 & 15 \\ 12 & 21 \end{bmatrix}$$

Now,

$$\mathbf{AB} = \begin{bmatrix} 2 & 5 \\ 4 & 1 \end{bmatrix} \cdot \begin{bmatrix} 24 & 15 \\ 12 & 21 \end{bmatrix} = \begin{bmatrix} 48+60 & 30+105 \\ 96+12 & 60+21 \end{bmatrix} = \begin{bmatrix} 108 & 135 \\ 108 & 81 \end{bmatrix}$$

Also,

$$\mathbf{BA} = \begin{bmatrix} 24 & 15 \\ 12 & 21 \end{bmatrix} \cdot \begin{bmatrix} 2 & 5 \\ 4 & 1 \end{bmatrix} = \begin{bmatrix} 48+60 & 120+15 \\ 24+84 & 60+21 \end{bmatrix} = \begin{bmatrix} 108 & 135 \\ 108 & 81 \end{bmatrix}$$

We see that $\mathbf{AB} = \mathbf{BA}$. This is not surprising since both are equal to $\mathbf{A}^3$.*
**Do Exercise 21.**  ■

## Properties of Matrix Multiplication

Another fact about matrix multiplication which is different from multiplication of numbers is that it is possible to have $\mathbf{AB} = \mathbf{0}$, even though neither $\mathbf{A}$ nor $\mathbf{B}$ is the zero matrix.

**EXAMPLE 7**  Find $\mathbf{AB}$ if

$$\mathbf{A} = \begin{bmatrix} 1 & -1 \\ -1 & 1 \end{bmatrix} \text{ and } \mathbf{B} = \begin{bmatrix} 2 & 2 \\ 2 & 2 \end{bmatrix}$$

**Solution**  We easily see that

$$\mathbf{AB} = \begin{bmatrix} 1 & -1 \\ -1 & 1 \end{bmatrix} \cdot \begin{bmatrix} 2 & 2 \\ 2 & 2 \end{bmatrix} = \begin{bmatrix} 1(2)+(-1)2 & 1(2)+(-1)2 \\ (-1)2+1(2) & (-1)2+1(2) \end{bmatrix}$$

$$= \begin{bmatrix} 0 & 0 \\ 0 & 0 \end{bmatrix}$$

Note that $\mathbf{AB} = \mathbf{0}$, although neither $\mathbf{A}$ nor $\mathbf{B}$ is a zero matrix.
**Do Exercise 23.**  ■

There are, however, some properties of matrices which are like the familiar properties of numbers.

---

*We may have $\mathbf{AB} = \mathbf{BA}$ even if each factor is not a power of the same matrix $\mathbf{C}$. (See Exercise 43.)

**Properties of Matrices**

1. Multiplication of matrices is *associative*. That is, if **A**, **B**, and **C** are $m \times n$, $n \times p$, and $p \times q$ matrices, respectively, then

$$\mathbf{A(BC)} = \mathbf{(AB)C}$$

2. Multiplication of matrices is *distributive over addition*. That is, if **A**, **B**, and **C** are matrices, **A** is $m \times n$, and both **B** and **C** are $n \times p$, then

$$\mathbf{A(B + C)} = \mathbf{AB} + \mathbf{AC}$$

In the preceding section, we defined the multiplicative identity of order $n$ which we denoted $\mathbf{I}_n$. We now state a third property of matrix multiplication which justifies the name of the matrix $\mathbf{I}_n$.

3. If **A** is an $m \times n$ matrix, the following are true:

$$\mathbf{AI}_n = \mathbf{A} \quad \text{and} \quad \mathbf{I}_m\mathbf{A} = \mathbf{A}$$

When we multiply **A** on the right by the identity matrix, we must choose $\mathbf{I}_n$ so that **A** is conformable to $\mathbf{I}_n$. For a similar reason, we must choose $\mathbf{I}_m$ when we multiply **A** on the left.

**EXAMPLE 8**    Let $\mathbf{A} = \begin{bmatrix} 2 & 5 & -3 \\ 7 & -2 & 6 \end{bmatrix}$. Find $\mathbf{I}_2\mathbf{A}$ and $\mathbf{AI}_3$.

**Solution**    The identity matrix $\mathbf{I}_2$ is $2 \times 2$ and **A** is $2 \times 3$. Hence, $\mathbf{I}_2$ is conformable to **A** for multiplication and the product matrix $\mathbf{I}_2\mathbf{A}$ is $2 \times 3$. We get

$$
\begin{aligned}
\mathbf{I}_2\mathbf{A} &= \begin{bmatrix} 1 & 0 \\ 0 & 1 \end{bmatrix} \cdot \begin{bmatrix} 2 & 5 & -3 \\ 7 & -2 & 6 \end{bmatrix} \\
&= \begin{bmatrix} 1(2)+0(7) & 1(5)+0(-2) & 1(-3)+0(6) \\ 0(2)+1(7) & 0(5)+1(-2) & 0(-3)+1(6) \end{bmatrix} \\
&= \begin{bmatrix} 2 & 5 & -3 \\ 7 & -2 & 6 \end{bmatrix} = \mathbf{A}
\end{aligned}
$$

The matrix **A** is $2 \times 3$ and the identity matrix $\mathbf{I}_3$ is $3 \times 3$. Therefore, **A** is conformable to $\mathbf{I}_3$ for multiplication and the product matrix $\mathbf{AI}_3$ is $2 \times 3$. We obtain

$$
\begin{aligned}
\mathbf{AI}_3 &= \begin{bmatrix} 2 & 5 & -3 \\ 7 & -2 & 6 \end{bmatrix} \cdot \begin{bmatrix} 1 & 0 & 0 \\ 0 & 1 & 0 \\ 0 & 0 & 1 \end{bmatrix} \\
&= \begin{bmatrix} 2(1)+5(0)+(-3)0 & 2(0)+5(1)+(-3)0 & 2(0)+5(0)+(-3)1 \\ 7(1)+(-2)0+6(0) & 7(0)+(-2)1+6(0) & 7(0)+(-2)0+6(1) \end{bmatrix} \\
&= \begin{bmatrix} 2 & 5 & -3 \\ 7 & -2 & 6 \end{bmatrix} = \mathbf{A}
\end{aligned}
$$

## Applications

EXAMPLE 9    A manufacturer builds regular, deluxe, and super sailboats. Each comes in two models, A and B. The following table gives the quantities the manufacturer plans to build in a certain month.

|         | Regular | Deluxe | Super |
|---------|---------|--------|-------|
| Model A | 5       | 4      | 3     |
| Model B | 6       | 3      | 2     |

The material needed for each type is given by the following table in appropriately chosen units.

|         | Wood | Paint | Canvas |
|---------|------|-------|--------|
| Regular | 5    | 4     | 6      |
| Deluxe  | 6    | 5     | 10     |
| Super   | 8    | 7     | 15     |

Finally, the cost per unit (in thousands of dollars) of each of the materials used is given in tabular form.

|        | Cost per unit |
|--------|---------------|
| Wood   | 2             |
| Paint  | .3            |
| Canvas | 3             |

**a.** Represent the given tables by matrices **Q**, **M**, and **C**, respectively.
**b.** Find the product **QM** and interpret the result.
**c.** Find the matrix **(QM)C** and interpret the result.

Solution    **a.** Using the values from the tables, we have

$$Q = \begin{bmatrix} 5 & 4 & 3 \\ 6 & 3 & 2 \end{bmatrix}, \ M = \begin{bmatrix} 5 & 4 & 6 \\ 6 & 5 & 10 \\ 8 & 7 & 15 \end{bmatrix}, \text{ and } C = \begin{bmatrix} 2 \\ .3 \\ 3 \end{bmatrix}$$

**b.**

$$QM = \begin{bmatrix} 5 & 4 & 3 \\ 6 & 3 & 2 \end{bmatrix} \cdot \begin{bmatrix} 5 & 4 & 6 \\ 6 & 5 & 10 \\ 8 & 7 & 15 \end{bmatrix}$$

$$= \begin{bmatrix} 25+24+24 & 20+20+21 & 30+40+45 \\ 30+18+16 & 24+15+14 & 36+30+30 \end{bmatrix}$$

$$= \begin{bmatrix} 73 & 61 & 115 \\ 64 & 53 & 96 \end{bmatrix}$$

Note that 73 is the dot product of row 1 of matrix **Q** by column 1 of matrix **M**. Since row 1 of matrix **Q** represents the A models and column 1 of matrix **M** represents the number of units of wood, 73 represents the total number of units of wood needed to build all the A models. Similarly, 61 units of paint and 115 units of canvas are needed to build all the A models. Also, to build all the B models, 64 units of wood, 53 units of paint, and 96 units of canvas are needed.

c.

$$(\mathbf{QM})\mathbf{C} = \begin{bmatrix} 73 & 61 & 115 \\ 64 & 53 & 96 \end{bmatrix} \cdot \begin{bmatrix} 2 \\ .3 \\ 3 \end{bmatrix}$$

$$= \begin{bmatrix} 146+18.3+345 \\ 128+15.9+288 \end{bmatrix} = \begin{bmatrix} 509.3 \\ 431.9 \end{bmatrix}$$

Since the cost per unit is given in thousands of dollars, 509.3 indicates that the total cost to build all the A models is $509,300 and the total cost to build all the B models is $431,900.

**Do Exercise 37.**                                                                     ■

To illustrate another application of matrix multiplication, we introduce the *basic principle of counting*, which is discussed in greater detail in Chapter 6. Suppose that an event (such as a flight from one city to another) can occur in $p$ different ways and once it has occurred in any of these, a second event (such as a connecting flight) can occur in $q$ different ways, then the two events can occur, in that order, in $p \cdot q$ different ways.

EXAMPLE 10    The following two tables give the number of daily nonstop flights between certain cities in the United States.

| From    To → ↓ | Chicago | Kansas City | Dallas |
|---|---|---|---|
| Seattle | 5 | 3 | 2 |
| San Francisco | 6 | 4 | 7 |
| Los Angeles | 7 | 3 | 9 |

| From    To → ↓ | Boston | New York |
|---|---|---|
| Chicago | 4 | 4 |
| Kansas City | 5 | 6 |
| Dallas | 3 | 5 |

Use matrix multiplication to find the total number of flights from each of the West Coast cities to each of the East Coast cities with exactly one overnight stop in either Chicago, Kansas City, or Dallas.

**Solution**   Let **W** and **E** be matrices representing the first and second tables, respectively. Then,

$$\mathbf{W} = \begin{bmatrix} 5 & 3 & 2 \\ 6 & 4 & 7 \\ 7 & 3 & 9 \end{bmatrix} \text{ and } \mathbf{E} = \begin{bmatrix} 4 & 4 \\ 5 & 6 \\ 3 & 5 \end{bmatrix}$$

The product **WE** is a $3 \times 2$ matrix $[c_{ij}]$. We first calculate $c_{11}$ to get

$$c_{11} = 5(4) + 3(5) + 2(3) = 20 + 15 + 6 = 41$$

On any day there are five flights from Seattle to Chicago and, on the next day, four flights from Chicago to Boston. Therefore, there are 20 flights from Seattle to Boston with one overnight stop in Chicago. Similarly, we see that there are 15 flights from Seattle to Boston with one overnight stop in Kansas City and 6 flights from Seattle to Boston with one overnight stop in Dallas. It is now clear that there are 41 flights from Seattle to Boston with exactly one overnight stop in Chicago, Kansas City, or Dallas and $c_{11} = 41$.

We complete the multiplication of the two matrices as shown.

$$\mathbf{WE} = \begin{bmatrix} 5 & 3 & 2 \\ 6 & 4 & 7 \\ 7 & 3 & 9 \end{bmatrix} \cdot \begin{bmatrix} 4 & 4 \\ 5 & 6 \\ 3 & 5 \end{bmatrix}$$

$$= \begin{bmatrix} 20+15+6 & 20+18+10 \\ 24+20+21 & 24+24+35 \\ 28+15+27 & 28+18+45 \end{bmatrix} = \begin{bmatrix} 41 & 48 \\ 65 & 83 \\ 70 & 91 \end{bmatrix}$$

Using the product matrix, we can write a table that gives the number of flights between the three western cities and the two eastern cities with exactly one overnight stop in Chicago, Kansas City, or Dallas.

| From   To → ↓ | Boston | New York |
|---|---|---|
| Seattle | 41 | 48 |
| San Francisco | 65 | 83 |
| Los Angeles | 70 | 91 |

**Do Exercise 39.**

We now introduce a procedure that we shall use in greater detail later.

**EXAMPLE 11**   Consider the system $\begin{cases} 2x + y = -3 \\ 5x + 3y = 4 \end{cases}$.

**a.** Write the system as one matrix equation.
**b.** Write the left side of this equation as a product of a $2 \times 2$ matrix and the column vector $\begin{bmatrix} x \\ y \end{bmatrix}$.

**Solution** **a.** We first write $\begin{bmatrix} 2x & + & y \\ 5x & + & 3y \end{bmatrix} = \begin{bmatrix} -3 \\ 4 \end{bmatrix}$.*

Since two matrices having the same dimensions are equal if, and only if, the corresponding elements are equal, the foregoing matrix equation is clearly equivalent to the original system of two equations.

**b.** $2x + y$ is the dot product of the vectors $[\ 2 \quad 1\ ]$ and $\begin{bmatrix} x \\ y \end{bmatrix}$, while $5x + 3y$ is

the dot product of the vectors $[\ 5 \quad 3\ ]$ and $\begin{bmatrix} x \\ y \end{bmatrix}$. We can factor the matrix on

the left side of the matrix equation of part **a** to get

$$\begin{bmatrix} 2 & 1 \\ 5 & 3 \end{bmatrix} \cdot \begin{bmatrix} x \\ y \end{bmatrix} = \begin{bmatrix} -3 \\ 4 \end{bmatrix}$$

**Do Exercise 31.** ■

In the preceding example, we wrote the linear system

$$\begin{cases} 2x + y = -3 \\ 5x + 3y = 4 \end{cases}$$

as a single matrix equation $\mathbf{AX} = \mathbf{B}$, where

$$\mathbf{A} = \begin{bmatrix} 2 & 1 \\ 5 & 3 \end{bmatrix}, \mathbf{X} = \begin{bmatrix} x \\ y \end{bmatrix}, \text{ and } \mathbf{B} = \begin{bmatrix} -3 \\ 4 \end{bmatrix}$$

The matrices **A**, **X**, and **B** are called the *coefficient matrix*, the *variable vector*, and the *value vector*, respectively. (The matrices **X** and **B** are 2 × 1, or 2-dimensional column vectors.)

In general, a system of *m* linear equations in *n* variables can be written as a single matrix equation

$$\mathbf{AX} = \mathbf{B}$$

Where **A** is an *m* × *n* coefficient matrix, **X** is an *n*-dimensional column vector whose entries are the variables of the system, and **B** is an *m*-dimensional value vector whose entries are the constants appearing on the right sides of the equations of the system.

**EXAMPLE 12** Write the system as a single matrix equation.

$$\begin{cases} 3w & + 2y - 3z = 5 \\ 5x + 3y + 5z = 2 \\ 2w + 4x - 6y + 2z = 7 \end{cases}$$

*Students often mistakenly think of the matrix on the left as a 2 × 2 matrix. However, each of the expressions $2x + y$ and $5x + 3y$ represents a single quantity and therefore that matrix has only one column. It is a 2 × 1 matrix.

**Solution** The coefficient matrix is $\mathbf{A}$ where

$$\mathbf{A} = \begin{bmatrix} 3 & 0 & 2 & -3 \\ 0 & 5 & 3 & 5 \\ 2 & 4 & -6 & 2 \end{bmatrix}$$

while the variable vector $\mathbf{X}$ and value vector $\mathbf{B}$ are defined by

$$\mathbf{X} = \begin{bmatrix} w \\ x \\ y \\ z \end{bmatrix} \text{ and } \mathbf{B} = \begin{bmatrix} 5 \\ 2 \\ 7 \end{bmatrix}$$

You should verify that $\mathbf{AX} = \mathbf{B}$.*

# Exercise Set 3.2

*In Exercises 1–5, find the products AB, BA, AC, CA, BC, and CB. If any of these cannot be found, state why.*

**1.** $\mathbf{A} = \begin{bmatrix} 1 & 3 & -2 \\ 0 & 2 & 5 \\ 2 & 4 & 0 \end{bmatrix}$, $\mathbf{B} = \begin{bmatrix} 2 & 7 \\ 3 & 5 \\ 7 & 2 \end{bmatrix}$,

$\mathbf{C} = \begin{bmatrix} -1 & 2 & 1 \\ 9 & 0 & 3 \end{bmatrix}$

**2.** $\mathbf{A} = \begin{bmatrix} 0 & -1 \\ 3 & 4 \\ 9 & 3 \\ 4 & 1 \end{bmatrix}$, $\mathbf{B} = \begin{bmatrix} 3 & 5 \\ 8 & 0 \\ 2 & 5 \\ 3 & 7 \end{bmatrix}$,

$\mathbf{C} = \begin{bmatrix} 4 & 6 & 3 \\ 3 & 0 & -1 \end{bmatrix}$

**3.** $\mathbf{A} = \begin{bmatrix} 2 & 5 \\ 1 & 0 \end{bmatrix}$, $\mathbf{B} = \begin{bmatrix} 3 & 5 & -2 \\ 9 & 3 & 0 \\ 0 & -1 & 3 \end{bmatrix}$,

$\mathbf{C} = \begin{bmatrix} 2 & 5 & 1 \end{bmatrix}$

**4.** $\mathbf{A} = \begin{bmatrix} 3 & 2 & 5 \\ 0 & 1 & 0 \\ 2 & 5 & 1 \\ 3 & -1 & 0 \end{bmatrix}$, $\mathbf{B} = \begin{bmatrix} 3 & 2 & 1 \\ 5 & 3 & 0 \end{bmatrix}$,

$\mathbf{C} = \begin{bmatrix} 3 & 1 & -8 \\ 0 & 2 & 5 \\ 3 & 2 & 0 \end{bmatrix}$

**5.** $\mathbf{A} = \begin{bmatrix} 5 & 0 & 8 & 1 \\ 5 & 1 & 0 & 2 \\ 0 & 2 & 5 & 7 \end{bmatrix}$, $\mathbf{B} = \begin{bmatrix} 9 & 5 & -2 \\ 0 & 3 & 2 \end{bmatrix}$,

$\mathbf{C} = \begin{bmatrix} 5 \\ 3 \\ 6 \end{bmatrix}$

*In Exercises 6–10, perform the indicated operations.*

**6.** $\begin{bmatrix} 2 & 3 & -4 \\ 3 & 0 & 2 \end{bmatrix} \cdot \left( \begin{bmatrix} 3 & 2 \\ 4 & 0 \\ 2 & -2 \end{bmatrix} + \begin{bmatrix} 5 & -2 \\ 3 & 1 \\ 7 & 5 \end{bmatrix} \right)$

*When writing a system of $m$ linear equations in $n$ variables in the form $\mathbf{AX} = \mathbf{B}$, make sure that the system has been written in such a way that the variables are in the same order in all the equations and that blank spaces appear in place of missing variables.

**7.** $4\begin{bmatrix} 3 & 1 & 7 \\ 0 & 4 & -2 \end{bmatrix}$

$\cdot \left( 2\begin{bmatrix} 1 & 2 & -3 \\ 0 & -1 & 5 \\ 7 & 0 & 2 \end{bmatrix} + 5\begin{bmatrix} 3 & 6 & 0 \\ 3 & 0 & 1 \\ 5 & 4 & 9 \end{bmatrix} \right)$

**8.** $-4\begin{bmatrix} 5 & 3 \\ 3 & 6 \\ 4 & -3 \end{bmatrix} \cdot \left( 3\begin{bmatrix} 5 & 3 & -1 \\ 6 & 0 & 9 \end{bmatrix} \right.$

$\left. + (-4)\begin{bmatrix} 3 & 4 & -2 \\ 7 & -4 & 6 \end{bmatrix} \right)$

**9.** $\left( \begin{bmatrix} 2 & 4 \\ 4 & 6 \\ 5 & -4 \end{bmatrix} + \begin{bmatrix} 3 & 5 \\ 6 & 7 \\ 0 & 3 \end{bmatrix} \right) \cdot \begin{bmatrix} 2 & 5 & 7 \\ 4 & 2 & 3 \end{bmatrix}$

**10.** $\left( 2\begin{bmatrix} 5 & 2 & 9 \\ 7 & 4 & 3 \\ 0 & 3 & 0 \end{bmatrix} + (-2)\begin{bmatrix} 3 & 1 & 3 \\ 2 & 4 & 0 \\ 3 & 1 & 5 \end{bmatrix} \right)$

$\cdot \begin{bmatrix} 2 & 4 & 7 & 1 \\ 3 & 8 & 3 & -2 \\ 2 & 7 & 9 & -5 \end{bmatrix}$

*In Exercises 11–20, find both products **AB** and **BA**.*

**11.** $\mathbf{A} = \begin{bmatrix} 2 & 5 & -1 \\ 4 & 0 & 7 \end{bmatrix}$ and $\mathbf{B} = \begin{bmatrix} 3 & -2 \\ 0 & 4 \\ 5 & 1 \end{bmatrix}$

**12.** $\mathbf{A} = \begin{bmatrix} 5 & -2 & 4 \\ 2 & 9 & 0 \end{bmatrix}$ and $\mathbf{B} = \begin{bmatrix} 0 & 2 \\ 4 & 1 \\ 6 & -1 \end{bmatrix}$

**13.** $\mathbf{A} = \begin{bmatrix} 3 & 1 & 7 & -9 \\ 0 & 4 & 2 & 3 \end{bmatrix}$ and $\mathbf{B} = \begin{bmatrix} 3 & 5 \\ 0 & -3 \\ 3 & 7 \\ 6 & 1 \end{bmatrix}$

**14.** $\mathbf{A} = \begin{bmatrix} 2 & 4 \\ 3 & 0 \end{bmatrix}$ and $\mathbf{B} = \begin{bmatrix} 3 & 1 \\ 1 & 5 \end{bmatrix}$

**15.** $\mathbf{A} = \begin{bmatrix} 5 & 2 \\ 7 & 3 \end{bmatrix}$ and $\mathbf{B} = \begin{bmatrix} 3 & -2 \\ -7 & 5 \end{bmatrix}$

**16.** $\mathbf{A} = \begin{bmatrix} 4 & 3 \\ 7 & 1 \end{bmatrix}$ and $\mathbf{B} = \begin{bmatrix} 2 & 7 \\ 8 & 4 \end{bmatrix}$

**17.** $\mathbf{A} = \begin{bmatrix} 1 & 2 & 3 \\ 4 & 5 & 6 \\ 7 & 8 & 9 \end{bmatrix}$ and $\mathbf{B} = \begin{bmatrix} 9 & 8 & 7 \\ 6 & 5 & 4 \\ 3 & 2 & 1 \end{bmatrix}$

**18.** $\mathbf{A} = \begin{bmatrix} 5 & -5 & 4 \\ 0 & 3 & 1 \\ 2 & -1 & 2 \end{bmatrix}$ and

$\mathbf{B} = \begin{bmatrix} 7 & 6 & -17 \\ 2 & 2 & -5 \\ -6 & -5 & 15 \end{bmatrix}$

**19.** $\mathbf{A} = \begin{bmatrix} 3 & 1 & 2 \\ 0 & 7 & -2 \\ 3 & 1 & 3 \end{bmatrix}$ and $\mathbf{B} = \begin{bmatrix} 0 & 2 & 0 \\ 3 & 6 & 1 \\ 2 & 1 & -6 \end{bmatrix}$

**20.** $\mathbf{A} = \begin{bmatrix} 1 & -4 & 0 & 1 \\ 5 & 0 & -5 & 4 \\ 0 & 1 & 3 & 1 \\ 2 & 2 & -1 & 2 \end{bmatrix}$ and

$\mathbf{B} = \begin{bmatrix} -28 & 35 & 34 & -73 \\ -1 & 1 & 1 & -2 \\ -8 & 10 & 10 & -21 \\ 25 & -31 & -30 & 65 \end{bmatrix}$

**21.** Find two $2 \times 2$ matrices $\mathbf{A}$ and $\mathbf{B}$ such that $\mathbf{AB} = \mathbf{BA}$.

**22.** Find two $3 \times 3$ matrices $\mathbf{A}$ and $\mathbf{B}$ such that $\mathbf{AB} = \mathbf{BA}$.

**23.** Find a $2 \times 2$ nonzero matrix $\mathbf{A}$ such that $\mathbf{AB} = 0$, where

$\mathbf{B} = \begin{bmatrix} 6 & 10 \\ 3 & 5 \end{bmatrix}$

**24.** Same as Exercise 23 with $\mathbf{B} = \begin{bmatrix} 8 & 18 \\ 4 & 9 \end{bmatrix}$.

**25.** Let $\mathbf{A} = \begin{bmatrix} 3 & 5 \\ 2 & -1 \\ 4 & 7 \end{bmatrix}$. Verify that $\mathbf{I}_3\mathbf{A} = \mathbf{AI}_2 = \mathbf{A}$.

**26.** Let $\mathbf{A} = \begin{bmatrix} 3 & 2 & 6 & 5 \\ 2 & 4 & 1 & 0 \end{bmatrix}$. Verify that $\mathbf{I}_2\mathbf{A} = \mathbf{AI}_4 = \mathbf{A}$.

**27.** Find the products $\mathbf{AB}$ and $\mathbf{BA}$ if
$\mathbf{A} = \begin{bmatrix} 7 & 5 \\ 4 & 3 \end{bmatrix}$ and $\mathbf{B} = \begin{bmatrix} 3 & -5 \\ -4 & 7 \end{bmatrix}$.

**28.** Same as Exercise 27 with $\mathbf{A} = \begin{bmatrix} 11 & 6 \\ 9 & 5 \end{bmatrix}$ and

$\mathbf{B} = \begin{bmatrix} 5 & -6 \\ -9 & 11 \end{bmatrix}$.

**29.** Same as Exercise 27 with
$\mathbf{A} = \begin{bmatrix} 3 & -5 & 8 \\ -4 & 7 & 13 \end{bmatrix}$ and $\mathbf{B} = \begin{bmatrix} 7 & 5 \\ 4 & 3 \\ 0 & 0 \end{bmatrix}$.

**30.** Same as Exercise 27 with

$$\mathbf{A} = \begin{bmatrix} 5 & 17 & 3 \\ 2 & 7 & 5 \end{bmatrix} \text{ and } \mathbf{B} = \begin{bmatrix} 7 & -17 \\ -2 & 5 \\ 0 & 0 \end{bmatrix}.$$

*In Exercises 31–35, write the linear system as a single matrix equation* $\mathbf{AX} = \mathbf{B}$, *where* $\mathbf{A}$ *is the coefficient matrix,* $\mathbf{X}$ *is the variable vector, and* $\mathbf{B}$ *is the value vector.*

**31.** $\begin{cases} 5x + 2y = -3 \\ 7x + 3y = \phantom{-}5 \end{cases}$   **32.** $\begin{cases} 7x + 5y = -2 \\ 4x + 3y = \phantom{-}8 \end{cases}$

**33.** $\begin{cases} 11x + 6y - 7 = 0 \\ 9x + 5y + 5 = 0 \end{cases}$   **34.** $\begin{cases} 5x - 5y + 4z = \phantom{-}3 \\ z + 3y = \phantom{-}1 \\ 2x - y + 2z = -2 \end{cases}$

**35.** $\begin{cases} x + w - 4y = -2 \\ 4w + 5x - 5z = \phantom{-}3 \\ y + 3z + w = \phantom{-}5 \\ 2x + 2y - z + 2w = -1 \end{cases}$

**36.** A manufacturer builds regular, deluxe, and super sofas. Each comes in two models, A and B. The following table gives the quantities he plans to build in a certain month.

|         | Regular | Deluxe | Super |
|---------|---------|--------|-------|
| Model A | 7       | 8      | 5     |
| Model B | 4       | 9      | 3     |

The material needed for each type is given by the following table in appropriately chosen units.

|         | Wood | Upholstery | Foam |
|---------|------|------------|------|
| Regular | 3    | 6          | 9    |
| Deluxe  | 4    | 8          | 10   |
| Super   | 5    | 11         | 12   |

Finally, the cost per unit (in dollars) of each of the materials used is given in tabular form.

|            | Cost per unit |
|------------|---------------|
| Wood       | 30            |
| Upholstery | 80            |
| Foam       | 15            |

a. Represent the tables by matrices $\mathbf{Q}$, $\mathbf{M}$, and $\mathbf{C}$, respectively.
b. Find the product $\mathbf{QM}$ and interpret the result.
c. Find the matrix $(\mathbf{QM})\mathbf{C}$ and interpret the result.

**37.** A builder can purchase materials from three different suppliers, but because of transportation costs she will buy all materials from the same supplier. The prices per unit (in dollars) charged by each of the suppliers are given in the following table.

|            | Bricks | Concrete | Lumber | Glass |
|------------|--------|----------|--------|-------|
| Supplier A | 8      | 5        | 9      | 3     |
| Supplier B | 7      | 6        | 8      | 2     |
| Supplier C | 9      | 5        | 7      | 2     |

The builder has contracted four different projects. The amount of materials required for each of the projects is given in another table.

|          | Project I | Project II | Project III | Project IV |
|----------|-----------|------------|-------------|------------|
| Bricks   | 1000      | 1500       | 1300        | 1400       |
| Concrete | 700       | 950        | 875         | 910        |
| Lumber   | 2200      | 1900       | 2100        | 1875       |
| Glass    | 800       | 950        | 875         | 650        |

Represent the first table by a matrix $\mathbf{P}$ and the second table by a matrix $\mathbf{R}$.

a. Find the product $\mathbf{PR}$ and interpret the result.
b. Which supplier should the contractor use for each of the four projects?

**38.** The following two tables give the number of daily nonstop flights between certain cities in the United States.

| From  To → | Minneapolis | Denver | Houston |
|------------|-------------|--------|---------|
| Philadelphia | 3         | 4      | 5       |
| Washington   | 6         | 5      | 8       |
| Miami        | 4         | 3      | 6       |

| From  To → | Portland | San Diego |
|------------|----------|-----------|
| Minneapolis | 3       | 5         |
| Denver      | 5       | 7         |
| Houston     | 4       | 6         |

Use matrix multiplication to find the total number of flights from each of the East Coast cities to each of the West Coast cities with exactly one overnight stop in either Minneapolis, Denver, or Houston.

**39.** The following two tables give the number of daily nonstop flights between certain cities in Canada.

| From  To → | Edmonton | Winnipeg | Calgary |
|------------|----------|----------|---------|
| Quebec     | 2        | 4        | 3       |
| Montreal   | 3        | 2        | 4       |
| Toronto    | 4        | 3        | 2       |

| From    To → ↓ | Vancouver | Victoria |
|---|---|---|
| Edmonton | 3 | 2 |
| Winnipeg | 4 | 3 |
| Calgary | 5 | 1 |

Use matrix multiplication to find the total number of flights from each of the eastern cities to each of the western cities with exactly one overnight stop in either Edmonton, Winnipeg, or Calgary.

**40.** Suppose four exams are given during a semester and a professor assigns weights of .2, .2, .2, and .4 to exams 1, 2, 3, and the final, respectively. Then, if a student receives scores of 80, 75, 82, and 90 on the four exams, that student's average is 83.4 since .2(80) + .2(75) + .2(82) + .4(90) = 16 + 15 + 16.4 + 36 = 83.4. Suppose that four mathematics professors teaching the same course assign different weights to the four exams during a semester according to the following table.

| Weight assignment | Exam I | Exam II | Exam III | Final Exam |
|---|---|---|---|---|
| Professor W | .1 | .1 | .1 | .7 |
| Professor X | .2 | .2 | .2 | .4 |
| Professor Y | .1 | .2 | .3 | .4 |
| Professor Z | .15 | .15 | .15 | .55 |

Five students who took the course received the scores displayed in the following table.

| Grades | Jeremy | Shirley | Kathy | Andre | Ted |
|---|---|---|---|---|---|
| Exam I | 80 | 85 | 90 | 65 | 70 |
| Exam II | 92 | 88 | 92 | 72 | 67 |
| Exam III | 87 | 89 | 88 | 81 | 72 |
| Final Exam | 92 | 90 | 85 | 83 | 99 |

Let **W** and **S** be matrices representing tables 1 and 2, respectively. Find the product **WS** and interpret the result.

**41.** Same as Exercise 40 with the following tables.

| Weight assignment | Exam I | Exam II | Exam III | Final Exam |
|---|---|---|---|---|
| Professor X | .15 | .15 | .2 | .5 |
| Professor Y | .3 | .3 | .3 | .1 |
| Professor Z | .25 | .2 | .25 | .3 |

| Grades | Anay | Janet | Alan | Carl | Dave | Burnett |
|---|---|---|---|---|---|---|
| Exam I | 87 | 92 | 90 | 78 | 98 | 25 |
| Exam II | 90 | 95 | 85 | 89 | 79 | 45 |
| Exam III | 88 | 96 | 95 | 94 | 97 | 32 |
| Final Exam | 79 | 98 | 92 | 93 | 100 | 51 |

**\*42.** Recall from Section 2.1 that if $A = \begin{bmatrix} a & b \\ c & d \end{bmatrix}$, then the determinant of **A** is denoted det**A** and is the number $ad - bc$. Show that if **A** and **B** are two 2 × 2 matrices, then $\det(AB) = \det A \cdot \det B$.

**\*43.** Let $A = \begin{bmatrix} 2 & 1 \\ 6 & 3 \end{bmatrix}$ and $B = \begin{bmatrix} 0 & 2 \\ 12 & 2 \end{bmatrix}$.

a. Show that $AB = BA$.

b. Use the result of Exercise 42 to show that **A** and **B** are not powers of the same 2 × 2 matrix **C**.

# 3.3 Solution of Linear Systems by Row Reduction

Much of the work that was done in Chapter 2 to solve a system of $n$ linear equations in $n$ variables can be simplified using matrix notation. The ideas are essentially the same, but the equations are written in a more compact form. We need only remember that the important data in a system of linear equations are the set of coefficients on the left sides of the equations and the set of constants on the right. In a matrix, the variables and the equal signs may be omitted if we are careful to place each coefficient in the position that indicates the variable it would be attached to.

## Augmented Matrix of a Linear System

**EXAMPLE 1**   Write a matrix that represents the following system of linear equations:

$$\begin{cases} 2x + 3z = -5 \\ 3y - 4z + 6x = 2 \\ x + 3z - 5y = 7 \end{cases}$$

**Solution**   We first write the system in such a way that the variables appear in the same order in all three equations. We make sure that all three variables appear in all three equations, using a zero coefficient where a variable is missing. We get

$$\begin{cases} 2x + 0y + 3z = -5 \\ 6x + 3y - 4z = 2 \\ x - 5y + 3z = 7 \end{cases}$$

We can now write the matrix

$$\left[\begin{array}{ccc|c} 2 & 0 & 3 & -5 \\ 6 & 3 & -4 & 2 \\ 1 & -5 & 3 & 7 \end{array}\right]$$

where it is understood that the elements of the first column are the x-coefficients, the elements of the second column are the y-coefficients, those of the third column are the z-coefficients, and those of the fourth column are the constant terms.
**Do Exercise 3.**                                                                       ■

A matrix such as this is usually called the *augmented matrix of the system*. It is customary to draw a vertical line where the equal signs would appear in the system of equations. Thus, if the matrix equation representing a linear system is $\mathbf{AX} = \mathbf{B}$, the augmented matrix of the system is $\mathbf{A} \mid \mathbf{B}$.

**EXAMPLE 2**   The following is the augmented matrix of a system of linear equations in $w$, $x$, $y$, and $z$. Write the system of equations.

$$\left[\begin{array}{cccc|c} 2 & -3 & 5 & 1 & -2 \\ 0 & 1 & 0 & 3 & 4 \\ 3 & 0 & 2 & -4 & 5 \\ 2 & 5 & 1 & 3 & 0 \end{array}\right]$$

**Solution**   The first, second, third, and fourth columns give the coefficients of $w$, $x$, $y$, and $z$, respectively, while the fifth column gives the constant terms on the right sides of the equations. The system is

$$\begin{cases} 2w - 3x + 5y + z = -2 \\ x + 3z = 4 \\ 3w + 2y - 4z = 5 \\ 2w + 5x + y + 3z = 0 \end{cases}$$

**Do Exercise 7.**                                                                       ■

Recall from Chapter 2 that given a system of linear equations, we can get an equivalent linear system if we multiply both sides of an equation by a nonzero constant or add to any equation one of the other equations. We can also interchange any two equations. In the examples in Chapter 2, we used these methods, sometimes in combination, to arrive at an equivalent linear system in the form

$$\begin{cases} x & = a \\ \quad y & = b \\ \quad\quad z = c \end{cases}$$

We found that the solution of the last system, and therefore of the original system, was the ordered triple $(a, b, c)$. In the same spirit, we shall perform "row operations" on the augmented matrix until we obtain an augmented matrix in the form

$$\begin{bmatrix} 1 & 0 & 0 & a \\ 0 & 1 & 0 & b \\ 0 & 0 & 1 & c \end{bmatrix}$$

## Elementary Row Operations

The following example will help clarify the technique.

**EXAMPLE 3**    Solve the following system as we did in Chapter 2, but this time simultaneously work the same steps on the augmented matrix of the system

$$\begin{cases} 2x + 9y = \phantom{-}3 \\ x + 5y = -2 \end{cases}$$

**Solution**

| Linear System | | Augmented Matrix |
|---|---|---|
| $\begin{cases} 2x + 9y = \phantom{-}3 \\ x + 5y = -2 \end{cases}$ | Interchange the two equations (rows). | $\begin{bmatrix} 2 & 9 & 3 \\ 1 & 5 & -2 \end{bmatrix}$ |
| $\begin{cases} x + 5y = -2 \\ 2x + 9y = \phantom{-}3 \end{cases}$ | Multiply the first equation (row) by $-2$ and add to the second equation (row). | $\begin{bmatrix} 1 & 5 & -2 \\ 2 & 9 & 3 \end{bmatrix}$ |
| $\begin{cases} x + 5y = -2 \\ \phantom{x}{-y} = \phantom{-}7 \end{cases}$ | Multiply the second equation (row) by $-1$. | $\begin{bmatrix} 1 & 5 & -2 \\ 0 & -1 & 7 \end{bmatrix}$ |
| $\begin{cases} x + 5y = -2 \\ \phantom{x+5}y = -7 \end{cases}$ | Multiply the second equation (row) by $-5$ and add to the first equation (row). | $\begin{bmatrix} 1 & 5 & -2 \\ 0 & 1 & -7 \end{bmatrix}$ |
| $\begin{cases} x \phantom{+5y} = \phantom{-}33 \\ \phantom{x+5}y = -7 \end{cases}$ | | $\begin{bmatrix} 1 & 0 & 33 \\ 0 & 1 & -7 \end{bmatrix}$ |

The solution of the system is the ordered pair $(33, -7)$.
**Do Exercise 9.**    ∎

**Procedure to Solve a Linear System of *n*
Equations in *n* Variables by Row Reduction**

**Step 1.**   Write the augmented matrix for the given system. If we perform any of the following row operations on the augmented matrix, we obtain an augmented matrix for an equivalent system.

  **a.** Interchange any two rows. The symbol $R_i \leftrightarrow R_j$ means "interchange the *i*th and *j*th rows."
  **b.** Multiply any row by a nonzero constant. The symbol $kR_i \to R_i$ means "the *i*th row is multiplied by the nonzero constant *k* to obtain the new *i*th row."
  **c.** Add to any row a constant multiple of another row. The symbol $kR_i + R_j \to R_j$ means "multiply the *i*th row by the constant *k* and add this product to the *j*th row to obtain the new *j*th row."
  **d.** We may combine some of these steps. The meaning of the symbol $mR_i + nR_j \to R_j$ should now be clear.

**Step 2.**   Perform a sequence of elementary row operations on the augmented matrix for the given system until we get an $n \times (n + 1)$ augmented matrix whose first *n* columns form the identity matrix $\mathbf{I}_n$. The solution of the given system can then be read from the $(n + 1)$st column of the last augmented matrix.

   We may perform more than one row operation in each step. However, we must clearly indicate to the right of the augmented matrix which operations we are performing. This is an essential part of the solution and serves two purposes:

1. In complicated cases it is virtually impossible for anyone to follow what was done unless the proper documentation is written along side the augmented matrices.
2. If you discover a wrong solution, you will have a much easier time in locating the error if each operation has been clearly described.

The format is such that the documentation written on the right side describes what was done to the augmented matrix directly to the left to get the next augmented matrix. The next examples will familiarize you with the form.

**EXAMPLE 4**   Use row reduction to solve the system $\begin{cases} 2x + 6y = 12 \\ 6x + 21y = -3 \end{cases}$.

**Solution**   The augmented matrix for the system is

$$\left[ \begin{array}{cc|c} 2 & 6 & 12 \\ 6 & 21 & -3 \end{array} \right]$$

We now perform a sequence of elementary row operations on this matrix.

$$\left[ \begin{array}{cc|c} 2 & 6 & 12 \\ 6 & 21 & -3 \end{array} \right] \qquad \begin{array}{l} \left(\tfrac{1}{2}\right)R_1 \to R_1 \\ \left(\tfrac{1}{3}\right)R_2 \to R_2 \end{array}$$

$$\begin{bmatrix} 1 & 3 & | & 6 \\ 2 & 7 & | & -1 \end{bmatrix} \qquad (-2)R_1 + R_2 \rightarrow R_2$$

$$\begin{bmatrix} 1 & 3 & | & 6 \\ 0 & 1 & | & -13 \end{bmatrix} \qquad (-3)R_2 + R_1 \rightarrow R_1$$

$$\begin{bmatrix} 1 & 0 & | & 45 \\ 0 & 1 & | & -13 \end{bmatrix}$$

The first two columns of the last augmented matrix form the identity matrix $\mathbf{I}_2$. Hence, we are done. The solution of the original system is the ordered pair $(45, -13)$ since the last augmented matrix represents the system

$$\begin{cases} x & = & 45 \\ & y = & -13 \end{cases}$$

**Do Exercise 13.**

**EXAMPLE 5**  Solve the following system by row reduction:

$$\begin{cases} 5x - 5y + 4z = 17 \\ \qquad 3y + z = 6 \\ 2x - y + 2z = 9 \end{cases}$$

**Solution**  The augmented matrix for the system is

$$\begin{bmatrix} 5 & -5 & 4 & | & 17 \\ 0 & 3 & 1 & | & 6 \\ 2 & -1 & 2 & | & 9 \end{bmatrix}$$

We perform elementary row operations on this augmented matrix until the part of the resulting matrix to the left of the vertical bar is the identity matrix $\mathbf{I}_3$. That is, until the last augmented matrix is in the form $\mathbf{I}_3 \mid \mathbf{C}$ where $\mathbf{C}$ is a 3-dimensional column vector. The basic idea is to get a 1 in the 1-1 position and 0's elsewhere in the first column, a 1 in the 2-2 position and 0's elsewhere in the second column, and a 1 in the 3-3 position and 0's elsewhere in the third column. There are many ways of doing this. We could start by multiplying row 1 by $\frac{1}{5}$ to get a 1 in the 1-1 position. However, this would result in fractions leading to awkward calculations. Another way to get a 1 in the 1-1 position is to multiply row 3 by $-2$ and add the result to row 1. This is the way we begin.

$$\begin{bmatrix} 5 & -5 & 4 & | & 17 \\ 0 & 3 & 1 & | & 6 \\ 2 & -1 & 2 & | & 9 \end{bmatrix} \qquad (-2)R_3 + R_1 \rightarrow R_1$$

$$\begin{bmatrix} 1 & -3 & 0 & | & -1 \\ 0 & 3 & 1 & | & 6 \\ 2 & -1 & 2 & | & 9 \end{bmatrix} \qquad (-2)R_1 + R_3 \rightarrow R_3$$

$$\begin{bmatrix} 1 & -3 & 0 & | & -1 \\ 0 & 3 & 1 & | & 6 \\ 0 & 5 & 2 & | & 11 \end{bmatrix} \qquad R_2 \leftrightarrow R_3$$

(Since we already have a 1 in column 3, we will work on that column.)

$$\begin{bmatrix} 1 & -3 & 0 & | & -1 \\ 0 & 5 & 2 & | & 11 \\ 0 & 3 & 1 & | & 6 \end{bmatrix} \qquad (-2)R_3 + R_2 \rightarrow R_2$$

$$\begin{bmatrix} 1 & -3 & 0 & | & -1 \\ 0 & -1 & 0 & | & -1 \\ 0 & 3 & 1 & | & 6 \end{bmatrix} \qquad \begin{array}{l} (-3)R_2 + R_1 \rightarrow R_1 \\ (3)R_2 + R_3 \rightarrow R_3 \\ (-1)R_2 \rightarrow R_2 \end{array}$$

$$\begin{bmatrix} 1 & 0 & 0 & | & 2 \\ 0 & 1 & 0 & | & 1 \\ 0 & 0 & 1 & | & 3 \end{bmatrix}$$

The first three columns of the last matrix form the identity matrix $I_3$. Hence, we are finished. The solution of the original system of equations is the ordered triple $(2, 1, 3)$. Verify that if we replace $x$, $y$, and $z$ in the original system by 2, 1, and 3, respectively, we indeed get three true statements.
**Do Exercise 15.**

**A CLOSER LOOK**   In the previous example, we worked on the first column, then on the third column to take advantage of the fact that a 1 already appeared in the third column. Basically, it does not matter in what order we work as long as we remember to avoid fractions as much as possible, and once we have a column, say the $j$th one, in the desired form ($a_{jj} = 1$ and all other elements $= 0$), to keep it in that form. That is, we should not add a nonzero multiple of the $j$th row to another row, because this operation would replace one of the 0's in the $j$th column by a nonzero number. On occasion, if the integers $m$ and $n$ appear in a column, it is advantageous to look for integers $p$ and $q$ so that $pm + qn = 1$. For example, if 5 and 7 appear in a column, $3(5) + (-2)7 = 1$. We illustrate this idea in the next example.

**EXAMPLE 6**   Use the method of row reduction to solve the system

$$\begin{cases} 13x + 7y + 17z = 0 \\ 15x + 5y + 14z = 2 \\ 37x \phantom{+ 5y} + 12z = 13 \end{cases}$$

**Solution**   The augmented matrix of the system is

$$\begin{bmatrix} 13 & 7 & 17 & | & 0 \\ 15 & 5 & 14 & | & 2 \\ 37 & 0 & 12 & | & 13 \end{bmatrix}$$

Since none of the entries in the first three columns is a 1 and since we wish to avoid fractions, we begin with column 2. The reason for this choice is the fact that one of the entries in the second column is a 0 and also that $3(5) + (-2)7 = 1$. Hence, we multiply row 1 by $-2$, row 2 by 3, and add the results to get the new row 2. We proceed with this first step. The explanation for each of the subsequent steps is given to the right of each augmented matrix.

$$\begin{bmatrix} 13 & 7 & 17 & | & 0 \\ 15 & 5 & 14 & | & 2 \\ 37 & 0 & 12 & | & 13 \end{bmatrix} \qquad (-2)R_1 + 3R_2 \rightarrow R_2$$

$$\begin{bmatrix} 13 & 7 & 17 & | & 0 \\ 19 & 1 & 8 & | & 6 \\ 37 & 0 & 12 & | & 13 \end{bmatrix} \qquad (-7)R_2 + R_1 \rightarrow R_1$$

$$\begin{bmatrix} -120 & 0 & -39 & | & -42 \\ 19 & 1 & 8 & | & 6 \\ 37 & 0 & 12 & | & 13 \end{bmatrix} \qquad \left(\tfrac{1}{3}\right)R_1 \rightarrow R_1$$

$$\begin{bmatrix} -40 & 0 & -13 & | & -14 \\ 19 & 1 & 8 & | & 6 \\ 37 & 0 & 12 & | & 13 \end{bmatrix} \qquad R_1 + R_3 \rightarrow R_3$$

$$\begin{bmatrix} -40 & 0 & -13 & | & -14 \\ 19 & 1 & 8 & | & 6 \\ -3 & 0 & -1 & | & -1 \end{bmatrix} \qquad \begin{array}{l} (-13)R_3 + R_1 \rightarrow R_1 \\ 8R_3 + R_2 \rightarrow R_2 \\ (-1)R_3 \rightarrow R_3 \end{array}$$

$$\begin{bmatrix} -1 & 0 & 0 & | & -1 \\ -5 & 1 & 0 & | & -2 \\ 3 & 0 & 1 & | & 1 \end{bmatrix} \qquad \begin{array}{l} (-5)R_1 + R_2 \rightarrow R_2 \\ 3R_1 + R_3 \rightarrow R_3 \\ (-1)R_1 \rightarrow R_1 \end{array}$$

$$\begin{bmatrix} 1 & 0 & 0 & | & 1 \\ 0 & 1 & 0 & | & 3 \\ 0 & 0 & 1 & | & -2 \end{bmatrix}$$

Since the first three columns of the last augmented matrix form the identity matrix $I_3$, we are finished. The solution of the given system is obtained from the last column of the last augmented matrix. We find that the solution is the ordered triple $(1, 3, -2)$. Verify that if we replace $x$, $y$, and $z$ in the given system by 1, 3, and $-2$, respectively, then we obtain three true statements.

**Do Exercise 19.**

## More General Linear Systems

Thus far we have only considered linear systems where the number of equations and variables are the same. We will now discuss briefly the more general case of a linear system of $m$ equations in $n$ variables where $m$ and $n$ need not be equal. We shall see that a linear system has either a unique solution, no solution, or an infinite number of solutions. In general, we proceed as we did in the previous examples until we get an augmented matrix which is in row-reduced echelon form.

**DEFINITION 3.10:**   An augmented matrix is said to be a *row-reduced echelon matrix* if all of the following conditions are met.

1. All rows consisting entirely of 0's are below all rows with nonzero entries;
2. In each row with nonzero entries, the first nonzero entry is a 1;
3. If a 1 is the first nonzero entry of a row and appears in column $j$, then all other entries in that column are 0's; and
4. The first nonzero entry in any row, except for the first row, is to the right of the first nonzero entry in each preceding row.

The variables corresponding to columns that contain a leading 1 are called *leading variables*.

The nature of the solution set of the linear system can be deduced from an augmented matrix which is in row-reduced echelon form. For example, if after performing a sequence of elementary row operations on a linear system we obtain a row-reduced echelon matrix that contains a row in which all entries except the last one are 0's, then the system has no solution; it is inconsistent. If this does *not* happen, then the system will have *at least one solution*. The reasons for this will be apparent in the following examples.

**EXAMPLE 7**   Solve the linear system

$$\begin{cases} x + 3y - z = 2 \\ 3x + 5y + 2z = 0 \\ 4x + 12y - 4z = 5 \end{cases}$$

**Solution**   The augmented matrix of the system is

$$\left[ \begin{array}{ccc|c} 1 & 3 & -1 & 2 \\ 3 & 5 & 2 & 0 \\ 4 & 12 & -4 & 5 \end{array} \right]$$

Since we have a 1 in the 1-1 position, we begin working with column 1.

$$\left[ \begin{array}{ccc|c} 1 & 3 & -1 & 2 \\ 3 & 5 & 2 & 0 \\ 4 & 12 & -4 & 5 \end{array} \right] \qquad \begin{array}{l} (-3)R_1 + R_2 \rightarrow R_2 \\ (-4)R_1 + R_3 \rightarrow R_3 \end{array}$$

$$\left[ \begin{array}{ccc|c} 1 & 3 & -1 & 2 \\ 0 & -4 & 5 & -6 \\ 0 & 0 & 0 & -3 \end{array} \right]$$

We need not go any further. All entries in the third row are 0's except for the last entry. We can conclude that the given system has no solution—it is inconsistent—

because the last row represents the equation

$$0x + 0y + 0z = -3$$

which clearly has no solution since any replacement of $x$, $y$, and $z$ by real numbers will yield 0 on the left side and $0 \neq -3$.

**Do Exercise 17.** ■

**EXAMPLE 8** Solve the system

$$\begin{cases} x - y = 3 \\ x + y = 7 \\ x - 2y = 1 \end{cases}$$

**Solution** The augmented matrix of the system is

$$\begin{bmatrix} 1 & -1 & | & 3 \\ 1 & 1 & | & 7 \\ 1 & -2 & | & 1 \end{bmatrix}$$

We perform the following elementary row operations.

$$\begin{bmatrix} 1 & -1 & | & 3 \\ 1 & 1 & | & 7 \\ 1 & -2 & | & 1 \end{bmatrix} \quad \begin{matrix} (-1)R_1 + R_2 \to R_2 \\ (-1)R_1 + R_3 \to R_3 \end{matrix}$$

$$\begin{bmatrix} 1 & -1 & | & 3 \\ 0 & 2 & | & 4 \\ 0 & -1 & | & -2 \end{bmatrix} \quad \left(\tfrac{1}{2}\right)R_2 \to R_2$$

$$\begin{bmatrix} 1 & -1 & | & 3 \\ 0 & 1 & | & 2 \\ 0 & -1 & | & -2 \end{bmatrix} \quad \begin{matrix} R_2 + R_1 \to R_1 \\ R_2 + R_3 \to R_3 \end{matrix}$$

$$\begin{bmatrix} 1 & 0 & | & 5 \\ 0 & 1 & | & 2 \\ 0 & 0 & | & 0 \end{bmatrix}$$

The last matrix is in row-reduced echelon form. All entries in the third row are 0's. This row represents the equation $0x + 0y = 0$, which is satisfied for all values of $x$ and $y$. This last row can thus be ignored and from the first two rows we see that $x = 5$ and $y = 2$. Hence, the system has a unique solution, the ordered pair $(5, 2)$.

**Do Exercise 33.** ■

It may be useful to think of this example geometrically. Each equation is linear in $x$ and $y$; hence, the graph of each equation is a straight line. Since the system has a unique solution, the three straight lines intersect at a single point.

**EXAMPLE 9**   Solve the following system of three linear equations in four variables $w$, $x$, $y$, and $z$.

$$\begin{cases} 3w + 6x + y + 18z = 9 \\ w + 2x + 5z = 2 \\ 2w + 4x + 3y + 19z = 13 \end{cases}$$

**Solution**   The augmented matrix of the system is

$$\begin{bmatrix} 3 & 6 & 1 & 18 & 9 \\ 1 & 2 & 0 & 5 & 2 \\ 2 & 4 & 3 & 19 & 13 \end{bmatrix}$$

We then perform a sequence of elementary row operations, giving the explanations on the right sides of the matrices.

$$\begin{bmatrix} 3 & 6 & 1 & 18 & 9 \\ 1 & 2 & 0 & 5 & 2 \\ 2 & 4 & 3 & 19 & 13 \end{bmatrix} \qquad R_1 \leftrightarrow R_2$$

$$\begin{bmatrix} 1 & 2 & 0 & 5 & 2 \\ 3 & 6 & 1 & 18 & 9 \\ 2 & 4 & 3 & 19 & 13 \end{bmatrix} \qquad \begin{array}{l}(-3)R_1 + R_2 \to R_2 \\ (-2)R_1 + R_3 \to R_3\end{array}$$

$$\begin{bmatrix} 1 & 2 & 0 & 5 & 2 \\ 0 & 0 & 1 & 3 & 3 \\ 0 & 0 & 3 & 9 & 9 \end{bmatrix} \qquad (-3)R_2 + R_3 \to R_3$$

$$\begin{bmatrix} 1 & 2 & 0 & 5 & 2 \\ 0 & 0 & 1 & 3 & 3 \\ 0 & 0 & 0 & 0 & 0 \end{bmatrix}$$

The last augmented matrix is in row-reduced echelon form. All entries in the third row are 0's. Thus the system is consistent.

Observe that $w$ and $y$ are leading variables. The nonleading variables, $x$ and $z$ in this case, end up being parameters* and the leading variables are expressed in terms of these parameters. The first two rows correspond to the equations

$$w + 2x + 5z = 2$$
$$y + 3z = 3$$

These can be written

$$w = 2 - 2x - 5z$$
$$y = 3 - 3z$$

---

*A parameter is an arbitrary constant in a mathematical expression that distinguishes specific cases. For example, in $y = mx + b$, $m$ and $b$ are parameters, and giving these certain values specifies which line is represented by the equation.

We can give the parameters $x$ and $z$ any values we wish, say $a$ and $b$, respectively, and obtain

$$w = 2 - 2a - 5b$$
$$y = 3 - 3b$$

Thus, the solution set of the given system is the set of all ordered quadruples $(2 - 2a - 5b, a, 3 - 3b, b)$, where $a$ and $b$ are arbitrary real numbers. We therefore have infinitely many solutions. For example, for $a = b = 0$ we get the solution $(2, 0, 3, 0)$ and for $a = 1$, $b = 2$ we get another solution $(-10, 1, -3, 2)$. You should check that these are indeed solutions of the original system.
**Do Exercise 35.**

In general, if the number of variables in a linear system is greater than the number of equations, the system will have infinitely many solutions provided that it is consistent. Later we shall see that the graph of a linear equation in three variables is a plane. You might find it helpful to consider a special case which illustrates the remark we just made. For example, if we have a linear system of two equations in three variables, then the graphs of these equations are two planes. Only one of the following three things can happen:

1. The two planes are parallel; they have no points in common, hence there are no solutions—the system is inconsistent.
2. The two planes coincide; they have infinitely many points in common, hence there are infinitely many solutions.
3. The intersection of the two planes is a straight line. Again, there are infinitely many common points and thus infinitely many solutions.

**EXAMPLE 10**    Solve the system

$$\begin{cases} 2x - 4y + 6z = 3 \\ 3x - 6y + 9z = 5 \end{cases}$$

**Solution**    The augmented matrix of the system is

$$\begin{bmatrix} 2 & -4 & 6 & \vert & 3 \\ 3 & -6 & 9 & \vert & 5 \end{bmatrix}$$

We proceed as we did in the previous examples.

$$\begin{bmatrix} 2 & -4 & 6 & \vert & 3 \\ 3 & -6 & 9 & \vert & 5 \end{bmatrix} \qquad (-3)R_1 + 2R_2 \to R_2$$

$$\begin{bmatrix} 2 & -4 & 6 & \vert & 3 \\ 0 & 0 & 0 & \vert & 1 \end{bmatrix}$$

All entries, except the last one, of the second row of this augmented matrix are 0's. Thus, the system has no solutions; it is inconsistent.
**Do Exercise 31.**

## Solving Several Linear Systems Simultaneously

In certain applications, it is necessary to solve several linear systems which differ only in the constants on the right sides of the equations. In such cases, the method of row reduction can be used to solve all systems simultaneously as in the following example.

**EXAMPLE 11** Consider the following four linear equations:

$$x + 3y = 15$$
$$x + 3y = -6$$
$$2x - y = -19$$
$$2x - y = 23$$

with graphs $L_1$, $L_2$, $L_3$, and $L_4$, respectively.

**a.** Show that $L_1$ is parallel to $L_2$ and that $L_3$ is parallel to $L_4$.
**b.** Find the vertices of the parallelogram formed by these four lines.

**Solution** **a.** We write the first two equations in the form

$$y = \left(-\tfrac{1}{3}\right)x + 5 \quad \text{and} \quad y = \left(-\tfrac{1}{3}\right)x - 2$$

The slopes of $L_1$ and of $L_2$ are equal, therefore $L_1$ is parallel to $L_2$. Similarly, we can show that the slopes of $L_3$ and $L_4$ are equal.

**b.** To find the four vertices of the parallelogram formed by these four lines, we must find the points of intersection of $L_1$ and $L_3$, $L_2$ and $L_4$, $L_1$ and $L_4$, and, $L_2$ and $L_3$. We must, therefore, solve the systems

$$\begin{cases} x + 3y = 15 \\ 2x - y = -19 \end{cases} \qquad \begin{cases} x + 3y = -6 \\ 2x - y = 23 \end{cases}$$

$$\begin{cases} x + 3y = 15 \\ 2x - y = 23 \end{cases} \qquad \begin{cases} x + 3y = -6 \\ 2x - y = -19 \end{cases}$$

Observe that the left sides of all four systems are identical. We can combine the four augmented matrices of these systems in a single matrix and apply the row operations to all four simultaneously. We have

$$\left[\begin{array}{rr|r|r|r|r} 1 & 3 & 15 & -6 & 15 & -6 \\ 2 & -1 & -19 & 23 & 23 & -19 \end{array}\right] \qquad (-2)R_1 + R_2 \rightarrow R_2$$

$$\left[\begin{array}{rr|r|r|r|r} 1 & 3 & 15 & -6 & 15 & -6 \\ 0 & -7 & -49 & 35 & -7 & -7 \end{array}\right] \qquad (-\tfrac{1}{7})R_2 \rightarrow R_2$$

$$\left[\begin{array}{rr|r|r|r|r} 1 & 3 & 15 & -6 & 15 & -6 \\ 0 & 1 & 7 & -5 & 1 & 1 \end{array}\right] \qquad (-3)R_2 + R_1 \rightarrow R_1$$

$$\left[\begin{array}{rr|r|r|r|r} 1 & 0 & -6 & 9 & 12 & -9 \\ 0 & 1 & 7 & -5 & 1 & 1 \end{array}\right] \qquad (-3)R_2 + R_1 \rightarrow R_1$$

The first two columns of the last matrix form the identity $\mathbf{I}_2$. Hence we are finished. The first three columns form the augmented matrix

$$\left[\begin{array}{rr|r} 1 & 0 & -6 \\ 0 & 1 & 7 \end{array}\right]$$

which indicates that the solution of the first system is $x = -6$ and $y = 7$. Therefore the point of intersection of $L_1$ and $L_3$ is the point $(-6, 7)$. Similarly, we find that the other three vertices of the parallelogram are $(9, -5)$, $(12, 1)$, and $(-9, 1)$. (See Figure 3.1.)

**Do Exercise 47.**

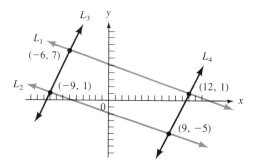

**FIGURE 3.1**

## Applications

**EXAMPLE 12 (Optional)**    A firm produces three commodities at four different plants. In the following matrix, $a_{ij}$ is the number of units of the $i$th commodity produced at the $j$th plant per day.

$$\begin{bmatrix} 3 & 4 & 5 & 2 \\ 2 & 3 & 2 & 1 \\ 0 & 2 & 2 & 3 \end{bmatrix}$$

The firm must fill orders of 202, 121, and 98 units of commodities 1, 2, and 3, respectively. How many days will it take each plant to produce enough commodities to fill the orders?

**Solution**    Let $w$, $x$, $y$, and $z$ be the number of days that plants 1, 2, 3, and 4, respectively, must produce a commodity. Then, the numbers of units of commodity 1 produced at plant 1 is $3w$, while plants 2, 3, and 4 produce $4x$, $5y$, and $2z$ units of commodity 1. Thus the total number of units of commodity 1 produced is

$$3w + 4x + 5y + 2z$$

Since 202 units of commodity 1 have been ordered, the first equation is

$$3w + 4x + 5y + 2z = 202$$

Similarly, the number of units of commodity 2 produced at plants 1, 2, 3, and 4 are $2w$, $3x$, $2y$, and $z$, respectively. Since 121 units of commodity 2 have been ordered, we get the second equation

$$2w + 3x + 2y + z = 121$$

The third equation is

$$2x + 2y + 3z = 98$$

and the linear system is

$$\begin{cases} 3w + 4x + 5y + 2z = 202 \\ 2w + 3x + 2y + \phantom{2}z = 121 \\ \phantom{2w + {}}2x + 2y + 3z = \phantom{20}98 \end{cases}$$

The augmented matrix of this system is

$$\left[\begin{array}{cccc|c} 3 & 4 & 5 & 2 & 202 \\ 2 & 3 & 2 & 1 & 121 \\ 0 & 2 & 2 & 3 & 98 \end{array}\right]$$

We now can perform the following elementary row operations.

$$\left[\begin{array}{cccc|c} 3 & 4 & 5 & 2 & 202 \\ 2 & 3 & 2 & 1 & 121 \\ 0 & 2 & 2 & 3 & 98 \end{array}\right] \qquad (-1)R_2 + R_1 \to R_1$$

$$\left[\begin{array}{cccc|c} 1 & 1 & 3 & 1 & 81 \\ 2 & 3 & 2 & 1 & 121 \\ 0 & 2 & 2 & 3 & 98 \end{array}\right] \qquad (-2)R_1 + R_2 \to R_2$$

$$\left[\begin{array}{cccc|c} 1 & 1 & 3 & 1 & 81 \\ 0 & 1 & -4 & -1 & -41 \\ 0 & 2 & 2 & 3 & 98 \end{array}\right] \qquad \begin{array}{c} (-1)R_2 + R_1 \to R_1 \\ (-2)R_2 + R_3 \to R_3 \end{array}$$

$$\left[\begin{array}{cccc|c} 1 & 0 & 7 & 2 & 122 \\ 0 & 1 & -4 & -1 & -41 \\ 0 & 0 & 10 & 5 & 180 \end{array}\right] \qquad \left(\tfrac{1}{10}\right)R_3 \to R_3$$

$$\left[\begin{array}{cccc|c} 1 & 0 & 7 & 2 & 122 \\ 0 & 1 & -4 & -1 & -41 \\ 0 & 0 & 1 & .5 & 18 \end{array}\right] \qquad \begin{array}{c} (-7)R_3 + R_1 \to R_1 \\ 4R_3 + R_2 \to R_2 \end{array}$$

$$\left[\begin{array}{cccc|c} 1 & 0 & 0 & -1.5 & -4 \\ 0 & 1 & 0 & 1 & 31 \\ 0 & 0 & 1 & .5 & 18 \end{array}\right]$$

The last matrix is in row-reduced echelon form. The leading variables are $w$, $x$, and $y$, and $z$ is a nonleading variable. If we let $z = a$, the last augmented matrix corresponds to the linear system

$$\begin{cases} w \phantom{+ x + y} - 1.5a = -4 \\ \phantom{w + {}} x \phantom{+ y} + a = \phantom{-}31 \\ \phantom{w + x + {}} y + .5a = \phantom{-}18 \end{cases}$$

Thus, we have

$$\begin{cases} w = 1.5a - 4 \\ x = 31 - a \\ y = 18 - .5a \\ z = a \end{cases}$$

There are infinitely many solutions. All ordered quadruples $(1.5a - 4, 31 - a, 18 - .5a, a)$, where $a$ is an arbitrary real number, are solutions of the linear system. However, we must consider only nonnegative values of $w$, $x$, $y$, and $z$ since those variables denote the number of days that plants 1, 2, 3, and 4 have to produce the three commodities to fill the orders. Therefore, $1.5a - 4 \geq 0$, $31 - a \geq 0$, $18 - .5a \geq 0$ and $a \geq 0$. The first, second, and third of the four inequalities are equivalent to $a \geq \frac{8}{3}$, $a \leq 31$, and $a \leq 36$, respectively. Since $0 < \frac{8}{3} < 31 < 36$, the four inequalities will be satisfied simultaneously if $\frac{8}{3} \leq a \leq 31$.

Although it is not necessary to do so, we may want our solutions to be integers. Since $1.5a$ and $.5a$ appear in two of the solutions, we choose an even integer between $\frac{8}{3}$ and 31 for the value of $a$. For example, if we let $a = 4$, then a feasible solution is $(2, 27, 16, 4)$, which means that the first, second, third, and fourth plants should produce the three commodities for 2, 27, 16, and 4 days, respectively. On the other hand, if we let $a = 20$, then we obtain another feasible solution, namely $(26, 11, 8, 20)$, and the production of plants 1, 2, 3, and 4 should last 26, 11, 8, and 20 days, respectively. You may find other feasible solutions to this problem.
**Do Exercise 57.**                                                                    ■

**EXAMPLE 13**   Suppose that when three commodities X, Y, and Z are priced at $x$, $y$, and $z$ dollars per unit, respectively, the quantities demanded are $D_x$, $D_y$, and $D_z$, respectively, (measured in thousands) and the quantities supplied are $S_x$, $S_y$, and $S_z$, respectively, (also measured in thousands). The demand and supply functions for these commodities are interrelated as expressed by

$$D_x = 1050 - 25x + 94y - 30z$$
$$S_x = -126 + 80x - 10y + 75z$$
$$D_y = 813 + 30x - 54y + 15z$$
$$S_y = -15 - 10x + 94y - 10z$$
$$D_z = 920 - 15x + 36y - 10z$$
$$S_z = -26 + 50x - 12y + 60z$$

Find the equilibrium prices and quantities for all three commodities.

**Solution**   Since at equilibrium supply equals demand, we obtain a system of three equations by setting $D_x = S_x$, $D_y = S_y$, and $D_z = S_z$. We obtain

$$\begin{cases} 1050 - 25x + 94y - 30z = -126 + 80x - 10y + 75z \\ 813 + 30x - 54y + 15z = -15 - 10x + 94y - 10z \\ 920 - 15x + 36y - 10z = -26 + 50x - 12y + 60z \end{cases}$$

The foregoing system may be written

$$\begin{cases} -105x + 104y - 105z = -1176 \\ 40x - 148y + 25z = -828 \\ -65x + 48y - 70z = -946 \end{cases}$$

We first write the augmented matrix of the system, then solve the system using row operations.

$$\left[\begin{array}{ccc|c} -105 & 104 & -105 & -1176 \\ 40 & -148 & 25 & -828 \\ -65 & 48 & -70 & -946 \end{array}\right] \quad \begin{array}{l} \left(\frac{1}{40}\right)R_2 \to R_2 \\ R_2 \leftrightarrow R_1 \end{array}$$

$$\left[\begin{array}{ccc|c} 1 & -3.7 & 0.625 & -20.7 \\ -105 & 104 & -105 & -1176 \\ -65 & 48 & -70 & -946 \end{array}\right] \quad \begin{array}{l} 105R_1 + R_2 \to R_2 \\ 65R_1 + R_3 \to R_3 \end{array}$$

$$\left[\begin{array}{ccc|c} 1 & -3.7 & 0.625 & -20.7 \\ 0 & -284.5 & -39.375 & -3349.5 \\ 0 & -192.5 & -29.375 & -2291.5 \end{array}\right] \quad \left(-\frac{1}{284.5}\right)R_2 \to R_2$$

$$\left[\begin{array}{ccc|c} 1 & -3.7 & 0.625 & -20.7 \\ 0 & 1 & \frac{315}{2276} & \frac{6699}{569} \\ 0 & -192.5 & -29.375 & -2291.5 \end{array}\right] \quad \begin{array}{l} (3.7)R_2 + R_1 \to R_1 \\ (192.5)R_2 + R_3 \to R_3 \end{array}$$

$$\left[\begin{array}{ccc|c} 1 & 0 & \frac{2588}{2276} & \frac{13008}{569} \\ 0 & 1 & \frac{315}{2276} & \frac{6699}{569} \\ 0 & 0 & \frac{-6220}{2276} & \frac{-14306}{569} \end{array}\right] \quad \left(\frac{-2276}{6220}\right)R_3 \to R_3$$

$$\left[\begin{array}{ccc|c} 1 & 0 & \frac{2588}{2276} & \frac{13008}{569} \\ 0 & 1 & \frac{315}{2276} & \frac{6699}{569} \\ 0 & 0 & 1 & \frac{92}{10} \end{array}\right] \quad \begin{array}{l} \left(\frac{-2588}{2276}\right)R_3 + R_1 \to R_1 \\ \left(\frac{-315}{2276}\right)R_3 + R_2 \to R_2 \end{array}$$

$$\left[\begin{array}{ccc|c} 1 & 0 & 0 & 12.4 \\ 0 & 1 & 0 & 10.5 \\ 0 & 0 & 1 & 9.2 \end{array}\right]$$

This matrix is in reduced-row echelon form. The equilibrium prices for products X, Y, and Z are $12.40, $10.50, and $9.20 per unit, respectively. The equilibrium quantity for product X is obtained by replacing $x$, $y$, and $z$ by 12.4, 10.5, and 9.2, respectively, in the equation giving $D_x$. (We could also use the equation giving $S_x$ since $D_x = S_x$.) We get $1050 - 25(12.4) + 94(10.5) - 30(9.2) = 1451$. The equilibrium quantity for product X is 1,451,000. Similarly, we find that the equilibrium quantities for products Y and Z are 756,000 and 1,020,000, respectively.
**Do Exercise 63.** ∎

## Homogeneous Linear Systems

We conclude this section with a brief discussion of a special type of system of equations.

> **DEFINITION 3.11:** A linear system, in which all constant terms are 0, is called a *homogeneous* system. The solution in which all variables are given the value 0 is called the *trivial solution*. Any other type of solution is called a *nontrivial solution*.

For example,

$$\begin{cases} 2x + 3y - 4z = 0 \\ 5x - 3y + 6z = 0 \\ 7x + 5y - 3z = 0 \end{cases}$$

is a homogeneous system and $(0, 0, 0)$ is the trivial solution for that system. In general, if a homogeneous linear system has more unknowns than equations, then nontrivial solutions can also be found as illustrated in the following example.

**EXAMPLE 14**   Solve the following homogeneous linear system:

$$\begin{cases} 3x + 2y - 5z = 0 \\ 2x + 3y + 5z = 0 \end{cases}$$

**Solution**   The augmented matrix of the system is

$$\begin{bmatrix} 3 & 2 & -5 & | & 0 \\ 2 & 3 & 5 & | & 0 \end{bmatrix}$$

We then perform a sequence of elementary row operations, giving the explanations on the right sides of the matrices.

$$\begin{bmatrix} 3 & 2 & -5 & | & 0 \\ 2 & 3 & 5 & | & 0 \end{bmatrix} \qquad (-1)R_2 + R_1 \rightarrow R_1$$

$$\begin{bmatrix} 1 & -1 & -10 & | & 0 \\ 2 & 3 & 5 & | & 0 \end{bmatrix} \qquad (-2)R_1 + R_2 \rightarrow R_2$$

$$\begin{bmatrix} 1 & -1 & -10 & | & 0 \\ 0 & 5 & 25 & | & 0 \end{bmatrix} \qquad \left(\tfrac{1}{5}\right)R_2 \rightarrow R_2$$

$$\begin{bmatrix} 1 & -1 & -10 & | & 0 \\ 0 & 1 & 5 & | & 0 \end{bmatrix} \qquad R_2 + R_1 \rightarrow R_1$$

$$\begin{bmatrix} 1 & 0 & -5 & | & 0 \\ 0 & 1 & 5 & | & 0 \end{bmatrix}$$

The last augmented matrix is in reduced-row echelon form. As expected, because we had more unknowns than equations, one column (the third) has no entry that can be the first nonzero entry of its row. So, in the corresponding system of equations, $z$ does not appear as the first variable in either equation. We give $z$ an arbitrary value, say $t$, and solve the system

$$\begin{cases} x - 5t = 0 \\ y + 5t = 0 \end{cases}$$

Thus, $x = 5t$ and $y = -5t$. The solution set of the system is $\{(5t, -5t, t) \mid t$ is any real number$\}$. The trivial solution is obtained by letting $t = 0$, and $(5, -5, 1)$ is obtained by letting $t = 1$. Other solutions may be found.
**Do Exercise 41.**

A procedure for solving a linear system by row reduction follows.

## Solving a Linear System by Row Reduction

**Step 1.** Write the augmented matrix for the given system.

**Step 2.** Perform a sequence of elementary row operations on the augmented matrix until you obtain an augmented matrix that is in row-reduced echelon form.

You now have exactly one of three possibilities.

1. There is no solution in the case where there is a row in which all entries except the last one are 0's.
2. There is a unique solution in the case where every variable is a leading variable.
3. There are infinitely many solutions in the case where there are no rows in which all entries except the last one are 0's, and at least one variable is a nonleading variable so that the leading variables are expressed in terms of at least one parameter.

# Exercise Set 3.3

*In Exercises 1–4, write the augmented matrix of the given linear system.*

**1.** $\begin{cases} 2x + 3y = -5 \\ 3x - 5y = 17 \end{cases}$

**2.** $\begin{cases} 3y + 12 = 6x \\ 5x - 15 = 2y \end{cases}$

**3.** $\begin{cases} 2y + 5z - 7x = 4 \\ 3x + 7z = 0 \\ 2z - 5y = 7 \end{cases}$

**4.** $\begin{cases} x + 2y - 5w = 3 \\ z + 2x - 3y = 7 \\ 5y + 2w = 3 \end{cases}$

*In Exercises 5–8, write a linear system corresponding to the given augmented matrix.*

**5.** $\begin{bmatrix} 2 & 0 & | & 3 \\ 5 & 3 & | & -2 \end{bmatrix}$

**6.** $\begin{bmatrix} 0 & 5 & 3 & | & 2 \\ 2 & 4 & 9 & | & 0 \\ -3 & 0 & 2 & | & 8 \end{bmatrix}$

**7.** $\begin{bmatrix} 2 & 0 & 3 & 4 & | & 7 \\ -3 & 2 & 5 & 0 & | & 2 \\ 2 & -1 & 0 & 4 & | & -3 \end{bmatrix}$

**8.** $\begin{bmatrix} \frac{1}{2} & 3 & \frac{5}{4} & 2 & 4 & | & -1 \\ 7 & 3 & 0 & 0 & 2 & | & \frac{1}{3} \\ 0 & 2 & \frac{1}{4} & 3 & \frac{3}{2} & | & 4 \\ -3 & 0 & 1 & \frac{1}{5} & 3 & | & 0 \end{bmatrix}$

*In Exercises 9–40, solve the given linear system. If the system is inconsistent, give the reason.*

**9.** $\begin{cases} 2x + y = 5 \\ 6x + 3y = 1 \end{cases}$

**10.** $\begin{cases} 6x + 17y = -1 \\ x + 3y = 0 \end{cases}$

**11.** $\begin{cases} 8x + 12y = 16 \\ 6x + 9y = 12 \end{cases}$

**12.** $\begin{cases} 3x + 7y = -1 \\ 2x + 5y = -1 \end{cases}$

**13.** $\begin{cases} 6x + 15y = 2 \\ 4x + 10y = 1 \end{cases}$

**14.** $\begin{cases} 15x + 21y = 6 \\ 2x + 6y = 4 \end{cases}$

**15.** $\begin{cases} 5x - 5y + 4z = 8 \\ 3y + z = 5 \\ 2x - y + 2z = 5 \end{cases}$

**16.** $\begin{cases} x - 3y = -7 \\ 2x + 2y + 3z = 5 \\ 2x - y + 2z = -2 \end{cases}$

**17.** $\begin{cases} x + 2y + 3z = 2 \\ 3x + 3y = 4 \\ 2x + y - 3z = 3 \end{cases}$

**18.** $\begin{cases} 7x + 6y - 17z = 49 \\ 2x + 2y - 5z = 15 \\ -6x - 5y + 15z = -42 \end{cases}$

**19.** $\begin{cases} 3x + 2y - 7z = 3 \\ 2x + 3y + 5z = 15 \\ 2x + 2y - 5z = 3 \end{cases}$

**20.** $\begin{cases} x - 2z = 2 \\ 3x + 2y - 7z = 11 \\ 4x + 5y - 10z = 20 \end{cases}$

**21.**
$$\begin{cases} 2x + y - 3z = 2 \\ 3x + 2y + 4z = 1 \\ 8x + 5y + 5z = 4 \end{cases}$$

**22.**
$$\begin{cases} -x - 2y - 2z = -10 \\ 4x + y + 5z = 2 \\ 2x - y + 2z = -7 \end{cases}$$

**23.**
$$\begin{cases} 3x - y + 3z = -1 \\ 2x + 2y + 3z = 8 \\ -3y - z = -10 \end{cases}$$

**24.**
$$\begin{cases} 2x + 3y + z = 4 \\ 9x + 11y + 4z = 11 \\ 6x + 4y + 2z = -2 \end{cases}$$

**25.**
$$\begin{cases} 2x - y + 2z = 14 \\ 4x + y + 5z = 37 \\ 11x - 8y + 10z = 69 \end{cases}$$

**26.**
$$\begin{cases} 2x + 5y + z = 1 \\ 3x + 4y + 5z = 0 \\ 8x + 13y + 11z = 2 \end{cases}$$

**27.**
$$\begin{cases} 7w - 8x - 5y + 6z = 27 \\ w - 4x + z = 7 \\ 3w - 11x + 3y + 4z = 22 \\ 4w - 6x - y + 4z = 18 \end{cases}$$

**28.**
$$\begin{cases} -w + x + y - 2z = -4 \\ -8w + 15x + 14y - 33z = -53 \\ w + x + y - 3z = -2 \\ -5w - x + 5z = -1 \end{cases}$$

**29.**
$$\begin{cases} x + 3y + z = 1 \\ 5w + x - 2y + 5z = 3 \\ w - 2x + 6y + 3z = 3 \\ 2w + 3x + 2y + 3z = 2 \end{cases}$$

**30.**
$$\begin{cases} 2w + 2x - y + 2z = 22 \\ 4w + 5x + y + 5z = 38 \\ 7w + 2x - 6y + 6z = 61 \\ 3w - 2x - y + 3z = 6 \end{cases}$$

**31.**
$$\begin{cases} 2x + 3y - z = 2 \\ 4x + 6y - 2z = 3 \end{cases}$$

**32.**
$$\begin{cases} 3x + y + 5z = 1 \\ 2x + 3y - 2z = 6 \end{cases}$$

**33.**
$$\begin{cases} x + 3y = -7 \\ 3x + y = 3 \\ 2x - 5y = 19 \end{cases}$$

**34.**
$$\begin{cases} x + 5y = 6 \\ 2x - y = 4 \\ 3x + 2y = 7 \end{cases}$$

**35.**
$$\begin{cases} 4w + 5x + 2y - 5z = -2 \\ 2w + y + 6z = 22 \\ 2w + 2x + y - z = 3 \end{cases}$$

**36.**
$$\begin{cases} -w - 2x - 2y + 3z = -20 \\ 4w + 5x + y - 2z = 16 \\ 2w + 2x - y + z = -5 \end{cases}$$

**37.**
$$\begin{cases} w + x - 3y = 0 \\ 2w + 2y + 3z = 13 \\ 8w + 3x - 4y + 8z = 34 \end{cases}$$

**38.**
$$\begin{cases} 3w - x + 3y + 2z = 2 \\ 2w + 2x + 3y - z = -2 \\ -3x - y + 3z = 5 \end{cases}$$

**39.**
$$\begin{cases} 3w + 7x + 2y - 7z = 21 \\ 2w + 4x + 2y - 5z = 17 \\ 2w + 3x + 3y + 5z = -10 \end{cases}$$

**40.**
$$\begin{cases} x - 3y = -7 \\ 2x + 2y + 3z = 13 \\ 2x - y + 2z = 3 \\ x + y - z = 4 \end{cases}$$

*In Exercises 41–45, find the solution set of the given homogeneous system.*

**41.**
$$\begin{cases} 4w + 5x + 2y - 5z = 0 \\ 2w + y + 6z = 0 \\ 2w + 2x + y - z = 0 \end{cases}$$

**42.**
$$\begin{cases} -w - 2x - 2y + 3z = 0 \\ 4w + 5x + y - 2z = 0 \\ 2w + 2x - y + z = 0 \end{cases}$$

**43.**
$$\begin{cases} w + x - 3y = 0 \\ 2w + 2y + 3z = 0 \\ 8w + 3x - 4y + 8z = 0 \end{cases}$$

**44.**
$$\begin{cases} 3w - x + 3y + 2z = 0 \\ 2w + 2x + 3y - z = 0 \\ -3x - y + 3z = 0 \end{cases}$$

**45.**
$$\begin{cases} 3w + 7x + 2y - 7z = 0 \\ 2w + 4x + 2y - 5z = 0 \\ 2w + 3x + 3y + 5z = 0 \end{cases}$$

*In Exercises 46–51, solve the linear systems of parts a, b, and c simultaneously.*

**46.**
$$\begin{cases} 2x - 3y = d_1 \\ -3x + 5y = d_2, \text{ where} \end{cases}$$
**a.** $d_1 = 3$, $d_2 = 0$; **b.** $d_1 = -2$, $d_2 = 5$; **c.** $d_1 = 4$, $d_2 = 7$.

**47.**
$$\begin{cases} 3x - 7y = d_1 \\ -2x + 5y = d_2, \text{ where} \end{cases}$$
**a.** $d_1 = 2$, $d_2 = 5$; **b.** $d_1 = -4$, $d_2 = 7$; **c.** $d_1 = 3$, $d_2 = 6$.

**48.**
$$\begin{cases} 3x + 5y = d_1 \\ 4x + 7y = d_2, \text{ where} \end{cases}$$
**a.** $d_1 = 0$, $d_2 = 4$; **b.** $d_1 = 11$, $d_2 = 6$; **c.** $d_1 = 3$, $d_2 = 2$.

49. $\begin{cases} x \quad - 2z = d_1 \\ 2x + 2y - 5z = d_2 \\ 4x + 5y - 10z = d_3, \text{ where} \end{cases}$
    **a.** $d_1 = 2, d_2 = 0, d_3 = 7$; **b.** $d_1 = 0, d_2 = 3, d_3 = 5$;
    **c.** $d_1 = 7, d_2 = -2, d_3 = 4$.

50. $\begin{cases} 3x + 2y - 7z = d_1 \\ 2x + 2y - 5z = d_2 \\ 2x + 3y + 5z = d_3, \text{ where} \end{cases}$
    **a.** $d_1 = 1, d_2 = 3, d_3 = 0$; **b.** $d_1 = 2, d_2 = 7, d_3 = 4$;
    **c.** $d_1 = 4, d_2 = -5, d_3 = 6$.

51. $\begin{cases} 5x - 5y + 4z = d_1 \\ \quad\quad 3y + z = d_2 \\ 2x - y + 2z = d_3, \text{ where} \end{cases}$
    **a.** $d_1 = 9, d_2 = 5, d_3 = 1$; **b.** $d_1 = 6, d_2 = 1, d_3 = 7$;
    **c.** $d_1 = 2, d_2 = 13, d_3 = -5$.

52. The Trebon Electronic Company manufactures VCRs. Their deluxe model requires 5 hours of assembly time and 2 hours of testing time, while their regular model requires 3 hours of assembly time and 1 hour of testing time. Assembly time and testing time available for production of VCRs vary from day to day according to the following schedule.

| Time available (in hours) | Monday | Tuesday | Wednesday | Thursday | Friday |
|---|---|---|---|---|---|
| For assembly | 68 | 66 | 70 | 71 | 69 |
| For testing | 26 | 25 | 27 | 27 | 26 |

For each day of the week, let $x$ and $y$ be the numbers of deluxe and regular models the company manufactures.

**a.** Write five linear systems of equations knowing that all available assembly and testing time is used each day.
**b.** Solve the five systems simultaneously.

53. A trucking company has two types of trucks, large and medium, that transport two kinds of cargo. The number of units that each type of truck can carry is given in the following table.

| Number of units of cargo transported per day | Truck type | |
|---|---|---|
| | Large | Medium |
| Kind A | 120 | 50 |
| Kind B | 80 | 30 |

The number of units of each kind of cargo the company must move each day of the week is given in the following table.

| | Monday | Tuesday | Wednesday | Thursday | Friday |
|---|---|---|---|---|---|
| Kind A | 750 | 820 | 730 | 1020 | 1040 |
| Kind B | 490 | 540 | 470 | 660 | 680 |

For each day of the week, let $x$ and $y$ be the numbers of large and medium trucks, respectively, the company schedules to transport the two commodities.

**a.** Write five linear systems of equations knowing that all units of cargo will have been transported each day.
**b.** Solve the five systems simultaneously.

54. The Companie de Bateaux de Quebec makes aluminum boats. Type A requires 7 hours of assembly time and 2 hours of painting time, while type B requires 4 hours of assembly time and 1 hour of painting time. Assembly and painting time varies from week to week each month. The times available in each department (in hours) is given in the table below.

| | Week 1 | Week 2 | Week 3 | Week 4 |
|---|---|---|---|---|
| Assembly time | 200 | 218 | 227 | 247 |
| Painting time | 55 | 60 | 62 | 68 |

Let $x$ and $y$ denote the numbers of type A and B boats, respectively, that the company will manufacture each week.

**a.** Write four linear systems of equations in $x$ and $y$ assuming that all assembly and painting time is used each week.
**b.** Solve the four systems simultaneously.

55. A television manufacturing company makes three models of television sets. The super model requires 5 hours of wiring time, 6 hours of assembly time, and 2 hours of testing time. The deluxe model requires 3 hours of wiring time, 4 hours of assembly time, and 1 hour of testing time. The regular model requires 2, 3, and 1 hours of wiring, assembly, and testing time, respectively. The number of hours available in each of the three departments varies from month to month. The number of hours available in the first three months of the year is given in the table below.

| | January | February | March |
|---|---|---|---|
| Wiring time | 690 | 675 | 785 |
| Assembly time | 910 | 900 | 930 |
| Testing time | 280 | 275 | 320 |

Let $x$, $y$, and $z$ denote the numbers of super, deluxe, and regular models, respectively, the company manufactures the first three months of the year.

**a.** Write three linear systems of equations assuming that all available time in all three departments is used.
**b.** Solve all three systems simultaneously.

56. An airline company flies three types of cargo planes: jumbo, large, and medium. On Mondays, Wednesdays, and Fridays

these planes transport three kinds of freight: kind A, kind B, and kind C. The number of units of freight that each plane is able to transport is given in the table below.

|  | Jumbo | Large | Medium |
|---|---|---|---|
| Kind A | 50 | 35 | 20 |
| Kind B | 60 | 40 | 35 |
| Kind C | 55 | 45 | 30 |

The number of units of each kind of cargo the company is committed to transport is given in the following table.

|  | Monday | Wednesday | Friday |
|---|---|---|---|
| Kind A | 405 | 470 | 505 |
| Kind B | 520 | 585 | 655 |
| Kind C | 510 | 580 | 615 |

Let $x$, $y$, and $z$ be the number of jumbo, large, and medium planes, respectively, the company must schedule to each of the days.

**a.** Write three systems of linear equations in $x$, $y$, and $z$, assuming that on each flight, each plane transports the maximum number of units of each kind of freight.

**b.** Solve the three systems simultaneously.

**57.** A manufacturer can produce three types of refrigerators called models A, B, and C. Each model takes a certain number of hours to assemble and a certain number of hours to test as shown in the following table.

|  | Model A | Model B | Model C |
|---|---|---|---|
| Assembly time (in hours) | 7 | 5 | 4 |
| Testing time | 2 | 1.5 | 1 |

Each day, the assembly and testing departments have a maximum of 285 and 78 hours available, respectively. How many refrigerators of each type can the company produce if it operates at full capacity? If the profit for each model is $50, $60, and $40, respectively, how many of each model should be produced and sold to maximize profit? What is the maximum profit?

**58.** Repeat Exercise 57 if each day the assembly and testing departments have a maximum of 304 and 85 hours available, respectively.

**59.** A company manufactures three models of vacuum cleaners: standard, regular, and deluxe. The production of these requires the use of two machines I and II. The time (in minutes) needed on each machine for each model is given in the following table.

|  | Model | | |
|---|---|---|---|
|  | Standard | Regular | Deluxe |
| Time on machine I | 20 | 30 | 35 |
| Time on machine II | 15 | 20 | 30 |

In a given week, machine I is available for 35 hours and machine II is available for 28 hours. How many of each model should be manufactured each week to have all the available time on each machine utilized? If the profit on each of the standard, regular, and deluxe models is $45, $56, and $60, respectively, how many of each model should be produced and sold in order to maximize profit? What is the maximum profit?

**60.** Repeat Exercise 59 if in a given week machine I is available for 33 hours and machine II is available for 27 hours.

**61.** A manufacturer can produce four types of inflatable rubber boats: small, medium, large, and super large. Each boat requires a certain number of hours from the cutting, assembly, and packaging departments as shown in the table.

|  | Small | Medium | Large | Super large |
|---|---|---|---|---|
| Cutting time (in hours) | 1 | 1.5 | 2 | 2.5 |
| Assembly time (in hours) | 1.5 | 2 | 2.5 | 3 |
| Packaging time (in hours) | .3 | .4 | .7 | .6 |

The maximum number of hours available each day in the cutting, assembly, and packaging departments are 295, 385, and 85, respectively. How many boats of each type can be produced each day if the plant operates at full capacity? If the profit on each model is $40, $50, $75, and $90, respectively, how many of each model should be produced and sold in order to maximize profit? What is the maximum profit?

**62.** Repeat Exercise 61 if the maximum number of hours available each day in the cutting, assembly, and packaging departments is 275, 385, and 93, respectively.

**63.** Suppose that when three commodities X, Y, and Z are priced at $x$, $y$, and $z$ dollars per unit, respectively, the quantities demanded are $D_x$, $D_y$, and $D_z$, respectively (measured in thousands), and the quantities supplied are $S_x$, $S_y$, and $S_z$, respectively (also measured in thousands). The demand and supply functions for these commodities are interrelated as expressed by

$D_x = 520 - 30x + 92y - 25z$

$S_x = -120 + 70x - 8y + 75z$

$D_y = 140 + 32x - 52y + 18z$

$S_y = -7 - 8x + 98y - 12z$

$D_z = 478 - 20x + 28y - 5z$

$S_z = -21 + 40x - 12y + 65z$

Find the equilibrium prices and quantities for all three commodities.

**64.** Same as Exercise 63 with the following equations:

$D_x = 150 - 10x + 85y - 17z$

$S_x = -8 + 90x - 15y + 83z$

$D_y = 280 + 42x - 20y + 81z$

$S_y = -17 - 8x + 120y - 19z$

$D_z = 310 - 17x + 51y - 12z$

$S_z = -44 + 63x - 9y + 88z$

**65.** Same as Exercise 63 with the following equations:

$D_x = 800 - 23x + 95y - 18z$

$S_x = -35 + 77x - 5y + 82z$

$D_y = 320 + 50x - 12y + 35z$

$S_y = -68 - 10x + 118y - 5z$

$D_z = 1310 - 10x + 42y - 20z$

$S_z = -112 + 80x - 8y + 100z$

---

# 3.4 Multiplicative Inverse

The following example illustrates the concept of the multiplicative inverse of matrices.

**EXAMPLE 1** Find the products **AB** and **BA** if

$$A = \begin{bmatrix} 2 & 1 \\ 5 & 3 \end{bmatrix} \text{ and } B = \begin{bmatrix} 3 & -1 \\ -5 & 2 \end{bmatrix}$$

**Solution** Since both matrices are $2 \times 2$, each is conformable to the other for multiplication. Both products are defined and are $2 \times 2$ matrices. We get

$$AB = \begin{bmatrix} 2 & 1 \\ 5 & 3 \end{bmatrix} \cdot \begin{bmatrix} 3 & -1 \\ -5 & 2 \end{bmatrix} = \begin{bmatrix} 2(3)+1(-5) & 2(-1)+1(2) \\ 5(3)+3(-5) & 5(-1)+3(2) \end{bmatrix}$$

$$= \begin{bmatrix} 1 & 0 \\ 0 & 1 \end{bmatrix} = I_2$$

$$BA = \begin{bmatrix} 3 & -1 \\ -5 & 2 \end{bmatrix} \cdot \begin{bmatrix} 2 & 1 \\ 5 & 3 \end{bmatrix} = \begin{bmatrix} 3(2)+(-1)5 & 3(1)+(-1)3 \\ (-5)2+2(5) & (-5)1+2(3) \end{bmatrix}$$

$$= \begin{bmatrix} 1 & 0 \\ 0 & 1 \end{bmatrix} = I_2$$

We have obtained the identity matrix $I_2$ for both products **AB** and **BA**. ∎

Recall that if $a$ and $b$ are numbers such that the product $ab$ is the number 1, then $b$ is said to be the multiplicative inverse of $a$ and we write $b = a^{-1}$. For example, .2 is the multiplicative inverse of 5 since $5(.2) = 1$. The number 0 has no multiplicative inverse since $0(b) = 0 \neq 1$ for all numbers $b$. This holds true for matrices as well.

**DEFINITION 3.12:** If **A** and **B** are $n \times n$ matrices and $AB = BA = I_n$, then we say that **B** is the *multiplicative inverse* of **A** and we write $B = A^{-1}$. Note that **A** is also the multiplicative inverse of **B** and $A = B^{-1}$.

Just as the number 0 has no multiplicative inverse, not all square matrices have multiplicative inverses, as you can see in the following example.

EXAMPLE 2　Show that the matrix $\begin{bmatrix} 2 & -3 \\ 6 & -9 \end{bmatrix}$ does not have an inverse.

Solution　Suppose that the given matrix had an inverse

$$\begin{bmatrix} a & b \\ c & d \end{bmatrix}$$

Then we would have

$$\begin{bmatrix} 2 & -3 \\ 6 & -9 \end{bmatrix} \cdot \begin{bmatrix} a & b \\ c & d \end{bmatrix} = \begin{bmatrix} 1 & 0 \\ 0 & 1 \end{bmatrix}$$

But

$$\begin{bmatrix} 2 & -3 \\ 6 & -9 \end{bmatrix} \cdot \begin{bmatrix} a & b \\ c & d \end{bmatrix} = \begin{bmatrix} 2a-3c & 2b-3d \\ 6a-9c & 6b-9d \end{bmatrix}$$

It follows that

$$\begin{bmatrix} 2a-3c & 2b-3d \\ 6a-9c & 6b-9d \end{bmatrix} = \begin{bmatrix} 1 & 0 \\ 0 & 1 \end{bmatrix}$$

The truth of the foregoing equality would imply that the system

$$\begin{cases} 2a - 3c = 1 \\ 6a - 9c = 0 \end{cases}$$

has a solution. But $6a - 9c = 0$ is equivalent to $2a - 3c = 0$. Since $2a - 3c$ cannot equal both 0 and 1, the given matrix does not have an inverse.
**Do Exercise 3.**　■

We shall discuss this further when we learn how to find the multiplicative inverse of a matrix.

> **DEFINITION 3.13:**　A square matrix that has an inverse is said to be *invertible* (or *nonsingular*). A square matrix that does not have an inverse is said to be a *singular* matrix.

If a square matrix has a multiplicative inverse, it is unique. To demonstrate that this is the case, suppose that **B** and **C** are both inverses of **A**. Then by definition we have

$$\mathbf{AB} = \mathbf{BA} = \mathbf{I}_n \quad \text{and} \quad \mathbf{AC} = \mathbf{CA} = \mathbf{I}_n$$

We write

$$\mathbf{C} = \mathbf{CI}_n = \mathbf{C(AB)} = \mathbf{(CA)B} = \mathbf{I}_n\mathbf{B} = \mathbf{B}$$

Thus, **B** and **C** are the same matrix.

**EXAMPLE 3** Find **AB** and **BA** if $\mathbf{A} = \begin{bmatrix} 7 & 2 & 5 \\ 3 & 1 & 2 \end{bmatrix}$ and $\mathbf{B} = \begin{bmatrix} 1 & -2 \\ -3 & 7 \\ 0 & 0 \end{bmatrix}$.

**Solution** Clearly, **A** is conformable to **B** for multiplication and the product **AB** is a 2 × 2 matrix. We get

$$\mathbf{AB} = \begin{bmatrix} 7 & 2 & 5 \\ 3 & 1 & 2 \end{bmatrix} \cdot \begin{bmatrix} 1 & -2 \\ -3 & 7 \\ 0 & 0 \end{bmatrix}$$

$$= \begin{bmatrix} 7(1)+2(-3)+5(0) & 7(-2)+2(7)+5(0) \\ 3(1)+1(-3)+2(0) & 3(-2)+1(7)+2(0) \end{bmatrix} = \begin{bmatrix} 1 & 0 \\ 0 & 1 \end{bmatrix} = \mathbf{I}_2$$

Since **B** is 3 × 2 and **A** is 2 × 3, **B** is conformable to **A** for multiplication and the product **BA** is a 3 × 3 matrix. We find that

$$\mathbf{BA} = \begin{bmatrix} 1 & -2 \\ -3 & 7 \\ 0 & 0 \end{bmatrix} \cdot \begin{bmatrix} 7 & 2 & 5 \\ 3 & 1 & 2 \end{bmatrix} = \begin{bmatrix} 1 & 0 & 1 \\ 0 & 1 & -1 \\ 0 & 0 & 0 \end{bmatrix}$$

Although **AB** is the identity matrix of order 2, **A** and **B** are not inverses of each other because **BA** is not an identity matrix. ∎

The previous example illustrates the fact that two matrices **A** and **B** are inverses of each other if, and only if, both matrices are *square* and $\mathbf{AB} = \mathbf{BA} = \mathbf{I}_n$.

## Procedure to Find the Inverse

There are several procedures to find the inverse of an invertible matrix. We shall introduce the first of these in the present section. Another will be discussed in Section 3.6.

**EXAMPLE 4** Find the inverse of the matrix $\begin{bmatrix} 7 & 5 \\ 4 & 3 \end{bmatrix}$ and check your result.

**Solution** Let the inverse be the matrix

$$\begin{bmatrix} x_1 & y_1 \\ x_2 & y_2 \end{bmatrix}$$

Then,

$$\begin{bmatrix} 7 & 5 \\ 4 & 3 \end{bmatrix} \cdot \begin{bmatrix} x_1 & y_1 \\ x_2 & y_2 \end{bmatrix} = \begin{bmatrix} 1 & 0 \\ 0 & 1 \end{bmatrix}$$

This can be written

$$\begin{bmatrix} 7x_1 + 5x_2 & 7y_1 + 5y_2 \\ 4x_1 + 3x_2 & 4y_1 + 3y_2 \end{bmatrix} = \begin{bmatrix} 1 & 0 \\ 0 & 1 \end{bmatrix}$$

This matrix equation is equivalent to the system

$$\begin{cases} 7x_1 + 5x_2 = 1 \\ 4x_1 + 3x_2 = 0 \\ 7y_1 + 5y_2 = 0 \\ 4y_1 + 3y_2 = 1 \end{cases}$$

We may consider the first two equations as a linear system in $x_1$ and $x_2$, and the last two equations as a linear system in $y_1$ and $y_2$. In so doing, the two systems have the same coefficient matrix. Thus, we may solve them simultaneously as we did in Section 3.3. (See Example 11 and Exercises 46–51 of that section.) The augmented matrix for both systems is

$$\left[\begin{array}{cc|c|c} 7 & 5 & 1 & 0 \\ 4 & 3 & 0 & 1 \end{array}\right]$$

We perform elementary row operations on this augmented matrix to solve the two systems simultaneously.

$$\left[\begin{array}{cc|c|c} 7 & 5 & 1 & 0 \\ 4 & 3 & 0 & 1 \end{array}\right] \qquad (-2)R_2 + R_1 \to R_1$$

$$\left[\begin{array}{cc|c|c} -1 & -1 & 1 & -2 \\ 4 & 3 & 0 & 1 \end{array}\right] \qquad \begin{array}{c} 4R_1 + R_2 \to R_2 \\ (-1)R_1 \to R_1 \end{array}$$

$$\left[\begin{array}{cc|c|c} 1 & 1 & -1 & 2 \\ 0 & -1 & 4 & -7 \end{array}\right] \qquad \begin{array}{c} R_2 + R_1 \to R_1 \\ (-1)R_2 \to R_2 \end{array}$$

$$\left[\begin{array}{cc|c|c} 1 & 0 & 3 & -5 \\ 0 & 1 & -4 & 7 \end{array}\right]$$

Thus, $x_1 = 3$, $x_2 = -4$, $y_1 = -5$, and $y_2 = 7$. The inverse of the given matrix is

$$\left[\begin{array}{cc} 3 & -5 \\ -4 & 7 \end{array}\right]$$

We can check our result by multiplication.

$$\left[\begin{array}{cc} 3 & -5 \\ -4 & 7 \end{array}\right] \cdot \left[\begin{array}{cc} 7 & 5 \\ 4 & 3 \end{array}\right] = \left[\begin{array}{cc} 3(7)+(-5)4 & 3(5)+(-5)3 \\ (-4)7+7(4) & (-4)5+7(3) \end{array}\right]$$

$$= \left[\begin{array}{cc} 1 & 0 \\ 0 & 1 \end{array}\right]$$

We leave it to you to check that

$$\left[\begin{array}{cc} 7 & 5 \\ 4 & 3 \end{array}\right] \cdot \left[\begin{array}{cc} 3 & -5 \\ -4 & 7 \end{array}\right] = \left[\begin{array}{cc} 1 & 0 \\ 0 & 1 \end{array}\right]$$

We can conclude that our result is correct.
**Do Exercise 5.**                                                                              ∎

EXAMPLE 5    Find the inverse of the matrix $\mathbf{A}$ if $\mathbf{A} = \begin{bmatrix} 5 & 2 & -6 \\ 7 & -3 & 4 \\ 3 & 5 & -12 \end{bmatrix}$.

**Solution**  Suppose that the inverse of the given matrix is

$$\begin{bmatrix} x_1 & y_1 & z_1 \\ x_2 & y_2 & z_2 \\ x_3 & y_3 & z_3 \end{bmatrix}$$

Then,

$$\begin{bmatrix} 5 & 2 & -6 \\ 7 & -3 & 4 \\ 3 & 5 & -12 \end{bmatrix} \cdot \begin{bmatrix} x_1 & y_1 & z_1 \\ x_2 & y_2 & z_2 \\ x_3 & y_3 & z_3 \end{bmatrix} = \begin{bmatrix} 1 & 0 & 0 \\ 0 & 1 & 0 \\ 0 & 0 & 1 \end{bmatrix}$$

where the $x_i$'s, $y_i$'s, and $z_i$'s need to be determined.

Performing the multiplication on the left, and equating each entry of the product to the corresponding entry of the matrix $\mathbf{I}_3$ yields the following three linear systems.

$$\begin{cases} 5x_1 + 2x_2 - 6x_3 = 1 \\ 7x_1 - 3x_2 + 4x_3 = 0 \\ 3x_1 + 5x_2 - 12x_3 = 0 \end{cases} \qquad \begin{cases} 5y_1 + 2y_2 - 6y_3 = 0 \\ 7y_1 - 3y_2 + 4y_3 = 1 \\ 3y_1 + 5y_2 - 12y_3 = 0 \end{cases}$$

and

$$\begin{cases} 5z_1 + 2z_2 - 6z_3 = 0 \\ 7z_1 - 3z_2 + 4z_3 = 0 \\ 3z_1 + 5z_2 - 12z_3 = 1 \end{cases}$$

The three linear systems have the same coefficient matrix; therefore, they can be solved simultaneously as follows.

$$\left[\begin{array}{ccc|ccc} 5 & 2 & -6 & 1 & 0 & 0 \\ 7 & -3 & 4 & 0 & 1 & 0 \\ 3 & 5 & -12 & 0 & 0 & 1 \end{array}\right] \qquad (-2)R_3 + R_1 \rightarrow R_1$$

$$\left[\begin{array}{ccc|ccc} -1 & -8 & 18 & 1 & 0 & -2 \\ 7 & -3 & 4 & 0 & 1 & 0 \\ 3 & 5 & -12 & 0 & 0 & 1 \end{array}\right] \qquad \begin{array}{l} 7R_1 + R_2 \rightarrow R_2 \\ 3R_1 + R_3 \rightarrow R_3 \\ (-1)R_1 \rightarrow R_1 \end{array}$$

$$\left[\begin{array}{ccc|ccc} 1 & 8 & -18 & -1 & 0 & 2 \\ 0 & -59 & 130 & 7 & 1 & -14 \\ 0 & -19 & 42 & 3 & 0 & -5 \end{array}\right] \qquad (-3)R_3 + R_2 \rightarrow R_2$$

Because we could not easily get a 1 in the 2-2 or 3-3 position without introducing fractions, we multiplied row 3 by $-3$ and added the result to row 2 to get smaller integers (still hoping to avoid fractions!).

$$\left[\begin{array}{ccc|ccc} 1 & 8 & -18 & -1 & 0 & 2 \\ 0 & -2 & 4 & -2 & 1 & 1 \\ 0 & -19 & 42 & 3 & 0 & -5 \end{array}\right] \qquad \left(\frac{-1}{2}\right)R_2 \rightarrow R_2$$

$$\left[\begin{array}{ccc|ccc} 1 & 8 & -18 & -1 & 0 & 2 \\ 0 & 1 & -2 & 1 & \frac{-1}{2} & \frac{-1}{2} \\ 0 & -19 & 42 & 3 & 0 & -5 \end{array}\right] \qquad \begin{array}{l} (-8)R_2 + R_1 \rightarrow R_1 \\ (19)R_2 + R_3 \rightarrow R_3 \end{array}$$

$$\left[\begin{array}{ccc|c|c|c} 1 & 0 & -2 & -9 & 4 & 6 \\ 0 & 1 & -2 & 1 & \frac{-1}{2} & \frac{-1}{2} \\ 0 & 0 & 4 & 22 & \frac{-19}{2} & \frac{-29}{2} \end{array}\right] \qquad \left(\tfrac{1}{4}\right)R_3 \rightarrow R_3$$

$$\left[\begin{array}{ccc|c|c|c} 1 & 0 & -2 & -9 & 4 & 6 \\ 0 & 1 & -2 & 1 & \frac{-1}{2} & \frac{-1}{2} \\ 0 & 0 & 1 & \frac{11}{2} & \frac{-19}{8} & \frac{-29}{8} \end{array}\right] \qquad \begin{array}{l} 2R_3 + R_1 \rightarrow R_1 \\ 2R_3 + R_2 \rightarrow R_2 \end{array}$$

$$\left[\begin{array}{ccc|c|c|c} 1 & 0 & 0 & 2 & \frac{-3}{4} & \frac{-5}{4} \\ 0 & 1 & 0 & 12 & \frac{-21}{4} & \frac{-31}{4} \\ 0 & 0 & 1 & \frac{11}{2} & \frac{-19}{8} & \frac{-29}{8} \end{array}\right]$$

Since the first three columns of the last augmented matrix form the identity matrix $\mathbf{I}_3$, we are done. We see that

$$x_1 = 2, \quad y_1 = \tfrac{-3}{4}, \quad z_1 = \tfrac{-5}{4}$$
$$x_2 = 12, \quad y_2 = \tfrac{-21}{4}, \quad z_2 = \tfrac{-31}{4}$$
$$x_3 = \tfrac{11}{2}, \quad y_3 = \tfrac{-19}{8}, \quad z_3 = \tfrac{-29}{8}$$

Hence,

$$\mathbf{A}^{-1} = \begin{bmatrix} 2 & \frac{-3}{4} & \frac{-5}{4} \\ 12 & \frac{-21}{4} & \frac{-31}{4} \\ \frac{11}{2} & \frac{-19}{8} & \frac{-29}{8} \end{bmatrix}$$

For convenience, we write

$$\mathbf{A}^{-1} = \left(\tfrac{1}{8}\right)\begin{bmatrix} 16 & -6 & -10 \\ 96 & -42 & -62 \\ 44 & -19 & -29 \end{bmatrix}$$

To check our result, we perform the following multiplication.

$$\left(\tfrac{1}{8}\right)\begin{bmatrix} 16 & -6 & -10 \\ 96 & -42 & -62 \\ 44 & -19 & -29 \end{bmatrix} \cdot \begin{bmatrix} 5 & 2 & -6 \\ 7 & -3 & 4 \\ 3 & 5 & -12 \end{bmatrix}$$
$$= \left(\tfrac{1}{8}\right)\begin{bmatrix} 8 & 0 & 0 \\ 0 & 8 & 0 \\ 0 & 0 & 8 \end{bmatrix} = \begin{bmatrix} 1 & 0 & 0 \\ 0 & 1 & 0 \\ 0 & 0 & 1 \end{bmatrix}$$

Similarly, we find that

$$\begin{bmatrix} 5 & 2 & -6 \\ 7 & -3 & 4 \\ 3 & 5 & -12 \end{bmatrix}\left(\tfrac{1}{8}\right)\begin{bmatrix} 16 & -6 & -10 \\ 96 & -42 & -62 \\ 44 & -19 & -29 \end{bmatrix}$$
$$= \left(\tfrac{1}{8}\right)\begin{bmatrix} 8 & 0 & 0 \\ 0 & 8 & 0 \\ 0 & 0 & 8 \end{bmatrix} = \begin{bmatrix} 1 & 0 & 0 \\ 0 & 1 & 0 \\ 0 & 0 & 1 \end{bmatrix}$$

Since in both cases the multiplications yielded the identity matrix $\mathbf{I}_3$, we conclude that our result is correct.

You should carry out the details of the two matrix multiplications.

**Do Exercise 11.**                                                                         ■

---

**A CLOSER LOOK**   In the third step of the sequence of elementary row operations, we could have obtained a 1 in the 2-2 position by multiplying row 2 by $\frac{-1}{59}$. However, this operation would have introduced cumbersome fractions. We could also have obtained a 1 in the 2-2 position by multiplying row 2 by 9, row 3 by $-28$, and adding the results to get the new row 2 since $9(-59) + (-28)(-19) = -531 + 532 = 1$. We did not use this approach, although it would have delayed the introduction of fractions, because we wished to illustrate the fact that at each step we have several options. We should always try to determine quickly which option leads to the simplest solution. In this case, it would be difficult to readily find integers $m$ and $n$ such that $m(-59) + n(-19) = 1$; although, as we have shown above, such integers exist.

---

In general, we use the following procedure to find the inverse of an $n \times n$ matrix $\mathbf{A}$. We start with the augmented matrix $\mathbf{A} \mid \mathbf{I}_n$. If a sequence of elementary row operations transforms this matrix into $\mathbf{I}_n \mid \mathbf{B}$, then the matrix $\mathbf{B}$ is $\mathbf{A}^{-1}$. However, if at any step we obtain all 0's in any row to the left of the vertical line, the matrix $\mathbf{A}$ does not have an inverse; it is singular.

**EXAMPLE 6**   Find the inverse, if it exists, of the matrix $\begin{bmatrix} 3 & 5 & 2 \\ 5 & 9 & -4 \\ 2 & 3 & 5 \end{bmatrix}$.

**Solution**

$$\left[\begin{array}{rrr|rrr} 3 & 5 & 2 & 1 & 0 & 0 \\ 5 & 9 & -4 & 0 & 1 & 0 \\ 2 & 3 & 5 & 0 & 0 & 1 \end{array}\right] \qquad (-1)R_3 + R_1 \rightarrow R_1$$

$$\left[\begin{array}{rrr|rrr} 1 & 2 & -3 & 1 & 0 & -1 \\ 5 & 9 & -4 & 0 & 1 & 0 \\ 2 & 3 & 5 & 0 & 0 & 1 \end{array}\right] \qquad \begin{array}{l} (-5)R_1 + R_2 \rightarrow R_2 \\ (-2)R_1 + R_3 \rightarrow R_3 \end{array}$$

$$\left[\begin{array}{rrr|rrr} 1 & 2 & -3 & 1 & 0 & -1 \\ 0 & -1 & 11 & -5 & 1 & 5 \\ 0 & -1 & 11 & -2 & 0 & 3 \end{array}\right] \qquad \begin{array}{l} (-1)R_2 + R_3 \rightarrow R_3 \\ (-1)R_2 \rightarrow R_2 \end{array}$$

$$\left[\begin{array}{ccc|ccc} 1 & 2 & -3 & 1 & 0 & -1 \\ 0 & 1 & -11 & 5 & -1 & -5 \\ 0 & 0 & 0 & 3 & -1 & -2 \end{array}\right]$$

We may stop. We have obtained all 0's to the left of the vertical line in the third row. We conclude that the given matrix does not have an inverse. It is singular.

**Do Exercise 15.**                                                                                                ■

EXAMPLE 7    Use the result of Example 5 to solve the system

$$\begin{cases} 5x + 2y - 6z = 8 \\ 7x - 3y + 4z = -24 \\ 3x + 5y - 12z = 16 \end{cases}^*$$

Solution    We first write the system as a matrix equation.

$$\begin{bmatrix} 5 & 2 & -6 \\ 7 & -3 & 4 \\ 3 & 5 & -12 \end{bmatrix} \cdot \begin{bmatrix} x \\ y \\ z \end{bmatrix} = \begin{bmatrix} 8 \\ -24 \\ 16 \end{bmatrix}$$

We then multiply both sides of the equation (on the *left* of each side) by the inverse of the coefficient matrix found in Example 5.

$$\frac{1}{8}\begin{bmatrix} 16 & -6 & -10 \\ 96 & -42 & -62 \\ 44 & -19 & -29 \end{bmatrix} \cdot \begin{bmatrix} 5 & 2 & -6 \\ 7 & -3 & 4 \\ 3 & 5 & -12 \end{bmatrix} \cdot \begin{bmatrix} x \\ y \\ z \end{bmatrix}$$

$$= \frac{1}{8}\begin{bmatrix} 16 & -6 & -10 \\ 96 & -42 & -62 \\ 44 & -19 & -29 \end{bmatrix} \cdot \begin{bmatrix} 8 \\ -24 \\ 16 \end{bmatrix}$$

After performing the multiplications we get

$$\begin{bmatrix} 1 & 0 & 0 \\ 0 & 1 & 0 \\ 0 & 0 & 1 \end{bmatrix} \cdot \begin{bmatrix} x \\ y \\ z \end{bmatrix} = \frac{1}{8}\begin{bmatrix} 112 \\ 784 \\ 344 \end{bmatrix}$$

That is,

$$\begin{bmatrix} x \\ y \\ z \end{bmatrix} = \begin{bmatrix} 14 \\ 98 \\ 43 \end{bmatrix}$$

---

*The procedure illustrated in this example is not as efficient as it may appear because it is usually as tedious to find the inverse of the coefficient matrix as it is to solve the linear system by the method described in Section 3.3.

The solution of the original linear system is the ordered triple $(14, 98, 43)$.
**Do Exercise 31.** ■

EXAMPLE 8   Find the inverse, if it exists, of the following matrix

$$\begin{bmatrix} 2 & 2 & -1 & 2 \\ 4 & 5 & 1 & 5 \\ 7 & 2 & -6 & 6 \\ 3 & -2 & -1 & 3 \end{bmatrix}$$

**Solution**   We proceed as follows.

$$\left[\begin{array}{cccc|cccc} 2 & 2 & -1 & 2 & 1 & 0 & 0 & 0 \\ 4 & 5 & 1 & 5 & 0 & 1 & 0 & 0 \\ 7 & 2 & -6 & 6 & 0 & 0 & 1 & 0 \\ 3 & -2 & -1 & 3 & 0 & 0 & 0 & 1 \end{array}\right] \quad \begin{array}{l} R_2 + R_1 \to R_1 \\ 6R_2 + R_3 \to R_3 \\ R_2 + R_4 \to R_4 \end{array}$$

$$\left[\begin{array}{cccc|cccc} 6 & 7 & 0 & 7 & 1 & 1 & 0 & 0 \\ 4 & 5 & 1 & 5 & 0 & 1 & 0 & 0 \\ 31 & 32 & 0 & 36 & 0 & 6 & 1 & 0 \\ 7 & 3 & 0 & 8 & 0 & 1 & 0 & 1 \end{array}\right] \quad R_2 \leftrightarrow R_3$$

$$\left[\begin{array}{cccc|cccc} 6 & 7 & 0 & 7 & 1 & 1 & 0 & 0 \\ 31 & 32 & 0 & 36 & 0 & 6 & 1 & 0 \\ 4 & 5 & 1 & 5 & 0 & 1 & 0 & 0 \\ 7 & 3 & 0 & 8 & 0 & 1 & 0 & 1 \end{array}\right] \quad (-1)R_1 + R_4 \to R_4$$

$$\left[\begin{array}{cccc|cccc} 6 & 7 & 0 & 7 & 1 & 1 & 0 & 0 \\ 31 & 32 & 0 & 36 & 0 & 6 & 1 & 0 \\ 4 & 5 & 1 & 5 & 0 & 1 & 0 & 0 \\ 1 & -4 & 0 & 1 & -1 & 0 & 0 & 1 \end{array}\right] \quad \begin{array}{l} (-7)R_4 + R_1 \to R_1 \\ (-36)R_4 + R_2 \to R_2 \\ (-5)R_4 + R_3 \to R_3 \end{array}$$

$$\left[\begin{array}{cccc|cccc} -1 & 35 & 0 & 0 & 8 & 1 & 0 & -7 \\ -5 & 176 & 0 & 0 & 36 & 6 & 1 & -36 \\ -1 & 25 & 1 & 0 & 5 & 1 & 0 & -5 \\ 1 & -4 & 0 & 1 & -1 & 0 & 0 & 1 \end{array}\right] \quad \begin{array}{l} (-5)R_1 + R_2 \to R_2 \\ (-1)R_1 + R_3 \to R_3 \\ R_1 + R_4 \to R_4 \\ (-1)R_1 \to R_1 \end{array}$$

$$\left[\begin{array}{cccc|cccc} 1 & -35 & 0 & 0 & -8 & -1 & 0 & 7 \\ 0 & 1 & 0 & 0 & -4 & 1 & 1 & -1 \\ 0 & -10 & 1 & 0 & -3 & 0 & 0 & 2 \\ 0 & 31 & 0 & 1 & 7 & 1 & 0 & -6 \end{array}\right] \quad \begin{array}{l} 35R_2 + R_1 \to R_1 \\ 10R_2 + R_3 \to R_3 \\ (-31)R_2 + R_4 \to R_4 \end{array}$$

$$\left[\begin{array}{cccc|cccc} 1 & 0 & 0 & 0 & -148 & 34 & 35 & -28 \\ 0 & 1 & 0 & 0 & -4 & 1 & 1 & -1 \\ 0 & 0 & 1 & 0 & -43 & 10 & 10 & -8 \\ 0 & 0 & 0 & 1 & 131 & -30 & -31 & 25 \end{array}\right]$$

Since the first four columns of the last augmented matrix form the identity matrix $I_4$, we are done. The inverse of the given matrix is

$$\begin{bmatrix} -148 & 34 & 35 & -28 \\ -4 & 1 & 1 & -1 \\ -43 & 10 & 10 & -8 \\ 131 & -30 & -31 & 25 \end{bmatrix}$$

You should check the correctness of the result by multiplying the given matrix by the matrix we just obtained to verify that the product is indeed the identity matrix $I_4$.*
**Do Exercise 21.**                                                                                             ■

## Application

**EXAMPLE 9**    A Seattle merchant owns three grocery stores and wishes to mix two kinds of nuts. Kind A is worth $4 a pound and kind B is worth $6 a pound. He would like to get 100 pounds of mixture to sell at $5.20 per pound for his central store, 120 pounds of mixture to sell at $4.80 per pound for his northern store, and 150 pounds of mixture to sell at $5.40 per pound for his southern store. How many pounds of each kind of nut must he mix for each of his three stores?

**Solution**    Let $q$ pounds be the quantity of mixture he wishes to obtain for a store and $c$ dollars per pound be the selling price of the mixture at that store. Thus, for the central store $q = 100$ and $c = \$5.20$, for the northern store $q = 120$ and $c = \$4.80$, and for the southern store $q = 150$ and $c = \$5.40$. Let $x$ pounds of kind A nuts and $y$ pounds of kind B nuts be the quantities he must mix in each store to get the desired result. Then, $x + y = q$. Since kind A costs $4 a pound, the cost of $x$ pounds is $\$4x$. Similarly, we find that the cost of $y$ pounds of kind B is $\$6y$. It follows that the cost of the mixture at any store is $\$(4x + 6y)$. But $q$ pounds of mixture at $c$ dollars a pound is worth $qc$ dollars. Thus, we have the following linear system in $x$ and $y$:

$$\begin{cases} x + \phantom{4}y = q \\ 4x + 6y = qc \end{cases}$$

The matrix equation for the system is

$$\begin{bmatrix} 1 & 1 \\ 4 & 6 \end{bmatrix} \cdot \begin{bmatrix} x \\ y \end{bmatrix} = \begin{bmatrix} q \\ qc \end{bmatrix}$$

We find the inverse of the coefficient matrix using row reduction.

$$\begin{bmatrix} 1 & 1 & | & 1 & 0 \\ 4 & 6 & | & 0 & 1 \end{bmatrix} \qquad (-4)R_1 + R_2 \rightarrow R_2$$

---

*In advanced texts, it is proved that if $\mathbf{A}$ and $\mathbf{B}$ are $n \times n$ matrices and $\mathbf{AB} = \mathbf{I}_n$, $\mathbf{BA} = \mathbf{I}_n$ also. Hence, when you have found the inverse $\mathbf{B}$ of some $n \times n$ matrix $\mathbf{A}$ and you wish to check the result, it is sufficient to verify through multiplication that $\mathbf{AB} = \mathbf{I}_n$. It is not necessary to check that $\mathbf{BA} = \mathbf{I}_n$ also.

$$\begin{bmatrix} 1 & 1 & | & 1 & 0 \\ 0 & 2 & | & -4 & 1 \end{bmatrix} \qquad (\tfrac{1}{2})R_2 \to R_2$$

$$\begin{bmatrix} 1 & 1 & | & 1 & 0 \\ 0 & 1 & | & -2 & \tfrac{1}{2} \end{bmatrix} \qquad (-1)R_2 + R_1 \to R_1$$

$$\begin{bmatrix} 1 & 0 & | & 3 & \tfrac{-1}{2} \\ 0 & 1 & | & -2 & \tfrac{1}{2} \end{bmatrix}$$

The inverse of the coefficient matrix is

$$\begin{bmatrix} 3 & \tfrac{-1}{2} \\ -2 & \tfrac{1}{2} \end{bmatrix}$$

We multiply both sides of the matrix equation (on the left of each side) by this inverse. We get

$$\begin{bmatrix} 3 & \tfrac{-1}{2} \\ -2 & \tfrac{1}{2} \end{bmatrix} \cdot \begin{bmatrix} 1 & 1 \\ 4 & 6 \end{bmatrix} \cdot \begin{bmatrix} x \\ y \end{bmatrix} = \begin{bmatrix} 3 & \tfrac{-1}{2} \\ -2 & \tfrac{1}{2} \end{bmatrix} \cdot \begin{bmatrix} q \\ qc \end{bmatrix}$$

Thus,

$$\begin{bmatrix} 1 & 0 \\ 0 & 1 \end{bmatrix} \cdot \begin{bmatrix} x \\ y \end{bmatrix} = \begin{bmatrix} 3q - (\tfrac{1}{2})qc \\ -2q + (\tfrac{1}{2})qc \end{bmatrix}$$

That is,

$$\begin{bmatrix} x \\ y \end{bmatrix} = q \begin{bmatrix} 3 - (\tfrac{1}{2})c \\ -2 + (\tfrac{1}{2})c \end{bmatrix}$$

Since $q = 100$ and $c = 5.2$ in the central store, we have

$$\begin{bmatrix} x \\ y \end{bmatrix} = 100 \begin{bmatrix} 3 - (\tfrac{1}{2})5.2 \\ -2 + (\tfrac{1}{2})5.2 \end{bmatrix} = 100 \begin{bmatrix} 3 - 2.60 \\ -2 + 2.60 \end{bmatrix} = \begin{bmatrix} 40 \\ 60 \end{bmatrix}$$

The merchant should mix 40 pounds of kind A nuts with 60 pounds of kind B nuts for his central store. Similarly, $q = 120$ and $c = \$4.80$ in the northern store. Therefore,

$$\begin{bmatrix} x \\ y \end{bmatrix} = 120 \begin{bmatrix} 3 - (\tfrac{1}{2})4.8 \\ -2 + (\tfrac{1}{2})4.8 \end{bmatrix} = 120 \begin{bmatrix} 3 - 2.4 \\ -2 + 2.4 \end{bmatrix} = \begin{bmatrix} 72 \\ 48 \end{bmatrix}$$

The merchant should mix 72 pounds of kind A nuts with 48 pounds of kind B nuts for his northern store.

Finally, for the southern store we have $q = 150$ and $c = 5.4$. Thus,

$$\begin{bmatrix} x \\ y \end{bmatrix} = 150 \begin{bmatrix} 3 - (\tfrac{1}{2})5.4 \\ -2 + (\tfrac{1}{2})5.4 \end{bmatrix} = 150 \begin{bmatrix} 3 - 2.7 \\ -2 + 2.7 \end{bmatrix} = \begin{bmatrix} 45 \\ 105 \end{bmatrix}.$$

The merchant should mix 45 pounds of kind A nuts with 105 pounds of kind B nuts for his southern store.

**Do Exercise 45.**

## Exercise Set 3.4

*In Exercises 1–4, show that the given matrix does not have an inverse.*

**1.** $\begin{bmatrix} 1 & 0 \\ 1 & 0 \end{bmatrix}$

**2.** $\begin{bmatrix} 2 & 1 \\ 6 & 3 \end{bmatrix}$

**3.** $\begin{bmatrix} 3 & 2 \\ 9 & 6 \end{bmatrix}$

**4.** $\begin{bmatrix} 5 & 7 \\ 20 & 28 \end{bmatrix}$

*In Exercises 5–24, find the inverse, if it exists, of the given matrix.*

**5.** $\begin{bmatrix} 5 & 7 \\ 2 & 3 \end{bmatrix}$

**6.** $\begin{bmatrix} 6 & 17 \\ 1 & 3 \end{bmatrix}$

**7.** $\begin{bmatrix} 9 & 13 \\ 2 & 3 \end{bmatrix}$

**8.** $\begin{bmatrix} -4 & 9 \\ 3 & -7 \end{bmatrix}$

**9.** $\begin{bmatrix} 3 & 8 \\ 5 & 12 \end{bmatrix}$

**10.** $\begin{bmatrix} 9 & 3 \\ 8 & 5 \end{bmatrix}$

**11.** $\begin{bmatrix} 5 & -5 & 4 \\ 0 & 3 & 1 \\ 2 & -1 & 2 \end{bmatrix}$

**12.** $\begin{bmatrix} 1 & -3 & 0 \\ 2 & 2 & 3 \\ 2 & -1 & 2 \end{bmatrix}$

**13.** $\begin{bmatrix} 7 & 6 & -17 \\ 2 & 2 & -5 \\ -6 & -5 & 15 \end{bmatrix}$

**14.** $\begin{bmatrix} 3 & 2 & -7 \\ 2 & 2 & -5 \\ 2 & 3 & -5 \end{bmatrix}$

**15.** $\begin{bmatrix} 1 & 0 & -2 \\ 2 & 5 & -6 \\ 4 & 5 & -10 \end{bmatrix}$

**16.** $\begin{bmatrix} -1 & -2 & -2 \\ 4 & 1 & 5 \\ 2 & -1 & 2 \end{bmatrix}$

**17.** $\begin{bmatrix} -7 & 1 & 1 \\ 3 & 2 & 3 \\ 5 & 1 & 2 \end{bmatrix}$

**18.** $\begin{bmatrix} 3 & -1 & 3 \\ 2 & 2 & 3 \\ 0 & -3 & -1 \end{bmatrix}$

**19.** $\begin{bmatrix} 2 & -1 & 7 \\ 3 & 9 & 0 \\ 5 & 0 & 2 \end{bmatrix}$

**20.** $\begin{bmatrix} 3 & 1 & 4 \\ -3 & -4 & 2 \\ -3 & 2 & -10 \end{bmatrix}$

**21.** $\begin{bmatrix} 7 & -8 & -5 & 6 \\ 1 & -4 & 0 & 1 \\ 3 & -11 & 3 & 4 \\ 4 & -6 & -1 & 4 \end{bmatrix}$

**22.** $\begin{bmatrix} -1 & 1 & 1 & -2 \\ -8 & 15 & 14 & -33 \\ 1 & 1 & 1 & -3 \\ -5 & -1 & 0 & 5 \end{bmatrix}$

**23.** $\begin{bmatrix} 0 & 1 & 3 & 1 \\ 5 & 1 & -2 & 5 \\ 1 & -2 & 6 & 3 \\ 2 & 3 & 2 & 3 \end{bmatrix}$

**24.** $\begin{bmatrix} 2 & 2 & -1 & 2 \\ 4 & 5 & 1 & 5 \\ 7 & 2 & -6 & 6 \\ 3 & -2 & -1 & 3 \end{bmatrix}$

*In Exercises 25–44, write the given linear system as a matrix equation and solve that equation by using the inverses found in Exercises 5–24.*

**25.** $\begin{cases} 5x + 7y = 2 \\ 2x + 3y = 5 \end{cases}$

**26.** $\begin{cases} 6x + 17y = -1 \\ x + 3y = 3 \end{cases}$

**27.** $\begin{cases} 9x + 13y = -3 \\ 2x + 3y = 5 \end{cases}$

**28.** $\begin{cases} -4x + 9y = 13 \\ 3x - 7y = -4 \end{cases}$

**29.** $\begin{cases} 3x + 8y = -1 \\ 5x + 12y = 17 \end{cases}$

**30.** $\begin{cases} 9x + 3y = 6 \\ 8x + 5y = 15 \end{cases}$

**31.** $\begin{cases} 5x - 5y + 4z = 3 \\ 3y + z = 1 \\ 2x - y + 2z = 5 \end{cases}$

**32.** $\begin{cases} x - 3y = -5 \\ 2x + 2y + 3z = -7 \\ 2x - y + 2z = 9 \end{cases}$

**33.** $\begin{cases} 7x + 6y - 17z = 0 \\ 2x + 2y - 5z = 3 \\ -6x - 5y + 15z = 7 \end{cases}$

**34.** $\begin{cases} 3x + 2y - 7z = 15 \\ 2x + 2y - 5z = -4 \\ 2x + 3y - 5z = 7 \end{cases}$

**35.** $\begin{cases} x + -2z = 3 \\ 2x + 5y - 6z = 0 \\ 4x + 5y - 10z = 5 \end{cases}$

**36.** $\begin{cases} -x - 2y - 2z = -4 \\ 4x + y + 5z = 11 \\ 2x - y + 2z = 5 \end{cases}$

**37.** $\begin{cases} -7x + y + z = 3 \\ 3x + 2y + 3z = 1 \\ 5x + y + 2z = 5 \end{cases}$

**38.** $\begin{cases} 3x - y + 3z = 12 \\ 2x + 2y + 3z = 0 \\ -3y - z = 17 \end{cases}$

**39.** $\begin{cases} 2x - y + 7z = 8 \\ 3x + 9y = 7 \\ 5x + 2z = 1 \end{cases}$

**40.** $\begin{cases} 3x + y + 4z = 9 \\ -3x - 4y + 2z = 19 \\ -3x + 2y - 10z = 3 \end{cases}$

**41.** $\begin{cases} 7w - 8x - 5y + 6z = 1 \\ w - 4x + z = 3 \\ 3w - 11x + 3y + 4z = 1 \\ 4w - 6x - y + 4z = 7 \end{cases}$

**42.**
$$\begin{cases} w + x + y - 2z = 0 \\ -8w + 15x + 14y - 33z = 2 \\ w + x + y - 3z = 3 \\ -5w - x + 5z = 0 \end{cases}$$

**43.**
$$\begin{cases} x + 3y + z = 8 \\ 5w + x - 2y + 5z = 3 \\ w - 2x + 6y + 3z = 5 \\ 2w + 3x + 2y + 3z = 7 \end{cases}$$

**44.**
$$\begin{cases} 2w + 2x - y + 2z = 6 \\ 4w + 5x + y + 5z = 2 \\ 7w + 2x - 6y + 6z = -1 \\ 3w - 2x - y + 3z = 1 \end{cases}$$

**45.** A Chicago merchant owns three grocery stores and wishes to mix two kinds of nuts. Kind A is worth \$3.60 a pound and kind B is worth \$4.60 a pound. She would like to get 200 pounds of mixture to sell at \$3.80 per pound for her central store, 300 pounds of mixture to sell at \$4 a pound for her northern store, and 250 pounds of mixture to sell at \$4.20 per pound for her southern store. How many pounds of each kind of nut must she mix for each of her three stores?

**46.** A Montreal confectioner owns three stores and wishes to mix candy worth \$4.25 per pound with candy worth \$5.25 per pound for each of the three stores. He wants to get 200 pounds of an assortment worth \$4.45 a pound for store A, 230 pounds of an assortment worth \$4.45 a pound for store B, and 250 pounds of an assortment worth \$5.05 a pound for store C. How much of each kind must he mix for each of the three stores?

**47.** A concert hall has 6000 seats. Each evening the manager divides the seats into two sections. The tickets are worth \$7 in section A and \$10 in section B. Assuming that all tickets can be sold, how many seats must he assign to each section on weekdays, on Saturdays, and on Sundays to get the following revenues: \$48,000 on each of the weekdays, \$57,000 on Saturdays, an \$51,000 on Sundays.

**48.** A football stadium has 60,000 seats. The manager divides the stadium into two sections for the exhibition games, the games the home team plays outside its division, and the games it plays within its division. The tickets are worth \$19 in section A and \$14 in section B. Assuming that all tickets can be sold, how many seats must he assign to each section on days of an exhibition game, a game outside the division, and a game within the division to bring in the following revenues: \$970,000 for each exhibition game, \$990,000 for each game outside the division, and \$1,120,000 for each game within the division.

**49.** Use the inverse of the coefficient matrix to solve the five systems obtained in Exercise 52, Section 3.3.

**50.** Use the inverse of the coefficient matrix to solve the five systems obtained in Exercise 53, Section 3.3.

**51.** Use the inverse of the coefficient matrix to solve the four systems obtained in Exercise 54, Section 3.3.

**52.** Use the inverse of the coefficient matrix to solve the three systems obtained in Exercise 55, Section 3.3.

**53.** Use the inverse of the coefficient matrix to solve the three systems obtained in Exercise 56, Section 3.3.

**54.** In the preceding section we have shown that a homogeneous system of linear equations has a nontrivial solution whenever the number of unknowns is larger than the number of equations. Show that if the number of equations is the same as the number of unknowns, the only solution of the system is the trivial solution whenever the coefficient matrix has an inverse.

**55.** Show that the only solution of the following homogeneous system is the trivial solution.

$$\begin{cases} 2x + 5y = 0 \\ 3x + 7y = 0 \end{cases}$$

**56.** Same as Exercise 55 for the system

$$\begin{cases} 3x + 2y - 3z = 0 \\ x + 2y - 3z = 0 \\ 4x + 3y + 3z = 0 \end{cases}$$

# 3.5 Determinants

In Section 2.1 we defined the determinant of a 2 × 2 matrix and showed that we could solve a linear system of two equations in two variables using a rule called Cramer's rule. In this section we wish to apply this concept to any square matrix. We shall describe how we assign to each $n \times n$ matrix a number called the determinant of that matrix. We shall do this inductively.*

## Cofactors

We begin by defining the determinant of the $1 \times 1$ matrix $[c]$ as the number $c$. We then define the cofactor of an entry of a square matrix.

> **DEFINITION 3.14:**    The *cofactor* $C_{ij}$ of the $i$-$j$ entry of a square matrix $\mathbf{A}$ is the number $(-1)^{i+j}$ times the determinant of the square matrix obtained from $\mathbf{A}$ by deleting its $i$th row and $j$th column.

(As of now we can obtain only the cofactors of the entries of a $2 \times 2$ matrix since we have not yet defined the determinant of square matrices other than $1 \times 1$ matrices.)

If we consider the $2 \times 2$ matrix

$$\mathbf{A} = \begin{bmatrix} a_{11} & a_{12} \\ a_{21} & a_{22} \end{bmatrix}$$

and let $C_{ij}$ denote the cofactor of each $a_{ij}$, we have

$$C_{11} = (-1)^{1+1}\det \begin{bmatrix} a_{11} & a_{12} \\ a_{21} & a_{22} \end{bmatrix} = \det[a_{22}] = a_{22}$$

$$C_{12} = (-1)^{1+2}\det \begin{bmatrix} a_{11} & a_{12} \\ a_{21} & a_{22} \end{bmatrix} = (-1)\det[a_{21}] = -a_{21}$$

$$C_{21} = (-1)^{2+1}\det \begin{bmatrix} a_{11} & a_{12} \\ a_{21} & a_{22} \end{bmatrix} = (-1)\det[a_{12}] = -a_{12}$$

$$C_{22} = (-1)^{2+2}\det \begin{bmatrix} a_{11} & a_{12} \\ a_{21} & a_{22} \end{bmatrix} = \det[a_{11}] = a_{11}$$

We now calculate the dot product of each row of the matrix $\mathbf{A}$ by the corresponding row of cofactors. We get

$$\begin{aligned}
[\, a_{11} \quad a_{12} \,] \cdot [\, C_{11} \quad C_{12} \,] &= [\, a_{11} \quad a_{12} \,] \cdot [\, a_{22} \quad -a_{21} \,] \\
&= a_{11}(a_{22}) + a_{12}(-a_{21}) \\
&= a_{11}a_{22} - a_{12}a_{21}
\end{aligned}$$

Also,

$$\begin{aligned}
[\, a_{21} \quad a_{22} \,] \cdot [\, C_{21} \quad C_{22} \,] &= [\, a_{21} \quad a_{22} \,] \cdot [\, -a_{12} \quad a_{11} \,] \\
&= a_{21}(-a_{12}) + a_{22}(a_{11}) \\
&= -a_{21}a_{12} + a_{22}a_{11} = a_{11}a_{22} - a_{12}a_{21}
\end{aligned}$$

---

*The process known as definition by induction consists of two parts: **a.** we define the determinant of a $1 \times 1$ matrix and **b.** assuming we have defined the determinant of an $n \times n$ matrix for fixed $n \geq 1$, we further define the determinant of a $(n + 1) \times (n + 1)$ matrix.

In both cases we get the same result. This common value of the two dot products is defined as the determinant of the $2 \times 2$ matrix **A**. This definition agrees with that given in Section 2.1. Thus,

$$\det\begin{bmatrix} a_{11} & a_{12} \\ a_{21} & a_{22} \end{bmatrix} = a_{11}a_{22} - a_{12}a_{21}$$

**EXAMPLE 1**   Calculate the determinant of the $2 \times 2$ matrix

$$\begin{bmatrix} 3 & -2 \\ 5 & 7 \end{bmatrix}$$

**Solution**   $\det\begin{bmatrix} 3 & -2 \\ 5 & 7 \end{bmatrix} = 3(7) - (-2)5$

$$= 21 + 10 = 31$$

**Do Exercise 1.**   ■

Now that we know how to calculate the determinant of a $2 \times 2$ matrix, we can find each of the cofactors of the nine entries of a $3 \times 3$ matrix.

**EXAMPLE 2**   Calculate the cofactors of the nine entries of the $3 \times 3$ matrix

$$\mathbf{A} = \begin{bmatrix} 3 & 5 & -2 \\ 2 & 1 & 4 \\ 6 & 3 & 8 \end{bmatrix}$$

**Solution**   The cofactor $C_{11}$ is found by multiplying $(-1)^{1+1}$ by the determinant of the $2 \times 2$ matrix obtained from **A** by deleting row 1 and column 1. Thus,

$$C_{11} = (-1)^{1+1}\det\begin{bmatrix} 3 & 5 & -2 \\ 2 & 1 & 4 \\ 6 & 3 & 8 \end{bmatrix}$$

$$= (-1)^{1+1}\det\begin{bmatrix} 1 & 4 \\ 3 & 8 \end{bmatrix}$$

$$= 1(8) - 4(3) = 8 - 12 = -4$$

Similarly, $C_{12}$ is obtained by multiplying $(-1)^{1+2}$ by the determinant of the $2 \times 2$ matrix obtained from **A** by deleting row 1 and column 2. Thus,

$$C_{12} = (-1)^{1+2}\det\begin{bmatrix} 3 & 5 & -2 \\ 2 & 1 & 4 \\ 6 & 3 & 8 \end{bmatrix}$$

$$= (-1)^{1+2}\det\begin{bmatrix} 2 & 4 \\ 6 & 8 \end{bmatrix}$$

$$= (-1)[2(8) - 4(6)] = (-1)[16 - 24]$$

$$= (-1)(-8) = 8$$

Similarly,

$$C_{13} = (-1)^{1+3} \det \begin{bmatrix} 3 & 5 & -2 \\ 2 & 1 & 4 \\ 6 & 3 & 8 \end{bmatrix}$$

$$= (-1)^{1+3} \det \begin{bmatrix} 2 & 1 \\ 6 & 3 \end{bmatrix}$$

$$= 2(3) - 1(6) = 6 - 6 = 0$$

$$C_{21} = (-1)^{2+1} \det \begin{bmatrix} 3 & 5 & -2 \\ 2 & 1 & 4 \\ 6 & 3 & 8 \end{bmatrix}$$

$$= (-1)^{2+1} \det \begin{bmatrix} 5 & -2 \\ 3 & 8 \end{bmatrix}$$

$$= (-1)[5(8) - (-2)3] = (-1)[40 + 6]$$

$$= (-1)46 = -46$$

$$C_{22} = (-1)^{2+2} \det \begin{bmatrix} 3 & 5 & -2 \\ 2 & 1 & 4 \\ 6 & 3 & 8 \end{bmatrix}$$

$$= (-1)^{2+2} \det \begin{bmatrix} 3 & -2 \\ 6 & 8 \end{bmatrix}$$

$$= 3(8) - (-2)6 = 24 + 12 = 36$$

$$C_{23} = (-1)^{2+3} \det \begin{bmatrix} 3 & 5 & -2 \\ 2 & 1 & 4 \\ 6 & 3 & 8 \end{bmatrix}$$

$$= (-1)^{2+3} \det \begin{bmatrix} 3 & 5 \\ 6 & 3 \end{bmatrix}$$

$$= (-1)[3(3) - 5(6)] = (-1)[9 - 30]$$

$$= (-1)(-21) = 21$$

$$C_{31} = (-1)^{3+1} \det \begin{bmatrix} 3 & 5 & -2 \\ 2 & 1 & 4 \\ 6 & 3 & 8 \end{bmatrix}$$

$$= (-1)^{3+1} \det \begin{bmatrix} 5 & -2 \\ 1 & 4 \end{bmatrix}$$

$$= 5(4) - (-2)1 = 20 + 2 = 22$$

$$C_{32} = (-1)^{3+2} \det \begin{bmatrix} 3 & 5 & -2 \\ 2 & 1 & 4 \\ 6 & 3 & 8 \end{bmatrix}$$

$$= (-1)^{3+2} \det \begin{bmatrix} 3 & -2 \\ 2 & 4 \end{bmatrix}$$

$$= (-1)[3(4) - (-2)2] = (-1)[12 + 4]$$
$$= (-1)16 = -16$$

$$C_{33} = (-1)^{3+3} \det \begin{bmatrix} 3 & 5 & -2 \\ 2 & 1 & 4 \\ 6 & 3 & 8 \end{bmatrix}$$

$$= (-1)^{3+3} \det \begin{bmatrix} 3 & 5 \\ 2 & 1 \end{bmatrix}$$

$$= 3(1) - 5(2) = 3 - 10 = -7$$

EXAMPLE 3   Calculate the dot product of each row of the matrix of the previous example by the corresponding row of cofactors and compare the three results.

Solution   We have

$$[\ a_{11} \quad a_{12} \quad a_{13}\ ] \cdot [\ C_{11} \quad C_{12} \quad C_{13}\ ] =$$
$$[\ 3 \quad 5 \quad -2\ ] \cdot [\ -4 \quad 8 \quad 0\ ] = 3(-4) + 5(8) + (-2)0$$
$$= -12 + 40 + 0 = 28$$

$$[\ a_{21} \quad a_{22} \quad a_{23}\ ] \cdot [\ C_{21} \quad C_{22} \quad C_{23}\ ] =$$
$$[\ 2 \quad 1 \quad 4\ ] \cdot [\ -46 \quad 36 \quad 21\ ] = 2(-46) + 1(36) + 4(21)$$
$$= -92 + 36 + 84 = 28$$

$$[\ a_{31} \quad a_{32} \quad a_{33}\ ] \cdot [\ C_{31} \quad C_{32} \quad C_{33}\ ] =$$
$$[\ 6 \quad 3 \quad 8\ ] \cdot [\ 22 \quad -16 \quad -7\ ] = 6(22) + 3(-16) + 8(-7)$$
$$= 132 - 48 - 56 = 28$$

Note that 28 was obtained all three times.

The fact that the same number was obtained for the three dot products is not a coincidence. It can be proved, but we shall not do it here, that if **A** is any 3 × 3 matrix, then the dot product of any row of **A** by the corresponding row of cofactors is equal to the dot product of any other row of **A** by the corresponding row of cofactors. This common value is, by definition, the determinant of the matrix **A**. Thus, from the results of Example 3, we can write

$$\det \begin{bmatrix} 3 & 5 & -2 \\ 2 & 1 & 4 \\ 6 & 3 & 8 \end{bmatrix} = 28$$

## Definition of Determinant

Now that we know how to find the determinant of a 3 × 3 matrix, we may define the determinant of a 4 × 4 matrix as the dot product of any row of that matrix by the corresponding row of cofactors. It can be proved that the result is independent of the choice of the row.

> **DEFINITION 3.15:**    In general, once we have defined the determinant of an $n \times n$ matrix, then the *determinant of an (n + 1) × (n + 1)* matrix is defined as the dot product of any row of that matrix by the corresponding row of cofactors.*

Again, it can be proved that this dot product is independent of the choice of the row.

It is not practical to use the definition of the determinant of a square matrix for computing its value. We already have seen that

$$\det \begin{bmatrix} a_{11} & a_{12} \\ a_{21} & a_{22} \end{bmatrix} = a_{11}a_{22} - a_{12}a_{21}$$

We now derive a formula for the determinant of a $3 \times 3$ matrix.

**EXAMPLE 4**    Use the definition of determinant to derive a formula for the determinant of a $3 \times 3$ matrix.

**Solution**    Consider an arbitrary $3 \times 3$ matrix $\mathbf{A}$, where

$$\mathbf{A} = \begin{bmatrix} a_{11} & a_{12} & a_{13} \\ a_{21} & a_{22} & a_{23} \\ a_{31} & a_{32} & a_{33} \end{bmatrix}$$

We first calculate the cofactors of each of the entries of row 1. We have

$$C_{11} = (-1)^{1+1} \det \begin{bmatrix} a_{11} & a_{12} & a_{13} \\ a_{21} & a_{22} & a_{23} \\ a_{31} & a_{32} & a_{33} \end{bmatrix}$$

$$= (-1)^{1+1} \det \begin{bmatrix} a_{22} & a_{23} \\ a_{32} & a_{33} \end{bmatrix} = a_{22}a_{33} - a_{23}a_{32}$$

---

* If $\mathbf{A} = \begin{bmatrix} a_{11} & a_{12} & a_{13} & \cdots & a_{1n} \\ a_{21} & a_{22} & a_{23} & \cdots & a_{2n} \\ \vdots & \vdots & \vdots & & \vdots \\ a_{n1} & a_{n2} & a_{n3} & \cdots & a_{nn} \end{bmatrix}$ is an $n \times n$ matrix, then the symbol

$\begin{vmatrix} a_{11} & a_{12} & a_{13} & \cdots & a_{1n} \\ a_{21} & a_{22} & a_{23} & \cdots & a_{2n} \\ \vdots & \vdots & \vdots & & \vdots \\ a_{n1} & a_{n2} & a_{n3} & \cdots & a_{nn} \end{vmatrix}$ is often used instead of det$\mathbf{A}$ to denote the determinant of the

matrix $\mathbf{A}$.

$$C_{12} = (-1)^{1+2} \det \begin{bmatrix} a_{11} & a_{12} & a_{13} \\ a_{21} & a_{22} & a_{23} \\ a_{31} & a_{32} & a_{33} \end{bmatrix}$$

$$= (-1)^{1+2} \det \begin{bmatrix} a_{21} & a_{23} \\ a_{31} & a_{33} \end{bmatrix} = (-1)^{1+2}(a_{21}a_{33} - a_{23}a_{31})$$

$$= -a_{21}a_{33} + a_{23}a_{31}$$

$$C_{13} = (-1)^{1+3} \det \begin{bmatrix} a_{11} & a_{12} & a_{13} \\ a_{21} & a_{22} & a_{23} \\ a_{31} & a_{32} & a_{33} \end{bmatrix}$$

$$= (-1)^{1+3} \det \begin{bmatrix} a_{21} & a_{22} \\ a_{31} & a_{32} \end{bmatrix} = a_{21}a_{32} - a_{22}a_{31}$$

Thus,

$$\det \mathbf{A} = [\, a_{11} \quad a_{12} \quad a_{13} \,] \cdot [\, C_{11} \quad C_{12} \quad C_{13} \,]$$

$$= a_{11}C_{11} + a_{12}C_{12} + a_{13}C_{13}$$

$$= a_{11}(a_{22}a_{33} - a_{23}a_{32}) + a_{12}$$

$$(-a_{21}a_{33} + a_{23}a_{31}) + a_{13}(a_{21}a_{32} - a_{22}a_{31})$$

$$= a_{11}a_{22}a_{33} - a_{11}a_{23}a_{32} - a_{12}a_{21}a_{33} + a_{12}a_{23}a_{31}$$

$$+ a_{13}a_{21}a_{32} - a_{13}a_{22}a_{31}.$$

There are six terms in the last line of the sum, three with a plus sign and three with a minus sign. Each term is a product of three factors, one from each row and each column. There are two mnemonic devices commonly used to remember the formula giving the determinant of the $3 \times 3$ matrix

$$\mathbf{A} = \begin{bmatrix} a_{11} & a_{12} & a_{13} \\ a_{21} & a_{22} & a_{23} \\ a_{31} & a_{32} & a_{33} \end{bmatrix}$$

These are illustrated schematically by the following diagrams. In each case, we multiply the three entries that are strung together by an arrow, assigning to each term the sign that is indicated at the tail of the corresponding arrow. In the second diagram, we duplicated columns 1 and 2 to the right of the matrix $\mathbf{A}$ to be able to extend the arrows.

First device

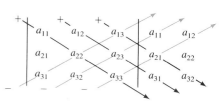

Second device

EXAMPLE 5   Find the determinant of the $3 \times 3$ matrix

$$\begin{bmatrix} 3 & -2 & 5 \\ 4 & 1 & 2 \\ 7 & 3 & -6 \end{bmatrix}$$

**Solution**   We use the first device to find which entries are multiplied together.

Thus,

$$\det \begin{bmatrix} 3 & -2 & 5 \\ 4 & 1 & 2 \\ 7 & 3 & -6 \end{bmatrix} = 3(1)(-6) + 4(3)(5) + 7(2)(-2)$$
$$-5(1)(7) - 2(3)(3) - (-6)(4)(-2)$$
$$= -18 + 60 - 28 - 35 - 18 - 48$$
$$= -87 \qquad \blacksquare$$

Unfortunately, there is no simple device to calculate the determinant of an $n \times n$ matrix when $n$ is larger than 3. (See Exercise 56.) Thus, we usually depend on the definition, as is illustrated in the following example.

EXAMPLE 6   Find the determinant of the matrix

$$\mathbf{A} = \begin{bmatrix} 3 & 5 & -1 & 6 \\ 2 & 0 & 3 & 0 \\ 5 & 2 & 5 & 2 \\ 7 & 1 & 2 & 7 \end{bmatrix}$$

**Solution**   Since two of the entries in row 2 are 0's, it is easiest to compute the dot product of row 2 by the corresponding row of cofactors. With the usual notation for cofactors, we get

$$\det \mathbf{A} = [\, 2 \quad 0 \quad 3 \quad 0 \,] \cdot [\, C_{21} \quad C_{22} \quad C_{23} \quad C_{24} \,]$$
$$= 2C_{21} + 0C_{22} + 3C_{23} + 0C_{24} = 2C_{21} + 3C_{23}$$

We do not need to calculate the cofactors $C_{22}$ and $C_{24}$. We find

$$C_{21} = (-1)^{2+1} \det \begin{bmatrix} 5 & -1 & 6 \\ 2 & 5 & 2 \\ 1 & 2 & 7 \end{bmatrix}$$

$$= (-1)[5 \cdot 5 \cdot 7 + (-1)2 \cdot 1 + 6 \cdot 2 \cdot 2$$
$$- 1 \cdot 5 \cdot 6 - 2 \cdot 2 \cdot 5 - 7 \cdot (-1)2]$$
$$= (-1)[175 - 2 + 24 - 30 - 20 + 14] = (-1)(161) = -161$$

Also,

$$C_{23} = (-1)^{2+3} \det \begin{bmatrix} 3 & 5 & 6 \\ 5 & 2 & 2 \\ 7 & 1 & 7 \end{bmatrix}$$
$$= (-1)[3 \cdot 2 \cdot 7 + 5 \cdot 2 \cdot 7 + 6 \cdot 1 \cdot 5$$
$$- 7 \cdot 2 \cdot 6 - 1 \cdot 2 \cdot 3 - 7 \cdot 5 \cdot 5]$$
$$= (-1)[42 + 70 + 30 - 84 - 6 - 175] = (-1)(-123) = 123$$

Thus,

$$\det\mathbf{A} = 2(-161) + 3(123) = -322 + 369$$
$$= 47$$

**Do Exercise 15.** ■

## Properties of Determinants

It is clear from the previous example that if we wish to calculate the determinant of a matrix using the definition, it is best to choose the row which has the most 0's. In fact, we can use elementary row operations to get a new matrix that has a row with all but one entry equal to 0 and whose determinant is related to the determinant of the original matrix. We will state the necessary properties of determinants to justify this process after the following definition.

> **DEFINITION 3.16:** If $\mathbf{A}$ is an $m \times n$ matrix, its *transpose* is the $n \times m$ matrix whose $i$-$j$ entry is the $j$-$i$ entry of the matrix $\mathbf{A}$. The rows of matrix $\mathbf{A}$ become the columns of the transpose and the columns of matrix $\mathbf{A}$ become the rows of the transpose, while preserving their order. The transpose of a matrix $\mathbf{A}$ is denoted $\mathbf{A}^T$.

**EXAMPLE 7** Find the transpose of the matrix $\mathbf{A}$ where

$$\mathbf{A} = \begin{bmatrix} 2 & 3 & 7 \\ 9 & 0 & 4 \end{bmatrix}$$

**Solution** Since $\mathbf{A}$ is a $2 \times 3$ matrix, its transpose is a $3 \times 2$ matrix. Furthermore, row 1 of $\mathbf{A}$ is column 1 of $\mathbf{A}^T$ and row 2 of $\mathbf{A}$ is column 2 of $\mathbf{A}^T$. We write

$$\mathbf{A}^T = \begin{bmatrix} 2 & 9 \\ 3 & 0 \\ 7 & 4 \end{bmatrix}$$

■

We now state (but do not prove) some basic properties of determinants which are useful for calculating the determinant of a square matrix.

### Basic Properties of Determinants

1. If $\mathbf{A}$ is a square matrix, then $\det\mathbf{A} = \det\mathbf{A}^T$.
2. If $\mathbf{A}$ is a square matrix, then $\det\mathbf{A}$ is equal to the dot product of any column of $\mathbf{A}$ by the corresponding column of cofactors.
3. If $\mathbf{A}$ is a square matrix and $\mathbf{B}$ is obtained from $\mathbf{A}$ by interchanging two rows (columns), then $\det\mathbf{B} = -\det\mathbf{A}$.
4. If $\mathbf{A}$ is a square matrix with two identical rows (columns), then $\det\mathbf{A} = 0$.
5. If $\mathbf{A}$ is a square matrix and $\mathbf{B}$ is obtained from $\mathbf{A}$ by multiplying one of the rows (columns) of $\mathbf{A}$ by a number $k$, then $\det\mathbf{B} = k \cdot \det\mathbf{A}$.
6. If $\mathbf{A}$ is a square matrix and $\mathbf{B}$ is obtained from $\mathbf{A}$ by replacing a row (column) of $\mathbf{A}$ by the sum of that row (column) and a constant multiple of another row (column) of $\mathbf{A}$, then $\det\mathbf{B} = \det\mathbf{A}$.

### Evaluation of Determinants

Using these properties repeatedly, we can get a new matrix with many entries which are 0, and whose determinant is a known multiple of that of the original matrix. We illustrate with the following example.

**EXAMPLE 8**    Evaluate the determinant of the matrix

$$\mathbf{A} = \begin{bmatrix} 3 & 5 & 7 & 8 \\ 7 & 10 & 3 & 2 \\ 2 & -5 & 8 & 3 \\ 2 & 15 & 9 & 5 \end{bmatrix}$$

**Solution**    Using Property 5, we can factor 5 out of column 2. Thus, we get

$$\det\mathbf{A} = \det\begin{bmatrix} 3 & 5 & 7 & 8 \\ 7 & 10 & 3 & 2 \\ 2 & -5 & 8 & 3 \\ 2 & 15 & 9 & 5 \end{bmatrix} = 5 \cdot \det\begin{bmatrix} 3 & 1 & 7 & 8 \\ 7 & 2 & 3 & 2 \\ 2 & -1 & 8 & 3 \\ 2 & 3 & 9 & 5 \end{bmatrix}$$

$$= 5 \cdot \det\begin{bmatrix} 3 & 1 & 7 & 8 \\ 1 & 0 & -11 & -14 \\ 5 & 0 & 15 & 11 \\ -7 & 0 & -12 & -19 \end{bmatrix}$$

where the last matrix was obtained by using Property 6 three times $((-2)R_1+R_2 \to R_2,\ R_1+R_3 \to R_3,\ (-3)R_1+R_4 \to R_4)$. Since column 2 of the last matrix has three 0's, we will find the determinant by calculating the dot product of column 2 by the corresponding column of cofactors. Since the 2-2, 3-2, and 3-4 entries of the last matrix are all 0, we need not calculate their corresponding cofactors. Therefore, we need to find $C_{12}$. We get

$$C_{12} = (-1)^{1+2} \det \begin{bmatrix} 1 & -11 & -14 \\ 5 & 15 & 11 \\ -7 & -12 & -19 \end{bmatrix}$$

$$= (-1)^{1+2} \det \begin{bmatrix} 1 & -11 & -14 \\ 0 & 70 & 81 \\ 0 & -89 & -117 \end{bmatrix}$$

$$= (-1)^{1+2}(-1)^{1+1} \det \begin{bmatrix} 70 & 81 \\ -89 & -117 \end{bmatrix}$$

$$= (-1)[70(-117) - 81(-89)] = (-1)(-981) = 981$$

Thus,

$$\det A = \det \begin{bmatrix} 3 & 5 & 7 & 8 \\ 7 & 10 & 3 & 2 \\ 2 & -5 & 8 & 3 \\ 2 & 15 & 9 & 5 \end{bmatrix} = 5 \cdot \det \begin{bmatrix} 3 & 1 & 7 & 8 \\ 7 & 2 & 3 & 2 \\ 2 & -1 & 8 & 3 \\ 2 & 3 & 9 & 5 \end{bmatrix}$$

$$= 5 \cdot \det \begin{bmatrix} 3 & 1 & 7 & 8 \\ 1 & 0 & -11 & -14 \\ 5 & 0 & 15 & 11 \\ -7 & 0 & -12 & -19 \end{bmatrix}$$

$$= 5(981) = 4905 \qquad \blacksquare$$

We now give another example to illustrate how the evaluation of a determinant can be done using the basic properties of determinants, as we did in the previous example, but writing the solution in a more compact way. We give the explanations on the right side using our earlier convention that $kR_i + R_j \to R_j$ means $k$ times row $i$ added to row $j$ gives the new row $j$, and $kC_i + C_j \to C_j$ means $k$ times column $i$ added to column $j$ gives the new column $j$.

**EXAMPLE 9** Evaluate the determinant of the matrix

$$\begin{bmatrix} 2 & 3 & 4 & 6 \\ 5 & 2 & 7 & -3 \\ 4 & 8 & -4 & 12 \\ 8 & 4 & -2 & 3 \end{bmatrix}$$

**Solution** Using the properties of determinants, we get

$$\det \begin{bmatrix} 2 & 3 & 4 & 6 \\ 5 & 2 & 7 & -3 \\ 4 & 8 & -4 & 12 \\ 8 & 4 & -2 & 3 \end{bmatrix} \qquad \text{Factoring } -4 \text{ out of row 3}$$

$$= (-4)\det \begin{bmatrix} 2 & 3 & 4 & 6 \\ 5 & 2 & 7 & -3 \\ -1 & -2 & 1 & -3 \\ 8 & 4 & -2 & 3 \end{bmatrix}$$

$(-4)R_3 + R_1 \rightarrow R_1$
$(-7)R_3 + R_2 \rightarrow R_2$
$2R_3 + R_4 \rightarrow R_4$

$$= (-4)\det \begin{bmatrix} 6 & 11 & 0 & 18 \\ 12 & 16 & 0 & 18 \\ -1 & -2 & 1 & -3 \\ 6 & 0 & 0 & -3 \end{bmatrix}$$

Calculating the dot product of column 1 by the corresponding column of cofactors and using the fact that three entries in that column are 0's.

$$= (-4)(1)C_{33}$$

$$= (-4)(1)(-1)^{3+3}\det \begin{bmatrix} 6 & 11 & 18 \\ 12 & 16 & 18 \\ 6 & 0 & -3 \end{bmatrix}$$

Factoring 6 out of column 1 and 3 out of column 3.

$$= (-4)(6)(3)\det \begin{bmatrix} 1 & 11 & 6 \\ 2 & 16 & 6 \\ 1 & 0 & -1 \end{bmatrix}$$

$C_1 + C_3 \rightarrow C_3$

$$= (-72)\det \begin{bmatrix} 1 & 11 & 7 \\ 2 & 16 & 8 \\ 1 & 0 & 0 \end{bmatrix}$$

Calculating the dot product of row 3 by the corresponding row of cofactors.

$$= (-72)(1)C_{31}$$

$$= (-72)\det \begin{bmatrix} 11 & 7 \\ 16 & 8 \end{bmatrix}$$

$$= (-72)[11 \cdot 8 - 16 \cdot 7]$$

$$= (-72)(88 - 112)$$

$$= (-72)(-24)$$

$$= 1728$$

**Do Exercise 15.**

## Cramer's Rule

We now expand Cramer's rule to solve linear systems of $n$ equations in $n$ variables.

### Cramer's Rule

Suppose a linear system of $n$ equations in $n$ variables is written as a matrix equation as follows:

$$\mathbf{AX} = \mathbf{B}$$

where

$$\mathbf{A} = [a_{ij}]$$

$$X = \begin{bmatrix} x_1 \\ x_2 \\ \vdots \\ x_n \end{bmatrix} \quad \text{and} \quad B = \begin{bmatrix} b_1 \\ b_2 \\ \vdots \\ b_n \end{bmatrix}$$

For each $i$, let $A_i$ be the matrix obtained from the coefficient matrix $A$ by replacing the $i$th column by the $n$-dimensional column vector $B$. The linear system has a unique solution $(c_1, c_2, \ldots, c_n)$ if, and only if, $\det A$ is nonzero. In that case,

$$c_i = \frac{\det A_i}{\det A}, \quad \text{for each } i = 1, 2, \ldots, n$$

**EXAMPLE 10**   Use Cramer's rule to solve the system

$$\begin{cases} 2x + 3y - 2z + 2 = 0 \\ z + 3x = 0 \\ 2y + 5x + 1 = 0 \end{cases}$$

**Solution**   We rearrange the terms of each equation so that all three variables are in the same order and eliminate any constant terms on the left sides of the equations.

$$\begin{cases} 2x + 3y - 2z = -2 \\ 3x \qquad + z = 0 \\ 5x + 2y \qquad = -1 \end{cases}$$

Thus, the coefficient matrix is

$$A = \begin{bmatrix} 2 & 3 & -2 \\ 3 & 0 & 1 \\ 5 & 2 & 0 \end{bmatrix}$$

The matrix $A_x$ is obtained from $A$ by replacing the first column (the $x$-coefficients) by the three-dimensional column vector $B$, where

$$B = \begin{bmatrix} -2 \\ 0 \\ -1 \end{bmatrix}$$

Thus,

$$A_x = \begin{bmatrix} -2 & 3 & -2 \\ 0 & 0 & 1 \\ -1 & 2 & 0 \end{bmatrix}$$

Similarly, we get

$$A_y = \begin{bmatrix} 2 & -2 & -2 \\ 3 & 0 & 1 \\ 5 & -1 & 0 \end{bmatrix}$$

and

$$A_z = \begin{bmatrix} 2 & 3 & -2 \\ 3 & 0 & 0 \\ 5 & 2 & -1 \end{bmatrix}$$

Thus,

$$\det A = \det \begin{bmatrix} 2 & 3 & -2 \\ 3 & 0 & 1 \\ 5 & 2 & 0 \end{bmatrix} = -1$$

$$\det A_x = \det \begin{bmatrix} -2 & 3 & -2 \\ 0 & 0 & 1 \\ -1 & 2 & 0 \end{bmatrix} = 1$$

$$\det A_y = \det \begin{bmatrix} 2 & -2 & -2 \\ 3 & 0 & 1 \\ 5 & -1 & 0 \end{bmatrix} = -2$$

$$\det A_z = \det \begin{bmatrix} 2 & 3 & -2 \\ 3 & 0 & 0 \\ 5 & 2 & -1 \end{bmatrix} = -3$$

It follows that,

$$x = \frac{\det A_x}{\det A} = \frac{1}{-1} = -1; \; y = \frac{\det A_y}{\det A} = \frac{-2}{-1} = 2; \; z = \frac{\det A_z}{\det A} = \frac{-3}{-1} = 3$$

The solution of the given linear system is the ordered triple $(-1, 2, 3)$.*
**Do Exercise 31.**

## Equation of a Line

We conclude this section with an interesting device to find the equation of a straight line through two given points.

EXAMPLE 11    **a.** Show that an equation of a straight line passing through the points $(a_1, b_1)$ and $(a_2, b_2)$ is

$$\det \begin{bmatrix} x & y & 1 \\ a_1 & b_1 & 1 \\ a_2 & b_2 & 1 \end{bmatrix} = 0$$

---

*In general, Cramer's rule is not the most efficient method of solving linear systems, except possibly in the case of two linear equations in two variables. However, Cramer's rule is important theoretically. For example, it allows us to determine that a linear system of $n$ equations in $n$ variables has a unique solution whenever the coefficient matrix has a nonzero determinant. In particular, the trivial solution is the only solution of a homogeneous system of $n$ linear equations in $n$ variables whenever the coefficient matrix has a nonzero determinant.

**b.** Use this result to find an equation of the line passing through the points (7, 3) and (5, 2).

**Solution**  **a.** If we replace $x$ by $a_1$ and $y$ by $b_1$ on the left side, rows 1 and 2 are identical. So, by Property 4, the determinant on the left is 0 and we conclude that $(a_1, b_1)$ is a solution of the equation. Thus, the point $(a_1, b_1)$ is on the graph of the equation. Similarly, we can show that the point $(a_2, b_2)$ is also on the graph of the equation. If we expand the left side of the equation, we get

$$A + Bx + Cy = 0$$

where $A$, $B$, and $C$ are constants obtained as follows:

$$A = \det \begin{bmatrix} a_1 & b_1 \\ a_2 & b_2 \end{bmatrix}, B = -\det \begin{bmatrix} 1 & b_1 \\ 1 & b_2 \end{bmatrix}, C = \det \begin{bmatrix} 1 & a_1 \\ 1 & a_2 \end{bmatrix}$$

Thus, we have verified that the given equation is linear, and that the two given points are on its graph.

**b.** An equation of the line through the points (7, 3) and (5, 2) is

$$\det \begin{bmatrix} x & y & 1 \\ 7 & 3 & 1 \\ 5 & 2 & 1 \end{bmatrix} = 0$$

Expanding the left side, we get

$$3x + 14 + 5y - 15 - 7y - 2x = 0$$

which simplifies to

$$x - 2y - 1 = 0$$

**Do Exercise 57.**

# Exercise Set 3.5

*In Exercises 1–16, evaluate the determinant of the given matrix.*

**1.** $\begin{bmatrix} 2 & 3 \\ 5 & 4 \end{bmatrix}$

**2.** $\begin{bmatrix} 3 & 4 \\ -2 & 5 \end{bmatrix}$

**3.** $\begin{bmatrix} 5 & 7 \\ 4 & 6 \end{bmatrix}$

**4.** $\begin{bmatrix} 5 & -1 \\ 2 & 6 \end{bmatrix}$

**5.** $\begin{bmatrix} 5 & 6 & 3 \\ 4 & 5 & -1 \\ 1 & 0 & 0 \end{bmatrix}$

**6.** $\begin{bmatrix} 3 & 7 & 5 \\ 2 & 5 & 1 \\ 1 & 3 & 3 \end{bmatrix}$

**7.** $\begin{bmatrix} 4 & -1 & 3 \\ 1 & 5 & -1 \\ 1 & 3 & -2 \end{bmatrix}$

**8.** $\begin{bmatrix} 1 & 4 & 1 \\ 2 & -1 & -2 \\ 1 & 1 & -1 \end{bmatrix}$

**9.** $\begin{bmatrix} 1 & -2 & 1 \\ 2 & -1 & -1 \\ 1 & 1 & 4 \end{bmatrix}$

**10.** $\begin{bmatrix} -1 & -2 & -3 \\ 2 & 4 & 1 \\ -1 & 2 & 0 \end{bmatrix}$

**11.** $\begin{bmatrix} 4 & -2 & 0 \\ 1 & 2 & 2 \\ 1 & 2 & 3 \end{bmatrix}$

**12.** $\begin{bmatrix} 2 & 1 & 0 \\ 1 & 0 & -1 \\ 0 & 1 & -1 \end{bmatrix}$

**13.** $\begin{bmatrix} 2 & -1 & 2 \\ 4 & 1 & 5 \\ 11 & -8 & 10 \end{bmatrix}$

**14.** $\begin{bmatrix} 6 & 4 & -14 \\ 2 & 2 & -5 \\ 2 & 3 & -5 \end{bmatrix}$

**15.** $\begin{bmatrix} 7 & -8 & -5 & 6 \\ 3 & -12 & 0 & 3 \\ 3 & -11 & 3 & 4 \\ 4 & -6 & -1 & 4 \end{bmatrix}$

**16.** $\begin{bmatrix} 2 & 2 & -5 & 2 \\ 4 & 5 & 5 & 5 \\ 7 & 2 & -30 & 6 \\ 3 & -2 & -5 & 3 \end{bmatrix}$

*In Exercises 17–20, solve the given equation for x and check your result.*

**17.** $\det \begin{bmatrix} 1 & x & -2 \\ 2 & 3 & 2 \\ 4 & 9 & -2 \end{bmatrix} = 0$

**18.** $\det \begin{bmatrix} 6x & 3 & 2 \\ 2 & x & 2 \\ -2 & -1 & 1 \end{bmatrix} = 0$

**19.** $\det \begin{bmatrix} x & 2x+1 & x \\ x^2 & x+1 & 3 \\ 1 & 0 & 0 \end{bmatrix} = -3$

**20.** $\det \begin{bmatrix} 2x+1 & 2x & -17 \\ 2 & x-1 & -5 \\ -2x & -5 & 15 \end{bmatrix} = 1$

*In Exercises 21–30, solve the given system using Cramer's rule.*

**21.** $\begin{cases} 3x + 2y = -2 \\ 7x + 5y = 3 \end{cases}$

**22.** $\begin{cases} 3x - 4y = -2 \\ -7x + 9y = 3 \end{cases}$

**23.** $\begin{cases} 5x - 3y = 7 \\ 2x + 7y = 11 \end{cases}$

**24.** $\begin{cases} 5x + 2y = -4 \\ 3x - 11y = -39 \end{cases}$

**25.** $\begin{cases} x + 13y = 34 \\ 2x - 3y = -19 \end{cases}$

**26.** $\begin{cases} 5x + 2y = 3 \\ 10x + 4y = -2 \end{cases}$

**27.** $\begin{cases} -4x + 3y = 15 \\ 3x - 7y = -16 \end{cases}$

**28.** $\begin{cases} 3x - 13y = 6 \\ 2x + 4y = 4 \end{cases}$

**29.** $\begin{cases} -5x + 7y = 64 \\ 3x + 5y = 26 \end{cases}$

**30.** $\begin{cases} 3x - 17y = -97 \\ 2x + 15y = 67 \end{cases}$

*Exercises 31–45: In each Exercise n ($31 \leq n \leq 45$), use Cramer's rule to solve the linear system of Exercise n − 30, Section 2.2.*

*Exercises 46–55: In each Exercise n ($46 \leq n \leq 55$), use Cramer's rule to solve the linear system of Exercise n − 18, Section 2.2.*

**56.** Two mnemonic devices used to calculate the determinant of a 3 × 3 matrix are given on page 253. Apply one of the devices to the matrix of Exercise 16 and show that it does not yield the correct value.

*In Exercises 57–62, write in determinant form an equation of the line passing through the two points whose coordinates are given. Then expand and simplify.*

**57.** (2, 3), (5, 6)

**58.** (2, −3), (−3, −4)

**59.** (1, 7), (3, −6)

**60.** (5, 1), (7, 8)

**61.** (4, −2), (−5, −3)

**62.** (2, 4), (5, −2)

*In Exercises 63–67, use the properties of determinants to find det B.*

**63.** **A** is a 2 × 2 matrix, det**A** = 3 and **B** = 5**A**.

**64.** **A** is a 3 × 3 matrix, det**A** = 7 and **B** = 3**A**.

**65.** **A** is a 4 × 4 matrix, det**A** = −2 and **B** = −3**A**.

**66.** det**A** = −8 and **B** is obtained by interchanging rows 2 and 3 of **A**.

**67.** det**A** = −6 and **B** = $\mathbf{A}^T$.

*It can be shown that if the vertices of a triangle have coordinates $(a_1, b_1)$, $(a_2, b_2)$ and $(a_3, b_3)$, then the area of the triangle is given by the absolute value of*

$$\left(\frac{1}{2}\right) \det \begin{bmatrix} a_1 & b_1 & 1 \\ a_2 & b_2 & 1 \\ a_3 & b_3 & 1 \end{bmatrix}$$

*In Exercises 68–71, use a determinant to find the area of the triangle whose vertices are the points with the given coordinates.*

**68.** (1, 2), (2, −3), (−2, −3)

**69.** (3, 1), (5, −2), (−3, −2)

**70.** (5, 3), (−2, 4), (2, −3)

**71.** (3, 7), (3, −1), (−4, −6)

## 3.6 Inverse of a Square Matrix Using Adjoints

In Section 3.4, we used elementary row operations to find the inverse of a nonsingular square matrix. In this section, we describe a different way of finding inverses. This way is often more efficient for small matrices (2 × 2 or 3 × 3), but is not very effective for larger matrices.

## Adjoint of a Square Matrix

In Example 2 of the preceding section we found the nine cofactors of the matrix

$$A = \begin{bmatrix} 3 & 5 & -2 \\ 2 & 1 & 4 \\ 6 & 3 & 8 \end{bmatrix}$$

We have tabulated the results. In this table the column headed $a_{ij}$ gives the entries of matrix $A$, while the column headed $C_{ij}$ gives the corresponding cofactors.

| $i$ | $j$ | $a_{ij}$ | $C_{ij}$ |
|-----|-----|----------|----------|
| 1 | 1 | 3 | $-4$ |
| 1 | 2 | 5 | 8 |
| 1 | 3 | $-2$ | 0 |
| 2 | 1 | 2 | $-46$ |
| 2 | 2 | 1 | 36 |
| 2 | 3 | 4 | 21 |
| 3 | 1 | 6 | 22 |
| 3 | 2 | 3 | $-16$ |
| 3 | 3 | 8 | $-7$ |

We have already observed that if we calculate the dot product of any row of matrix $A$ by the corresponding row of cofactors, we obtain the determinant of matrix $A$.

$$[ \ 3 \quad 5 \quad -2 \ ] \cdot [ \ -4 \quad 8 \quad 0 \ ] = 28$$
$$[ \ 2 \quad 1 \quad 4 \ ] \cdot [ \ -46 \quad 36 \quad 21 \ ] = 28$$
$$[ \ 6 \quad 3 \quad 8 \ ] \cdot [ \ 22 \quad -16 \quad -7 \ ] = 28$$

Suppose now we calculate the dot product of any row by a row of cofactors corresponding to some different row. For example,

$$[ \ a_{11} \quad a_{12} \quad a_{13} \ ] \cdot [ \ C_{21} \quad C_{22} \quad C_{23} \ ]$$
$$= [ \ 3 \quad 5 \quad -2 \ ] \cdot [ \ -46 \quad 36 \quad 21 \ ]$$
$$= 3(-46) + 5(36) + (-2)21 = -138 + 180 + -42 = 0$$

You can verify that the following are also true.

$$[ \ a_{11} \quad a_{12} \quad a_{13} \ ] \cdot [ \ C_{31} \quad C_{32} \quad C_{33} \ ]$$
$$= [ \ 3 \quad 5 \quad -2 \ ] \cdot [ \ 22 \quad -16 \quad -7 \ ] = 0$$
$$[ \ a_{21} \quad a_{22} \quad a_{23} \ ] \cdot [ \ C_{11} \quad C_{12} \quad C_{13} \ ]$$
$$= [ \ 2 \quad 1 \quad 4 \ ] \cdot [ \ -4 \quad 8 \quad 0 \ ] = 0$$
$$[ \ a_{21} \quad a_{22} \quad a_{23} \ ] \cdot [ \ C_{31} \quad C_{32} \quad C_{33} \ ]$$
$$= [ \ 2 \quad 1 \quad 4 \ ] \cdot [ \ 22 \quad -16 \quad -7 \ ] = 0$$

$$[\; a_{31} \quad a_{32} \quad a_{33}\; ] \cdot [\; C_{11} \quad C_{12} \quad C_{13}\; ]$$
$$= [\; 6 \quad 3 \quad 8\; ] \cdot [\; -4 \quad 8 \quad 0\; ] = 0$$

$$[\; a_{31} \quad a_{32} \quad a_{33}\; ] \cdot [\; C_{21} \quad C_{22} \quad C_{23}\; ]$$
$$= [\; 6 \quad 3 \quad 8\; ] \cdot [\; -46 \quad 36 \quad 21\; ] = 0$$

Consider the matrix $\mathbf{C}$ whose entries are the cofactors of $\mathbf{A}$.

$$\mathbf{C} = \begin{bmatrix} C_{11} & C_{12} & C_{13} \\ C_{21} & C_{22} & C_{23} \\ C_{31} & C_{32} & C_{33} \end{bmatrix} = \begin{bmatrix} -4 & 8 & 0 \\ -46 & 36 & 21 \\ 22 & -16 & -7 \end{bmatrix}$$

Then,

$$\mathbf{C}^T = \begin{bmatrix} -4 & -46 & 22 \\ 8 & 36 & -16 \\ 0 & 21 & -7 \end{bmatrix}$$

The first row of matrix $\mathbf{C}$ is the first column of $\mathbf{C}^T$, the second row of matrix $\mathbf{C}$ is the second column of $\mathbf{C}^T$, and the third row of matrix $\mathbf{C}$ is the third column of $\mathbf{C}^T$. Recalling that in matrix multiplication the $i$-$j$ entry of the product is the dot product of the $i$th row of the first factor by the $j$th column of the second factor, we see that the $i$-$j$ entry of the product $\mathbf{A}\mathbf{C}^T$ is the dot product of the $i$th row of $\mathbf{A}$ by the $j$th column of $\mathbf{C}^T$. That is to say, it is the dot product of the $i$th row of $\mathbf{A}$ by the $j$th row of $\mathbf{C}$. It follows from the preceding results that the $i$-$j$ entry of the product $\mathbf{A}\mathbf{C}^T$ is 28 when $i = j$ and 0 when $i \neq j$. Therefore,

$$\mathbf{A}\mathbf{C}^T = \begin{bmatrix} 28 & 0 & 0 \\ 0 & 28 & 0 \\ 0 & 0 & 28 \end{bmatrix} = 28 \begin{bmatrix} 1 & 0 & 0 \\ 0 & 1 & 0 \\ 0 & 0 & 1 \end{bmatrix}$$

DEFINITION 3.17:   The *adjoint* of an $n \times n$ matrix $\mathbf{A}$ is denoted adj$\mathbf{A}$ and is the transpose of the $n \times n$ matrix $\mathbf{C}$ whose $i$-$j$ entry is the cofactor of the $i$-$j$ entry of the matrix $\mathbf{A}$ for each $i = 1, 2, \ldots , n$ and each $j = 1, 2, \ldots , n$.

EXAMPLE 1   Find the adjoint of the matrix

$$\mathbf{A} = \begin{bmatrix} 1 & 3 & -5 \\ 6 & 1 & 2 \\ 4 & -3 & 7 \end{bmatrix}$$

**Solution**  The calculations are given in the table.

| $i$ | $j$ | $a_{ij}$ | Calculations | Cofactor $C_{ij}$ |
|---|---|---|---|---|
| 1 | 1 | 1 | $(-1)^{1+1} \det \begin{bmatrix} 1 & 2 \\ -3 & 7 \end{bmatrix} = 7 + 6 =$ | 13 |
| 1 | 2 | 3 | $(-1)^{1+2} \det \begin{bmatrix} 6 & 2 \\ 4 & 7 \end{bmatrix} = -(42 - 8) =$ | $-34$ |
| 1 | 3 | $-5$ | $(-1)^{1+3} \det \begin{bmatrix} 6 & 1 \\ 4 & -3 \end{bmatrix} = -18 - 4 =$ | $-22$ |
| 2 | 1 | 6 | $(-1)^{2+1} \det \begin{bmatrix} 3 & -5 \\ -3 & 7 \end{bmatrix} = -(21 - 15) =$ | $-6$ |
| 2 | 2 | 1 | $(-1)^{2+2} \det \begin{bmatrix} 1 & -5 \\ 4 & 7 \end{bmatrix} = 7 + 20 =$ | 27 |
| 2 | 3 | 2 | $(-1)^{2+3} \det \begin{bmatrix} 1 & 3 \\ 4 & -3 \end{bmatrix} = -(-3 - 12) =$ | 15 |
| 3 | 1 | 4 | $(-1)^{3+1} \det \begin{bmatrix} 3 & -5 \\ 1 & 2 \end{bmatrix} = 6 + 5 =$ | 11 |
| 3 | 2 | $-3$ | $(-1)^{3+2} \det \begin{bmatrix} 1 & -5 \\ 6 & 2 \end{bmatrix} = -(2 + 30) =$ | $-32$ |
| 3 | 3 | 7 | $(-1)^{3+3} \det \begin{bmatrix} 1 & 3 \\ 6 & 1 \end{bmatrix} = 1 - 18 =$ | $-17$ |

Thus, the cofactor matrix **C** is

$$\begin{bmatrix} C_{11} & C_{12} & C_{13} \\ C_{21} & C_{22} & C_{23} \\ C_{31} & C_{32} & C_{33} \end{bmatrix} = \begin{bmatrix} 13 & -34 & -22 \\ -6 & 27 & 15 \\ 11 & -32 & -17 \end{bmatrix}$$

By definition, adj**A** = **C**$^T$. Thus,

$$\text{adj}\mathbf{A} = \begin{bmatrix} 13 & -6 & 11 \\ -34 & 27 & -32 \\ -22 & 15 & -17 \end{bmatrix}$$

**EXAMPLE 2**   **a.** Calculate the product **A**(adj**A**) for the matrix **A** of Example 1.
**b.** Find det**A** for the same matrix **A**.

**Solution**   **a.** **A** adj**A** $= \begin{bmatrix} 1 & 3 & -5 \\ 6 & 1 & 2 \\ 4 & -3 & 7 \end{bmatrix} \begin{bmatrix} 13 & -6 & 11 \\ -34 & 27 & -32 \\ -22 & 15 & -17 \end{bmatrix}$

$$= \begin{bmatrix} 13 - 102 + 110 & -6 + 81 - 75 & 11 - 96 + 85 \\ 78 - 34 - 44 & -36 + 27 + 30 & 66 - 32 - 34 \\ 52 + 102 - 154 & -24 - 81 + 105 & 44 + 96 - 119 \end{bmatrix}$$

$$= \begin{bmatrix} 21 & 0 & 0 \\ 0 & 21 & 0 \\ 0 & 0 & 21 \end{bmatrix}$$

**b.** Note that 21 was obtained by multiplying each row of matrix **A** by the corresponding row of cofactors. Hence, $\det\mathbf{A} = 21$. ∎

## Finding the Inverse Using Adjoints

We now give the following theorems.

**THEOREM 3.1:*** If **A** is any $n \times n$ matrix, then

$$\mathbf{A}(\text{adj}\mathbf{A}) = (\text{adj}\mathbf{A})\mathbf{A} = (\det\mathbf{A})\mathbf{I}_n$$

**THEOREM 3.2:** If **A** and **B** are $n \times n$ matrices, then

$$\det(\mathbf{AB}) = (\det\mathbf{A})(\det\mathbf{B})$$

**THEOREM 3.3:*** An $n \times n$ matrix **A** is invertible if, and only if, $\det\mathbf{A} \neq 0$. In that case,

$$\mathbf{A}^{-1} = \frac{1}{\det\mathbf{A}} \, \text{adj}\mathbf{A}$$

**EXAMPLE 3**    Find $\mathbf{A}^{-1}$ if $\mathbf{A} = \begin{bmatrix} 5 & -8 \\ -4 & 7 \end{bmatrix}$.

**Solution**    We see that **A** is invertible since

$$\det\mathbf{A} = 5(7) - (-4)(-8) = 35 - 32 = 3$$

---

*See Exercises 25–28 for the proofs of Theorems 3.1 and 3.3. The proof of Theorem 3.2 is too technical at this level and may be found in more advanced books on matrix algebra.

Also, the cofactors are

$$C_{11} = (-1)^{1+1}(7) = 7$$
$$C_{12} = (-1)^{1+2}(-4) = 4$$
$$C_{21} = (-1)^{2+1}(-8) = 8$$
$$C_{22} = (-1)^{2+2}(5) = 5$$

Therefore, the cofactor matrix is

$$\begin{bmatrix} 7 & 4 \\ 8 & 5 \end{bmatrix}$$

and the adjoint is

$$\begin{bmatrix} 7 & 8 \\ 4 & 5 \end{bmatrix}$$

It follows that

$$\mathbf{A}^{-1} = \left(\frac{1}{3}\right)\begin{bmatrix} 7 & 8 \\ 4 & 5 \end{bmatrix} = \begin{bmatrix} \frac{7}{3} & \frac{8}{3} \\ \frac{4}{3} & \frac{5}{3} \end{bmatrix}$$

**Do Exercise 9.**

**EXAMPLE 4**   Find $\mathbf{A}^{-1}$ for the matrix $\mathbf{A}$ of Example 1.

**Solution**   We found in Example 2 that $\det\mathbf{A} = 21$; hence, it is nonzero. It follows that the matrix $\mathbf{A}$ is invertible. We also found in Example 1 that

$$\text{adj}\mathbf{A} = \begin{bmatrix} 13 & -6 & 11 \\ -34 & 27 & -32 \\ -22 & 15 & -17 \end{bmatrix}$$

Thus,

$$\mathbf{A}^{-1} = \left(\frac{1}{21}\right)\begin{bmatrix} 13 & -6 & 11 \\ -34 & 27 & -32 \\ -22 & 15 & -17 \end{bmatrix}$$

**Do Exercise 13.**

### Procedure to Find the Inverse of an $n \times n$ Matrix A Using the Adjoint

**Step 1.**   Calculate $\det\mathbf{A}$ by finding the dot product of any row of $\mathbf{A}$ by the corresponding row of cofactors. Recall that a cofactor $C_{ij}$ is the product $(-1)^{i+j}$ times the determinant of the matrix obtained from $\mathbf{A}$ by deleting row $i$ and column $j$. If $\det\mathbf{A} = 0$, the matrix does not have an inverse. Otherwise we proceed to Step 2.

**Step 2.**   For each $i = 1, 2, \ldots, n$ and $j = 1, 2, \ldots, n$ find the cofactor $C_{ij}$, if not already found in Step 1.

**Step 3.**   Find the matrix adj$\mathbf{A}$. Recall that this matrix is the transpose of the cofactor matrix which may be written from the results of Steps 1 and 2.

**Step 4.**   The inverse $\mathbf{A}^{-1}$ is obtained by dividing each entry of adj$\mathbf{A}$ found in Step 3 by det$\mathbf{A}$ found in Step 1.

## Exercise Set 3.6

*Exercises 1–24: Do Exercises 1–24 of Section 3.4 using the procedure summarized at the end of this section.*

*25. Prove that if two rows of a square matrix $\mathbf{A}$ are identical, then det$\mathbf{A} = 0$. (*Hint:* Suppose that row $i$ and row $j$ are identical. Let $\mathbf{B}$ be the matrix obtained from $\mathbf{A}$ by interchanging rows $i$ and $j$. Then, $\mathbf{A} = \mathbf{B}$ and det$\mathbf{B} = -$det$\mathbf{A}$.)

*26. Prove that if $\mathbf{A}$ is an $n \times n$ matrix, then the dot product of any row by the row of cofactors corresponding to a different row is 0. (*Hint:* Suppose $i \neq j$. We want to show that

$(a_{i1} \; a_{i2} \; . \; . \; . \; a_{in}) \cdot (C_{j1} \; C_{j2} \; . \; . \; . \; C_{jn}) = 0$. Obtain a matrix $\mathbf{B}$ from $\mathbf{A}$ by replacing row $j$ by row $i$. Then matrix $\mathbf{B}$ has two identical rows and therefore its determinant is 0 by Exercise 25. Now find det$\mathbf{B}$ by calculating the dot product of row $j$ of matrix $\mathbf{B}$ by the corresponding row of cofactors.)

*27. Prove Theorem 3.1 (*Hint:* Use the result of Exercise 26 and the definition of determinant.)

*28. Prove Theorem 3.3 (*Hint:* Use Theorems 3.1 and 3.2.)

## 3.7  Applications in Economics and Cryptography: Input-Output Analysis and Coded Messages

### Input-Output Models: The Closed System

Matrix multiplication and the multiplicative inverse of a matrix provide tools to solve some important problems in economics. We begin with a discussion of *input-output models*, which were first introduced in the late 1940s by the American economist Wassily Leontief. (In 1973, Leontief received the Nobel Prize in Economics.) The purpose of the models is to allow economists to predict the production levels needed to meet anticipated demands.

Leontief made the assumption that the model representing the production of all industries in an economy is a linear system. The simplest model is the *closed* system where the output of each industry serves only as input to all other industries within the system and no other inputs are needed. An *open* system consists of a productive segment made up of a certain number of industries, and a nonproductive (consumer) segment creating a demand for commodities over and above their uses for production. It is assumed that whatever is produced is consumed. We begin with a simple example of a closed system.

EXAMPLE 1   Suppose that in a simple society there are three types of individuals: the carpenters who build all housing, the farmers who grow all the food, and the tailors who make all the clothing. Each type of individual will consume goods produced not only by the other types but also by his own type. Carpenters must have shelters, farmers must eat, and tailors must wear clothes. Assume that all commodities produced are consumed and that the fraction of each commodity consumed by each is given in the following table.

|  | Housing | Food | Clothing |
|---|---|---|---|
| Carpenters | 20% | 30% | 30% |
| Farmers | 45% | 40% | 25% |
| Tailors | 35% | 30% | 45% |

Is it possible, under these assumptions, to have *interior equilibrium*? That is, is it possible for consumption and production to be equal?

**Solution**  Let us measure consumption and production in dollars and suppose that the carpenters, farmers, and tailors have total incomes of $x$, $y$, and $z$ dollars per year, respectively. From the table, it is clear that the carpenters spend $\$.20x$ on housing, $\$.30y$ on food, and $\$.30z$ on clothing. Since we assume that they spend exactly the amount they earn, we have the equation

$$.2x + .3y + .3z = x$$

Similarly, using the fact that the farmers spend $\$.45x$, $\$.40y$, and $\$.25z$ on housing, food, and clothing, respectively, we get a second equation

$$.45x + .4y + .25z = y$$

The third equation is obtained similarly, giving the system

$$\begin{cases} .2x + .3y + .3z = x \\ .45x + .4y + .25z = y \\ .35x + .3y + .45z = z \end{cases}$$

This system is equivalent to

$$\begin{cases} -.8x + .3y + .3z = 0 \\ .45x - .6y + .25z = 0 \\ .35x + .3y - .55z = 0 \end{cases}$$

We see that the ordered triple $(0, 0, 0)$ is a solution of this system. However, the determinant of the coefficient matrix is also zero. Hence, this solution is not unique. We seek some solution $(a, b, c)$ where $a$, $b$, and $c$ are positive. We use the method of row reduction discussed in Section 3.3.

$$\begin{bmatrix} -.8 & .3 & .3 & | & 0 \\ .45 & -.6 & .25 & | & 0 \\ .35 & .3 & -.55 & | & 0 \end{bmatrix} \quad R_1 + R_2 \rightarrow R_2$$

$$\begin{bmatrix} -.8 & .3 & .3 & | & 0 \\ -.35 & -.3 & .55 & | & 0 \\ .35 & .3 & -.55 & | & 0 \end{bmatrix} \quad \begin{matrix} R_2 + R_3 \rightarrow R_3 \\ R_2 + R_1 \rightarrow R_1 \end{matrix}$$

$$\begin{bmatrix} -1.15 & 0 & .85 & \bigg| & 0 \\ -.35 & -.3 & .55 & \bigg| & 0 \\ 0 & 0 & 0 & \bigg| & 0 \end{bmatrix} \qquad \left(\frac{-1}{1.15}\right)R_1 \rightarrow R_1$$

$$\begin{bmatrix} 1 & 0 & \frac{-17}{23} & \bigg| & 0 \\ -.35 & -.3 & .55 & \bigg| & 0 \\ 0 & 0 & 0 & \bigg| & 0 \end{bmatrix} \qquad .35R_1 + R_2 \rightarrow R_2$$

$$\begin{bmatrix} 1 & 0 & \frac{-17}{23} & \bigg| & 0 \\ 0 & -.3 & \frac{67}{230} & \bigg| & 0 \\ 0 & 0 & 0 & \bigg| & 0 \end{bmatrix} \qquad \left(\frac{-1}{.3}\right)R_2 \rightarrow R_2$$

$$\begin{bmatrix} 1 & 0 & \frac{-17}{23} & \bigg| & 0 \\ 0 & 1 & \frac{-67}{69} & \bigg| & 0 \\ 0 & 0 & 0 & \bigg| & 0 \end{bmatrix}$$

The last augmented matrix represents the system

$$\begin{cases} x = \left(\frac{17}{23}\right)z \\ y = \left(\frac{67}{69}\right)z \end{cases}$$

We have an infinite number of solutions. If we let $z = 69c$, where $c$ is a positive constant, then $x = \left(\frac{17}{23}\right)69c = 51c$ and $y = \left(\frac{67}{69}\right)69c = 67c$. Thus, the set of solutions is the set of ordered triples $(51c, 67c, 69c)$, where $c$ is an arbitrary positive number. Therefore, the carpenters', farmers', and tailors' incomes are in the ratios 51:67:69.

■

## Input-Output Models: The Open System

We now discuss the more interesting open Leontief system.

**EXAMPLE 2**    Assume that we have a simple economy with only two industries I and II. To produce 1 unit of its own product, industry I needs inputs of .52 units of its own product and .44 units of the product of industry II. Industry II needs inputs of .25 units of the product of industry I and .25 units of its own product to produce 1 unit of its own product. The commodities that are not used internally by the industries themselves are called *final demand*. These goods are produced for the consumer. Find the production necessary to provide for final demands of

**a.** 32 units of the product of industry I and 54 units of the product of industry II.
**b.** 60 units of the product of industry I and 48 units of the product of industry II.

**Solution**    **a.** The inputs are $x$ and $y$ units of the products of industries I and II, respectively. Of the $x$ units produced by industry I, $.52x$ units are used by industry I, $.25y$ units are used by industry II, and 32 units are provided for final demand. These three quantities added must give the total output of industry I, which is $x$ units. Thus,

$$.52x + .25y + 32 = x$$

Similarly, of the $y$ units produced by industry II, $.44x$ units are used by industry I, $.25y$ units are used by industry II, and 54 units are provided for final demand. These three quantities added must give the total output of industry II, which is $y$ units. Therefore,

$$.44x + .25y + 54 = y$$

Writing these two equations together gives the linear system

$$\begin{cases} .52x + .25y + 32 = x \\ .44x + .25y + 54 = y \end{cases}$$

Although this system can easily be solved, we proceed as follows to illustrate the general model that we wish to discuss. We first write the system as a matrix equation

$$\mathbf{BX} + \mathbf{D} = \mathbf{X}$$

where

$$\mathbf{B} = \begin{bmatrix} .52 & .25 \\ .44 & .25 \end{bmatrix}$$

$$\mathbf{X} = \begin{bmatrix} x \\ y \end{bmatrix} \quad \text{and} \quad \mathbf{D} = \begin{bmatrix} 32 \\ 54 \end{bmatrix}$$

We get the following sequence of equivalent matrix equations:

$$\mathbf{BX} + \mathbf{D} = \mathbf{X}$$
$$\mathbf{X} - \mathbf{BX} = \mathbf{D}$$
$$(\mathbf{I}_2 - \mathbf{B})\mathbf{X} = \mathbf{D}$$
$$(\mathbf{I}_2 - \mathbf{B})^{-1}(\mathbf{I}_2 - \mathbf{B})\mathbf{X} = (\mathbf{I}_2 - \mathbf{B})^{-1}\mathbf{D}$$
$$\mathbf{I}_2\mathbf{X} = (\mathbf{I}_2 - \mathbf{B})^{-1}\mathbf{D}$$
$$\mathbf{X} = (\mathbf{I}_2 - \mathbf{B})^{-1}\mathbf{D} \qquad (*)$$

Now,

$$\mathbf{I}_2 - \mathbf{B} = \begin{bmatrix} 1 & 0 \\ 0 & 1 \end{bmatrix} - \begin{bmatrix} .52 & .25 \\ .44 & .25 \end{bmatrix} = \begin{bmatrix} .48 & -.25 \\ -.44 & .75 \end{bmatrix}$$

We easily find that

$$(\mathbf{I}_2 - \mathbf{B})^{-1} = \begin{bmatrix} 3 & 1 \\ 1.76 & 1.92 \end{bmatrix}$$

Thus,

$$\mathbf{X} = \begin{bmatrix} 3 & 1 \\ 1.76 & 1.92 \end{bmatrix} \begin{bmatrix} 32 \\ 54 \end{bmatrix}$$

$$= \begin{bmatrix} 3(32)+1(54) \\ 1.76(32)+1.92(54) \end{bmatrix} = \begin{bmatrix} 150 \\ 160 \end{bmatrix}$$

We conclude that industry I should produce 150 units while industry II should produce 160 units.

**b.** We need only use the demand matrix $\mathbf{D} = \begin{bmatrix} 60 \\ 48 \end{bmatrix}$ in equation (*) to get

$$\mathbf{X} = \begin{bmatrix} 3 & 1 \\ 1.76 & 1.92 \end{bmatrix} \begin{bmatrix} 60 \\ 48 \end{bmatrix}$$

$$= \begin{bmatrix} 3(6)+1(48) \\ 1.76(60)+1.92(48) \end{bmatrix} = \begin{bmatrix} 228 \\ 197.76 \end{bmatrix}$$

We conclude that industry I should produce 228 units and industry II should produce 197.76 units.
**Do Exercise 1.**                                                                                         ▪

## Input-Output Models Applied

We now generalize the ideas presented in this last example. Assume that we have an economy consisting of $n$ industries, each one of which produces a single commodity. Assume further that these are interdependent; that is, each uses some of the others' output in order to operate. In addition, each industry produces some goods for the consumer (nonproductive) segment of the economy that are called *final demands*. We assume that production of one unit of the $j$th commodity (output) requires a quantity $b_{ij}$ of the $i$th commodity (input). These provide the entries of the $n \times n$ matrix

$$\mathbf{B} = \begin{bmatrix} b_{11} & b_{12} & \cdots & b_{1n} \\ b_{21} & b_{22} & \cdots & b_{2n} \\ \vdots & & & \\ b_{n1} & b_{n2} & \cdots & b_{nn} \end{bmatrix}$$

where the entries of the $j$th column give the input requirements for production of one unit of the $j$th commodity. That is to say, if the unit is the dollar, then $b_{ij}$ gives the dollar value that industry $j$ must purchase from industry $i$ to produce one dollar worth of its own commodity. The matrix $\mathbf{B}$ is often called the *technological coefficient matrix* or the *input-output matrix*. (The matrix

$$\begin{bmatrix} .52 & .25 \\ .44 & .25 \end{bmatrix}$$

in the previous example is a technological coefficient matrix).*
    Now, for $i = 1, 2, \ldots, n$, let $x_i$ dollars be the total output of the $i$th industry and let $d_i$ dollars be the demand for the product of the $i$th industry by the nonproductive segment (final demand). The $x_i$ dollars output of the $i$th industry is divided up as the $n$ inputs used by the $n$ industries and the final demand. The first industry requires $b_{i1}x_1$

---

*Each entry in a technological coefficient matrix is between 0 and 1, and the sum of the entries in each column is never greater than 1. In fact, in the preceding example, the sum of the entries in columns 1 and 2 were .96 and .5, respectively. The "missing" .04 and .5 in the two columns is attributed to external sources of supply. We will illustrate this in Example 3 where primary inputs will be involved.

dollars of the output of the $i$th industry since it requires $b_{i1}$ dollar of that industry for each dollar it produces and its total output is $x_1$. Similarly, the second industry requires $b_{i2}x_2$ dollars of the output of the $i$th industry, and so on. Since the interindustry demand for the product of the $i$th industry added to the final (consumer) demand for that product must equal the total output of that industry, we have

$$\underbrace{b_{i1}x_1 + b_{i2}x_2 + b_{i3}x_3 + \cdots + b_{in}x_n} + \underbrace{d_i} = x_i$$

Output of the $i$th industry used by    Final demand.
the $n$ industries.

The foregoing is true for $i = 1, 2, 3, \ldots, n$. Thus, we have the linear system

$$\begin{cases} b_{11}x_1 + b_{12}x_2 + b_{13}x_3 + \cdots + b_{1n}x_n + d_1 = x_1 \\ b_{21}x_1 + b_{22}x_2 + b_{23}x_3 + \cdots + b_{2n}x_n + d_2 = x_2 \\ \quad\vdots \\ b_{n1}x_1 + b_{n2}x_2 + b_{n3}x_3 + \cdots + b_{nn}x_n + d_n = x_n \end{cases}$$

Recalling that

$$\mathbf{B} = \begin{bmatrix} b_{11} & b_{12} & \cdots & b_{1n} \\ b_{21} & b_{22} & \cdots & b_{2n} \\ \vdots & & & \\ b_{n1} & b_{n2} & \cdots & b_{nn} \end{bmatrix} \quad \text{and letting } \mathbf{X} = \begin{bmatrix} x_1 \\ x_2 \\ \vdots \\ x_n \end{bmatrix}, \mathbf{D} = \begin{bmatrix} d_1 \\ d_2 \\ \vdots \\ d_n \end{bmatrix}$$

we may write the system as a single matrix equation

$$\mathbf{BX} + \mathbf{D} = \mathbf{X}$$

where $\mathbf{B}$ is the technological coefficient matrix, $\mathbf{X}$ is the *total output vector*, and $\mathbf{D}$ is the *demand vector* of the consumer segment (final demand).

We can solve for the vector $\mathbf{X}$ using the methods of matrix algebra as follows:

$$\mathbf{D} = \mathbf{X} - \mathbf{BX}$$
$$= (\mathbf{I}_n - \mathbf{B})\mathbf{X}$$

Thus, if the matrix $\mathbf{I}_n - \mathbf{B}$ is invertible, we get

$$\mathbf{X} = (\mathbf{I}_n - \mathbf{B})^{-1}\mathbf{D} \quad *$$

The following example will illustrate a situation in which we are given the industry inputs and the final demands in tabular form. In this example, the industries will be using *primary inputs* which include such inputs as labor, raw material, and land. The primary inputs used by each industry are given in the last row of the table.

---

*In more advanced books, it is proved that if all entries of the matrix $\mathbf{B}$ are nonnegative and if for each $j = 1, 2, \ldots, n$ the sum of all the entries in the $j$th column is less than 1, then the matrix $\mathbf{I}_n - \mathbf{B}$ is invertible and all the entries of its inverse are nonnegative.

EXAMPLE 3    Suppose we have a simple economy consisting of three industries: agriculture, housing, and manufacturing. The interaction between these industries is given in the following table, where the units are in millions of dollars per year.

|  | Agriculture | Housing | Manufacturing | Final demands | Total output |
|---|---|---|---|---|---|
| Agriculture | 3000 | 4050 | 2840 | 2110 | 12,000 |
| Housing | 3600 | 2700 | 5680 | 1520 | 13,500 |
| Manufacturing | 1800 | 4050 | 2840 | 5510 | 14,200 |
| Primary inputs | 3600 | 2700 | 2840 |  |  |

Assuming that the technological coefficient matrix remains constant over the periods of time under consideration, find the outputs needed from each industry if the final demands change to the following, where the units are in millions of dollars per year.

**a.** 774 for agriculture, 1290 for housing, and 1032 for manufacturing.
**b.** 516 for agriculture, 774 for housing, and 1548 for manufacturing.
**c.** 1806 for agriculture, 2322 for housing, and 2838 for manufacturing.

Solution    **a.** We must first find the technological coefficient matrix. If we add the entries in each of the first three columns of the table we obtain 12,000, 13,500, and 14,200, which are the total outputs of agriculture, housing, and manufacturing, respectively. This agrees with the fact that for each industry, the total value of the output is equal to the sum of the values of all inputs, and all outputs are used by either the producing segment or the nonproducing segment (final demand) of the economy.

Agriculture uses 3000, 3600, and 1800 millions of dollars per year of agriculture, housing, and manufacturing, respectively, to produce 12,000 millions of dollars per year. Therefore, to produce 1 dollar per year, it will use .25, .3, and .15 dollars per year of agriculture, housing, and manufacturing, respectively, since $\frac{3000}{12,000} = .25$, $\frac{3600}{12,000} = .3$ and $\frac{1800}{12,000} = .15$.

Similarly, we find that to produce 1 dollar per year, housing uses .3, .2, and .3 dollars of agriculture, housing, and manufacturing, respectively, since $\frac{4050}{13,500} = .3$, $\frac{2700}{13,500} = .2$, and $\frac{4050}{13,500} = .3$.

Finally, we find that to produce 1 dollar per year, manufacturing uses .2, .4, and .2 dollars of agriculture, housing, and manufacturing, respectively, since $\frac{2840}{14,200} = .2$, $\frac{5680}{14,200} = .4$, and $\frac{2840}{14,200} = .2$.

Hence, the technological coefficient matrix is

$$\mathbf{B} = \begin{bmatrix} .25 & .3 & .2 \\ .3 & .2 & .4 \\ .15 & .3 & .2 \end{bmatrix}$$

As in the preceding example, the output vector is given by

$$\mathbf{X} = (\mathbf{I}_3 - \mathbf{B})^{-1}\mathbf{D}$$

where

$$\mathbf{D} = \begin{bmatrix} 774 \\ 1290 \\ 1032 \end{bmatrix}$$

$$\mathbf{I}_3 - \mathbf{B} = \begin{bmatrix} 1 & 0 & 0 \\ 0 & 1 & 0 \\ 0 & 0 & 1 \end{bmatrix} - \begin{bmatrix} .25 & .3 & .2 \\ .3 & .2 & .4 \\ .15 & .3 & .2 \end{bmatrix}$$

$$= \begin{bmatrix} .75 & -.3 & -.2 \\ -.3 & .8 & -.4 \\ -.15 & -.3 & .8 \end{bmatrix}$$

Since each entry of matrix $\mathbf{B}$ is nonnegative and the sum of the three elements of each column of that matrix is less than 1, we know that $\mathbf{I}_3 - \mathbf{B}$ is invertible and that the entries of its inverse are nonnegative. We find the nine cofactors as follows:

| $i$ | $j$ | $C_{ij}$ |
|---|---|---|
| 1 | 1 | $(-1)^{1+1}[.8(.8) - (-.3)(-.4)] = .52$ |
| 1 | 2 | $(-1)^{1+2}[(-.3)(.8) - (-.15)(-.4)] = .3$ |
| 1 | 3 | $(-1)^{1+3}[(-.3)(-.3) - (-.15)(.8)] = .21$ |
| 2 | 1 | $(-1)^{2+1}[(-.3)(.8) - (-.3)(-.2)] = .3$ |
| 2 | 2 | $(-1)^{2+2}[.75(.8) - (-.15)(-.2)] = .57$ |
| 2 | 3 | $(-1)^{2+3}[.75(-.3) - (-.15)(-.3)] = .27$ |
| 3 | 1 | $(-1)^{3+1}[(-.3)(-.4) - (.8)(-.2)] = .28$ |
| 3 | 2 | $(-1)^{3+2}[.75(-.4) - (-.3)(-.2)] = .36$ |
| 3 | 3 | $(-1)^{3+3}[.75(.8) - (-.3)(-.3)] = .51$ |

To find the determinant of $\mathbf{I}_3 - \mathbf{B}$, we calculate the dot product of its first row by the corresponding row of cofactors. We get

$$(.75 \quad -.3 \quad -.2) \cdot (.52 \quad .3 \quad .21) =$$
$$.75(.52) + (-.3)(.3) + (-.2)(.21) = 0.258$$

Therefore,

$$\det(\mathbf{I}_3 - \mathbf{B}) = 0.258$$

Also,

$$\mathrm{adj}(\mathbf{I}_3 - \mathbf{B}) = \begin{bmatrix} .52 & .3 & .21 \\ .3 & .57 & .27 \\ .28 & .36 & .51 \end{bmatrix}^T$$

$$= \begin{bmatrix} .52 & .3 & .28 \\ .3 & .57 & .36 \\ .21 & .27 & .51 \end{bmatrix}$$

Thus,

$$(\mathbf{I}_3 - \mathbf{B})^{-1} = (1/0.258)\begin{bmatrix} .52 & .3 & .28 \\ .3 & .57 & .36 \\ .21 & .27 & .51 \end{bmatrix}$$

$$= \begin{bmatrix} \frac{520}{258} & \frac{300}{258} & \frac{280}{258} \\ \frac{300}{258} & \frac{570}{258} & \frac{360}{258} \\ \frac{210}{258} & \frac{270}{258} & \frac{510}{258} \end{bmatrix}$$

It follows that

$$\mathbf{X} = \begin{bmatrix} \frac{520}{258} & \frac{300}{258} & \frac{280}{258} \\ \frac{300}{258} & \frac{570}{258} & \frac{360}{258} \\ \frac{210}{258} & \frac{270}{258} & \frac{510}{258} \end{bmatrix} \cdot \begin{bmatrix} 774 \\ 1290 \\ 1032 \end{bmatrix}$$

$$= \begin{bmatrix} 4180 \\ 5190 \\ 4020 \end{bmatrix}$$

Agriculture, housing, and manufacturing must produce 4180, 5190, and 4020 millions of dollars per year, respectively.

**b.** When the final demand vector is

$$\mathbf{D} = \begin{bmatrix} 516 \\ 774 \\ 1548 \end{bmatrix}$$

the output vector is obtained as follows:

$$\mathbf{X} = \begin{bmatrix} \frac{520}{258} & \frac{300}{258} & \frac{280}{258} \\ \frac{300}{258} & \frac{570}{258} & \frac{360}{258} \\ \frac{210}{258} & \frac{270}{258} & \frac{510}{258} \end{bmatrix} \cdot \begin{bmatrix} 516 \\ 774 \\ 1548 \end{bmatrix}$$

$$= \begin{bmatrix} 3620 \\ 4470 \\ 4290 \end{bmatrix}$$

Hence, the production of agriculture, housing, and manufacturing must be 3620, 4470, and 4290 millions of dollars per year, respectively.

**c.** Finally, when the final demand vector is

$$\mathbf{D} = \begin{bmatrix} 1806 \\ 2322 \\ 2838 \end{bmatrix}$$

the output vector is obtained as follows:

$$\mathbf{X} = \begin{bmatrix} \frac{520}{258} & \frac{300}{258} & \frac{280}{258} \\ \frac{300}{258} & \frac{570}{258} & \frac{360}{258} \\ \frac{210}{258} & \frac{270}{258} & \frac{510}{258} \end{bmatrix} \cdot \begin{bmatrix} 1806 \\ 2322 \\ 2838 \end{bmatrix}$$

$$= \begin{bmatrix} 9420 \\ 11,190 \\ 9510 \end{bmatrix}$$

We conclude that the production of agriculture, housing, and manufacturing must be 9420, 11,190, and 9510 millions of dollars per year, respectively.

**Do Exercise 17.**                                                                                    ■

If an economist wishes to calculate the output vectors $\mathbf{X}$ for a number of different final demand vectors $\mathbf{D}$, then it is more convenient to first calculate $(\mathbf{I}_n - \mathbf{B})^{-1}$ and to use the formula $\mathbf{X} = (\mathbf{I}_n - \mathbf{B})^{-1}\mathbf{D}$ than it is to use the row reduction method to solve for $\mathbf{X}$ for each $\mathbf{D}$.

## Encoding and Decoding (Optional)

Another interesting application of the inverse of a matrix is encoding and decoding secret messages. The art of putting messages into code or cipher is called *cryptography* and dates back to the ancient Greeks. Since World War II, matrices and their inverses have been used effectively to encode and decode messages. The following example illustrates the technique.

**EXAMPLE 4**   Use the matrix $\mathbf{A}$ to encode the message "Mathematics can be fun!" where

$$\mathbf{A} = \begin{bmatrix} 5 & 2 & -6 \\ 7 & -3 & 4 \\ 3 & 5 & -12 \end{bmatrix}$$

**Solution**   We first assign a number to each letter of the message. Let $a = 1$, $b = 2$, $c = 3$, . . . , $z = 26$. Let the number 27 correspond to a blank space and the number 28 correspond to the exclamation mark. Thus,

M  A  T  H  E  M  A  T  I  C  S  –  C  A  N  –  B  E  –  F  U  N  !
13  1  20  8  5  13  1  20  9  3  19  27  3  1  14  27  2  5  27  6  21  14  28

Since we are using a $3 \times 3$ matrix to encode the message, we separate the numbers into groups of three and write these as column vectors

$$\begin{bmatrix} 13 \\ 1 \\ 20 \end{bmatrix} \begin{bmatrix} 8 \\ 5 \\ 13 \end{bmatrix} \begin{bmatrix} 1 \\ 20 \\ 9 \end{bmatrix} \begin{bmatrix} 3 \\ 19 \\ 27 \end{bmatrix} \begin{bmatrix} 3 \\ 1 \\ 14 \end{bmatrix} \begin{bmatrix} 27 \\ 2 \\ 5 \end{bmatrix} \begin{bmatrix} 27 \\ 6 \\ 21 \end{bmatrix} \begin{bmatrix} 14 \\ 28 \\ 27 \end{bmatrix}$$

Note that we entered a 27 as the third entry of the last vector since there are 23 letters, which is not a multiple of 3. Because 27 corresponds to a blank space, the message is not altered by this entry. We now multiply each of these vectors, on the left, by the given matrix **A**. We get

$$\begin{bmatrix} 5 & 2 & -6 \\ 7 & -3 & 4 \\ 3 & 5 & -12 \end{bmatrix} \begin{bmatrix} 13 \\ 1 \\ 20 \end{bmatrix} = \begin{bmatrix} -53 \\ 168 \\ -196 \end{bmatrix}$$

$$\begin{bmatrix} 5 & 2 & -6 \\ 7 & -3 & 4 \\ 3 & 5 & -12 \end{bmatrix} \begin{bmatrix} 8 \\ 5 \\ 13 \end{bmatrix} = \begin{bmatrix} -28 \\ 93 \\ -107 \end{bmatrix}$$

$$\begin{bmatrix} 5 & 2 & -6 \\ 7 & -3 & 4 \\ 3 & 5 & -12 \end{bmatrix} \begin{bmatrix} 1 \\ 20 \\ 9 \end{bmatrix} = \begin{bmatrix} -9 \\ -17 \\ -5 \end{bmatrix}$$

$$\begin{bmatrix} 5 & 2 & -6 \\ 7 & -3 & 4 \\ 3 & 5 & -12 \end{bmatrix} \begin{bmatrix} 3 \\ 19 \\ 27 \end{bmatrix} = \begin{bmatrix} -109 \\ 72 \\ -220 \end{bmatrix}$$

$$\begin{bmatrix} 5 & 2 & -6 \\ 7 & -3 & 4 \\ 3 & 5 & -12 \end{bmatrix} \begin{bmatrix} 3 \\ 1 \\ 14 \end{bmatrix} = \begin{bmatrix} -67 \\ 74 \\ -154 \end{bmatrix}$$

$$\begin{bmatrix} 5 & 2 & -6 \\ 7 & -3 & 4 \\ 3 & 5 & -12 \end{bmatrix} \begin{bmatrix} 27 \\ 2 \\ 5 \end{bmatrix} = \begin{bmatrix} 109 \\ 203 \\ 31 \end{bmatrix}$$

$$\begin{bmatrix} 5 & 2 & -6 \\ 7 & -3 & 4 \\ 3 & 5 & -12 \end{bmatrix} \begin{bmatrix} 27 \\ 6 \\ 21 \end{bmatrix} = \begin{bmatrix} 21 \\ 255 \\ -141 \end{bmatrix}$$

$$\begin{bmatrix} 5 & 2 & -6 \\ 7 & -3 & 4 \\ 3 & 5 & -12 \end{bmatrix} \begin{bmatrix} 14 \\ 28 \\ 27 \end{bmatrix} = \begin{bmatrix} -36 \\ 122 \\ -142 \end{bmatrix}$$

The message is transmitted as "−53, 168, −196, −28, 93, −107, −9, −17, −5, −109, 72, −220, −67, 74, −154, 109, 203, 31, 21, 255, −141, −36, 122, −142." The person who receives the message knows the matrix **A** that was used to encode the message and proceeds to separate the numbers into groups of three, writes these groups as column vectors, and multiplies each of these vectors—on the left—by $\mathbf{A}^{-1}$, thus getting back the original set of vectors. Then the message is deciphered by replacing each 1 by "a," each 2 by "b," and so on.
**Do Exercise 25.**    ■

**EXAMPLE 5**    The matrix of the preceding example was used to encode the message: "−90, 141, −235, 73, −42, 138, 109, 183, 43, −137, 114, −290, −128, 125, −283." Decode the message.

**Solution**    Since the matrix is a $3 \times 3$ matrix, we separate these numbers into groups of three and write these groups as column vectors as follows:

$$\begin{bmatrix} -90 \\ 141 \\ -235 \end{bmatrix} \quad \begin{bmatrix} 73 \\ -42 \\ 138 \end{bmatrix} \quad \begin{bmatrix} 109 \\ 183 \\ 43 \end{bmatrix} \quad \begin{bmatrix} -137 \\ 114 \\ -290 \end{bmatrix} \quad \begin{bmatrix} -128 \\ 125 \\ -283 \end{bmatrix}$$

The inverse of the matrix that was used to encode the message was found in Example 5, Section 3.4. It is

$$\mathbf{A}^{-1} = \left(\frac{1}{8}\right) \begin{bmatrix} 16 & -6 & -10 \\ 96 & -42 & -62 \\ 44 & -19 & -29 \end{bmatrix}$$

We now multiply each of the vectors we obtained by this inverse. We get

$$\left(\tfrac{1}{8}\right) \begin{bmatrix} 16 & -6 & -10 \\ 96 & -42 & -62 \\ 44 & -19 & -29 \end{bmatrix} \begin{bmatrix} -90 \\ 141 \\ -235 \end{bmatrix} = \begin{bmatrix} 8 \\ 1 \\ 22 \end{bmatrix} = \begin{bmatrix} H \\ A \\ V \end{bmatrix}$$

$$\left(\tfrac{1}{8}\right) \begin{bmatrix} 16 & -6 & -10 \\ 96 & -42 & -62 \\ 44 & -19 & -29 \end{bmatrix} \begin{bmatrix} 73 \\ -42 \\ 138 \end{bmatrix} = \begin{bmatrix} 5 \\ 27 \\ 1 \end{bmatrix} = \begin{bmatrix} E \\ [] \\ A \end{bmatrix}$$

$$\left(\tfrac{1}{8}\right) \begin{bmatrix} 16 & -6 & -10 \\ 96 & -42 & -62 \\ 44 & -19 & -29 \end{bmatrix} \begin{bmatrix} 109 \\ 183 \\ 43 \end{bmatrix} = \begin{bmatrix} 27 \\ 14 \\ 9 \end{bmatrix} = \begin{bmatrix} [] \\ N \\ I \end{bmatrix}$$

$$\left(\tfrac{1}{8}\right) \begin{bmatrix} 16 & -6 & -10 \\ 96 & -42 & -62 \\ 44 & -19 & -29 \end{bmatrix} \begin{bmatrix} -137 \\ 114 \\ -290 \end{bmatrix} = \begin{bmatrix} 3 \\ 5 \\ 27 \end{bmatrix} = \begin{bmatrix} C \\ E \\ [] \end{bmatrix}$$

$$\left(\tfrac{1}{8}\right) \begin{bmatrix} 16 & -6 & -10 \\ 96 & -42 & -62 \\ 44 & -19 & -29 \end{bmatrix} \begin{bmatrix} -128 \\ 125 \\ -283 \end{bmatrix} = \begin{bmatrix} 4 \\ 1 \\ 25 \end{bmatrix} = \begin{bmatrix} D \\ A \\ Y \end{bmatrix}$$

Thus, the message is HAVE A NICE DAY.
**Do Exercise 21.**    ■

   In spite of the fact that this type of encoding and decoding is simple to do, it is nevertheless difficult to break. In practice, long messages are divided into large groups which are written as $n$-dimensional column vectors where $n$ may be as large as 30. Then each vector is multiplied, on the left, by an $n \times n$ invertible matrix $\mathbf{A}$, to encode the message. The message can be decoded by writing it as a group of $n$-dimensional column vectors and multiplying each of these, on the left, by $\mathbf{A}^{-1}$. Finding the inverse of a large matrix requires a huge amount of time if done longhand. However, government agencies involved in encoding and decoding have access to computers, which greatly facilitates this process.

# Exercise Set 3.7

*In Exercises 1–10, a technological coefficient matrix for a simple economy is given. In each case, find the necessary output by each industry to meet the projected final demands.*

1. $\begin{bmatrix} .25 & .50 \\ .40 & .40 \end{bmatrix}$

   a. 240 units from industry I and 360 units from industry II.
   b. 320 units from industry I and 450 units from industry II.
   c. 380 units from industry I and 290 units from industry II.

2. $\begin{bmatrix} .40 & .68 \\ .25 & .30 \end{bmatrix}$

   a. 550 units from industry I and 460 units from industry II.
   b. 480 units from industry I and 560 units from industry II.
   c. 395 units from industry I and 470 units from industry II.

3. $\begin{bmatrix} .60 & .40 \\ .35 & .40 \end{bmatrix}$

   a. 580 units from industry I and 490 units from industry II.
   b. 620 units from industry I and 600 units from industry II.
   c. 710 units from industry I and 850 units from industry II.

4. $\begin{bmatrix} .30 & .40 \\ .625 & .50 \end{bmatrix}$

   a. 890 units from industry I and 750 units from industry II.
   b. 750 units from industry I and 645 units from industry II.
   c. 900 units from industry I and 865 units from industry II.

5. $\begin{bmatrix} .20 & .46 \\ .50 & .40 \end{bmatrix}$

   a. 780 units from industry I and 820 units from industry II.
   b. 690 units from industry I and 850 units from industry II.
   c. 950 units from industry I and 670 units from industry II.

6. $\begin{bmatrix} .20 & .30 & .40 \\ .30 & .20 & .30 \\ .20 & .40 & .20 \end{bmatrix}$

   a. 680, 780, and 850 units from industries I, II, and III, respectively.
   b. 590, 705, and 574 units from industries I, II, and III, respectively.
   c. 470, 680, and 725 units from industries I, II, and III, respectively.

7. $\begin{bmatrix} .30 & .25 & .30 \\ .20 & .40 & .20 \\ .40 & .30 & .20 \end{bmatrix}$

   a. 550, 800, and 470 units from industries I, II, and III, respectively.
   b. 650, 745, and 475 units from industries I, II, and III, respectively.
   c. 470, 750, and 840 units from industries I, II, and III, respectively.

8. $\begin{bmatrix} .25 & .30 & .40 \\ .30 & .20 & .10 \\ .40 & .40 & .40 \end{bmatrix}$

   a. 450, 540, and 630 units from industries I, II, and III, respectively.
   b. 650, 805, and 580 units from industries I, II, and III, respectively.
   c. 870, 970, and 850 units from industries I, II, and III, respectively.

9. $\begin{bmatrix} .10 & .40 & .30 \\ .50 & .25 & .20 \\ .20 & .30 & .40 \end{bmatrix}$

   a. 380, 540, and 730 units from industries I, II, and III, respectively.
   b. 670, 890, and 500 units from industries I, II, and III, respectively.
   c. 600, 785, and 435 units from industries I, II, and III, respectively.

10. $\begin{bmatrix} .25 & .10 & .20 \\ .40 & .20 & .30 \\ .30 & .40 & .40 \end{bmatrix}$

    a. 860, 870, and 580 units from industries I, II, and III, respectively.
    b. 490, 825, and 745 units from industries I, II, and III, respectively.

c. 970, 880, and 925 units from industries I, II, and III, respectively.

*In Exercises 11–20, a table describes the interaction between sectors of a simple hypothetical economy. In each of these, find the output needed by each sector to meet the anticipated final demands. The units are in millions of dollars per year.*

**11.**

|  | Industry I | Industry II | Final demands | Total output |
|---|---|---|---|---|
| Industry I | 455 | 1050 | 315 | 1820 |
| Industry II | 728 | 840 | 532 | 2100 |
| Primary inputs | 637 | 210 | | |

a. The final demands change to 450 for industry I and 640 for industry II. What are the new primary input requirements?

b. The final demands change to 650 for industry I and 780 for industry II. What are the new primary input requirements?

c. The final demands change to 760 for industry I and 860 for industry II. What are the new primary input requirements?

**12.**

|  | Industry I | Industry II | Final demands | Total output |
|---|---|---|---|---|
| Industry I | 1280 | 1904 | 16 | 3200 |
| Industry II | 800 | 840 | 1160 | 2800 |
| Primary inputs | 1120 | 56 | | |

a. The final demands change to 50 for industry I and 940 for industry II. What are the new primary input requirements?

b. The final demands change to 120 for industry I and 1780 for industry II. What are the new primary input requirements?

c. The final demands change to 180 for industry I and 1260 for industry II. What are the new primary input requirements?

**13.**

|  | Industry I | Industry II | Final demands | Total output |
|---|---|---|---|---|
| Industry I | 1512 | 720 | 288 | 2520 |
| Industry II | 882 | 720 | 198 | 1800 |
| Primary inputs | 126 | 360 | | |

a. The final demands change to 250 for industry I and 340 for industry II. What are the new primary input requirements?

b. The final demands change to 380 for industry I and 580 for industry II. What are the new primary input requirements?

c. The final demands change to 460 for industry I and 420 for industry II. What are the new primary input requirements?

**14.**

|  | Industry I | Industry II | Final demands | Total output |
|---|---|---|---|---|
| Industry I | 750 | 1640 | 110 | 2500 |
| Industry II | 1562.5 | 2050 | 487.5 | 4100 |
| Primary inputs | 187.5 | 410 | | |

a. The final demands change to 100 for industry I and 540 for industry II. What are the new primary input requirements?

b. The final demands change to 150 for industry I and 870 for industry II. What are the new primary input requirements?

c. The final demands change to 60 for industry I and 950 for industry II. What are the new primary input requirements?

**15.**

|  | Industry I | Industry II | Final demands | Total output |
|---|---|---|---|---|
| Industry I | 652 | 1886 | 722 | 3260 |
| Industry II | 1630 | 1640 | 830 | 4100 |
| Primary inputs | | 574 | | |

a. The final demands change to 850 for industry I and 740 for industry II. What are the new primary input requirements?

b. The final demands change to 920 for industry I and 710 for industry II. What are the new primary input requirements?

c. The final demands change to 670 for industry I and 840 for industry II. What are the new primary input requirements?

**16.**

|  | Industry I | Industry II | Industry III | Final demands | Total output |
|---|---|---|---|---|---|
| Industry I | 720 | 1230 | 1280 | 370 | 3600 |
| Industry II | 1080 | 820 | 960 | 1240 | 4100 |
| Industry III | 720 | 1640 | 640 | 200 | 3200 |
| Primary inputs | 1080 | 410 | 320 | | |

a. The final demand for industry I products increases by 20%; for industry II products, it decreases by 10%; and for industry III products, it increases by 50%.

**b.** The final demands change to 650 for industry I, 1100 for industry II, and 510 for industry III.

**17.**

|  | Industry I | Industry II | Industry III | Final demands | Total output |
|---|---|---|---|---|---|
| Industry I | 1560 | 1025 | 1380 | 1235 | 5200 |
| Industry II | 1040 | 1640 | 920 | 500 | 4100 |
| Industry III | 2080 | 1230 | 920 | 370 | 4600 |
| Primary inputs | 520 | 205 | 1380 | | |

**a.** The final demand for industry I products increases by 10%; for industry II products, it decreases by 15%; and for industry III products, it increases by 30%.

**b.** The final demands change to 950 for industry I, 850 for industry II, and 720 for industry III.

**18.**

|  | Industry I | Industry II | Industry III | Final demands | Total output |
|---|---|---|---|---|---|
| Industry I | 1675 | 1200 | 2920 | 905 | 6700 |
| Industry II | 2010 | 800 | 73 | 460 | 4000 |
| Industry III | 2680 | 1600 | 2920 | 100 | 7300 |
| Primary inputs | 335 | 400 | 730 | | |

**a.** The final demand for industry I products decreases by 20%; for industry II products, it increases by 10%; and for industry III products, it increases by 70%.

**b.** The final demands change to 980 for industry I, 870 for industry II, and 320 for industry III.

**19.**

|  | Industry I | Industry II | Industry III | Final demands | Total output |
|---|---|---|---|---|---|
| Industry I | 420 | 1760 | 1140 | 880 | 4200 |
| Industry II | 2100 | 1100 | 760 | 440 | 4400 |
| Industry III | 840 | 1320 | 1520 | 120 | 3800 |
| Primary inputs | 840 | 220 | 380 | | |

**a.** The final demand for industry I products increases by 15%; for industry II products, it decreases by 20%; and for industry III products, it increases by 80%.

**b.** The final demands change to 750 for industry I, 1250 for industry II, and 480 for industry III.

**20.**

|  | Industry I | Industry II | Industry III | Final demands | Total output |
|---|---|---|---|---|---|
| Industry I | 1630 | 623 | 1490 | 2777 | 6520 |
| Industry II | 2608 | 1246 | 2235 | 141 | 6230 |
| Industry III | 1956 | 2492 | 2980 | 22 | 7450 |
| Primary inputs | 326 | 1869 | 745 | | |

**a.** The final demand for industry I products decreases by 30%; for industry II products, it increases by 70%; and for industry III products, it increases by 90%.

**b.** The final demands change to 3250 for industry I, 380 for industry II, and 95 for industry III.

**21.** The matrix

$$\begin{bmatrix} 1 & 0 & 2 \\ 2 & 1 & 4 \\ 3 & 5 & 5 \end{bmatrix}$$

was used to encode a message. Suppose that you received the message: "46, 110, 207, 61, 141, 251, 35, 79, 138, 66, 159, 306." Decode that message.

**22.** The matrix

$$\begin{bmatrix} 1 & 0 & 0 \\ 3 & 1 & 2 \\ 2 & 0 & 1 \end{bmatrix}$$

was used to encode a message. Suppose that you received the message: "5, 49, 20, 15, 124, 57, 12, 57, 30, 5, 60, 19, 20, 105, 49, 19, 122, 57, 8, 75, 34, 20, 101, 58, 27, 117, 62, 1, 71, 29, 25, 132, 71, 27, 117, 62, 9, 63, 29." Decode that message.

**23.** The matrix

$$\begin{bmatrix} 1 & 0 & 1 \\ 0 & 1 & 1 \\ 2 & 0 & 3 \end{bmatrix}$$

was used to encode a message. Suppose that you received the message: "40, 41, 101, 31, 52, 89, 10, 27, 25, 45, 52, 117, 29, 26, 83." Decode that message.

**24.** Use the matrix of Exercise 21 to encode the following message:

DO NOT DRINK AND DRIVE

**25.** Use the matrix of Exercise 22 to encode the following message:

EDUCATION IS A PRIVILEGE

**26.** Use the matrix of Exercise 23 to encode the following message:

RACQUETBALL IS FUN

# 3.8    Markov Chains (Optional)

In this section we begin the study of sequences of events, where at any stage an element of chance is present. Such a sequence is called a *stochastic process*. This sequence may be such that at any stage an outcome is independent of all previous outcomes. In this case, it is a sequence of *Bernoulli trials*. For example, a sequence of coin tosses or a sequence of rolls of a die are sequences of Bernoulli trials. However, in general, each outcome in a sequential process depends on what has happened previously. For example, the price of a stock on a certain day depends on the behavior of that stock on preceding days. Similarly, the results of an election may be affected by the results of previous elections. The simplest case occurs when at any stage the outcome depends only on the outcome at the immediately preceding stage and is not affected by earlier outcomes. Such a process is called a *Markov Chain*. Andrei Markov (1856–1922) was a Russian mathematician who has been given much of the credit for introducing the systematic study of stochastic processes.

## Introductory Terminology

We begin with a simple discussion and we assume that you know some of the basic terminology of games of chance.

Suppose that private educators have found that 15% of high school students attend private schools, while the remaining 85% attend public school. They wish to increase the percentage of students attending private schools, so they hire a team of trained recruiters. These recruiters carry out a campaign and collect the following information. Among all students who are attending private schools this year, 82% will attend private schools again the following year, while 18% will switch to public schools. On the other hand, among all students who are attending public schools this year, 6% will attend private schools the following year, while 94% will remain in public schools. Let us calculate what percentages of students will attend private schools and public schools the following year, starting with 15% attending private schools and 85% attending public schools.

If there are only 100 students attending school this year, 15 would be in private schools, while 85 would be in public schools. Since 82% of the students attending private schools remain in private schools, of the 15 who are attending private schools, 12.3 would still attend private schools. (Do not be be concerned with the .3. Remember that we are dealing with percentages!) However, of the 85 who are now attending public schools, 5.1 would switch to private schools, since 85(.06) = 5.1. Thus, of the 100 students who are in school this year, 17.4 will be in private schools the following year since 12.3 + 5.1 = 17.4. We conclude that 17.4% of all students will be attending private schools the following year. Note that

$$(.15)(.82) + (.85)(.06) = .174$$

is the dot product of the vectors

$$[ \ .15 \quad .85 \ ] \text{ and } \begin{bmatrix} .82 \\ .06 \end{bmatrix}$$

Similarly, of these 100 students, 82.6 will be in public schools the following year, since $15(.18) + 85(.94) = 82.6$. We conclude that 82.6% of all students will be attending public schools the following year. Again, we note that

$$(.15)(.18) + (.85)(.94) = .826$$

is the dot product of the vectors

$$[ .15 \quad .85 ] \text{ and } \begin{bmatrix} .18 \\ .94 \end{bmatrix}$$

Combining these results, we write

$$[ .15 \quad .85 ] \begin{bmatrix} .82 & .18 \\ .06 & .94 \end{bmatrix} = [ .174 \quad .826 ]$$

The vectors $[ .15 \quad .85 ]$ and $[ .174 \quad .826 ]$ are called *state vectors*, while the matrix

$$\begin{bmatrix} .82 & .18 \\ .06 & .94 \end{bmatrix}$$

is called the *transition matrix* of the system. Note that the entries of the first row add up to 1, which is to be expected since these entries give the percentages of students from private schools who will be attending private schools and public schools, respectively, the following year. Of course, we are assuming that no student drops out of school.

For the same reason, the entries of the second row of the transition matrix, as well as the entries of both state vectors add up to 1. (See Exercises 37 and 38.) Also, all entries of the transition matrix and of any state vector must be nonnegative. It should now be clear that if the transition matrix remains the same, we can calculate how the percentages change from year to year simply by multiplying on the right the present state vector by the transition matrix to obtain the new state vector.

Thus, the state vectors after 1, 2, 3, . . . , 13 years are obtained as follows:

$$\mathbf{V}_1 = \mathbf{V}_0\mathbf{T} = [ .1500 \quad .8500 ] \begin{bmatrix} .82 & .18 \\ .06 & .94 \end{bmatrix} = [ .1740 \quad .8260 ]$$

$$\mathbf{V}_2 = \mathbf{V}_1\mathbf{T} = [ .1740 \quad .8260 ] \begin{bmatrix} .82 & .18 \\ .06 & .94 \end{bmatrix} = [ .1922 \quad .8078 ]$$

$$\mathbf{V}_3 = \mathbf{V}_2\mathbf{T} = [ .1922 \quad .8078 ] \begin{bmatrix} .82 & .18 \\ .06 & .94 \end{bmatrix} = [ .2061 \quad .7939 ]$$

$$\mathbf{V}_4 = \mathbf{V}_3\mathbf{T} = [ .2061 \quad .7939 ] \begin{bmatrix} .82 & .18 \\ .06 & .94 \end{bmatrix} = [ .2166 \quad .7834 ]$$

$$\mathbf{V}_5 = \mathbf{V}_4\mathbf{T} = [ .2166 \quad .7834 ] \begin{bmatrix} .82 & .18 \\ .06 & .94 \end{bmatrix} = [ .2246 \quad .7754 ]$$

$$\mathbf{V}_6 = \mathbf{V}_5\mathbf{T} = [ .2246 \quad .7754 ] \begin{bmatrix} .82 & .18 \\ .06 & .94 \end{bmatrix} = [ .2307 \quad .7693 ]$$

$$\mathbf{V}_7 = \mathbf{V}_6\mathbf{T} = [ .2307 \quad .7693 ] \begin{bmatrix} .82 & .18 \\ .06 & .94 \end{bmatrix} = [ .2353 \quad .7647 ]$$

$$\mathbf{V}_8 = \mathbf{V}_7\mathbf{T} = [\ .2353 \quad .7647\ ] \begin{bmatrix} .82 & .18 \\ .06 & .94 \end{bmatrix} = [\ .2388 \quad .7612\ ]$$

$$\mathbf{V}_9 = \mathbf{V}_8\mathbf{T} = [\ .2388 \quad .7612\ ] \begin{bmatrix} .82 & .18 \\ .06 & .94 \end{bmatrix} = [\ .2415 \quad .7585\ ]$$

$$\mathbf{V}_{10} = \mathbf{V}_9\mathbf{T} = [\ .2415 \quad .7585\ ] \begin{bmatrix} .82 & .18 \\ .06 & .94 \end{bmatrix} = [\ .2435 \quad .7565\ ]$$

$$\mathbf{V}_{11} = \mathbf{V}_{10}\mathbf{T} = [\ .2435 \quad .7565\ ] \begin{bmatrix} .82 & .18 \\ .06 & .94 \end{bmatrix} = [\ .2451 \quad .7549\ ]$$

$$\mathbf{V}_{12} = \mathbf{V}_{11}\mathbf{T} = [\ .2451 \quad .7549\ ] \begin{bmatrix} .82 & .18 \\ .06 & .94 \end{bmatrix} = [\ .2463 \quad .7537\ ]$$

$$\mathbf{V}_{13} = \mathbf{V}_{12}\mathbf{T} = [\ .2463 \quad .7537\ ] \begin{bmatrix} .82 & .18 \\ .06 & .94 \end{bmatrix} = [\ .2472 \quad .7528\ ]$$

## Steady-State Vector

Note that we have rounded off the results to four decimal places. We can see that there was not much change in the last five state vectors. In fact, it appears that these vectors seem to be approaching the vector [ .25    .75 ]. If we use this vector as a state vector, we get

$$[\ .2500 \quad .7500\ ] \begin{bmatrix} .82 & .18 \\ .06 & .94 \end{bmatrix} = [\ .2500 \quad .7500\ ]$$

We see that the transition of the vector [ .25    .75 ] leads to the same state vector. We call this vector the *steady-state vector*.

The following example illustrates an interesting characteristic of Markov chains.

**EXAMPLE 1**  Using the same transition matrix as in the prior discussion, find the first seven state vectors if initially 21% of the students attended private schools and 79% attended public schools.

**Solution**  The initial state vector is $\mathbf{V}_0 = [\ .21 \quad .79\ ]$. Thus,

$$\mathbf{V}_1 = \mathbf{V}_0\mathbf{T} = [\ .21 \quad .79\ ] \begin{bmatrix} .82 & .18 \\ .06 & .94 \end{bmatrix} = [\ .2196 \quad .7804\ ]$$

Continuing, we get the following results:

$$\mathbf{V}_2 = \mathbf{V}_1\mathbf{T} = [\ .2269 \quad .7731\ ]$$
$$\mathbf{V}_3 = \mathbf{V}_2\mathbf{T} = [\ .2324 \quad .7676\ ]$$
$$\mathbf{V}_4 = \mathbf{V}_3\mathbf{T} = [\ .2366 \quad .7634\ ]$$
$$\mathbf{V}_5 = \mathbf{V}_4\mathbf{T} = [\ .2398 \quad .7602\ ]$$
$$\mathbf{V}_6 = \mathbf{V}_5\mathbf{T} = [\ .2422 \quad .7578\ ]$$
$$\mathbf{V}_7 = \mathbf{V}_6\mathbf{T} = [\ .2441 \quad .7559\ ]$$

Again, we see that the state vectors seem to stabilize at [ .25    .75 ]. This is not a coincidence. It follows from Theorem 3.4.

**Do Exercise 25.**

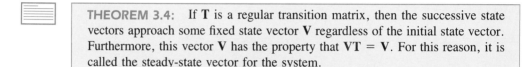

**DEFINITION 3.18:**    A transition matrix **T** is said to be *regular* if for some positive integer $n$, all entries of the $n$th power of **T** are positive.

**THEOREM 3.4:**    If **T** is a regular transition matrix, then the successive state vectors approach some fixed state vector **V** regardless of the initial state vector. Furthermore, this vector **V** has the property that $\mathbf{VT} = \mathbf{V}$. For this reason, it is called the steady-state vector for the system.

It is useful to be able to find the steady-state vector without calculating a succession of state vectors and guessing the steady state from the resulting sequence. We illustrate how this can be done in the next two examples.

**EXAMPLE 2**    Suppose that the team of recruiters hired by the private educators want to predict the percentages of students attending private and public schools, respectively, after a number of years, without calculating the different state vectors. How would they proceed?

**Solution**    The transition matrix is

$$\begin{bmatrix} .82 & .18 \\ .06 & .94 \end{bmatrix}$$

Assume that the steady-state vector is $[\, x \quad y \,]$. Then, we have

$$[\, x \quad y \,] \begin{bmatrix} .82 & .18 \\ .06 & .94 \end{bmatrix} = [\, x \quad y \,]$$

This matrix equation is equivalent to the system

$$\begin{cases} .82x + 0.6y = x \\ .18x + .94y = y \end{cases}$$

This system can be written

$$\begin{cases} -.18x + .06y = 0 \\ .18x - .06y = 0 \end{cases}$$

These two equations are obviously equivalent. Hence, we keep only one of them and write it with the equation $x + y = 1$ (which must be satisfied since the vector $[\, x \quad y \,]$ is a state vector), to obtain the system

$$\begin{cases} x + y = 1 \\ -.18x + .06y = 0 \end{cases}$$

Using row reduction to solve the system, we get

$$\begin{bmatrix} 1 & 1 & | & 1 \\ -.18 & .06 & | & 0 \end{bmatrix} \qquad .18R_1 + R_2 \rightarrow R_2$$

$$\begin{bmatrix} 1 & 1 & | & 1 \\ 0 & .24 & | & .18 \end{bmatrix} \qquad \left(\tfrac{1}{.24}\right)R_2 \rightarrow R_2$$

$$\begin{bmatrix} 1 & 1 & | & 1 \\ 0 & 1 & | & .75 \end{bmatrix} \qquad (-1)R_2 + R_1 \rightarrow R_1$$

$$\begin{bmatrix} 1 & 0 & | & .25 \\ 0 & 1 & | & .75 \end{bmatrix}$$

Thus, $x = .25$ and $y = .75$. We conclude that the steady-state vector of the system is [ .25   .75 ]. The recruiters could have predicted that, after a number of years, enrollments would stabilize to 25% of the student population attending private schools and 75% attending public schools.

**Do Exercise 21.**

**EXAMPLE 3**  The price of a certain stock varies from day to day. On any given day, the price may increase, remain unchanged, or decrease. Furthermore, the change on any given day affects the change on the following day. The transition matrix is described by the following table.

|  |  | **Tomorrow's change** | | |
|---|---|---|---|---|
|  |  | **Increase** | **Unchanged** | **Decrease** |
| **Today's change** | Increase | 35% | 40% | 25% |
|  | Unchanged | 25% | 60% | 15% |
|  | Decrease | 35% | 40% | 25% |

The interpretation of the first row of the table is that if the stock increased today, there is a 35% chance that it will increase tomorrow, a 40% chance that its value will remain unchanged tomorrow, and a 25% chance that its value will decrease tomorrow. The second and third rows have similar interpretations.

**a.** If the value of the stock went up today, what are the chances that its value will increase, remain the same, or decrease in four days?

**b.** Find the steady-state vector and interpret the result.

**Solution**   **a.** The transition matrix is $\mathbf{T} = \begin{bmatrix} .35 & .4 & .25 \\ .25 & .6 & .15 \\ .35 & .4 & .25 \end{bmatrix}$.

Since the value of the stock went up today, the chance of its value going up is 100%, while the chances of its value remaining the same or decreasing are 0%. Thus, the initial state vector is $\mathbf{V}_0 = [\,1 \quad 0 \quad 0\,]$. If we let $\mathbf{V}_1, \mathbf{V}_2, \mathbf{V}_3,$ and $\mathbf{V}_4$ be the state vectors after 1, 2, 3, and 4 days, respectively, we have

$$\mathbf{V}_1 = \mathbf{V}_0\mathbf{T} = [\,1 \quad 0 \quad 0\,] \begin{bmatrix} .35 & .4 & .25 \\ .25 & .6 & .15 \\ .35 & .4 & .25 \end{bmatrix} = [\,.35 \quad .4 \quad .25\,]$$

Similarly, we get

$$V_2 = V_1T = [ \ .31 \quad .48 \quad .21 \ ]$$
$$V_3 = V_2T = [ \ .302 \quad .496 \quad .202 \ ]$$
$$V_4 = V_3T = [ \ .3004 \quad .4992 \quad .2004 \ ]$$

Considering the three entries of the state vector $V_4$, we conclude that after 4 days, the value of the stock has a 30.04% chance of increasing, a 49.92% chance of remaining the same, and a 20.04% chance of decreasing.

**b.** Let $[ \ x \quad y \quad z \ ]$ be the steady-state vector. We know such a vector exists because the transition matrix $T$ is regular. Then we have $x + y + z = 1$, with $x$, $y$, and $z$ nonnegative.

$$[ \ x \quad y \quad z \ ] \begin{bmatrix} .35 & .4 & .25 \\ .25 & .6 & .15 \\ .35 & .4 & .25 \end{bmatrix} = [ \ x \quad y \quad z \ ]$$

This matrix equation is equivalent to the linear system

$$\begin{cases} .35x + .25y + .35z = x \\ .40x + .60y + .40z = y \\ .25x + .15y + .25z = z \end{cases}$$

This system, with the added equation $x + y + z = 1$, may be written

$$\begin{cases} x + \quad y + \quad z = 1 \\ -.65x + .25y + .35z = 0 \\ .40x - .40y + .40z = 0 \\ .25x + .15y - .75z = 0 \end{cases}$$

The solution is $x = .3$, $y = .5$, and $z = .2$. The steady-state vector is $[ \ .3 \quad .5 \quad .2 \ ]$. This means that after a certain length of time, the behavior of the stock will stabilize so that each day, the chances for its value to increase, remain the same, or decrease will be 30%, 50%, 20%, respectively.

**Do Exercise 31.** ■

## Nonregular Transition Matrix

We conclude this section with an example to illustrate that if the transition matrix fails to be regular, then a steady-state vector may not exist.

**EXAMPLE 4**  Consider the nonregular transition matrix $T$ where $T = \begin{bmatrix} 0 & 1 \\ 1 & 0 \end{bmatrix}$.

Show that if we start with a state vector $[ \ a \quad b \ ]$ with $a \neq b$, then $[ \ a \quad b \ ]T^n$ does not approach a limiting vector.

**Solution**  Note that

$$[ \ a \quad b \ ] \begin{bmatrix} 0 & 1 \\ 1 & 0 \end{bmatrix} = [ \ b \quad a \ ] \quad \text{and} \quad [ \ b \quad a \ ] \begin{bmatrix} 0 & 1 \\ 1 & 0 \end{bmatrix} = [ \ a \quad b \ ]$$

Thus, $[\,a\quad b\,]\mathbf{T} = [\,b\quad a\,]$, $[\,a\quad b\,]\mathbf{T}^2 = [\,a\quad b\,]$, $[\,a\quad b\,]\mathbf{T}^3 = [\,b\quad a\,]$, $[\,a\quad b\,]\mathbf{T}^4 = [\,a\quad b\,]$, $\cdots$. In general, $[\,a\quad b\,]\mathbf{T}^n = [\,a\quad b\,]$, when $n$ is even and $[\,a\quad b\,]\mathbf{T}^n = [\,b\quad a\,]$, when $n$ is odd. Since $a \neq b$, $[\,a\quad b\,]\mathbf{T}^n$ does not approach a limiting steady-state vector.  ■

## Exercise Set 3.8

*In Exercises 1–6, determine whether the given vector is a state vector. If a vector is not a state vector, explain why.*

**1.** $\begin{bmatrix} \frac{1}{3} & \frac{1}{6} & \frac{1}{6} \end{bmatrix}$  **2.** $\begin{bmatrix} \frac{1}{4} & \frac{1}{2} & \frac{1}{4} \end{bmatrix}$

**3.** $[\,.7\quad .6\quad -.3\,]$  **4.** $[\,1\quad 2\quad 5\quad 6\,]$

**5.** $[\,.15\quad .25\quad .6\,]$

**6.** $[\,.1\quad .23\quad .42\quad .25\,]$

*In Exercises 7–12, determine whether the given matrix is a transition matrix of some system. If a matrix is not a transition matrix, explain why.*

**7.** $\begin{bmatrix} .3 & .4 \\ .7 & .6 \end{bmatrix}$  **8.** $\begin{bmatrix} .2 & .5 & .3 \\ .4 & .1 & .5 \end{bmatrix}$

**9.** $\begin{bmatrix} .4 & .8 & -.2 \\ .3 & .4 & .3 \\ .5 & .1 & .4 \end{bmatrix}$  **10.** $\begin{bmatrix} \frac{1}{3} & \frac{1}{6} & \frac{1}{2} \\ \frac{1}{5} & \frac{11}{20} & \frac{1}{4} \\ 0 & 1 & 0 \end{bmatrix}$

**11.** $\begin{bmatrix} \frac{1}{5} & \frac{1}{3} & \frac{7}{15} \\ \frac{2}{7} & \frac{1}{4} & \frac{13}{28} \\ \frac{2}{3} & \frac{1}{9} & \frac{2}{9} \end{bmatrix}$  **12.** $\begin{bmatrix} .2 & .3 & .5 \\ .5 & .1 & .4 \\ .9 & .05 & .05 \\ .8 & .15 & .05 \end{bmatrix}$

*In Exercises 13–18, determine whether the given transition matrix is regular. If a transition matrix is not regular, explain why.*

**13.** $\begin{bmatrix} .2 & .8 \\ 1 & 0 \end{bmatrix}$  **14.** $\begin{bmatrix} 1 & 0 \\ 0 & 1 \end{bmatrix}$

**15.** $\begin{bmatrix} 0 & 1 \\ 1 & 0 \end{bmatrix}$  **16.** $\begin{bmatrix} 0 & 1 \\ .3 & .7 \end{bmatrix}$

**17.** $\begin{bmatrix} .2 & .4 & .4 \\ .1 & .7 & .2 \\ .6 & 0 & .4 \end{bmatrix}$  **18.** $\begin{bmatrix} 0 & 0 & 1 \\ 1 & 0 & 0 \\ 0 & 1 & 0 \end{bmatrix}$

*In Exercises 19–24, a transition matrix of a system is given. The matrices are regular since all entries are positive. In each case find the steady-state vector.*

**19.** $\begin{bmatrix} .3 & .7 \\ .3 & .7 \end{bmatrix}$  **20.** $\begin{bmatrix} .675 & .325 \\ .175 & .825 \end{bmatrix}$

**21.** $\begin{bmatrix} .78 & .22 \\ .18 & .82 \end{bmatrix}$  **22.** $\begin{bmatrix} .595 & .405 \\ .095 & .905 \end{bmatrix}$

**23.** $\begin{bmatrix} .15 & .1 & .75 \\ .25 & .3 & .45 \\ .3125 & .55 & .1375 \end{bmatrix}$

**24.** $\begin{bmatrix} .3 & .4 & .3 \\ .16 & .78 & .06 \\ .2 & .1 & .7 \end{bmatrix}$

*In Exercises 25–30, a transition matrix $T$ of a system is given along with an initial state vector $V_0$. These transition matrices are regular since they have positive entries. In each case find a sequence $V_0, V_1, V_2, V_3, \ldots, V_n$ of state vectors long enough so that you can guess the steady-state vector $V$. Then check your answer by showing that $VT = V$.*

**25.** $\mathbf{T} = \begin{bmatrix} .91 & .09 \\ .03 & .97 \end{bmatrix}$; $\mathbf{V}_0 = [\,.15\quad .85\,]$

**26.** $\mathbf{T} = \begin{bmatrix} .4 & .6 \\ .15 & .85 \end{bmatrix}$; $\mathbf{V}_0 = [\,.3\quad .7\,]$

**27.** $\mathbf{T} = \begin{bmatrix} .585 & .415 \\ .085 & .915 \end{bmatrix}$; $\mathbf{V}_0 = [\,.25\quad .75\,]$

**28.** $\mathbf{T} = \begin{bmatrix} .715 & .285 \\ .215 & .785 \end{bmatrix}$; $\mathbf{V}_0 = [\,.35\quad .65\,]$

**29.** $\mathbf{T} = \begin{bmatrix} .3 & .5 & .2 \\ .2 & .7 & .1 \\ .1 & .4 & .5 \end{bmatrix}$; $\mathbf{V}_0 = [\,.35\quad .52\quad .13\,]$

**30.** $\mathbf{T} = \begin{bmatrix} .575 & .3 & .125 \\ .2 & .3 & .5 \\ .4 & .5 & .1 \end{bmatrix}$; $\mathbf{V}_0 = [\,.35\quad .52\quad .13\,]$

**31.** The price of a certain stock varies from day to day. On any given day, the price may increase, remain unchanged, or decrease. Furthermore, the change on any given day affects the change on the following day. The transition matrix is described by the following table.

**Tomorrow's change**

|  | Increase | Unchanged | Decrease |
|---|---|---|---|
| Increase | 25% | 30% | 45% |
| Today's change   Unchanged | 30% | 40% | 30% |
| Decrease | 55% | 20% | 25% |

**a.** If the value of the stock went up on January 5, what are the chances that its value will increase, remain the same, or decrease on January 9 of the same year?

**b.** If the value of the stock was unchanged on March 23, what are the chances that its value will increase, remain the same, or decrease on March 28 of the same year?

**c.** If the value of the stock went down on September 23, what are the chances that its value will increase, remain the same, or decrease on September 29 of the same year?

**32.** Same as Exercise 31 with the following table.

**Tomorrow's change**

|  | Increase | Unchanged | Decrease |
|---|---|---|---|
| Increase | 15% | 20% | 65% |
| Today's change   Unchanged | 25% | 50% | 25% |
| Decrease | 45% | 15% | 40% |

**a.** If the value of the stock went up on April 27, what are the chances that its value will increase, remain the same, or decrease on April 30 of the same year?

**b.** If the value of the stock was unchanged on November 24, what are the chances that its value will increase, remain the same, or decrease on November 29 of the same year?

**c.** If the value of the stock went down on June 4, what are the chances that its value will increase, remain the same, or decrease on June 10 of the same year?

**33.** Assume that a nation is controlled by three main political parties, A, B, and C. The result of each election depends somewhat on which party is in control at the time of the election. The following transition matrix describes the chances that the nation will be controlled by one of the parties after an election, given the possible different parties in control at the time of the election.

**In control after the election**

|  | A | B | C |
|---|---|---|---|
| A | 65% | 20% | 15% |
| In control now   B | 25% | 60% | 15% |
| C | 35% | 25% | 40% |

Given that the elections take place every three years and that party A won in 1987, what are the chances that

**a.** party B will take control in 1993?
**b.** party A will take control in 1996?
**c.** party C will take control in 1999?

**34.** Same as Exercise 33 if the transition matrix is described by the following table.

**In control after the election**

|  | A | B | C |
|---|---|---|---|
| A | 63% | 17% | 20% |
| In control now   B | 30% | 55% | 15% |
| C | 40% | 20% | 40% |

Given that the elections take place every four years and that party A won in 1988, what are the chances that

**a.** party B will take control in 1996?
**b.** party A will take control in 2000?
**c.** party C will take control in 2004?

**35.** The spring weather conditions in Seattle are influenced from day to day by the weather conditions on the previous day. The transition matrix is described by the following table.

**Tomorrow**

|  | Rainy | Cloudy | Sunny |
|---|---|---|---|
| Rainy | 60% | 30% | 10% |
| Today   Cloudy | 30% | 55% | 15% |
| Sunny | 20% | 20% | 60% |

Given that it is sunny on a Monday, what are the chances that

**a.** it will be sunny on Thursday of the same week?
**b.** it will be cloudy on Friday of the same week?
**c.** it will be rainy on Saturday of the same week?

**36.** The summer weather conditions in Quebec are influenced from day to day by the weather conditions on the previous day. The transition matrix is described by the following table.

**Tomorrow**

|  | Rainy | Cloudy | Sunny |
|---|---|---|---|
| Rainy | 40% | 30% | 30% |
| Today   Cloudy | 30% | 45% | 25% |
| Sunny | 10% | 15% | 75% |

Given that it is rainy on a Monday, what are the chances that

a. it will be sunny on Thursday of the same week?
b. it will be cloudy on Friday of the same week?
c. it will be rainy on Saturday of the same week?

37. Given that $\mathbf{V} = [\ x \quad y\ ]$ is a state vector and that

$$\mathbf{T} = \begin{bmatrix} a & b \\ c & d \end{bmatrix}$$

is a transition matrix of a system, show that $\mathbf{VT}$ is a state vector.

38. Given that $\mathbf{V} = [\ x \quad y \quad z\ ]$ is a state vector and that

$$\mathbf{T} = \begin{bmatrix} a & b & c \\ d & e & f \\ g & h & i \end{bmatrix}$$

is a transition matrix of a system, show that $\mathbf{VT}$ is a state vector.

39. Show that if a transition matrix has a 1 on its main diagonal, it cannot be regular.

# 3.9 Chapter Review

**IMPORTANT SYMBOLS AND TERMS**

$\mathbf{A}_i$ [3.5]
$a_{ij}$ [3.1]
$[a_{ij}]m \times n$ [3.1]
$\mathbf{A}^T$ [3.5]
DetA [3.5]
Additive identity [3.1]
Adjoint [3.6]
Associative [3.2]
Augmented matrix [3.3]
Basic principle of counting [3.2]
Bernoulli trials [3.8]
Closed system [3.7]
Coefficient matrix [3.2]
Cofactor [3.5]
Conformable [3.2]
Cryptography [3.7]
Decoding [3.7]
Demand vector [3.7]
Determinant [3.5]

Diagonal element [3.1]
Difference [3.1]
Dimensions [3.1]
Distributive [3.2]
Dot product (inner product) [3.2]
Elementary row operation [3.3]
Encoding [3.7]
Final demands [3.7]
$i$-$j$ entry [3.1]
Input-Output model [3.7]
Interior equilibrium [3.7]
Invertible (nonsingular) [3.4]
Leading variable [3.3]
Matrix [3.1]
Multiplicative identity [3.1]
Multiplicative inverse [3.2]
$n$-dimensional column vector [3.1]

$n$-dimensional row vector [3.1]
Open Leontief system [3.7]
Product [3.2]
Row-reduced echelon matrix [3.3]
Scalar matrix [3.1]
Singular [3.4]
Square matrix [3.1]
Stochastic process [3.8]
Technological coefficient matrix (input-output matrix) [3.7]
Total output vector [3.7]
Transpose [3.5]
Zero matrix [3.1]
Scalar multiplication [3.1]
Sum [3.1]
Value vector [3.2]
Variable vector [3.2]

**SUMMARY**

A rectangular array of numbers enclosed within brackets is called a matrix. We use the notation $[\mathbf{a}_{ij}]_{m \times n}$ to denote a matrix with $m$ rows, $n$ columns, and the number $\mathbf{a}_{ij}$ in its $i$th row and $j$th column. A $1 \times n$ matrix is called an $n$-dimensional row vector, an $n \times 1$ matrix is called an $n$-dimensional column vector. A zero matrix is a matrix all of whose entries are zero, and is called an additive identity. If $\alpha$ is a scalar (number), and $\mathbf{A}$ is an $m \times n$ matrix, $\alpha\mathbf{A}$ is the $m \times n$ matrix $\mathbf{B}$ such that $b_{ij} = \alpha a_{ij}$ for all $i,j$. Two matrices may be added or subtracted, if and only if they have the same dimensions. Addition of matrices is commutative, and associative. The dot product of two $n$-dimensional vectors is the sum of the $n$ products obtained by multiplying corresponding entries of the two vectors. If $\mathbf{A}$ is an $m \times n$ matrix and $\mathbf{B}$ is an $n \times p$ matrix, the product $\mathbf{AB}$ is the $m \times p$ matrix $\mathbf{C}$ whose $i$-$j$ entry is the dot product of the $i$th row of $\mathbf{A}$ by the $j$th column of $\mathbf{B}$, for all $i,j$. Matrix multiplication is associative and distributive over addition. However, multiplication is not commutative. Also it is possible to have

$\mathbf{AB} = 0$ where neither $\mathbf{A}$ nor $\mathbf{B}$ is the zero matrix. Because of this, we may have $\mathbf{AB} = \mathbf{AC}$ even though $\mathbf{B} \neq \mathbf{C}$. The $n \times n$ matrix $[\mathbf{e}_{ij}]$ where $\mathbf{e}_{ij} = 1$ if $i = j$, and $0$ if $i \neq j$ is called the multiplicative identity of order $n$ and is denoted $\mathbf{I}_n$. If $\mathbf{A}$ and $\mathbf{B}$ are $n \times n$ matrices and $\mathbf{AB} = \mathbf{BA} = \mathbf{I}_n$, then we say that $\mathbf{B}$ is the multiplicative inverse of $\mathbf{A}$ and write $\mathbf{B} = \mathbf{A}^{-1}$. Furthermore, we say that $\mathbf{A}$ and $\mathbf{B}$ are invertible (nonsingular). The system

$$\begin{cases} a_{11}x_1 + a_{12}x_2 + \cdots + a_{1n}x_n = b_1 \\ a_{21}x_1 + a_{22}x_2 + \cdots + a_{2n}x_n = b_2 \\ \qquad\qquad\vdots \\ a_{m1}x_1 + a_{m2}x_2 + \cdots + a_{mn}x_n = b_m \end{cases}$$

may be written $\mathbf{AX} = \mathbf{B}$ where

$$\mathbf{A} = \begin{bmatrix} a_{11} & a_{12} & \cdots & a_{1n} \\ a_{21} & a_{22} & \cdots & a_{2n} \\ & & \vdots & \\ a_{m1} & a_{m2} & \cdots & a_{mn} \end{bmatrix} \quad \mathbf{X} = \begin{bmatrix} x_1 \\ x_2 \\ \vdots \\ x_n \end{bmatrix} \text{ and } \mathbf{B} = \begin{bmatrix} b_1 \\ b_2 \\ \vdots \\ b_m \end{bmatrix}$$

The matrices $\mathbf{A}$, $\mathbf{X}$, and $\mathbf{B}$ are called the coefficient matrix, the variable vector, and the value vector, respectively. If $m = n$ and $\mathbf{A}$ is invertible, we may find $\mathbf{A}^{-1}$ and multiply both sides of $\mathbf{AX} = \mathbf{B}$ (on the left) by $\mathbf{A}^{-1}$ to get $\mathbf{X} = \mathbf{A}^{-1}\mathbf{B}$, from which the solution of the original system may be read. The augmented matrix of the system of $m$ linear equations in $n$ variables

$$\begin{cases} a_{11}x_1 + a_{12}x_2 + \cdots + a_{1n}x_n = b_1 \\ a_{21}x_1 + a_{22}x_2 + \cdots + a_{2n}x_n = b_2 \\ \qquad\qquad\vdots \\ a_{m1}x_1 + a_{m2}x_2 + \cdots + a_{mn}x_n = b_m \end{cases}$$

is the $m \times (n + 1)$ matrix

$$\begin{bmatrix} a_{11} & a_{12} & \cdots & a_{1n} & | & b_1 \\ a_{21} & a_{22} & \cdots & a_{2n} & | & b_2 \\ & & \vdots & & & \\ a_{m1} & a_{m2} & \cdots & a_{mn} & | & b_m \end{bmatrix}$$

and is written in compact form $\mathbf{A}|\mathbf{B}$. Given the augmented matrix of a linear system, we get the augmented matrix of an equivalent linear system if we perform any of the following row operations: interchange row $i$ and row $j (R_i \leftrightarrow R_j)$; multiply row $i$ by the nonzero constant $k$ to obtain the new row $i (kR_i \to R_i)$; multiply row $i$ by the constant $k$ and add the result to row $j$ to obtain the new row $j (kR_i + R_j \to R_j)$; and multiply row $i$ by the constant $k_1$ and add the result to $k_2$ times row $j$, where $k_2$ is a nonzero constant, to obtain the new row $j (k_1R_i + k_2R_j \to R_j)$. We perform these operations until we obtain a matrix in row-reduced echelon form, from which we can determine the solutions of the system.

To find the multiplicative inverse of an $n \times n$ matrix $\mathbf{A}$ we start with the $n \times 2n$ matrix $\mathbf{A} \mid \mathbf{I}_n$. If a sequence of elementary row operations transforms this matrix into $\mathbf{I}_n \mid \mathbf{B}$, then the matrix $\mathbf{B}$ is the inverse of $\mathbf{A}$. However, if at any step we obtain all 0's in any row to the left of the vertical line, the matrix $\mathbf{A}$ does not have an inverse, it is singular.

The determinant of a square matrix is a number that can easily be calculated for $2 \times 2$ and $3 \times 3$ matrices but, in general, not for larger matrices.

To solve a system of $n$ linear equations in $n$ variables using Cramer's rule we proceed as follows.

**Step 1.** Arrange all variables so that corresponding variables are underneath each other and make sure all constants are on the right of the equal signs in all $n$ equations.

**Step 2.** Write the system as a single matrix equation $\mathbf{AX} = \mathbf{B}$, where $\mathbf{A}$ is the coefficient matrix

$$\mathbf{X} = \begin{bmatrix} x_1 \\ x_2 \\ \cdot \\ \cdot \\ \cdot \\ x_n \end{bmatrix} \text{ and } \mathbf{B} = \begin{bmatrix} b_1 \\ b_2 \\ \cdot \\ \cdot \\ \cdot \\ b_n \end{bmatrix}$$

**Step 3.** For each $i \in \{1, 2, \ldots, n\}$, let $\mathbf{A}_i$ be the matrix obtained by replacing the $i$th column of $\mathbf{A}$ by the $n$-dimensional vector $\mathbf{B}$.

**Step 4.** Calculate $\det \mathbf{A}$. If $\det \mathbf{A} = 0$, either the system has no solution, or infinitely many solutions, which we can determine by using the row reduction method. If $\det \mathbf{A} \neq 0$, calculate $\det \mathbf{A}_i$ for each $i \in \{1, 2, \ldots, n\}$. The system has a unique solution $(c_1, c_2, \ldots, c_n)$, where for each $i$, $c_i = \det \mathbf{A}_i / \det \mathbf{A}$. We can also find the multiplicative inverse of a nonsingular square matrix $\mathbf{A}$ using determinants.

In an open Leontief system, we have an economy consisting of $n$ industries, each of which produces a single commodity. Each uses some of the others' output in order to operate. In addition, each industry produces some goods for the consumer (nonproductive) segment of the economy. We assume that production of one unit of the $j$th commodity (output) requires a quantity $b_{ij}$ of the $i$th commodity (input), the $n \times n$ matrix $\mathbf{B} = [b_{ij}]$ is called the technological coefficient matrix (input-output matrix). We let $x_i$ dollars be the total output of the $i$th industry and $d_i$ dollars be the demand for the product of the $i$th industry by the nonproductive segment. Then, we have

$$\mathbf{BX} + \mathbf{D} = \mathbf{X}$$

where,

$$\mathbf{X} = \begin{bmatrix} x_1 \\ x_2 \\ \cdot \\ \cdot \\ \cdot \\ x_n \end{bmatrix} \text{ and } \mathbf{D} = \begin{bmatrix} d_1 \\ d_2 \\ \cdot \\ \cdot \\ \cdot \\ d_n \end{bmatrix}$$

The $n$-dimensional vectors $\mathbf{X}$ and $\mathbf{D}$ are called the total output vector and the demand vector, respectively. Given the technological matrix $\mathbf{B}$ and the demand vector $\mathbf{D}$, we can find the production necessary to meet the final demands by solving the matrix equation $\mathbf{BX} + \mathbf{D} = \mathbf{X}$. We get

$$\mathbf{X} = (\mathbf{I}_n - \mathbf{B})^{-1}\mathbf{D}$$

Matrix multiplication can also be used to encode and decode messages.

A stochastic process is a sequence of events where at each stage an element of chance is present. If at any stage an outcome is independent of all previous outcomes, we have a sequence of Bernoulli trials. If at any stage the outcome depends only on the outcome of the immediately preceding stage and is not affected by earlier outcomes, the process is called a Markov chain. Given that outcome $O_i$ has occurred at some trial, the chance that outcome $O_j$ will occur on the next trial is $p_{ij}$, where $0 \le p_{ij} \le 1$, the $n \times n$ matrix $[p_{ij}]$ is called the transition matrix. For each $i$, we have $\sum_{j=1}^{n} p_{ij} = 1$, since at any stage one of the $n$ possible outcomes must occur.

If it is given that at the initial trial, the chance of outcome $O_i$ occurring is $p_i$, the vector $[\, p_1 \quad p_2 \quad \cdots \quad p_n \,]$ is called the initial state vector. We must have $0 \le p_i \le 1$ for each $i$ and $p_1 + p_2 + \cdots + p_n = 1$. If we let $V_0$ denote the initial state vector and $T$ denote the transition matrix, then $V_1 = V_0 T$ is the state vector for the first trial, $V_2 = V_1 T$ is the state vector for the second trial, $V_3 = V_2 T$ is the state vector for the third trial, and so on. For the $k$th trial, the $j$th entry of the vector $V_k$ gives the chance that the $j$th outcome will occur. A transition matrix $T$ is said to be regular if for some positive integer $n$, all entries of the $n$th power of $T$ are positive. If $T$ is a regular transition matrix, then the successive state vectors approach some fixed state vector $V$ regardless of the initial vector. Furthermore, this vector has the property that $VT = V$. If $T$ is a regular $n \times n$ transition matrix, we can find the steady state vector as follows. We let $X = [\, x_1, x_2, \ldots, x_n \,]$ so that $XT = X$. Also, we have $x_1 + x_2 + \cdots + x_n = 1$. The last two equalities yield a system of linear equations which can be solved for $x_1, x_2, \ldots, x_n$.

SAMPLE EXAM QUESTIONS

1. Write the $2 \times 3$ matrix $[a_{ij}]$ if $a_{ij} = 2i - 3j$ for all $i,j$.

2. Write the $3 \times 2$ matrix $[b_{ij}]$ if $b_{ij} = (-1)^{i+j}(2i + j^2)$ for all $i,j$.

3. Write the $4 \times 4$ matrix $[c_{ij}]$ if

$$c_{ij} = \begin{cases} 0 & \text{if } i = j \\ -i^2 + 3j & \text{whenever } i \ne j. \end{cases}$$

4. Perform the indicated operations.

a. $4\begin{bmatrix} 2 & 3 \\ -1 & 5 \end{bmatrix}$

b. $-3\begin{bmatrix} 4 & 5 & 0 \\ -3 & 5 & 7 \\ -2 & 1 & 9 \end{bmatrix} + 5\begin{bmatrix} 2 & 7 & 1 \\ -5 & 3 & 9 \\ -3 & 0 & 4 \end{bmatrix}$

5. For each equality, find the value of the variables which make the matrix equation a true statement.

a. $\begin{bmatrix} x & 4 \\ -1 & 7 \end{bmatrix} = \begin{bmatrix} 3 & y \\ -1 & z \end{bmatrix}$

b. $\begin{bmatrix} x & 2 & 3 \\ -5 & y & 3 \\ -2 & 6 & z \end{bmatrix} = \begin{bmatrix} 6 & 2 & 3 \\ -5 & 2 & 3 \\ -2 & 6 & 3 \end{bmatrix}$

c. $\begin{bmatrix} x+2 & 3 & 5 \\ -9 & y-3 & 3 \\ 4 & 7 & z+4 \end{bmatrix} = \begin{bmatrix} 2x-1 & 3 & 5 \\ -9 & 1+2y & 3 \\ 4 & 7 & 5z \end{bmatrix}$

6. Let $\mathbf{A} = \begin{bmatrix} 2 & 4 & 0 \\ 3 & 1 & 2 \\ 4 & 9 & 3 \end{bmatrix}$, $\mathbf{B} = \begin{bmatrix} 3 & 1 \\ 4 & 7 \\ 2 & 5 \end{bmatrix}$, $\mathbf{C} = \begin{bmatrix} 1 & 0 & 5 \\ 4 & 2 & 8 \end{bmatrix}$.

Find the following if possible. If not possible, state why.

a. The product $\mathbf{AB}$.
b. The product $\mathbf{CA}$.
c. The product $\mathbf{AC}$.
d. The product $\mathbf{CB}$.
e. The product $\mathbf{BC}$.
f. The inverse $\mathbf{A}^{-1}$.
g. The product $\mathbf{BA}$.
h. The inverse $\mathbf{B}^{-1}$.

7. a. Find two $2 \times 2$ matrices $\mathbf{A}$ and $\mathbf{B}$ such that $\mathbf{AB} \neq \mathbf{BA}$.
   b. Find two $2 \times 2$ matrices $\mathbf{A}$ and $\mathbf{B}$ such that $\mathbf{AB} = \mathbf{BA}$, where neither $\mathbf{A}$ nor $\mathbf{B}$ is the identity or the zero matrix.

8. Write the following system of linear equations as a single matrix equation:

$$\begin{cases} 5w + 2x + 7y - 3z = 2 \\ -3w + 6x + 3y - 4z = 7 \\ 2w + 5x + 7y - 5z = 2 \end{cases}$$

9. Solve each of the following linear systems by row reduction.

a. $\begin{cases} 3x + 4y - 5z = 4 \\ -4x - 2y + 5z = 6 \end{cases}$

b. $\begin{cases} 5x + 3y - 4z = -5 \\ -6x + 4y - 3z = -25 \\ x + 4y - 6z = -20 \end{cases}$

10. Solve the linear systems of parts a, b, and c simultaneously.

$$\begin{cases} 2x + 5y = d_1 \\ 7x + 17y = d_2 \end{cases}$$

a. $d_1 = 5$, $d_2 = -3$
b. $d_1 = 3$, $d_2 = -7$
c. $d_1 = -2$, $d_2 = 6$

11. Solve the linear systems of parts a, b, and c simultaneously.

$$\begin{cases} 2x + 6y + 3z = d_1 \\ 2x + 7y + 5z = d_2 \\ x + 3y + 2z = d_3 \end{cases}$$

a. $d_1 = 3$, $d_2 = -4$, $d_3 = 7$
b. $d_1 = 0$, $d_2 = -3$, $d_3 = 5$
c. $d_1 = -5$, $d_2 = 1$, $d_3 = -3$

12. Find the inverse of each of the following matrices, if it exists.

a. $\begin{bmatrix} 9 & 11 \\ 4 & 5 \end{bmatrix}$

b. $\begin{bmatrix} 9 & 3 \\ 12 & 4 \end{bmatrix}$

c. $\begin{bmatrix} 2 & 4 & 3 \\ 6 & 9 & 8 \\ 1 & 2 & 2 \end{bmatrix}$ (*Hint:* Use the adjoint.)

d. $\begin{bmatrix} 2 & 4 & 1 \\ 3 & 7 & 3 \\ 5 & 10 & 3 \end{bmatrix}$

e. $\begin{bmatrix} 3 & 2 & 2 \\ 9 & 3 & 5 \\ 6 & 1 & 3 \end{bmatrix}$

13. Solve the following systems. First write the given system as a matrix equation and then multiply both sides by the inverse of the coefficient matrix.

a. $\begin{cases} 9x + 11y = 3 \\ 4x + 5y = -2 \end{cases}$

b. $\begin{cases} 2x + 4y + 3z = -3 \\ 6x + 9y + 8z = 6 \\ x + 2y + 2z = 18 \end{cases}$ (Use the result of Exercise 12c.)

c. $\begin{cases} 2x + 4y + z = -3 \\ 3x + 7y + 3z = 2 \\ 5x + 10y + 3z = 4 \end{cases}$ (Use the result of Exercise 12d.)

14. Evaluate each of the following determinants:

a. $\det \begin{bmatrix} 5 & 2 \\ 3 & 4 \end{bmatrix}$

b. $\det \begin{bmatrix} 1 & 2 & 3 \\ 3 & 1 & 6 \\ 5 & 2 & 7 \end{bmatrix}$

c. $\det \begin{bmatrix} 3 & 0 & 6 & 9 \\ 5 & -1 & 9 & 14 \\ 4 & 2 & 6 & -2 \\ 7 & 5 & 10 & -1 \end{bmatrix}$ (*Hint:* Use the basic properties of determinants.)

15. Solve each of the following systems using Cramer's rule.

a. $\begin{cases} 2x + 5y = -11 \\ 3x - 2y = 12 \end{cases}$

b. $\begin{cases} 2x + 3y - 4z = -12 \\ 3x - 2y + 5z = 13 \\ 5x + y - 3z = -15 \end{cases}$

16. Assume that we have a simple economy with only two industries I and II. Assume that the technological coefficient matrix for that economy is

$$\begin{bmatrix} .5 & .5 \\ .3 & .2 \end{bmatrix}$$

where the entries of the first column are the amounts industry I uses from its own industry and from industry II to produce one unit of its own commodity, while the second column gives the amounts industry II uses from industry I and from its own industry to produce one unit of its own commodity. Find the production necessary to provide for final demands of

**a.** 250 units of the product of industry I and 300 units of the product of industry II.
**b.** 400 units of the product of industry I and 500 units of the product of industry II.

**17.** Suppose we have a simple economy consisting of three industries: manufacturing, housing, and agriculture. The interaction between these industries is shown in the following table, where the units are in millions of dollars per year.

|  | Agriculture | Housing | Manufacturing | Final demands | Total output |
|---|---|---|---|---|---|
| Agriculture | 3300 | 3100 | 3960 | 640 | 11,000 |
| Housing | 2200 | 4960 | 2640 | 2600 | 12,400 |
| Manufacturing | 4400 | 3720 | 2640 | 2440 | 13,200 |
| Primary inputs | 1100 | 620 | 3960 |  |  |

Assuming that the technological coefficient matrix remains constant over the period under consideration, find the outputs needed from each industry if the final demands change to the following, where the units are in millions of dollars per year.

**a.** 660 for agriculture, 960 for housing, and 564 for manufacturing.
**b.** 910 for agriculture, 1043 for housing, and 665 for manufacturing.
**c.** 940 for agriculture, 1500 for housing, and 1680 for manufacturing.

**18.** Use the matrix $\begin{bmatrix} 3 & 1 \\ 5 & 2 \end{bmatrix}$ to encode the message: THIS EXAM IS HARD.

**19.** The matrix of Exercise 18 was used to encode a message. Then the message was sent to you to decode during your exam. This is what you received: "36, 65, 49, 83, 93, 159, 66, 111, 60, 109." What did the message say?

**20.** Which of the following is a state vector? If a vector is not a state vector, explain why.

**a.** [ .5    .7    −.2 ]         **b.** [ .4    0    .6 ]
**c.** [ .13    .27    .6 ]        **d.** [ .15    .25    −.2    .8 ]
**e.** [ .3    .2    1.5    .6 ]

**21.** Which of the following is a transition matrix of a system? If a matrix is not a transition matrix, explain why.

**a.** $\begin{bmatrix} .2 & .6 \\ .8 & .4 \end{bmatrix}$         **b.** $\begin{bmatrix} .3 & .7 \\ .6 & .4 \end{bmatrix}$

**c.** $\begin{bmatrix} .1 & .3 & .6 \\ .2 & .5 & .3 \end{bmatrix}$         **d.** $\begin{bmatrix} .2 & .1 & .7 \\ .4 & .3 & .3 \\ .1 & .1 & .8 \end{bmatrix}$

22. Which of the following transition matrices are regular? If a transition matrix is not regular, explain why.

a. $\begin{bmatrix} .1 & .9 \\ 0 & 1 \end{bmatrix}$

b. $\begin{bmatrix} .4 & .6 \\ 1 & 0 \end{bmatrix}$

c. $\begin{bmatrix} .1 & .3 & .6 \\ .2 & 0 & .8 \\ .4 & .5 & .1 \end{bmatrix}$

23. The matrix $\begin{bmatrix} .74 & .26 \\ .24 & .76 \end{bmatrix}$ is the transition matrix of a system. Find the steady-state vector of that system.

24. The matrix **T** is the transition matrix of a system and the vector $v_0$ is the initial state vector. Find the state vectors $v_1$ and $v_2$.

$\begin{bmatrix} .87 & .13 \\ .21 & .79 \end{bmatrix}$; $v_0 = [\ .32 \quad .68\ ]$

25. The fall weather conditions in Montreal are influenced from day to day by the weather conditions on the previous day. The transition matrix is described by the following table.

|  |  | Tomorrow | | |
|---|---|---|---|---|
|  |  | **Rainy** | **Cloudy** | **Sunny** |
|  | Rainy | 55% | 35% | 10% |
| **Today** | Cloudy | 50% | 45% | 5% |
|  | Sunny | 15% | 25% | 60% |

Given that it is cloudy on a Monday, what are the chances that

a. it will be sunny on Wednesday of the same week?
b. it will be cloudy on Thursday of the same week?
c. it will be rainy on Friday of the same week?

26. A manufacturer builds regular, deluxe, and super sailboats. Each comes in two models, A and B. The following table gives the quantities which are scheduled to be built in a certain month.

|  | **Regular** | **Deluxe** | **Super** |
|---|---|---|---|
| Model A | 6 | 6 | 3 |
| Model B | 5 | 2 | 4 |

The material needed for each type is given by the following table in appropriate units:

|  | **Wood** | **Paint** | **Canvas** |
|---|---|---|---|
| Regular | 7 | 3 | 5 |
| Deluxe | 8 | 4 | 10 |
| Super | 9 | 6 | 17 |

Finally, the cost per unit (in thousands of dollars) of each of the materials used is given in tabular form.

|  | Cost per unit |
|---|---|
| Wood | 3 |
| Paint | .2 |
| Canvas | 3 |

a. Represent the given tables by matrices $Q$, $M$, and $C$, respectively.
b. Calculate the product $QM$ and interpret the result.
c. Calculate the product $(QM)C$ and interpret the result.

27. The Bonnechance Company manufactures slot machines. Each deluxe model requires 8 hours of assembly time and 3 hours of testing time, while each regular model requires 5 hours of assembly time and 2 hours of testing time. The time available for assembly and testing varies according to the following schedule:

| Time available (in hours) | Monday | Tuesday | Wednesday | Thursday | Friday |
|---|---|---|---|---|---|
| For assembly | 70 | 75 | 64 | 75 | 62 |
| For testing | 20 | 30 | 28 | 33 | 25 |

For each day of the week, let $x$ and $y$ denote the numbers of deluxe and regular models the company manufactures.

a. Write five linear systems of equations knowing that all available time for assembly and testing is used each day.
b. Solve the five systems simultaneously.

# Linear Programming

**4**

Of the many applications of mathematics in business and economics, linear programming is perhaps the most useful. We are often faced with the task of maximizing or minimizing a function subject to constraints. For example, a farmer may have a number of acres on which to grow corn and peanuts. He may already have accepted orders that will require him to use a given number of those acres for corn and a given number for peanuts. He may also be required to follow regulations concerning the number of acres he may use to grow corn or peanuts. All these conditions represent constraints or limitations on how the farmer may grow his crops. If $x$ and $y$ are the number of acres he will use for corn and peanuts, respectively, then his profit may also be expressed in terms of $x$ and $y$. The problem then is to maximize profit and yet have $x$ and $y$ satisfy all the constraints. The profit, expressed in terms of $x$ and $y$, is often called the *objective function*, while $x$ and $y$ are called the *decision variables*. It is useful to introduce the following definition.

> **DEFINITION 4.1:**    The set of points $(x, y)$ for which $x$ and $y$ satisfy all the constraints is called the *region of feasible solutions*.

If the constraints and the objective function are expressed as linear inequalities and a linear equation, respectively, then we have a *linear programming problem*. In this chapter, we shall first introduce a *graphical method* for solving linear programming problems involving only two variables. In practice, however, most linear programming problems involve many variables and many constraints. Consequently, we shall also discuss the *simplex method*, which provides, with the help of computers, an efficient way of solving these problems.

# 4.1  The Geometric Approach

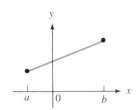

**FIGURE 4.1**

A linear function of one variable defined on a closed bounded interval $[a, b]$ will assume its maximum and minimum values at the endpoints of that interval. It is easy to see this geometrically since the graph of a linear equation is a straight line. When we draw the portion of that line over the interval $[a, b]$, the highest and lowest points of the graph will be endpoints. (See Figure 4.1.)

Geometrically there are two ways of looking at a linear programming problem with two variables. We shall discuss both ways briefly before giving specific examples. As we shall see, just as the graph of a function of one variable is often a curve in the coordinate plane, the graph of a function of two variables is often a surface in three-dimensional space. And just as the graph of a linear function of one variable is a straight line, the graph of a linear function of two variables is a plane.

## Bounded Regions

To help you understand the concept, visualize a house with a polygonal base and flat but slanted roof. The highest point of the roof will be at one corner of the house and the lowest point will be at another corner. (See Figure 4.2.) For the same reason,

**FIGURE 4.2**

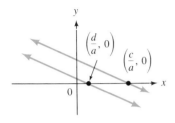

**FIGURE 4.3**

when the set of points $(x, y)$, which satisfy all the constraints of a given problem (region of feasible solutions), happens to be a polygon and its interior and the objective function is linear (defined by $z = ax + by + c$ where $a$, $b$, $c$ are constants), then the maximum and minimum values of $z$ over that polygon and its interior will occur at the corners of the polygon.

Thus, to find the maximum and minimum values of $z$ when the point $(x, y)$ is on the polygon, or in its interior, we need only evaluate the expression $ax + by + c$ at each corner of the polygon. Of all the numbers that are obtained, the largest is the maximum value of the function and the smallest is its minimum value.

For a slightly different way of looking at a linear programming problem geometrically, we will recall a few facts about linear equations in two variables. First, the equation $ax + by = c$, where $a$, $b$, and $c$ are real numbers, is called a linear equation in two variables. If at least $a$ or $b$ is not 0, its graph is a straight line intersecting the $x$-axis at $\left(\frac{c}{a}, 0\right)$ if $a \neq 0$ and the $y$-axis at $\left(0, \frac{c}{b}\right)$ if $b \neq 0$. The slope of the graph of that equation is $\frac{-a}{b}$ if $b \neq 0$. Since two lines with slopes $m_1$ and $m_2$ are parallel if, and only if, $m_1 = m_2$, the graphs of $ax + by = c$ and $ax + by = d$ are parallel since both have slope $\frac{-a}{b}$, when $b \neq 0$, and both are parallel to the $y$-axis when $b = 0$. It is also true that the graph of $ax + by = c$ is farther away from the origin than the graph of $ax + by = d$ if $|d| < |c|$, since these lines intersect the $x$-axis at $\left(\frac{c}{a}, 0\right)$ and $\left(\frac{d}{a}, 0\right)$. (See Figure 4.3.)

EXAMPLE 1    Sketch the graphs of $2x + 3y = 6$ and $2x + 3y = 12$. The two lines are parallel. Which graph is farther away from the origin and why?

Solution    Let $L_1$ and $L_2$ be the graphs of the first and second equations, respectively. The slope of each line is $\frac{-2}{3}$. Hence, the two lines are parallel. (See Figure 4.4.)

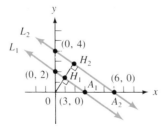

**FIGURE 4.4**

The $x$-intercept of $L_1$ is $(3, 0)$ and that of $L_2$ is $(6, 0)$. Thus, $L_2$ is farther away from the origin than $L_1$. You can see in Figure 4.4 that the right triangles $OH_1A_1$ and $OH_2A_2$ are similar. The hypotenuse $OA_2$ is longer than the hypotenuse $OA_1$, so $OH_2$ must be longer than $OH_1$.    ■

At this point, you should review Section 2.4 on systems of linear inequalities. Recall that the graph of a single linear inequality in two variables is a half-plane and therefore, the graph of a system of linear inequalities is the intersection of half-planes. If this intersection is bounded (that is, enclosed in a circle centered at the origin), then it is called a *bounded convex polygonal region*.

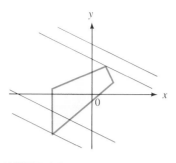

**FIGURE 4.5**

Suppose we have $z = ax + by + c$ where $a$, $b$, and $c$ are real numbers and we consider only values of $x$ and $y$ so that $(x, y)$ is in a bounded convex polygonal region. Then, as claimed earlier, the largest and smallest values of $z$ will occur at corners of that region. To see this, we need only replace $z$ by several constants $c_1$, $c_2$, $c_3$, ... , $c_n$ and observe that the graphs of all equations $ax + by + c = c_i$ are parallel (see Figure 4.5) and the graphs with large values for $|c_i|$ will be the farthest away from the origin. Some of these lines will have points in common with the convex polygonal region. Some will not because $|c_i|$ is either too large or too small. As you can see from Figure 4.5, the lines corresponding to the largest and smallest values of $|c_i|$ which still have points in common with the convex polygonal region are those that touch the region at a corner.

**EXAMPLE 2**  Find the maximum and minimum values of $z$ if $z = 5x + 2y + 1$ and $x$ and $y$ are subject to the following constraints:

$$3x + 2y \leq 210$$
$$x - y \leq 20$$
$$x - 3y \geq -150$$
$$x \geq 0 \text{ and } y \geq 0$$

**Solution**  We sketch the system

$$\begin{cases} 3x + 2y \leq 210 \\ x - y \leq 20 \\ x - 3y \geq -150 \\ x \geq 0 \text{ and } y \geq 0 \end{cases}$$

as in Section 2.4. (See Figure 4.6.) The graph is a bounded convex polygonal region with vertices (corners) $(0, 0)$, $(20, 0)$, $(50, 30)$, $(30, 60)$, and $(0, 50)$.

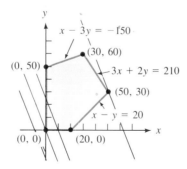

**FIGURE 4.6**

We evaluate $5x + 2y + 1$ at each vertex:

at $(0, 0)$, $z = 5(0) + 2(0) + 1 = 0 + 0 + 1 = 1$,

at $(20, 0)$, $z = 5(20) + 2(0) + 1 = 100 + 0 + 1 = 101$,

at $(50, 30)$, $z = 5(50) + 2(30) + 1 = 250 + 60 + 1 = 311$,

at $(30, 60)$, $z = 5(30) + 2(60) + 1 = 150 + 120 + 1 = 271$, and

at $(0, 50)$, $z = 5(0) + 2(50) + 1 = 0 + 100 + 1 = 101$.

Of the five numbers obtained, 311 is the largest and 1 is the smallest. Thus, the maximum value of $z$ is 311; it occurs at (50, 30). The minimum value of $z$ is 1; it occurs at (0, 0).
**Do Exercise 9.**

EXAMPLE 3    Find the maximum value of $z$ if $z = 3x + 2y + 6$, where $x$ and $y$ are subject to the constraints of Example 2.

Solution    The set of feasible solutions is the same as in Example 2 and is drawn in Figure 4.7.

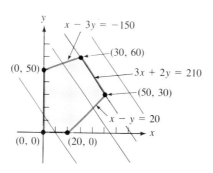

**FIGURE 4.7**

We evaluate $z$ at each of the six corners:

at (0, 0), $z = 3(0) + 2(0) + 6 = 0 + 0 + 6 = 6$,
at (20, 0), $z = 3(20) + 2(0) + 6 = 60 + 0 + 6 = 66$,
at (50, 30), $z = 3(50) + 2(30) + 6 = 150 + 60 + 6 = 216$,
at (30, 60), $z = 3(30) + 2(60) + 6 = 90 + 120 + 6 = 216$, and
at (0, 50), $z = 3(0) + 2(50) + 6 = 0 + 100 + 6 = 106$.

The largest number obtained is 216. Thus, the maximum value of $z$ is 216. However, it occurs at both corners (50, 30) and (30, 60). In that case, the maximum value of $z$ is attained not only at these two points, but at every point of the segment joining these two corners. For example, the point (40, 45) is on that segment and $3(40) + 2(45) + 6 = 120 + 90 + 6 = 216$. The lines whose equations are $c = 3x + 2y + 6$, for different values of $c$, are parallel to the edge joining the endpoints (50, 30) and (30, 60). Therefore the largest value of $c$ for which such a line intersects the region of feasible solutions is 216. In this case, the intersection of the line and the polygonal region is one of the edges of the region. (See Figure 4.7.)
**Do Exercise 11.**

EXAMPLE 4    A farmer has 150 acres of land on which he may grow corn and wheat. His costs to plant the wheat and the corn are $20 per acre and $10 per acre, respectively. He has at most $2000 to spend for planting his crops. He already has accepted orders which require that he plant at least 10 acres of wheat and 20 acres of corn. His profit per acre is $150 for wheat and $120 for corn. How many acres of each should he plant to maximize his profit?

**Solution**   Let $x$ denote the number of acres of wheat and $y$ the number of acres of corn that he should plant. Since he has only 150 acres of land available,

$$x + y \leq 150$$

His total cost (measured in dollars) of planting the wheat will be $20x$ and that of planting the corn will be $10y$. Hence, his total cost will be $(20x + 10y)$. Since he has at most $2000 available, we have

$$20x + 10y \leq 2000$$

which is equivalent to

$$2x + y \leq 200$$

Because of the orders he has accepted, he must grow at least 10 acres of wheat and 20 acres of corn. Therefore, it is also necessary that $x \geq 10$ and $y \geq 20$. His total profit will be $150x from the wheat and $120y from the corn. If $z$ dollars is his total profit,

$$z = 150x + 120y$$

Our problem is to maximize $z$ subject to the constraints

$$\begin{cases} x + y \leq 150 \\ 2x + y \leq 200 \\ \quad\;\; x \geq 10 \\ \quad\;\; y \geq 20 \end{cases}$$

The graph of this system is shown in Figure 4.8. The vertex (50, 100) is obtained by solving the system

$$\begin{cases} x + y = 150 \\ 2x + y = 200 \end{cases}$$

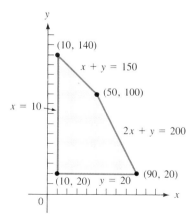

**FIGURE 4.8**

The other vertices are obtained similarly. We must now evaluate $z$ at each of the four corners of the region of feasible solutions shaded in Figure 4.8.

$$\text{At } (10, 20), z = 150(10) + 120(20) = 1500 + 2400 = 3900,$$
$$\text{at } (90, 20), z = 150(90) + 120(20) = 13,500 + 2400 = 15,900,$$
$$\text{at } (50, 100), z = 150(50) + 120(100) = 7500 + 12,000 = 19,500,$$
$$\text{and at } (10, 140), z = 150(10) + 120(140) = 1500 + 16,800 = 18,300.$$

The largest of these four values of $z$ is 19,500, which occurs when $x = 50$ and $y = 100$. We conclude that the farmer should grow 50 acres of wheat and 100 acres of corn to realize a maximum profit of \$19,500.
**Do Exercise 19.**                                                                              ▪

## Unbounded Regions

It should be clear that if the region of feasible solutions is a *bounded* convex polygonal region, then a linear function will always assume both a maximum and a minimum value and each will always occur at a vertex. But what happens when the region of feasible solutions is unbounded? It may help to first look at the case of a linear function of one variable defined on the unbounded interval $[a, \infty)$. If the graph has positive slope, then the linear function has a minimum at the left endpoint of the interval and it has no maximum. (See Figure 4.9a.) If the graph has negative slope, then the function has a maximum at the left endpoint of the interval and it has no minimum. (See Figure 4.9b.)

**FIGURE 4.9**                                   (a)                                   (b)

 Similarly, when we consider $z = ax + by + c$ and the region of feasible solutions is unbounded, then $z$ may, or may not, have a maximum or a minimum value (see Exercise 17). However, if a maximum or a minimum exists, then it will occur at one of the corners of the region of feasible solutions.

**EXAMPLE 5**  Find an optimum (maximum or minimum) value of $z$, if it exists, where

$$z = 2x + 3y + 13$$

and

$$4x - y + 10 \geq 0$$
$$x - 10y + 50 \leq 0$$
$$x + y \geq 60$$

**Solution**  The region of feasible solutions is shown in Figure 4.10. This region is unbounded and it has two corners, (10, 50), and (50, 10).

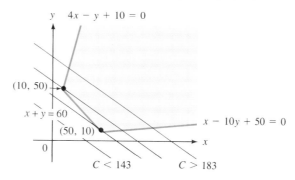

**FIGURE 4.10**

We wish to find a maximum or minimum value of $z$, if it exists. First we evaluate $z$ at the two corners. At (10, 50), we get $z = 183$ and at (50, 10) we get $z = 143$. We may be tempted to state that 143 is the minimum value of $z$ and 183 is the maximum value of $z$. But let us not be too hasty. Consider the graphs of the equations

$$2x + 3y + 13 = 183 \quad \text{and} \quad 2x + 3y + 13 = 143$$

The two lines are parallel and they intersect the region of feasible solutions at the points (10, 50) and (50, 10), respectively. It is clear from Figure 4.10 that if we draw the graph of $2x + 3y + 13 = c$, the graph will not intersect the region of feasible solutions if $c < 143$, but will always intersect the region if $c > 183$. Therefore $z$ has no maximum value. It has a minimum value, however, which is 143. This occurs at (50, 10).

**Do Exercise 13.**

In summary, when we wish to solve geometrically a linear programming problem with two variables, we proceed as follows:

**Step 1.**  Graph the system of linear inequalities that defines all constraints.

**Step 2.**  Find the coordinates of each corner of the region by solving the appropriate pairs of linear equations.

**Step 3.**  If the region of feasible solutions is bounded and the objective function is nonconstant, then both a maximum and a minimum value of the objective function will exist. In this case, evaluate the objective function at each corner of the region. The largest number obtained is the maximum value of the objective function and the smallest is its minimum. If an optimum value occurs at two corners, then it occurs at every point on the line segment joining these two corners.

**Step 4.**   If the region of feasible solutions is unbounded, evaluate the objective function $z = ax + by + c$ at each corner of the region. If $c_1$ and $c_2$ are the largest and smallest numbers obtained, respectively, then if a maximum exists, it will be $c_1$; if a minimum exists, it will be $c_2$. Replace $z$ by $c_1$ and by $c_2$ and draw the graphs of the two equations. The graphs will intersect the region of feasible solutions at certain corners. Then draw the graph of an equation $c_3 = ax + by + c$ where $c_3$ is larger than $c_1$. If this graph does not intersect the region of feasible solutions, then $c_1$ is the maximum value of $z$. If this graph does intersect the region, then $z$ has no maximum. Similarly, draw the graph of an equation $c_4 = ax + by + c$ where $c_4$ is less than $c_2$. If this graph does not intersect the region of feasible solutions, then the minimum value of $z$ exists and is $c_2$. If this graph does intersect the region, then $z$ has no minimum.

# Exercise Set 4.1

*In Exercises 1–6, a region of feasible solutions is shown and an equation defining an objective function is given. In each case, find the maximum and minimum value of $z$.*

**1.** $z = 2x + 5y + 3$

**2.** $z = 3x + 2y - 5$

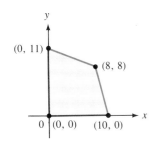

**5.** $z = .25x + .45y - .75$

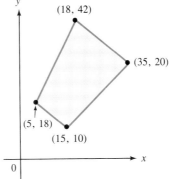

**6.** $z = -4x + 7y - 9$

**3.** $z = 5x + 3y - 7$

**4.** $z = 4x + 7y + 5$

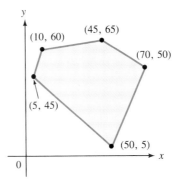

**7.** Find the maximum and minimum values of $z$ if $z = 4x + 7y$ and

$$3x + 5y \leq 180$$
$$2x - 3y + 70 \geq 0$$
$$x \geq 0 \text{ and } y \geq 0$$

**8.** Find the maximum and minimum values of $z$ if $z = 3x + 5y$ and

$$2x + 7y \leq 285$$
$$3x - 5y + 115 \geq 0$$
$$x \geq 0 \text{ and } y \geq 0$$

**9.** Find the maximum and minimum values of $z$ if $z = 5x - 13y$ and

$$5x + 13y \leq 359$$
$$4x - 5y + 67 \geq 0$$
$$x \geq 5 \text{ and } y \geq 12$$

**10.** Find the maximum and minimum values of $z$ if $z = -3x + 5y$ and

$$5x + 9y \leq 363$$
$$3x - 7y + 179 \geq 0$$
$$x \geq 7 \text{ and } y \geq 20$$

**11.** Find the maximum and minimum values of $z$ if $z = 2x + 2y - 13$ and

$$2x - y \leq 20$$
$$x + y \leq 220$$
$$-2x + y \leq 70$$
$$x \geq 20 \text{ and } y \geq 80$$

**12.** Find the maximum and minimum values of $z$ if $z = 7x - 3y + 25$ and

$$8x - 5y + 340 \geq 0$$
$$x + y \leq 120$$
$$2x + y + 40 \geq 0$$
$$x \leq 100 \text{ and } y \geq -20$$

**13.** If it exists, find the maximum or minimum value of $z$ if $z = 2x + 5y - 7$ and

$$x - 2y \leq 10$$
$$-2x + y \leq 10$$
$$x \geq 10 \text{ and } y \geq 5$$

**14.** If it exists, find the maximum or minimum value of $z$ if $z = -5x + 3y + 12$ and

$$x - y \leq 50$$
$$x + y \leq 150$$
$$x - y + 50 \geq 0$$

**15.** If possible, find the maximum or minimum value of $z$ if $z = 4x - 5y + 150$ and

$$x - 7y + 50 \leq 0$$
$$3x + 4y \geq 100$$
$$4x - 3y + 200 \geq 0$$

**16.** If possible, find the maximum or minimum value of $z$ if $z = 2x + y + 10$ and

$$x - y + 50 \geq 0$$
$$3x - y + 50 \geq 0$$
$$2x + 3y \geq 40$$
$$x - 2y + 15 \leq 0$$

**17.** Let $z = x - y$ and

$$x \geq 0$$
$$y \geq 0$$

Show that $z$ has neither a maximum nor a minimum value.

*Exercises 18–30 were given in Exercise Set 2.4 without an objective function to optimize. Refer to the corresponding exercises to obtain the regions of feasible solutions.*

**18.** If the profit is $350 per acre of peanuts and $280 per acre of corn, write the profit $z$ as a function of $x$ and $y$. How many acres of each should the farmer plant to maximize his profit?

**19.** Suppose that the profit is $370 per acre of wheat and $290 per acre of barley. Write the profit $z$ as a function of $x$ and $y$. How many acres of each should the farmer grow to maximize her profit?

**20.** Suppose that the company makes a $120 profit on each model A and a $95 profit on each model B. Express the total profit per month $z$ as a function of $x$ and $y$. How many of each model should be produced each month to maximize the profit?

**21.** Assume that the profit per deluxe model is $350 and that per standard model is $275. Express the total profit per month $z$ as a function of $x$ and $y$. How many of each model should be produced per month to maximize the profit?

**22.** Suppose that regular and premium gasoline sell for $1.05 per gallon and $1.15 per gallon, respectively. Express the revenue per day $z$ as a function of $x$ and $y$. How many gallons of each type should be sold per day in order to maximize the revenue?

**23.** Suppose that stock A sells at $80 a share and stock B sells at $70 a share. Write the total cost $z$ as a function of $x$ and $y$. How many shares of each kind should the investor buy to minimize her total cost?

**24.** Suppose that stock A sells at $120 a share and stock B sells at $95 a share. Write the cost $z$ as a function of $x$ and $y$. How many shares of each kind should the investor buy to minimize his total cost?

**25.** If the church makes a profit of $10 per doll and $25 per blanket, how many dolls and blankets should Shirley make per month to maximize the profit?

**26.** If the handyman makes a profit of $3 per basket and $6 per vase, how many baskets and vases should he make per week to maximize his profit?

**27.** If the handyman makes a profit of $5 per basket and $4 per vase, how many baskets and vases should he make per week to maximize his profit?

**28.** If the charge per night is $90 per first-class room and $55 per ordinary room, how many rooms of each type should there be to maximize the revenue?

**29.** If the charge per night is $95 per first-class room and $50 per ordinary room, how many rooms of each type should there be to maximize the revenue?

**30.** If type I food costs $3 per pound and type II costs $3.75 per pound, how many pounds of each kind should be used per week to minimize cost?

# 4.2 The Three-Dimensional Coordinate System

In Section 4.1, we discussed how a linear programming problem with two variables can be solved geometrically. However, in the applications, we usually have many variables and many constraints. It is necessary in such cases to use a method other than the geometric one. To describe this method effectively, we will restrict ourselves, at the start, to problems in which there are two or three decision variables. In doing so, we will be able to justify geometrically some of the steps we must take in an algebraic method. Then we shall be able to expand the method to cases with many decision variables and many constraints. First, however, we need to briefly introduce the three-dimensional Cartesian coordinate system. A more detailed presentation will be given in Section 11.1.

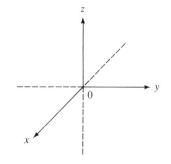

**FIGURE 4.11**

## The System of Coordinates

A three-dimensional coordinate system is obtained by adding a third axis to the familiar $xy$-Cartesian plane. This third axis is usually called the $z$-axis and is perpendicular to the $xy$-plane through the origin. The $xyz$-coordinate system is pictured in Figure 4.11. This system is often called a right-handed system because, if a right hand is placed so that the index finger points in the positive direction of the $x$-axis and the middle finger points in the direction of the positive $y$-axis, then the thumb points in the positive direction of the $z$-axis. (See Figure 4.12.)

Given an ordered $(a, b, c)$ of real numbers, the *graph* of that ordered triple in the coordinate system can be described as follows. We find the points on the $x$ and $y$ axes

FIGURE 4.12

FIGURE 4.13

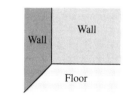

FIGURE 4.14

at directed distances $a$ and $b$ units from the origin, respectively, as we do when we locate $(a, b)$ in the $xy$-plane. Then we find the point on the $z$-axis at a directed distance $c$ from the origin. Now through each of the three points we construct a plane perpendicular to the axis on which the point lies. (See Figure 4.13.)

These three planes intersect at a unique point $P$, which is the graph of the ordered triple $(a, b, c)$. We also say that $a$, $b$, $c$, are the *coordinates* of the point $P$: $a$ is the $x$-coordinate or first coordinate, $b$ is the $y$-coordinate or second coordinate, and $c$ is the $z$-coordinate or third coordinate. Note that any point on the $x$-axis has coordinates $(x, 0, 0)$, any point on the $y$-axis has coordinates $(0, y, 0)$, and any point on the $z$-axis has coordinates $(0, 0, z)$. Similarly, points on the $xy$-plane have coordinates $(x, y, 0)$, points on the $xz$-plane have coordinates $(x, 0, z)$, and points on the $yz$-plane have coordinates $(0, y, z)$. The three coordinate planes divide the three-dimensional space into eight unbounded regions called *octants*. The region that consists of all points with coordinates $(a, b, c)$, where $a$, $b$, and $c$ are positive, is called the *first octant*. The other seven octants are not given special names. It is easy to visualize a three-dimensional coordinate system using two walls and the floor of a room. (See Figure 4.14.)

**EXAMPLE 1**   Plot the points $(3, 4, 2)$ and $(3, 2, -4)$.

**Solution**   The point $(3, 4, 2)$ is shown in Figure 4.15a and the graph of $(3, 2, -4)$ is shown in Figure 4.15b.
**Do Exercise 1.**

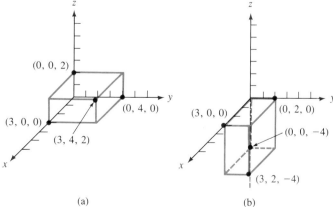

FIGURE 4.15                                    (a)                                    (b)

## Graphs of Linear Equations

The graph of an equation $ax + by = c$, where $a$, $b$, and $c$ are real numbers with at least one of the numbers $a$ and $b \neq 0$, is a straight line in the $xy$-plane. Similarly, if $a$, $b$, $c$, and $d$ are real numbers, with at least one of the numbers $a$, $b$, and $c \neq 0$, the graph of the equation $ax + by + cz = d$ is a plane in the three-dimensional coordinate system. We find the intercepts of that plane with the three axes by letting two of the variables $x$, $y$, and $z$ equal 0 and solve for the third variable. Since three noncollinear points determine a plane, it is usually sufficient to find the three intercepts of the plane with the three axes in order to sketch the plane whose equation is given.

**EXAMPLE 2**    Sketch the graph of $2x + 3y + 5z = 60$.

**Solution**    To find the $x$-intercept, replace $y$ and $z$ by 0 and solve for $x$. We get

$$2x = 60, \text{ so that } x = 30$$

The $x$-intercept is the point $(30, 0, 0)$.
  To find the $y$-intercept, replace $x$ and $z$ by 0 and solve for $y$. We obtain

$$3y = 60, \text{ so that } y = 20$$

The $y$-intercept is the point $(0, 20, 0)$.
  The $z$-intercept is found by replacing $x$ and $y$ by 0 and solving for $z$. We get

$$5z = 60, \text{ so that } z = 12$$

The $z$-intercept is $(0, 0, 12)$.
  We now plot $(30, 0, 0)$, $(0, 20, 0)$, and $(0, 0, 12)$. These points are the three vertices of a first octant triangle. (See Figure 4.16.) This triangle is part of the plane which is the graph of $2x + 3y + 5z = 60$.
**Do Exercise 5.**

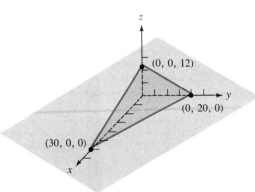

**FIGURE 4.16**

The plane whose equation is $ax + by + cz = d$ divides the three-dimensional space into three subsets, the plane itself and two half-spaces. One of these half-spaces is the graph of the inequality $ax + by + cz < d$, while the other is the graph of $ax + by + cz > d$. For example, in Figure 4.16, the half-space bounded by the plane and which contains the origin is the graph of $2x + 3y + 5z < 60$, since $2(0) + 3(0) + 5(0) < 60$ is true. The other half-space is the graph of $2x + 3y + 5z > 60$.

In general, two planes coincide, are parallel, or intersect along a straight line. In the latter case, a third plane may pass through that line, be parallel to it, or intersect it at a single point. To find the intersection of three planes, we need only solve the system of three linear equations whose graphs are the given planes.

**EXAMPLE 3**   **a.** Find the intersection of the planes $P_1$, $P_2$, and $P_3$, whose equations are,

$$
\begin{aligned}
x - y + 2z &= 6 \\
-x + 3y + 4z &= 12 \\
4x + 6y - 7z &= 24
\end{aligned}
$$

respectively.

**b.** Sketch the region in the first octant bounded by these three planes and the three coordinate planes.

**Solution**   **a.** We must solve the linear system

$$
\begin{cases}
x - y + 2z = 6 \\
-x + 3y + 4z = 12 \\
4x + 6y - 7z = 24
\end{cases}
$$

We use any of the techniques of Chapters 2 and 3 to obtain $x = 5$, $y = 3$, and $z = 2$. The point $(5, 3, 2)$ is, therefore, the point of intersection of the three planes.

**b.** It is easy to verify that $(6, 0, 0)$ is the $x$-intercept of both $P_1$ and $P_3$, $(0, 4, 0)$ is the $y$-intercept of both $P_2$ and $P_3$, and $(0, 0, 3)$ is the $z$-intercept of both $P_1$ and $P_2$. Hence, $(6, 0, 0)$, $(0, 0, 3)$, and $(5, 3, 2)$ are the vertices of a triangle that is part of the plane $P_1$. The points $(0, 4, 0)$, $(0, 0, 3)$, and $(5, 3, 2)$ are the vertices of a triangle that is part of the plane $P_2$. The points $(6, 0, 0)$, $(0, 4, 0)$, and $(5, 3, 2)$ are the vertices of a triangle that is part of the plane $P_3$. These triangles are sketched in Figure 4.17. They form part of the boundary of the region in the first octant bounded by the three given planes and by the three coordinate planes.

**Do Exercise 9.**

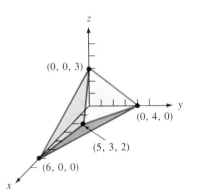

**FIGURE 4.17**

As we shall see in the next section, this region is the graph of the system

$$\begin{cases} x - y + 2z \leq 6 \\ -x + 3y + 4z \leq 12 \\ 4x + 6y - 7z \leq 24 \\ x \geq 0,\ y \geq 0,\ z \geq 0 \end{cases}$$

If we have a set of $n$ linear equations where each is in the two variables $x$ and $y$ and is the graph of a straight line in the $xy$-plane it may be desirable to find all possible points of intersection of pairs of these lines. Since each of these points is the intersection of two lines, we must solve all systems of two equations that can be obtained from the given set of $n$ equations. The number of systems is the same as the number of subsets, each with two members, that we obtain from a set with $n$ members.

Similarly, if we have a set of $n$ linear equations where each is in the three variables $x$, $y$, and $z$ and is the graph of a plane in the $xyz$-coordinate system, it may be desirable to find all possible points of intersection of triples of these planes. Since each of these points is the intersection of three planes, we must solve all systems of three equations that can be obtained from the given set of $n$ equations. The number of systems is the same as the number of subsets, each with three members, that we obtain from a set with $n$ members.

We shall show in Chapter 6 that the number of subsets, each with $k$ elements, that can be obtained from a set of $n$ elements is $\frac{n!}{k!(n-k)!}$, where the symbol $n!$ (read "$n$ factorial") represents the product of the first $n$ positive integers. For example, $5! = 5 \cdot 4 \cdot 3 \cdot 2 \cdot 1 = 120$.

**EXAMPLE 4**    We have a set of six planes. Assume that the maximum number of points of intersection we can get is finite. What is that number?

**Solution**    We have six linear equations. The coordinates of each point of intersection of three planes are obtained by solving a system of three linear equations. Each system is obtained by choosing three equations from the given set of six equations. Hence, the number of systems obtained is the same as the number of subsets, each with three members, that can be obtained from a set with six members. This number is 20 since

$$\frac{6!}{3!(6-3)!} = \frac{6!}{3!3!}$$

$$= \frac{6(5)(4)}{3!}$$

$$= \frac{120}{6} = 20$$

 Note that 20 is the maximum number of points of intersection that we can have. Several of the systems may have identical solutions. In that case, the number of points of intersection will be less than 20. This will happen if more than three planes pass through the same point. Similarly, some of the systems may have no solution; for example, if two of the planes are parallel. Again, this will result in less than 20 points of intersection. On the other hand, if we had not assumed that the number of points of intersection is finite, it is possible for all six planes to intersect along the same line. In that case, the number of points of intersection is infinite.

**Do Exercise 21.**                                                              ■

---

## Exercise Set 4.2

**1.** Plot the points **a.** $(2, 4, 7)$, **b.** $(-3, 4, 5)$, and **c.** $(2, -3, 4)$.

**2.** Plot the points **a.** $(3, 5, 1)$, **b.** $(-2, 5, 7)$, and **c.** $(3, -4, 6)$.

**3.** Plot the points **a.** $(-2, 3, 6)$, **b.** $(2, -4, -5)$, and **c.** $(7, -5, 8)$.

**4.** Sketch the graph of the equation $2x + 3y + 4z = 24$.

**5.** Sketch the graph of the equation $5x + 2y + 3z = 60$.

**6.** Sketch the graph of the equation $x + 2y - 3z = 18$.

**7.** Sketch the graph of the equation $-x + 5y + 3z = 30$.

**8.** Sketch the graph of the equation $-3x + 4y - 3z = 36$.

*In Exercises 9–13, the equations of three planes $P_1$, $P_2$, and $P_3$ are given. In each case:*

*a. Find the intersection of the three planes.*

*b. Sketch the region in the first octant bounded by the three planes and the three coordinate planes.*

**9.**
$$9x - 7y + 3z = 9$$
$$-11x + 6y + 4z = 12$$
$$8x + 4y - 5z = 8$$

**10.**
$$6x - 7y + 12z = 12$$
$$-14x + 3y + 12z = 12$$
$$6x + 3y - 8z = 12$$

**11.**
$$20x - 38y + 25z = 100$$
$$-x + y + z = 4$$
$$24x + 30y - 33z = 120$$

**12.**
$$21x - 37y + 9z = 63$$
$$-6x + 7y + z = 7$$
$$x + 3y - z = 3$$

**13.**
$$7x - 10y + 14z = 42$$
$$-13x + 12y + 20z = 60$$
$$5x + 6y - 13z = 30$$

*In Exercises 14–21, assume that the maximum number of points of intersection is finite.*

**14.** Five lines have been drawn in a plane. What is the maximum number of points of intersection of pairs of these lines?

**15.** Same as Exercise 14 with six lines.

**16.** Same as Exercise 14 with seven lines.

**17.** Same as Exercise 14 with eight lines.

**18.** Four planes have been drawn in the three-dimensional space. What is the maximum finite number of points of intersection of triples of these planes?

**19.** Same as Exercise 18 with five planes.

**20.** Same as Exercise 18 with seven planes.

**21.** Same as Exercise 18 with nine planes.

---

## 4.3 Introduction to the Simplex Method

In Section 4.1, the linear programming problem was discussed geometrically. However, this approach is cumbersome unless there are only two decision variables and relatively few constraints. In practice, there are usually many decision variables and constraints. In such cases, we use the *simplex method*. The name of this method comes from a geometric object. This object is analogous to the convex polygonal region introduced earlier and, in more advanced work, is called a simplex. We need not concern ourselves with this object now, but we should understand that many of its properties are similar to those of a convex polygonal region so that the techniques used to solve problems with only two decision variables may be extended to problems with many variables.

Interest in linear programming developed during World War II when military forces wanted to find efficient ways to bring supplies to the front. It was then that the simplex method was developed. Today, it has far-reaching applications in the fields of business and economics.

As stated, there are often many decision variables; we use symbols such as $x_1$, $x_2$, $x_3$, $x_4$, ... , $x_n$ to represent them. The variable names are convenient when computers are used to solve linear programming problems.

We will at first restrict our discussion to the so-called *standard maximum linear programming problems*. In these problems, all constraints are in the form

$$a_{i1}x_1 + a_{i2}x_2 + a_{i3}x_3 + \cdots + a_{in}x_n \leq b_i$$

where $x_1$, $x_2$, $x_3$, ... , $x_n$ are decision variables and $a_{i1}$, $a_{i2}$, $a_{i3}$, ... , $a_{in}$, and $b_i$ are constants and

1. All variables are nonnegative.
2. All constant terms $b_i$ are nonnegative.
3. The objective function is expressed in the form

$$z = c_1x_1 + c_2x_2 + c_3x_3 + \cdots + c_nx_n$$

where $c_1$, $c_2$, $c_3$, ... , $c_n$ are constants.
4. We are trying to maximize the objective function.

We will discuss minimization briefly at the end of this chapter.

When we considered problems with only two decision variables, we noted that the graph of $z = ax + by + c$ is a plane and asked you to visualize a house with a slanted flat roof over a convex polygonal floor. We argued that such a house would clearly have its highest point at the corner. To find this point, we could measure the height of the house at each of its corners; the largest of the numbers would be the highest point. Similarly, when we wanted to find the maximum value of the objective function, we evaluated that function at each corner of the region of feasible solutions and concluded that the largest number obtained was the maximum value of the function. A disadvantage of this method is that it is necessary to find the coordinates of each corner of the region of feasible solutions.

To understand the idea behind the simplex method, let us go back to visualizing the house with a slanted flat roof. Imagine that a squirrel is running along the edges of the roof and he has decided to reach the highest point. He will surely succeed if he adopts the following strategy. Starting at a corner, he runs along the edge of the roof from that corner provided he is going uphill by doing so. If both edges starting at that corner are going up, he chooses the steeper edge. He keeps doing this, from corner to corner, until he reaches a corner where no edge is going up. He clearly has reached the highest point. The advantage of this strategy is that he will likely reach the highest point without having tested each corner. The idea behind the simplex method is similar to the squirrel's strategy.

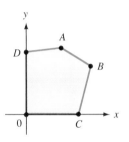

**FIGURE 4.18**

We will illustrate the technique geometrically and first restrict the discussion to cases where the number of decision variables is two or three. Therefore, the set of feasible solutions is a convex polygonal region in the case of two variables and a convex polyhedral region in the case of three variables. We shall say that two corners of such a region are *adjacent* if they are connected by an edge of the region. In Figure 4.18, $A$ and $B$ are adjacent, while $A$ and $C$ are not. In Figure 4.19, $A$ and $C$ are adjacent while $A$ and $B$ are not.

The simplex method is a procedure that allows us to move from one corner of the region of feasible solutions to an adjacent corner in such a way that the value of the objective function is increased at each step, until we either find the maximum value of the function or see that no maximum exists. The following theorem, stated without proof, justifies this procedure.

**FIGURE 4.19**

**THEOREM 4.1:**  Let $A$ be a corner of the region of feasible solutions of a linear programming problem. If the value of the objective function is not maximum at $A$, then there is an edge of the region of feasible solutions, starting at $A$, along which the values of the objective function increase as we move away from $A$.

Our goal is to develop an algebraic way to identify corners of the region of feasible solutions and also to find a way to move from one corner to an adjacent corner.

## Slack Variables

The first step is to replace the inequalities by equations through the introduction of new variables called *slack variables*. For example, suppose that a farmer has 200 acres of land on which he wishes to grow wheat and barley. Let $x$ and $y$ represent the numbers of acres he uses for wheat and barley, respectively. Then

$$x + y \leq 200$$

Since he may not use all of his land for these two grains, let $s$ represent the number of acres on which he plants neither wheat nor barley. Then

$$x + y + s = 200$$

since each of the 200 acres is used for wheat, barley, or neither. Of course, $s$ is 0 if the farmer uses all of his land for wheat and barley, but $s$ will always be nonnegative. The variable $s$ represents the "slack," the acreage not used for the two grains.

> **DEFINITION 4.2:** If each constraint which is described by an inequality
>
> $$a_{i1}x_1 + a_{i2}x_2 + a_{i3}x_3 + \cdots + a_{in}x_n \leq b_i$$
>
> is converted to an equation
>
> $$a_{i1}x_1 + a_{i2}x_2 + a_{i3}x_3 + \cdots + a_{in}x_n + s_i = b_i,$$
>
> where $s_i \geq 0$, then each $s_i$ is called a *slack variable*.

**EXAMPLE 1** Sketch the graph of the inequality $2x + 3y \leq 12$. Then introduce a variable $s$ to change the inequality to an equation. Finally, discuss geometrically the situation when **a.** $s > 0$, **b.** $s = 0$, and **c.** $s < 0$.

**Solution** The graph is shown in Figure 4.20.

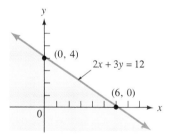

**FIGURE 4.20**

We introduce the variable $s$ by writing

$$2x + 3y + s = 12$$

As we know, the graph of $2x + 3y = 12$ is a line $L$ which divides the plane into three subsets: the open half-plane containing the origin (the shaded region in Figure 4.20), the line $L$, and the other open half-plane (not shaded in Figure 4.20). The coordinates of each point in the first half-plane satisfy the strict inequality $2x + 3y < 12$, those of each point on the line satisfy the equation $2x + 3y = 12$, and those of each point in the nonshaded half-plane satisfy the strict inequality $2x + 3y > 12$. Consequently, suppose that $(a, b, c)$ satisfies

$$2x + 3y + s = 12$$

That is, suppose that

$$2a + 3b + c = 12$$

Then,

**a.** If $c > 0$, $2a + 3b < 12$ and the point $(a, b)$ is in the shaded half-plane in Figure 4.20.

**b.** If $c = 0$, $2a + 3b = 12$ and the point $(a, b)$ is on the line $L$.
**c.** If $c < 0$, $2a + 3b > 12$ and the point $(a, b)$ is in the nonshaded half-plane.
**Do Exercise 1.**   ■

In Example 1, the variable $s$ was not called a slack variable since we allowed it to assume negative values. However, this example and the next two examples illustrate that a slack variable must be nonnegative for the corresponding point to be in the region of feasible solutions.

**EXAMPLE 2**   Give a discussion similar to that of Example 1 for $3x + 5y + 2z \le 30$.

**Solution**   The graph of $3x + 5y + 2z = 30$ is a plane $P$, part of which is shown in Figure 4.21.

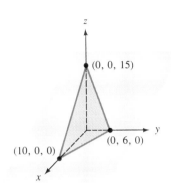

**FIGURE 4.21**

This plane divides the three-dimensional space into three subsets: the open half-space containing the origin, the plane $P$, and the open half-space not containing the origin. The coordinates of each point in the first half-space satisfy the strict inequality $3x + 5y + 2z < 30$, those of each point of the plane $P$ satisfy the equation $2x + 5y + 2z = 30$, and those of each point in the other half-space satisfy the strict inequality $3x + 5y + 2z > 30$. Converting the inequality to an equation by introducing the variable $s$, we get

$$3x + 5y + 2z + s = 30$$

Suppose now that $(a, b, c, d)$ is a solution of the foregoing equation. That is, suppose that $3a + 5b + 2c + d = 30$ is true. Then,

**a.** If $d > 0$, $3a + 5b + 2c < 30$ and the point $(a, b, c)$ is in the half-space containing the origin.
**b.** If $d = 0$, $3a + 5b + 2c = 30$ and the point $(a, b, c)$ is on the plane $P$.
**c.** If $d < 0$, $3a + 5b + 2c > 30$ and the point $(a, b, c)$ is in the half-space not containing the origin.
**Do Exercise 7.**   ■

**EXAMPLE 3**   **a.** Sketch the graph of the system

$$x + 2y \le 8$$
$$x + y \le 6$$
$$x \ge 0, \, y \ge 0$$

**b.** Introduce variables to convert the first two inequalities to equations.

**c.** Discuss the relationship between the values of the variables and the position of points in the coordinate plane.

**Solution**    **a.** Let $L_1$ be the graph of $x + 2y = 8$ and $L_2$ be the graph of $x + y = 6$. The ordered pair $(4, 2)$ is the solution of the system

$$\begin{cases} x + 2y = 8 \\ x + \phantom{2}y = 6 \end{cases}$$

Hence, the point $(4, 2)$ is the intersection of the lines $L_1$ and $L_2$. Each of these lines divides the plane into half-planes. The two half-planes containing the origin are the graphs of the inequalities

$$x + 2y < 8 \quad \text{and} \quad x + y < 6.$$

The half-plane to the right of the $y$-axis is the graph of the inequality $x > 0$ and the half-plane above the $x$-axis is the graph of the inequality $y > 0$. Thus, the intersection of these four half-planes, together with its boundary, is the graph of the given system of inequalities shown as the shaded region in Figure 4.22.

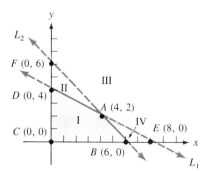

**FIGURE 4.22**

The corners (vertices) $A$, $B$, $C$, and $D$ of the convex polygonal region have coordinates $(4, 2)$, $(6, 0)$, $(0, 0)$, and $(0, 4)$, respectively.

**b.** We introduce the variables $s$ and $t$ so that the first two inequalities are converted to equations.

$$x + 2y + s = 8$$
$$x + \phantom{2}y + t = 6.$$

**c.** For convenience, let $E(8, 0)$ be the $x$-intercept of $L_1$ and $F(0, 6)$ be the $y$-intercept of $L_2$. (See Figure 4.22.) The lines $L_1$ and $L_2$ divide the first quadrant into four regions labeled I, II, III, and IV. Region I is the interior of the convex polygonal region $ABCD$; region II is the interior of triangle $ADF$; region III is the unbounded set to the right of the $y$-axis, the lines $L_1$ and $L_2$, and above the $x$-axis; and region IV is the interior of triangle $ABE$. Now suppose that $(a, b, c, d)$ is a solution of the system

$$\begin{cases} x + 2y + s = 8 \\ x + \phantom{2}y + t = 6 \end{cases}$$

That is, suppose that

$$a + 2b + c = 8 \quad \text{and} \quad a + b + d = 6.$$

Then, using the same reasoning as in Example 1, we see that the point $(a, b)$ will be in different regions depending on the following conditions:

1. In region I if $a$, $b$, $c$, and $d$ are positive.
2. In region II if $a$, $b$, and $d$ are positive while $c$ is negative.
3. In region III if $a$ and $b$ are positive while $c$ and $d$ are negative.
4. In region IV if $a$, $b$, and $c$ are positive while $d$ is negative.
5. On the boundary of region I if one or two of $a$, $b$, $c$, and $d$ are 0 while the others are positive. For example, if both $c$ and $d$ are 0, the point will be on both $L_1$ and $L_2$, so it must be point $A$.

You should consider other cases. However, for the purpose of what will follow, we wish to stress that given the linear system

$$\begin{cases} x + 2y + s = 8 \\ x + y + t = 6 \end{cases}$$

and a solution $(a, b, c, d)$ of that system, we can associate a point $(a, b)$ in the coordinate plane and locate it vis-a-vis the lines $L_1$, $L_2$, the $x$-axis, and the $y$-axis, depending on the signs of $a$, $b$, $c$, and $d$. We emphasize that $(a, b)$ is *inside* the convex polygonal region $ABCD$ if $a$, $b$, $c$, and $d$ are all positive, and that $(a, b)$ is at a corner (vertex) of the region if two of the four numbers $a$, $b$, $c$, and $d$ are 0. Specifically, $A$ is associated with the solution $(4, 2, 0, 0)$, $B$ with the solution $(6, 0, 2, 0)$, $C$ with the solution $(0, 0, 8, 6)$, and $D$ with the solution $(0, 4, 0, 2)$. Also $E$ is associated with the solution $(8, 0, 0, -2)$ and $F$ is associated with the solution $(0, 6, -4, 0)$.
**Do Exercise 13.**                                                        ■

## Vertices of the Region of Feasible Solutions

The next example is lengthy and detailed; we will omit all but the essential steps. It is important that you understand the role of slack variables in analyzing a system of linear inequalities. You are urged to become familiar with the steps taken.

EXAMPLE 4

Consider the system $\begin{cases} y + z \le 5 \\ 5x + 7y - 3z \le 35 \\ 15x + y + 21z \le 105 \\ x \ge 0,\ y \ge 0,\ z \ge 0 \end{cases}$

Introduce slack variables to convert the first three inequalities to equations. Use the values of these slack variables to describe geometrically the graph of the given system of inequalities and sketch the graph.

**Solution**    We introduce three slack variables $r$, $s$, and $t$ and write

$$
\begin{aligned}
y + \phantom{5x + 7}z + r &= 5 \\
5x + 7y - \phantom{15}3z + s &= 35 \\
15x + \phantom{7}y + 21z + t &= 105 \\
x \geq 0,\ y \geq 0,\ z \geq 0,\ r \geq 0,\ s \geq 0,\ t \geq 0
\end{aligned}
$$

Now consider the equations

$$
\begin{aligned}
y + \phantom{5x + 7}z &= 5 & (1) \\
5x + 7y - \phantom{15}3z &= 35 & (2) \\
15x + \phantom{7}y + 21z &= 105 & (3) \\
x &= 0 & (4) \\
y &= 0 & (5) \\
z &= 0 & (6)
\end{aligned}
$$

The graph of each equation is a plane in three dimensions. Let $P_i$ denote the graph of equation $(i)$ for $i = 1, 2, 3, 4, 5$, and 6. Note that $P_4$ is the $yz$-plane, $P_5$ is the $xz$-plane, and $P_6$ is the $xy$-plane. To find the points of intersection of these planes, we solve linear systems of three equations taken from the six equations. For example, $(4, 3, 2)$ is the solution of the system

$$
\begin{cases}
y + \phantom{5x + 7}z = 5 & (1) \\
5x + 7y - \phantom{15}3z = 35 & (2) \\
15x + \phantom{7}y + 21z = 105 & (3)
\end{cases}
$$

Hence, the point $(4, 3, 2)$ is the intersection of the planes $P_1$, $P_2$, and $P_3$. Since we form each system by taking three equations from a set of six equations, the number of different systems we can obtain is 20 since $\frac{6!}{3!(6 - 3)!} = \frac{6!}{3!3!} = 20$. The 20 subsets of the set $\{1, 2, 3, 4, 5, 6\}$, each with three elements, are $\{1, 2, 3\}$, $\{1, 2, 4\}$, $\{1, 2, 5\}$, $\{1, 2, 6\}$, $\{1, 3, 4\}$, $\{1, 3, 5\}$, $\{1, 3, 6\}$, $\{1, 4, 5\}$, $\{1, 4, 6\}$, $\{1, 5, 6\}$, $\{2, 3, 4\}$, $\{2, 3, 5\}$, $\{2, 3, 6\}$, $\{2, 4, 5\}$, $\{2, 4, 6\}$, $\{2, 5, 6\}$, $\{3, 4, 5\}$, $\{3, 4, 6\}$, $\{3, 5, 6\}$, $\{4, 5, 6\}$. With each of these subsets, we associate a linear system of three equations taken from the original set of six equations. For example, the system

$$
\begin{cases}
y + \phantom{5x + 7}z = 5 & (1) \\
5x + 7y - \phantom{15}3z = 35 & (2) \\
15x + \phantom{7}y + 21z = 105 & (3)
\end{cases}
$$

is associated with the subset $\{1, 2, 3\}$, since it is made up of equations (1), (2), and (3); and the system

$$
\begin{cases}
y + \phantom{5x + 7}z = 5 & (1) \\
5x + 7y - \phantom{15}3z = 35 & (2) \\
x = 0 & (4)
\end{cases}
$$

is associated with the subset $\{1, 2, 4\}$, since it is made up of equations (1), (2), and (4). The table gives the solutions of the 20 linear systems of three equations.

| System | Solution |
|--------|----------|
| {1, 2, 3} | (4, 3, 2) |
| {1, 2, 4} | (0, 5, 0) |
| {1, 2, 5} | (10, 0, 5) |
| {1, 2, 6} | (0, 5, 0) |
| {1, 3, 4} | (0, 0, 5) |
| {1, 3, 5} | (0, 0, 5) |
| {1, 3, 6} | $\left(\frac{20}{3}, 5, 0\right)$ |
| {1, 4, 5} | (0, 0, 5) |
| {1, 4, 6} | (0, 5, 0) |
| {1, 5, 6} | no solution |
| {2, 3, 4} | $\left(0, 7, \frac{14}{3}\right)$ |
| {2, 3, 5} | (7, 0, 0) |
| {2, 3, 6} | (7, 0, 0) |
| {2, 4, 5} | $\left(0, 0, \frac{-35}{3}\right)$ |
| {2, 4, 6} | (0, 5, 0) |
| {2, 5, 6} | (7, 0, 0) |
| {3, 4, 5} | (0, 0, 5) |
| {3, 4, 6} | (0, 105, 0) |
| {3, 5, 6} | (7, 0, 0) |
| {4, 5, 6} | (0, 0, 0) |

Notice that there are duplications. For example, (0, 5, 0) is the solution of the systems represented by {1, 2, 4}, {1, 2, 6}, {1, 4, 6}, and {2, 4, 6}. This is because the planes $P_1$, $P_2$, $P_4$, and $P_6$ intersect at (0, 5, 0). Although the steps used to solve each of the 20 systems were left out, it is clear that this process is time consuming. In the future we will not need to find the solutions of all 20 systems. Remember that we are interested only in graphing the system of inequalities

$$\begin{cases} y + z \le 5 \\ 5x + 7y - 3z \le 35 \\ 15x + y + 21z \le 105 \\ x \ge 0, y \ge 0, z \ge 0 \end{cases}$$

Now for each of the 20 solutions in the table to the left, we calculate the values of the variables $r$, $s$, and $t$, so that $(x, y, z, r, s, t)$ is a solution of the system

$$\begin{cases} y + z + r = 5 \\ 5x + 7y - 3z + s = 35 \\ 15x + y + 21z + t = 105 \end{cases}$$

The results are given in the following table.

| System | Solution |
|--------|----------|
| {1, 2, 3} | (4, 3, 2, 0, 0, 0) |
| {1, 2, 4} | (0, 5, 0, 0, 0, 100) |
| {1, 2, 5} | (10, 0, 5, 0, 0, −150) |
| {1, 2, 6} | (0, 5, 0, 0, 0, 100) |
| {1, 3, 4} | (0, 0, 5, 0, 50, 0) |
| {1, 3, 5} | (0, 0, 5, 0, 50, 0) |
| {1, 3, 6} | $\left(\frac{20}{3}, 5, 0, 0, \frac{-100}{3}, 0\right)$ |
| {1, 4, 5} | (0, 0, 5, 0, 50, 0) |
| {1, 4, 6} | (0, 5, 0, 0, 0, 100) |
| {1, 5, 6} | no solution |
| {2, 3, 4} | $\left(0, 7, \frac{14}{3}, \frac{-20}{3}, 0, 0\right)$ |
| {2, 3, 5} | (7, 0, 0, 5, 0, 0) |
| {2, 3, 6} | (7, 0, 0, 5, 0, 0) |
| {2, 4, 5} | $\left(0, 0, \frac{-35}{3}, \frac{50}{3}, 0, 350\right)$ |
| {2, 4, 6} | (0, 5, 0, 0, 0, 100) |
| {2, 5, 6} | (7, 0, 0, 5, 0, 0) |
| {3, 4, 5} | (0, 0, 5, 0, 50, 0) |
| {3, 4, 6} | (0, 105, 0, −100, 0, 0) |
| {3, 5, 6} | (7, 0, 0, 5, 0, 0) |
| {4, 5, 6} | (0, 0, 0, 5, 35, 105) |

Recall that $x$, $y$, $z$, $r$, $s$, and $t$ are required to be nonnegative. Thus, among the ordered sextuples listed, we can reject all those with a negative entry. Each of the ordered sextuples listed has at least three 0's. Recall the significance of each 0. Consider, for example, the sextuple (4, 3, 2, 0, 0, 0). This means that $x = 4$, $y = 3$, $z = 2$, $r = 0$, $s = 0$, and $t = 0$. The fact that $r = 0$ indicates that the point (4, 3, 2) is on the plane $P_1$. That point is also on the plane $P_2$ since $s = 0$, and on the plane $P_3$ since $t = 0$. Since $x = 4$, $y = 3$, and $z = 2$ satisfy the system of linear inequalities, we conclude that the point is on the graph of that system. In fact, it is a corner of the graph since it is the intersection of three of the planes which bound the graph. We now list all five of the ordered sextuples that have no negative entry. They are $(4, 3, 2, 0, 0, 0)$, $(0, 5, 0, 0, 0, 100)$, $(0, 0, 5, 0, 50, 0)$, $(7, 0, 0, 5, 0, 0)$, and $(0, 0, 0, 5, 35, 105)$. Thus, the points $(4, 3, 2)$, $(0, 5, 0)$, $(0, 0, 5)$, $(7, 0, 0)$, and $(0, 0, 0)$ are the corners of the graph of the given system of inequalities. Let us label these points $A$, $B$, $C$, $D$, and $E$, respectively. The graph is the polyhedron whose faces are the triangles $ABD$, $ABC$, $ADC$, $BCE$, $BDE$, and $CDE$. (See Figure 4.23.) This polyhedron has nine edges.

**Do Exercise 27.**

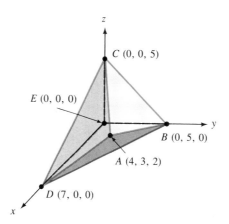

**FIGURE 4.23**

In the next section, we shall introduce a good technique to identify the corners of the graph of a system of linear inequalities. We shall also show how to move from any corner to an adjacent corner.

# Exercise Set 4.3

*In Exercises 1–12, sketch the graph of the given inequality. Then introduce a variable s to convert the inequality to an equation. Finally, discuss geometrically the situation when a. $s > 0$, b. $s = 0$, and c. $s < 0$.*

**1.** $3x + y \le 15$

**2.** $x + 6y \le 18$

**3.** $2x + 3y \le 12$

**4.** $2x + 5y \le 30$

**5.** $2x + y \le 14$

**6.** $4x + 3y \le 24$

**7.** $4x + 5y + 4z \le 40$

**8.** $12x + 8y + 9z \le 72$

**9.** $4x + y + 4z \le 16$

**10.** $10x + 6y + 5z \le 90$

**11.** $2x + 2y + 3z \le 12$

**12.** $2x + 4y + 5z \le 30$

*In Exercises 13–28, a system of linear inequalities is given.*

*a. Sketch the graph of the system.*
*b. Introduce variables to convert the inequalities that have more than one variable to equations.*
*c. Discuss the relationship between the values of the variables and the position of points in the appropriate coordinate systems.*

**13.**  $x + y \le 7$
$3x + y \le 15$
$x \ge 0, y \ge 0$

**14.** $x + y \le 8$
$x + 6y \le 18$
$x > 0, y \ge 0$

**15.** $\begin{aligned} x + 2y &\leq 7 \\ 2x + 3y &\leq 12 \\ x \geq 0, y &\geq 0 \end{aligned}$

**16.** $\begin{aligned} 2x + 5y &\leq 25 \\ 3x + 2y &\leq 21 \\ x \geq 0, y &\geq 0 \end{aligned}$

**17.** $\begin{aligned} 2x + 5y &\leq 30 \\ 2x + y &\leq 14 \\ x \geq 0, y &\geq 0 \end{aligned}$

**18.** $\begin{aligned} 7x + 2y &\leq 105 \\ x + y &\leq 20 \\ x \geq 0, y &\geq 0 \end{aligned}$

**19.** $\begin{aligned} x + y &\leq 10 \\ x + 2y &\leq 17 \\ 3x + 2y &\leq 27 \\ x \geq 0, y &\geq 0 \end{aligned}$

**20.** $\begin{aligned} x + y &\leq 17 \\ 2x + y &\leq 26 \\ 4x + 3y &\leq 58 \\ x \geq 0, y &\geq 0 \end{aligned}$

**21.** $\begin{aligned} 2x + y &\leq 18 \\ x + 2y &\leq 21 \\ 4x + 5y &\leq 72 \\ x \geq 0, y &\geq 0 \end{aligned}$

**22.** $\begin{aligned} 8x + 9y &\leq 260 \\ 5x + 3y &\leq 131 \\ x + 2y &\leq 50 \\ x \geq 0, y &\geq 0 \end{aligned}$

**23.** $\begin{aligned} 4x + 5y + 4z &\leq 40 \\ 12x + 9y + 8z &\leq 96 \\ x \geq 0, y \geq 0, z &\geq 0 \end{aligned}$

**24.** $\begin{aligned} 12x + 8y + 9z &\leq 72 \\ x + y + z &\leq 7 \\ x \geq 0, y \geq 0, z &\geq 0 \end{aligned}$

**25.** $\begin{aligned} 4x + y + 4z &\leq 16 \\ 2x + 3y + 7z &\leq 18 \\ x \geq 0, y \geq 0, z &\geq 0 \end{aligned}$

**26.** $\begin{aligned} 10x + 6y + 5z &\leq 90 \\ 8x + 12y + 13z &\leq 108 \\ x \geq 0, y \geq 0, z &\geq 0 \end{aligned}$

**27.** $\begin{aligned} 2x + y + z &\leq 10 \\ 2x + 2y + 3z &\leq 14 \\ x \geq 0, y \geq 0, z &\geq 0 \end{aligned}$

**28.** $\begin{aligned} 2x + 4y + 5z &\leq 30 \\ 4x + 4y + 5z &\leq 40 \\ x \geq 0, y \geq 0, z &\geq 0 \end{aligned}$

# 4.4 Moving from Corner to Corner

Now we shall discuss algebraically how to identify corners of the region of feasible solutions of a linear programming problem and how to go from any corner to an adjacent corner. This is the final step before describing the simplex method in Section 4.5. Again, because we wish our presentation to have a geometric flavor, we shall restrict our examples to cases with two or three variables. However, the procedure can be generalized to any number of variables.

## Basic Feasible Solutions

In the preceding section, we showed that by introducing slack variables we could convert all constraints, except the nonnegative ones, to equations. In doing so, we always obtain a system of linear equations with more variables than equations. In general, as we saw in Section 3.3, such a system has infinitely many solutions. In the preceding section, we also observed that given a solution of the linear system obtained by introducing slack variables, we could locate a point in the appropriate coordinate system in relation to the boundary of the region of feasible solutions. In particular, we could tell whether a point was outside the region, on the boundary, a vertex, or inside the region. Since we are primarily interested in the vertices of the region of feasible solutions, recall that in each example given so far, if an ordered $n$-tuple was a solution of the linear system obtained by introducing slack variables and it represented a vertex of the region of feasible solutions, then the number of 0's in the $n$-tuple was at least as large as the number of decision variables in the problem. This is not a coincidence. It derives from the following theorem, which we state without proof.

THEOREM 4.2: Suppose that in a linear programming problem we have $m$ decision variables $x_1, x_2, \ldots, x_m$, subject to $n$ constraints, and suppose that we introduce $n$ slack variables $s_1, s_2, \ldots, s_n$, thus obtaining a system of $n$ equations in $m + n$ variables. Suppose further that an optimum solution exists. Then, there exists a solution $(a_1, a_2, \ldots, a_m, b_1, b_2, \ldots, b_n)$ of the linear system where at least $m$ of the numbers $a_1, a_2, \ldots, a_m, b_1, b_2, \ldots, b_n$ are 0 with the objective function attaining its optimum value at $(a_1, a_2, \ldots, a_m)$.

You need not memorize the foregoing theorem, but you should remember the following.

In trying to locate where the objective function in a linear programming problem attains its maximum, look for solutions of the linear system of equations obtained by introducing slack variables, where the number of 0's is at least as large as the number of decision variables and all other values are nonnegative.

At this point, you may wish to review Section 3.3, where you learned to solve a system of linear equations by row reduction working on the augmented matrix of the system. The following example will illustrate the concepts we need to remember.

**EXAMPLE 1** You are given a system of three equations in the five variables $x_1, x_2, x_3, x_4$, and $x_5$. After performing a sequence of elementary row operations on the augmented matrix of the system, you obtain

$$\left[\begin{array}{ccccc|c} 2 & 0 & 1 & 3 & 0 & 4 \\ -7 & 1 & 0 & 8 & 0 & -3 \\ 4 & 0 & 0 & 9 & 1 & 9 \end{array}\right]$$

Find three solutions of the system.

**Solution** This augmented matrix represents

$$\begin{cases} 2x_1 & + x_3 + 3x_4 & = 4 \\ -7x_1 + x_2 & + 8x_4 & = -3 \\ 4x_1 & + 9x_4 + x_5 & = 9 \end{cases}$$

Since $x_3, x_2$, and $x_5$ appear only in the first, second, and third equations, respectively, we can solve for these variables in terms of $x_1$ and $x_4$ as follows:

$$\begin{cases} x_3 = 4 - 2x_1 - 3x_4 \\ x_2 = -3 + 7x_1 - 8x_4 \\ x_5 = 9 - 4x_1 - 9x_4 \end{cases}$$

We may now give any values to $x_1$ and $x_4$ and obtain corresponding values for $x_3, x_2$, and $x_5$.

We begin with the easiest and let $x_1 = x_4 = 0$. We get

$$x_3 = 4 - 2(0) - 3(0) = 4$$
$$x_2 = -3 + 7(0) - 8(0) = -3$$
$$x_5 = 9 - 4(0) - 9(0) = 9$$

The corresponding solution is $(0, -3, 4, 0, 9)$.

We let $x_1 = 0$ and $x_4 = 1$ and get

$$x_3 = 4 - 2(0) - 3(1) = 1$$
$$x_2 = -3 + 7(0) - 8(1) = -11$$
$$x_5 = 9 - 4(0) - 9(1) = 0$$

The corresponding solution is $(0, -11, 1, 1, 0)$.

Finally, we let $x_1 = 1$ and $x_4 = 0$ and obtain

$$x_3 = 4 - 2(1) - 3(0) = 2$$
$$x_2 = -3 + 7(1) - 8(0) = 4$$
$$x_5 = 9 - 4(1) - 9(0) = 5$$

A third solution is $(1, 4, 2, 0, 5)$.

**Do Exercise 3.**

**EXAMPLE 2**   You are given a system of three equations in the six variables $x_1$, $x_2$, $x_3$, $x_4$, $x_5$, and $x_6$. After performing a sequence of elementary row operations on the augmented matrix of the system, you obtain

$$\begin{bmatrix} 3 & 0 & 2 & 1 & 6 & 0 & | & 7 \\ 2 & 0 & 7 & 0 & 9 & 1 & | & 8 \\ -1 & 1 & 3 & 0 & 7 & 0 & | & -3 \end{bmatrix}$$

Find an obvious solution of the system.

**Solution**   Note that columns 4, 6, and 2, in that order, form the $3 \times 3$ identity matrix. Thus, the entries of the last column give the values of $x_4$, $x_6$, and $x_2$ provided that the other variables are given the value 0. An obvious solution is $(0, -3, 0, 7, 0, 8)$.

**Do Exercise 5.**

> **DEFINITION 4.3:**   In general, if we have a linear system of $m$ equations in $n$ variables where $n > m$, and if we have performed a sequence of elementary row operations on the augmented matrix of the system to obtain a matrix in which $m$ of the columns, written in some order, form the identity matrix $\mathbf{I}_m$, then the variables corresponding to those columns are called *basic variables*. The other $(n - m)$ variables are called the *nonbasic variables*.

If we give the value 0 to all nonbasic variables, the values of the basic variables may be read directly from the last column of the last augmented matrix. We now show

how a matrix can be used to identify the corners of a region of feasible solutions of a linear programming problem and how to move from any corner to an adjacent corner. For the time being, we consider only the constraints of the problem, rather than the objective function, focusing instead on the region of feasible solutions.

## Entering and Leaving Variables

**EXAMPLE 3**   Consider the system

$$x + 2y \le 10$$
$$x + y \le 6$$
$$2x + y \le 11$$
$$x \ge 0,\ y \ge 0$$

**a.** Sketch the graph of the system.
**b.** Introduce slack variables to convert the first three inequalities to equations.
**c.** Use the augmented matrix of the linear system obtained in part **b** to identify one corner of the graph of part **a** and use elementary row operations on that matrix to move from corner to corner.

**Solution**   **a.** The graph is shown in Figure 4.24.

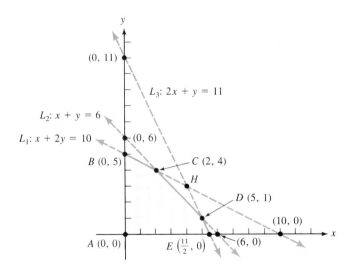

**FIGURE 4.24**

**b.** We introduce slack variables $r$, $s$, and $t$ by writing

$$\begin{cases} x + 2y + r \phantom{+ s + t} = 10 \\ x + y \phantom{+2} + s \phantom{+ t} = 6 \\ 2x + y \phantom{+2 + s} + t = 11 \end{cases}$$
$$x \ge 0,\ y \ge 0,\ r \ge 0,\ s \ge 0,\ t \ge 0$$

**c.** The augmented matrix of the system is

$$\begin{bmatrix} 1 & 2 & 1 & 0 & 0 & | & 10 \\ 1 & 1 & 0 & 1 & 0 & | & 6 \\ 2 & 1 & 0 & 0 & 1 & | & 11 \end{bmatrix}$$

Note that columns 3, 4, and 5 that give the $r$, $s$, and $t$ coefficients, respectively, form the identity matrix $I_3$. Thus, we begin by selecting $r$, $s$, and $t$ as basic variables and $x$, $y$ as nonbasic variables. We say that $\{r, s, t\}$ is the *basis*. Then we let $x = y = 0$, and from the last column of the matrix we read $r = 10$, $s = 6$, and $t = 11$. The first solution is $(0, 0, 10, 6, 11)$. This is a feasible solution since all entries are nonnegative. We call it a *basic feasible solution*. It corresponds to the corner of the graph of part **a** that has coordinates $(0, 0)$. This corner has been labeled $A$ in Figure 4.24.

We would like to move to corner $B$. We see that $A$ is the intersection of the $x$ and $y$ axes. Now we wish to move along the $y$-axis. Thus, $x$ must remain 0 and $y$ must become positive. That is, $x$ will still be a nonbasic variable, while $y$ will become a basic variable. We say that $y$ is the *entering variable* and we indicate this on the matrix by drawing an arrow below the column of $y$ coefficients as indicated in the next matrix. That column is called the *pivot column*.

|  | Basis | $x$ | $y$ | $r$ | $s$ | $t$ | Values of basic variables |
|---|---|---|---|---|---|---|---|
| ← | $r$ | 1 | ② | 1 | 0 | 0 | 10 |
|  | $s$ | 1 | 1 | 0 | 1 | 0 | 6 |
|  | $t$ | 2 | 1 | 0 | 0 | 1 | 11 |
|  |  |  | ↑ |  |  |  |  |

Since $y$ is entering the basis, one present element of the basis must leave. Should it be $r$, $s$, or $t$? We go to the point $B$ and we see that $B$ is not only on the $y$-axis, but also on the line $L_1$, the graph of $x + 2y = 10$. Thus, $r$ must be 0 since the first equation of our system was $x + 2y + r = 0$. That is, $r$ must become nonbasic. We say that $r$ is the *leaving variable* (or *departing variable*) and indicate this with an arrow pointing out to the left of the matrix by the row representing $r$. That row is called the *pivot row*. The number 2 which is in the pivot column and in the pivot row is called the *pivot element* and is circled. Now we perform elementary row operations on the augmented matrix of the system to convert the pivot element to the number 1 and all other elements in the pivot column to 0. We proceed as follows:

| Basis | $x$ | $y$ | $r$ | $s$ | $t$ | Values of basic variables |  |
|---|---|---|---|---|---|---|---|
| $r$ | 1 | 2 | 1 | 0 | 0 | 10 | $\left(\frac{1}{2}\right)R_1 \to R_1$ |
| $s$ | 1 | 1 | 0 | 1 | 0 | 6 |  |
| $t$ | 2 | 1 | 0 | 0 | 1 | 11 |  |

| Basis | $x$ | $y$ | $r$ | $s$ | $t$ | Values of basic variables | |
|---|---|---|---|---|---|---|---|
| $r$ | $\frac{1}{2}$ | 1 | $\frac{1}{2}$ | 0 | 0 | 5 | |
| $s$ | 1 | 1 | 0 | 1 | 0 | 6 | $-R_1 + R_2 \rightarrow R_2$ |
| $t$ | 2 | 1 | 0 | 0 | 1 | 11 | $-R_1 + R_3 \rightarrow R_3$ |

| Basis | $x$ | $y$ | $r$ | $s$ | $t$ | Values of basic variables |
|---|---|---|---|---|---|---|
| $y$ | $\frac{1}{2}$ | 1 | $\frac{1}{2}$ | 0 | 0 | 5 |
| $s$ | $\frac{1}{2}$ | 0 | $\frac{-1}{2}$ | 1 | 0 | 1 |
| $t$ | $\frac{3}{2}$ | 0 | $\frac{-1}{2}$ | 0 | 1 | 6 |

The basis is now $\{y, s, t\}$ and the values of the basic variables can be read from the last column: $y = 5$, $s = 1$, and $t = 6$. The nonbasic variables are $x$ and $r$ and are both 0. The solution of the linear system is $(0, 5, 0, 1, 6)$. The first two entries give us $(0, 5)$, which are the coordinates of the corner $B$. Now we wish to move along the line $L_1$ and go to adjacent corner $C$. Since we are staying on $L_1$, the value of $r$ must remain 0; that is, $r$ will remain nonbasic. Since we are leaving the $y$-axis, $x$ will be positive. That is, $x$ will enter the basis. We indicate this with an arrow below the matrix under the column of $x$ coefficients. Which of the variables $y$, $s$, or $t$ must leave the basis? Since the point $C$ is on the line $L_2$ which is the graph of the equation $x + y = 6$, we must have $s = 0$ because the second equation of our system was $x + y + s = 6$. Thus, $s$ must become nonbasic and will be the leaving variable. We indicate this with an arrow to the left of the augmented matrix by the $s$ row. (See the following matrix.)

| | Basis | $x$ | $y$ | $r$ | $s$ | $t$ | Values of basic variables |
|---|---|---|---|---|---|---|---|
| | $y$ | $\frac{1}{2}$ | 1 | $\frac{1}{2}$ | 0 | 0 | 5 |
| $\leftarrow$ | $s$ | $\left(\frac{1}{2}\right)$ | 0 | $\frac{-1}{2}$ | 1 | 0 | 1 |
| | $t$ | $\frac{3}{2}$ | 0 | $\frac{-1}{2}$ | 0 | 1 | 6 |
| | | $\uparrow$ | | | | | |

Thus, the pivot element is in column 1, row 2 and is circled. Now we must perform elementary row operations to convert the pivot element to 1 and all other entries in the pivot column to 0.

| Basis | $x$ | $y$ | $r$ | $s$ | $t$ | Values of basic variables | |
|---|---|---|---|---|---|---|---|
| $y$ | $\frac{1}{2}$ | 1 | $\frac{1}{2}$ | 0 | 0 | 5 | |
| $s$ | $\frac{1}{2}$ | 0 | $\frac{-1}{2}$ | 1 | 0 | 1 | $2R_2 \rightarrow R_2$ |
| $t$ | $\frac{3}{2}$ | 0 | $\frac{-1}{2}$ | 0 | 1 | 6 | |

| Basis | $x$ | $y$ | $r$ | $s$ | $t$ | Values of basic variables | |
|---|---|---|---|---|---|---|---|
| $y$ | $\frac{1}{2}$ | 1 | $\frac{1}{2}$ | 0 | 0 | 5 | $\left(\frac{-1}{2}\right)R_2 + R_1 \to R_1$ |
| $s$ | 1 | 0 | $-1$ | 2 | 0 | 2 | |
| $t$ | $\frac{3}{2}$ | 0 | $\frac{-1}{2}$ | 0 | 1 | 6 | $\left(\frac{-3}{2}\right)R_2 + R_3 \to R_3$ |

| Basis | $x$ | $y$ | $r$ | $s$ | $t$ | Values of basic variables |
|---|---|---|---|---|---|---|
| $y$ | 0 | 1 | 1 | $-1$ | 0 | 4 |
| $x$ | 1 | 0 | $-1$ | 2 | 0 | 2 |
| $t$ | 0 | 0 | 1 | $-3$ | 1 | 3 |

Note that $r$ and $s$ are the nonbasic variables and we set them equal to 0. From the last column, we read $y = 4$, $x = 2$, and $t = 3$. Thus, the solution of the linear system is $(2, 4, 0, 0, 3)$. The first two entries give us the coordinates $(2, 4)$ of $C$. Since the other entries are nonnegative, we have a feasible solution to the given system of inequalities. Since two of the entries in $(2, 4, 0, 0, 3)$ are 0, we have a basic feasible solution.

We leave it as an exercise for you to proceed from $C$ to $D$ and then from $D$ to $E$, and from $E$ back to $A$.

**Do Exercise 13.** ■

## Selecting the Leaving Variable

In the preceding example, we depended on the graph of the system of inequalities to select the entering and leaving variables each time we moved from one corner to an adjacent corner. However, we are striving for a way to do this without a diagram. The way the entering variable is selected will be discussed in the next section. We shall now show how the leaving variable should be selected once the entering variable has been chosen.

Let us go back for a moment to the previous example where we wanted to go from corner $B$ to corner $C$. Suppose that we decided that $t$ should be the leaving variable. Then $t$ would have become nonbasic and its value would have been 0. We would have obtained the point $H$ on $L_3$. However, the point $H$ is on the side of the line $L_2$ which does not contain the origin, so the value of $s$ would have been negative and the solution would not have been a feasible solution. The underlying principle is as follows: once an entering variable has been chosen, the leaving variable should be chosen so that after performing elementary row operations to convert the pivot element to a 1 and all other elements in the pivot column to 0, *all entries in the last column of the resulting matrix should be nonnegative*. We use this principle to arrive at a systematic way of choosing the leaving variable.

Suppose that we have an $m \times n$ augmented matrix, with $m < (n - 1)$ such that $m$ of the first $(n - 1)$ columns, arranged in the proper order, form the identity matrix $\mathbf{I}_m$ and suppose further that all entries in the last column are nonnegative. If we have selected the $j$th variable to be the entering variable, the $j$th column is the pivot column.

In the following matrix we write both the pivot column and the last column, which at this stage gives the values of the basic variables, and we omit all other columns.

$$\begin{bmatrix} & a_{1j} & & b_1 \\ & a_{2j} & & b_2 \\ \cdots & a_{3j} & \cdots & b_3 \\ & \vdots & & \vdots \\ & a_{mj} & & b_m \end{bmatrix}$$

We wonder whether $a_{kj}$ is a good candidate for the pivot element. If it were, we would perform elementary row operations on the matrix to convert $a_{kj}$ to a 1 and all other entries in the pivot column to 0. Our first step might be to multiply row $k$ by $1/a_{kj}$. By doing this, the corresponding entry in the last column becomes $b_k/a_{kj}$. This already puts some restrictions on $a_{kj}$. It cannot be 0. Furthermore, it should not be negative since the quotient $b_k/a_{kj}$ gives the value of the new basic variable and that value must be nonnegative (remember that $b_k$ is nonnegative). Thus, as a first step, we note that the pivot element should be positive. But what do we do if we have several positive entries in the pivot column? Suppose that both $a_{pj}$ and $a_{kj}$ are positive. We choose $a_{kj}$ over $a_{pj}$ for the pivot element provided that by converting $a_{kj}$ to 1 and $a_{pj}$ to 0 through elementary row operations, the corresponding entries in the last column will remain nonnegative. Clearly, $a_{pj}$ will be converted to a 0 by multiplying row $k$ by $-a_{pj}/a_{kj}$ and adding the result to row $p$ to obtain the new row $p$. The new entry in row $p$, last column will be $b_p - (a_{pj}/a_{kj})b_k$. This last element must be nonnegative. Thus, we have

$$b_p - (a_{pj}/a_{kj})b_k \geq 0$$

Multiplying both sides by $a_{kj}$ (which is legal since we know $a_{kj}$ is positive), we get

$$b_p a_{kj} - b_k a_{pj} \geq 0$$

Equivalently,

$$b_p a_{kj} \geq b_k a_{pj}$$

or

$$b_p/a_{pj} \geq b_k/a_{kj}$$

---

### Procedure for Finding the Pivot

Having chosen the entering variable, from which we get the pivot column, we determine the pivot as follows. We divide each positive entry of the pivot column into the corresponding entry in the last column. If $b_k/a_{kj}$ is the smallest quotient obtained, $a_{kj}$ is chosen as the pivot element. Note that $b_k/a_{kj}$ is necessarily nonnegative since all $b_i$'s are nonnegative. In case of a tie between $b_k/a_{kj}$ and $b_q/a_{qj}$, either $a_{kj}$ or $a_{qj}$ may be chosen.

EXAMPLE 4    Consider the system of inequalities of Example 4, Section 4.3.

$$y + z \leq 5$$
$$5x + 7y - 3z \leq 35$$
$$15x + y + 21z \leq 105$$
$$x \geq 0, y \geq 0, z \geq 0$$

Find three vertices of the graph of this system using the method just described.

**Solution**    We first introduce slack variables $r$, $s$, and $t$ and write

$$\begin{cases} y + z + r & = 5 \\ 5x + 7y - 3z + s & = 35 \\ 15x + y + 21z + t = 105 \end{cases}$$

The augmented matrix is

| Basis | $x$ | $y$ | $z$ | $r$ | $s$ | $t$ | Values of basic variables |
|-------|-----|-----|-----|-----|-----|-----|---------------------------|
| $r$ | 0 | 1 | 1 | 1 | 0 | 0 | 5 |
| $s$ | 5 | 7 | -3 | 0 | 1 | 0 | 35 |
| $t$ | 15 | 1 | 21 | 0 | 0 | 1 | 105 |

We choose $r$, $s$, and $t$ as our initial basic variables since the $r$, $s$, and $t$ columns form the identity matrix $\mathbf{I}_3$. Thus, $x$, $y$, and $z$ are nonbasic variables and we set them equal to 0. The first basic feasible solution is (0, 0, 0, 5, 35, 105). The first three entries of this sextuple give us the coordinates (0, 0, 0) of the origin, which is a vertex of the graph of the original system of inequalities. Suppose that we wish to have $y$ as the entering variable. All three entries in the $y$ column are positive. We divide each entry into the corresponding element in the last column. We get $\frac{5}{1} = 5$, $\frac{35}{7} = 5$, and $\frac{105}{1} = 105$. It is a tie between the first two quotients. We may choose either 1 or 7 to be the pivot element. We choose 1, since that choice will lead to simpler calculations. This means that $r$ will be the leaving variable. We perform elementary row operations as follows to convert the other entries in the $y$ column to 0.

| Basis | $x$ | $y$ | $z$ | $r$ | $s$ | $t$ | Values of basic variables | |
|-------|-----|-----|-----|-----|-----|-----|---------------------------|---|
| ← $r$ | ⓪ | 1 | 1 | 1 | 0 | 0 | 5 | $(-7)R_1 + R_2 \rightarrow R_2$ |
| $s$ | 5 | 7 | -3 | 0 | 1 | 0 | 35 | $(-1)R_1 + R_3 \rightarrow R_3$ |
| $t$ | 15 | 1 | 21 | 0 | 0 | 1 | 105 | |

| Basis | $x$ | $y$ | $z$ | $r$ | $s$ | $t$ | Values of basic variables |
|-------|-----|-----|-----|-----|-----|-----|---------------------------|
| $y$ | 0 | 1 | 1 | 1 | 0 | 0 | 5 |
| $s$ | 5 | 0 | -10 | -7 | 1 | 0 | 0 |
| $t$ | 15 | 0 | 20 | -1 | 0 | 1 | 100 |

We see from the last column that $y = 5$, $s = 0$, and $t = 100$. The other variables are nonbasic and we set them equal to 0. We get the basic feasible solution $(0, 5, 0, 0, 0, 100)$. The first three entries give us the coordinates $(0, 5, 0)$ of a second vertex of the graph of the original system.

Next we choose $z$ to be the entering variable. Then the pivot column is the $z$ column. Only two of its entries are positive. We divide them into the corresponding entries of the last column. We get $\frac{5}{1} = 5$ and $\frac{100}{20} = 5$. We have a tie again and choose the 1 to be the pivot, which means that $y$ will be the leaving variable, and we proceed.

|  | Basis | $x$ | $y$ | $z$ | $r$ | $s$ | $t$ | Values of basic variables |
|---|---|---|---|---|---|---|---|---|
| ← | $y$ | 0 | ① | 1 | 1 | 0 | 0 | 5 |
|  | $s$ | 5 | 0 | $-10$ | $-7$ | 1 | 0 | 0 |
|  | $t$ | 15 | 0 | 20 | $-1$ | 0 | 1 | 100 |

$$10R_1 + R_2 \to R_2$$
$$(-20)R_1 + R_3 \to R_3$$

| Basis | $x$ | $y$ | $z$ | $r$ | $s$ | $t$ | Values of basic variables |
|---|---|---|---|---|---|---|---|
| $z$ | 0 | 1 | 1 | 1 | 0 | 0 | 5 |
| $s$ | 5 | 10 | 0 | 3 | 1 | 0 | 50 |
| $t$ | 15 | $-20$ | 0 | $-21$ | 0 | 1 | 0 |

From this, we read the values of the basic variables $z = 5$, $s = 50$, and $t = 0$. The other variables are nonbasic and have the value 0. The solution is $(0, 0, 5, 0, 50, 0)$. The first three entries give the coordinates $(0, 0, 5)$ of another vertex of the graph of the original system.

**Do Exercise 17.** ■

## Unbounded Region of Feasible Solutions

In the preceding two examples, we saw that a pivot column was determined by the choice of an entering variable. We then considered only the positive entries of that column as candidates for the pivot element. The following example should help shed light on the significance of having a pivot column whose entries are all negative.

EXAMPLE 5 Consider the system

$$
\begin{aligned}
-2x + y &\le 1 \\
x - y &\le 10 \\
5x - 8y &\le 25 \\
x \ge 0, \, y &\ge 0
\end{aligned}
$$

**a.** Sketch the graph of the system.

**b.** Introduce slack variables to convert the first three inequalities to equations, then move from corner to corner of the graph using the method introduced in this section.

**Solution**   **a.** The graph is shown in Figure 4.25. The graphs of the equations $-2x + y = 1$, $x - y = 10$, and $5x - 8y = 25$ have been labeled $L_1$, $L_2$, and $L_3$, respectively. The four corners of the graph have been denoted $A$, $B$, $C$, and $D$.

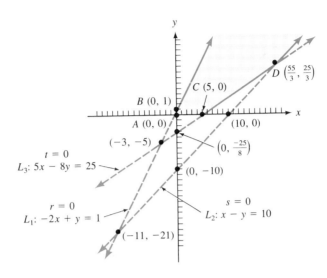

**FIGURE 4.25**

**b.** We introduce the slack variables $r$, $s$, and $t$ and write

$$\begin{cases} -2x + y + r \qquad\quad = 1 \\ \quad\; x - y \qquad + s \qquad = 10 \\ \quad\; 5x - 8y \qquad\quad + t = 25 \end{cases}$$

The augmented matrix is

$$\begin{bmatrix} -2 & 1 & 1 & 0 & 0 & | & 1 \\ 1 & -1 & 0 & 1 & 0 & | & 10 \\ 5 & -8 & 0 & 0 & 1 & | & 25 \end{bmatrix}$$

Since the third, fourth, and fifth columns form $\mathbf{I}_3$, we select $r$, $s$, and $t$ as our initial basis. The values of the basic variables are read from the last column. They are $r = 1$, $s = 10$, and $t = 25$. The nonbasic variables are $x$ and $y$ and are given the value 0. Thus, a basic feasible solution is $(0, 0, 1, 10, 25)$. The first two entries give the coordinates $(0, 0)$ of the origin, which is a corner of the region of feasible solutions. It has been labeled $A$ in Figure 4.25. Suppose that we select $y$ as the entering variable as indicated by the arrow below the $y$ column in the augmented matrix below.

|  | Basis | $x$ | $y$ | $r$ | $s$ | $t$ | Values of basic variables |
|---|---|---|---|---|---|---|---|
| ← | $r$ | $-2$ | ① | 1 | 0 | 0 | 1 |
|  | $s$ | 1 | $-1$ | 0 | 1 | 0 | 10 |
|  | $t$ | 5 | $-8$ | 0 | 0 | 1 | 25 |

The only positive entry in the pivot column is the 1 in the $r$ row. Hence, $r$ will be the leaving variable as indicated by the arrow on the left of the matrix. Therefore 1 is the pivot element and it is circled. We must perform elementary row operations on this augmented matrix so that the pivot element is converted to a 1 (in this case it already is 1), and then all other entries in the pivot column are converted to 0. From here on, we shall not document on the right side the type of elementary row operations we are performing as we have done earlier, since we shall always proceed in the same way. That is, once $a_{kj}$ has been selected as the pivot element, we shall multiply row $k$ by $1/a_{kj}$ to get a new row $k$. This will convert $a_{kj}$ to 1. Also for each $p \neq k$, we shall multiply row $k$ by $-a_{pj}/a_{kj}$ and add the result to row $p$ to get a new row $p$. This will convert $a_{pj}$ to 0 as required, since $a_{pj} + (-a_{pj}/a_{kj})a_{kj} = a_{pj} - a_{pj} = 0$.

|        | Basis | $x$ | $y$ | $r$ | $s$ | $t$ | Values of basic variables |
|--------|-------|-----|-----|-----|-----|-----|---------------------------|
| $\leftarrow$ | $r$ | $-2$ | ① | 1 | 0 | 0 | 1 |
|        | $s$ | 1 | $-1$ | 0 | 1 | 0 | 10 |
|        | $t$ | 5 | $-8$ | 0 | 0 | 1 | 25 |

|  | Basis | $x$ | $y$ | $r$ | $s$ | $t$ | Values of basic variables |
|--|-------|-----|-----|-----|-----|-----|---------------------------|
|  | $y$ | $-2$ | 1 | 1 | 0 | 0 | 1 |
|  | $s$ | $-1$ | 0 | 1 | 1 | 0 | 11 |
|  | $t$ | $-11$ | 0 | 8 | 0 | 1 | 33 |

The new basis is $\{y, s, t\}$. The values of the basic variables are $y = 1$, $s = 11$, and $t = 33$. The nonbasic variables are $x$ and $r$ and we give them the value 0. The new feasible solution is $(0, 1, 0, 11, 33)$. The first two entries give the coordinates $(0, 1)$ of corner $B$ of the graph of the given system. The nonbasic variables are $x$ and $r$. Therefore, one of these two must be the next entering variable. If we choose $r$, the basis will be $\{r, s, t\}$, which was the initial basis, and we will be back at the initial corner $A$. So, we choose $x$ instead. But that choice makes the first column the pivot column and all entries in that column are negative. Why is this so? Notice in Figure 4.25 that point $B$ is the only corner of the region of feasible solutions that is on line $L_1$ and the region of feasible solutions is unbounded. Thus, if we wish to keep $r = 0$, that is to keep $r$ as a nonbasic variable and enter $x$ as the new basic variable, we cannot choose an entering variable in a way which will keep us within the region of feasible solutions. The following will illustrate this fact.

Suppose we choose $s$ as the leaving variable, then $-1$ becomes the pivot element and we proceed as shown.

|  | Basis | $x$ | $y$ | $r$ | $s$ | $t$ | Values of basic variables |
|--|-------|-----|-----|-----|-----|-----|---------------------------|
|  | $y$ | $-2$ | 1 | 1 | 0 | 0 | 1 |
| $\leftarrow$ | $s$ | $-1$ | 0 | 1 | 1 | 0 | 11 |
|  | $t$ | $-11$ | 0 | 8 | 0 | 1 | 33 |

| Basis | $x$ | $y$ | $r$ | $s$ | $t$ | Values of basic variables |
|-------|-----|-----|-----|-----|-----|-----|
| $y$ | 0 | 1 | $-1$ | $-2$ | 0 | $-21$ |
| $x$ | 1 | 0 | $-1$ | $-1$ | 0 | $-11$ |
| $t$ | 0 | 0 | $-3$ | $-11$ | 1 | $-88$ |

The corresponding solution is $(-11, -21, 0, 0, -88)$. This is not a feasible solution since several of the values of the variables are negative. The first two entries give the coordinates $(-11, -21)$ of the point of intersection of the lines $L_1$ and $L_2$. You will find it instructive to rework part **b** of this example choosing, at the beginning, $x$ instead of $y$ as the entering variable. This should give the point $C$ with coordinates $(5, 0)$. The next time, choose $y$ as the entering variable, thus obtaining the point $D$ with coordinates $\left(\frac{55}{3}, \frac{25}{3}\right)$. Since point $D$ is the only corner of the region of feasible solutions which is on line $L_2$, we will have the same situation as we had when we reached point $B$.

**Do Exercise 25.**

## Procedure for Moving from Corner to Corner

Given a set of "≤" constraints for a linear programming problem, we identify the basic feasible solutions as follows:

**Step 1.** Convert each "≤" constraint to an equation by introducing a slack variable for each constraint.

**Step 2.** Write the augmented matrix of the system of equations obtained in Step 1. The columns representing the slack variables introduced in Step 1 form an identity matrix. Thus, we choose the slack variables as our initial basis and their values are read from the last column. All other variables are given the value 0. This gives us our initial basic feasible solution.

**Step 3.** Having chosen one of the nonbasic variables to become a basic variable (the entering variable), which at this point we do arbitrarily, the column representing that variable becomes the pivot column. Divide all positive entries of the pivot column into the corresponding entry of the last column. If $b_k/a_{kj}$ is the smallest quotient obtained, $a_{kj}$ is chosen as the pivot element. In case of a tie between $b_k/a_{kj}$ and $b_p/a_{pj}$, either $a_{kj}$ or $a_{pj}$ may be chosen.

**Step 4.** Perform elementary row operations on the augmented matrix to convert the pivot element to a 1 and all other entries in the pivot column to 0. From now on we shall refer to this process as *pivoting*. The new basis consists of the original basis with one of its members replaced by the entering variable. The values of the basic variables are read from the last column and the nonbasic variables are given the value 0 to get the next basic feasible solution.

**Step 5.** Choose another entering variable and return to Step 3.

## Exercise Set 4.4

*In Exercises 1–6, the given matrix has been obtained by performing elementary row operations on the augmented matrix of a linear system of equations. Give the solution that can be read directly from this matrix.*

1. $\begin{bmatrix} 2 & 0 & -3 & 1 & 7 & | & 5 \\ 3 & 1 & 9 & 0 & 6 & | & -3 \end{bmatrix}$

2. $\begin{bmatrix} 3 & 1 & 7 & 2 & 0 & | & 9 \\ 5 & 0 & 8 & 1 & 1 & | & 7 \end{bmatrix}$

3. $\begin{bmatrix} 0 & 3 & 1 & 5 & 0 & 8 & | & -2 \\ 1 & 7 & 0 & 9 & 0 & 2 & | & 13 \\ 0 & -3 & 0 & 4 & 1 & 9 & | & 17 \end{bmatrix}$

4. $\begin{bmatrix} 2 & 3 & 0 & 0 & 5 & 1 & | & 4 \\ 9 & -2 & 1 & 0 & 3 & 0 & | & -5 \\ -8 & 5 & 0 & 1 & 6 & 0 & | & 7 \end{bmatrix}$

5. $\begin{bmatrix} 0 & 0 & 3 & 1 & 7 & 0 & | & 5 \\ 1 & 0 & 8 & 0 & 6 & 0 & | & -2 \\ 0 & 1 & 9 & 0 & 8 & 0 & | & 3 \\ 0 & 0 & 1 & 0 & 6 & 1 & | & 13 \end{bmatrix}$

6. $\begin{bmatrix} 6 & 1 & 7 & 0 & 3 & 0 & 0 & | & 8 \\ 4 & 0 & -9 & 0 & 6 & 1 & 0 & | & 7 \\ 5 & 0 & 3 & 1 & 9 & 0 & 0 & | & 6 \\ -4 & 0 & 6 & 0 & -5 & 0 & 1 & | & 8 \end{bmatrix}$

*In Exercises 7–24, a system of linear inequalities is given. In each case, find the coordinates of the indicated number of corners of the graph of the system using the method described in this section. Do not sketch the graphs. For Exercise n, where $7 \le n \le 22$, use the system of linear inequalities of Exercise $n + 6$, Exercise Set 4.3.*

7. Three corners.

8. Three corners.

9. Three corners.

10. Three corners.

11. Three corners.

12. Three corners.

13. Four corners.

14. Four corners.

15. Four corners.

16. Four corners.

17. Four corners.

18. Four corners.

19. Four corners.

20. Four corners.

21. Four corners.

22. Four corners.

23.
$$x + y + z \le 15$$
$$x + y \le 10$$
$$4x - 3y \le 12$$
$$-3x + 4y \le 12$$
$$x \ge 0, y \ge 0, z \ge 0$$
Five corners.

24.
$$2x + 3y + 5z \le 60$$
$$-y + z \le 4$$
$$3y - z \le 24$$
$$2y + 3z \le 27$$
$$x \ge 0, y \ge 0, z \ge 0$$
Five corners.

*In Exercises 25–28, show, without sketching the graphs, that the region of feasible solutions of the given set of constraints is unbounded.*

25.
$$-3x + y \le 3$$
$$x - y \le 5$$
$$2x - y \le 15$$
$$x \ge 0, y \ge 0$$

26.
$$-x + y \le 10$$
$$x - 3y \le 9$$
$$x - 2y \le 10$$
$$x \ge 0, y \ge 0$$

27.
$$-4x + y \le 6$$
$$-3x + y \le 7$$
$$x - 4y \le 5$$
$$x - 3y \le 6$$
$$x \ge 0, y \ge 0$$

28.
$$-10x + 3y \le 30$$
$$-7x + 3y \le 39$$
$$3x - 5y \le 21$$
$$7x - 4y \le 72$$
$$x \ge 0, y \ge 0$$

# 4.5 The Simplex Method

We are ready for the simplex method. In Section 4.3, we learned how to convert the set of "≤" constraints to a linear system of equations by using slack variables. In Section 4.4, we used the augmented matrix of the system to identify corners of the region of feasible solutions and to move from any corner to an adjacent corner. It is a simple matter now to incorporate the objective function into the system.

## Description of the Simplex Method

Since the objective function is defined by

$$z = c_1x_1 + c_2x_2 + c_3x_3 + \cdots + c_nx_n$$

and since the linear equations obtained from the constraints have all their variables on the left side, we first write

$$z - c_1x_1 - c_2x_2 - c_3x_3 - \cdots - c_nx_n = 0$$

We then add this equation to the linear system obtained from the "≤" constraints by introducing slack variables. We then proceed exactly as in the preceding section treating $z$ as a basic variable throughout. In this manner, we can read the value of the objective function in the right column as we move from corner to corner. It is useful to have the following definition.

> **DEFINITION 4.4:** Performing the necessary row operations to convert the pivot element to 1 and the other entries in the pivot column to 0 is called *pivoting*.

**EXAMPLE 1** Find the maximum value of the objective function $z = 5x_1 + 4x_2$ subject to the constraints

$$x_1 + x_2 \leq 5$$
$$3x_1 + 2x_2 \leq 12$$
$$x_1 \geq 0, x_2 \geq 0$$

**Solution** We introduce slack variables $s_1$ and $s_2$ and write the constraints as

$$x_1 + x_2 + s_1 = 5$$
$$3x_1 + 2x_2 + s_2 = 12$$
$$x_1 \geq 0, x_2 \geq 0, s_1 \geq 0, s_2 \geq 0$$

We now write the objective function as

$$z - 5x_1 - 4x_2 = 0$$

Adding this to these two equations, we get

$$\begin{cases} x_1 + x_2 + s_1 = 5 \\ 3x_1 + 2x_2 + s_2 = 12 \\ z - 5x_1 - 4x_2 = 0 \end{cases}$$

The augmented matrix is

$$\begin{bmatrix} 0 & 1 & 1 & 1 & 0 & 5 \\ 0 & 3 & 2 & 0 & 1 & 12 \\ 1 & -5 & -4 & 0 & 0 & 0 \end{bmatrix}$$

It is clear that columns 4, 5, and 1, in that order, form the identity $I_3$. Thus, we let $\{s_1, s_2, z\}$ be our initial basis. From this point on, we will display the augmented matrix and variables in a slightly different format. This is often called a *tableau*. The initial tableau of the problem is

| Basis | $z$ | $x_1$ | $x_2$ | $s_1$ | $s_2$ | Values of basic variables |
|-------|-----|-------|-------|-------|-------|---------------------------|
| $s_1$ | 0 | 1 | 1 | 1 | 0 | 5 |
| $s_2$ | 0 | 3 | 2 | 0 | 1 | 12 |
| $z$ | 1 | −5 | −4 | 0 | 0 | 0 |

We use a horizontal line to separate the coefficients of the objective function from those of the constraints. We give $x_1$ and $x_2$ the value 0 since they are nonbasic variables. Consequently, the initial value of the objective function is 0, which can be read in the lower right corner of the tableau. From the tableau, we see that the initial basic feasible solution is (0, 0, 5, 12). The first two entries give (0, 0), the origin, as a corner of the region of feasible solutions. Since the objective function is $z = 5x_1 + 4x_2$, its value will increase if either $x_1$ or $x_2$ becomes positive. Since the coefficient of $x_1$ is larger than that of $x_2$, the value of $z$ will increase more per unit increase of $x_1$; thus, it seems reasonable to use $x_1$ as the entering variable. (See Example 2.) Because the largest coefficient in the system is 5, that of $x_1$, the bottom entry in the $x_1$ column is −5, which is the *most negative* entry in the last row of the tableau. Since $\frac{12}{3}$ is less than $\frac{5}{1}$, we select $s_2$ as the leaving variable. Thus, 3 is the pivot element and is circled in the initial tableau.

**Tableau I**

| Basis | $z$ | $x_1$ | $x_2$ | $s_1$ | $s_2$ | Values of basic variables |
|-------|-----|-------|-------|-------|-------|---------------------------|
| $s_1$ | 0 | 1 | 1 | 1 | 0 | 5 |
| ← $s_2$ | 0 | ③ | 2 | 0 | 1 | 12 |
| $z$ | 1 | −5 | −4 | 0 | 0 | 0 |

↑

The numbers in the bottom row, except for the last one, are called *indicators* because they are used to determine the entering variable. Pivoting, we get the next tableau.

**Tableau II**

| Basis | $z$ | $x_1$ | $x_2$ | $s_1$ | $s_2$ | Values of basic variables |
|-------|-----|-------|-------|-------|-------|---------------------------|
| $s_1$ | 0 | 0 | $\frac{1}{3}$ | 1 | $\frac{-1}{3}$ | 1 |
| $x_1$ | 0 | 1 | $\frac{2}{3}$ | 0 | $\frac{1}{3}$ | 4 |
| $z$ | 1 | 0 | $\frac{-2}{3}$ | 0 | $\frac{5}{3}$ | 20 |

We see from the tableau that $x_1 = 4$, $x_2 = 0$ (since it is a nonbasic variable), $s_1 = 1$, $s_2 = 0$ (since it is the other nonbasic variable), and $z = 20$. This means that our second basic feasible solution is $(4, 0, 1, 0)$. The first two entries give the coordinates $(4, 0)$ of another corner of the region of feasible solutions, and the value of the objective function at that corner is 20. The last row of the tableau indicates that

$$z - \left(\tfrac{2}{3}\right)x_2 + \left(\tfrac{5}{3}\right)s_2 = 20$$

which we can write

$$z = \left(\tfrac{2}{3}\right)x_2 - \left(\tfrac{5}{3}\right)s_2 + 20$$

The values of $x_2$ and $s_2$ are 0, so $z = 20$. Clearly the value of $z$ will increase if $s_2$ remains 0 and $x_2$ becomes positive. Thus, we choose $x_2$ as our entering variable. The bottom entry in the $x_2$ column is $\tfrac{-2}{3}$ which is the most negative indicator. Since $1/(1/3)$ is less than $4/(2/3)$, we choose $s_1$ to be the leaving variable and $\tfrac{1}{3}$ becomes the pivot element as indicated in the next tableau.

**Tableau III**

| | Basis | $z$ | $x_1$ | $x_2$ | $s_1$ | $s_2$ | Values of basic variables |
|---|---|---|---|---|---|---|---|
| $\leftarrow$ | $s_1$ | 0 | 0 | $\boxed{\tfrac{1}{3}}$ | 1 | $\tfrac{-1}{3}$ | 1 |
| | $x_1$ | 0 | 1 | $\tfrac{2}{3}$ | 0 | $\tfrac{1}{3}$ | 4 |
| | $z$ | 1 | 0 | $\tfrac{-2}{3}$ | 0 | $\tfrac{5}{3}$ | 20 |

Pivoting, we obtain the next tableau.

**Tableau IV**

| Basis | $z$ | $x_1$ | $x_2$ | $s_1$ | $s_2$ | Values of basic variables |
|---|---|---|---|---|---|---|
| $x_2$ | 0 | 0 | 1 | 3 | $-1$ | 3 |
| $x_1$ | 0 | 1 | 0 | $-2$ | 1 | 2 |
| $z$ | 1 | 0 | 0 | 2 | 1 | 22 |

All indicators in the bottom row are nonnegative. Thus, the value of $z$ cannot be increased. To see this, note that the last row of the tableau indicates that

$$z + 2s_1 + s_2 = 22$$

Since $s_1$ and $s_2$ are 0 (they are nonbasic variables), the value of $z$ is 22, which could have been read from the tableau. However, if we write

$$z = -2s_1 - s_2 + 22$$

it is clear that the value of $z$ will decrease if either $s_1$ or $s_2$ is made positive while the other remains 0. Thus, neither $s_1$ nor $s_2$ can be the entering variable. But these were the only two choices. Thus, we are through, and 22 is the maximum value of the objective function. It occurs at $(2, 3)$.

**Do Exercise 3.**

The $z$ column was not changed throughout the problem. It never does since we wish $z$ to remain a basic variable so that its value may be read from the tableau at each step. Consequently, that column will be omitted in future tableaus.

### Procedure of the Simplex Method

**Step 1.**   Convert the "$\leq$" constraints to equations by introducing slack variables. Write the objective function in the form

$$z - c_1x_1 - c_2x_2 - c_3x_3 - \cdots - c_nx_n = 0$$

**Step 2.**   For the system obtained in Step 1 (with the objective equation written last), write the associated tableau, leaving out the $z$-column. This is the initial tableau. By choosing the slack variables to be the basic variables, we obtain the initial basic feasible solution at the origin. The value of the objective function appears in the lower right corner of the tableau.

**Step 3.**   Determine whether the maximum value of the objective function has been attained by examining the bottom row of the tableau for negative indicators. If there are no negative indicators, the maximum has been attained. Its value is the entry in the lower right corner of the tableau. Otherwise, choose as the entering variable the one corresponding to the most negative indicator in the bottom row.

**Step 4.**   The entering variable determines the pivot column. Divide each positive entry in the pivot column into its corresponding element in the last column (the current values of the basic variables). Choose as the leaving variable the one corresponding to the smallest nonnegative quotient. In case of a tie, choose either.

**Step 5.**   The leaving (departing) variable determines the pivot row. The element in the pivot row and the pivot column is the pivot element. If the pivot element is $a_{kj}$, multiply row $k$ by $1/a_{kj}$ to obtain the new row $k$, and for each $p \neq k$, add to row $p$ row $k$ multiplied by $(-a_{pj}/a_{kj})$ to obtain the new row $p$. In this manner, the pivot element has been converted to 1 and all other entries of the pivot column have been converted to 0. Performing Step 5 is referred to as pivoting. Return to Step 3.

## A Remark on the Choice of the Entering Variable

The manner in which we choose the entering variable does not necessarily give the best choice. It gives the best choice for each increase of *one unit* of the entering variable. Recall the example in Section 4.3 where we asked you to visualize a house with a slanted flat roof and a squirrel trying to get to the highest point by running along the edges of the roof from corner to corner. His strategy was to go along the steeper edge when he was at a corner from which both edges were going uphill. This does not guarantee that the next corner he will reach will be higher than the one he would have reached if he had made the other choice because the steeper edge could be much shorter than the other edges. However, since our squirrel is nearsighted, he cannot see how long the edges are and he must make a choice. Hence, it is reasonable for him to choose the steeper edge although it may not be the best choice. Our situation is similar. In Step 3, we make the most reasonable choice of the entering variable because this is the best we can do. The next example should illustrate this point.

**EXAMPLE 2**  Consider the objective function $z = 5x + 10y$ subject to the constraints

$$
\begin{aligned}
-x + y &\le 5 \\
-x + 2y &\le 15 \\
x + 7y &\le 120 \\
x \ge 0,\ y &\ge 0
\end{aligned}
$$

  **a.** Find the maximum value of $z$ using the simplex method.
  **b.** Repeat part **a** except that on the first choice of the entering variable make the "wrong" choice and select $x$ as the entering variable instead of $y$.
  **c.** Explain geometrically why the solution in **b** is simpler than the solution in **a**.

**Solution**  **a.** We first introduce slack variables $r$, $s$, and $t$ and we write the objective function with all its variables on the left.

$$
\left\{
\begin{aligned}
-x + y + r &\qquad\qquad = 5 \\
-x + 2y &\quad + s \qquad = 15 \\
x + 7y &\qquad\quad + t = 120 \\
z - 5x - 10y &\qquad\qquad = 0
\end{aligned}
\right.
$$

We then set up the associated tableau which represents the initial solution of the problem, leaving out the $z$-column.

**Tableau I**

| Basis | $x$ | $y$ | $r$ | $s$ | $t$ | Values of basic variables |
|-------|-----|-----|-----|-----|-----|---------------------------|
| $r$ ← | $-1$ | ① | 1 | 0 | 0 | 5 |
| $s$ | $-1$ | 2 | 0 | 1 | 0 | 15 |
| $t$ | 1 | 7 | 0 | 0 | 1 | 120 |
| $z$ | $-5$ | $-10$ | 0 | 0 | 0 | 0 |

The most negative indicator in the bottom row is $-10$. Thus, we select $y$ as the entering variable, as indicated by the arrow at the bottom of the $y$ column. Since $\frac{5}{1}$ is less than both $\frac{15}{2}$ and $\frac{120}{7}$, we select $r$ as the leaving variable. The pivot element is circled. It is already 1. Pivoting, we obtain the next tableau.

**Tableau II**

| Basis | $x$ | $y$ | $r$ | $s$ | $t$ | Values of basic variables |
|-------|-----|-----|-----|-----|-----|---------------------------|
| $y$ | $-1$ | 1 | 1 | 0 | 0 | 5 |
| $s$ ← | ① | 0 | $-2$ | 1 | 0 | 5 |
| $t$ | 8 | 0 | $-7$ | 0 | 1 | 85 |
| $z$ | $-15$ | 0 | 10 | 0 | 0 | 50 |

Thus, $y = 5$, $s = 5$, $t = 85$, $z = 50$, and $x = r = 0$. The second basic feasible solution is $(0, 5, 0, 5, 85)$ and the value of $z$ is 50 at that point. Since there is a negative indicator in the bottom row, the value of $z$ can be improved. We choose $x$ as the entering variable. The $x$ column becomes the pivot column. Since $\frac{5}{1}$ is less than $\frac{85}{8}$, we choose $s$ as the leaving variable. The pivot element is circled. It is already 1. We need only convert the other entries in the pivot column to 0.

**Tableau III**

| Basis | $x$ | $y$ | $r$ | $s$ | $t$ | Values of basic variables |
|---|---|---|---|---|---|---|
| $y$ | 0 | 1 | $-1$ | 1 | 0 | 10 |
| $x$ | 1 | 0 | $-2$ | 1 | 0 | 5 |
| $t$ | 0 | 0 | ⑨ | $-8$ | 1 | 45 |
| $z$ | 0 | 0 | $-20$ | 15 | 0 | 125 |

The next basic feasible solution is $(5, 10, 0, 0, 45)$. The value of the objective function is 125. Since there is a negative indicator in the last row, the value of $z$ can be improved. The most negative indicator is $-20$; hence, we select $r$ as our next entering variable. Note that 9 is the only positive entry in the pivot column. Thus, we choose $t$ as the leaving variable. The pivot element is circled. Pivoting again, we get the next tableau.

**Tableau IV**

| Basis | $x$ | $y$ | $r$ | $s$ | $t$ | Values of basic variables |
|---|---|---|---|---|---|---|
| $y$ | 0 | 1 | 0 | $\frac{1}{9}$ | $\frac{1}{9}$ | 15 |
| $x$ | 1 | 0 | 0 | $\frac{-7}{9}$ | $\frac{2}{9}$ | 15 |
| $r$ | 0 | 0 | 1 | $\frac{-8}{9}$ | $\frac{1}{9}$ | 5 |
| $z$ | 0 | 0 | 0 | $\frac{-25}{9}$ | $\frac{20}{9}$ | 225 |

The corresponding basic feasible solution is $(15, 15, 5, 0, 0)$, and the value of $z$ is 225. There is still a negative indicator in the bottom row but it is the only one. Consequently, it is the most negative and we choose $s$ as the entering variable. Since $\frac{1}{9}$ is the only positive entry in the pivot column, it must be the pivot. Pivoting, we get

**Tableau V**

| Basis | $x$ | $y$ | $r$ | $s$ | $t$ | Values of basic variables |
|---|---|---|---|---|---|---|
| $s$ | 0 | 9 | 0 | 1 | 1 | 135 |
| $x$ | 1 | 7 | 0 | 0 | 1 | 120 |
| $r$ | 0 | 8 | 1 | 0 | 1 | 125 |
| $z$ | 0 | 25 | 0 | 0 | 5 | 600 |

There are no negative indicators in the bottom row. Thus, the value of the objective function cannot be improved. The maximum value is 600. It occurs at (120, 0).

**b.** Let us consider the initial tableau again.

**Tableau I**

| Basis | $x$ | $y$ | $r$ | $s$ | $t$ | Values of basic variables |
|---|---|---|---|---|---|---|
| $r$ | $-1$ | 1 | 1 | 0 | 0 | 5 |
| $s$ | $-1$ | 2 | 0 | 1 | 0 | 15 |
| $t$ | ①  | 7 | 0 | 0 | 1 | 120 |
| $z$ | $-5$ | $-10$ | 0 | 0 | 0 | 0 |

The most negative indicator in the bottom row is $-10$. For this reason, we chose $y$ as our original entering variable. Suppose we had decided not to follow the usual procedure and had chosen $x$ as our first entering variable. Since 1 is the only positive entry in the $x$ column, it would have to be the pivot element. Therefore, $t$ would have to be the leaving variable. Converting the other entries in the pivot column to 0, we would have obtained

**Tableau II**

| Basis | $x$ | $y$ | $r$ | $s$ | $t$ | Values of basic variables |
|---|---|---|---|---|---|---|
| $r$ | 0 | 8 | 1 | 0 | 1 | 125 |
| $s$ | 0 | 9 | 0 | 1 | 1 | 135 |
| $x$ | 1 | 7 | 0 | 0 | 1 | 120 |
| $z$ | 0 | 25 | 0 | 0 | 5 | 600 |

Since there are no negative indicators in the bottom row, we have reached the maximum value of the objective function. This value is 600 as obtained in part **a**.

**c.** Note that we obtained the maximum value in part **b** much faster than in part **a**. Why did this happen? It will be easier to understand if we solve the same problem using the geometric approach, as we did in Section 4.1. The graph of the region of feasible solutions is shown in Figure 4.26.

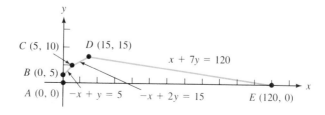

**FIGURE 4.26**

The vertices of the region have been found to be the points $A$, $B$, $C$, $D$, and $E$ with coordinates $(0, 0)$, $(0, 5)$, $(5, 10)$, $(15, 15)$, and $(120, 0)$, respectively. The value of the objective function at each of these points is found as follows:

$$\text{at } (0, 0), z = 0; \text{ at } (0, 5), z = 50; \text{ at } (5, 10), z = 125;$$
$$\text{at } (15, 15), z = 225; \text{ and at } (120, 0), z = 600.$$

We again see that the maximum value is 600 and is attained at $E$ with coordinates $(120, 0)$. Using the initial tableau, the initial basic solution was the origin. The maximum value of the objective function occurs at the adjacent corner $E$. By choosing $x$ as the entering variable, we get to the corner $E$ in one step. However, by choosing $y$ as the first entering variable, we go through the corners $B$, $C$, and $D$ before getting to $E$. Note that the values of $z$ which we calculated at these corners are 50, 125, 225, and 600, respectively, and these are the values which appeared in the lower right corners of the second, third, fourth, and fifth tableaus. By going from $A$ to $B$, we increased $y$ by only 5 units and the objective function by 50 even though we increased the objective function by 10 for each unit increase in $y$. However, by going from $A$ to $E$, although we increased the objective function by only 5 for each unit increase in $x$, we increased $x$ by 120 units and the objective function by 600. ∎

In spite of what happened in the preceding example, in problems with many decision variables and many constraints, the choice of the entering variable will be based on the most negative indicator in the bottom row of the tableau. We must do this because in such cases we do not have the benefit of a graph or the simplicity of the situation in the foregoing example to decide otherwise. Clearly, the number of steps necessary to solve a linear programming problem using the simplex method is much less than the number of steps necessary when we use the method of Section 4.3. Using that method, we had to solve many systems of linear equations and many of the solutions obtained were not feasible. Therefore, the simplex method is much more efficient.

**EXAMPLE 3**    Maximize $z = 2x_1 + 3x_2 + 5x_3 + 2x_4$ subject to

$$
\begin{aligned}
x_1 - x_2 + x_3 + 2x_4 &\leq 3 \\
x_1 + 2x_2 + x_3 + x_4 &\leq 11 \\
2x_1 + x_2 + 4x_4 &\leq 10 \\
x_1 \geq 0, x_2 \geq 0, x_3 \geq 0, x_4 &\geq 0
\end{aligned}
$$

**Solution**    We introduce slack variables $s_1$, $s_2$, and $s_3$ and write

$$
\begin{aligned}
x_1 - x_2 + x_3 + 2x_4 + s_1 &= 3 \\
x_1 + 2x_2 + x_3 + x_4 + s_2 &= 11 \\
2x_1 + x_2 + 4x_4 + s_3 &= 10 \\
x_1 \geq 0, x_2 \geq 0, x_3 \geq 0, x_4 \geq 0, s_1 \geq 0, s_2 \geq 0, s_3 &\geq 0
\end{aligned}
$$

We now write the objective function as

$$z - 2x_1 - 3x_2 - 5x_3 - 2x_4 = 0$$

The initial tableau is

**Tableau I**

| Basis | $x_1$ | $x_2$ | $x_3$ | $x_4$ | $s_1$ | $s_2$ | $s_3$ | Values of basic variables |
|-------|-------|-------|-------|-------|-------|-------|-------|---------------------------|
| $s_1$ ← | 1 | −1 | ① | 2 | 1 | 0 | 0 | 3 |
| $s_2$ | 1 | 2 | 1 | 1 | 0 | 1 | 0 | 11 |
| $s_3$ | 2 | 1 | 0 | 4 | 0 | 0 | 1 | 10 |
| $z$ | −2 | −3 | −5 | −2 | 0 | 0 | 0 | 0 |

The most negative indicator in the bottom row is $-5$. Thus, we select $x_3$ as the entering variable. Since $\frac{3}{1}$ is less than $\frac{11}{1}$, $s_1$ is the leaving variable and the circled 1 is the pivot element. Pivoting, we get

**Tableau II**

| Basis | $x_1$ | $x_2$ | $x_3$ | $x_4$ | $s_1$ | $s_2$ | $s_3$ | Values of basic variables |
|-------|-------|-------|-------|-------|-------|-------|-------|---------------------------|
| $x_3$ | 1 | −1 | 1 | 2 | 1 | 0 | 0 | 3 |
| $s_2$ ← | 0 | ③ | 0 | −1 | −1 | 1 | 0 | 8 |
| $s_3$ | 2 | 1 | 0 | 4 | 0 | 0 | 1 | 10 |
| $z$ | 3 | −8 | 0 | 8 | 5 | 0 | 0 | 15 |

Since there is a negative indicator in the bottom row, we choose $x_2$ as our next entering variable. Since $\frac{8}{3}$ is less than $\frac{10}{1}$, $s_2$ is selected as the leaving variable. The circled 3 is the pivot element. Pivoting, we get

**Tableau III**

| Basis | $x_1$ | $x_2$ | $x_3$ | $x_4$ | $s_1$ | $s_2$ | $s_3$ | Values of basic variables |
|-------|-------|-------|-------|-------|-------|-------|-------|---------------------------|
| $x_3$ | 1 | 0 | 1 | $\frac{5}{3}$ | $\frac{2}{3}$ | $\frac{1}{3}$ | 0 | $\frac{17}{3}$ |
| $x_2$ | 0 | 1 | 0 | $\frac{-1}{3}$ | $\frac{-1}{3}$ | $\frac{1}{3}$ | 0 | $\frac{8}{3}$ |
| $s_3$ | 2 | 0 | 0 | $\frac{13}{3}$ | $\frac{1}{3}$ | $\frac{-1}{3}$ | 1 | $\frac{22}{3}$ |
| $z$ | 3 | 0 | 0 | $\frac{16}{3}$ | $\frac{7}{3}$ | $\frac{8}{3}$ | 0 | $\frac{109}{3}$ |

There are no negative indicators in the bottom row. Thus, the maximum value of the objective function has been attained. It is $\frac{109}{3}$ and occurs at $\left(0, \frac{8}{3}, \frac{17}{3}, 0\right)$ since $x_2 = \frac{8}{3}$, $x_3 = \frac{17}{3}$, and $x_1 = x_4 = 0$.
**Do Exercise 13.**

To appreciate the effectiveness of the simplex method, you are urged to compare this example to Example 4, Section 4.3. If we had to find all vertices of the region in that example and evaluate the objective function at each vertex, the work would have been overwhelming.

## Checking Your Calculations

**A CLOSER LOOK** Suppose that in a maximization problem in standard form, there are $m$ "$\leq$" constraints and therefore $m$ slack variables are introduced. In any tableau used in the problem, the $m$ entries in the bottom row and in the columns corresponding to the slack variables form an $m$-dimensional row vector. If we calculate the dot product of that vector in the $kth$ tableau with the $m$-dimensional column vector formed by the values of these $m$ variables read in the last column of the *initial tableau*, the result should always give the value of the objective function in the $kth$ tableau. For instance, in the first tableau of the previous example, the three-dimensional column vector formed by the initial values of the slack variables (read in the last column)

was $\begin{bmatrix} 3 \\ 11 \\ 10 \end{bmatrix}$. The three-dimensional row vectors formed by the numbers read in the

bottom row and in the columns corresponding to the slack variables $s_1$, $s_2$, and $s_3$ in the second and third tableaus are $[\ 5 \quad 0 \quad 0\ ]$ and $\left[\ \frac{7}{3} \quad \frac{8}{3} \quad 0\ \right]$, respectively. Note that

$$[\ 5 \quad 0 \quad 0\ ] \cdot \begin{bmatrix} 3 \\ 11 \\ 10 \end{bmatrix} = 3(5) + 11(0) + 10(0) = 15$$

and 15 is in the lower right corner of the second tableau. Similarly,

$$\left[\ \frac{7}{3} \quad \frac{8}{3} \quad 0\ \right] \cdot \begin{bmatrix} 3 \\ 11 \\ 10 \end{bmatrix} = 3\left(\frac{7}{3}\right) + 11\left(\frac{8}{3}\right) + 10(0) = \frac{109}{3}$$

and $\frac{109}{3}$ is in the lower right corner of the third tableau. (You may find this useful in checking your steps when solving a maximization problem in standard form by the simplex method.) If the solution requires several tableaus and the arithmetic is troublesome, then for each tableau, calculate the dot product of the row vector (which is formed by the entries in the bottom row and the columns corresponding to the slack variables) and the column vector (which is formed using the initial values of the slack variables in the initial tableau). The result should be equal to the value of the objective function that appears in the lower right corner. If it is equal, there is no guarantee that the work is correct, but if it is not, then you know there is an error in the solution.

## An Application

**EXAMPLE 4** Jack bought 7 acres of land on which he wishes to grow wheat, barley, and corn. The cost of planting one acre is $120 for wheat, $80 for barley, and $90 for corn. Jack has only $720 available for planting. The anticipated profit per acre is $1000 for wheat,

$800 for barley, and $400 for corn. How many acres of each should he plant to maximize his profit?

**Solution**  Let $x$, $y$, and $z$ denote the numbers of acres of wheat, barley, and corn, respectively, that Jack will plant. Then the total number of acres planted will be $x + y + z$. We have

$$x + y + z \leq 7$$

The planting costs (measured in dollars) will be $120x$ for wheat, $80y$ for barley, and $90z$ for corn. Thus, the total cost of planting is $(120x + 80y + 90z)$. Since Jack has only $720 available, we have

$$120x + 80y + 90z \leq 720$$

or

$$12x + 8y + 9z \leq 72$$

Jack will make $1000x$ profit on the wheat, $800y$ profit on the barley, and $400z$ profit on the corn. Thus, his total profit will be $\$(1000x + 800y + 400z)$. The problem is how to maximize $w$ where

$$w = 1000x + 800y + 400z$$

subject to

$$x + \ y + \ z \leq \ 7$$
$$12x + 8y + 9z \leq 72$$
$$x \geq 0, y \geq 0, z \geq 0$$

We begin as usual and write

$$x + \ y + \ z + r \ \ \ \ = \ 7$$
$$12x + 8y + 9z \ \ \ \ + s = 72$$

We then write the objective function with all variables on the left.

$$w - 1000x - 800y - 400z = 0$$

The initial tableau is

**Tableau I**

| Basis | $x$ | $y$ | $z$ | $r$ | $s$ | Values of basic variables |
|---|---|---|---|---|---|---|
| $r$ | 1 | 1 | 1 | 1 | 0 | 7 |
| $s$ | (12) | 8 | 9 | 0 | 1 | 72 |
| $w$ | $-1000$ | $-800$ | $-400$ | 0 | 0 | 0 |

Since $-1000$ is the most negative indicator in the bottom row, we choose $x$ as the entering variable. We then choose $s$ as the leaving variable, since $\frac{72}{12}$ is less than $\frac{7}{1}$. Thus, 12 is the circled pivot element. We pivot to get the second tableau.

**Tableau II**

| Basis | $x$ | $y$ | $z$ | $r$ | $s$ | Values of basic variables |
|---|---|---|---|---|---|---|
| $r$ | 0 | $\left(\frac{1}{3}\right)$ | $\frac{1}{4}$ | 1 | $\frac{-1}{12}$ | 1 |
| $x$ | 1 | $\frac{2}{3}$ | $\frac{3}{4}$ | 0 | $\frac{1}{12}$ | 6 |
| $w$ | 0 | $\frac{-400}{3}$ | 350 | 0 | $\frac{250}{3}$ | 6000 |

Since we have a negative indicator in the bottom row, the value of the objective function, which is now 6000, may be improved. We choose $y$ as the entering variable. Comparing $1/(1/3) = 3$ to $6/(2/3) = 9$, we choose $r$ to be the leaving variable. Thus, $\frac{1}{3}$ is the new pivot element. Again, we pivot to obtain the third tableau.

**Tableau III**

| Basis | $x$ | $y$ | $z$ | $r$ | $s$ | Values of basic variables |
|---|---|---|---|---|---|---|
| $y$ | 0 | 1 | $\frac{3}{4}$ | 3 | $\frac{-1}{4}$ | 3 |
| $x$ | 1 | 0 | $\frac{1}{4}$ | $-2$ | $\frac{1}{4}$ | 4 |
| $w$ | 0 | 0 | 450 | 400 | 50 | 6400 |

Since there are no negative indicators in the bottom row, the value of the objective function cannot be improved. It is 6400 and occurs when $x = 4$, $y = 3$, and $z = 0$. Thus, Jack should plant 4 acres of wheat, 3 acres of barley, and no corn.
**Do Exercise 29.** ∎

## Unbounded Feasible Solution Set in the Optimizing Direction

We conclude this section with a brief discussion of cases where we have an unbounded feasible solution set in the optimizing direction and where we have multiple optimum solutions. Suppose, as in the next example, we have a linear programming problem that gives rise to a tableau in which an entering variable has been selected and all entries above the indicator are negative or 0. Then, the linear programming problem has an unbounded feasible solution set in the optimizing direction.

**EXAMPLE 5** Maximize $z = 3x + 4y$ subject to

$$-x + y \leq 5$$
$$-3x + 5y \leq 35$$
$$x \geq 0, y \geq 0$$

**Solution**  As usual we get

$$-x + y + s_1 = 5$$
$$-3x + 5y + s_2 = 35$$
$$z - 3x - 4y = 0$$

The initial tableau is

**Tableau I**

| Basis | $x$ | $y$ | $s_1$ | $s_2$ | Values of basic variables |
|-------|-----|-----|-------|-------|---------------------------|
| $s_1$ | $-1$ | ① | 1 | 0 | 5 |
| $s_2$ | $-3$ | 5 | 0 | 1 | 35 |
| $z$ | $-3$ | $-4$ | 0 | 0 | 0 |

The most negative indicator is $-4$. Thus, $y$ is the entering variable. Comparing ratios, we see that $s_1$ is the leaving variable. Pivoting, we get

**Tableau II**

| Basis | $x$ | $y$ | $s_1$ | $s_2$ | Values of basic variables |
|-------|-----|-----|-------|-------|---------------------------|
| $y$ | $-1$ | 1 | 1 | 0 | 5 |
| $s_2$ | ② | 0 | $-5$ | 1 | 10 |
| $z$ | $-7$ | 0 | 4 | 0 | 20 |

Since $-7$ is the only negative indicator, we select $x$ as the next entering variable. The only positive entry in the pivot column is 2. Thus, the leaving variable is $s_2$. Pivoting, we get

**Tableau III**

| Basis | $x$ | $y$ | $s_1$ | $s_2$ | Values of basic variables |
|-------|-----|-----|-------|-------|---------------------------|
| $y$ | 0 | 1 | $\frac{-3}{2}$ | $\frac{1}{2}$ | 10 |
| $x$ | 1 | 0 | $\frac{-5}{2}$ | $\frac{1}{2}$ | 5 |
| $z$ | 0 | 0 | $\frac{-27}{2}$ | $\frac{7}{2}$ | 55 |

The only negative indicator is $\frac{-27}{2}$. Thus, $s_1$ is the entering variable and the $s_1$ column is the pivot column. Note that all entries in the pivot column are negative. Thus, the region of feasible solutions is unbounded in the optimizing direction and $z$ has no maximum value. You would find it instructive to solve this example geometrically.
**Do Exercise 3.**

## Multiple Optimum Solutions

We suggested earlier that you visualize a house with a convex polygonal base and a slanted flat roof to see that such a house would have its highest point at a corner. If two adjacent corners of the roof are the same altitude, then every point on the edge connecting these two corners are at that altitude. Similarly, if in a linear programming problem the objective function attains its maximum value at two adjacent vertices of the region of feasible solutions, then it attains its maximum value at each point on the edge connecting these two vertices. In the case where we have $n$ decision variables, if the function attains its maximum value at $(x_1, x_2, x_3, \ldots, x_n)$ and also at $(y_1, y_2, y_3, \ldots, y_n)$ then, if for each $i \in \{1, 2, \ldots, n\}$ we let $z_i = (1 - t)x_i + ty_i$, where $0 \le t \le 1$, the function attains its maximum value at $(z_1, z_2, z_3, \ldots, z_n)$ also. (See Exercises 38, 39, and 40.)

Suppose now that we have a linear programming problem that creates a tableau in which there are no negative indicators. In this case, an optimum solution has been reached. But what if one or more of the indicators corresponding to nonbasic variables are 0? Then, by using one of these nonbasic variables as an entering variable, we may get a different basic feasible solution. This would give the same value of the objective function, since in pivoting, the bottom row would remain unchanged. In such a case, both basic feasible solutions, together with every point on the edge joining them, would yield the maximum value of the objective function as is illustrated in the next example.

**EXAMPLE 6**    Maximize $z = 10x + 6y$ subject to

$$x + y \le 10$$
$$5x + 3y \le 36$$
$$2x + y \le 14$$
$$x \ge 0, y \ge 0$$

**Solution**    As usual, we write

$$x + y + s_1 = 10$$
$$5x + 3y + s_2 = 36$$
$$2x + y + s_3 = 14$$
$$z - 10x - 6y = 0$$
$$x \ge 0, y \ge 0$$

The initial tableau is

**Tableau I**

| Basis | $x$ | $y$ | $s_1$ | $s_2$ | $s_3$ | Values of basic variables |
|-------|-----|-----|-------|-------|-------|---------------------------|
| $s_1$ | 1 | 1 | 1 | 0 | 0 | 10 |
| $s_2$ | 5 | 3 | 0 | 1 | 0 | 36 |
| $s_3$ | ② | 1 | 0 | 0 | 1 | 14 |
| $z$ | $-10$ | $-6$ | 0 | 0 | 0 | 0 |

Since $-10$ is the most negative indicator, $x$ is the entering variable. Comparing ratios, we see that $s_3$ is the leaving variable. Pivoting, we get

**Tableau II**

| Basis | $x$ | $y$ | $s_1$ | $s_2$ | $s_3$ | Values of basic variables |
|---|---|---|---|---|---|---|
| $s_1$ | 0 | $\frac{1}{2}$ | 1 | 0 | $\frac{-1}{2}$ | 3 |
| $\leftarrow$ $s_2$ | 0 | $\left(\frac{1}{2}\right)$ | 0 | 1 | $\frac{-5}{2}$ | 1 |
| $x$ | 1 | $\frac{1}{2}$ | 0 | 0 | $\frac{1}{2}$ | 7 |
| $z$ | 0 | $-1$ | 0 | 0 | 5 | 70 |

Since $-1$ is the only negative indicator, $y$ is the next entering variable. Comparing ratios, we see that $s_2$ is the leaving variable. Pivoting, we get

**Tableau III**

| Basis | $x$ | $y$ | $s_1$ | $s_2$ | $s_3$ | Values of basic variables |
|---|---|---|---|---|---|---|
| $\leftarrow$ $s_1$ | 0 | 0 | 1 | $-1$ | $\textcircled{2}$ | 2 |
| $y$ | 0 | 1 | 0 | 2 | $-5$ | 2 |
| $x$ | 1 | 0 | 0 | $-1$ | 3 | 6 |
| $z$ | 0 | 0 | 0 | 2 | 0 | 72 |

There are no negative indicators in the bottom row, so the optimum solution has been reached. The maximum value of $z$ is 72 and it occurs when $x = 6$, $y = 2$, $s_1 = 2$, and $s_2 = s_3 = 0$. That is, it occurs at $(6, 2)$. Since $s_3$ is a nonbasic variable and the indicator in the $s_3$ column is 0, we shall check for multiple solutions. We treat $s_3$ as an entering variable. Comparing ratios, we see that $s_1$ is the leaving variable. Pivoting, we get

**Tableau IV**

| Basis | $x$ | $y$ | $s_1$ | $s_2$ | $s_3$ | Values of basic variables |
|---|---|---|---|---|---|---|
| $s_3$ | 0 | 0 | $\frac{1}{2}$ | $\frac{-1}{2}$ | 1 | 1 |
| $y$ | 0 | 1 | $\frac{5}{2}$ | $\frac{-1}{2}$ | 0 | 7 |
| $x$ | 1 | 0 | $\frac{-3}{2}$ | $\frac{1}{2}$ | 0 | 3 |
| $z$ | 0 | 0 | 0 | 2 | 0 | 72 |

The bottom row is the same as in the preceding tableau. We have, therefore, no negative indicators and we still have an optimum solution. The maximum value of $z$ is 72 and occurs when $x = 3$, $y = 7$, $s_1 = s_2 = 0$, and $s_3 = 1$. That is, it occurs at $(3, 7)$. Note that $(3, 7)$ and $(6, 2)$ are two adjacent vertices of the region of feasible solutions. Since $z$ has the value 72 at both points, it will have the value 72 at every point on the edge connecting these two vertices. That is, $z$ will have the value 72 at each $(x, y)$ where

$$x = (1 - t)3 + 6t, \text{ and } y = (1 - t)7 + 2t, 0 \le t \le 1$$

When $t = 0$, we get the endpoint $(3, 7)$ and when $t = 1$, we obtain the other endpoint $(6, 2)$. When $t = \frac{1}{2}$ we get the point $\left(\frac{9}{2}, \frac{9}{2}\right)$. We verify that, as expected, at that point the value of $z$ is 72 by calculating

$$z = 10\left(\frac{9}{2}\right) + 6\left(\frac{9}{2}\right) = 45 + 27 = 72$$

**Do Exercise 17.**

■

# Exercise Set 4.5

*In Exercises 1–37, use the simplex method to solve the given problems.*

**1.** Maximize $z = 2x + 4y$ subject to

$$x + 5y \le 75$$
$$2x + 5y \le 90$$
$$x + y \le 33$$
$$x \ge 0, y \ge 0$$

**2.** Maximize $z = 3x + 3y$ subject to

$$x + 5y \le 22$$
$$3x + 7y \le 42$$
$$x \ge 0, y \ge 0$$

**3.** Maximize $z = 3x + 2y$ subject to

$$-2x + 3y \le 3$$
$$-x + 4y \le 8$$
$$x \ge 0, y \ge 0$$

**4.** Maximize $z = 20x + 5y$ subject to

$$x - y \le 2$$
$$x + 2y \le 16$$
$$x + y \le 10$$
$$x \ge 0, y \ge 0$$

**5.** Maximize $z = 5x + 5y$ subject to

$$x - y \le 60$$
$$8x + 5y \le 600$$
$$2x + y \le 90$$
$$-x + y \le 60$$
$$x \ge 0, y \ge 0$$

**6.** Maximize $z = 3x + 5y$ subject to

$$x - y \le 5$$
$$7x - 5y \le 49$$
$$x \ge 0, y \ge 0$$

**7.** Maximize $z = 14x + 24y$ subject to

$$7x + 3y \le 210$$
$$x + 7y \le 140$$
$$2x + 5y \le 118$$
$$x \ge 0, y \ge 0$$

**8.** Maximize $z = 12x + 15y$ subject to

$$x + 3y \le 30$$
$$4x + 5y \le 64$$
$$2x + y \le 26$$
$$x \ge 0, y \ge 0$$

**9.** Maximize $w = 8x + 9y + 6z$ subject to

$$4x + 5z \le 210$$
$$x + y \le 48$$
$$x \le 45$$
$$x \ge 0, y \ge 0, z \ge 0$$

**10.** Maximize $w = 2x - 24y + 8z$ subject to

$$4x + 3y - z \le 7$$
$$x - y - z \le 7$$
$$-x - y + z \le 14$$
$$x \ge 0, y \ge 0, z \ge 0$$

**11.** Maximize $w = 2x + 4y + 12z$ subject to

$$x + y + 2z \le 14$$
$$y + 4z \le 12$$
$$x \ge 0, y \ge 0, z \ge 0$$

**12.** Maximize $w = 5x + y + z$ subject to

$$x + y + z \le 66$$
$$-x + 3y + 2z \le 72$$
$$2x + y + 3z \le 108$$
$$x \ge 0, y \ge 0, z \ge 0$$

**13.** Maximize $w = x + y + z$ subject to

$$x + 2y + z \le 168$$
$$2x + y + z \le 280$$
$$x + y + 3z \le 210$$
$$x \ge 0, y \ge 0, z \ge 0$$

**14.** Maximize $w = 12x + 10y + 2z$ subject to

$$3x + 9y - 2z \leq 10$$
$$2x + 4y - z \leq 4$$
$$-4x + y + z \leq 6$$
$$x \geq 0, y \geq 0, z \geq 0$$

**15.** Maximize $w = 10x + 25y + 15z$ subject to

$$x + 4y + 2z \leq 200$$
$$4x + 6y + 5z \leq 750$$
$$x \geq 0, y \geq 0, z \geq 0$$

**16.** Maximize $w = 3x + 6y + 3z$ subject to

$$2x + y + 3z \leq 24$$
$$x + 2y \leq 12$$
$$2x + z \leq 8$$
$$x \geq 0, y \geq 0, z \geq 0$$

**17.** Maximize $w = 8x + 2y + 3z$ subject to

$$8x + 10y - 9z \leq 80$$
$$30x - 25y + 60z \leq 300$$
$$-22x + 35y + 56z \leq 280$$
$$x \geq 0, y \geq 0, z \geq 0$$

**18.** Maximize $w = 10x + 20y + 15z$ subject to

$$x + 5y + 2z \leq 80$$
$$x + 3y + 4z \leq 60$$
$$x \geq 0, y \geq 0, z \geq 0$$

**19.** Maximize $w = 3x + 6y - 3z$ subject to

$$x + y + z \leq 40$$
$$2x + y - 3z \leq 20$$
$$x - y + 2z \leq 30$$
$$x \geq 0, y \geq 0, z \geq 0$$

**20.** Maximize $w = 2x + 24y + 4z$ subject to

$$x + 5y + 2z \leq 420$$
$$2x + y + 3z \leq 980$$
$$x + 7y \leq 560$$
$$x \geq 0, y \geq 0, z \geq 0$$

**21.** Maximize $w = 3x + 4y + 5z$ subject to

$$3x + 2y \leq 60$$
$$3x + 2y + 3z \leq 72$$
$$2y + 3z \leq 48$$
$$x \geq 0, y \geq 0, z \geq 0$$

**22.** Maximize $z = 3x_1 + 2x_3$ subject to

$$x_1 + x_2 \leq 16$$
$$x_1 - 2x_2 \leq 28$$
$$x_3 - 2x_4 \leq 8$$
$$x_3 + x_4 \leq 20$$
$$x_1 \geq 0, x_2 \geq 0, x_3 \geq 0, x_4 \geq 0$$

**23.** Maximize $z = 5x_1 + 10x_2 + 5x_3 + 25x_4$ subject to

$$3x_1 + x_2 + 2x_3 + x_4 \leq 10$$
$$x_1 + 2x_2 + x_3 + x_4 \leq 5$$
$$x_1 \geq 0, x_2 \geq 0, x_3 \geq 0, x_4 \geq 0$$

**24.** Maximize $z = 2x_1 + 3x_2 - x_3 + 5x_4$ subject to

$$x_1 + x_2 - x_3 + 2x_4 \leq 8$$
$$2x_1 - x_2 + x_3 + x_4 \leq 7$$
$$3x_1 - x_2 + 2x_3 + x_4 \leq 9$$
$$x_1 \geq 0, x_2 \geq 0, x_3 \geq 0, x_4 \geq 0$$

**25.** A farmer has 100 acres of land on which to grow tomatoes and peas. The cost of planting each acre of tomatoes is $50. Each acre of peas will cost $25 to plant. The farmer's budget for planting is $6250. If the anticipated profits are $100 per acre of peas and $150 per acre of tomatoes, how many acres of each should be planted to maximize profit? What is the maximum profit?

**26.** A farmer has 25 acres on which to grow apples and pears. Because of an agreement, he cannot grow more than 20 acres of apples and 15 acres of pears. If the profit per acre is $600 for apples and $400 for pears, how many acres of each should be planted to maximize profits? What is the maximum profit?

**27.** A farmer has 100 acres on which he wishes to grow oranges and lemons. He has been advised not to grow more than 50 acres of lemons. The cost of harvesting is $150 per acre of oranges and $300 per acre of lemons. He has only $18,000 budgeted for harvesting. If the profit per acre is $500 for oranges and $250 for lemons, how many acres of each should be planted to maximize profit? What is the maximum profit?

**28.** A company makes two gadgets, A and B. Each gadget A requires 1 hour on machine I, 1 hour on machine II, 2 hours on machine III, and 1 hour on machine IV. Each gadget B requires 2 hours on machine I, 1 hour on machine II, 1 hour on machine III, and 10 hours on machine IV. The time available per week for the production of these two gadgets is 48 hours on machine I, 40 hours on machine II, 50 hours on machine III, and 200 hours on machine IV. If the profit on each gadget A is $300 and on each gadget B it is $200, and if the company can sell as many gadgets as it produces, how many of each gadget should be produced per week to maximize profit? What is the maximum profit?

**29.** A company makes products A, B, and C. The times required for each unit of these products on each of the three machines are given in the following table.

|            | Product A | Product B | Product C |
|------------|-----------|-----------|-----------|
| Machine I  | 3 hours   | 5 hours   | 3 hours   |
| Machine II | 3 hours   | 2 hours   | 0 hour    |
| Machine III| 1 hour    | 2 hours   | 0 hour    |

The amount of time available per day on each type of machine is 450 hours on machine I, 60 hours on machine II, and 120 hours on machine III. If the profits per unit of products A, B, and C are $70, $100, and $40, respectively, how many units of each product should be manufactured per day to maximize profit? What is the maximum profit?

**30.** A camera company makes two models of movie cameras. The deluxe model takes 3 hours to assemble, 6 minutes to test, and 16 minutes to pack. The regular model takes 4 hours to assemble, 3 minutes to test, and 4 minutes to pack. The number of work hours available per day for assembling, testing, and packing are 300, 5, and $\frac{32}{3}$, respectively. If the profit per deluxe model is $400 and that per regular model is $150, and if the company can sell as many cameras as it can produce, how many deluxe and how many regular models should be produced per day to maximize profit? What is the maximum profit?

**31.** A farmer has 40 acres of land on which he wishes to plant peanuts, wheat, and barley. The cost of planting each acre is $100 for peanuts, $50 for wheat, and $100 for barley. The cost of harvesting each acre is $300 for peanuts, $200 for wheat, and $100 for barley. The farmer has $3000 available for planting and $8000 available for harvesting. If the profits per acre are $100 for peanuts, $300 for wheat, and $400 for barley, how many acres of each should be planted to maximize profit? What is the maximum profit?

**32.** A company manufactures products A, B, and C, each of which is made from raw materials I, II, and III. The unit requirements for each of the three products are tabulated.

| | **Each Unit of Product Requires** | | |
|-----------|-----------------|------------------|------------------|
|           | Raw material I | Raw material II | Raw material III |
| Product A | 1 unit          | 1 unit           | 1 unit           |
| Product B | 3 units         | 6 units          | 9 units          |
| Product C | 6 units         | 3 units          | 10 units         |

The maximum quantities available each day are 960 units of raw material I, 1800 units of raw material II, and 2740 units of raw material III. The cost to manufacture each unit of products A, B, and C are $5, $37, and $120, respectively. The company can sell as many units as it can produce at the following prices per unit: $13 for A, $69 for B, and $168 for C. How many units per day of each product should be manufactured to maximize profit? What is the maximum profit?

**33.** A company has 1000 pounds of peanuts, 1200 pounds of cashew nuts, and 800 pounds of Brazil nuts. It wishes to package three varieties of mixed nuts. The quantities (in pounds) of each kind of nut required to make each pound of each variety of mixed nuts are given in the following table.

|             | **Variety I** | **Variety II** | **Variety III** |
|-------------|---------------|----------------|-----------------|
| Peanuts     | $\frac{1}{2}$ | $\frac{1}{3}$  | 0               |
| Cashew nuts | $\frac{1}{2}$ | $\frac{1}{3}$  | $\frac{1}{2}$   |
| Brazil nuts | 0             | $\frac{1}{3}$  | $\frac{1}{2}$   |

Variety I sells for $3 per pound, variety II sells for $4 per pound, and variety III sells for $5 per pound. How many pounds of each variety should be mixed to maximize revenue? What is the maximum revenue?

**34.** A candy manufacturer has 320 pounds of chocolate, 180 pounds of fruit, and 240 pounds of nuts. She produces three types of candy. The quantities (in pounds) required to make each box of each type are given in the following table.

|           | **Type A**    | **Type B**    | **Type C**    |
|-----------|---------------|---------------|---------------|
| Chocolate | $\frac{1}{4}$ | $\frac{1}{2}$ | 1             |
| Fruit     | $\frac{1}{2}$ | $\frac{1}{4}$ | $\frac{1}{4}$ |
| Nuts      | $\frac{1}{4}$ | $\frac{1}{2}$ | $\frac{3}{4}$ |

If she can sell each box of type A for $5, each box of type B for $6, and each box of type C for $10, how many boxes of each should she make to maximize revenue? What is the maximum revenue?

**35.** A company manufactures products A, B, and C. The numbers of hours of machine time and of finishing time necessary to produce each unit of each product are given in the following table.

|  | Machine Time | Finishing Time |
|---|---|---|
| Product A | 1 | 1 |
| Product B | 2 | 3 |
| Product C | 3 | 5 |

The number of work hours of machine time and of finishing time available each day are 50 and 60, respectively. The profits per unit realized from A, B, and C are $10, $20, and $40, respectively. How many units of each should be manufactured each day to maximize profit? What is the maximum profit?

**36.** A company manufactures small boats. Each type requires materials as indicated in the following table.

|  | Type A | Type B | Type C |
|---|---|---|---|
| Aluminum | 2 units | 3 units | 6 units |
| Wood | 3 units | 4 units | 1 unit |
| Paint | 3 units | 5 units | 2 units |

Each week, the company has available 600 units of aluminum, 480 units of wood, and 240 units of paint. If the profit per boat is $300 for A, $500 for B, and $300 for C, and if the company can sell all the boats it can produce, how many boats of each type should be produced each week to maximize profit? What is the maximum profit?

**37.** A contractor has 75 acres of land, $30,000,000 in capital, and 800,000 labor hours available to build houses. The amounts of resources required for each type of house are given in the following table.

|  | Type A | Type B | Type C |
|---|---|---|---|
| Land | .25 acre | .25 acre | 1 acre |
| Capital | $50,000 | $150,000 | $100,000 |
| Labor hours | 6000 | 4000 | 2000 |

If the contractor's profit is $20,000 on each A house, $40,000 on each B house, and $60,000 on each C house, how many houses of each type should he build to maximize profit? What is the maximum profit?

**38.** Suppose that the function defined by

$$z = f(x, y) = mx + ny,$$

where $m$ and $n$ are constants, assumes the same value at $(x_1, y_1)$ and $(x_2, y_2)$. That is,

$$mx_1 + ny_1 = mx_2 + ny_2 = A$$

Show that if $x = (1 - t)x_1 + tx_2$ and $y = (1 - t)y_1 + ty_2$, where $0 \le t \le 1$, then $f(x, y) = A$ also. The points $(x, y)$ are the points on the line segment joining $(x_1, y_1)$ and $(x_2, y_2)$. (*Hint:* Replace $x$ and $y$ by $(1 - t)x_1 + tx_2$ and $(1 - t)y_1 + ty_2$, respectively, in the right side of the equation $f(x, y) = mx + ny$, and simplify.)

**39.** Suppose that the function defined by

$$w = f(x, y, z) = mx + ny + pz$$

where $m$, $n$, and $p$ are constants, assumes the same value at $(x_1, y_1, z_1)$ and $(x_2, y_2, z_2)$. That is,

$$mx_1 + ny_1 + pz_1 = mx_2 + ny_2 + pz_2 = A$$

Show that if $x = (1 - t)x_1 + tx_2$, $y = (1 - t)y_1 + ty_2$, and $z = (1 - t)z_1 + tz_2$ where $0 \le t \le 1$, then $f(x, y, z) = A$ also. The points $(x, y, z)$ are the points on the line segment joining $(x_1, y_1, z_1)$ and $(x_2, y_2, z_2)$. (See the hint in Exercise 38.)

**40.** Let $Z = f(x_1, x_2, \ldots, x_n) = c_1x_1 + c_2x_2 + \cdots + c_nx_n$, where $c_1, c_2, \ldots, c_n$ are constants and suppose that $Z$ has the same value at $(a_1, a_2, \ldots, a_n)$ and at $(b_1, b_2, \ldots, b_n)$. That is, suppose that

$$c_1a_1 + c_2a_2 + \cdots + c_na_n$$
$$= c_1b_1 + c_2b_2 + \cdots + c_nb_n = A$$

Let

$$z_1 = (1 - t)a_1 + tb_1$$
$$z_2 = (1 - t)a_2 + tb_2$$
$$\vdots \qquad \vdots \qquad \vdots$$
$$z_n = (1 - t)a_n + tb_n$$

where $0 \le t \le 1$. Show that $f(z_1, z_2, \ldots, z_n) = A$ also. (See the hint in Exercise 38.)

# 4.6 Minimization and Duality (Optional)

In the preceding section we studied only maximization problems in standard form. In this section, we discuss minimization problems in standard form. This form differs from that of maximums only in the fact that all "$\le$" inequalities are reversed and the coefficients in the equation defining the objective function must be nonnegative. We shall solve a minimization problem in standard form by applying the simplex method to a corresponding maximization problem in standard form. We shall first show how to obtain the corresponding maximization problem, then illustrate the technique with examples. We conclude with two examples that will help show why the technique works.

## The Dual Problem

Recall the definition of the transpose of a matrix from Section 3.5. If $\mathbf{A}$ is an $m \times n$ matrix, its transpose $\mathbf{A}^T$ is the $n \times m$ matrix whose $i$-$j$ entry is the $j$-$i$ entry of $\mathbf{A}$. That is, the $i$th row of $\mathbf{A}^T$ is the $i$th column of $\mathbf{A}$. For example, if

$$\mathbf{A} = \begin{bmatrix} 1 & 5 & 9 \\ 0 & 3 & 4 \end{bmatrix}$$

then

$$\mathbf{A}^T = \begin{bmatrix} 1 & 0 \\ 5 & 3 \\ 9 & 4 \end{bmatrix}$$

Consider the following minimization problem. Minimize $z = c_{11}x_1 + c_{12}x_2 + \cdots + c_{1n}x_n$ where all the coefficients $c_{ij}$ are nonnegative and the variables $x_i$ are subject to

$$\begin{aligned}
a_{11}x_1 + a_{12}x_2 + \cdots + a_{1n}x_n &\geq b_1 \\
a_{21}x_1 + a_{22}x_2 + \cdots + a_{2n}x_n &\geq b_2 \\
&\vdots \\
a_{m1}x_1 + a_{m2}x_2 + \cdots + a_{mn}x_n &\geq b_m \\
x_1 \geq 0, \; x_2, \geq 0, \; \ldots, \; x_n &\geq 0
\end{aligned}$$

All $b_i$'s need not be nonnegative. We write the first $m$ inequalities as equalities, *without introducing additional variables*. The equality defining the objective function may be written on the last line with the $z$ on the right side of the equal sign as follows:

$$\begin{aligned}
a_{11}x_1 + a_{12}x_2 + \cdots + a_{1n}x_n &= b_1 \\
a_{21}x_1 + a_{22}x_2 + \cdots + a_{2n}x_n &= b_2 \\
&\vdots \\
a_{m1}x_1 + a_{m2}x_2 + \cdots + a_{mn}x_n &= b_m \\
c_{11}x_1 + c_{12}x_2 + \cdots + c_{1n}x_n &= z
\end{aligned}$$

The augmented matrix of the system is

$$\begin{bmatrix} a_{11} & a_{12} & \cdots & a_{1n} & b_1 \\ a_{21} & a_{22} & \cdots & a_{2n} & b_2 \\ \cdot & \cdot & & \cdot & \cdot \\ \cdot & \cdot & & \cdot & \cdot \\ \cdot & \cdot & & \cdot & \cdot \\ a_{m1} & a_{m2} & \cdots & a_{mn} & b_m \\ c_{11} & c_{12} & \cdots & c_{1n} & z \end{bmatrix}$$

Finally, we write the transpose of the matrix.

$$\begin{bmatrix} a_{11} & a_{21} & \cdots & a_{m1} & c_{11} \\ a_{12} & a_{22} & \cdots & a_{m2} & c_{12} \\ \cdot & \cdot & & \cdot & \cdot \\ \cdot & \cdot & & \cdot & \cdot \\ \cdot & \cdot & & \cdot & \cdot \\ a_{1n} & a_{2n} & \cdots & a_{mn} & c_{1n} \\ b_1 & b_2 & \cdots & b_m & z \end{bmatrix}$$

Using this matrix, we write the following maximization problem, which is called the *dual* of our original minimization problem.

Maximize $Z = b_1 y_1 + b_2 y_2 + \cdots + b_m y_m$, where the $y_j$'s are subject to

$$a_{11} y_1 + a_{21} y_2 + \cdots + a_{m1} y_m \le c_{11}$$
$$a_{12} y_1 + a_{22} y_2 + \cdots + a_{m2} y_m \le c_{12}$$
$$\begin{matrix} \cdot & \cdot & & \cdot & \cdot \\ \cdot & \cdot & & \cdot & \cdot \\ \cdot & \cdot & & \cdot & \cdot \end{matrix}$$
$$a_{1n} y_1 + a_{2n} y_2 + \cdots + a_{mn} y_m \le c_{1n}$$
$$y_1 \ge 0, y_2 \ge 0, \ldots, y_m \ge 0$$

The original minimization problem is called the *primal* problem. We used different names for the variables in the dual to emphasize that it is a different problem.

**EXAMPLE 1**   Write the dual of the following primal minimization problem.
Minimize $z = 3x + 5y$, where $x$ and $y$ are subject to

$$2x + 5y \ge \phantom{-}4$$
$$3x + 7y \ge -2$$
$$x - 6y \ge \phantom{-}3$$
$$x \ge 0, y \ge 0$$

**Solution**   We change the first three constraint inequalities to equations and write

$$2x + 5y = \phantom{-}4$$
$$3x + 7y = -2$$
$$x - 6y = \phantom{-}3$$
$$3x + 5y = \phantom{-}z$$

The augmented matrix of this system is

$$\begin{bmatrix} 2 & 5 & 4 \\ 3 & 7 & -2 \\ 1 & -6 & 3 \\ 3 & 5 & z \end{bmatrix}$$

The transpose is

$$\begin{bmatrix} 2 & 3 & 1 & 3 \\ 5 & 7 & -6 & 5 \\ 4 & -2 & 3 & z \end{bmatrix}$$

The last row of the transpose gives the coefficients of the equation defining the new objective function, while the first two rows give the coefficients of the new constraint inequalities. The dual problem is the following.

Maximize $Z = 4u - 2v + 3w$, where $u$, $v$, and $w$ are subject to

$$2u + 3v + w \leq 3$$
$$5u + 7v - 6w \leq 5$$
$$u \geq 0, v \geq 0, w \geq 0$$

As required, the constants 3 and 5 of the first two inequalities are nonnegative. This results from the fact that these quantities were the coefficients of the equation defining the original objective function, which in a minimization problem in standard form are nonnegative.                                        ■

## Solving Minimization Problems in Standard Form

The following theorem, which is proved in more advanced texts, allows us to solve a minimization problem in standard form by solving the dual problem.

> **THEOREM 4.3:** The objective function $z$ of a minimization problem in standard form assumes a minimum value if, and only if, the objective function $Z$ of the dual problem assumes a maximum value. Furthermore, the minimum value of $z$ is equal to the maximum value of $Z$. The coordinates of the point where the objective function $z$ attains its minimum value are read in the final tableau of the dual problem: in the bottom row and in the columns corresponding to the slack variables used to solve the dual problem.

**EXAMPLE 2**   Minimize $z = 54x + 22y$ subject to

$$9x + y \geq 144$$
$$x + y \geq 120$$
$$x \geq 0, y \geq 0$$

**Solution**   The matrix of the primal problem is

$$\begin{bmatrix} 9 & 1 & 144 \\ 1 & 1 & 120 \\ 54 & 22 & z \end{bmatrix}$$

The transpose is

$$\begin{bmatrix} 9 & 1 & 54 \\ 1 & 1 & 22 \\ 144 & 120 & z \end{bmatrix}$$

This is the matrix of the dual problem. Thus, we can formulate the dual problem as follows. Maximize $Z = 144u + 120v$ subject to

$$9u + v \le 54$$
$$u + v \le 22$$
$$u \ge 0, v \ge 0$$

Using slack variables in the usual way, we obtain the system

$$\begin{cases} 9u + v + r & = 54 \\ u + v + s & = 22 \\ Z - 144u - 120v & = 0 \end{cases}$$

The initial tableau for this system is given with the entering and leaving variables and the pivot element as indicated.

**Tableau I**

| Basis | $u$ | $v$ | $r$ | $s$ | Values of basic variables |
|---|---|---|---|---|---|
| $r$ | ⑨ | 1 | 1 | 0 | 54 |
| $s$ | 1 | 1 | 0 | 1 | 22 |
| $Z$ | $-144$ | $-120$ | 0 | 0 | 0 |

Pivoting, we get

**Tableau II**

| Basis | $u$ | $v$ | $r$ | $s$ | Values of basic variables |
|---|---|---|---|---|---|
| $u$ | 1 | $\frac{1}{9}$ | $\frac{1}{9}$ | 0 | 6 |
| $s$ | 0 | $\frac{8}{9}$ | $\frac{-1}{9}$ | 1 | 16 |
| $Z$ | 0 | $-104$ | 16 | 0 | 864 |

The entering and leaving variables and the pivot element are indicated above. Pivoting, we obtain

**Tableau III**

| Basis | $u$ | $v$ | $r$ | $s$ | Values of basic variables |
|-------|-----|-----|-----|-----|---------------------------|
| $u$ | 1 | 0 | $\frac{1}{8}$ | $\frac{-1}{8}$ | 4 |
| $v$ | 0 | 1 | $\frac{-1}{8}$ | $\frac{9}{8}$ | 18 |
| $Z$ | 0 | 0 | 3 | 117 | 2736 |

All indicators in the bottom row are nonnegative, so an optimum solution has been reached. The maximum value of $Z$ is 2736. It occurs when $u = 4$ and $v = 18$. Therefore, the minimum value of $z$ in the primal problem is 2736. It occurs at (3, 117) since 3 and 117 appear in the bottom row and in the columns corresponding to the slack variables $r$ and $s$ of the final tableau. Hence, the minimum value of $z$ occurs when $x = 3$ and $y = 117$.
**Do Exercise 1.**

The next example illustrates the fact that in a minimization problem in standard form, the constants on the right of the constraint inequalities may be negative, with the exception of those expressing the nonnegativity of the decision variables.

**EXAMPLE 3**   Minimize $z = 4x + 7y$ subject to

$$x + y \geq 14$$
$$x - 3y \leq 6$$
$$x \geq 0, y \geq 0$$

**Solution**   We must first multiply the second inequality by $-1$ to change the "$\leq$" to "$\geq$."
Minimize $z$ if $z = 4x + 7y$ subject to

$$x + y \geq 14$$
$$-x + 3y \geq -6$$
$$x \geq 0, y \geq 0$$

The problem is now in standard form. The matrix of the primal problem is

$$\begin{bmatrix} 1 & 1 & 14 \\ -1 & 3 & -6 \\ 4 & 7 & z \end{bmatrix}$$

The transpose is

$$\begin{bmatrix} 1 & -1 & 4 \\ 1 & 3 & 7 \\ 14 & -6 & z \end{bmatrix}$$

This is the matrix of the dual problem. Thus, we can formulate the dual problem as follows. Maximize $Z = 14u - 6v$ subject to

$$u - v \le 4$$
$$u + 3v \le 7$$
$$u \ge 0, v \ge 0$$

Using slack variables, we obtain

$$\begin{cases} u - v + r & = 4 \\ u + 3v & + s = 7 \\ Z - 14u + 6v & = 0 \end{cases}$$

The initial tableau is now given with the entering and leaving variables and the pivot element indicated.

**Tableau I**

| Basis | $u$ | $v$ | $r$ | $s$ | Values of basic variables |
|-------|-----|-----|-----|-----|---------------------------|
| $\leftarrow$ $r$ | ① | $-1$ | $1$ | $0$ | $4$ |
| $s$ | $1$ | $3$ | $0$ | $1$ | $7$ |
| $Z$ | $-14$ | $6$ | $0$ | $0$ | $0$ |

Pivoting, we get

**Tableau II**

| Basis | $u$ | $v$ | $r$ | $s$ | Values of basic variables |
|-------|-----|-----|-----|-----|---------------------------|
| $u$ | $1$ | $-1$ | $1$ | $0$ | $4$ |
| $\leftarrow$ $s$ | $0$ | ④ | $-1$ | $1$ | $3$ |
| $Z$ | $0$ | $-8$ | $14$ | $0$ | $56$ |

The entering and leaving variables and the pivot element are indicated. Pivoting, we get

**Tableau III**

| Basis | $u$ | $v$ | $r$ | $s$ | Values of basic variables |
|-------|-----|-----|-----|-----|---------------------------|
| $u$ | $1$ | $0$ | $\frac{3}{4}$ | $\frac{1}{4}$ | $\frac{19}{4}$ |
| $v$ | $0$ | $1$ | $\frac{-1}{4}$ | $\frac{1}{4}$ | $\frac{3}{4}$ |
| $Z$ | $0$ | $0$ | $12$ | $2$ | $62$ |

All indicators in the bottom row are nonnegative. Hence, the maximum of $Z$ has been reached. It is 62 and occurs when $u = \frac{19}{4}$ and $v = \frac{3}{4}$. Therefore, the minimum value of $z$ in the primal problem is 62. It occurs at $(12, 2)$ since 12 and 2 appear in the bottom row and in the columns corresponding to the slack variables $r$ and $s$ of the final tableau. Hence, the minimum value of $z$ occurs when $x = 12$ and $y = 2$.
**Do Exercise 21.**                                                                    ■

**EXAMPLE 4**    John must supplement his daily intake of vitamins A, B, and C. He should add at least 24 units of vitamin A, 7 units of vitamin B, and 18 units of vitamin C. John can buy vitamin pills at store S and at store P. The store S pills cost \$15 per bottle of 100 pills, while the store P pills cost \$18 per bottle of 200 pills. The vitamin content of each type of pill is given in the following table.

|  | Vitamin A | Vitamin B | Vitamin C |
|---|---|---|---|
| Store S pill | 6 units | 2 units | 3 units |
| Store P pill | 4 units | 1 unit | 6 units |

How many pills from each store must he take each day to minimize his cost?

**Solution**    Let $x$ and $y$ denote the numbers of pills he takes daily from store S and store P, respectively. The cost per day of the store S pills is \$$.15x$ since the cost per pill is given by $\frac{15}{100} = .15$. Also, the cost per day of the store P pills is \$$.09y$ since the cost per pill is given by $\frac{18}{200} = .09$. Hence, his total daily cost is \$$(.15x + .09y)$. This is the quantity we wish to minimize. There are some constraints. John will get $6x$ units of vitamin A from the $x$ pills from store S and $4y$ units of vitamin A from the $y$ pills from store P. Therefore, his total intake of vitamin A from these pills is $(6x + 4y)$ units. Since this intake must be at least 24 units, we write

$$6x + 4y \geq 24$$

Similarly, we get

$$2x + y \geq 7 \quad \text{and} \quad 3x + 6y \geq 18$$

Simplifying these inequalities, we can state the minimization problem in standard form as follows:
   Minimize $z = .15x + .09y$ subject to

$$3x + 2y \geq 12$$
$$2x + \phantom{2}y \geq \phantom{1}7$$
$$\phantom{2}x + 2y \geq \phantom{1}6$$
$$x \geq 0, y \geq 0$$

The matrix of the primal problem is

$$\begin{bmatrix} 3 & 2 & 12 \\ 2 & 1 & 7 \\ 1 & 2 & 6 \\ .15 & .09 & z \end{bmatrix}$$

The transpose is

$$\begin{bmatrix} 3 & 2 & 1 & .15 \\ 2 & 1 & 2 & .09 \\ 12 & 7 & 6 & z \end{bmatrix}$$

This is the matrix of the dual problem. Thus, we can formulate the dual problem as follows.

Maximize $Z = 12u + 7v + 6w$ subject to

$$3u + 2v + w \le .15$$
$$2u + v + 2w \le .09$$
$$u \ge 0, v \ge 0, w \ge 0$$

Using slack variables, we get

$$\begin{cases} 3u + 2v + w + r & = .15 \\ 2u + v + 2w & + s = .09 \\ Z - 12u - 7v - 6w & = 0 \end{cases}$$

The initial tableau is given below with the entering and leaving variables and the pivot element as indicated.

**Tableau I**

| Basis | $u$ | $v$ | $w$ | $r$ | $s$ | Values of basic variables |
|---|---|---|---|---|---|---|
| $r$ | 3 | 2 | 1 | 1 | 0 | .15 |
| $s$ | ② | 1 | 2 | 0 | 1 | .09 |
| $Z$ | −12 | −7 | −6 | 0 | 0 | 0 |

Pivoting, we obtain

**Tableau II**

| Basis | $u$ | $v$ | $w$ | $r$ | $s$ | Values of basic variables |
|---|---|---|---|---|---|---|
| $r$ | 0 | $\tfrac{1}{2}$ | −2 | 1 | $\tfrac{-3}{2}$ | .015 |
| $u$ | 1 | $\tfrac{1}{2}$ | 1 | 0 | $\tfrac{1}{2}$ | .045 |
| $Z$ | 0 | −1 | 6 | 0 | 6 | .54 |

The entering and leaving variables and the pivot element are indicated. Pivoting, we get

**Tableau III**

| Basis | $u$ | $v$ | $w$ | $r$ | $s$ | Values of basic variables |
|-------|-----|-----|-----|-----|-----|---------------------------|
| $v$   | 0   | 1   | $-4$ | 2  | $-3$ | .03 |
| $u$   | 1   | 0   | 3   | $-1$ | $-1$ | .03 |
| $Z$   | 0   | 0   | 2   | 2   | 3   | .57 |

All indicators in the bottom row are nonnegative. The maximum value of $Z$ has been reached. It is .57 and occurs at $u = .03$ and $v = .03$. Therefore, the minimum value of $z$ in the primal problem is .57. It occurs at the point $(2, 3)$ since 2 and 3 appear in the bottom row and in the columns corresponding to the slack variables $r$ and $s$ of the final tableau. Thus, the minimum value of $z$ occurs when $x = 2$ and $y = 3$. We conclude that John should take two of the store S pills and three of the store P pills per day to minimize his cost. His daily cost is $.57.
**Do Exercise 23.**                                                                    ■

## Warning About Simplifications in the Dual Problem

Students are usually encouraged to simplify equations and inequalities before solving them. This is good practice because in general simplication does make subsequent calculations easier. When we use the method described in this section, it is correct to simplify a constraint inequality in the primal problem by multiplying it by a positive number. However, *a constraint inequality should not be multiplied by a positive number in a dual maximization problem if that maximization problem is used to solve a minimization problem.* The next example is given to illustrate this difficulty.

**EXAMPLE 5**    Minimize $z = 12x + 16y$ subject to

$$x + 2y \geq 40$$
$$3x + 2y \geq 72$$
$$x \geq 0, y \geq 0$$

**Solution**    The matrix of the primal problem is

$$\begin{bmatrix} 1 & 2 & 40 \\ 3 & 2 & 72 \\ 12 & 16 & z \end{bmatrix}$$

The transpose is

$$\begin{bmatrix} 1 & 3 & 12 \\ 2 & 2 & 16 \\ 40 & 72 & z \end{bmatrix}$$

This is the matrix of the dual problem. Thus, we can formulate the dual problem as follows:

Maximize $Z = 40u + 72v$ subject to

$$u + 3v \le 12$$
$$2u + 2v \le 16$$
$$u \ge 0, v \ge 0$$

At this point, we might be tempted to multiply both sides of the second inequality by $\frac{1}{2}$. However, we should not do so. Instead, we continue by using slack variables to obtain the system

$$\begin{cases} u + 3v + r & = 12 \\ 2u + 2v & + s = 16 \\ Z - 40u - 72v & = 0 \end{cases}$$

The initial tableau for this system showing the entering and leaving variables and the pivot element is

**Tableau I**

| Basis | $u$ | $v$ | $r$ | $s$ | Values of basic variables |
|---|---|---|---|---|---|
| $r$ | 1 | ③ | 1 | 0 | 12 |
| $s$ | 2 | 2 | 0 | 1 | 16 |
| $Z$ | $-40$ | $-72$ | 0 | 0 | 0 |

Pivoting, we obtain

**Tableau II**

| Basis | $u$ | $v$ | $r$ | $s$ | Values of basic variables |
|---|---|---|---|---|---|
| $v$ | $\frac{1}{3}$ | 1 | $\frac{1}{3}$ | 0 | 4 |
| $s$ | $\frac{4}{3}$ | 0 | $\frac{-2}{3}$ | 1 | 8 |
| $Z$ | $-16$ | 0 | 24 | 0 | 288 |

The entering and leaving variables and the pivot element are indicated. Pivoting, we obtain

**Tableau III**

| Basis | $u$ | $v$ | $r$ | $s$ | Values of basic variables |
|---|---|---|---|---|---|
| $v$ | 0 | 1 | $\frac{1}{2}$ | $\frac{-1}{4}$ | 2 |
| $u$ | 1 | 0 | $\frac{-1}{2}$ | $\frac{3}{4}$ | 6 |
| $Z$ | 0 | 0 | 16 | 12 | 384 |

All indicators in the bottom row are nonnegative. Hence, the maximum value of $Z$ has been reached. It is 384 and occurs when $u = 6$ and $v = 2$. Therefore, the minimum value of $z$ in the primal problem is also 384. It occurs at (16, 12). Hence, the minimum value of $z$ occurs when $x = 16$ and $y = 12$. Suppose that we had simplified the constraint inequality in the dual maximization problem as follows:

$$u + 3v \leq 12$$
$$2u + 2v \leq 16$$
$$u \geq 0, v \geq 0$$

Multiplying both sides of the second inequality by $\frac{1}{2}$ we would obtain

$$u + 3v \leq 12$$
$$u + v \leq 8$$
$$u \geq 0, v \geq 0$$

We use slack variables as usual to get the system

$$\begin{cases} u + 3v + r \phantom{+ s} = 12 \\ u + v \phantom{+ 3r} + s = 8 \\ Z - 40u - 72v \phantom{+ r + s} = 0 \end{cases}$$

The initial tableau showing the entering and leaving variables and the pivot element is

**Tableau I**

| Basis | $u$ | $v$ | $r$ | $s$ | Values of basic variables |
|---|---|---|---|---|---|
| $r$ | 1 | ③ | 1 | 0 | 12 |
| $s$ | 1 | 1 | 0 | 1 | 8 |
| $Z$ | $-40$ | $-72$ | 0 | 0 | 0 |

Pivoting, we obtain

**Tableau II**

| Basis | $u$ | $v$ | $r$ | $s$ | Values of basic variables |
|---|---|---|---|---|---|
| $v$ | $\frac{1}{3}$ | 1 | $\frac{1}{3}$ | 0 | 4 |
| $s$ | $\frac{2}{3}$ | 0 | $\frac{-1}{3}$ | 1 | 4 |
| $Z$ | $-16$ | 0 | 24 | 0 | 288 |

The entering and leaving variables and the pivot element are indicated. Pivoting, we obtain

**Tableau III**

| Basis | $u$ | $v$ | $r$ | $s$ | Values of basic variables |
|---|---|---|---|---|---|
| $v$ | 0 | 1 | $\frac{1}{2}$ | $\frac{-1}{4}$ | 2 |
| $u$ | 1 | 0 | $\frac{-1}{2}$ | $\frac{3}{2}$ | 6 |
| $Z$ | 0 | 0 | 16 | 24 | 384 |

All indicators in the bottom row are nonnegative. Hence, the maximum value of $Z$ is 384. It occurs when $u = 6$ and $v = 2$. This is the same answer that we obtained earlier for the *dual maximization problem*. Reading the bottom line of the final tableau, we conclude that the minimum value of $z$ in the primal problem is 384. It occurs at $(16, 24)$. Hence, the minimum value of $z$ occurs when $x = 16$ and $y = 24$. But this answer is *wrong*. The minimum value of $z$ is 384. But it occurs when $x = 16$ and $y = 12$. You are urged to give some thought to the reason why simplifying a constraint inequality does not alter the answer to the dual problem, but does alter the answer to the primal problem.   ■

## Relationship Between a Primal Minimization Problem and its Dual

The last two examples are given to help you see why the solution of a primal minimization problem in standard form and that of its dual are equal.

**A CLOSER LOOK**

**EXAMPLE 6**   The Rouquier brothers own a company that manufactures products A and B. The profit from product A is $5280 per unit, while the profit from product B is $1650 per unit. The manufacturing of these products requires the use of two machines. The time required on each machine for each product is given in the following table.

| | Number of Hours Required to Produce One Unit of | |
|---|---|---|
| | Product A | Product B |
| Machine I | 11 | 2 |
| Machine II | 3 | 1 |

The time available on each machine per year is 2310 hours on machine I and 990 hours on machine II. Assuming that every unit of each product made can be sold, use the simplex method to determine how many units of each product should be manufactured each year to maximize the yearly profit. What is the maximum yearly profit?

**Solution**   Let $x$ and $y$ denote the numbers of units per year of products A and B, respectively. The profit realized from $x$ units of A is $\$5280x$ and that from $y$ units of B is $\$1650y$. Hence, the total profit is $\$(5280x + 1650y)$. We wish to maximize this quantity. The number of hours used on machine I is $11x$ in the manufacturing of A and $2y$ hours in the manufacturing of B. Hence, the total number of hours machine I is used is $11x + 2y$. Since the total number of hours available on machine I is 2310, we have

$$11x + 2y \leq 2310$$

Similarly,

$$3x + y \leq 990$$

Our problem is to maximize $z = 5280x + 1650y$ subject to

$$11x + 2y \leq 2310$$
$$3x + \phantom{2}y \leq \phantom{2}990$$
$$x \geq 0, y \geq 0$$

The initial tableau is

**Tableau I**

| Basis | $x$ | $y$ | $r$ | $s$ | Values of basic variables |
|-------|-----|-----|-----|-----|---------------------------|
| $r$ | ⑪ | 2 | 1 | 0 | 2310 |
| $s$ | 3 | 1 | 0 | 1 | 990 |
| $z$ | −5280 | −1650 | 0 | 0 | 0 |

Pivoting, we get

**Tableau II**

| Basis | $x$ | $y$ | $r$ | $s$ | Values of basic variables |
|-------|-----|-----|-----|-----|---------------------------|
| $x$ | 1 | $\frac{2}{11}$ | $\frac{1}{11}$ | 0 | 210 |
| $s$ | 0 | ⑤/11 | $\frac{-3}{11}$ | 1 | 360 |
| $z$ | 0 | −690 | 480 | 0 | 1,108,800 |

Pivoting, we get

**Tableau III**

| Basis | x | y | r | s | Values of basic variables |
|-------|---|---|---|---|---------------------------|
| x | 1 | 0 | $\frac{1}{5}$ | $\frac{-2}{5}$ | 66 |
| y | 0 | 1 | $\frac{-3}{5}$ | $\frac{11}{5}$ | 792 |
| z | 0 | 0 | 66 | 1518 | 1,655,280 |

There are no negative indicators in the bottom row, so an optimum solution has been reached. The maximum profit per year is $1,655,280 reached at $x = 66$ and $y = 792$. The Rouquier company should manufacture 66 units of A and 792 units of B per year.

∎

**EXAMPLE 7**   One of the Rouquier brothers of Example 6 decides that instead of making the two products, they could rent the machine time to another company. Suppose that machine I is rented for $u$ dollars per hour and machine II is rented for $v$ dollars per hour.

**a.** Express the amount of rent the company would receive per year in terms of $u$ and $v$.

**b.** Write the inequalities that must be satisfied for this alternative to be at least as profitable as making the two products.

**c.** Since the Rouquier company is competing with other companies, they must try to minimize the amount of rent they charge while still making a profit. Give a geometric solution to find the minimum total amount of rent they receive each year and how much per hour is charged for each machine.

**Solution**   **a.** Since machine I is rented for 2310 hours at $u$ dollars per hour, the yearly rental for machine I is $2310u$. Similarly, machine II is rented for 990 hours at $v$ dollars per hour, therefore, the yearly rental for machine II is $990v$. The total yearly income from the rental is $(2310u + 990v)$.

**b.** In producing one unit of product A, machine I is used 11 hours and machine II is used 3 hours. The value of the rental for 11 hours of machine I is $11u$. For the 3 hours of machine II, it is $3v$. The value of the rental to produce one unit of product A is $(11u + 3v)$. This amount must be at least as large as the profit the Rouquier brothers would make by producing one unit of product A; otherwise, they should make the product rather than rent the machines. We have

$$11u + 3v \geq 5280$$

Similarly, in producing one unit of product B, machine I is used 2 hours and machine II is used 1 hour. The value of the rental for these 2 hours of machine I is $2u$. For the 1 hour of machine II it is $v$. The value of the rental to produce one unit of product B is $(2u + v)$. This amount must be at least as large as the profit the Rouquier brothers would make by producing one unit of product

B; otherwise, they should make the product rather than rent the machines. We have

$$2u + v \geq 1650$$

**c.** The sum received for the rental per year is $z$ dollars, where $z = 2310u + 990v$. Thus, to be competitive, the Rouquier brothers must minimize $z$ subject to

$$11u + 3v \geq 5280$$
$$2u + \phantom{3}v \geq 1650$$
$$u \geq 0, v \geq 0$$

The geometric solution to this minimization problem appears in Figure 4.27. The region of feasible solutions is shaded. The region of feasible solutions is unbounded. It has three vertices: $(0, 1760)$, $(66, 1518)$, and $(825, 0)$. We evaluate $z$ at each vertex.

At $(0, 1760)$, $z = 2310(0) + 990(1760) = 1{,}742{,}400$,

at $(66, 1518)$, $z = 2310(66) + 990(1518) = 1{,}655{,}280$,

and at $(825, 0)$, $z = 2310(825) + 990(0) = 1{,}905{,}750$.

The smallest of these three values is $1{,}655{,}280$. If $c < 1{,}655{,}280$, the graph of the equation $2310u + 990v = c$ does not intersect the region of feasible solutions. Therefore, the minimum value of $z$ is $1{,}655{,}280$ and it occurs when $u = 66$ and $v = 1518$. Therefore, the Rouquier company should charge \$66 per hour for machine I and \$1518 per hour for machine II. The minimum total amount they can receive for the rent and be as well off as if they made the two products is \$1,655,280. Note that this amount is equal to the maximum profit they can make if they do, in fact, make the two products. Observe that the maximization problem of Example 6 is the dual of the minimization problem of Example 7.

**Do Exercise 34.** ∎

$(0, 1760)$

$(66, 1518)$

$(825, 0)$

$2310u + 990v = c$
$c < 1{,}655{,}280$

**FIGURE 4.27**

### Procedure to Solve a Minimization Problem in Standard Form

**Step 1.** Be sure the problem is in standard form.

**Step 2.** Find the dual maximization problem by

**a.** Writing the given "$\geq$" inequalities as equations together with the equation defining the objective function, which has the dependent variable on the right side and appears last.

**b.** Writing the augmented matrix of the system obtained in **a**.

**c.** Writing the transpose of the matrix obtained in **b**.

**d.** Writing the system of equations corresponding to the matrix obtained in **c**.

**e.** Replacing all equal signs by "$\leq$" in all but the last equation.

**f.** *Not* simplifying any of the constraint inequalities in the dual maximization problem just obtained.

**Step 3.**  Use the simplex method to get the final tableau of the maximization problem obtained in Step 2.

**Step 4.**  Read the solution to the original minimization problem as follows. The minimum value of the objective function is found in the lower right corner of the final tableau obtained in Step 3. This minimum value occurs at a point whose coordinates are read in the bottom row and in the columns corresponding to the slack variables of the same final tableau.

# Exercise Set 4.6

*In Exercises 1–33, use the method described in the procedure at the end of this section to solve a minimization problem.*

**1.** Minimize $z = 45x + 27y$ subject to

$$x + y \geq 45$$
$$3x + y \geq 81$$
$$x \geq 0, y \geq 0$$

**2.** Minimize $z = 12x + 16y$ subject to

$$x + 2y \geq 40$$
$$3x + 2y \geq 72$$
$$x \geq 0, y \geq 0$$

**3.** Minimize $z = 57x + 133y$ subject to

$$2x + 5y \geq 210$$
$$5x + 3y \geq 240$$
$$x \geq 0, y \geq 0$$

**4.** Minimize $z = 10x + 4y$ subject to

$$-x + y \geq 2$$
$$11x - 4y \geq 55$$
$$x \geq 0, y \geq 0$$

**5.** Minimize $z = 27x + 20y$ subject to

$$4x + 5y \geq 260$$
$$3x + y \geq 96$$
$$x \geq 0, y \geq 0$$

**6.** Minimize $z = 10x + 18y$ subject to

$$x + y \geq 15$$
$$x + 3y \geq 21$$
$$x \geq 0, y \geq 0$$

**7.** Minimize $z = 16x + 25y$ subject to

$$2x + y \geq 7$$
$$x + y \geq 5$$
$$2x + 5y \geq 16$$
$$x \geq 0, y \geq 0$$

**8.** Minimize $z = 1974x + 1470y$ subject to

$$7x + 3y \geq 210$$
$$x + 7y \geq 140$$
$$2x + 5y \geq 118$$
$$x \geq 0, y \geq 0$$

**9.** Minimize $w = 210x + 48y + 45z$ subject to

$$4x + y + z \geq 8$$
$$y \geq 9$$
$$5x \geq 6$$
$$x \geq 0, y \geq 0, z \geq 0$$

**10.** Minimize $w = 7x + 7y + 14z$ subject to

$$4x + y - z \geq 2$$
$$3x - y - z \geq -24$$
$$-x - y + z \geq 8$$
$$x \geq 0, y \geq 0, z \geq 0$$

**11.** Minimize $w = 66x + 72y + 108z$ subject to

$$x - y + 2z \geq 5$$
$$x + 3y + z \geq 1$$
$$x + 2y + 3z \geq 1$$
$$x \geq 0, y \geq 0, z \geq 0$$

**12.** Minimize $w = 168x + 280y + 210z$ subject to

$$x + 2y + z \geq 14$$
$$2x + y + z \geq 14$$
$$x + y + 3z \geq 14$$
$$x \geq 0, y \geq 0, z \geq 0$$

**13.** Minimize $z = 200x + 750y$ subject to

$$x + 4y \geq 20$$
$$4x + 6y \geq 50$$
$$2x + 5y \geq 30$$
$$x \geq 0, y \geq 0$$

**14.** Minimize $w = 24x + 12y$ subject to

$$2x + y + 2z \geq 3$$
$$x + 2y \geq 6$$
$$3x + z \geq 3$$
$$x \geq 0, y \geq 0, z \geq 0$$

**15.** Minimize $z = 80x + 60y$ subject to

$$x + y \geq 10$$
$$5x + 3y \geq 20$$
$$2x + 4y \geq 15$$
$$x \geq 0, y \geq 0$$

**16.** Minimize $w = 40x + 20y + 30z$ subject to

$$x + 2y + z \geq 3$$
$$x + y - z \geq 6$$
$$-x + 3y - 2z \leq 3$$
$$x \geq 0, y \geq 0, z \geq 0$$

**17.** Minimize $w = 420x + 910y + 560z$ subject to

$$x + 2y + \phantom{5}z \geq 2$$
$$5x + \phantom{2}y + 7z \geq 24$$
$$2x + 3y \phantom{+ 7z} \geq 4$$
$$x \geq 0,\, y \geq 0,\, z \geq 0$$

**18.** Minimize $w = 60x + 72y + 48z$ subject to

$$x + y \phantom{+ 7z} \geq 3$$
$$x + y + \phantom{3}z \geq 6$$
$$3y + 3z \geq 15$$
$$x \geq 0,\, y \geq 0,\, z \geq 0$$

**19.** Minimize $z = 16x_1 + 28x_2 + 8x_3 + 20x_4$ subject to

$$x_1 + \phantom{2}x_2 \phantom{+ 8x_3} \geq 3$$
$$x_1 - 2x_2 \phantom{+ 8x_3} \geq 0$$
$$x_3 + x_4 \geq 2$$
$$-2x_3 + x_4 \geq 0$$
$$x_1 \geq 0,\, x_2 \geq 0,$$
$$x_3 \geq 0,\, x_4 \geq 0$$

**20.** Minimize $z = 10x + 5y$ subject to

$$3x + \phantom{2}y \geq 5$$
$$x + 2y \geq 10$$
$$2x + \phantom{2}y \geq 5$$
$$x + \phantom{2}y \geq 25$$
$$x \geq 0,\, y \geq 0$$

**21.** Minimize $w = 8x + 7y + 9z$ subject to

$$x + 2y + 3z \geq 2$$
$$x - \phantom{2}y - \phantom{2}z \geq 3$$
$$x - \phantom{2}y - 2z \leq 1$$
$$2x + \phantom{2}y + \phantom{2}z \geq 5$$
$$x \geq 0,\, y \geq 0,\, z \geq 0$$

**22.** A publishing company is organizing a fishing trip for its 200 employees. Small and large boats are available at costs of $300 and $1200 each, respectively. A small boat can take six employees and requires one skipper, while a large boat can take 20 employees but requires one skipper and two baiters. The charter company has, at most, 31 people available who can be skippers or baiters. How many small boats and large boats should the publishing company charter to minimize the cost? What is the minimum cost?

**23.** A politician hired a consultant who advised him that for each dollar spent on radio advertising, 10 Republicans and 30 Democrats can be reached; while for each dollar spent on television advertising, 20 Republicans and 20 Democrats can be reached. The politician wishes to reach at least 700,000 Republicans and 1,200,000 Democrats. Find the amounts the politician should spend on television and radio advertising to reach his goal at a minimum cost. What is the minimum cost?

**24.** A Jesuit university found that for each dollar spent advertising in a nonreligious publication, 40 Catholic and 80 non-Catholic families can be reached. For each dollar spent on advertising in a Catholic publication, 100 Catholic and 30 non-Catholic families can be reached. The university wants to reach at least 330,000 Catholic and 235,000 non-Catholic

families. How much should be spent on advertising in each publication to reach the goal at a minimum cost? What is the minimum cost?

**25.** A corporation produces chemicals X, Y, and Z at sites A, B, and C. The daily production of each chemical at each of the sites is given in the following table, where the units are tons.

|  | Site A | Site B | Site C |
|---|---|---|---|
| Chemical X | 6 | 3 | 3 |
| Chemical Y | 5 | 5 | 5 |
| Chemical Z | 4 | 8 | 4 |

The corporation has received orders of 270 tons of X, 300 tons of Y, and 288 tons of Z. The daily cost of operation for each site is $30,000, $36,000, and $27,000, respectively. How many days should the corporation operate each site to be able to fill the orders at a minimum operating cost? What is the minimum operating cost?

**26.** A dietitian has to plan a special diet for athletes using two types of food, A and B. The numbers of units of iron, calcium, vitamin A, and cholesterol contained in each ounce of each type of food are given in the following table.

|  | Food A | Food B |
|---|---|---|
| Iron | 18 | 9 |
| Calcium | 24 | 20 |
| Vitamin A | 7 | 35 |
| Cholesterol | 9 | 5 |

The daily minimum requirements for an athlete are 180 units of iron, 320 units of calcium, and 210 units of vitamin A. How many ounces of each type of food should be used daily to minimize the intake of cholesterol? What is the minimum intake of cholesterol per day?

**27.** A dietitian in a school cafeteria has to plan a diet for school children using foods X, Y, and Z. The numbers of units of calcium, vitamin B, iron, and cholesterol in each ounce of each type of food are given in the following table.

|  | Food X | Food Y | Food Z |
|---|---|---|---|
| Calcium | 15 | 10 | 10 |
| Vitamin B | 8 | 6 | 2 |
| Iron | 10 | 6 | 2 |
| Cholesterol | 16 | 8 | 4 |

The minimum daily requirements for each child are 160 units of calcium, 56 units of vitamin B, and 48 units of iron. How many ounces of each of the three types of food should be mixed for each child to minimize the daily intake of cholesterol? What is the minimum daily intake of cholesterol?

**28.** A dietician in the local hospital is planning meals for patients using foods I, II, and III. The amounts of protein and calcium, and the costs per unit of each type of food are given in the following table.

| | Type I | Type II | Type III |
|---|---|---|---|
| Protein | 8g | 4g | 20g |
| Calcium | 80mg | 20mg | 20mg |
| Cost | 40c | 12c | 40c |

The minimum requirements for each patient are 80 grams of protein and 600 milligrams of calcium. How many units of each type of food should be used for each patient to minimize the total cost of food per patient per day? What is that minimum cost?

**29.** A dairy company pasteurizes milk at three different sites: Arlington, Bellingham, and Mount Vernon. The daily production of 2% and 4% milk (in thousands of gallons) as well as the daily cost of operation are given for each site in the table.

| | 2% | 4% | Daily cost in dollars |
|---|---|---|---|
| Arlington | 10 | 20 | 20,000 |
| Bellingham | 10 | 10 | 14,000 |
| Mount Vernon | 20 | 10 | 24,000 |

Each month, the company must fill orders of 630,000 gallons of 2% and 360,000 gallons of 4% milk. How many days per month should each site operate to minimize the total operating cost per month? What is the minimum total monthly operating cost?

**30.** The dean of the business school at a local university is faced with the following problem. University policies dictate that during an academic year, each assistant professor teaches 12 credits, each associate professor teaches 10 credits, and each full professor teaches 8 credits, and that the number of assistant professors be at least twice the number of full professors. On the average, each assistant professor publishes one research paper per year, each associate professor publishes two research papers per year, and each full professor publishes three research papers per year. The enrollment in the school creates a need of at least 470 credits taught per academic year. Because of accreditation, the school's faculty is expected to publish a total of at least 80 research papers per academic year. The average salaries per academic year for assistant professors, associate professors, and full professors are $20,000, $33,000, and $45,000, respectively. How many professors should be at each rank to minimize the teaching budget of the school? What is the minimum teaching budget?

**31.** An electronics company makes its VCRs at sites A and B and has two distributors I and II. Site A can manufacture at most 750 VCRs per week, while site B can manufacture at

most 600 VCRs a week. Distributors I and II must receive at least 450 and 750 VCRs each week, respectively. The transportation costs per VCR from each site to each distributor are given in the following table.

| From | To | |
|---|---|---|
| | Distributor I | Distributor II |
| Site A | $6.00 | $12.00 |
| Site B | $15.00 | $9.00 |

How many VCRs should be shipped each week from each site to each distributor to minimize the cost of transportation? What is the minimum cost?

**32.** A company manufactures its refrigerators at plants A and B and has two distributors I and II. Plant A can manufacture at most 48 refrigerators per week, while plant B can manufacture at most 100 refrigerators a week. Distributors I and II must receive at least 60 and 80 refrigerators each week, respectively. The transportation costs per refrigerator from each plant to each distributor are given in the following table.

| From | To | |
|---|---|---|
| | Distributor I | Distributor II |
| Plant A | $20.00 | $25.00 |
| Plant B | $30.00 | $25.00 |

How many refrigerators should be shipped each week from each plant to each distributor to minimize the total cost of transportation? What is the minimum cost?

**33.** An RV manufacturer builds its motor homes at factories A and B and has two distributors I and II. Factory A can manufacture at most 25 motor homes a month, while factory B can manufacture 25 motor homes a month at most. Distributors I and II must receive at least 20 and 30 motor homes each month, respectively. The transportation costs per motor home from each factory to each distributor are given in the following table.

| From | To | |
|---|---|---|
| | Distributor I | Distributor II |
| Factory A | $200.00 | $250.00 |
| Factory B | $225.00 | $175.00 |

How many motor homes should be shipped each week from each factory to each distributor to minimize the cost of transportation? What is that minimum cost?

**34.** Verify that the maximization problem of Example 6 of this section is the dual of the minimization problem of Example 7. Also check that the coordinates of the point where the function of Example 7 attains its minimum appear in the bottom row of the final tableau of Example 6.

# 4.7  The Big *M* Method

In the preceding sections, we presented methods for solving maximization and minimization problems in standard form. For a linear programming problem in standard form, we always begin with a basic feasible solution in which all decision variables are 0. However, for a maximization problem which is not in standard form, such a basic feasible solution may not exist. In this section, we show how the simplex method may be used in such cases. We begin with a simple example which we first solve geometrically. We shall use this example to illustrate the idea behind the new method.

**EXAMPLE 1**    Maximize $P = 3x + 2y$ subject to

$$x - y \le 1$$
$$x + y \le 9$$
$$2x + y \ge 5$$
$$x \ge 0, y \ge 0$$

**Solution**    We first graph the equations

$$x - y = 1$$
$$x + y = 9$$
$$2x + y = 5$$

(See Figure 4.28.) We then shade the region that represents the set of feasible solutions. The vertices of this region are (2, 1), (5, 4), (0, 9), and (0, 5). We find that

$P = 8$ at (2, 1) since $3(2) + 2(1) = 8$
$P = 23$ at (5, 4) since $3(5) + 2(4) = 23$
$P = 18$ at (0, 9) since $3(0) + 2(9) = 18$
$P = 10$ at (0, 5) since $3(0) + 2(5) = 10$

Hence, the maximum value of $P$ is 23 and occurs at (5, 4).    ∎

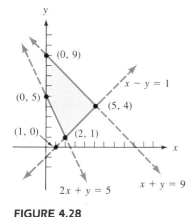

**FIGURE 4.28**

## The Modified Problem

We cannot start the simplex method as we did in the standard problem by letting the decision variables be 0, since, as we see in Figure 4.28, the point (0, 0) is not in the region of feasible solutions. We must find some way to get to one of the vertices of the region of feasible solutions. As we often do in mathematics, we modify our problem so that the techniques already available may apply. We first introduce slack variables to change the "≤" constraints to equalities. We obtain

$$x - y + s_1 = 1, \text{ where } s_1 \ge 0$$

and

$$x + y + s_2 = 9, \text{ where } s_2 \ge 0$$

We now introduce a *surplus variable* to change the "≥" constraint to an equality. At this time, it serves our purpose best to name the surplus variable $z$, and we write

$$2x + y - z = 5, \text{ where } z \geq 0$$

If we modify our problem, thinking of $z$ as a new decision variable, so that the constraints are

$$x - y \leq 1$$
$$x + y \leq 9$$
$$2x + y - z \leq 5$$
$$x \geq 0, \, y \geq 0, \, z \geq 0$$

we have a problem in standard form. The region of feasible solutions of the modified problem is sketched in Figure 4.29, where in the $xy$-plane we also show the region of feasible solutions of the original problem. Note that the edge connecting the points $A$ and $B$ belongs to both regions. Adding slack variables to change the "≤" to equalities, we obtain

$$x - y + s_1 = 1$$
$$x + y + s_2 = 9$$
$$2x + y - z + s_3 = 5$$
$$x \geq 0, \, y \geq 0, \, z \geq 0, \, s_1 \geq 0, \, s_2 \geq 0, \, s_3 \geq 0$$

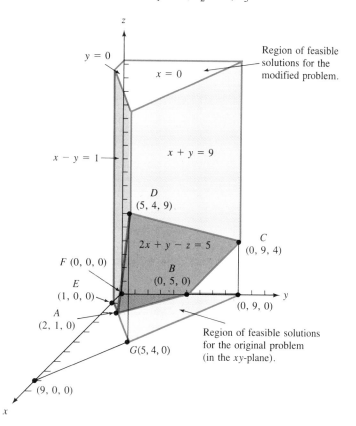

**FIGURE 4.29**

The slack variable $s_3$ introduced here is called an *artificial variable* for the original problem. In Figure 4.29, the region of feasible solutions for the modified problem is an unbounded region and its vertices are the points $A, B, C, D, E, F$, with coordinates $(2, 1, 0)$, $(0, 5, 0)$, $(0, 9, 4)$, $(5, 4, 9)$, $(1, 0, 0)$, and $(0, 0, 0)$, respectively. The points $A$ and $B$ and the projections of the points $C$ and $D$ onto the $xy$-plane are vertices of the region of feasible solutions of the original problem. For this reason, it is desirable for the optimum solution of the modified problem to occur at one of the four points $A, B, C,$ or $D$. That is, to occur on the plane whose equation is $2x + y - z = 5$. Hence, *it is desirable for the value of the artificial variable $s_3$ to be 0*, since the constraint $2x + y - z + s_3 = 5$ must be satisfied.

At this point, we cannot expect the original objective function to attain its maximum value at one of the points $A, B, C,$ or $D$. We must, therefore, modify the objective function also. Since we wish to have the new objective function attain its maximum value at a point where the value of the artificial variable is 0, we impose a very large "penalty," $M$, for each unit value of the artificial variable. That is, we subtract $s_3 M$ from the value of the objective function where M is a very large positive number. The modified objective function is defined by

$$P_M = 3x + 2y - s_3 M$$

It is clear that if $P_M$ has a maximum value, it must be at a point where $s_3 = 0$.

We now solve the modified problem using the simplex method. We shall then show how this solution is related to the solution of the original problem. Finally, we shall describe the general method to solve linear programming problems that are not in standard form.

**EXAMPLE 2**    Use the simplex method to solve the modified problem. That is, maximize $P_M$ if $P_M = 3x + 2y - s_3 M$, where

$$x - y + s_1 = 1$$
$$x + y + s_2 = 9$$
$$2x + y - z + s_3 = 5$$
$$x \geq 0, y \geq 0, z \geq 0, s_1 \geq 0, s_2 \geq 0, s_3 \geq 0$$

**Solution**    We proceed as we did in Section 4.5. We write the equation defining the objective function as

$$P_M - 3x - 2y + s_3 M = 0$$

The first tableau is

**Tableau I**

| Basis | $x$ | $y$ | $z$ | $s_1$ | $s_2$ | $s_3$ | Values of basic variables |
|-------|-----|-----|-----|-------|-------|-------|---------------------------|
| $s_1$ | 1 | $-1$ | 0 | 1 | 0 | 0 | 1 |
| $s_2$ | 1 | 1 | 0 | 0 | 1 | 0 | 9 |
| $s_3$ | 2 | 1 | $-1$ | 0 | 0 | 1 | 5 |
| $P_M$ | $-3$ | $-2$ | 0 | 0 | 0 | $M$ | 0 |

Here $x = y = z = 0$, $s_1 = 1$, $s_2 = 9$, and $s_3 = 5$. Note first that $(0, 0, 0)$ is a feasible solution for the modified problem (See Figure 4.29), but the corresponding point $(0, 0)$ is not a feasible solution for the original problem. We observe that the value of the new objective function is $-5M$ since $3(0) + 2(0) - 5(M) = -5M$. This is not the value that is in the lower right corner of the tableau. We modify this by changing the $M$ at the bottom of the $s_3$ column to 0. $((-M)R_3 + R_4 \rightarrow R_4)$. We obtain

**Tableau II**

| Basis | $x$ | $y$ | $z$ | $s_1$ | $s_2$ | $s_3$ | Values of basic variables |
|---|---|---|---|---|---|---|---|
| $s_1$ | ① | $-1$ | 0 | 1 | 0 | 0 | 1 |
| $s_2$ | 1 | 1 | 0 | 0 | 1 | 0 | 9 |
| $s_3$ | 2 | 1 | 1 | 0 | 0 | 1 | 5 |
| $P_M$ | $-2M - 3$ | $-M - 2$ | $M$ | 0 | 0 | 0 | $-5M$ |

Since $M$ is a very large positive number, $-2M - 3$ is the most negative indicator in the bottom row of the tableau. Thus, we choose $x$ as the entering variable. The quotient $\frac{1}{1}$ is smaller than both quotients $\frac{9}{1}$ and $\frac{5}{2}$. So we choose $s_1$ as the leaving variable. Pivoting, we obtain

**Tableau III**

| Basis | $x$ | $y$ | $z$ | $s_1$ | $s_2$ | $s_3$ | Values of basic variables |
|---|---|---|---|---|---|---|---|
| $x$ | 1 | $-1$ | 0 | 1 | 0 | 0 | 1 |
| $s_2$ | 0 | 2 | 0 | $-1$ | 1 | 0 | 8 |
| $s_3$ | 0 | ③ | $-1$ | $-2$ | 0 | 1 | 3 |
| $P_M$ | 0 | $-3M - 5$ | $M$ | $2M + 3$ | 0 | 0 | $-3M + 3$ |

Here $x = 1$, $y = z = s_1 = 0$, $s_2 = 8$, and $s_3 = 3$. The point $E$ with coordinates $(1, 0, 0)$ is a corner of the region of feasible solutions for the modified problem but is not in the region of feasible solutions of the original problem. (See Figures 4.28 and 4.29, where in Figure 4.28 the point $E$ has coordinates $(1, 0)$.) The most negative indicator in the bottom row is $-3M - 5$. We choose $y$ as the entering variable and $s_3$ as the leaving variable since $\frac{3}{3}$ is less than $\frac{8}{2}$. Pivoting, we obtain

**Tableau IV**

| Basis | $x$ | $y$ | $z$ | $s_1$ | $s_2$ | $s_3$ | Values of basic variables |
|---|---|---|---|---|---|---|---|
| $x$ | 1 | 0 | $\frac{-1}{3}$ | $\frac{1}{3}$ | 0 | $\frac{1}{3}$ | 2 |
| $s_2$ | 0 | 0 | $\frac{2}{3}$ | $\frac{1}{3}$ | 1 | $\frac{-2}{3}$ | 6 |
| $y$ | 0 | 1 | $\frac{-1}{3}$ | $\frac{-2}{3}$ | 0 | $\frac{1}{3}$ | 1 |
| $P_M$ | 0 | 0 | $\frac{-5}{3}$ | $\frac{-1}{3}$ | 0 | $M + \frac{5}{3}$ | 8 |

So, $x = 2$, $y = 1$, $z = 0$, $s_1 = 0$, $s_2 = 6$, and $s_3 = 0$. The point $A$ with coordinates $(2, 1, 0)$ is a corner of the region of feasible solutions of the modified problem. (See Figure 4.29.) It is also a corner of the region of feasible solutions of the original problem. In the $xy$-plane the point $A$ has coordinates $(2, 1)$. We have reached a point which is on the common edge of both regions. The value of the artificial variable $s_3$ is 0. Therefore, we no longer need to carry the $s_3$ column in succeeding tableaus. We delete that column and obtain the simplified tableau.

**Tableau V**

| Basis | $x$ | $y$ | $z$ | $s_1$ | $s_2$ | Values of basic variables |
|---|---|---|---|---|---|---|
| $x$ | 1 | 0 | $\frac{-1}{3}$ | $\frac{1}{3}$ | 0 | 2 |
| $s_2$ | 0 | 0 | $\left(\frac{2}{3}\right)$ | $\frac{1}{3}$ | 1 | 6 |
| $y$ | 0 | 1 | $\frac{-1}{3}$ | $\frac{-2}{3}$ | 0 | 1 |
| $P_M$ | 0 | 0 | $\frac{-5}{3}$ | $\frac{-1}{3}$ | 0 | 8 |

Clearly, $z$ and $s_2$ should be entering and leaving variables, respectively. Pivoting, we obtain

**Tableau VI**

| Basis | $x$ | $y$ | $z$ | $s_1$ | $s_2$ | Values of basic variables |
|---|---|---|---|---|---|---|
| $x$ | 1 | 0 | 0 | $\frac{1}{2}$ | $\frac{1}{2}$ | 5 |
| $z$ | 0 | 0 | 1 | $\frac{1}{2}$ | $\frac{3}{2}$ | 9 |
| $y$ | 0 | 1 | 0 | $\frac{-1}{2}$ | $\frac{1}{2}$ | 4 |
| $P_M$ | 0 | 0 | 0 | $\frac{1}{2}$ | $\frac{5}{2}$ | 23 |

Here $x = 5$, $y = 4$, $z = 9$, and $s_1 = s_2 = 0$. The point $D$ with coordinates $(5, 4, 9)$ is a corner of the region of feasible solutions of the modified problem. Also its projection onto the $xy$-plane is the point $G$, it has coordinates $(5, 4)$ and is a corner of the region of feasible solutions of the original problem. There are no negative indicators in the bottom row. Thus, the maximum value of the objective function has been attained. It is read in the lower right corner of the final tableau. The maximum value of $P_M$ is 23. It turns out that 23 is also the maximum value of the objective function of the original problem. To see this, suppose that we could find feasible values of $x$, $y$, $z$, $s_1$, and $s_2$, which would be solutions of the original system and would yield a value of $P$ larger than 23. Using these values along with $s_3 = 0$, we would get a value of $P_M$ larger than 23, contradicting the fact that we found the maximum value of $P_M$ to be 23. ■

The previous example serves as an illustration of the steps that we follow to solve a linear programming problem that is not in standard form using the Big *M* Method.

### Procedure: The Big *M* Method

**Step 1.** Make sure that all constraints have nonnegative constants on the right side. If necessary, multiply both sides of a constraint by $-1$, remembering to reverse the sense of the inequality sign if the constraint is an inequality.

**Step 2.** Add a slack variable $s_i$ on the left side of each "$\leq$" constraint to obtain an equality.

**Step 3.** Subtract a surplus variable $s_j$ and add an artificial variable $a_i$ on the left side of each "$\geq$" constraint to obtain an equality.

**Step 4.** Add an artificial variable $a_j$ on the left side of each equality constraint.

**Step 5.** For each artificial variable $a_i$ which was added, subtract $Ma_i$ from the right side of the equation defining the objective function $P$. Use the same constant $M$ for all artificial variables. Rename the objective function $P_M$ to obtain the modified problem. *All constraint, slack, surplus, and artificial variables must be nonnegative.*

**Step 6.** Form the preliminary simplex tableau for the modified problem obtained in the first five steps.

**Step 7.** Eliminate the $M$'s in the bottom row of the preliminary simplex tableau from all columns which correspond to artificial variables by *adding all the rows corresponding to artificial variables, multiplying the result by* $-M$, *and adding the product to the bottom row*. At this stage, all artificial variables are basic variables. In this way, we obtain the initial simplex tableau.

**Step 8.** Solve the modified problem using the simplex method on the tableau obtained in Step 7. A tableau may be simplified as soon as all artificial variables have the value 0. We can then delete all columns corresponding to the artificial variables and continue until the solution of the modified problem is complete.

**Step 9.** The solution of the original problem is related to that of the modified problem as follows:

   **a.** If the modified problem has no solution, neither does the original problem.
   **b.** If we obtain a solution of the modified problem where one of the artificial variables has a positive value, the original problem has no solution.
   **c.** If we obtain a solution of the modified problem where all artificial variables are 0, then the solution of the original problem is obtained by deleting all artificial variables in the solution of the modified problem.

We now give several examples to illustrate the Big *M* Method.

**EXAMPLE 3** Formulate the modified problem for the following, but do not solve the modified problem. Maximize $P = 2x + 3y - 4z$ subject to

$$2x + 5y - 7z \leq 2$$
$$x - 2y - 3z \leq -4$$
$$x + 4y + 5z = 20$$
$$2x + 3y - z = -3$$
$$x \geq 0, y \geq 0, z \geq 0$$

**Solution** In order to have positive constants on the right sides of the first four constraints, multiply both sides of the second inequality by $-1$, reversing the sense of the inequality. Also multiply both sides of the second equality by $-1$. The constraints are written

$$2x + 5y - 7z \leq 2$$
$$-x + 2y + 3z \geq 4$$
$$x + 4y + 5z = 20$$
$$-2x - 3y + z = 3$$
$$x \geq 0, y \geq 0, z \geq 0$$

Now add a slack variable $s_1$ to the left side of the first inequality. Subtract a surplus variable $s_2$ from, and add an artificial variable $a_1$ to, the left side of the second inequality. Also add artificial variables $a_2$ and $a_3$ to the left sides of the first and second equalities, respectively. Write the first four constraints as equalities. We obtain

$$2x + 5y - 7z + s_1 \qquad = 2$$
$$-x + 2y + 3z - s_2 + a_1 = 4$$
$$x + 4y + 5z \qquad + a_2 = 20$$
$$-2x - 3y + z \qquad + a_3 = 3$$
$$x \geq 0, y \geq 0, z \geq 0, s_1 \geq 0, s_2 \geq 0, a_1 \geq 0, a_2 \geq 0, a_3 \geq 0$$

We subtract $(a_1 + a_2 + a_3)M$ from the right side of the equation defining the objective function and rename the objective function $P_M$. We get

$$P_M = 2x + 3y - 4z - a_1M - a_2M - a_3M$$

The modified problem is to maximize $P_M$ subject to

$$2x + 5y - 7z + s_1 \qquad = 2$$
$$-x + 2y + 3z - s_2 + a_1 = 4$$
$$x + 4y + 5z \qquad + a_2 = 20$$
$$-2x - 3y + z \qquad + a_3 = 3$$
$$x \geq 0, y \geq 0, z \geq 0, s_1 \geq 0, s_2 \geq 0, a_1 \geq 0, a_2 \geq 0, a_3 \geq 0$$

**EXAMPLE 4**  Maximize $P = 5x + 8y + z$ subject to

$$2x + 3y + z \leq 14$$
$$4x + 3y + z \geq 18$$
$$x + 2y + z = 9$$
$$x \geq 0, y \geq 0, z \geq 0$$

**Solution**  First we introduce a slack variable $s_1$ in the first constraint, a surplus variable $s_2$ and an artificial variable $a_1$ in the second constraint, and an artificial variable $a_2$ in the third constraint. Then we subtract $M(a_1 + a_2)$ from the right side of the equality defining the objective function $P$. We rename the objective function $P_M$ and obtain the modified problem. Maximize $P_M = 5x + 8y + z - a_1M - a_2M$ subject to

$$2x + 3y + z + s_1 \qquad\quad = 14$$
$$4x + 3y + z - s_2 + a_1 = 18$$
$$x + 2y + z \qquad\quad + a_2 = 9$$
$$x \geq 0, y \geq 0, z \geq 0, s_1 \geq 0, s_2 \geq 0, a_1 \geq 0, \text{ and } a_2 \geq 0$$

The equation defining the objective function is written

$$P_M - 5x - 8y - z + a_1M + a_2M = 0$$

The preliminary simplex tableau for the modified problem is

**Tableau I**

| Basis | $x$ | $y$ | $z$ | $s_1$ | $s_2$ | $a_1$ | $a_2$ | Values of basic variables |
|-------|-----|-----|-----|-------|-------|-------|-------|---------------------------|
| $s_1$ | 2 | 3 | 1 | 1 | 0 | 0 | 0 | 14 |
| $a_1$ | 4 | 3 | 1 | 0 | −1 | 1 | 0 | 18 |
| $a_2$ | 1 | 2 | 1 | 0 | 0 | 0 | 1 | 9 |
| $P_M$ | −5 | −8 | −1 | 0 | 0 | $M$ | $M$ | 0 |

Now add rows 2 and 3, corresponding to the artificial variables and multiply the result by $-M$. Add the product to row 4 to obtain the initial simplex tableau for the modified problem.

**Tableau II**

| Basis | $x$ | $y$ | $z$ | $s_1$ | $s_2$ | $a_1$ | $a_2$ | Values of basic variables |
|-------|-----|-----|-----|-------|-------|-------|-------|---------------------------|
| $s_1$ | 2 | 3 | 1 | 1 | 0 | 0 | 0 | 14 |
| $a_1$ | 4 | 3 | 1 | 0 | −1 | 1 | 0 | 18 |
| ← $a_2$ | 1 | ② | 1 | 0 | 0 | 0 | 1 | 9 |
| $P_M$ | $-5M - 5$ | $-5M - 8$ | $-2M - 1$ | 0 | $M$ | 0 | 0 | $-27M$ |

↑

The most negative indicator is $-5M - 8$. Therefore, $y$ is the entering variable. Since $\frac{9}{2}$ is less than $\frac{18}{3}$ and $\frac{14}{3}$, $a_2$ is the leaving variable. Pivoting, we obtain

### Tableau III

| | Basis | $x$ | $y$ | $z$ | $s_1$ | $s_2$ | $a_1$ | $a_2$ | Values of basic variables |
|---|---|---|---|---|---|---|---|---|---|
| $\leftarrow$ | $s_1$ | $\tfrac{1}{2}$ | 0 | $\tfrac{-1}{2}$ | 1 | 0 | 0 | $\tfrac{-3}{2}$ | $\tfrac{1}{2}$ |
| | $a_1$ | $\tfrac{5}{2}$ | 0 | $\tfrac{-1}{2}$ | 0 | $-1$ | 1 | $\tfrac{-3}{2}$ | $\tfrac{9}{2}$ |
| | $y$ | $\tfrac{1}{2}$ | 1 | $\tfrac{1}{2}$ | 0 | 0 | 0 | $\tfrac{1}{2}$ | $\tfrac{9}{2}$ |
| | $P_M$ | $\tfrac{-5M}{2} - 1$ | 0 | $\tfrac{M}{2} + 3$ | 0 | $M$ | 0 | $\tfrac{5M}{2} + 4$ | $\tfrac{-9M}{2} + 36$ |

The only negative indicator is $\frac{-5M}{2} - 1$. Thus, $x$ is the entering variable. Since $\frac{1}{2}/\frac{1}{2}$ is less than $\frac{9}{2}/\frac{5}{2}$ and $\frac{9}{2}/\frac{1}{2}$, $s_1$ is the leaving variable. Pivoting, we obtain

### Tableau IV

| | Basis | $x$ | $y$ | $z$ | $s_1$ | $s_2$ | $a_1$ | $a_2$ | Values of basic variables |
|---|---|---|---|---|---|---|---|---|---|
| | $x$ | 1 | 0 | $-1$ | 2 | 0 | 0 | $-3$ | 1 |
| $\leftarrow$ | $a_1$ | 0 | 0 | 2 | $-5$ | $-1$ | 1 | ⑥ | 2 |
| | $y$ | 0 | 1 | 1 | $-1$ | 0 | 0 | 2 | 4 |
| | $P_M$ | 0 | 0 | $-2M + 2$ | $5M + 2$ | $M$ | 0 | $-5M + 1$ | $-2M + 37$ |

The most negative indicator is $-5M + 1$. Thus, $a_2$ is the entering variable. Since $\frac{2}{6}$ is less than $\frac{4}{2}$, $a_1$ is the leaving variable. Pivoting, we obtain

### Tableau V

| | Basis | $x$ | $y$ | $z$ | $s_1$ | $s_2$ | $a_1$ | $a_2$ | Values of basic variables |
|---|---|---|---|---|---|---|---|---|---|
| | $x$ | 1 | 0 | 0 | $\tfrac{-1}{2}$ | $\tfrac{-1}{2}$ | $\tfrac{1}{2}$ | 0 | 2 |
| $\leftarrow$ | $a_2$ | 0 | 0 | $\tfrac{1}{3}$ | $\tfrac{-5}{6}$ | $\tfrac{-1}{6}$ | $\tfrac{1}{6}$ | 1 | $\tfrac{1}{3}$ |
| | $y$ | 0 | 1 | $\tfrac{1}{3}$ | $\tfrac{2}{3}$ | $\tfrac{1}{3}$ | $\tfrac{-1}{3}$ | 0 | $\tfrac{10}{3}$ |
| | $P_M$ | 0 | 0 | $\tfrac{-M + 5}{3}$ | $\tfrac{5M + 17}{6}$ | $\tfrac{M + 1}{6}$ | $\tfrac{5M - 1}{6}$ | 0 | $\tfrac{-M + 110}{3}$ |

Since $\frac{-M + 5}{3}$ is the only negative indicator, $z$ is the entering variable. Comparing ratios, we see that $a_2$ should be the leaving variable. Pivoting, we get

**Tableau VI**

| Basis | $x$ | $y$ | $z$ | $s_1$ | $s_2$ | $a_1$ | $a_2$ | Values of basic variables |
|-------|-----|-----|-----|-------|-------|-------|-------|---------------------------|
| $x$   | 1   | 0   | 0   | $\frac{-1}{2}$ | $\frac{-1}{2}$ | $\frac{1}{2}$ | 0   | 2  |
| $z$   | 0   | 0   | 1   | $\frac{-5}{2}$ | $\frac{-1}{2}$ | $\frac{1}{2}$ | 3   | 1  |
| $y$   | 0   | 1   | 0   | $\frac{3}{2}$  | $\frac{1}{2}$  | $\frac{-1}{2}$ | $-1$ | 3  |
| $P_M$ | 0   | 0   | 0   | 7 | 1 | $M - 1$ | $M - 5$ | 35 |

All indicators are positive. Thus, 35 is the maximum value of $P_M$. It occurs where $x = 2$, $y = 3$, $z = 1$, and $s_1 = s_2 = a_1 = a_2 = 0$. Since both artificial variables are 0, 35 is also the maximum value of $P$. It occurs at the point $(2, 3, 1)$.
**Do Exercise 5.**   ■

**EXAMPLE 5**   Maximize $P = 5x + 6y$ subject to

$$2x + 3y \le 6$$
$$4x + 5y \ge 20$$
$$x \ge 0,\ y \ge 0$$

**Solution**   Add a slack variable $s_1$ to the left side of the first constraint, subtract a surplus variable $s_2$ and add an artificial variable $a_1$ to the left side of the second constraint to change these constraints to equalities. Next, subtract $a_1 M$ from the right side of the equation defining the objective function and rename the objective function $P_M$ to get the modified problem. Maximize $P_M = 5x + 6y - a_1 M$ subject to

$$2x + 3y + s_1 \qquad = 6$$
$$4x + 5y - s_2 + a_1 = 20$$
$$x \ge 0,\ y \ge 0,\ s_1 \ge 0,\ s_2 \ge 0,\ a_1 \ge 0$$

We write the equation defining the objective function as

$$P_M - 5x - 6y + a_1 M = 0$$

The preliminary tableau for the modified problem is

**Tableau I**

| Basis | $x$ | $y$ | $s_1$ | $s_2$ | $a_1$ | Values of basic variables |
|-------|-----|-----|-------|-------|-------|---------------------------|
| $s_1$ | 2   | 3   | 1     | 0     | 0     | 6  |
| $a_1$ | 4   | 5   | 0     | $-1$  | 1     | 20 |
| $P_M$ | $-5$ | $-6$ | 0   | 0     | $M$   | 0  |

Multiply row 2 by $-M$ and add the result to row 3 to change the bottom entry of the column corresponding to $a_1$ to 0. We obtain the initial simplex tableau for the modified problem.

**Tableau II**

| Basis | $x$ | $y$ | $s_1$ | $s_2$ | $a_1$ | Values of basic variables |
|---|---|---|---|---|---|---|
| $s_1$ ← | 2 | ③ | 1 | 0 | 0 | 6 |
| $a_1$ | 4 | 5 | 0 | $-1$ | 1 | 20 |
| $P_M$ | $-4M - 5$ | $-5M - 6$ | 0 | $M$ | 0 | $-20M$ |

Clearly $-5M - 6$ is the most negative indicator, so $y$ is the entering variable. Since $\frac{6}{3}$ is less than $\frac{20}{5}$ we choose $s_1$ to be the leaving variable. Hence 3 is the pivot. Pivoting, we obtain

**Tableau III**

| Basis | $x$ | $y$ | $s_1$ | $s_2$ | $a_1$ | Values of basic variables |
|---|---|---|---|---|---|---|
| $y$ ← | $\left(\frac{2}{3}\right)$ | 1 | $\frac{1}{3}$ | 0 | 0 | 2 |
| $a_1$ | $\frac{2}{3}$ | 0 | $\frac{-5}{3}$ | $-1$ | 1 | 10 |
| $P_M$ | $\frac{-2M}{3} - 1$ | 0 | $\frac{5M}{3} + 2$ | $M$ | 0 | $-10M + 12$ |

↑

Since $(-2M/3) - 1$ is the only negative indicator, $x$ is the entering variable and, comparing ratios, $y$ should be the leaving variable. Pivoting, we obtain

**Tableau IV**

| Basis | $x$ | $y$ | $s_1$ | $s_2$ | $a_1$ | Values of basic variables |
|---|---|---|---|---|---|---|
| $x$ | 1 | $\frac{3}{2}$ | $\frac{1}{2}$ | 0 | 0 | 3 |
| $a_1$ | 0 | $-1$ | $-2$ | $-1$ | 1 | 8 |
| $P_M$ | 0 | $M + \frac{3}{2}$ | $2M + \frac{5}{2}$ | $M$ | 0 | $-8M + 15$ |

All indicators in the bottom row are nonnegative. Thus, the simplex procedure terminates with $x = 3$, $y = s_1 = s_2 = 0$, $a_1 = 8$, and $P_M = -8M + 15$. Since the artificial variable $a_1$ has a positive value, the original problem has no solution. **Do Exercise 3.** ■

## Solving a Minimization Problem

In order to locate the point where a function attains its minimum value, it suffices to locate the point where the negative of the function attains its maximum value. For example, if a function $f$ attains its minimum value at $(x_0, y_0, z_0)$, then $f(x_0, y_0, z_0) \leq f(x, y, z)$ for all $(x, y, z)$ in the domain of $f$. Consequently, $-f(x_0, y_0, z_0) \geq -f(x, y, z)$ for all $(x, y, z)$ in the domain of $-f$. That is, $-f$ attains its maximum at $(x_0, y_0, z_0)$. Furthermore, $\min f = f(x_0, y_0, z_0) = -[-f(x_0, y_0, z_0)] = -\max(-f)$.

**EXAMPLE 6**    Minimize $C = 6x + 4y + 4z$ subject to

$$5x + 7y + 4z \leq 65$$
$$5x + 5y + 4z \geq 57$$
$$x + 2y + 4z = 25$$
$$x \geq 0, \, y \geq 0, \, z \geq 0$$

**Solution**    Let $P = -C = -6x - 4y - 4z$ and try to find the maximum value of $P$ subject to the same constraints. First, add a slack variable $s_1$ to the left side of the first constraint changing it to an equality. Change the second constraint to an equality by subtracting a surplus variable $s_2$ and adding an artificial variable $a_1$ to its left side. Then, add an artificial variable $a_2$ to the left side of the equality constraint. Subtract $(a_1 + a_2)M$ from the right side of the equality defining $P$ and rename the objective function to get the modified problem. Maximize $P_M = -6x - 4y - 4z - a_1M - a_2M$ subject to

$$5x + 7y + 4z + s_1 \qquad\qquad = 65$$
$$5x + 5y + 4z \quad - s_2 + a_1 \quad = 57$$
$$x + 2y + 4z \qquad\qquad + a_2 = 25$$
$$x \geq 0, \, y \geq 0, \, z \geq 0, \, s_1 \geq 0, \, s_2 \geq 0, \, a_1 \geq 0, \, a_2 \geq 0$$

Write the equation defining the objective function in the following form

$$P_M + 6x + 4y + 4z + a_1M + a_2M = 0$$

The preliminary tableau for the modified problem is

**Tableau I**

| Basis | $x$ | $y$ | $z$ | $s_1$ | $s_2$ | $a_1$ | $a_2$ | Values of basic variables |
|-------|-----|-----|-----|-------|-------|-------|-------|---------------------------|
| $s_1$ | 5 | 7 | 4 | 1 | 0 | 0 | 0 | 65 |
| $a_1$ | 5 | 5 | 4 | 0 | -1 | 1 | 0 | 57 |
| $a_2$ | 1 | 2 | 4 | 0 | 0 | 0 | 1 | 25 |
| $P_M$ | 6 | 4 | 4 | 0 | 0 | $M$ | $M$ | 0 |

Now add rows 2 and 3, corresponding to the artificial variables. Then, multiply the result by $-M$ and add the product to row 4 to obtain the initial simplex tableau for the modified problem.

**Tableau II**

| Basis | $x$ | $y$ | $z$ | $s_1$ | $s_2$ | $a_1$ | $a_2$ | Values of basic variables |
|---|---|---|---|---|---|---|---|---|
| $s_1$ | 5 | 7 | 4 | 1 | 0 | 0 | 0 | 65 |
| $a_1$ | 5 | 5 | 4 | 0 | $-1$ | 1 | 0 | 57 |
| $a_2$ | 1 | 2 | ④ | 0 | 0 | 0 | 1 | 25 |
| $P_M$ | $-6M + 6$ | $-7M + 4$ | $-8M + 4$ | 0 | $M$ | 0 | 0 | $-82M$ |

Since the most negative indicator is $-8M + 4$, $z$ is the entering variable. Comparing $\frac{25}{4}$, $\frac{57}{4}$, and $\frac{65}{4}$, $\frac{25}{4}$ is the smallest. Thus, $a_2$ is the leaving variable. Pivoting, we obtain

**Tableau III**

| Basis | $x$ | $y$ | $z$ | $s_1$ | $s_2$ | $a_1$ | $a_2$ | Values of basic variables |
|---|---|---|---|---|---|---|---|---|
| $s_1$ | 4 | 5 | 0 | 1 | 0 | 0 | $-1$ | 40 |
| $a_1$ | ④ | 3 | 0 | 0 | $-1$ | 1 | $-1$ | 32 |
| $z$ | $\frac{1}{4}$ | $\frac{1}{2}$ | 1 | 0 | 0 | 0 | $\frac{1}{4}$ | $\frac{25}{4}$ |
| $P_M$ | $-4M + 5$ | $-3M + 2$ | 0 | 0 | $M$ | 0 | $2M - 1$ | $-32M - 25$ |

Since the most negative indicator is $-4M + 5$, $x$ is the entering variable. Comparing ratios, we see that $a_1$ should be the leaving variable. Pivoting, we get

**Tableau IV**

| Basis | $x$ | $y$ | $z$ | $s_1$ | $s_2$ | $a_1$ | $a_2$ | Values of basic variables |
|---|---|---|---|---|---|---|---|---|
| $s_1$ | 0 | 2 | 0 | 1 | 1 | $-1$ | 0 | 8 |
| $x$ | 1 | $\frac{3}{4}$ | 0 | 0 | $\frac{-1}{4}$ | $\frac{1}{4}$ | $\frac{-1}{4}$ | 8 |
| $z$ | 0 | $\frac{5}{16}$ | 1 | 0 | $\frac{1}{16}$ | $\frac{-1}{16}$ | $\frac{5}{16}$ | $\frac{17}{4}$ |
| $P_M$ | 0 | $\frac{-7}{4}$ | 0 | 0 | $\frac{5}{4}$ | $M - \frac{5}{4}$ | $M + \frac{1}{4}$ | $-65$ |

Here, $x = 8$, $y = 0$, $z = \frac{17}{4}$, $s_1 = 8$, and $s_2 = a_1 = a_2 = 0$. Since both artificial variables are equal to 0, delete the two columns corresponding to these variables to obtain the simplified tableau.

**Tableau V**

|  | Basis | $x$ | $y$ | $z$ | $s_1$ | $s_2$ | Values of basic variables |
|---|---|---|---|---|---|---|---|
| ← | $s_1$ | 0 | ② | 0 | 1 | 1 | 8 |
|  | $x$ | 1 | $\frac{3}{4}$ | 0 | 0 | $\frac{-1}{4}$ | 8 |
|  | $z$ | 0 | $\frac{5}{16}$ | 1 | 0 | $\frac{1}{16}$ | $\frac{17}{4}$ |
|  | $P_M$ | 0 | $\frac{-7}{4}$ | 0 | 0 | $\frac{5}{4}$ | $-65$ |

↑

Since $\frac{-7}{4}$ is the only negative indicator, $y$ is the entering variable. Comparing ratios, we find that $s_1$ is the leaving variable. Pivoting, we obtain

**Tableau VI**

| Basis | $x$ | $y$ | $z$ | $s_1$ | $s_2$ | Values of basic variables |
|---|---|---|---|---|---|---|
| $y$ | 0 | 1 | 0 | $\frac{1}{2}$ | $\frac{1}{2}$ | 4 |
| $x$ | 1 | 0 | 0 | $\frac{-3}{8}$ | $\frac{-5}{8}$ | 5 |
| $z$ | 0 | 0 | 1 | $\frac{-5}{32}$ | $\frac{-3}{32}$ | 3 |
| $P_M$ | 0 | 0 | 0 | $\frac{7}{8}$ | $\frac{17}{8}$ | $-58$ |

Since all indicators in the bottom row are nonnegative, an optimum solution has been reached. The maximum value of $P_M$ is $-58$ and occurs when $x = 5$, $y = 4$, $z = 3$, and $s_1 = s_2 = a_1 = a_2 = 0$. Thus, the maximum value of $P$ is also $-58$. Recalling that $P = -C$, the minimum value of $C$ is 58, since $-(-58) = 58$. This minimum value occurs at the point (5, 4, 3).

**Do Exercise 11.**                                                                    ■

## An Application

EXAMPLE 7    An oil refinery's facilities allow the blending of, at most, 9 million gallons per week. Some antiknock ingredient must be used. Each million gallons of grade A gasoline requires 4 pounds of the ingredient, each million gallons of grade B gasoline requires 5 pounds of the ingredient, while each million gallons of grade C gasoline requires only 2 pounds. Only 36 pounds of the ingredient is available each week. Because of market conditions, the sum of twice the output of grade A and half the output of grade B must exceed the output of grade C by at least 6 million gallons. Suppose that the refinery makes profits of 11 cents per gallon on grade A gasoline, 14 cents per gallon on grade B gasoline, and 6 cents per gallon on grade C gasoline. How many gallons of each grade of gasoline should be blended per week to maximize the weekly profit? What is the maximum weekly profit?

**Solution**   Let $x$, $y$, and $z$ be the numbers of millions of gallons of grade A, grade B, and grade C gasoline, respectively, that the refinery will blend each week. Since the facilities allow the blending of at most 9 million gallons per week, we have

$$x + y + z \leq 9$$

The amounts of antiknock ingredient required is $4x$ pounds for the $x$ million gallons of grade A, $5y$ pounds for the $y$ million gallons of grade B, and $2z$ pounds for the $z$ million gallons of grade C. Since only 36 pounds of the ingredient are available per week we have

$$4x + 5y + 2z \leq 36$$

Because of market conditions, $2x + \frac{1}{2}y$ must exceed $z$ by at least 6. Thus,

$$2x + \frac{1}{2}y - z \geq 6$$

or

$$4x + y - 2z \geq 12$$

The profits on grade A gasoline, grade B gasoline, and grade C gasoline, in tens of thousands of dollars, will be $11x$, $14y$, and $6z$, respectively. Thus, the total profit $P$, also in tens of thousands of dollars, is given by

$$P = 11x + 14y + 6z$$

The problem is to maximize $P$, subject to

$$\begin{aligned}
x + y + z &\leq 9 \\
4x + 5y + 2z &\leq 36 \\
4x + y - 2z &\geq 12 \\
x \geq 0, \, y \geq 0, \, z &\geq 0
\end{aligned}$$

We change the first three inequalities to equalities by adding a slack variable $s_1$ to the left side of the first inequality, adding a slack variable $s_2$ to the left side of the second inequality, and subtracting a surplus variable $s_3$ and adding an artificial variable $a_1$ to the left side of the third inequality. We then subtract $a_1 M$ from the right side of the equality defining the objective function, where $M$ is a very large positive number, and rename the objective function $P_M$ to get the modified problem.

Maximize $P_M = 11x + 14y + 6z - a_1 M$ subject to

$$\begin{aligned}
x + y + z + s_1 &= 9 \\
4x + 5y + 2z + s_2 &= 36 \\
4x + y - 2z - s_3 + a_1 &= 12 \\
x \geq 0, \, y \geq 0, \, z \geq 0, \, s_1 \geq 0, \, s_2 \geq 0, \, s_3 \geq 0, \, a_1 &\geq 0
\end{aligned}$$

Write the equation defining the objective function as

$$P_M - 11x - 14y - 6z + a_1 M = 0$$

Then write the preliminary simplex tableau for the modified problem.

**Tableau I**

| Basis | $x$ | $y$ | $z$ | $s_1$ | $s_2$ | $s_3$ | $a_1$ | Values of basic variables |
|-------|-----|-----|-----|-------|-------|-------|-------|---------------------------|
| $s_1$ | 1 | 1 | 1 | 1 | 0 | 0 | 0 | 9 |
| $s_2$ | 4 | 5 | 2 | 0 | 1 | 0 | 0 | 36 |
| $a_1$ | 4 | 1 | $-2$ | 0 | 0 | $-1$ | 1 | 12 |
| $P_M$ | $-11$ | $-14$ | $-6$ | 0 | 0 | 0 | $M$ | 0 |

Now multiply row 3, corresponding to the artificial variable $a_1$, by $-M$ and add the result to row 4 to obtain the initial simplex tableau for the modified problem.

**Tableau II**

| Basis | $x$ | $y$ | $z$ | $s_1$ | $s_2$ | $s_3$ | $a_1$ | Values of basic variables |
|-------|-----|-----|-----|-------|-------|-------|-------|---------------------------|
| $s_1$ | 1 | 1 | 1 | 1 | 0 | 0 | 0 | 9 |
| $s_2$ | 4 | 5 | 2 | 0 | 1 | 0 | 0 | 36 |
| $a_1$ | ④ | 1 | $-2$ | 0 | 0 | $-1$ | 1 | 12 |
| $P_M$ | $-4M - 11$ | $-M - 14$ | $2M - 6$ | 0 | 0 | $M$ | 0 | $-12M$ |

Since $-4M - 11$ is the most negative indicator, $x$ is the entering variable. Comparing ratios, we see that $a_1$ is the leaving variable. Pivoting, we obtain

**Tableau III**

| Basis | $x$ | $y$ | $z$ | $s_1$ | $s_2$ | $s_3$ | $a_1$ | Values of basic variables |
|-------|-----|-----|-----|-------|-------|-------|-------|---------------------------|
| $s_1$ | 0 | $\frac{3}{4}$ | $\frac{3}{2}$ | 1 | 0 | $\frac{1}{4}$ | $\frac{-1}{4}$ | 6 |
| $s_2$ | 0 | 4 | 4 | 0 | 1 | 1 | $-1$ | 24 |
| $x$ | 1 | $\frac{1}{4}$ | $\frac{-1}{2}$ | 0 | 0 | $\frac{-1}{4}$ | $\frac{1}{4}$ | 3 |
| $P_M$ | 0 | $\frac{-45}{4}$ | $\frac{-23}{2}$ | 0 | 0 | $\frac{-11}{4}$ | $M + \frac{11}{4}$ | 33 |

Here, $x = 3$, $y = 0$, $z = 0$, $s_1 = 6$, $s_2 = 24$, and $s_3 = a_1 = 0$. Since the value of the artificial variable is 0, delete the column corresponding to the artificial variable to obtain the simplified tableau.

**Tableau IV**

| | Basis | $x$ | $y$ | $z$ | $s_1$ | $s_2$ | $s_3$ | Values of basic variables |
|---|---|---|---|---|---|---|---|---|
| ← | $s_1$ | 0 | $\frac{3}{4}$ | $\left(\frac{3}{2}\right)$ | 1 | 0 | $\frac{1}{4}$ | 6 |
| | $s_2$ | 0 | 4 | 4 | 0 | 1 | 1 | 24 |
| | $x$ | 1 | $\frac{1}{4}$ | $\frac{-1}{2}$ | 0 | 0 | $\frac{-1}{4}$ | 3 |
| | $P_M$ | 0 | $\frac{-45}{4}$ | $\frac{-23}{2}$ | 0 | 0 | $\frac{-11}{4}$ | 33 |

↑

Since $\frac{-23}{2}$ is the most negative indicator in the bottom row, $z$ is the entering variable. Comparing ratios, we see that $s_1$ is the leaving variable. Pivoting, we obtain

**Tableau V**

| | Basis | $x$ | $y$ | $z$ | $s_1$ | $s_2$ | $s_3$ | Values of basic variables |
|---|---|---|---|---|---|---|---|---|
| | $z$ | 0 | $\frac{1}{2}$ | 1 | $\frac{2}{3}$ | 0 | $\frac{1}{6}$ | 4 |
| ← | $s_2$ | 0 | $(2)$ | 0 | $\frac{-8}{3}$ | 1 | $\frac{1}{3}$ | 8 |
| | $x$ | 1 | $\frac{1}{2}$ | 0 | $\frac{1}{3}$ | 0 | $\frac{-1}{6}$ | 5 |
| | $P_M$ | 0 | $\frac{-11}{2}$ | 0 | $\frac{23}{3}$ | 0 | $\frac{-5}{6}$ | 79 |

↑

Since $\frac{-11}{2}$ is the most negative indicator in the bottom row, $y$ is the entering variable. Comparing ratios, we see that $s_2$ is the leaving variable. Pivoting, we obtain

**Tableau VI**

| Basis | $x$ | $y$ | $z$ | $s_1$ | $s_2$ | $s_3$ | Values of basic variables |
|---|---|---|---|---|---|---|---|
| $z$ | 0 | 0 | 1 | $\frac{4}{3}$ | $\frac{-1}{4}$ | $\frac{1}{12}$ | 2 |
| $y$ | 0 | 1 | 0 | $\frac{-4}{3}$ | $\frac{1}{2}$ | $\frac{1}{6}$ | 4 |
| $x$ | 1 | 0 | 0 | 1 | $\frac{-1}{4}$ | $\frac{-1}{4}$ | 3 |
| $P_M$ | 0 | 0 | 0 | $\frac{1}{3}$ | $\frac{11}{4}$ | $\frac{1}{12}$ | 101 |

All indicators in the bottom row are nonnegative. Thus, an optimum solution has been reached. The maximum value of $P_M$ is 101 and occurs when $x = 3$, $y = 4$, $z = 2$, and $s_1 = s_2 = s_3 = a_1 = 0$. Since the value of the artificial variable is 0, 101 is also the maximum value of the original objective function $P$ and occurs when $x = 3$, $y = 4$, and $z = 2$. Thus, the refinery should blend 3 million gallons of type A gasoline, 4 million gallons of type B gasoline, and 2 million gallons of type C gasoline per week to maximize the weekly profit. The weekly profit will be $1,010,000, since $101(10,000) = 1,010,000$.

**Do Exercise 19.**

■

# Exercise Set 4.7

*Use the Big M Method to solve all problems. If for some problem a simpler technique is available, use that technique only to check the answer you obtained with the Big M Method.*

**1.** Maximize $P = 3x + 2y$ subject to

$$2x + 3y \leq 36$$
$$-x + y \geq 2$$
$$x \geq 0, y \geq 0$$

**2.** Maximize $P = 2x + y$ subject to

$$x - y + 50 \geq 0$$
$$3x - y + 50 \geq 0$$
$$2x + 3y \geq 40$$
$$x - 2y + 15 \leq 0$$
$$x \geq 0, y \geq 0$$

**3.** Maximize $P = 3x + 2y$ subject to

$$2x + 3y \geq 36$$
$$x + 4y \leq 8$$
$$x \geq 0, y \geq 0$$

**4.** Maximize $P = 4x + 3y$ subject to

$$2x - y \leq -3$$
$$-x + 2y \geq 12$$
$$x + y \leq 18$$
$$x \geq 0, y \geq 0$$

**5.** Maximize $P = 4x + 4y + 3z$ subject to

$$2x + 4y + 3z \geq 48$$
$$2x + y + 3z \leq 45$$
$$2x + 6y + z \leq 56$$
$$x + 2y + 3z = 30$$
$$x \geq 0, y \geq 0, z \geq 0$$

**6.** Minimize $P$ where $P$ is the function of Exercise 5 subject to the same constraints.

**7.** Maximize $P = 3x + 2y + 5z$ subject to

$$-4x + y - 3z \leq -29$$
$$6x + 3y \leq 39$$
$$5x + y + 6z \leq 61$$
$$x + y + z = 12$$
$$x \geq 0, y \geq 0, z \geq 0$$

**8.** Minimize $C = x + 5y + 3z$ subject to the constraints of Exercise 7.

**9.** Minimize $C = 2x + 3y + 5z$ subject to the constraints of Exercise 7.

**10.** Maximize $P = 6x + 11y + 12z$ subject to

$$2x + 3y + 4z \leq 34$$
$$5x + 5y + 8z \geq 64$$
$$x + 2y + 4z = 26$$
$$x \geq 0, y \geq 0, z \geq 0$$

**11.** Minimize $C = 9x + 4y + 4z$ subject to the constraints of Exercise 10.

**12.** Maximize $P = 5x + 6y + 3z$ subject to

$$3x + 2y + 3z \leq 33$$
$$7x + 3y + 6z \geq 62$$
$$2x + y + 3z = 25$$
$$x \geq 0, y \geq 0, z \geq 0$$

**13.** Minimize $C = 9x + 2y + 3z$ subject to the constraints of Exercise 12.

**14.** Maximize $P = 4x + 3y + 5z$ subject to

$$-13x + 33y + 4z \leq 183$$
$$16x + 24y - 13z \leq 219$$
$$9x - 4y + 8z \leq 121$$
$$2x + 3y + z \geq 30$$
$$x \geq 0, y \geq 0, z \geq 0$$

**15.** Minimize $C = 4x + 3y + 5z$ subject to the constraints of Exercise 14.

**16.** Maximize $P = 11x + 5y + z$ subject to

$$3x + 4y + z \leq 35$$
$$3x + 2y + z \leq 29$$
$$5x + 4y + z \geq 49$$
$$2x + 3y + z = 25$$
$$x \geq 0, y \geq 0, z \geq 0$$

**17.** A farmer has 18 acres on which he may grow peanuts and corn. Because of market conditions the acreage for corn must exceed twice the acreage for peanuts by at least 3 acres and twice the acreage for corn must exceed the acreage for peanuts by at least 12 acres. Let $x$ and $y$ denote the numbers of acres allocated to peanuts and corn, respectively. If his profit is $400 per acre of peanuts and $300 per acre of corn, write his profit $P$ as a function of $x$ and $y$. How many acres of each should he plant to maximize his profit?

**18.** A farmer has 800 acres on which she may grow wheat and barley. She has accepted orders for 150 acres of wheat and 100 acres of barley. Moreover, she must follow regulations that require that four times the acreage for barley must exceed the acreage for wheat by at least 200 acres. Let $x$ and $y$ denote the numbers of acres allocated to wheat and barley, respectively. Suppose that her profit is \$370 per acre of wheat and \$290 per acre of barley. Write her total profit $P$ as a function of $x$ and $y$. How many acres of each should she grow to maximize profit?

**19.** An electronics company makes two models of VCRs. Model A takes 2 hours to assemble and 15 minutes to test, while model B takes 3 hours to assemble and 30 minutes to test. A maximum of 60,000 hours per month are available for assembly, while a maximum of 8000 hours a month are available for testing. Market studies indicate that 15 times the monthly production of model B should exceed the monthly production of the A model by at least 3000. Let $x$ and $y$ denote the numbers of A and B models, respectively, the company can make per month. Suppose that the company makes a \$120 profit on each model A and a \$95 profit on each model B. Express the total profit per month $P$ as a function of $x$ and $y$. How many of each model should be produced each month to maximize the profit per month?

**20.** A camera company makes two models of movie cameras. The deluxe model takes 3 hours to assemble and 10 minutes to test, while the standard model takes 2 hours to assemble and 6 minutes to test. A maximum of 38,000 hours per month are available for assembly, while a maximum of 2000 hours a month are available for testing. Because of market conditions, the management has been advised that three times the monthly production of the standard model should exceed the monthly production of deluxe models by at least 6000. Let $x$ and $y$ denote the numbers of deluxe and standard models, respectively, the company can make per month. Assume that the profit per deluxe model is \$450 and that per standard model is \$175. Express the total profit per month $P$ as a function of $x$ and $y$. How many of each model should be produced per month to maximize the profit per month?

**21.** An owner of a gasoline station can buy at most 1600 gallons of gasoline per day from his supplier. On a certain day he estimates that his regular customers will purchase at least 500 gallons of regular gasoline and at least 200 gallons of premium gasoline. He also knows from experience that each day five times the number of gallons of premium sold exceeds the number of gallons of regular sold by at least 2000. Let $x$ and $y$ denote, respectively, the numbers of gallons of regular and premium gasoline he can sell per day. Suppose that regular and premium gasoline sell for \$1.05 per gallon and \$1.15 per gallon, respectively. Express the

revenue per day $R$ dollars as a function of $x$ and $y$. How many gallons of each type should be sold per day to maximize his revenue per day?

**22.** An investor wishes to buy 300 shares of common stock. Her broker recommends stocks A, B, and C and suggests that to reduce the risk of a fluctuating market, twice the number of shares of stock A plus the number of shares of stock B should not exceed 275. He also recommends that twice the number of shares of stock A plus the number of shares of stock C should not exceed 325 and that seven times the number of shares of stock A should exceed the number of shares of stock C by at least 125. Let $x$, $y$, and $z$ represent the numbers of shares of stocks A, B, and C, respectively, that the investor will buy. Suppose that stock A sells at \$80 a share, stock B sells at \$50 a share, and stock C sells at \$70 a share. Write the total cost $C$ as a function of $x$, $y$, and $z$. How many shares of each kind should the investor buy to minimize her total cost?

**23.** An investor wishes to buy 500 shares of common stock. His broker recommends stocks A, B, and C and suggests that to reduce the risk of a fluctuating market, the number of shares of stock A plus the number of shares of stock B should not exceed 400. She also recommends that the number of shares of stock A should exceed that of stock B by less than 100 and that twice the number of shares of stock A should exceed the number of shares of stock C by at least 100. Let $x$, $y$, and $z$ represent the numbers of shares of stocks A, B, and C, respectively, that the investor will buy. Suppose that stock A sells at \$20 a share, stock B sells at \$35 a share, and stock C sells at \$25 a share. Write the total cost $C$ as a function of $x$, $y$, and $z$. How many shares of each kind should the investor buy to minimize his total cost?

**24.** Henri's motel has 120 rooms and a restaurant that can seat 45 people. During the tourist season, the motel will be full each day. Some of the rooms may be furnished to accommodate travelers who like to go first class, while some of the remaining rooms will be furnished to satisfy the second-class guests. The remaining rooms will be furnished for third-class guests. The manager of the motel knows from experience that 60% of the first-class guests, 30% of the second-class guests, and 10% of the third-class guests will eat in the motel restaurant. Because of market conditions, not more than 90 rooms should be furnished for first- and second-class guests. If the manager furnishes $x$ rooms for first-class guests, $y$ rooms for second-class guests, and $z$ rooms for third-class guests, and if the charge per night is \$90 per first-class room, \$55 per second-class room, and \$30 per third-class room, write the revenue per night $R$ as a function of $x$, $y$, and $z$. How many rooms of each type should there be to maximize the revenue per night?

25. A nursing home dietician is planning menus for patients. She wishes to provide a diet which has a minimum of 120 units of fat, 60 units of carbohydrates, and 90 units of proteins, and a maximum of 70 units of cholesterol per patient per month. These goals can be met using two types of food. Type I contains 3 units of fats, 2 units of carbohydrates, 6 units of proteins, and 2 units of cholesterol per pound, while each pound of type II contains 12 units of fats, 4 units of carbohydrates, 3 units of proteins, and 3 units of cholesterol. If the dietician plans to use $x$ pounds of type I and $y$ pounds of type II food per patient per month, and if type I food costs $3 per pound and type II costs $3.75 per pound, write the cost $C$ per patient per month as a function of $x$ and $y$. How many pounds of each kind of food should be used per month to minimize the cost?

26. An oil refinery's facilities allow the blending of at most 11 million gallons per week. Some antiknock ingredient must be used. Each million gallons of grade A gasoline requires 6 pounds of the ingredient, each million gallons of grade B gasoline requires 4 pounds of the ingredient, while each million gallons of grade C gasoline requires only 2 pounds. Only 38 pounds of the ingredient are available each week. Because of market conditions, the sum of the output of grade A and twice the output of grade B must exceed the output of grade C by at least 1 million gallons. Suppose that the refinery makes profits of $.15 per gallon on grade A gasoline, $.12 per gallon on grade B gasoline, and $.08 per gallon on grade C gasoline. How many gallons of each grade of gasoline should be blended per week to maximize the weekly profit? What is this profit?

27. The Tallgrass Company makes motors for its lawn mowers at two different sites, A and B. Site A can make at most 900 motors a week while site B can make at most 500 motors a week. The motors are shipped to two assembly plants I and II. The plants have already accepted orders so that plant I must assemble at least 600 lawn mowers a week and plant II must assemble at least 700 lawn mowers a week. The shipping costs per motor are as follows:

From site A to plant I: $8

From site A to plant II: $6

From site B to plant I: $7

From site B to plant II: $5

Let $x_1$, $x_2$, $x_3$, and $x_4$ denote the numbers of motors shipped from site A to plant I, from site A to plant II, from site B to plant I, and from site B to plant II, respectively. Write the total shipping cost $C as a function of $x_1$, $x_2$, $x_3$, and $x_4$. What weekly shipping schedule will minimize the total shipping cost?

28. The chairperson of the mathematics department must decide how many professors, associate professors, and assistant professors the department will have. The administration requires that the number of assistant professors must exceed the sum of the numbers of professors and associate professors by at least nine. Exactly 202 sections of mathematics courses must be taught each academic year and the instructional budget available is $1,015,000. On the average, professors earn $50,000, associate professors earn $45,000, and assistant professors earn $25,000. The department must have a good research publication record to satisfy accreditation requirements. On the average, professors publish five research papers and teach five sections a year, associate professors publish four research papers and teach six sections a year, and assistant professors publish .5 research paper and teach seven sections a year. How many professors, associate professors, and assistant professors should be in the department to maximize the number of research publications per year?

# 4.8  Chapter Review

**IMPORTANT SYMBOLS AND TERMS**

$C(n,k) = n!/[k!(n - k)!]$ [4.2]

$n!$

Adjacent corners [4.3]

Artificial variable [4.7]

Basic feasible solution [4.4]

Basic variables [4.4]

Basis [4.4]

Big $M$ Method [4.7]

Bounded [4.1]

Convex polygonal region [4.1]

Coordinates [4.2]

Decision variable [4.1]

Dual maximization problem [4.6]

Entering variable [4.4]

First octant [4.2]

Graphical method [4.1]

Graph of an ordered triple of real numbers $(a, b, c)$ [4.2]

Indicators [4.5]

Intersection of planes [4.2]

$k$th tableau [4.5]

Leaving variable [4.4]

Linear programming [4.1]

Modified problem [4.7]

Multiple optimum solutions [4.5]

$n$ factorial
Nonbasic variables [4.4]
Objective function [4.1],
  [4.5]
Octant [4.2]
Pivot column [4.4]
Pivot element [4.4]
Pivoting [4.4]

Pivot row [4.4]
Preliminary simplex tableau
  [4.7]
Primal problem [4.6]
Right-handed system [4.2]
Set of feasible solutions
  [4.1]
Simplex method [4.3]

Slack variables [4.3]
Surplus variable [4.7]
Tableau [4.4]
Three-dimensional Cartesian
  coordinate system [4.2]
Unbounded [4.1]
Unbounded sets of feasible
  solutions [4.5]

**SUMMARY**

In a linear programming problem, we either maximize or minimize a linear function, called the objective function, subject to linear constraints which are expressed as inequalities. It is possible to use the graphical method to solve a linear programming problem involving only two variables, but we use the simplex method for solving problems with more variables. Given a problem, we first form a mathematical model by introducing some decision variables and writing a linear objective function. We translate the constraints of the problem as linear inequalities and/or equations and write inequalities which show that all constraints are nonnegative. If the objective function is $z = ax + by + d$, and we wish to solve the problem graphically, we proceed as follows:

**Step 1.** We sketch the graph of the system of linear constraints to get the region of feasible solutions. If this region is empty, the problem has no solution. Otherwise, we find the corners of the region by solving the appropriate systems of equations.

**Step 2.** If the region of feasible solutions is bounded, then both a maximum and minimum value of the objective function exist. In that case, we evaluate the objective function at each corner of the region. The largest of the numbers obtained is the maximum value of the objective function, and the smallest is its minimum. If an optimum value occurs at two corners, then it occurs at every point on the segment joining the two corners.

**Step 3.** If the region of feasible solutions is unbounded, evaluate the objective function at each corner of the region. If $c_1$ and $c_2$ are the largest and smallest numbers obtained, respectively, then if a maximum exists, it will be $c_1$, and if a minimum exists, it will be $c_2$. We sketch the graph of $d = ax + by + c$ for some $d > c_1$. If this graph intersects the region of feasible solutions, the objective function has no maximum. Otherwise $c_1$ is the maximum. We proceed similarly with some $d < c_2$ to check for a minimum.

The simplex method is used if there are more than two decision variables.
If the problem is a maximization problem in standard form, use the procedure on page 342.
If the problem is a minimization problem in standard form, use the procedure on page 372.
If the problem is not in standard form, use the Big $M$ Method procedure on page 381.

**SAMPLE EXAM QUESTIONS**

1. Suppose the vertices of the set of feasible solutions for a linear programming problem are $(1, 5)$, $(3, 8)$, $(5, 4)$, and $(2, 1)$ and the objective function is $z = 3x + 2y$. Find the maximum and minimum value of $z$.

2. Same as Exercise 1 if the vertices are $(2, 4)$, $(4, 7)$, $(10, 5)$, $(8, 3)$, and $(3, 1)$ and the objective function is $z = 5x - 2y$.

3. Same as Exercise 1 if the vertices are $(1, 1, 1)$, $(4, 3, 5)$, $(5, 2, 3)$, and $(7, 4, 6)$ and the objective function is $z = 2x + 5y + 3z$.

*Solve Exercises 4, 5, and 6 geometrically.*

4. Find the maximum and minimum value of $z = 3x + 11y$ subject to

$$2x + \phantom{5}y \le 24$$
$$2x + 5y \le 40$$
$$x \ge 0 \text{ and } y \ge 0$$

5. Find the maximum and minimum value of $z = 3x + 4y$ subject to

$$x + 7y \ge \phantom{0}17$$
$$x + \phantom{7}y \ge \phantom{00}5$$
$$-4x + 5y \le \phantom{00}8$$
$$11x + 5y \le 115$$
$$x \ge 0 \text{ and } y \ge 0$$

6. Find an optimum value of $z = 2x + 3y$ subject to

$$x + 7y \ge 17$$
$$x + \phantom{7}y \ge \phantom{0}5$$
$$-4x + 5y \le \phantom{0}8$$
$$x \ge 0 \text{ and } y \ge 0$$

7. Plot the points **a.** $(-1, 2, 3)$, **b.** $(2, 4, 5)$, and **c.** $(-2, 3, 5)$.

8. Sketch the graph of the equation $3x + 5y + 2z = 120$.

9. Suppose that five lines have been drawn in a plane and that the number of points of intersection is finite. What is the maximum number of points of intersection?

10. Suppose that five planes have been drawn in a three-dimensional Cartesian coordinate system and that the number of points of intersection is finite. What is the maximum number of points of intersection?

11. The equations

$$-10x + 12y + \phantom{0}9z = 72$$
$$18x + 15y - \phantom{0}4z = 90$$
$$16x - \phantom{0}7y + 10z = 80$$

are the equations of planes $P_1$, $P_2$, and $P_3$, respectively.

**a.** Find the intersection of the three planes.
**b.** Sketch the region in the first octant bounded by the three planes and the three coordinate planes.
**c.** Using the information of **a** and **b**, find the maximum value of $w = 2x + 5y + 3z$ subject to

$$-10x + 12y + \phantom{0}9z \le 72$$
$$18x + 15y - \phantom{0}4z \le 90$$
$$16x - \phantom{0}7y + 10z \le 80$$
$$x \ge 0, \ y \ge 0, \text{ and } z \ge 0$$

**12.** The following has been obtained by performing elementary row operations on the augmented matrix of a linear system of equations in the variables $x_1, x_2, \ldots, x_7$. This system has infinitely many solutions. Give the solution which can be read directly from the given matrix.

$$\begin{bmatrix} 3 & -5 & 0 & 1 & 4 & 0 & -6 & 7 \\ -1 & 4 & 1 & 0 & 7 & 0 & 9 & -5 \\ 4 & 3 & 0 & 0 & -2 & 1 & 7 & 3 \end{bmatrix}$$

**13.** For each of the following systems of linear inequalities, find the coordinates of the indicated number of corners of the graph of the system by introducing slack variables and obtaining a sequence of tableaux. Start with the origin. Do not sketch the graphs.

**a.**   $2x + y \le 24$

$2x + 5y \le 40$

$x \ge 0$ and $y \ge 0$, (3 corners)

**b.**   $x + y \le 15$

$5x + 3y \le 55$

$5x + 2y \le 50$

$x \ge 0$ and $y \ge 0$, (4 corners)

**c.**   $-10x + 12y + 9z \le 72$

$18x + 15y - 4z \le 90$

$16x - 7y + 10z \le 80$

$x \ge 0, y \ge 0$, and $z \ge 0$, (4 corners)

*Solve each of the following using the simplex method.*

**14.** Find the maximum value of $z = 2x + 7y$ subject to

$2x + y \le 24$

$2x + 5y \le 40$

$x \ge 0$ and $y \ge 0$

**15.** Find the maximum value of $z = 3x + 8y$ subject to

$-2x + y \le 2$

$-2x + 3y \le 14$

$x \ge 0$ and $y \ge 0$

**16.** Find the maximum value of $z = 2x + y$ subject to

$x + y \le 15$

$5x + 3y \le 55$

$5x + 2y \le 50$

$x \ge 0$ and $y \ge 0$

**17.** Maximize $w = 2x + 5y + 3z$ subject to

$-10x + 12y + 9z \le 72$

$18x + 15y - 4z \le 90$

$16x - 7y + 10z \le 80$

$x \ge 0, y \ge 0$, and $z \ge 0$

**18.** Maximize $z = 6x + 9y$ subject to

$3x + 7y \le 49$

$2x + 3y \le 26$

$x + y \le 12$

$x \ge 0$ and $y \ge 0$

**19.** Maximize $w = 15x + 3y + 3z$
subject to

$$-x + 3y + 2z \leq 216$$
$$x + y + z \leq 198$$
$$2x + y + 3z \leq 324$$
$$x \geq 0,\ y \geq 0,\ z \geq 0$$

**20.** Minimize $z = 9x + 7y$ subject to

$$3x + 4y \geq 120$$
$$9x + 2y \geq 180$$
$$x \geq 0,\ y \geq 0$$

**21.** Minimize $z = 5x + 2y$ subject to

$$3x + y \geq 50$$
$$x + y \geq 30$$
$$x + 4y \geq 60$$
$$x \geq 0,\ y \geq 0$$

**22.** Minimize $w = 33x + 54y + 36z$
subject to

$$x + 2y - z \geq 10$$
$$x + y + 3z \geq 2$$
$$x + 3y + 2z \geq 2$$
$$x \geq 0,\ y \geq 0,\ z \geq 0$$

**23.** A company makes gadgets A and B. The production of each requires time on machines I, II, and III. Gadget A requires 2 hours on I, 1 hour on II, and 1 hour on III; gadget B requires 1 hour on I, 1 hour on II, and 2 hours on III. The time available on these machines per week is 200 hours on I, 140 hours on II, and 240 hours on III. The profit on each A is $90, and the profit on each B is $120. The company can sell all it produces. How many of each should it make per week to maximize profit? What is that profit?

**24.** A company builds three types of sailboats. The deluxe model requires 3 units of wood, 2 units of canvas, and 1 unit of paint; the super model requires 5 units of wood, 1 unit of canvas, and 2 units of paint; and the regular model requires 2 units of wood, 1 unit of canvas, and 1 unit of paint. Each month, the company has, at most, 69 units of wood, 24 units of canvas, and 21 units of paint available. If the profits are $4000 per deluxe model, $2000 per super model, and $3000 per regular model, and if the company can sell all the boats it can build each month, how many of each model should be built each month to maximize profit? What is this profit?

**25.** A dairy company processes milk at two locations. Its Arlington plant produces 50 gallons of 4% and 100 gallons of 2% milk per hour and costs $168 per hour to operate, while the Marysville plant produces 100 gallons of 4% and 150 gallons of 2% milk per hour and costs $288 per hour to operate. The company needs a total daily production of at least 1200 gallons of 4% and 2250 gallons of 2% milk. How many hours per day should each plant operate to minimize cost?

**26.** Use the Big $M$ Method to maximize $z = 5x + 4y$ subject to

$$3x + 8y \leq 120$$
$$3x + 2y \leq 46$$
$$3x - 5y \leq -10$$
$$x \geq 0,\ y \geq 0$$

**27.** Use the Big $M$ Method to maximize $w = 2x + 5y + 3z$ subject to

$$4x - y + 3z \geq 29$$
$$x + y + z = 12$$
$$6x + 3y \leq 39$$
$$5x + y + 6z \leq 61$$
$$x \geq 0,\ y \geq 0,\ z \geq 0$$

**28.** Use the Big $M$ Method to minimize $w = 3x - 2y + 7z$ subject to the constraints of Exercise 27.

**29.** Use the Big $M$ Method to maximize $w = 7x + 4y + z$ subject to the constraints of Exercise 27.

# 5

# Sequences and Mathematics of Finance

Often, events occur sequentially. That is, there will be a first event, then a second, then a third, and so on. In this chapter we describe a type of function that serves as a mathematical model for solving problems involving sequential outcomes or events. Special business applications will also be discussed.

# 5.1 Sequences, Arithmetic and Geometric Progressions

> **DEFINITION 5.1:** Suppose $n$ is a positive integer and for each $i$ in $\{1, 2, 3, \ldots, n\}$ we associate a real number $a_i$. Then, the list $a_1, a_2, a_3, \ldots, a_n$ is called a *finite sequence*. If for each $j$ in $\{1, 2, 3, \ldots, n, \ldots\}$ we associate a real number $b_j$, then the list $b_1, b_2, b_3, \ldots, b_n, \ldots$ is called an *infinite sequence*.

In other words, a finite sequence is a function with domain $\{1, 2, \ldots, n\}$ and an infinite sequence is a function with domain $\{1, 2, 3, \ldots, n, \ldots\}$. If $s$ is the name of a sequence, we write $s_i$ instead of the usual function notation $s(i)$. The function values are called the *terms* of the sequence. Thus, $s_1$ is called the first term of the sequence, $s_2$ is the second term, and so on. The methods used to describe sequences are the same as those used to describe functions.

**EXAMPLE 1** The sequence $s$ is defined by $s_n = \frac{2}{n + 1}$. Find the first four terms of that sequence.

**Solution** The first term is 1, since $s_1 = \frac{2}{1 + 1} = 1$. The second term is $s_2 = \frac{2}{2 + 1} = \frac{2}{3}$; the third term is $\frac{1}{2}$; and the fourth term is $\frac{2}{5}$.

**Do Exercise 3.**  ■

**EXAMPLE 2** The sequence $t$ is defined by $t_n = (-1)^{[n/2]}n^2$. Find the first five terms of the sequence.

**Solution** Recall that if $x$ is a real number, $[x]$ denotes the largest integer less than or equal to $x$. Thus,

$$t_1 = (-1)^{[1/2]}1^2 = (-1)^0 \cdot 1 = 1$$
$$t_2 = (-1)^{[2/2]}2^2 = (-1)^1 \cdot 4 = -4$$
$$t_3 = (-1)^{[3/2]}3^2 = (-1)^1 \cdot 9 = -9$$
$$t_4 = (-1)^{[4/2]}4^2 = (-1)^2 \cdot 16 = 16$$
$$t_5 = (-1)^{[5/2]}5^2 = (-1)^2 \cdot 25 = 25$$

**Do Exercise 7.**  ■

**EXAMPLE 3** In Section 1.4, it was shown that if $P$ dollars in an account earns interest at an annual rate $r$ (written as a decimal) compounded $k$ times per year, then the amount $A_n$ dollars in the account at the end of the $n$th conversion period is

$$A_n = P(1 + i)^n$$

where $i = \frac{r}{k}$. If \$1000 is deposited in an account at 8% interest compounded quarterly, find the amount in the account at the end of each of the first six conversion periods.

**Solution** The interest per conversion period is .02 since $\frac{.08}{4} = .02$. We get the following amounts in the account:

$$A_1 = 1000(1 + .02)^1 = 1020$$

$$A_2 = 1000(1 + .02)^2 = 1040.40$$

$$A_3 = 1000(1 + .02)^3 \doteq 1061.21$$

$$A_4 = 1000(1 + .02)^4 \doteq 1082.43$$

$$A_5 = 1000(1 + .02)^5 \doteq 1104.08$$

$$A_6 = 1000(1 + .02)^6 \doteq 1126.16$$

These results were obtained with a calculator and were rounded to two decimal places.
**Do Exercise 31.** ■

We will briefly discuss two types of sequences which occur so frequently that they have been given special names. They are called *arithmetic progressions* and *geometric progressions*.

## Arithmetic Progressions

An arithmetic progression is a sequence that starts with an arbitrary number as the first term and then the same constant is added to obtain successive terms. For example, 2, 5, 8, 11, 14, and 17 are the first six terms of an arithmetic progression. We started with 2 as our first term and kept adding 3 to obtain successive terms.

> **DEFINITION 5.2:** A sequence in the form $a, a+d, a+2d, \ldots$ is called an *arithmetic progression* (abbreviated A.P.) with *first term a* and *common difference d*.

**EXAMPLE 4** The first term of an arithmetic progression is 6 and its common difference is 2. Find the first six terms of this A.P.

**Solution** It is given that $a_1 = 6$. Thus, $a_2 = 6 + 2 = 8$, $a_3 = 8 + 2 = 10$, $a_4 = 10 + 2 = 12$, $a_5 = 12 + 2 = 14$, and $a_6 = 14 + 2 = 16$.
**Do Exercise 9.** ■

If an A.P. has first term $a$ and common difference $d$, then $a_1 = a$, $a_2 = a + d$, $a_3 = a_2 + d = (a + d) + d = a + 2d$, $a_4 = a_3 + d = (a + 2d) + d = a + 3d$ and, in general,

$$a_k = a + (k - 1)d$$

**EXAMPLE 5**    The 21st and 35th terms of an A.P. are 64 and 106, respectively. Find the first three terms.

**Solution**    We have

$$a_{21} = a + (21 - 1)d = a + 20d = 64$$

and

$$a_{35} = a + (35 - 1)d = a + 34d = 106$$

Subtracting these two equalities, we obtain

$$(a + 34d) - (a + 20d) = 106 - 64$$

That is,

$$14d = 42$$

and

$$d = \frac{42}{14} = 3$$

Replacing $d$ by 3 in the expression for $a_{21}$, we obtain

$$a + 20 \cdot 3 = 64$$

so that $a = 64 - 60 = 4$. It follows that $a_1 = 4$, $a_2 = 4 + 3 = 7$, and $a_3 = 7 + 3 = 10$.

**Do Exercise 13.**    ■

Now we wish to derive a formula for the sum $S_n$ of the first $n$ terms of an A.P. with first term $a$ and common difference $d$. We first write

$$S_n = a + (a + d) + (a + 2d) + \cdots + (a + (n - 2)d) + (a + (n - 1)d)$$

Writing this sum in reverse order, we get

$$S_n = (a + (n - 1)d) + (a + (n - 2)d) + \cdots + (a + 2d) + (a + d) + a$$

Each of the sums on the right of the equalities has $n$ terms. When we add these equalities, we have a sum of $n$ identical terms.

$$2S_n = [2a + (n - 1)d] + [2a + (n - 1)d] + \cdots + [2a + (n - 1)d]$$
$$= n[2a + (n - 1)d]$$

Thus,

$$S_n = \frac{n[2a + (n - 1)d]}{2}$$

If we note that $2a + (n - 1)d = a + [a + (n - 1)d] = a_1 + a_n$, we obtain a simpler formula

$$S_n = \frac{n(a_1 + a_n)}{2}$$

Since $(a_1 + a_n)/2$ is the average of the first and $n$th terms and $n$ is the number of terms in the sum, we may state the following:

> To find the sum of the first $n$ terms of an A.P., we need only calculate the average of the first and $n$th terms and multiply this average by $n$, the number of terms.

**EXAMPLE 6**   Find the sum of the first 50 terms of the A.P. of Example 5.

**Solution**   We found that the first term $a_1$ is 4 and the common difference $d$ is 3. Thus, the 50th term is 151 since $4 + (50 - 1)3 = 4 + 49 \cdot 3 = 4 + 147 = 151$. The average of the 1st and 50th terms is $\frac{4 + 151}{2} = 77.5$. Hence, the sum of the first 50 terms is $50 \cdot 77.5 = 3875$.

**Do Exercise 19.**   ■

## Geometric Progressions

A geometric progression is a sequence where each successive term is obtained by multiplying the preceding term by the same constant. For example, 3, 6, 12, 24, 48, 96 are the first six terms of a geometric progression. We start with 3 as the first term and keep multiplying by 2 to obtain successive terms.

> **DEFINITION 5.3:**   A sequence in the form $a, ar, ar^2, \ldots$ is called a *geometric progression* (abbreviated G.P.) with *first term $a$* and *common ratio $r$*.

**EXAMPLE 7**   The first term of a G.P. is 5 and its common ratio is $-2$. Find the second, third, fourth, and fifth terms of this G.P.

**Solution**   The second term is

$$a_2 = a_1 \cdot r = 5 \cdot (-2) = -10$$

Similarly

$$a_3 = a_2 \cdot r = (-10) \cdot (-2) = 20$$
$$a_4 = a_3 \cdot r = (20) \cdot (-2) = -40$$
$$a_5 = a_4 \cdot r = (-40) \cdot (-2) = 80$$

**Do Exercise 37.**   ■

If a G.P. has first term $a$ and common ratio $r$, then $a_1 = a$, $a_2 = ar$, $a_3 = a_2r = (ar)r = ar^2$, $a_4 = a_3r = (ar^2)r = ar^3$ and, in general,

$$a_k = ar^{k-1}$$

**EXAMPLE 8**   Suppose that the 10th and 13th terms of a G.P. are 39,366 and $-1,062,882$, respectively. Find its first three terms.

**Solution**   We have

$$a_{10} = ar^{10-1} = ar^9 = 39,366 \tag{1}$$

and

$$a_{13} = ar^{13-1} = ar^{12} = -1,062,882 \tag{2}$$

Dividing the second equality by the first, we obtain

$$r^3 = -27$$

Thus, $r = -3$. Replacing $r$ by $-3$ in (1), we get $a(-3)^9 = 39,366$. Thus,

$$-19,683a = 39,366$$

and

$$a = -2$$

The first three terms are $-2$, $6$, and $-18$ since $(-2)(-3) = 6$ and $6(-3) = -18$.
**Do Exercise 41.**   ■

In applications, it is often necessary to add the first $n$ terms of a G.P. If a G.P. has first term $a$ and common ratio $r$, then the sum of the first $n$ terms $S_n$ is

$$S_n = a + ar + ar^2 + ar^3 + \cdots + ar^{n-2} + ar^{n-1} \tag{1}$$

Multiplying both sides by $r$, we obtain

$$S_n r = ar + ar^2 + ar^3 + ar^4 + \cdots + ar^{n-1} + ar^n \tag{2}$$

Subtracting the second equality from the first, we get

$$S_n - S_n r = a - ar^n$$

Thus,

$$S_n(1 - r) = a(1 - r^n)$$

and if $1 - r \neq 0$, we have

$$S_n = \frac{a(1 - r^n)}{1 - r}$$

In the case that $1 - r = 0$, $r = 1$, and the G.P. is the constant sequence $a$, $a$, $a$, . . . , the sum of the first $n$ terms is $na$.

**EXAMPLE 9**   The first term of a G.P. is 5 and the common ratio is 2. Find the sum of the first 20 terms.

**Solution**   We use the formula

$$S_n = \frac{a(1 - r^n)}{1 - r}$$

with $a = 5$, $r = 2$, and $n = 20$. We obtain

$$S_{20} = \frac{5(1 - 2^{20})}{1 - 2}$$

$$= \frac{5(1 - 1,048,576)}{-1}$$

$$= 5,242,875$$

The sum of the first 20 terms of the G.P. is 5,242,875.
**Do Exercise 47.**

The next three examples illustrate how arithmetic and geometric progressions occur in applications.

## Applications

**EXAMPLE 10**    Office equipment was purchased for $30,000 and is assumed to have a scrap value of $12,000 after 6 years. Determine the value of the equipment $V_k$ after $k$ years, $0 < k \le 6$, assuming that the value is depreciated linearly.

**Solution**    Since the depreciation is linear, each year the value of the equipment will be $3000 less than the preceding year since $\frac{30,000 - 12,000}{6} = 3000$. Thus the value $V_1$ at the end of the first year is given by

$$V_1 = 30,000 - 3000 = 27,000$$

The values $V_2$, $V_3$, $V_4$, $V_5$, and $V_6$ are obtained as follows:

$$V_2 = 27,000 - 3000 = 24,000$$
$$V_3 = 24,000 - 3000 = 21,000$$
$$V_4 = 21,000 - 3000 = 18,000$$
$$V_5 = 18,000 - 3000 = 15,000$$
$$V_6 = 15,000 - 3000 = 12,000$$

Therefore, 30,000, 27,000, 24,000, 21,000, 18,000, 15,000, and 12,000 are the first seven terms of an A.P. whose first term is 30,000 and common difference is 3000.
**Do Exercise 27.**

**EXAMPLE 11**    The value of a building which cost its owner $200,000 is declining by 10% per year. What will the building be worth at the end of the $k$th year for $k = 1, 2, 3, 4,$ and 5?

**Solution**    Let $V_k$ be the value of the building at the end of the $k$th year. The value declined by 10% the first year. Thus, at the end of the first year, the value of the building is 90% of the original value.

$$V_1 = 200,000(.9) = 180,000$$

At the end of the first year, the building is worth $180,000. Similarly, we get

$$V_2 = 180,000(.9) = 162,000$$
$$V_3 = 162,000(.9) = 145,800$$
$$V_4 = 145,800(.9) = 131,220$$
$$V_5 = 131,220(.9) = 118,098$$

At the end of the second, third, fourth, and fifth years, the value of the building will be $162,000, $145,800, $131,220, and $118,098, respectively. Therefore, 200,000, 180,000, 162,000, 145,800, 131,220, and 118,098 are the first six terms of a G.P. whose first term is 200,000 and common ratio is .9.
**Do Exercise 29.**                                                                                ■

**EXAMPLE 12**    Mary's bank pays a nominal rate of interest of 6% compounded monthly. She deposits $200 on the first of each month. How much is in Mary's account immediately after the sixth deposit?

**Solution**    We shall use the formula $A_n = P(1 + i)^n$ to determine the value of each deposit immediately after she makes the sixth deposit. We have $i = \frac{.06}{12} = .005$. Immediately after the sixth deposit, the value of the sixth deposit is still $200. The fifth deposit has been in the bank 1 month; thus, its value is $200(1 + .005) = $200(1.005)$. The fourth deposit has been in the bank 2 months; thus, its value is $200(1 + .005)^2 = $200(1.005)^2$. The values of the third, second, and first deposits are $200(1.005)^3$, $200(1.005)^4$, and $200(1.005)^5$, respectively. (See Figure 5.1.) Therefore, the balance in Mary's account is the sum

$$200 + 200(1.005) + 200(1.005)^2 + 200(1.005)^3 + 200(1.005)^4 + 200(1.005)^5$$
$$= \frac{200(1 - 1.005^6)}{1 - 1.005} \doteq 1215.10$$

This is the sum of the first six terms of a G.P. with first term 200 and common ratio 1.005. This result was obtained with a calculator and was rounded to two decimal places. The balance in Mary's account is $1215.10.
**Do Exercise 57.**

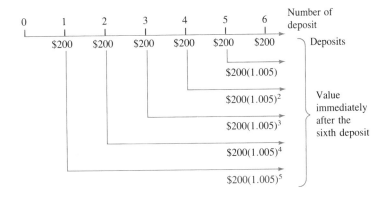

**FIGURE 5.1**

# Exercise Set 5.1

*In Exercises 1–8, find the first five terms of each of the sequences.*

**1.** $s_n = \dfrac{2}{2n - 1}$

**2.** $t_n = \dfrac{3}{2n + 1}$

**3.** $a_n = \dfrac{-2}{n^2 + 1}$

**4.** $s_n = \dfrac{(-1)^n}{3n - 2}$

**5.** $t_n = \dfrac{(-1)^{n+1}n}{n + 1}$

**6.** $a_n = \dfrac{2n - 1}{2n + 1}$

**7.** $s_n = \dfrac{(-1)^{[n/3]}(n + 3)}{3n - 1}$

**8.** $t_n = \dfrac{n^2 + n + 1}{n^2 + n - 1}$

**9.** The first term of an A.P. is $-4$ and its common difference is 3. Find the first ten terms.

**10.** The first term of an A.P. is 12 and its common difference is 5. Find the first five terms.

**11.** The first term of an A.P. is $-5$ and its common difference is $-2$. Find the first eight terms.

**12.** The second term of an A.P. is 4 and its common difference is 4. Find the first ten terms.

**13.** The 20th and 30th terms of an A.P. are 73 and 113, respectively. Find the first five terms.

**14.** The 32nd and 43rd terms of an A.P. are 95 and 128, respectively. Find the first seven terms.

**15.** The 16th and 36th terms of an A.P. are $-20$ and $-60$, respectively. Find the first five terms.

**16.** The 17th and 51st terms of an A.P. are 76 and 246, respectively. Find the first six terms.

**17.** The 19th and 63rd terms of an A.P. are 46 and 178, respectively. Find the first eight terms.

**18.** Find the sum of the first 100 terms of the A.P. of Exercise 9.

**19.** Find the sum of the first 70 terms of the A.P. of Exercise 10.

**20.** Find the sum of the first 80 terms of the A.P. of Exercise 11.

**21.** Find the sum of the first 90 terms of the A.P. of Exercise 12.

**22.** Find the sum of the first 60 terms of the A.P. of Exercise 13.

**23.** Find the sum of the first 100 terms of the A.P. of Exercise 14.

**24.** Find the sum of the first 200 terms of the A.P. of Exercise 15.

**25.** Find the sum of the first 90 terms of the A.P. of Exercise 16.

**26.** Find the sum of the first 300 terms of the A.P. of Exercise 17.

**27.** Office equipment was purchased for $25,000 and is assumed to have a scrap value of $5000 after 8 years. Determine the value of the equipment $V_k$ after $k$ years, $0 < k \le 8$, assuming that the value is depreciated linearly.

**28.** A truck was purchased for $35,000 and is assumed to have a trade-in value of $8000 after 10 years. Determine the value of the truck $V_k$ after $k$ years, $0 < k \le 10$, assuming that the value is depreciated linearly.

**29.** The value of a building which cost its owner $300,000 is declining by 5% per year. What will the building be worth at the end of the $k$th year for $k = 1, 2, 3, 4,$ and 5?

**30.** The value of a machine which cost its owner $60,000 is declining by 8% per year. What will the machine be worth at the end of the $k$th year for $k = 1, 2, 3, 4,$ and 5?

**31.** An amount of $500 is deposited in a bank that advertises a nominal interest rate of 6% compounded monthly. Find the amount at the end of each of the first six conversion periods.

**32.** If $300 is deposited in a bank that advertises a nominal interest rate of 8% compounded quarterly, what is the amount at the end of each of the first six conversion periods?

**33.** An amount of $800 is deposited in a bank that advertises a nominal interest rate of 8% compounded monthly. Find the amount at the end of each of the first six conversion periods.

**34.** If $1500 is deposited in a bank that advertises a nominal interest rate of 9% compounded quarterly, what is the amount at the end of each of the first six conversion periods?

**35.** An amount of $600 is deposited in a bank that advertises a nominal interest rate of 7% compounded monthly. Find the amount at the end of each of the first six conversion periods.

**36.** The first term $a_1$ of a G.P. is 3 and its common ratio $r$ is 4. Find the second, third, fourth, and fifth terms.

**37.** The first term $b_1$ of a G.P. is 2 and its common ratio $r$ is 5. Find the second, third, fourth, and fifth terms.

**38.** The first term $c_1$ of a G.P. is $-3$ and its common ratio $r$ is 2. Find the second, third, fourth, and fifth terms.

**39.** The first term $t_1$ of a G.P. is 512 and its common ratio $r$ is $\frac{1}{2}$. Find the second, third, fourth, and fifth terms.

**40.** The first term $a_1$ of a G.P. is 243 and its common ratio $r$ is $\frac{-1}{3}$. Find the second, third, fourth, and fifth terms.

**41.** The fifth and seventh terms of a G.P. are 324 and 2916, respectively. Find its first three terms.

**42.** The eighth and tenth terms of a G.P. are $-384$ and $-1536$, respectively. Find its first three terms.

**43.** The seventh and tenth terms of a G.P. are $-1458$ and $-39366$, respectively. Find its first three terms.

**44.** The 8th and 11th terms of a G.P. are 8 and 1, respectively. Find its first three terms.

**45.** The fifth and eighth terms of a G.P. are 9 and $\frac{1}{3}$, respectively. Find its first three terms.

**46.** Find the sum of the first 20 terms of the G.P. of Exercise 36.

**47.** Find the sum of the first 20 terms of the G.P. of Exercise 37.

**48.** Find the sum of the first 50 terms of the G.P. of Exercise 38.

**49.** Find the sum of the first 30 terms of the G.P. of Exercise 39.

**50.** Find the sum of the first 15 terms of the G.P. of Exercise 40.

**51.** Find the sum of the first 25 terms of the G.P. of Exercise 41.

**52.** Find the sum of the first 42 terms of the G.P. of Exercise 42.

**53.** Find the sum of the first 27 terms of the G.P. of Exercise 43.

**54.** Find the sum of the first 17 terms of the G.P. of Exercise 44.

**55.** Find the sum of the first 28 terms of the G.P. of Exercise 45.

**56.** A bank advertises a nominal rate of interest of 8% compounded monthly. Kris deposits $300 on the first of each month. How much will be in his account immediately after the tenth deposit?

**57.** A bank advertises a nominal rate of interest of 6% compounded monthly. Kathy deposits $250 on the first of each month. How much will be in her account immediately after the 15th deposit?

**58.** A bank advertises a nominal rate of interest of 7% compounded monthly. Steven deposits $150 on the first of each month. How much will be in his account immediately after the fifth deposit?

**59.** A bank advertises a nominal rate of interest of 6% compounded monthly. Barbara deposits $400 on the first of each month. How much will be in her account immediately before the tenth deposit?

**60.** A bank advertises a nominal rate of interest of 7% compounded monthly. Mike deposits $500 on the first of each month. How much will be in his account immediately before the tenth deposit?

**61.** A bank advertises a nominal rate of interest of 6% compounded monthly. Carol deposits $450 on the first of each month. How much will be in her account immediately before the 23rd deposit?

**62.** Suppose that each bacterium in a culture divides into three bacteria every hour and that there are five bacteria at the start. How many bacteria will be present after 8 hours?

**63.** Suppose that a dropped Ping-Pong ball always rebounds half the height it falls. If it is dropped from a height of 64 inches, how far will it have traveled when it hits the floor for the tenth time?

**64.** One-fourth of a container full of alcohol is replaced by water and the solution is mixed thoroughly. Then one-fourth of that solution is replaced by water and the new solution is mixed thoroughly. If this is done repeatedly, what fraction of the original volume of alcohol is left after the tenth replacement?

**65.** A man has signed a 12-year contract which gives him a $30,000 yearly salary the first year and a 6% raise each year after that. How much will he have earned when his contract expires?

**66.** Suppose that a new car is worth $18,000 and that at the end of each year its value depreciates by 15% of the value it had at the beginning of that year. What will the car be worth at the end of 8 years?

**67.** Suppose that a company made you the following offer. Your monthly salary would either be $15,000 or each month you would be paid 1 cent the first day, 2 cents the second day, 4 cents the third day, 8 cents the fourth day, and so on for each of the month's 21 working days. Which offer would you choose? Justify your answer.

**68.** Suppose that you are playing blackjack at your favorite casino. You are playing at a table where a minimum bet of $1 is required. You bet $1 on the first hand. If you lose, you bet $2 on the next hand. Each time you lose you double your bet on the next hand. How much will you have won if you lose $k$ straight hands, but win the $(k + 1)$st hand?

**69.** The casino of Exercise 68 has a table limit of $1 minimum and $500 maximum for bets for each hand. Explain the danger of using the strategy of Exercise 68.

# 5.2 Annuities

Everyone is familiar with financial transactions involving a sequence of equal payments made periodically. For example, you may have: deposited money in a savings account; or borrowed money and repaid the sum, plus interest, in a certain number of payments; or bought life insurance and made yearly payments. Whenever a transaction involves a sequence of equal payments made periodically, it is called an *annuity*. This section is devoted to defining terms, deriving formulas, and giving examples involving annuities. First, we shall give three examples to illustrate the basic ideas involved. Try to do the problems in this section, applying the techniques used in these examples, until you thoroughly understand the basic ideas. The other examples will use formulas to get the answers more efficiently. In most examples, the computations will be made using a calculator. Review the section "How to Use a Calculator," p. xxvii.

Recall the following two basic formulas.

1. If a sum of $P$ dollars earns compound interest at a rate $i$ (written as a decimal) per compounding period, then the total amount $A_n$ dollars after $n$ compounding periods is given by the formula

$$A_n = P(1 + i)^n$$

2. The sum of the first $n$ terms of a geometric progression is

$$a + ar + ar^2 + ar^3 + \cdots + ar^{n-1} = \frac{a(1 - r^n)}{1 - r}$$

provided that $r \neq 1$. Most of the problems of this section can be solved using these two formulas.

## Simple Annuities

**EXAMPLE 1**   On the first of each month Mary deposits $200 in a bank that pays 6% interest compounded monthly. How much is in Mary's account immediately after the 20th deposit? How much interest is earned?

**Solution**   The rate of interest per conversion period is .005 since $\frac{.06}{12} = .005$. Immediately after the 20th deposit, the value of that deposit is still $200. The 19th deposit has been in the bank 1 month, so its value is $200(1 + .005)$. The 18th deposit has been in the bank 2 months, its value is $200(1 + .005)^2$. In general, the $k$th deposit has been in the bank $(20 - k)$ months, and its value is $200(1 + .005)^{20-k}$. Thus, the value of the first deposit is $200(1 + .005)^{19}$. (See Figure 5.2.) The total value (deposits plus interest) is given by

$$200 + 200(1.005) + 200(1.005)^2 + \cdots + 200(1.005)^{19}$$

This is the sum of the first 20 terms of a geometric progression with $a = 200$ and $r = 1.005$. It is equal to

$$\frac{200[1 - (1.005)^{20}]}{1 - 1.005} = \frac{200(1 - 1.10489558)}{-.005} = 4195.8232$$

Therefore, Mary has $4195.82 in her savings account. She has made 20 deposits of $200 each for a total of $4000. Since $4195.82 - 4000 = 195.82$, $195.82 is the earned interest.
**Do Exercise 17.**

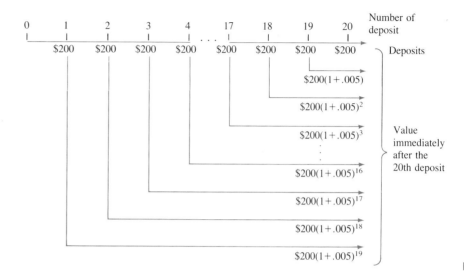

**FIGURE 5.2**

The foregoing is an example of a *simple annuity* because the interval between payments coincides with the conversion period at which the interest is being paid. The $4195.82 represents the *amount* (or *future value*) of the annuity.

### Sinking Fund

**EXAMPLE 2**     When John entered college, his parents decided to reward him upon graduation with a trip to Europe. They estimate the trip will cost $8000. They wish to make equal deposits on the first of each month, starting October 1, 1992, in a bank that pays a nominal rate of 6% interest compounded monthly. The last deposit will be June 1, 1996, at which time John will graduate. How much should they deposit each month so that the value of the annuity will be $8000?

**Solution**     John's parents will make 45 deposits of $D$ dollars each. We must find the value of $D$ which will make the sum of all deposits plus interest equal to $8000. As in Example 1, $i = .005$. On June 1, 1991, immediately after the 45th deposit, the value of that deposit will still be $D$ dollars. At that time, the 44th deposit will have earned interest for 1 month, so its value will be $D(1 + .005)$, and the 43rd deposit will have earned interest for 2 months and its value will be $D(1 + .005)^2$. In general, the $k$th deposit will have earned interest for $(45 - k)$ months and its value will be $D(1 + .005)^{45-k}$. (See Figure 5.3). The total value of all deposits plus interest will be

$$D + D(1.005) + D(1.005)^2 + D(1.005)^3 + \cdots + D(1.005)^{44}$$

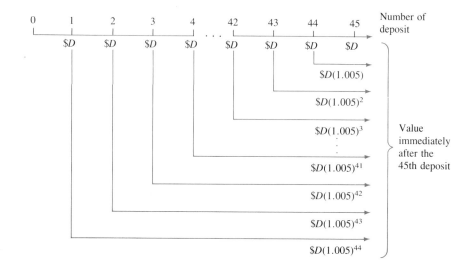

**FIGURE 5.3**

This is the sum of the first 45 terms of a geometric progression with $a = D$ and $r = 1.005$. Its value is

$$\frac{D[1 - (1.005)^{45}]}{1 - 1.005} \doteq \frac{D(1 - 1.25162082)}{-.005} = (50.324164)D$$

This must equal $8000, so

$$(50.324164)D \doteq 8000$$

and

$$D \doteq 8000/50.324164$$

$$\doteq 158.9693572$$

Thus, John's parents will deposit $158.97 each month.
**Do Exercise 19.**

Example 2 illustrates what is often called a *sinking fund*. A sinking fund is a fund into which equal payments are made at regular intervals in order to accumulate (deposits plus interest) a definite amount of money by a specified date. Often, sinking funds are established by companies to meet future replacement costs of equipment.

## Deferred Annuity Certain

**EXAMPLE 3**   On the day of their daughter's birth, Mr. and Mrs. Wise decided to set aside a sum of money to provide for the daughter's college education. They wish to make a single deposit in a bank that pays a rate of interest of 12% compounded yearly to provide a payment of $9000 on each of the daughter's 18th, 19th, 20th, and 21st birthdays. How much should they deposit?

**Solution**    It is easier to think of the single deposit as the sum of four deposits $A_1$, $A_2$, $A_3$, and $A_4$, made at the same time where $A_1$ plus interest will provide the first $9000, $A_2$ plus interest will provide the second $9000, and so on. On the daughter's 18th birthday, $A_1$ will have earned interest for 18 years. Thus, its value will be $A_1(1 + .12)^{18}$. Since $A_1$ provides the first $9000, we must have $A_1(1 + .12)^{18} = 9000$. Thus,

$$A_1 = 9000(1.12)^{-18}$$

Similarly, we find that

$$A_2 = 9000(1.12)^{-19}$$
$$A_3 = 9000(1.12)^{-20}$$
$$A_4 = 9000(1.12)^{-21} \qquad \text{(See Figure 5.4)}$$

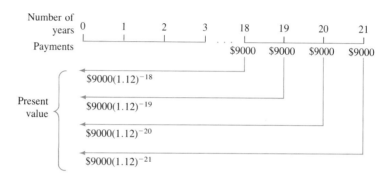

**FIGURE 5.4**

The total, single deposit is given by

$$9000(1.12)^{-18} + 9000(1.12)^{-19} + 9000(1.12)^{-20} + 9000(1.12)^{-21}$$

This is the sum of the first four terms of a geometric progression with $a = 9000(1.12)^{-18}$ and $r = 1.12^{-1}$. Its value is

$$\frac{9000(1.12)^{-18}[1 - (1.12^{-1})^4]}{1 - 1.12^{-1}} = \frac{9000(1.12)^{-17}(1 - (1.12)^{-4})}{1.12 - 1}$$

$$\doteq \frac{(1310.799068)(0.364481921)}{.12}$$

$$\doteq 3981.35$$

Thus, on the day of their daughter's birth, Mr. and Mrs. Wise should deposit $3981.35 in the bank.
**Do Exercise 21.**                                                                                        ■

Example 3 is an illustration of what is called a *deferred annuity certain*. It is "deferred" because the payments start at a future date, and it is "certain" because the payments are not contingent upon some uncertain event, such as the daughter being alive. In case of death, the payments would go to the heirs. The sum $3981.35 is the *present value* of this annuity.

The first three examples illustrate different transactions called annuities. We shall now formally define the basic terms even though some of them already have been introduced. We shall then derive several formulas and give examples to demonstrate how they can be used.

## Annuities: Basic Terms Defined

> **DEFINITION 5.4:**  An *ordinary annuity* is a series of equal payments to be paid, or received, at the end of consecutive periods of equal length. If the payments are made at the beginning of each period, the annuity is called an *annuity due*.
>
> An annuity is said to be *simple* if the interval of time between successive payments coincides with the conversion period at which interest is being paid. Otherwise, it is called a *general annuity*.
>
> The *payment period* is the interval of time at the end (or beginning) of which each payment is to be made.
>
> The *term* of an annuity is the time from the beginning of the first payment period to the end of the last payment period.
>
> A *deferred annuity* (or *intercepted annuity*) is an annuity for which the first payment is not made until more than one payment period has elapsed. It is an *immediate annuity* if the term begins immediately.
>
> An annuity is said to be *certain* if its term is fixed.
>
> A *contingent annuity* is an annuity for which the term depends on some uncertain event such as the death of an individual. In particular, a *temporary annuity* is an annuity payable over a given period, provided the recipient continues to live through that period, otherwise terminating at death. A *last survivor annuity* is an annuity to be paid until the last of two or more survivors' lives end. A *perpetuity* is an annuity whose payments continue forever.
>
> A *sinking fund* is an annuity that is established for the purpose of accumulating a sum of money to pay an obligation at a future designated date. Often sinking funds are established by companies to provide money for anticipated replacement costs of equipment.
>
> The *amount* (*future value*) of an annuity is the sum of money which would be accumulated by the end of the term if each of the payments were invested at a given rate of compound interest at the time of payment until the end of the term. The *accumulated value* of an annuity at a certain date is the sum of the compound amounts of all payments to that date.
>
> An amount deposited today that yields a given sum in the future is called the *present value* of the sum.
>
> The *present value of an annuity* is the sum of the present values of all the payments at the beginning of the term of the annuity.

## Derivations

We now derive formulas for the amount and present value of an annuity. In these, $R$, $i$, and $n$ will denote the periodic payment, the interest rate per period (written as a decimal), and the number of periods, respectively.

Let $S$ dollars be the amount of an ordinary annuity. Since the annuity is ordinary, the payments are made at the end of each period. Therefore the value of the $n$th payment, at the time of the $n$th payment, is $R$ dollars; the $(n - 1)$st payment has earned interest for one period, so its value is $\$R(1 + i)$; the $(n - 2)$nd payment has earned interest for two periods, so its value is $\$R(1 + i)^2$; and so on. In general, the $k$th payment has earned interest for $(n - k)$ periods, and its value is $\$R(1 + i)^{n-k}$. (See Figure 5.5.) Thus,

$$S = R + R(1 + i) + R(1 + i)^2 + \cdots + R(1 + i)^{n-1}$$

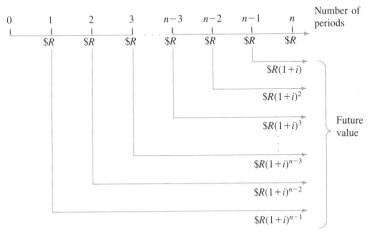

**FIGURE 5.5**

Since the expression on the right is the sum of the first $n$ terms of a G.P. with $a = R$ and $r = 1 + i$, we write

$$S = \frac{R[1 - (1 + i)^n]}{1 - (1 + i)} = R \cdot \frac{(1 + i)^n - 1}{i} \qquad (1)$$

The expression $[(1 + i)^n - 1]/i$ is often abbreviated $s_{\overline{n}|i}$ (read: "$s$ angle $n$ at $i$"). (See Table 5.1, p. 424.) Thus, the amount $S$ of an ordinary annuity is given by

$$S = R \cdot s_{\overline{n}|i} \qquad (2)$$

Now let $A$ dollars be the present value of an ordinary annuity. Let $A_k$ dollars be the present value of the $k$th payment. Since the payments are made at the end of each period, $A_1$ will have earned interest for one period, hence we have $A_1(1 + i) = R$ and $A_1 = \frac{R}{(1 + i)}$. Similarly, $A_2$ will have earned interest for two periods and we have $A_2(1 + i)^2 = R$, so that $A_2 = R/[(1 + i)^2]$. In general, it is easy to see that $A_k = R/[(1 + i)^k]$. (See Figure 5.6.) Thus,

$$A = \frac{R}{1 + i} + \frac{R}{(1 + i)^2} + \cdots + \frac{R}{(1 + i)^n}$$

This is the sum of the first $n$ terms of a G.P. whose first term and common ratio are $\frac{R}{(1 + i)}$ and $\frac{1}{(1 + i)}$, respectively. Therefore,

$$A = \frac{\dfrac{R}{1 + i}\left[1 - \dfrac{1}{(1 + i)^n}\right]}{1 - \dfrac{1}{1 + i}}$$

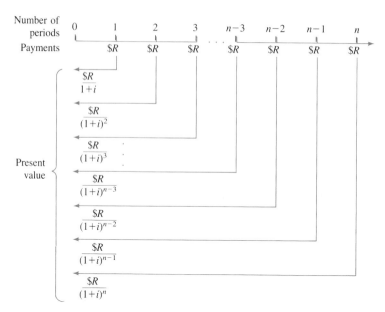

**FIGURE 5.6**

Multiplying the numerator and denominator of this fraction by $(1 + i)$, we obtain

$$A = R \cdot \frac{1 - (1 + i)^{-n}}{i} \tag{3}$$

The fractional quantity is often abbreviated $a_{\overline{n}|i}$ (read: "$a$ angle $n$ at $i$"). (See Table 5.2, p. 425.) Thus, we have

$$A = R \cdot a_{\overline{n}|i}* \tag{4}$$

---

*We derived the present value of an ordinary annuity by adding $A_1, A_2, \ldots, A_n$ where each $A_k$ was the value of the $k$th payment at the beginning of the first period. We could have argued instead, that the present value $A$, deposited as a lump sum together with the compound interest for $n$ periods, should yield the same amount at the end of the $n$th period. Thus, we should have

$$A(1 + i)^n = S$$

and

$$A = S(1 + i)^{-n}$$

We found earlier that

$$S = R \cdot \frac{(1 + i)^n - 1}{i}$$

Therefore,

$$A = R \cdot \frac{(1 + i)^n - 1}{i}(1 + i)^{-n}$$

$$A = R \cdot \frac{1 - (1 + i)^{-n}}{i}$$

This is the same formula we obtained earlier.

Now we wish to calculate the amount of an annuity due. Rather than derive another formula, we can proceed as follows. Payments are usually made at the beginning of each period, but assume instead that the term of the annuity started one period earlier and make an extra payment at the end of the term. Then, the first payment is made at the end of the first period, the second payment is made at the end of the second period, and so on. The extra payment made at the end will then be the $(n + 1)$st payment made at the end of the $(n + 1)$st period. We, therefore, have changed the annuity to an ordinary annuity with $(n + 1)$ periods. We can calculate the amount of this new annuity using Formula (1) (see the following annuity formulas), then subtract the extra payment which is made at the end of the $n$th period to obtain the amount of the original annuity.

Similarly, suppose we wish to calculate the present value of an annuity due. Since the payments are made at the beginning of the period, the second payment is made at the beginning of the second period which can also be considered the end of the first period. Thus, if the first payment was not made, we would have an ordinary annuity with $(n - 1)$ periods. We could calculate the present value of that annuity using Formula (3) and add to the result the value of the first payment which we assumed was not made.

### Annuity Formulas

*Amount of an ordinary annuity:*

$$S = R \cdot \frac{(1 + i)^n - 1}{i} \tag{1}$$

$$S = R \cdot s_{\overline{n}|i} \tag{2}$$

*Present value of an ordinary annuity:*

$$A = R \cdot \frac{1 - (1 + i)^{-n}}{i} \tag{3}$$

$$A = R \cdot a_{\overline{n}|i} \tag{4}$$

## Applications

**EXAMPLE 4**   John deposits $500 at the end of every 3 months for a period of 4 years in a bank that pays a nominal rate of 12% interest compounded quarterly. What is the amount in his account at the end of the 4 years? How much interest did he earn?

**Solution**   The rate of interest per period is .03 since $\frac{.12}{4} = .03$. In 4 years, there are 16 periods. The payments are made at the end of each period, so we use Formula (1) for an ordinary annuity. We obtain

$$S = 500 \left[ \frac{(1 + .03)^{16} - 1}{.03} \right] \doteq 10{,}078.44$$

Thus, at the end of the fourth year, John will have $10,078.44 in his account. We used a calculator to obtain this result. John made 16 deposits of $500 each for a total of $8000. Therefore, the interest earned was $2,078.44.
**Do Exercise 23.**

**EXAMPLE 5** Betty deposits $300 at the beginning of each month for a period of 4 years in a bank that pays a nominal rate of 12% interest compounded monthly. What is the amount in her account at the end of the 4 years?

**Solution** We must find the amount of an annuity due (since the payments are made at the beginning of each month). We have $R = 300$, $n = 48$, and $i = .01$ $\left(\text{since } \frac{.12}{12} = .01\right)$. We cannot use Formula (2) yet because the annuity is not ordinary. We change the problem by assuming that Betty makes an extra payment at the end of the fourth year, thus making 49 payments. We further assume that the term of the annuity started 1 month before she made the first payment, so that all 49 payments are made at the end of the periods. We have changed the problem to that of an ordinary annuity, and now Formula (2) can be used. We get

$$S = 300 \cdot s_{\overline{49}|.01}$$

$$\doteq 300(62.83483)$$

$$\doteq 18,850.449 \quad *$$

Since Betty did not, in fact, make the 49th payment, we must subtract the amount of that payment and conclude that after 4 years the amount in Betty's account is $18,550.45 since $18,850.45 - 300 = 18,550.45$. We have rounded the answer to the nearest cent.

**Do Exercise 25.** (See Exercise 45.) ■

**EXAMPLE 6** Roy wishes to purchase an annuity that will pay his daughter Lisa $1000 at the end of each month for the next 4 years. If the rate of interest is 12% compounded monthly, how much must he pay for this annuity?

**Solution** This annuity is ordinary since the payments are made at the end of each period. Thus, we wish to find the present value of an ordinary annuity and we use Formula (4), where $R = 1000$, $n = 48$, and $i = .01$ (since $\frac{.12}{12} = .01$). We get

$$A = 1000 \cdot a_{\overline{48}|.01}$$

$$\doteq 1000(37.973958)$$

$$\doteq 37,973.96 \quad **$$

Roy will have to pay $37,973.96 for this annuity.
**Do Exercise 27.** ■

**EXAMPLE 7** Peter won $2,400,000 at a state lottery. The state must pay $120,000 at the beginning of each of the next 20 years and purchases an annuity for that purpose. What is the present value of this annuity if the rate of interest is 8% compounded annually?

---

*The value of $s_{\overline{49}|.01}$ was obtained from Table 5.1 on page 425. We found the row with 49 on the left (since $n = 49$) and the column with .01 at the top (since $i = .01$). The number in that row and that column was 62.83483, the value of $s_{\overline{49}|.01}$.

**The value of $a_{\overline{48}|.01}$ was obtained using Table 5.2 on page 426.

**Solution**  We will use Formula (4). Since the payments are made at the beginning of each period, the annuity is due. If we assume that the first payment is not made, then the second payment becomes the first payment made at the end of the first year, the third payment becomes the second payment made at the end of the second year, and so on. We have changed the annuity to an ordinary annuity where $R = 120,000$, $n = 19$, and $i = .08$. The present value of this annuity is

$$S = 120,000 \frac{1-(1 + .08)^{-19}}{.08}$$

$$\doteq 120,000(9.60359920)$$

$$\doteq 1,152,431.904$$

We must add to this value the first payment which was made at the beginning of the first year. Hence, the present value of the original annuity due is $1,272,431.90 since $1,152,431.90 + $120,000 = $1,272,431.90.
**Do Exercise 29.** (See Exercise 46.)  ■

**EXAMPLE 8**  On her 30th birthday, Norma established a retirement fund by paying $200 a month to a life insurance company that pays 9% interest compounded monthly. If she retires on her 65th birthday, at which time she makes her last payment, and if she wishes to make equal withdrawals from the retirement fund until she reaches her 80th birthday, the first withdrawal being made 1 month after her 65th birthday, how much will she get each month from the retirement fund?

**Solution**  Norma established an ordinary annuity since the payments are made at the end of each period (the first payment was made 1 month after the beginning of the term of the annuity). The amount of this annuity on her 65th birthday may be calculated using Formula (1) with $R = 200$, $i = .0075$ (since $\frac{.09}{12} = .0075$), and $n = 420$ (since $35 \cdot 12 = 420$). We get

$$S = 200 \cdot \frac{(1 + .0075)^{420} - 1}{.0075}$$

$$\doteq 200(2,941.78448)$$

$$\doteq 588,356.896$$

When Norma retires, she essentially purchases an ordinary annuity whose present value is $588,356.90. Since she wishes to make monthly withdrawals until she reaches her 80th birthday, there will be 180 payments of $R$ dollars each. Using Formula (3), we get

$$588,356.90 = R \cdot \frac{1 - (1 + .0075)^{-180}}{.0075}$$

$$\doteq R(98.59340886)$$

Therefore,

$$R \doteq \frac{588,356.90}{98.59340886}$$

$$\doteq 5967.507431$$

Rounding the answer to the nearest cent, we conclude that Norma will receive $5967.51 a month. While Norma paid only $84,000 into the fund ($200 \cdot 420 = 84,000$) she will withdraw $1,074,151.80 (since $5967.51 \cdot 180 = 1,074,151.80$). The difference represents the interest earned.
**Do Exercise 31.**                                                                                    ■

**EXAMPLE 9**    A roofing company establishes a sinking fund that will provide $200,000 to replace its trucks in 5 years. It purchases an ordinary annuity and agrees to make equal payments every 6 months. How much is each payment if the rate of interest is 8% compounded semiannually?

**Solution**    Since the company purchased an ordinary annuity, we may use Formula (2) where $S = 200,000$, $R$ is to be found, $n = 10$, and $i = .04$. We get

$$200,000 = R \cdot \frac{(1 + .04)^{10} - 1}{.04}$$

$$\doteq R \cdot (12.006107)$$

Thus,

$$R \doteq \frac{200,000}{12.006107} \doteq 16,658.189$$

Thus, the company should pay $16,658.19 at the end of every 6 months for a period of 5 years.
**Do Exercise 33.**                                                                                    ■

**EXAMPLE 10**    Ralph would like to have $120,000 in his savings account 15 years from now. He is only able to deposit $200 at the end of each month for the next 4 years in a bank that pays a nominal interest rate of 12% compounded monthly. How much must each of his monthly deposits be in the last 11 years if he is to reach his goal?

**Solution**    Since the payments are made at the end of each month, the amount Ralph will have in his account in 4 years may be obtained using Formula (2). We get

$$S = 200(s_{\overline{48}|.01})$$

$$\doteq 200(61.222604)$$

$$\doteq 12,244.5208$$

This amount will continue to earn interest in the next 11 years. If Ralph did not make any additional deposits, the amount he would have in 11 more years would be given by

$$A = 12,244.5208(1 + .01)^{132}$$

$$\doteq 12,244.5208(3.71895856)$$

$$\doteq 45,536.86544$$

Rounding to the nearest cent, we see that the additional amount Ralph must accumulate is $74,463.13 since $120,000 - 45,536.87 = 74,463.13$.

Using Formula (1) with $n = 132$, $i = .01$, and $S = 74,463.13$, we get

$$74,463.13 = R \cdot \frac{(1 + .01)^{132} - 1}{.01}$$

$$\doteq R \cdot (271.895856)$$

Therefore,

$$R \doteq \frac{74,463.13}{271.895856}$$

$$\doteq 273.866366$$

Rounding to the nearest cent, Ralph should deposit $273.87 at the end of each month for the next 11 years.
**Do Exercise 35.**                                                                          ◾

**EXAMPLE 11**  Upon retirement, Fran used her life savings of $150,000 to purchase an annuity. The life insurance company from which she purchased the annuity offers 8% interest compounded semiannually and she is to receive one of 40 equal payments every 6 months starting in 6 months. How much will she receive every 6 months?

**Solution**  The payment being made at the end of each period indicates that the annuity is ordinary. Thus, we may use Formula (3) where $A = 150,000$, $R$ is to be found, $n = 40$, and $i = .04$. We get

$$150,000 = R \cdot \frac{1 - (1 + .04)^{-40}}{.04}$$

$$\doteq R(19.79277389)$$

$$R \doteq \frac{150,000}{19.79277389}$$

$$\doteq 7578.52$$

Fran will receive $7578.52 at the end of each 6-month period for the next 20 years.
**Do Exercise 37.**                                                                          ◾

We conclude with an example of a special type of annuity called a *perpetuity*.

## Perpetuities

**EXAMPLE 12**  Suppose someone wishes to establish a fund, earning 8% interest compounded annually, for the purpose of maintaining a family crypt. The maintenance cost is $300 per year with the first payment due immediately. How much should be paid now to purchase an annuity that will take care of the perpetual maintenance of the family crypt?

**Solution**  The first $300 would have to be paid now. Then, an amount earning $300 in interest each year should be deposited so that each year the interest could be withdrawn to take care of the maintenance of the crypt, keeping the principal amount in the bank.

This principal should be \$P where P(.08) = 300. Thus, $P = \frac{300}{.08} = 3750$. Adding the first payment, we obtain 4050 since 3750 + 300 = 4050. The cost of this perpetuity is \$4050.

**Do Exercise 41.**                                                                                    ∎

# Exercise Set 5.2

*In Exercises 1–6, find the value of the given symbol using a calculator.*

**1.** $(1 + .02)^{23}$

**2.** $(1 + .03)^{42}$

**3.** $(1 + .04)^{26}$

**4.** $(1 + .05)^{-17}$

**5.** $(1 + .02)^{-26}$

**6.** $(1 + .06)^{-9}$

*In Exercises 7–15, find the value of the given symbol using the tables on pages 424–426.*

**7.** $s_{\overline{10}|.0025}$

**8.** $s_{\overline{20}|.005}$

**9.** $s_{\overline{15}|.0075}$

**10.** $s_{\overline{32}|.01}$

**11.** $s_{\overline{20}|.0125}$

**12.** $a_{\overline{17}|.0025}$

**13.** $a_{\overline{12}|.005}$

**14.** $a_{\overline{34}|.0075}$

**15.** $a_{\overline{17}|.01}$

*In Exercises 16–21, use only the formulas that were used in Examples 1, 2, and 3 and do the computations with a calculator.*

**16.** Steven deposits \$150 in his savings account on the first of each month. Steven's bank pays 8% interest compounded monthly. How much is in Steven's account immediately after the 25th deposit? How much is earned interest?

**17.** Julie deposits \$300 in her savings account on the first of each month. Julie's bank pays 6% interest compounded monthly. How much is in Julie's account immediately after the 40th deposit? How much is earned interest?

**18.** When Gwen entered college, her parents decided that they would reward her upon graduation with a trip to Japan. They estimate the trip will cost \$6000. They wish to make equal deposits on the first of each month, starting December 1, 1990, in a bank that pays 9% interest compounded monthly. The last deposit will be made on June 1, 1994. How much should they deposit each month?

**19.** When Mike entered college, his parents decided that they would reward him upon graduation with a trip to India. They estimate the trip will cost \$9000. They wish to make equal deposits on the first of each month, starting March 31, 1990, in a bank that pays 12% interest compounded monthly. The last deposit will be made on December 31, 1994. How much should they deposit each month?

**20.** On the day of their son's birth, Mr. and Mrs. Planer decided to set aside a sum of money to provide for his college education. They wish to make a single deposit in a bank that pays 9% interest compounded yearly in order to provide a payment of \$8500 on each of the son's 18th, 19th, 20th, and 21st birthdays. How much should they deposit?

**21.** On the day of their daughter's birth, Mr. and Mrs. Bernard decided to set aside a sum of money to provide for the daughter's college education. They wish to make a single deposit in a bank that pays 8% interest compounded yearly in order to provide a payment of \$9500 on each of the daughter's 18th, 19th, 20th, and 21st birthdays. How much should they deposit?

*In Exercises 22–39, you may use Formulas 1–4 and a calculator. If, for a problem, the values of n and i appear in the tables on pages 424–426, you may use the tables for that problem.*

**22.** Lisa deposits \$450 at the end of every 6 months for a period of 7 years in a bank that pays 7% interest compounded semiannually. What is the amount in her account at the end of 7 years? How much interest did she earn?

**23.** Louis deposits \$250 at the end of each month for a period of 6 years in a bank that pays 6% interest compounded monthly. What is the amount in his account at the end of 6 years? How much interest did he earn?

**24.** Shirley deposits \$250 at the beginning of each month for a period of 3 years in a bank that pays 9% interest compounded monthly. What is the amount in her account at the end of 3 years? How much interest did she earn?

**25.** Peter deposits \$350 at the beginning of each quarter for a period of 5 years in a bank that pays 8% interest compounded quarterly. What is the amount in his account at the end of 5 years? How much interest did he earn?

**26.** Ted wishes to purchase an annuity that will pay his son Kris \$750 at the end of each month for the next 3 years starting in 1 month. If the rate of interest is 6% compounded monthly, how much must Ted pay for this annuity?

**27.** Dorothy wishes to purchase an annuity that will pay her son Mark \$950 at the end of each quarter for the next 6 years starting in 3 months. If the rate of interest is 8% compounded quarterly, how much must Dorothy pay for this annuity?

**28.** What is the present value of an annuity due that will pay $15,000 per year for a period of 15 years if the rate of interest is 7% compounded annually?

**29.** What is the present value of an annuity due that will pay $500 per month for a period of 4 years if the rate of interest is 9% compounded monthly?

**30.** On his 25th birthday, Charles established a retirement fund by paying $250 a month to a life insurance company that pays 6% interest compounded monthly. If he retires on his 65th birthday, at which time he will make his last payment, and if he wishes to make equal withdrawals from the retirement fund until he reaches his 75th birthday, the first withdrawal being made 1 month after his 65th birthday, how much will he get each month from the fund?

**31.** On her 35th birthday, Louise established a retirement fund by paying $350 a month to a life insurance company that pays 9% interest compounded monthly. If she retires on her 65th birthday, at which time she will make her last payment, and if she wishes to make equal withdrawals from the retirement fund until she reaches her 75th birthday, the first withdrawal being made 1 month after her 65th birthday, how much will she get each month from the fund?

**32.** A trucking company wishes to establish a sinking fund that will provide $350,000 to replace its trucks in 4 years. For this purpose, it purchases an ordinary annuity and will make equal payments every quarter. How much is each payment if the rate of interest is 9% compounded quarterly?

**33.** A manufacturer wishes to establish a sinking fund that will provide $500,000 to replace some of its machinery 5 years from now. For this purpose, it purchases an ordinary annuity and will make equal payments every 6 months. How much is each payment if the rate of interest is 7% compounded semiannually?

**34.** Steven would like to have $250,000 in his savings account 20 years from now. He will be able to deposit $150 at the end of each month for the next 5 years in a bank that pays 6% interest compounded monthly. How much must each of his monthly deposits be in the last 15 years if he is to reach his goal?

**35.** Kathy would like to have $150,000 in her savings account 12 years from now. She will be able to deposit $100 at the end of each month for the next 4 years in a bank that pays 12% interest compounded monthly. How much must each of her monthly deposits be in the last 8 years if she is to reach her goal?

**36.** Upon retirement, George used his life savings of $125,000 to purchase an annuity. The life insurance company from which he purchased the annuity pays 6% interest compounded monthly and he is to receive one of 240 equal payments every month, starting in 1 month. How much will he receive each month?

**37.** On her 62nd birthday, Gwen used her life savings of $200,000 to purchase an annuity. The life insurance company from which she purchased the annuity pays 9% interest compounded monthly. Gwen is to receive equal monthly payments starting in 1 month, until she receives the last payment on her 75th birthday. How much will she receive each month?

**38.** John has been depositing $150 on the first of each month over the last 15 years. The interest was 9% compounded monthly for the first 5 years, but it decreased to 6% compounded monthly for the last 10 years. How much was in his account immediately after the 180th deposit?

**39.** Rose has been depositing $250 on the first of each month over the last 25 years. The interest was 6% compounded monthly for the first 8 years, but it increased to 9% compounded monthly for the last 17 years. How much was in her account immediately after the 300th deposit?

**40.** Suppose that Mrs. Tradition wishes to establish a fund, earning 7% compounded annually, to maintain a family crypt. The maintenance cost is $250 per year with the first payment due immediately. How much must she pay now to purchase an annuity that will take care of the perpetual maintenance of the crypt?

**41.** Suppose that Mr. Goodfellow wishes to establish a fund, earning 8% compounded annually, to award a scholarship to the student majoring in business at his alma mater. The maintenance of the scholarship will cost $9500 a year with the first payment due immediately. How much should be paid now to purchase an annuity that will take care of the perpetual maintenance of the scholarship?

**42.** Prove the formula
$$\frac{1}{a_{\overline{n}|i}} = \frac{1}{s_{\overline{n}|i}} + i$$

**43.** Prove that the amount of an annuity due is given by
$$S = \frac{R[(1+i)^{n+1} - 1 - i]}{i} = R(S_{\overline{n+1}|i} - 1)$$

**44.** Prove that the present value of an annuity due is given by
$$A = \frac{R[1 + i - (1+i)^{-n+1}]}{i} = R(1 + a_{\overline{n+1}|i})$$

**45.** Solve Example 5 of this section using the formula of Exercise 43.

**46.** Solve Example 7 of this section using the formula of Exercise 44.

**47.** Do Exercise 24 using the formula of Exercise 43.

**48.** Do Exercise 25 using the formula of Exercise 43.

**49.** Do Exercise 28 using the formula of Exercise 44.

**50.** Do Exercise 29 using the formula of Exercise 44.

TABLE 5.1
AMOUNT OF ANNUITY
OF 1 PER PERIOD

$$S_{\overline{n}|i} = \frac{(1 + i)^n - 1}{i}$$

| $n$ | $\frac{1}{4}\%$ | $\frac{1}{2}\%$ | $\frac{3}{4}\%$ | $1\%$ | $1\frac{1}{4}\%$ |
|---|---|---|---|---|---|
| 1 | 1.000000 | 1.000000 | 1.000000 | 1.000000 | 1.000000 |
| 2 | 2.002499 | 2.005000 | 2.007500 | 2.010000 | 2.012500 |
| 3 | 3.007505 | 3.015025 | 3.022556 | 3.030100 | 3.037656 |
| 4 | 4.015023 | 4.030100 | 4.045225 | 4.060401 | 4.075627 |
| 5 | 5.025060 | 5.050251 | 5.075564 | 5.101005 | 5.126572 |
| 6 | 6.037623 | 6.075502 | 6.113631 | 6.152015 | 6.190654 |
| 7 | 7.052717 | 7.105880 | 7.159483 | 7.213535 | 7.268037 |
| 8 | 8.070347 | 8.141409 | 8.213179 | 8.285670 | 8.358888 |
| 9 | 9.090523 | 9.182116 | 9.274778 | 9.368527 | 9.463374 |
| 10 | 10.113249 | 10.228027 | 10.344339 | 10.462212 | 10.581666 |
| 11 | 11.138532 | 11.279167 | 11.421921 | 11.566834 | 11.713936 |
| 12 | 12.166377 | 12.335563 | 12.507586 | 12.682502 | 12.860361 |
| 13 | 13.196793 | 13.397241 | 13.601393 | 13.809327 | 14.021115 |
| 14 | 14.229784 | 14.464227 | 14.703403 | 14.947421 | 15.196379 |
| 15 | 15.265359 | 15.536549 | 15.813679 | 16.096895 | 16.386334 |
| 16 | 16.303521 | 16.614231 | 16.932281 | 17.257864 | 17.591163 |
| 17 | 17.344280 | 17.697302 | 18.059273 | 18.430442 | 18.811052 |
| 18 | 18.387640 | 18.785789 | 19.194717 | 19.614747 | 20.046190 |
| 19 | 19.433609 | 19.879718 | 20.338678 | 20.810894 | 21.296767 |
| 20 | 20.482192 | 20.979116 | 21.491218 | 22.019003 | 22.562977 |
| 21 | 21.533398 | 22.084012 | 22.652402 | 23.239193 | 23.845014 |
| 22 | 22.587230 | 23.194432 | 23.822295 | 24.471585 | 25.143077 |
| 23 | 23.643699 | 24.310404 | 25.000962 | 25.716301 | 26.457365 |
| 24 | 24.702807 | 25.431957 | 26.188469 | 26.973463 | 27.788082 |
| 25 | 25.764564 | 26.559116 | 27.384883 | 28.243198 | 29.135433 |
| 26 | 26.828975 | 27.691912 | 28.590269 | 29.525630 | 30.499626 |
| 27 | 27.896046 | 28.830372 | 29.804696 | 30.820886 | 31.880871 |
| 28 | 28.965785 | 29.974524 | 31.028231 | 32.129095 | 33.279382 |
| 29 | 30.038200 | 31.124396 | 32.260943 | 33.450386 | 34.695374 |
| 30 | 31.113295 | 32.280018 | 33.502900 | 34.784890 | 36.129066 |
| 31 | 32.191078 | 33.441419 | 34.754172 | 36.132739 | 37.580679 |
| 32 | 33.271555 | 34.608626 | 36.014828 | 37.494066 | 39.050438 |
| 33 | 34.354734 | 35.781669 | 37.284939 | 38.869006 | 40.538568 |
| 34 | 35.440620 | 36.960577 | 38.564576 | 40.257696 | 42.045300 |
| 35 | 36.529221 | 38.145380 | 39.853810 | 41.660273 | 43.570866 |
| 36 | 37.620543 | 39.336107 | 41.152714 | 43.076876 | 45.115502 |
| 37 | 38.714594 | 40.532788 | 42.461359 | 44.507645 | 46.679446 |
| 38 | 39.811380 | 41.735452 | 43.779819 | 45.952721 | 48.262939 |
| 39 | 40.910909 | 42.944129 | 45.108168 | 47.412248 | 49.866225 |
| 40 | 42.013185 | 44.158850 | 46.446479 | 48.886371 | 51.489553 |

**TABLE 5.1**
**AMOUNT OF ANNUITY**
**OF 1 PER PERIOD**

$$S_{\overline{n}|i} = \frac{(1 + i)^n - 1}{i}$$

| n | $\frac{1}{4}\%$ | $\frac{1}{2}\%$ | $\frac{3}{4}\%$ | $1\%$ | $1\frac{1}{4}\%$ |
|---|---|---|---|---|---|
| 41 | 43.118218 | 45.379644 | 47.794828 | 50.375234 | 53.133172 |
| 42 | 44.226012 | 46.606543 | 49.153289 | 51.878987 | 54.797337 |
| 43 | 45.336577 | 47.839575 | 50.521938 | 53.397776 | 56.482304 |
| 44 | 46.449918 | 49.076773 | 51.900853 | 54.931754 | 58.188332 |
| 45 | 47.566043 | 50.324167 | 53.290109 | 56.481072 | 59.915686 |
| 46 | 48.684957 | 51.575788 | 54.689785 | 58.045882 | 61.664632 |
| 47 | 49.806669 | 52.833667 | 56.099958 | 59.626341 | 63.435440 |
| 48 | 50.931185 | 54.097835 | 57.520707 | 61.222604 | 65.228383 |
| 49 | 52.058514 | 55.368324 | 58.952113 | 62.834830 | 67.043738 |
| 50 | 53.188659 | 56.645166 | 60.394253 | 64.463178 | 68.881784 |

**TABLE 5.2**
**AMOUNT OF ANNUITY**
**OF 1 PER PERIOD**

$$A_{\overline{n}|i} = \frac{1 - (1 + i)^{-n}}{i}$$

| n | $\frac{1}{4}\%$ | $\frac{1}{2}\%$ | $\frac{3}{4}\%$ | $1\%$ | $1\frac{1}{4}\%$ |
|---|---|---|---|---|---|
| 1 | 0.997506 | 0.995025 | 0.992556 | 0.990099 | 0.987654 |
| 2 | 1.992524 | 1.985099 | 1.977723 | 1.970395 | 1.963115 |
| 3 | 2.985061 | 2.970248 | 2.955556 | 2.940985 | 2.926534 |
| 4 | 3.975123 | 3.950496 | 3.926110 | 3.901965 | 3.878058 |
| 5 | 4.962716 | 4.925867 | 4.889439 | 4.853431 | 4.817835 |
| 6 | 5.947846 | 5.896385 | 5.845597 | 5.795476 | 5.746010 |
| 7 | 6.930519 | 6.862074 | 6.794637 | 6.728194 | 6.662725 |
| 8 | 7.910741 | 7.822960 | 7.736613 | 7.651677 | 7.568124 |
| 9 | 8.888520 | 8.779064 | 8.671576 | 8.566017 | 8.462344 |
| 10 | 9.863860 | 9.730412 | 9.599579 | 9.471304 | 9.345525 |
| 11 | 10.836767 | 10.677027 | 10.520674 | 10.367628 | 10.217803 |
| 12 | 11.807249 | 11.618933 | 11.434912 | 11.255077 | 11.079311 |
| 13 | 12.775310 | 12.556152 | 12.342345 | 12.133740 | 11.930184 |
| 14 | 13.740957 | 13.488708 | 13.243022 | 13.003702 | 12.770552 |
| 15 | 14.704197 | 14.416626 | 14.136994 | 13.865052 | 13.600545 |
| 16 | 15.665033 | 15.339926 | 15.024312 | 14.717873 | 14.420291 |
| 17 | 16.623475 | 16.258633 | 15.905024 | 15.562251 | 15.229918 |
| 18 | 17.579525 | 17.172769 | 16.779180 | 16.398268 | 16.029548 |
| 19 | 18.533192 | 18.082357 | 17.646829 | 17.226008 | 16.819307 |
| 20 | 19.484480 | 18.987420 | 18.508019 | 18.045552 | 17.599315 |
| 21 | 20.433396 | 19.887980 | 19.362798 | 18.856982 | 18.369694 |
| 22 | 21.379946 | 20.784060 | 20.211214 | 19.660379 | 19.130562 |
| 23 | 22.324136 | 21.675682 | 21.053314 | 20.455820 | 19.882036 |
| 24 | 23.265970 | 22.562867 | 21.889145 | 21.243386 | 20.624233 |
| 25 | 24.205456 | 23.445639 | 22.718754 | 22.023155 | 21.357268 |
| 26 | 25.142599 | 24.324019 | 23.542188 | 22.795203 | 22.081252 |
| 27 | 26.077405 | 25.198029 | 24.359492 | 23.559607 | 22.796298 |
| 28 | 27.009879 | 26.067691 | 25.170711 | 24.316442 | 23.502517 |
| 29 | 27.940030 | 26.933025 | 25.975892 | 25.065784 | 24.200016 |
| 30 | 28.867859 | 27.794055 | 26.775079 | 25.807707 | 24.888905 |

TABLE 5.2
AMOUNT OF ANNUITY
OF 1 PER PERIOD

$$A_{\overline{n}|i} = \frac{1 - (1 + i)^{-n}}{i}$$

| $n$ | $\frac{1}{4}\%$ | $\frac{1}{2}\%$ | $\frac{3}{4}\%$ | $1\%$ | $1\frac{1}{4}\%$ |
|---|---|---|---|---|---|
| 31 | 29.793376 | 28.650802 | 27.568317 | 26.542284 | 25.569289 |
| 32 | 30.716584 | 29.503285 | 28.355649 | 27.269588 | 26.241273 |
| 33 | 31.637490 | 30.351527 | 29.137121 | 27.989691 | 26.904961 |
| 34 | 32.556099 | 31.195550 | 29.912775 | 28.702665 | 27.560455 |
| 35 | 33.472417 | 32.035373 | 30.682655 | 29.408579 | 28.207857 |
| 36 | 34.386451 | 32.871018 | 31.446804 | 30.107504 | 28.847266 |
| 37 | 35.298205 | 33.702505 | 32.205264 | 30.799509 | 29.478781 |
| 38 | 36.207685 | 34.529856 | 32.958079 | 31.484662 | 30.102500 |
| 39 | 37.114898 | 35.353091 | 33.705289 | 32.163032 | 30.718518 |
| 40 | 38.019848 | 36.172230 | 34.446937 | 32.834685 | 31.326932 |
| 41 | 38.922541 | 36.987293 | 35.183064 | 33.499688 | 31.927834 |
| 42 | 39.822983 | 37.798302 | 35.913711 | 34.158107 | 32.521317 |
| 43 | 40.721180 | 38.605275 | 36.638919 | 34.810007 | 33.107474 |
| 44 | 41.617136 | 39.408234 | 37.358729 | 35.455452 | 33.686394 |
| 45 | 42.510859 | 40.207198 | 38.073180 | 36.094507 | 34.258167 |
| 46 | 43.402353 | 41.002187 | 38.782312 | 36.727235 | 34.822881 |
| 47 | 44.291624 | 41.793221 | 39.486166 | 37.353698 | 35.380623 |
| 48 | 45.178676 | 42.580320 | 40.184780 | 37.973958 | 35.931479 |
| 49 | 46.063518 | 43.363502 | 40.878194 | 38.588077 | 36.475535 |
| 50 | 46.946152 | 44.142788 | 41.566445 | 39.196116 | 37.012874 |

## 5.3 Mortgages

Installment buying is of great interest since almost everyone has been, or will be, involved in making purchases where it is necessary, because of the size of the original price, to pay only part of the price (the down payment) and pay off the balance by making payments at regular intervals. The purchase of a car or a home are examples of this type of financial transaction. Essentially, when a lending institution lends a sum of money to a person to be repaid with interest in equal payments at regular intervals, the lending institution purchases an annuity from that person. If the payments are made at the end of each period, the annuity is ordinary. Furthermore, if the interval of time between successive payments coincides with the conversion period at which the interest is being paid, the annuity is also simple and we can use Formula (3) of the preceding section.

$$A = R \cdot \frac{1 - (1 + i)^{-n}}{i}$$

where $A$ dollars is the present value of the annuity, $R$ dollars is the amount of each payment, $i$ (written as a decimal) is the rate of interest per conversion period, and $n$ is the number of periods (payments). Thus, $A$ dollars is the amount that the lending institution lends to the purchaser at the time of the purchase. When making such a purchase, the buyer is concerned about the portion of each payment that is interest and the portion that is being repaid on the principal (original loan). The treatment of annuities in the preceding section did not answer the following questions: how much of the loan remains to be paid after $k$ payments? how much of the $k$th payment is an interest payment? how much is used to repay the loan? We shall derive formulas to answer these and related questions.

## Derivation of Formulas

Suppose that a lending institution lends $A$ dollars to an individual who will repay the loan plus interest by making $n$ monthly payments of $R$ dollars each, the first payment being made after 1 month. The rate of interest is $i$ (written as a decimal) compounded monthly on the unpaid balance. Then, the relationships between $A$, $R$, $n$, and $i$ are given by Formula (3) of the preceding section.

Let $A_k$ dollars ($k = 0, 1, 2, \ldots, n$) be the unpaid balance on the loan after the $k$th payment, where $A_0 = A$ and $A_n = 0$. Let $B_k$ dollars ($k = 1, 2, 3, \ldots, n$) be the amount of the $k$th payment used to repay the principal. Let $I_k$ dollars ($k = 1, 2, 3, \ldots, n$) be the amount of the $k$th payment used for interest payment. Then, $B_k + I_k = R$ for each $k$.

We wish to derive formulas for $B_k$ and $I_k$. A simple way to derive these is to make use of the formulas we derived in the preceding section. (A more direct approach may be found in the interesting treatment of installment buying in [2], at the end of this section.)

Immediately after the $k$th payment has been made, $(n - k)$ payments remain to be made. Thus, the principal that remains to be paid is the present value of these $(n - k)$ payments. Using Formula (3) of the previous section, we find the present value to be

$$A_k = R \cdot \frac{1 - (1 + i)^{-(n-k)}}{i} = R \cdot \frac{1 - (1 + i)^{k-n}}{i}$$

It is easy to see that the amount of the $k$th payment used to repay the principal is the difference between what was left to be paid after the $(k-1)$st payment and what is left to be paid after the $k$th payment. That is, $B_k = A_{k-1} - A_k$. Therefore,

$$B_k = R \cdot \frac{1 - (1 + i)^{k-1-n}}{i} - R \cdot \frac{1 - (1 + i)^{k-n}}{i}$$

$$= R \cdot \frac{(1 + i)^{k-n} - (1 + i)^{k-1-n}}{i}$$

$$= R \cdot \frac{(1 + i)^{k-1-n}[(1 + i) - 1]}{i} = R \cdot (1 + i)^{k-1-n}$$

and

$$I_k = R - B_k$$

$$= R - R \cdot (1 + i)^{k-1-n} = R[1 - (1 + i)^{k-1-n}]$$

We now derive a formula that will enable us to find the number of payments necessary to pay off a certain loan. Start with the formula

$$A = R \cdot \frac{1 - (1 + i)^{-n}}{i}$$

This yields

$$(1 + i)^n = \frac{R}{R - Ai}$$

Taking the natural logarithm of both sides, we get

$$n\ln(1 + i) = \ln(R) - \ln(R - Ai)$$

and

$$n = \frac{\ln(R) - \ln(R - Ai)}{\ln(1 + i)}$$

### Summary of Related Formulas

*Total amount of the loan:*

$$A = R \cdot \frac{1 - (1 + i)^{-n}}{i} \tag{1}$$

*Unpaid balance on the loan after the kth payment:*

$$A_k = R \cdot \frac{1 - (1 + i)^{k-n}}{i} \tag{2}$$

*Amount of the principal repaid in the kth payment:*

$$B_k = R \cdot (1 + i)^{k-1-n} \tag{3}$$

*Amount of interest paid in the kth payment:*

$$I_k = R[1 - (1 + i)^{k-1-n}] \tag{4}$$

*Number of payments necessary to pay off the loan:*

$$n = \frac{\ln(R) - \ln(R - Ai)}{\ln(1 + i)} \tag{5}$$

$A$ = Amount of the loan, in dollars.
$R$ = Payment per period, in dollars.
$i$ = Rate of interest per period, written as a decimal.
$n$ = Number of periods (payments).
$A_k$ = Unpaid balance on the loan, after the $k$th payment, in dollars.
$B_k$ = Amount of the principal paid in the $k$th payment, in dollars.
$I_k$ = Amount of interest paid in the $k$th payment, in dollars.

**A CLOSER LOOK**   When someone borrows money and agrees to repay the loan plus interest by making equal payments at regular intervals, it is understood that the interest is being charged on the unpaid balance of the principal. We have based the derivation of the five formulas on the concept of annuity since the lending institution does, in fact, purchase an annuity from the borrower. We should verify that the results we obtain are consistent with the concept of installment buying.

1. The interest paid in the $(k + 1)$st payment should be the interest earned by the unpaid balance after the $k$th payment. That is, we should have

   $$I_{k+1} = A_k \cdot i$$

   Replacing $k$ by $k + 1$ in Formula (4), we get

   $$I_{k+1} = R[1 - (1 + i)^{k-n}]$$

   Using Formula (2), we obtain

   $$A_k \cdot i = R \cdot \frac{1 - (1 + i)^{k-n}}{i} \cdot i$$

   $$= R[1 - (1 + i)^{k-n}]$$

   Thus, $I_{k+1} = A_k \cdot i$ since both quantities are equal to $R[1 - (1 + i)^{k-n}]$.

2. The unpaid balance on the loan after the $k$th payment $A_k$ should be equal to the total amount of the loan $A$, minus the sum of the amounts of the principal paid in each of the first $k$ payments. That is, the following should be true:

   $$A_k = A - (B_1 + B_2 + \cdots + B_k) \qquad \text{(See Exercise 19.)}$$

3. $B_1$, the amount of principal repaid in the first payment, is $R(1 + i)^{-n}$, which is the present value of the $n$th payment. Similarly, $B_2$, the amount of principal repaid in the second payment, is $R(1 + i)^{1-n}$, which is the present value of the $(n - 1)$st payment. In general, $B_k$, the amount of principal repaid in the $k$th payment, is equal to the present value of the $(n - k + 1)$st payment, since both of these quantities are equal to $R(1 + i)^{k-1-n}$.

4. See Exercise 21.

## Applications

**EXAMPLE 1**   Mr. and Mrs. Beauman bought a house for $150,000. They made a 20% down payment, with the balance amortized by a 25-year mortgage at a rate of 9% compounded monthly.

   **a.** Find the amount of their monthly mortgage payment.
   **b.** Find the amount of interest paid in the 100th payment.
   **c.** Find the amount of principal paid in the 100th payment.
   **d.** Find the amount of the mortgage the Beaumans will have paid in 15 years.
   **e.** Find the total amount of interest the Beaumans will have paid once the mortgage is paid off.

**Solution**   **a.** The down payment was $30,000 since 150,000(0.2)= 30,000. Thus, the mortgage is $120,000. Use Formula (1) with $A = 120{,}000$, $i = .0075$ (since $\frac{.09}{12} = .0075$), $n = 300$ (since $25 \cdot 12 = 300$), and $R$ is to be found. We get

$$120{,}000 = R \cdot \frac{1 - (1 + .0075)^{-300}}{.0075}$$

$$\doteq R(119.1616222)$$

Thus,

$$R \doteq \frac{120{,}000}{119.1616222}$$

$$\doteq 1007.035636$$

Rounding the answer to the nearest cent, we conclude that the Beaumans will have a $1007.04 monthly mortgage payment.

**b.** The amount of interest paid in the 100th payment may be found using Formula (4) with $R = 1007.04$, $i = .0075$, $k = 100$, and $n = 300$. We get

$$I_{100} = 1007.04[1 - (1 + .0075)^{100-1-300}]$$

$$\doteq 1007.04(.777287813)$$

$$\doteq 782.7599192$$

Rounding to the nearest cent, the interest paid in the 100th payment is $782.76.

**c.** Using Formula (3) to find the amount of principal repaid in the 100th payment, with $R = 1007.04$, $i = .0075$, $k = 100$, and $n = 300$. We obtain

$$B_{100} = 1007.04(1 + .0075)^{100-1-300}$$

$$\doteq 1007.04(.222712187)$$

$$\doteq 224.2800809$$

Thus, the amount of interest paid in the 100th payment is $224.28. Note that, as expected, $224.28 + 782.76 = 1007.04$.

**d.** The amount of principal the Beaumans have paid after 15 years is the total amount of the mortgage minus the unpaid balance after 15 years. To find the unpaid balance after 15 years use Formula (2) with $R = 1007.04$, $i = .0075$, $n = 300$, and $k = 180$ (since $15 \cdot 12 = 180$).

$$A_{180} = 1007.04 \cdot \frac{1 - (1 + .0075)^{180-300}}{.0075}$$

$$\doteq 1007.04(78.94169267)$$

$$\doteq 79{,}497.44218$$

Rounding to the nearest cent, $A_{180} = 79{,}497.44$. Thus, after 15 years, the Beaumans will have paid $40,502.56 of the principal since $120{,}000 - 79{,}497.44 = 40{,}502.56$.

**e.** Since the Beaumans are making monthly payments of $1007.04 for a period of 25 years, when the mortgage is paid off they will have paid a total of $302,112.00

since $12 \cdot 25 \cdot 1007.04 = 302,112$. Therefore, the total amount they paid in interest is \$182,112 since $302,112 - 120,000 = 182,112$.

**Do Exercise 3.** ∎

**EXAMPLE 2**    Mr. and Mrs. Dougherty have a combined take-home pay of \$4187 per month. A mortgage company will lend them an amount which will yield a maximum monthly payment of 25% of their take-home pay. If the rate of interest is 12% compounded monthly, the loan is to be amortized over a period of 25 years, and a 20% down payment is required, what is the maximum value of a house the Doughertys can purchase?

**Solution**    The maximum monthly payment is given by

$$\$4187(.25) = \$1046.75$$

Since the loan is to be amortized over a period of 25 years, the number of monthly payments will be 300. The interest rate per month is .01 since $\frac{.12}{12} = .01$. Using Formula (1), we get

$$A = 1046.75 \frac{1 - (1 + .01)^{-300}}{.01}$$

$$\doteq 1046.75(94.94655125)$$

$$\doteq 99,385.30252$$

Rounding to the nearest cent, we conclude that the Doughertys may borrow up to \$99,385.30. But a 20% down payment is required; therefore, if $V$ dollars is the maximum value of a house they can purchase, we must have

$$V - (.2)V = 99,385.30$$

That is,

$$(.8)V = 99,385.30$$

and

$$V = \frac{99,385.3}{.8} \doteq 124,231.625$$

Thus, the maximum value of a house the Doughertys can purchase is approximately \$124,230.

**Do Exercise 9.** ∎

**EXAMPLE 3**    Ann bought a new car for \$12,500. She is required to make a 20% down payment. She borrows the balance from her credit union. Her debt is to be amortized over a period of 5 years at 9% interest compounded monthly.

**a.** How much is each monthly payment?

**b.** How much of the 30th payment will pay interest?

**c.** How much of the 30th payment is loan payment?

**d.** If Ann decides to sell her car after 3 years, how much should she get to be able to pay the balance of her loan?

**Solution** **a.** Ann must make a $2500 down payment since $12,500(.20) = 2500$. Therefore, she will borrow $10,000 from her credit union. Use Formula (1) with $A = 10,000$, $n = 60$ (since $5 \cdot 12 = 60$), and $i = .0075$ (since $\frac{.09}{12} = .0075$).

$$10,000 = R \cdot \frac{1 - (1 + .0075)^{-60}}{.0075}$$

$$\doteq R(48.17337347)$$

Therefore,

$$R \doteq \frac{10,000}{48.17337347} \doteq 207.5835525$$

Rounding to the nearest cent, Ann's payments are $207.58 per month.

**b.** To find the amount of interest paid in the 30th payment, we use Formula (4) with $k = 30$, $R = 207.58$, $i = .0075$, and $n = 60$. We get

$$I_{30} = 207.58[1 - (1 + .0075)^{30-1-60}]$$

$$= 207.58(0.206762384)$$

$$= 42.91973567$$

Thus, $42.92 is paid in interest in the 30th payment.

**c.** Using part **b**, the amount of the loan paid in the 30th payment is $164.66 since $207.58 - 42.92 = 164.66$.

**d.** In 3 years, Ann will have made 36 payments. The remaining balance is found using Formula (2) with $k = 36$ and the values of $R$, $i$, and $n$ as in part **b**.

$$A_{36} = 207.58 \cdot \frac{1 - (1 + .0075)^{36-60}}{.0075}$$

$$= 207.58(21.88914613)$$

$$= 4,543.748954$$

Ann should sell her car for at least $4543.75 to be able to pay off the loan.
**Do Exercise 13.** ∎

**EXAMPLE 4** Upon graduation from college, Roy had accumulated a debt of $7600. No interest was charged while he was in college, but the lending institution charges 9% interest compounded monthly starting at graduation. Roy wishes to repay his debt as quickly as possible but can only afford payments of $200 per month. How long will it take him to repay his debt?

**Solution** We shall use Formula (5) with $R = 200$, $A = 7600$, and $i = .0075$ since $\frac{.09}{12} = .0075$. We get

$$n = \frac{\ln(200) - \ln[200 - 7600(.0075)]}{\ln(1 + .0075)}$$

$$= 44.89722564$$

Hence, it will take Roy 45 months to repay his student loan. (The last payment will be a bit less than $200.)
**Do Exercise 15.** ∎

# Exercise Set 5.3

1. Mr. and Mrs. Sherman bought a lot for $36,500. They made a 20% down payment, with the balance amortized by a 25-year mortgage at 7.25% interest compounded monthly.

   a. Find the amount of their monthly mortgage payment.
   b. Find the amount of interest paid in the 150th payment.
   c. Find the amount of principal paid in the 150th payment.
   d. Find the amount of the mortgage the Shermans will have paid in 12 years.
   e. Find the total amount of interest the Shermans will have paid once the mortgage is paid off.

2. Mr. and Mrs. Larsen bought a house for $200,000. They made a 40% down payment, with the balance amortized by a 25-year mortgage at 9% interest compounded monthly.

   a. Find the amount of their monthly mortgage payment.
   b. Find the amount of interest paid in the 200th payment.
   c. Find the amount of principal paid in the 200th payment.
   d. Find the amount of the mortgage the Larsens will have paid in 17 years.
   e. Find the total amount of interest the Larsens will have paid once the mortgage is paid off.

3. Mr. and Mrs. Bateau bought a boat for $12,500. They made a 20% down payment, with the balance amortized by a 5-year mortgage at 9% interest compounded monthly.

   a. Find the amount of their monthly mortgage payment.
   b. Find the amount of interest paid in the 25th payment.
   c. Find the amount of principal paid in the 25th payment.
   d. Find the amount of the mortgage the Bateaus will have paid in 3 years.
   e. Find the total amount of interest the Bateaus will have paid once the mortgage is paid off.

4. Charles bought a car for $10,000. He made a 20% down payment, with the balance amortized by a 3-year loan at 12% interest compounded monthly.

   a. Find the amount of his monthly payment.
   b. Find the amount of interest paid in the 22nd payment.
   c. Find the amount of principal paid in the 22nd payment.
   d. Find the amount of the loan Charles will have paid in 2 years.
   e. Find the total amount of interest Charles will have paid once the loan is paid off.

5. Mr. and Mrs. Kelley bought a house for $120,000. They paid 25% down, with the balance amortized by a 20-year mortgage at 12% interest compounded monthly.

   a. Find the amount of their monthly mortgage payment.
   b. Find the amount of interest paid in the 125th payment.

   c. Find the amount of principal paid in the 125th payment.
   d. Find the amount of the mortgage the Kelleys will have paid in 13 years.
   e. Find the total amount of interest the Kelleys will have paid once the mortgage is paid off.

6. Mr. and Mrs. Nelson bought a house for $190,000. They paid 25% down, with the balance amortized by a mortgage at 12% interest compounded monthly. Find the amount of their monthly mortgage payment if the loan is to be paid off in a. 40 years, b. 30 years, c. 25 years, d. 20 years, e. 15 years, and f. 10 years. Compare the monthly payments.

7. Same as Exercise 6 if the house was bought for $210,000 and the rate of interest was 9% compounded monthly.

8. Mr. and Mrs. Leva have a combined take-home pay of $5180 per month. A mortgage company will lend them an amount which will yield a maximum monthly payment of 20% of their take-home pay. If the rate of interest is 11% compounded monthly, the loan is to be amortized over a period of 30 years, and a 20% down payment is required, what is the maximum value of a house the Levas can purchase?

9. Mr. and Mrs. Chang have a combined take-home pay of $5212 per month. A mortgage company will lend them an amount which will yield a maximum monthly payment of 25% of their take-home pay. If the rate of interest is 11.5% compounded monthly, the loan is to be amortized over a period of 20 years, and a 15% down payment is required, what is the maximum value of a house the Changs can purchase?

10. Dave has a take-home pay of $3214 per month. A bank will lend him an amount which will yield a maximum monthly payment of 15% of his take-home pay toward the purchase of a boat. If the rate of interest is 12% compounded monthly, the loan is to be amortized over a period of 10 years, and a 30% down payment is required, what is the maximum value of a boat Dave can purchase?

11. Mr. and Mrs. Udeen have a combined take-home pay of $45,228 per year. A mortgage company will lend them an amount which will yield a maximum monthly payment of 25% of their take-home pay. If the rate of interest is 10% compounded monthly, the loan is to be amortized over a period of 35 years, and a 15% down payment is required, what is the maximum value of a house the Udeens can purchase?

12. Jerry bought a car for $12,500. He is required to make a 20% down payment. He borrows the balance from his credit union. His debt is to be amortized over a period of 5 years at 9% interest compounded monthly.

**a.** How much is each monthly payment?

**b.** How much of the 23rd payment will pay interest?

**c.** How much of the 23rd payment is loan payment?

**d.** If Jerry decides to trade his car after 2 years, how much must he pay the credit union to pay off his loan, assuming there is no penalty for early payment?

**13.** Kathy bought a new car for $12,000. She makes a one-third down payment. She borrows the balance from her bank. Her debt is to be amortized over a period of 3 years at 12% interest compounded monthly.

**a.** How much is each monthly payment?

**b.** How much of the 19th payment will pay interest?

**c.** How much of the 19th payment is loan payment?

**d.** If Kathy decides to sell her car after the 25th payment, how much should she get to be able to pay the balance of her loan from the proceeds?

**14.** Mr. and Mrs. Galbraith bought a new home for $120,000. They made a 25% down payment and borrowed the balance from a mortgage company. Their debt is to be amortized over a period of 20 years at 12% interest compounded monthly.

**a.** How much is each monthly payment?

**b.** How much of the 50th payment will pay interest?

**c.** How much of the 50th payment is principal payment?

**d.** If the Galbraiths decide to sell their home after 10 years and succeed in selling it for $130,000, how much cash will be available to them (neglecting the costs of the transaction)?

**15.** Upon graduation from college, Lisa had accumulated a debt of $9800. No interest was charged while she was in college, but the lending institution charges 9% interest compounded monthly starting at graduation. Lisa wishes to repay her debt as quickly as possible, but can afford payments of only $250 per month. How long will it take to repay her debt?

**16.** To build a new hot tub in his backyard, Bob borrowed $9850. The lending institution charges 12% interest compounded

monthly. Bob wishes to repay his debt as quickly as possible, but can afford payments of only $375 a month. How long will it take to repay his debt?

**17.** To start a small business, Sharon had accumulated a debt of $57,600. No interest was charged until the business started, but the lending institution charged 12% interest compounded monthly as soon as the business was in operation. Sharon wished to repay her debt as quickly as possible and decided that she could pay $1200 a month. How long will it take to repay her debt?

**18.** Mr. and Mrs. Esser borrowed $15,000 to remodel their home. The lending institution charges 11% interest compounded monthly. The Essers wish to repay their debt as quickly as possible but can afford payments of only $380 a month. How long will it take to repay their debt?

**19.** Verify that the unpaid balance on a loan after the $k$th payment $A_k$ is equal to the total amount of the loan $A$ minus the sum of the amounts of the principal paid in each of the first $k$ payments. That is, verify that

$$A_k = A - (B_1 + B_2 + B_3 + \cdots + B_k)$$

(*Hint:* Use Formula (2) on the left for $A_k$ and Formulas (1) and (3) on the right for $A$ and each of the $B_i$'s; then, simplify using the formula for the sum of the first $n$ terms of a G.P.)

**20.** Formula (5) gives the number of payments necessary to pay off a loan in terms of $\ln(R)$, $\ln(R - Ai)$, and $\ln(1 + i)$. Clearly $R$ and $1 + i$ are positive, so $\ln(R)$ and $\ln(1 + i)$ are defined. Verify that $R - Ai > 0$. (*Hint:* Use Formula (1) and replace $A$ by $R[1 - (1 + i)^{-n}]/i$.)

**21.** The total amount of interest paid on a loan is the total amount paid ($\$nR$) minus the original amount of the loan ($\$A$). It is also the sum of the interest amounts paid with each of the $n$ payments. That is

$$nR - A = I_1 + I_2 + I_3 + \cdots + I_n$$

Verify this fact using Formulas (1) and (4) of this section.

For those interested in pursuing these applications further, the following readings will be helpful.

1. John D. Baildon. Arithmetic Progressions and the Consumer, *The College Mathematics Journal*, Vol. 16, No. 5 (1985), 395–397.

2. Ann D. Holley, A Question of Interest, *The Two-Year College Mathematics Journal*, Vol. 9, No. 2 (1978), 81–83.

3. Morris Morduchow, Discrete and Continuous Compounding, *The American Mathematical Monthly*, Vol. 92, No. 10 (1985), 734–735.

4. Stanley G. Wayment, Another Question of Interest, *The Two-Year College Mathematics Journal*, Vol. 11, No. 4 (1980), 252–254.

# 5.4 Chapter Review

SUMMARY

A finite sequence is a function whose domain is the set $\{1, 2, \ldots, n\}$; an infinite sequence is a function whose domain is the set $\{1, 2, 3, \ldots, n, \ldots\}$. The sequence $a, a + d, a + 2d, a + 3d, \ldots$ is called an arithmetic progression. The $k$th term of an A.P. is

$$s_k = a + (k - 1)d$$

and the sum of the first $n$ terms is

$$S_n = \frac{n[2a + (n - 1)d]}{2} = \frac{n(s_1 + s_n)}{2}$$

The sequence $a, ar, ar^2, ar^3, \ldots$ is called a geometric progression. The $k$th term of a G.P. is

$$s_k = ar^{k-1}$$

and the sum of the first $n$ terms is

$$S_n = \frac{a(1 - r^n)}{1 - r} \quad \text{whenever } 1 - r \neq 0$$

If interest is compounded at regular time intervals, and if $P$ dollars is the principal (present value), $i$ is the rate of interest per conversion period (written as a decimal), $n$ is the number of conversion periods, and $S_n$ is the compound amount (future value) after $n$ conversion periods, then

$$S_n = P(1 + i)^n \quad \text{and} \quad P = S_n(1 + i)^{-n}$$

Whenever a transaction involves a sequence of equal payments made periodically, this transaction is called an annuity. In an *ordinary annuity*, each payment is made at the end of a payment period and in an *annuity due* each payment is made at the beginning of a payment period.

The basic formulas for ordinary annuities are

$$S = R \cdot \frac{(1 + i)^n - 1}{i} = R \cdot s_{\overline{n}|i}$$

$$A = R \cdot \frac{1 - (1 + i)^{-n}}{i} = R \cdot a_{\overline{n}|i}$$

and for an annuity due:

$$S = R(s_{\overline{n+1}|i} - 1)$$

$$A = R(1 + a_{\overline{n-1}|i})$$

where $R$ = amount of each payment, $n$ = number of payments, $i$ = rate of interest per period (written as a decimal), $S$ = amount (future value) of the annuity, and $A$ = the present value of the annuity.

A sinking fund is an annuity which is established for the express purpose of accumulating a sum of money to pay an obligation at a future designated date.

Suppose that a lending institution lends $A$ dollars, at the periodic rate of interest $i$ (written as a decimal) on the unpaid balance, to be repaid in $n$ periodic payments of $R$ dollars each and each payment is made at the end of each period. Let $A_k$ dollars be the unpaid balance on the loan after the $k$th payment, $B_k$ dollars be the portion of the $k$th payment used to repay the principal, and $I_k$ dollars be the portion of the $k$th payment used to pay interest on the unpaid balance. Then

$$A = R \cdot \frac{1 - (1 + i)^{-n}}{i}$$

$$A_k = R \cdot \frac{1 - (1 + i)^{k-n}}{i}$$

$$B_k = R \cdot (1 + i)^{k-1-n}$$

$$I_k = R[1 - (1 + i)^{k-1-n}]$$

$$n = \frac{\ln(R) - \ln(R - Ai)}{\ln(1 + i)}$$

**SAMPLE EXAM QUESTIONS**

1. Find the first five terms of each of the sequences defined by

   a. $s_n = \dfrac{4}{3n + 1}$

   b. $s_n = (-1)^n \dfrac{2n}{n + 3}$

   c. $s_n = \dfrac{n^2 + 2n + 3}{2n^2 + n - 2}$

2. If $700 is deposited in a bank that pays 6% interest compounded monthly, find the amount in that account at the end of each of the first 6 months.

3. Let $s$ be an arithmetic progression whose first term is $-3$ and whose common difference is 3.

   a. Find the first four terms of this A.P.

   b. Find the sum of the first 200 terms.

**4.** The 35th and 57th terms of an A.P. are 20 and 31, respectively.

    **a.** Find the first five terms of this A.P.
    **b.** Find the sum of the first 1000 terms.

**5.** Let $s$ be a G.P. whose first term is 4 and whose common ratio is $-2$.

    **a.** Find the second, third, and fourth terms.
    **b.** Find the sum of the first 20 terms.

**6.** The seventh and tenth terms of a G.P. are 192 and 1536, respectively.

    **a.** Find the first five terms.
    **b.** Find the sum of the first 20 terms.

**7.** A bank pays 6% interest compounded monthly. Wendy deposits $325 the first of each month. How much is in her account immediately after the 54th deposit?

**8.** Suppose that one-third of a container full of pure alcohol is replaced by water and the solution is mixed thoroughly. Then, one-third of the solution is replaced by water and mixed thoroughly. If this is done repeatedly, what fraction of the original amount of alcohol is left after the 20th replacement?

**9.** Suppose that a new car is worth $20,000 and that at the end of each year its value depreciates 17% of the value it had at the beginning of that year. What will the car be worth at the end of 10 years?

**10.** Tim deposits $500 on the first of each month in a bank that pays 7% interest compounded monthly. How much is in his account immediately before the 40th deposit?

**11.** Jeremy deposits $300 on the first of each month in a bank that pays 6.5% interest compounded monthly. How much is in Jeremy's account immediately after the 30th deposit?

**12.** When Lisa entered college, her parents decided that they would reward her upon graduation with a trip to France. They estimate the trip will cost $10,000. They wish to make equal deposits on the first of each month, starting October 1, 1990, in a bank that pays 6% interest compounded monthly. The last deposit will be June 1, 1994, at which time Lisa will graduate. How much should they deposit each month?

**13.** On the day of their son's birth, Mr. and Mrs. Tarte decided to set aside a sum of money to provide for his college education. They wish to make a single deposit in a bank that pays 9% interest compounded yearly in order to provide a payment of $12,000 on each of the son's 18th, 19th, 20th, and 21st birthdays. How much should they deposit?

**14.** Janet deposits $200 at the beginning of each month in a bank that pays 9% interest compounded monthly. How much will be in her account at the end of 5 years?

**15.** John deposits $350 at the end of every 3 months for a period of 5 years in a bank that pays 8% interest compounded quarterly. What is the final amount in his account?

**16.** Mary wishes to purchase an annuity that will pay her son $800 at the end of each month for the next 6 years. If the rate of interest is 8% compounded monthly, how much must she pay for this annuity?

**17.** What is the present value of an annuity due that will pay $15,000 per year for a period of 20 years if the rate of interest is 9% compounded annually?

**18.** Nancy won $500,000 at the Washington State Lottery. After taxes, she is to receive $20,000 a year for the next 20 years with the first payment being made immediately. The State of Washington purchases an annuity that will provide these payments. If the rate of interest is 8% compounded annually, what is the cost of the annuity?

**19.** On his 25th birthday, John established a retirement fund by paying $300 per month to a life insurance company that pays 8% interest compounded monthly, the first payment being made 1 month after his birthday. If he retires on his 65th birthday, at which time he makes the last payment, and if he wishes to make equal withdrawals until he reaches his 75th birthday, the first withdrawal being made 1 month after his 65th birthday, how much will he get each month from the retirement fund?

**20.** A company establishes a sinking fund that will provide $300,000 to update its machinery in 6 years. For this purpose, it purchases an ordinary annuity and will make equal payments every 3 months. How much is each payment if the rate of interest is 8% compounded quarterly?

**21.** Judy would like to have $200,000 in her savings account 20 years from now. She will be able to deposit $250 at the end of each month for the next 5 years in a bank that pays 9% interest compounded monthly. How much must she deposit monthly in the last 15 years if she is to reach her goal?

**22.** Mr. and Mrs. Bernard purchased a house for $240,000. They made a 30% down payment, with the balance amortized by a 20-year mortgage at 11.5% interest compounded monthly.

   **a.** Find the amount of their monthly mortgage payment.
   **b.** Find the amount of interest in the 50th payment.
   **c.** Find the amount of principal in the 50th payment.
   **d.** Find the amount of the mortgage the Bernards will have paid in 13 years.
   **e.** Find the total amount of interest the Bernards will have paid once the mortgage is paid off.

**23.** Mr. and Mrs. Sherman have a combined take-home pay of $5540 per month. A mortgage company will lend them an amount that will yield a maximum monthly payment of 25% of their take-home pay. If the rate of interest is 12% compounded monthly, the loan is to be amortized over a period of 20 years, and a 20% down payment is required, what is the maximum value of a house the Shermans can purchase?

**24.** Gwen bought a new car for $15,000. She made a 25% down payment and borrowed the balance from her credit union. Her debt is to be amortized over a period of 5 years at 12% interest compounded monthly, the first payment being made one month from the date of purchase.

   **a.** How much is each monthly payment?
   **b.** How much of the 20th payment will pay interest?
   **c.** How much of the 20th payment is loan payment?
   **d.** If Gwen sells her car immediately after the 24th payment, how much should she get to pay the balance of the loan?

**25.** Upon graduation from college, Pierre had accumulated a debt of $12,500 in financial aid. No interest was charged while he was in school, but the lending institution charges 12% interest compounded monthly starting at graduation. Pierre wishes to repay his debt as promptly as possible, but can pay only $250 a month. How long will it take to pay his debt?

**26.** Kathy establishes a fund earning 9% compounded annually to maintain a family crypt. The maintenance cost is $350 per year with the first payment due immediately. How much must be paid to purchase an annuity that will take care of the perpetual maintenance of the family crypt?

# Probability

**6**

Almost everyone has some idea of the meaning of probability, and the word *probably* is (probably!) one of the most widely used words in English. In fact, because of the various connotations of *probability*, there are differences of opinion about probability theory even among mathematicians and philosophers. Nevertheless, the basic laws and calculations are almost universal. The purpose of this chapter is to familiarize you with the fundamental ideas and to show how the theory may be derived from a few simple axioms. Although our goal is to give you the ability to apply the acquired knowledge in your major field ("probably" business or economics), many of the examples and exercises will involve tossing coins, rolling dice, and drawing cards from a standard deck or marbles from an urn. There are two reasons for this. The first is that such experiments are simple and unambiguous; therefore, we will be able to concentrate on the mathematical ideas we are trying to illustrate. The second is that we are assuming most of you are using this book in a course that is a prerequisite to business and economics courses; as a consequence, you are "probably" not yet ready to handle concepts that come later in these courses.

You are urged to review Sections 0.2 and 0.3 on sets and Venn diagrams before beginning the study of this chapter.

# 6.1  Introduction

We are often forced to make decisions on probable knowledge and incomplete evidence. Things we often call facts are actually a mixture of physical data and inferences about them. Let us, for example, examine what surgeons mean when they tell a patient who has undergone eye surgery that his chances of recovery are about three out of four. This means that out of a great many patients who have undergone the same type of surgery, approximately 75% have recovered. But can the truth of the statement be tested? Suppose the patient does not recover. Does this mean that the doctor's statement was false? No, because on the average 25% of the patients are not expected to recover. It means only that the present patient was one of the unlucky ones. In spite of the fact that there is no satisfactory way of testing the validity of the statement, we feel it is still significant. In fact, each day most of us are willing to entrust our lives on actions based on probability considerations. The fact that someone can purchase over $100,000 worth of flight insurance for only $5 makes you feel that traveling by air is extremely safe since insurance companies base their rates on probability.

Often, probabilities are assigned to events that cannot be duplicated exactly. These are called *subjective probabilities* and are assigned on the basis of repeated trials and personal interpretation of these trials. In statistical studies, *success* means that the event under consideration actually occurred. The *relative frequency* of success is the ratio of the number of successes to the number of trials. We can now formulate a definition of probability.

**DEFINITION 6.1:**    Suppose that in $n$ trials of an experiment there are $s_n$ successes and that the sequence $s_n/n$ of relative frequencies of successes approaches a number as $n$ gets larger and larger. Then this number is defined to be the *probability* of success in a single trial.

In practice, the number of trials that can possibly be performed is always finite. Therefore, the probability is approximated by the relative frequency for the largest number of trials that have been performed for the experiment. The following example illustrates this definition of probability.

**EXAMPLE 1**   A company is making light bulbs. Its quality control team tested light bulbs chosen at random among those produced in the last 6 months. The following table gives the relative frequencies for different numbers of bulbs tested. The results listed in this table are cumulative. For example, 5 of the first 100 bulbs were defective and 6 of the next 100 were defective; thus, the second line shows that 11 of 200 were defective.

| Number of bulbs tested $n$ | Number of defective bulbs $s_n$ | Relative frequency $\dfrac{s_n}{n}$ |
|---|---|---|
| 100 | 5 | .05 |
| 200 | 11 | .055 |
| 300 | 18 | .06 |
| 500 | 29 | .058 |
| 1000 | 55 | .055 |
| 2000 | 105 | .0525 |
| 3000 | 162 | .054 |
| 4000 | 214 | .0535 |
| 5000 | 266 | .0532 |
| 6000 | 318 | .053 |

What is the probability that a light bulb selected at random is defective?

**Solution**   Since the results are cumulative and the largest number of trials is 6000, we conclude that if a light bulb is chosen randomly, the probability that it is defective is approximately .053.

**Do Exercise 1.**   ■

## Uniform Sample Space

Often, especially in games of chance, we can define probability without collecting data on relative frequencies. We illustrate with an example.

**EXAMPLE 2**   Suppose two dice are rolled.* What are the chances that the sum of the faces is five?

**Solution**   When two dice are rolled, the set of possible outcomes consists of all the different ways that the two dice may come to rest. If we let $x$ and $y$ denote the numbers that come up on the first die and the second die, respectively, then an outcome is represented by the ordered pair $(x, y)$, where $x$ and $y$ are members of the set $\{1, 2, 3, 4, 5, 6\}$.

---

*Unless otherwise stated, we tacitly assume that the dice are "fair." This means only that they are not biased to favor certain outcomes.

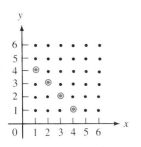

**FIGURE 6.1**

Since there are 6 choices for $x$ and 6 choices for $y$, there are 36 ordered pairs. These ordered pairs are represented in Figure 6.1. With "fair" dice, each of the 36 outcomes occurs equally often. Thus, the chance of any one of them occurring is $\frac{1}{36}$. We are interested not in just one ordered pair $(x, y)$, but in those for which $x + y = 5$. There are four such outcomes: $(1, 4)$, $(2, 3)$, $(3, 2)$, and $(4, 1)$. They are the circled points in Figure 6.1. It is reasonable to say that if two fair dice are rolled, the chances that the sum of the faces is 5 is $\frac{1}{9}$ since $\frac{4}{36} = \frac{1}{9}$.

**Do Exercise 5.**    ∎

Since we are often interested in a set of possible outcomes, we must decide what the chances are that one or more of these outcomes actually will occur. We have the following definition.

> **DEFINITION 6.2:**   The set of all possible outcomes of an experiment is called the *sample space* for that experiment. Each outcome is a *point* in the sample space and each subset of the sample space is an *event*. If all outcomes are equally likely, we say that they are *equiprobable* and the sample space is *uniform*. In that case, the *probability* of an event is the quotient of the number of elements in the event by the number of elements in the sample space.

In this book, we consider mostly finite sample spaces. Symbolically, if $n(A)$ denotes the number of elements in a set $A$, we write $p(E) = \frac{n(E)}{n(S)}$ where $S$ is a uniform sample space, $E$ is an event (a subset of $S$), and $p(E)$ denotes the probability of the event $E$.*

Since $E$ is a subset of $S$, we will always have

$$0 \leq n(E) \leq n(S)$$

from which it follows that

$$0 \leq \frac{n(E)}{n(S)} \leq \frac{n(S)}{n(S)}$$

and

$$0 \leq p(E) \leq 1$$

In Example 3, the sample space consisted of 36 points. An event can be a singleton (a subset with just one point) or it can be a subset with several points (for example, the event that the sum of the faces is 5 is the subset $\{(1, 4), (2, 3), (3, 2), (4, 1)\}$). Since the empty set is a subset of any set, it is always an event. It is called the *impossible event*. If $S$ is a sample space and $E$ is a subset of $S$, then $S\backslash E$ (the complement of $E$ in $S$, also denoted $E'$) represents the event that $E$ does not occur. If $E_1$ and $E_2$

---

*In one sense, the preceding definition of probability is circular because the expression "equally likely" involves the notion of probability. However, this technical difficulty should not cause a problem in understanding the concepts.

are subsets of $S$, then $E_1 \cup E_2$ is the event that either $E_1$ or $E_2$ (or both) occur, and $E_1 \cap E_2$ is the event that both $E_1$ and $E_2$ occur. If $S$ is a sample space and $E$ is an event, it is clear that $n(E) + n(E') = n(S)$ since every member of the sample space must be either in $E$ or in $E'$ and no outcome is in both $E$ and $E'$. Thus,

$$\frac{n(E) + n(E')}{n(S)} = \frac{n(S)}{n(S)}$$

and

$$\frac{n(E)}{n(S)} + \frac{n(E')}{n(S)} = 1$$

Therefore,

$$p(E) + p(E') = 1$$

**EXAMPLE 3**   Two dice are rolled. Represent the following situations in terms of events and calculate the probability of each event.

   **a.** The sum of the faces is 3.
   **b.** The sum of the faces is 7.
   **c.** The sum of the faces is 7 or 3.
   **d.** The sum of the faces is 7 and 3.
   **e.** The sum of the faces is not 3.

**Solution**   The sample space is the set of all ordered pairs $(x, y)$ where $x$ and $y$ are members of $\{1, 2, 3, 4, 5, 6\}$. Therefore, the sample space has 36 members.

   **a.** The sum of the faces is 3 if, and only if, the outcome is a member of the event $E_3 = \{(1, 2), (2, 1)\}$. Since there are 2 members in $E_3$ and 36 members in the sample space, the probability that the sum of the numbers is 3 is $\frac{1}{18}$ since $\frac{2}{36} = \frac{1}{18}$. We write

$$p(E_3) = \frac{1}{18}$$

   **b.** The sum of the faces is 7 if, and only if, the outcome is a member of the event $E_7 = \{(1, 6), (2, 5), (3, 4), (4, 3), (5, 2), (6, 1)\}$. Since there are 6 members in $E_7$, the probability that the sum is 7 is $\frac{1}{6}$ since $\frac{6}{36} = \frac{1}{6}$. We write

$$p(E_7) = \frac{1}{6}$$

   **c.** The sum of the faces is 7 or 3 if, and only if, the outcome is a member of the event $E_7 \cup E_3 = \{(1, 6,), (2, 5), (3, 4), (4, 3), (5, 2), (6, 1), (1, 2), (2, 1)\}$. Since there are 8 members in $E_7 \cup E_3$, the probability that the sum is 7 or 3 is $\frac{2}{9}$ since $\frac{8}{36} = \frac{2}{9}$. We write

$$p(E_7 \cup E_3) = \frac{2}{9}$$

**d.** The sum of the faces is 7 and 3 if, and only if, the outcome is a member of the event $E_7 \cap E_3$. But $E_7 \cap E_3 = \emptyset$, which means that this event is the impossible event. The number of elements in $\emptyset$ is 0. Hence, the probability that the sum is 7 and 3 is 0 since $\frac{0}{36} = 0$. We write

$$p(E_7 \cap E_3) = 0$$

**e.** The sum of the faces is not 3 if, and only if, the outcome is a member of the event $E_3'$. But

$$p(E_3) + p(E_3') = 1$$

Therefore,

$$\frac{1}{18} + p(E_3') = 1$$

and the probability that the sum of the numbers is not 3 is $\frac{17}{18}$. We write

$$p(E_3') = \frac{17}{18}$$

**Do Exercise 11.**                                                    ■

## Giving Odds

Often, probabilities are given in terms of *odds*, especially in games of chance. The odds in favor of an event $E$ is defined to be the ratio $\frac{p(E)}{p(E')}$. Thus, if the odds in favor of event $E$ are $m$ to $n$, with $mn \neq 0$, then $\frac{p(E)}{p(E')} = \frac{m}{n}$. Equivalently, $\frac{p(E')}{p(E)} = \frac{n}{m}$. Adding 1 to both sides of this equality we get

$$\frac{p(E')}{p(E)} + 1 = \frac{n}{m} + 1$$

or

$$\frac{p(E) + p(E')}{p(E)} = \frac{n + m}{m}$$

Using the fact that $p(E) + p(E') = 1$, we obtain

$$\frac{1}{p(E)} = \frac{n + m}{m}$$

and so

$$p(E) = \frac{m}{n + m}$$

Also,

$$p(E') = 1 - p(E) = 1 - \frac{m}{n + m} = \frac{n + m - m}{n + m} = \frac{n}{n + m}$$

**EXAMPLE 4**   If the odds in favor of UCLA going to the next Rose Bowl are 2 to 13, what is the probability that UCLA will go to the next Rose Bowl?

**Solution**   Replacing $m$ by 2 and $n$ by 13 in the formula

$$p(E) = \frac{m}{n + m}$$

we obtain

$$p(E) = \frac{2}{13 + 2} = \frac{2}{15}$$

The probability that UCLA will play in the next Rose Bowl is $\frac{2}{15}$.
**Do Exercise 15.**

**EXAMPLE 5**   Suppose the probability of an event $E$ is 0.21. What are the odds in favor of that event?

**Solution**   When $p(E) = 0.21$, $p(E') = 0.79$ since $1 - 0.21 = 0.79$. Hence, the ratio $\frac{0.21}{0.79} = \frac{21}{79}$ gives the odds in favor of event $E$. The odds in favor of event $E$ are 21 to 79.
**Do Exercise 17.**

## Biased Sample Space

In many cases, the outcomes of an experiment are not equally likely; however, we can still assign probabilities to the members of a finite sample space $E$ according to the following axioms:

**Axiom 1:** If $E = \{O_1, O_2, O_3, \ldots, O_n\}$, then for each $i$, $0 \le p(O_i) \le 1$, where $p(O_i)$ is the probability of outcome $O_i$.
**Axiom 2:** $p(O_1) + p(O_2) + p(O_3) + \cdots + p(O_n) = 1$.

**EXAMPLE 6**   Six teams are playing in a basketball tournament. The probabilities of winning the tournament for each of the first five teams is given in the following table.

| Team | Probability of winning the tournament |
|------|----------------------------------------|
| A | .12 |
| B | .21 |
| C | .08 |
| D | .35 |
| E | .09 |

What is the probability that team F wins the tournament?

**Solution**   Let $p(F)$ denote the probability that team F wins the tournament. By Axiom 2, we have

$$.12 + .21 + .08 + .35 + .09 + p(\text{F}) = 1$$

Thus,

$$.85 + p(\text{F}) = 1$$

and

$$p(\text{F}) = 1 - .85$$
$$= .15$$

The probability that team F wins the tournament is 0.15.
**Do Exercise 21.**

# Exercise Set 6.1

*The data given in the tables of Exercises 1–3 are cumulative, as were the data given in the table of Example 1.*

**1.** An electronics company is testing new picture tubes that it manufactured in the last 12 months. Its quality control division randomly chose tubes manufactured during that period and came up with the following results.

| Number of tubes tested<br>$n$ | Number of defective tubes<br>$s_n$ |
|---|---|
| 100 | 3 |
| 200 | 5 |
| 300 | 8 |
| 400 | 11 |
| 500 | 15 |
| 600 | 17 |
| 700 | 22 |
| 800 | 25 |
| 900 | 29 |
| 1000 | 32 |
| 1100 | 35 |
| 1200 | 38 |
| 1300 | 40 |
| 1400 | 44 |
| 1500 | 47 |

**a.** Calculate the relative frequency for each of the values of $n$.

**b.** Estimate the probability that a tube selected at random will be defective.

**2.** A manufacturer is testing the reliability of its new electric razor. Its quality control division randomly chose razors manufactured during the past year and came up with the following results.

| Number of razors tested<br>$n$ | Number of defective razors<br>$s_n$ |
|---|---|
| 100 | 4 |
| 200 | 9 |
| 300 | 11 |
| 400 | 17 |
| 500 | 22 |
| 600 | 23 |
| 700 | 29 |
| 800 | 34 |
| 900 | 36 |
| 1000 | 41 |
| 1100 | 47 |
| 1200 | 49 |
| 1300 | 55 |
| 1400 | 57 |
| 1500 | 62 |

**a.** Calculate the relative frequency for each of the values of $n$.

**b.** Estimate the probability that a razor selected at random will be defective.

**3.** A company is manufacturing and testing a new hair dryer. Its quality control division randomly tested dryers manufactured during the past 6 months and came up with the following results.

| Number of dryers tested $n$ | Number of defective dryers $s_n$ |
|:---:|:---:|
| 200 | 4 |
| 400 | 9 |
| 600 | 11 |
| 800 | 17 |
| 1000 | 21 |
| 1200 | 26 |
| 1400 | 30 |
| 1600 | 34 |
| 1800 | 35 |
| 2000 | 39 |

a. Calculate the relative frequency for each of the values of $n$.

b. What is the best estimate of the probability that a dryer selected at random will be defective?

*In Exercises 4–7, suppose two fair dice are rolled.*

4. What are the chances that the sum of the faces is 4?

5. What are the chances that the sum of the faces is 6?

6. What are the chances that the sum of the faces is 7?

7. What are the chances that the sum of the faces is 8?

8. Three fair dice are rolled. What are the chances that the sum of the faces is 4?

9. Three fair dice are rolled. What are the chances that the sum of the faces is 5?

10. Two fair dice are rolled. Represent the following situations in terms of events and calculate the probability of each event.

   a. The sum of the faces is 4.
   b. The sum of the faces is 6.
   c. The sum of the faces is 4 or 6.
   d. The sum of the faces is 4 and 6.
   e. The sum of the faces is not 4.

11. Two fair dice are rolled. Represent the following situations in terms of events and calculate the probability of each event.

   a. The sum of the faces is 8.
   b. The sum of the faces is 9.
   c. The sum of the faces is 8 or 9.
   d. The sum of the faces is 8 and 9.
   e. The sum of the faces is not 8.

12. A fair coin is tossed twice. Since a coin comes up heads (H) or tails (T), each possible outcome is an ordered pair where each entry is either H or T. Thus the sample space has four members.

   a. Describe the sample space.

b. Represent the following situations in terms of events and calculate the probability of each event.
   (i)   Exactly one toss comes up heads.
   (ii)  At least one toss comes up heads.
   (iii) Both are heads.
   (iv)  They are different.

13. A fair coin is tossed three times. Since a coin comes up heads (H) or tails (T), each possible outcome is an ordered triple where each entry is either H or T. Thus the sample space has eight members.

   a. Describe the sample space.
   b. Represent each of the following situations in terms of events and calculate the probability of each event.
   (i)   Exactly one toss comes up heads.
   (ii)  At least one toss comes up heads.
   (iii) All three tosses are heads.
   (iv)  At least two of the tosses are heads.

14. If the odds in favor of the Chicago Bears going to the next Super Bowl are 3 to 10, what is the probability that the Bears will go to the next Super Bowl?

15. If the odds in favor of the Boston Celtics winning the next NBA championship are 4 to 20, what is the probability that the Celtics will not win the next NBA championship?

16. If the odds in favor of a stock going up in value tomorrow are 5 to 3, what is the probability that its value will increase tomorrow?

17. If the probability that the Seattle Seahawks will win the next Super Bowl is 0.08, what are the odds in favor of the Seahawks winning the next Super Bowl?

18. If the probability that the Montreal Expos win the next World Series is 0.06, what are the odds in favor of the Expos winning the next World Series?

19. If the probability that John wins the next racquetball championship at his club is 0.35, what are the odds of John winning the championship?

20. If the probability that Betty wins the next golf tournament at her club is 0.24, what are the odds of Betty winning the tournament?

21. Five players are entered in a racquetball tournament. The probabilities that players A, B, C, or D win the tournament are 0.23, 0.13, 0.10, and 0.35, respectively. What is the probability that player E wins the tournament?

22. Eight players are entered in a table tennis tournament. The probabilities that players A, B, C, D, E, F, and G win the tournament are 0.13, 0.03, 0.07, 0.11, 0.06, 0.09, and 0.12, respectively. Show that player H is a heavy favorite to win the tournament.

# 6.2 Counting Techniques

In the preceding section, we explained how a probability function can be defined when the sample space is finite. That is, in a uniform sample space with $n$ elements, if $E$ is an event with $m$ elements, then the probability of $E$ is $p(E) = \frac{m}{n}$. The examples given were so simple that we could list the elements of the sample space. In many applications, however, the sample spaces and events are large and it is necessary to be able to find the number of elements in these sets without actually listing them. Imagine, for example, that you wanted to know how many people were attending a certain show, given that the theater was full. If the theater had 75 rows with 50 seats each, you would not enumerate all the people but would conclude that 3750 people were attending the show since $75(50) = 3750$. We devote this section to a brief introduction to techniques that allow us to determine the number of elements in sets without actually listing them.

## Tree Diagrams

**EXAMPLE 1**   Suppose we draw one marble from an urn containing a red, a green, and a white marble and then flip a coin. Describe the sample space.

**Solution**   We will use a scheme called a *tree diagram*. (See Figure 6.2.) Let $R$, $G$, and $W$ denote the red, green, and white marbles, respectively; and let $H$ and $T$ denote heads and tails, respectively.

**FIGURE 6.2**

The set of combined outcomes has six members: $(R, H)$, $(R, T)$, $(G, H)$, $(G, T)$, $(W, H)$, and $(W, T)$. The first element of each pair represents the marble drawn and the second element represents the way the coin landed. There are three possible outcomes when we first draw a marble and two possible outcomes when we flip a coin; and $3 \cdot 2 = 6$.
**Do Exercise 17.**                                                                                       ■

**EXAMPLE 2**   A coin is tossed three times. Use a tree diagram to find all possible combined outcomes.

**Solution**   We let $H$ and $T$ represent heads and tails, respectively, and draw the following tree diagram. (See Figure 6.3.) Following the branches of the tree diagram, we get the following eight possible combined outcomes of heads and tails: $(H, H, H)$, $(H, H,$

$T$), $(H, T, H)$, $(H, T, T)$, $(T, H, H)$, $(T, H, T)$, $(T, T, H)$, $(T, T, T)$. Note that there are two possible outcomes with each of the three tosses and that $2 \cdot 2 \cdot 2 = 8$.
**Do Exercise 15.**

First toss   Second toss   Third toss   Combined outcomes

Start

H — H — H   (H, H, H)
H — T   (H, H, T)
H — T — H   (H, T, H)
T   (H, T, T)
T — H — H   (T, H, H)
T   (T, H, T)
T — H   (T, T, H)
T   (T, T, T)

**FIGURE 6.3**

Although a tree diagram is a convenient way to list combined outcomes when we perform a sequence of experiments in order, it is not practical when the number of outcomes in an experiment is large. In such cases, we use the multiplication principle that we introduced earlier. For convenience, we state this principle again.

Suppose that we perform an experiment in two parts. The first part has $m$ possible outcomes and, regardless of which of these occurs, the second part has $n$ possible outcomes. Then, when the two parts of the experiment are performed, in that order, there are $mn$ possible outcomes. This principle can be generalized as follows.

> **Fundamental Principle of Counting**  Suppose that we perform an experiment in $k$ parts. The first part has $n_1$ possible outcomes; regardless of which of these occurs, the second part has $n_2$ possible outcomes; regardless of which of these occurs, the third part has $n_3$ possible outcomes; and so on. Then when the $k$ parts of the experiment are performed, in that order, the number of possible outcomes is $n_1 \cdot n_2 \cdots \cdots n_k$.

**EXAMPLE 3**   A license plate is made up of 3 of the 26 letters of the alphabet followed by 3 of the 10 digits. How many different license plates can we have if no letter or digit is repeated?

**Solution**   There are 26 ways to pick one letter. Then, since we don't allow repetition, there are 25 ways to pick a second letter. The third letter can be picked in 24 ways. Similarly, there are 10 ways to pick the first digit, 9 ways to pick the next digit, and 8 ways to pick the last digit. Therefore we can have 11,232,000 different license plates since

$$26 \cdot 25 \cdot 24 \cdot 10 \cdot 9 \cdot 8 = 11,232,000$$

We now derive some formulas that will be useful in calculating the number of elements in a set. For some of these formulas, we will need to recall the meaning of the symbol $n!$.

If $n$ is a positive integer, the product $n(n - 1)(n - 2) \cdots (3)(2)(1)$ is abbreviated $n!$ (read: "$n$ factorial").

For example,

$$3! = 3(3 - 1)(3 - 2) = 3 \cdot 2 \cdot 1 = 6$$

and

$$5! = 5(5 - 1)(5 - 2)(5 - 3)(5 - 4) = 5 \cdot 4 \cdot 3 \cdot 2 \cdot 1 = 120$$

As a consequence of this definition, we see that if $k$ is a positive integer, then

$$(k + 1)! = (k + 1)k!$$

Although we have not defined $0!$, if we replace $k$ by 0 in the foregoing formula, we obtain

$$1! = 1(0!)$$

That is,

$$1 = 0!$$

Therefore, we define $0!$ to be equal to 1. With this value attached to $0!$, the formula

$$(k + 1)! = (k + 1)k!$$

is true for any nonnegative integer $k$.

## Permutations and Combinations

**DEFINITION 6.3:**    Suppose we have a set of $n$ distinct objects from which we select $r$ members. The subset obtained is called a *combination of n things taken r at a time*. If the members of this subset are arranged in a definite order, then the ordered subset obtained is called a *permutation of n things taken r at a time*.

If $n$ and $r$ are integers such that $0 \leq r \leq n$, the symbol $C(n, r)$ denotes the number of combinations of $n$ things taken $r$ at a time and the symbol $P(n, r)$ denotes the number of permutations of $n$ things taken $r$ at a time. In the next example, we will show how to calculate $P(n, r)$ and $C(n, r)$ for small values of $n$ and $r$.

**EXAMPLE 4**    Find the values of $P(7, 3)$ and $C(7, 3)$.

**Solution**    Given a set with seven members, to form a permutation with three members, we must fill three places without using the same object twice. There are seven ways to fill the first place. Once this has been done, there are six ways to fill the second place. When the first two places have been filled, there are five ways to fill the third place. Using the multiplication principle, the total number of distinct three-member permutations that can be formed is 210 since $7 \cdot 6 \cdot 5 = 210$. Thus,

$$P(7, 3) = 210$$

Note that, multiplying and dividing by 4!, we obtain

$$P(7, 3) = 7 \cdot 6 \cdot 5 = \frac{7 \cdot 6 \cdot 5 \cdot 4 \cdot 3 \cdot 2 \cdot 1}{4 \cdot 3 \cdot 2 \cdot 1} = \frac{7!}{4!} = \frac{7!}{(7-3)!}$$

To obtain any one of the 210 permutations of seven things taken three at a time, we do two things in order. We pick a subset with three members. There are $C(7, 3)$ ways to do this. We then must arrange that subset in some order. There are three ways to fill the first place, two ways to fill the second place, and only one way to fill the third place. Thus, for each of the subsets with three members, we can make six ordered subsets since $3 \cdot 2 \cdot 1 = 6$. Therefore, using the multiplication principle, there are $C(7, 3) \cdot 6$ ways to get a permutation of seven things taken three at a time. Since we already know that this number is 210, we can write

$$C(7, 3) \cdot 6 = 210$$

We get

$$C(7, 3) = \frac{210}{6} = 35$$

We could have written

$$C(7, 3) = \frac{P(7, 3)}{3!} = \frac{7!}{(7-3)!(3!)}$$ ∎

Using exactly the same reasoning as in the previous example, we get the following formulas.

If $n$ and $r$ are integers with $0 \le r \le n$, then

$$P(n, r) = \frac{n!}{(n-r)!}$$

and

$$C(n, r) = \frac{n!}{(n-r)!r!}$$

If $r = n$, then

$$P(n, n) = \frac{n!}{(n-n)!} = \frac{n!}{0!} = \frac{n!}{1} = n!$$

Thus, there are $n!$ permutations of $n$ things taken $n$ at a time.

**EXAMPLE 5**   In how many ways can the positions of president, vice president, and secretary be filled, in that order, in a club of 15 members if no member can hold more than one position?

**Solution** The number of ways the positions can be filled is the number of permutations of 15 things take 3 at a time, namely

$$P(15, 3) = \frac{15!}{(15 - 3)!} = \frac{15 \cdot 14 \cdot 13 \cdot 12!}{12!} = 15 \cdot 14 \cdot 13 = 2730$$

**Do Exercise 27.** ■

**EXAMPLE 6** How many lines are determined by five points, no three of which are collinear?*

**Solution** Since two distinct points determine a line, we are interested in finding how many two-point subsets there are in the given set. That number is

$$C(5, 2) = \frac{5!}{(5 - 2)!2!} = \frac{5!}{3!2!} = \frac{5 \cdot 4 \cdot 3!}{3!2!} = \frac{20}{2} = 10$$

Therefore, ten lines are determined by the five points.
**Do Exercise 29.** ■

 It is important to remember that when determining the number of ways a subset of a given set may be selected, we use permutations if the subset is ordered, and if the subset is not ordered, we use combinations.**

**EXAMPLE 7** A club membership is composed of 42 women and 37 men. How many different committees of seven members can be formed with the following restrictions:

**a.** The committee must have four women and three men.
**b.** Women must have the majority on the committee.

**Solution** **a.** In forming a committee of four women and three men, we do two things. We first choose 4 women among the 42 women. Since the members chosen need not be ordered, we use combinations. The number of ways we can choose the four women is $C(42, 4)$. We then choose 3 men among the 37 male members. Again, these need not be ordered. Thus, we use combinations to find that the number of ways we can choose the three men is $C(37, 3)$. By the fundamental principle of counting, the number of ways we can choose the four women and the three men is $C(42, 4) \cdot C(37, 3)$. Using the formula to calculate $C(n, r)$, we get

$$C(42, 4) \cdot C(37, 3) = \frac{42!}{(42 - 4)!4!} \cdot \frac{37!}{(37 - 3)!3!}$$

$$= \frac{42 \cdot 41 \cdot 40 \cdot 39 \cdot 38!}{38!4!} \cdot \frac{37 \cdot 36 \cdot 35 \cdot 34!}{34!3!}$$

---

*Points are collinear if they lie on the same straight line.
**Each time we select a subset $A$ with $r$ elements, given a set $S$ with $n$ elements, we get a corresponding subset, the complement of $A$ in $S$ with $n - r$ elements. Therefore, the number of subsets with $r$ elements is the same as the number of subsets with $n - r$ elements. That is,

$$C(n, r) = C(n, n - r)$$

This equality can be verified using the formula for $C(n, r)$. (See Exercise 19.)

$$= \frac{42 \cdot 41 \cdot 40 \cdot 39}{4!} \cdot \frac{37 \cdot 36 \cdot 35}{3!}$$

$$= 111,930 \cdot 7770$$

$$= 869,696,100$$

Therefore, 869,696,100 different committees of four women and three men can be formed. Whether we choose the women first or the men is irrelevant since

$$C(42, 4) \cdot C(37, 3) = C(37, 3) \cdot C(42, 4)$$

**b.** In a committee of seven members, the women will have the majority if there are four, five, six, or seven women on that committee. We already know that we can form 869,696,100 committees with four women and three men. In exactly the same manner, we calculate the number of committees with five women and two men. We get

$$C(42, 5) \cdot C(37, 2) = 850,668 \cdot 666 = 566,544,888$$

The number of committees with six women and one man is

$$C(42, 6) \cdot C(37, 1) = 5,245,786 \cdot 37 = 194,094,082$$

The number of committees with seven women and no men is, of course, $C(42, 7) = 26,978,328$. To find the total number of different committees with a majority of women, we simply add the results we just obtained to get

$$869,696,100 + 566,544,888 + 194,094,082 + 26,978,328 = 1,657,313,398$$

Therefore, 1,657,313,398 committees of seven members with a majority of women can be formed.

**Do Exercise 37.**                                                                      ■

In discussing permutations of $n$ things taken $r$ at a time, we have assumed that the $n$ things were distinct. We may encounter situations where we have a set of objects not all of which are distinguishable from one another. Suppose we have a set of $n$ elements consisting of only two types of objects. There are $n_1$ indistinguishable objects of type I and $n_2$ indistinguishable objects of type II with $n_1 + n_2 = n$. If all $n$ objects were distinct, there would be $n!$ permutations. However, some of these permutations are indistinguishable. Let $P(n; n_1, n_2)$ denote the number of distinguishable permutations. Assume that we have labeled the objects of type I and type II so that all objects are now distinguishable. Then with each of the original indistinguishable permutations, we can form $n_1!n_2!$ distinguishable permutations because the objects of type I can be arranged in $n_1!$ ways and those of type II can be arranged in $n_2!$ ways. Thus, the total number of distinguishable permutations is $P(n; n_1, n_2) \cdot n_1! \cdot n_2!$. Consequently,

$$P(n; n_1, n_2) \cdot n_1! \cdot n_2! = n!$$

and

$$P(n; n_1, n_2) = \frac{n!}{n_1! \cdot n_2!}$$

This formula may be generalized for cases where we have sets with more than two subsets of like elements.

Suppose we have a set with $n$ elements and $k$ subsets whose union is the given set. The first subset is of type I and has $n_1$ indistinguishable elements, the second subset is of type II and has $n_2$ indistinguishable elements, and the $k$th subset is of type $k$ and has $n_k$ indistinguishable elements. Consequently, $n_1 + n_2 + \cdots + n_k = n$.

The number $P(n; n_1, n_2, \ldots, n_k)$ of distinct permutations is given by

$$P(n; n_1, n_2, \ldots, n_k) = \frac{n!}{n_1! n_2! \ldots n_k!}$$

**EXAMPLE 8**    A store has three identical A television sets, four identical B television sets, and two identical C television sets. At the beginning of the Christmas sale, the store manager puts these nine sets on display by arranging them against one of the walls in the store. In how many different ways can he do this?

**Solution**    Since he has nine sets, of which the three A sets are indistinguishable, the four B sets are indistinguishable, and the two C sets are also indistinguishable, the number of distinct ways that the manager can arrange the nine sets against a wall is 1260 since

$$P(9; 3, 4, 2) = \frac{9!}{3!4!2!} = 1260$$

**Do Exercise 23.**                                                                                 ■

## Set Partitioning

We conclude this section with a brief discussion of set partitioning. For example, when four players are playing bridge and a hand is dealt using all 52 cards, an ordered partition $(H_1, H_2, H_3, H_4)$ of the deck of cards is obtained where each $H_i$ has 13 members. Suppose that you want to know how many different hands can be dealt. To deal with such problems, we need a formula that will allow us to find the number of distinct ordered partitions of a set.

**DEFINITION 6.4:**   Given a set $S$ with $n$ members, an *ordered partition* of $S$ is an ordered collection of subsets $H_1, H_2, H_3, \ldots, H_k$ of $S$ such that each member of $S$ is in exactly one of the subsets $H_i$.

Suppose that $H_1$ has $r_1$ members, $H_2$ has $r_2$ members, $H_3$ has $r_3$ members, . . . , and $H_k$ has $r_k$ members, so that $r_1 + r_2 + r_3 + \cdots + r_k = n$. If we let

$C(n; r_1, r_2, \ldots, r_k)$ denote the number of such ordered partitions, then it can be shown that

$$C(n; r_1, r_2, \ldots, r_k) = \frac{n!}{r_1! r_2! \cdots r_n!} \quad *$$

**EXAMPLE 9**   In how many different ways can 15 salespersons be assigned to 3 departments if 5 must be assigned to department A, 7 to department B, and 3 to department C?

**Solution**   Any such assignment is an ordered partition $(H_A, H_B, H_C)$ of the set of 15 salespersons where $H_A$ has five elements, $H_B$ has seven elements, and $H_C$ has three elements. The number of ordered partitions is 360,360 since

$$\frac{15!}{5!7!3!} = 360{,}360$$

**Do Exercise 41.**   ◼

---

# Exercise Set 6.2

*In Exercises 1–14, evaluate the given expression.*

**1.** 7!

**2.** $\dfrac{12!}{8!}$

**3.** $\dfrac{15!}{9!3!3!}$

**4.** $P(5, 3)$

**5.** $P(6, 6)$

**6.** $P(13, 6)$

**7.** $C(5, 3)$

**8.** $C(6, 2)$

**9.** $C(13, 7)$

**10.** $C(7, 2)$ and $C(7, 5)$

**11.** $C(10, 3)$ and $C(10, 7)$

**12.** $C(20, 7)$ and $C(20, 13)$

**13.** $P(10; 6, 4)$

**14.** $P(20; 4, 5, 3, 8)$

**15.** A coin is tossed four times. With the help of a tree diagram, list all possible combined outcomes.

**16.** A coin is tossed twice, then a die is rolled once. With the help of a tree diagram, list all possible combined outcomes.

**17.** A coin is tossed three times. Then a marble is drawn from an urn containing a red marble, a green marble, and a blue marble. With the help of a tree diagram, list all possible combined outcomes.

**18.** An urn contains marbles numbered 1, 2, 3, and 4. A marble is drawn, then a coin is tossed twice. With the help of a tree diagram, list all possible combined outcomes.

**19.** Using the formula

$$C(n, k) = \frac{n!}{(n - k)!k!}$$

show that if $n$ and $r$ are integers with $0 \le r \le n$, then

$$C(n, r) = C(n, n - r)$$

**20.** How many permutations of the three letters x, y, and z are there? List all of them.

**21.** How many permutations of the letters a, b, c, and d are there? List all of them.

**22.** How many distinct letter arrangements can be made using all the letters in the word *Mississippi*?

---

*Note the similarity of this formula to the formula giving the number of distinguishable permutations of $n$ objects when some of these objects are indistinguishable. Also if $k = 2$, then $r_2 = n - r_1$ and the foregoing formula may be written

$$C(n; r_1, r_2) = \frac{n!}{r_1!(n - r_1)!}$$

This is precisely the formula giving the number of combinations of $n$ things taken $r_1$ at a time.

23. Repeat Exercise 22 with the word *mathematics*.

24. Repeat Exercise 22 with the word *statistics*.

25. How many different signals, each consisting of seven flags placed in a horizontal line, can be formed using a set of three indistinguishable blue flags and four indistinguishable red flags?

26. Repeat Exercise 25 with a set of four indistinguishable blue flags, five indistinguishable red flags, and six indistinguishable green flags.

27. In how many ways can the positions of president, vice president, and secretary be filled in a club of 13 members if no member can hold more than one position?

28. How many lines are determined by seven points, no three of which are collinear?

29. Repeat Exercise 28 with ten points.

30. How many planes are determined by eight points, no four of which are coplanar* and no three of which are collinear? (*Hint:* Each three points determine a plane.)

31. Repeat Exercise 30 with 12 points.

32. How many permutations of the letters of the word *hypersonic* are there? If you could write one of these every 5 seconds, how long would it take you to list them all?

33. A foreman is in charge of 12 machines and has 12 workers to assign to these machines, one worker per machine. In how many different ways can this be done? Suppose that the foreman wishes to make the assignments in the best possible way and thinks about each of the possible assignments for two seconds. How long would it take him? Does the answer surprise you?

34. Repeat Exercise 33 with 15 machines and 15 workers.

35. Six soldiers are selected from a squad of eight for patrol duty. In how many ways can this be done?

36. How many committees of five can be formed from a group of six Republicans and three Democrats if each committee is to have at least three Republicans?

37. A club membership is composed of 18 women and 20 men. How many committees of five can be formed if **a.** exactly three committee members must be men? **b.** men must have the majority on the committee?

38. A basketball coach has three centers, six forwards, and five guards available. She must field a team made up of one center, two forwards, and two guards. In how many ways can this be done?

39. Repeat Exercise 38 if there are two centers, seven forwards, and six guards available.

40. In how many ways can ten cooks be assigned to three kitchens if five cooks are needed for kitchen I, two cooks are needed for kitchen II, and three cooks are needed for kitchen III?

41. In how many distinct ways can 5-card hands be dealt to 6 poker players from a standard 52-card deck? (*Hint:* Number the players 1, 2, 3, 4, 5, and 6, then each hand corresponds to an ordered partition $(H_1, H_2, H_3, H_4, H_5, H_6, B)$ of the deck into seven subsets where each $H_i$ is the hand player $i$ holds and $B$ is the set of cards left after the hand is dealt.)

42. A publishing company has 25 salespersons. In how many ways can they be assigned to four territories if eight are assigned to the northern territory, seven are assigned to the eastern territory, six are assigned to the western territory, and four are assigned to the southern territory?

# 6.3  Conditional Probability

When we consider a probability problem for which the sample space is finite and all outcomes are equally likely, we need only to be able to calculate the number of elements in the sample space and the number of elements in an event to find the probability of that event. The preceding section contains formulas that enable us to find these numbers without actually listing all the outcomes. Mathematicians have formulated another method. There are ways to express a complex probability problem in terms of simpler ones. We briefly describe some of these techniques.

Frequently, an event is simply a combination of events. For example, if one die is thrown twice, the experiment is to throw it once, and then, after this is done, to throw

---

*Points that lie on the same plane are said to be coplanar.

it again. Each throw is independent of the other. This is not always the case, however, with sequential events. To best formulate the basic techniques of this section, we must distinguish between events that are dependent, independent, or mutually exclusive. Before giving the formal definitions, we give some examples.

Suppose one card is drawn from a standard deck of cards and then another is drawn without replacing the first. What is the probability that both are spades? Clearly, if the first card is a spade, then the probability of the second card being a spade is $\frac{12}{51}$ (since there are 12 spades left among 51 cards). On the other hand, if the first card is not a spade, then the probability of the second being a spade is $\frac{13}{51}$. We see that the probability of the second card being a spade depends on the outcome of the first draw and therefore the two events are dependent.

However, suppose a card is drawn, replaced in the deck, and then the deck is shuffled before a second card is drawn. Then, the probability that the second card is a spade is $\frac{13}{52}$, regardless of the outcome of the first drawing; the two events are independent.

Suppose a single card is drawn from a deck. The event "it is a heart" and the event "it is a club" are mutually exclusive since one cannot happen if the other does. However, the event "it is a heart" and the event "it is a red card" are alternative outcomes of a single draw that are not mutually exclusive since they can both happen.

## Conditional Probability

In the first example, in which two cards were drawn without replacement, we saw that the probability that the second card is a spade given that the first was a spade is $\frac{12}{51}$. For two events $A$ and $B$, $p(A \mid B)$ denotes the probability that $A$ occurs given that $B$ does occur. Thus, if $U$ is the sample space, $p(A \mid U)$ is synonymous to $p(A)$. Recall that if $C$ is a finite set, $n(C)$ denotes the number of elements in $C$. For example, if $X = \{2, 4, 5\}$, then $n(X) = 3$. Now suppose that $U$ is a finite sample space in which all outcomes are equally likely and let $A$ and $B$ be events in $U$. Then, to calculate $p(A \mid B)$, we do not consider points in $A \backslash B$ because, for any of these outcomes, $B$ failed to occur. Thus, we consider only points in $A \cap B$. (See Figure 6.4.) Since $B$ does occur, it plays the role of the sample space and we have

$$p(A \mid B) = \frac{n(A \cap B)}{n(B)}$$

But,

$$p(B) = \frac{n(B)}{n(U)}$$

Therefore,

$$p(A \mid B) \cdot p(B) = \frac{n(A \cap B)}{n(B)} \cdot \frac{n(B)}{n(U)} = \frac{n(A \cap B)}{n(U)}$$

and

$$p(A \mid B) \cdot p(B) = p(A \cap B)$$

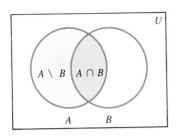

**FIGURE 6.4**

**A CLOSER LOOK**

This formula was derived when we had a finite uniform sample space. Suppose we have a finite sample space $U$ in which the outcomes are not necessarily equally likely. Let

$$U = \{o_1, o_2, o_3, \ldots, o_n\}$$

If $p(o_i)$ denotes the probability that outcome $o_i$ occurs, we have

$$0 \leq p(o_i) \leq 1$$

for each $i$, and

$$p(o_1) + p(o_2) + p(o_3) + \cdots + p(o_n) = 1$$

Now suppose that $A$ and $B$ are events in $U$. Again let $p(A \mid B)$ denote the probability that event $A$ occurs given that event $B$ has occurred. We must assign new probabilities to the original outcomes. We assign the probability 0 to any member of $U$ that is not in $B$ since it is given that $B$ has occurred and, therefore, any outcome that is not in $B$ cannot occur. Consequently, if $o_i$ is not in $B$, we let $p(o_i \mid B) = 0$. We also treat the event $B$ as our new sample space. It is natural to assign the conditional probabilities to elements of $B$ in the same ratio as their probabilities were assigned in the original sample space. That is, if $o_r$ and $o_s$ are in $B$, then

$$\frac{p(o_r \mid B)}{p(o_s \mid B)} = \frac{p(o_r)}{p(o_s)}$$

Without loss of generality, we may assume that the elements of $U$ have been labeled so that $B = \{o_1, o_2, o_3, \ldots, o_q\}$ and $U \backslash B = \{o_{q+1}, o_{q+2}, o_{q+3}, \ldots, o_n\}$. Then, for any integer $i$ between 1 and $q$, inclusive, we have

$$\frac{p(B)}{p(o_i)} = \frac{p(o_1) + p(o_2) + p(o_3) + \cdots + p(o_q)}{p(o_i)}$$

$$= \frac{p(o_1)}{p(o_i)} + \frac{p(o_2)}{p(o_i)} + \frac{p(o_3)}{p(o_i)} + \cdots + \frac{p(o_q)}{p(o_i)}$$

$$= \frac{p(o_1 \mid B)}{p(o_i \mid B)} + \frac{p(o_2 \mid B)}{p(o_i \mid B)} + \frac{p(o_3 \mid B)}{p(o_i \mid B)} + \cdots + \frac{p(o_q \mid B)}{p(o_i \mid B)}$$

$$= \frac{p(o_1 \mid B) + p(o_2 \mid B) + p(o_3 \mid B) + \cdots + p(o_q \mid B)}{p(o_i \mid B)} = \frac{1}{p(o_i \mid B)}$$

We have shown that $p(B)/p(o_i) = 1/p(o_i \mid B)$. Consequently, we have $p(o_i \mid B) = p(o_i)/p(B)$. That is, the conditional probability of any outcome $o_i$ in $B$ is obtained by multiplying the original probability of $o_i$ by $1/p(B)$. Note the following:

1. We have assumed that $p(o_i) \neq 0$. In the case that $p(o_i) = 0$, we let $p(o_i \mid B) = 0$.

2. We have used the fact that since $B = \{o_1, o_2, \ldots, o_q\}$,

$$p(o_1) + p(o_2) + p(o_3) + \cdots + p(o_q) = p(B)$$

3. We have also used the fact that since we treat $B$ as our new sample space, then

$$p(o_1 \mid B) + p(o_2 \mid B) + p(o_3 \mid B) + \cdots + p(o_q \mid B) = 1$$

Again let us assume that the elements of $U$ have been labeled so that $A \cap B = \{o_1, o_2, \ldots, o_m\}$. When we calculate the conditional probability $p(A \mid B)$, we must add the conditional probabilities of those outcomes which are in both $A$ and $B$ (remember $B$ is the new sample space). Thus,

$$p(A \mid B) = p(o_1 \mid B) + p(o_2 \mid B) + p(o_3 \mid B) + \cdots + p(o_m \mid B)$$

$$= \frac{p(o_1)}{p(B)} + \frac{p(o_2)}{p(B)} + \frac{p(o_3)}{p(B)} + \cdots + \frac{p(o_m)}{p(B)}$$

$$= \frac{p(o_1) + p(o_2) + p(o_3) + \cdots + p(o_m)}{p(B)}$$

$$= \frac{p(A \cap B)}{p(B)}$$

and

$$p(A \mid B) \cdot p(B) = p(A \cap B)$$

Note that this is exactly the same relationship as we obtained when all outcomes in $U$ were equally likely. Therefore, we have the following definition.

> **DEFINITION 6.5:**    Suppose that $A$ and $B$ are events in a sample space $U$. Then the *conditional probability* of $A$ given that $B$ has occurred, denoted $p(A \mid B)$, is defined by
>
> $$p(A \mid B) = \frac{p(A \cap B)}{p(B)}, \text{ provided that } p(B) \neq 0$$
>
> Similarly,
>
> $$p(B \mid A) = \frac{p(A \cap B)}{p(A)}, \text{ provided that } p(A) \neq 0$$

**EXAMPLE 1**    Suppose a bag contains six red marbles and four blue marbles, and a second bag contains four red marbles and five blue marbles. A marble is drawn from the first bag and placed in the second. The second bag is shaken and a marble is drawn from it. What is the probability that a red marble was drawn each time?

**Solution**    An outcome is drawing one marble and then another. It can be represented by an ordered pair $(x, y)$. Let

$$A = \{(x, y) \mid (x, y) \in U \text{ and } x \text{ is a red marble}\}$$

and

$$B = \{(x, y) \mid (x, y) \in U \text{ and } y \text{ is a red marble}\}$$

Then the probability that a red marble was drawn each time is obtained as follows:

$$p(A \cap B) = p(A) \cdot p(B \mid A)$$
$$= \frac{6}{10} \cdot \frac{5}{10} = \frac{3}{10}$$

We know $p(A) = \frac{6}{10}$ since six of the ten marbles in the first bag were red. Also $p(B \mid A) = \frac{5}{10}$, since if the first marble drawn is red and placed in the second bag, then five of the ten marbles in that bag are red.

**Do Exercise 1.** ■

**EXAMPLE 2** The probability that a freshman takes an English course is .62. The probability of taking a mathematics course is .35 and the probability of taking both is .23.

    **a.** What is the probability that a freshman will take an English course given that a mathematics course is also taken by that student?

    **b.** What is the probability that a freshman will take a mathematics course given that an English course is also taken by that student?

**Solution** Let $E$ denote the event that the student takes an English course and $M$ be the event that the student takes a mathematics course. It is given that $p(E) = .62$, $p(M) = .35$, and $p(E \cap M) = .23$.

    **a.** The probability that the student takes an English course given that a mathematics course is taken is $p(E \mid M)$; the value for which is obtained as follows:

$$p(E \mid M) = \frac{p(E \cap M)}{p(M)} = \frac{.23}{.35} \doteq .6571$$

    **b.** The probability that the student takes a mathematics course given that an English course is taken is $p(M \mid E)$, the value for which is obtained as follows:

$$p(M \mid E) = \frac{p(E \cap M)}{p(E)} = \frac{.23}{.62} \doteq .371.$$

**Do Exercise 3a, b.** ■

## Independent Events

**DEFINITION 6.6:** If $A$ and $B$ are events and $p(A) = p(A \mid B)$, we say that $A$ and $B$ are *independent events*. In that case, we have

$$p(A \cap B) = p(A) \cdot p(B)$$

**EXAMPLE 3** An urn contains three white marbles and one black marble. A marble is drawn from the urn and replaced, the urn is shaken, and a marble is drawn again. What is the probability that both marbles drawn are white?

**Solution**  Since before the second drawing the original situation was restored, the two events are independent. Each time, the probability of drawing a white marble is $\frac{3}{4}$. Therefore the probability of drawing a white marble both times is $\frac{3}{4} \cdot \frac{3}{4} = \frac{9}{16}$.

**Do Exercise 5.**

## Mutually Exclusive Events

**DEFINITION 6.7:**  If $U$ is a sample space and $A$ and $B$ are disjoint subsets of $U$, then $A$ and $B$ are said to be *mutually exclusive events*. (See Figure 6.5.)

**FIGURE 6.5**

If $A$ and $B$ are mutually exclusive events, then

$$p(A \cup B) = p(A) + p(B)$$

**EXAMPLE 4**  Suppose a card is drawn at random from a standard deck of cards. What is the probability that it is either a king or a queen?

**Solution**  The probability of drawing a king is $\frac{4}{52} = \frac{1}{13}$. The probability of drawing a queen is also $\frac{1}{13}$. Hence, the probability of drawing a king or a queen is $\frac{1}{13} + \frac{1}{13} = \frac{2}{13}$, since those two events are mutually exclusive.

**Do Exercise 9.**

If two events $A$ and $B$ are not mutually exclusive, and we calculate the probability that one or the other will occur, we use the formula

$$p(A \cup B) = p(A) + p(B) - p(A \cap B)$$

To verify this, suppose the outcomes have been labeled in such a way that

$$A \cap B = \{c_1, c_2, \cdots, c_k\}$$
$$A = \{c_1, c_2, \cdots, c_k, a_1, a_2, \cdots, a_m\}$$
$$B = \{c_1, c_2, \cdots, c_k, b_1, b_2, \cdots, b_n\}$$

Then,

$$A \cup B = \{c_1, c_2, \cdots, c_k, a_1, a_2, \cdots, a_m, b_1, b_2, \cdots, b_n\}$$

Thus,

$$p(A \cup B) = p(c_1) + p(c_2) + \cdots + p(c_k) + p(a_1)$$
$$+ p(a_2) + \cdots + p(a_m) + p(b_1) + p(b_2) + \cdots + p(b_n)$$
$$p(A) = p(c_1) + p(c_2) + \cdots + p(c_k)$$
$$+ p(a_1) + p(a_2) + \cdots + p(a_m)$$
$$p(B) = p(c_1) + p(c_2) + \cdots + p(c_k) + p(b_1) + p(b_2) + \cdots + p(b_n)$$
$$p(A \cap B) = p(c_1) + p(c_2) + \cdots + p(c_k)$$

It is now easy to verify that

$$p(A \cup B) = p(A) + p(B) - p(A \cap B)$$

**EXAMPLE 5**  A card is drawn from a standard deck of cards. What is the probability that it is either a spade or a king?

**Solution**  Here the sample space is the set of 52 cards. Let $A = \{x \mid x$ is a spade$\}$ and $B = \{x \mid x$ is a king$\}$. Then clearly $A \cap B \neq \emptyset$. In fact, $A \cap B$ has exactly one member, the king of spades. Thus, $p(A \cap B) = \frac{1}{52}$, $p(A) = \frac{13}{52} = \frac{1}{4}$, and $p(B) = \frac{4}{52} = \frac{1}{13}$. It follows that

$$p(A \cup B) = \frac{1}{4} + \frac{1}{13} - \frac{1}{52} = \frac{13 + 4 - 1}{52} = \frac{16}{52} = \frac{4}{13}$$

**Do Exercise 11.**                                                            ■

**EXAMPLE 6**  What is the probability that the freshman of Example 2 takes both English and mathematics given that at least one of the two is taken?

**Solution**  We first find the probability that the student takes at least one of the two subjects. We get

$$p(E \cup M) = p(E) + p(M) - p(E \cap M) = .62 + .35 - .23 = .74$$

Thus,

$$p(E \cap M \mid E \cup M) = \frac{p(E \cap M)}{p(E \cup M)} = \frac{.23}{.74} \doteq .3108$$

**Do Exercise 3c.**                                                           ■

## Sequence of Dependent Events

The following theorem allows us to solve problems in which we have a sequence of $k$ dependent events.

> **THEOREM 6.1:**  Suppose we have a sequence $E_1, E_2, \ldots, E_k$ of $k$ dependent events. The probability of the first event $E_1$ occurring is $p_1$. After the first event has occurred, the probability of the second event $E_2$ occurring is $p_2$; after these two events have occurred, the probability of the third event $E_3$ occurring is $p_3$, and so on, then the probability that all $k$ events occur, in that order, is $p_1 p_2 p_3 \cdots p_k$.

When applying this theorem to certain problems, the number of events that are considered failures may be much smaller than the number of events that are considered successes. In such cases, it is simpler to first calculate the probability of failure, then subtract the result from 1 to get the probability of success.

**EXAMPLE 7**  Suppose 23 people have been randomly selected. What is the probability that at least two of them celebrate their birthdays the same day?

**Solution**  If "at least two of the people celebrate their birthdays the same day" is success, failure is "no two people celebrate their birthdays the same day." That is, all 23 people have birthdays falling on different days of the year. We assume that each year has 365 days and that it is equally likely for a person to be born on any of these days.

Let us name these people $P_1, P_2, P_3, \ldots, P_{23}$. The 23 people will celebrate their birthdays on different days if the following sequence of events occurs. $P_1$ celebrates his/her birthday on some day of the year; this has probability 1. Once this happens, $P_2$ celebrates his/her birthday on any of the other 364 days and this event has probability $\frac{364}{365}$. Now $P_3$ must celebrate his/her birthday on any day of the year except the days that $P_1$ and $P_2$ celebrate their birthdays, and the probability of this event is $\frac{363}{365}$. Continuing in this manner, and using Theorem 6.1, we find that the probability that no 2 of the 23 people celebrate their birthdays the same day is

$$\frac{365}{365} \cdot \frac{364}{365} \cdot \frac{363}{365} \cdot \ldots \cdot \frac{343}{365} \doteq .4927$$

Therefore, the probability that at least 2 of the 23 people celebrate their birthdays the same day is .5073 since $1 - .4927 = .5073$. Does this result surprise you? **Do Exercise 15.** ■

It is often useful in a probability problem to represent the situation pictorially. Venn diagrams and probability trees may be used, as illustrated in the next three examples.

## Venn Diagrams

**EXAMPLE 8**  A sociologist theorized that couples with religious affiliations had a better chance to remain married than those without affiliations. To test his theory, he conducted a study of 1000 randomly selected couples who had married ten years previously. He let $U$ be the universe (the set of married couples he surveyed), $R$ be the set of all married couples who had religious affiliations, and $M$ be the set of all couples who were still married. He then drew the following Venn diagram. (See Figure 6.6.)

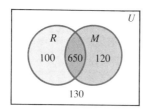

**FIGURE 6.6**

a. Find the probability that a couple with a religious affiliation is still married.
b. Find the probability that a couple who is still married has a religious affiliation.
c. Find the probability that a couple with a religious affiliation is divorced.

**Solution**  a. We must find the probability that a couple is still married given that it has a religious affiliation. That is, we must find $p(M \mid R)$. From the Venn diagram, we readily see that

$$n(U) = 1000$$
$$n(M \cap R) = 650$$
$$n(M) = 770$$
$$n(R) = 750$$

Thus,

$$p(M \cap R) = \frac{650}{1000} = .65$$

$$p(R) = \frac{75}{1000} = .75$$

It follows that

$$p(M \mid R) = \frac{p(M \cap R)}{p(R)} = \frac{.65}{.75} \doteq .8667$$

Thus, if we know that a couple has a religious affiliation, there is a probability of .8667 that they are still married.

b. We must find $p(R \mid M)$. Since $p(R \mid M) = \frac{p(R \cap M)}{p(M)}$, we first calculate $p(M)$. From the Venn diagram we see that

$$n(M) = 770$$

Hence,

$$p(M) = \frac{770}{1000} = .77$$

Thus,

$$p(R \mid M) = \frac{p(R \cap M)}{p(M)} = \frac{.65}{.77} \doteq .8442$$

Therefore, if we know that a couple is still married, the probability that the couple has a religious affiliation is .8442.

c. Since $M$ represents the set of all couples who are still married, $M'$ represents the set of all couples who are divorced. (Here, we assume that all couples studied are still married or divorced, and that none of the people studied has died.) From the Venn diagram we see that $n(M') = 230$ since $130 + 100 = 230$. Thus,

$$p(M') = \frac{230}{1000} = .23$$

We also see that

$$n(M' \cap R) = 100$$

Thus,

$$p(M' \cap R) = \frac{100}{1000} = .1$$

Hence,

$$p(M' \mid R) = \frac{p(M' \cap R)}{p(R)} = \frac{.1}{.75} \doteq .1333$$

Therefore, if a couple has a religious affiliation, the probability that they are divorced is .1333.

**Do Exercise 21.** ◼

## Probability Trees

Another device that is helpful in picturing the possible outcomes in a probability problem is the probability tree. A probability tree is used to represent a multiphase process and consists of chance nodes from which branches are extended. Each chance node represents a phase of the process and each possible outcome of that phase is represented by a branch extending from the node. The probability of each outcome is written by the corresponding branch connecting adjacent nodes. A sequence of branches starting at the first node and ending at a tip is called a *path*. The paths are mutually exclusive and represent all possible outcomes of the process. Thus, the set of all possible paths represents the sample space and the probability of each outcome in the sample space is obtained by multiplying the probabilities written by the connecting branches of the corresponding path. (See Figure 6.7.)

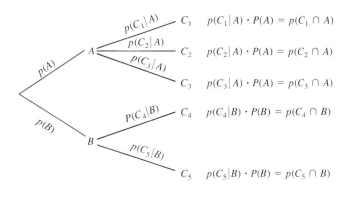

**FIGURE 6.7**

**EXAMPLE 9**   An urn contains three red, five white, and two blue marbles. Two marbles are drawn, one after the other, without replacement. What is the probability that one blue and one white marble were drawn?

**Solution**   A probability tree showing the different possible outcomes is drawn in Figure 6.8. In the tree, $R$, $W$, and $B$ represent the events "drawing a red marble," "drawing a white marble," and "drawing a blue marble," respectively. Since there are ten marbles in the urn and all outcomes are equally likely, $p(R) = .3$, $p(W) = .5$, and $p(B) = .2$. Since the first marble is not replaced, there are only nine marbles in the urn when the second marble is drawn. If the first marble drawn is red, only two red marbles remain in the urn, so that $p(R$ on second draw $\mid R$ on first draw$) = \frac{2}{9}$. The other conditional probabilities are computed similarly and the results are shown on the probability tree.

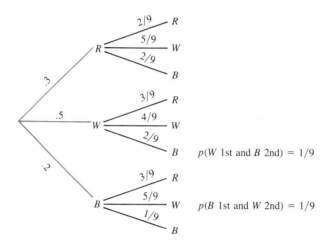

**FIGURE 6.8**

Since we were asked to find the probability that one blue and one white marble were drawn, we need only show the probability of a white marble being drawn on the first draw and a blue marble being drawn on the second draw, as well as the probability of a blue marble being drawn on the first draw and a white marble being drawn on the second draw. These probabilities are calculated as follows:

$$p(W \text{ 1st and } B \text{ 2nd}) = p(B \text{ 2nd} \mid W \text{ 1st}) \cdot p(W \text{ 1st}) = \frac{2}{9}(.5) = \frac{1}{9}$$

and

$$p(B \text{ 1st and } W \text{ 2nd}) = p(W \text{ 2nd} \mid B \text{ 1st}) \cdot p(B \text{ 1st}) = \frac{5}{9}(.2) = \frac{1}{9}$$

These two events are mutually exclusive, so the final probability is the sum of these two probabilities. That is, $p(\text{one } W \text{ and one } B) = \frac{1}{9} + \frac{1}{9} = \frac{2}{9}$.
**Do Exercise 23.**   ∎

   Note that in the preceding example, we showed only the necessary probabilities on the probability tree. Often, some of the paths will be shorter than others because we may leave out branches that are not relevant to the particular problem. We illustrate this in the following example.

**EXAMPLE 10**   Suppose that the Boston Celtics and Los Angeles Lakers are playing in the finals of the NBA championship and that the Lakers win the first two games in Boston. Games 3, 4, and 5 are scheduled in Los Angeles, while games 6 and 7, if necessary, will be played in Boston. The two teams are evenly matched and the only advantage is the home court. For each game, the probability of the home team winning that game is .6. What is the probability that the Celtics will win the championship in spite of losing the first two games at home?

**Solution**   We represent the possible outcomes in a probability tree. An $L$ by a node indicates that the Lakers won that game, while a $C$ indicates that the Celtics were victorious. Also .6 written by a branch indicates the probability of the home team winning, while .4 written by a branch indicates the probability of the visiting team winning. Note that as soon as the Lakers win two more games, the champion is determined and we end the path. Similarly, as soon as the Celtics win four games we end the path. (See Figure 6.9.) For example, the top path has only two branches, indicating that the Lakers won in four straight games. The probability of this is .36 since $(.6)(.6) = 0.36$. (Remember that games 3 and 4 are played in Los Angeles.) The lowest path indicates that the Celtics won games 3, 4, 5, and 6. The occurrence of this outcome has probability 0.0384, since the Celtics are the visiting team for games 3, 4, and 5 and the home team for game 6 and $(.4)(.4)(.4)(.6) = 0.0384$. Since we wish to find the probability of the Celtics winning the championship, we consider only paths that end with a $C$.

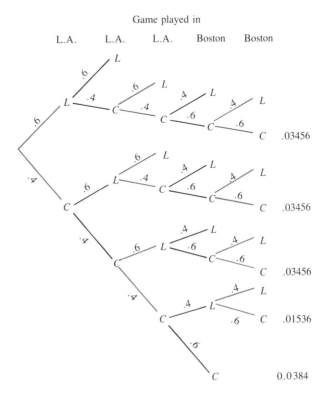

FIGURE 6.9

We calculated the probabilities of all outcomes represented by those paths and showed the results on the probability tree. For example, the first path that ends with a $C$ represents the outcome where the Lakers won game 3 and the Celtics won the last four games. This outcome has probability 0.03456 since $(.6)(.4)(.4)(.6)(.6) = 0.03456$. To find the probability of the Celtics winning the championship, we add the probabilities indicated by the five paths that terminate with a $C$, since these represent the outcome "the Celtics win" and they are mutually exclusive. We get

$$0.03456 + 0.03456 + 0.03456 + 0.01536 + 0.0384 = 0.15744$$

Therefore, the probability of the Celtics winning the championship after losing the first two games at home is 0.15744.
**Do Exercise 29.** ∎

## Infinite Sample Spaces

Although we have considered so far only sample spaces with a finite number of members, there are instances where we may wish to discuss the possibility of having infinitely many outcomes. We illustrate this situation with the following example.

**EXAMPLE 11** Steven, Michael, and Kris take turns tossing a coin, in that order. The first one to throw tails wins. What is the probability that Michael wins?

**Solution** The results of the tosses can be represented by ordered $n$-tuples. A successful outcome ("Michael wins") is represented by an $n$-tuple where the first $n - 1$ entries are $H$, and the last entry representing a toss by Michael is a $T$. For example, $(H, T)$, $(H, H, H, H, T)$, $(H, H, H, H, H, H, H, T)$, represent successful outcomes. The probability of the first of these is $\frac{1}{4}$ since $\frac{1}{2} \cdot \frac{1}{2} = \frac{1}{4}$, the probability of the second is $\frac{1}{32}$ since $\frac{1}{2} \cdot \frac{1}{2} \cdot \frac{1}{2} \cdot \frac{1}{2} \cdot \frac{1}{2} = \frac{1}{32}$, and so on. It is clear that, theoretically, we must consider infinitely many outcomes that are mutually exclusive. Adding the probabilities of all these outcomes we obtain

$$\frac{1}{2^2} + \frac{1}{2^5} + \frac{1}{2^8} + \frac{1}{2^{11}} + \cdots$$

We will see in Chapter 9 that this expression is a geometric series, and that its value can be found by using the formula

$$a + ar + ar^2 + ar^3 + \cdots = \frac{a}{1 - r}$$

whenever $|r| < 1$. In this series, $a = \frac{1}{4}$ and $r = \frac{1}{8}$. Thus,

$$\frac{1}{2^2} + \frac{1}{2^5} + \frac{1}{2^8} + \frac{1}{2^{11}} + \cdots = \frac{\frac{1}{4}}{1 - \frac{1}{8}} = \frac{2}{7}$$

Thus, the probability that Michael wins is $\frac{2}{7}$.
**Do Exercise 33.** ∎

# Exercise Set 6.3

**1.** A bag contains five red marbles and seven blue marbles. A second bag contains three red marbles and four blue marbles. A marble is drawn from the first bag and placed in the second. The second bag is shaken and a marble is drawn from it. What is the probability that a red marble was drawn each time?

**2.** An urn contains six white marbles and three green marbles. A second urn contains seven white marbles and six green marbles. A marble is drawn from the first urn and placed in the second. The second urn is shaken and a marble is drawn from it. What is the probability that a red marble was drawn each time?

**3.** The probability that a freshman takes an English course is .73. The probability of taking a mathematics course is .27 and the probability of taking both is .18. What is the probability that a freshman will take

   **a.** an English course given that a mathematics course is also taken by that student?
   **b.** a mathematics course given that an English course is also taken by that student?
   **c.** both English and mathematics given that at least one of the two is taken by that student?

**4.** The probability that a sophomore takes a philosophy course is .21. The probability of taking a sociology course is .28 and the probability of taking both is .12. What is the probability that a sophomore will take

   **a.** a philosophy course given that a sociology course is also taken by that student?
   **b.** a sociology course given that a philosophy course is also taken by that student?
   **c.** both philosophy and sociology given that at least one of the two is taken by that student?

**5.** A bag contains five red marbles and two blue marbles. A marble is drawn from the bag and replaced, the bag is shaken, and a marble is drawn again. What is the probability that both marbles drawn are red?

**6.** Repeat Exercise 5 without replacement.

**7.** An urn contains eight white marbles and five black marbles. A marble is drawn from the urn and replaced, the urn is shaken, and a marble is drawn again. What is the probability that both marbles drawn are white?

**8.** Repeat Exercise 7 without replacement.

**9.** A card is drawn at random from a standard deck of cards. What is the probability that it is either an ace or a jack?

**10.** A card is drawn at random from a standard deck of cards. What is the probability that it is either a ten or a face card?

**11.** A card is drawn from a standard deck of cards. What is the probability that it is either a heart or a queen?

**12.** A card is drawn from a standard deck of cards. What is the probability that it is either a spade or a face card?

**13.** Suppose two dice are rolled. What is the probability that the sum of the numbers that come up is either even or less than seven?

**14.** Suppose two dice are rolled. What is the probability that the sum of the numbers that come up is either odd or less than nine?

**15.** Suppose that ten people have been randomly selected. What is the probability that at least two of them celebrate their birthdays the same day? (*Hint:* See Example 7 and use a calculator.)

**16.** Repeat Exercise 15 for 15 people.

**17.** Repeat Exercise 15 for 17 people.

**18.** Repeat Exercise 15 for 20 people.

**19.** Repeat Exercise 15 for 30 people.

**20.** Repeat Exercise 15 for 40 people.

**21.** A sociologist theorized that people who do not drink alcoholic beverages have a better chance to complete their college education than those who do. To test her theory, she conducted a study of 2000 randomly selected people in the 23-to-25 age group. She let $U$ be the universe (the set of people she surveyed), $A$ be the set of all people who abstain from drinking alcoholic beverages, and $C$ be the set of all people who completed their college education. She then drew the following Venn diagram.

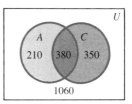

Find the probability that

   **a.** a person who does not drink alcoholic beverages completes a college education.
   **b.** a person who completes a college education does drink alcoholic beverages.

**c.** a person who does not drink alcoholic beverages does not complete a college education.

Did the results of the survey agree with the theory?

**22.** A sociologist theorized that teenagers who participate in sports have a better chance of not smoking than those who do not participate. To support his theory, he conducted a study of 1500 randomly selected young people who had just turned 20. He let $U$ be the universe (the set of young people he surveyed), $A$ be the set of all people who participate in some sport, and $N$ be the set of all people who do not smoke. He then drew the following Venn diagram.

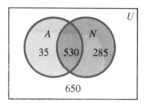

Find the probability that

**a.** a teenager who participates in sports does not smoke.
**b.** a teenager who does not smoke participates in sports.
**c.** a teenager who participates in sports smokes.

Did the results of the survey agree with the theory?

*Draw a probability tree to solve Exercises 23–29.*

**23.** An urn contains five red, three white, and seven blue marbles. Two marbles are drawn, one after the other, without replacement. What is the probability that one red and one white marble were drawn?

**24.** A bag contains four green, seven red, and one blue marble. Two marbles are drawn, one after the other, without replacement. What is the probability that one green and one red marble were drawn?

**25.** Suppose that the Seattle Supersonics and Milwaukee Bucks are playing in the finals of the NBA championship and that the Bucks win the first two games in Seattle. Games 3, 4, and 5 are scheduled in Milwaukee, and games 6 and 7, if necessary, will be played in Seattle. The two teams are evenly matched and the only advantage is the home court. For each game, the probability of the home team winning that game is .55. What is the probability that the Supersonics will win the championship in spite of losing the first two games at home?

**26.** Repeat Exercise 25 if the teams split the first two games in Seattle.

**27.** Suppose that the Montreal Expos and Detroit Tigers are playing in the World Series and that the Expos win the first two games in Detroit. Games 3, 4, and 5 are scheduled in Montreal, and games 6 and 7, if necessary, will be played in Detroit. The probabilities of Montreal winning games 3, 4, 5, 6, and 7 are .65, .55, .6, .35, and .3, respectively. What is the probability that the Tigers will win the World Series in spite of losing the first two games at home?

**28.** Repeat Exercise 27 if the two teams split the first two games in Detroit.

**29.** Repeat Exercise 27 if the Tigers win the first two games at home.

**30.** Suppose that the New York Rangers and Quebec Nordiques are playing for the Stanley Cup and that the Nordiques win the first two games in New York. Games 3, 4, and 5 are scheduled in Quebec, and games 6 and 7, if necessary, will be played in New York. The two teams are evenly matched and the only advantage is the home arena. For each game, the probability of the home team winning that game is .62. What is the probability that the Rangers will win the Stanley Cup in spite of losing the first two games at home?

**31.** Repeat Exercise 30 if the teams split the first two games in New York.

**32.** In Example 10 of this section, we found the probability that the Boston Celtics would win the championship after losing the first two games was 0.15744. Using a tree diagram, find the probabilities that

**a.** the Lakers will win in four games.
**b.** the Lakers will win in five games.
**c.** the Lakers will win in six games.
**d.** the Lakers will win in seven games.
**e.** the Lakers will win the championship.

Add your answer in part **e** to 0.15744 and verify that this sum is 1. Why?

**33.** Betty, Shirley, and Fran take turns tossing a coin, in that order. The first one to throw heads wins. What is the probability that Betty wins?

**34.** In Exercise 33, what is the probability that Fran wins?

**35.** Jeremy and André take turns rolling a die, in that order. The first to roll a 6 wins. What is the probability that Jeremy wins?

**36.** In Exercise 35, what is the probability that André wins?

**37.** Julie, Gwen, and Laura take turns rolling a die. The first one to roll a number less than 3 wins. What is the probability that Julie wins?

**38.** In Exercise 37, what is the probability that Gwen wins?

**39.** In Exercise 37, what is the probability that Laura wins?

# 6.4 Bayes' Formula

We begin by considering a sample space $S$ and $n$ mutually exclusive events $E_1$, $E_2$, $E_3$, . . . , $E_n$, such that each outcome in the sample space is in one of the events $E_i$. In other words, $E_i \cap E_j = \emptyset$ if $i \neq j$, and

$$E_1 \cup E_2 \cup E_3 \cup \cdots \cup E_n = U.$$

Let $A$ be an event such that $p(A) \neq 0$. Bayes' formula* provides a simple tool for computing the conditional probability of each event $E_i$ given $A$, from the conditional probability of $A$ given each event $E_i$, and the unconditional probability of each $E_i$.

### Derivation of Bayes' Formula

We have

$$P(E_1) + P(E_2) + P(E_3) + \cdots + P(E_n) = 1$$

Let $i$ be any integer between 1 and $n$, inclusive. Then,

$$p(E_i \mid A) \cdot p(A) = p(E_i \cap A)$$

and

$$p(A \mid E_i) \cdot p(E_i) = p(A \cap E_i)$$

But, $E_i \cap A = A \cap E_i$. Thus, $p(E_i \cap A) = p(A \cap E_i)$, from which it follows that

$$p(E_i \mid A) \cdot p(A) = p(A \mid E_i) \cdot p(E_i)$$

Since $p(A) \neq 0$, we get

$$p(E_i \mid A) = \frac{p(A \mid E_i) \cdot p(E_i)}{p(A)}$$

Note that,

$$A = (A \cap E_1) \cup (A \cap E_2) \cup (A \cap E_3) \cup \cdots \cup (A \cap E_n)$$

(See the Venn diagram in Figure 6.10.)

Since the events $A \cap E_i$, $1 \leq i \leq n$ are mutually exclusive, we have

$$p(A) = p(A \cap E_1) + p(A \cap E_2) + p(A \cap E_3) + \cdots + p(A \cap E_n)$$

It follows that

$$p(E_i \mid A) = \frac{p(A \mid E_i) \cdot p(E_i)}{p(A)}$$

$$= \frac{p(A \mid E_i) \cdot p(E_i)}{p(A \cap E_1) + p(A \cap E_2) + p(A \cap E_3) + \cdots + p(A \cap E_n)}$$

**FIGURE 6.10**

---

*This formula was first derived by the English Presbyterian minister Thomas Bayes and was first published in 1763 although Bayes died in 1761.

But, for each $j$ we have $p(A \cap E_j) = p(A \mid E_j) \cdot P(E_j)$. Hence,

$$p(E_i \mid A) = \frac{p(A \mid E_i) \cdot p(E_i)}{p(A \mid E_1) \cdot p(E_1) + p(A \mid E_2) \cdot p(E_2) + \cdots + p(A \mid E_n) \cdot p(E_n)}$$

## Applications

Bayes' formula may appear difficult to memorize. In actuality, problems involving this formula can be done using a probability tree as we shall illustrate in the following example.

**EXAMPLE 1**     A novice golfer must hit a certain shot. He has eight different clubs in his bag, only one of which is the right club for this particular shot. The probability that he hits a good shot is .25 if he uses the right club, but only .12 if he uses a wrong club. Suppose he picks a club randomly from his bag and hits a good shot. What is the probability that he selected the right club?

**Solution**     Let $G$ and $B$ represent the events "hit a good shot" and "hit a bad shot," respectively. Let $R$ and $W$ represent the events "selecting the right club" and "selecting the wrong club," respectively. Since he is just as likely to select one club as any other (remember he is a novice), we have $p(R) = \frac{1}{8} = .125$ and $p(W) = \frac{7}{8} = .875$. It is given that $p(G \mid R) = .25$ and that $p(G \mid W) = .12$. Hence, $p(B \mid R) = .75$ since $1 - .25 = .75$, and $p(B \mid W) = .88$ since $1 - .12 = .88$. We are asked to find $p(R \mid G)$.
Using Bayes' formula, we know that

$$p(R \mid G) = \frac{p(G \mid R) \cdot p(R)}{p(G \mid R) \cdot p(R) + p(G \mid W) \cdot p(W)}$$

Thus,

$$p(R \mid G) = \frac{.25(.125)}{.25(.125) + .12(.875)} \doteq .2294$$

Thus, if it is known that the novice hit a good shot, the probability that he had selected the right club for that shot is approximately .2294. We draw a probability tree for this problem. (See Figure 6.11.)

**FIGURE 6.11**

We see from the diagram that the numerator of the expression giving $p(R \mid G)$ is the product $p(G \mid R)p(R) = .25(.125) = .03125$. This is the product of the probabilities on the branches in the path to $G$ through $R$. On the other hand, the denominator of that expression is $p(G \mid R)p(R) + p(G \mid W)p(W) = .25(.125) + .12(.875) = .13625$. This is the sum of all products of the probabilities on the branches of the paths leading to $G$.

**Do Exercise 1.**

Therefore, instead of using the formula

$$p(E_i \mid A) = \frac{p(A \mid E_i) \cdot p(E_i)}{p(A \mid E_1) \cdot p(E_1) + p(A \mid E_2) \cdot p(E_2) + \cdots + p(A \mid E_n) \cdot p(E_n)}$$

to solve this type of problem, we can use the following procedure.

**STEP 1.** Draw a probability tree for the problem, indicating by each branch of the tree $p(E_i)$ and $p(A \mid E_i)$ for $i = 1, 2, \ldots, n$.

**STEP 2.** Calculate the product of the probabilities on the branches of each path leading to $A$.

**STEP 3.** Add all products obtained in Step 2.

**STEP 4.** To calculate $p(E_i \mid A)$, divide the product of the probabilities on the branches of the path leading to $A$ through $E_i$ (this is one of the products calculated in Step 2) by the sum obtained in Step 3.

**EXAMPLE 2** Three machines are used in manufacturing a certain gadget. Machines A, B, and C are used to produce 30%, 45%, and 25% of the total output, respectively. It is known that 2% of the gadgets made using machine A are defective, 4% of the gadgets made using machine B are defective, and 3% of the gadgets made using machine C are defective. A gadget is selected at random from the total output. If it is defective, what is the probability that it was made using machine B?

**Solution** **Step 1.** We begin by drawing a probability tree (see Figure 6.12), where $A$, $B$, and $C$ represent the events "produced using machine A," "produced using machine B," and "produced using machine C," respectively; $D$ represents the event "the gadget is defective" and $D'$ represents the event "the gadget is not defective."

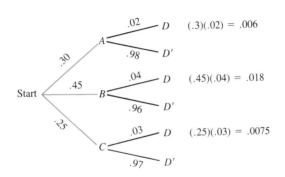

**FIGURE 6.12**

**Step 2.** Since it is given that the gadget chosen at random is defective, we calculate each of the products of the probabilities on the branches of each path leading to $D$.

For the first path we get $.3(.02) = .006$.
For the second path we get $.45(.04) = .018$
For the third path we get $.25(.03) = .0075$.

**Step 3.** Now we add all products obtained in Step 2.

$$.006 + .018 + .0075 = .0315$$

**Step 4.** Since we wish to find the probability that the defective gadget was produced using machine B, we consider the second path since it leads to $D$ through $B$. The product of the probabilities of the branches in that path was obtained in Step 2. It is .018. To find the desired probability, we divide .018 by the sum obtained in Step 3. We get

$$\frac{.018}{.0315} \doteq .57143$$

where the result has been rounded off to five decimal places. Thus, the probability that the gadget was made using machine B, given that it is defective, is approximately .57143.

**Do Exercise 9.**    ■

Sometimes we are faced with the problem of finding probabilities of experiments that have several stages. We illustrate this with an example.

**EXAMPLE 3**    Suppose that an urn contains three fair coins and a coin with tails on both sides. A coin is selected randomly from the urn and it is tossed.

**a.** Tails comes up. What is the probability that the coin selected was fair?
**b.** The coin is tossed again, and again tails comes up. Calculate the probability that the coin selected was fair.
**c.** The coin is tossed a third time and this time heads comes up. What is the probability that the coin selected was a fair coin?

**Solution**    **a. Step 1.** We first draw a probability tree where $F$ represents the event "the coin selected was fair," $F'$ represents the event "the coin selected has two tails," $H$ represents the event "heads came up" and $T$ represents the event "tails came up." We indicate the probability of each of these by the corresponding branch on the tree. (See Figure 6.13.)

**FIGURE 6.13**

**Step 2.** We calculate the product of the probabilities on the branches of each path leading to $T$. We get, for the first path leading to $T$, $.75(.5) = .375$, and for the second path leading to $T$, $.25(1) = .25$.

**Step 3.** We now add the products obtained in Step 2. We get .375 + .25 = .625.

**Step 4.** We divide .375, which is the product of the probabilities on the branches of the path leading to $T$ through $F$, by the sum obtained in Step 3. We get $\frac{.375}{.625} = .6$.

Therefore, the probability that the coin selected was fair, given that the toss came up tails, is .6.

**b.** We use the result of part **a**. The probability $p(F \mid T)$ was found to be .6. Consequently, $p(F' \mid T) = .4$. These probabilities can now be used for the second stage. We go through the same steps as in part **a**.

**Step 1.** We first draw a probability tree and indicate the new probabilities on the branches of the tree. (See Figure 6.14.)

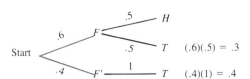

**FIGURE 6.14**

**Step 2.** We calculate the product of the probabilities on the branches of each path leading to $T$. We get, for the first path leading to $T$, $.6(.5) = .3$, and for the second path leading to $T$, $.4(1) = .4$.

**Step 3.** We now add the products obtained in Step 2. We get $.3 + .4 = .7$.

**Step 4.** We divide .3, which is the product of the probabilities on the branches of the path leading to $T$ through $F$, by the sum obtained in Step 3. We get $\frac{.3}{.7} = \frac{3}{7}$. The probability that the coin selected was fair, given that the tosses came up tails both times, is $\frac{3}{7}$.

**c.** If at any stage the coin is tossed and heads comes up, then we know that a fair coin was selected since it could not be the coin with tails on both sides. In that case, the probability that a fair coin was selected is 1.

**Do Exercise 19.**  ■

---

## Exercise Set 6.4

**1.** A novice golfer must hit a certain shot. He has six different clubs in his bag, only one of which is the right club for this particular shot. The probability that he hits a good shot is .23 if he uses the right club, but only .1 if he uses a wrong club. Suppose that he picks a club randomly from his bag and hits a good shot. What is the probability that he selected the right club?

**2.** Repeat Exercise 1 if the golfer has five clubs in his bag and the probability that he hits a good shot is .26 if he uses the right club and .11 if he uses a wrong club.

**3.** In Exercise 1, what is the probability that the golfer selected the right club given that the shot he hit was a bad shot?

**4.** Urn 1 contains four green marbles and three red marbles, and urn 2 contains five red marbles and seven red marbles.

A coin is tossed. If heads comes up, urn 1 is selected and a marble is drawn from that urn. If tails comes up, urn 2 is selected and a marble is drawn. Suppose the marble drawn is red.

**a.** What is the probability that the first urn was selected?

**b.** What is the probability that the second urn was selected?

**5.** Repeat Exercise 4 if urn 1 has 3 green marbles and 10 red marbles, and urn 2 has 13 green marbles and 9 red marbles.

**6.** Urn 1 contains five black marbles and seven white marbles, and urn 2 contains six black marbles and ten white marbles. A die is rolled. If a 2 comes up, urn 1 is selected and a marble is drawn from that urn. If any other number comes up, urn 2 is selected and a marble is drawn. Suppose the marble drawn is white.

**a.** What is the probability that the first urn was selected?
**b.** What is the probability that the second urn was selected?

7. Urn 1 contains four yellow marbles and six blue marbles, and urn 2 contains seven yellow marbles and nine blue marbles. A die is rolled. If a number greater than 4 comes up, urn 1 is selected and a marble is drawn from that urn. If any other number comes up, urn 2 is selected and a marble is drawn. Suppose the marble drawn is blue.

   **a.** What is the probability that the first urn was selected?
   **b.** What is the probability that the second urn was selected?

8. Urn 1 contains five red marbles and six white marbles, urn 2 contains seven red marbles and nine blue marbles, and urn 3 contains ten red marbles and three blue marbles. A die is rolled. If an even number comes up, urn 1 is selected and a marble is drawn from that urn. If a 1 or a 3 comes up, urn 2 is selected and a marble is drawn. If a 5 comes up, urn 3 is selected and a marble is drawn. Suppose the marble drawn is red.

   **a.** What is the probability that the first urn was selected?
   **b.** What is the probability that the second urn was selected?
   **c.** What is the probability that the third urn was selected?
   **d.** If a white marble is drawn, what is the probability that urn 1 was selected?

9. Three machines are used in manufacturing a certain gadget. Machines A, B, and C are used to produce 20%, 65%, and 15% of the total output, respectively. It is known that 3% of the gadgets made using machine A are defective, 5% of the gadgets made using machine B are defective, and 4% of the gadgets made using machine C are defective. A gadget is selected at random from the total output. If it is defective, what is the probability that it was made using machine B?

10. A company manufactures car batteries at three plants. Plant A produces 850 batteries a day, plant B produces 630 batteries a day, and plant C produces 520 batteries a day. Experience indicates that 4% of the batteries produced at plant A are defective, 5% of the batteries produced at plant B are defective, and 2% of those produced at plant C are defective. Each day, all batteries produced are shipped to the company's warehouse. Suppose that a battery randomly selected in the warehouse is found to be defective. What is the probability that it was produced at **a.** plant A? **b.** plant B? **c.** plant C?

11. A company manufactures VCRs at four plants. Plant A produces 350 VCRs a week, plant B produces 420 VCRs a week, plant C produces 330 VCRs a week, and plant D produces 400 VCRs a week. Experience indicates that 2% of the VCRs produced at plant A are defective, 4% of the VCRs produced at plant B are defective, 3% of those pro-

duced at plant C are defective, and 1.5% of those produced at plant D are defective. Each week, all VCRs produced are shipped to the company's warehouse. Suppose that a VCR randomly selected in the warehouse is found to be defective. What is the probability that it was produced at **a.** plant A? **b.** plant B? **c.** plant C? **d.** plant D?

12. In Exercise 10, suppose that a battery randomly selected in the warehouse is not defective. What is the probability that it was produced at **a.** plant A? **b.** plant B? **c.** plant C?

13. In Exercise 11, suppose that a VCR randomly selected in the warehouse is not defective. What is the probability that it was produced at **a.** plant A? **b.** plant B? **c.** plant C? **d.** plant D?

14. The director of an MBA program decided that applicants to the program will be required to take a qualifying exam in mathematics. After several years of administering the qualifying test, it was found that 70% of all applicants completed the program and that 85% of those who completed the program had passed the test, while only 12% of those who did not complete the program had passed the test. Suppose an applicant has passed the test. What is the probability that this applicant will complete the program?

15. Repeat Exercise 14 with 60% of all applicants completing the program, 82% of those completing the program passed the test, and only 14% of those not completing the program passed the test.

16. It is known that .02% of the people in a city have AIDS. A new test is tried and it is found to give a positive reaction to 45% of the people who have the disease and to 15% of the people who do not have the disease. A certain person's test is positive. What is the probability that this person has the disease?

17. It is known that 2.6% of the people in the United States are diabetic. A new glucose test is tried and it is found to give a positive reaction to 90% of the people who are diabetic and to 5% of the people who are not diabetic. A certain person's test is positive. What is the probability that this person is diabetic?

18. Suppose that an urn contains four fair coins and a coin with tails on both sides. A coin is selected randomly from the urn and it is tossed.

   **a.** Tails comes up. What is the probability that the coin selected was fair?
   **b.** The coin is tossed again, and again tails comes up. Calculate the probability that the coin selected was fair.
   **c.** The coin is tossed a third time and this time heads comes up. What is the probability that the coin selected was a fair coin?

**19.** Suppose that an urn contains six fair coins and two coins with heads on both sides. A coin is selected randomly from the urn and it is tossed.

**a.** Heads comes up. What is the probability that the coin selected was fair?

**b.** The coin is tossed again, and again heads comes up. Calculate the probability that the coin selected was fair.

**c.** The coin is tossed a third time and this time tails comes up. What is the probability that the coin selected was a fair coin?

---

# 6.5 Expected Value

In the 17th century, the Chevalier de Méré, a gambling enthusiast, asked a great French mathematician, Blaise Pascal (1623–1662), how the stakes should be divided in an unfinished game of chance. Subsequently, Pascal corresponded with another great French mathematician, Pierre de Fermat (1601–1665), about the problem raised by the Chevalier's request. The resulting series of letters laid the foundation for probability theory. It is therefore natural that we use simple games of chance to illustrate some of the ideas involved. However, you should keep in mind that probability theory is an important subject with wide applications both in applied and pure mathematics.

## Expectation

We begin with the following example.

**EXAMPLE 1**  In drawing a single card from a standard deck, you win $5 if you draw an ace, $2 if you draw a face card, and 25 cents if any other card is drawn. It costs $1 to be allowed to draw a card. Would you expect to win money in the long run by playing this game?

**Solution**  Suppose you play the game 520 times. Since the probability of drawing an ace is $\frac{4}{52}$, you would expect to draw an ace approximately 40 times, and win approximately $200 doing so. Similarly, there are 12 face cards in the deck; thus, the probability of drawing a face card is $\frac{12}{52}$ and you would expect to draw a face card approximately 120 times, winning approximately $240 doing so. Finally, you would expect to draw some other card approximately 360 times (since $520 - (40 + 120) = 360$), and win approximately $90 dollars doing so. Thus, if you played 520 times, you would win approximately $530. However, it would cost you $520 dollars to play 520 games. Thus, you would expect to win about $10 dollars if you played 520 games or an average of $\frac{1}{52}$ dollar per game (slightly less than 2 cents per game).

**Do Exercise 1.**                                                                                         ∎

Of course, we know that things will not always happen exactly as expected. In the preceding example, when the game is played 520 times, we may not have exactly 40 aces drawn or 120 face cards drawn. However, if the game is played a great number of times, it is likely that the average gain per game will be slightly less than 2 cents. We used the 520 games to illustrate the basic idea.

Suppose we present the example in a different way. Each time we draw an ace we win $5, but we pay $1 to play, so the net payoff is only $4. Each time we draw a face card we win $2, but it cost $1; thus, the net payoff is only $1. Each time we draw any other card we win 25 cents, but the cost is still $1, so we lose 75 cents and the net payoff is −$.75. If we play 520 games, we can calculate the average payoff as follows:

$$\frac{40(4) + 120(1) + 360(-.75)}{520} = 4 \cdot \frac{40}{520} + 1 \cdot \frac{120}{520} + (-.75) \cdot \frac{360}{520}$$

$$= 4 \cdot \frac{1}{13} + 1 \cdot \frac{3}{13} + (-.75) \cdot \frac{9}{13}$$

$$= 4p(A) + 1p(F) + (-.75)p(T)$$

where $A$ is the event "drawing an ace," $F$ is the event "drawing a face card," and $T$ is the event "drawing any other card."

> **DEFINITION 6.8:** If a game has $n$ outcomes, the probabilities of which are $p_1, p_2, \ldots, p_n$, respectively, and if the net payoff is the amount $A_i$ if the $i$th outcome occurs, where $(1 \leq i \leq n)$, then the *expectation* for the game is the sum
>
> $$p_1A_1 + p_2A_2 + \cdots + p_nA_n.$$

This concept is useful when deciding whether or not to enter a gambling game. Since the expectation represents approximately the average winnings per trial, provided the game is played many times, then the decision to enter the game should be made only if the expectation is positive. This, of course, does not apply to the purchase of insurance, where it is customary for the insured to have a negative expectation. The purpose of insurance is not to win money, but to spread the cost of accidents among the insured.

## Probability Distribution

To expand on this, we will introduce the concept of random variables. Basically, a random variable is a variable whose values are real numbers that depend on a random process.

> **DEFINITION 6.9:** A *random variable* is a function $X$ whose domain is a sample space and whose range is a set of real numbers. If the range is finite, or if it can be put in a one-to-one correspondence with the set of positive integers, then $X$ is said to be a *discrete random variable*. If the range is an interval of real numbers, the random variable is said to be *continuous*.

Suppose, for example, a die is rolled twice and we represent an outcome by an ordered pair $(x, y)$, where $x$ is the number that comes up on the first roll and $y$ is the number that comes up on the second roll. Then, the sample space $S$ has 36 members.

We may define a random variable $X$ by $X(x, y) = x + y$. For example, $X(1, 3) = 4$, $X(2, 5) = 7$, and so on. Then, the range of $X$ is the set $\{2, 3, 4, 5, 6, 7, 8, 9, 10, 11, 12\}$ and $X$ is a discrete random variable. Now suppose we keep tossing a fair coin until tails comes up and $X$ is the number of tosses it takes to get tails. In this case, the sample space is $\{T, HT, HHT, HHHT, HHHHT, HHHHHT, \ldots\}$. Then, $X(T) = 1, X(HT) = 2, X(HHT) = 3$, and so on. Again, $X$ is a discrete random variable. On the other hand, if a traffic light changes every 90 seconds and $X$ seconds is the time a randomly selected motorist must wait at a red light, then the value of $X$ is in the interval $[0, 90]$ and therefore $X$ is a continuous random variable.

Note that for each value of a random variable we have a corresponding event in the sample space. For example, if $X(x, y) = x + y$, where $x$ and $y$ represent the numbers that came up on two rolls of a die, then

$$\{(x, y) \mid X(x, y) = 5\} = \{(1, 4), (2, 3), (3, 2), (4, 1)\}$$

so that the event $E = \{(1, 4), (2, 3), (3, 2), (4, 1)\}$ corresponds to the value 5 of the random variable. Consequently, we can assign to each value of a random variable a probability that the random variable will take that value. In this example, since the outcomes in a roll of a fair die are equally likely, $p(E) = \frac{4}{36} = \frac{1}{9}$. So, $p(X = 5) = p(E) = \frac{1}{9}$, where $p(X = 5)$ denotes the probability that the random variable assumes the value 5. In general, the following definition applies.

> **DEFINITION 6.10:**  Suppose $S$ is a sample space with a probability function $p$ and $X$ is a discrete random variable with domain $S$. For each value of $x$ in the range of $X$, let
>
> $$p(X = x) = p(E), \text{ where } E = \{s \mid s \in S \text{ and } X(s) = x\}$$
>
> The *probability function* (or *probability distribution*) of $X$ is the function $f$, defined by $f(x) = p(X = x)$ for each $x$ in the range of $X$.

**EXAMPLE 2**  A fair coin is tossed three times and $X$ is the number of tails that come up. Determine the probability distribution of $X$.

**Solution**  In this experiment, the sample space has eight elements. Each of these is an ordered triple. For example, $(T, T, H)$ represents the outcome "tails came up the first two tosses, then heads came up." The number of tails that can come up is 0, 1, 2, or 3. Thus, the range of the random variable $X$ is the set $\{0, 1, 2, 3\}$. This set is the domain of the probability distribution.

The event corresponding to $X = 0$ is $\{(H, H, H)\}$. Since all outcomes are equally likely, the probability of that event is $\frac{1}{8}$, and $f(0) = P(X = 0) = \frac{1}{8}$. The event corresponding to $X = 1$ is $\{(T, H, H), (H, T, H), (H, H, T)\}$. The probability of that event is $\frac{3}{8}$, and $f(1) = P(X = 1) = \frac{3}{8}$. The event corresponding to $X = 2$ is $\{(T, T, H), (T, H, T), (H, T, T)\}$. The probability of that event is $\frac{3}{8}$, and $f(2) = P(X = 2) = \frac{3}{8}$. The event corresponding to $X = 3$ is $\{(T, T, T)\}$. The probability of that event is $\frac{1}{8}$, and $f(3) = P(X = 3) = \frac{1}{8}$.

The results are summarized in the following *probability table*.

| $x$ | $f(x) = P(X = x)$ |
|-----|-------------------|
| 0   | $\frac{1}{8}$     |
| 1   | $\frac{3}{8}$     |
| 2   | $\frac{3}{8}$     |
| 3   | $\frac{1}{8}$     |

It should be noted that

$$f(0) + f(1) + f(2) + f(3) = \frac{1}{8} + \frac{3}{8} + \frac{3}{8} + \frac{1}{8} = 1$$

and that $0 \leq f(x) \leq 1$ for each $x \in \{0, 1, 2, 3\}$. This follows from the fact that the events corresponding to $X = 0$, $X = 1$, $X = 2$, and $X = 3$ are mutually exclusive and their union is the sample space.
**Do Exercise 5.**                                                                           ■

In general, if $X$ is a discrete random variable with domain $S$ and range $D$, and $f$ is the probability distribution of $X$, then **a.** The domain of $f$ is $D$, **b.** $0 \leq f(x) \leq 1$, for all $x$ in $D$, and **c.** $\sum_{x \in D} f(x) = 1$.

**EXAMPLE 3**  A fair coin is tossed until heads comes up. Let $X$ be the number of tosses it takes to get heads. Determine the probability distribution of $X$.

**Solution**  The sample space is {H, TH, TTH, TTTH, TTTTH, . . .}. Since the coin is fair, we have $p(\text{H}) = \frac{1}{2}$, $p(\text{TH}) = \left(\frac{1}{2}\right)\left(\frac{1}{2}\right) = \frac{1}{4}$, $p(\text{TTH}) = \left(\frac{1}{2}\right)\left(\frac{1}{2}\right)\left(\frac{1}{2}\right) = \frac{1}{8}$, $p(\text{TTTH}) = \left(\frac{1}{2}\right)\left(\frac{1}{2}\right)\left(\frac{1}{2}\right)\left(\frac{1}{2}\right) = \frac{1}{16}$; and so on. If $X$ is the number of tosses required to get heads, the range of $X$ is {1, 2, 3, 4, . . .}. Thus, the domain of the probability distribution $f$ is also {1, 2, 3, 4, . . .}. We have

$$f(1) = P(X = 1) = p(\text{H}) = \frac{1}{2}$$

$$f(2) = P(X = 2) = p(\text{TH}) = \frac{1}{4}$$

$$f(3) = P(X = 3) = p(\text{TTH}) = \frac{1}{8}$$

$$f(4) = P(X = 4) = p(\text{TTTH}) = \frac{1}{16}$$

$$\vdots$$

In general, $f(n) = 1/2^n$.

We see that

$$f(1) + f(2) + f(3) + f(4) + \cdots = \frac{1}{2} + \frac{1}{4} + \frac{1}{8} + \frac{1}{16} + \cdots$$

$$= \frac{\frac{1}{2}}{1 - \frac{1}{2}} = 1$$

which agrees with our earlier statement about probability distributions.  ■

Probability distributions for continuous random variables will be defined in Chapter 10.

## Histogram

The probability distribution of a discrete random variable is often represented pictorially by a *probability histogram*. The probability histogram for the random variable of Example 2 is shown in Figure 6.15.

$f(x) = P(X = x)$

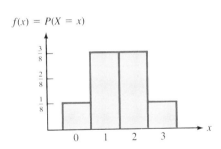

**FIGURE 6.15**

The rectangle over each value of $x$ has width 1 and height $f(x) = P(X = x)$. The sum of the areas of the rectangles is 1 square unit.

## Mathematical Expectation

Returning to Example 1, we can define the value of a discrete random variable to be the net payoff (winnings − cost). The range of $X$ is $\{-.75, 1, 4\}$. The probability distribution function is defined as follows:

$$f(-.75) = P(X = -.75) = p(\text{drawing a card other than an ace or a face card})$$

$$= \frac{9}{13}$$

$$f(1) = P(X = 1) = p(\text{drawing a face card}) = \frac{3}{13}$$

$$f(4) = P(X = 4) = p(\text{drawing an ace}) = \frac{1}{13}$$

The expectation for that game was calculated as follows:

$$(-.75)\left(\frac{9}{13}\right) + (1)\left(\frac{3}{13}\right) + (4)\left(\frac{1}{13}\right) = \frac{.25}{13} = \frac{1}{52} \qquad (*)$$

But, $\frac{9}{13} = f(-.75)$, $\frac{3}{13} = f(1)$, and $\frac{1}{13} = f(4)$. So, equality (*) may be written

$$(-.75)f(-.75) + (1)f(1) + (4)f(4) = \frac{1}{52}$$

This equality suggests the following definition.

> **DEFINITION 6.11:**    Suppose $S$ is a finite sample space and $X$ is a discrete random variable with probability distribution $f$ defined on the range $\{x_1, x_2, x_3, \ldots, x_n\}$ of $X$. The *mathematical expectation* of $X$ is given by
>
> $$E(X) = x_1 f(x_1) + x_2 f(x_2) + x_3 f(x_3) + \cdots + x_n f(x_n)$$

**EXAMPLE 4**    Calculate the mathematical expectation of the random variable of Example 2.

**Solution**    The values of $X$ were the numbers of tails that can come up when a fair coin is tossed three times. Therefore, the range of $X$ is $\{0, 1, 2, 3\}$. We found in Example 2 that $f(0) = \frac{1}{8}$, $f(1) = \frac{3}{8}$, $f(2) = \frac{3}{8}$, and $f(3) = \frac{1}{8}$. Thus, the mathematical expectation of the random variable is

$$E(X) = 0\left(\frac{1}{8}\right) + 1\left(\frac{3}{8}\right) + 2\left(\frac{3}{8}\right) + 3\left(\frac{1}{8}\right) = \frac{12}{8} = 1.5$$

**Do Exercise 11.**    ∎

This result should not surprise you since we would expect tails to come up about half of the time; thus for three tosses, we expect the average number of tails to be 1.5. Note that for no outcome of the experiment do we have 1.5 tails coming up. The number 1.5 is simply an average, just as when an instructor announces that the class average on a certain test is 73.52, probably no student actually scored 73.52 on that exam.

**EXAMPLE 5**    The annual premium that a Seattle homeowner must pay an insurance company to protect her house, worth $150,000, against total destruction by fire is $125. From experience, the company has determined that the probability of a house being totally destroyed by fire in that neighborhood is 0.00025. Calculate the expected gain per policy to the company.

**Solution**    Let $X$ be the gain to the company per policy. If the house is not destroyed by fire, the company gains $125. On the other hand, if the house is destroyed by fire, the company loses $149,875 since $150,000 - 125 = 149,875$. Therefore, the value of $X$ is 125 in the first case and $-149,875$ in the second case. If $f$ is the probability distribution of $X$, we have

$$f(-149,875) = P(X = -149,875) = 0.00025$$

and

$$f(125) = P(X = 125) = 0.99975$$

The expected value of $X$ is found as follows:

$$E(X) = -149,875f(-149,875) + 125f(125)$$
$$= -149,875(0.00025) + 125(0.99975) = 87.5$$

On the average, the insurance company will gain $\$87.50$ per policy.
**Do Exercise 13.**                                                          ■

## Variance and Standard Deviation

Since the mathematical expectation $E(X)$ of a random variable $X$ is the average value of $X$ in the long run, it is a measure of the central tendency of $X$. You have undoubtedly seen cases of two exams where the averages were identical, but in one exam all scores were close to the average and in the other exam there were a few high grades and a few low grades, but not many near the average. Thus, the average alone does not tell us anything about the dispersion of the grades. We need to have some way to measure the dispersion of the values of a random variable. The variance, often denoted $VAR(X)$ or $\sigma^2$ (read: "sigma squared"), and the standard deviation are two such measures. Basically, the variance of a discrete random variable $X$ is the mathematical expectation of the squares of the deviations of the values of the random variable $X$ from its expected value $E(X)$.

**DEFINITION 6.12:**   If $X$ is a random variable whose range is $\{x_1, x_2, x_3, \ldots, x_n\}$, $f$ is the probability distribution of $X$, and $E(X)$ is the mathematical expecation of $X$, then the *variance* of $X$ is

$$\sigma^2 = [x_1 - E(X)]^2 f(x_1) + [x_2 - E(X)]^2 f(x_2) + \cdots + [x_n - E(X)]^2 f(x_n)$$

In applications, it is sometimes difficult to use variance as a measure of dispersion of a random variable because the values of the variance and the variable are expressed in different units. For example, if $X$ represents a number of feet, the variance would represent a number of square feet. To alleviate this difficulty, we use the standard deviation.

**DEFINITION 6.13:**   The *standard deviation* of a random variable is the non-negative square root of its variance.

**EXAMPLE 6**   Suppose both random variables $X$ and $Y$ have range $\{1, 2, 3, 4\}$. Suppose $f$ and $g$ are the probability distributions of $X$ and $Y$, respectively, defined by the following probability tables.

| $x$ | $f(x) = P_1(X = x)$ |
|-----|---------------------|
| 1   | .3                  |
| 2   | .4                  |
| 3   | .2                  |
| 4   | .1                  |

| $y$ | $g(y) = P_2(Y = y)$ |
|-----|---------------------|
| 1   | .55                 |
| 2   | .1                  |
| 3   | .05                 |
| 4   | .3                  |

**a.** Show that both $X$ and $Y$ have the same mathematical expectation.
**b.** Find the variance and standard deviation of $X$.
**c.** Find the variance and standard deviation of $Y$.
**d.** Draw a probability histogram for $X$ and for $Y$.

**Solution**   **a.** We first find $E(X)$.

$$E(X) = 1 \cdot f(1) + 2 \cdot f(2) + 3 \cdot f(3) + 4 \cdot f(4)$$
$$= 1(.3) + 2(.4) + 3(.2) + 4(.1) = 2.1$$

We now find $E(Y)$.

$$E(Y) = 1 \cdot g(1) + 2 \cdot g(2) + 3 \cdot g(3) + 4 \cdot g(4)$$
$$= 1(.55) + 2(.1) + 3(.05) + 4(.3) = 2.1$$

Thus, $E(X) = E(Y)$.
**b.** The variance of $X$ is found as follows:

$$\text{VAR}(X) = (1 - 2.1)^2(.3) + (2 - 2.1)^2(.4) + (3 - 2.1)^2(.2)$$
$$+ (4 - 2.1)^2(.1) = .89$$

Therefore, the standard deviation of $X$ is

$$\sigma = \sqrt{\text{VAR}(X)} = \sqrt{.89} \doteq .943398$$

**c.** The variance of $Y$ is found as follows:

$$\text{VAR}(Y) = (1 - 2.1)^2(.55) + (2 - 2.1)^2(.1) + (3 - 2.1)^2(.05)$$
$$+ (4 - 2.1)^2(.3) = 1.79$$

The standard deviation of $Y$ is

$$\sigma = \sqrt{\text{VAR}(Y)} = \sqrt{1.79} \doteq 1.337909$$

**d.** The histograms for $X$ and $Y$ are shown in Figure 6.16.
**Do Exercise 21.**

**FIGURE 6.16**

As we pointed out earlier, the variance is a rough measure of the spread of values of the random variable. In fact, if VAR($X$) is small and each of the terms used to calculate its value is nonnegative, then each of these terms must be small. Therefore if for any $i$, $|x_i - E(X)|$ is large, then the corresponding $f(x_i)$ must be small. In other words, when the variance is small, large deviations of $X$ from the mean $E(X)$ are not likely. On the other hand, when the variance is large, it is likely that not all values of $X$ lie near the mean.

**A CLOSER LOOK**    The formula

$$\sigma^2 = [x_1 - E(X)]^2 f(x_1) + [x_2 - E(X)]^2 f(x_2) + \cdots + [x_n - E(X)]^2 f(x_n)$$

which defines the variance, can be simplified to a form that is often easier to use. This simplified formula is given in the following theorem.

**THEOREM 6.2:**    Suppose $X$ is a random variable whose range is $\{x_1, x_2, x_3, \ldots, x_n\}$, $f$ is the probability distribution of $X$, and $E(X)$ is the mathematical expected value of $X$ (the mean). Then the variance of $X$ is given by

$$\text{VAR}(X) = \sigma^2 = \left[\sum_{i=1}^{n} x_i^2 f(x_i)\right] - [E(X)]^2 = E(X^2) - [E(X)]^2$$

**Proof:** (Optional) If you are reading this proof, you should be familiar with the sigma notation introduced in Section 1.7. If you are not familiar with this notation but you want to know how the formula is obtained, you should work Exercises *24, *25, and *26 at the end of this section. In the proof, we shall make use of the following facts.

$$E(X) = \sum_{i=1}^{n} x_i f(x_i) \quad \text{(by definition)}$$

and

$$\sum_{i=1}^{n} f(x_i) = 1 \quad \text{(since } f \text{ is a probability density function)}$$

We proceed as follows.

$$VAR(X) = \sum_{i=1}^{n} [x_i - E(X)]^2 f(x_i)$$

$$= \sum_{i=1}^{n} \{x_i^2 - 2x_i E(X) + [E(X)]^2\} f(x_i)$$

$$= \sum_{i=1}^{n} \{x_i^2 f(x_i) - 2x_i E(X) f(x_i) + [E(X)]^2 f(x_i)\}$$

$$= \sum_{i=1}^{n} x_i^2 f(x_i) - \sum_{i=1}^{n} 2x_i E(X) f(x_i) + \sum_{i=1}^{n} [E(X)]^2 f(x_i)$$

$$= \sum_{i=1}^{n} x_i^2 f(x_i) - 2E(X) \sum_{i=1}^{n} x_i f(x_i) + [E(X)]^2 \sum_{i=1}^{n} f(x_i)$$

$$= \left[ \sum_{i=1}^{n} x_i^2 f(x_i) \right] - 2[E(X)]^2 + [E(X)]^2$$

$$= \left[ \sum_{i=1}^{n} x_i^2 f(x_i) \right] - [E(X)]^2 = E(X^2) - [E(X)]^2$$

**EXAMPLE 7** Find the variance of the random variable $X$ of Example 6 using the preceding formula.

**Solution** It was found in Example 6 that $E(X) = 2.1$. Using this formula and the values of $x_i$ and $f(x_i)$ given in the probability table of Example 6, we obtain

$$VAR(X) = [1^2(.3) + 2^2(.4) + 3^2(.2) + 4^2(.1)] - (2.1)^2$$
$$= 5.3 - 4.41 = 0.89$$

As expected, this is the result obtained in Example 6. ∎

---

# Exercise Set 6.5

1. In drawing a single card from a standard deck, you win $10 if you draw an ace, $5 if you draw a face card, and 50 cents if any other card is drawn. It costs $2 to be allowed to draw a card. Would you expect to win money in the long run by playing this game? Approximately how much would you expect to win or lose if you played 200 times?

2. In a throw of two dice, you win $3 if the sum of the numbers that come up is 7 or 11, and lose $1 if the sum is anything else. Would you expect to win money in the long run by playing this game? Approximately how much would you expect to win or lose if you played 100 times?

3. You are given a bag containing six black coins and three white coins and you are offered the following choice.

   a. Draw one coin. If it is white, you win $5; if it is black, you lose $1.
   b. Draw two coins. If both are white, you win $10; if they are of different color, you lose $2.

   Assuming that you want to play one of the two games, which game should you play?

4. A fair coin was tossed four times and $X$ is the number of heads that came up. Determine the probability distribution of $X$.

5. Repeat Exercise 4 if the coin is tossed five times.

6. Two fair dice are rolled. Let $X$ be the sum of the two numbers that come up. Determine the probability distribution of $X$.

7. A fair die is rolled until a six comes up. Let $X$ represent the number of rolls it takes for a six to come up. Determine the probability distribution of $X$.

8. A bag contains five black marbles and eight white marbles. You draw a marble randomly. If it is black, you stop; otherwise, you replace the marble in the bag, shake the bag, and draw a marble again. If it is black, you stop; otherwise, you repeat the process until you draw a black marble. Let $X$ be the number of times you must draw a marble to get a black one. Determine the probability distribution of $X$.

9. Calculate the mathematical expectation of the random variable of Exercise 4.

10. Calculate the mathematical expectation of the random variable of Exercise 5.

11. Calculate the mathematical expectation of the random variable of Exercise 6.

12. The annual premium that a San Francisco homeowner must pay an insurance company to protect his house, worth $275,000, against total destruction by fire is $325. From experience, the company has determined that the probability of a house being totally destroyed by fire in that neighborhood is 0.00018. Calculate the expected gain per policy to the company.

13. Repeat Exercise 12 if the annual premium is $360, the house is worth $325,000, and the probability of the house being totally destroyed by fire is 0.00025.

14. Repeat Exercise 12 if the annual premium is $280, the house is worth $275,000, and the probability of the house being totally destroyed by fire is 0.0006.

15. An insurance company offers the following medical plan to university employees. It will pay $150 a day for hospitalization up to a maximum of 7 days per year. The annual premium for this policy is $50. From experience with similar groups, the company has estimated that the probabilities that an employee is hospitalized for exactly 1 day, exactly 2 days, exactly 3 days, exactly 4 days, exactly 5 days, exactly 6 days, and more than 6 days are 0.0015, 0.0025, 0.0035, 0.005, 0.0065, 0.008, and 0.02, respectively. Calculate the expected gain per policy to the company.

16. The insurance company of Exercise 15 wishes to have an expected gain of $25 per policy and has hired an actuary to calculate the necessary annual premium to reach that goal. What should the annual premium be?

17. The Accounting Club at a small university is holding a raffle to raise money for its activities. Five hundred tickets at $1.50 each have been sold. The first prize is $200, the second prize is $100, and the third prize is $50. Let $X$ be the random variable that represents the net gain on one ticket. Determine the expected value of $X$.

18. Suppose a game consists of tossing a fair coin three times. The number of dollars you receive is equal to the square of the number of tails that came up. If you must pay $3 to play the game, is the game a fair game? (A fair game is defined to be a game for which the expected value per play is $0.)

19. Calculate the variance of the random variable $Y$ of Example 6 of this section using the formula

$$\text{VAR}(X) = \sigma^2 = \left[ \sum_{i=1}^{n} x_i^2 f(x_i) \right] - [E(X)]^2$$

20. Suppose both random variables $X$ and $Y$ have range $\{2, 4, 6, 8\}$. Suppose $f$ and $g$ are probability distributions of $X$ and $Y$, respectively, defined by the following probability tables.

| $x$ | $f(x) = P_1(X = x)$ | | $y$ | $g(y) = P_2(Y = y)$ |
|---|---|---|---|---|
| 2 | .4 | | 2 | .5 |
| 4 | .3 | | 4 | .2 |
| 6 | .2 | | 6 | .1 |
| 8 | .1 | | 8 | .2 |

a. Show that both $X$ and $Y$ have the same mathematical expectation.

b. Find the variance and standard deviation of $X$.

c. Find the variance and standard deviation of $Y$.

d. Draw a probability histogram for $X$ and $Y$.

21. Suppose both random variables $X$ and $Y$ have range $\{1, 3, 5, 7, 9\}$. Suppose $f$ and $g$ are probability distributions of $X$ and $Y$, respectively, defined by the following probability tables.

| $x$ | $f(x) = P_1(X = x)$ | | $y$ | $g(y) = P_2(Y = y)$ |
|---|---|---|---|---|
| 1 | .1 | | 1 | .15 |
| 3 | .1 | | 3 | .25 |
| 5 | .3 | | 5 | .05 |
| 7 | .2 | | 7 | .05 |
| 9 | .3 | | 9 | .5 |

a. Show that both $X$ and $Y$ have the same mathematical expectation.

b. Find the variance and standard deviation of $X$.

c. Find the variance and standard deviation of $Y$.

d. Draw a probability histogram for $X$ and for $Y$.

22. Calculate the variance and standard deviation of the random variable $X$ of Exercise 4.

23. Repeat Exercise 22 for the random variable of Exercise 6.

*24. The range of a random variable $X$ is $\{x_1, x_2\}$ and $f$ is the probability distribution of $X$. Show that $\text{VAR}(X) = x_1^2 f(x_1) + x_2^2 f(x_2) - [E(X)]^2$. (*Hint:* By definition $\text{VAR}(X) = [x_1 - E(X)]^2 f(x_1) + [x_2 - E(X)]^2 f(x_2)$. Expand $[x_1 - E(X)]^2$ and $[x_2 - E(X)]^2$, rearrange the terms and use the fact that $E(X) = x_1 f(x_1) + x_2 f(x_2)$ and that $f(x_1) + f(x_2) = 1$.)

**\*25.** The range of a random variable $X$ is $\{x_1, x_2, x_3\}$ and $f$ is the probability distribution of $X$. Show that $\mathrm{VAR}(X) = x_1^2 f(x_1) + x_2^2 f(x_2) + x_3^2 f(x_3) - [E(X)]^2$. (*Hint:* See Exercise \*24.)

**\*26.** The range of a random variable $X$ is $\{x_1, x_2, x_3, x_4\}$ and $f$ is the probability distribution of $X$. Show that $\mathrm{VAR}(X) = x_1^2 f(x_1) + x_2^2 f(x_2) + x_3^2 f(x_3) + x_4^2 f(x_4) - [E(X)]^2$. (*Hint:* See Exercise \*24.)

# 6.6　The Binomial Distribution

In this section, we shall discuss a discrete probability distribution that is connected with repeated, but independent, trials of an experiment where only two outcomes are possible. There are many applications in which such experiments take place. When a television network conducts a survey, either a person is or is not watching a certain program; when the quality control of a manufacturing company tests an item randomly selected, it either is or is not defective, and so on.

## The Binomial Theorem

The terms in the expansion of a power of a binomial are useful in describing the probability distributions of random variables. We begin with a brief discussion of the binomial theorem. You undoubtedly recall, from elementary algebra, at least the first few of the following formulas. (See also the "Getting Started" section.)

$$(a + b)^1 = a + b$$
$$(a + b)^2 = a^2 + 2ab + b^2$$
$$(a + b)^3 = a^3 + 3a^2b + 3ab^2 + b^3$$
$$(a + b)^4 = a^4 + 4a^3b + 6a^2b^2 + 4ab^3 + b^4$$
$$(a + b)^5 = a^5 + 5a^4b + 10a^3b^2 + 10a^2b^3 + 5ab^4 + b^5$$

and so on.

We observe the following for $n = 1, 2, 3, 4,$ and $5$:

1. The expansion of $(a + b)^n$ has $n + 1$ terms.
2. The first term is $a^n$ and the last is $b^n$.
3. After the first term of the expansion, in each successive term, the exponent assigned to $b$ increases by one from 0 to $n$, while that assigned to $a$ decreases by one, from $n$ to 0. Thus the sum of the exponents for each term is $n$.
4. If we multiply the coefficient of any term, except the last, by the exponent assigned to $a$ in that term and then divide by one more than the exponent assigned to $b$, we get the coefficient of the next term.
5. The coefficients of terms equidistant from the ends of the expansion are equal.
6. Recalling that $C(n, r) = \frac{n!}{(n - r)!r!}$, we can verify that the coefficient of each of the terms involving $a^{n-r}b^r$ is $C(n, r)$.

The observations we just made for $n = 1, 2, 3, 4,$ and $5$ are special cases of the following general theorem. (See Exercise \*41.)

> **THEOREM 6.3 (Binomial Theorem):** If $n$ is a positive integer and $a$, $b$ are real (or complex) numbers, then $(a + b)^n = C(n, 0)a^n + C(n, 1)a^{n-1}b^1 + C(n, 2)a^{n-2}b^2 + \cdots + C(n, n - 1)a^1b^{n-1} + C(n, n)b^n$. Using the sigma notation,
>
> $$(a + b)^n = \sum_{r=0}^{n} C(n, r)a^{n-r}b^r$$

**EXAMPLE 1**    Write the expansion of $(a + b)^7$.

**Solution**    The expansion will have eight terms. The first and last terms are $a^7$ and $b^7$, respectively. Without coefficients, the terms are $a^7$, $a^6b^1$, $a^5b^2$, $a^4b^3$, $a^3b^4$, $a^2b^5$, $a^1b^6$, and $b^7$. The coefficients are $C(7, 0)$, $C(7, 1)$, $C(7, 2)$, $C(7, 3)$, $C(7, 4)$, $C(7, 5)$, $C(7, 6)$, and $C(7, 7)$. These coefficients are easily calculated using the formula

$$C(n, r) = \frac{n!}{(n - r)!r!}$$

They are 1, 7, 21, 35, 35, 21, 7, 1. As an example, we show one of the calculations.

$$C(7, 3) = \frac{7!}{(7 - 3)!3!} = \frac{7(6)(5)(4!)}{4!3!} = \frac{7(6)(5)}{3(2)(1)} = 35$$

Thus,

$$(a + b)^7 = a^7 + 7a^6b + 21a^5b^2 + 35a^4b^3 + 35a^3b^4 + 21a^2b^5 + 7ab^6 + b^7$$

**Do Exercise 1.**    ■

## Bernoulli Trials

We are now ready to begin the discussion of the main topic of this section. Repeated independent trials of an experiment are said to be *Bernoulli trials* if there are only two possible outcomes for each trial and if their probabilities remain the same throughout the $n$ trials of the experiment. Jakob (Jacques) Bernoulli (1654–1705) was a member of a Swiss family of mathematicians and scientists. He was one of the first people who undertook the systematic study of probability problems related to experiments with only two possible outcomes.

In working with Bernoulli experiments, it is customary to refer to one of the two outcomes as a *success*, denoted $S$, and to the other as a *failure*, denoted $F$. If we let the probability of success be $p$ and the probability of failure be $q$, we have

$$P(S) + P(F) = p + q = 1$$

For example, if we roll a fair die and ask for the probability of a 5 showing, we may consider the outcome "a 5 showing" a success and the outcome "any of the numbers 1, 2, 3, 4, or 6 showing" a failure, then $P(S) = \frac{1}{6}$ and $P(F) = \frac{5}{6}$.

Suppose a Bernoulli trial is repeated $n$ times. We may want to calculate the probability that exactly $r$ of these trials were successes, where $0 \leq r \leq n$.

In applying our results to real-life situations, there will be very few cases where the conditions described above are perfectly satisfied. However, in most cases, the assumption that the conditions are satisfied will not affect the results. For example, suppose 30% of a population of 15,000 is watching a football game on a Sunday afternoon. If the television station calls one of these people, the probability that this person is watching the game will be $\frac{3}{10}$, but the probability that the second person called is watching will be $\frac{4500}{14,999}$, or $\frac{4499}{14,999}$, depending on whether the first person called was, or was not, watching the game. Clearly, these numbers are near, but not equal, to .3. For all practical purposes, we may assume that the probability remains .3 as long as the number of people called remains small. This again illustrates that when mathematics is used in real-life applications, the mathematical model only approximates the situation. This creates no difficulty as long as we are careful in choosing a model that best fits the problem at hand.

## Binomial Distribution

In working with Bernoulli trials, we shall use the following rule.

Suppose that there are $n$ independent trials of an experiment and that the probability of outcome $o_i$ in the $i$th trial is $p(o_i)$. Then, the probability of the event $(o_1, o_2, \ldots, o_n)$ occurring is given by

$$p(o_1, o_2, \ldots, o_n) = p(o_1)p(o_2) \cdots p(o_n)$$

The next two examples illustrate the basic idea behind a binomial distribution.

**EXAMPLE 2**    Suppose a coin is weighted so that the probability that tails comes up on any toss is $\frac{3}{4}$. The coin is tossed five times. What is the probability that heads will come up on the second and fourth tosses and tails will come up on the first, third, and fifth tosses?

**Solution**    For convenience, success means the result of the toss is tails and failure means the coin came up heads. Then $p(S) = \frac{3}{4}$ and $p(F) = \frac{1}{4}$ since $1 - \frac{3}{4} = \frac{1}{4}$. Furthermore, let $S_i$ and $F_j$ denote success and failure in the $i$th and $j$th toss, respectively. We want to find the probability of $(S_1, F_2, S_3, F_4, S_5)$. This probability is found as follows:

$$p(S_1, F_2, S_3, F_4, S_5) = p(S_1)p(F_2)p(S_3)p(F_4)p(S_5)$$
$$= \left(\frac{3}{4}\right)\left(\frac{1}{4}\right)\left(\frac{3}{4}\right)\left(\frac{1}{4}\right)\left(\frac{3}{4}\right) = \left(\frac{3}{4}\right)^3\left(\frac{1}{4}\right)^2 = \frac{27}{1024}$$

**Do Exercise 3.**    ∎

Before considering the next example, we should note that since multiplication is commutative, we would have obtained exactly the same result had we been asked to calculate the probability that tails will come up on the second, fourth, and fifth tosses and heads will come up on the first and third tosses.

**EXAMPLE 3**    If we use the coin of Example 2 and toss it five times, what is the probability that tails will come up exactly three times?

**Solution**    For convenience, success means that tails came up exactly three times. Then, the outcome $(S_1, F_2, S_3, F_4, S_5)$ is a success, but so are the outcomes $(F_1, S_2, S_3, F_4, S_5)$ and $(S_1, F_2, F_3, S_4, S_5)$. Each of these outcomes has probability $\left(\frac{3}{4}\right)^3\left(\frac{1}{4}\right)^2$, as was found in Example 2. We must consider all possible outcomes for which we have exactly three $S$'s and two $F$'s. Note that in $(S_1, F_2, S_3, F_4, S_5)$ the $S$'s have subscripts 1, 3, and 5, while in $(F_1, S_2, S_3, F_4, S_5)$ they have subscripts 2, 3, and 5, and in $(S_1, F_2, F_3, S_4, S_5)$, the subscripts for $S$ are 1, 4, and 5. Therefore, with each successful outcome, we can associate a subset of three elements chosen from the set $\{1, 2, 3, 4, 5\}$. We know, however, that the number of such subsets is $C(5, 3)$, which is easily found to be 10. Thus, we have exactly ten outcomes in a successful event and each has probability $\left(\frac{3}{4}\right)^3\left(\frac{1}{4}\right)^2$. We could add the probabilities, but the ten terms of that sum are equal, so we obtain the same result by multiplying the common term by 10. Therefore, the probability that exactly three tails come up when the weighted coin is tossed five times is $\frac{135}{512}$, since $10\left(\frac{3}{4}\right)^3\left(\frac{1}{4}\right)^2 = C(5, 3)\left(\frac{3}{4}\right)^3\left(\frac{1}{4}\right)^2 = \frac{135}{512}$. **Do Exercise 9.**

Because the number of outcomes in a successful event corresponds to a coefficient of the expansion of a power of a binomial, an experiment such as the one in the preceding example is called a *binomial experiment* and the probability distribution of the number of successes is called a *binomial distribution*.

The result of the preceding example can be generalized to any number $n$ of Bernoulli trials. This generalization is stated in the following theorem.

> **THEOREM 6.4:**    Suppose that in a binomial experiment the probabilities of success and failure in any trial are $p$ and $q$, respectively. Let $X$ be the number of successes in $n$ independent trials. Then, the probability distribution $f$ of $X$ is given by
>
> $$f(x) = P(X = x) = C(n, x)p^x q^{n-x}$$
>
> where $p + q = 1$ and $x$ is an integer such that $0 \leq x \leq n$.

Any random variable with this probability distribution is called a *binomial random variable* and is said to have a *binomial distribution.**

**EXAMPLE 4**    Evaluate $b\left(3; 7, \frac{1}{3}\right)$.

---

*In many textbooks on probability, the symbol $b(x; n, p)$ is used to denote $C(n, x)p^x q^{n-x}$.

**Solution**   We have

$$b\left(3; 7, \frac{1}{3}\right) = C(7, 3)\left(\frac{1}{3}\right)^3\left(\frac{2}{3}\right)^{7-3}$$

$$= \frac{7!}{4!3!}\left(\frac{1}{3}\right)^3\left(\frac{2}{3}\right)^4 = 35 \cdot \frac{1}{27} \cdot \frac{16}{81} = \frac{560}{2187}$$

**Do Exercise 15.**

**EXAMPLE 5**   Suppose $X$ is a binomial random variable with $p = \frac{1}{4}$ and $n = 4$. Find the probability distribution of $X$.

**Solution**   We have $q = 1 - \frac{1}{4} = \frac{3}{4}$. We shall use the formula

$$f(x) = P(X = x) = C(n, x)p^x q^{n-x}$$

with $n = 4$ and $x = 0, 1, 2, 3, 4$.
    We get

$$f(0) = P(X = 0) = C(4, 0)\left(\frac{1}{4}\right)^0\left(\frac{3}{4}\right)^4 = \frac{4!}{4!0!} \cdot 1 \cdot \frac{81}{256} = \frac{81}{256}$$

$$f(1) = P(X = 1) = C(4, 1)\left(\frac{1}{4}\right)^1\left(\frac{3}{4}\right)^3 = \frac{4!}{3!1!} \cdot \frac{1}{4} \cdot \frac{27}{64} = \frac{27}{64}$$

$$f(2) = P(X = 2) = C(4, 2)\left(\frac{1}{4}\right)^2\left(\frac{3}{4}\right)^2 = \frac{4!}{2!2!} \cdot \frac{1}{16} \cdot \frac{9}{16} = \frac{27}{128}$$

$$f(3) = P(X = 3) = C(4, 3)\left(\frac{1}{4}\right)^3\left(\frac{3}{4}\right)^1 = \frac{4!}{1!3!} \cdot \frac{1}{64} \cdot \frac{3}{4} = \frac{3}{64}$$

$$f(4) = P(X = 4) = C(4, 4)\left(\frac{1}{4}\right)^4\left(\frac{3}{4}\right)^0 = \frac{4!}{0!4!} \cdot \frac{1}{256} \cdot 1 = \frac{1}{256}$$

Note that

$$f(0) + f(1) + f(2) + f(3) + f(4) = \frac{81}{256} + \frac{27}{64} + \frac{27}{128} + \frac{3}{64} + \frac{1}{256}$$

$$= \frac{81 + 108 + 54 + 12 + 1}{256} = \frac{256}{256} = 1$$

This result is expected since $f$ is a probability distribution.
**Do Exercise 17.**

    As we have seen earlier, it is sometimes advantageous when we wish to find the probability of an event $E$, to compute the probability of the complement $E'$ and then use the fact that $P(E) + P(E') = 1$, as in the following example.

**EXAMPLE 6**   A fair die is rolled ten times. Find the probability of a five coming up at least twice.

**Solution**   Let $X$ be the number of times that a five comes up. Then $X$ is a discrete random variable with a binomial distribution. Here $n = 10, p = \frac{1}{6}$, and $q = \frac{5}{6}$. The computations will be simplified if we first find the probability that the number of times a five came up is less than two.

$$P(X < 2) = P(X = 0) + P(X = 1)$$

$$= C(10, 0)\left(\frac{1}{6}\right)^0\left(\frac{5}{6}\right)^{10} + C(10, 1)\left(\frac{1}{6}\right)^1\left(\frac{5}{6}\right)^9$$

$$= \frac{10!}{10!0!}\left(\frac{1}{6}\right)^0\left(\frac{5}{6}\right)^{10} + \frac{10!}{9!1!}\left(\frac{1}{6}\right)^1\left(\frac{5}{6}\right)^9$$

$$= 1 \cdot 1 \cdot \frac{5^{10}}{6^{10}} + 10 \cdot \frac{1}{6} \cdot \frac{5^9}{6^9} = \left(\frac{5}{6}\right)^{10}(1 + 2) \doteq .48452$$

Therefore, the probability that a five comes up at least twice in ten rolls of a fair die is .51548 since $1 - .48452 = .51548$. The answer has been rounded off to five decimal places.

**Do Exercise 21.**    ■

The formulas used to calculate the mathematical expectation and variance of a binomial random variable are simple.

> **THEOREM 6.5:**  Suppose that in a binomial experiment the probabilities of success and failure in any trial are $p$ and $q$, respectively. Let $X$ be the number of successes in $n$ independent trials. Then, the mathematical expectation (mean), variance, and standard deviation of $X$ are given by $E(X) = np$, $\text{VAR}(X) = npq$, $\sigma = \sqrt{npq}$, respectively.

**A CLOSER LOOK**  The proof of the first of these formulas is given for those of you who feel at ease with manipulations involving the sigma notation and factorials. It may be omitted without any loss of understanding of the concepts. The proof of the formula for the variance is more involved and is left as an exercise. (See Exercise **40.)

*Proof:* (Optional) The possible values of $X$ are $0, 1, 2, \cdots, n$. Thus, if $f$ is the probability distribution of $X$, we have

$$E(X) = 0 \cdot f(0) + 1 \cdot f(1) + 2 \cdot f(2) + \cdots + n \cdot f(n)$$

But in this case, $f(k) = b(k; n, p) = C(n, k)p^k q^{n-k}$. Thus,

$$E(X) = 0 + \sum_{k=1}^{n} kb(k; n, p) = \sum_{k=1}^{n} kC(n, k)p^k q^{n-k}$$

$$= \sum_{k=1}^{n} k\frac{n!}{(n-k)!k!}p^k q^{n-k}$$

$$= np \sum_{k=1}^{n} \frac{(n-1)!}{(n-k)!(k-1)!}p^{k-1}q^{n-k}$$

$$= np(C(n-1, 0)q^{n-1} + C(n-1, 1)q^{n-2}p^1 + C(n-1, 2)q^{n-3}p^2$$
$$+ \cdots + C(n-1, n-2)q^1 p^{n-2} + C(n-1, n-1)p^{n-1})$$
$$= np(q + p)^{n-1} = np(1)^{n-1} = np$$

**EXAMPLE 7**    Find the mean, variance, and standard deviation for the binomial random variable of Example 5.

**Solution**    Using the formula given in Theorem 6.5 with $n = 4$, $p = \frac{1}{4}$, and $q = \frac{3}{4}$, we get

$$E(X) = 4\left(\frac{1}{4}\right) = 1$$

$$\text{VAR}(X) = 4\left(\frac{1}{4}\right)\left(\frac{3}{4}\right) = \frac{3}{4}$$

and

$$\sigma = \sqrt{\text{VAR}(X)} = \sqrt{\frac{3}{4}} = \frac{\sqrt{3}}{2}$$

**Do Exercise 33.**    ∎

Although it is not necessary to do so, since we already have obtained the desired results, we shall calculate these quantities using the results of Example 5 to compare the efficiency of the two methods.

In Example 5, the probability distribution was $f(0) = \frac{81}{256}$, $f(1) = \frac{27}{64}$, $f(2) = \frac{27}{128}$, $f(3) = \frac{3}{64}$, and $f(4) = \frac{1}{256}$. Thus,

$$E(X) = 0 \cdot \frac{81}{256} + 1 \cdot \frac{27}{64} + 2 \cdot \frac{27}{128} + 3 \cdot \frac{3}{64} + 4 \cdot \frac{1}{256}$$

$$= \frac{108 + 108 + 36 + 4}{256} = 1$$

$$\text{VAR}(X) = 0^2 f(0) + 1^2 f(1) + 2^2 f(2) + 3^2 f(3) + 4^2 f(4) - E(X)^2$$

$$= 0^2 \cdot \frac{81}{256} + 1^2 \cdot \frac{27}{64} + 2^2 \cdot \frac{27}{128} + 3^2 \cdot \frac{3}{64} + 4^2 \cdot \frac{1}{256} - 1^2$$

$$= \frac{108 + 216 + 108 + 16 - 256}{256} = \frac{3}{4}$$

As we expect, the results are the same.

## Testing Small Samples

One of the important applications of binomial experiments is in making decisions about whether to accept or reject a large number of items on the basis of evidence on a small sample. We conclude this section with an example to illustrate this type of problem.

**EXAMPLE 8** An auto supply company has a sampling scheme to inspect large lots of incoming car batteries. Twenty batteries are to be tested and the lot rejected if three or more are found to be defective. Suppose that a lot contains exactly 4% defective batteries. What is the probability that the lot will be accepted?

**Solution** Let $X$ be the number of defective batteries found. Then the lot will be accepted if $X < 3$. Thus, we must find $P(X < 3)$. But, $P(X < 3) = P(X = 0) + P(X = 1) + P(X = 2)$. Note that the probability of the first battery tested being defective is .04. Strictly speaking, the probability that the second battery tested is defective depends on whether or not the first battery tested was defective. However, since we were told that the lot of batteries was large, we may assume that for each of the 20 batteries tested, the probability that it is defective remains .04. Hence, $p = .04$, $q = .96$ (since $1 - .04 = .96$), and $n = 20$. It follows that

$$
\begin{aligned}
P(X < 3) &= P(X = 0) + P(X = 1) + P(X = 2) \\
&= C(20, 0)(.04)^0(.96)^{20} + C(20, 1)(.04)^1(.96)^{19} \\
&\qquad + C(20, 2)(.04)^2(.96)^{18} \\
&= 1 \cdot 1 \cdot (.96)^{20} + 20(.04)(.96)^{19} + 190(.04)^2(.96)^{18} \\
&= (.96)^{18}[(.96)^2 + 20(.04)(.96) + 190(.04)^2] \\
&\doteq .4796(1.9936) \doteq .95614
\end{aligned}
$$

Therefore, the probability that a lot is accepted is .95614. Consequently, the probability that a lot is rejected is .04386 since $1 - .95614 = .04386$.
**Do Exercise 37.**

# Exercise Set 6.6

**1.** Write the expansion of $(a + b)^6$.

**2.** Write the expansion of $(2x + 3y)^8$.

*In Exercises 3–14, assume that the trials are independent, although in some cases this is not actually true. However, in each case, the values of p and q remain nearly constant so that this assumption is justified.*

**3.** Suppose a coin is weighted so that the probability that heads comes up on any toss is $\frac{5}{7}$. The coin is tossed five times. What is the probability that heads will come up on the second and fifth tosses, and tails will come up on the first, third, and fourth tosses?

**4.** Suppose a coin is weighted so that the probability that tails comes up on any toss is $\frac{8}{13}$. The coin is tossed six times. What is the probability that heads will come up on the second and fourth tosses, and tails will come up on the first, third, fifth, and sixth tosses?

**5.** Suppose a die is weighted so that the probability that a six comes up on any roll is $\frac{1}{4}$. The die is rolled five times. What is the probability that a six comes up only on the second and fourth rolls?

**6.** Suppose a die is weighted so that the probability that a two comes up on any roll is $\frac{4}{13}$. The die is rolled seven times. What is the probability that a two comes up only on the first, second, and fourth rolls?

**7.** Suppose the coin of Exercise 3 is tossed five times. What is the probability that exactly three tails will come up?

**8.** Repeat Exercise 7 if the coin is tossed eight times and we want exactly five tails to come up.

**9.** Suppose the coin of Exercise 4 is tossed eight times. What is the probability that exactly three heads will come up?

**10.** Repeat Exercise 9 if the coin is tossed ten times and we want exactly six tails to come up.

11. The die of Exercise 5 is rolled seven times. What is the probability that a six comes up exactly four times?

12. Repeat Exercise 11 if the die is rolled nine times and we want the probability of a six coming up exactly five times.

13. The die of Exercise 6 is rolled eight times. What is the probability that a two comes up exactly four times?

14. Repeat Exercise 13 if the die is rolled 11 times and we want the probability of a two coming up exactly four times.

15. Evaluate the following:

    **a.** $b(2; 8, .2)$     **b.** $b(3; 10, .3)$     **c.** $b(9; 13, .4)$

16. Evaluate the following:

    **a.** $b(3; 9, .6)$     **b.** $b(6; 11, .2)$     **c.** $b(7; 14, .1)$

17. Suppose $X$ is a random variable with $p = \frac{1}{5}$ and $n = 4$.

    **a.** Find the probability distribution of $X$.
    **b.** Draw a probability histogram for $X$.

18. Repeat Exercise 17 with $p = \frac{2}{3}$ and $n = 5$.

19. Repeat Exercise 17 with $p = \frac{3}{5}$ and $n = 6$.

20. A fair die is rolled 12 times. Find the probability that a six comes up at least twice.

21. A fair die is rolled 20 times. Find the probability that a one comes up at least three times.

22. Suppose a basketball player hits 76% of the free throws he attempts. If he attempts eight free throws in a game, what is the probability that **a.** he will hit exactly six free throws? **b.** he will hit at least six free throws?

23. Suppose a basketball player hits 80% of the free throws she attempts. If she attempts eight free throws in a game, what is the probability that **a.** she will hit exactly seven free throws? **b.** she will hit at least seven free throws?

24. In a field goal kicking contest, John is successful on 55% of his attempts and he needs at least six successful kicks in order to win. He has ten attempts left in the contest. What is the probability that he will win?

25. If there are six children in a family, what is the probability that at least two are boys? (Assume that the probability that a child is a boy is $\frac{1}{2}$.)

26. If there are seven children in a family, what is the probability that at least four are girls? (Assume that the probability that a child is a girl is $\frac{1}{2}$.)

27. A company wins a contract on 65% of the bids it submits. If the company plans to bid on ten projects next year and if it must win at least four of the contracts to have a successful year, what is the probability that next year will be successful?

28. Mr. Nonbright, who is a student in a mathematics course, is taking a true-false examination which has 20 questions.

If he guesses the answer for each question, what is the probability that he gets **a.** all twenty correct? **b.** at least eighteen correct? **c.** at least fifteen correct?

29. Repeat Exercise 28 if the test is multiple choice and five answers are given for each question, only one of which is correct.

30. What is the expected score of the student in Exercise 28?

31. What is the expected score of the student in Exercise 29?

32. Compute the expected value (mean), variance, and standard deviation of the binomial random variable with $n = 8$ and $p = .3$.

33. Repeat Exercise 32 with $n = 12$ and $p = .4$.

34. Repeat Exercise 32 with $n = 15$ and $p = .2$.

35. Repeat Exercise 32 with $n = 18$ and $p = .3$.

36. Repeat Exercise 32 with $n = 20$ and $p = .35$.

37. A hardware store has a sampling scheme to inspect large lots of incoming light bulbs. Fifteen light bulbs are to be tested and the lot rejected if two or more are found to be defective. Suppose that a lot contains exactly 3% defective light bulbs. What is the probability that the lot will be accepted?

38. The quality control department of a large chain of computer stores established the following sampling scheme to inspect large lots of incoming computers. Fifteen computers would be tested and the lot rejected if two or more were found to be defective. Suppose that a lot contains exactly 3% defective computers. What is the probability that the lot will be accepted?

*39. The lines in the proof of Theorem 6.4 have been numbered. Give a justification for claiming equality between

    **a.** lines (1) and (2)     **b.** lines (2) and (3)
    **c.** lines (3) and (4)     **d.** lines (4) and (5)

$$E(X) = 0 + \sum_{k=1}^{n} kb(k; n, p) = \sum_{k=1}^{n} kC(n, k)p^k q^{n-k} \qquad (1)$$

$$= \sum_{k=1}^{n} k \frac{n!}{(n-k)!k!} p^k q^{n-k} \qquad (2)$$

$$= np \sum_{k=1}^{n} \frac{(n-1)!}{(n-k)!(k-1)!} p^{k-1} q^{n-k} \qquad (3)$$

$$= np(C(n-1, 0)q^{n-1} +$$
$$C(n-1, 1)q^{n-2}p^1$$
$$+ C(n-1, 2)q^{n-3}p^2 +$$
$$\cdots + C(n-1, n-2)q^1 p^{n-2}$$
$$+ C(n-1, n-1)p^{n-1}) \qquad (4)$$

$$= np(q + p)^{n-1} = np(1)^{n-1} = np \qquad (5)$$

**\*\*40.** Prove the formula $\text{VAR}(X) = npq$ which appeared in Theorem 6.5. (*Hint:* Start with $\text{VAR}(X) = [0^2 f(0) + 1^2 f(1) + 2^2 f(2) + \cdots + n^2 f(n)] - E(X)^2 = [0^2 f(0) + 1^2 f(1) + 2^2 f(2) + \cdots + n^2 f(n)] - (np)^2$. Then replace each $f(k)$ by $C(n, k)p^k q^{n-k}$. Factor $np$ out of the sum in brackets and simplify each term. The sum in brackets should reduce to $[1 \cdot b(0; n - 1, p) + 2 \cdot b(1; n - 1, p) + 3 \cdot b(2; n - 1, p) + \cdots + n \cdot b(n - 1; n - 1, p)]$. Each term can be written as the sum of two terms. For example, $3 \cdot b(2; n - 1, p)$ can be written $2 \cdot b(2; n - 1, p) +$ $b(2; n - 1, p)$. After doing this, rearrange the terms and group them into two sums so that one sum is the mean of a binomial random variable for an experiment with $n - 1$ trials each with probability of success $p$ (thus equal to $(n - 1)p$), and the other sum is the sum of all values of a probability distribution (thus equal to 1). Then simplify and use the fact that $1 - p = q$.)

**\*41.** Using mathematical induction, prove the binomial theorem.

# 6.7 More on Markov Chains

Markov chains were introduced in Section 3.8 because they provided an interesting application of matrix multiplication. However, this introduction was limited by the fact that we had not discussed probability prior to that introduction. In this section, we give a few more examples involving Markov chains and we shall use some of the results we obtained elsewhere in this chapter. You are urged to read Section 3.8 before proceeding. However, we shall state again the main concepts.

## Review of Definitions

We have seen that if we have a finite sequence of independent trials with possible outcomes $O_1, O_2, \ldots, O_n$, where for each $i$ the probability of $O_i$ is $p_i$, then the probability of $(O_1, O_2, \ldots, O_n)$ is the product $p_1 p_2 \cdots p_n$. In the theory of Markov chains, we generalize this result slightly by making the outcome of any trial, except the first, dependent on the outcome of the immediately preceding trial, and only on that outcome. Thus, to every ordered pair of outcomes $(O_i, O_j)$, there corresponds a conditional probability $p_{ij}$. That is, given that $O_i$ has occurred at some trial, the probability that $O_j$ will occur at the next trial is $p_{ij}$. In addition, we must be given the probability $p_k$ of the outcome $O_k$ occurring at the initial trial. The vector $[\, p_1 \quad p_2 \quad \cdots \quad p_n \,]$ is called the *initial-state vector*. We must have $p_k \geq 0$ for each $k$, and $p_1 + p_2 + \cdots + p_n = 1$, since the $p_k$'s are probabilities and one of the outcomes $O_1, O_2, \ldots, O_n$ must occur. The matrix

$$\begin{bmatrix} p_{11} & p_{12} & p_{13} & \cdots & p_{1n} \\ p_{21} & p_{22} & p_{23} & \cdots & p_{2n} \\ p_{31} & p_{32} & p_{33} & \cdots & p_{3n} \\ \cdot & \cdot & \cdot & & \cdot \\ \cdot & \cdot & \cdot & & \cdot \\ \cdot & \cdot & \cdot & & \cdot \\ p_{n1} & p_{n2} & p_{n3} & \cdots & p_{nn} \end{bmatrix}$$

is called the *transition matrix*. If the possible outcomes of any trial are $O_1, O_2, \ldots, O_n$, and if during the experiment $O_i$ has just occurred in the latest trial, we say that the experiment is in the state $O_i$. If outcome $O_j$ occurs in the next trial, we say that the experiment has gone through the *transition* from state $O_i$ to state $O_j$. In the transition matrix, all $p_{ij}$'s must be nonnegative since they are probabilities, and for any $i$, the sum $p_{i1} + p_{i2} + \cdots + p_{in} = 1$, since when the experiment is in state $O_i$, it must be in one of the states $O_1, O_2, \ldots, O_n$ in the next trial. A *state vector* is a vector [ $p_1$   $p_2$   $p_3$   $\cdots$   $p_n$ ] where each entry $p_i$ gives the probability of the system being in state $O_i$. Again, we must have $p_k \geq 0$ for each $k$, and $p_1 + p_2 + \cdots + p_n = 1$.

The examples that follow are given to clarify the preceding definitions.

**EXAMPLE 1** Professor Meany, who teaches a mathematics course for business students, gives frequent pop quizzes. Members of a fraternity have studied the pattern he follows. They have established that if he gives a quiz on a certain day, the probability that he will give a quiz the next class day is 0.1. On the other hand, if he does not give a quiz, then the probability that he will give a quiz the next class day is 0.3. He never gives a quiz the first week of the semester, but the probability that he gives a quiz on the first day of the second week is 0.4.

**a.** Find the probability that he will give a quiz on the second day of the second week.
**b.** Find the probability that he will not give a quiz the second day of the second week.

**Solution**   **a.** We first solve the problem using a probability tree. We let $Q_i$ represent the outcome "he gives a quiz on the $i$th day of the second week" and $Q_i'$ the outcome "he does not give a quiz on the $i$th day of the second week." We were given $p(Q_1) = 0.4$, hence $p(Q_1') = .6$ since $1 - .4 = .6$. Similarly, it was given that $p(Q_2 \mid Q_1) = 0.1$ and $p(Q_2 \mid Q_1') = 0.3$. Consequently, $p(Q_2' \mid Q_1) = 0.9$ and $p(Q_2' \mid Q_1') = 0.7$. We have drawn a probability tree showing this information in Figure 6.17.

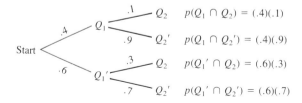

**FIGURE 6.17**

Therefore,

$$p(Q_2) = p(Q_1 \cap Q_2) + p(Q_1' \cap Q_2)$$
$$= (0.4)(0.1) + (0.6)(0.3) = 0.22$$

The probability that he will give a quiz the second day of the second week is 0.22.
**b.** We can obtain the probability that he will not give a quiz the second day of the second week immediately since $1 - .22 = .78$. However, using the probability tree, we get

$$p(Q_2') = p(Q_1 \cap Q_2') + p(Q_1' \cap Q_2') = (0.4)(0.9) + (0.6)(0.7) = 0.78$$

as expected. ∎

If we were asked to find the probability that Professor Meany gives a quiz on the fifth day of the second week, using a probability tree would be inefficient because the number of branches that would have to be drawn would be too large. In this case, it is best to use matrix multiplication as we now illustrate.

**EXAMPLE 2** Repeat Example 1 by first writing the appropriate transition matrix and initial-state vector and then showing that the first-state vector is the product of the initial-state vector by the transition matrix.

**Solution** If we think of "he gives a quiz" as outcome $O_1$ and "he does not give a quiz" as outcome $O_2$, then the probability $p_{11}$ is 0.1. We know this because the probability that he gives a quiz one day if he gave a quiz the previous day is 0.1. Also, $p_{12} = 0.9$, since we must have $p_{11} + p_{12} = 1$. Similarly, $p_{21} = 0.3$ since we know that if he does not give a quiz one day, the probability that he gives a quiz the next day is 0.3. Thus, $p_{22}$ is 0.7 because $p_{21} + p_{22} = 1$. So, the transition matrix is

$$\begin{bmatrix} .1 & .9 \\ .3 & .7 \end{bmatrix}$$

For the initial-state vector, it was given that on the first day of the second week, the probability that "he gives a quiz" is 0.4. Thus, $p_1 = 0.4$. Consequently, $p_2 = 0.6$ since we must have $p_1 + p_2 = 1$. Therefore, the initial-state vector is [ .4   .6 ]. If we multiply the initial-state vector (which is a $1 \times 2$ matrix) by the $2 \times 2$ transition matrix, we will get a $1 \times 2$ matrix (a vector). The first entry of that vector is the dot product of [ .4   .6 ] by the first column of the transition matrix. It is calculated as follows:

$$(.4)(.1) + (.6)(.3) = .22$$

Compare this calculation to that of $p(Q_2)$ in Example 1. Similarly, the second entry of the product is the dot product of [ .4   .6 ] by the second column of the transition matrix. It is calculated as follows:

$$(.4)(.9) + (.6)(.7) = .78$$

Compare this calculation to that of $p(Q_2')$ in Example 1. We can summarize the calculations as follows:

$$[ .4 \quad .6 ]\begin{bmatrix} .1 & .9 \\ .3 & .7 \end{bmatrix} = [ .22 \quad .78 ]$$

The second-state vector is [ .22   .78 ]. ∎

**EXAMPLE 3** What is the probability that Professor Meany of Example 1 will give a pop quiz on the following days.

**a.** On the third day of the second week?
**b.** On the fourth day of the second week?
**c.** On the fifth day of the second week?

**Solution** **a.** We wish to get a third-state vector. We simply multiply the second-state vector by the transition matrix.

$$[ .22 \quad .78 ]\begin{bmatrix} .1 & .9 \\ .3 & .7 \end{bmatrix} = [ .256 \quad .744 ]$$

Thus, the probability that Professor Meany will give a pop quiz on the third day of the second week is 0.256.

**b.** We now multiply the third-state vector by the transition matrix to get the fourth-state vector. We get

$$[\ .256 \quad .744\ ] \begin{bmatrix} .1 & .9 \\ .3 & .7 \end{bmatrix} = [\ .2488 \quad .7512\ ]$$

We see that the probability that the professor will give a pop quiz on the fourth day of the second week is 0.2488.

**c.** We get the fifth-state vector by multiplying the fourth-state vector by the transition matrix.

$$[\ .2488 \quad .7512\ ] \begin{bmatrix} .1 & .9 \\ .3 & .7 \end{bmatrix} = [\ .25024 \quad .74976\ ]$$

The probability that he gives a quiz on the fifth day of the second week is 0.25024.

∎

## Steady-State Vector

In the preceding example, the fourth- and fifth-state vectors seem to be approaching the vector $[\ .25 \quad .75\ ]$. In fact, calculating the sixth- and seventh-state vectors, we get $[\ .249952 \quad .750048\ ]$ and $[\ .2500096 \quad .7499904\ ]$, respectively. Multiplying the vector $[\ .25 \quad .75\ ]$ by the transition matrix, we get

$$[\ .25 \quad .75\ ] \begin{bmatrix} .1 & .9 \\ .3 & .7 \end{bmatrix} = [\ .25 \quad .75\ ]$$

We see that the transition of the state vector $[\ .25 \quad .75\ ]$ leads to the same state vector. This vector is called the *steady-state vector*. In the preceding example, the fact that the steady-state vector is $[\ .25 \quad .75\ ]$ means that after a few days, the probability that Professor Meany gives a pop quiz on an arbitrary day will be 0.25. **Do Exercise 15.**

Recall that a transition matrix **T** is said to be regular if for some positive integer $n$, all entries of the $n$th power of **T** are positive. As we observed in Section 3.8, if **T** is a regular transition matrix, then the successive state vectors approach some fixed state vector **v** regardless of the initial-state vector. (See also Theorem 6.6.)

**EXAMPLE 4**    Let $\mathbf{T} = \begin{bmatrix} .1 & 0 & .9 \\ .2 & .3 & .5 \\ .6 & .2 & .2 \end{bmatrix}$ be a transition matrix.

**a.** Show that **T** is regular.

**b.** Find the steady-state vector.

**Solution**    **a.**

$$T^2 = \begin{bmatrix} .1 & 0 & .9 \\ .2 & .3 & .5 \\ .6 & .2 & .2 \end{bmatrix} \cdot \begin{bmatrix} .1 & 0 & .9 \\ .2 & .3 & .5 \\ .6 & .2 & .2 \end{bmatrix}$$

$$= \begin{bmatrix} .55 & .18 & .27 \\ .38 & .19 & .43 \\ .22 & .10 & .68 \end{bmatrix}$$

All entries of $T^2$ are positive. Hence $T$ is regular.

**b.** Since the transition matrix is regular, we know that a steady-state vector exists. Let $[\, x \quad y \quad z \,]$ be this vector. Then, we must have

$$[\, x \quad y \quad z \,] \cdot \begin{bmatrix} .1 & 0 & .9 \\ .2 & .3 & .5 \\ .6 & .2 & .2 \end{bmatrix} = [\, x \quad y \quad z \,]$$

Performing the multiplication on the left, we get a three-dimensional row vector that is equal to the vector on the right. Because corresponding entries of these vectors are equal, we obtain the system

$$\begin{cases} .1x + .2y + .6z = x \\ \qquad .3y + .2z = y \\ .9x + .5y + .2z = z \end{cases}$$

which may be written

$$\begin{cases} -.9x + \quad .2y + .6z = 0 \quad\quad\quad (1) \\ \qquad\quad -.7y + .2z = 0 \quad\quad\quad (2) \\ .9x + \quad .5y - .8z = 0 \quad\quad\quad (3) \end{cases}$$

Adding equations (1) and (2) and multiplying the resulting equation by $-1$ gives equation (3). We may omit that equation. Since the steady-state vector is a probability vector, we have $x + y + z = 1$. Therefore, we must solve the system

$$\begin{cases} -.9x + \quad .2y + .6z = 0 \\ \qquad\quad -.7y + .2z = 0 \\ x + \quad y + \quad z = 1 \end{cases}$$

The solution is $x = \frac{230}{635}$, $y = \frac{90}{635}$, and $z = \frac{315}{635}$. The steady-state vector is

$$\mathbf{v} = \begin{bmatrix} \dfrac{230}{635} & \dfrac{90}{635} & \dfrac{315}{635} \end{bmatrix} \doteq [\, .3622 \quad .1417 \quad .4961 \,]$$

Note that the three entries are between 0 and 1, as they should be since they are probabilities. You are encouraged to verify that the sum of the three entries in the vector $\mathbf{v}$ is 1, and that the product $\mathbf{v}T$ is $\mathbf{v}$.

**Do Exercise 5.**

In many cases, we have a fixed transition matrix $T$ for a Markov chain and an initial-state vector $\mathbf{v}_0$. We are then asked to find a state vector $\mathbf{v}_k$ for the $k$th trial of the experiment. In such cases, we proceed to find $\mathbf{v}_1$ by calculating the product $\mathbf{v}_0 T$, then we find $\mathbf{v}_2$ by calculating the product $\mathbf{v}_1\mathbf{T}$, and so on, until we find $\mathbf{v}_k$ by calculating the product $\mathbf{v}_{k-1}\mathbf{T}$. In other cases, we have a fixed transition matrix but we wish to find state vectors $\mathbf{v}_k$ using different initial-state vectors. For this, it is best to first find the $k$th power $\mathbf{T}^k$ of $\mathbf{T}$ and use the fact that $\mathbf{v}_k = \mathbf{v}_0\mathbf{T}^k$. Once $\mathbf{T}^k$ has been found, we need only multiply any initial-state vector $\mathbf{v}_0$ by $\mathbf{T}^k$ to find the corresponding state vector $\mathbf{v}_k$. The matrix $\mathbf{T}^k$ is often called the *k-step transition matrix* for the Markov chain.

**EXAMPLE 5**   The transition matrix for a Markov process is

$$\mathbf{T} = \begin{bmatrix} .2 & .8 \\ .6 & .4 \end{bmatrix}$$

Find the state vector for the fifth trial if the initial state vector is

**a.** [ 1    0 ]     **b.** [ 0    1 ]     **c.** [ .4    .6 ]
**d.** [ .2    .8 ]     **e.** [ .3    .7 ]     **f.** [ .9    .1 ]

**Solution**   **a.** If we let $\mathbf{v}_0$ be the initial-state vector for the first trial, then $\mathbf{v}_1 = \mathbf{v}_0\mathbf{T}$ is the state vector for the second trial, $\mathbf{v}_2 = \mathbf{v}_0\mathbf{T}^2$ is the state vector for the third trial, $\mathbf{v}_3 = \mathbf{v}_0\mathbf{T}^3$ is the state vector for the fourth trial, and $\mathbf{v}_4 = \mathbf{v}_0\mathbf{T}^4$ is the state vector for the fifth trial. We first find $\mathbf{T}^4$ as follows:

$$\mathbf{T}^2 = \begin{bmatrix} .2 & .8 \\ .6 & .4 \end{bmatrix} \cdot \begin{bmatrix} .2 & .8 \\ .6 & .4 \end{bmatrix} = \begin{bmatrix} .52 & .48 \\ .36 & .64 \end{bmatrix}$$

$$\mathbf{T}^4 = \begin{bmatrix} .52 & .48 \\ .36 & .64 \end{bmatrix} \cdot \begin{bmatrix} .52 & .48 \\ .36 & .64 \end{bmatrix} = \begin{bmatrix} .4432 & .5568 \\ .4176 & .5824 \end{bmatrix}$$

(We used the fact that $\mathbf{T}^4 = \mathbf{T}^2 \cdot \mathbf{T}^2$.) Now, when $\mathbf{v}_0 = [\ 1\quad 0\ ]$,

$$\mathbf{v}_4 = [\ 1\quad 0\ ] \cdot \begin{bmatrix} .4432 & .5568 \\ .4176 & .5824 \end{bmatrix} = [\ .4432\quad .5568\ ]$$

**b.** When $\mathbf{v}_0 = [\ 0\quad 1\ ]$,

$$\mathbf{v}_4 = [\ 0\quad 1\ ] \cdot \begin{bmatrix} .4432 & .5568 \\ .4176 & .5824 \end{bmatrix} = [\ .4176\quad .5824\ ]$$

**c.** When $\mathbf{v}_0 = [\ .4\quad .6\ ]$,

$$\mathbf{v}_4 = [\ .4\quad .6\ ] \cdot \begin{bmatrix} .4432 & .5568 \\ .4176 & .5824 \end{bmatrix} = [\ .42784\quad .57216\ ]$$

**d.** When $\mathbf{v}_0 = [\ .2\quad .8\ ]$,

$$\mathbf{v}_4 = [\ .2\quad .8\ ] \cdot \begin{bmatrix} .4432 & .5568 \\ .4176 & .5824 \end{bmatrix} = [\ .42272\quad .57728\ ]$$

**e.** When $\mathbf{v}_0 = [\ .3\quad .7\ ]$,

$$\mathbf{v}_4 = [\ .3\quad .7\ ] \cdot \begin{bmatrix} .4432 & .5568 \\ .4176 & .5824 \end{bmatrix} = [\ .42528\quad .57472\ ]$$

**f.** When $v_0 = [\ .9 \quad .1\ ]$,

$$v_4 = [\ .9 \quad .1\ ] \cdot \begin{bmatrix} .4432 & .5568 \\ .4176 & .5824 \end{bmatrix} = [\ .44064 \quad .55936\ ]$$ ■

**EXAMPLE 6**  Let **T** be the transition matrix of Example 5.

**a.** Find $T^7$.
**b.** What do you notice?
**c.** Find the steady-state vector for the matrix **T**.
**d.** Comment.

**Solution**  **a.** We have already found $T^2$ and $T^4$ in the previous example. Thus,

$$T^7 = T^4 \cdot T^2 \cdot T = \begin{bmatrix} .4276352 & .5723648 \\ .4292736 & .5707264 \end{bmatrix}$$

**b.** Note that if we round off the entries to two decimal places, the two rows of the matrix $T^7$ would be identical.
**c.** Let the steady-state vector be $[\ x \quad y\ ]$. Then,

$$[\ x \quad y\ ] \cdot \begin{bmatrix} .2 & .8 \\ .6 & .4 \end{bmatrix} = [\ x \quad y\ ]$$

and

$$[\ .2x + .6y \quad .8x + .4y\ ] = [\ x \quad y\ ]$$

This matrix equation is equivalent to the system

$$\begin{cases} .2x + .6y = x \\ .8x + .4y = y \end{cases}$$

which may be written

$$\begin{cases} -.8x + .6y = 0 \\ .8x - .6y = 0 \end{cases}$$

Clearly, these two equations are equivalent, we need to keep only one of them. We must also have $x + y = 1$ since $[\ x \quad y\ ]$ is a probability vector. We now solve the system

$$\begin{cases} .8x - .6y = 0 \\ x + y = 1 \end{cases}$$

The solution is easily found to be $x = \frac{3}{7}$ and $y = \frac{4}{7}$. Thus, the steady-state vector is $\left(\frac{3}{7} \quad \frac{4}{7}\right) \doteq [\ .42857 \quad .57143\ ]$.
**d.** We see that the steady-state vector is nearly equal to each of the two rows of the matrix $T^7$.

**Do Exercise 27.** ■

The result of the preceding example is a special case of what always happens with regular transition matrices. We state the following theorem without proof.

> **THEOREM 6.6:** Suppose **T** is a regular transition matrix. Then,
>
> 1. **T** has a unique steady-state vector **v** all of whose entries are positive.
> 2. If **W** is a square matrix, all of whose rows are equal to **v**, the powers **T**, $\mathbf{T}^2$, $\mathbf{T}^3$, $\mathbf{T}^4$, . . . approach **W**.
> 3. If **u** is an arbitrary state vector, the sequence of vectors $\mathbf{uT}$, $\mathbf{uT}^2$, $\mathbf{uT}^3$, $\mathbf{uT}^4$, . . . approaches the steady-state vector **v**.

# Exercise Set 6.7

*In Exercises 1–6, a transition matrix for a Markov chain is given. In each exercise, show that the given matrix is regular and find the steady-state vector.*

**1.** $\begin{bmatrix} .3 & .7 \\ 1 & 0 \end{bmatrix}$
    **2.** $\begin{bmatrix} .2 & .8 \\ 1 & 0 \end{bmatrix}$

**3.** $\begin{bmatrix} .75 & .25 \\ 1 & 0 \end{bmatrix}$
    **4.** $\begin{bmatrix} .3 & .2 & .5 \\ .1 & .9 & 0 \\ .4 & .5 & .1 \end{bmatrix}$

**5.** $\begin{bmatrix} .3 & 0 & .7 \\ 0 & .2 & .8 \\ .4 & .1 & .5 \end{bmatrix}$
    **6.** $\begin{bmatrix} .2 & 0 & .8 \\ 0 & .3 & .7 \\ .4 & .6 & 0 \end{bmatrix}$

*In Exercises 7 and 8, show that the given transition matrix is not regular.*

**7.** $\begin{bmatrix} .3 & .7 \\ 0 & 1 \end{bmatrix}$
    **8.** $\begin{bmatrix} 1 & 0 & 0 \\ .2 & .7 & .1 \\ .3 & .6 & .1 \end{bmatrix}$

*In Exercises 9, 10, and 11, determine the values of the variables that will transform the given matrix into a transition matrix for a Markov chain.*

**9.** $\begin{bmatrix} .2 & .7 & x \\ .3 & y & .5 \\ z & .2 & .4 \end{bmatrix}$
    **10.** $\begin{bmatrix} .6 & x & 2y \\ y & 3x & .3 \\ x & 2y & z \end{bmatrix}$

**11.** $\begin{bmatrix} 2x & 3y & .2 \\ z & .4 & 3x \\ .4 & 4x & y \end{bmatrix}$

**12.** The transition matrix for a Markov process is

$$\mathbf{T} = \begin{bmatrix} .7 & .3 \\ .6 & .4 \end{bmatrix}$$

and $\mathbf{v}_0 = [\ .4 \quad .6\ ]$, $\mathbf{w}_0 = [\ .2 \quad .8\ ]$ are initial-state vectors. Do the following:

**a.** Perform the multiplication $\mathbf{v}_0\mathbf{T}$ and interpret the result with the help of a probability tree.
**b.** Perform the multiplication $\mathbf{w}_0\mathbf{T}$ and interpret the result with the help of a probability tree.
**c.** Find $\mathbf{w}_0\mathbf{T}^2$ and interpret the result.
**d.** Find $\mathbf{w}_0\mathbf{T}^3$ and interpret the result.

**13.** Repeat Exercise 12 with

$$\mathbf{T} = \begin{bmatrix} .3 & .2 & .5 \\ .1 & .7 & .2 \\ .4 & .5 & .1 \end{bmatrix}$$

$\mathbf{v}_0 = [\ .2 \quad .4 \quad .4\ ]$ and $\mathbf{w}_0 = [\ .5 \quad .3 \quad .2\ ]$.

**14.** Professor Tuffy, who teaches a mathematics course for business students, gives frequent pop quizzes. Members of a sorority have studied the pattern she follows. They have established that if she gives a quiz on a certain day, the probability that she will give a quiz the next day is 0.08. On the other hand, if she does not give a quiz, then the probability that she will give a quiz the next day is 0.27. She never gives a quiz the first three days of the quarter, but the probability that she gives a quiz on the fourth day of the quarter is 0.35. Find the following probabilities.

**a.** She will give a quiz on the fifth day of the quarter.
**b.** She will not give a quiz on the fifth day of the quarter.

**c.** She will give a quiz on the sixth day of the quarter.

**d.** She will not give a quiz on the sixth day of the quarter.

**e.** She will give a quiz on the tenth day of the quarter.

**f.** She will not give a quiz on the tenth day of the quarter.

**15.** A group attempting to fight cancer in a large city began an active campaign against smoking. For each of several months a survey was conducted. On the basis of their study, they found they could predict that 85% of adults who smoke in any given month will continue to do so in the next month, and only 2.5% of adults who do not smoke in any given month will do so the next month.

**a.** If initially 25% of the adult population smoked, what percentage will still be smoking after 3 months?

**b.** If the campaign is continued over a long period of time, what percentage of the adult population will continue smoking?

**16.** Assume that a nation is controlled by three political parties: X, Y, and Z. The results at each election depend somewhat on which party is in control at the time of the election. If party X is in control now, the probabilities that parties X, Y, Z will be in control after the next elections are 0.4, 0.35, and 0.25, respectively. If party Y is in control now, the probabilities that parties X, Y, Z will be in control after the next elections are 0.2, 0.45, and 0.35, respectively. If party Z is in control now, the probabilities that parties X, Y, Z will be in control after the next elections are 0.3, 0.35, and 0.35, respectively. Given that the elections take place every four years and that party Y won in 1988, what is the probability that **a.** party X will take control in 1996? **b.** party Y will take control in 2000? **c.** party Z will take control in 2004?

**17.** A new transportation system has been established in Seattle and the city is conducting a survey over several months to establish the worth of that system. The survey found that the people traveling to work either use the new system, drive their own car, or car pool. If a person uses the new system now, the probabilities that he/she continues to use the new system, uses his/her own car, or car pools the next month are 0.8, 0.15, and 0.05, respectively. If a person uses his/her own car now, the probabilities that he/she switches to the new system, continues to use his/her own car, or car pools the next month are 0.3, 0.6, and 0.1, respectively. Finally, if a person car pools now, the probabilities that he/she uses the new system, uses his/her own car, or continues to car pool the next month are 0.3, 0.05, and 0.65, respectively. The first month the new system went into operation, 20% of the working people used the new system, 75% used their own cars, and 5% car pooled.

**a.** What percentages of the work force will be using each type of transportation after 3 months?

**b.** After the system has been in operation for a long time, what percentages of the work force will be using each type of transportation?

**18.** Sally is taking a mathematics class at 8:00 A.M. and has difficulty arriving on time. Suppose the events that she is on time or late on consecutive days form a Markov chain. If she is on time on a given day, then the probability that she is on time the next day is 0.4. On the other hand, if she is late on a given day, then the probability that she will be late again on the next day is 0.1.

**a.** Suppose Sally was on time the first day of class, what is the probability that she will be on time on the fifth day of class?

**b.** Suppose Sally was late the first day of class, what is the probability that she will be on time on the fifth day of class?

**19.** Repeat Exercise 18a if we know only that the probability she is late on the first day of class is 0.2.

**20.** Repeat Exercise 18a if we know only that the probability she is on time the first day of class is 0.65.

**21.** A car insurance company classifies its insured motorists into three categories. Those who have not had an accident during the previous year are in the $A_0$ category, those who had exactly one accident during the previous year are in the $A_1$ category, while those who had more than one accident during the previous year are in the $A_+$ category. At the end of each year, all insured motorists are reclassified based only on their record during the preceding year. Long studies by the company have shown that if an insured motorist is in $A_0$ in any given year, then the probabilities that he/she will be in $A_0$, $A_1$, or $A_+$ the following year are 0.85, 0.1, and 0.05, respectively. If an insured motorist is in $A_1$ in any given year, then the probabilities that he/she will be in $A_0$, $A_1$, or $A_+$ the following year are 0.9, 0.08, and 0.02, respectively. Finally, if an insured motorist is in $A_+$ in any given year, then the probabilities that he/she will be in $A_0$, $A_1$, or $A_+$ the following year are 0.5, 0.3, and 0.2, respectively. Suppose that initially the company insured only motorists who belonged in $A_0$.

**a.** Find the probability that after 3 years, an insured motorist randomly selected belongs to $A_0$.

**b.** Find the probability that after 4 years, an insured motorist randomly selected belongs to $A_1$.

**c.** Find the probability that after 5 years, an insured motorist randomly selected belongs to $A_+$.

**22.** Suppose we play the following game. We have three dice which are weighted. When the blue die is rolled, the probability that a 1 or 2 comes up is 0.2, the probability that a 3 or 4 comes up is 0.3, and the probability that a 5 or 6 comes up is 0.5. For the white die, the probabilities are 0.3, 0.4, and 0.3, respectively; for the red die, they are 0.1, 0.5, and 0.4, respectively. The color die that is rolled is determined by the number that came up on the previous roll. That is, if a 1 or 2 comes up, use the blue die on the next roll; if a 3 or 4 comes up, use the white die on the next roll; and if a 5 or 6 comes up, use the red die on the next roll.

**a.** What is the probability that we roll the blue die on the fifth roll if the blue die was used on the first roll?

**b.** What is the probability that we roll the red die on the fifth roll if the white die was used on the first roll?

**c.** What is the probability that we roll the white die on the fifth roll if the first roll was determined as follows: Draw a card from a standard deck. If it is an ace, roll the blue die; if it is a face card, roll the white die; if it is any other card, roll the red die.

**23.** Let $\mathbf{T}$ be the regular matrix of Exercise 1. Find the matrix $\mathbf{W}$ that the sequence $\mathbf{T}, \mathbf{T}^2, \mathbf{T}^3, \mathbf{T}^4, \ldots$ approaches.

**24.** Repeat Exercise 23 for the matrix of Exercise 2.

**25.** Repeat Exercise 23 for the matrix of Exercise 3.

**26.** Repeat Exercise 23 for the matrix of Exercise 4.

**27.** Repeat Exercise 23 for the matrix of Exercise 5.

**28.** Repeat Exercise 23 for the matrix of Exercise 6.

**29.** The matrix of Exercise 7 is not regular. Does it have a steady-state vector?

**30.** The matrix of Exercise 8 is not regular. Does it have a steady-state vector?

**\*31.** Let $\mathbf{v} = [\; a_1 \quad a_2 \quad a_3 \quad \cdots \quad a_n \;]$ be an $n$-dimensional row vector such that $a_1 + a_2 + a_3 + \cdots + a_n = 1$. (The $a_i$'s need not be nonnegative since they are not necessarily probabilities.) Let $\mathbf{W}$ be the $n \times n$ matrix with all rows equal to $\mathbf{v}$. Show that the product $\mathbf{vW}$ is equal to $\mathbf{v}$.

# 6.8 Chapter Review

**IMPORTANT SYMBOLS AND TERMS**

$b(x; n, p)$ [6.6]
$C(n, r)$ [6.2]
$C(n; r_1, r_2, \ldots, r_k)$ [6.2]
$n!$ [6.2]
$p(A \mid B)$ [6.3]
$P(n, r)$ [6.2]
$P(n; n_1, n_2, \ldots, n_k)$ [6.2]
$p(E)$ [6.1]
$P(X = x)$ [6.5]
$\sigma$ [6.5]
$\sigma = \sqrt{\mathrm{VAR}(X)}$ [6.5]
Bayes' formula [6.4]
Bernoulli trials [6.6]
Binomial distribution [6.6]
Binomial experiment [6.6]
Binomial random variable [6.6]
Binomial theorem [6.6]
Combination of $n$ things taken $r$ at a time [6.2]
Conditional probability [6.3]
Continuous random variable [6.5]
Discrete random variable [6.5]

Equiprobable [6.1]
Event [6.1]
Expectation [6.5]
Fundamental principle of counting [6.2]
Impossible event [6.1]
Independent events [6.3]
Infinite sample space [6.3]
Initial-state vector [6.7]
$k$-step transition matrix [6.7]
Markov chain [6.7]
Mathematical expectation (mean) [6.5], [6.6]
Multiplication principle [6.2]
Mutually exclusive events [6.3]
$n$ factorial [6.2]
Odds [6.1]
Ordered partition [6.2]
Outcome [6.1]
Permutation of $n$ things taken $r$ at a time [6.2]
Probability function (probability distribution) [6.5]

Probability histogram [6.5]
Probability of success [6.1]
Probability table [6.5]
Probability tree [6.3]
Random variable [6.5]
Regular transition matrix [6.7]
Relative frequency [6.1]
Sample space [6.1]
Standard deviation [6.5], [6.6]
State vector [6.7]
Steady-state vector [6.7]
Subjective probabilities [6.1]
Success [6.1]
Transition matrix [6.7]
Tree diagram [6.2]
Uniform sample space [6.1]
Variance [6.6]
Variance of a discrete random variable [6.5]

SUMMARY    An event is a subset of a sample space (the set of all possible outcomes of an experiment). If $S$, $E$, $n(S)$, and $n(E)$ denote a sample space, an event, the number of elements in the sample space, and the number of elements in the event, respectively, and if the outcomes are equally likely, then the probability of the event $E$ is

$$p(E) = \frac{n(E)}{n(S)}$$

The odds in favor of an event $E$ is the ratio $\frac{p(E)}{p(E')}$.

Suppose we perform $k$ experiments, in order, and there are $n_1, n_2, \ldots, n_k$ possible outcomes, respectively, for these $k$ experiments. Then, the number of combined outcomes performed in the given order is $n_1 \cdot n_2 \cdots \cdots n_k$. A subset with $r$ elements chosen from a set of $n$ distinct objects is called a combination of $n$ things taken $r$ at a time. If the members of this subset are arranged in a definite order, then the ordered subset is called a permutation of $n$ things taken $r$ at a time. If $n$ and $r$ are integers such that $0 \leq r \leq n$, the symbols $C(n, r)$ and $P(n, r)$ denote the numbers of combinations and permutations, respectively, of $n$ things taken $r$ at a time. We can calculate these numbers using the formulas

$$C(n, r) = \frac{n!}{r!(n-r)!} \text{ and } P(n, r) = \frac{n!}{(n-r)!}$$

If we have a set with $n$ elements composed of $k$ subsets, the first subset being type 1 and having $n_1$ indistinguishable elements, the second subset being type 2 and having $n_2$ indistinguishable elements, and the $k$th subset being type $k$ and having $n_k$ indistinguishable elements, then the number of distinct permutations of these $n$ elements (taken $n$ at a time) is denoted $P(n; n_1, n_2, \ldots, n_k)$ and is given by $P(n; n_1, n_2, \ldots, n_k) = n!/[n_1!n_2! \ldots n_k!]$.

Given a set $S$ with $n$ members, an ordered partition of $S$ is an ordered collection $H_1$, $H_2, \ldots, H_n$ of subsets of $S$ such that each member of $S$ is in exactly one of the subsets $H_i$. Suppose $H_1$ has $r_1$ members, $H_2$ has $r_2$ members, $\ldots$, $H_k$ has $r_k$ members, and let $C(n; r_1, r_2, \ldots, r_k)$ denote the number of ordered partitions, then $C(n; r_1, r_2, \ldots, r_k) = n!/[r_1!r_2! \ldots r_k!]$.

If $A$ and $B$ are events, the conditional probability of $A$, given that $B$ has occurred, is denoted $p(A \mid B)$ and is

$$p(A \mid B) = \frac{p(A \cap B)}{p(B)}, \text{ provided that } p(B) \neq 0.$$

If $A$ and $B$ are events and $p(A) = p(A \mid B)$, we say that $A$ and $B$ are independent. If $A$ and $B$ are disjoint, we say that they are mutually exclusive. In this case, $p(A \cup B) = p(A) + p(B)$. If two events are not mutually exclusive, we may use the formula $p(A \cup B) = p(A) + p(B) - p(A \cap B)$.

If we have a sequence $E_1, E_2, \ldots, E_k$ of $k$ dependent events, where the probability of the first event $E_1$ occurring is $p_1$ and, after the first event has occurred, the probability of the second event occurring is $p_2$, and so forth, then the probability that all $k$ events occur, in that order, is $p_1 \cdot p_2 \cdot p_3 \cdots \cdots p_k$.

If $E_1, E_2, \ldots, E_n$ are mutually exclusive events whose union is $U$ and $A$ is an event such that $p(A) \neq 0$, then Bayes' formula states that

$$p(E_i \mid A) = \frac{p(A \mid E_i) \cdot p(E_i)}{p(A \mid E_1) \cdot p(E_1) + p(A \mid E_2) \cdot p(E_2) + \cdots + p(A \mid E_n) \cdot p(E_n)}$$

If a game has $n$ outcomes, the probabilities of which are $p_1, p_2, \ldots,$ and $p_n$, respectively, and if the net payoff to a person is the amount $A_i$ if the $i$th outcome occurs, where $1 \leq i \leq n$, then the person's expectation for that game is the sum $p_1 \cdot A_1 + p_2 \cdot A_2 + \cdots + p_n \cdot A_n$.

If $S$ is a sample space with a probability function $P$ and $X$ is a discrete random variable with domain $S$, the probability distribution of $X$ is the function $f$, whose domain is the range of $X$, defined by

$$f(x) = P(X = x) = P(E) \text{ where } E = \{s \mid s \in S \text{ and } X(s) = x\}$$

A probability distribution is often represented by a probability histogram. If $S$ is a finite sample space and $X$ is a discrete random variable with probability distribution $f$ defined on the range $\{x_1, x_2, \ldots, x_n\}$ of $X$, then the mathematical expectation of $X$ is

$$E(x) = x_1 \cdot f(x_1) + x_2 \cdot f(x_2) + x_3 \cdot f(x_3) + \cdots + x_n \cdot f(x_n)$$

Its variance (denoted $\text{VAR}(X)$ or $\sigma^2$) is

$$\sigma^2 = [x_1 - E(X)]^2 f(x_1) + [x_2 - E(X)]^2 f(x_2) + \cdots + [x_n - E(X)]^2 f(x_n)$$

$$= \left[ \sum_{i=1}^{n} x_i^2 f(x_i) \right] - [E(X)]^2 = E(X^2) - [E(X)]^2$$

and its standard deviation is $\sigma = \sqrt{\text{VAR}(X)}$.

The Binomial Theorem states that if $n$ is a positive integer and $a$ and $b$ are real (or complex) numbers, then

$$(a + b)^n = \sum_{i=0}^{n} C(n, r)a^{n-r}b^r$$

If $X$ is the number of successes in $n$ independent trials of a binomial experiment, then the probability distribution of $X$ is given by

$$f(x) = P(X = x) = C(n, x)p^x q^{n-x}$$

where $p + q = 1$ and $x$ is an integer such that $0 \leq x \leq n$. The mathematical expectation (mean), variance, and standard deviation of $X$ are given by

$$E(X) = np, \quad \text{VAR}(X) = npq, \quad \text{and} \quad \sigma = \sqrt{npq}$$

respectively.

If in a Markov chain it is given that outcome $O_i$ has occurred at some trial and that the probability that outcome $O_j$ will occur at the next trial is $p_{ij}$, then the $n \times n$ marix $[p_{ij}]$ is called the transition matrix. If it is given that at the initial trial the probability of outcome $O_i$ occurring is $p_i$, then the vector $[\, p_1 \quad p_2 \quad \cdots \quad p_n \,]$ is called the initial-state vector.

If we let $\mathbf{v}_0$ and $\mathbf{T}$ denote the initial-state vector and transition matrix, respectively, then $\mathbf{v}_1 = \mathbf{v}_0\mathbf{T}$ is the state vector for the first trial, $\mathbf{v}_2 = \mathbf{v}_1\mathbf{T}$ is the state vector for the second trial, and so on. For the $k$th trial, the $j$th entry of the vector $\mathbf{v}_k$ gives the probability that the $j$th outcome will occur. A transition matrix $\mathbf{T}$ is said to be regular if for some positive integer $n$ all entries of the $n$th power of $\mathbf{T}$ are positive. If $\mathbf{T}$ is a regular transition matrix, then the successive state vectors approach some fixed vector $\mathbf{v}$ regardless of the initial vector. This vector has the property that $\mathbf{v}\mathbf{T} = \mathbf{v}$.

**SAMPLE EXAM QUESTIONS**

1. Evaluate the expressions

    **a.** $5!$     **b.** $\dfrac{12!}{10!}$     **c.** $\dfrac{17!}{14!5!3!}$     **d.** $P(7, 3)$

    **e.** $P(8, 8)$     **f.** $C(12, 3)$     **g.** $P(15; 3, 5, 7)$     **h.** $C(18; 4, 6, 8)$

2. A die is rolled and then a coin is tossed twice. With the help of a tree diagram, list all the possible outcomes.

3. How many distinct letter arrangements can be made using all the letters of the word *Massachusetts*?

4. A basketball team has three centers, three power forwards, three small forwards, and four guards. The coach must choose one center, one power forward, one small forward, and two guards to form his starting team. In how many ways can he do it?

5. How many committees of six can be formed from a group of seven Republicans and six Democrats if each committee is to have at least three Democrats?

6. In how many ways can 5-card hands be dealt to five poker players from a standard 52-card deck?

7. A store has two identical A television sets, three identical B television sets, and five identical C television sets. At the end of a sale, the store manager displays these sets by arranging them against one of the walls in her store. In how many distinguishable ways can she do this?

8. A foreman is in charge of eight machines and has eight workers, one worker per machine. In how many ways can this be done? Suppose the foreman thinks about each of the possible ways for one second. How long will it take him to reach a decision?

9. A fair coin is tossed four times.

   a. Describe the sample space. (It is a set of ordered quadruples.)
   b. Represent the following situations in terms of events and calculate the probability of each event: (i) exactly one toss results in tails; (ii) at least one toss results in tails; (iii) at least two of the tosses result in tails; (iv) all four tosses result in tails.

10. The odds in favor of the Seattle Supersonics winning the next NBA championship are 2 to 488. What is the probability that Seattle will win the next championship?

11. The probability that Lendl wins the next U.S. Open Tennis Championship is .32. What are the odds in favor of Lendl winning?

12. Shirley, Betty, Norma, Fran, Barbara, Carol, and Kathy are the only entries in a golf tournament. The probabilities that Shirley, Betty, Norma, Fran, and Kathy win are .11, .05, .09, .23, and .32, respectively. What is the probability that Barbara or Carol wins the tournament?

13. Suppose an urn contains five black and four white marbles, and a second urn contains three black and six white marbles. A marble is drawn from the first urn and placed in the second urn. The second urn is shaken and a marble is drawn from it. What is the probability that a black marble was drawn each time?

14. The probability that a freshman takes an English course is .7, the probability that she takes a mathematics course is .25, and the probability that she takes both is .18. What is the probability that a freshman takes **a.** an English course given that she takes a mathematics course? **b.** a mathematics course given that she takes an English course? **c.** both English and mathematics given that she takes at least one of the two subjects?

15. Suppose that a bag contains five red chips and three blue chips. A chip is drawn from the bag and replaced, the bag is shaken, and a chip is drawn again. What is the probability that both chips drawn were blue?

16. A card is drawn from a standard deck of cards.
    a. What is the probability that it is either a king or an ace?
    b. What is the probability that it is either a diamond or a queen?

17. Suppose that five people have been asked to randomly choose a number between 1 and 50, inclusive. What is the probability that no two of them chose the same number?

18. An urn contains five black, three white, and four yellow marbles. Two marbles are drawn, one after the other, without replacing the first before the second is drawn. With the help of a probability tree, find the probability that one black and one yellow marble were drawn.

19. Gwen, Julie, and Marian take turns rolling a fair die, in that order. The first one to roll a 6 wins. What is the probability that Marian wins?

20. A sociologist theorized that people who do not drink alcohol have a better chance to be nonsmokers than those who do. To support his theory, he conducted a study of 1000 randomly selected people in the 18-to-23 age group. He let $U$ be the set of people he studied (the universe), $A$ be the set of people who abstain from drinking alcohol, and $N$ be the set of nonsmokers. He then drew the following Venn diagram.

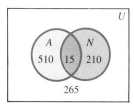

   a. Find the probability that a person who does not drink alcohol is a nonsmoker.
   b. Find the probability that a person who is a nonsmoker does drink alcohol.
   c. Find the probability that a person who does not drink alcohol is a smoker.

21. A novice enters a marksmanship competition and has the choice of six different rifles, only one of which is the right one for a shot from a certain distance. The probability that he hits a bull's-eye is .12 if he uses the right rifle and only .03 if he uses a wrong rifle. Suppose he picks a rifle randomly and hits a bull's-eye. What is the probability that he selected the right rifle?

22. Three machines are used in manufacturing a certain golf ball. Machines I, II, and III are used to manufacture 32%, 43%, and 25% of the total output, respectively. It is known that 1.5% of the balls manufactured using machine I are defective, 2.1% of the balls manufactured using machine II are defective, and 3.7% of the balls manufactured using machine III are defective. A ball is selected at random from the total output and is found to be defective. What is the probability that it was made using machine III?

23. It costs $2 to roll a pair of fair dice. If a sum of 2 or 12 comes up, you are paid $10; if a sum of 3 or 11 comes up, you are paid $5; and if a sum of 4 or 10 comes up, you are paid $3. You lose if any other sum comes up. If you played this game many times, how much would you expect to win or lose per game, on the average?

24. Suppose that a fair coin is tossed four times and that $X$ is the number of heads that came up.

   a. Determine the probability distribution of $X$.
   b. Draw a probability histogram for $X$.
   c. Calculate the mathematical expection of $X$.

25. The annual premium that a Boston homeowner must pay to an insurance company to protect her house, worth $280,000, against total destruction by fire is $300. From experience, the company has determined that the probability of a house being totally destroyed by fire in any one year is .00021. Calculate the company's expected gain.

26. Suppose that a random variable $X$ has range $\{1, 2, 3, 4, 5\}$ and that its probability distribution is defined by the following probability table.

| $x$ | $f(x) = P(X = x)$ |
|---|---|
| 1 | .2 |
| 2 | .25 |
| 3 | .1 |
| 4 | .3 |
| 5 | .15 |

a. Find the variance and standard deviation of $X$.

b. Draw a probability histogram for $X$.

27. Suppose a coin is weighted so that the probability that heads comes up on any toss is .6. The coin is tossed four times. What is the probability that heads comes up exactly two times?

28. The coin of Exercise 27 is tossed five times. What is the probability that heads comes up
a. exactly three times? b. exactly four times? c. at least three times?

29. Suppose that $X$ is a binomial random variable with $p = .3$ and $n = 3$.

a. Find the probability distribution of $X$.

b. Find the mean, variance, and standard deviation of $X$.

30. A fair die is rolled 12 times. Find the probability that a three came up at least twice.

31. A hardware store has a sampling scheme to inspect large lots of incoming light bulbs. Twenty-five light bulbs are to be tested and the lot is rejected if four or more are found to be defective. Suppose a lot contains exactly 5% defective light bulbs. What is the probability that the lot will be rejected?

32. Show that the given transition matrix is not regular.

$$\begin{bmatrix} .2 & .8 \\ 0 & 1 \end{bmatrix}$$

33. Determine the values of the variables that will transform the given matrix into a transition matrix for a Markov chain.

$$\begin{bmatrix} .1 & .3 & x \\ .4 & y & .2 \\ z & .5 & .3 \end{bmatrix}$$

34. Let $\mathbf{A} = \begin{bmatrix} .6 & .4 \\ 1 & 0 \end{bmatrix}$.

a. Show that $\mathbf{A}$ is a regular transition matrix for a Markov chain.

b. Find the steady-state vector.

35. An organization attempting to fight heart disease in a large city began an active campaign against smoking. For each of several months a survey is conducted. The survey found that 78% of those who smoke in any given month will continue to do so the next month, and only 2% of those who do not smoke in any given month will smoke the next month.

a. If initially 20% of the adult population of that city smoked, what percentage will still be smoking after 4 months?

b. If the campaign is continued over a long period, what percentage of the adult population will be smoking after many months?

# The Derivative

**7**

There are two important problems for which we need calculus. We need to use calculus to find the *instantaneous* speed of a moving object and to determine the slope of a *tangent line* to a curve at a certain point. When a motorist is issued a speeding ticket, the average speed for the trip may be under the speed limit, but the speed at some particular instant was not. In both problems, and in many others, we consider a ratio in the form $\frac{f(x) - f(a)}{x - a}$ (the average rate of change of the function $f$) and then evaluate it as $x$ takes on values closer and closer to $a$. In fact, we find the limit of that ratio. We will see that this limit is the derivative of the function $f$ at $a$. In this chapter, we will not only give you a feel for what the derivative of a function is, but also derive formulas for easily finding derivatives of functions we studied in Chapter 1. First, however, you must understand the concepts of rate of change and limits.

## 7.1   Rate of Change

### Increments

When we consider a variable quantity, an *increment* is a change in the value of that quantity. To calculate an increment, we subtract the initial value from the terminal value of the quantity. For instance, to find an increment in your bank account, you subtract the original amount from the final amount in the account. Note that an increment can be a negative number!

**EXAMPLE 1**   The Joneses had \$5000 in their savings account on January 1, 1988, and \$7000 on January 1, 1990. Find the increment in their savings between these dates.

**Solution**   The initial amount was \$5000 and the final amount is \$7000. Since $7000 - 5000 = 2000$, the increment is \$2000.
**Do Exercise 1.**                                                                                            ∎

If the name of a variable is $x$, we use the symbol $\Delta x$ (read: "delta $x$") to denote an increment (a change) in $x$. Suppose we have a function $f$ and $y = f(x)$. If the value of $x$ changes from $x_1$ to $x_2$, then

$$\Delta x = x_2 - x_1$$

and, letting $y_1 = f(x_1)$ and $y_2 = f(x_2)$

$$\Delta y = y_2 - y_1$$

**EXAMPLE 2**   Let $y = f(x) = -x^2 + 2x + 1$. Find $\Delta x$ and $\Delta y$ as $x$ changes from 2 to 4.

**Solution**   The initial value $x_1$ is 2 and the terminal value $x_2$ is 4. Therefore,

$$\Delta x = 4 - 2 = 2$$

Also,

$$y_1 = f(x_1) = f(2) = -2^2 + 2(2) + 1 = 1$$

and

$$y_2 = f(x_2) = f(4) = -4^2 + 2(4) + 1 = -7$$

Thus,

$$\Delta y = y_2 - y_1 = -7 - 1 = -8$$

Note that the increment in $y$ is negative.

**Do Exercise 5.**    ■

**EXAMPLE 3**    If an object is released from rest and falls freely under the influence of gravity, the distance $d$ feet traveled by the object in $t$ seconds is given by

$$d = 16t^2$$

Find $\Delta t$ and $\Delta d$ for the following intervals of time.

**a.** From $t_1 = 2$ to $t_2 = 6$ seconds.
**b.** From $t_1 = 2$ to $t_2 = 5$ seconds.
**c.** From $t_1 = 2$ to $t_2 = 4$ seconds.
**d.** From $t_1 = 2$ to $t_2 = 3$ seconds.
**e.** From $t_1 = 2$ to $t_2 = 2.1$ seconds.

**Solution**    **a.** Here, $\Delta t = 6 - 2 = 4$. Also $d_1 = 16(2^2) = 64$ and $d_2 = 16(6^2) = 576$. Thus,

$$\Delta d = 576 - 64 = 512$$

**b.** In this case, $\Delta t = 5 - 2 = 3$ and

$$\Delta d = 16(5^2) - 16(2^2)$$
$$= 400 - 64 = 336$$

**c.** We have $\Delta t = 4 - 2 = 2$ and

$$\Delta d = 16(4^2) - 16(2^2)$$
$$= 256 - 64 = 192$$

**d.** Here, $\Delta t = 3 - 2 = 1$ and

$$\Delta d = 16(3^2) - 16(2^2)$$
$$= 144 - 64 = 80$$

**e.** Finally, we get $\Delta t = 2.1 - 2 = 0.1$ and

$$\Delta d = 16(2.1^2) - 16(2^2)$$
$$= 70.56 - 64 = 6.56$$

**Do Exercise 13.**    ■

## Average Speed

The average speed of a moving object is found by dividing the distance traveled by the amount of time the object was moving. If $d = f(t)$ gives the distance traveled by an object at instant $t$, then $\Delta t = t_2 - t_1$ gives the amount of time the object was moving from instants $t_1$ to $t_2$ and $\Delta d = f(t_2) - f(t_1)$ gives the distance traveled by that object in that time interval. Thus, $\frac{\Delta d}{\Delta t}$ is the average speed of the object from time $t = t_1$ to $t = t_2$.

**EXAMPLE 4**   For parts **a** through **e** of Example 3, find the average speed of the freely falling object.

**Solution**   **a.** In this case, the amount of time $\Delta t$ is 4 seconds and the distance traveled $\Delta d$ is 512 feet. Thus, $\frac{\Delta d}{\Delta t} = \frac{512}{4} = 128$. So, the average speed is 128 ft/sec.

**b.** Here we have $\frac{\Delta d}{\Delta t} = \frac{336}{3} = 112$. Therefore the average speed is 112 ft/sec.

**c.** We have $\frac{\Delta d}{\Delta t} = \frac{192}{2} = 96$. The average speed is 96 ft/sec.

**d.** Here $\frac{\Delta d}{\Delta t} = \frac{80}{1} = 80$ and the average speed is 80 ft/sec.

**e.** In this case $\frac{\Delta d}{\Delta t} = \frac{6.56}{0.1} = 65.6$. The average speed is 65.6 ft/sec.

**Do Exercise 19.**                                                                                   ■

Although we have not yet defined instantaneous speed of a moving object, consider the following question. The falling object in the previous example was moving at a certain speed at instant $t_1 = 2$ seconds. We obtained the average speeds of the object over five time intervals, all starting at $t_1 = 2$ seconds. Among the five answers in Example 4, which one best approximates the speed of the object at instant $t_1 = 2$ seconds?

## Slope of a Secant Line

In Section 1.3, we saw that if $(x_1, y_1)$ and $(x_2, y_2)$ are two distinct points on a straight line, the slope of that line is

$$m = \frac{y_2 - y_1}{x_2 - x_1}$$

whenever $x_2 \neq x_1$. If $f$ is a function, this ratio gives the slope of the secant line through the points $(x_1, f(x_2))$ and $(x_1, f(x_2))$.

**EXAMPLE 5**   Let $f(x) = x^2$.

**a.** Find the slope of the secant line through $(1, 1)$ and $(3, 9)$.

**b.** Find the slope of the secant line through $(1, 1)$ and $(1 + h, f(1 + h))$ where $h \neq 0$.

**Solution**   **a.** Since $f(1) = 1^2 = 1$ and $f(3) = 3^2 = 9$, the points $(1, 1)$ and $(3, 9)$ are on the graph of $f$. The slope of the secant line is 4 since $\frac{9 - 1}{3 - 1} = 4$.

**b.** The slope of the line through $(1, 1)$ and $(1 + h, f(1 + h))$ is

$$\frac{f(1 + h) - 1}{(1 + h) - 1} = \frac{(1 + h)^2 - 1}{h} = \frac{1 + 2h + h^2 - 1}{h}$$

$$= \frac{h(2 + h)}{h} = 2 + h \qquad (\text{since } h \neq 0).$$

If $h = 2$, $(1 + h, f(1 + h)) = (3, 9)$ and $2 + h = 4$. So, the answer in part **a** is consistent with the general case. The graphs of $y = x^2$ and both secant lines are shown in Figure 7.1.

**Do Exercise 33.**                                                                                   ■

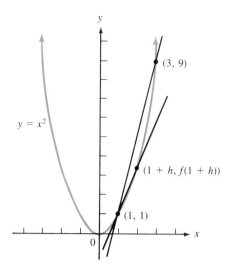

**FIGURE 7.1**

## Average Rate of Change

Using the increment notation, the slope of a secant line may be written

$$m = \frac{\Delta y}{\Delta x}$$

In general, this ratio gives the average rate of change of the variable $y$ with respect to the variable $x$. That is, the average number of units of change in the value of $y$ per unit of change in the value of $x$.

> **DEFINITION 7.1:** If the interval $[x_1, x_2]$ is a subset of the domain of a function $f$, the *average rate of change* of $f$ with respect to $x$ over this interval is defined to be the ratio
>
> $$\frac{\Delta y}{\Delta x} = \frac{f(x_2) - f(x_1)^*}{x_2 - x_1}$$

**EXAMPLE 6** A sporting equipment manufacturer knows that the cost (in dollars) per month to build $q$ rowing machines is given by $C(q) = 30{,}000 + 75q$. The revenue obtained from these $q$ rowing machines (also in dollars) is given by $R(q) = 400q - 0.4q^2$, where $0 \le q \le 500$. If the manufacturer decides to change the production level from 300 machines to 400 machines per month, what are the corresponding increments in cost, revenue, and profit? Find the average rate of change in each of these quantities.

*To calculate the average rate of change of $f$ over the interval $[x_1, x_2]$, it is *necessary* for the whole interval to be a subset of the domain of $f$. For example, if $f(x) = \frac{1}{x-2}$, the average rate of change of $f$ over the interval $[1, 4]$ cannot be calculated because 2 is in the interval but is not in the domain of $f$.

**Solution**  To find the profit, we subtract the cost from the revenue.

$$P(q) = R(q) - C(q)$$
$$= (400q - 0.4q^2) - (30{,}000 + 75q)$$
$$= -0.4q^2 + 325q - 30{,}000$$

The increment in production is $\Delta q = 400 - 300 = 100$. Now,

$$\Delta C = C(400) - C(300)$$
$$= [30{,}000 + 75(400)] - [30{,}000 + 75(300)]$$
$$= 60{,}000 - 52{,}500 = 7500$$

and

$$\Delta R = R(400) - R(300)$$
$$= [400(400) - 0.4(400^2)] - [400(300) - 0.4(300^2)]$$
$$= 96{,}000 - 84{,}000 = 12{,}000$$

Also,
$$\Delta P = P(400) - P(300)$$
$$= [-0.4(400^2) + 325(400) - 30{,}000] - [-0.4(300^2) + 325(300) - 30{,}000]$$
$$= 36{,}000 - 31{,}500 = 4500$$

The increments in cost, revenue, and profit are \$7500, \$12,000, and \$4500, respectively. Since $\frac{\Delta C}{\Delta q} = \frac{7500}{100} = 75$, the average rate of change in cost is \$75 per rowing machine. Similarly, we find that the average rate of change in revenue and in profit are \$120 and \$45 per rowing machine, respectively.
**Do Exercise 27.**

# Exercise Set 7.1

**1.** The Shermans had \$3200 in their savings account on February 1, 1987, and \$5600 on March 1, 1990. Find the increment in their savings between these dates.

**2.** The Doughertys had \$5700 in their savings account on September 1, 1986, and \$3200 on November 1, 1990. Find the increment in their savings between these dates.

*In Exercises 3–11, find $\Delta x$ and $\Delta y$ as x changes from the value $x_1$ to the value $x_2$.*

**3.** $y = f(x) = -5x + 2$, $x_1, = 2$, $x_2 = 8$

**4.** $y = f(x) = 3x^2 + 4x + 2$, $x_1 = -2$, $x_2 = 3$

**5.** $y = f(x) = -x^2 + 6x + 3$, $x_1 = 4$, $x_2 = -2$

**6.** $y = g(x) = \dfrac{1}{x + 2}$, $x_1 = 0$, $x_2 = 3$

**7.** $y = h(x) = \sqrt{x + 2}$, $x_1 = 2$, $x_2 = 7$

**8.** $y = f(x) = \dfrac{-x}{\sqrt{x + 3}}$, $x_1 = 6$, $x_2 = 1$

**9.** $y = g(x) = 5 \ln x$, $x_1 = e$, $x_2 = e^3$

**10.** $y = h(x) = 2^x$, $x_1 = 3$, $x_2 = 5$

**11.** $y = g(x) = -3^x$, $x_1 = 2$, $x_2 = -1$

*In Exercises 12–16, use the formula of Example 3 to find $\Delta t$ and $\Delta d$ for the given time intervals.*

**12.** From $t_1 = 1$ to $t_2 = 5$

**13.** From $t_1 = 3$ to $t_2 = 6$

**14.** From $t_1 = 5$ to $t_2 = 5.1$

**15.** From $t_1 = 4$ to $t_2 = 4.01$

**16.** From $t_1 = 3$ to $t_2 = 3 + h$

**17.** In the time interval of Exercise 12.

**18.** In the time interval of Exercise 13.

**19.** In the time interval of Exercise 14.

**20.** In the time interval of Exercise 15.

**21.** In the time interval of Exercise 16.

*In Exercises 22–26, find $\Delta z$, $\Delta w$, and $\frac{\Delta w}{\Delta z}$.*

**22.** $w = f(z) = z^2 + 1$, $z_1 = -1$, $z_2 = 3$

**23.** $w = g(z) = -z^2 + 2z + 3$, $z_1 = -2$, $z_2 = 3$

**24.** $w = h(z) = \dfrac{1}{z + 3}$, $z_1 = 2$, $z_2 = 4$

**25.** $w = f(z) = z^2 + 3z - 1$, $z_1 = 2$, $z_2 = 2 + a$

**26.** $w = g(z) = 2z^2 - 5z + 1$, $z_1 = a$, $z_2 = a + b$

**27.** A toy manufacturer knows that the cost (in dollars) per month to make $q$ dolls is given by $C(q) = 10,000 + 12q$. The revenue generated from these $q$ dolls (also in dollars) is given by $R(q) = 60q - 0.03q^2$. The manufacturer decides to change the production level from 600 dolls per month to 800 dolls per month. Find the corresponding increments in cost, revenue, and profit. Also find the average rate of change per doll for each of these quantities.

**28.** If the price of peanuts is $p$ dollars per pound, the number of pounds people are willing to buy per month is given by the demand equation

$$D(p) = 2000 - 20p^2, \text{ for } 0 \le p \le 10$$

Find the average rate of change in demand when the price changes from \$4 per pound to \$6 per pound.

**29.** The average person learning to type will type $W(t)$ words per minute after $t$ weeks of learning where

$$W(t) = 80\left(1 - \frac{10}{t^2}\right) \text{ provided that } 4 \le t \le 10$$

Find the average rate of change in the number of words per minute for a change in time from 5 to 7 weeks.

**30.** Find the average rate of change of the area of a circle as the radius changes from 3 inches to 5 inches.

**31.** A projectile is fired straight up. Its height $d$ feet above the ground $t$ seconds later is given by the formula

$$d = -16t^2 + 2567t$$

Find the average speed from instant $t_1 = 1$ to $t_2 = 3$ seconds.

**32.** The size of a population on an island is given by $P(t) = 5000\,(2^{t/10})$ where $t$ is measured in years and $10 \le t \le 40$. Find the average rate of change of the population from time $t_1 = 20$ to $t_2 = 30$ years.

*In Exercises 33–37, find the slope of the secant line passing through the points $(x_1, f(x_1))$ and $(x_2, f(x_2))$.*

**33.** $y = f(x) = x^2 + 3x + 1$

    **a.** $x_1 = 2$, $x_2 = 4$
    **b.** $x_1 = 2$, $x_2$, $= 2 + h$, where $h \neq 0$

**34.** $y = f(x) = -x^2 + 4x + 3$

    **a.** $x_1 = -1$, $x_2 = 2$
    **b.** $x_1 = -1$, $x_2 = -1 + h$, where $h \neq 0$

**35.** $y = f(x) = \sqrt{x + 2}$

    **a.** $x_1 = 7$, $x_2 = 23$
    **b.** $x_1 = 7$, $x_2 = 7 + h$, where $h \neq 0$

**36.** $y = f(x) = \dfrac{1 + x}{2 + \sqrt{x}}$

    **a.** $x_1 = 4$, $x_2 = 9$
    **b.** $x_1 = 4$, $x_2 = 4 + h$, where $h \neq 0$

**37.** $y = f(x) = x + 2^x$

    **a.** $x_1 = -1$, $x_2 = 1$
    **b.** $x_1 = -1$, $x_2 = -1 + h$, where $h \neq 0$

# 7.2    Intuitive Description of Limit

## Instantaneous Speed

In Example 4 of the preceding section, we calculated the average speed of a freely falling object as it traveled from time $t_1 = 2$ seconds to $t_2 = 2 + h$ seconds, where $h$ took on the values of five positive numbers. Let us do the same for the general value of $h$. Recall that $d = 16t^2$.

In this case, we get

$$\Delta t = t_2 - t_1 = (2 + h) - 2 = h$$

and

$$\Delta d = 16(2 + h)^2 - 16(2^2)$$
$$= (64 + 64h + 16h^2) - 64$$
$$= 64h + 16h^2$$
$$= 16h(4 + h)$$

Therefore,

$$\frac{\Delta d}{\Delta t} = \frac{16h(4 + h)}{h} = 16(4 + h) \qquad \text{(provided } h \neq 0).$$

Therefore, the average speed during the given time interval is $16(4 + h)$ ft/sec. Note that we can get the five answers we obtained in the preceding section by replacing $h$ successively by 4, 3, 2, 1, and 0.1.

How fast was the object moving at instant $t = 2$ seconds? Since it is reasonable to assume that the speed does not change very much over a small time interval, we can get a good approximation by considering values of $16(4 + h)$ where $|h|$ is small. When $h$ is very close to 0, $4 + h$ is near 4 and $16(4 + h)$ is near 64. We, therefore, conclude that the speed of the falling object at instant $t = 2$ seconds was 64 ft/sec. Observe that the value of the average speed over each interval $[2, 2 + h]$ is within a specified distance of 64, provided $h$ is sufficiently close to 0, and the specified distance can be made as small as we please.* For example, if we wish the average speed to be within 0.00001 of 64, then we require

$$|16(4 + h) - 64| < 0.00001$$

which is equivalent to

$$\frac{-0.00001}{16} < h < \frac{0.00001}{16}$$

So, if we choose $h$ so that $|h| < \frac{0.00001}{16}$, the average speed of the falling object over the time interval $[2, 2 + h]$ will be within 0.00001 ft/sec of 64 ft/sec. However, we cannot allow $h = 0$, since in calculating the average speed over the interval $[2, 2 + h]$, *we required $h \neq 0$.*

## Tangent Lines

In Euclidean geometry, a tangent line to a circle is a line that intersects the circle at exactly one point. However, this definition is not satisfactory for other curves. It is possible for a line to intersect a curve at exactly one point, although we would not consider this line a tangent. (See Figure 7.2a.) It is also possible that a line, which

---

*Recall that if $a$ and $b$ are real numbers, $|a - b|$ gives the distance between $a$ and $b$ on the number line.

we consider to be a tangent to a curve, intersects that curve at more than one point. (See Figure 7.2b.)

**FIGURE 7.2**

(a)                          (b)

> **DEFINITION 7.2:**   The *tangent line* to a curve $C$ at a point $P$ is the line, in the limiting position if it exists, of the secant lines through the fixed point $P$ on $C$, and a variable point $Q$ on $C$, as $Q$ moves along $C$ closer and closer to $P$. (See Figure 7.3.)

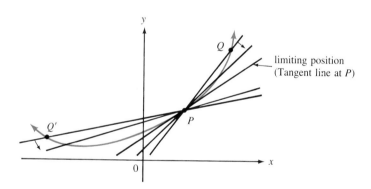

**FIGURE 7.3**

This is not a rigorous definition since we have not yet defined what we mean by a limit or "limiting position." However, it does describe intuitively what we would consider a tangent line to be.

Let us return to Example 5, part **b**, of the preceding section. The curve $C$ was the graph of $y = f(x) = x^2$. We found that the slope of the secant line through the points $(1, 1)$ and $(1 + h, f(1 + h))$ was $2 + h$, with $h \neq 0$. Let $(1, 1)$ and $(1 + h, f(1 + h))$ be the fixed point $P$ and the variable point $Q$, respectively. (See Figure 7.4.) As $Q$, with coordinates $(1 + h, (1 + h)^2)$, gets closer and closer to $P$, with coordinates $(1, 1)$, the values of $h$ get closer and closer to 0. So, the values $2 + h$ of the slopes of the secant lines will be within any specified distance of 2, no matter how small, provided $h$ is sufficiently close to 0. For example, if we wish the slope of a secant line to be within 0.0000001 of 2, we need only require that

$$|(2 + h) - 2| < 0.0000001$$

which will be true if we choose $h$ such that $|h| < 0.0000001$. Therefore, it is reasonable to state that the slope of the tangent line at $(1, 1)$ is 2. Again, keep in mind that we have required $h \neq 0$.

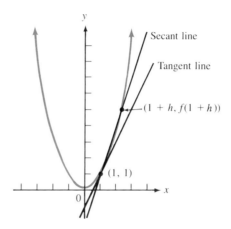

**FIGURE 7.4**

**EXAMPLE 1**   Let $y = f(x) = x^2 + 3x - 5$. Find the slope of the tangent line to the graph of $f$ at $(2, 5)$

**Solution**   The point $(2, 5)$ is on the graph of $f$ since

$$f(2) = 2^2 + 3 \cdot 2 - 5 = 5$$

Consider the secant line through the point $(2, 5)$ and another point $(x, f(x))$ on the graph of $f$. The slope of that secant line is

$$m_x = \frac{f(x) - 5}{x - 2}$$

But,

$$\frac{f(x) - 5}{x - 2} = \frac{(x^2 + 3x - 5) - 5}{x - 2}$$

$$= \frac{x^2 + 3x - 10}{x - 2} = \frac{(x - 2)(x + 5)}{x - 2}$$

$$= x + 5 \quad (\text{provided } x \neq 2)$$

If $(x, f(x))$ gets closer and closer to $(2, 5)$, the corresponding secant lines get closer and closer to the tangent line, and $x$ takes on values that are closer and closer to 2. Therefore, the slope of the secant lines take on values that are closer and closer to 7. In fact, the values $x + 5$ of the slopes of the secant lines will be within any specified distance (no matter how small) of 7, provided $x$ is sufficiently close to 2 and $x \neq 2$. For example, if we desire the slope of a secant line to be within 0.00000003 of 7, then we write

$$|(x + 5) - 7| < 0.00000003$$

which is true provided that $|x - 2| < 0.00000003$. Again, we cannot allow $x = 2$, since $x \neq 2$ was required earlier. See Figure 7.5.

**FIGURE 7.5**
**Do Exercise 33.**

## Limit of a Function

The above examples illustrate the limit concept which we now define.

> **DEFINITION 7.3:** Suppose that the number $c$ belongs to some interval $I$, and $f$ is a function defined in this interval except possibly at $c$. We say that the *limit of $f(x)$ as $x$ tends to $c$* (approaches $c$) is the number $L$ and we write
>
> $$\lim_{x \to c} f(x) = L$$
>
> provided the following is true. The values of $f(x)$ are within a specified distance of $L$, as long as $x$ is sufficiently close to, but distinct from, $c$. The specified distance can be as small as we please.

Notice that we are not concerned with the value of $f(x)$ when $x$ equals $c$, but only with those values when $x$ is close to $c$. As we have seen in the earlier examples, we must often find $\lim_{x \to c} f(x)$ even though the function $f$ is not defined at $c$. In general, however, $f(c)$ may or may not be defined.

**EXAMPLE 2**  Suppose $f(x) = 2x + 3$. Find $\lim_{x \to 4} f(x)$.

**Solution**  To find the value of $f(x)$, we multiply the value of $x$ by 2 then add 3 to the product. Thus if the value of $x$ is close to 4, the value of $2x$ is close to 8 and the value of $f(x)$ is close to 11. We conclude that

$$\lim_{x \to 4} f(x) = 11$$

**Do Exercise 1.**                                                                              ◼

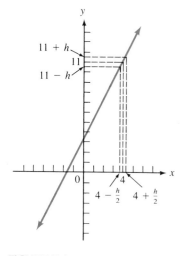

**FIGURE 7.6**

To give a formal justification for the result in Example 2, we proceed as follows: We want $f(x)$ to be within a specified distance of 11. Let $h$ be any positive number. Note that $h$ can be as small as we wish. Next, we show that

$$|(2x + 3) - 11| < h$$

whenever $x$ is close enough to 4 and $x \neq 4$. This inequality is equivalent to

$$|2x - 8| < h$$

and to

$$|x - 4| < \frac{h}{2}$$

Thus, if we choose any real number $x$ such that

$$0 < |x - 4| < \frac{h}{2}$$

then we will have

$$|(2x + 3) - 11| < h$$

We have shown that $f(x)$ is arbitrarily close to 11 (within a very small distance $h$) by taking $x$ sufficiently close to 4 (within a distance $\frac{h}{2}$.) (See Figure 7.6.)

In Example 2, both $\lim_{x \to 4} f(x)$ and $f(4)$ are equal to 11. That is, $\lim_{x \to 4} f(x) = f(4)$. Although this does happen often, as we shall see in Section 7.3, it is not always the case. It is important to remember that *in dealing with limits, we let x approach c but we keep x ≠ c.*

We state the following theorem without proof. As we shall see in the examples, you will often be able to evaluate the limits of functions using your intuition. However, if asked to justify your steps, you *must* invoke the properties of limits stated in the theorem.

> **THEOREM 7.1:**   Suppose $a$, $b$, and $c$ are constants; $r$ is a rational number; $\lim_{x \to c} f(x) = L$; and $\lim_{x \to c} g(x) = M$. Then
>
> 1. $\lim_{x \to c} a = a$.
> 2. $\lim_{x \to c} (ax + b) = ac + b$.
> 3. $\lim_{x \to c} [f(x) + g(x)] = L + M$.
> 4. $\lim_{x \to c} [f(x) - g(x)] = L - M$.
> 5. $\lim_{x \to c} [f(x) \cdot g(x)] = L \cdot M$.
> 6. $\lim_{x \to c} \dfrac{f(x)}{g(x)} = \dfrac{L}{M}$   (provided that $M \neq 0$).
> 7. $\lim_{x \to 4} [f(x)]^r = L^r$   (provided that $[f(x)]^r$ is defined when $x$ is near $c$).

**EXAMPLE 3**   Evaluate each of the following limits.

  **a.** $\lim_{x \to 2} (3x + 2)$

  **b.** $\lim_{x \to 3} \dfrac{4x + 1}{3x - 4}$

  **c.** $\lim_{x \to -2} (x^2 + 3x - 4)$

  **d.** $\lim_{x \to 2} (2x + 5)^{3/2}$

**Solution**   **a.** $\lim_{x \to 2} (3x + 2) = 3 \cdot 2 + 2 = 8$   (by Property 2)

  **b.** It is easy to see that $\lim_{x \to 3} (4x + 1) = 13$ and $\lim_{x \to 3} (3x - 4) = 5$. Therefore,

$$\lim_{x \to 3} \frac{4x + 1}{3x - 4} = \frac{13}{5} \quad \text{(by Property 6)}$$

  **c.** When $x$ is close to $-2$, $x^2$ and $3x$ are close to 4 and $-6$, respectively. Hence, $x^2 + 3x - 4$ is close to $-6$ since $4 + (-6) - 4 = -6$. Thus,

$$\lim_{x \to -2} (x^2 + 3x - 4) = -6$$

If asked to justify your steps, you should proceed as follows.

$$\lim_{x \to -2} (x^2 + 3x - 4) = \lim_{x \to -2} x^2 + \lim_{x \to -2} (3x - 4) \quad \text{(by Property 3)}$$
$$= (\lim_{x \to -2} x)(\lim_{x \to -2} x) + \lim_{x \to -2} (3x - 4) \quad \text{(by Property 5)}$$
$$= (-2)(-2) + (3(-2) - 4) \quad \text{(by Property 2)}$$
$$= 4 - 10 = -6$$

**d.** Clearly $\lim_{x \to 2}(2x + 5) = 9$. Thus,

$$\lim_{x \to 2}[(2x + 5)^{3/2}] = 9^{3/2} = (3^2)^{3/2} = 3^3 = 27 \quad \text{(by Property 7)}$$

**Do Exercise 9.**

From now on, we will evaluate limits without referring to their properties. You should, as you read the examples, be aware of which properties are being used.

We have seen in part **b** of the previous example that

$$\lim_{x \to 3} \frac{4x + 1}{3x - 4} = \frac{13}{5}$$

We know that whenever $x - 3 \neq 0$, we have

$$\frac{4x + 1}{3x - 4} = \frac{(4x + 1)(x - 3)}{(3x - 4)(x - 3)}$$

Thus, the following must be true:

$$\lim_{x \to 3} \frac{(4x + 1)(x - 3)}{(3x - 4)(x - 3)} = \lim_{x \to 3} \frac{4x + 1}{3x - 4} = \frac{13}{5}$$

Using the approach we took earlier to evaluate $\lim_{x \to 3} \frac{(4x + 1)(x - 3)}{(3x - 4)(x - 3)}$, we would have obtained $\frac{0}{0}$, which is not a number. Whenever this happens, we must try to write the fraction in a different but equivalent form that will enable us to calculate the limit.

**EXAMPLE 4**    Evaluate $\lim_{x \to 2} (x^2 + x - 6)/(x^2 - 3x + 2)$.

**Solution**    Note that $\lim_{x \to 2}(x^2 + x - 6) = \lim_{x \to 2}(x^2 - 3x + 2) = 0$. This being the case, we first try to simplify the fraction.

$$\frac{x^2 + x - 6}{x^2 - 3x + 2} = \frac{(x - 2)(x + 3)}{(x - 2)(x - 1)} = \frac{x + 3}{x - 1} \quad \text{(provided } x \neq 2\text{)}.$$

Thus,

$$\lim_{x \to 2} \frac{x^2 + x - 6}{x^2 - 3x + 2} = \lim_{x \to 2} \frac{x + 3}{x - 1} = \frac{5}{1} = 5$$

The graph of $y = f(x) = (x^2 + x - 6)/(x^2 - 3x + 2)$ is shown in Figure 7.7. Note that $(2, 5)$ is not on the graph since the function $f$ is not defined at 2. However, we see on the graph that if $x$ is near 2, then the corresponding $y$ is close to 5. In fact, if we choose $y$ within a specified distance of 5, then we can take $x$ sufficiently close to 2 so that $f(x) = y$.

**Do Exercise 11.**

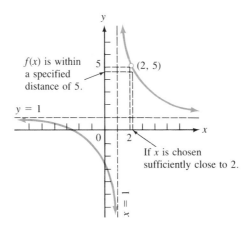

$f(x)$ is within a specified distance of 5.

$y = 1$

If $x$ is chosen sufficiently close to 2.

**FIGURE 7.7**

*Suggestion:* Using a calculator, compute the value of the fraction

$$\frac{x^2 + x - 6}{x^2 - 3x + 2}$$

for $x = 1.9$, $x = 1.99$, $x = 1.999$, $x = 2.1$, $x = 2.01$, and $x = 2.001$.

Note that if $f(x)$ and $g(x)$ are polynomials and if

$$\lim_{x \to a} f(x) = \lim_{x \to a} g(x) = 0$$

then $x - a$ is a factor of both $f(x)$ and $g(x)$ since $f(a) = g(a) = 0$. (See Section 0.6.) This fact is often helpful in factoring the numerator and denominator of a rational function for the purpose of simplification.

In certain cases, however, we will want to evaluate $\lim_{x \to a} \frac{f(x)}{g(x)}$ where $\lim_{x \to a} f(x) = \lim_{x \to a} g(x) = 0$, but $f(x)$ or $g(x)$, or both, are not polynomials.*

The case where $f(x)$ or $g(x)$, or both, contain radicals can often be treated by the process of rationalization. Recall that to rationalize either $a - b$ or $a + b$ where $a$ or $b$, or both, is a radical of order 2, we use the fact that $(a - b)(a + b) = a^2 - b^2$ because $a^2 - b^2$ will be free of radicals.

**EXAMPLE 5** Evaluate each of the following limits.

**a.** $\lim\limits_{x \to 3} \dfrac{\sqrt{x + 1} - 2}{x - 3}$

**b.** $\lim\limits_{x \to 2} \dfrac{x - 2}{3 - \sqrt{2x + 5}}$

**c.** $\lim\limits_{x \to 3} \dfrac{\sqrt{4 - x} - 1}{3 - \sqrt{2x + 3}}$

---

*In more advanced texts, a technique to handle such cases is introduced (L'Hospital's Rule). In this book, however, we shall not concern ourselves with this technique, but we will give other examples in which algebraic manipulations of the fraction are helpful.

**Solution** If we attempted to use Property 6 of Theorem 7.1 to evaluate these limits, we would get a meaningless expression in all three cases. Thus, in each case, we rationalize the part of the fraction that contains a radical.

a. $\dfrac{\sqrt{x+1}-2}{x-3} = \dfrac{(\sqrt{x+1}-2)(\sqrt{x+1}+2)}{(x-3)(\sqrt{x+1}+2)}$

$$= \dfrac{(x+1)-4}{(x-3)(\sqrt{x+1}+2)}$$

$$= \dfrac{x-3}{(x-3)(\sqrt{x+1}+2)}$$

$$= \dfrac{1}{\sqrt{x+1}+2} \quad \text{(provided } x \neq 3)$$

Thus,

$$\lim_{x \to 3} \dfrac{\sqrt{x+1}-2}{x-3} = \lim_{x \to 3} \dfrac{1}{\sqrt{x+1}+2} = \dfrac{1}{4}$$

b. $\dfrac{x-2}{3-\sqrt{2x+5}} = \dfrac{(x-2)(3+\sqrt{2x+5})}{(3-\sqrt{2x+5})(3+\sqrt{2x+5})}$

$$= \dfrac{(x-2)(3+\sqrt{2x+5})}{9-(2x+5)}$$

$$= \dfrac{(x-2)(3+\sqrt{2x+5})}{4-2x}$$

$$= \dfrac{(x-2)(3+\sqrt{2x+5})}{-2(x-2)}$$

$$= \dfrac{3+\sqrt{2x+5}}{-2} \quad \text{(provided } x \neq 2)$$

Thus,

$$\lim_{x \to 2} \dfrac{x-2}{3-\sqrt{2x+5}} = \lim_{x \to 2} \dfrac{3+\sqrt{2x+5}}{-2} = \dfrac{6}{-2} = -3$$

c. We multiply the numerator and denominator of the fraction by $(\sqrt{4-x}+1)$ to rationalize $(\sqrt{4-x}-1)$, and by $(3+\sqrt{2x+3})$ to rationalize $(3-\sqrt{2x+3})$. We obtain

$$\dfrac{\sqrt{4-x}-1}{3-\sqrt{2x+3}} = \dfrac{(\sqrt{4-x}-1)(\sqrt{4-x}+1)(3+\sqrt{2x+3})}{(3-\sqrt{2x+3})(\sqrt{4-x}+1)(3+\sqrt{2x+3})}$$

$$= \dfrac{[(4-x)-1](3+\sqrt{2x+3})}{[9-(2x+3)](\sqrt{4-x}+1)}$$

$$= \dfrac{(3-x)(3+\sqrt{2x+3})}{(6-2x)(\sqrt{4-x}+1)}$$

$$= \dfrac{(3-x)(3+\sqrt{2x+3})}{2(3-x)(\sqrt{4-x}+1)}$$

$$= \frac{3 + \sqrt{2x + 3}}{2(\sqrt{4 - x} + 1)} \quad \text{(provided } x \neq 3\text{)}$$

Thus,

$$\lim_{x \to 3} \frac{\sqrt{4 - x} - 1}{3 - \sqrt{2x + 3}} = \lim_{x \to 3} \frac{3 + \sqrt{2x + 3}}{2(\sqrt{4 - x} + 1)} = \frac{6}{4} = \frac{3}{2}$$

**Do Exercises 19 and 23.**                                                                  ■

> In evaluating $\lim_{x \to a} \frac{f(x)}{g(x)}$, where $\lim_{x \to a} f(x) = b$ and $\lim_{x \to a} g(x) = c$, if $b = 0$, or $c = 0$, or both, it is useful to keep in mind that
>
> $$\frac{b}{c} \text{ is } \begin{cases} \text{equal to 0 if } b = 0 \text{ and } c \neq 0 \\ \text{undefined if } b \neq 0 \text{ and } c = 0 \\ \text{an indeterminate form when } b = 0 \text{ and } c = 0. \end{cases}$$
>
> In the latter two cases, the expression $\frac{b}{c}$ is meaningless. In the indeterminate case we must attempt to simplify the fraction *before* using the properties of limits in Theorem 7.1.

In Examples 4 and 5 we illustrated indeterminate forms. We conclude with the following example.

**EXAMPLE 6**   Evaluate each of the following limits.

**a.** $\lim_{x \to 2} \dfrac{x - 2}{x + 3}$

**b.** $\lim_{x \to 1} \dfrac{x + 2}{x - 1}$

**Solution**   **a.** Here, $\lim_{x \to 2} (x - 2) = 0$ and $\lim_{x \to 2} (x + 3) = 5$. Thus,

$$\lim_{x \to 2} \frac{x - 2}{x + 3} = \frac{0}{5} = 0$$

**b.** In this case, $\lim_{x \to 1} (x + 2) = 3$ and $\lim_{x \to 1} (x - 1) = 0$. Therefore, $\lim_{x \to 1} \frac{x + 2}{x - 1}$ is undefined.

**Do Exercises 13 and 17.**                                                                  ■

# Exercise Set 7.2

*In Exercises 1–27, evaluate the limits. Also, in each case, use a calculator to evaluate the function, whose limit was found, at four values of the independent variable chosen near its limiting value.*

**1.** $\lim_{x \to 2} (5x - 2)$

**2.** $\lim_{x \to -3} (2x + 1)$

**3.** $\lim_{x \to 0} (2^x + 5)$

**4.** $\lim_{x \to 1} (3x + 1)^4$

**5.** $\lim_{x \to 2} (3x + 2)^{5/3}$

**6.** $\lim_{x \to -3} (-2x^2 + 5x + 10)$

**7.** $\lim_{x \to 2} (x^2 + 6x + 7)$

**8.** $\lim_{x \to 1} \sqrt{x^2 + 6x + 2}$

**9.** $\lim_{x \to 1} \dfrac{2x + 1}{3x - 2}$

**10.** $\lim_{t \to 2} \dfrac{5t + 1}{2t - 7}$

**11.** $\lim_{y \to 2} \dfrac{y^2 + y - 6}{y^2 - 4}$

**12.** $\lim_{x \to 1} \dfrac{x^2 + 5x - 6}{x^2 + 2x - 3}$

**13.** $\lim\limits_{u\to 3} \dfrac{u^2-9}{u^2+5u+2}$

**14.** $\lim\limits_{x\to 2} \dfrac{x^3-8}{x^2-x-2}$

**15.** $\lim\limits_{x\to 1/2}[(6x+3)(8x-5)]$

**16.** $\lim\limits_{x\to e}[(3x+2)\ln x]$

**17.** $\lim\limits_{x\to 4} \dfrac{\sqrt{x}-2}{x-4}$

**18.** $\lim\limits_{x\to 3} \dfrac{\sqrt{x+6}-3}{x-3}$

**19.** $\lim\limits_{x\to -1} \dfrac{\sqrt{5+x}-2}{x+1}$

**20.** $\lim\limits_{x\to 2} \dfrac{x-2}{2-\sqrt{x+2}}$

**21.** $\lim\limits_{x\to 5} \dfrac{5-x}{\sqrt{2x-1}-3}$

**22.** $\lim\limits_{t\to -2} \dfrac{t+2}{3-\sqrt{7-t}}$

**23.** $\lim\limits_{x\to 1} \dfrac{\sqrt{x+3}-2}{3-\sqrt{x+8}}$

**24.** $\lim\limits_{x\to 2} \dfrac{3-\sqrt{2x+5}}{\sqrt{8x+9}-5}$

**25.** $\lim\limits_{x\to 4} \dfrac{x-\sqrt{3x+4}}{x^2-16}$

**\*26.** $\lim\limits_{x\to 8} \dfrac{x^{1/3}-2}{x-8}$ (*Hint:* Use the fact that $(a-b)(a^2+ab+b^2)=a^3-b^3$.)

**\*27.** $\lim\limits_{x\to 4} \dfrac{(6x+3)^{1/3}-3}{x-(3x+52)^{1/3}}$ (See the hint in Exercise 26.)

**28.** Let $f(x)=x^2+1$. Find $\lim\limits_{x\to 2}\dfrac{f(x)-f(2)}{x-2}$.

**29.** Let $g(x)=x^2+2x+5$. Find $\lim\limits_{h\to 0}\dfrac{g(3+h)-g(3)}{h}$.

**30.** Let $f(x)=\sqrt{x}$. Find $\lim\limits_{h\to 0}\dfrac{f(4+h)-f(4)}{h}$.

**31.** A projectile is fired straight up. After $t$ seconds, its distance from the ground is $d$ feet where $d=-16t^2+192t$. Find its average speed from instant $t_1=4$ seconds to instant

$t_2=(4+h)$ seconds. What is its instantaneous speed at instant 4 seconds?

**32.** A projectile is fired straight up. After $t$ seconds, its distance from the ground is $d$ feet where $d=-16t^2+384t$. Find its average speed from instant $t_1=t$ seconds to instant $t_2=t+h$ seconds. What is its instantaneous speed at instant $t$ seconds? At what instant will the instantaneous speed be 0? How far is the projectile above the ground at that instant? Will it go any higher?

**33.** Let $y=f(x)=x^2+x+3$.

  **a.** Verify that the point $(2,9)$ is on the graph of $f$.

  **b.** Find the slope $m_h$ of the secant line through the points $(2,9)$ and $(2+h,f(2+h))$.

  **c.** Evaluate $\lim\limits_{h\to 0} m_h$. Let this value be $m$.

  **d.** Find the equation of the line passing through $(2,9)$ and with slope $m$.

  **e.** On the same coordinate plane, carefully sketch the graphs of the equation $y=f(x)$ and the equation found in part **d**.

*In Exercises 34–39, a function $f$ and a number $c$ are given. In each case, find $\lim\limits_{h\to 0}\dfrac{f(c+h)-f(c)}{h}$.*

**34.** $f(x)=5x+3,\ c=2$

**35.** $f(x)=-3x+2,\ c=4$

**36.** $f(x)=x^2+3,\ c=-1$

**37.** $f(x)=2x^2+3x+1,\ c=4$

**38.** $f(x)=\sqrt{x},\ c=9$

**39.** $f(x)=\dfrac{15}{x},\ c=3$

# 7.3 Continuity and One-Sided Limits

## Continuous Functions

In the preceding section we encountered functions with the following property:

$$\lim_{x\to a} f(x)=f(a)$$

Any such function is said to be *continuous at a*. If a function is continuous at each point of its domain, we simply say that it is *continuous*.

The statement $\lim_{x \to a} f(x) = f(a)$ describes three properties of the function $f$.

1. $f(a)$ is defined. That is, $a$ is in the domain of $f$.
2. $\lim_{x \to a} f(x)$ exists.
3. The two numbers $\lim_{x \to a} f(x)$ and $f(a)$ are equal.

If any of these three properties is not satisfied, then the function is not continuous at $a$. For example, is the function $f$ defined by $f(x) = (x^2 - 9)/(x - 3)$ continuous at 3? The answer is no, since 3 is not in the domain of $f$. We need not check Properties 2 and 3.

Suppose in playing a lottery game you paid \$3 for a chance to press a button on an electronic device that selected a real number between 1 and 9, inclusive. If an integer $n$ is selected, you win \$$3^{n+2}$; otherwise, you lose. If \$$p(x)$ is the payoff when the number $x$ is selected,

$$p(x) = \begin{cases} -3 & \text{if } x \text{ is not an integer (since you lost what you paid)} \\ 3^{x+2} - 3 & \text{if } x \in \{1, 2, 3, 4, 5, 6, 7, 8, 9\} \text{ (since the payoff is the amount you win minus the amount you paid to play)} \end{cases}$$

For instance, $p(4) = 3^{4+2} - 3 = 3^6 - 3 = 729 - 3 = 726$. Now if $x$ is sufficiently close to 4, but $x \neq 4$, then $x$ is not an integer and $p(x) = -3$. Thus, $\lim_{x \to 4} p(x) = -3$. Since $-3 \neq 726$, $\lim_{x \to 4} p(x) \neq p(4)$. Thus the function $p$ is not continuous at 4.

Here are some other examples.

**EXAMPLE 1** Discuss the continuity of each function at the given point $c$.

**a.** $f(x) = \begin{cases} 0 & \text{if } x = 0 \\ \dfrac{1}{x} & \text{if } x \neq 0 \end{cases}$  $c = 0$

**b.** $f(x) = \begin{cases} 2 & \text{if } x = 3 \\ \dfrac{x^2 - 9}{x - 3} & \text{if } x \neq 3 \end{cases}$  $c = 3$

**c.** $f(x) = \dfrac{x^2 - 1}{x - 1}$,  $c = 1$

**d.** $f(x) = \begin{cases} 4 & \text{if } x = 2 \\ \dfrac{x^2 - 4}{x - 2} & \text{if } x \neq 2 \end{cases}$  $c = 2$

**Solution** **a.** Here, $f(0) = 0$ but $\lim_{x \to 0} f(x)$ does not exist. Thus the function $f$ is not continuous at 0.

**b.** $f(3) = 2$. So we check Property 2.

$$\lim_{x \to 3} f(x) = \lim_{x \to 3} \frac{x^2 - 9}{x - 3}$$

$$= \lim_{x \to 3} \frac{(x - 3)(x + 3)}{x - 3} = \lim_{x \to 3} (x + 3) = 6$$

Property 2 is satisfied. Since $6 \neq 2$, $\lim_{x \to 3} f(x) \neq f(3)$ and Property 3 is not satisfied. Thus the function $f$ is not continuous at 3.

**c.** Here, $f(1)$ is not defined. Thus, we need not check any further. The function $f$ is not continuous at 1.

**d.** We see that $f(2) = 4$. So Property 1 is satisfied. We now check Property 2.

$$\lim_{x \to 2} f(x) = \lim_{x \to 2} \frac{x^2 - 4}{x - 2}$$

$$= \lim_{x \to 2} \frac{(x - 2)(x + 2)}{x - 2} = \lim_{x \to 2} (x + 2) = 4$$

Property 2 is satisfied. We see that $\lim_{x \to 2} f(x) = f(2)$, since both quantities are equal to 4, and Property 3 is satisfied. Thus the function $f$ is continuous at 2. The discontinuities of the functions of Example 1 are illustrated in Figure 7.8. Note the jumps in the graphs of the functions of parts **a** and **b** and the hole in the graph of the function of part **c**.

**Do Exercises 1, 3, and 5.**

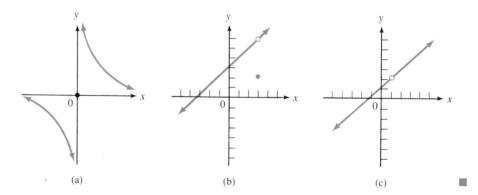

**FIGURE 7.8**

(a)          (b)          (c)

Most of the functions studied in Chapter 2 are continuous. However, if $h$ is a rational function and $h(x) = f(x)/g(x)$, where $f(x)$ and $g(x)$ are polynomial functions, then $h$ is not continuous at any $c$ for which $g(c) = 0$. However, it is continuous at every other point. In graphing a function, it is important to know whether it is continuous. The graph of a continuous function has no breaks, so it can be drawn without lifting the pencil.

**EXAMPLE 2**    Using Theorem 7.1, show that the polynomial $p(x) = 3x^2 + 5x - 4$ is continuous.

**Solution**    Let $a$ be an arbitrary real number. Then

$$\lim_{x \to a}(3x^2 + 5x - 4) = \lim_{x \to a} 3x^2 + \lim_{x \to a}(5x - 4) \quad \text{(by Property 3)}$$

$$= (\lim_{x \to a} 3x)(\lim_{x \to a} x) + \lim_{x \to a}(5x - 4) \quad \text{(by Property 5)}$$

$$= 3a \cdot a + (5 \cdot a - 4) \quad \text{(by Property 2)}$$

$$= 3a^2 + 5a - 4$$

Also,

$$p(a) = 3a^2 + 5a - 4$$

Therefore,

$$\lim_{x \to a}(3x^2 + 5x - 4) = p(a)$$

The polynomial is continuous at $a$. Since the real number $a$ was arbitrary, the polynomial is continuous everywhere. (See Exercises *29 and **30.)
**Do Exercise 23.**    ∎

## Applications of Continuity

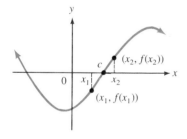

**FIGURE 7.9**

Recall that one of the ways to solve a polynomial equation $p(x) = 0$ is to graph the equation $y = p(x)$ and find the abscissa of each intersection of the graph with the $x$-axis. If it is found that for some real numbers $x_1$ and $x_2$, $f(x_1)$ and $f(x_2)$ have different signs, then $f(c) = 0$ for some $c$ between $x_1$ and $x_2$. A rigorous proof of this fact requires concepts not covered at this level. However, you may see the geometric reason for it. The points $(x, f(x_1))$ and $(x_2, f(x_2))$ are on opposite sides of the $x$-axis. Since the graph of a continuous function has no breaks, it must cross the $x$-axis between $x_1$ and $x_2$. (See Figure 7.9.)

**EXAMPLE 3**    Show that the equation $x^3 - x^2 - 11x + 3 = 0$ has a solution between 3 and 4. Then approximate that solution to one decimal place.

**Solution**    Let $p(x) = x^3 - x^2 - 11x + 3$. Then

$$p(3) = 3^3 - 3^2 - 11 \cdot 3 + 3 = -12 \quad \text{and}$$

$$p(4) = 4^3 - 4^2 - 11 \cdot 4 + 3 = 7$$

Since $p(3) < 0$ and $p(4) > 0$, the polynomial equation $p(x) = x^3 - x^2 - 11x + 3 = 0$ has a solution between 3 and 4. We find that $p(3.5) = -4.875$. Therefore, the solution is between 3.5 and 4. After several similar calculations, we find that $p(3.73) = -0.047783$ and $p(3.74) = 0.186024$. Since $p(3.73) < 0$ and $p(3.74) > 0$, the solution is between 3.73 and 3.74. Thus, 3.7 is a solution correct to one decimal place.
**Do Exercise 27.**    ∎

EXAMPLE 4    Sketch the graph of $y = f(x) = \dfrac{3}{x-2}$.

Solution    Observe that $f$ is a rational function which is continuous everywhere except at $x = 2$. So the only break of the graph is at $x = 2$. We calculate several values of $f(x)$ and obtain

| $x$ | $-2$ | $-1$ | $0$ | $1$ | $3$ | $4$ | $5$ | $6$ |
|-----|------|------|-----|-----|-----|-----|-----|-----|
| $f(x)$ | $\dfrac{-3}{4}$ | $-1$ | $\dfrac{-3}{2}$ | $-3$ | $3$ | $\dfrac{3}{2}$ | $1$ | $\dfrac{3}{4}$ |

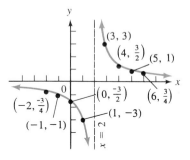

We plot the points $\left(-2, \frac{-3}{4}\right)$, $(-1, -1)$, $\left(0, \frac{-3}{2}\right)$, $(1, -3)$, $(3, 3)$, $\left(4, \frac{3}{2}\right)$, $(5, 1)$, and $\left(6, \frac{3}{4}\right)$.

Since we know the graph has only one break, we draw a smooth curve through the points that are to the left of $x = 2$, and another smooth curve through the points that are to the right of $x = 2$. (See Figure 7.10.)

**FIGURE 7.10**

**Do Exercise 25.**                                                                                                  ◼

## Left- and Right-Hand Limits

A good example of the next concept we wish to discuss is postage. Let $p(x)$ cents be the postage necessary to mail a letter weighing $x$ ounces. Then, if $0 < x \leq 1$, $p(x) = 25$, and if $1 < x \leq 2$, $p(x) = 45$. If $x$ is less than 1 but is sufficiently close to 1, the corresponding $p(x)$ is 25; however, if $x$ is sufficiently close to 1 but larger than 1, then the value of $p(x)$ is 45. We say that the left-hand limit of $p(x)$ as $x$ approaches 1 is 25, while the right-hand limit is 45. We write

$$\lim_{x \to 1^-} p(x) = 25 \quad \text{and} \quad \lim_{x \to 1^+} p(x) = 45$$

**DEFINITION 7.4:**   Let $c$ be a number in some interval $I$, and $f$ be a function defined on $I$ except possibly at $c$. We say that the *left-hand limit of $f$ at $c$ is $L$* and we write

$$\lim_{x \to c^-} f(x) = L$$

provided the following is true. The values of $f(x)$ are within a specified distance of $L$, as long as $x$ is sufficiently close to, but less than, $c$. The specified distance may be as small as we wish.

The right-hand limit is defined similarly.

Both the names of the limits and the notation have a geometric flavor. The first limit is called the left-hand limit because, when $x < c$, $x$ is to the left of $c$ on the number line. The second limit is called the right-hand limit because $x$ is to the right of $c$ whenever $x > c$.

Note that for a function $f$, $\lim_{x \to a} f(x)$ exists if, and only if, both $\lim_{x \to a^-} f(x)$ and $\lim_{x \to a^+} f(x)$ exist and are equal.

Recall that if $x$ is a real number, $[x]$ denotes the greatest integer that is less than or equal to $x$. This idea is used frequently in pricing merchandise. When an item is priced at $4.99, the merchant hopes that the customer thinks of the price as $4 ($[4.99] = 4$) rather than $5. We now consider another example of one-sided limits.

**EXAMPLE 5**   Sketch the graph of $f(x) = \left[\frac{x}{2}\right]$ for $-4 \le x \le 4$.

a. Find $\lim_{x \to -2^-} f(x)$.

b. $\lim_{x \to -2^+} f(x)$.

**Solution**   Note that $f(-4) = \left[\frac{-4}{2}\right] = [-2] = -2, f(-3.5)$
$= \left[\frac{-3.5}{2}\right] = [-1.75] = -2, f(-3) = \left[\frac{-3}{2}\right]$
$= [-1.5] = -2, f(-2) = \left[\frac{-2}{2}\right] = [-1] = -1$, and so on.
The graph is illustrated in Figure 7.11.

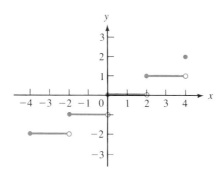

**FIGURE 7.11**

From the graph, we see that if $x$ is close to, but less than, $-2$, $f(x) = -2$. Thus $\lim_{x \to 2^-} f(x) = -2$. On the other hand, if $x$ is larger than, but close to, $-2$, then $f(x) = -1$. Thus, $\lim_{x \to -2^+} f(x) = -1$.

**Do Exercise 11.**                                                                                    ∎

**EXAMPLE 6**   For the function $f$, whose graph is sketched in Figure 7.12, find the following.

a. $\lim_{x \to 2^-} f(x)$

b. $\lim_{x \to 2^+} f(x)$

c. $f(2)$

**Solution**   a. Choose $x$ to the left of 2, but very close to 2, on the $x$-axis. From that point, draw a perpendicular to the $x$-axis. This perpendicular intersects the graph at the point

**FIGURE 7.12**

$(x, f(x))$. From this point, draw a perpendicular to the y-axis. This perpendicular intersects the y-axis at the value of $f(x)$ which we see is near 3. Thus
$$\lim_{x \to 2^-} f(x) = 3.$$

**b.** Similarly, starting with x near 2 but to the right of 2, we see that $\lim_{x \to 2^+} f(x) = 5$.

**c.** Clearly, $f(2) = 4$.

Note that for the function of this example, $\lim_{x \to 2} f(x)$ does not exist since the left-hand and right-hand limits are not equal. Therefore, this function is not continuous at 2.

**Do Exercise 19.**

■

---

# Exercise Set 7.3

*In Exercises 1–10, discuss the continuity of the function at the given point c.*

**1.** $f(x) = \dfrac{x^2 - 16}{x + 4}$,   $c = -4$

**2.** $f(x) = \begin{cases} 3 & \text{if } x = -5 \\ \dfrac{x^2 - 25}{x + 5} & \text{if } x \neq -5 \end{cases}$   $c = -5$

**3.** $g(t) = \begin{cases} 2 & \text{if } t = 1 \\ \dfrac{t^2 + t - 2}{t^2 - 3t + 2} & \text{if } t \neq 1 \end{cases}$   $c = 1$

**4.** $h(u) = \begin{cases} u + 1 & \text{if } u < 1 \\ u^2 + 2 & \text{if } u \geq 1 \end{cases}$   $c = 1$

**5.** $f(x) = \begin{cases} \dfrac{9}{7} & \text{if } x = 2 \\ \dfrac{x^2 + 5x - 14}{x^2 + 3x - 10} & \text{if } x \neq 2 \end{cases}$   $c = 2$

**6.** $g(x) = \begin{cases} 10 & \text{if } x = 2 \\ \dfrac{x^3 - 8}{x - 2} & \text{if } x \neq 2 \end{cases}$   $c = 2$

**7.** $h(s) = \begin{cases} 0 & \text{if } s = 9 \\ \dfrac{\sqrt{s} - 3}{s - 9} & \text{if } s \neq 9 \end{cases}$   $c = 9$

**8.** $f(t) = \begin{cases} 10 & \text{if } t = 5 \\ \dfrac{1}{t - 5} & \text{if } t \neq 5 \end{cases}$   $c = 5$

**9.** $g(t) = \begin{cases} 5 & \text{if } t = -2 \\ \dfrac{t^2 + 3t + 2}{t^2 + 4t + 4} & \text{if } t \neq -2 \end{cases}$   $c = -2$

**10.** $h(x) = \begin{cases} \pi & \text{if } x = 4 \\ \dfrac{\sqrt{x} - 2}{3 - \sqrt{x + 5}} & \text{if } x \neq 4 \end{cases}$   $c = 4$

*In Exercises 11–17, sketch the graphs of the functions over the given intervals and find both the left-hand and right-hand limits at the indicated point c.*

**11.** $f(x) = \left[\dfrac{x}{3}\right]$, $-6 \leq x \leq 6$, $c = -3$

**12.** $g(x) = \left[\dfrac{x}{4}\right]$, $-8 \leq x \leq 8$, $c = 4$

**13.** $h(t) = [t^2]$, $-3 \leq t \leq 3$, $c = -2$

**14.** $f(x) = \begin{cases} x + 1 & \text{if } x < 1 \\ 2x + 3 & \text{if } x > 1 \end{cases}$   $-1 \leq x \leq 3$, $c = 1$

**15.** $g(x) = \begin{cases} 2x - 1 & \text{if } x < 2 \\ 3x - 3 & \text{if } x > 2 \end{cases}$   $-1 \leq x \leq 5$, $c = 2$

**16.** $h(u) = \begin{cases} [u] & \text{if } u < 3 \\ 5 & \text{if } u = 3 \\ u^2 + 1 & \text{if } u > 3 \end{cases}$   $-1 \leq u \leq 5$, $c = 3$

**17.** $f(x) = \begin{cases} \dfrac{\sqrt{x} - 3}{x - 9} & \text{if } x < 9 \\ 3 & \text{if } x = 9 \\ x - 2 & \text{if } x > 9 \end{cases}$   $1 \leq x \leq 12$, $c = 9$

*In Exercises 18–21, the graph of a function f is sketched. Find* **a.** $\lim\limits_{x \to c^-} f(x)$, **b.** $\lim\limits_{x \to c^+} f(x)$, *and* **c.** $f(c)$ *for the given value of c.*

**18.**

$c = 3$

**19.**

$c = -2$

**20.**

$c = 2$

**21.**

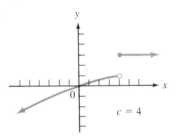

$c = 4$

**22.** Let $p(x) = 2x^3 + 5x^2 - 6x + 3$. Use the properties of limits to show that $p$ is continuous.

**23.** Same as Exercise 22 for $p(x) = -3x^4 + 3x^2 - 8x + 4$.

**24.** Same as Exercise 22 for $p(x) = 5x^5 + 2x^3 + 6x^2 - 5x + 9$.

**25.** Sketch the graph of $y = \dfrac{2}{x + 3}$.

**26.** Sketch the graph of $y = \dfrac{-4}{x - 1}$.

**27.** Show that $x^3 - 4x^2 - 5x + 14 = 0$ has a solution between 4 and 5. Approximate that solution to one decimal place.

**28.** Show that $2x^3 - 5x^2 - 6x + 4 = 0$ has a solution between $-2$ and $-1$. Approximate that solution to one decimal place.

**\*29.** Use the properites of limits to prove that any polynomial function is continuous. (*Hint:* Start with $p(x) = a_0 x^n + a_1 x^{n-1} + \cdots + a_{n-1}x + a_n$ and use the properties of limits to show that $\lim\limits_{x \to a}(a_0 x^n + a_1 x^{n-1} + \cdots + a_{n-1}x + a_n) = p(a)$.)

**\*\*30.** Use mathematical induction to give another proof of the continuity of polynomials. (*Hint:* If $p(x)$ is a polynomial of degree $n > 1$, by the Factor Theorem, there is a polynomial $q(x)$ of degree $n - 1$ such that $p(x) - p(a) = (x - a)q(x)$. Use this and the induction hypothesis to show that $\lim\limits_{x \to a}(p(x) - p(a)) = 0$, and so on.)

# 7.4 The Derivative

We are now ready to discuss one of the most important concepts in calculus—the derivative. We will spend the remaining part of the chapter describing how to find derivatives of the elementary functions presented thus far. In the next chapter, we will present basic applications of derivatives.

In Sections 7.1 and 7.2, we discussed the concepts of rate of change and limits. In the applications, it is often useful to consider the limit of a rate of change of a function with respect to its independent variable, as the increment in the independent variable approaches 0. The instantaneous velocity of a moving object could serve as one example of a derivative. However, we again choose to use the concept of a tangent line (see Section 7.2) to introduce the definition because this geometric approach may help you visualize, and therefore understand, one of the most important definitions of elementary calculus.

## Slope of a Tangent Line

Suppose we were asked to draw a line tangent to the graph of a function $f$ at the point $(a, f(a))$. Using a ruler, we would draw a line through the points $P$ and $Q$ with coordinates $(a, f(a))$ and $(a + h, f(a + h))$, respectively. (See Figure 7.13.) We would then rotate the ruler about the point $P$, making sure that the point $Q$ would approach the point $P$. Note that as we do this, the values of $h$ approach 0 and the slope $m_h$ of each secant line through $(a, f(a))$ and $(a + h, f(a + h))$ is

$$m_h = \frac{f(a + h) - f(a)}{(a + h) - a} = \frac{f(a + h) - f(a)}{h}$$

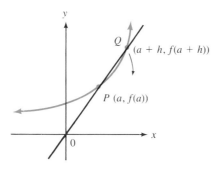

**FIGURE 7.13**

If it was possible to actually do so, we would stop the rotation of the ruler at the exact instant where $(a + h, f(a + h))$ coincides with $(a, f(a))$. Since, as we rotate the ruler very slightly, the slope of the secant line does not change very much, we would expect that the values of $\frac{f(a + h) - f(a)}{h}$, for $h$ sufficiently close to 0, are good approximations of the slope of the tangent line.

**DEFINITION 7.5:**   The *slope of the tangent line* at $(a, f(a))$ is

$$\lim_{h \to 0} \frac{f(a + h) - f(a)}{h}$$

provided this limit exists.

## The Derivative

As we shall see later, many concepts in the physical and social sciences involve limits such as the one defining the slope of the tangent line. It is therefore useful to give this limit a special name and study it.

**DEFINITION 7.6:**  Suppose $f$ is a function defined on an open interval containing the point $x$ and $\lim\limits_{h \to 0} \dfrac{f(x + h) - f(x)}{h}$ exists. Then the value of that limit is called the *derivative of $f$ at $x$* and is denoted $f'(x)$. When $f'(x)$ exists, we say that *$f$ is differentiable at $x$*. If a function $f$ is differentiable at each point of its domain, we simply say that *$f$ is differentiable*.

Note that given a function $f$, for each fixed value of $x$ in the domain of $f$ at which $f$ is differentiable, we have a unique value $f'(x)$ of the derivative. Hence, $f'$ is a function whose domain is the following subset of the domain of $f$:

$$\{x \mid f \text{ is differentiable at } x\}$$

The process of finding the derivative of a function is called *differentiation*. When we perform this operation, it is important to keep in mind which of the variables is the independent variable. We will always state that we differentiate with respect to $x$, if $x$ is the independent variable, or with respect to $t$, if $t$ is the independent variable. Often, we shall use the capital letter $D$ to denote the operation of differentiation, and a subscript to indicate the independent variable. For example $D_u$ and $D_x$ will be used to denote differentiation with respect to $u$ and with respect to $x$, respectively.

If $f$ is a differentiable function and $y = f(x)$, then each of the symbols $f'(x)$, $y'$, $\dfrac{dy}{dx}$, $\dfrac{df(x)}{dx}$, $D_x y$, and $D_x f(x)$ denotes the derivative of $f$ with respect to $x$.*

## Calculation of the Derivative

Although we shall soon learn formulas to facilitate finding derivatives, it is important that you practice finding derivatives using the definition.

Given a function $f$, the basic steps to calculate $f'(a)$ are the following:

**STEP 1.**  Calculate $f(a + h)$.

**STEP 2.**  Find $f(a)$.

**STEP 3.**  Find $f(a + h) - f(a)$.

**STEP 4.**  Calculate $\dfrac{f(a + h) - f(a)}{h}$ and simplify.

(We may assume that $h \neq 0$ since we will take the limit as $h$ tends to 0; hence, we will not allow $h = 0$.)

**STEP 5.**  Evaluate $\lim\limits_{h \to 0} \dfrac{f(a + h) - f(a)}{h}$. If this limit exists, its value is $f'(a)$.

---

*The notation $\dfrac{dy}{dx}$ for the derivative of $y = f(x)$ was introduced by Gottfried Wilhelm Leibniz (1646–1716).

EXAMPLE 1    Let $f(x) = x^2 + 2x + 3$. Find $f'(2)$.

Solution    **Step 1.** $f(2 + h) = (2 + h)^2 + 2(2 + h) + 3 = 4 + 4h + h^2 + 4 + 2h + 3 = 11 + 6h + h^2$

**Step 2.** $f(2) = 2^2 + 2(2) + 3 = 11$

**Step 3.** $f(2 + h) - f(2) = (11 + 6h + h^2) - 11$

$$= 6h + h^2 = h(6 + h)$$

**Step 4.** $\dfrac{f(2 + h) - f(2)}{h} = \dfrac{h(6 + h)}{h} = 6 + h$    (since $h \neq 0$)

**Step 5.** $\lim\limits_{h \to 0} \dfrac{f(2 + h) - f(2)}{h} = \lim\limits_{h \to 0} (6 + h) = 6$

Therefore $f'(2) = 6$.

**Do Exercise 3.** ∎

EXAMPLE 2    Let $g(x) = \sqrt[3]{x}$.

**a.** Find $g'(x)$.
**b.** Find the domain of $g'$.

Solution    **a.** With practice, you will be able to condense the five steps as follows:

$$g'(x) = \lim_{h \to 0} \frac{g(x + h) - g(x)}{h}$$

$$= \lim_{h \to 0} \frac{\sqrt[3]{x + h} - \sqrt[3]{x}}{h}$$

If we tried to evaluate this limit using our intuition, we would obtain an indeterminate form. Therefore, we must rationalize the numerator of the fraction. Since the radicals are of order 3, we use the fact that

$$a^3 - b^3 = (a - b)(a^2 + ab + b^2)$$

Thus, we multiply the numerator and denominator of the fraction by $[(\sqrt[3]{x + h})^2 + (\sqrt[3]{x + h})(\sqrt[3]{x}) + (\sqrt[3]{x})^2]$. We obtain

$$\frac{\sqrt[3]{x + h} - \sqrt[3]{x}}{h} = \frac{(\sqrt[3]{x + h} - \sqrt[3]{x})\,[(\sqrt[3]{x + h})^2 + (\sqrt[3]{x + h})(\sqrt[3]{x}) + (\sqrt[3]{x})^2]}{h[(\sqrt[3]{x + h})^2 + (\sqrt[3]{x + h})(\sqrt[3]{x}) + (\sqrt[3]{x})^2]}$$

$$= \frac{(\sqrt[3]{x + h})^3 - (\sqrt[3]{x})^3}{h[(\sqrt[3]{x + h})^2 + (\sqrt[3]{x + h})(\sqrt[3]{x}) + (\sqrt[3]{x})^2]}$$

$$= \frac{h}{h[(\sqrt[3]{x + h})^2 + (\sqrt[3]{x + h})(\sqrt[3]{x}) + (\sqrt[3]{x})^2]}$$

$$= \frac{1}{(\sqrt[3]{x + h})^2 + (\sqrt[3]{x + h})(\sqrt[3]{x}) + (\sqrt[3]{x})^2} \quad \text{(since } h \neq 0\text{)}$$

Thus,

$$= \lim_{h \to 0} \frac{\sqrt[3]{x+h} - \sqrt[3]{x}}{h} = \lim_{h \to 0} \frac{1}{(\sqrt[3]{x+h})^2 + (\sqrt[3]{x+h})(\sqrt[3]{x}) + (\sqrt[3]{x})^2}$$

$$= \frac{1}{(\sqrt[3]{x})^2 + (\sqrt[3]{x})^2 + (\sqrt[3]{x})^2}$$

$$= \frac{1}{3(\sqrt[3]{x})^2} = \frac{1}{3x^{2/3}}$$

and

$$g'(x) = \frac{1}{3x^{2/3}}$$

**b.** Since this last fraction is defined for all values of $x$ except 0, the domain of $g'$ is the set of all real numbers except 0.

**Do Exercise 13.**

## Applications of Differentiation

**EXAMPLE 3**   Let $y = f(x) = x^2 + 1$.

**a.** Verify that $(-2, 5)$ is on the graph of $f$.
**b.** Find the slope of the tangent line to the graph of $f$ at $(-2, 5)$.
**c.** Find the equation of the tangent line.

**Solution**   **a.** When $x = -2$, $y = f(-2) = (-2)^2 + 1 = 4 + 1 = 5$. Thus $(-2, 5)$ is on the graph of $f$.

**b.** To find the slope of the tangent line, we first find the slope $m_h$ of the line through $(-2, 5)$ and $(-2 + h, f(-2 + h))$ for any real number $h \neq 0$. We proceed as follows.

$$m_h = \frac{f(-2 + h) - f(-2)}{(-2 + h) - (-2)}$$

$$= \frac{[(-2 + h)^2 + 1] - 5}{h}$$

$$= \frac{4 - 4h + h^2 + 1 - 5}{h} = \frac{-4h + h^2}{h}$$

$$= -4 + h \quad (\text{since } h \neq 0)$$

Thus the slope of the tangent line is $\lim_{h \to 0} (-4 + h)$. That is, $m = -4$.

**c.** Using the point-slope formula of a line, we find that the equation of the tangent line is

$$y - 5 = -4[x - (-2)]$$

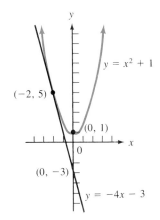

**FIGURE 7.14**

This equation can be written

$$y = -4x - 3$$

The graphs of the function $f$ and the tangent line are illustrated in Figure 7.14.
**Do Exercise 17.** ■

Note that in the preceding example, the steps taken to find the slope of the tangent line at $(-2, 5)$ were identical to those we would have taken to find $f'(-2)$. Thus, the slope of the tangent line at $(-2, f(-2))$ is $f'(-2)$.

**EXAMPLE 4**    A particle is moving on a directed line. Let $s = f(t) = 16t^2 + 128t + 50$, where $s$ feet is the directed distance of the particle from the origin at instant $t$ seconds. Find the instantaneous velocity of the particle at instant $t = 3$ seconds.

**Solution**    We first find the average velocity from instant $t = 3$ seconds to instant $t = 3 + h$ seconds, as at the beginning of Section 7.2.
Here, $\Delta t = t_2 - t_1 = (3 + h) - 3 = h$. Also,

$$\begin{aligned}
\Delta s = s_2 - s_1 &= f(3 + h) - f(3) \\
&= [16(3 + h)^2 + 128(3 + h) + 50] - [16(3^2) + 128(3) + 50] \\
&= 224h + 16h^2 = 16h(14 + h)
\end{aligned}$$

Therefore, the average velocity from instant 3 seconds to instant $(3 + h)$ seconds is $16(14 + h)$ ft/sec since

$$\begin{aligned}
\frac{\Delta s}{\Delta t} &= \frac{16h(14 + h)}{h} \\
&= 16(14 + h) \quad \text{(provided } h \neq 0)
\end{aligned}$$

The instantaneous velocity of the moving particle at instant $t = 3$ seconds is 224 ft/sec since $\lim_{h \to 0} 16(14 + h) = 224$.
Again, note that this limit is in fact the value of $f'(3)$.
**Do Exercise 21.** ■

## Continuity and Differentiability

If $f(x) = |x|$, then $\lim_{x \to 0} f(x) = f(0)$. Therefore $f$ is continuous at 0. However, it is not differentiable at 0. To see this, we find

$$\begin{aligned}
\lim_{h \to 0^-} \frac{f(0 + h) - f(0)}{h} &= \lim_{h \to 0^-} \frac{|h| - 0}{h} \\
&= \lim_{h \to 0^-} \frac{-h}{h} \quad \text{(since } h < 0) \\
&= \lim_{h \to 0^-} (-1) = -1
\end{aligned}$$

On the other hand,

$$\lim_{h \to 0^+} \frac{f(0 + h) - f(0)}{h} = \lim_{h \to 0^+} \frac{|h| - 0}{h}$$

$$= \lim_{h \to 0^+} \frac{h}{h} \quad \text{(since } h > 0\text{)}$$

$$= \lim_{h \to 0^+} 1 = 1$$

So

$$\lim_{h \to 0^-} \frac{f(0 + h) - f(0)}{h} \neq \lim_{h \to 0^+} \frac{f(0 + h) - f(0)}{h}$$

and $f'(0)$ does not exist. (See Figure 7.15.)

See Exercises 27–30 for other examples of functions that are continuous at $c$, but not differentiable at $c$.

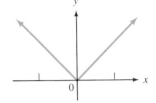

**FIGURE 7.15**

---

THEOREM 7.2:   If a function is differentiable at $c$, then it is continuous at $c$.

---

***Proof (Optional):*** In the definition of the derivative, $f'(c) = \lim\limits_{h \to 0} \frac{f(c + h) - f(c)}{h}$, let $c + h = x$. Then, $x \to c$ as $h \to 0$, $h = x - c$, and the definition may be written

$$f'(c) = \lim_{x \to c} \frac{f(x) - f(c)}{x - c}$$

If $x \neq c$, we write

$$f(x) - f(c) = (x - c)\frac{f(x) - f(c)}{x - c}$$

so that

$$\lim_{x \to c}\left[f(x) - f(c)\right] = \lim_{x \to c}\left[(x - c)\frac{f(x) - f(c)}{x - c}\right]$$

$$= \lim_{x \to c}(x - c)\lim_{x \to c}\frac{f(x) - f(c)}{x - c}$$

$$= 0 \cdot f'(c) \quad \text{(since the derivative exists)}$$

$$= 0$$

Thus,

$$\lim_{x \to c} f(x) = \lim_{x \to c}[f(x) - f(c) + f(c)]$$

$$= \lim_{x \to c}[f(x) - f(c)] + \lim_{x \to c} f(c) = 0 + f(c) = f(c)$$

We have shown that

$$\lim_{x \to c} f(x) = f(c)$$

Thus $f$ is continuous at $c$.

Theorem 7.2 tells us that if a function is differentiable at $c$, then it *must* be continuous at $c$. Equivalently, it tells us that if a function is not continuous at $c$, then it *cannot* be differentiable at $c$. For example, the function $f(x) = \frac{x}{x-3}$ is not continuous at 3 (it is not defined at 3), so it is not differentiable at 3. Also, the function $f(x) = \left[\frac{x}{2}\right]$ of Example 5, Section 7.3, is defined but not continuous at $-2$. So it is not differentiable at $-2$. However, if a function is continuous at $c$, it may or may not be differentiable at $c$. We have seen in Section 7.3 that polynomial functions are continuous. (See also Exercises 29 and 30, Section 7.3.) In fact, they are differentiable.

> **THEOREM 7.3:**   If $f$ is a polynomial function, then $f$ is differentiable.

*Proof (Optional):* Let $f$ be a polynomial function, $c$ be an arbitrary real number, and $g(x) = f(x) - f(c)$. Then, $g(x)$ is a polynomial and $g(c) = 0$. By the Factor Theorem, $x - c$ is a factor of $g(x)$. So, $f(x) - f(c) = (x - c)h(x)$ for some polynomial $h(x)$. Therefore,

$$\lim_{x \to c} \frac{f(x) - f(c)}{x - c} = \lim_{x \to c} \frac{(x - c)h(x)}{x - c}$$

$$= \lim_{x \to c} h(x) \quad \text{(since } x \neq c\text{)}$$

$$= h(c) \quad \text{(since } h \text{ is continuous)}$$

We have shown that $\lim\limits_{x \to c} \dfrac{f(x) - f(c)}{x - c}$ exists. Since $c$ was arbitrary, $f$ is differentiable.*

# Exercise Set 7.4

*In Exercises 1–7, find $f'(c)$ for the given value of $c$.*

**1.** $f(x) = 3x + 2$, $c = 2$

**2.** $f(x) = x^2 + 7x - 10$, $c = -3$

**3.** $f(x) = x^3 + x + 1$, $c = 1$

**4.** $f(x) = \dfrac{2}{x}$, $c = 2$

**5.** $f(x) = \sqrt{x}$, $c = 4$

**6.** $f(x) = \sqrt[3]{x + 1}$, $c = 7$

**7.** $f(x) = \dfrac{3}{\sqrt{x + 1}}$, $c = 8$

*In Exercises 8–15, find $\dfrac{dy}{dx}$.*

**8.** $y = f(x) = 6$

**9.** $y = g(x) = 5x + 3$

**10.** $y = h(x) = -3x + 2$

**11.** $y = f(x) = x^2 + 5x + 13$

**12.** $y = g(x) = -3x^2 + 6x + 1$

**13.** $y = h(x) = \sqrt{3x + 1}$

**14.** $y = f(x) = \sqrt[3]{x^4}$

**15.** $y = g(x) = \dfrac{2}{\sqrt{x + 5}}$

*In Exercises 16–20, a function and a point are given.*

   *a. Verify that the point is on the graph of the function.*
   *b. Sketch the graph of the function.*
   *c. Find the slope of the tangent line at the given point.*

---

*See Section 7.5 for a simpler proof of this theorem. See also Exercises 41 and 42, Exercise Set 7.5.

*d.* Find the equation of the tangent line and draw the tangent line on the graph sketched in part *b*.

**16.** $f(x) = x^2 + 3x + 1$, $(-1, -1)$

**17.** $f(x) = x^2 + 5x - 3$, $(2, 11)$

**18.** $f(x) = -\sqrt{x}$, $(4, -2)$

**19.** $g(x) = \sqrt[3]{x^2 + 4}$, $(2, 2)$

**20.** $h(x) = \dfrac{6}{\sqrt{x + 6}}$, $(3, 2)$

*In Exercises 21–26, a particle is moving along a directed line. At instant t seconds its directed distance from the origin is s feet. Find the instantaneous velocity at the given instant.*

**21.** $s = t^2 + 3t + 7$, $t = 4$ seconds

**22.** $s = -t^2 + 5t + 8$, $t = 3$ seconds

**23.** $s = 16t^2 + 64t + 70$, $t = 2$ seconds

**24.** $s = 4t^3 + 3t^2 + 4t + 2$, $t = 2$ seconds

**25.** $s = \dfrac{120}{\sqrt{t + 5}}$, $t = 4$ seconds

**26.** $s = \sqrt[3]{3t + 12}$, $t = 5$ seconds

*In Exercises 27–30, a function f and a point c are given.*

*a.* Find $\lim\limits_{x \to c^-} f(x)$.

*b.* Find $\lim\limits_{x \to c^+} f(x)$.

*c.* Find $f(c)$.

*d.* Is $f$ continuous at $c$?

*e.* Find $\lim\limits_{h \to 0^-} \dfrac{f(c + h) - f(c)}{h}$.

*f.* Find $\lim\limits_{h \to 0^+} \dfrac{f(c + h) - f(c)}{h}$.

*g.* Is $f$ differentiable at $c$?

**\*27.** $f(x) = \begin{cases} x - 2 & \text{if } x \le 2 \\ 2 - x & \text{if } x > 2 \end{cases}$   $c = 2$

**\*28.** $f(x) = \begin{cases} x^2 + 5x - 3 & \text{if } x \le 1 \\ -x^2 + x + 3 & \text{if } x > 1 \end{cases}$   $c = 1$

**\*29.** $f(x) = \begin{cases} x^3 + x + 1 & \text{if } x \le 2 \\ x^2 + 3x + 1 & \text{if } x > 2 \end{cases}$   $c = 2$

**\*30.** $f(x) = \begin{cases} x^2 - 5x - 3 & \text{if } x \le 6 \\ \sqrt{x + 3} & \text{if } x > 6 \end{cases}$   $c = 6$

## 7.5  Basic Differentiation Formulas

Even in simple cases, the process of finding the derivative from the definition can be tedious. To facilitate our work, we now derive five basic formulas to help us find derivatives of many functions in a straightforward way.

### Some Basic Differentiation Formulas

The following theorem is stated, but the proof will be given only for some special cases at the end of the section.

> **THEOREM 7.4:**   If $c$ and $r$ are any real numbers, and $f$ and $g$ are differentiable functions, then
>
> 1. $D_x c = 0$
> 2. $D_x(f(x) + g(x)) = f'(x) + g'(x)$
> 3. $D_x(f(x) - g(x)) = f'(x) - g'(x)$
> 4. $D_x c f(x) = c f'(x)$
> 5. $D_x x^r = r x^{r-1}$   \*

---

\*In the statement of the theorem, we used the $D_x$ notation on the left side of each formula and the prime notation on the right side of some of the formulas. This should not cause any difficulty once you familiarize yourself with the different notations for derivatives.

It will be useful for you to learn these formulas in words. That is,

1. The derivative of a constant is 0.
2. and 3. The derivative of the sum (difference) of two differentiable functions is the sum (difference) of their derivatives.
4. The derivative of the product of a constant by a differentiable function is the product of the constant by the derivative of the function.
5. The derivative, with respect to $x$, of the $r$th power of $x$ is the product of $r$ by the $(r - 1)$th power of $x$.

Following are several examples to illustrate the use of this theorem. We justify some of the steps by referring to which formula of Theorem 7.4 we are using.

**EXAMPLE 1**    Find $f'(x)$ for each of the following functions.

**a.** $f(x) = 8$
**b.** $f(x) = 3x^5$
**c.** $f(x) = \sqrt[3]{x^2}$
**d.** $f(x) = 2x^2 + 5x + 1$
**e.** $f(x) = \dfrac{x^2 + 2}{\sqrt{x}}$

**Solution**    **a.** $f'(x) = D_x 8 = 0$    (by Formula 1; the derivative of a constant is 0)

**b.** $f'(x) = D_x 3x^5$

$\qquad = 3D_x x^5$    (by Formula 4)

$\qquad = 3(5x^{5-1})$    (by Formula 5)

$\qquad = 15x^4$

**c.** $f'(x) = D_x \sqrt[3]{x^2} = D_x x^{2/3}$

$\qquad = \dfrac{2}{3} x^{2/3 - 1}$    (by Formula 5)

$\qquad = \dfrac{2}{3} x^{-1/3}$

$\qquad = \dfrac{2}{3x^{1/3}}$

**d.** $f'(x) = D_x(2x^2 + 5x + 1)$

$\qquad = D_x(2x^2) + D_x(5x) + D_x 1$    (by Formula 2)

$\qquad = 2D_x x^2 + 5D_x x + 0$    (by Formulas 4 and 1)

$\qquad = 2(2x^{2-1}) + 5(1x^{1-1})$    (by Formula 5)

$\qquad = 4x^1 + 5x^0 = 4x + 5$

**e.** $f'(x) = D_x \dfrac{x^2 + 2}{\sqrt{x}} = D_x \dfrac{x^2 + 2}{x^{1/2}}$

$\qquad = D_x(x^{3/2} + 2x^{-1/2})$

$\qquad = D_x x^{3/2} + D_x(2x^{-1/2})$    (by Formula 2)

$\qquad = D_x x^{3/2} + 2D_x x^{-1/2}$    (by Formula 4)

$$= \frac{3}{2}x^{3/2-1} + 2\left(\frac{-1}{2}x^{-1/2-1}\right) \quad \text{(by Formula 5)}$$

$$= \frac{3}{2}x^{1/2} - x^{-3/2} = \frac{3}{2}x^{1/2} - \frac{1}{x^{3/2}}$$

**Do Exercises 1, 5, and 9.**

Note that in parts **c** and **e** of the preceding example, we changed the form of the expressions defining the function in order to use the differentiation formulas.

With practice, you will be able to condense the solutions of differentiation problems to a small number of steps, as illustrated in Example 2.

**EXAMPLE 2**   Find $f'(x)$ for the following functions.

**a.** $f(x) = 4x^3 + 5x^2 - 10x + 6$

**b.** $f(x) = \dfrac{2x^2 + 3x - 2}{\sqrt[3]{x}}$

**Solution**   **a.** $f'(x) = D_x(4x^3 + 5x^2 - 10x + 6)$

$$= 4(3x^2) + 5(2x) - 10(1) + 0 = 12x^2 + 10x - 10$$

**b.** $f'(x) = D_x\left(\dfrac{2x^2 + 3x - 2}{\sqrt[3]{x}}\right)$

$$= D_x\left(2x^{5/3} + 3x^{2/3} - 2x^{-1/3}\right)$$

$$= 2\left(\frac{5}{3}x^{5/3-1}\right) + 3\left(\frac{2}{3}x^{2/3-1}\right) - 2\left(\frac{-1}{3}x^{-1/3-1}\right)$$

$$= \frac{10}{3}x^{2/3} + 2x^{-1/3} + \frac{2}{3}x^{-4/3} = \frac{10}{3}x^{2/3} + \frac{2}{x^{1/3}} + \frac{2}{3x^{4/3}}$$

**Do Exercise 15.**

## Applications

**EXAMPLE 3**   Find the equation of the line tangent to the graph of the function $f$ at $(3, 11)$, where $y = f(x) = 2x^2 - 3x + 2$.

**Solution**   We use the fact that the slope of the tangent line to the graph of a differentiable function is the value of the derivative at the point of tangency. Thus, we first find $f'(x)$.

$$f'(x) = D_x(2x^2 - 3x + 2)$$

$$= D_x(2x^2) - D_x(3x) + D_x 2 = 2D_x x^2 - 3D_x x + 0$$

$$= 2(2x^{2-1}) - 3(1x^{1-1}) + 0 = 4x^1 - 3x^0 = 4x - 3$$

To find the slope of the tangent line at $(3, 11)$, we evaluate $f'(x)$ at the point of tangency; that is, when $x = 3$. Thus, $m = f'(3) = 4(3) - 3 = 9$. Using the point-slope form of the equation of a line, we find that the equation of the tangent line is $y - 11 = 9(x - 3)$, which can be written

$$y = 9x - 16$$

**Do Exercise 19.**

When we wish to find the slope of the tangent line to the graph of a function $f$ at the point $(a, f(a))$, we must be sure to evaluate the derivative $f'$ at $a$. Sometimes students wrongly assume that the slope is $f'(x)$ and obtain $y - f(a) = f'(x)(x - a)$ for the equation of the tangent line. A moment of reflection should convince you that, in general, this equation cannot be the equation of the tangent line since it is not linear!

**EXAMPLE 4**   Find the instantaneous velocity of the particle of Example 4 of the preceding section at instant $t = 3$ seconds.

**Solution**   Recall that $s = f(t) = 16t^2 + 128t + 50$. We shall use the fact that the instantaneous velocity is obtained by evaluating $f'(t)$ at the given instant. Note that the independent variable is $t$; thus, we differentiate with respect to $t$.

$$f'(t) = D_t(16t^2 + 128t + 50)$$
$$= D_t(16t^2) + D_t(128t) + D_t 50$$
$$= 16D_t t^2 + 128D_t t + 0$$
$$= 16(2t^{2-1}) + 128(1t^{1-1})$$
$$= 32t^1 + 128t^0 = 32t + 128$$

Therefore, the velocity at instant $t = 3$ seconds is 224 ft/sec since $f'(3) = 32(3) + 128 = 224$.

**Do Exercise 23.**

Compare this solution to that of Example 4, Section 7.4. For an application of differentiation to be discussed in greater detail in Chapter 8, see Exercises 28 and 29.

## Partial Proof of Theorem 7.4 (Optional)

**A CLOSER LOOK**   You should refer to Theorem 7.1 in Section 7.2 for the justification of some of the steps.

1. Here, we let $K(x) = c$. Then

$$D_x c = K'(x)$$

$$= \lim_{h \to 0} \frac{K(x + h) - K(x)}{h} = \lim_{h \to 0} \frac{c - c}{h}$$

$$= \lim_{h \to 0} \frac{0}{h} = 0$$

2. $D_x(f(x) + g(x)) = \lim\limits_{h \to 0} \dfrac{[f(x + h) + g(x + h)] - [f(x) + g(x)]}{h}$

$$= \lim\limits_{h \to 0}\left[\dfrac{f(x + h) - f(x)}{h} + \dfrac{g(x + h) - g(x)}{h}\right]$$

$$= \lim\limits_{h \to 0} \dfrac{f(x + h) - f(x)}{h} + \lim\limits_{h \to 0} \dfrac{g(x + h) - g(x)}{h}$$

$$= f'(x) + g'(x)$$

3. The proof is entirely analogous to that of Formula 2.

4. $D_x cf(x) = \lim\limits_{h \to 0} \dfrac{cf(x + h) - cf(x)}{h}$

$$= \lim\limits_{h \to 0} c\dfrac{f(x + h) - f(x)}{h}$$

$$= (\lim\limits_{h \to 0} c) \cdot \lim\limits_{h \to 0}\dfrac{f(x + h) - f(x)}{h} = cf'(x)$$

5. The proof where $r$ is a positive integer is not difficult using the formula

$$a^n - b^n = (a - b)(a^{n-1} + a^{n-2}b + a^{n-3}b^2 + \cdots + ab^{n-2} + b^{n-1})$$

The proof where $r$ is a rational number will be given in Section 7.8 and the proof where $r$ is any real constant will be given in Section 7.10. Here, we give the proof for the special case $r = 4$. The general case where $r$ is an arbitrary positive integer is analogous. Let $p(x) = x^4$. Then

$$D_x x^4 = \lim\limits_{h \to 0} \dfrac{(x + h)^4 - x^4}{h}$$

$$= \lim\limits_{h \to 0} \dfrac{[(x + h) - x][(x + h)^3 + (x + h)^2 x + (x + h)x^2 + x^3]}{h}$$

$$= \lim\limits_{h \to 0} \dfrac{h[(x + h)^3 + (x + h)^2 x + (x + h)x^2 + x^3]}{h}$$

$$= \lim\limits_{h \to 0}[(x + h)^3 + (x + h)^2 x + (x + h)x^2 + x^3] \quad \text{(Since } h \neq 0.\text{)}$$

$$= x^3 + x^2(x) + x(x^2) + x^3 = x^3 + x^3 + x^3 + x^3 = 4x^3$$

In Section 7.4, we stated and proved that polynomial functions are differentiable. Here is an alternate proof. A polynomial is an algebraic sum of terms in the form $a_{n-r}x^r$, where $a_{n-r}$ is a constant and $r$ is a nonnegative integer. Each of these terms is differentiable. Thus, the sum is differentiable.

# Exercise Set 7.5

*In Exercises 1–17, differentiate the given functions.*

**1.** $f(x) = 6$

**2.** $g(x) = e$

**3.** $h(x) = \log_7 3$

**4.** $f(x) = \pi^4$

**5.** $f(t) = 2t^2 + 3t + 1$

**6.** $g(s) = 5s^2 + 6s - e^5$

**7.** $h(x) = 3x^4 + 10x^2 - 6x + 3$

**8.** $f(u) = \pi u^3 + \frac{4}{3}u^2 - 6u + 1$

**9.** $g(x) = \sqrt{x^3}$

**10.** $h(t) = \frac{t^2}{2}$

**11.** $f(s) = (s^2 + 1)^3$ (*Hint:* Expand $(s^2 + 1)^3$ first.)

**12.** $g(x) = \frac{x^2 + 5x + 2}{\sqrt[3]{x}}$

**13.** $h(t) = (t^2 + 1)(3t^3 + 2)$

**14.** $f(u) = \frac{2u^2 + 3u - 6}{\sqrt{u}}$

**15.** $f(x) = \frac{5x^{4/3} + 3x^{2/3} + 5}{2x^{2/5}}$

**16.** $g(t) = \frac{5}{\sqrt[3]{t^2}}$

**17.** $h(u) = \frac{4}{\sqrt[5]{u^3}}$

*In Exercises 18–22, find the equation of the line tangent to the graph of the given function at the indicated point of the graph.*

**18.** $y = f(x) = x^2 + 8x - 3$, $(2, 17)$

**19.** $y = g(x) = -x^2 + 3x + 4$, $(-2, -6)$

**20.** $y = f(x) = x^3 + 3x^2 + 2x + 1$, $(1, 7)$

**21.** $y = g(x) = \frac{x^2 + 5x - 10}{\sqrt{x}}$, $(4, 13)$

**22.** $y = f(x) = (x^2 + 1)(3x^2 - 4)$, $(3, 230)$

*In Exercises 23–27, find the instantaneous velocity of a particle moving along a directed line at the given instant where s is the directed distance, in feet, from the origin to the particle at instant t seconds.*

**23.** $s = 5t^2 + 6t + 10$, at $t = 2$ seconds

**24.** $s = 16t^2 - 64t + 100$, at $t = 12$ seconds

**25.** $s = 5\sqrt[3]{t^2}$, at $t = 8$ seconds

**26.** $s = (t^3 + 1)(2t^2 - 3)$, at $t = 2$ seconds

**27.** $s = \frac{t^2 + 7t - 2}{\sqrt{t}}$, at $t = 4$ seconds

**28.** If $C(q)$ is the cost of producing $q$ units of a product, then $C$ is called the cost function. Often, the domain of this function is the set of positive integers. However, even in those cases, the formula defining it may yield a differentiable function defined over the set of positive real numbers. In that case, the derivative $C'$ is called the *marginal cost function*. The value of $C'(n)$, where $n$ is a positive integer, is interpreted by some economists as an approximation of the cost of producing the $n$th unit, and by others as an approximation of the cost of producing the $(n+1)$st unit. We can use either of the two interpretations.

Suppose the cost $C(q)$ (in dollars) of producing $q$ racquetball racquets is given by

$$C(q) = 10{,}000 + 150q^{2/3}$$

**a.** Find the marginal cost function.

**b.** Find the approximate cost of producing the 1000th racquet.

**c.** Find the exact cost of producing the 1000th racquet.

**29.** Suppose the cost $C(q)$ (in dollars) of producing $q$ television sets is given by

$$C(q) = 20{,}000 + 150q - 2q^2 + .01q^3$$

**a.** What is the marginal cost function?

**b.** What is the approximate cost of producing the 100th television set? (See Exercise 28.)

**c.** Find the exact cost of producing the 100th television set.

**30.** When an object is thrown vertically upward, from ground level, with an initial velocity $v_0$ ft/sec, its height $s$ feet above the ground at instant $t$ seconds later is given by

$$s = -16t^2 + v_0 t$$

Experience tells us that the speed of the object decreases until the instantaneous velocity is 0; at that point, the object begins to fall toward the ground again.

Suppose an object is thrown vertically upward, from ground level, with an initial velocity of 288 feet per second.

**a.** At what instant will its instantaneous velocity be 0?

**b.** Find the maximum height reached by the object.

**c.** At what instant will the instantaneous velocity be 96 feet per second?

**d.** At what instant will the object reach the ground?

**31.** Same as Exercise 30 with an initial velocity of 384 ft/sec.

**32.** Same as Exercise 30 with an initial velocity of 480 ft/sec.

**33.** Same as Exercise 30 with an initial velocity of 1664 ft/sec.

*In Exercises 34–37, an equation is given. Find the points on the graph of each equation where the lines tangent to the graph are parallel to the x-axis.*

**34.** $y = 2x^3 + 3x^2 - 12x + 4$

**35.** $y = 2x^3 + 3x^2 - 36x + 10$

**36.** $y = x^3 + 3x^2 - 45x + 20$

**37.** $y = x^4 - 32x + 15$

**38.** Find the point on the graph of $y = x^2 + 6x + 1$ where the slope of the tangent line is $\frac{-2}{3}$.

**39.** Find the point on the graph of $y = x^2 - 5x + 2$ where the tangent line is parallel to the graph of $3x + 5y = 7$.

**40.** Find the point on the graph of $y = -x^2 + 5x - 3$ where the tangent line is perpendicular to the graph of $5x + 7y = 3$.

**41.** Let $f(x) = 3x^2 + 6x - 2$.

   **a.** Find $f(3)$.
   **b.** Find $h(x)$ such that $h(x)(x - 3) = f(x) - f(3)$.
   **c.** Find $f'(x)$.
   **d.** Evaluate $h(3)$ and $f'(3)$ and compare your answers.

**42.** Let $f(x) = 2x^4 - 5x^2 + 3x + 7$.

   **a.** Find $f(-2)$.
   **b.** Find $h(x)$ such that $h(x)(x + 2) = f(x) - f(-2)$.
   **c.** Find $f'(x)$.
   **d.** Evaluate $h(-2)$ and $f'(-2)$ and compare your answers.

# 7.6  Basic Differentiation Formulas (Continued)

## The Product Rule

In the preceding section, we derived five basic differentiation formulas. We saw that the derivative of the sum (difference) of two differentiable functions is the sum (difference) of their derivatives. We also saw that the derivative of a constant multiplied by a differentiable function is equal to the constant multiplied by the derivative of that function. You might, therefore, be tempted to conjecture that the derivative of the product of two differentiable functions is the product of their derivatives. However, this is false, as is illustrated by the following example.

**EXAMPLE 1**  Let $f(x) = x^2$ and $g(x) = x^3$. Show that

$$D_x[f(x) \cdot g(x)] \neq f'(x) \cdot g'(x)$$

**Solution**  $f(x) \cdot g(x) = x^2 \cdot x^3 = x^5$

Therefore,

$$D_x[f(x) \cdot g(x)] = D_x x^5 = 5x^4$$

Also,

$$f'(x) = D_x x^2 = 2x$$

and

$$g'(x) = D_x x^3 = 3x^2$$

Thus,

$$f'(x) \cdot g'(x) = (2x)(3x^2) = 6x^3$$

We see that

$$D_x[f(x) \cdot g(x)] \neq f'(x) \cdot g'(x), \text{ since } 5x^4 \neq 6x^3$$

∎

We now wish to derive a formula to enable us to differentiate the product of two differentiable functions.

> **THEOREM 7.5 (Product Rule):**   If $f$ and $g$ are two differentiable functions, their product is a differentiable function and
>
> $$D_x[f(x) \cdot g(x)] = f(x) \cdot g'(x) + g(x) \cdot f'(x)$$

**Proof (Optional):** By definition,

$$D_x[f(x) \cdot g(x)] = \lim_{h \to 0} \frac{f(x + h) \cdot g(x + h) - f(x) \cdot g(x)}{h}$$

We write

$$\lim_{h \to 0} \frac{f(x + h) \cdot g(x + h) - f(x) \cdot g(x)}{h}$$

$$= \lim_{h \to 0} \frac{f(x + h) \cdot g(x + h) - f(x + h) \cdot g(x) + f(x + h) \cdot g(x) - f(x) \cdot g(x)}{h}$$

$$= \lim_{h \to 0} \left[ f(x + h) \frac{g(x + h) - g(x)}{h} + g(x) \frac{f(x + h) - f(x)}{h} \right]$$

$$= \lim_{h \to 0} f(x + h) \lim_{h \to 0} \frac{g(x + h) - g(x)}{h} + \lim_{h \to 0} g(x) \lim_{h \to 0} \frac{f(x + h) - f(x)}{h}$$

$$= f(x) \cdot g'(x) + g(x) \cdot f'(x)$$

Note that we used the fact that $\lim_{h \to 0} f(x + h) = f(x)$. This step can be justified by stating that differentiable functions are continuous (see Section 7.4).

In words, Theorem 7.5 can be stated as follows.

**Product Rule**

The derivative of the product of two differentials functions is equal to the product of the first factor by the derivative of the second factor, plus the product of the second factor by the derivative of the first factor.

**EXAMPLE 2**   Let $f(x) = (x^2 + 5x + 2)(x^3 + 3x - 2)$. Find $f'(x)$ in two different ways.

**Solution**  Using the Product Rule of differentiation (Theorem 7.5), we get

$$f'(x) = D_x[(x^2 + 5x + 2)(x^3 + 3x - 2)]$$
$$= (x^2 + 5x + 2)D_x(x^3 + 3x - 2) + (x^3 + 3x - 2)D_x(x^2 + 5x + 2)$$
$$= (x^2 + 5x + 2)(3x^2 + 3) + (x^3 + 3x - 2)(2x + 5)$$
$$= 5x^4 + 20x^3 + 15x^2 + 26x - 4$$

If we write $f(x)$ as a polynomial by multiplying first, we get

$$f(x) = (x^2 + 5x + 2)(x^3 + 3x - 2) = x^5 + 5x^4 + 5x^3 + 13x^2 - 4x - 4$$

Thus,

$$f'(x) = D_x(x^5 + 5x^4 + 5x^3 + 13x^2 - 4x - 4)$$
$$= 5x^4 + 20x^3 + 15x^2 + 26x - 4$$

Note that the two answers are identical, as expected.
**Do Exercise 9.** ∎

## The Quotient Rule

Next we state and prove a theorem that will enable us to differentiate quotients of differentiable functions.

> **THEOREM 7.6 (Quotient Rule):**  Let $f$ and $g$ be differentiable functions. Then, whenever $g(x) \neq 0$, the quotient $f/g$ is a differentiable function and
>
> $$D_x \frac{f(x)}{g(x)} = \frac{g(x)f'(x) - f(x)g'(x)}{[g(x)]^2}$$

***Proof (Optional):*** It is simpler to first derive a formula for $D_x \dfrac{1}{g(x)}$.

$$D_x \frac{1}{g(x)} = \lim_{h \to 0} \frac{\dfrac{1}{g(x+h)} - \dfrac{1}{g(x)}}{h} = \lim_{h \to 0} \frac{g(x) - g(x+h)}{h \cdot g(x+h) \cdot g(x)}$$

$$= \lim_{h \to 0} \left[ \frac{g(x+h) - g(x)}{h} \cdot \frac{-1}{g(x+h) \cdot g(x)} \right]$$

$$= \lim_{h \to 0} \frac{g(x+h) - g(x)}{h} \cdot \lim_{h \to 0} \frac{-1}{g(x+h) \cdot g(x)}$$

$$= g'(x) \frac{-1}{[g(x)]^2} = \frac{-g'(x)}{[g(x)]^2}$$

Thus,

$$D_x \frac{f(x)}{g(x)} = D_x \left[ f(x) \cdot \frac{1}{g(x)} \right] = f(x) \cdot D_x \frac{1}{g(x)} + \frac{1}{g(x)} \cdot D_x f(x) \quad \text{(by Theorem 7.5)}$$

$$= f(x) \cdot \frac{-g'(x)}{[g(x)]^2} + \frac{1}{g(x)} \cdot f'(x) = \frac{g(x)f'(x) - f(x)g'(x)}{[g(x)]^2}$$

EXAMPLE 3   Find $\dfrac{dy}{dx}$ if $y = \dfrac{x^2 + 5x + 6}{x^2 + 4}$.

Solution   $\dfrac{dy}{dx} = D_x \dfrac{x^2 + 5x + 6}{x^2 + 4}$

$$= \frac{(x^2 + 4)D_x(x^2 + 5x + 6) - (x^2 + 5x + 6)D_x(x^2 + 4)}{(x^2 + 4)^2}$$

$$= \frac{(x^2 + 4)(2x + 5) - (x^2 + 5x + 6)(2x)}{(x^2 + 4)^2}$$

$$= \frac{(2x^3 + 5x^2 + 8x + 20) - (2x^3 + 10x^2 + 12x)}{(x^2 + 4)^2}$$

$$= \frac{-5x^2 - 4x + 20}{(x^2 + 4)^2}$$

**Do Exercise 15.**

In differentiating $kf(x)$ or $\dfrac{f(x)}{k}$ where $k$ is constant and $f$ is a differentiable function, you should not use the Product Rule or the Quotient Rule. It would not be wrong to do so, but it would be as silly as what an old farmer did when he wanted to know how many cows he had—he proceeded to count the legs and divided his answer by four! The following example illustrates what should be done in such cases.

EXAMPLE 4   Find $\dfrac{dy}{dx}$ for the following equations.

**a.** $y = 5(x^2 + 3)$
**b.** $y = x^4/4$

Solution   **a.** $\dfrac{dy}{dx} = D_x 5(x^2 + 3)$

$$= 5D_x(x^2 + 3) = 5(2x) = 10x$$

**b.** $\dfrac{dy}{dx} = D_x \dfrac{x^4}{4}$

$$= D_x\left(\frac{1}{4} \cdot x^4\right) = \frac{1}{4} \cdot D_x x^4 = \frac{1}{4} \cdot 4x^3 = x^3$$

## An Application

EXAMPLE 5   From experience, a manufacturer of Ping-Pong paddles has learned that the cost $C(q)$ (in dollars) per day to manufacture $q$ paddles is given by

$$C(q) = \frac{5q^3 + 4500q^2 + 375q + 117{,}500}{q^2 + 75}$$

**a.** Find the marginal cost function.
**b.** Find the approximate cost to produce the 76th paddle. (See Exercise 28 of the preceding section.)

**Solution** **a.** By definition, the marginal cost function is the derivative of the cost function. Thus,

$$C'(q) = D_q \frac{5q^3 + 4500q^2 + 375q + 117{,}500}{q^2 + 75}$$

$$= \frac{(q^2 + 75)D_q(5q^3 + 4500q^2 + 375q + 117{,}500) - (5q^3 + 4500q^2 + 375q + 117{,}500)D_q(q^2 + 75)}{(q^2 + 75)^2}$$

$$= \frac{(q^2 + 75)(15q^2 + 9000q + 375) - (5q^3 + 4500q^2 + 375q + 117{,}500)(2q)}{(q^2 + 75)^2}$$

$$= \frac{(15q^4 + 9000q^3 + 1500q^2 + 675{,}000q + 28{,}125) - (10q^4 + 9000q^3 + 750q^2 + 235{,}000q)}{(q^2 + 75)^2}$$

$$= \frac{5q^4 + 750q^2 + 440{,}000q + 28{,}125}{(q^2 + 75)^2}$$

**b.** The approximate cost of producing the 76th unit is given by $C'(75)$ or by $C'(76)$. We calculate $C'(75)$.

$$C'(75) = \frac{5(75^4) + 750(75^2) + 440{,}000(75) + 28{,}125}{(75^2 + 75)^2} \doteq 6.0157$$

Therefore, the approximate cost of producing the 76th paddle is $6.02.

Note that the total cost of producing 76 paddles is $C(76)$ and the total cost of producing 75 paddles is $C(75)$. Hence, the exact cost of producing the 76th paddle is given by $C(76) - C(75)$, which is found to be 5.9961.

Thus, rounding this to $6, we see that by using the derivative (marginal cost function), we obtained an answer with a 0.3% error.

**Do Exercise 21.** ◼

# Exercise Set 7.6

*In Exercises 1–4, use the given functions to illustrate the fact that $D_x[f(x) \cdot g(x)] \neq f'(x) \cdot g'(x)$.*

**1.** $f(x) = 5$ and $g(x) = 3x$

**2.** $f(x) = 2x$ and $g(x) = 5x^2$

**3.** $f(x) = 3x^2$ and $g(x) = 5x^4$

**4.** $f(x) = \sqrt{x}$ and $g(x) = \sqrt[3]{x^2}$

*In Exercises 5–8, use the given functions to illustrate the fact that $D_x \dfrac{f(x)}{g(x)} \neq \dfrac{f'(x)}{g'(x)}$.*

**5.** $f(x) = x^3$ and $g(x) = x^2$

**6.** $f(x) = 5x^6$ and $g(x) = 3x^2$

**7.** $f(x) = 2x^2$ and $g(x) = 4x^5$

**8.** $f(x) = \sqrt[3]{x^2}$ and $g(x) = \sqrt[4]{x^3}$

*In Exercises 9–14, find $\dfrac{dy}{dx}$ in two different ways.*

**9.** $y = (x^2 + 11x + 3)(x^4 + 5)$

**10.** $y = (3x^3 + 10x + 1)(2x^2 + 7)$

**11.** $y = (5x^3 + 3x + 8)(2x^4 + 5)$

**12.** $y = (2x^5 - 3x^2 + 2)(5x^3 + 2x^2 - 5x + 7)$

**13.** $y = (x^2 + 3x - 1)(\sqrt{x} + 3)$

**14.** $y = \left(\sqrt[3]{x^2} + 1\right)\left(\sqrt[4]{x^3} + 6\right)$

*In Exercises 15–20, find $\dfrac{dy}{dx}$.*

**15.** $y = \dfrac{x^2 + 1}{x^4 + 3}$

**16.** $y = \dfrac{x^3 + 3x - 21}{x^2 + 4x + 1}$

**17.** $y = \dfrac{2x^5 - 4}{3x^2 + 10}$

**18.** $y = \dfrac{5x^2 + 8x - 4}{\sqrt{x} + 6}$

**19.** $y = \dfrac{\sqrt{x^3} + 3}{\sqrt{x} - 2}$

**20.** $y = \dfrac{\sqrt[4]{x^3} + \sqrt{x}}{5 - \sqrt[3]{x^2}}$

**21.** A manufacturer found that the cost $C(q)$ (in dollars) of producing $q$ units of a commodity is given by

$$C(q) = 20q + \dfrac{60,000}{\sqrt{q} + 10}$$

**a.** Find the marginal cost function.
**b.** Find the approximate cost of producing the 101st unit. (*Hint:* See Exercise 28, Section 7.5.)

**22.** A television manufacturer knows that the cost $C(q)$ (in dollars) of building $q$ deluxe television sets is given by

$$C(q) = \dfrac{2800q^{5/2} + 2800q^{1/2} + 172,800}{q^2 + 1}$$

**a.** Find the marginal cost function.
**b.** Find the approximate cost of producing the 50th set. (*Hint:* See Exercise 28, Section 7.5.)

*Marginal revenue and marginal profit are also defined as rates of change. That is, if the revenue and profit functions are R and P, respectively, then the marginal revenue and the marginal profit functions are R' and P', respectively. Recall that profit is obtained by subtracting cost from revenue.*

*In Exercises 23–25, the cost $\$C(q)$ and revenue $\$R(q)$ when $q$ units of a commodity are produced and sold are given.*

**a.** *Find the profit function.*
**b.** *Find the marginal cost function.*
**c.** *Find the marginal revenue function.*
**d.** *Find the marginal profit function.*

**23.** $C(q) = \dfrac{100q^2 + 600q}{q + 3}$

$R(q) = \dfrac{200q^2 + 500q}{q + 2}$

**24.** $C(q) = 6000 + q\left(300 + \dfrac{6000}{q + 50}\right),\ 0 \le q \le 200$

$R(q) = \dfrac{-q^3}{50} + q^2 + 600q,\ 0 \le q \le 200$

**25.** $C(q) = 1000 + 50q + \dfrac{30,000q}{q^2 + 500},\ 0 \le q \le 100$

$R(q) = \dfrac{-q^4 + 100q^3 - q^2 + 100q}{1250},\ 0 \le q \le 100$

**26.** A population of rare birds on an island is slowly decreasing. The number $P(t)$ of birds (where $t$ is the number of years a count has been kept of that population) is given by $P(t) = (31,000 + 3t)/(t^2 + 31)$. Find the rate of change of the population when $t = 13$.

**27. a.** Use the Quotient Rule to find $D_x\, x^{-10}$. (*Hint:* Write $x^{-10} = 1/x^{10}$.)
**b.** If $n$ is a positive integer, use the Quotient Rule to derive the formula $D_x\, x^{-n} = -nx^{-n-1}$.

## 7.7  The Chain Rule

Recall that the derivative of a function has been defined as the instantaneous rate of change of that function with respect to the independent variable. Suppose $f$ and $g$ are differentiable functions, can we conclude that the composition $g \circ f$ is differentiable? If it is, how can the derivative $D_x[(f \circ g)(x)]$ be obtained? (See Section 1.1 for the definition of composition of functions.)

### The Chain Rule

Let us introduce variables and write $y = g(u)$ and $u = f(x)$ so that $y = (g \circ f)(x) = g(f(x)) = g(u)$. (See Figure 7.16.)

**FIGURE 7.16**

Note that $y$ is the dependent variable for both the functions $g$ and $g \circ f$, and $x$ is the independent variable for both $f$ and $g \circ f$, while $u$ is the dependent variable for the function $f$ but also the independent variable for the function $g$. Therefore, when we differentiate $f$, it is with respect to $x$; when we differentiate $g$, it must be with respect to $u$; and when we differentiate $g \circ f$ it is again with respect to $x$.

Assuming for the time being that $g \circ f$ is differentiable, the three derivatives we can obtain are $\frac{dy}{du}$, $\frac{du}{dx}$, and $\frac{dy}{dx}$. Although these three symbols do not denote fractions, if we treat them as such, we get the following relationship:

$$\left(\frac{dy}{du}\right)\left(\frac{du}{dx}\right) = \left(\frac{dy}{dx}\right)$$

It is interesting that this equality is in fact correct. It is called the *Chain Rule*. The formal statement of this relationship between derivatives of composite functions is stated in the following theorem. The proof of this theorem, which is valid in all cases, is beyond the scope of this book. However, we shall give a proof that is correct for most of the standard functions we encounter in the applications.

> **THEOREM 7.7 (Chain Rule):**   Let $f$ be a function that is differentiable at $x$, and $g$ be a function differentiable at $f(x)$. Then the composition $g \circ f$ is differentiable at $x$ and $D_x[(g \circ f)(x)] = g'(f(x))f'(x)$.

**Proof (Optional):** Note first that on the right side of the formula, the prime notation indicates differentiation. Keep in mind that $f'(x)$ is the derivative of $f$ with respect to $x$, evaluated at $x$, while $g'(f(x))$ is the value of the derivative of $g$, with respect to $u$, evaluated at $f(x)$. Now we write

$$\frac{(g \circ f)(x + h) - (g \circ f)(x)}{h} = \frac{g(f(x + h) - g(f(x))}{h}$$

$$= \frac{g(f(x + h)) - g(f(x))}{f(x + h) - f(x)} \cdot \frac{f(x + h) - f(x)}{h}$$

Of course, this is valid provided $f(x + h) - f(x) \neq 0$. Now, let $k = f(x + h) - f(x)$, so that $f(x + h) = f(x) + k$. Then we can write

$$\frac{(g \circ f)(x + h) - (g \circ f)(x)}{h} = \frac{g(f(x) + k) - g(f(x))}{k} \cdot \frac{f(x + h) - f(x)}{h}$$

Since $f$ is differentiable at $x$, it is continuous at $x$. Thus, $\lim_{h \to 0} f(x + h) = f(x)$, from which we see that as $h$ goes to 0, $k$ also goes to 0, and we get

$$D_x[(g \circ f)(x)] = \lim_{h \to 0} \frac{(g \circ f)(x + h) - (g \circ f)(x)}{h}$$

$$= \lim_{h \to 0}\left[\frac{g(f(x) + k) - g(f(x))}{k} \cdot \frac{f(x + h) - f(x)}{h}\right]$$

$$= \lim_{k \to 0} \frac{g(f(x) + k) - g(f(x))}{k} \cdot \lim_{h \to 0} \frac{f(x + h) - f(x)}{h}$$

$$= g'(f(x)) \cdot f'(x).^*$$

**EXAMPLE 1**    Find $\dfrac{dy}{dx}$ if $y = (x^2 + 1)^{10}$.

**Solution**    Let $u = x^2 + 1$, so that $y = u^{10}$. It follows that

$$\frac{du}{dx} = 2x \text{ and } \frac{dy}{du} = D_u u^{10} = 10u^9$$

Thus,

$$\frac{dy}{dx} = \frac{dy}{du} \cdot \frac{du}{dx} = 10u^9(2x)$$

$$= 10(x^2 + 1)^9(2x) = 20x(x^2 + 1)^9$$

**Do Exercise 1.**

**EXAMPLE 2**    Find $f'(x)$ if $f(x) = \sqrt[3]{x^4 + x^2 + 5}$.

**Solution**    We write

$$f(x) = \sqrt[3]{x^4 + x^2 + 5} = (x^4 + x^2 + 5)^{1/3}$$

Now let $u = x^4 + x^2 + 5$, so that $y = f(x) = u^{1/3}$. Differentiating $u$ with respect to $x$, and differentiating $y$ with respect to $u$, we get

$$\frac{du}{dx} = 4x^3 + 2x \text{ and } \frac{dy}{du} = \frac{1}{3}u^{1/3-1} = \frac{1}{3}u^{-2/3} = \frac{1}{3u^{2/3}}$$

Thus,

$$\frac{dy}{dx} = \frac{dy}{du} \cdot \frac{du}{dx} = \frac{1}{3u^{2/3}}(4x^3 + 2x) = \frac{1}{3(x^4 + x^2 + 5)^{2/3}}(4x^3 + 2x)$$

$$= \frac{4x^3 + 2x}{3\sqrt[3]{(x^4 + x^2 + 5)^2}}$$

**Do Exercise 5.**

## The Generalized Power Rule

We often need to differentiate power functions of the type $[f(x)]^r$, where $f$ is a differentiable function and $r$ is a constant. It is therefore useful to derive a formula to combine both the Power Rule and the Chain Rule.

---

*The reason this proof is not valid in the general case is that the condition $f(x + h) - f(x) \neq 0$, which we required at the beginning of the proof, may not hold even in a small interval containing $x$. However, in most of the applications, this difficulty does not arise. Therefore, for all practical purposes, our simpler proof will suffice.

**THEOREM 7.8 (Generalized Power Rule):**   If $f$ is a differentiable function, $r$ is a constant, and $y = [f(x)]^r$, then

$$\frac{dy}{dx} = r[f(x)]^{r-1} \cdot f'(x)$$

**Proof:** Let $u = f(x)$, so that $y = u^r$. Thus,

$$\frac{du}{dx} = f'(x) \text{ and } \frac{dy}{du} = ru^{r-1}$$

Therefore,

$$\frac{dy}{dx} = \frac{dy}{du} \cdot \frac{du}{dx} = ru^{r-1} \cdot f'(x) = r[f(x)]^{r-1} \cdot f'(x)$$

This formula is often written as

$$D_x u^r = ru^{r-1} \cdot D_x u$$

and is called the Generalized Power Rule.

**EXAMPLE 3**   Find $\dfrac{dy}{dx}$ in each of the following cases.

**a.** $y = (x^3 + 3)^{10}$
**b.** $y = (x^2 + 3x + 5)^3$
**c.** $y = \dfrac{(x^2 + 3x + 5)^3}{(x^3 + 3)^{10}}$

**Solution**   **a.** Use the Generalized Power Rule with $u = x^3 + 3$ and $r = 10$ to get

$$\frac{dy}{dx} = D_x(x^3 + 3)^{10} = 10(x^3 + 3)^{10-1}D_x(x^3 + 3)$$

$$= 10 (x^3 + 3)^9(3x^2) = 30x^2(x^3 + 3)^9$$

**b.** Using the same formula with $u = x^2 + 3x + 5$ and $r = 3$, we get

$$\frac{dy}{dx} = D_x(x^2 + 3x + 5)^3 = 3(x^2 + 3x + 5)^{3-1}D_x(x^2 + 3x + 5)$$

$$= 3(x^2 + 3x + 5)^2(2x + 3) = (6x + 9)(x^2 + 3x + 5)^2$$

**c.** Using the Quotient Rule, we get

$$\frac{dy}{dx} = D_x \frac{(x^2 + 3x + 5)^3}{(x^3 + 3)^{10}}$$

$$= \frac{(x^3 + 3)^{10}D_x(x^2 + 3x + 5)^3 - (x^2 + 3x + 5)^3D_x(x^3 + 3)^{10}}{[(x^3 + 3)^{10}]^2}$$

Now we substitute the results obtained in parts **a** and **b** to get

$$\frac{dy}{dx} = \frac{(x^3 + 3)^{10}(6x + 9)(x^2 + 3x + 5)^2 - (x^2 + 3x + 5)^3 \, 30x^2(x^3 + 3)^9}{(x^3 + 3)^{20}}$$

$$= \frac{3(x^3 + 3)^9(x^2 + 3x + 5)^2[(x^3 + 3)(2x + 3) - (x^2 + 3x + 5)10x^2]}{(x^3 + 3)^{20}}$$

$$= \frac{3(x^2 + 3x + 5)^2[(2x^4 + 3x^3 + 6x + 9) - (10x^4 + 30x^3 + 50x^2)]}{(x^3 + 3)^{11}}$$

$$= \frac{3(x^2 + 3x + 5)^2(-8x^4 - 27x^3 - 50x^2 + 6x + 9)}{(x^3 + 3)^{11}}$$

**Do Exercise 9.**                                                      ∎

**EXAMPLE 4**   Use the General Power Rule and the Product Rule to derive the Quotient Rule which was given in Theorem 7.6 of Section 7.6.

**Solution**   To write the quotient $\dfrac{f(x)}{g(x)}$ as a product, we proceed as follows.

$$\frac{f(x)}{g(x)} = f(x) \cdot \frac{1}{g(x)} = f(x) \cdot [g(x)]^{-1}$$

Thus,

$$D_x \frac{f(x)}{g(x)} = D_x[f(x) \cdot [g(x)]^{-1}]$$

$$= f(x) \cdot D_x[g(x)]^{-1} + [g(x)]^{-1} \cdot D_x f(x)$$

$$= f(x) \cdot (-1)[g(x)]^{-1-1} \cdot g'(x) + [g(x)]^{-1} \cdot f'(x)$$

$$= f(x) \cdot \frac{-g'(x)}{[g(x)]^2} + \frac{1}{g(x)} \cdot f'(x) = \frac{g(x) \cdot f'(x) - f(x) \cdot g'(x)}{[g(x)]^2}$$   ∎

**EXAMPLE 5**   Suppose $f$ is a differentiable function and $f'(x) = \dfrac{1}{x^2 + 1}$. Let $y = g(x) = f(x^3 + 2)$. Find $g'(x)$.

**Solution**   Let $u = x^3 + 2$, so that $y = f(u)$. Now, by the Chain Rule, we have

$$\frac{dy}{dx} = \frac{dy}{du} \cdot \frac{du}{dx} = f'(u)D_x u = \frac{1}{u^2 + 1}D_x(x^3 + 2)$$

$$= \frac{1}{(x^3 + 2)^2 + 1}D_x(x^3 + 2) = \frac{1}{(x^6 + 4x^3 + 4) + 1}3x^2 = \frac{3x^2}{x^6 + 4x^3 + 5}$$

**Do Exercise 37.**                                                     ∎

In the remaining part of this chapter, and in future chapters, we shall need to use the differentiation formulas frequently. It is therefore useful to begin to list and to number the formulas we have derived thus far.

**Summary of Differentiation Formulas**

In the following formulas, $k$ and $r$ are constants, $f$ and $g$ are differentiable functions, $u = f(x)$, and $v = g(x)$.

1. $D_x k = 0$   (derivative of a constant)
2. $D_x(u + v) = D_x u + D_x V$   (the Sum Rule)
3. $D_x(u - v) = D_x u - D_x v$   (the Difference Rule)
4. $D_x(ku) = kD_x u$
5. $D_x(uv) = uD_x v + vD_x u$   (the Product Rule)
6. $D_x \dfrac{u}{v} = \dfrac{vD_x u - uD_x v}{v^2}$   (the Quotient Rule)
7. If $y = g(f(x))$, then $\dfrac{dy}{dx} = g'(f(x)) \cdot f'(x)$. Equivalently, if $y = g(u)$ and $u = f(x)$, then

$$\frac{dy}{dx} = \frac{dy}{du} \cdot \frac{du}{dx} \quad \text{(the Chain Rule)}$$

8. $D_x u^r = ru^{r-1} \cdot D_x u$   (the Generalized Power Rule)

# Exercise Set 7.7

*In Exercises 1–20, find the derivative of the given function with respect to the appropriate independent variable.*

**1.** $f(x) = (x^2 + 6x + 3)^{40}$

**2.** $g(x) = (x^3 + 5x - 4)^{10}$

**3.** $h(x) = \dfrac{5}{(x^2 + 3x + 1)^{13}}$

**4.** $f(t) = \sqrt[5]{t^2 + 1}$

**5.** $g(u) = \sqrt[4]{(u^2 + 5u + 1)^5}$

**6.** $h(s) = \dfrac{7}{\sqrt[3]{s^2 + 5s + 7}}$

**7.** $f(x) = (x^3 + 1)^4(x^4 + 3x^2 + 2)^3$

**8.** $g(t) = (t^2 + 1)^6(t^3 + 3t^2 + 5)^5$

**9.** $h(u) = \dfrac{(u^2 + 1)^4}{(u^3 + 2u^2 + 3)^5}$

**10.** $g(s) = \dfrac{(2s^3 + 3s^2 + 10)^4}{(s^2 + 1)^5}$

**11.** $f(t) = \dfrac{\sqrt[3]{t^2 + 1}}{(t^3 + 4)^2}$

**12.** $g(x) = \dfrac{\sqrt[4]{x^2 + 4}}{\sqrt{x^4 + 1}}$

**13.** $h(t) = \sqrt[3]{1 + t}$

**14.** $f(x) = \dfrac{(x^2 + 3)^5(x^4 + 3x + 7)^2}{\sqrt{x^6 + 8}}$

**15.** $g(t) = \left(\dfrac{t^2 + 3}{t^4 + 1}\right)^{12}$

**16.** $h(u) = (1 + \sqrt{u})^{50}$

**17.** $f(s) = \left(\dfrac{3s + 1}{\sqrt{s + 2}}\right)^{20}$

**18.** $g(x) = [(x^2 + 3)^4(\sqrt{x} + 1)^5]^{10}$

**19.** $h(y) = \left(y^2 + \dfrac{1}{\sqrt{y}}\right)^{17}$

**20.** $f(t) = \sqrt{9 - t^2}$

*In Exercises 21–25, find the equation of the line tangent to the graph of the given equation at the indicated point.*

**21.** $y = (x^2 + 1)^5$, at $(1, 32)$

**22.** $y = (x^3 + 7)^{10}$, at $(-2, 1)$

**23.** $y = \dfrac{(x^2 + 1)^4}{(x^3 + x + 29)^5}$, at $(-3, -10{,}000)$

**24.** $y = \sqrt{x^3 + 10}$, at $(-1, 3)$

**25.** $y = \dfrac{(x^3 + x^2 + x + 1)^3}{\sqrt{x^2 + 1}}$, at $(-1, 0)$

**26.** Recall that $y = \sqrt{25 - x^2}$ is the equation of a semicircle with center at the origin and radius 5.

**a.** Find the slope of the line tangent to the semicircle at the point $(-3, 4)$

**b.** Find the slope of the line through the origin and the point $(-3, 4)$.

**c.** Using your results from parts **a** and **b**, verify that the tangent line at the point $(-3, 4)$ is perpendicular to the radius from the origin to the point of tangency $(-3, 4)$.

**27.** Same as Exercise 26 for $y = -\sqrt{169 - x^2}$ and the point $(12, -5)$.

**28.** A particle is moving along a directed line. Its distance $d$ (in meters) from the origin at instant $t$ (in minutes) is given by

$$d = \sqrt[3]{t^2 + 15}$$

Find the instantaneous velocity when $t = 7$ minutes.

**29.** An object is falling downward subject to gravity and other forces. Its distance $d$ (in feet) from the starting point after $t$ seconds is given by

$$d = 16t^2 - 81\sqrt[5]{t^2 + 18}$$

Find the instantaneous velocity at instant 15 seconds.

*In Exercises 30–33, the cost $C(q)$ (in dollars) of producing $q$ units of a commodity is given.*

*a. Find the marginal cost function.*

*b. Find the approximate cost of producing the ath unit where a is given in each case. (Hint: See Exercise 28, Section 7.5.)*

**30.** $C(q) = 2000 + q\sqrt[3]{206 + 1000/q}$, $a = 100$

**31.** $C(q) = 50,000 + 30q - \sqrt[5]{q^2 + 625}$, $a = 50$

**32.** $C(q) = 3500 + \left[\dfrac{400q^4 + 100,000q^2 + 6,250,000}{q^3}\right]^{1/2}$, $a = 50$

**33.** $C(q) = 10,000 + q\left(10 + \dfrac{q^2}{q^3 + 50}\right)^3$, $a = 10$

**34.** The number of rabbits on a small island $t$ months after they were brought onto the island is $P(t)$, where

$$P(t) = 50\left(1 + \dfrac{t^2}{1000}\right)^3$$

Find the instantaneous rate of change of that rabbit population when $t = 30$.

**35.** When a new employee is hired by a company, the employee will produce $P(x)$ items after $x$ weeks of work, where $P(x)$ is given by

$$P(x) = 60 - \dfrac{60}{(x^2 + 1)^{.1}}$$

Find the rate of change of production on the 12th week of employment.

**36.** Suppose $f$ is a differentiable function, $f'(x) = \dfrac{2}{x}$, and let $y = g(x) = f(x^2 + 9)$. Find $g'(x)$.

**37.** Suppose $h$ is a differentiable function, $h'(z) = \dfrac{1}{z + 1}$, and let $y = f(t) = h(t^4 + 3)$. Find $f'(t)$.

**38.** Suppose $g$ is a differentiable function, $g'(x) = \dfrac{1}{2\sqrt{x}}$, and let $y = f(x) = g(9x^2)$. Find $f'(x)$.

# 7.8 Implicit Differentiation

## Functions Defined Implicitly

We introduced the concept of function in Chapter 1. Essentially, a function is a rule that associates with each element of one set (the domain), one, and only one, member of another set (the range). Frequently in mathematics, the domain and range are sets of numbers and the rule is described by a formula written as $y = f(x)$. In this case, we say that the function $f$ is defined *explicitly*. At other times, however, an equation in two variables, say $x$ and $y$, is given, but it is not in the form $y = f(x)$. In that case, for each value of one of the variables, one or more values of the other variable may correspond. Thus such an equation may describe one or more functions. Any function defined in this manner is said to be defined *implicitly*.

EXAMPLE 1    Show that the equation $x^2 + y^2 = 16$ defines two functions with independent variable $x$.

Solution    Solving for $y$, we get

$$y^2 = 16 - x^2$$

and

$$y = \sqrt{16 - x^2}, \text{ or } y = -\sqrt{16 - x^2}$$

Thus, one function is defined by $y = f_1(x) = \sqrt{16 - x^2}$ and the other by $y = f_2(x) = -\sqrt{16 - x^2}$.

Note that the graph of $x^2 + y^2 = 16$ is a circle with the center at the origin and radius 4. The graph of $f_1$ is the upper semicircle, and the graph of $f_2$ is the lower semicircle. (See Figure 7.17.)

**Do Exercise 1.**

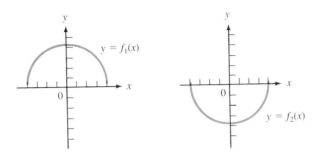

**FIGURE 7.17**

## Implicit Differentiation

We may not always be able to solve for $y$ in terms of $x$. In fact, there are applications where it is not essential to obtain a formula for $y$ in terms of $x$. Instead, the value of the derivative at certain points must be obtained. It is possible to accomplish this goal using a technique called *implicit differentiation*. The idea is simple. If an equation is given, say, in $x$ and $y$, and we know that there is a differentiable function $f$ which is defined implicitly by this equation, then we need only differentiate both sides with respect to the independent variable. We then use all the relevant formulas, keeping in mind that if $x$ is the independent variable then $D_x x = 1$, but in general $D_x y \neq 1$. Thus, we must remember that $D_x x^n = n x^{n-1}$ while $D_x y^n \neq n y^{n-1}$ because, by the Chain Rule, we have $D_x y^n = n y^{n-1} D_x y$. Upon completing the differentiation, we solve for $D_x y$.

EXAMPLE 2    The equation $x^2 y^3 + x^2 = y^4 + 3x^3 + 2$ defines a function $f$ with $y = f(x)$. Find $D_x y$.

Solution    First, we replace $y$ by $f(x)$ in the equation. We get

$$x^2[f(x)]^3 + x^2 = [f(x)]^4 + 3x^3 + 2$$

Next, we differentiate both sides of the equation with respect to $x$.

$$D_x\{x^2[f(x)]^3 + x^2\} = D_x\{[f(x)]^4 + 3x^3 + 2\}$$

Hence,

$$D_x(x^2[f(x)]^3) + D_x x^2 = D_x[f(x)]^4 + D_x 3x^3 + D_x 2$$

Using the appropriate differentiation formula for each term, we obtain

$$x^2 D_x[f(x)]^3 + [f(x)]^3 D_x x^2 + 2x = 4[f(x)]^3 f'(x) + 3(3x^2) + 0$$

Continuing the differentiation, we get

$$x^2(3[f(x)]^2 f'(x)) + [f(x)]^3(2x) + 2x = 4[f(x)]^3 f'(x) + 9x^2$$

We must solve for $f'(x)$. Thus, we bring all terms that involve $f'(x)$ to the left side and all other terms to the right side to get

$$3x^2[f(x)]^2 f'(x) - 4[f(x)]^3 f'(x) = 9x^2 - 2x[f(x)]^3 - 2x$$

Factoring $f'(x)$ on the left side, we obtain

$$\{3x^2[f(x)]^2 - 4[f(x)]^3\}f'(x) = 9x^2 - 2x[f(x)]^3 - 2x$$

If $3x^2[f(x)]^2 - 4[f(x)]^3 \neq 0$, we may divide both sides of this equation by that quantity to get

$$f'(x) = \frac{9x^2 - 2x[f(x)]^3 - 2x}{3x^2[f(x)]^2 - 4[f(x)]^3} = \frac{9x^2 - 2xy^3 - 2x*}{3x^2 y^2 - 4y^3}$$

**Do Exercise 5.**                                                                      ∎

**EXAMPLE 3 (Optional)**    We have stated that if $x > 0$ and $r$ is any real number, then $D_x x^r = rx^{r-1}$. This formula is certainly valid if $r$ is a rational number. Use implicit differentiation to give a proof of the formula in this case.

**Solution**    If $r$ is rational, there exist integers $m$ and $n$ such that $r = \frac{m}{n}$. Let $y = x^r$. Thus $y = x^{m/n}$. Raising both sides of the equality to the $n$th power, we get

$$y^n = x^m$$

Differentiating both sides with respect to $x$, we may use the formula $D_x u^r = ru^{r-1}D_x u$ since the formula has been proved when $r$ is an integer. Thus,

$$ny^{n-1}D_x y = mx^{m-1}D_x x = mx^{m-1}$$

It follows that

$$D_x y = \frac{mx^{m-1}}{ny^{n-1}}$$

Replacing $y$ by $x^{m/n}$, we get

$$D_x y = \frac{m}{n} \cdot \frac{x^{m-1}}{(x^{m/n})^{n-1}} = \frac{m}{n} \cdot \frac{x^{m-1}}{x^{(m/n)(n-1)}}$$

---

*You are encouraged to perform implicit differentiation without first replacing $y$ by $f(x)$, after having worked through a few examples.

$$= \frac{m}{n} \cdot \frac{x^{m-1}}{x^{m-m/n}}$$

$$= \frac{m}{n} \cdot x^{(m-1)-(m-m/n)} = \frac{m}{n} \cdot x^{m/n-1} = rx^{r-1} \qquad \blacksquare$$

**EXAMPLE 4**    The equation $x^2 + y^5 = x^3y^4 + 5x - 15$ defines a function $f$ whose graph passes through the point $(2, -1)$. Find the equation of the line tangent to the graph at that point.

**Solution**    When we replace $x$ by 2 and $y$ by $-1$, we get a true statement since

$$2^2 + (-1)^5 = (2^3)(-1)^4 + 5(2) - 15$$

simplifies to

$$3 = 3$$

Therefore, $(2, -1)$ is indeed on the graph of the equation. While there may be more than one function defined by the given equation, we are asked to find the equation of the line tangent to the graph of the function whose graph passes through $(2, -1)$. Thus, we must evaluate the derivative of that function at that particular point. We use implicit differentiation to find $\frac{dy}{dx}$.

Differentiating both sides of the equation with respect to $x$, we get

$$D_x(x^2 + y^5) = D_x(x^3y^4 + 5x - 15)$$

Using the differentiation formulas, we obtain

$$2x + 5y^4 \frac{dy}{dx} = x^3 D_x y^4 + y^4 D_x x^3 + 5 - 0$$

That is,

$$2x + 5y^4 \frac{dy}{dx} = x^3 \left( 4y^3 \frac{dy}{dx} \right) + y^4(3x^2) + 5$$

Solving for $\frac{dy}{dx}$, we get

$$5y^4 \frac{dy}{dx} - 4x^3 y^3 \frac{dy}{dx} = 3x^2 y^4 - 2x + 5$$

Factoring out $\frac{dy}{dx}$ on the left side, we have

$$(5y^4 - 4x^3 y^3)\frac{dy}{dx} = 3x^2 y^4 - 2x + 5$$

and

$$\frac{dy}{dx} = \frac{3x^2 y^4 - 2x + 5}{5y^4 - 4x^3 y^3},$$

provided that $5y^4 - 4x^3 y^3 \neq 0$

To find the slope of the tangent line, we evaluate $\frac{dy}{dx}$ at $(2, -1)$. Thus, we replace $x$ by 2 and $y$ by $-1$ in the last equation and get

$$m = \frac{3(2^2)(-1)^4 - 2(2) + 5}{5(-1)^4 - 4(2^3)(-1)^3} = \frac{13}{37}$$

Thus, the tangent line has slope $\frac{13}{37}$ and passes through $(2, -1)$. (See Figure 7.18.) Using the point-slope form of the equation of a line, we write

$$y - (-1) = \frac{13}{37}(x - 2)$$

and

$$y = \frac{13}{37}x - \frac{63}{37}$$

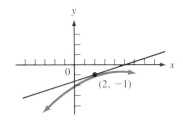

**FIGURE 7.18**

**Do Exercise 13.**

**EXAMPLE 5**    In Example 1, we showed that the equation $x^2 + y^2 = 16$ defines implicitly two functions $f_1$ and $f_2$ such that $y = f_1(x) = \sqrt{16 - x^2}$ and $y = f_2(x) = -\sqrt{16 - x^2}$.

**a.** Find $\frac{dy}{dx}$ implicitly.
**b.** Find $f_1'(x)$.
**c.** Find $f_2'(x)$.
**d.** Explain why we have only one answer for $\frac{dy}{dx}$ in part **a** although there are two functions defined by the equation and therefore two derivatives.

**Solution**    **a.** Differentiating both sides of the equation with respect to $x$, we get

$$D_x(x^2 + y^2) = D_x 16$$

That is,

$$2x + 2yD_x y = 0$$

Solving for $D_x y$, we get

$$2yD_x y = -2x$$

and

$$D_x y = \frac{-x}{y} \quad \text{(provided that } y \neq 0)$$

**b.** $f_1(x) = \sqrt{16 - x^2} = (16 - x^2)^{1/2}$

Thus,

$$f_1'(x) = D_x(16 - x^2)^{1/2} = \frac{1}{2}(16 - x^2)^{1/2-1} \cdot D_x(16 - x^2)$$

$$= \frac{1}{2}(16 - x^2)^{-1/2} \cdot (-2x) = \frac{-x}{(16 - x^2)^{1/2}} = \frac{-x}{\sqrt{16 - x^2}}$$

**c.** Similarly, we obtain

$$f_2(x) = -\sqrt{16 - x^2} = -(16 - x^2)^{1/2}$$

Thus,

$$f_2'(x) = D_x[-(16 - x^2)^{1/2}] = \frac{-1}{2}(16 - x^2)^{1/2-1} \cdot D_x(16 - x^2)$$

$$= \frac{-1}{2}(16 - x^2)^{-1/2} \cdot (-2x) = \frac{x}{(16 - x^2)^{1/2}} = \frac{x}{\sqrt{16 - x^2}}$$

Note that in part **a**, the answer is $D_x y = \frac{-x}{y}$. Thus, the derivative is described not only in terms of $x$ but also in terms of $y$. Furthermore, for each value of $x$ between $-4$ and $4$, there are two values of $y$. The first is $y = f_1(x) = \sqrt{16 - x^2}$. Thus, for that value of $y$, we get $D_x y = \frac{-x}{\sqrt{16 - x^2}}$, which is the answer obtained in part **b** for $f_1'(x)$. The other value of $y$ is $f_2(x) = -\sqrt{16 - x^2}$. Replacing $y$ by that value in the equation $D_x y = \frac{-x}{y}$, we get

$$D_x y = \frac{-x}{-\sqrt{16 - x^2}} = \frac{x}{\sqrt{16 - x^2}}$$

which is the answer obtained in part **c** for $f_2'(x)$. Therefore our three answers are consistent.
**Do Exercise 16.**

### Procedure for Implicit Differentiation

If an equation in two variables, say $x$ and $y$, is given and we are told that this equation defines a differentiable function $f$ such that $y = f(x)$, then we proceed as follows to find $\frac{dy}{dx}$.

**STEP 1.** Differentiate both sides of the equation with respect to $x$.

**STEP 2.** Use all the relevant formulas of differentiation, being careful to use the Chain Rule and keeping in mind that $D_x y \neq 1$.

**STEP 3.** Solve for $D_x y$.

**STEP 4.** If we need to find the value of $D_x y$ for a function $f$ whose graph passes through a point $(a, b)$, simply replace $x$ by $a$ and $y$ by $b$ in the answer obtained in Step 3.

# Exercise Set 7.8

1. Show that there are two functions $f_1$ and $f_2$ with independent variable $x$ defined implicitly by the equation $y^2 + 5y - x^2 = 0$. (*Hint:* Use the quadratic formula.)

2. Same as Exercise 1 with equation $y^2 + 3xy - 4 = 0$.

3. Same as Exercise 1 with equation $y^2 - 4x^2 = 0$.

4. Show that the equation

$$y^3 - 2xy^2 - 5x^2y + 6x^3 = 0$$

defines three functions with independent variable $x$. (*Hint:* $(y - x)$ is a factor of the left side of the equation.)

*In Exercises 5–12, find $D_x y$ using implicit differentiation.*

**5.** $x^3 + x^2 y + 5 = 3x^2 y^2 - 7y + 6x$

**6.** $x^2 y^4 + 10 = 3x^4 y^3 + x^2 - 5$

**7.** $5x^2 y^4 + 6x = 8x^3 + 7y^2 - 10$

**8.** $e^3 x^2 + 5xy^3 = e^2 y^5 + 6xy + \pi$

**9.** $(\log_5 3) x^4 + 6y^7 = e^6 x^2 y^4 + \pi^2$

**10.** $\sqrt[3]{x^2} + \sqrt[5]{y^2} = xy + 5$

**11.** $x\sqrt[4]{y^3} + y^2 \sqrt[7]{x^3} = x^2 y + 7$

**12.** $\dfrac{y}{\sqrt[3]{x^2}} + x^2 = \dfrac{x^3}{\sqrt[9]{y^3}} + 5x + 10$

*In Exercises 13–15, an equation and a point are given. **a.** Verify that the point is on the graph of the equation. **b.** Find the equation of the line tangent to the graph at the given point.*

**13.** $x^2 + 3x^2 y = 5x^2 y^3 + 140$, $(2, -2)$

**14.** $2x^3 + 3y^2 = 5x^2 y + y^3 - 8$, $(-1, 2)$

**15.** $\sqrt[3]{x^2 + 4} + y^2 = \sqrt{y^2 + 3} + 3y - 2$, $(2, 1)$

**16.** Find $\frac{dy}{dx}$ implicitly for the equation of Exercise 1. Then find $f_1'(x)$ and $f_2'(x)$ where $f_1(x)$ and $f_2(x)$ were found in Exercise 1. Show that the three answers are consistent.

**17.** Same as Exercise 16 for the equation of Exercise 2.

**18.** Same as Exercise 16 for the equation of Exercise 3.

**19.** Find $\frac{dy}{dx}$ implicitly for the equation of Exercise 4. Then find $f_1'(x)$, $f_2'(x)$, and $f_3'(x)$ where $f_1(x)$, $f_2(x)$, and $f_3(x)$ were found in Exercise 4. Show that the four answers are consistent.

# 7.9 Derivatives of Exponential and Logarithmic Functions

## The Constant $e$

The exponential and logarithmic functions were introduced in Sections 1.4 and 1.5, respectively. We now undertake the derivation of formulas that will allow us to differentiate these functions.

Suppose $b$ is a positive real number and $b \neq 1$. Then the exponential function with base $b$ is defined by

$$y = f(x) = b^x$$

Now,

$$f'(x) = \lim_{h \to 0} \frac{f(x + h) - f(x)}{h} = \lim_{h \to 0} \frac{b^{(x + h)} - b^x}{h}$$

$$= \lim_{h \to 0} \frac{b^x b^h - b^x}{h}$$

$$= \lim_{h \to 0} \frac{b^x (b^h - 1)}{h}$$

$$= (\lim_{h \to 0} b^x)\left(\lim_{h \to 0} \frac{b^h - 1}{h}\right)$$

$$= b^x \lim_{h \to 0} \frac{b^h - 1}{h}$$

provided this last limit exists. (In more advanced texts, it is shown that this limit does exist. Furthermore, it is proved that $\lim_{h \to 0} \dfrac{e^h - 1}{h} = 1$.) Recall that the number $e$ is an irrational number introduced in Chapter 1 and that $e \doteq 2.71828$.

## The Derivative of $e^x$

If we choose $e$ as the base of our exponential function, we get

$$f'(x) = D_x e^x = e^x \lim_{h \to 0} \frac{e^h - 1}{h} = e^x \cdot 1 = e^x$$

It is indeed an interesting fact that the exponential function is its own derivative. We now state the following theorem.

> **THEOREM 7.9:**   If $f$ is a differentiable function and $y = e^{f(x)}$, then
> $D_x e^{f(x)} = e^{f(x)} f'(x)$.

**Proof:** Let $u = f(x)$ so that $y = e^u$. Then

$$\frac{du}{dx} = f'(x) \text{ and } \frac{dy}{du} = e^u$$

By the Chain Rule, we get

$$\frac{dy}{dx} = \frac{dy}{du} \cdot \frac{du}{dx} = e^u \cdot f'(x) = e^{f(x)} f'(x)$$

This formula is often written

$$D_x e^u = e^u \cdot D_x u$$

and is called the formula for the derivative of $e$ to a function.

**EXAMPLE 1**   Find $\dfrac{dy}{dx}$ in each of the following cases.

**a.** $y = e^{x^2 + 3x + 1}$
**b.** $y = (x^2 + 5)^3 e^{5x + 6}$
**c.** $y = \dfrac{e^{x^3}}{x^4 + 3}$

**Solution**   **a.** We use the formula $D_x e^u = e^u D_x u$, where $u = x^2 + 3x + 1$.

Thus,

$$\frac{dy}{dx} = D_x e^{x^2 + 3x + 1} = e^{x^2 + 3x + 1} D_x (x^2 + 3x + 1)$$

$$= e^{x^2 + 3x + 1}(2x + 3) = (2x + 3)e^{x^2 + 3x + 1}$$

**b.** We first use the Product Rule (Formula 5, Section 7.7).

$$\frac{dy}{dx} = D_x[(x^2 + 5)^3 e^{5x + 6}] = (x^2 + 5)^3 D_x e^{5x + 6} + e^{5x + 6} D_x (x^2 + 5)^3$$

Next, we use the formula $D_x e^u = e^u \cdot D_x u$ and the General Power Rule (Formula 8, Section 7.7). We get

$$\frac{dy}{dx} = (x^2 + 5)^3 e^{5x+6} D_x(5x + 6) + e^{5x+6} 3(x^2 + 5)^{3-1} D_x(x^2 + 5)$$

$$= (x^2 + 5)^3 e^{5x+6}(5) + e^{5x+6} 3(x^2 + 5)^2(2x)$$

$$= (x^2 + 5)^2 e^{5x+6}[5(x^2 + 5) + 6x] = (x^2 + 5)^2 e^{5x+6}(5x^2 + 6x + 25)$$

$$= (x^2 + 5)^2 (5x^2 + 6x + 25)e^{5x+6}.$$

**c.** We first use the Quotient Rule (Formula 6, Section 7.7).

$$\frac{dy}{dx} = D_x \frac{e^{x^3}}{x^4 + 3} = \frac{(x^4 + 3)D_x e^{x^3} - e^{x^3} D_x(x^4 + 3)}{(x^4 + 3)^2}$$

$$= \frac{(x^4 + 3)e^{x^3} D_x x^3 - e^{x^3} 4x^{4-1}}{(x^4 + 3)^2}$$

$$= \frac{(x^4 + 3)e^{x^3} 3x^2 - e^{x^3} 4x^3}{(x^4 + 3)^2}$$

$$= \frac{e^{x^3}(3x^6 + 9x^2 - 4x^3)}{(x^4 + 3)^2} = \frac{x^2(3x^4 - 4x + 9)e^{x^3}}{(x^4 + 3)^2}$$

**Do Exercises 3, 7, and 13.**

■

Students who are beginning their calculus studies often have difficulty remembering the difference between $D_x e^u$ and $D_x e^x$. The correct formulas are

$$D_x e^u = e^u \cdot D_x u \text{ and } D_x e^x = e^x$$

To avoid difficulties, a beginner might always want to use the first of the two formulas to get

$$D_x e^x = e^x \cdot D_x x = e^x \cdot 1 = e^x$$

demonstrating that the second formula is a special case of the first one.

### The Derivative of $b^{f(x)}$

Using a method that we introduce in the following section, it can be proved that if $b$ is a positive number and $f$ is a differentiable function, then

$$D_x b^{f(x)} = b^{f(x)}(\ln b)f'(x)$$

Since $\ln e = 1$, we may apply this formula to differentiate $e^{f(x)}$ to get

$$D_x e^{f(x)} = e^{f(x)}(\ln e)f'(x)$$

$$= e^{f(x)}(1)f'(x) = e^{f(x)}f'(x)$$

Therefore the formula used to differentiate $e^{f(x)}$ is a special case of the formula used to differentiate $b^{f(x)}$. Differentiation of functions involving exponents is often confusing to the beginner. It is extremely important to keep in mind that in

$$D_x u^r = r u^{r-1} D_x u \qquad (*)$$

the base $u$ is a variable and the exponent $r$ is a constant. On the other hand, in

$$D_x b^u = b^u(\ln b)D_x u \qquad (**)$$

the base $b$ is constant while the exponent $u$ is variable. For example, to find $D_x(x^3 + 3)^{10}$, we should use (*) and to find $D_x 2^{x^2+3x+1}$, we should use (**). We shall soon learn to differentiate $f(x)^{g(x)}$ where both the base and exponent are variables.

## The Derivative of the Logarithmic Function

We proceed as follows.*

$\quad y = \ln x$ is equivalent to $e^y = x$

Differentiate both sides of the equation to obtain

$$D_x e^y = D_x x$$

That is,

$$e^y D_x y = 1$$

Thus,

$$D_x y = \frac{1}{e^y} = \frac{1}{x}$$

It can be shown that if $f$ is a differentiable function, then $D_x |f(x)| = \frac{|f(x)|}{f(x)} \cdot f'(x)$ whenever $f(x) \neq 0$ (see Exercise 45). Using this result, we can prove the following theorem.

> **THEOREM 7.10:**  Let $f$ be a differentiable function. Then whenever $f(x) \neq 0$, we have
>
> $$D_x \ln|f(x)| = \frac{f'(x)}{f(x)}$$

**Proof:** Let $u = |f(x)|$ and $y = \ln u$. Then $u > 0$, so that $\ln u$ is defined. Furthermore, $\frac{dy}{du} = \frac{1}{u}$ and

$$D_x \ln|f(x)| = \frac{dy}{dx} = \frac{dy}{du} \cdot \frac{du}{dx}$$

$$= \frac{1}{u} \cdot \frac{|f(x)|}{f(x)} \cdot f'(x) = \frac{1}{u} \cdot \frac{u}{f(x)} \cdot f'(x) = \frac{f'(x)}{f(x)}$$

---

*In Chapter 1, we saw that the logarithmic function is the inverse of the exponential function. In more advanced texts, a formula giving the relationship between the derivative of a differentiable one-to-one function and its inverse is derived. For our purpose, it is sufficient to assume that the derivative of the logarithmic function exists and to derive the formula using implicit differentiation.

This result is often written as

$$D_x \ln|u| = \frac{D_x u}{u}$$

Note that no absolute value sign appears on the right side of the formula.

**EXAMPLE 2** Find $\frac{dy}{dx}$ in each case.

    **a.** $y = \ln|x^3 + 3x + 6|$
    **b.** $y = (x^2 + 1)\ln(x^2 + 1)$
    **c.** $y = \dfrac{\ln|x^3 + 10|}{x^3 + 10}$

**Solution**   **a.** We use the formula $D_x\ln|u| = D_x u/u$, where $u = x^3 + 3x + 6$. We get

$$\frac{dy}{dx} = D_x\ln|x^3 + 3x + 6|$$

$$= \frac{D_x(x^3 + 3x + 6)}{x^3 + 3x + 6} = \frac{3x^2 + 3}{x^3 + 3x + 6}$$

**b.** We first use the Product Rule.

$$\frac{dy}{dx} = (x^2 + 1)D_x\ln(x^2 + 1) + [\ln(x^2 + 1)]D_x(x^2 + 1)$$

Now we use the formula $D_x\ln|u| = D_x u/u$, where $u = x^2 + 1$, to get

$$\frac{dy}{dx} = (x^2 + 1)\frac{D_x(x^2 + 1)}{x^2 + 1} + [\ln(x^2 + 1)]2x$$

$$= 2x + [\ln(x^2 + 1)]2x$$

$$= 2x[1 + \ln(x^2 + 1)]$$

**c.** Using the Quotient Rule, we get

$$\frac{dy}{dx} = \frac{(x^3 + 10)D_x\ln|x^3 + 10| - [\ln|x^3 + 10|]D_x(x^3 + 10)}{(x^3 + 10)^2}$$

$$= \frac{(x^3 + 10) D_x(x^3 + 10)/(x^3 + 10) - [\ln|x^3 + 10|]3x^2}{(x^3 + 10)^2}$$

$$= \frac{3x^2 - [\ln|x^3 + 10|]3x^2}{(x^3 + 10)^2}$$

$$= \frac{3x^2[1 - \ln|x^3 + 10|]}{(x^3 + 10)^2}$$

**Do Exercises 17, 21, and 25.**

**EXAMPLE 3** The equation $3x^2 + e^y = 4y^3\ln|x| + 5$ defines a differentiable function $f$ where $y = f(x)$. Use implicit differentiation to find $\frac{dy}{dx}$.

**Solution** We first differentiate both sides of the equation with respect to $x$.

$$D_x(3x^2 + e^y) = D_x(4y^3 \ln|x| + 5)$$

Using the appropriate differentiation formulas, we get

$$6x + e^y \frac{dy}{dx} = 4y^3 D_x \ln|x| + [\ln|x|]D_x 4y^3 + 0$$

and

$$6x + e^y \frac{dy}{dx} = 4y^3\left(\frac{1}{x}\right) + [\ln|x|]4(3y^2)\frac{dy}{dx}$$

Solving for $\frac{dy}{dx}$, we get

$$e^y \frac{dy}{dx} - 12y^2[\ln|x|]\frac{dy}{dx} = \frac{4y^3}{x} - 6x$$

Factoring $\frac{dy}{dx}$, we obtain

$$(e^y - 12y^2\ln|x|)\frac{dy}{dx} = \frac{4y^3 - 6x^2}{x}$$

and

$$\frac{dy}{dx} = \frac{4y^3 - 6x^2}{x(e^y - 12y^2\ln|x|)} \quad \text{(provided } e^y - 12y^2\ln|x| \neq 0)$$

**Do Exercise 39.** ∎

**EXAMPLE 4** Recall that $\log_b|u| = \frac{\ln|u|}{\ln b}$. (See Section 1.5.) Use this fact to find $\frac{dy}{dx}$ if $y = \log_5|3x + 2|$.

**Solution** Write $\log_5|3x + 2| = \frac{\ln|3x + 2|}{\ln 5} = \frac{1}{\ln 5}\ln|3x + 2|$. Then

$$\frac{dy}{dx} = D_x \log_5|3x + 2| = D_x \frac{1}{\ln 5}\ln|3x + 2| = \frac{1}{\ln 5}D_x\ln|3x + 2|$$

$$= \frac{1}{\ln 5} \cdot \frac{D_x(3x + 2)}{3x + 2} = \frac{3}{(\ln 5)(3x + 2)}.$$

**Do Exercise 31.** ∎

### More Differentiation Formulas

At the end of Section 7.7, we listed the basic differentiation formulas we derived up to that point. These formulas were numbered 1–8. We continue numbering our new formulas for future reference.

9. $D_x e^u = e^u \cdot D_x u$
10. $D_x b^u = b^u(\ln b)D_x u$
11. $D_x \ln|u| = \dfrac{D_x u}{u}$

## Exercise Set 7.9

*In Exercises 1–30, find $\dfrac{dy}{dx}$.*

**1.** $y = e^{x^5}$

**2.** $y = e^{\sqrt{x}}$

**3.** $y = e^{2x^2+5x+1}$

**4.** $y = e^{x^3+5x/2+3}$

**5.** $y = e^{(x^2+5x+1)^2}$

**6.** $y = (x^3 + 1)e^{5x}$

**7.** $y = (x^2 + 3x + 7)e^{x^2+3}$

**8.** $y = (x^4 + 3x + 2)^2 e^{x^4+5}$

**9.** $y = \sqrt{x^2 + 7}e^{x^2+7}$

**10.** $y = \sqrt[3]{x^4 + 3}e^{6x+2}$

**11.** $y = \dfrac{e^{5x}}{x^2}$

**12.** $y = \dfrac{e^{3x^2+1}}{x^3 + x + 2}$

**13.** $y = \dfrac{x^4 + 2}{e^{5x}}$

**14.** $y = \dfrac{x^2 + 6x + 2}{e^{7x}}$

**15.** $y = \dfrac{\sqrt{x^2 + 1}}{e^{3x+2}}$

**16.** $y = \dfrac{\sqrt[3]{x^2}}{e^{x^2+3}}$

**17.** $y = \ln(x^6 + x^2 + 1)$

**18.** $y = \ln|x^3 + 2x^2 + 5x + 1|$

**19.** $y = \ln|x^5 + 3x^2 - 6|$

**20.** $y = x^3\ln(x^2 + 7)$

**21.** $y = (x^4 + 6x + 1)\ln|x^3 - 1|$

**22.** $y = \sqrt{x^2 + 8}\,\ln|5x + 1|$

**23.** $y = \sqrt[3]{x^2 + 6}\,\ln(x^2 + 6)$

**24.** $y = \dfrac{\ln|x^3 + 5x - 2|}{x^4 + 10}$

**25.** $y = \dfrac{\ln(5x^4 + 2)}{x^5 + 2x + 1}$

**26.** $y = \dfrac{\ln|x^3 + x - 6|}{\sqrt{x^2 + 8}}$

**27.** $y = \dfrac{\ln(x^4 + 1)}{\ln(x^2 + 2)}$

**28.** $y = (e^{x^2} + 1)^5[\ln(x^2 + 4) + 2]^2$

**29.** $y = \ln(x^2 + 2) + \ln|x^4 - 3| - e^{x^2}(3 + \ln|5x|)^2$

**30.** $y = 5x^e + e^{5x} + (x^2 + 1)^e + e^{x^2+1} + e^2$

*Recall that $\log_b|u| = \dfrac{\ln|u|}{\ln b}$ (see Section 1.5). Use this fact to find $\dfrac{dy}{dx}$ in Exercises 31–35.*

**31.** $y = \log_3|5x + 1|$

**32.** $y = \log_5|x^3 + 5x + 6|$

**33.** $y = \log|x^2 + 5x + 2|$

**34.** $y = \log_7|x^5 + 3x^2 + 6x - 2|$

**35.** $y = (x^2 + 6)\log_6|x^3 + 3x + 7|$

*In Exercises 36–41, the given equation defines a differentiable function f where $y = f(x)$. In each case, use implicit differentiation to find $\frac{dy}{dx}$.*

**36.** $e^{xy} + x^2 = y^2 + x^3 + 2$

**37.** $e^{x^2} + e^{x^3y^2} = 10x^2y^4 + 11$

**38.** $\ln(x^2y^4) + 5x^2 = 10x^2y^5 + 10$

**39.** $\ln(x^6y^2) + e^{xy} = 3x^2y^3$

**40.** $\ln \sqrt[5]{x^2y^6} = 10x^2y^4 + 17$

**41.** $5x^2y^3 + 10 = 5\ln \sqrt[7]{x^2y^\pi} + x^2 + 2$

*In Exercises 42–44, find the equation of the line tangent to the graph of the given equation at the given point.*

**42.** $e^{x^2y} + 3x^2y = y^3 - 7$, $(0, 2)$

**43.** $\ln(x^2y^4) = \dfrac{x + y^2}{e} + 2$, $(e, \sqrt{e})$

**44.** $y = 5e^{x^2} + 3x + 1$, $(0, 6)$

**45.** Suppose that $f$ is a differentiable function and that $f(x) \neq 0$. Show that $D_x|f(x)| = \dfrac{|f(x)|}{f(x)}f'(x)$.

(*Hint:* Use the following facts: **a.** $|f(x)| = f(x)$ whenever $f(x) > 0$; **b.** $|f(x)| = -f(x)$ whenever $f(x) < 0$.)

**46.** The size $P(t)$ of a population in year $t$ is given by the equation

$$P(t) = 5000e^{5t/2000}$$

Find the rate of change of the population in the year 1990.

**47.** A strain of bacteria is being grown in a culture. Its size $P(t)$ after $t$ hours is given by the formula

$$P(t) = 2000e^{(.2)t}$$

Find the rate of change of the populaton when $t = 10$.

*Before working Exercises 48–51, see Exercise 28, Section 7.5. In each of these exercises, the cost $C(q)$ (in dollars) to produce $q$ units of a commodity is given. A positive integer $a$ is also given.*

a. Find the marginal cost function.
b. Find the approximate cost of producing the $a^{th}$ unit.

**48.** $C(q) = 50,000 + 10qe^{-3q/1000}$, $a = 100$

**49.** $C(q) = 1000 + 75qe^{(-q^2+100)/50,000}$, $a = 100$

**50.** $C(q) = 5000 + \dfrac{10q\ln(q^2 + 1)}{e^{.05q}}$, $a = 10$

**51.** $C(q) = e^{.002q}(-85,000q + 60,000,000)$, $a = 150$

**52.** The amount $A$ of a radioactive substance present at time $t$ is given by the equation

$$A = 500e^{-t^2/100}$$

where $t$ is measured in hours and $A$ is measured in grams. Find **a.** the amount of substance at $t = 10$ and **b.** the rate of change of the amount of substance at $t = 10$.

**53.** An object is moving along a directed line. Its distance $d$ meters from the origin at instant $t$ minutes is given by the equation

$$d = 500 + \ln\sqrt{t^3 + 9}$$

Find the velocity of the object at instant $t = 6$ minutes.

**54.** Determine the interval over which $f'(x) > 0$ if

$$f(x) = (x - 2)e^x$$

(*Hint:* $e^x > 0$ for all $x$.)

**55.** Determine the intervals over which $f'(x) > 0$ if

$$f(x) = (x^2 - x - 1)e^x$$

**56.** Determine the interval over which $f'(x) > 0$ if

$$f(x) = (x^2 + 3x - 4)e^{2x}$$

# 7.10 Logarithmic Differentiation

## Review of the Laws of Logarithms

Until recently, the labor involved in many numerical computations was considerably simplifed by the logarithmic function. This function can change a multiplication to addition, a division to subtraction, raising to a power to multiplication, and extracting a root to division, as shown in the following formulas.

$$\log_b|xy| = \log_b|x| + \log_b|y|$$

(The logarithm of a *product* equals the *sum* of the logarithms of its factors.)

$$\log_b\left|\frac{x}{y}\right| = \log_b|x| - \log_b|y|$$

(The logarithm of a *quotient* equals the logarithm of the dividend *minus* the logarithm of the divisor.)

$$\log_b|x|^r = r\log_b|x|$$

(The logarithm of a *power* equals the exponent of the power *multiplied* by the logarithm of the base of the power.)

$$\text{lob}_b\sqrt[n]{|x|} = \frac{\log_b|x|}{n}$$

(The logarithm of a *root* of a quantity equals the logarithm of that quantity *divided* by the index of the radical.)

## Logarithmic Differentiation

With the advent of calculators, logarithms are no longer used for the purpose of arithmetical computations. However, the logarithmic function is extremely useful in differentiating functions that involve products, quotients, powers, and radicals since it is indeed simpler to differentiate sums and differences than it is to differentiate products and quotients. We illustrate the method with the following example.

EXAMPLE 1   Find $f'(x)$ if

$$f(x) = \frac{(x^3 + 1)^5 \sqrt[3]{x^3 + 3x + 1}}{(2x^2 - 5x + 1)^3}$$

**Solution**   Whenever $f(x) \neq 0$, $|f(x)| > 0$ and $\ln |f(x)|$ is defined. Thus, we first take the absolute value of both sides of the given equation to obtain

$$|f(x)| = \left| \frac{(x^3 + 1)^5 \sqrt[3]{x^3 + 3x + 1}}{(2x^2 - 5x + 1)^3} \right|$$

$$= \frac{|x^3 + 1|^5 |x^3 + 3x + 1|^{1/3}}{|2x^2 - 5x + 1|^3}$$

Thus,

$$\ln |f(x)| = \ln \frac{|x^3 + 1|^5 |x^3 + 3x + 1|^{1/3}}{|2x^2 - 5x + 1|^3}$$

$$= \ln |x^3 + 1|^5 + \ln |x^3 + 3x + 1|^{1/3} - \ln |2x^2 - 5x + 1|^3$$

$$= 5\ln |x^3 + 1| + \frac{1}{3}\ln |x^3 + 3x + 1| - 3\ln |2x^2 - 5x + 1|$$

Now, differentiating both sides of this equality with respect to $x$, we get

$$\frac{f'(x)}{f(x)} = 5\frac{D_x(x^3 + 1)}{x^3 + 1} + \frac{1}{3}\frac{D_x(x^3 + 3x + 1)}{x^3 + 3x + 1} - 3\frac{D_x(2x^2 - 5x + 1)}{2x^2 - 5x + 1}$$

$$= 5\frac{3x^2}{x^3 + 1} + \frac{1}{3}\frac{3x^2 + 3}{x^3 + 3x + 1} - 3\frac{4x - 5}{2x^2 - 5x + 1}$$

$$= \frac{15x^2}{x^3 + 1} + \frac{x^2 + 1}{x^3 + 3x + 1} - \frac{12x - 15}{2x^2 - 5x + 1}$$

Muliltplying both sides by $f(x)$, we obtain

$$f'(x) = f(x)\left( \frac{15x^2}{x^3 + 1} + \frac{x^2 + 1}{x^3 + 3x + 1} - \frac{12x - 15}{2x^2 - 5x + 1} \right)$$

That is,

$$f'(x) = \frac{(x^3 + 1)^5 \sqrt[3]{x^3 + 3x + 1}}{(2x^2 - 5x + 1)^3}\left( \frac{15x^2}{x^3 + 1} + \frac{x^2 + 1}{x^3 + 3x + 1} - \frac{12x - 15}{2x^2 - 5x + 1} \right)$$

**Do Exercise 5.**

    The technique used in the preceding example is called *logarithmic differentiation*.
    Recall that in dealing with arbitrary real exponents, we require the base to be a positive number to avoid difficulties with imaginary numbers. We now generalize the formula $D_x x^n = nx^{n-1}$, which was proved earlier for the case where $n$ is an integer and also where $n$ is a rational number.

EXAMPLE 2   Prove that if $x > 0$ and $r$ is an arbitrary real number, then

$$D_x x^r = rx^{r-1}$$

**Solution**   We let $y = x^r$. Thus, $y > 0$ and

$$\ln y = \ln x^r = r\ln x$$

Assuming that $\dfrac{dy}{dx}$ exists, we differentiate both sides with respect to $x$, to obtain

$$D_x\ln y = D_x(r\ln x)$$

or

$$\frac{D_x y}{y} = rD_x \ln x = r \cdot \frac{1}{x}$$

Multiplying both sides by $y$, we get

$$D_x y = y\left(\frac{r}{x}\right) = x^r\left(\frac{r}{x}\right) = rx^{r-1} \qquad *$$

In the preceding section we introduced the formula

$$D_x b^u = b^u(\ln b)D_x u$$

However, if you are willing to use logarithmic differentiation, you need not memorize this formula.   ◼

**EXAMPLE 3**   Find $D_x y$ if $y = 2^{x^2+3x+5}$.

**Solution**   Clearly, $y > 0$. Thus we can take the natural logarithm of both sides of the given equation to get

$$\ln y = \ln 2^{x^2+3x+5} = (x^2 + 3x + 5)\ln 2$$

Differentiating both sides with respect to $x$, we obtain

$$\frac{D_x y}{y} = (\ln 2)D_x(x^2 + 3x + 5) \quad \text{(since } \ln 2 \text{ is a constant)}$$

Thus,

$$\frac{D_x y}{y} = (\ln 2)(2x + 3)$$

Multiplying both sides by $y$, we get

$$D_x y = y(\ln 2)(2x + 3)$$

That is,

$$D_x y = 2^{x^2+3x+5}(\ln 2)(2x + 3)$$

**Do Exercise 11.**   ◼

---

*Example 2 could have been solved as follows. If $x > 0$, we can write

$$x = e^{\ln x}$$

So, $x^r = (e^{\ln x})^r = e^{r\ln x}$ and $D_x x^r = D_x e^{r\ln x} = e^{r\ln x}(D_x r\ln x) = x^r \cdot r \cdot \dfrac{1}{x} = rx^{r-1}$.

EXAMPLE 4　In a calculus exam, a student was asked to find $\frac{dy}{dx}$ if $y = e^{x^2}/(x^3 + 3x + 5)$. The student could not recall the Product Rule and the Quotient Rule, but knew the logarithmic function well  Show how the student was able to find the correct answer.

**Solution**　The student probably wrote

$$|y| = \left| \frac{e^{x^2}}{x^3 + 3x + 5} \right| = \frac{e^{x^2}}{|x^3 + 3x + 5|}$$

Therefore,

$$\ln|y| = \ln \frac{e^{x^2}}{|x^3 + 3x + 5|}$$

$$= \ln e^{x^2} - \ln|x^3 + 3x + 5| = x^2 \ln e - \ln|x^3 + 3x + 5|$$

$$= x^2 - \ln|x^3 + 3x + 5| \quad \text{(since } \ln e = 1\text{)}$$

Differentiating both sides with respect to $x$ we get

$$D_x \ln|y| = D_x(x^2 - \ln|x^3 + 3x + 5|)$$

and

$$\frac{D_x y}{y} = 2x - \frac{D_x(x^3 + 3x + 5)}{x^3 + 3x + 5} = 2x - \frac{3x^2 + 3}{x^3 + 3x + 5}$$

Multiplying both sides by $y$,

$$D_x y = y\left(2x - \frac{3x^2 + 3}{x^3 + 3x + 5}\right)$$

or

$$D_x y = \frac{e^{x^2}}{x^3 + 3x + 5}\left(2x - \frac{3x^2 + 3}{x^3 + 3x + 5}\right)$$

■

We remind you once more that in differentiating $u^v$, we use the formula $D_x u^r = ru^{r-1}D_x u$ in the case where $u$ is variable and $v$ is constant, and we use $D_x b^v = b^v(\ln b)D_x v$ in the case where $u$ is constant and $v$ is variable. But what do we do if both $u$ and $v$ are variables? Use logarithmic differentiation!

EXAMPLE 5　Differentiate $f$ if $f(x) = (x^4 + 1)^{3x}$.

**Solution**　Note that $x^4 + 1 > 0$ for all $x$. Thus, $f(x) > 0$. Taking the natural logarithm of both sides, we obtain

$$\ln f(x) = \ln(x^4 + 1)^{3x}$$

$$= 3x \ln(x^4 + 1)$$

Differentiating both sides with respect to $x$, we get

$$D_x \ln f(x) = D_x[3x\ln(x^4 + 1)]$$

Using Formula 11 (Section 7.9) on the left side and the Product Rule on the right side, we obtain

$$\frac{f'(x)}{f(x)} = 3x \cdot D_x \ln(x^4 + 1) + [\ln(x^4 + 1)]D_x 3x$$

Continuing to use the appropriate formulas,

$$\frac{f'(x)}{f(x)} = 3x \frac{D_x(x^4 + 1)}{x^4 + 1} + [\ln(x^4 + 1)]3 = 3x \frac{4x^3}{x^4 + 1} + 3\ln(x^4 + 1)$$

Thus,

$$\frac{f'(x)}{f(x)} = \frac{12x^4}{x^4 + 1} + 3\ln(x^4 + 1)$$

Multiplying both sides of this equation by $f(x)$, we get

$$f'(x) = f(x)\left[\frac{12x^4}{x^4 + 1} + 3\ln(x^4 + 1)\right]$$

That is,

$$f'(x) = (x^4 + 1)^{3x}\left[\frac{12x^4}{x^4 + 1} + 3\ln(x^4 + 1)\right]$$

**Do Exercise 17.**

■

When performing logarithmic differentiation, you must be able to use the laws of logarithms to justify your steps at all times. Be aware, for example, that in general

$$\ln|f(x) + g(x)| \neq \ln|f(x)| + \ln|g(x)|$$

(See Exercise 30.)

### Steps to Follow in Performing Logarithmic Differentiation

In differentiating a function $f$ that involves products, quotients, powers, or radicals, it is often advantageous to use logarithmic differentiation.
   This method is valid only when $f(x) \neq 0$.

**STEP 1.**   Take the absolute value of both sides of the equation defining $f(x)$.

**STEP 2.**   Take the natural logarithm of both sides.

**STEP 3.**   Use the laws of logarithms on the right side to change it to an algebraic sum of terms in the form $k_i \ln|g_i(x)|$.

**STEP 4.**   Differentiate both sides with respect to $x$.

**STEP 5.**   Obtain $f'(x)$ by multiplying both sides by $f(x)$.

# Exercise Set 7.10

*In Exercises 1–21, use logarithmic differentiation to find $\dfrac{dy}{dx}$.*

**1.** $y = (x^2 + 1)(x^3 + 3x + 2)(x^4 + x^2 + 1)$

**2.** $y = \dfrac{e^{5x}(x^4 + 1)}{x^2 + 3x + 2}$

**3.** $y = (x^3 + 5x^2 + 6x + 2)^2 \sqrt{x^3 + 1}$

**4.** $y = (x^4 + 3x^2 - 6)^5 e^{x^2+1} \sqrt[5]{x^3 + 3x^2 + 10}$

**5.** $y = \dfrac{(3x^5 - 4x^2 + 6)^4(x^2 - 6x + 7)^3}{(x^2 + 1)^{1/3}(x^3 - 4x^2 + 6)^2}$

**6.** $y = \dfrac{\sqrt[4]{x^4 + x^2 + 1}\sqrt[3]{5x^2 - 6x + 3}}{(x^3 + 6)^2}$

**7.** $y = \dfrac{e^{x^2}(x^3 + 1)^2}{(x^4 + 3)^2(x^3 + 1)^2}$

**8.** $y = \dfrac{\sqrt{e^{2x} + 1}(x^2 + 10x - 1)^2}{(x^2 + 1)^5(4x + 3)^3}$

**9.** $y = \dfrac{5}{(x^4 + 1)(x^2 + 6)^2 \sqrt[3]{x^3 + x^2 + x + 1}}$

**10.** $y = 2^{x^3+5x^2+6x+1}$

**11.** $y = 5^{5x^2-6x+10}$

**12.** $y = \pi^{6x^3+2x^2-5x+6}$

**13.** $y = 4^{\sqrt{x^2+3}}$

**14.** $y = \dfrac{1}{5^{x^2}}$

**15.** $y = (x^2)(2^x)$

**16.** $y = (x^4 + x^2 + 3)^{x^2+2}$

**17.** $y = (e^x + 4)^{x^2}$

**18.** $y = (5x^2 + 6)^{\sqrt{x^2+4}}$

**19.** $y = (x^6 + 3x^2 + 5)^{e^{3x}}$

**20.** $y = (e^{3x} + e^{2x} + e^x + 1)^{\ln(x^2+5)}$

**21.** $y = [(x^2 + 1)^{5x+2}]^{x^2+6}$

**22.** Suppose an object is moving on a directed line and its distance $d$ feet from the origin at instant $t$ seconds is given by

$$d = \dfrac{(t^2 + 1)e^{5t}}{\sqrt{t^3 + t^2 + 1}}$$

Find its velocity at instant $t = 4$ seconds.

**23.** Find the equation of the tangent line to the graph of

$$y = \dfrac{108x(x^4 + 1)^2 \sqrt[3]{x^2 + x + 2}}{(x^2 + 2)^3 \sqrt{x^8 + x^4 + 8x + 1}}$$

at the point $(2, 34)$.

**24.** The cost $C(q)$ (in dollars) to produce $q$ units of a commodity is given by the equation

$$C(q) = 6000 + q(q^2 + 10)^{-3q/100 + 1}$$

**a.** Find the marginal cost function.
**b.** Find the approximate cost to produce the 101st unit.
(*Hint:* See Exercise 28, Section 7.5.)

**25.** If $f_1, f_2, \ldots , f_n$ are differentiable functions and $u_1 = f_1(x)$, $u_2 = f_2(x), \ldots , u_n = f_n(x)$, use logarithmic differentiation to show that

$$D_x(u_1 u_2 \ldots u_n) = \sum_{i=1}^{n} (u_1 u_2 \ldots u_n)\left(\dfrac{D_x u_i}{u_i}\right)$$

**26.** Use the result of Exercise 25 to find $D_x y$ for

**a.** $y = (x^2 + 5x + 7)e^{x^2}\ln(x^2 + 3)$
**b.** $y = (x^3 + 5)^2 \sqrt{x^4 + 7}\,(e^{5x} + 1)^{10}(5^{x^2})$

**27.** It has been said that two wrongs don't make a right. But sometimes it does—as illustrated by the following. On a calculus exam, students were asked to find $\frac{dy}{dx}$ where $y = (x^2 + 1)^{5x}$. One student wrongly used the formula $D_x u^r = ru^{r-1}D_x u$, with $u = x^2 + 1$ and $r = 5x$ and claimed that

$$\dfrac{dy}{dx} = 5x(x^2 + 1)^{5x-1}D_x(x^2 + 1)$$

$$= 5x(x^2 + 1)^{5x-1}(2x) = 10x^2 (x^2 + 1)^{5x-1}$$

Of course this answer is wrong. Another student thought he should use the formula $D_x b^u = b^u(\ln b)D_x u$, with $b = x^2 + 1$ and $u = 5x$ and got

$$\dfrac{dy}{dx} = (x^2 + 1)^{5x}[\ln(x^2 + 1)]D_x(5x)$$

$$= 5(x^2 + 1)^{5x}[\ln(x^2 + 1)]$$

Again, this answer is wrong. The correct way to find $D_x y$ is to use logarithmic differentiation. Show that if you add the two wrong answers you get the correct answer.

28. Suppose $f$ and $g$ are differentiable functions, $u = f(x)$ and $v = g(x)$, and $f(x) > 0$ for all $x$. Use logarithmic differentiation to show that

$$D_x u^v = v(u^{v-1})(D_x u) + u^v(\ln u)(D_x v)$$

(*Hint:* See Exercise 27.)

29. Use the result of Exercise 28 to find $D_x y$ if

$$y = (x^4 + 3)^{x^2}$$

*30. In general, the statement

$$\ln|f(x) + g(x)| = \ln|f(x)| + \ln|g(x)|$$

is false.

a. Give three examples of functions $f$ and $g$ for which the statement is false.
b. Show, however, that if $x^2 - 1 \neq 0$, $f(x) = x^2$, and $g(x) = x^2/(x^2 - 1)$, then the statement is true.
c. What relationship between $f(x)$ and $g(x)$ must be satisfied for the statement to be true?

## 7.11 Higher Order Derivatives

### The Second Derivative

We have seen that the instantaneous velocity of a moving object may be found by evaluating the derivative of the distance function at a given instant. Thus, if $s$ is the distance traveled by a moving object at instant $t$ and $s = f(t)$, then the velocity $v$ is given by the equation

$$v = f'(t)$$

Often we hear that a certain car has good acceleration. What is meant by that statement? It simply indicates that the speed of the car can be increased rapidly. Thus, acceleration measures a rate of change of velocity and, since we use derivatives to calculate instantaneous rates of change, it is natural to define *acceleration* as the derivative, with respect to $t$, of the velocity. Thus,

$$a = D_t v$$

Since $v = f'(t)$, the acceleration is the derivative of the derivative of $f$. This is but one of the many examples of what is called a *second derivative*.

### Higher Order Derivatives

In general, if we start with a differentiable function, its derivative is a function. If the derivative is differentiable, we can differentiate it and get another function (the second derivative). If the second derivative is differentiable, we can differentiate again to get the third derivative, and so on. We may keep differentiating each derivative as long as we have a differentiable function at each stage.

EXAMPLE 1    Let $f(x) = x^3 + 2x^2 + 5x + 1$. Find the first, second, third, and fourth derivatives of $f$.

**Solution**   $f(x) = x^3 + 2x^2 + 5x + 1.$

$$f'(x) = D_x(x^3 + 2x^2 + 5x + 1)$$
$$= 3x^2 + 4x + 5$$
$$f''(x) = D_x f'(x) = D_x(3x^2 + 4x + 5) = 6x + 4$$
$$f'''(x) = D_x f''(x) = D_x(6x + 4) = 6$$
$$f''''(x) = D_x\, 6 = 0$$

**Do Exercise 3.**                                                                                     ■

## Notation

We have seen that if $y = f(x)$, where $f$ is a differentiable function, then each of the following symbols denotes the (first) derivative of $f$ with respect to $x$:

$$y', f'(x), D_x y, D_x f(x), \frac{dy}{dx}, \text{ and } \frac{df(x)}{dx}$$

Corresponding to each type of symbol, we have symbols that denote the second, third, . . . , $n$th derivatives of $f$. We list those in the following table.

| First | Second | Third | $n$th (derivative) |
|-------|--------|-------|--------------------|
| $y'$ | $y''$ | $y'''$ | $y^{(n)}$ |
| $f'(x)$ | $f''(x)$ | $f'''(x)$ | $f^{(n)}(x)$ |
| $D_x y$ | $D_x^2 y$ | $D_x^3 y$ | $D_x^n y$ |
| $D_x f(x)$ | $D_x^2 f(x)$ | $D_x^3 f(x)$ | $D_x^n f(x)$ |
| $\dfrac{dy}{dx}$ | $\dfrac{d^2 y}{dx^2}$ | $\dfrac{d^3 y}{dx^3}$ | $\dfrac{d^n y}{dx^n}$ |
| $\dfrac{df(x)}{dx}$ | $\dfrac{d^2 f(x)}{dx^2}$ | $\dfrac{d^3 f(x)}{dx^3}$ | $\dfrac{d^n f(x)}{dx^n}$ |

Note the position of 2 in $d^2y/dx^2$. It is above the $d$ in the top term and above the $x$ in the bottom term. The reason for this peculiarity is the notation used for differentiating with respect to $x$. Recall that either $\frac{d}{dx}$ or $D_x$ is used for that purpose. Thus, if we differentiate $y$ with respect to $x$, we get $\frac{d}{dx}(y) = \frac{dy}{dx}$. If we differentiate $\frac{dy}{dx}$ with respect to $x$, we get $\frac{d}{dx}\left(\frac{dy}{dx}\right)$ which we can condense to $d^2y/dx^2$. Also note the use of parentheses in the symbols $y^{(n)}$ and $f^{(n)}(x)$. This is to avoid confusing these symbols with $y^n$ and $f^n(x)$, which denote the $n$th power of $y$ and $f(x)$, respectively. Finally, it is sometimes convenient to write $f^{(0)}(x)$ for $f(x)$. Be careful not to treat the superscript indicating the order of a derivative as an exponent. For example, $D_x^2 x^3 \neq (D_x x^3)^2$. To prove this, observe that $D_x^2 x^3 = D_x 3x^2 = 6x$ and $(D_x x^3)^2 = (3x^2)^2 = 9x^4$.

**EXAMPLE 2**   Find the first, second, and third derivatives of the functions defined by the following equations.

**a.** $y = e^{x^2}$
**b.** $s = \ln(t^2 + 1)$

**Solution**    We shall use the different notations to help you become familiar with all of them.

**a.** $y' = D_x e^{x^2} = e^{x^2} D_x x^2 = e^{x^2}(2x) = 2xe^{x^2}$

$$\frac{d^2y}{dx^2} = D_x(2xe^{x^2}) = 2x(D_x e^{x^2}) + e^{x^2}(D_x 2x)$$

$$= 2xe^{x^2}(2x) + e^{x^2}(2) = 2e^{x^2}(2x^2 + 1)$$

$$D_x^3 y = D_x\left(\frac{d^2y}{dx^2}\right) = D_x\,(2e^{x^2}(2x^2 + 1))$$

$$= 2e^{x^2}D_x(2x^2 + 1) + (2x^2 + 1)D_x 2e^{x^2} = 2e^{x^2}(4x) + (2x^2 + 1)2e^{x^2}(2x)$$

$$= 4xe^{x^2}[2 + (2x^2 + 1)] = 4xe^{x^2}(2x^2 + 3)$$

**b.** $s = \ln(t^2 + 1)$.

Here the independent variable is $t$. Thus, we differentiate with respect to $t$. We get

$$\frac{ds}{dt} = D_t \ln(t^2 + 1) = \frac{D_t(t^2 + 1)}{t^2 + 1} = \frac{2t}{t^2 + 1}$$

Therefore,

$$D_t^2 s = D_t\left(\frac{2t}{t^2 + 1}\right) = \frac{(t^2 + 1)D_t(2t) - 2tD_t(t^2 + 1)}{(t^2 + 1)^2}$$

$$= \frac{(t^2 + 1)(2) - 2t(2t)}{(t^2 + 1)^2} = \frac{-2t^2 + 2}{(t^2 + 1)^2}$$

Differentiating again, we get

$$s''' = D_t \frac{-2t^2 + 2}{(t^2 + 1)^2}$$

$$= \frac{(t^2 + 1)^2 D_t(-2t^2 + 2) - (-2t^2 + 2)D_t(t^2 + 1)^2}{[(t^2 + 1)^2]^2}$$

$$= \frac{(t^2 + 1)^2(-4t) - (-2t^2 + 2)2(t^2 + 1)(2t)}{(t^2 + 1)^4}$$

$$= \frac{-4t(t^2 + 1)[(t^2 + 1) + (-2t^2 + 2)]}{(t^2 + 1)^4}$$

$$= \frac{-4t(-t^2 + 3)}{(t^2 + 1)^3} = \frac{4t(t^2 - 3)}{(t^2 + 1)^3}$$

**Do Exercises 5 and 9.**    ■

## Higher Order Derivatives Found Implicitly

We have shown that when a differentiable function is defined implicitly, it is possible to find the derivative without having an explicit formula for the function itself. Expanding on the technique, it is also possible to find the values of higher derivatives. We illustrate this idea with the following example.

**EXAMPLE 3** The equation $x^3 + y^3 = 5x + 4$ defines a three-times-differentiable function $f$ such that $y = f(x)$. Find the values of the first, second, and third order derivatives at $(1, 2)$.

**Solution** Differentiating both sides of the given equation with respect to $x$, we get

$$D_x(x^3 + y^3) = D_x(5x + 4)$$

That is,

$$3x^2 + 3y^2 y' = 5 \tag{1}$$

from which we get

$$y' = \frac{5 - 3x^2}{3y^2} \quad \text{(provided } y \neq 0)$$

To evaluate $y'$ at $(1, 2)$, replace $x$ by 1 and $y$ by 2 in the preceding equation. We obtain

$$y' = \frac{5 - 3(1^2)}{3(2)^2} = \frac{1}{6}$$

Differentiating both sides of (1) with respect to $x$, we get

$$D_x(3x^2 + 3y^2 y') = D_x 5$$

or

$$6x + 3y^2 y'' + y'(6yy') = 0$$

or

$$6x + 3y^2 y'' + 6y(y')^2 = 0 \tag{2}$$

Thus, if $y \neq 0$, we have

$$y'' = \frac{-6x - 6y(y')^2}{3y^2} = \frac{-2x - 2y(y')^2}{y^2} \tag{3}$$

To evaluate $y''$ at $(1, 2)$, we replace $x$ by 1, $y$ by 2, and $y'$ by $\frac{1}{6}$ in the preceding equation. We get

$$y'' = \frac{-2(1) - 2(2)(1/6)^2}{2^2} = \frac{-19}{36}$$

Differentiating both sides of (2) with respect to $x$, we get

$$D_x(6x + 3y^2 y'' + 6y(y')^2) = D_x 0$$

and

$$6 + 3y^2 y''' + y'' 6yy' + 6y2(y')y'' + (y')^2)6y' = 0$$

Thus,

$$3y^2 y''' = -6 - 18yy'y'' - 6(y')^3$$

so that

$$y''' = \frac{-2 - 6yy'y'' - 2(y')^3}{y^2} \quad \text{(provided } y \neq 0)$$

To evaluate $y'''$ at $(1, 2)$, replace $x$ by 1, $y$ by 2, $y'$ by $\frac{1}{6}$, and $y''$ by $\frac{-19}{36}$ in the preceding equation. We get

$$y''' = \frac{-2 - 6(2)(1/6)(-19/36) - 2(1/6)^3}{2^2} = \frac{-103}{432}$$

If we wish to obtain a formula for $y''$ in terms of $x$ and $y$, we simply replace $y'$ by $(5 - 3x^2)/3y^2$ in equation (3).
**Do Exercise 21.**   ∎

## Applications

EXAMPLE 4    An object is moving along a directed line. The directed distance $s$ meters from the origin to the object at instant $t$ seconds is given by

$$s = 3t^3 + 52t^2 - 6t + 10$$

**a.** Find the instantaneous velocity of the object at instant 5 seconds.
**b.** Find the acceleration of the object at instant 5 seconds.

Solution    **a.** $s = f(t) = 3t^3 + 52t^2 - 6t + 10$. Thus, the velocity at instant $t$ is given by

$$v = f'(t) = D_t(3t^3 + 52t^2 - 6t + 10) = 9t^2 + 104t - 6$$

At instant 5 seconds, we get

$$v = 9(5^2) + 104(5) - 6 = 225 + 520 - 6 = 739$$

At instant 5 seconds, the velocity is 739 meters per second.
**b.** By definition, the acceleration $a$ is given by

$$a = \frac{dv}{dt} = f''(t)$$

Therefore,

$$a = f''(t) = D_t(9t^2 + 104t - 6) = 18t + 104$$

At instant 5 seconds, we have

$$a = 18(5) + 104 = 90 + 104 = 194$$

At instant 5 seconds, the acceleration is 194 meters per second per second (abbreviated 194 m/sec²).
**Do Exercise 15.**   ∎

EXAMPLE 5    A furniture manufacturer found that the cost $C(q)$ (in dollars) to produce $q$ deluxe sofas is given by

$$C(q) = 0.02q^3 - 12q^2 + 3600q + 10,000$$

a. Find the marginal cost function.
b. Approximate the cost of making the 151st sofa.
c. Find the second derivative $C''(q)$.
d. Evaluatae $C''(150)$ and interpret the result.

(See Exercise 28, Section 7.5.)

Solution    a. The marginal cost function is $C'$. Thus we get

$$C'(q) = D_q(0.02q^3 - 12q^2 + 3600q + 10,000) = 0.06q^2 - 24q + 3600$$

b. The approximate cost of making the 151st sofa is obtained by evaluating the marginal cost function at 150.

$$C'(150) = 0.06(150^2) - 24(150) + 3600$$
$$= 1350 - 3600 + 3600 = 1350$$

Therefore the cost of producing the 151st sofa is approximately $1350.
c. $C''(q) = D_q(0.06q^2 - 24q + 3600) = 0.12q - 24$.
d. $C''(150) = 0.12(150) - 24 = -6$.
This means that at the level of production of 150 sofas, an increase in production of 1 sofa would cause a decrease in the marginal cost of approximately $6.
**Do Exercise 25.**

# Exercise Set 7.11

*In Exercises 1–14, an equation that defines a function explicitly is given. Find the first, second, and third derivatives with respect to the independent variable.*

**1.** $y = 2x^2 + 5x + 10$

**2.** $y = 3x^2 - 6x + 4$

**3.** $y = x^5 + 3x^2 - 6x + 7$

**4.** $y = x^2 + 1$

**5.** $u = e^{t^4}$

**6.** $w = \ln(t^4 + t^2 + 1)$

**7.** $t = ue^{5u}$

**8.** $s = \dfrac{t^2 + 1}{e^t}$

**9.** $y = (x^2 + 1)\ln(x^2 + 1)$

**10.** $y = \dfrac{x}{x + 1}$

**11.** $z = 2\sqrt[3]{t^2}$

**12.** $u = s^2 e^{5s}$

**13.** $w = \dfrac{v^2 + 1}{v^4 + 1}$

**14.** $y = 5\sqrt[3]{x^2} + 2\sqrt[5]{x^3}$

*In Exercises 15–18, the position function of a particle moving on a directed line is given, where s is measured in centimeters and t is in seconds. **a.** Find the velocity at the indicated instant. **b.** Find the acceleration at the indicated instant.*

**15.** $s = 2t^3 + 5t^2 - 6t + 7$, $t = 5$ seconds

**16.** $s = 5t^4 - 3t^2 + 13$, $t = 10$ seconds

**17.** $s = 5e^{-t} + 5t^2 + 7$, $t = 3$ seconds

**18.** $s = 7t\ln(t^2 + 1) + 4$, $t = 6$ seconds

*In Exercises 19–24, each equation defines a three-times-differentiable function f such that $y = f(x)$. In each case, find the values of the first, second, and third derivatives at the given point.*

**19.** $x^3 + y^3 = 7$, $(-1, 2)$

**20.** $x^2 + xy + y^2 = 7$, $(1, 2)$

**21.** $x^2y^3 + y^3 = 5x^4 - 3$, (1, 1)

**22.** $e^{x^2} + e^{3y} = 9$, (0, ln 2)

**23.** $\ln(x^2 + y^2) = 3x^5$, (0, 1)

**24.** $\sqrt[3]{x^2} + \sqrt[5]{y^2} = 5x + 6y - 29$, (8, −1)

*Before working Exercises 25–27, see Exercise 28, Section 7.5.*

**25.** A radio company knows that if $C(q)$ is the cost of producing $q$ radios, then

$$C(q) = 20,000 + 100qe^{(-0.001)q} \text{ for } 0 < q \leq 1000.$$

  **a.** Find the marginal cost function.
  **b.** Find the approximate cost of producing the 501st radio.
  **c.** Find $C''(q)$.
  **d.** Evaluate $C''(500)$ and interpret your result.

**26.** The total cost of producing $q$ electronic modules manufactured by a company is $C(q)$, where

$$C(q) = \frac{10,000q + 2000}{q + 1}, \text{ if } 0 \leq q \leq 50$$

  **a.** Find the marginal cost function.
  **b.** Estimate the cost of producing the 20th module.
  **c.** Find $C''(q)$.
  **d.** Evaluate $C''(19)$ and give an interpretation of your answer.

**27.** A sporting goods company has found that the cost $C(q)$ (in dollars) of manufacturing $q$ racquetball racquets is given by

$$C(q) = 2000 + 10q \ln \frac{q^2 + 10,000}{10,000}$$

where $0 \leq q \leq 200$

  **a.** Find the marginal cost function.
  **b.** Approximate the cost of manufacturing the 101st racquet.
  **c.** Find $C''(q)$.
  **d.** Evaluate $C''(100)$ and give an interpretation of your result.

**\*28.** Recall from basic algebra that

$$(a + b)^2 = a^2 + 2ab + b^2$$
$$(a + b)^3 = a^3 + 3a^2b + 3ab^2 + b^3$$
$$(a + b)^4 = a^4 + 4a^3b + 6a^2b^2 + 4ab^3 + b^4$$

Note that the coefficients on the right sides of the formulas are

$$1 \quad 2 \quad 1$$
$$1 \quad 3 \quad 3 \quad 1$$
$$1 \quad 4 \quad 6 \quad 4 \quad 1$$

Notice also that each row begins and ends with a 1 and that each of the other numbers in each row (starting with row 2) is the sum of the two numbers immediately above it. For example, the next row should be 1 5 10 10 5 1. This gives us the coefficients for

$$(a + b)^5 = a^5 + 5a^4b + 10a^3b^2 + 10a^2b^3 + 5ab^4 + b^5.$$

The pattern above is part of the triangle known as Pascal's triangle. (See the section "Getting Started.")

  **a.** Write the next row of Pascal's Triangle.
  **b.** Write the expansion of $(a + b)^6$.
  **c.** Let $f$ and $g$ be functions that have at least six derivatives: $u = f(x)$, $v = g(x)$, and $w = uv$. Verify that

$$\frac{d^2w}{dx^2} = uv'' + 2u'v' + u''v$$

  Note that the coefficients are 1, 2, and 1.
  **d.** Verify that

$$\frac{d^3w}{dx^3} = uv''' + 3u'v'' + 3u''v' + u'''v$$

  Note that the coefficients are 1, 3, 3, and 1.
  **e.** Discuss, but do not prove, the formulas for

$$\frac{d^4w}{dx^4}, \frac{d^5w}{dx^5}, \text{ and } \frac{d^6w}{dx^6}$$

*In Exercises 29–32, use the results of Exercise 28 to find the indicated derivatives.*

**\*29.** Find $\dfrac{d^3w}{dx^3}$ where $y = x^5e^{5x}$.

**\*30.** Find $\dfrac{d^4w}{dx^4}$ where $y = (x^3 + x^2 + x + 1)\ln|x + 1|$.

**\*31.** Find $\dfrac{d^5w}{dx^5}$ where $y = x^2e^{x^2}$.

**\*32.** Find $\dfrac{d^6w}{dx^6}$ where $y = e^{10x}\ln|x + 1|$.

# 7.12 Chapter Review

IMPORTANT SYMBOLS
AND TERMS

$df(x)/dx$ [7.4]
$D_x y$ [7.4]
$D_x f(x)$ [7.4]
$D_x^2 y$ [7.11]
$D_x^3 y$ [7.11]
$D_x^n y$ [7.11]
$D_x^2 f(x)$ [7.11]
$D_x^3 f(x)$ [7.11]
$D_x^n f(x)$ [7.11]
$d^2 y/dx^2$ [7.11]
$d^3 y/dx^3$ [7.11]
$d^n y/dx^n$ [7.11]
$d^2 f(x)/dx^2$ [7.11]
$d^3 f(x)/dx^3$ [7.11]
$d^n f(x)/dx^n$ [7.11]
$dy/dx$ [7.4]
$f'(x)$ [7.4]
$f''(x)$ [7.11]
$f'''(x)$ [7.11]

$f^{(n)}(x)$ [7.11]
$$\lim_{h \to 0} \frac{f(x+h) - f(x)}{h} =$$
$f'(x)$ [7.4]
$\lim_{x \to a} f(x) = L$ [7.2]
$\lim_{x \to a^-} f(x)$ [7.3]
$\lim_{x \to a^+} f(x)$ [7.3]
$y'$ [7.4]
$y''$ [7.11]
$y'''$ [7.11]
$\Delta x$ [7.1]
Acceleration [7.11]
Average rate of change [7.1]
Continuous [7.3]
Cost function [7.5]
Chain Rule [7.7]
Derivative [7.4]

Differentiable [7.4]
Differentiation [7.4]
Discontinuous [7.3]
Function defined implicitly
   [7.8]
Higher order derivative
   [7.11]
Implicit differentiation [7.8]
Increment [7.1]
Left-hand limit [7.3]
Limit [7.2]
Logarithmic differentiation
   [7.10]
Marginal cost function [7.5]
Right-hand limit [7.3]
Second derivative [7.11]
Slope [7.4]
Tangent line [7.2]

SUMMARY

If a function $f$ is defined by $y = f(x)$, the average rate of change of $f$ with respect to $x$ is the ratio $\frac{\Delta y}{\Delta x}$, where $\Delta x = x_2 - x_1$ and $\Delta y = f(x_2) - f(x_1)$.

If $f$ is a function and the values of $f(x)$ are within a specified distance of the number $L$, provided that $x$ is sufficiently close to $c$ and $x \neq c$, then we say that $\lim_{x \to c} f(x) = L$. The specified distance may be as small as we wish. If $\lim_{x \to c} f(x) = L$ and $\lim_{x \to c} g(x) = M$, then

1. $\lim_{x \to c} [f(x) \pm g(x)] = L \pm M$.

2. $\lim_{x \to c} [f(x) \cdot g(x)] = L \cdot M$.

3. $\lim_{x \to c} \dfrac{f(x)}{g(x)} = \dfrac{L}{M}$, provided $M \neq 0$.

4. $\lim_{x \to c} [f(x)]^r = L^r$, provided $[f(x)]^r$ and $L^r$ are defined.

Also if $k$ is constant and $n$ is a positive integer, then

5. $\lim_{x \to c} k = k$.

6. $\lim_{x \to c} x^n = c^n$.

If $p(x)$ is a polynomial, then

7. $\lim_{x \to c} p(x) = p(c)$.

This means that polynomial functions are continuous.

We may evaluate limits intuitively. In cases where we obtain a ratio $\frac{a}{b}$, we must consider the following cases. If $a = 0$, and $b \neq 0$, the limit is 0. If $a \neq 0$ and $b = 0$, the limit is not defined. If both $a$ and $b$ are 0, then we must try to simplify the fraction, either by factoring or by rationalization when radicals are involved. To evaluate left-hand and right-hand limits as $x$

approaches $c$, we evaluate the function at values of $x$ which are close to $c$ but less than $c$ for the left-hand limit, or greater than $c$ for the right-hand limit. The derivative $f'(x)$ is

$$\lim_{\Delta x \to 0} \frac{f(x + \Delta x) - f(x)}{\Delta x}$$

provided this limit exists. The derivative gives the instantaneous rate of change of the function. If $y = f(x)$ and $f'(a)$ exists, then $f'(a)$ is the slope of the tangent line to the graph of $f$ at $(a, f(a))$. If a function is differentiable at $a$, then it is continuous at $a$. However, a continuous function may fail to be differentiable. If $s = f(t)$ is the equation of motion of a particle along a directed line, $f'(t)$ gives the instantaneous velocity at instant $t$, and $f''(t)$ gives the acceleration at instant $t$. In economics, the adjective "marginal" used with the name of a function indicates the derivative of that function. For example, the "marginal cost function" is the derivative $C'(q)$ of the cost function $C(q)$. The number $C'(a)$ is interpreted as the approximate cost to produce the $a$th (or $(a + 1)$st) unit. The same remarks apply to the marginal revenue and marginal profit functions.

The basic differentiation formulas are as follows, where $k$ and $c$ are constants and $u$ and $v$ are differentiable functions of $x$.

$$D_x(k) = 0 \tag{1}$$

$$D_x(u + v) = D_x u + D_x v \tag{2}$$

$$D_x(u - v) = D_x u - D_x v \tag{3}$$

$$D_x(ku) = kD_x u \tag{4}$$

$$D_x(uv) = uD_x v + vD_x u \tag{5}$$

$$D_x\left(\frac{u}{v}\right) = \frac{vD_x u - uD_x v}{v^2} \tag{6}$$

If $y = g(f(x))$, where $f$ is differentiable at $x$, and $g$ is differentiable at $f(x)$, then

$$\frac{dy}{dx} = g'(f(x))f'(x) \tag{7}$$

Equivalently, if $y = g(u)$ and $u = f(x)$, then

$$\frac{dy}{dx} = \frac{dy}{du} \cdot \frac{du}{dx} \tag{7'}$$

$$D_x u^c = cu^{c-1}D_x u \tag{8}$$

$$D_x e^u = e^u D_x u \tag{9}$$

$$D_x b^u = b^u(\ln b)D_x u \tag{10}$$

$$D_x \ln|u| = \frac{D_x u}{u} \tag{11}$$

You must be careful when differentiating $u^v$ with respect to $x$. If the base $u$ is variable and the exponent $v$ is constant, use Formula 8; if the base $u$ is constant and the exponent $v$ is variable, use Formula 9 or 10 depending on the value of the base. (Note that Formula 9 is a special case of formula 10 since $\ln e = 1$.) If both $u$ and $v$ are variables, use logarithmic differentiation. Logarithmic differentiation may also be used to find $f'(x)$ when the expression defining $f(x)$ consists of products, quotients, or powers. We first take the absolute value of both sides to get $|y| = |f(x)|$. We then take the natural logarithm of both sides $\ln|y| = \ln|f(x)|$ and use the laws of logarithms to change products to sums, quotients to differences, and powers to products in the expression $\ln|f(x)|$. We finally differentiate both sides with respect to $x$ and solve for $\frac{dy}{dx}$. If an equation in $x$ and $y$ defines $y$ implicitly as a differentiable function of $x$, we may obtain $\frac{dy}{dx}$ as follows.

Differentiate both sides with respect to $x$. Apply all basic formulas to perform the differentiation, being careful to remember that $D_x y \neq 1$ so that $D_x y^c = cy^{c-1}D_x y$, $D_x e^y = e^y D_x y$, and so on. Then solve for $\frac{dy}{dx}$.

Given a function $f$, if we differentiate once, we get the first derivative $f'$; if we differentiate again (differentiate $f'$), we get the second derivative $f''$, and so on.

**SAMPLE EXAM QUESTIONS**

1. Let $f(x) = x^2 + 5x - 4$. Find $\Delta x$ and $\Delta y$ as **a.** $x$ increases from 3 to 6 and **b.** $x$ decreases from 1 to $-3$. In each case, also find the average rate of change of $f$ with respect to $x$.

2. A particle moves along a directed line. Its directed distance $s$ meters from the origin at instant $t$ seconds is given by the formula $s = g(t) = 3t^3 + 2t^2 + 3t - 5$. Find $\Delta t$ and $\Delta s$ and $t$ increases from **a.** 3 to 5 seconds, **b.** 3 to 3.5 seconds, and **c.** 3 to 3.1 seconds. In each case, also find the average rate of change of $g$ with respect to $t$. Find the derivative $\frac{ds}{dt}$. Evaluate the derivative at $t = 3$ and compare your answer to the average rate of change of $g$ as $t$ increases from 3 to 3.1.

3. Find the average rate of change of the area of a circle as the radius increases from 6 to 9 centimeters.

4. A manufacturer knows that the cost per month to manufacture $q$ units of a certain commodity is $\$C(q)$, where $C(q) = 5000 + 5q^{0.75}$. The revenue $\$R(q)$ generated from these $q$ units is given by $R(q) = 3.3q - 0.0003q^2$. The manufacturer decides to increase the production level from 1296 units per month to 4096 units per month.

   **a.** Find the corresponding increments in production, cost, revenue, and profit.
   **b.** Find the corresponding average rates of change in cost, revenue, and profit.
   **c.** Find the marginal cost when 4096 units are produced and interpret the result.
   **d.** Find the marginal revenue when 4096 units are produced and interpret the result.
   **e.** Find the marginal profit when 4096 units are produced and interpret the result.

5. Let $y = f(x) = x^2 + 3x - 2$.

   **a.** Find the slope of the line passing through the points $(2, f(2))$ and $(4, f(4))$.
   **b.** Find the slope of the line passing through the points $(2, f(2))$ and $(3, f(3))$.
   **c.** Find the slope of the line passing through the points $(2, f(2))$ and $(2.1, f(2.1))$.
   **d.** Find the slope of the tangent line at the point $(2, f(2))$.
   **e.** Find the equation of the tangent line at $(2, f(2))$.

6. Evaluate each of the following limits.

   **a.** $\lim\limits_{x \to 3}(2x + 1)$

   **b.** $\lim\limits_{x \to -1}(3x^2 + 4x - 2)$

   **c.** $\lim\limits_{x \to 3}(3x + 7)^{5/2}$

   **d.** $\lim\limits_{x \to 5}\dfrac{x - 5}{x + 2}$

   **e.** $\lim\limits_{x \to -2}\dfrac{x + 3}{x + 2}$

   **f.** $\lim\limits_{x \to 1}\dfrac{x^2 + 3x - 4}{x^2 - 1}$

   **g.** $\lim\limits_{x \to 5}\dfrac{\sqrt{4 + x} - 3}{x - 5}$

**h.** $\lim\limits_{x \to 3} \dfrac{x - 3}{2 - \sqrt{7 - x}}$

**7.** Let $f(x) = x^2 + 7x - 4$. Find

**a.** $\lim\limits_{x \to 3} \dfrac{f(x) - f(3)}{x - 3}$

**b.** $\lim\limits_{x \to -1} \dfrac{f(x) - f(-1)}{x + 1}$

**c.** $f'(x)$ (Using any of the formulas 1–11.)

**d.** Evaluate $f'(3)$ and $f'(-1)$ and compare these values to your answers in parts **a** and **b**.

**8.** Let $f$ be defined as follows:

$$f(x) = \begin{cases} \dfrac{x^2 - 16}{x - 4} & \text{if } x \neq 4 \\ 10 & \text{if } x = 4 \end{cases}$$

Is $f$ continuous at 4?

**9.** Let $g$ be defined as follows:

$$g(x) = \begin{cases} x + 2 & \text{if } x < 1 \\ -x + 3 & \text{if } x \geq 1 \end{cases}$$

Find **a.** $\lim\limits_{x \to 1^-} g(x)$ and **b.** $\lim\limits_{x \to 1^+} g(x)$. Is $g$ continuous at 1?

**10.** Sketch the graph of the function $h$ if $h(x) = \left[\frac{x}{5}\right]$ for $-15 \leq x \leq 15$ and find **a.** $\lim\limits_{x \to 10^-} h(x)$

and **b.** $\lim\limits_{x \to 10^+} h(x)$.

**11.** Define the following terms or expressions:

**a.** The function $f$ is *continuous at* $c$.

**b.** The *derivative* of a function $g$ at $b$.

**12.** Let the function $f$ be defined by $f(x) = -x^2 + x - 3$. Using the definition of derivative, not the formulas, calculate $f'(3)$.

**13.** Let the function $g$ be defined by $g(x) = \sqrt{x}$. Using the definition of derivative, not the formulas, calculate $g'(9)$.

**14.** Differentiate with respect to the appropriate variable. (You may use logarithmic differentiation.)

**a.** $y = \pi^2$

**b.** $x = 4t^3 - 5t^2 + 7t - 5$

**c.** $r = \sqrt{s + 1}$

**d.** $y = (x^2 + 5)(e^{3x + 2})$

**e.** $z = \dfrac{u^3 - 3u + 2}{\ln(u^2 + 9)}$

**f.** $d = 5^{x^2 + 2}$

**g.** $y = e^{\ln 5x}$

**h.** $y = (x^2 + 1)^{3x+2}$

**i.** $y = \dfrac{(x^2 + 3)\sqrt{1 - 3x}}{(x^3 + 5)^{1/4}}$

**j.** $y = \sqrt{\dfrac{(x^2 + 1)\sqrt{4x + 7}}{e^{x^4 + 4x + 2}}}$

**k.** $r = \dfrac{4}{t^3}$

**l.** $y = \dfrac{-3}{\sqrt{x^3}}$

**m.** $y = \dfrac{(x^4 + 9)\sqrt{x^6 + 5}}{(4x^6 + 4x^2 + 1)e^{x^2+3}}$

**n.** $z = \dfrac{e^u + e^{-u}}{2}$

**o.** $z = \dfrac{u^2 + 1}{e^u + e^{-u}}$

**p.** $y = 2x^e + 5e^x + (x^3 + e)^e + e^{x^3+e} + e^4$

**15.** The equation $x^2y^3 + 5xy^2 + x^2 = 6 + 5e^y$ defines a differentiable function $f$ such that $y = f(x)$. Find $\frac{dy}{dx}$.

**16. a.** verify that the point $(1, 2)$ is on the graph of the equation

$$x^3 + y^4 = 5e^{(x-1)}y + 12$$

**b.** Find the equation of the line tangent to the graph of the equation at that point.

**17.** Find the value of the indicated higher derivative at the given point. You need not simplify the expression you obtain before making the substitution to evaluate.

**a.** $y = x^4 + 3x^3 - 5x^2 + 9x - 6$, $\dfrac{d^3y}{dx^3}$, $(1, 2)$

**b.** $y = \sqrt{x+1}$, $y''$, $(8, 3)$

**c.** $x^2 - 3xy + y^2 = 11$, $y''$, $(-1, 2)$

**d.** $y = 2e^{x^2}$, $D_x^3y$, $(0, 2)$

**e.** $y = (x + 1)\ln(x + 1)$, $D_x^3y$, $(e - 1, e)$

**f.** $y = e^{2x}$, $y^{(10)}$, $(0, 1)$

**g.** $y = 5x^{50} + 30x^{31} + 5x^{12} - 6x^2 + 5x - 3$, $y^{(51)}$, $(0, -3)$

**18.** The equation $x^2 + y^2 = 169$ defines a three-times-differentiable function $f$ such that $y = f(x)$. Find the values of the first, second, and third derivatives at the point $(5, -12)$.

**19.** Same as Exercise 18 with equation

$$x^2e^{y-1} + x^3y^2 = 5x + 2$$

and point $(2, 1)$.

**20.** Suppose $f$ is a differentiable function, $f'(x) = 7/x^3$, and let $y = g(x) = f(3x + 1)$. Find $g'(x)$.

**21.** Suppose $g$ is a differentiable function, $g'(x) = 5/\sqrt[3]{x^2}$, and let $y = f(x) = g(x^2 + 1)$. Find $f'(x)$.

# Applications of Derivatives

8

Since the value of the derivative of a function $f$ at $c$ gives the instantaneous rate of change of that function at $c$, it is often the case that derivatives are useful in many situations that involve a rate of change. For example, velocity, acceleration, the slope of a tangent line, as well as marginal cost and marginal revenue, are instantaneous rates of change of certain functions and therefore can be evaluated using derivatives. Examples of these were given in the preceding chapter so that you might see the application of calculus to fields such as economics (marginal analysis), physical sciences (velocity and acceleration), psychology (rate of learning), and sociology (rate of change of population). In this chapter, we continue to discuss applications, including the concept of optimization—one of the most useful applications of derivatives. But first, we shall discuss marginal analysis in more detail.

# 8.1 Marginal Analysis

## Marginal Cost and Marginal Revenue

Marginal analysis was introduced in Exercise 28, Section 7.5. Generally speaking, in economics the word *marginal* translates to *derivative*. For example, the marginal cost, marginal revenue, and marginal profit functions are the derivatives of the cost, revenue, and profit functions, respectively.

When we say that the cost of producing $q$ units of a certain commodity is $C(q)$ dollars, the domain of the function is usually, but not always, a set of nonnegative integers. However, even in cases where the domain is the set of positive integers, the formula that describes $C(q)$ defines a differentiable function on an interval of real numbers. The derivative of that function, as we know, is called the *marginal cost function*. Let us analyze the significance of the value of the marginal cost function geometrically at a positive integer $n$. Let $C$ be the cost function. We draw the graph of $C$ over an interval of real numbers, even though the domain of $C$ may consist of only nonnegative integers. (See Figure 8.1.)

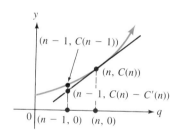

**FIGURE 8.1**

The number $C'(n)$ is the slope of the tangent line to the graph of $y = C(q)$ at the point $(n, C(n))$. The equation of the tangent line is

$$y - C(n) = C'(n)(q - n)$$

Replace $q$ by $n - 1$ in this equation and get

$$y - C(n) = C'(n)((n - 1) - n) = -C'(n)$$

Therefore

$$y = C(n) - C'(n)$$

Thus the point $(n - 1, C(n) - C'(n))$ is on the tangent line. On the other hand, the point $(n - 1, C(n - 1))$ is on the graph of the cost function. Note in Figure 8.1 that the line perpendicular to the $x$-axis passing through $(n - 1, 0)$ intersects both the graph of the cost function and the tangent line at two close points. The absolute value of the difference between the ordinates of these two points is generally small.

Consequently,

$$(C(n) - C'(n)) - C(n - 1) = (C(n) - C(n - 1)) - C'(n)$$

is small in absolute value.

We conclude that in general, $C'(n)$ is a good approximation of $C(n) - C(n - 1)$. Now note that $C(n)$ and $C(n - 1)$ are the costs of producting $n$ and $n - 1$ units, respectively. The difference $C(n) - C(n - 1)$ is the cost of producing the $n$th unit, a fact that we stated and used in Section 7.5, Exercises 28 and 29. Using a similar argument, it can be shown that $C'(n)$ also approximates the cost of producing the $(n + 1)$st unit.

The same idea may be used when working with the revenue and profit functions.

The use of derivatives to approximate functional values will be discussed in greater detail later in this chapter.

**EXAMPLE 1**   The Mason Rubber Company has made a study of the relationship between the cost $C(q)$ (in dollars), revenue $R(q)$ (in dollars), and number $q$ of rubber boats it manufactures each month. The equations are

$$C(q) = 2000 + 1000 \sqrt[3]{q^2 + 1}$$

and

$$R(q) = -1.5q^2 + 600q$$

where $0 \leq q \leq 200$.

**a.** Find the profit function.
**b.** Find the marginal cost function.
**c.** Find the marginal revenue function.
**d.** Find the marginal profit function.
**e.** Evaluate $C'(100)$ and interpret your result.
**f.** Evaluate $R'(100)$ and interpret your result.
**g.** Evaluate $P'(100)$ and interpret your result.

**Solution**   **a.** The profit is obtained by subtracting the cost from the revenue. Thus

$$\begin{aligned} P(q) &= R(q) - C(q) \\ &= (-1.5q^2 + 600q) - (2000 + 1000 \sqrt[3]{q^2 + 1}) \\ &= -1.5q^2 + 600q - 1000 \sqrt[3]{q^2 + 1} - 2000 \end{aligned}$$

**b.** By definition, the marginal cost function is the derivative $C'$. Therefore,

$$\begin{aligned} C'(q) &= D_q(2000 + 1000(q^2 + 1)^{1/3}) \\ &= 0 + 1000\left(\frac{1}{3}\right)(q^2 + 1)^{1/3-1}D_q(q^2 + 1) \\ &= \frac{1000}{3}(q^2 + 1)^{-2/3}2q = \frac{2000q}{3(q^2 + 1)^{2/3}} \end{aligned}$$

**c.** The marginal revenue function is the derivative $R'$. Therefore, we differentiate $R$ and obtain

$$R'(q) = D_q(-1.5q^2 + 600q) = -3q + 600$$

**d.** The marginal profit function is the derivative of the profit function. Since

$$P(q) = R(q) - C(q)$$

we have

$$P'(q) = R'(q) - C'(q)$$

$$= (-3q + 600) - \frac{2000q}{3(q^2 + 1)^{2/3}}$$

$$= \frac{(1800 - 9q)(q^2 + 1)^{2/3} - 2000q}{3(q^2 + 1)^{2/3}}$$

**e.** $C'(100) = \dfrac{2000(100)}{3(100^2 + 1)^{2/3}} \doteq 143.62.$

The cost of producing the 100th rubber boat is approximately $143.62.
**f.** $R'(100) = -3(100) + 600 = 300.$

The revenue from the 100th boat is approximately $300.
**g.** $P'(100) = R'(100) - C'(100)$

$$= 300 - 143.62 = 156.38$$

The profit from the 100th boat is approximately $156.38.
**Do Exercise 1.**

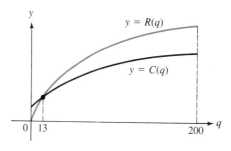

**FIGURE 8.2**

Note that in the preceding example the company would operate at a loss if it manufactured fewer than 13 boats. (See Figure 8.2.)

## Marginal Average Cost

In general, when $q$ units are produced, the average cost $A(q)$ is given by

$$A(q) = \frac{C(q)}{q}$$

Accordingly, the marginal average cost function is the derivative $A'$. Thus,

$$A'(q) = D_q\frac{C(q)}{q} = \frac{qC'(q) - C(q)}{q^2}$$

$$= \frac{C'(q)}{q} - \frac{C(q)}{q^2} = \frac{1}{q}\left[C'(q) - \frac{C(q)}{q}\right]$$

$$= \frac{1}{q}[C'(q) - A(q)]$$

This equality shows that the value of the marginal average cost function at $q$ is the average of the difference between the marginal cost and the average cost when $q$ units are produced. It approximates the change in the average cost if production is increased from $q - 1$ to $q$ units or from $q$ to $q + 1$ units.

**EXAMPLE 2** Find the average cost function and the marginal average cost function for the rubber boats produced in Example 1. Evaluate these functions for $q = 100$ and interpret your results.

**Solution** The cost function of Example 1 was defined by

$$C(q) = 2000 + 1000 \sqrt[3]{q^2 + 1}$$

Thus the average cost function is defined by

$$A(q) = \frac{2000 + 1000 \sqrt[3]{q^2 + 1}}{q}$$

To find the marginal average cost function, we need only differentiate both sides of this equality. However, since we already have $C'(q)$ (from Example 1) and $A(q)$, we use the formula

$$A'(q) = \frac{1}{q}(C'(q) - A(q))$$

and get

$$A'(q) = \frac{1}{q}\left(\frac{2000q}{3(q^2 + 1)^{2/3}} - \frac{2000 + 1000 \sqrt[3]{q^2 + 1}}{q}\right)$$

$$= \frac{2000q^2 - 6000(q^2 + 1)^{2/3} - 3000(q^2 + 1)}{3q^2(q^2 + 1)^{2/3}}$$

$$= \frac{-1000q^2 - 6000(q^2 + 1)^{2/3} - 3000}{3q^2(q^2 + 1)^{2/3}}$$

Now,

$$A(100) = \frac{2000 + 1000 \sqrt[3]{100^2 + 1}}{100} \doteq 235.45$$

Therefore, if the cost of production per boat had been constant at \$235.45, then the total cost to build the first 100 boats would be the same as it was when the cost varied as production changed.

$$A'(100) = \frac{-1000(100^2) - 6000(100^2 + 1)^{2/3} - 3000}{3(100^2)(100^2 + 1)^{2/3}}$$

$$\doteq -0.9183124$$

An increase in production of one unit (from 99 to 100) will *decrease* the average cost by approximately 92 cents.
**Do Exercise 9.**

---

# Exercise Set 8.1

**1.** If $q$ motorcycles are built by the Romero Cycle Company, the cost of production $C(q)$ (in dollars) and the revenue $R(q)$ (also in dollars) are given by

$$C(q) = 14,000 + 13,428\sqrt{q}$$

and

$$R(q) = -4q^2 + 2400q$$

where $0 \le q \le 300$.

**a.** Find the profit function.
**b.** Find the marginal cost function.
**c.** Find the marginal revenue function.
**d.** Find the marginal profit function.
**e.** Find the approximate cost of producing the 144th motorcycle.
**f.** Find the approximate revenue from the 144th motorcycle.
**g.** Find the approximate profit derived from the 144th motorcycle.

**2.** The owner of a small winery has found that if $q$ bottles of wine are produced by the winery each day, then the cost $C(q)$ and revenue $R(q)$, in dollars, are given by

$$C(q) = 100 + 150 \sqrt[3]{q}$$

and

$$R(q) = -0.12q^2 + 24q$$

where $0 \le q \le 100$. Answer the same questions as in Exercise 1. For parts **e**, **f**, and **g**, do the calculations for the 64th bottle.

**3.** Experience shows that when $q$ type A pocket calculators are produced by a manufacturer, the cost of production $C(q)$ (in dollars), the revenue $R(q)$ (also in dollars), and $q$ are related by the equations

$$C(q) = -0.25q^2 + 25q + 600$$

and

$$R(q) = 100qe^{(-0.02)q}$$

where $0 \le q \le 50$. Answer the same questions as in Exercise 1. For parts **e**, **f**, and **g**, do the computations for the 40th calculator.

**4.** The manager of a small toy company knows that the cost $C(q)$ (in dollars) of producing $q$ handmade dolls is

$$C(q) = -0.3q^2 + 60q + 630$$

and that the corresponding revenue $R(q)$ (also in dollars) is

$$R(q) = q(45 + \ln(q + 1))$$

where $0 \le q \le 100$. Answer the same questions as in Exercise 1. For parts **e**, **f**, and **g**, do the calculations for the 74th doll.

**5.** The Triplenight company found that if it manufactures $q$ racquetball racquets, the corresponding cost and revenue (in dollars) for $0 \le q \le 2500$ are given by

$$C(q) = 5000 + 1000\sqrt{q}$$

and

$$R(q) = -0.0125q^2 + 62.5q$$

respectively. Answer the same questions as in Exercise 1. For parts **e**, **f**, and **g**, do the calculations for the 1600th racquet.

**6.** An athletic equipment supplier knows that if $q$ dozen Ping-Pong balls are produced, the revenue generated and the cost incurred (in dollars) are given by

$$R(q) = q\left(3.5 + \frac{1}{50}\ln(q + 1)\right)$$

and

$$C(q) = -0.0007q^2 + 2.8q + 200$$

respectively, provided $0 \le q \le 1000$. Answer the same questions as in Exercise 1. For parts **e**, **f**, and **g**, do the calculations for the 799th dozen.

**7.** The Walker Electrical Company determined that the cost $C(q)$ and revenue $R(q)$ (in dollars) corresponding to a production of $q$ deluxe batteries are

$$C(q) = 30q + 60,000e^{-.001q}$$

and

$$R(q) = -0.05q^2 + 150q$$

respectively, where $0 \leq q \leq 1000$. Answer the same questions as in Exercise 1. For parts **e**, **f**, and **g**, do the calculations for the 300th battery.

**8.** A computer manufacturer found that the cost $C(q)$ and revenue $R(q)$ (in dollars) corresponding to a production of $q$ units of the new PEAR computer are

$$C(q) = 50,000 + \frac{1000q^2}{2q + 1}$$

and

$$R(q) = \frac{-q^3}{6} + 30q^2$$

respectively, where $0 \leq q \leq 100$. Answer the same questions as in Exercise 1. For parts **e**, **f**, and **g**, do the calculations for the 37th computer.

*In Exercises 9–16, perform the following.*

*a. Find the marginal average cost function.*
*b. Compute the value of the marginal average cost function for the given value of q and interpret your result.*

**9.** Use the cost function of Exercise 1, $q = 200$.

**10.** Use the cost function of Exercise 2, $q = 80$.

**11.** Use the cost function of Exercise 3, $q = 35$.

**12.** Use the cost function of Exercise 4, $q = 90$.

**13.** Use the cost function of Exercise 5, $q = 2000$.

**14.** Use the cost function of Exercise 6, $q = 800$.

**15.** Use the cost function of Exercise 7, $q = 750$.

**16.** Use the cost function of Exercise 8, $q = 75$.

**17.** The marketing division of a company found that the relationship between the number $q$ units of its product the consumers are willing to buy each month and the price $p$ (in dollars) per unit is

$$p = -0.001q^2 + 640$$

provided that $0 \leq q \leq 450$.

**a.** Find the revenue function.
**b.** Find the marginal revenue function.
**c.** Evaluate the marginal revenue function at $q = 400$ and interpret the result.

**18.** Suppose the relationship between the price $p$ (in dollars) per unit of a commodity and the quantity $q$ units the consumers are willing to buy per week is

$$p = 0.00025q^2 - 0.05q + 200$$

provided that $0 \leq q \leq 800$.

**a.** Find the revenue function.
**b.** Find the marginal revenue function.
**c.** Evaluate the marginal revenue function at $q = 600$ and interpret the result.

**19.** Suppose that when the price of a commodity is $50 per unit a store will sell 400 units per week. If the price is reduced to $40 per unit, the same store will sell 480 units per week.

**a.** Find the demand equation, assuming it is linear.
**b.** Find the revenue $R$ in terms of $q$.
**c.** Find the marginal revenue function.
**d.** Evaluate the marginal revenue function at $q = 350$ and interpret the result.

**20.** Marketing studies have shown that the demand equation for a product is linear. A store can sell 60 units per day when the price is $10 per unit and only 45 units per day when the price is increased to $13 per unit.

**a.** Find the demand equation.
**b.** Find the revenue $R$ in terms of $q$.
**c.** Find the marginal revenue function.
**d.** Find the value of the marginal revenue function at $q = 50$ and interpret the result.

*In Exercises 21–26, find the equation of the tangent line to the graph of the given equation at the indicated point.*

**21.** $y = 2x^2 + 5x + 6$, $(-2, 4)$

**22.** $y = (x^3 + 2)^{10}$, $(-1, 1)$

**23.** $y = \dfrac{x^2 - 3x + 6}{\sqrt{x}}$, $(4, 5)$

**24.** $y = x \ln x$, $(e, e)$

**25.** $y = 2e^{x^2}$, $(0, 2)$

**26.** $x^2 + 2x^3y^2 + y^4 = 9 - 20y$, $(-2, 1)$

*In Exercises 27–30, a particle is moving along a directed line. Its directed distance from the origin, at instant t seconds, is s feet. Find the velocity and acceleration of the particle for the given value of t.*

**27.** $s = 5t^2 + 7t + 7$, $t = 2$

**28.** $s = \sqrt[5]{t^2 + 194}$, $t = 7$

**29.** $s = (5t^2 + 6t - 4)e^t$, $t = 2$

**30.** $s = (t^3 + 1) \ln(t^2 + 1)$, $t = 4$

# 8.2 Elasticity

## Definition of Elasticity

The derivative has been used to evaluate the rate of change of a function with respect to its independent variable. It can also be used to measure the relationship between the percentage changes in two variables through a concept called *elasticity*. Suppose $f$ is a differentiable function and $y = f(x)$. As we have seen earlier, an increment $\Delta x$ in the independent variable will create a corresponding increment $\Delta y$ in the dependent variable, where

$$\Delta y = f(x + \Delta x) - f(x)$$

To find the percentage changes in $x$ and $y$, we multiply each ratio $\frac{\Delta x}{x}$ and $\frac{\Delta y}{y}$ by 100. For example, if

$$y = f(x) = x^2$$

then when $x$ changes values from 5 to 5.2,

$$\Delta x = 5.2 - 5 = 0.2$$

and

$$\Delta y = (5.2)^2 - 5^2 = 27.04 - 25 = 2.04$$

Thus the percentage increase in $x$ is 4% since $\left(\frac{0.2}{5}\right)100 = 4$, and the corresponding percentage increment in $y$ is 8.16% since $\left(\frac{2.04}{25}\right)100 = 8.16$.

To study the relationship between the percentage changes in variables, it is useful to consider their ratios. Thus we can write

$$\frac{\frac{\Delta y}{y}(100)}{\frac{\Delta x}{x}(100)} = \frac{x}{y} \cdot \frac{\Delta y}{\Delta x}$$

Taking the limit of this expression as $\Delta x$ tends to (approaches) 0, we get

$$\lim_{\Delta x \to 0}\left(\frac{x}{y} \cdot \frac{\Delta y}{\Delta x}\right) = \frac{x}{y} \cdot \frac{dy}{dx} = \frac{x}{y} \cdot f'(x)$$

This last expression occurs frequently, so it has been given a special name and is usually represented by the Greek letter $\eta$ (eta).

**DEFINITION 8.1:**  If $f$ is a differentiable function and $y = f(x)$, then the *elasticity of y with respect to x* is defined to be

$$\eta = \frac{x}{y} \cdot f'(x) = \frac{x}{y} \cdot \frac{dy}{dx}$$

Since $f'(x) = \lim_{\Delta x \to 0} \frac{\Delta y}{\Delta x}$, the values of $f'(x)$ and $\frac{\Delta y}{\Delta x}$ are close when $\Delta x$ is near 0. Therefore, $\eta$ approximates the value of $\frac{x}{y} \cdot \frac{\Delta y}{\Delta x}$ when $\Delta x$ is small. But we have seen that $\frac{x}{y} \cdot \frac{\Delta y}{\Delta x}$ is equal to the ratio between the percentage changes in $y$ and $x$, respectively; therefore, the value of $\eta$ approximates this ratio also. Note, for example, that for $y = f(x) = x^2$, when $x$ changes from 5 to 5.2, the corresponding values of $y$ change from 25 to 27.04. Thus a 4% increase in the value of $x$ causes an 8.16% increase in the value of $y$. The ratio between these percentages is 2.04 since $\frac{8.16}{4} = 2.04$. Now, $f'(x) = D_x x^2 = 2x$. The elasticity is 2 since $\frac{5}{25} f'(5) = \frac{5}{25}(10) = 2$. Note that $2 \doteq 2.04$. Thus, the elasticity is approximately equal to the ratio between the percentage changes in the dependent and independent variables.

### Elasticity of Demand

Clearly, there is a relationship between the number $q$ of units of a product that consumers are willing to buy and the price $p$ per unit. In general, this relationship is written as $p = f(q)$ and is called a *demand equation*. For our present purpose, it is more convenient to write $q = D(p)$, where $D$ is called a *demand function*.

> **DEFINITION 8.2:** Assuming that $D$ is a differentiable function, the *elasticity of demand* is defined to be
> $$\eta = \frac{p}{q} \cdot D'(p) = \frac{p}{q} \cdot \frac{dq}{dp}$$

**EXAMPLE 1** The demand equation for a commodity is

$$q = -p^2 - 10p + 2475 \qquad \text{(provided } 0 \le p \le 45\text{)}$$

**a.** Find the elasticity of demand when $p = 25$.
**b.** Find the percentage change in price when the price increases from $25 to $27.
**c.** Find the corresponding percentage change in demand.
**d.** Calculate the value of the ratio between the answers in parts **b** and **c**.
**e.** Compare the answers in part **d** to that of part **a**.

**Solution** **a.** Since $q = f(p) = -p^2 - 10p + 2475$, we have

$$f'(p) = -2p - 10$$

Thus,

$$f'(25) = -2(25) - 10 = -60$$

When $p = 25$, $q = 1600$ since $-25^2 - 10(25) + 2475 = 1600$. Therefore the elasticity of demand is given by

$$\eta = \frac{25}{1600}(-60) = -0.9375$$

**b.** When the price increases from $25 to $27, the increment is $2, an increase of 8%, since $\left(\frac{2}{25}\right)(100) = 8$.
**c.** We know that $f(25) = 1600$. We calculate

$$f(27) = -27^2 - 10(27) + 2475 = 1476$$

The demand decreases by 124 units since $1476 - 1600 = -124$. So the percentage decrease in demand is 7.75% since $\left(\frac{-124}{1600}\right)(100) = -7.75$.

**d.** The ratio between the percentages is $\frac{-7.75}{8} = -0.96875$.

**e.** Note that the actual value of the ratio between the percentage changes in demand and price ($-0.96875$) is close to the value of the elasticity of demand ($-0.9375$).

**Do Exercise 3.**

In the future, we shall only calculate the value of the elasticity of demand to estimate the value of the ratio between the percentage changes in demand and price. Then we shall use this result to approximate the percentage change in one variable given the percentage change in the other variable.

**EXAMPLE 2**    On the basis of market studies, a publisher found that the quantity $q$ of a book the public is willing to buy and the price $p$ (in dollars) per book are related by the equation

$$q = -10p^2 - 500p + 50,000$$

where $0 \leq p \leq 50$.

**a.** Find the elasticity of demand if $p = 30$.

**b.** Using the result of part **a**, estimate the percentage decrease in demand if there is a price increase of 60 cents per book.

**Solution**    Since $q = f(p) = -10p^2 - 500p + 50,000$, we get

$$f'(p) = -20p - 500$$

Thus

$$f'(30) = -20(30) - 500 = -1100$$

Also, when $p = 30$, $q = 26,000$ since $f(30) = 26,000$. Therefore, the elasticity of demand when $p = 30$ is

$$\eta = \frac{30}{26,000}(-1100) \doteq -1.2692$$

If the price is increased by 60 cents (0.6 dollars), the percentage increase is 2% since $\left(\frac{0.6}{30}\right)(100) = 2$. Thus,

$$\frac{\text{percentage decrease in demand}}{2} \doteq -1.2692$$

It follows that when the price per book is increased to $30.60, the percentage decrease in demand is approximately 2.5384% since $(-1.2692)(2) = -2.5384$.

**Do Exercise 11.**

We shall see later that when a differentiable function is decreasing on an open interval, its derivative cannot be positive on that interval. In general, the elasticity of demand is a negative number, since the demand function is a decreasing function.*

---

*In some textbooks, the elasticity of demand is defined to be $\left|\frac{p}{q}f'(p)\right|$. We do not follow this convention.

### Three Cases of Demand Elasticity

**Case 1.** $\eta < -1$. Here, a percentage increase in price will result in a greater percentage decrease in demand and a percentage decrease in price will cause a greater percentage increase in demand. In this case, we say that the *demand is elastic*.

**Case 2.** $\eta = -1$. This means that a percentage increase in price will cause an equal percentage decrease in demand and a percentage decrease in price will create an equal percentage increase in demand. This is referred to as *unit elasticity*.

**Case 3.** $-1 < \eta < 0$. Clearly this indicates that a percentage increase in price will result in a smaller percentage decrease in demand and a percentage decrease in price will create a smaller percentage increase in demand. In this case, we say that the *demand is inelastic*.

Note that in Example 2, we found the elasticity of demand to be $-1.2692$. In that case, the demand was elastic.

## Exercise Set 8.2

*In Exercises 1–8, an equation relating the number q of items of a commodity consumers are willing to buy and the price p (in dollars) per unit is given. Calculate the elasticity of demand at the indicated price.*

**1.** $q = -p^2 + 400$ for $0 \le p \le 20$, at $p = 15$.

**2.** $q = -p^2 - 20p + 1500$ for $0 \le p \le 30$, at $p = 20$.

**3.** $q = 3p^4 - 200p^3 + 6p^2 - 600p + 6,265,000$ for $0 \le p \le 50$, at $p = 30$.

**4.** $q = p^3 - 15p^2 - 600p + 10,000$ for $0 \le p \le 20$, at $p = 10$.

**5.** $q = p^3 - 4800p + 128,000$ for $0 \le p \le 40$, at $p = 10$.

**6.** $q = 100(50 - p)e^{-.01p}$ for $0 \le p \le 50$, at $p = 25$.

**7.** $q = 100\sqrt{169 - p^2}$ for $0 \le p \le 13$, at $p = 5$.

**8.** $q = 50,000\sqrt[4]{4096 - p}$ for $0 \le p \le 4096$, at $p = 3471$.

**9.** For the demand equation of Exercise 1, approximate the percentage decrease in demand when the price is increased from \$15 to \$15.30.

**10.** For the demand equation in Exercise 2, approximate the percentage increase in demand as the price is decreased from \$20 to \$19.

**11.** In Exercise 3, approximate the percentage decrease in demand when the price is increased by 3%.

**12.** In Exercise 4, approximate the percentage increase in demand when the price is decreased by 2%.

**13.** In Exercise 5, approximate the percentage increase in price if demand is decreased by 2%.

**14.** In Exercise 6, approximate the percentage decrease in price if demand is increased by 3%.

**15.** For the demand function of Exercise 7, approximate the percentage change in revenue when the price is increased from \$5 to \$5.20. (*Hint:* Recall that $R = pq$.)

**16.** For the demand function of Exercise 8, approximate the percentage change in revenue when the price is decreased by 3%.

*In Exercises 17–24, determine whether the demand is elastic, inelastic, or has unit elasticity for the given demand function and values of p.*

**17.** $q = -p^2 + 400$
  **a.** $p = 5$, **b.** $p = 10$, **c.** $p = 18$.

**18.** $q = -p^2 - 20p + 1500$
  **a.** $p = 5$, **b.** $p = 10$, **c.** $p = 15$.

**19.** $q = 3p^4 - 200p^3 + 6p^2 - 600p + 6,265,000$
  **a.** $p = 10$, **b.** $p = 20$, **c.** $p = 40$.

**20.** $q = p^3 - 15p^2 - 600p + 10,000$
  **a.** $p = 5$, **b.** $p = 8$, **c.** $p = 12$.

**21.** $q = p^3 - 4800p + 128,000$

    **a.** $p = 10$, **b.** $p = 20$, **c.** $p = 30$.

**22.** $q = 100(50 - p)e^{-0.01p}$

    **a.** $p = 15$, **b.** $p = 30$, **c.** $p = 45$.

**23.** $q = 100\sqrt{169 - p^2}$

    **a.** $p = 3$, **b.** $p = 6$, **c.** $p = 9$.

**24.** $q = 50,000 \sqrt[4]{4096 - p}$

    **a.** $p = 1000$, **b.** $p = 2000$, **c.** $p = 3000$.

**25.** The demand equation for a commodity is

$$q = -p^2 + 5p + 48,$$

where $3 \le p \le 9$. Determine the values of $p$ for which the demand is elastic, inelastic, and has unit elasticity.

**26.** Same as Exercise 25 for the demand equation

$$q = -p^3 - 2p^2 - 5p + 1736,$$

where $0 \le p \le 11$.

**27.** Suppose $f$ is a differentiable function and $y = f(x)$. Show

that the elasticity of $y$ with respect to $x$ is equal to $D_x\ln|y|/D_x\ln|x|$, provided that $xy \neq 0$.

**28.** If $q = D(p)$ gives the demand in terms of price, then the revenue $R(q)$ is given by $R(q) = qp$.

    **a.** Show that the marginal revenue function is given by

$$R'(q) = p\left(1 + \frac{1}{\eta}\right)$$

    **b.** Using your result from part **a**, show that the marginal revenue is positive when the demand is elastic and negative when the demand is inelastic. (*Hint:* differentiate both sides of the equation $R(q) = qp$ with respect to $q$ and use the fact that $\frac{dq}{dp} = 1/(dp/dq)$.)

**29.** Suppose that $q = D(p)$ is a demand equation and that for $p = p_0$, we have unit elasticity. Thus $\eta = -1$, and a percentage increase in price is balanced by an approximately equal percentage decrease in demand. Will the total revenue remain approximately the same? Justify your answer.

**30.** Suppose an increase in the price of $a\%$ causes a decrease in the demand of $b\%$. How should $a$ and $b$ be related for the total revenue to remain the same?

# 8.3 Optimization

In this section, we begin the study of optimization. "Think geometrically" in studying this concept, for although we shall state several formal definitions and theorems to justify results, the ideas are quite simple if we consider them graphically. We begin with geometric illustrations of the concepts we shall need.

## Increasing and Decreasing Functions

If $g(t)$ dollars is the price per gallon of regular gasoline where $t$ is the number of weeks since January 1, 1990, and the graph of $g$ is sketched in Figure 8.3, you conclude that the price of gasoline has been increasing during the period shown.

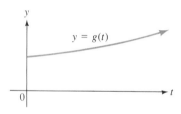

**FIGURE 8.3**

In general, if the graph of a function is rising as we go from left to right, the function is increasing. In Figure 8.4, we see that if $x_1 < x_2$, then $f(x_1) < f(x_2)$.

**FIGURE 8.4**

**DEFINITION 8.3:** A function $f$ is *strictly increasing on* an interval $I$ if for any two numbers $x_1$ and $x_2$ in that interval, $f(x_1) < f(x_2)$ whenever $x_1 < x_2$.

**FIGURE 8.5**

Figure 8.5 represents the rate of unemployment for some period of time. We see that this rate increased part of the time, but was also constant for some of the time. However, it never decreased. This is an example of an increasing (nondecreasing) function.

**DEFINITION 8.4:** A function $f$ is *increasing (nondecreasing)* on an interval $I$ if for any two numbers $x_1$ and $x_2$ in that interval, $f(x_1) \leq f(x_2)$ whenever $x_1 < x_2$.

Suppose now $q = D(p)$ is the demand equation for a commodity. The graph of the function $D$ is shown in Figure 8.6. We see that the graph is falling as we move from left to right. This is because, as we would expect, the quantity demanded decreases as the price increases.

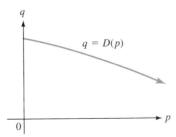

**FIGURE 8.6**

In general, we have the following definition.

> **DEFINITION 8.5:**    A function $f$ is *strictly decreasing* on an interval $I$ if for any two numbers $x_1$ and $x_2$ in that interval, $f(x_1) > f(x_2)$ whenever $x_1 < x_2$.

The percentage of a population who smoke, during some time period, is represented in Figure 8.7. From the graph, you would conclude that part of the time the percentage decreased and part of the time it remained constant—but, it never increased. This is an example of a decreasing (nonincreasing) function.

**FIGURE 8.7**

> **DEFINITION 8.6:**    A function $f$ is *decreasing* (*nonincreasing*) on an interval $I$ if for any two numbers $x_1$ and $x_2$ in that interval, $f(x_1) \geq f(x_2)$ whenever $x_1 < x_2$.

## Local Extrema of a Function*

Suppose a company realizes $P(q)$ dollars profit if it produces and sells $q$ units of a commodity in a month. The graph of the function $P$ is sketched in Figure 8.8. The graph has a high point when $q = q_0$. So the profit is maximum when $q_0$ units are made and sold each month. Clearly, knowing the value of $q_0$ will be useful to the company!

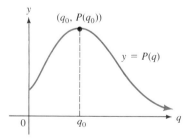

**FIGURE 8.8**

---

*We refer to either a maximum or a minimum value of a function as an extremum (plural: extrema).

As another example, suppose $C(x)$ dollars is the cost of fencing a rectangular region of given area, where $x$ feet is the length of one of the sides of the rectangle. The graph of the function $C$ is shown in Figure 8.9. The graph has a low point when $x = x_0$, so the cost is minimum when one of the dimensions of the rectangle is $x_0$ feet. It would, of course, be advantageous to the fence builder to know the value of $x_0$.

Often, the graph of a function over some interval will have more than one low or high point. (See Figure 8.10.) To help you understand the next definitions, imagine that a nearsighted bug is climbing the graph and reaches point $A$. He might think he is at the top of the graph, because he can only see as far as the two vertical lines drawn in Figure 8.10. The interval $(a, b)$ determined on the $x$-axis by these vertical lines is a neighborhood of the point $x_0$, and $A$ is the highest point of the graph over that neighborhood.

**FIGURE 8.9**

**FIGURE 8.10**

**DEFINITION 8.7:**   A *neighborhood* of a number $c$ is an open interval containing $c$. In particular, the interval $(c - h, c + h)$ is a neighborhood of $c$.

**DEFINITION 8.8:**   A function $f$ has a *local (relative) maximum* at $c$ if there is a positive number $h$ such that for each $x$ in the domain of $f$ for which $c - h < x < c + h$, it is true that $f(x) \leq f(c)$.

**DEFINITION 8.9:**   A function $f$ has a *local (relative) minimum* at $c$ if there is a positive number $h$ such that for each $x$ in the domain of $f$ for which $c - h < x < c + h$, it is true that $f(c) \leq f(x)$.

(See point $B$ on the graph of Figure 8.10.)

## Absolute Extrema of a Function

If the bug who climbed the graph of Figure 8.10 had reached point $C$, he would be at the highest point of the graph over the entire interval. This illustrates the following definition.

> **DEFINITION 8.10:** A function $f$ has an *absolute maximum at d* if for every $x$ in the domain of $f$ it is true that $f(x) \le f(d)$.

> **DEFINITION 8.11:** A function $f$ has an *absolute minimum at d* provided that $f(d) \le f(x)$ for all $x$ in the domain of $f$.

(See point $B$ on the graph of Figure 8.10.)

**EXAMPLE 1**   The graph of a function $f$ is shown in Figure 8.11. Locate the relative and absolute extrema of the function. What are these extrema?

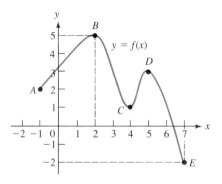

**FIGURE 8.11**

**Solution**   Clearly, $B$ and $E$ are the highest and lowest points of the graph, respectively. The absolute maximum of $f$ occurs at 2. Its value is $f(2) = 5$. Similarly, the absolute minimum occurs at 7 and its value is $-2$. From the graph, we also see that $f$ has a local minimum at $-1$ (its value is 2) and another at 4 (its value is 1). Furthermore, $f$ has a local maximum at 5 and its value is 3. ∎

   Observe that we give the *location* of an extremum and then its value. It would be wrong, for example, to say that 2 is the absolute maximum of the function $f$ of Example 1; 2 is the *location* of the absolute maximum, 5 *is* the absolute maximum.

## The Sign of the Derivative and the Behavior of a Function

Note that in general, if a function is strictly increasing over an interval $I$ and if its graph has a tangent line at each of its points, then each tangent line cannot be falling to the right; therefore, the graph of the function has a nonnegative slope. (See Figure 8.12.) Thus, the derivative will be nonnegative on the interval.

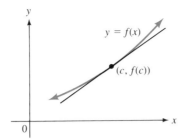

**FIGURE 8.12**

Similarly, if a function is strictly decreasing over an interval and its graph has a tangent line at each of its points, then each tangent line cannot be rising to the right and therefore the graph of the function has a nonpositive slope. It follows that the derivative of the function is nonpositive on that interval. (See Figure 8.13.)

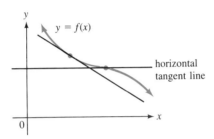

**FIGURE 8.13**

These ideas can be formalized in Theorem 8.1. The proof of parts 1, 2, and 5 will be given in the next section as optional reading.

**THEOREM 8.1:** Suppose a function $f$ is differentiable on an open interval $I$, then the following are true.

1. If $f'(x) > 0$ for all $x$ in $I$, then the function $f$ is strictly increasing on $I$.
2. If $f'(x) < 0$ for all $x$ in $I$, then the function $f$ is strictly decreasing on $I$.
3. If $f$ is increasing on $I$, then $f'(x) \geq 0$ for all $x$ in $I$.
4. If $f$ is decreasing on $I$, then $f'(x) \leq 0$ for all $x$ in $I$.
5. If $f'(x) = 0$ for all $x$ in $I$, then $f$ is constant on $I$.

**EXAMPLE 2**  Let the function $f$ be defined by

$$y = f(x) = 2x^3 + 3x^2 - 12x - 5$$

**a.** Find the intervals over which $f$ is increasing.
**b.** Find the intervals over which $f$ is decreasing.
**c.** Sketch the graph of $f$.

**Solution** We must determine the intervals over which $f'(x) > 0$ and those over which $f'(x) < 0$. Thus, we must first find the derivative $f'(x)$. We get

$$f'(x) = D_x(2x^3 + 3x^2 - 12x - 5)$$
$$= 6x^2 + 6x - 12 = 6(x^2 + x - 2) = 6(x + 2)(x - 1)$$

The function will be increasing when $f'(x) > 0$ and decreasing when $f'(x) < 0$.

We can therefore tabulate the behavior of $f'(x)$ in the same way as we did earlier when we solved quadratic inequalities. Since $f'$ is continuous, it will take the value 0 as it changes from positive to negative values. But $f'(x) = 0$ if, and only if, $x = -2$ or $x = 1$. Thus, on the first line of the table, we write the values $-2$ and 1 for $x$, with $-2$ to the left of 1 since $-2 < 1$.

On the second line of the table, we have a 0 underneath $-2$ and 1 because $f'(-2) = f'(1) = 0$.

| $x$ | | $-2$ | | 1 | |
|---|---|---|---|---|---|
| $f'(x)$ | $+$ | 0 | $-$ | 0 | $+$ |
| $f(x)$ | $\nearrow$ | | $\searrow$ | | $\nearrow$ |

Since $f'(x) = 6(x + 2)(x - 1)$ and the factors $(x + 2)$ and $(x - 1)$ are of degree 1, the sign of the derivative changes as $x$ increases through $-2$ and through 1. It is sufficient to check the sign of $f'(0)$. We find $f'(0) = -12$. Since 0 is between $-2$ and 1 in the first line and $f'(0) < 0$, we enter a minus sign in the second line between the two zeros. We now enter plus signs to the left of the first zero and to the right of the second zero because we know that the sign of the derivative changes as $x$ increases through $-2$ and through 1. Finally, on the third line we show arrows rising to the right where $f$ is increasing, which is where $f'(x) > 0$, and we have an arrow falling to the right where $f$ is decreasing, which is where $f'(x) < 0$. Considering the values of $x$ in the first line above the three arrows of the third line, we can now answer the questions.

**a.** The function $f$ is increasing on the interval $(-\infty, -2)$ and on the interval $(1, \infty)$.
**b.** The function is decreasing on the interval $(-2, 1)$.
**c.** Since the function $f$ is continuous, we know its graph has no jump. We calculate the functional values $f(-2)$ and $f(1)$, because $f'(-2) = f'(1) = 0$ and therefore the tangent lines at $(-2, f(-2))$ and $(1, f(1))$ will be horizontal. Now,

$$f(-2) = 2(-2)^3 + 3(-2)^2 - 12(-2) - 5 = 15$$

and

$$f(1) = 2(1)^3 + 3(1)^2 - 12(1) - 5 = -12$$

Thus the points $(-2, 15)$ and $(1, -12)$ are on the graph of $f$. Since $f(0) = -5$, the point $(0, -5)$ is also on the graph. We now have enough information to sketch a rough graph of the function. (See Figure 8.14.)

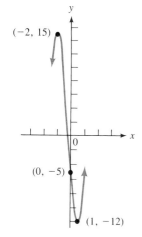

**FIGURE 8.14**

**Do Exercise 9.**

EXAMPLE 3    Find the intervals over which the graph of the function $g$ is rising, falling, if

$$g(x) = xe^{2x}$$

Solution    We want to find the intervals over which $g'(x)$ is positive and over which it is negative. We first differentiate $g(x)$ to get

$$g'(x) = D_x(xe^{2x}) = xD_xe^{2x} + e^{2x}D_xx$$
$$= xe^{2x}(2) + e^{2x}(1) = (2x + 1)e^{2x}$$

Since $e^{2x} > 0$ for all values of $x$, the sign of $g'(x)$ will be the same as that of the factor $2x + 1$. Note that

$$2x + 1 = 0 \text{ when } x = \frac{-1}{2}$$

and

$$2x + 1 < 0 \text{ when } x < \frac{-1}{2}$$

Furthermore,

$$2x + 1 > 0 \text{ when } x > \frac{-1}{2}$$

Therefore, $g'(x)$ is positive on the interval $\left(\frac{-1}{2}, \infty\right)$ and $g'(x)$ is negative on the interval $\left(-\infty, \frac{-1}{2}\right)$. It follows that the graph of $g$ is rising on the interval $\left(\frac{-1}{2}, \infty\right)$ and falling on the interval $\left(-\infty, \frac{-1}{2}\right)$. (See Figure 8.15.)

**Do Exercise 17.**

**FIGURE 8.15**

The graph shown has the point $\left(-\frac{1}{2}, \frac{-1}{2e}\right)$ labeled.

EXAMPLE 4    The graph of the *derivative* of a function $f$ is shown in Figure 8.16. Assume that this function has a continuous second derivative.

a. Identify the intervals over which the function $f$ is increasing.
b. Identify the intervals over which the function $f$ is decreasing.
c. Find the $x$-coordinate of each point on the graph of $y = f(x)$ where the tangent to the graph is horizontal.

**FIGURE 8.16**

d. Identify the intervals over which $f''(x) \geq 0$.
e. Identify the intervals over which $f''(x) \leq 0$.
f. Find the values of $x$ for which $f''(x) = 0$.

**Solution**    **a.** The graph of $y = f'(x)$ is above the $x$-axis on the intervals $(-1, 2)$ and $(5, \infty)$. Thus, $f'(x) > 0$ on these intervals. It follows that the function $f$ is strictly increasing on the intervals $(-1, 2)$ and $(5, \infty)$.

**b.** The graph of $y = f'(x)$ is below the $x$-axis on the intervals $(-\infty, -1)$ and $(2, 5)$. Thus, $f'(x) < 0$ on these intervals. The function $f$ is strictly decreasing on the intervals $(-\infty, -1)$ and $(2, 5)$.

**c.** The graph of $y = f'(x)$ crosses the $x$-axis at $x = -1$, $x = 2$, and $x = 5$. Thus $f'(-1) = f'(2) = f'(5) = 0$. Since the value of the derivative gives the slope of the tangent line, the slope of the tangent line at each of the points $(-1, f(-1))$, $(2, f(2))$, and $(5, f(5))$ is 0. Consequently, these tangent lines are parallel to the $x$-axis.

**d.** The graph of $f'$ is rising on the intervals $(-\infty, 1)$ and $(3, \infty)$. The derivative of $f'$ must be nonnegative on these intervals. That is, $f''(x) \geq 0$ on the intervals $(-\infty, 1)$ and $(3, \infty)$.

**e.** The graph of $f'$ is falling on the interval $(1, 3)$. The derivative of $f'$ must be nonpositive on this interval. That is, $f''(x) \leq 0$ on the interval $(1, 3)$.

**f.** Since $f''(x)$ changes sign as $x$ increases through 1 and since $f''$ is continuous, $f''(1) = 0$. Similarly, we argue that $f''(3) = 0$.

**Do Exercises 25–30.**

---

# Exercise Set 8.3

*In Exercises 1–4, the graph of a function f is given.*

*a. Find the intervals over which f is increasing.*
*b. Find the intervals over which f is decreasing.*
*c. Find the points where the graph has horizontal tangents.*
*d. Find the local maxima.*
*e. Find the local minima.*
*f. Find the absolute maximum.*
*g. Find the absolute minimum.*

**1.**

**2.**

**3.**

**4.**

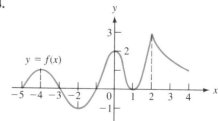

*In Exercises 5–14, a function f is defined.*

*a. Identify the intervals over which f is increasing.*
*b. Identify the intervals over which f is decreasing.*

c. Sketch the graph of the function.

d. Draw the horizontal tangent lines.

5. $f(x) = x^2 - 6x + 3$

6. $f(x) = x^2 - 3x + 1$

7. $f(x) = x^2 - 16$

8. $f(x) = x^3 + 3x^2 - 9x - 15$

9. $f(x) = 2x^3 - 3x^2 - 12x + 10$

10. $f(x) = x^3 - 3x + 2$

11. $f(x) = 2x^3 + 3x^2 - 72x + 100$

12. $f(x) = x^3 - 27x + 25$

13. $f(x) = x^4 - 32x + 45$

14. $f(x) = 3x^4 - 8x^3 - 6x^2 + 24x + 7$

In Exercises 15–24, a function g is defined.

a. Identify the intervals over which $g'(x) > 0$.

b. Identify the intervals over which $g'(x) < 0$.

c. Find the values of x for which $g'(x) = 0$.

d. Find the values of x in the domain of g for which $g'(x)$ is not defined.

15. $g(x) = e^{x^2}$

16. $g(x) = (x + 1)e^{3x}$

17. $g(x) = (2x + 4)e^{x/2}$

18. $g(x) = (2x - 3)e^{x^2}$

19. $g(x) = (x + 1)\ln|x + 1|$

20. $g(x) = \ln|(x + 1)(x + 3)|$

21. $g(x) = \sqrt[3]{x}$

22. $g(x) = \sqrt[5]{x - 1}$

23. $g(x) = \dfrac{2x + 3}{3x - 4}$

24. $g(x) = \dfrac{4x + 3}{2(x^2 + 1)}$

Exercises 25–30 refer to the following graph of the derivative $f'$. Assume that the function f has a continuous second derivative.

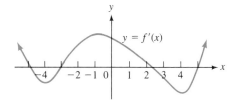

25. Identify the intervals over which the function f is increasing.

26. Identify the intervals over which the function f is decreasing.

27. Find the value of the abscissa of each point on the graph of $y = f(x)$ where the tangent to the graph is horizontal.

28. Identify the intervals over which $f''(x) \geq 0$.

29. Identify the intervals over which $f''(x) \leq 0$.

30. Find the values of x for which $f''(x) = 0$.

31. The Sherman Company found that the profit P, in thousands of dollars, from the sale of q thousand cans of racquetball balls is $P = -q^3 + 9q^2 - 15q - 9$, where $0 \leq q \leq 6$.

a. Find the intervals over which profit is decreasing.

b. Find the intervals over which profit is increasing.

c. Over which interval does the company operate at a loss?

32. The Schiavo Television Company found that the profit P, in thousands of dollars, it could make from the sale of q thousand color television sets is

$$P = -q^3 + 33q^2 - 120q - 252$$

for $0 \leq q \leq 28$.

a. Find the intervals over which profit is decreasing.

b. Find the intervals over which profit is increasing.

c. Over which interval does the company operate at a loss?

33. Market studies have shown that the relationship between the number q of units (in thousands) of a product the consumers are willing to buy and the price p (in dollars) per unit is given by

$$p = -q^2 + 48$$

provided $0 \leq q \leq 6$.

a. Find the total revenue function.

b. Find the intervals over which the revenue function is increasing.

c. Find the intervals over which the revenue function is decreasing.

34. The manager of a delicatessen knows the number q of cans (in thousands) of escargots the public is willing to buy and the price p (in dollars) per can are related by

$$p = -q^2 - 2q + 15$$

where $0 \leq q \leq 3$.

a. Show that the function defined by this equation is decreasing on the interval [0, 3].

b. Find the revenue function.

c. Determine the intervals over which the revenue function is increasing and over which it is decreasing.

35. The Udeen Electrical Company found that the relationship between the price p per deluxe flashlight and the number q of flashlights sold per week is

$$p = (10 - q)e^{-.1q}$$

for $0 \leq q \leq 10$.

**a.** Show that this function is decreasing on the interval [0, 10].

**b.** Find the revenue function.

**c.** Determine the intervals over which the revenue function is increasing and over which it is decreasing.

**36.** For the average human being, the ability to learn a language at age $t$ years old is given by the equation

$$L(t) = 10te^{-.05t}$$

for $2 \leq t \leq 70$, where $L(t)$ is approximately the number of new words one can learn per day. Find the age intervals where the ability to learn is increasing and where it is decreasing.

**37.** Let the function $f$ be defined by

$$f(x) = x^7 + 2x^5 + 3x^3 + 5x - 7.$$

**a.** Show that $f'(x) \geq 5$ for all $x$, and therefore $f'(x) > 0$ for all $x$.

**b.** Using the result of part **a**, show that $f$ is a one-to-one function.

# 8.4 The Mean Value Theorem (Optional)

In this section, we give the proof of Theorem 8.1 stated in the preceding section. In general, business and social sciences students are interested in the applications of mathematics to their fields of study rather than in the theoretical considerations. However, some of you have a keen interest in mathematics. The theorems and proofs in this section are intended for you.

The following example will prepare you for the next theorem.

**EXAMPLE 1**   Let $f$ be defined on the interval [2, 5] by the equation $f(x) = x^2 + 1$. Find a point on the graph of $f$ where the tangent line is parallel to the line that passes through the points $(2, f(2))$ and $(5, f(5))$.

**Solution**   Recall that two lines with slopes $m_1$ and $m_2$ are parallel if, and only if, $m_1 = m_2$. The line that passes through the two points has slope $\frac{f(5) - f(2)}{5 - 2}$. Let $(c, f(c))$ be the point on the graph where the tangent line is parallel to the line through $(2, f(2))$ and $(5, f(5))$. The slope of that tangent line is the value of the derivative at $c$. Therefore

$$f'(c) = \frac{f(5) - f(2)}{5 - 2}$$

But

$$f'(x) = D_x(x^2 + 1) = 2x$$

Thus

$$f'(c) = 2c$$

We get

$$2c = \frac{f(5) - f(2)}{5 - 2} = \frac{(5^2 + 1) - (2^2 + 1)}{3} = 7$$

Therefore,

$$c = \frac{7}{2}$$

Note that $2 < \frac{7}{2} < 5$, thus the point $\left(\frac{7}{2}, f\left(\frac{7}{2}\right)\right)$ is on the graph of $f$. Now $f\left(\frac{7}{2}\right) =$ $\left(\frac{7}{2}\right)^2 + 1 = \frac{53}{4}$. The point $\left(\frac{7}{2}, \frac{53}{4}\right)$ is the point we were asked to locate. (See Figure 8.17.)

**Do Exercise 3.**

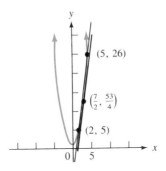

**FIGURE 8.17**

The preceding example illustrates an important theorem called the *Mean Value Theorem.**

**THEOREM 8.2 (MEAN VALUE THEOREM):** If a function $f$ is continuous on a closed bounded interval $[a, b]$ and differentiable on the open interval $(a, b)$, then there is at least one point $c$ in $(a, b)$ such that

$$f'(c) = \frac{f(b) - f(a)}{b - a}$$

This theorem is easy to understand geometrically. (See Figure 8.18.)

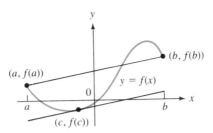

**FIGURE 8.18**

If we could draw all tangent lines to the graph of the equation $y = f(x)$, it is intuitively clear that there is at least one tangent line (there may be more than one) that is parallel to the line passing through the endpoints of the graph. But if $(c, f(c))$

---

*The proof of the *Mean Value Theorem* is beyond the scope of this book, but may be found in most standard calculus textbooks.

is the point of tangency, then $f'(c)$ is the slope of the tangent line. However, two parallel lines have equal slopes. Therefore,

$$f'(c) = \frac{f(b) - f(a)}{b - a}$$

(For another interpretation of the Mean Value Theorem, see Exercises 13 and 14 at the end of this section.)

Using the Mean Value Theorem, it is easy to prove parts 1, 2, and 5 of Theorem 8.1, of the preceding section. For convenience, we state these parts of the theorem again.

Suppose a function $f$ is differentiable on an open interval $I$.

1. If $f'(x) > 0$ for all $x$ in $I$, then $f$ is strictly increasing on $I$.
2. If $f'(x) < 0$ for all $x$ in $I$, then $f$ is strictly decreasing on $I$.
3. If $f'(x) = 0$ for all $x$ in $I$, then $f$ is constant on $I$.

***Proof:***

1. Let $x_1$ and $x_2$ be in $I$ with $x_1 < x_2$. By the Mean Value Theorem, there is a number $c$ between $x_1$ and $x_2$ (hence in $I$) such that

$$f'(c) = \frac{f(x_2) - f(x_1)}{x_2 - x_1}$$

Since $f'(x) > 0$ for all $x$ in $I$, $f'(c) > 0$. Thus

$$\frac{f(x_2) - f(x_1)}{x_2 - x_1} > 0$$

We know that $x_2 - x_1 > 0$ (since $x_1 < x_2$). Thus, $f(x_2) - f(x_1)$ is also positive. We conclude that $f(x_1) < f(x_2)$. We have proved that $x_1 < x_2$ implies $f(x_1) < f(x_2)$. Therefore the function $f$ is strictly increasing.
2. The proof for this part is analogous and is therefore omitted.
3. Let $x_1$ be a fixed point in $I$ and let $x$ be any other point in $I$. By the Mean Value Theorem, there is a point $c$ between $x_1$ and $x$ such that

$$f'(c) = \frac{f(x) - f(x_1)}{x - x_1}$$

But, since $f'(x) = 0$ for all $x$ in $I$, $f'(c) = 0$. Thus

$$\frac{f(x) - f(x_1)}{x - x_1} = 0$$

from which we conclude that

$$f(x) - f(x_1) = 0$$

and

$$f(x) = f(x_1)$$

Therefore, $f(x) = f(x_1)$ for all $x$ in $I$ and we conclude that $f$ is a constant function on $I$.

## Exercise Set 8.4

*In Exercises 1–9, a function f is defined and numbers a and b are given.*

*a. In each case, find all numbers c between a and b such that*

$$f'(c) = \frac{f(b) - f(a)}{b - a}$$

*b. In each case, draw the graph of the function f over the interval [a, b]. Also, draw the line through the points (a, f(a)) and (b, f(b)) and then draw the parallel tangent line(s).*

**1.** $f(x) = x^2 + 5$; $a = 3$, $b = 7$
**2.** $f(x) = x^2 + x + 2$; $a = 2$, $b = 5$
**3.** $f(x) = \sqrt{x}$; $a = 4$, $b = 9$
**4.** $f(x) = \ln x$; $a = 1$, $b = e$
**5.** $f(x) = \dfrac{1}{x}$; $a = 4$, $b = 9$
**6.** $f(x) = \dfrac{x + 2}{x - 3}$; $a = 4$, $b = 8$
**7.** $f(x) = \sqrt[3]{x}$; $a = 1$, $b = 8$
**8.** $f(x) = x^3 - 3x^2 + 3x + 3$; $a = -1$, $b = 3$
**9.** $f(x) = x^3 + 6x^2 + 12x + 11$; $a = -2 - \sqrt{3}$, $b = -2 + \sqrt{3}$

**10.** Prove that if $a$ and $b$ are real numbers and $a \neq b$, then $a^2 + ab + b^2 > 0$. (*Hint:* Only the case $ab < 0$ is not obvious. In this case, note that $a^2 + ab + b^2 = (a + b)^2 - ab$.)

**11.** Let $f(x) = x^3$.

   **a.** Prove that $f$ is strictly increasing. (*Hint:* Use the definition, the formula $a^3 - b^3 = (a - b)(a^2 + ab + b^2)$, and the result of Exercise 10.)
   **b.** Show that $f'(0) = 0$.

This exercise shows that in part 3 of Theorem 8.1 of the preceding section, we could only conclude that $f'(x) \geq 0$, but not that $f'(x) > 0$.

**12.** Let $f(x) = x^{2/3}$.

   **a.** Show that there is no number $c$ between $-1$ and $1$ such that

$$f'(c) = \frac{f(1) - f(-1)}{1 - (-1)}$$

   **b.** Draw the graph of $f$ over the interval $[-1, 1]$ and explain why part **a** does not contradict the conclusion of the Mean Value Theorem.

**13.** A particle is moving along a directed line. Its directed distance $d$ feet from the origin at instant $t$ seconds is

$$d = t^2 + 5t - 4$$

   **a.** Find the average speed of the particle from instant $t_1 = 2$ seconds to instant $t_2 = 6$ seconds.
   **b.** Show that there is an instant in that time interval at which the instantaneous velocity is exactly the same as the average velocity.

**14.** A motorist who frequently travels on a certain highway buys a radar detector. Two police officers are stationed on the side of the highway at points $A$ and $B$ which are 40 miles apart. As the motorist passes point $A$, he is traveling at a speed of 50 mph (remember: he knows the police officer is there). The police officer radios to point $B$ that the motorist has just passed point $A$. Forty minutes later, the motorist passes point $B$, again traveling 50 mph because he knows the radar is being used. The police officer stops the motorist and she tells him: "You have traveled 40 miles in 40 minutes, which is an average speed of 60 mph. I took calculus when I was in college and I know that at some instant in the last 40 minutes you were traveling at a speed of 60 mph, which is over the 55 mph limit. Therefore, I am issuing this ticket. Have a good day, Sir." The police officer went back to her station satisfied that her college education was useful after all! Explain why the police officer could be sure that the motorist had traveled 60 mph at some instant.

## 8.5  The First Derivative Test

Local (relative) maximum and minimum as well as absolute maximum and minimum of a function were defined in Section 8.3. In this section, we introduce a technique to find these extrema. (Recall that we use the term extremum to denote either a maximum or a minimum.)

## Critical Values of a Function

When the graph of a function is drawn as in Figure 8.19, it is easy to locate its extrema. The function $f$ of Figure 8.19 has local maxima at $x_1$ and $x_3$, and local minima at $a$, $x_2$, and $b$. Furthermore, the absolute maximum occurs at $x_3$ and the absolute minimum is at $x_2$. Again, we emphasize the terminology. For example, we say that the function $f$ has a local maximum *at* $x_1$, and the maximum value *is* $f(x_1)$. The first statement gives the location of the maximum, the second gives its value.

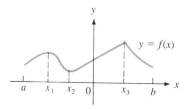

**FIGURE 8.19**

We wish to locate the relative extrema of a function without having the graph as a visual aid. We first state the following theorem.

---

**THEOREM 8.3:** Suppose a function $f$ is defined on an open interval $I$, $x_1$ is in $I$, and $f'(x_1)$ exists.

1. If $f'(x_1) > 0$, then there is a neighborhood of $x_1$, say $(x_1 - h, x_1 + h)$, such that $x_1 - h < s < x_1 < t < x_1 + h$ implies $f(s) < f(x_1) < f(t)$.
2. If $f'(x_1) < 0$, then there is a neighborhood of $x_1$, say $(x_1 - h, x_1 + h)$, such that $x_1 - h < s < x_1 < t < x_1 + h$ implies $f(s) > f(x_1) > f(t)$.

---

(See Figure 8.20.)

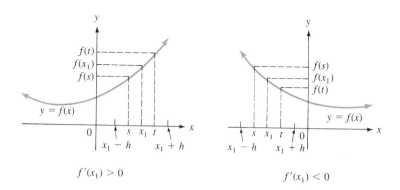

**FIGURE 8.20**

The details of the proof of Theorem 8.3 are left as an exercise. The basic idea in part 1 is the fact that $f'(x_1) > 0$ implies the existence of a small interval about the point $x_1$ on which the ratio $(f(x) - f(x_1))/(x - x_1)$ is positive, from which we can deduce that $f(x) - f(x_1))$ and $x - x_1$ must both be positive, or both be negative. Using this fact, we can complete the proof.

Now suppose a function $f$ is defined on an open interval $I$ and $f$ has a local extremum at some point $x_1$ of $I$. Then by Theorem 8.3, we cannot have $f'(x_1) > 0$ (otherwise, there would be an interval $(x_1 - h, x_1 + h)$ such that $x_1 - h < s < x_1 < t < x_1 + h$ implies $f(s) < f(x_1) < f(t)$, and $f(x_1)$ would not be a local extremum). Similarly, we cannot have $f'(x_1) < 0$. Thus, there are only two possibilities left. Either $f'(x_1) = 0$ or $f'(x_1)$ is not defined. Therefore we give the following definition.

**DEFINITION 8.12:**   Suppose that a function $f$ is defined on some interval $I$ and that for some $c$ in $I$, which is not an endpoint of that interval, either $f'(c) = 0$ or $f'(c)$ is not defined. Then $c$ is called a *critical value* of the function $f$ and $(c, f(c)$ is called a *critical point*.

The discussion preceding the definition proves the following theorem.

**THEOREM 8.4:**   Suppose a function $f$ is defined on an interval $I$ and $f$ has a local extremum at $c$. Then either $c$ is an endpoint of the interval $I$, or $c$ is a critical value of the function.

It is important to note that the theorem does *not* state that if $c$ is a critical value of the function $f$, then the function necessarily has a local extremum at $c$. What it does state is that if $f$ has a local extremum at $c$, then $c$ must be a critical value of $f$ or an endpoint of the interval on which $f$ is defined.

Therefore in trying to locate the extrema of a function, we need not be concerned with points that are not critical values or endpoints of the interval of definition of that function. The first step will always be to find all critical values of the function.

Now suppose that $x_1$ is a critical value of the function $f$. Figure 8.21**a–i** illustrates some of the different cases we could have. In Figure 8.21**a–d**, the graph has a horizontal tangent line at $(x_1, f(x_1))$; thus $f'(x_1) = 0$. Note that in **a**, we have a maximum since the function is increasing to the left of $x_1$ and decreasing to the right. In **b**, we have a minimum since $f$ changes from decreasing to increasing at $x_1$. In **c** and **d**, we have neither a maximum nor a minimum since in **c** the graph keeps falling to the right and in **d** it keeps rising to the right as it passes through $(x_1, f(x_1))$.

In Figure 8.21**e–g**, the derivative is not defined at $x_1$ and we have a maximum in **e**, a minimum in **f**, and neither in **g** for the same reasons as for **c** and **d**.

In **h** and **i**, the function is not continuous at $x_1$, so $f'(x_1)$ does not exist. Note that in **h**, the graph is rising to the left of $x_1$ and falling to the right. Thus we might expect a maximum, but we have neither a maximum nor a minimum. In **i**, the graph is falling to the left of $x_1$ and rising to the right. Therefore we would expect a local minimum, but instead we have a local maximum. To avoid these difficulties, we shall require the functions to be continuous.

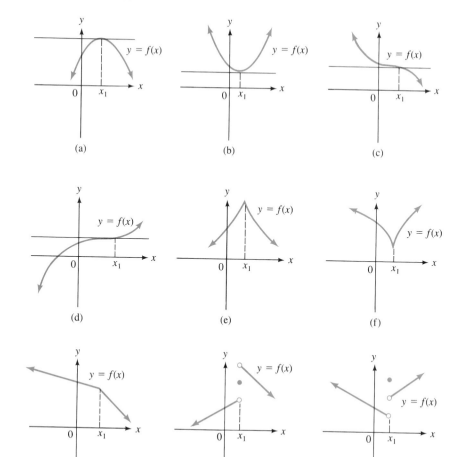

**FIGURE 8.21**

## The First Derivative Test

The following theorem is usually referred to as the *First Derivative Test*.

**THEOREM 8.5 (First Derivative Test):** Assume that a function $f$ is continuous on some interval $I$ and that $c$ is a critical value of $f$ in $I$. Let $h$ be a small enough positive number so that $c$ is the only critical value of $f$ in the neighborhood $(c - h, c + h)$, and $(c - h, c + h) \subset I$. Let $J = (c - h, c + h)$. Then:

**a.** $f$ has a local minimum at $c$ if $f'(x) < 0$ when $c - h < x < c$, and $f'(x) > 0$ when $c < x < c + h$.

**b.** $f$ has a local maximum at $c$ if $f'(x) > 0$ when $c - h < x < c$, and $f'(x) < 0$ when $c < x < c + h$.

**c.** *f* does not have a local extremum at *c* if $f'(x) > 0$ for all *x* in *J* with $x \neq c$, or if $f'(x) < 0$ for all *x* in *J* with $x \neq c$. That is, *f* does not have a local extremum at *c* if $f'(x)$ does not change sign as *x* goes from the interval $(c - h, c)$ to the interval $(c, c + h)$.

(See Figure 8.22.)

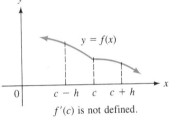

$f'(c) = 0$

Local maximum at *c* if
$f'(x) > 0$ on $(c - h, c)$
and $f'(x) < 0$ on $(c, c + h)$.

$f'(c)$ is not defined.

Local minimum at *c* if
$f'(x) < 0$ on $(c - h, c)$
and $f'(x) > 0$ on $(c, c + h)$.

$f'(c) = 0$

Neither a local maximum
nor minimum at *c* if
$f'(x) > 0$ on both $(c - h, c)$
and $(c, c + h)$.

$f'(c)$ is not defined.

Neither a local maximum
nor minimum at *c* if
$f'(x) < 0$ on both $(c - h, c)$
and $(c, c + h)$.

**FIGURE 8.22**

As an illustration of the ideas involved, imagine a blind person riding in a car. If that person could feel the car traveling uphill then downhill, he or she would know that the car passed through a high point of the highway.

Essentially, the sign of the derivative $f'(x)$ indicates whether the graph goes uphill or downhill. Therefore, without actually seeing the picture, we can deduce the right conclusion in each case.

## Endpoint Extrema

The only other fact to remember is that an extremum can occur only at a critical point or at an endpoint of the interval over which the function is defined.

> **THEOREM 8.6:**   Let $f$ be a continuous function defined on a closed bounded interval $[a, b]$. Then:
>
> **a.** $f$ has a local maximum at the endpoint $a$ if there is some $x_1$ in the interval $(a, b)$ such that $f'(x) < 0$ for each $x$ in $(a, x_1)$.
> **b.** $f$ has a local minimum at the endpoint $a$ if there is some $x_1$ in the interval $(a, b)$ such that $f'(x) > 0$ for each $x$ in $(a, x_1)$.
> **c.** $f$ has a local maximum at the endpoint $b$ if there is some $x_2$ in the interval $(a, b)$ such that $f'(x) > 0$ for each $x$ in $(x_2, b)$.
> **d.** $f$ has a local minimum at the endpoint $b$ if there is some $x_2$ in the interval $(a, b)$ such that $f'(x) < 0$ for each $x$ in $(x_2, b)$.

(See Figure 8.23.)

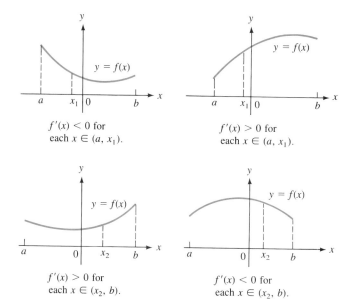

**FIGURE 8.23**

## Examples

**EXAMPLE 1**   The function $f$ is defined on the closed interval $[-4, 3]$ by the equation

$$f(x) = 2x^3 + 3x^2 - 12x + 7$$

Locate all relative extrema of $f$ on that interval.

**Solution**   Since $f(x)$ is a polynomial, it is differentiable. Therefore the only critical values will be the solutions of the equation $f'(x) = 0$.
   Now,

$$f'(x) = D_x(2x^3 + 3x^2 - 12x + 7) = 6x^2 + 6x - 12 = 6(x + 2)(x - 1)$$

The solutions of the equation

$$6(x + 2)(x - 1) = 0$$

are $-2$ and $1$. Thus the only critical values of $f$ are $-2$ and $1$, and both are in the defined interval of the function. We make a table to determine the intervals on which $f'(x) > 0$ and those on which $f'(x) < 0$. In the first line of the table, we enter the values of $x$ that are the endpoints of the interval over which $f$ is defined and the critical values of $f$. We write them in increasing order.

| $x$ | $-4$ | | $-2$ | | $1$ | | $3$ |
|---|---|---|---|---|---|---|---|
| $f'(x)$ | | $+$ | $0$ | $-$ | $0$ | $+$ | |
| $f(x)$ | | $\nearrow$ | | $\searrow$ | | $\nearrow$ | |

The explanations for the other entries in the table are the same as for those values tabulated in Section 8.3, Example 2.

We see that $f'(x) > 0$ on the intervals $(-4, -2)$ and $(1, 3)$. Thus $f$ is increasing on these intervals. Also, $f'(x) < 0$ on the interval $(-2, 1)$; consequently, $f$ is decreasing on that interval. We conclude that the function has a local minimum at the left endpoint $-4$, a local maximum at $-2$, another local minimum at $1$, and a local maximum at the right endpoint $3$.
**Do Exercise 17.**   ■

Recall that $f(u)$ and $f(v)$ are the absolute minimum and absolute maximum, respectively, of a function $f$ provided that $f(u) \leq f(x) \leq f(v)$ for all $x$ in the domain of $f$.

In general, a function may or may not have an absolute maximum or an absolute minimum. However, in advanced mathematics texts, it is proved that if a function $f$ is continuous on a closed bounded interval $[a, b]$, then it does have both absolute extrema. Since an absolute extremum is also a local extremum, it must occur at a critical value of $f$ or at an endpoint of the interval. To find the absolute extrema, we first find all critical values of $f$ in the open interval $(a, b)$. Then we evaluate the function at each of these values and at the endpoints $a$ and $b$. Out of all the numbers obtained, the smallest is the absolute minimum and the largest is the absolute maximum.

**EXAMPLE 2**   Find the absolute extrema of the function of Example 1.

**Solution**   The critical values of $f$ in the open interval $(-4, 3)$ are $-2$ and $1$. (See Example 1.) Evaluating the function at each of these values and at the endpoints $-4$ and $3$, we get

$$f(-4) = -25, f(-2) = 27, f(1) = 0, \text{ and } f(3) = 52$$

Among the numbers $-25$, $27$, $0$, and $52$, the smallest is $-25$ and the largest is $52$. Thus the absolute minimum is $-25$ and occurs at $-4$, and the absolute maximum is $52$, which occurs at $3$.
**Do Exercise 29.**   ■

EXAMPLE 3    Find the critical values* of the functions defined by the following equations.

    **a.** $f(x) = (x - 1)^{1/3}$

    **b.** $g(x) = \dfrac{1}{(x - 1)^{1/3}}$

Solution    **a.** We must first find $f'(x)$. Differentiating, we get

$$f'(x) = D_x(x - 1)^{1/3} = \frac{1}{3}(x - 1)^{1/3-1}D_x(x - 1)$$

$$= \frac{1}{3(x - 1)^{2/3}}$$

Clearly, $f'(1)$ is not defined. Since $f(1) = 0$, 1 is in the domain of $f$. Therefore 1 is a critical value of $f$. We see that the equation $1/3(x - 1)^{2/3} = 0$ has no solution. Thus 1 is the only critical value of $f$.

    **b.** $g'(x) = D_x\dfrac{1}{(x - 1)^{1/3}} = D_x(x - 1)^{-1/3}$

$$= \frac{-1}{3}(x - 1)^{-1/3-1}D_x(x - 1) = \frac{-1}{3}(x - 1)^{-4/3} = \frac{-1}{3(x - 1)^{4/3}}$$

Again $g'(1)$ is not defined. However, in this case $g(1)$ is not defined either. Therefore, 1 is not a critical value of $g$ since it is not in the domain of $g$.
**Do Exercises 3 and 5.**    ■

EXAMPLE 4    A function $f$ is defined by

$$f(x) = x^{1/3}e^{x/6}$$

Find its extrema.

Solution    The domain of $f$ is the set of all real numbers. Thus, the only points that we need to consider are the critical values of $f$. We first find the derivative. We get

$$f'(x) = D_x(x^{1/3}e^{x/6})$$

$$= x^{1/3}D_xe^{x/6} + e^{x/6}D_xx^{1/3} = x^{1/3}e^{x/6}\left(\frac{1}{6}\right) + e^{x/6}\left(\frac{1}{3}\right)x^{-2/3}$$

$$= \frac{e^{x/6}(x + 2)}{6x^{2/3}}$$

It is clear that $f'(x) = 0$ only when $x = -2$, and that $f'(0)$ is not defined. Since 0 is in the domain of $f$, it is a critical value. Thus the only critical values are $-2$ and 0. Before making a table, we note that $e^{x/6} > 0$ for all values of $x$, and that $x^{2/3} = (x^{1/3})^2$ and is therefore positive for all values of $x$ except $x = 0$. Consequently, $f'(x)$ changes sign only when $x + 2$ does. That is, as $x$ increases through $-2$. We enter the critical numbers $-2$ and 0 on the first line of the table. The $u$ in line 2 indicates that $f'(x)$ is undefined when $x = 0$. We find that $f'(1) > 0$. So we enter a

*It must be emphasized that the critical values of a function are in the domain of that function. If for some function $f$, $f'(c)$ is not defined and $c$ is not in the domain of $f$, then $c$ is not a critical value of $f$.

plus sign to the right of the $u$ in the second line of the table. Since $f'(x)$ changes sign only at $x = -2$ we enter a plus sign between 0 and $u$ on the second line and a minus sign to the left of 0. The arrows on the third line indicate where $f$ is increasing, decreasing.

| $x$ | | $-2$ | | | 0 | |
|---|---|---|---|---|---|---|
| $f'(x)$ | $-$ | 0 | $+$ | $u$ | $+$ | |
| $f(x)$ | $\searrow$ | | $\nearrow$ | | $\nearrow$ | |

Using the last line of the table, we see that the function $f$ has a local minimum at $-2$, but it has neither a maximum nor a minimum at 0.

As we stated earlier, unless a function is continuous on a closed bounded interval, it may not have an absolute extremum. In this example, the function has an absolute minimum at $-2$ but it has no absolute maximum.

**Do Exercise 21.**

**EXAMPLE 5** Locate the relative extrema of the function $f$ if

$$f(x) = (x^2 + 5x + 18)(x - 2)^{1/3}.$$

**Solution** We first find the derivative.

$$\begin{aligned} f'(x) &= D_x[(x^2 + 5x + 18)(x - 2)^{1/3}] \\ &= (x^2 + 5x + 18)D_x(x - 2)^{1/3} + (x - 2)^{1/3}D_x(x^2 + 5x + 18) \\ &= (x^2 + 5x + 18)\frac{1}{3}(x - 2)^{-2/3} + (x - 2)^{1/3}(2x + 5) \\ &= \frac{(x^2 + 5x + 18) + 3(x - 2)(2x + 5)}{3(x - 2)^{2/3}} \\ &= \frac{7x^2 + 8x - 12}{3(x - 2)^{2/3}} = \frac{(x + 2)(7x - 6)}{3(x - 2)^{2/3}}. \end{aligned}$$

It is now clear that $f'(-2) = f'\left(\frac{6}{7}\right) = 0$, and that $f'$ is not defined at 2. Since $-2, \frac{6}{7}$, and 2 are in the domain of $f$, all three numbers are critical values of $f$. We enter these three values on the first line of the table.

| $x$ | | $-2$ | | $\dfrac{6}{7}$ | | 2 | |
|---|---|---|---|---|---|---|---|
| $f'(x)$ | $+$ | 0 | $-$ | 0 | $+$ | $u$ | $+$ |
| $f(x)$ | $\nearrow$ | | $\searrow$ | | $\nearrow$ | | $\nearrow$ |

On the second line for $f'(x)$, we enter 0 below $-2$ and $\frac{6}{7}$, and $u$ below 2 to indicate that $f'(x) = 0$ when $x = -2$ and when $x = \frac{6}{7}$, and that $f'(x)$ is not defined when $x = 2$. Since $(x - 2)^{2/3} = [(x - 2)^{1/3}]^2 \geq 0$, $f'(x)$ changes sign only as $x + 2$ and

$7x - 6$ do. That is, as $x$ increases through $-2$ and again through $\frac{6}{7}$. We find that $f'(0) < 0$. Since 0 is between $-2$ and $\frac{6}{7}$ on the first line, we enter a minus sign between the two zeros on the second line. Remembering that $f'(x)$ changes sign only at $-2$ and at $\frac{6}{7}$, we enter a plus sign to the left of the first 0, a plus sign between 0 and $u$, and another plus sign to the right of $u$. We use arrows to indicate where $f(x)$ is increasing (decreasing) on the third line of the table. We conclude that $f$ has a local maximum at $-2$, a local minimum at $\frac{6}{7}$, and has neither a maximum nor a minimum at 2.

**Do Exercise 27.**

### Procedure for Finding the Extrema of a Continuous Function $f$

**STEP 1.** Find $f'(x)$.

**STEP 2.** Solve the equation $f'(x) = 0$

**STEP 3.** Determine all values of $x$ for which $f'(x)$ is not defined.

**STEP 4.** Make a table to determine the intervals on which $f'(x) > 0$ and those on which $f'(x) < 0$. On the first line of the table, write, in increasing order, the *solutions* of the equation $f'(x) = 0$ obtained in Step 2 that are in the domain of $f$, the *values* of $x$ obtained in Step 3 that are in the domain of $f$, and the *endpoints* of the interval on which $f$ is defined, if appropriate. In the second line, under each critical number $c$ entered in the first line, write 0 if $f'(c) = 0$, and $u$ if $f'(c)$ is undefined. In each open interval with endpoints that are two entries on the first line of the table and which contains no critical value of $f$, choose an arbitrary number $d$. Evaluate $f'(d)$. We need not calculate its exact value since we are only interested in whether $f'(d)$ is positive or negative. If $f'(d) > 0$, then $f'(x) > 0$ for all $x$ on the interval and we show this on the second line of the table by entering a "+" between the endpoints of the interval. Similarly if $f'(d) < 0$, we enter a "−" between the endpoints of the interval. (See also Example 7, Section 0.8 for a slightly more efficient procedure.)

**STEP 5.** On the last line of the table, draw arrows going up where $f'(x) > 0$ and arrows going down where $f'(x) < 0$.

**STEP 6.** Using the last line of the table as a guide, determine for each entry on the top line of the table (critical values and endpoints of the interval on which $f$ is defined) whether the function has a local maximum, a local minimum, or neither at that entry.

**STEP 7.** In the case where $f$ is continuous on a closed bounded interval $[a, b]$, evaluate $f$ at all critical values in the open interval $(a, b)$ and at the endpoints $a$ and $b$. Among the values so obtained, the smallest is the absolute minimum and the largest is the absolute maximum.

# Exercise Set 8.5

*In Exercises 1–13, find the critical values of the given functions.*

**1.** $f(x) = x^2 + 5x - 6$

**2.** $f(x) = -x^2 + 6x - 3$

**3.** $g(x) = 5x^{4/5}$

**4.** $g(x) = 3(x - 5)^{2/3}$

**5.** $h(x) = 2(x + 1)^{-2/3}$

**6.** $f(t) = t^3 + 3t^2 - 24t + 10$

**7.** $f(x) = 2x^3 + 9x^2 - 108x + 30$

**8.** $g(x) = 3xe^{5x}$

**9.** $g(x) = x^2 e^{3x}$

**10.** $g(x) = \ln|x + 2|$

**11.** $h(x) = x^3 + 6x^2 + 9x + 4$, where $-2 \le x \le 2$

**12.** $h(x) = xe^{x/5}$, where $-3 \le x \le 5$

**13.** $f(x) = (x + 1)\ln|x + 1|$

*In Exercises 14–27, find all local extrema of the function defined by the given equation.*

**14.** $f(x) = x^2 + 4x - 5$

**15.** $f(x) = -x^2 + 6x + 10$

**16.** $g(x) = x^3 + 3x^2 - 9x + 5$

**17.** $g(x) = x^3 + 6x^2 - 36x + 7$, where $-4 \le x \le 4$

**18.** $h(x) = 4(x - 3)^{2/3}$

**19.** $h(x) = x^3 + 3x^2 - 24x + 10$, where $-5 \le x \le 1$

**20.** $h(x) = 2x^3 + 9x^2 - 108x + 30$, where $-4 \le x \le 6$

**21.** $f(t) = 4te^{3t}$

**22.** $f(u) = u^2 e^{4u}$

**23.** $g(x) = x\ln|x + 1|$

**24.** $g(x) = \dfrac{x^2 + 1}{x + 2}$

**25.** $h(t) = \dfrac{t^2 + t + 2}{t - 1}$

**26.** $h(x) = \dfrac{3x^2 + 10x + 14}{x - 1}$

**27.** $f(x) = (x^2 + 3x + 11) \sqrt[3]{x - 2}$

*In Exercises 28–37, find the absolute extrema (if there are any) of the given functions.*

**28.** $f(x) = x^2 + 6x - 10$

**29.** $f(x) = x^2 + 4x + 4$, $-1 \le x \le 2$

**30.** $g(x) = x^3 + 3x^2 - 24x + 10$, $-5 \le x \le 4$

**31.** $g(x) = 2x^3 + 9x^2 - 108x + 30$, $-4 \le x \le 5$

**32.** $h(x) = xe^{4x}$

**33.** $h(x) = x^2 e^{3x}$

**34.** $f(x) = \ln|x + 1|$, $0 \le x \le 5$

**35.** $f(x) = xe^{-x/4}$, $-1 \le x \le 6$

**36.** $g(x) = \sqrt{x^2 + x + 4}$

**37.** $f(x) = (x^2 + 3x + 11) \sqrt[3]{x - 2}$, $-2 \le x \le 3$

**\*38.** Part of the proof of part 1 of Theorem 8.3 was given in this section. Complete the details of the proof.

**\*39.** Prove part 2 of Theorem 8.3.

**40.** Let the functions $f$, $g$, and $h$ be defined by $f(x) = x^4$, $g(x) = -x^4$, and $h(x) = x^3$, respectively. Using the First Derivative Test show the following.

  **a.** The function $f$ has a local minimum at $0$.
  **b.** The function $g$ has a local maximum at $0$.
  **c.** The function $h$ has neither a maximum nor a minimum at $0$.

**\*41.** If $C(q)$ dollars is the total cost of producing $q$ units of a commodity, then

$$C(q) = F + V(q)$$

where $F$ is the *fixed cost* (a constant) and $V(q)$ is the *total variable cost*.

  **a.** Show that if the average variable cost is minimum at $q_0$, then the marginal cost and the average variable cost are equal at $q_0$.
  **b.** Show that if the average total cost is minimum at $q_0$, then the marginal cost is equal to the average total cost at $q_0$.

**\*42.** Do this exercise only if you know the content of optional Section 8.4. Prove Theorem 8.6 of this section. (*Hint:* Use the Mean Value Theorem.)

# 8.6 The Second Derivative Test

There is another test which is often simpler to use than the First Derivative Test that you learned in the preceding section. It also has a nice geometric interpretation.

## Concavity

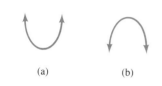

(a)          (b)

**FIGURE 8.24**

A curve is said to be concave up if it opens upward as in Figure 8.24**a**, and it is concave down if it opens downward as in Figure 8.24**b**. A more formal definition considers the tangent lines, as in Figure 8.25**a**. As the point of tangency moves to the right, the slopes of the tangent lines are increasing, which means that the first derivative is increasing. On the other hand, in Figure 8.25**b** when the graph is concave down, as the point of tangency moves to the right, the slopes of the tangent lines are decreasing, which indicates that the first derivative of the function is decreasing. Thus we have the following definition.

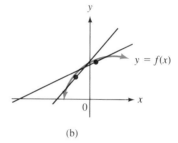

(a)                                (b)

**FIGURE 8.25**

The slope of the tangent line increases as the point of tangency moves to the right.

The slope of the tangent line decreases as the point of tangency moves to the right.

**DEFINITION 8.13:**   The graph of a function $f$ is said to be *concave up* (*concave down*) on an interval $I$ provided $f'$ is increasing (decreasing) on $I$.

By Theorem 8.1 of Section 8.3, if $f''(x) > 0$ ($f''(x) < 0$) on an interval, then $f'$ is increasing (decreasing) on that interval and we have the following theorem.

**THEOREM 8.7:**   Suppose a function $f$ is twice differentiable on an interval $I$.

1. If $f''(x) > 0$ for all $x$ in $I$, then the graph of $y = f(x)$ is concave up on $I$.
2. If $f''(x) < 0$ for all $x$ in $I$, then the graph of $y = f(x)$ is concave down on $I$.

## The Second Derivative Test

We may now state the next theorem, which is usually referred to as the *Second Derivative Test*.

**THEOREM 8.8:**    Suppose $c$ is a critical value of a function $f$.

1. If $f''(c) > 0$, then the function $f$ has a local minimum at $c$.
2. If $f''(c) < 0$, then $f$ has a local maximum at $c$.
3. If $f''(c) = 0$, no conclusion can be drawn.

The proof of this theorem will not be given.

To illustrate part 3 of the theorem, consider the functions $f$, $g$, and $h$ defined by $f(x) = x^4$, $g(x) = -x^4$, and $h(x) = x^3$. As an exercise, you may show that 0 is a critical value of all three functions and that $f''(0) = g''(0) = h''(0) = 0$. However, $f$ has a local minimum at 0, $g$ has a local maximum at 0, and $h$ has neither a maximum nor a minimum at 0. (See Exercise 40, Section 8.5.) Therefore the Second Derivative Test is inconclusive in that case.

The Second Derivative Test is often simpler to use than the First Derivative Test because we need only evaluate the second derivative at the critical values of the function; whereas in the case of the First Derivative Test, we need to determine the sign of the first derivative on intervals. Therefore, in cases where it is easy to obtain a formula for the second derivative—as with polynomials—the second derivative test is by far the simpler test.

On the other hand, the Second Derivative Test cannot be used at a critical value $c$ where $f''(c) = 0$. Also, it cannot be used at those critical values of $f$ where $f'(c)$ is not defined because $f''(c)$ is not defined either.

Sometimes, the details in differentiating the first derivative to obtain the second derivative can be more troublesome than using the First Derivative Test. In those cases, good judgment would suggest using the First Derivative Test.

We introduced the notion of concavity at the beginning of this section. Note that concavity was defined over an interval, while in the Second Derivative Test, we used only the value of the second derivative at the critical value $c$. However, almost always in the applications, the second derivative is continuous. Thus if $f''(c) > 0$, then, necessarily, there is an open interval containing $c$ on which $f''(x) > 0$. Similarly, if $f''(c) < 0$, then $f''(x) < 0$ for all $x$ in some open interval containing $c$. These facts make it easy to visualize the idea behind the Second Derivative Test.

If $f''(c) > 0$ and $f''$ is continuous at $c$, then, near $c$, the graph of $f$ is concave up and we see that $f$ has a local minimum at $c$. (See Figure 8.26a.) On the other hand, if $f''(c) < 0$ and $f''$ is continuous at $c$, then there is an interval containing $c$ over which the graph of $f$ is concave down. Therefore $f$ has a local maximum at $c$. (See Figure 8.26b.)

(a)

(b)

**FIGURE 8.26**

Some students use a mnemonic device to recall the Second Derivative Test. Since people who have a positive attitude have a smiling face (see Figure 8.27**a**), and those with a negative attitude have a sad face (see Figure 8.27**b**) they remember, by analogy, that a function with a positive second derivative has a "smiling" concave-up graph, while one with a negative second derivative has a "sad" or concave-down graph. (See Figure 8.28**a, b**.)

(a)          (b)

**FIGURE 8.27**

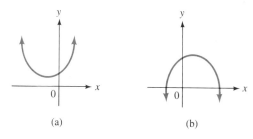

**FIGURE 8.28**          (a)          (b)

## Examples

We are now ready to give some examples.

**EXAMPLE 1**  Using the Second Derivative Test, locate the relative extrema of the function

$$f(x) = x^3 - 6x^2 + 9x + 4$$

**Solution**  Since $f(x)$ is a polynomial, the derivative exists for all $x$. Therefore the only critical values are the solutions of $f'(x) = 0$. Differentiating, we get

$$f'(x) = 3x^2 - 12x + 9 = 3(x - 3)(x - 1)$$

Thus, $f'(x) = 0$ when $x = 1$ or $x = 3$. The critical values are 1 and 3.
Differentiating again, we get

$$f''(x) = 6x - 12$$

We now evaluate $f''(x)$ at the critical values. We get

$$f''(1) = 6(1) - 12 = -6$$

and

$$f''(3) = 6(3) - 12 = 6$$

Since $f''(1) < 0$, $f$ has a local maximum at 1, and $f''(3) > 0$ implies that $f$ has a local minimum at 3.
**Do Exercise 17.**                                                                          ∎

**EXAMPLE 2**  Using the Second Derivative Test, find the absolute extrema of $f$ if

$$f(x) = (x^2 - 8x + 16)e^x, \text{ where } 0 \le x \le 3$$

**Solution**   We need to find the critical values of $f$ that are in the open interval $(0, 3)$. Differentiating, we get

$$f'(x) = (x^2 - 8x + 16)D_x e^x + e^x D_x(x^2 - 8x + 16)$$
$$= (x^2 - 8x + 16)e^x + e^x(2x - 8) = (x^2 - 6x + 8)e^x$$

The critical values must be solutions of

$$(x^2 - 6x + 8)e^x = 0$$

Factoring, we get

$$(x - 2)(x - 4)e^x = 0$$

Since $e^x > 0$ for all $x$, the only solutions are 2 and 4. Only one of these is in the open interval $(0, 3)$. Thus, the only critical value of $f$ is 2.

We must find the second derivative. Differentiating again,

$$f''(x) = (x^2 - 6x + 8)e^x + e^x(2x - 6)$$
$$= (x^2 - 4x + 2)e^x$$

We now evaluate $f''(x)$ at 2. We get

$$f''(2) = -2e^2$$

Since $f''(2) < 0$, the function $f$ has a local maximum at 2.

To find the absolute extrema, we must evaluate the function at the critical value 2 and also at the endpoints of the interval $[0, 3]$.

We get

$$f(0) = [0^2 - 8(0) + 16]e^0 = 16$$
$$f(2) = [2^2 - 8(2) + 16]e^2 = 4e^2$$
$$f(3) = [3^2 - 8(3) + 16]e^3 = e^3$$

Since $e < 4$, we have $e^3 < 4e^2$. Using a calculator, we find that $e^3 \doteq 20.0855$. Therefore, $16 < e^3 < 4e^2$. The absolute minimum is 16 and it occurs at 0. The absolute maximum is $4e^2$ and it occurs at 2.

**Do Exercise 37.**

Observe that in the preceding example, the function $f$ had only one critical point in the interval $(0, 3)$ and the relative maximum at that point was also the absolute maximum of $f$ on $[0, 3]$. This is a special case of the following theorem whose proof is left as an exercise.

> **THEOREM 8.9:**   If $f$ is a continuous function on an interval and has only one critical value in that interval, then a relative extremum at that value is also an absolute extremum.

## Procedure to Find the Extrema of a Continuous Function Using the Second Derivative Test

**STEP 1.** Find $f'(x)$.

**STEP 2.** Solve the equation $f'(x) = 0$. The solutions that are in the domain of $f$ are critical values of $f$.

**STEP 3.** Determine all values of $x$ for which $f'(x)$ is not defined. Those values that are in the domain of $f$ are also critical values of $f$. However, since the Second Derivative Test cannot be used at these points, use the First Derivative Test.

**STEP 4.** Differentiate $f'(x)$ obtained in Step 1 to get $f''(x)$.

**STEP 5.** For each critical value $c$ obtained in Step 2, evaluate $f''(c)$. (Note that you need not carry out the computation completely. As soon as it is clear whether $f''(c)$ is positive or negative, you may stop.)

    **a.** If $f''(c) > 0$, $f$ has a local minimum at $c$.
    **b.** If $f''(c) < 0$, $f$ has a local maximum at $c$.
    **c.** If $f''(c) = 0$, use the First Derivative Test.

To find the local extrema at the endpoints of the interval on which $f$ is defined, and to find the absolute extrema, proceed as in the First Derivative Test.

---

# Exercise Set 8.6

*Exercises 1–37. In Exercises 1–37 of Section 8.5, p. 625, use the Second Derivative Test, whenever possible, to find all local extrema of the functions defined.*

**\*38.** In Example 1 of this section, we saw that $f''(1) = -6$ and $f''(3) = 6$. Thus $f''(3) = -f''(1)$. In general, is it true that if $f(x) = ax^3 + bx^2 + cx + d$ is a polynomial of degree 3 and if $x_1$ and $x_2$ are two critical values of the function $f$, then $f''(x_1) = -f''(x_2)$? Justify your answer.

**\*39.** Suppose $f$ is a continuous function on an interval $I$, $x_1$ and $x_2$ are in $I$, and $f$ has no critical values in the open interval $(x_1, x_2)$.

    **a.** If $f(x_1) < f(x_2)$, then $f$ is increasing on the interval $[x_1, x_2]$.
    **b.** If $f(x_1) > f(x_2)$, then $f$ is decreasing on the interval $[x_1, x_2]$.

Give a geometric argument to justify this claim. (*Hint:* see Theorem 8.4, Section 8.5. This may sometimes be used in cases where a function $f$ has critical values $x_1$, $x_2$, and $x_3$, but no critical values in the intervals $(x_1, x_2)$ and $(x_2, x_3)$. If the Second Derivative Test can be used at $x_1$ and $x_3$ but not at $x_2$, we can determine whether $f$ has an extremum at $x_2$ by comparing the functional values $f(x_1)$, $f(x_2)$, and $f(x_3)$. In this case, we need not use the First Derivative test at $x_2$.)

**\*40.** Using the Second Derivative Test only and the result of Exercise 39\*, identify the extrema of the function $f$ defined by

$$f(x) = x^{2/3} e^{-x^2}.$$

**\*\*41.** If a function $f$ is not continuous at $c$, then it is possible for $f$ to be decreasing to the left of $c$ and increasing to its right, and yet not have a minimum at $c$. (See Figure 8.21i, Section 8.5.)

a. Give a geometric argument to show that this cannot happen if $f$ is continuous.

b. Prove that if $f''(c)$ exists, then there is a neighborhood of $c$ over which $f$ is continuous. (*Hint:* First show that $f$ is differentiable in some neighborhood of $c$ using the fact that $f''(c) = \lim\limits_{x \to c} \frac{f'(x) - f'(c)}{x - c}$.)

**42. Let $f(x) = x^4$. Show that $f'$ is increasing on the interval $(-1, 1)$ and, therefore, the graph of $f$ is concave up on that interval. However, $f''(0)$ is not positive. Does this contradict the statement of Theorem 8.7 in the preceding section? Justify your answer.

# 8.7 Applications

We now have the tools to find the extrema of a function. In the applications, there are many situations where we need to maximize or minimize a certain variable. Two examples of such applications are minimizing the cost of building an object or maximizing the profit in producing and selling a commodity.

In general, in an application of optimization, we will have a situation where there will be several variable quantities involved. The question is what should the value of one of the variables be in order to maximize or minimize the value of one of the other variables?

### Procedure for Solving Applied Optimization Problems

**STEP 1.** The variable whose value is to be found can serve as the independent variable. Call it $x$, or any other symbol you wish to use for the independent variable. Be sure to indicate the units.

**STEP 2.** The variable that is to be maximized, or minimized, should be chosen as the dependent variable. Call it $y$, or any symbol you wish different from the symbol used in Step 1.

**STEP 3.** Whenever possible, draw a picture and label it.

**STEP 4.** Write down all formulas involved in the problem. For example: total cost = (average variable cost)(number of units) + fixed cost.

**STEP 5.** Using the formulas of Step 4, write down all variable quantities in terms of $x$ and $y$, or the symbols used in Steps 1 and 2.

**STEP 6.** Using the information given, write down an equation relating $x$ and $y$. If necessary, solve for $y$ to get an equation in the form $y = f(x)$. This equation should involve only the two variables $x$ and $y$. If any other variable is present, eliminate it using the results of Step 5. Be sure to specify the domain of the function, because our interest is in finding an absolute extremum on that domain.

**STEP 7.** In this step, find the absolute maximum or minimum value of the function $f$ on its domain using any of the techniques you learned in Sections 8.5 and 8.6. Recall that if the domain of a function is an interval, the absolute extrema may occur at the endpoints of the interval. (See Exercise 23.) In particular, in a case where the function has only one critical value, use

a. Theorem 8.9, Section 8.6

or

**b.** The result of Exercise 39* of Section 8.6. That is, if $c$ is the only critical value of a function $f$ on an open interval $(a, b)$ and that function is continuous on the closed interval $[a, b]$, then

(i) $f$ has a maximum at $c$ in the case where $f(a) < f(c)$ and $f(b) < f(c)$.

(ii) $f$ has a minimum at $c$ in the case where $f(c) < f(a)$ and $f(c) < f(b)$.

In other cases, recall that if the domain is a closed bounded interval and $f$ is continuous, you need not determine the behavior of $f$ at each critical value. Simply evaluate $f$ at each critical value and at each endpoint. Choose the largest answer if you seek a maximum and the smallest if you seek a minimum.

In some instances, the domain of the function is a set of positive integers; for the purpose of differentiation, extend the domain to an interval of positive real numbers. If you find that the absolute extremum of such a function occurs at $c$ where $c$ is not an integer, it is customary to evaluate the function at $i$ and at $i+1$ where $i$ is the largest integer less than $c$, provided of course that these two integers are in the domain of the function. Then take the larger of these two functional values if you seek a maximum and the smaller one if you want a minimum. There are functions for which this practice may yield wrong results (see Exercises 29, 30, and 31). However, in the applications, these types of functions do not usually occur and you may proceed as suggested. It should also be noted that when you are doing a problem in which you are asked to find the number of units of a commodity (such as television sets or VCRs), that will yield a maximum profit, maximum revenue, or minimum cost, and so on, a noninteger answer may make sense. For example, producing 212.3 units per day may be interpreted as producing 2123 units in 10 days.

**STEP 8.** Write down the right conclusion and make sure you have answered the question being asked.

**EXAMPLE 1** We wish to enclose with a fence, a rectangular region of 12,250 square feet next to a wall. Only three sides of the rectangle need to be fenced. The fencing for one side perpendicular to the wall costs \$30/ft, while the fencing for the other two sides costs \$20/ft. What are the dimensions of the rectangle that would minimize cost? Find the minimum cost.

**Solution** **Step 1.** Let $x$ feet be the length of the side that is to be fenced at a cost of \$30/ft. Note that $0 < x$.

**Step 2.** Let $y$ dollars be the total cost of the fence.

**Step 3.** The region is drawn in Figure 8.29.

**Step 4.** The relevant formulas in this problem are the following:

**a.** Area = (length)(width).

**b.** Total cost = cost of one type of fence + cost of the other type of fence.

**c.** Cost of one type of fence = (cost per unit)(number of units).

Wall

x feet    Area of region = 12,250 square feet

←——Other dimension of rectangle——→

FIGURE 8.29

**Step 5.** Using formula (a), we see that

$$12{,}250 = x \text{ (other dimension of the rectangle)},$$

or

$$\text{other dimension of the rectangle} = \frac{12{,}250}{x}$$

Thus the length of the $20/ft fencing is

$$\left(\frac{12{,}250}{x} + x\right) \text{ feet}$$

Its cost is

$$20\left(\frac{12{,}250}{x} + x\right) \text{ dollars}$$

The cost of the $30/ft fencing is $30x$ dollars.

**Step 6.** The total cost is

$$y = C(x) = 30x + 20\left(\frac{12{,}250}{x} + x\right) = 50x + \frac{245{,}000}{x}$$

where $x > 0$.

**Step 7.** We shall use the Second Derivative Test. Differentiating, we get

$$C'(x) = D_x\left(50x + \frac{245{,}000}{x}\right) = 50 - 245{,}000x^{-2}$$

$$= \frac{50x^2 - 245{,}000}{x^2}$$

Note that $C'(0)$ is not defined. However, 0 is not in the domain of the function $C$, so it is not a critical value. We must solve $C'(x) = 0$. We have

$$\frac{50x^2 - 245{,}000}{x^2} = 0, \text{ where } x > 0$$

Therefore,

$$50x^2 - 245{,}000 = 0$$

or

$$x^2 = 4900$$

Thus

$$x = -70 \text{ or } x = 70$$

Only 70 is in the domain of $C$, so 70 is the only critical value. Differentiate again to find $C''(x)$.

$$C''(x) = D_x(50 - 245{,}000x^{-2}) = -245{,}000(-2)x^{-3} = \frac{490{,}000}{x^3}$$

Obviously, $C''(70) > 0$, since $490{,}000/70^3 > 0$. (Note that we did not need to calculate the exact value of $C''(70)$.) We conclude that $C$ has a local minimum at 70. Since 70 is the only critical value of $C$, $C(70)$ is in fact the absolute minimum of the function $C$.

**Step 8.** The side being fenced with \$30/ft fencing should be 70 feet long. The other side should be 175 feet long since $\frac{12{,}250}{70} = 175$.
The minimum cost is determined by

$$C(70) = 50(70) + \frac{245{,}000}{70} = 7000$$

Thus, the minimum cost is \$7000.
**Do Exercise 3.**                                                                 ■

EXAMPLE 2   A small university charged \$141 per credit hour during the 1990–1991 academic year and its students registered for a total of 180,000 credit hours. The president of the university, who had taken a calculus course for business students while he was in college, wanted to raise the tuition. Studies showed that an increase in tuition of $x$ dollars per credit hour would cause a decrease in enrollment of $100x^2$ in the total number of credit hours taught, provided $0 \le x \le 40$. He calculated the increase in cost per credit hour that would create a maximum total revenue for the following year. What were the results of his calculations?

Solution   **Step 1.** Let $x$ dollars be the increase per credit hour. Recall that $0 \le x \le 40$.
**Step 2.** Let $y$ dollars be the total revenue.
**Step 3.** No diagram is necessary in this case.
**Step 4.** The relevant formulas are

**a.** Total revenue = (cost per credit hour)(total number of credit hours).
**b.** New cost per credit hour = original cost per credit hour + increase per credit hour.
**c.** New number of credit hours = original number of credit hours − decrease in the number of credit hours.

**Step 5.** New cost per credit hour = $(141 + x)$ dollars. New number of credit hours taught = $180{,}000 - 100x^2$.
**Step 6.** Using formula (a), he got

$$y = R(x) = (141 + x)(180{,}000 - 100x^2)$$
$$= -100x^3 - 14{,}100x^2 + 180{,}000x + 25{,}380{,}000$$

**Step 7.** Since $R(x)$ is a polynomial, it was easy to use the Second Derivative Test. Thus, differentiating $R(x)$ twice, the president got

$$D_x(-100x^3 - 14,100x^2 + 180,000x + 25,380,000)$$
$$= -300x^2 - 28,200x + 180,000$$

and

$$R''(x) = -600x - 28,200$$

To find the critical values of $R$, he set $R'(x) = 0$ and solved the resulting equation

$$-300x^2 - 28,200x + 180,000 = 0$$

Dividing both sides by $-300$,

$$x^2 + 94x - 600 = 0$$
$$(x - 6)(x + 100) = 0$$

The solutions are $-100$ and $6$. Obviously, $-100$ is not in the domain of $R$. Thus the only critical value is $6$. Now,

$$R''(6) = -600(6) - 28,200 = -31,800$$

Since $R''(6) < 0$, the function $R$ has a relative maximum at $6$. Since $6$ is the only critical value in the interval $[0, 40]$, by Theorem 8.9, Section 8.6, $R(6)$ is in fact the absolute maximum of the revenue function.*
**Step 8.** The increase in tuition should be \$6 per credit hour to maximize the total revenue.
**Do Exercise 13.**                                                                    ■

On occasion, the formula obtained in Step 6 for $f(x)$ will involve radicals. In such cases, the differentiation and the details in carrying out Step 7 may be tedious. It is then useful to use the following theorem.

---

**THEOREM 8.10:**   Suppose a function $f$ is defined on an interval $I$ and $f(x) > 0$ for all $x$ in $I$, and let $g(x) = [f(x)]^2$. Then:

1. $f$ has a relative maximum at $x_0$ if, and only if, $g$ has a relative maximum at $x_0$.
2. $f$ has a relative minimum at $x_0$ if, and only if, $g$ has a relative minimum at $x_0$.

---

The proof is simple and is left as an exercise. It uses the fact that if $b$ and $c$ are two nonnegative numbers, then $b \le c$ if, and only if, $b^2 \le c^2$. (See Exercise 38.) The following example will illustrate this idea.

---

*Since $R$ is continuous on the closed interval $[0, 40]$ and $6$ is the only critical value in the open interval $(0, 40)$, we could have evaluated $R(0)$, $R(6)$, and $R(40)$ instead of using the Second Derivative Test. Clearly, $R(0) = 25,380,000$, $R(6) = 25,930,800$, and $R(40) = 3,620,000$. Of these three functional values, $25,930,000$ is the largest. Therefore it is the absolute maximum of the function $R$ on the interval $[0, 40]$.

**EXAMPLE 3** Find the point on the graph of the equation $y = x^2 + 2x + 3$ that is closest to the point $(14, 9)$.

**Solution** **Step 1.** Let $x$ be the abscissa of a point on the graph of the equation.
**Step 2.** Let $d$ units be the distance between the point $(14, 9)$ and that point.

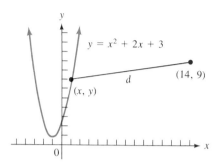

**FIGURE 8.30**

**Step 3.** The graph of the equation, the point $(14, 9)$, and the distance $d$ are illustrated in Figure 8.30.
**Step 4.** Recall that the distance between $(x_1, y_1)$ and $(x_2, y_2)$ is

$$d = \sqrt{(x_2 - x_1)^2 + (y_2 - y_1)^2}$$

**Step 5.** When the abscissa of a point on the graph of the equation is $x$, the ordinate is $x^2 + 2x + 3$.
**Step 6.** The distance between points $(14, 9)$ and $(x, x^2 + 2x + 3)$ is

$$d = f(x) = \sqrt{(x - 14)^2 + ((x^2 + 2x + 3) - 9)^2}$$
$$= \sqrt{(x - 14)^2 + (x^2 + 2x - 6)^2}$$

where $x$ is any real number.
**Step 7.** We wish to find the minimum value of $f(x)$. Since the formula that gives $f(x)$ involves a radical of order 2, we consider instead the function $g$ defined by

$$g(x) = [f(x)]^2 = (x - 14)^2 + (x^2 + 2x - 6)^2$$

Since $f$ will have a minimum value at $x_0$ if, and only if, $g$ has a minimum value at $x_0$, we need only locate the minimum value of $g$. We shall use the Second Derivative Test.

Differentiating, we get

$$g'(x) = 2(x - 14) + 2(x^2 + 2x - 6)(2x + 2)$$
$$= 4x^3 + 12x^2 - 14x - 52$$

Differentiating a second time, we get

$$g''(x) = 12x^2 + 24x - 14$$

To find the critical values of $g$, we set $g'(x) = 0$ and solve the resulting equation

$$4x^3 + 12x^2 - 14x - 52 = 0$$

Factoring,

$$(x - 2)(4x^2 + 20x + 26) = 0$$

Therefore,

$$x - 2 = 0 \quad \text{or} \quad 4x^2 + 20x + 26 = 0$$

The second equation has a negative discriminant since $b^2 - 4ac = 20^2 - 4(4)(26) = -16$. Thus it has no real solution. Therefore, 2 is the only critical value of $g$. Evaluating $g''(2)$, we obtain

$$g''(2) = 12(2^2) + 24(2) - 14 = 82$$

Since $g''(2) > 0$, $g$ has a relative minimum at 2. The domain of $g$ is the set of all real numbers and 2 is the only critical value, so $g$ has an absolute minimum at 2. (See Exercise 51.)

**Step 8.** The function $f$ also has an absolute minimum at 2. We have $f(2) = 2^2 + 2(2) + 3 = 11$. Thus the point $(2, 11)$ is the closest point on the parabola to the point $(14, 9)$. (See Exercise 27.)

**Do Exercise 32.**    ■

**EXAMPLE 4**  The Landau Radio Electronics Company conducted a market study and found that the relationship between the price $p$ (in dollars) per Walkman and the number $q$ of Walkmans consumers are willing to buy is given by

$$p = -0.00007q^2 - 0.04q + 200 \text{ provided that } 0 \le q \le 1400.$$

It estimated that the cost of production of Walkmans is $15,000 plus $20 per unit. How many Walkmans should be manufactured to maximize profit?

**Solution**  **Step 1.** Let $q$ be the number of Walkmans produced.
**Step 2.** Let $y$ dollars be the profit.
**Step 3.** No diagram is necessary in this case.
**Step 4.** The relevant formulas in this problem are

**a.** Profit = revenue − cost.
**b.** Revenue = (price per unit)(number of units sold).
**c.** Total cost = (average variable cost)(number of units made) + fixed cost.

**Step 5.** By formula (b), the revenue function is given by

$$R(q) = (-0.00007q^2 - 0.04q + 200)q = -0.00007q^3 - 0.04q^2 + 200q$$

where $0 \le q \le 1400$.
  By formula (c), the cost function is

$$C(q) = 20q + 15,000$$

**Step 6.** By formula (a), the profit function is

$$P(q) = (-0.00007q^3 - 0.04q^2 + 200q) - (20q + 15,000)$$
$$= -0.00007q^3 - 0.04q^2 + 180q - 15,000$$

where $0 \le q \le 1400$.
**Step 7.** We wish to locate the absolute maximum of $P$ on the closed interval $[0, 1400]$. Since $P(q)$ is a polynomial, it will be easy to use the Second Derivative Test.

Differentiating, we get

$$P'(q) = -0.00021q^2 - 0.08q + 180$$

and

$$P''(q) = -0.00042q - 0.08$$

Setting $P'(q) = 0$, we get

$$-0.00021q^2 - 0.08q + 180 = 0$$

Using the quadratic formula, we obtain

$$q = \frac{0.08 \pm \sqrt{(-0.08)^2 - 4(-0.00021)(180)}}{2(-0.00021)} = \frac{0.08 \pm \sqrt{0.1576}}{-0.00042}$$

That is,

$$q \doteq -1135.6873 \text{ or } q \doteq 754.7349$$

Obviously, 754.7349 is the only critical value in the interval [0, 1400]. Also, $P''(754.7349) = (-0.00042)(754.7349) - 0.08$ is negative. (Note that we did not need to carry out the calculations.)

Thus $P(q)$ has a local maximum when $q = 754.7349$. Since 754.7349 is the only critical value in [0, 1400], by Theorem 8.9 Section 8.6, $P$ has an absolute maximum at 754.7349. However, this number is not an integer, so we evaluate $P(754)$ and $P(755)$ to get

$$P(754) = -0.00007(754^3) - 0.04(754^2) + 180(754) - 15,000 = 67,973.0855$$

and

$$P(755) = -0.00007(755^3) - 0.04(755^2) + 180(755) - 15,000 = 67,973.1787$$

**Step 8.** The maximum profit will be realized if 755 Walkmans are produced since $P(755)$ is larger than $P(754)$. Note that we had only one critical value in the domain of the function $P$; therefore we can be sure that evaluating $P$ at the integer 755 will indeed give us the maximum profit. (See Exercise 31.)
**Do Exercise 33.** ∎

The last example illustrates an interesting application of optimization. In general, a store must keep merchandise in inventory at a certain cost and is charged a certain amount each time an order is placed. If the store keeps a small inventory, the cost of maintaining inventory will be smaller but more orders will have to be placed so the cost of ordering will be larger, and vice versa. The problem then is to determine the size of each order that will minimize the total cost of maintaining inventory and ordering.

EXAMPLE 5    The C.L.E.A.N. Store sells 8640 cases of soap per year. The manager of the store buys all of the soap from a single supplier and is charged $96 per order, regardless of the quantity ordered. The manager has estimated that the cost of maintaining inventory is $3.20 per case per year. How many cases of soap should be ordered each time to minimize the total cost of ordering and maintaining inventory?

Solution    **Step 1.** Let $x$ be the number of cases requested each time an order is placed. Clearly, $x > 0$.

**Step 2.** Let $y$ dollars be the total cost of ordering and maintaining inventory.

**Step 3.** Figure 8.31 illustrates the size of inventory on the vertical axis and the horizontal axis indicates time. Note that an $i$ on this axis represents the time of arrival of the $(i + 1)$st order. We are assuming that the soap is being consumed at a constant rate. The diagram shows that just before the arrival of an order, the size of inventory is near 0; at the time of arrival, it is $x$ cases; and on the average, it is $\frac{x}{2}$ cases.

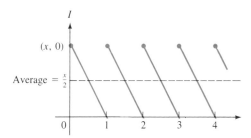

**FIGURE 8.31**

**Step 4.** The formulas involved are

**a.** Total cost = cost of ordering + cost of maintaining inventory.

**b.** Cost of ordering = (charge per order)(number of orders).

**c.** Number of orders = $\dfrac{\text{total demand per year}}{\text{number requested with each order}}$.

**d.** Cost of maintaining inventory = (cost per case)(average number of cases on hand).

**e.** Average number of cases on hand

$$= \frac{\text{number of units purchased with each order}}{2}$$

(See Figure 8.31.)

**Step 5.** By formula (c), the number of orders placed per year is $\frac{8640}{x}$. Thus, by formula (b) the cost of ordering is $96 \cdot \frac{8640}{x} = \frac{829,440}{x}$ dollars. By formula (e), the average number of cases on hand is $\frac{x}{2}$. Thus, by formula (d) the cost of maintaining inventory is $(3.20)\frac{x}{2} = 1.6x$ dollars. By formula (a), the total cost is $\left(\frac{829,440}{x} + 1.6x\right)$ dollars.

**Step 6.** Let $y = C(x) = \frac{829,440}{x} + 1.6x$.

**Step 7.** We wish to minimize $C(x)$. We find that

$$C'(x) = \frac{-829,440}{x^2} + 1.6$$

Therefore $C'(0)$ is not defined. However, 0 is not in the domain of $C$. Thus it is not a critical value. We set $C'(x) = 0$ and get

$$\frac{-829,440}{x^2} + 1.6 = 0$$

and

$$\frac{1.6x^2 - 829{,}440}{x^2} = 0$$

The solutions of this equation are $\pm720$. Only 720 is in the domain of $C$; thus 720 is the only critical value of $C$.

We find that

$$C''(x) = D_x\left(\frac{-829{,}440}{x^2} + 1.6\right) = \frac{1{,}658{,}880}{x^3}$$

Clearly, $C''(720) > 0$. Thus $C$ has a local minimum at 720. Since 720 is the only critical value of $f$, $C$ has in fact an absolute minimum at 720.

**Step 8.** With each order, the manager should request 720 cases of soap.

**Do Exercise 47.**  ∎

## Exercise Set 8.7

**1.** A farmer has 400 feet of fence available and wishes to enclose a rectangular region. What dimensions of the rectangle would maximize the area?

**2.** A rectangular box with open top is to be made from a piece of cardboard 24 by 9 inches by cutting out small squares in the four corners and folding up the edges. What dimensions of the box would maximize its volume?

**3.** A topless box with a square bottom is to be built so that its volume is 64 cubic feet. The bottom is to be made with material costing $6/ft$^2$, while the four rectangular sides are to be made with material costing $1.50/ft$^2$. What dimensions of the box would minimize the cost of material?

**4.** A rectangular sign is to contain 150 square inches of printed matter. The margins at the top and bottom must be 1.5 inches each. Each margin at the side must be 1 inch. What dimensions of the sign would minimize its area?

**5.** A fence needs to enclose a rectangular region of 1200 square feet. The fencing along three sides will cost $9/ft while the fencing along the fourth side will cost $15/ft. What dimensions of the rectangle are required for the fencing to be most economical?

**6.** An ecologist is conducting a research project on breeding pheasants in captivity. She first must construct suitable pens. She wants a rectangular region with two additional fences down the center as shown in the diagram. Find the maximum area that may be enclosed with 4400 meters of fencing.

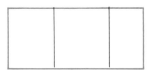

**7.** A company manufactures a top-open rectangular trash bin. Each bin must be 7 feet high and have a volume of 1764 cubic feet. The material for the front and back rectangular sides costs $8/ft$^2$, the material for the two rectangular sides costs $14/ft$^2$, and the material for the rectangular bottom costs $20/ft$^2$. What dimensions of the bin would minimize the cost of material?

**8.** A rectangular pen is being constructed to hold foxes and rabbits. The wall between the foxes and the rabbits must be stronger than the outer fence. The wall costs $18/ft and the fence costs $6/ft. The total area enclosed must be 16,000 square feet. What dimensions would minimize the cost?

**9.** A rectangular garden is to be enclosed with a fence. One side of the garden is against an existing wall so that no fence is needed along that side. The material to be used costs $5/ft for the side parallel to the wall and $8/ft for the other two sides. Only $320 is available for the material for the fence. If all the money available is used, what dimensions of the garden would maximize the area?

**10.** A piece of wire 40 cm long is to be bent to form a right angle. Where should the vertex of the angle be to minimize the distance between the two loose ends?

11. A rectangular region of 1260 square feet is to be surrounded on three sides by a brick wall and on one side by a fence. The brick wall costs $35/ft to build, while the fencing costs $15/ft. Find the dimensions of the rectangle that will minimize the cost.

12. A food cannery must can beans in cylindrical containers each having a volume of 160 cubic inches. What are the dimensions of each can that would minimize its surface area?

13. The manager of a peach orchard is trying to decide when to harvest. If he has the peaches picked now, the average yield per tree will be 96 pounds, which he can sell for 90 cents per pound. Experience shows that the yield per tree will increase about 4 pounds per week, while the price per pound will decrease about 3 cents per week. When should the peaches be picked to maximize the income per tree? What is the maximum income per tree?

14. The members of the Oregon Ducklings, a Little League baseball team, have been collecting aluminum cans to help pay the cost of bats and balls. They already have 150 pounds of aluminum for which they can get 80 cents a pound now. If they continue to collect, they can get 5 more pounds per day, but the price per pound will decrease 2 cents per day. Knowing that they can make only one transaction, how many days should they wait to maximize their income? What is the maximum income?

15. The manager of a delicatessen can buy a can of escargots for $5.50. She knows that if she sets the price at $x$ dollars per can, $-x^2 + 144$ cans will be sold per month, provided that $0 \le x \le 12$. What price per can would maximize the profit per month?

16. A Washington State University plant physiologist has shown that when potato roots are treated with carbon dioxide, the number of potatoes set on a plant can be increased by as many as ten times the normal number. However, the more potatoes set on a plant, the smaller the average size of each potato. Suppose an untreated potato plant sets 5 potatoes each weighing 195 grams on the average, and that for each additional potato set on the plant, the average weight of the potato drops by 3 grams. Find the number of potatoes that should be set on each plant to maximize the total weight of the potatoes produced.

17. A pipeline is to be run to a resort community which is on an island 16 miles from the closest point on the shoreline. The water source for the pipe is 30 miles down the shore. If it costs $500/mi to lay the pipe through water and $300/mi to lay the pipe on land, at what point on the shoreline should the pipe turn into the water to minimize cost? What is the total cost? (See the diagram.)

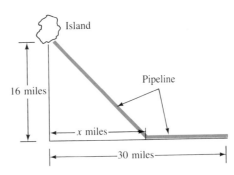

18. The owner of a trout farm purchased 4620 rainbow trout from a hatchery and placed them in a small private lake to grow. The average weight of a young rainbow trout is .08 pounds and each gains, on the average, .05 lb/wk. However, on the average, 75 fish die each week. When should the owner harvest the lake to maximize the total weight of the trout?

19. A farmer uses his 500 acres of land to grow two kinds of grain. Grain A is sold at the market and grain B is used to feed his herd of 500 cattle. An acre planted with grain A will yield 240 bushels of grain which sell for $3 a bushel. Feeding the cattle with the yield of $x$ acres of grain B will result in each steer weighing $300 + 3x - 0.003x^2$ pounds. The steer sell for $1.20 per pound. How many acres of each crop should be planted to maximize revenue?

20. A 150-room hotel in Reno will rent all of its rooms if it charges $40 a room. However, from experience, the manager knows that for each $2 increase per room, three fewer rooms will be rented per night. What rent per room would maximize the revenue per night?

21. A tour charges groups of individuals $300 per person, minus $3 for each extra person over a group size of 40. It costs $2000 plus $120 per person to organize and conduct the tour. How many individuals should take the tour to maximize profit?

22. Two straight highways are perpendicular. A motorist traveling south on one of them at 48 mph passes through the intersection at 1:00 P.M. At that instant, a motorist traveling west on the other highway at 36 mph is 50 miles east of the intersection. At what time will they be closest to each other?

23. A truck consumes gasoline at the rate of $(3 + x^2/600)$ gallons per hour when it travels at a speed of $x$ miles per hour. The cost of gasoline is $1.25 per gallon and a truck driver earns $10 per hour. The minimum and maximum legal speeds for trucks are 30 and 55 mph respectively. Find the speed that will make a 600-mile trip most economical.

24. Find the point on the graph of the equation $y^2 = 4x$ that is closest to **a.** the point $(-1, 0)$, **b.** the point $(1, 0)$, **c.** the point $(2, 0)$, and **d.** the point $(3, 0)$.

25. Same as Exercise 23, but assume the truck is in a country where the legal speeds are the same but gasoline is more expensive—$4.20 per gallon—and labor is cheaper—the truck driver earns only $4.90 an hour.

26. **a.** Find the point on the graph of the equation $y = x^2 - 3x + 2$ that is closest to the point $(6, 1)$. **b.** Show that the line through the point $(6, 1)$ and the point you found in part **a** is perpendicular to the tangent line to the graph of the equation at that point.

27. Show that the line through points $(2, 11)$ and $(14, 9)$ is perpendicular to the tangent line to the graph of $y = x^2 + 2x + 3$ at $(2, 11)$. (*Hint:* See Example 3.)

28. The Reid Electronic Company decided to produce a new Walkman. The cost of manufacturing $q$ thousand Walkmans is given by

$$C(q) = 20{,}000 + 50{,}000q$$

where $C(q)$ is in dollars and $0 \le q \le 14$. Studies of teenage buying habits showed that the relationship between the price $p$ dollars per Walkman and the quantity $q$ thousand Walkmans that will be sold is given by

$$p = -2q^2 - 2q + 500$$

where $0 \le q \le 14$. Find the number of Walkmans that should be produced to maximize profit.

29. Suppose $f$ is a differentiable function defined on an interval $I$ and we wish to find the maximum and minimum value of $f$ on the set of integers in $I$. We usually locate the absolute maximum and minimum of $f$ on $I$. If this occurs at $c$ and $c$ is not an integer, then we evaluate $f(i)$ and $f(i + 1)$ where $i$ is the largest integer less than $c$. We then take the larger of these two values if we seek a maximum and the smaller one if we want a minimum. This may give us a wrong answer as illustrated in this and the next exercise. The graph of a function $f$ is sketched below. Using this graph, find the following.

**a.** The point $c$ where the function $f$ assumes its absolute maximum.
**b.** The largest integer $i$ less than $c$.
**c.** The functional values $f(i)$ and $f(i + 1)$.
**d.** An integer $k$ in the domain of $f$ such that $f(k)$ is larger than both $f(i)$ and $f(i + 1)$.

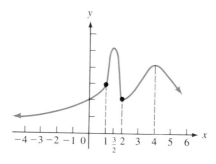

*30. Same as Exercise 29 for the function $f$ defined by

$$f(x) = \begin{cases} 6\sqrt{1 - 4\left(x - \dfrac{1}{2}\right)^2}, & \text{if } 0 \le x \le 1 \\ -\sqrt{1 - (x - 2)^2}, & \text{if } 1 \le x \le 3 \\ \sqrt{25 - (x - 8)^2}, & \text{if } 3 \le x \le 13. \end{cases}$$

(*Hint:* Sketch the graph of $f$. It is the union of a half-ellipse and two semicircles.)

*31. Explain why, if $f$ is a continuous function defined on a closed bounded interval $[a, b]$ and $f$ has only one critical point in the open interval $(a, b)$, then the difficulty illustrated in Exercises 29 and 30 cannot occur.

32. Show that $(2, 3)$ is the closest point on the graph of $y = x^3 - x^2 - x + 1$ to the point $(9, 2)$.

33. The Nelson Electronics Company is introducing a new deluxe model of personal computers. The cost of production is estimated to be a fixed cost of $4,500,000 plus $500 per computer produced. Market studies showed that the demand equation for this type of PC is

$$p = D(q) = -0.00005q^2 - 0.3q + 8000$$

where $p$ is the price per PC in dollars and $q$ is the number of PCs the consumers will buy at that price, with $0 \le q \le 9000$. How many PCs should be produced to maximize profit?

34. The C.O.L.D. Ski Equipment Company wishes to introduce a new line of deluxe ski boots. The cost $C(q)$ (in dollars) to manufacture $q$ pairs of these boots is

$$C(q) = e^{-0.0032q}(0.3q^2 + 200q + 20{,}000)$$

provided that $0 \le q \le 300$. Market studies showed that the relationship between the price $p$ dollars per pair of boots and the number $q$ of pairs the consumers will buy is given by

$$p = (0.5q + 500)e^{-0.0032q}, \text{ for } 0 \le q \le 300$$

How many pairs of boots should be produced to maximize profit?

**35.** The L.E.G. Cycling Co. found that if it built $q$ ten-speed bicycles per month where $0 \leq q \leq 100$, the cost of production will be $(5000 + 2000 \cdot \ln(q + 1))$ dollars. Market studies showed that the demand equation is linear and that each month 100 bicycles can be sold at a price of $200 each, while only 20 will be sold if the price is $520 per bike. How many bicycles should be manufactured each month to maximize the profit?

**36.** An electronics company found that the cost of manufacturing $q$ VCRs is given by

$$C(q) = 35,000 + 500q - q^2$$

for $0 \leq q \leq 130$. It also found that the relationship between the price $p$ dollars per VCR and the number $q$ VCRs consumers are willing to buy is given by

$$p = -0.02q^2 - 0.6q + 1500 \quad \text{for} \quad 0 \leq q \leq 130$$

How many VCRs should the company manufacture to maximize profit?

**37.** The Mason Boat and Waterski Company found that the cost function to manufacture its new type of speed boat would be linear with a fixed cost of $22,500,000 plus $1500 per boat as long as the production was kept under 15,000 boats. Market studies showed that the demand equation for the boats is

$$p = -0.000025q^2 - 0.25q + 15,000$$

provided that $0 \leq q \leq 15,000$. How many boats should be manufactured to maximize profit?

**\*38.** Prove Theorem 8.10 of the preceding section.

**39.** The Vino Alonzo Winery knows that the cost of producing $q$ thousand bottles of its 1991 wine is $C(q)$ thousand dollars where

$$C(q) = 0.0005q^3 + 3q + 8$$

provided that $0 \leq q \leq 40$. How many bottles should be produced to minimize the average cost of production per bottle?

**40.** When a truck travels at a speed of $x$ miles per hour, where $0 < x \leq 55$, it will burn $(3 + x^2/500)$ gallons of gasoline per hour. If the gasoline costs $2 per gallon and the truck driver earns $4 per hour, at what speed should the truck travel to minimize the cost per mile?

**41.** The M.D.Y. Land Development Company purchased a piece of land in Kent, Washington, and wanted to erect a building on that land. The cost of the lot was $343,000. The manager of the company broke down the cost of construction into two types. The first type would be the same for each floor

and he estimated that it would cost $15,000 per floor. The second type would increase with the height of the building and is estimated to be $7000 times the square of the number of floors for each floor. How many stories should the building have to minimize the average cost per floor?

**42.** The M.D.Y. Development Company purchased a lot in downtown Kent for $2,000,000. It plans to erect an office building on that lot. The cost of building the first floor is $200,000, that of building the second floor is $210,000, that of building the third floor is $220,000, and so on. Knowing that the net annual income for each story is $55,300, how many stories should the building be to return the maximum rate of interest?

**43.** A man on an island 16 miles north of a straight shoreline must reach a point 30 miles east of the closest point on the shore to the island. If he can row at a speed of 3 mph and jog at a speed of 5 mph, where should he land on the shore in order to reach his destination as soon as possible?

**44.** A ship lies 10 miles from a straight shoreline and 15 miles farther down the shore, another ship is anchored 20 miles off shore. A boat leaves the first ship to land some passengers on shore and then proceeds to the second ship. What is the length of the shortest distance the boat can travel? Give two solutions to this problem: one using calculus, another using geometry.

**45.** A Norman glass window consists of a rectangle with a semicircle on its top side, the diameter of the semicircle coinciding with the top side of the rectangle. **a.** What dimensions of the window would let in the most light if the perimeter is 20 feet? **b.** If the semicircle is made up of stained glass admitting only three-fourths the amount of light clear glass does for the same area, answer the same question as in part **a.**

**46.** An Indian teepee has the shape of a right circular cone and is made by tying together at the top poles 20 feet long and stretching buffalo skin over them. What dimensions of the teepee would give the maximum volume?

**47.** A hardware store sells 1620 cases of a certain type of light bulb per year. The manager of the store buys all of these from a single supplier and is charged $66 per order, regardless of the quantity ordered. The manager has estimated that the cost of maintaining inventory is $1.65 per case per year. How many cases should be ordered each time to minimize the total cost of ordering and maintaining inventory?

**48.** An auto supplies store sells 1960 cases of motor oil per year. The manager of the store buys all of the motor oil from a single supplier and is charged $85 per order, regardless of the quantity ordered. The manager has estimated that the cost of maintaining inventory is $4.25 per case per year.

How many cases of this motor oil should be ordered each time to minimize the total cost of ordering and maintaining inventory?

49. The manager of a seafood restaurant has calculated that during the noon hour on any weekday, the restaurant can make a profit of up to $792. However, if she does not have enough employees, patrons will be lost. She estimated that if she has only $x$ employees working during that hour, there will be a loss in profit of $\frac{792}{x+1}$ dollars. Each employee is paid $5.50 per hour. How many employees should be working during the noon hour to minimize the cost? (*Hint:* Treat a loss in profit as cost.)

50. Let $f$ be a differentiable function and suppose that, using the same technique as in Example 3, $(c, d)$ has been found to be the point on the graph of $y = f(x)$ closest to a given point $(a, b)$ not on the graph. Assume that $a \neq c$ and $b \neq d$. Prove that the line passing through points $(a, b)$ and $(c, d)$ is perpendicular to the tangent line to the graph of $y = f(x)$ at $(c, d)$.

**51. Suppose $f$ is a continuous function defined on the set of real numbers and $c$ is the only critical value of $f$. Prove that if $f$ has a local extremum at $c$, then this extremum is an absolute extremum.

# 8.8 Curve Sketching

It is useful to be able to sketch the graph of a function. Usually, a rough sketch is sufficient. We may only be interested in the general shape of the curve and in the location of significant points, such as the local extrema. A knowledge of derivatives is useful for this purpose. In this section, we illustrate how to sketch the graphs of a polynomial and of a simple rational function. In the next section, we further discuss rational and other functions.

## Graphing Polynomials

**EXAMPLE 1**    Sketch the graph of

$$y = f(x) = x^3 + 3x^2 - 9x - 11$$

**Solution**    We use the derivative to determine the intervals over which $f$ is increasing, decreasing, and to locate the relative extrema. We shall use the second derivative to determine the intervals over which the graph is concave up, down. Differentiating, we get

$$f'(x) = 3x^2 + 6x - 9 = 3(x + 3)(x - 1)$$

Setting $f'(x) = 0$,

$$3(x + 3)(x - 1) = 0$$

The solutions are $-3$ and $1$. These are the critical values of $f$. We now make a table, as in Section 8.3, to determine the intervals over which $f'(x)$ is positive, negative. We found that $f'(0) = -9$. So we entered a minus sign between the two zeros in the second row of the table. Since the sign of $f'(x)$ changes as $x$ increases through $-3$ and through $1$, we entered plus signs to the left of the first 0 and to the right of the second 0. The table indicates that the function is increasing on the intervals $(-\infty, -3)$ and $(1, \infty)$. It is decreasing on $(-3, 1)$. We conclude that $f$ has a local maximum at $-3$ and a local minimum at $1$.

| $x$ | | $-3$ | | $1$ | |
|---|---|---|---|---|---|
| $f'(x)$ | $+$ | $0$ | $-$ | $0$ | $+$ |
| $f(x)$ | ↗ | | ↘ | | ↗ |

We found that $f(-3) = 16$, and $f(1) = -16$. So $(-3, 16)$ and $(1, -16)$ are critical points. Differentiating $f'(x)$, we get

$$f''(x) = 6x + 6 = 6(x + 1)$$

Clearly, $f''(x) = 0$ if $x = -1$, $f''(x) > 0$ if $x > -1$, and $f''(x) < 0$ if $x < -1$. Thus, the graph of $y = f(x)$ is concave up (smiling!) on the interval $(-1, \infty)$ and concave down (sad!) on the interval $(-\infty, -1)$. Since $f(-1) = 0$, the point $(-1, 0)$ is on the graph of $f$. This is a special point of the graph since on one side, the graph is concave one way, and on the other side, it is concave the opposite way. This point is called a *point of inflection*.

It is convenient, because it is easy, to find the $y$-intercept of the graph. We replace $x$ by 0 in the equation defining the function and get $f(0) = -11$. Thus the point $(0, -11)$ is the $y$-intercept. To find the $x$-intercepts, we replace $y$ by 0 in the original equation and solve for $x$. We get

$$0 = x^3 + 3x^2 - 9x - 11 = (x + 1)(x^2 + 2x - 11)$$

Thus,

$$x + 1 = 0 \quad \text{or} \quad x^2 + 2x - 11 = 0$$

Solving these equations (the second one using the quadratic formula), we find that the $x$-intercepts are $(-1 - 2\sqrt{3}, 0)$, $(-1, 0)$, and $(-1 + 2\sqrt{3}, 0)$. Often, finding the $x$-intercepts is troublesome. In such cases, we sketch the graph without plotting the $x$-intercepts. We can instead plot a few other points of the graph.

Before sketching the graph, we tabulate the information we have obtained and plot the points $(-1 - 2\sqrt{3}, 0)$, $(-3, 16)$, $(-1, 0)$, $(0, -11)$, $(1, -16)$, and $(-1 + 2\sqrt{3}, 0)$.

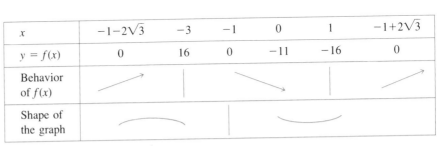

| $x$ | $-1-2\sqrt{3}$ | $-3$ | $-1$ | $0$ | $1$ | $-1+2\sqrt{3}$ |
|---|---|---|---|---|---|---|
| $y = f(x)$ | $0$ | $16$ | $0$ | $-11$ | $-16$ | $0$ |
| Behavior of $f(x)$ | ↗ | | ↘ | | | ↗ |
| Shape of the graph | ⌢ | | | ⌣ | | |

Using the information displayed in the table, we now draw a smooth continuous curve through the points we have plotted. (See Figure 8.32.)

Note that $(-3, 16)$ and $(1, -16)$ are the local high and low points of the graph. We may think of these points as critical points of the graph because they occur when $f'(x) = 0$. If $f(x)$ is a polynomial of degree $n$, then $f'(x)$ is a polynomial of degree $n - 1$. Thus, the equation $f'(x) = 0$ has at most $n - 1$ real solutions and the graph

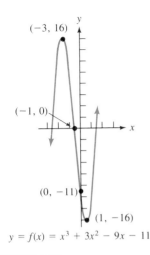

$(-3, 16)$

$(-1, 0)$

$(0, -11)$

$(1, -16)$

$y = f(x) = x^3 + 3x^2 - 9x - 11$

**FIGURE 8.32**

of $y = f(x)$ has at most $n - 1$ critical points. (See Exercises 28 and 29.) Also note that the point $(-1, 0)$ is the point of inflection.
**Do Exercise 5.** ■

## Points of Inflection

> **DEFINITION 8.14:** A *point of inflection* of the graph of a function $f$ is a point $(c, f(c))$ for which there is a positive number $h$ such that $f$ is continuous on the interval $(c - h, c + h)$, and the graph of $f$ is concave one way over the interval $(c - h, c)$ and concave the opposite way over the interval $(c, c + h)$.

As $x$ increases in the interval $(c - h, c + h)$, the point $(x, f(x))$ moves along the graph and passes through the point of inflection. The sign of $f''(x)$ must go from positive to negative, or vice versa. Therefore if $f''$ is continuous, then $f''(c) = 0$. That is, the vanishing of $f''(x)$ (if $f''$ is continuous) is necessary for a point of inflection. However, it is not a sufficient condition, because, as $x$ increases through a number $c$, it is possible to have $f''(c) = 0$ and $f''(x)$ not change signs. For example, $(0, 0)$ is *not* a point of inflection of the graph of $f(x) = x^4$, even though $f''(0) = 0$. In cases where $f''$ is not continuous, $f''(c) = 0$ is not a necessary condition for a point of inflection. For example, you may verify that $(0, 0)$ is a point of inflection of the graph of $f(x) = x^{5/3}$, and that $f''(0)$ is not defined.

**EXAMPLE 2**  Find the points of inflection for the graphs of each of the following.

**a.** $f(x) = x^4 + 2x^3 - 12x^2 + 48x + 5$
**b.** $g(x) = x^6 + 5x^4 - 7$
**c.** $h(x) = x^{9/5}$

**Solution**  **a.** Differentiating, we get

$$f'(x) = 4x^3 + 6x^2 - 24x + 48$$

and

$$f''x) = 12x^2 + 12x - 24$$

Since $f''$ is a polynomial function, it is continuous. Thus the condition $f''(x) = 0$ is necessary for a point of inflection. Solving

$$12x^2 + 12x - 24 = 12(x - 1)(x + 2) = 0$$

we get

$$x = -2, \text{ or } x = 1$$

We make the table

| $x$ | | $-2$ | | $1$ | |
|---|---|---|---|---|---|
| $f''(x)$ | $+$ | $0$ | $-$ | $0$ | $+$ |

and conclude that the graph is concave up on $(-\infty, -2)$, concave down on $(-2, 1)$, and concave up on $(1, \infty)$. So the points $(-2, f(-2))$ and $(1, f(1))$ are points of inflection.

**b.** Differentiating, we get

$$g'(x) = 6x^5 + 20x^3$$

and

$$g''(x) = 30x^4 + 60x^2$$

Since $g''$ is a polynomial, it is continuous. Therefore the condition $g''(x) = 0$ is necessary for a point of inflection. The only real solution of $30x^4 + 60x^2 = 0$ is 0. However, when $x \neq 0$, $30x^4 + 60x^2 > 0$. So $g''(x)$ does not change sign and the graph of $g$ has no point of inflection.

**c.** Differentiating, we get

$$h'(x) = \frac{9}{5}x^{4/5}$$

and

$$h''(x) = \frac{36}{25}x^{-1/5} = \frac{36}{25x^{1/5}}$$

Therefore, $h''(x) < 0$ when $x < 0$, and $h''(x) > 0$ when $x > 0$. Furthermore, $h$ is continuous. It follows that the point $(0, h(0))$ is a point of inflection of the graph of $h$. Notice that $h''(0)$ is not defined.
**Do Exercises 31, 33, and 35.**

## Limits at Infinity

Recall that a rational function is the quotient of two polynomial functions. It is often more tedious to sketch the graph of a rational function than to sketch the graph of a polynomial function. One of the reasons for this is that if

$$R(x) = \frac{f(x)}{g(x)}$$

then the function $R$ is not continuous at any $x$ for which $g(x) = 0$. We begin with a simple example.

**EXAMPLE 3**   Sketch the graph of $f$ if $y = f(x) = \dfrac{1}{x}$.

**Solution**   Note that 0 is not in the domain of $f$. However, if $x > 0$, but sufficiently close to 0, then $\frac{1}{x}$ will be positive and large. The corresponding point $(x, y)$ will be in the first quadrant, close to the $y$-axis but high. Further, if $x > 0$ and large, then $\frac{1}{x}$ will be positive but close to 0. Thus the corresponding point $(x, y)$ will be in the first quadrant, near the $x$-axis but far to the right. Obviously, the point $(1, 1)$ is on the graph. Using this information, we sketch the part of the graph over the interval $(0, \infty)$. (See Figure 8.33.)

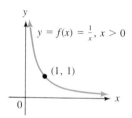

**FIGURE 8.33**

If $x$ is near 0 but negative, $\frac{1}{x}$ is negative but large in absolute value. Thus the corresponding point $(x, y)$ is in the third quadrant, near the $y$-axis but low. On the other hand, if $x$ is negative and large in absolute value, $\frac{1}{x}$ is negative but near 0. Thus, the corresponding point $(x, y)$ is in the third quadrant, near the $x$-axis and far to the left. Using the fact that $(-1, -1)$ is on the graph, we sketch the part of the graph over the interval $(-\infty, 0)$. (See Figure 8.34.)

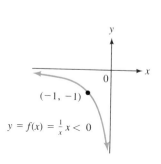

**FIGURE 8.34**

**FIGURE 8.35**

Now we can draw the graph of $y = \frac{1}{x}$, combining the sketches of Figures 8.33 and 8.34. (See Figure 8.35.)

The $y$-axis is a vertical asymptote and the $x$-axis is a horizontal asymptote of the preceding graph.

**DEFINITION 8.15:** A line $L$ is an *asymptote* for a curve $C$ if the perpendicular distance from a moving point on $C$ to the line $L$ approaches 0 as the distance from the point to the origin gets larger and larger without bound.

We will discuss asymptotes in detail in the next section; however, here we must discuss a new type of limit. First, recall the symbol $\infty$ which is read "infinity." This is not a number. It is used in the notation for limits. Thus, $x \to \infty$ will mean that $x$ increases without bound, and $x \to -\infty$ indicates that $x$ decreases without bound. We now give the following definitions.

**DEFINITION 8.16:** Let $f$ be a function. If the values of $f(x)$ are within a specified distance of $L$, provided $x$ is sufficiently large, we say that the *limit of $f(x)$ as $x$ tends to infinity is $L$*, and we write

$$\lim_{x \to \infty} f(x) = L$$

The specified distance may be arbitrarily small.

In other words, the values of $f(x)$ get arbitrarily close to $L$ as $x$ increases without bound.

DEFINITION 8.17: If the values of $f(x)$ are within a specified distance of $M$, provided $x$ is sufficiently small (that is, negative and sufficiently large in absolute value), we say that the *limit of $f(x)$ as $x$ tends to minus infinity* is $M$, and we write

$$\lim_{h \to -\infty} f(x) = M$$

That is, the values of $f(x)$ get arbitrarily close to $M$ as $x$ decreases without bound. These are illustrated in Figures 8.36 **a, b**.

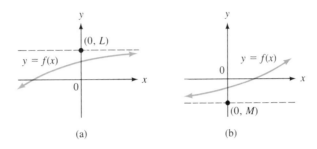

**FIGURE 8.36**                    (a)                              (b)

EXAMPLE 4    Let $f(x) = \dfrac{2x^2 + 5x - 1}{x^2 + 3x + 11}$.

**a.** Find $\lim\limits_{x \to \infty} f(x)$.

**b.** Find $\lim\limits_{x \to -\infty} f(x)$.

**Solution**    If $x \neq 0$, we have

$$\frac{2x^2 + 5x - 1}{x^2 + 3x + 11} = \frac{\dfrac{2x^2 + 5x - 1}{x^2}}{\dfrac{x^2 + 3x + 11}{x^2}}$$

$$= \frac{2 + \dfrac{5}{x} - \dfrac{1}{x^2}}{1 + \dfrac{3}{x} + \dfrac{11}{x^2}}$$

As the value of $x$ increases without bound, the values of $5/x$, $1/x^2$, $3/x$, and $11/x^2$ are all near 0. We conclude that

$$\lim_{x \to \infty} \frac{2 + \dfrac{5}{x} - \dfrac{1}{x^2}}{1 + \dfrac{3}{x} + \dfrac{11}{x^2}} = \frac{2}{1} = 2$$

Thus, $\lim\limits_{x \to \infty} f(x) = 2$.

**b.** The same argument shows that

$$\lim_{x \to -\infty} f(x) = 2, \text{ also.}$$

**Do Exercise 11.**

## Infinite Limits

A frequently occurring type of discontinuity of a function is where the absolute value of the function increases without bound as $x$ approaches a number $a$. We illustrate the different possibilities geometrically in Figure 8.37. In Figure 8.37**a**, we see that if $x$ is close to, but larger than, a number $a$, then $f(x)$ is positive and large.

(a)

(b)

(c)

(d)

**FIGURE 8.37**

**DEFINITION 8.18:** Let $f$ be a function. If the values of $f(x)$ are larger than any preassigned number, provided $x$ is sufficiently close to, but larger than $a$, we say that the *limit of $f(x)$ as $x$ approaches $a$ from the right is infinity* and we write

$$\lim_{x \to a+} f(x) = \infty$$

The preassigned number may be as large as we please.

We call this limit a right-hand limit because when $x$ is larger than $a$, it is to the right of $a$ on the number scale.

Similarly, in Figure 8.37**d**, we see that if $x$ is close to, but less than $a$, then $f(x)$ is negative and large in absolute value. We write

$$\lim_{x \to a^-} f(x) = -\infty$$

(Read: "limit of $f(x)$ as $x$ approaches $a$ from the left is minus infinity.") The meaning of Figures 8.37**b** and 8.37**c** should be clear. We have not written the definitions of the limits illustrated in Figures 8.37**b**, **c**, **d** because the definitions are analogous to the definition of $\lim_{x \to a^+} f(x) = \infty$.

As in the case of finite limits, we ask you to depend on intuition rather than on formal definitions.

**EXAMPLE 5**   Let $g(x) = \dfrac{x + 2}{x - 1}$.

    **a.** Find $\lim\limits_{x \to 1^+} g(x)$.

    **b.** Find $\lim\limits_{x \to 1^-} g(x)$.

**Solution**   **a.** If $x$ is near 1, but greater than 1, then $x - 1 > 0$ and $x - 1$ is close to 0. Furthermore, $x + 2$ is near 3; therefore $x + 2$ is positive. Since both $x + 2$ and $x - 1$ are positive, $g(x)$ is positive. However, $x + 2$ is near 3 and $x - 1$ is near 0; thus $\frac{x + 2}{x - 1}$ is large. We conclude that

$$\lim_{x \to 1^+} g(x) = \infty$$

    **b.** If $x$ is near to, but less than 1, $x - 1$ is near 0 but negative. However, $x + 2$ is still near 3; therefore it is still positive. Thus $\frac{x + 2}{x - 1}$ is negative but large in absolute value. We conclude that

$$\lim_{x \to 1^-} g(x) = -\infty$$

Therefore, near 1, the graph of the function $g$ will look as illustrated in Figure 8.38.

We shall draw the complete graph of this function in the next section.

**Do Exercise 21.**

**FIGURE 8.38**

**EXAMPLE 6**   Let $f(x) = \dfrac{(x - 2)(x + 3)}{(x - 5)(x + 6)}$.

    **a.** Find $\lim\limits_{x \to -6^-} f(x)$.

    **b.** Find $\lim\limits_{x \to -6^+} f(x)$.

    **c.** Find $\lim\limits_{x \to 5^-} f(x)$.

    **d.** Find $\lim\limits_{x \to 5^+} f(x)$.

**Solution**   It is convenient to make a table to determine the intervals over which $f(x)$ is positive and negative. When $f(x)$ is undefined at $c$, we indicate this on the table by writing a $u$ in the last line of the table under the number $c$.

| $x$ | | $-6$ | | $-3$ | | $2$ | | $5$ | |
|---|---|---|---|---|---|---|---|---|---|
| $x-2$ | $-$ | $-$ | $-$ | $-$ | $-$ | $0$ | $+$ | $+$ | $+$ |
| $x+3$ | $-$ | $-$ | $-$ | $0$ | $+$ | $+$ | $+$ | $+$ | $+$ |
| $x-5$ | $-$ | $-$ | $-$ | $-$ | $-$ | $-$ | $-$ | $0$ | $+$ |
| $x+6$ | $-$ | $0$ | $+$ | $+$ | $+$ | $+$ | $+$ | $+$ | $+$ |
| $f(x)$ | $+$ | $u$ | $-$ | $0$ | $+$ | $0$ | $-$ | $u$ | $+$ |

**a.** Now we argue as follows. When $x < -6$, but close to $-6$, $f(x)$ is positive (see the table). But $x + 6$ is near 0. Therefore, $(x - 5)(x + 6)$ is near 0 while $(x - 2)(x + 3)$ is not. Thus $\frac{(x - 2)(x + 3)}{(x - 5)(x + 6)}$ is large and therefore $\lim\limits_{x \to -6^-} f(x) = \infty$.

   Observe that we could have been more precise and state that when $x$ is near $-6$, $(x - 2)(x + 3)$ is near $(-8)(-3) = 24$, but we did not need to do so. Remember we are using our intuition!

**b.** When $x > -6$, but near $-6$, $f(x)$ is negative (see the table). As in part **a**, $(x - 2)(x + 3)$ is not near 0, while $(x - 5)(x + 6)$ is near 0. Thus $\frac{(x - 2)(x + 3)}{(x - 5)(x + 6)}$ is large in absolute value, but negative. We conclude that

$$\lim_{x \to -6^+} f(x) = -\infty$$

By similar arguments, we see that

**c.** $\lim\limits_{x \to 5^-} f(x) = -\infty$

and

**d.** $\lim\limits_{x \to 5^+} f(x) = \infty$.

Because of these results, the graph of $f$ near $-6$ and 5 will look as illustrated in Figure 8.39. We shall complete the graph of this function in the next section.
**Do Exercise 27.**

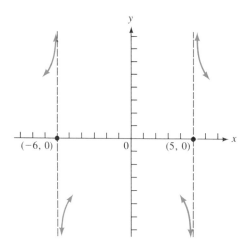

**FIGURE 8.39**

# Exercise Set 8.8

*In Exercises 1–10, sketch the graph of the given polynomial.*

**1.** $f(x) = x^2 + 6x - 2$

**2.** $g(x) = -x^2 + 8x + 3$

**3.** $h(x) = 2x^2 - 12x + 5$

**4.** $f(x) = -3x^2 + 24x - 7$

**5.** $g(x) = x^3 + 3x^2 - 24x + 5$

**6.** $h(x) = x^3 + 6x^2 - 15x + 4$

**7.** $f(x) = -x^3 - 3x^2 + 9x + 6$

**8.** $g(x) = x^3 - 12x + 7$

**9.** $h(x) = x^4 - 4x + 3$

**10.** $f(x) = x^5 - 80x + 60$

*In Exercises 11–17, find* **a.** $\lim_{x \to \infty} f(x)$ *and* **b.** $\lim_{x \to -\infty} f(x)$.

**11.** $f(x) = \dfrac{x^2 + 3x - 2}{3x^2 - 5x + 10}$

**12.** $f(x) = \dfrac{5x^2 + 6x - 3}{3x^2 - 8x + 2}$

**13.** $f(x) = \dfrac{-x^2 + 7x + 4}{x^2 + 8x - 6}$

**14.** $f(x) = \dfrac{6x^2 - 3x + 4}{2x^2 + 5x - 2}$

**15.** $f(x) = \dfrac{4x^3 - 2x^2 + 5x + 1}{2x^3 + 6x - 7}$

**16.** $f(x) = \dfrac{2x^3 - 5x^2 + 6}{4x^2 + 7x - 3}$

**17.** $f(x) = \dfrac{4x^2 - 5x + 7}{6x^3 + 3x^2 - 5}$

*In Exercises 18–27, find* **a.** $\lim_{x \to c^-} f(x)$ *and* **b.** $\lim_{x \to c^+} f(x)$ *for the given function and the given* $c$. *In some of the exercises, more than one value of* $c$ *is given. Do both part* **a** *and* **b** *for each value of* $c$.

**18.** $f(x) = \dfrac{5}{x - 1}$, $c = 1$

**19.** $f(x) = \dfrac{-2}{x + 3}$, $c = -3$

**20.** $f(x) = \dfrac{5x + 2}{x - 4}$, $c = 4$

**21.** $f(x) = \dfrac{5 - 3x}{x + 2}$, $c = -2$

**22.** $f(x) = \dfrac{(x + 1)(x + 3)}{x - 7}$, $c = 7$

**23.** $f(x) = \dfrac{x + 2}{(x - 1)^2}$, $c = 1$

**24.** $f(x) = \dfrac{2x - 3}{(x - 3)(x + 2)}$; $c = -2, c = 3$

**25.** $f(x) = \dfrac{x + 8}{(x + 1)(x - 5)}$; $c = -1, c = 5$

**26.** $f(x) = \dfrac{(x + 1)(x - 5)}{(x - 2)(x + 4)}$; $c = -4, c = 2$

**27.** $f(x) = \dfrac{(5 - x)(2x + 3)}{(2 - x)(7 - x)}$; $c = 2, c = 7$

**28.** Let $f(x) = x^3 - 3x^2 + 3x - 1$. Sketch the graph of $f$. Since the degree of this polynomial is 3, we expect at most two local high and low points on the graph. How many are there?

**29.** Let $g(x) = x^4 - 8x^3 + 24x^2 - 32x + 16$. Show that 2 is the only critical value of this function. Sketch the graph of $g$.

*In Exercises 30–36, find the points of inflection for the graphs of the given functions.*

**30.** $f(x) = x^4 - 4x^3 - 18x^2 + 12x + 7$

**31.** $g(x) = x^8 + 6x^4 - 4x + 2$

**32.** $h(x) = x^{13/7}$

**33.** $f(x) = x^4 - 4x^3 - 48x^2 + 36x + 3$

**34.** $g(x) = x^4 + 6x^2 + 5$

**35.** $h(x) = x^{17/9}$

**36.** $f(x) = x^4 + 2x^3 - 12x^2 + 48x + 5$

**37.** Let $f(x) = x^4 + x$

    **a.** Show that $f''(0) = 0$.

    **b.** Show that $(0, 0)$ is not a point of inflection of the graph of $f$.

    **c.** Show that $f$ does not have an extremum at 0.

    **d.** Sketch the graph of $f$ over the interval $[-1, 1]$.

# 8.9  More on Curve Sketching

In the previous section, we introduced new types of limits; namely, $\lim_{x \to a} f(x) = \pm\infty$, $\lim_{x \to a^+} f(x) = \pm\infty$, $\lim_{x \to \infty} f(x) = L$, and $\lim_{x \to -\infty} f(x) = L$. In this section, we discuss the geometric significance of these types of limits and use this information to sketch graphs of certain functions.

## Special Types of Asymptotes

Asymptotes were introduced in the preceding section. For the graph of a function $f$, there are three types of asymptotes, which are illustrated in Figure 8.40.

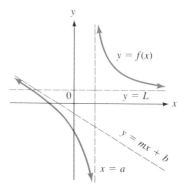

**FIGURE 8.40**

The basic rule to find two of these asymptotes is given in Theorem 8.11, stated without proof.

> **THEOREM 8.11:**    If $f$ is a function and either $\lim\limits_{x \to a^-} f(x) = \pm\infty$ or $\lim\limits_{x \to a^+} f(x) = \pm\infty$, then the graph of the equation $x = a$ is a vertical asymptote. If $\lim\limits_{x \to -\infty} f(x) = L$ or $\lim\limits_{x \to \infty} f(x) = L$, then the graph of $y = L$ is a horizontal asymptote.

**EXAMPLE 1**   Sketch the graph of the function of Example 5 in Section 8.8.

**Solution**   The function was defined by

$$g(x) = \frac{x + 2}{x - 1}$$

In the preceding section, we established that $\lim\limits_{x \to 1^-} g(x) = -\infty$ and $\lim\limits_{x \to 1^+} g(x) = \infty$. Thus the graph of $x = 1$ is a vertical asymptote. We drew two pieces of the graph of $g$ accordingly. (See Figure 8.38 in the preceding section.)
   Observe that if $x \neq 0$,

$$\frac{x + 2}{x - 1} = \frac{1 + \dfrac{2}{x}}{1 - \dfrac{1}{x}}$$

Since $\frac{2}{x}$ and $\frac{1}{x}$ tend to 0 as $x$ tends to $\infty$ or $-\infty$, we have

$$\lim\limits_{x \to \infty} g(x) = 1$$

and

$$\lim_{x \to -\infty} g(x) = 1$$

Therefore the graph of $y = 1$ is a horizontal asymptote. We find that $g(0) = -2$. Thus the point $(0, -2)$ is the $y$-intercept. It is easy to see that $g(-2) = 0$ and therefore the point $(-2, 0)$ is the $x$-intercept.

Differentiating, we get

$$g'(x) = \frac{-3}{(x - 1)^2}$$

and

$$g''(x) = \frac{6}{(x - 1)^3}$$

Note that $g'(1)$ is not defined. However, 1 is not in the domain of $g$ and therefore it is not a critical value of $g$.

| $x$ | | 1 | |
|---|---|---|---|
| $x - 1$ | $-$ | 0 | $+$ |
| $(x - 1)^2$ | $+$ | 0 | $+$ |
| $(x - 1)^3$ | $-$ | 0 | $+$ |
| $g'(x) = \dfrac{-3}{(x - 1)^2}$ | $-$ | $u$ | $-$ |
| Behavior of $g$ | ↘ | | ↘ |
| $g''(x) = \dfrac{6}{(x - 1)^3}$ | $-$ | $u$ | $+$ |
| Shape of the graph | ⌢ | | ⌢ |

We have entered the necessary information in the table. Note that we have entered the behavior of $g$ on the sixth line, immediately below the line giving the sign of the derivative $g'$, because whether the function is increasing or decreasing is determined by the sign of the first derivative. We have also entered information on the shape of the graph on the eighth line, immediately below the line giving the sign of the second derivative $g''$, because whether the graph is concave up or down is determined by the sign of the second derivative. Also note that although the graph is concave down to the left of 1, and concave up to the right of 1, it does not have a point of inflection at 1 because the function is not continuous at 1. Using this information, we can now complete the graph of $g$. (See Figure 8.41.)

 Observe that the function $g$ is not decreasing on its domain since, for example, $g(0) < g(2)$. However, it is decreasing over each interval contained in the domain of $g$. (The interval $[-1, 3]$ is not contained in the domain of $g$ since 1 is in the interval but not in the domain of $g$.)

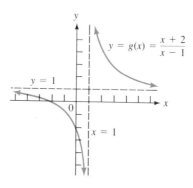

**FIGURE 8.41**

**Do Exercise 1.**

**EXAMPLE 2**   Sketch the graph of the function $f$ defined in Example 6, Section 8.8.

**Solution**   The function was defined by $f(x) = \dfrac{(x-2)(x+3)}{(x-5)(x+6)}$

We have seen in the preceding section that $\lim\limits_{x \to -6^-} f(x) = \infty$, $\lim\limits_{x \to -6^+} f(x) = -\infty$, $\lim\limits_{x \to 5^-} f(x) = -\infty$, and $\lim\limits_{x \to 5^+} f(x) = \infty$. Thus the graphs of $x = -6$ and $x = 5$ are vertical asymptotes. When $x \neq 0$, we can write

$$\frac{(x-2)(x+3)}{(x-5)(x+6)} = \frac{\left(1 - \dfrac{2}{x}\right)\left(1 + \dfrac{3}{x}\right)}{\left(1 - \dfrac{5}{x}\right)\left(1 + \dfrac{6}{x}\right)},$$

and since as $x$ tends to $\infty$ and as $x$ tends to $-\infty$, the four quantitites $\frac{2}{x}, \frac{3}{x}, \frac{5}{x}$, and $\frac{6}{x}$ all tend to 0 and have

$$\lim_{x \to -\infty} f(x) = 1 \text{ and } \lim_{x \to \infty} f(x) = 1$$

Thus the graph of $y = 1$ is a horizontal asymptote. At this time, it is useful to sketch the asymptotes—pieces of the graph—using the information we already have, including the x-intercepts $(2, 0)$ and $(-3, 0)$, and the y-intercept $\left(0, \frac{1}{5}\right)$. (See Figure 8.42.)

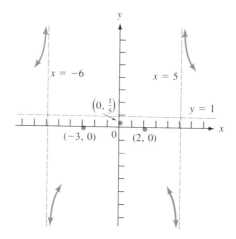

**FIGURE 8.42**

Differentiating $f(x)$ and simplifying the result, we get

$$f'(x) = \frac{-48x - 24}{(x - 5)^2(x + 6)^2}$$

Clearly, $f'(x) = 0$ only if $-48x - 24 = 0$. That is, only if $x = \frac{-1}{2}$. Thus, $\frac{-1}{2}$ is a critical value of the function $f$. It is the only critical value since $-6$ and $5$ are not in the domain of $f$. We now tabulate the information.

| $x$ | | $-6$ | | $\dfrac{-1}{2}$ | | $5$ | |
|---|---|---|---|---|---|---|---|
| $-48x - 24$ | $+$ | $+$ | $+$ | $0$ | $-$ | $-$ | $-$ |
| $(x - 5)^2(x + 6)^2$ | $+$ | $0$ | $+$ | $+$ | $+$ | $0$ | $+$ |
| $f'(x)$ | $+$ | $u$ | $+$ | $0$ | $-$ | $u$ | $-$ |
| $f(x)$ | $\nearrow$ | | $\nearrow$ | | $\searrow$ | | $\searrow$ |

Thus $f$ has a local maximum at $\frac{-1}{2}$. We calculate $f\left(\frac{-1}{2}\right)$ and obtain $f\left(\frac{-1}{2}\right) = \frac{25}{121}$. Therefore the point $\left(\frac{-1}{2}, \frac{25}{121}\right)$ is a local high point of the graph.

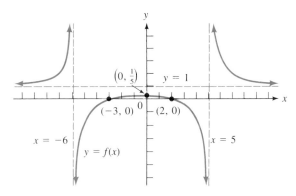

**FIGURE 8.43**

Now we plot the local high point and, using the information from the table, we complete the graph started in Figure 8.42. (See Figure 8.43.)*
**Do Exercise 3.** ∎

**EXAMPLE 3**   Sketch the graph of the function $h$ if

$$y = h(x) = \frac{x^2 + 5x - 6}{x + 1}$$

**Solution**   Note that

$$\frac{x^2 + 5x - 6}{x + 1} = \frac{(x - 1)(x + 6)}{x + 1}$$

---

*We did not use the second derivative to determine where the graph was concave up (and down) because the differentiation would have been tedious and the shape of the graph could be determined from the information we already had.

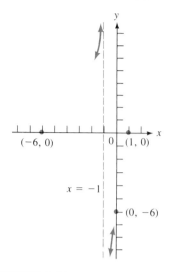

**FIGURE 8.44**

Therefore, $h(-6) = 0$ and $h(1) = 0$. It follows that the points $(-6, 0)$ and $(1, 0)$ are the $x$-intercepts of the graph. Also $h(0) = -6$ and $h(-1)$ is not defined. Therefore, the point $(0, -6)$ is the $y$-intercept. We make a table to determine where the function is positive, negative.

| $x$ | | $-6$ | | $-1$ | | $1$ | |
|---|---|---|---|---|---|---|---|
| $h(x)$ | $-$ | $0$ | $+$ | $u$ | $-$ | $0$ | $+$ |

Using the table, we see that if $x < -1$, but is close to $-1$, then $h(x) > 0$ and large (since the divisor $x + 1$ is near 0). Thus $\lim\limits_{x \to -1^-} h(x) = \infty$. Similarly, we see that $\lim\limits_{x \to -1^+} h(x) = -\infty$. Therefore the graph of $x = -1$ is a vertical asymptote. Using the information we have gathered so far, it is useful to draw pieces of the graph, including the intercepts, as shown in Figure 8.44.

Since the degree of $x^2 + 5x - 6$ is 2, and that of $x + 1$ is 1,

$$\lim_{x \to -\infty} \frac{x^2 + 5x - 6}{x + 1} = -\infty$$

and

$$\lim_{x \to \infty} \frac{x^2 + 5x - 6}{x + 1} = \infty.$$

 There is no horizontal asymptote. Is there an oblique asymptote? We might be tempted to proceed as we did in previous examples and state that if $x \neq 0$,

$$h(x) = \frac{x^2 + 5x - 6}{x + 1} = \frac{x + 5 - \dfrac{6}{x}}{1 + \dfrac{1}{x}}$$

so that as $x$ tends to infinity or minus infinity, $\frac{6}{x}$ and $\frac{1}{x}$ approach 0, $h(x)$ approaches $x + 5$, we conclude that the graph of $y = x + 5$ is an oblique asymptote. However, this is *wrong*! (See Exercise 17.) The correct way to proceed is as follows. Using division, we get

$$\frac{x^2 + 5x - 6}{x + 1} = x + 4 - \frac{10}{x + 1}$$

Therefore,

$$\frac{x^2 + 5x - 6}{x + 1} - (x + 4) = \frac{-10}{x + 1}$$

If we let $y = x + 4$, this equality can be written

$$h(x) - y = \frac{-10}{x + 1}$$

Since $\lim\limits_{x\to\pm\infty} \dfrac{-10}{x+1} = 0$, we see that when $|x|$ is large, $h(x) - y$ is close to 0. Thus, the graph of $y = h(x)$ is close to the graph of $y = x + 4$ and the latter is an oblique asymptote.

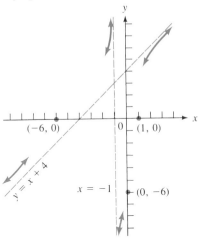

**FIGURE 8.45**

We now draw this asymptote and two more pieces of the graph. (See Figure 8.45.)
Differentiating, we find that

$$h'(x) = D_x \frac{x^2 + 5x - 6}{x + 1} = D_x\left[x + 4 - \frac{10}{x + 1}\right] = 1 + \frac{10}{(x + 1)^2}$$

and

$$h''(x) = D_x\left[1 + \frac{10}{(x + 1)^2}\right] = \frac{-20}{(x + 1)^3}$$

It is now clear that $h'(x) > 0$ for all values of $x$ except $-1$ since $h'(-1)$ is not defined. It follows that the function $h$ is increasing on the intervals $(-\infty, -1)$ and $(-1, \infty)$. Furthermore, we see that $h''(x) > 0$ on $(-\infty, -1)$ and $h''(x) < 0$ on $(-1, \infty)$. Thus the graph of the function $h$ is concave up on $(-\infty, -1)$ and concave down on $(-1, \infty)$. We can now complete the graph of the equation $y = h(x)$. (See Figure 8.46.)
**Do Exercise 7.**

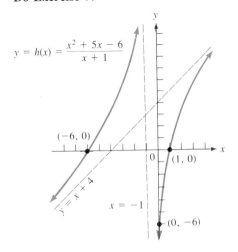

**FIGURE 8.46**

The next two examples illustrate functions whose definitions involve the exponential and the logarithmic functions.

**EXAMPLE 4**  Sketch the graph of the function $f$ if

$$y = f(x) = xe^{-x}$$

**Solution**  The domain of this function is the set of all real numbers. We first find the first and second derivatives.

$$f'(x) = xe^{-x}(-1) + e^{-x} = (1 - x)e^{-x}$$

and

$$f''(x) = (x - 2)e^{-x}$$

Clearly, 1 is a critical value of $f$. Since $e^{-x} > 0$ for all values of $x$, it is easy to see that $f'(x) > 0$ when $x < 1$ and $f'(x) < 0$ when $x > 1$. Thus the function is increasing on the interval $(-\infty, 1)$ and it is decreasing on the interval $(1, \infty)$. Consequently, it has a local, and absolute, maximum at 1. The point $(1, \frac{1}{e})$ is the highest point of the graph. The graph passes through the origin since $f(0) = 0$. Also, $f''(x) < 0$ when $x < 2$ and $f''(x) > 0$ when $x > 2$. Consequently, the graph is concave down over $(-\infty, 2)$, concave up over $(2, \infty)$, and the point $(2, 2/e^2)$ is a point of inflection. Since $e^x$ gets large much faster than $x$ does, $\lim\limits_{x \to \infty} f(x) = \lim\limits_{x \to \infty} x/e^x = 0$. Thus the $x$-axis is an asymptote. The graph is sketched in Figure 8.47.
**Do Exercise 13.**

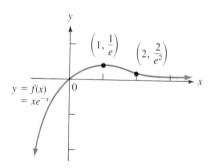

**FIGURE 8.47**

**EXAMPLE 5**  Sketch the graph of the function $g$ if

$$y = g(x) = x\ln x$$

**Solution**  The domain of $g$ is the set of all positive real numbers. There is no $y$-intercept, but the point $(1, 0)$ is the $x$-intercept since $g(1) = 0$. Differentiating, we get

$$g'(x) = xD_x\ln x + (\ln x)D_x x = 1 + \ln x$$

and

$$g''(x) = \frac{1}{x}$$

To find the critical values of $g$, set $g'(x) = 0$ to obtain

$$1 + \ln x = 0$$

or

$$\ln x = -1$$

Thus, $x = e^{\ln x} = e^{-1}$. It follows that $e^{-1}$ is a critical value of $g$. Since ln is an increasing function, and $g'(x) = 1 + \ln x$, it is easy to see that $g'(x) < 0$ when $x < e^{-1}$ and $g'(x) > 0$ when $x > e^{-1}$. Thus the function is decreasing on the interval $(0, e^{-1})$ and increasing on the interval $(e^{-1}, \infty)$. It follows that $g$ has a local, and absolute, minimum at $e^{-1}$. We find that $g(e^{-1}) = -e^{-1}$. So the point $(e^{-1}, -e^{-1})$ is the lowest point of the graph. Clearly, $g''(x) > 0$ for all $x > 0$. Hence, the graph of the function $g$ is concave up over its entire domain. In more advanced texts, it is shown that $\lim\limits_{x \to 0^+} x \ln x = 0$ and therefore the graph gets closer and closer to the origin as $x$ approaches 0 from the right. Of course, we could have drawn the same conclusion using intuition since the function is decreasing on $(0, e^{-1})$ and the point $(e^{-1}, -e^{-1})$ is close to the $x$-axis. It also is intuitively clear that as $x$ tends to infinity, $g(x)$ increases without bound. Thus $\lim\limits_{x \to \infty} g(x) = \infty$. The graph is sketched in Figure 8.48.

**Do Exercise 15.** ∎

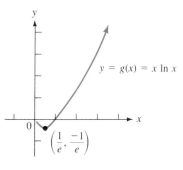

$y = g(x) = x \ln x$

$\left(\dfrac{1}{e}, \dfrac{-1}{e}\right)$

**FIGURE 8.48**

## Symmetry

We first state the following rules.

> If $f$ is a function and $f(-x) = f(x)$ for all $x$ in its domain, then its graph is symmetric with respect to the $y$-axis.
> If $f(-x) = -f(x)$ for all $x$ in the domain of $f$, then the graph is symmetric with respect to the origin.

This information can reduce the time spent sketching a graph. If we know that a graph is symmetric about the $y$-axis or the origin, we can sketch only half of it and obtain the other half by symmetry.

**EXAMPLE 6** Sketch the graph of the function $f$ if

$$y = f(x) = \frac{2x^2}{x^2 - 9}$$

**Solution** Checking for symmetry, we find that

$$f(-x) = \frac{2(-x)^2}{(-x)^2 - 9} = \frac{2x^2}{x^2 - 9} = f(x)$$

The graph is symmetric with respect to the $y$-axis. We shall sketch only the part of the graph over the interval $[0, \infty)$ and obtain the other part by symmetry. It is clear that $f(0) = 0$. Thus the origin is the $y$-intercept. It is also the $x$-intercept. It is easy to show that $\lim\limits_{x\to\infty} f(x) = 2$ and therefore the graph of $y = 2$ is a horizontal asymptote. Also, we find that $\lim\limits_{x\to3^-} f(x) = -\infty$ and $\lim\limits_{x\to3^+} f(x) = \infty$. Thus the graph of $x = 3$ is a vertical asymptote. We find that

$$f'(x) = \frac{-36x}{(x^2 - 9)^2}$$

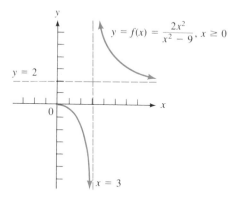

FIGURE 8.49

Clearly, $f'(x) < 0$ when $x > 0$ except at 3 where $f'$ is not defined. Therefore the function $f$ is decreasing on the intervals $(0, 3)$ and $(3, \infty)$. Using this information, we sketch half of the graph of $f$. (See Figure 8.49.) The other half is obtained by symmetry. (See Figure 8.50).
**Do Exercise 9.**

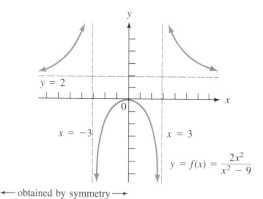

FIGURE 8.50

$\longleftarrow$ obtained by symmetry $\longrightarrow$

### Procedure for Sketching the Graph of a Function

**STEP 1.** Test for symmetry with respect to the $y$-axis and with respect to the origin. In the case of symmetry, sketch the graph over the interval $[0, \infty)$ and obtain the other half by symmetry.

**STEP 2.**   Find the y-intercept by replacing $x$ by 0 in the equation $y = f(x)$. Find the x-intercepts by replacing $y$ by 0 and solving the equation $f(x) = 0$. If this is impossible or too tedious, find a few other points by replacing $x$ by several numbers.

**STEP 3.**   Find all asymptotes. If the function is a rational function—that is, $y = R(x) = \frac{f(x)}{g(x)}$ where $f(x)$ and $g(x)$ are polynomials—then:

    **a.** If $g(c) = 0$ and $f(c) \neq 0$, the graph of $x = c$ is a vertical asymptote. Find $\lim\limits_{x \to c^-} R(x)$ and $\lim\limits_{x \to c^+} R(x)$. These limits are used to determine the location of the graph near the asymptote and away from the origin.

    **b.** If degree $f(x)$ = degree $g(x)$, then $\lim\limits_{x \to \pm\infty} \frac{f(x)}{g(x)} = a_0/b_0$, where $a_0$ and $b_0$ are the leading coefficients of $f(x)$ and $g(x)$, respectively. In this case, the graph of $y = a_0/b_0$ is a horizontal asymptote.

    **c.** If degree $f(x)$ < degree $g(x)$, then $\lim\limits_{x \to \pm\infty} \frac{f(x)}{g(x)} = 0$ and the x-axis is a horizontal asymptote.

    **d.** If degree $f(x)$ > degree $g(x)$, there is no horizontal asymptote. However, if degree $f(x)$ is one more than degree $g(x)$, there is an oblique asymptote found as follows.

        Divide $f(x)$ by $g(x)$ to get $\frac{f(x)}{g(x)} = mx + b + \frac{c}{g(x)}$ where $m$, $b$, and $c$ are real numbers, then the graph of $y = mx + b$ is an oblique asymptote. At this point, it is useful to draw the asymptotes and to sketch pieces of the graph near the asymptotes and away from the origin. Also plot all points found in Step 2.

**STEP 4.**   Use the first derivative to determine where the graph is rising, falling.

**STEP 5.**   Determine all critical values of the function and plot the corresponding points of the graph.

**STEP 6.**   If it is not too tedious, use the second derivative to determine where the graph is concave up and concave down. Find and plot the points of inflection.

**STEP 7.**   Complete the sketch of the graph that was started in Step 3.

---

# Exercise Set 8.9

*In Exercises 1–16, sketch the graph of the given equation.*

**1.** $y = f(x) = \dfrac{x - 3}{x + 2}$

**2.** $y = g(x) = \dfrac{2x - 5}{4 - x}$

**3.** $y = h(x) = \dfrac{(x - 3)(x - 4)}{(x - 1)(x - 6)}$

**4.** $y = f(x) = \dfrac{(x - 6)(x + 1)}{(x - 1)(x - 4)}$

**5.** $y = g(x) = \dfrac{(x - 4)(x - 1)}{(x - 3)^2}$

**6.** $y = h(x) = \dfrac{(x - 5)(x - 1)}{2x - 1}$

**7.** $y = f(x) = \dfrac{(x - 6)(x - 3)}{x - 2}$

**8.** $y = g(x) = \dfrac{3x^2 + 1}{x^2 - 16}$

**9.** $y = h(x) = \dfrac{(x - 1)(x + 7)}{x^2 + 1}$

*10. $y = f(x) = \dfrac{(x + 2)(7x - 74)}{(x + 1)(x - 7)}$

*11. $y = g(x) = \dfrac{(x + 3)(2x + 1)}{(x - 2)(x + 5)}$

12. $y = h(x) = xe^x$

13. $y = f(x) = xe^{-3x}$

14. $y = g(x) = (x + 1)\ln(x + 1)$

15. $y = h(x) = (x + 1)\ln|x + 1|$

16. $y = (x^2 + 1)\ln(x^2 + 1)$

17. Let

$$y = h(x) = (x^2 + 5x - 6)/(x + 1)$$

If $x \neq 0$, we can write

$$(x^2 + 5x - 6)/(x + 1) = \left(x + 5 - \tfrac{6}{x}\right)/\left(1 + \tfrac{1}{x}\right)$$

If $|x|$ is large, then $\tfrac{6}{x}$ and $\tfrac{1}{x}$ are close to 0. Thus $h(x)$ is close to $x + 5$. Show that this conclusion is false. (*Hint:* Show that

$$|(x^2 + 5x - 6)/(x + 1) - (x + 5)|$$

does *NOT* become arbitrarily small as $x$ tends to $\infty$.)

**18. Suppose $f(x) = a_0x^3 + a_1x^2 + a_2x + a_3$ and $g(x) = b_0x^3 + b_1x^2 + b_2x + b_3$ where $a_0 \neq 0$ and $b_0 \neq 0$, let $R(x) = \dfrac{f(x)}{g(x)}$. To find the $\lim\limits_{x \to \infty} R(x)$, we first divide numerator and denominator by $x^3$ and write

$$R(x) = \dfrac{a_0 + \dfrac{a_1}{x} + \dfrac{a_2}{x^2} + \dfrac{a^3}{x^3}}{b_0 + \dfrac{b_1}{x} + \dfrac{b_2}{x^2} + \dfrac{b_3}{x^3}}$$

Then using the fact that $a_1/x$, $a_2/x^2$, $a_3/x^3$, $b_1/x$, $b_2/x^2$, and $b_3/x^3$ all tend to 0 as $x$ tends to $\infty$, we claim that $\lim\limits_{x \to \infty} R(x) = a_0/b_0$. Prove that this claim is correct. (*Hint:* Show that $\lim\limits_{x \to \infty}\left(\dfrac{f(x)}{g(x)} - a_0/b_0\right) = 0$ using the fact that if $r(x)$ and $s(x)$ are polynomials and degree $r(x) <$ degree $s(x)$, then $\lim\limits_{x \to \infty} \dfrac{r(x)}{s(x)} = 0$.) Compare this Exercise to Exercise 17.

19. To appreciate the significance of the preceding two exercises, use a calculator to find **a.** $\dfrac{5.0001}{2.0001} - \dfrac{5}{2}$ and **b.** $\dfrac{9,000,000.0001}{2.0001} - \dfrac{9,000,000}{2}$. Compare the sizes of the two results.

20. Sketch the graph of $y = 9x^{8/3}$

21. Sketch the graph of $y = 50x^{7/5}$

# 8.10  Differentials

We have seen that the value of the marginal cost function at a positive integer $k$ is used by economists as an estimate of the cost of producing the $k$th (and by others the $(k + 1)$st) unit. Similarly, the values of the marginal revenue function and the marginal profit function at $k$ approximate the revenue and profit, respectively, derived from the $k$th (or $(k + 1)$st) unit. We shall expand on this idea to approximate the value of the increment $\Delta y$ of a function as the value of the independent variable changes from $x$ to $x + \Delta x$.

## Differentials

Recall that the derivative of a function $f$ is defined by

$$f'(x) = \lim_{\Delta x \to 0} \frac{f(x + \Delta x) - f(x)}{\Delta x}$$

provided this limit exists. We know that $\Delta y = f(x + \Delta x) - f(x)$. Thus we can write

$$f'(x) = \lim_{\Delta x \to 0} \frac{\Delta y}{\Delta x}$$

This means that if $\Delta x$ is close to 0, then $f'(x)$ is close to $\frac{\Delta y}{\Delta x}$, or

$$\Delta y \doteq f'(x)\Delta x$$

Recall that the symbol "$\doteq$" denotes "approximately equal to." The expression on the right is called a differential and is denoted $dy$.

> **DEFINITION 8.19:** If a function $f$ is differentiable at $x$, its *differential* is denoted $dy$ and is defined by the equation
>
> $$dy = f'(x)dx$$
>
> where $dx$ is any real number. It is also useful to call $dx$ the *differential of $x$*.

Note that the value of $dy$ depends on both the value of $x$ and the value of $dx$. Also, if $dx = \Delta x$ and $\Delta x$ is small, then $\Delta y \doteq dy$. This idea can be illustrated as follows. (See Figure 8.51.)

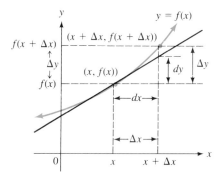

**FIGURE 8.51**

We draw the tangent to the graph of $y = f(x)$ at the point $(x, f(x))$. The slope of the tangent line to the graph at that point is $f'(x)$. But this is also equal to $\frac{\text{rise}}{\text{run}}$, which on the diagram is shown as $\frac{dy}{dx}$. Therefore the value of the differential $dy$ is the rise of the tangent line as the independent variable changes from $x$ to $x + \Delta x$. Note that on the diagram, the exact value of $\Delta y$ is close to the value of $dy$. These two quantities would be even closer if the value of $dx$ was taken closer to 0. Thus, this geometric illustration agrees with our previous claim that for small values of $dx$, $dy \doteq \Delta y$.

## Approximations Using Differentials

Often, it is easier to calculate the value of $dy$ than it is to find the exact value of $\Delta y$— a fact that is frequently used in the applications. The notation $dy$ and $dx$ used to define differentials was chosen to be consistent with the notation $\frac{dy}{dx}$ used for the derivative. From now on, we shall use $\frac{dy}{dx}$ either to denote a derivative or to denote the quotient of two differentials.

EXAMPLE 1   Let the function $f$ be defined by

$$y = f(x) = x^2 + 5x + 1$$

**a.** Find the exact value of $\Delta y$ as $x$ changes from $x = 3$ to $x = 3.2$.
**b.** Find the value of the differential $dy$ when $x = 3$ and $dx = 0.2$.
**c.** Compare the two answers.

Solution   **a.** $\Delta y = f(3.2) - f(3)$
$$= [(3.2)^2 + 5(3.2) + 1] - (3^2 + 5(3) + 1)$$
$$= 27.24 - 25 = 2.24$$

**b.** $f(x) = D_x(x^2 + 5x + 1) = 2x + 5$.
Thus

$$dy = f'(x)dx = (2x + 5)dx$$

When $x = 3$ and $dx = 0.2$, we get

$$dy = (2(3) + 5)(0.2) = (11)(0.2) = 2.2.$$

**c.** We see that the value of $dy$ is 2.2, which is close to the exact value of $\Delta y$ which was found to be 2.24.
**Do Exercise 3.**                                                                     ■

In the applications, we are often confronted with the task of evaluating a function $f$ at a number $b$. It may be the case that the value of that function at another number $c$, close to $b$, is already known or can readily be found. In such cases, we may use differentials to get an approximation for the value of $f(b)$. We proceed as follows.

**Procedure to Approximate Functional Values Using Differentials**

**STEP 1.**   Evaluate $f(c)$.

**STEP 2.**   Find $f'(x)$.

**STEP 3.**   Evaluate the differential $dy$ where $dy = f'(c)dx$ with $dx = b - c$.

**STEP 4.**   Since $\Delta y = f(b) - f(c)$, we have

$$f(b) = f(c) + \Delta y \doteq f(c) + dy$$

The value of this last expression can be obtained by adding the results of Steps 1 and 3.

EXAMPLE 2   Approximate $\sqrt[3]{997}$ using differentials.

Solution   Note that $\sqrt[3]{1000} = 10$. To make use of this fact, let

$$f(x) = \sqrt[3]{x} = x^{1/3}$$

Differentiating, we get

$$f'(x) = D_x x^{1/3} = \frac{1}{3x^{2/3}} = \frac{1}{3(\sqrt[3]{x})^2}$$

If we let $dx = 997 - 1000 = -3$ and $x = 1000$, we have

$$dy = f'(1000)(-3) = \frac{1}{3(\sqrt[3]{1000})^2}(-3) = \frac{-1}{100} = -0.001$$

Thus

$$\Delta y \doteq -0.01 \quad \text{and} \quad f(997) \doteq f(1000) + (-0.01)$$

That is,

$$f(997) \doteq 10 + (-0.01) = 9.99$$

Note that the value of $\sqrt[3]{997}$, obtained with a calculator, is 9.98998998.
**Do Exercise 9.** ■

## Estimating Errors Using Differentials

Differentials can also be used to estimate errors in measurements.

**DEFINITION 8.20:** The *relative error* in a measurement is the ratio $\frac{E}{A}$, where $E$ is the error and $A$ is the measurement. The *percentage error* is the relative error multiplied by 100.

**EXAMPLE 3** The diameter of a bowling ball is measured to be 8.6 inches with a possible maximum error of 1%. Estimate the maximum relative error and the maximum percentage error in volume.

**Solution** The radius is 4.3 inches. Therefore in absolute value, the maximum possible error in the measurement of the radius is .043 since $(4.3)(.01) = .043$. The formula for the volume of a sphere is

$$V = \frac{4}{3}\pi r^3$$

Thus,

$$\frac{dV}{dr} = D_r\left(\frac{4}{3}\pi r^3\right) = 4\pi r^2$$

and

$$dV = 4\pi r^2 dr$$

Here we have $r = 4.3$ and $|dr| \leq 0.043$. Thus

$$|dV| \leq 4\pi(4.3)^2(0.043) = 3.18028\pi$$

The maximum possible error in volume is approximately $3.18028\pi$ cubic inches. But the volume calculated with $r = 4.3$ is $106.0093333\pi$ cubic inches. Therefore the maximum relative error is approximately $\frac{3.18028\pi}{106.0093333\pi} = 0.03$, and the maximum percentage error is approximately 3%.

**Do Exercise 15.**                                                          ■

EXAMPLE 4   The admissions director of a private university estimates that 8000 students will enroll at the university the following year. He believes his estimate is correct within a maximum percentage error of 4%. The revenue function is defined by

$$y = R(x) = 5000x + 1{,}000{,}000 + 10{,}000\sqrt[3]{x}$$

where $x$ is the number of students enrolled and $R(x)$ is in dollars. Approximate the maximum percentage error in revenue.

Solution   With a student enrollment of 8000, the revenue is \$41,200,000 since

$$R(8000) = 5000(8000) + 1{,}000{,}000 + 10{,}000\sqrt[3]{8000} = 41{,}200{,}000$$

But the possible error in the enrollment estimate can be as large as 4%. This represents 320 students since $(8000)(0.04) = 320$. Therefore we have $|dx| \leq 320$.

Now,

$$R'(x) = D_x(5000x + 1{,}000{,}000 + 10{,}000\sqrt[3]{x})$$

$$= 5000 + \frac{10{,}000}{3x^{2/3}}$$

Calculating the approximation of $|dy|$ with $x = 8000$ and $|dx| \leq 320$, we get

$$|dy| \leq \left(5000 + \frac{10{,}000}{3(8000)^{2/3}}\right)(320) \doteq 1{,}602{,}666.667$$

In absolute value, the relative maximum error is given by

$$\frac{1{,}602{,}666.667}{41{,}200{,}000} \doteq 0.0389$$

The maximum percentage error in revenue is 3.89% since $(0.0389)(100) = 3.89$.

**Do Exercise 21.**                                                          ■

# Exercise Set 8.10

*In Exercises 1–7, an equation is given. A value of x and a value of dx = Δx are also given. In each case, find **a**. the exact value of Δy, **b**. the value of dy, and **c**. compare your two answers.*

**1.** $y = x^2 + 6x - 3$; $x = 4$, $dx = 0.2$

**2.** $y = -x^2 + 3x + 5$; $x = -3$, $dx = 0.3$

**3.** $y = x^3 + x^2 - 5x + 1$; $x = 3$, $dx = 0.1$

**4.** $y = x^4 + 2x^2 + 1$; $x = 3$, $dx = 0.1$

**5.** $y = \sqrt{x}$; $x = 225$, $dx = 0.3001$

**6.** $y = \sqrt[4]{x}$; $x = 256$, $dx = -4$

**7.** $y = \ln x$; $x = 75$, $dx = -2$

*In Exercises 8–13, use differentials to approximate the value of the given quantity.*

**8.** $\sqrt{102}$

**9.** $\sqrt[3]{515}$

**10.** $\sqrt[4]{621}$

**11.** $\sqrt[5]{245}$

**12.** $(10.02)^5$

**13.** $(9.9)^6$

**14.** The side of a square has been measured to be 15 inches with a maximum possible error of 2%. If the area of the square has been calculated using $s = 15$, estimate the maximum relative and percentage errors.

**15.** The radius of a circular circus ring has been measured to be 30 feet with a possible maximum error of 3%. If the circumference and area of the ring have been calculated using $r = 30$, estimate the maximum relative and percentage errors in **a.** the circumference and **b.** the area.

**16.** The side of a cube has been measured to be 8 cm with a possible maximum error of 0.1 cm. Estimate the maximum relative and percentage errors in volume if that volume has been calculated using $s = 8$.

**17.** The diameter of a sphere has been measured to be 50 inches with a maximum possible error of 0.5 inch. If the volume has been calculated using $d = 50$, approximate the maximum relative and percentage errors in volume.

**18.** A hollow cube is made up of 3/4-inch plywood. Each side of the cube is 8 feet. Approximate the volume of plywood in two different ways. (One of the ways should use differentials.)

**19.** A standard racquetball is 2 inches in diameter. If an increase in temperature of 15 degrees causes an increase in radius of 0.0225 inch, use differentials to approximate the corresponding increase in volume.

**20.** The Veranzi Electronics Co. established that the production and sales of $q$ VCRs would produce a profit of $p(q)$ dollars where

$$p(q) = -0.02q^3 + 0.4q^2 + 1000q - 35,000$$

provided $0 \leq q \leq 130$. The manager of the company estimated that 100 VCRs would be produced and sold by the company in the following month and calculated the expected profit accordingly. If his estimate is correct up to a maximum error of 5%, approximate the maximum relative and percentage errors in profit.

**21.** The profit function in Exercise 33, Section 8.7 was defined by

$$P(q) = -0.00005q^3 - 0.3q^2 + 7500q - 4,500,000$$

where $0 \leq q \leq 9000$. The sales manager predicted that 6000 PCs would be manufactured and sold in the first year of production. Assuming that his estimate is correct up to a maximum error of 3%, approximate the maximum relative and percentage errors in profit.

**22.** The profit function found in Exercise 28, Section 8.7 was defined by the equation

$$p(q) = -2000q^3 - 2000q^2 + 450,000q - 20,000$$

where $q$ is the number, in thousands, of Walkmans produced and sold by the Reid Electronics Company, $p(q)$ is in dollars, and $0 \leq q \leq 14$. The sales manager predicted that 7000 Walkmans would be manufactured and sold during the coming year. Assuming that the prediction is accurate within a maximum error of 2%, approximate the maximum relative and percentage errors in profit.

**23.** The profit function found in Exercise 37, Section 8.7 was given by

$$P(q) = -0.000025q^3 - 0.25q^2 + 13,500q - 22,500,000$$

where $0 \leq q \leq 15,000$. The president of the Mason Boat and Waterski Company predicted that 10,000 boats would be manufactured and sold by her company during the coming year. If her prediction is correct within a maximum error of 4%, approximate the maximum percentage error in profit.

# 8.11 Related Rates (Optional)

We have seen that one of the many interpretations of the derivative of a function is the instantaneous rate of change of that function with respect to its independent variable. In particular, if the independent variable is time, we use derivatives to find the velocity and acceleration of a moving object (see Example 4, Section 7.11). There are many cases where two or more dependent variables are related and all have the same independent variable. This independent variable often, but not always, is time. For example, in the process of inflating a ball, the radius $r$ and the volume $V$ are related by the formula

$$V = \frac{4}{3}\pi r^3$$

and each variable changes with time. As the process of inflating the ball goes on, both the radius and volume are increasing. The rate of change of one of the variables may be known, and we are asked to find the other.

In this type of application, use the procedure described below.

---

### Procedure for Solving Related Rates Problems

**STEP 1.**   In the problem at hand, identify among all variables, those whose rates of change are given and the one whose rate of change is to be found.

**STEP 2.**   Assign a symbol to each variable. It is common practice to use the first letter of the name of the variable; for example, $V$ for volume, $r$ for radius, and so on.

**STEP 3.**   Write an equation relating all the variables. Once this has been done, using the given information, known formulas, or any other legitimate means, eliminate all variables in the equation except those identified in Step 1. This step may not be needed if there are no extraneous variables.

**STEP 4.**   Differentiate both sides of the equation with respect to the independent variable, which often is time. Usually you need to use the Chain Rule or implicit differentiation (see Section 7.8).

**STEP 5.**   In the resulting equation, substitute the numerical data supplied in the problem.

**STEP 6.**   Solve the equation obtained in Step 5 to get the value of the rate of change you were asked to find.

---

 You should be careful to follow these steps in the correct order. A common mistake is to perform Step 5 before Step 4. A moment of reflection will show that this practice leads to the following difficulty. When a variable is replaced by a number, you get a constant. Then if you differentiate that constant, the result is 0.

**EXAMPLE 1**   A rock is thrown into a lake producing a circular ripple. The radius of the ripple is increasing at the rate of 10 ft/sec. How fast is the area inside the ripple increasing when the radius is 25 feet?

**Solution**   We are given the rate of change of the radius and we are asked to find the rate of change of the area.

Let $r$ feet be the radius and $A$ square feet be the area at time $t$. Thus

$$A = \pi r^2$$

Differentiating both sides of the equation with respect to $t$, we get

$$\frac{dA}{dt} = D_t(\pi r^2) = 2\pi r \frac{dr}{dt}$$

It is given that $\frac{dr}{dt} = 10$ and we are asked to find $\frac{dA}{dt}$ when $r = 25$. We substitute this data in the equation and get

$$\frac{dA}{dt} = 2\pi(25)(10) = 500\pi$$

The area is increasing at the rate of $500\pi$ square feet per second.
**Do Exercise 1.**                                                            ∎

**EXAMPLE 2**   A spherical balloon is being inflated at the rate of 5 cubic feet per second. How fast is the radius of the balloon increasing 20 seconds after the start?

**Solution**   We are given the rate of change of the volume and we are asked to find the rate of change of the radius. Let $r$ feet be the radius and $V$ cubic feet be the volume at instant $t$. Thus

$$V = \frac{4}{3}\pi r^3$$

Differentiating both sides of this equation with respect to $t$, we get

$$\frac{dV}{dt} = \frac{4}{3}\pi(3r^2)\frac{dr}{dt} = 4\pi r^2 \frac{dr}{dt}$$

It is given that $\frac{dV}{dt} = 5$. In order to use the preceding formula, we must know the value of $r$. Since the balloon is being inflated at the rate of 5 ft³/sec, in 20 seconds its volume is 100 cubic feet. Thus

$$100 = \frac{4}{3}\pi r^3$$

Therefore

$$r = \left(\frac{300}{4\pi}\right)^{1/3}$$

Substituting these values in the formula for $\dfrac{dV}{dt}$, we get

$$5 = 4\pi\left[\left(\frac{300}{4\pi}\right)^{1/3}\right]^2 \frac{dr}{dt} = (600^{2/3})(\pi^{1/3})\frac{dr}{dt}$$

Solving for $\dfrac{dr}{dt}$, we get

$$\frac{dr}{dt} = \frac{5}{(600^{2/3})(\pi^{1/3})} \doteq 0.048$$

The radius is increasing at the approximate rate of 0.048 ft/sec.
**Do Exercise 7.**                                                            ∎

EXAMPLE 3    The inside of a glass is shaped like an inverted right circular cone with a radius at the top of 2 inches and a height of 6 inches. The glass is set on a table and a child is sipping a milkshake out of it using a straw. When the depth of the milkshake left in the glass is 3 inches, the boy is sipping at the rate of 0.5 in.$^3$/sec. How fast is the depth of the milkshake decreasing at that time?

Solution    We are given the rate of change of volume and we are asked to find the rate of change of depth.

At any instant $t$, the shape of the milkshake left in the glass is that of an inverted right circular cone. Let $r$ inches be its radius, $h$ inches be its height, and $V$ cubic inches be its volume. Thus,

$$V = \frac{1}{3}\pi r^2 h$$

We must eliminate the extraneous variable $r$. Using the similar triangles in Figure 8.52, we see that

$$\frac{r}{2} = \frac{h}{6}$$

Therefore,

$$r = \frac{h}{3}$$

FIGURE 8.52

Substituting $\frac{h}{3}$ for $r$ in the formula for the volume of a cone, we obtain

$$V = \frac{1}{3}\pi\left(\frac{h}{3}\right)^2 h = \frac{\pi}{27}h^3$$

Differentiating both sides of this equation with respect to $t$, we get

$$\frac{dV}{dt} = \frac{\pi}{9}h^2\frac{dh}{dt}$$

The child is sipping the milkshake at the rate of 0.5 in.$^3$/sec. Since the volume is decreasing, we write $\frac{dV}{dt} = -0.5$. The given value of $h$ is 3. Substituting this data in the equation, we obtain

$$-0.5 = \frac{\pi}{9}(3^2)\frac{dh}{dt} = \pi\frac{dh}{dt}$$

Solving for $\frac{dh}{dt}$, we get

$$\frac{dh}{dt} = \frac{-0.5}{\pi} \doteq -0.159155$$

The fact that $\frac{dh}{dt}$ is negative is consistent with the obvious fact that the depth of the milkshake is decreasing. We conclude that when the depth of the milkshake is 3 inches, it is decreasing at the rate of approximately 0.16 in./sec.

**Do Exercise 15.**

EXAMPLE 4   Experience has shown a company that the cost of producing $q$ thousand pocket cal-culators is given by

$$C = C(q) = -0.25q^2 + 25q + 600$$

where $0 \leq q \leq 50$ and $C(q)$ is in thousands of dollars. When the level of production is 30,000 calculators, the production is increasing at the rate of 2000 per month. Find the corresponding rate of change in cost.

Solution   It is given that $\frac{dq}{dt} = 2$ (since $q$ is in thousands) and we are asked to find $\frac{dC}{dt}$. The relationship between $C$ and $q$ is given as

$$C = -0.25q^2 + 25q + 600$$

Differentiating both sides of this equation with respect to $t$, we obtain

$$\frac{dC}{dt} = (-0.5q + 25)\frac{dq}{dt}$$

Substituting in the given data ($q = 30$ and $\frac{dq}{dt} = 2$), we get

$$\frac{dC}{dt} = [(-0.5)(30) + 25](2) = 20$$

The cost of production is increasing at the rate of $20,000 per month.   ■
**Do Exercise 19.**


EXAMPLE 5   A particle is moving on the right half of a coordinate plane. Its position $(x, y)$ at instant $t$ is given by the equation

$$x^2y^2 - 5x = y^4 + 5$$

The abscissa is decreasing at the rate of 2 units per second when the ordinate is $-1$. How fast is the ordinate changing at that instant?

Solution   Since the abscissa is decreasing, we write $\frac{dx}{dt} = -2$. We are asked to find $\frac{dy}{dt}$. The relationship between $x$ and $y$ is given by

$$x^2y^2 - 5x = y^4 + 5$$

Differentiating both sides of this equation with respect to $t$, we get

$$x^2D_ty^2 + y^2D_tx^2 - 5\frac{dx}{dt} = 4y^3\frac{dy}{dt} + 0$$

Therefore,

$$x^2(2y)\frac{dy}{dt} + y^2(2x)\frac{dx}{dt} - 5\frac{dx}{dt} = 4y^3\frac{dy}{dt}$$

Equivalently,

$$(2xy^2 - 5)\frac{dx}{dt} = (4y^3 - 2x^2y)\frac{dy}{dt}$$

It is given that $\frac{dx}{dt} = -2$ and $y = -1$. We must find the value of $x$. We replace $y$ by $-1$ in the original equation and get

$$x^2(-1)^2 - 5x = (-1)^4 + 5$$

That is,

$$x^2 - 5x - 6 = 0$$

The solutions of this equation are $-1$ and $6$. Since the particle is moving in the right half-plane, $x$ must be positive and we choose $x = 6$. We now use all the data available in the equation relating $\frac{dx}{dt}$ and $\frac{dy}{dt}$ to get

$$[2(6)(-1)^2 - 5](-2) = [4(-1)^3 - 2(6)^2(-1)]\frac{dy}{dt}$$

Solving for $\frac{dy}{dt}$, we get

$$\frac{dy}{dt} = \frac{-7}{34}$$

Since $\frac{dy}{dt}$ is negative, we know that the ordinate is decreasing. We conclude that $y$ is decreasing at the rate of $\frac{7}{34}$ units per second.

**Do Exercise 17.**                                                               ■

---

## Exercise Set 8.11

**1.** A police officer who is positioned 400 feet north of an intersection notices a red sportscar traveling west through the intersection. He decides to take a radar reading and determines that the car is now 500 feet from him and that this distance is increasing at the rate of 66 ft/sec. What is the speed of the red car at that instant in miles per hour?

**2.** A motorist is 40 miles east of an intersection and is traveling east away from it at a speed of 45 mph. Another motorist is 30 miles south of that intersection and is traveling north toward it at a speed of 55 mph. At what rate is the distance between them changing?

**3.** A young child is flying a kite horizontally 120 feet above the ground. The child lets out 2.5 feet of string per second. If we assume that there is no sag in the string, at what speed is the kite moving when there is 130 feet of string out?

**4.** A 25-foot metal ladder is leaning against a vertical wall. The horizontal floor is slightly slippery and the bottom of the ladder slips away from the wall at the rate of 0.2 in./ sec. How fast is the top of the ladder sliding down against the wall when it is 20 feet high?

**5.** A paper cone of radius 3.5 inches and height 7 inches is being used as a filter to make coffee by pouring boiling water over coffee grounds which have been placed in the filter. If coffee filters through the pointed part of the cone at the constant rate of 0.75 in³/sec, how fast is a bubble on the surface of the water falling when it is 4 inches from the pointed part of the filter?

**6.** A bag of sand that is being lifted vertically has a small hole in it and sand is falling out on the ground to form a conical pile whose height is half its radius. If the volume of the pile is increasing at the rate of 2 in³/sec, how fast is the radius increasing when the volume of the pile is $36\pi$ in³?

**7.** A ball is batted along the first base line at a constant speed of 80 mph. How fast is its distance from third base increasing when it is 120 feet from home plate? (*Hint:* A baseball diamond is a square with sides 90 feet long.)

**8.** On a dark night, a thin man 6 feet tall walks away from a lamp post 24 feet high at the rate of 5 mph. How fast is the end of his shadow moving?

**9.** In Exercise 8, how fast is the shadow lengthening?

**10.** A light stands 40 feet from a house and 30 feet from the straight path leading from the street to the house. A man walks on the path toward the house at a speed of 4 feet per second. How fast is his shadow moving on the wall when he is 10 feet from the house?

11. A kite moves horizontally 120 feet high at a rate of 4 mph directly away from the young girl flying it. How fast is the girl letting out the string when 130 feet of string is out? Assume there is no sag in the string.

12. A train crosses a street on a track 40 feet above the ground at a speed of 25 feet per second. At that instant, a car 80 feet up the street is approaching at a speed of 30 feet per second. Find the rate of change of the distance between the car and the train 2 seconds later.

13. A balloon has a small leak and loses air at the constant rate of 3 ft³/min. How fast is the radius of the balloon decreasing when it is 20 feet long?

14. A tank is in the shape of a right circular cone, vertex down. Its radius is 5 feet and its height is 15 feet. Water is running into the tank at the rate of 1.5 ft³/min. A champagne cork floats on the surface of the water. How fast is the cork rising when the water is 10 feet deep?

15. Using the same data as in Exercise 14, but with the vertex of the tank up, answer the same question.

16. A particle is moving in a coordinate plane on the circle whose equation is $x^2 + y^2 = 2500$. If the value of $x$ is decreasing at the rate of 4 units per second when the value of $y$ is 30 feet and the particle is in the second quadrant, how fast is the value of $y$ changing at that instant?

17. A particle is moving on the graph of the equation $y = x^4 + 2x^3 - 5x^2 - 85$. If the abscissa is increasing at the constant rate of 2 units per second, how fast is the value of $y$ increasing, or decreasing, when $x = 3$?

18. A rock is dropped from the top of a cliff overlooking a river. Five seconds later, another rock is dropped from the same location. How fast is the distance between the two rocks increasing when they are 720 feet apart? (*Hint:* The distance function for a free-falling object is defined by $d = 16t^2$ where $d$ is in feet and $t$ is in seconds.)

19. The total cost $C(q)$, in thousands of dollars, to manufacture $q$ thousand radios is given by the equation $C(q) = 4 +$

$21q - 2q^2$, provided $0 \le q \le 5$. When the level of production is 3000 radios, production is increasing at the rate of 50 radios per month. Find the corresponding rate of change in cost.

20. The total cost $C(q)$, in thousands of dollars, to manufacture $q$ thousand VCRs is given by the equation $C(q) = 100 + 200q - q^3$, provided $0 \le q \le 8$. When the level of production is 5000 VCRs, production is increasing at the rate of 50 VCRs per week. Find the corresponding rate of increase in cost.

21. Referring to Exercise 1, Section 8.1, suppose that when the production level is 225 motorcycles, the production is decreasing at the rate of 10 motorcycles per month. Find the corresponding rates of change in **a.** cost, **b.** revenue, and **c.** profit.

22. Suppose that the owner of the small winery of Exercise 2, Section 8.1, decides to increase production by 6 bottles per day when the level of production is 64 bottles. Find the corresponding rates of change in **a.** cost, **b.** revenue, and **c.** profit.

23. Referring to Exercise 3, Section 8.1, suppose that when the production level is 30 pocket calculators, production is increased at the rate of 3 calculators per day. Find the corresponding rates of change in **a.** cost, **b.** revenue, and **c.** profit.

24. If the Triplenight Company of Exercise 5, Section 8.1, increases production at the rate of 50 racquets per day when the level of production reaches 1600 racquets, what are the corresponding rates of change in **a.** cost, **b.** revenue, and **c.** profit.

25. Suppose that in Exercise 33, Section 8.7, the demand is increasing at the rate of 100 personal computers per year when the level of demand is 6000. At what rate is the revenue changing assuming that the Nelson Electronics Co. will adjust its price to the demand?

# 8.12 Chapter Review

| IMPORTANT SYMBOLS AND TERMS | | | |
|---|---|---|---|
| | $dx$ [8.10] | $\underset{x \to a^+}{\text{Lim}}\, f(x) = -\infty$ [8.8] | Absolute minimum [8.3] |
| | $dy$ [8.10] | $\underset{x \to a^+}{\text{Lim}}\, f(x) = \infty$ [8.8] | Asymptote [8.9] |
| | $\underset{x \to \infty}{\text{Lim}}\, f(x) = L$ [8.8] | $\underset{x \to a^-}{\text{Lim}}\, f(x) = -\infty$ [8.8] | Concave down [8.6] |
| | $\underset{x \to -\infty}{\text{Lim}}\, f(x) = L$ [8.8] | Absolute extrema [8.3] | Concave up [8.6] |
| | $\underset{x \to a^+}{\text{Lim}}\, f(x) = \infty$ [8.8] | Absolute maximum [8.3] | Critical point [8.5] |
| | | | Critical value [8.5] |

## SUMMARY

In economics, "marginal" translates to "derivative." The value of the marginal cost function at the integer $n$, where $n \geq 1$, approximates the cost of the $n$th (or $(n + 1)$st) unit of the product. Similar statements can be made about the marginal revenue and marginal profit functions. If $f$ is a differentiable function and $y = f(x)$, the elasticity of $y$ with respect to $x$ is defined by

$$\eta = \frac{x}{y} \cdot \frac{dy}{dx}$$

When the increment in $x$ is small, $\eta$ approximates the ratio between the percentage changes in $y$ and $x$, respectively. Usually, the graph of the demand equation for a commodity has negative slope, so $\frac{dq}{dp}$ is negative. In this case, $\eta$ is a negative number. If $\eta < -1$, a percentage increase in price results in a greater percentage decrease in demand; the demand is elastic. If $\eta = -1$, then a percentage increase (decrease) in price will cause an approximately equal percentage decrease (increase) in demand; we have unit elasticity. If $-1 < \eta < 0$, then a percentage increase (decrease) in price will create a smaller percentage decrease (increase) in demand; the demand is inelastic.

If a function $f$ is differentiable on an interval $I$, then

**a.** $f'(x) > 0$ for all $x$ in $I$ implies that $f$ is strictly increasing on $I$.
**b.** $f'(x) < 0$ for all $x$ in $I$ implies that $f$ is strictly decreasing on $I$.
**c.** $f'(x) = 0$ for all $x$ in $I$ implies $f$ is constant on $I$.
**d.** $f''(x) > 0$ for all $x$ in $I$ implies the graph of $f$ is concave up over $I$.
**e.** $f''(x) < 0$ for all $x$ in $I$ implies the graph of $f$ is concave down over $I$.

If $f$ is continuous on an interval containing $c$, and $f''(x)$ changes sign as $x$ increases through $c$, then $(c, f(c))$ is a point of inflection of the graph of $f$.

If a function $f$ is defined on some interval $I$ and for some number $c$, which is not an endpoint of that interval, either $f'(c) = 0$ or $f'(c)$ is not defined, then $c$ is called a critical value of $f$. If $f$ has a local extremum at $c$, then either $c$ is an endpoint of the interval or $c$ is a critical value of the function.

Let $f$ be a continuous function on some interval $I$. To locate the relative extrema of $f$ we proceed as follows. First find all critical values of $f$ in $I$. If $c$ is a critical value of $f$, we may use one of several methods to determine whether or not $f$ has an extremum at $c$.

*First Derivative Test:* Consider a neighborhood $(c - h, c + h)$ of $c$ such that $c$ is the only critical value of $f$ in that interval.

1. If $f'(x) > 0$ on $(c - h, c)$ and $f'(x) < 0$ on $(c, c + h)$, then $f$ has a local maximum at $c$.
2. If $f'(x) < 0$ on $(c - h, c)$ and $f'(x) > 0$ on $(c, c + h)$, then $f$ has a local minimum at $c$.
3. If $f'(x)$ has the same sign on $(c - h, c)$ and $(c, c + h)$, then $f$ has neither a local maximum nor a local minimum at $c$.

*Second Derivative Test:* Evaluate $f''(c)$.

1. If $f''(c) < 0$, the function has a local maximum at $c$.
2. If $f''(c) > 0$, the function has a local minimum at $c$.
3. If $f''(c) = 0$, the test is inconclusive.

You may also do the following: If $c$ is the only critical value of $f$ on the interval $(c - h, c + h)$, evaluate $f(c - h)$, $f(c)$, and $f(c + h)$.

1. If $f(c)$ is between $f(c - h)$ and $f(c + h)$, then $f$ has neither a maximum nor a minimum at $c$.
2. If $f(c) < \min\{f(c - h), f(c + h)\}$, then $f$ has a local minimum at $c$.
3. If $f(c) > \max\{f(c - h), f(c + h)\}$, then $f$ has a local maximum at $c$.

If the domain of definition of $f$ has endpoints, you must also check these for extrema.

If $f$ is continuous on a closed bounded interval $[a, b]$, then it has both an absolute maximum and an absolute minimum. To find these, first find all critical values of $f$ in the open interval $(a, b)$. Then evaluate $f$ at each of these and at the endpoints $a$ and $b$. Among all values you obtained, the smallest is the absolute minimum of $f$ on $[a, b]$ and the largest is the absolute maximum of $f$ on $[a, b]$. If the domain of $f$ is not a bounded interval, $f$ may or may not have an absolute maximum or minimum. The correct conclusion can be drawn from geometric considerations.

In the applications, we generally wish to maximize or minimize a certain quantity. The general procedure is to let the dependent variable, say $y$, represent the quantity which is to be maximized (minimized), and to let the independent variable, say $x$, represent the quantity to be found. Using the information in the problem, we derive an expression for $y = f(x)$, making sure to specify the domain of $f$. Then we use any of the techniques to find the absolute extremum of $f$ on its domain.

If you wish to sketch the graph of $y = f(x)$, use the first derivative to determine the intervals over which the graph is rising (falling). You can also locate the relative extrema of the function. Using the second derivative, find the intervals over which the graph is concave up (down). Also find the points of inflection. Check for symmetry, find all asymptotes, and the $x$ and $y$-intercepts, before sketching the graph as outlined in the summary at the end of Section 8.9.

If a function $f$ is differentiable at $x$, its differential is denoted $dy$ and is defined by

$$dy = f'(x)dx$$

where $dx$ is any real number and is called the differential of $x$. If $dx = \Delta x$ and $\Delta x$ is small, then $dy \doteq \Delta y$. This fact may be used if the value of a function $f$ at $c$ is to be approximated and its value at $a$ is known. We let $dx = \Delta x = c - a$, and we calculate $dy = f'(a)dx$. Then we use the fact that $f(c) \doteq f(a) + dy$. The differential can also be used to estimate errors in measurement.

In certain problems, there are two or more dependent variables related by some equation and having a common independent variable. Some instantaneous rate(s) of change with respect to the independent variable is known and an unknown rate of change must be found. Write an equation relating the dependent variables. Differentiate both sides of the equation with respect to the independent variable. In the resulting equation, substitute the numerical data supplied in the problem. Solve the equation obtained to get the value of the rate of change you were asked to find.

1. A company has found that the cost $C(q)$ (in dollars) to manufacture $q$ units of its product is given by

$$C(q) = 20,000 + 60 \sqrt{q} + 50$$

a. Find the marginal cost.
b. Use your result of part a to estimate the cost of manufacturing the 175th unit.

2. A company has found that if it prices its product at $18 per unit, 10,000 units will be produced and sold; and if the price is $12 per unit, 40,000 will be produced and sold.

a. Derive the demand equation $p = D(q)$, assuming it is linear.
b. Find the revenue $R(q)$ (in dollars) in terms of the quantity $q$ units produced and sold.
c. Find the marginal revenue.
d. What is the approximate revenue derived from the 12,000th unit?
e. How many units should be produced and sold in order to maximize the revenue?

3. A company found that the cost $C(q)$ (in dollars) of manufacturing and selling $q$ units per month of a commodity is given by

$$C(q) = -q^2 + 980q + 20,000$$

where $0 \le q \le 490$ since a maximum of 490 units can be produced and sold. It is also known that 50 units can be sold per month at a price of $1140 per unit, and 200 units per month can be sold if the price is $960 per unit. Let $p = D(q)$ be the demand equation.

a. Assuming that the demand equation is linear, find $D(q)$.
b. Find the revenue function.
c. Find the profit function.
d. At what level of production will the company start making a profit?
e. Find the marginal profit.
f. Approximate the profit derived from the 300th unit produced and sold.
g. At what level of production per month will the profit be maximum?
h. What is the average cost per unit if $q$ units are produced per month?
i. What is the marginal average cost?

4. The demand equation for a commodity is

$$q = -p^2 - 25p + 7500$$

provided that $0 \le p \le 75$. In this equation, $q$ is the number of units the consumers are willing to buy when the price is $p$ dollars per unit.

a. Find the elasticity of demand when $p = 40$.
b. Find the percentage change in price when the price increases from $40 to $43.
c. Find the corresponding percentage change in demand.
d. Calculate the value of the ratio between your answers in parts b and c.
e. Compare the answer in part d to that of part a.

5. For the demand equation of Exercise 4, determine whether the demand is elastic, inelastic, or has unit elasticity when a. $p = 20$ and b. $p = 50$.

6. Determine the intervals over which the given functions are increasing, decreasing.

a. $f(x) = x^3 + 6x^2 + 9x - 10$
b. $g(x) = (x^2 + 3x - 9)e^{-x}$
c. $h(x) = (x + 1)\ln(x + 1)$

7. Determine over which intervals the graphs of the given functions are concave up or concave down. Also find all points of inflection.

    a. $f(x) = 2x^3 + 9x^2 - 60x + 13$
    b. $g(x) = (x^2 + 5x + 6)e^{-x}$
    c. $h(x) = 6x^{1/3}$

8. Find all critical values of each of the following functions.

    a. $f(x) = x^3 - 15x^2 + 63x - 7$
    b. $g(x) = \dfrac{2x + 1}{x - 1}$
    c. $h(x) = (x - 3)^{1/5}$
    d. $p(x) = (x - 3)^{-1/5}$

9. Let $f(x) = x^3 - 3x^2 - 45x + 10$. Using the second derivative test, locate all relative extrema of the function $f$.

10. let $f(x) = (2x^2 - 10x + 5)e^{-2x}$. Using the first derivative test, locate all relative extrema of the function $f$.

11. Let $f(x) = x^3 + 9x^2 - 21x + 5$. Find the absolute maximum and absolute minimum of $f$ on the interval $[0, 3]$.

12. Let $g(x) = (x - 8)^{2/3}$. Find the absolute maximum and absolute minimum of $g$ on the interval $[7, 16]$.

13. Suppose that a function $f$ is twice differentiable. The graph of its first derivative is shown in the accompanying figure.

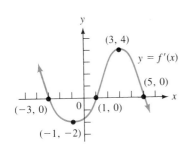

    a. Identify the intervals over which the function $f$ is increasing.
    b. Identify the intervals over which the function $f$ is decreasing.
    c. Identify the intervals over which the graph of the function $f$ is concave down.
    d. Identify the intervals over which the graph of the function $f$ is concave up.
    e. Find the abscissa of each point of the graph where the tangent to the graph is parallel to the $x$-axis.
    f. Find the abscissa of each point of inflection of the graph of the function $f$.

14. Let $f$ be defined on the interval $[1, 5]$ by the equation $f(x) = x^2 + x - 3$. Find the point on the graph of the function $f$ where the tangent line is parallel to the line passing through the points $(1, f(1))$ and $(5, f(5))$. What is the equation of that tangent line?

15. What is $dy$ if

    a. $y = x^2 + 5x - 6$
    b. $y = t^2 e^{3t}$
    c. $y = \dfrac{\ln(x^4 + 3)}{x^4 + 3}$

16. Let $y = 3x^2 - 5x + 6$. Suppose $x = -2$ and $\Delta x = dx = 0.2$.

    a. Find the exact value of $\Delta y$.
    b. Find the value of $dy$.
    c. Compare your two answers.

17. Using differentials, approximate $\sqrt{623}$. (*Hint:* $\sqrt{625} = 25$.)

18. The diameter of a sphere has been measured to be 60 centimeters with a maximum possible error of $\pm 0.3$ cms. If the volume has been calculated using 60 for the diameter, approximate

the relative and percentage errors in volume. (*Hint:* The volume $V$ of a sphere is given by the formula $V = \frac{4}{3}\pi r^3$.)

19. A police officer located 500 feet south of an intersection notices a sports car traveling east through the intersection. He decides to take a radar reading and determines that the car is now 1300 feet from him and that this distance is increasing at the rate of 96 ft/sec. What is the speed of the sports car in miles per hour at that instant?

20. Let $y = f(x) = x^3 + 3x^2 - 9x - 14$.

    **a.** Determine the intervals over which the graph of $f$ is rising, falling.
    **b.** Determine the intervals over which the graph of $f$ is concave up, concave down.
    **c.** Find the points of inflection.
    **d.** Find the relative extrema of $f$.
    **e.** Sketch the graph of $f$.

21. Same as Exercise 20 if $y = f(x) = (4x^2 - 2x + 2)e^{-x}$.

22. Let $y = \frac{3x - 6}{x + 2}$. Find all asymptotes, the $x$- and $y$-intercepts of the graph of this equation, and sketch the graph.

23. Same as Exercise 22 if $y = \dfrac{3(x - 4)(x - 7)}{(x - 5)(3x - 13)}$.

24. Same as Exercise 22 if $y = \dfrac{4x^2 + 5x - 26}{x + 3}$.

25. Same as Exercise 22 if $y = \dfrac{4x^2 + 2}{x^2 - 4}$.

26. Sketch the graph of $y = xe^{-5x}$.

27. Sketch the graph of $y = (x + 4)\ln(x + 4)$.

28. A fence is to be built to enclose a rectangular region of 5600 square meters. The fencing along three sides will cost $16 per meter, while the fencing along the fourth side will cost $12 a meter. What dimensions are required for the fencing to be the most economical?

29. A company has found that if it sells each of the cameras it produces for $p$ dollars, the number that can be produced and sold in any given month is $q$ thousands where

$$p = q^2 - 36q + 432$$

The maximum number that the company can produce and sell each month is 18,000. Find the revenue $R(q)$ (in dollars) per month. How many cameras should be produced and sold each month to maximize the revenue?

30. Find the point on the graph of the equation $y = x^2 + 3x + 2$ that is closest to the point $(9, 11)$. If your answer is the point $(a, b)$, verify that the tangent line to the graph of that equation at the point $(a, b)$ is perpendicular to the line that passes through $(9, 11)$ and $(a, b)$.

31. The number of credit hours taught in a year at a small university was 225,000 when tuition was $150 per credit hour. The president of the university decided that the revenue for the following year must be increased by increasing the tuition. She was told by the finance committee that an increase of tuition of $x$ dollars per credit hour would cause a reduction of $98x^2 + 4794$ in the number of credit hours taught. How much should tuition be increased per credit hour to maximize the revenue for the following year?

# Integration

**9**

We began the study of calculus by introducing derivatives and some applications of differentiation. This is a branch of calculus called *differential calculus*. We now introduce another part of calculus called *integral calculus*.

# 9.1 Antidifferentiation

## Antiderivatives

We saw earlier that if the position function of a particle moving along a directed line is known, then we can find the instantaneous velocity function of that particle by differentiation. There may be cases where the velocity function is known and we wish to find the position function. Therefore the problem is to find a function given its derivative. This process is called *antidifferentiation*. Students who have mastered the technique of finding derivatives can perform elementary antidifferentiation in the same manner as students who have mastered the operation of multiplication can perform simple division. We formally give the following definition.

> **DEFINITION 9.1:**   If $f$ and $g$ are two functions defined on an interval $I$ and $f'(x) = g(x)$ for every $x$ in $I$, then $f$ is said to be an *antiderivative* of $g$.

Note that we used the indefinite article "an" rather than the definite article "the," which suggests that there is more than one antiderivative. We will show that if a function $g$ has an antiderivative, then it has infinitely many antiderivatives; any two antiderivatives of the same function differ at most by a constant.

**EXAMPLE 1**    Let $f(x) = 6e^{3x}$, $g(x) = 2e^{3x}$, $h(x) = 2e^{3x} + 5$, and . Show that the functions $g$, $h$, and $i$ are antiderivatives of $f$.

**Solution**    Differentiating $g$, we get

$$g'(x) = D_x 2e^{3x} = 2D_x e^{3x} = 2e^{3x}(3) = 6e^{3x} = f(x)$$

Thus, $g$ is an antiderivative of $f$.
  Since $h(x) = g(x) + 5$,

$$h'(x) = D_x(g(x) + 5) = g'(x) + D_x 5 = f(x) + 0 = f(x)$$

We conclude that $h$ is also an antiderivative of $f$.
  Similarly,

$$i(x) = g(x) + \frac{\pi}{6}$$

Thus

$$i'(x) = D_x\left(g(x) + \frac{\pi}{6}\right) = g'(x) + D_x\left(\frac{\pi}{6}\right) = f(x) + 0 = f(x)$$

It follows that $i$ is an antiderivative of $f$. At this point, it should be clear that if $C$ is any constant and $j(x) = 2e^{3x} + C$, then $j$ is an antiderivative of $f$ also. Thus, $f$ has infinitely many antiderivatives.

**Do Exercise 1.**

## Indefinite Integrals

*Notation*: We use the symbol $\int f(x)\, dx$ to represent all antiderivatives of $f(x)$. We write

$$\int f(x)\, dx = g(x) + C \text{ if, and only if, } g'(x) = f(x)$$

This symbol is called an *indefinite integral*. Also, $f(x)$, $\int$, and $C$, are called the *integrand*, *integral sign*, and *constant of integration*, respectively. Note that the symbol $\int \ldots dx$ indicates antidifferentiation with respect to $x$, just as $D_x$ indicates differentiation with respect to $x$. Observe that the function to be antidifferentiated is placed between the integral sign $\int$ and the $dx$.

For instance, the result of Example 1 can be written

$$\int 6e^{3x}\, dx = 2e^{3x} + C$$

It is essential to write the constant of integration $C$ when performing antidifferentiation. For example, it is wrong to write

$$\int 6e^{3x}\, dx = 2e^{3x}$$

because the left side represents *all* antiderivatives of $6e^{3x}$, while the right side represents only one of them.

## Finding Antiderivatives by Guessing

Later in this chapter, we shall develop techniques and formulas commonly used when performing antidifferentiation. In many cases, however, antiderivatives can be readily obtained using trial and error. It is important for you to gain some proficiency using this approach. We first recall the basic differentiation formulas. In the following, $k$, $r$, and $b$ are constants with $b > 0$ and $b \neq 1$; $u$ and $v$ are differentiable functions of $x$.

1. $D_x k = 0$
2. $D_x(u + v) = D_x u + D_x v$
3. $D_x(u - v) = D_x u - D_x v$
4. $D_x(ku) = kD_x u$
5. $D_x x = 1$
6. $D_x u^r = ru^{r-1}D_x u$
7. $D_x(uv) = uD_x v + vD_x u$
8. $D_x \dfrac{u}{v} = \dfrac{vD_x u - uD_x v}{v^2}$

9. $D_x e^u = e^u D_x u$

10. $D_x b^u = b^u (\ln b) D_x u$

11. $D_x \ln|u| = \dfrac{D_x u}{u}$

When confronted with a problem of antidifferentiation, we shall try to identify one of the formulas by looking at the right side of the formula. For example, $D_x e^u = e^u D_x u$ tells us not only that $e^u D_x u$ is the derivative of $e^u$, but also that $e^u$ is an antiderivative of $e^u D_x u$. Using the differentiation formulas, we can often guess an antiderivative of a given function, up to a constant factor. We can then determine the value of the constant factor that will make our guess the correct answer. We illustrate this method with some examples.

**EXAMPLE 2**   Find $\displaystyle\int 5x^3 \, dx$.

**Solution**   Considering the right side of Formula 6, we guess that an antiderivative of $5x^3$ could be $kx^4$, where $k$ is a constant. This guess will be correct if

$$D_x kx^4 = 5x^3$$

We find that $D_x kx^4 = 4kx^3$. Thus we must have

$$4kx^3 = 5x^3$$

It follows that $4k = 5$ and $k = \frac{5}{4}$. Therefore an antiderivative of $5x^3$ is $5x^4/4$. So $5x^4/4 + C$, where $C$ is any constant, represents all antiderivatives of $5x^3$, and we write

$$\int 5x^3 \, dx = \frac{5x^4}{4} + C$$

**Do Exercise 13.**

**EXAMPLE 3**   Find $\displaystyle\int e^{x^2} x \, dx$.

**Solution**   Considering the right side of Formula 9, because it is the only one that contains a power of $e$ and $e^{x^2}x$ resembles $e^u D_x u$, we guess that an antiderivative of $e^{x^2}x$ could be $ke^{x^2}$, where $k$ is a constant. Our guess will be correct provided that

$$D_x ke^{x^2} = e^{x^2}x$$

We find that $D_x ke^{x^2} = 2ke^{x^2}x$. Thus, we must have

$$2ke^{x^2}x = e^{x^2}x$$

It follows that $2k = 1$ and $k = \frac{1}{2}$. Therefore, an antiderivative of $e^{x^2}x$ is $\frac{1}{2}e^{x2}$. Hence, $\frac{1}{2}e^{x^2} + C$, where $C$ is any constant, represents all antiderivatives of $e^{x^2}x$, and we write

$$\int e^{x^2}x \, dx = \frac{1}{2}e^{x^2} + C$$

**Do Exercise 19.**

With a little practice, you should be able to find antiderivatives by trial and error without going through the steps of writing a constant factor whose value is to be found. We illustrate with the following:

**EXAMPLE 4**  Find $\displaystyle\int 2(x^3 + 4)^{10}x^2 \, dx$.

**Solution**  We consider the right side of Formula 6 because $(x^3 + 4)^{10}$ represents a function to a constant power, and Formula 6 is the only formula involving a function to a constant power. We see that $2(x^3 + 4)^{10}x^2$ resembles $ru^{r-1}D_xu$, where $r - 1 = 10$ and $u = x^3 + 4$. We guess that an antiderivative might be $(x^3 + 4)^{11}$. Remember, this is only a guess! Now,

$$D_x(x^3 + 4)^{11} = 11(x^3 + 4)^{10}(3x^2) = 33(x^3 + 4)^{10}x^2$$

We do not want the derivative to be $33(x^3 + 4)^{10}x^2$; we want it to be $2(x^3 + 4)^{10}x^2$. Thus, we divide our guess by 33 (to cancel the unwanted factor) and multiply it by 2 (to introduce the wanted factor). Thus,

$$\int 2(x^3 + 4)^{10}x^2 \, dx = \frac{2}{33}(x^3 + 4)^{11} + C$$

where the constant $C$ is added to get all antiderivatives of $2(x^3 + 4)^{10}x^2$. **Do Exercise 25.** ■

You must be careful when using trial and error to multiply and divide your trial solutions by *constants* only. (See Exercise 46, Section 9.2.)

## Applications

**EXAMPLE 5**  A particle is moving along a directed line. Its velocity at instant $t$ seconds is $v(t)$ m/sec, where $v(t)$ is given by

$$v(t) = \frac{t}{t^2 + 1}$$

At instant 2 seconds, the particle is $-3$ meters from the origin. Find its position function.

**Solution**  Let $s$ be the position function. We know that

$$s'(t) = v(t) = \frac{t}{t^2 + 1}$$

To find $s(t)$, we antidifferentiate with respect to $t$. That is, we must find

$$\int \frac{t}{t^2 + 1} \, dt$$

Since $D_t(t^2 + 1) = 2t$, we see that $t/(t^2 + 1)$ resembles $D_xu/u$, the right side of Formula 11. Thus, we guess that an antiderivative of $t/(t^2 + 1)$ could be $\ln|t^2 + 1|$. But,

$$D_t \ln|t^2 + 1| = \frac{D_t(t^2 + 1)}{t^2 + 1} = \frac{2t}{t^2 + 1}$$

Since the integrand was $t/(t^2 + 1)$, we do not want the factor 2. We cancel it by dividing our guess by 2. We conclude that

$$\int \frac{t}{t^2 + 1}\, dt = \frac{1}{2}\ln|t^2 + 1| + C$$

where $C$ is any constant. Thus

$$s(t) = \frac{1}{2}\ln(t^2 + 1) + C$$

for some value of $C$. To find the value of $C$, we use the fact that $s(2) = -3$. Calculating the value of $s(2)$, we obtain

$$s(2) = \frac{1}{2}\ln(2^2 + 1) + C = \frac{1}{2}\ln(5) + C = \ln\sqrt{5} + C$$

Therefore

$$-3 = \ln\sqrt{5} + C$$

and

$$C = -3 - \ln\sqrt{5}$$

Thus the position function of the particle is

$$s(t) = \frac{1}{2}\ln(t^2 + 1) - 3 - \ln\sqrt{5}$$

where $t$ is in seconds and $s(t)$ is in meters.
**Do Exercise 33.**

■

**EXAMPLE 6**   A firm knows that the marginal cost function at a production level of $q$ units of a commodity is given by

$$M_c(q) = q^2 - 2q + 10$$

If the fixed cost is $35,000, find the cost function.

**Solution**   By definition, the marginal cost function is the derivative of the cost function. Thus,

$$C'(q) = M_c(q) = q^2 - 2q + 10$$

To find $C(q)$, we must antidifferentiate $q^2 - 2q + 10$. We have

$$C(q) = \int(q^2 - 2q + 10)dq = \frac{1}{3}q^3 - q^2 + 10q + K$$

where $K$ is some constant. The right side of this equality was obtained by first guessing an antiderivative, $q^3 - q^2 + q + K$, and finding the coefficients to change the guess to a correct antiderivative, as was done in Example 5. The fixed cost is $35,000. Therefore the cost at a production level of 0 units is $35,000, and we get

$$C(0) = \frac{1}{3}(0)^3 - 0^2 + 10(0) + K = 35,000$$

We conclude that $K = 35,000$. The cost function is

$$C(q) = \frac{1}{3}q^3 - q^2 + 10q + 35,000$$

**Do Exercise 41.**

## A Basic Theorem

We conclude this section by justifying a claim we made earlier.

> **THEOREM 9.1:**  Suppose $f$ and $g$ are two antiderivatives of the same function $h$ on some interval $I$. Then there is a constant $C$ such that
>
> $$f(x) - g(x) = C$$
>
> for every $x$ in $I$.

**Proof:** Let $F(x) = f(x) - g(x)$. Then

$$F'(x) = f'(x) - g'(x) = h(x) - h(x) = 0$$

for every $x$ in $I$. By Theorem 8.1, Section 8.3, we must have a constant $C$ such that $F(x) = C$ for every $x$ in $I$. Thus $f(x) - g(x) = C$.

We have seen that if a function $f$ is any antiderivative of a function $h$, then the function $g$ defined by $g(x) = f(x) + C$, where $C$ is any constant, is also an antiderivative of $h$. We just proved that if $f$ and $g$ are any two antiderivatives of a function $h$, then $f(x) = g(x) + C$. We conclude that if $f$ is any antiderivative of $h$, then $f(x) + C$ (where $C$ is any constant) represents all antiderivatives of $h$.

---

# Exercise Set 9.1

*In Exercises 1–11, verify that the function g is an antiderivative of the function f.*

**1.** $f(x) = 6x^2 + 4x - 3$; $g(x) = 2x^3 + 2x^2 - 3x + 5$

**2.** $f(x) = \sqrt{x}$; $g(x) = \frac{2}{3}\sqrt{x^3} + 10$

**3.** $f(x) = 5x(x^2 + 1)^{15}$; $g(x) = \frac{5}{32}(x^2 + 1)^{16} + 4$

**4.** $f(x) = x^4 \sqrt[3]{x^5 + 1}$; $g(x) = \frac{3}{20}\sqrt[3]{(x^5 + 1)^4} + e^5$

**5.** $f(x) = 3e^{5x} + 2x + 3$; $g(x) = \frac{3}{5}e^{5x} + x^2 + 3x + 5$

**6.** $f(x) = 5x^2e^{x^3}$; $g(x) = \frac{5}{3}e^{x^3} + 7$

**7.** $f(x) = \frac{x^2}{x^3 + 1}$; $g(x) = \frac{1}{3}\ln|x^3 + 1| + 15$

**8.** $f(x) = x3^{x^2}$; $g(x) = \frac{3^{x^2}}{2\ln 3} + \pi$

**9.** $f(x) = \frac{e^x(x - 1)}{x^2}$; $g(x) = \frac{e^x}{x} + 18$

**10.** $f(x) = xe^x$; $g(x) = e^x(x - 1) + 13$.

**11.** $f(x) = \ln|x|$; $g(x) = x(\ln|x| - 1) - 3$

*In Exercises 12–21, find the antiderivatives using the technique illustrated in Examples 2 and 3.*

**12.** $\displaystyle\int 3x^5 \, dx$

**13.** $\displaystyle\int (2x^3 + 5) \, dx$

**14.** $\displaystyle\int 3e^{10x} \, dx$

**15.** $\displaystyle\int \frac{t^3}{t^4 + 1} \, dt$

**16.** $\int e^{-2u} \, du$

**17.** $\int 3(x^4 + 1)^{50} x^3 \, dx$

**18.** $\int x^2 \sqrt{x^3 + 1} \, dx$

**19.** $\int 7s^3 e^{s^4} \, ds$

**20.** $\int \left( \dfrac{2}{\sqrt{x}} + e^{2x} \right)$

**21.** $\int \dfrac{x^2 + 1}{x^3 + 3x + 1} \, dx$

*In Exercises 22–30, find the antiderivatives using the technique illustrated in Example 4.*

**22.** $\int (3x^4 + 5x^3 + 10x^2 + 6) \, dx$

**23.** $\int (2t^2 \sqrt{t} + 6t + 1) \, dt$

**24.** $\int (xe^{3x^2} + 5x + 1) \, dx$

**25.** $\int (x^2 + 1)^{7/3} x \, dx$

**26.** $\int \dfrac{2x^2}{\sqrt{x^3 + 1}} \, dx$ (Hint: Try $(x^3 + 1)^{1/2}$.)

**27.** $\int \dfrac{5x^3}{x^4 + 1} \, dx$ (Hint: Try $\ln(x^4 + 1)$.)

**28.** $\int 3u 8^{u^2} \, du$ (Hint: Try $8^{u^2}$.)

**29.** $\int e^{t^2 + 2t + 5}(t + 1) \, dt$

**30.** $\int \dfrac{3xe^{3x} - e^{3x}}{4x^2} \, dx$ $\left( \text{Hint: Try } \dfrac{e^{3x}}{2x}. \right)$

*In Exercises 31–35, the velocity function of a particle moving along a directed line is given. The value of the position function at a certain instant is also given. Derive the formula for the position function of the particle. Time is measured in seconds and distance in feet.*

**31.** $v(t) = 32t + 15$; $s(0) = 20$

**32.** $v(t) = -48t + 5$; $s(2) = 15$

**33.** $v(t) = 6t^2 + 4t + 1$; $s(1) = -5$

**34.** $v(t) = e^{0.001t}$; $s(0) = 10$

**35.** $v(t) = \dfrac{t^2 + 1}{t^3 + 3t + 1}$; $s(0) = 50$

*In Exercises 36–40, the acceleration function of a particle moving along a directed line is given. The values of the velocity and position functions at a certain instant are also given. Derive the formulas for the velocity and position functions of the particle. Time is measured in seconds and distance in meters.*

**36.** $a(t) = 15$; $v(0) = 5$, $s(0) = 13$

**37.** $a(t) = -12t + 3$; $v(3) = 15$, $s(3) = 25$

**38.** $a(t) = 3t^2 + 6t + 8$; $v(1) = 6$, $s(1) = -15$

**39.** $a(t) = e^{0.03t}$; $v(0) = 5$, $s(0) = 100$

**40.** $a(t) = \dfrac{5}{\sqrt{t + 6}}$; $v(3) = 10$, $s(3) = 15$

*In Exercises 41–43, the marginal cost function at a production level of q units of a commodity is given. In each case the fixed cost is also given. Find the cost function.*

**41.** $M_c(q) = 0.01q^2 - 6q + 1000$, provided $0 \le q \le 300$; \$45,000.

**42.** $M_c(q) = 0.01q^2 - 7q + 5000$, provided $0 \le q \le 350$; \$50,000.

**43.** $M_c(q) = 3e^{-0.002q}$; \$35,000.

**44.** If $\int f(x) \, dx = 2x^4 - 5x^3 + 7x^2 + C$, then what is $f(x)$?

**45.** If $\int g(t) \, dt = 2t^2 \sqrt{t} + 6t + C$, then what is $g(t)$?

## 9.2  The U-Substitution

In the preceding section we introduced antiderivatives and we showed that, in simple cases, antiderivatives may be obtained by trial and error. However, to do this you must be able to reduce each indefinite integral to a form that appears in one of several standard antidifferentiation formulas. This section is devoted to a technique that is frequently used to accomplish this task.

### Rule of Substitution

Recall that $D_x e^{x^2} = 2e^{x^2} x$. Thus, $e^{x2}$ is an antiderivative of $2e^{x2}x$ and

$$\int 2e^{x^2} x \, dx = e^{x^2} + C \tag{1}$$

This equality does not appear in any of the standard tables of antidifferentiation. Instead, the following is given:

$$\int e^u \, du = e^u + C \tag{2}$$

You must be able to recognize that equalities (1) and (2) are equivalent.

In general, suppose that $u = f(x)$ where $f$ is a differentiable function of $x$, and let

$$F(x) = e^{f(x)}$$

Then, by the Chain Rule,

$$F'(x) = D_x e^{f(x)} = e^{f(x)} f'(x)$$

Thus, $e^{f(x)} f'(x)$ is the derivative of $F(x)$ and, consequently, $F(x)$ is an antiderivative of $e^{f(x)} f'(x)$. We write

$$\int e^{f(x)} f'(x) \, dx = e^{f(x)} + C \tag{3}$$

Note that equality (3) reduces to equality (2) if, in (3), we substitute $u$ for $f(x)$ and $du$ for $f'(x) \, dx$. This process is a special case of a technique frequently called the *u-substitution*. The validity of that technique stems from the following theorem.

THEOREM 9.2: Suppose the function $G$ is an antiderivative of a function $g$ and $u = f(x)$ where $f$ is a differentiable function of $x$. Then letting $f'(x) \, dx = du$, we have

$$\int g(f(x)) f'(x) \, dx = \int g(u) \, du = G(u) + C$$

This theorem can easily be proved using the Chain Rule. It is usually stated as follows.

**Rule of Substitution.** If in an indefinite integral, we replace $f(x)$ by $u$, then we must replace $f'(x) \, dx$ by $du$. The resulting integral must be free of the variable $x$.

Note the following:

1. As a mnemonic device, you should recall that if $u = f(x)$, then $\frac{du}{dx} = f'(x)$. Thus it is natural to let $du = f'(x) \, dx$ whenever we substitute $u$ for $f(x)$.
2. The process of antidifferentiation is often called integration.
3. If $f$ is a differentiable function, then $\int f'(x) \, dx = f(x) + C$.
4. If $f$ is a function which has an antiderivative, then $D_x \int f(x) \, dx = f(x)$, where the left side indicates that we are differentiating any of the infinitely many functions represented by $\int f(x) \, dx$.

## Some Standard Integration Formulas

We now list a few integration formulas. In the following, $k$, $n$, and $C$ are constants; $f$ and $g$ are functions that have antiderivatives.*

1. $\int (f(u) + g(u))\, du = \int f(u)\, du + \int g(u)\, du$
2. $\int (f(u) - g(u))\, du = \int f(u)\, du - \int g(u)\, du$
3. $\int kf(u)\, du = k \int f(u)\, du$
4. $\int u^n\, du = \dfrac{u^{n+1}}{n+1} + C$ (provided $n \neq -1$)
5. $\int u^{-1}\, du = \int \frac{1}{u}\, du = \ln|u| + C$
6. $\int e^u\, du = e^u + C$
7. $\int b^u\, du = \dfrac{b^u}{\ln b} + C$

To prove these formulas, we need only differentiate the right side of each equality with respect to $u$ and check that the derivative is in fact the integrand which appears on the left side. (See Exercises 34–40.)

Note also that in a formula such as the first one, each side represents infinitely many functions.

Success in applying the $u$-substitution method depends on your ability to choose the part of the integrand that should be replaced by $u$. This will come with the practice of working many examples. The important fact to keep in mind is that if you replace $f(x)$ by $u$ in an integrand, then $f'(x)\, dx$ must be present in the integrand *up to a constant factor*. The fact that, according to Formula 3, constant factors can be moved freely across an integral sign is useful in obtaining, in the integrand, the exact expression representing $du$.

Suppose we have an integral of the form

$$\int F(f(x))ng(x)\, dx$$

to evaluate, and we notice that $f'(x) = kg(x)$ where $n$ and $k$ are constants. Then it is useful to let $u = f(x)$ and $du = f'(x)\, dx = kg(x)\, dx$. At this point, we can proceed in one of two ways.

*First method:* Since $du = kg(x)\, dx$, we have

$$\frac{1}{k}\, du = g(x)\, dx$$

and

$$\frac{n}{k}\, du = ng(x)\, dx$$

Thus we may replace $ng(x)\, dx$ by $\frac{n}{k}\, du$ in the integrand, as well as $f(x)$ by $u$. We get

$$\int F(f(x))ng(x)\, dx = \int F(u)\frac{n}{k}\, du = \frac{n}{k} \int F(u)\, du$$

---

*You will later come to see that not all functions have antiderivatives.

*Second method:* We first write

$$\int F(f(x))ng(x)\ dx = n \int F(f(x))g(x)\ dx$$

We need to have $kg(x)\ dx$, which we replace by $du$, in the integrand. Thus, we multiply the integrand by the constant $k$ and, to compensate, we divide the integral by $k$. We get

$$n \int F(f(x))g(x)\ dx = \frac{n}{k} \int F(\underbrace{f(x)}_{u})\underbrace{kg(x)\ dx}_{du}$$

$$= \frac{n}{k} \int F(u)\ du$$

The two ways to adjust the integrand that we have just described are essentially identical; it is a matter of taste as to which one you wish to use.

It is important to remember, however, that although you can move a constant factor across the integral sign, it is wrong to move a variable across the integral sign. (See Exercise 45.) Furthermore, in making a substitution, you must always be sure to replace an expression by some other expression that you know is equal to it. Finally, you will always want to have only one variable in the integrand after the substitutions have been made.

The following examples have been selected to illustrate the $u$-substitution technique.

**EXAMPLE 1**  Find $\int (x^3 + 1)^4 x^2\ dx$.

**Solution**  Since the integrand involves $(x^3 + 1)$ raised to the fourth power, the formula that comes to mind is

$$\int u^n\ du = \frac{u^{n+1}}{n+1} + C$$

with $n = 4$. Thus we must let $u = x^3 + 1$, so that $\frac{du}{dx} = 3x^2$ and $du = 3x^2\ dx$.

Looking at the integral at hand, we notice that the constant factor 3 is missing. Therefore, we multiply the integrand by 3 to get $3x^2\ dx$ and we compensate by dividing the integral by 3. We get

$$\int (x^3 + 1)^4 x^2\ dx = \frac{1}{3} \int \underbrace{(x^3 + 1)^4}_{u} \underbrace{3x^2\ dx}_{du} = \frac{1}{3} \int u^4\ du$$

Then, using Formula 4 with $n = 4$, we get

$$\int (x^3 + 1)^4 x^2\ dx = \frac{1}{3} \int u^4\ du$$

$$= \frac{1}{3} \left[ \frac{u^{4+1}}{4+1} + C \right] = \frac{1}{3} \cdot \frac{u^5}{5} + \frac{1}{3}C$$

$$= \frac{1}{15} u^5 + K = \frac{1}{15}(x^3 + 1)^5 + K$$

where in the last step we replaced $u$ by $x^3 + 1$, because we wanted our answer to be a function of $x$, and the constant $\frac{1}{3}C$ by the constant $K$.

*Check:* $D_x\left[\dfrac{1}{15}(x^3 + 1)^5 + K\right] = \dfrac{1}{15} \cdot 5(x^3 + 1)^4(3x^2) + 0 = (x^3 + 1)^4x^2.$

Since this last expression is the original integrand, the answer is correct.
**Do Exercise 1.**

**EXAMPLE 2**   Find $\displaystyle\int e^{x^3+3x+7}2(x^2 + 1)\, dx.$

**Solution**   Since the integrand involves the exponential function to the base $e$, we try to use the formula $\int e^u\, du = e^u + C$.

Comparing the left side of this formula with the integral we wish to evaluate, we let $u = x^3 + 3x + 7$. Thus

$$\frac{du}{dx} = 3x^2 + 3 = 3(x^2 + 1)$$

and

$$du = 3(x^2 + 1)\, dx$$

Now we write

$$\int e^{x^3+3x+7}2(x^2 + 1)\, dx = 2\int e^{x^3+3x+7}(x^2 + 1)\, dx$$

$$= \frac{2}{3}\int e^{x^3+3x+7}3(x^2 + 1)\, dx$$

where we multiplied the integrand by 3 to get an expression equal to $du$ and divided the integral by 3 to compensate. Thus,

$$\int e^{x^3+3x+7}2(x^2 + 1)\, dx = \frac{2}{3}\int e^{x^3+3x+7}3(x^2 + 1)\, dx = \frac{2}{3}\int e^u\, du$$

$$= \frac{2}{3}(e^u + C) = \frac{2}{3}e^u + \frac{2}{3}C = \frac{2}{3}e^{x^3+3x+7} + K$$

*Check:* $D_x\left(\dfrac{2}{3}e^{x^3+3x+7} + K\right) = \dfrac{2}{3}[e^{x^3+3x+7}(3x^2 + 3)] + 0 = e^{x^3+3x+7}2(x^2 + 1).$

This last expression is the original integrand, so the answer is correct.
**Do Exercise 17.**

**EXAMPLE 3**   Find $\displaystyle\int \frac{2x^4}{\sqrt[3]{x^5 + 1}}\, dx.$

**Solution**   We first write the integral in the following form:

$$\int (x^5 + 1)^{-1/3}2x^4\, dx$$

It seems natural to use Formula 4. Thus we let

$$u = x^5 + 1$$

Therefore,

$$\frac{du}{dx} = 5x^4 \text{ and } du = 5x^4 \, dx$$

which means

$$x^4 \, dx = \frac{1}{5} \, du \text{ and } 2x^4 \, dx = \frac{2}{5} \, du$$

Thus

$$\int (x^5 + 1)^{-1/3} 2x^4 \, dx = \int u^{-1/3} \cdot \frac{2}{5} \, du = \frac{2}{5} \int u^{-1/3} \, du$$

$$= \frac{2}{5} \left[ \frac{u^{-1/3+1}}{-\frac{1}{3} + 1} + C \right] = \frac{3}{5} u^{2/3} + \frac{2C}{5}$$

$$= \frac{3}{5}(x^5 + 1)^{2/3} + K = \frac{3}{5} \sqrt[3]{(x^5 + 1)^2} + K$$

$$\text{Check: } D_x \left[ \frac{3}{5} \sqrt[3]{(x^5 + 1)^2} + K \right] = D_x \left[ \frac{3}{5}(x^5 + 1)^{2/3} + K \right]$$

$$= \frac{3}{5} \cdot \frac{2}{3}(x^5 + 1)^{2/3 - 1}(5x^4) + 0$$

$$= (x^5 + 1)^{-1/3} 2x^4 = \frac{2x^4}{\sqrt[3]{x^5 + 1}}$$

The last expression is the original integrand. The answer is correct.    ■

**EXAMPLE 4**    Find $\int e^{5x} 2x \, dx$.

**Solution**    As in Example 2, the integrand involves the exponential function to the base $e$; therefore we would want to use Formula 6 again. In this case, we let $u = 5x$, from which we get $\frac{du}{dx} = 5$ and $du = 5 \, dx$. Note that the unwanted factor in the integrand is $x$, which is not a constant. Therefore, we cannot use the $u$-substitution technique for this problem.

In the next section, we introduce a technique called integration by parts which will enable us to find integrals such as the one in this example. For now, we return to the trial-and-error technique. We guess that an antiderivative of $e^{5x} 2x$ is of the form $e^{5x} f(x)$. Because

$$D_x \, e^{5x} f(x) = e^{5x} f'(x) + f(x) e^{5x}(5) = e^{5x}(5f(x) + f'(x))$$

we should have

$$e^{5x} 2x = e^{5x}(5f(x) + f'(x))$$

from which we deduce that

$$2x = 5f(x) + f'(x)$$

Since the left side of this equality is a linear function, we try $f(x) = ax + b$ where $a$ and $b$ are constants. But $f(x) = ax + b$ implies that $f'(x) = a$. Thus we must have

$$2x = 5(ax + b) + a = 5ax + (a + 5b)$$

This will be an identity if, and only if,

$$5a = 2 \text{ and } a + 5b = 0$$

Solving the system

$$\begin{cases} 5a & = 2 \\ a + 5b = 0 \end{cases}$$

we get

$$a = \frac{2}{5} \text{ and } b = \frac{-2}{25}$$

Thus

$$\int e^{5x}2x \, dx = e^{5x}\left(\frac{2}{5}x - \frac{2}{25}\right) + C$$

$$Check: D_x \left[ e^{5x}\left(\frac{2}{5}x - \frac{2}{25}\right) + C \right] = e^{5x}\frac{2}{5} + \left(\frac{2}{5}x - \frac{2}{25}\right)e^{5x}(5) + 0$$

$$= e^{5x}\left[\frac{2}{5} + 5\left(\frac{2}{5}x - \frac{2}{25}\right)\right]$$

$$= e^{5x}\left[\frac{2}{5} + 2x - \frac{2}{5}\right] = e^{5x}2x$$

The answer is correct.
**Do Exercise 23.** (See Exercises 45 and 46.)

# Exercise Set 9.2

*In Exercises 1–24, evaluate the given integral using a u-substitution and check your answer by differentiation.*

**1.** $\int (x^3 + 5)^{10}x^2 \, dx$

**2.** $\int \sqrt[4]{x^3 + 8} \, x^2 \, dx$

**3.** $\int \frac{x^3}{(x^4 + 3)^5} \, dx$

**4.** $\int \frac{x^4}{\sqrt{x^5 + 6}} \, dx$

**5.** $\int (x^2 + 2x + 5)^{10}(x + 1) \, dx$

**6.** $\int (x^4 + 8x + 6)^6 \, 3(x^3 + 2) \, dx$

**7.** $\int \sqrt[3]{x^2 + 6x + 1}(5x + 15) \, dx$

**8.** $\int \frac{2x^2 + 8x + 6}{\sqrt[5]{x^3 + 6x^2 + 9x + 1}} \, dx$

**9.** $\int \frac{7x^4}{x^5 + 4x} \, dx$

**10.** $\int \frac{x + 2}{x^2 + 4x + 5} \, dx$

**11.** $\int \frac{e^{2x}}{e^{2x} + 1} \, dx$ (*Hint:* Let $u = e^{2x} + 1$.)

**12.** $\int \frac{1}{x \ln|x|} \, dx$ (*Hint:* Let $u = \ln|x|$.)

**13.** $\int (e^{5x} + 1)^{12}2e^{5x} \, dx$

**14.** $\int \frac{5e^{3x}}{\sqrt{e^{3x} + 2}} \, dx$

**15.** $\int e^{6x} \, dx$

**16.** $\int e^{x^5}3x^4 \, dx$

**17.** $\int e^{x^3 + 6x + 10}(x^2 + 2) \, dx$

**18.** $\int e^{x^3+6x^2+15x+4}(2x^2 + 8x + 10)\, dx$

**19.** $\int \dfrac{e^{\sqrt{x}}}{\sqrt{x}}\, dx$

**20.** $\int \dfrac{e^{2x}}{(e^{2x} + 1)\ln(e^{2x} + 1)}\, dx$ (*Hint:* Let $u = \ln(e^{2x} + 1)$.)

**21.** $\int 2^{x^2}x\, dx$

**22.** $\int 5^{x^2+4x+3}(x + 2)\, dx$

**23.** $\int 5xe^{3x}\, dx$ (*Hint:* Try $(ax + b)e^{3x}$ where $a$ and $b$ are constants. See Example 4.)

**24.** $\int x^2 e^{5x}\, dx$ (*Hint:* Try $(ax^2 + bx + c)e^{5x}$ where $a$, $b$, and $c$ are constants. See Example 4.)

*In Exercises 25–29, the velocity function of a particle moving along a directed line is given. The value of the position function at a certain instant is also given. Derive the formula for the position function. Time is measured in seconds and distance in centimeters.*

**25.** $v(t) = (t^2 + 1)^{1/2}3t,\ s(0) = 10$

**26.** $v(t) = (t^3 + 6t + 5)^{1/3}(t^2 + 2),\ s(1) = 5$

**27.** $v(t) = e^{-t^2}3t,\ s(0) = 4$

**28.** $v(t) = \dfrac{3t^3}{t^4 + 5},\ s(0) = 8$

**29.** $v(t) = 5^{t^2+1}(4t),\ s(0) = 25$

*In Exercises 30–33, the marginal cost function at a production level of q units of a commodity is given. In each case the fixed cost is also given. Find the cost function.*

**30.** $M_c(q) = \dfrac{2q^3 + 1}{\sqrt[5]{(q^4 + 2q + 3)^4}},\ \$50{,}000$

**31.** $M_c(q) = \dfrac{q + 3}{\sqrt[3]{(q^2 + 6q + 1)^2}},\ \$30{,}000$

**32.** $M_c(q) = \dfrac{e^{0.001q}}{(e^{0.001q} + 5)^{11/10}},\ \$10{,}000$

**33.** $M_c(x) = \dfrac{2q + 1}{q^2 + q + 5},\ \$40{,}000$

**34.** Verify Formula 1 by differentiating the right side.

**35.** Verify Formula 2.    **36.** Verify Formula 3.

**37.** Verify Formula 4.    **38.** Verify Formula 5.

**39.** Verify Formula 6.    **40.** Verify Formula 7.

*In Exercises 41–44, the acceleration function of a particle moving along a directed line is given. The values of the velocity and position functions at a certain instant are also given. Derive the formula for the position function. Time is measured in seconds and distance in feet.*

**41.** $a(t) = e^{4t};\ v(0) = 12,\ s(0) = 25$

**42.** $a(t) = e^{t/3};\ v(0) = 6,\ s(0) = -35$

**43.** $a(t) = te^{3t};\ v(0) = 8,\ s(0) = 40$ (*Hint:* See Example 4 and Exercises 23, 24.)

**44.** $a(t) = 5t^2 e^{5t};\ v(0) = 45,\ s(0) = 20$ (*Hint:* See Example 4 and Exercises 23, 24.)

**45.** A student attempted to find the integral of Example 4 as follows. He let $u = 5x$, $du = 5\, dx$, $dx = \frac{du}{5}$ and wrote

$$\int e^{5x}2x\, dx = 2x \int e^{5x}\, dx = 2x \int e^u \frac{du}{5}$$

$$= \frac{2x}{5} \int e^u\, du = \frac{2x}{5}e^u + C = \frac{2x}{5}e^{5x} + C$$

Show that this answer is wrong. (*Hint:* Differentiate the function he obtained.)

**46.** A student attempted to find the integral of Example 4 as follows. She guessed that the answer must be of the form $ke^{5x}$ and wrote

$$D_x ke^{5x} = ke^{5x}(5) = 5ke^{5x} = e^{5x}2x$$

She solved for $k$, obtained $k = \frac{2x}{5}$, and concluded that

$$\int e^{5x}2x\, dx = \frac{2x}{5}e^{5x} + C$$

Show that her answer is incorrect and explain where she made her error.

# 9.3  Integration by Parts

In the preceding two sections, we used differentiation formulas to obtain integration formulas. In this section, we introduce a new method based on the product rule of differentiation.

## Integration by Parts Formula

If $f$ and $g$ are two differentiable functions, then their product is also a differentiable function and

$$D_x(f(x)g(x)) = f(x)g'(x) + g(x)f'(x)$$

Antidifferentiating both sides with respect to $x$, we obtain

$$f(x)g(x) + C = \int [f(x)g'(x) + g(x)f'(x)]\, dx$$

$$= \int f(x)g'(x)\, dx + \int g(x)f'(x)\, dx$$

The last expression can be written

$$\int f(x)g'(x)\, dx = f(x)g(x) + C - \int g(x)f'(x)\, dx$$

Since the integral on the right side represents the sum of an antiderivative of $g(x)f'(x)$ and a constant, we usually omit the $C$ in writing the equality. This equality is called the *integration by parts formula*.

$$\int f(x)g'(x)\, dx = f(x)g(x) - \int g(x)f'(x)\, dx$$

## When to Use Integration by Parts

We now briefly discuss the conditions under which you should attempt to use this formula.

Suppose we have an integral $\int H(x)\, dx$ to evaluate and none of the methods we have already discussed can be used. Suppose further that it is possible to write $H$ as a product of a function $f$ and a derivative $g'$ of a function $g$. Then we can write

$$\int H(x)\, dx = \int f(x)g'(x)\, dx$$

and, using the integration by parts formula, we have

$$\int H(x)\, dx = f(x)g(x) - \int g(x)f'(x)\, dx$$

Thus the difficulty of evaluating $\int H(x)\, dx$ has been replaced by the task of finding the integral $\int g(x)f'(x)\, dx$. It may appear that the latter integral is easier to evaluate than the original one. Sometimes, two or more applications of the formula are necessary to obtain an integral that can be easily evaluated. (See Example 4.)

## Hints on How to Use Integration by Parts

The difficulty in using integration by parts is in choosing the functions $f$ and $g'$. You will become proficient in making that choice after working a number of examples. For the beginner, we suggest the following guidelines.

1. Since it is easier to differentiate than it is to integrate, choose $g'(x)$ in such a way that $g(x)$ can easily be obtained by integration.
2. The integral obtained on the right side should be simpler than the original integral on the left. (See Exercise 35.)
3. If the original integrand does not appear to be the product of two functions, we can let $g'(x) = 1$. (See Example 2.)
4. If we introduce the substitutions $u = f(x)$, $du = f'(x)\,dx$, $v = g(x)$, and $dv = g'(x)\,dx$, the integration by parts formula takes the following form, which many students find easier to memorize,

$$\int u\,dv = uv - \int v\,du$$

## Examples

The examples that follow have been chosen to illustrate the integration by parts method, beginning with a repeat of Example 4 of the last section.

**EXAMPLE 1**    Evaluate $\displaystyle\int 2xe^{5x}\,dx$.

**Solution**    We let $f(x) = 2x$ and $g'(x) = e^{5x}$. Thus, we can easily see that $f'(x) = 2$ and $g(x) = \frac{1}{5}e^{5x} + C$. Using the integration by parts formula, we get

$$\int 2xe^{5x}\,dx = \underbrace{(2x)}_{\substack{\uparrow \\ f(x)}}\underbrace{\left(\frac{1}{5}e^{5x} + C\right)}_{\substack{\uparrow \\ g(x)}} - \int \underbrace{\left(\frac{1}{5}e^{5x} + C\right)}_{\substack{\uparrow \\ g(x)}}\underbrace{2}_{\substack{\uparrow \\ f'(x)}}\,dx$$

$$= \frac{2}{5}xe^{5x} + 2Cx - \int \left(\frac{2}{5}e^{5x} + 2C\right)\,dx$$

The last integral can be evaluated by trial and error, or using a $u$-substitution. We obtain

$$\int 2xe^{5x}\,dx = \frac{2}{5}xe^{5x} + 2Cx - \left(\frac{2}{25}e^{5x} + 2Cx\right) + K$$

$$= \frac{2}{5}xe^{5x} - \frac{2}{25}e^{5x} + K = e^{5x}\left(\frac{2x}{5} - \frac{2}{25}\right) + K$$

**Do Exercise 1.**    ■

Notice that the constant $C$ does not appear in the final answer. This is not a coincidence. In using the method of integration by parts, the constant $C$, which arises in integrating $g'(x)$, always cancels out and does not appear in the final answer. (See Exercise 34.) Therefore, any antiderivative of $g'(x)$ may be used and, frequently, we choose the value 0 for the constant $C$. However, in certain cases, it is more advantageous to use a different value for $C$. (See Example 3.)

**EXAMPLE 2** Evaluate $\int \ln|x|\, dx$.

**Solution** Let $u = \ln|x|$ and $dv = (1)\, dx = dx$. Then $du = \frac{1}{x}\, dx$ and $v = x + C$. Since we know that the constant $C$ will cancel out of the final answer, we give it the value 0 and let $v = x$. Thus

$$\underbrace{\int \ln|x|}_{u}\underbrace{dx}_{dv} = \underbrace{(\ln|x|)}_{u}\underbrace{x}_{v} - \int \underbrace{x}_{v}\underbrace{\left(\frac{1}{x}\, dx\right)}_{du}$$

$$= x\ln|x| - \int 1\, dx = x\ln|x| - x + K$$

$$= x(\ln|x| - 1) + K$$

**Do Exercise 9.**

**EXAMPLE 3** Evaluate $\int 2x \ln|x + 1|\, dx$.

**Solution** We have two alternatives. Either let $u = 2x$ and $dv = \ln|x + 1|\, dx$, or $u = \ln|x + 1|$ and $dv = 2x\, dx$. The second choice seems correct since if we let $dv = \ln|x + 1|\, dx$, $v$ cannot readily be found. Thus, we select the second alternative and get

$$du = \frac{1}{x + 1}\, dx \text{ and } v = x^2 + C$$

At this point, we know that the constant $C$ will cancel out of the final answer. Therefore we may choose any value we wish for $C$. We let $C = -1$, so that $v = x^2 - 1 = (x - 1)(x + 1)$. As we shall soon see, this choice will lead to a useful simplification. We have

$$\int 2x \ln|x + 1|\, dx = \int \underbrace{(\ln|x + 1|)}_{u}\underbrace{2x\, dx}_{dv}$$

$$= \underbrace{(\ln|x + 1|)}_{u}\underbrace{(x^2 - 1)}_{v} - \int \underbrace{(x^2 - 1)}_{v}\underbrace{\frac{1}{x + 1}\, dx}_{du}$$

$$= (x^2 - 1)\ln|x + 1| - \int (x - 1)(x + 1)\frac{1}{x + 1}\, dx$$

$$= (x^2 - 1)\ln|x + 1| - \int (x - 1)\, dx$$

$$= (x^2 - 1)\ln|x + 1| - \left(\frac{x^2}{2} - x\right) + K$$

$$= (x^2 - 1)\ln|x + 1| - \frac{x^2}{2} + x + K$$

Note that if we had chosen to let $C = 0$ instead of $-1$, we would have obtained

$$\int 2x \ln|x + 1|\, dx = x^2 \ln|x + 1| - \int \frac{x^2}{x + 1}\, dx$$

This last integral is not as easily evaluated as $\int (x - 1)\, dx$, which was obtained in our example. It requires performing division to get

$$\frac{x^2}{x + 1} = x - 1 + \frac{1}{x + 1}$$

so that

$$\int \frac{x^2}{x + 1}\, dx = \int \left( x - 1 + \frac{1}{x + 1} \right) dx$$

$$= \frac{x^2}{2} - x + \ln|x + 1| + K$$

The final answer would have been the same, but the details would have been much more tedious.
**Do Exercise 11.**                                                                   ■

## Repeated Applications of Integration by Parts

Sometimes it is necessary to apply integration by parts several times.

**EXAMPLE 4**    Using integration by parts, evaluate

$$\int x^2 e^{3x}\, dx$$

**Solution**    Let $u = x^2$ and $dv = e^{3x}\, dx$. Then $du = 2x\, dx$ and $v = \frac{1}{3}e^{3x}$. (Here we choose $C = 0$.) It follows that

$$\int \underbrace{x^2}_{u}\, \underbrace{e^{3x}\, dx}_{dv} = \underbrace{x^2}_{u} \underbrace{\left( \frac{1}{3}e^{3x} \right)}_{v} - \int \underbrace{\frac{1}{3}e^{3x}}_{v} \underbrace{(2x\, dx)}_{du}$$

$$= \frac{1}{3}x^2 e^{3x} - \frac{2}{3} \int x e^{3x}\, dx \tag{*}$$

Although the last integral is simpler than the original integral, its evaluation also requires integration by parts. Thus we consider it separately. We let $u = x$, $dv = e^{3x}\, dx$ so that $du = dx$ and $v = \frac{1}{3}e^{3x}$. We get

$$\int \underbrace{x}_{u}\, \underbrace{e^{3x}\, dx}_{dv} = \underbrace{x}_{u} \underbrace{\left( \frac{1}{3}e^{3x} \right)}_{v} - \int \underbrace{\frac{1}{3}e^{3x}}_{v} \underbrace{dx}_{du}$$

$$= \frac{1}{3}x e^{3x} - \frac{1}{3} \int e^{3x}\, dx = \frac{1}{3}x e^{3x} - \frac{1}{3}\left( \frac{1}{3}e^{3x} \right) + C$$

$$= \frac{1}{3}x e^{3x} - \frac{1}{9}e^{3x} + C$$

Using this result in equality (*), we obtain

$$\int x^2 e^{3x}\, dx = \frac{1}{3}x^2 e^{3x} - \frac{2}{3}\int xe^{3x}\, dx$$

$$= \frac{1}{3}x^2 e^{3x} - \frac{2}{3}\left[\frac{1}{3}xe^{3x} - \frac{1}{9}e^{3x} + C\right]$$

$$= e^{3x}\left(\frac{1}{3}x^2 - \frac{2}{9}x + \frac{2}{27}\right) + K$$

where $K$ is a constant.
**Do Exercise 27.**

# Exercise Set 9.3

*In Exercises 1–25, use integration by parts to evaluate the given integral.*

**1.** $\int 3xe^{8x}\, dx$

**2.** $\int 5xe^{6x}\, dx$

**3.** $\int (5x + 6)e^{2x}\, dx$

**4.** $\int \frac{4x}{e^{3x}}\, dx$

**5.** $\int (3x + 1)e^{4x}\, dx$

**6.** $\int e^{\sqrt{x}}\, dx$ (*Hint:* First substitute $y = \sqrt{x}$.)

**7.** $\int x^5 e^{x^3}\, dx$ (*Hint:* First substitute $y = x^3$.)

**8.** $\int \ln|x + 3|\, dx$

**9.** $\int \ln|x^3|\, dx$

**10.** $\int (\ln|x|)^2\, dx$

**11.** $\int 2x \ln|x + 3|\, dx$

**12.** $\int (x + 3)\ln|x|\, dx$

**13.** $\int \frac{\ln|x|}{x^3}\, dx$

**14.** $\int 5x \ln(x^2)\, dx$

**15.** $\int \frac{\ln|x|}{\sqrt{x}}\, dx$

**16.** $\int x\, 2^x\, dx$

**17.** $\int x\, 3^x\, dx$

**18.** $\int x^2\ln|x + 5|\, dx$

**19.** $\int x^3\ln|x + 4|\, dx$

**20.** $\int x\sqrt{x + 2}\, dx$ (*Hint:* Let $u = x$ and $dv = (x + 2)^{1/2}\, dx$.)

**21.** $\int x\sqrt[3]{x + 2}\, dx$ (*Hint:* Let $u = x$ and $dv = (x + 2)^{1/3}\, dx$.)

**22.** $\int x(x + 5)^{100}\, dx$

**23.** $\int \frac{x}{\sqrt{x + 4}}\, dx$

**24.** $\int x^3\sqrt{5 - x^2}\, dx$

**25.** $\int x^2(x + 6)^{20}\, dx$

**26.** Use integration by parts to derive the following reduction formula:

$$\int x^n e^{ax}\, dx = \frac{1}{a}x^n e^{ax} - \frac{n}{a}\int x^{n-1} e^{ax}\, dx$$

*In Exercises 27–30, use the reduction formula obtained in Exercise 26 to evaluate the given integral. The formula must be used more than once in each exercise.*

**27.** $\int x^2 e^{10x}\, dx$

**28.** $\int x^3 e^{5x}\, dx$

**29.** $\int x^4 e^{-2x}\, dx$

**30.** $\int x^5 e^{4x}\, dx$

**31.** The marginal cost function at a production level of $q$ units of a commodity is given by

$$M_c(q) = (-q^2 + 10{,}000)e^{-0.1q}$$

provided that $0 \le q \le 100$. The fixed cost is $15,000. Find the cost function.

**32.** The marginal revenue function for a product is given by

$$M_r(q) = (-0.0001q^2 + 4)\ln|q + 2|$$

where $q$ is the number of units sold, provided that $0 \le q \le 150$. Find the revenue function. (*Hint:* $R(0) = 0$.)

**33.** A particle moves along a directed line. Its acceleration at instant $t$ seconds is $a$ ft/sec$^2$ where $a = te^{-2t}$.

   **a.** Find the velocity function if the velocity at instant 0 is 25 ft/sec.

   **b.** Find the position function if at instant 5 seconds the particle is 25 feet from the origin.

**34.** Clearly the function $g$ is an antiderivative of the function $g'$. Thus $g(x) + C$ represents all antiderivatives of $g'(x)$. Show that if we replace each $g(x)$ by $g(x) + C$ in the integration by parts formula, the constant $C$ will cancel out.

**35.** Suppose that in trying to evaluate $\int 2xe^{5x}\, dx$ of Example 1, we let $u = e^{5x}$ and $dv = 2x\, dx$. Show that this substitution would lead to an integral that is more complicated than the original one.

---

# 9.4   Integration Tables

We have discussed the process of finding antiderivatives by trial and error, by making a $u$-substitution, and by using the method of integration by parts. There are many techniques of integration available, and most of the standard calculus textbooks written for students majoring in the sciences devote many sections discussing them. In many integration problems, a great amount of labor and ingenuity are required, even of those who have mastered all these techniques. Often, the most effective method of evaluating an integral is using tables.

## How to Use Integration Tables

Many integration tables have been published, some more extensive than others. A relatively short table of integrals can be found in the appendix. To make it easy to use, the formulas have been classified under certain headings. For example, all integrands involving exponentials and logarithms appear under the same heading, all those containing $(a^2 \pm u^2)$ as a factor appear under another heading, and so forth. In many instances, a $u$-substitution must be made to transform the given integral into one that appears in the tables. The examples in this section have been selected to illustrate the use of tables for evaluating integrals.

**EXAMPLE 1**   Find $\int \dfrac{1}{9 - x^2}\, dx$.

**Solution**   Since $9 - x^2 = 3^2 - x^2$, we look for the appropriate formula under the heading "Integrals containing $a^2 \pm u^2$." We find

$$\int \frac{1}{a^2 - u^2}\, du = \frac{1}{2a} \ln \left| \frac{u + a}{u - a} \right|$$

Note that the constant of integration $C$ is not included in the tables.

Comparing our integral to the left side of the formula, we see that $a = 3$ and $u = x$. Thus we replace every $a$ by 3 and every $u$ by $x$ in the formula. We obtain

$$\int \frac{1}{3^2 - x^2}\, dx = \frac{1}{2(3)} \ln \left| \frac{x + 3}{x - 3} \right| + C = \frac{1}{6} \ln \left| \frac{x + 3}{x - 3} \right| + C$$

**Do Exercise 1.**

**EXAMPLE 2**    Evaluate $\int \dfrac{5x}{\sqrt{x^4 + 9}}\, dx.$

**Solution**    We cannot find in our table a heading for "Integrals containing $u^4 + a^2$." However, we do have a heading "Integrals containing $u^2 + a^2$." Thus we let $u = x^2$ so that $du = 2x\, dx$. We proceed with the $u$-substitution and write

$$\int \frac{5x}{\sqrt{x^4 + 9}}\, dx = 5 \int \frac{x}{\sqrt{x^4 + 9}}\, dx = \frac{5}{2} \int \frac{1}{\sqrt{(x^2)^2 + 3^2}}\, 2x\, dx$$

$$= \frac{5}{2} \int \frac{1}{\sqrt{u^2 + 3^2}}\, du$$

We now refer to the tables and find the following formula:

$$\int \frac{1}{\sqrt{u^2 \pm a^2}}\, du = \ln\left| u + \sqrt{u^2 \pm a^2}\right|$$

Comparing our integral to the left side of the formula, we see that $a = 3$ and we must select the plus sign. Thus,

$$\int \frac{5x}{\sqrt{x^4 + 9}}\, dx = \frac{5}{2} \int \frac{1}{\sqrt{u^2 + 3^2}}\, du$$

$$= \frac{5}{2} \ln\left| u + \sqrt{u^2 + 3^2}\right| + C = \frac{5}{2} \ln(x^2 + \sqrt{x^4 + 9}) + C$$

In the last step, we replaced $u$ by $x^2$ and used parentheses instead of absolute value signs since we know that $x^2 + \sqrt{x^4 + 9}$ is positive.
**Do Exercise 11.**    ∎

## Reduction Formulas

For certain integrals, it is necessary to use more than one formula, or to use the same formula several times. We illustrate this in the next example.

**EXAMPLE 3**    Find $\int x^2 e^{5x}\, dx.$

**Solution**    Under the heading "Integrals containing exponentials and logarithms" we find the formula

$$\int u^n e^{au}\, du = \frac{1}{a} u^n e^{au} - \frac{n}{a} \int u^{n-1} e^{au}\, du$$

Comparing the left side of the formula to the integral we were given, we see that $n = 2$, $a = 5$, and $u = x$. Thus

$$\int x^2 e^{5x}\, dx = \frac{1}{5} x^2 e^{5x} - \frac{2}{5} \int x e^{5x}\, dx \tag{*}$$

The last integral can be evaluated using the same formula, this time with $n = 1$. We get

$$\int xe^{5x} \, dx = \frac{1}{5}xe^{5x} - \frac{1}{5}\int x^0 e^{5x} \, dx$$

$$= \frac{1}{5}xe^{5x} - \frac{1}{5}\int e^{5x} \, dx = \frac{1}{5}xe^{5x} - \frac{1}{25}e^{5x} + C$$

Replacing $\int xe^{5x} \, dx$ by the last expression in equality (*), we get

$$\int x^2 e^{5x} \, dx = \frac{1}{5}x^2 e^{5x} - \frac{2}{5}\left[\frac{1}{5}xe^{5x} - \frac{1}{25}e^{5x} + C\right]$$

$$= \frac{1}{5}x^2 e^{5x} - \frac{2}{25}xe^{5x} + \frac{2}{125}e^{5x} + K$$

$$= e^{5x}\left(\frac{x^2}{5} - \frac{2x}{25} + \frac{2}{125}\right) + K$$

**Do Exercise 13.**

It is often useful to first manipulate the integrand algebraically to obtain a sum of integrals, some that can be evaluated using tables, and the others using known techniques. For example, in Exercise 18, it is useful to first change $(25 - 16x^2)^{3/2}$ to $(25 - 16x^2)(25 - 16x^2)^{1/2}$ and in Exercise 20, you should first multiply the numerator and denominator of the integrand by $e^{-4x^2}$.

# Exercise Set 9.4

*In Exercises 1–30, evaluate the integral using the table of integrals. In some of the exercises, you may have to perform a u-substitution first.*

**1.** $\int \dfrac{1}{\sqrt{x^2 - 25}} \, dx$

**2.** $\int \dfrac{x^2}{(x^2 + 36)^{3/2}} \, dx$

**3.** $\int \sqrt{x^2 + 16} \, dx$

**4.** $\int \dfrac{1}{x\sqrt{25 - x^2}} \, dx$

**5.** $\int e^x\sqrt{e^{2x} + 16} \, dx$

**6.** $\int \dfrac{1}{x^2(3x + 4)} \, dx$

**7.** $\int x\sqrt{5x + 8} \, dx$

**8.** $\int \dfrac{1}{x\sqrt{7x + 3}} \, dx$

**9.** $\int \dfrac{1}{49 - x^2} \, dx$

**10.** $\int \sqrt{e^{2x} + 25e^{4x}} \, dx$

**11.** $\int \dfrac{1}{5 + 3e^x} \, dx$

**12.** $\int \dfrac{1}{x(5x^3 + 4)} \, dx$

**13.** $\int x^2 e^{10x} dx$

**14.** $\int \dfrac{1}{2x^2 + 5x - 3} \, dx$

**15.** $\int \dfrac{\ln^5|x|}{x} \, dx$

**16.** $\int \dfrac{\sqrt{x^2 + 16}}{x} \, dx$

**17.** $\int \dfrac{3}{25 - 4x^2} \, dx$

**\*18.** $\int \dfrac{(25 - 16x^2)^{3/2}}{x} \, dx$

**19.** $\int x^5 e^{x^2} dx$

**\*20.** $\int \dfrac{6x}{5 + 3e^{4x^2}} \, dx$

**21.** $\int x^2\sqrt{4x^2 - 25} \, dx$

**22.** $\int (16x^2 + 25)^{3/2} \, dx$

**23.** $\int \dfrac{e^x}{(e^{2x} - 25)^{3/2}} \, dx$

**24.** $\int \dfrac{\sqrt{\ln^2|x| + 49}}{x \ln|x|} \, dx$

**25.** $\int x^3(9x^2 + 64)^{3/2} \, dx$

**26.** $\int x^3(4x^2 - 25)^{3/2} \, dx$

**27.** $\int x^3\sqrt{25x^2 - 121} \, dx$

**28.** $\int 4x^3\sqrt{16x^2 + 25} \, dx$

**29.** $\int 8x^7 \sqrt{x^4 + 16} \, dx$

**30.** $\int e^{6x} \sqrt{e^{2x} + 25} \, dx$

**31.** $\int x(x^4 + 4x^2 + 13)^{3/2} \, dx$ (*Hint:* First complete a square, then make a $u$-substitution.)

**32.** $\int \dfrac{1}{4 - x^2 + 3x} \, dx$ (*Hint:* First complete a square, then make a $u$-substitution.)

**33.** $\int \dfrac{1}{36 - x^2 + 5x} \, dx$ (*Hint:* First complete a square, then make a $u$-substitution.)

**34.** $\int x(x^4 + 10x^2 + 16)^{3/2} \, dx$ (*Hint:* First complete a square, then make a $u$-substitution.)

**35.** $\int \dfrac{x(x^4 + 6x^2 - 16)^{1/2}}{(x^2 + 3)^2} \, dx$ (*Hint:* First complete a square, then make a $u$-substitution.)

---

# 9.5 Guessing Again (Optional)

## When to Use Trial and Error

In the preceding section, we stated that often the most effective method of evaluating an integral is by using tables. However, in Example 3, it was necessary to use the formulas several times and the details of the procedure were tedious. A formula such as

$$\int u^n e^{au} \, du = \frac{1}{a} u^n e^{au} - \frac{n}{a} \int u^{n-1} e^{au} \, du$$

is called a *reduction formula*. Note that the integrand on the right side has $u^{n-1}$ as a factor, while $u^n$ is a factor of the integrand on the left. The power of $u$ is reduced by one each time this formula is used. Consequently, if we start with $\int u^n e^{au} \, du$ where $n$ is a positive integer, we must use the reduction formula $n$ times to arrive at an integral of the form $\int e^{au} \, du$, which can be easily evaluated. If $n$ is larger than 2, the solution is lengthy and awkward. In such cases, it is often best to find an antiderivative by trial and error. We need only know what type of function an antiderivative must be.

Suppose, for example, that $\int f(x)e^{ax} \, dx$ is to be evaluated. We guess that some antiderivative is of the form $q(x)e^{ax}$. Thus we must have

$$D_x(q(x)e^{ax}) = f(x)e^{ax}$$

But,

$$D_x(q(x)e^{ax}) = q(x)e^{ax}a + e^{ax}q'(x)$$
$$= e^{ax}(aq(x) + q'(x))$$

Therefore, necessarily,

$$e^{ax}(aq(x) + q'(x)) = f(x)e^{ax}$$

which yields

$$aq(x) + q'(x) = f(x)$$

In the case where $f(x)$ is a polynomial of degree $n$, we may assume that $q(x)$ is also a polynomial of degree $n$ because $q'(x)$ will be a polynomial of degree $n - 1$ and therefore $aq(x) + q'(x)$ will be of degree $n$ as required. Thus if we wish to evaluate $\int f(x)e^{ax}\ dx$ where $f(x)$ is a polynomial of degree $n$, we use as a trial antiderivative $q(x)e^{ax}$ where $q(x)$ is a polynomial of degree $n$ also.

## Examples

**EXAMPLE 1**   Using trial and error, evaluate the integral of Example 3 of the preceding section.

**Solution**   To find $\int x^2e^{5x}\ dx$, we assume that an antiderivative is $(ax^2 + bx + c)e^{5x}$ where $a$, $b$, and $c$ are constants to be determined. We must have

$$D_x[(ax^2 + bx + c)e^{5x}] = x^2e^{5x}$$

That is,

$$(ax^2 + bx + c)e^{5x}(5) + e^{5x}(2ax + b) = x^2e^{5x}$$

or

$$e^{5x}[5ax^2 + (2a + 5b)x + (b + 5c)] = x^2e^{5x}$$

This equality is an identity if, and only if,

$$5ax^2 + (2a + 5b)x + (b + 5c) = x^2$$

is an identity. Using the fact that two polynomials are identical if, and only if, corresponding coefficients are equal, we set

$$\begin{cases} 5a & = 1 \\ 2a + 5b & = 0 \\ b + 5c & = 0 \end{cases}$$

This system of equations is easily solved. We obtain $a = \frac{1}{5}$, $b = \frac{-2}{25}$, and $c = \frac{2}{125}$. Therefore

$$\int x^2e^{5x}\ dx = \left(\frac{x^2}{5} - \frac{2x}{25} + \frac{2}{125}\right)e^{5x} + C$$

**Do Exercise 11.**   ∎

**EXAMPLE 2**   Evaluate $\displaystyle\int (5x^3 + 13x^2 - 21x + 25)e^{5x}\ dx$.

**Solution**   Since the integrand is the product of $e^{5x}$ and a polynomial of degree 3, we assume that an antiderivative is $p(x)e^{5x}$ where $p(x)$ is a polynomial of degree 3. Thus,

$$\int (5x^3 + 13x^2 - 21x + 25)e^{5x}\ dx = (ax^3 + bx^2 + cx + d)e^{5x} + K$$

where $a$, $b$, $c$, and $d$ are constants to be determined and $K$ is an arbitrary constant of integration. We must have

$$D_x[(ax^3 + bx^2 + cx + d)e^{5x} + K] = (5x^3 + 13x^2 - 21x + 25)e^{5x}$$

However,

$$D_x[(ax^3 + bx^2 + cx + d)e^{5x} + K]$$
$$= (ax^3 + bx^2 + cx + d)e^{5x}(5) + e^{5x}(3ax^2 + 2bx + c) + 0$$
$$= e^{5x}[5ax^3 + (3a + 5b)x^2 + (2b + 5c)x + (c + 5d)]$$

Thus,

$$e^{5x}[5ax^3 + (3a + 5b)x^2 + (2b + 5c)x + (c + 5d)]$$
$$= (5x^3 + 13x^2 - 21x + 25)e^{5x}$$

Since the polynomial factors on each side must be identical, we obtain the system

$$\begin{cases} 5a & = & 5 \\ 3a + 5b & = & 13 \\ 2b + 5c & = & -21 \\ c + 5d & = & 25 \end{cases}$$

We easily get the solution $a = 1$, $b = 2$, $c = -5$, and $d = 6$. Therefore

$$\int (5x^3 + 13x^2 - 21x + 25)e^{5x} \, dx = (x^3 + 2x^2 - 5x + 6)e^{5x} + K$$

**Do Exercise 17.** ■

---

**EXAMPLE 3**  Evaluate $\int x^5 2^{x^2} \, dx$.

**Solution**  We first perform a $u$-substitution to reduce the power of the variable. We let $u = x^2$ so that $du = 2x \, dx$. Now we write

$$\int x^5 2^{x^2} \, dx = \frac{1}{2} \int (x^2)^2 2^{x^2} \, 2x \, dx = \frac{1}{2} \int u^2 2^u \, du$$

Since $u^2$ is a polynomial of degree 2, we try

$$\int u^2 2^u \, du = (Au^2 + Bu + C)2^u + K$$

where $A$, $B$, and $C$ are constants to be determined. We must have

$$D_u[(Au^2 + Bu + C)2^u + K] = u^2 2^u$$

or

$$(Au^2 + Bu + C)2^u(\ln 2) + 2^u(2Au + B) = u^2 2^u$$

Therefore

$$[(A\ln 2)u^2 + (2A + B\ln 2)u + (B + C\ln 2)]2^u = u^2 2^u$$

Equating the corresponding coefficients of the polynomial factors on each side, we obtain

$$\begin{cases} (\ln 2)A & = 1 \\ 2A + (\ln 2)B & = 0 \\ B + (\ln 2)C & = 0 \end{cases}$$

This system yields

$$A = \frac{1}{\ln 2}, \; B = \frac{-2}{(\ln 2)^2}, \text{ and } C = \frac{2}{(\ln 2)^3}$$

Thus

$$\int x^5 2^{x^2} \, dx = \frac{1}{2} \int u^2 2^u \, du$$

$$= \frac{1}{2} \left( \frac{u^2}{\ln 2} - \frac{2u}{(\ln 2)^2} + \frac{2}{(\ln 2)^3} \right) 2^u + K$$

$$= \left( \frac{x^4}{2 \ln 2} - \frac{x^2}{(\ln 2)^2} + \frac{1}{(\ln 2)^3} \right) 2^{x^2} + K$$

**Do Exercise 13.**                                                                                    ■

## Integrals of the Form $\int p(x)(a^2 - x^2)^r \, dx$

**A CLOSER LOOK**   A polynomial is said to be *odd* if each of its terms is of an odd degree. Similarly, a polynomial is said to be *even* if each of its terms is of an even degree. For example, $3x^5 + 2x^3 - 5x$ is an odd polynomial and $5x^6 + 4x^2 - 7$ is an even polynomial. It can be shown that if $p(x)$ is an odd polynomial of degree $2n - 1$, and if $r$ is a constant, then

$$\int p(x)(a^2 - x^2)^r \, dx = q(x)(a^2 - x^2)^r + C$$

where $q(x)$ is an even polynomial of degree $2n$. We illustrate how this fact can be used with the following example.

**EXAMPLE 4**   Using trial and error, evaluate $\int x^3 (25 - x^2)^{3/2} \, dx$.

**Solution**   Since in the integrand of $\int x^3 (25 - x^2)^{3/2} \, dx$, $x^3$ is an odd polynomial of degree 3, an antiderivative will be $q(x)(25 - x^2)^{3/2}$ where $q(x)$ is an even polynomial of degree 4. That is, an antiderivative must be $(ax^4 + bx^2 + c)(25 - x^2)^{3/2}$, where $a$, $b$, and $c$ are constants to be determined. We must have

$$D_x[(ax^4 + bx^2 + c)(25 - x^2)^{3/2}] = x^3(25 - x^2)^{3/2}$$

But,

$$D_x[(ax^4 + bx^2 + c)(25 - x^2)^{3/2}]$$

$$= (ax^4 + bx^2 + c)\frac{3}{2}(25 - x^2)^{1/2}(-2x) + (25 - x^2)^{3/2}(4ax^3 + 2bx)$$

$$= (25 - x^2)^{1/2}[(-3x)(ax^4 + bx^2 + c) + (25 - x^2)(4ax^3 + 2bx)]$$

$$= (25 - x^2)^{1/2}[-7ax^5 + (100a - 5b)x^3 + (50b - 3c)x]$$

This last expression must be identical to the integrand, which we write as follows:

$$x^3(25 - x^2)^{3/2} = (25 - x^2)^{1/2}(25 - x^2)x^3 = (25 - x^2)^{1/2}(-x^5 + 25x^3)$$

Therefore

$$(25 - x^2)^{1/2}[-7ax^5 + (100a - 5b)x^3 + (50b - 3c)x]$$
$$= (25 - x^2)^{1/2}(-x^5 + 25x^3)$$

It follows that

$$-7ax^5 + (100a - 5b)x^3 + (50b - 3c)x = -x^5 + 25x^3$$

Equating corresponding coefficients we obtain the system

$$\begin{cases} -7a & = -1 \\ 100a - 5b & = 25 \\ 50b - 3c & = 0 \end{cases}$$

This system yields

$$a = \frac{1}{7}, b = \frac{-15}{7}, \text{ and } c = \frac{-250}{7}$$

Thus

$$\int x^3(25 - x^2)^{3/2} \, dx = \left(\frac{x^4}{7} - \frac{15x^2}{7} - \frac{250}{7}\right)(25 - x^2)^{3/2} + K$$

**Do Exercise 23.** (See Exercise *28.)

We conclude this section with the following observation.

In general, it is good practice to differentiate the answer in an integration problem as a check. One advantage of the method of trial and error is that, in one sense, the checking is part of the method. For instance, to find $\int (x^3 + 1)^{20}x^2 \, dx$, we guess that an antiderivative must be of the form $k(x^3 + 1)^{21}$, where $k$ is a constant. Thus we must have

$$D_x k(x^3 + 1)^{21} = (x^3 + 1)^{20}x^2$$

That is,

$$21k(x^3 + 1)^{20}(3x^2) = (x^3 + 1)^{20}x^2$$

which yields

$$63k = 1$$

or

$$k = \frac{1}{63}$$

Therefore

$$\int (x^3 + 1)^{20}x^2 \, dx = \frac{1}{63}(x^3 + 1)^{21} + C$$

At this point, there is no need to differentiate the answer to check if it is correct, since this has already been done.

## Exercise Set 9.5

*In Exercises 1–27, use trial and error to evaluate the given integral.*

**1.** $\displaystyle\int (x^4 + 3)^{25} 6x^3 \, dx$

**2.** $\displaystyle\int (x^3 + 3x^2 + 3x + 1)^{50}(x^2 + 2x + 1) \, dx$

**3.** $\displaystyle\int \sqrt[3]{x^5 + 4}\, 3x^4 \, dx$

**4.** $\displaystyle\int \frac{5x}{(x^2 + 1)^3} \, dx$

**5.** $\displaystyle\int \frac{2x^2 + 4}{\sqrt[3]{(x^3 + 6x + 1)^2}} \, dx$

**6.** $\displaystyle\int \frac{5x}{x^2 + 1} \, dx$

**7.** $\displaystyle\int (e^{2x} + 1)^{50} 4e^{2x} \, dx$

**8.** $\displaystyle\int \frac{4e^{3x}}{(e^{3x} + 2)^5} \, dx$

**9.** $\displaystyle\int xe^{3x} \, dx$

**10.** $\displaystyle\int x^2 e^{4x} \, dx$

**11.** $\displaystyle\int x^3 e^{-3x} \, dx$

**12.** $\displaystyle\int x3^x \, dx$

**13.** $\displaystyle\int x^2 5^x \, dx$

**14.** $\displaystyle\int x^3 6^x \, dx$

**15.** $\displaystyle\int (4x^2 + 22x + 9)e^{4x} \, dx$

**16.** $\displaystyle\int (6x^2 + 8x + 19)e^{6x} \, dx$

**17.** $\displaystyle\int (3x^3 + 9x^2 + 19x + 8)e^{3x} \, dx$

**18.** $\displaystyle\int (2x^3 + x^2 - 14x + 7)e^{-2x} \, dx$

**19.** $\displaystyle\int x^3(25 - x^2)^{5/2} \, dx$

**20.** $\displaystyle\int x^3(49 - x^2)^{3/2} \, dx$

**21.** $\displaystyle\int x^5(16 - x^2)^{3/4} \, dx$

**22.** $\displaystyle\int (x^5 + 4x^3 + 3x)(4 - x^2)^{1/2} \, dx$

**23.** $\displaystyle\int (x^3 - 3x)(16 - x^2)^{5/2} \, dx$

**24.** $\displaystyle\int (2x^5 + 5x^3 - 6x)(25 - x^2)^{1/5} \, dx$

**25.** $\displaystyle\int (36x^3 - 149x)(16 - x^2)^{5/2} \, dx$

**26.** $\displaystyle\int (22x^3 - 274x)(25 - x^2)^{1/5} \, dx$

**27.** $\displaystyle\int (10x^3 + 17x)(4 - x^2)^{1/2} \, dx$

**\*28.** Find the integral of Example 4 by first performing the following algebraic manipulations.

$$
\begin{aligned}
x^3(25 - x^2)^{3/2} &= (x^3 - 25x + 25x)(25 - x^2)^{3/2} \\
&= (x^3 - 25x)(25 - x^2)^{3/2} + 25x(25 - x^2)^{3/2} \\
&= -x(25 - x^2)(25 - x^2)^{3/2} + 25x(25 - x^2)^{3/2} \\
&= -x(25 - x^2)^{5/2} + 25x(25 - x^2)^{3/2}
\end{aligned}
$$

## 9.6 Tabular Integration (Optional)

We have seen that integrals such as $\int x^n e^{ax} \, dx$ may be evaluated using integration by parts, or using a reduction formula, repeatedly. In either case, the algebraic details can be tedious when $n > 2$ and we suggested that in the case where $p(x)$ is a polynomial of degree $n$, the evaluation of $\int p(x)e^{ax} \, dx$ can best be accomplished by trial and error.

There is another method which is extremely efficient in some cases. This method is called *tabular integration* and is a generalization of the technique of integration by parts.

### Generalization of the Integration by Parts Formula

> **THEOREM 9.3:**  Suppose $f$ and $g$ have continuous derivatives of order 2. Then
> $$\int f(x)g''(x) \, dx = f(x)g'(x) - f'(x)g(x) + \int f''(x)g(x) \, dx$$

**Proof:** Differentiating the right side of the equality stated in the theorem, we get

$$D_x[f(x)g'(x) - f'(x)g(x) + \int f''(x)g(x)\, dx]$$
$$= f(x)g''(x) + f'(x)g'(x) - [f'(x)g'(x) + f''(x)g(x)] + f''(x)g(x)$$
$$= f(x)g''(x) + f'(x)g'(x) - f'(x)g'(x) - f''(x)g(x) + f''(x)g(x)$$
$$= f(x)g''(x)$$

Since the last function we obtained is the integrand in the equality stated in the theorem, the proof is complete.

Observe that the continuity of the derivatives is needed to ensure that the functions involved can be antidifferentiated. This fact will be discussed later. We now expand on the foregoing generalization of the integration by parts technique.

**THEOREM 9.4:** Suppose $f$ and $g$ have continuous derivatives of order $n + 1$. Then

$$\int f(x)g^{(n+1)}(x)\, dx = f(x)g^{(n)}(x) - f'(x)g^{(n-1)}(x) + f''(x)g^{(n-2)}(x) - \cdots$$
$$+ (-1)^n f^{(n)}(x)g(x) + (-1)^{(n+1)} \int f^{(n+1)}(x)g(x)\, dx$$

The proof is analogous to that of Theorem 9.3 and is left as an exercise.

This result can be remembered easily if it is written in tabular form as illustrated below.

The integral $\int f(x)g^{(n+1)}(x)\, dx$ is equal to the algebraic sum of the products of the functions connected by arrows, assigning each product "+" and "−" alternately as indicated over each of the arrows, and $(-1)^{(n+1)}$ times the integral of the product of the bottom two functions as indicated below the horizontal arrow at the bottom.

| $k$ | $f^{(k)}(x)$ | $g^{(n+1-k)}(x)$ |
|-----|--------------|------------------|
| 0 | $f(x)$ | $g^{(n+1)}(x)$ |
| 1 | $f'(x)$ | $g^{(n)}(x)$ |
| 2 | $f''(x)$ | $g^{(n-1)}(x)$ |
| 3 | $f'''(x)$ | $g^{(n-2)}(x)$ |
| $\vdots$ | $\vdots$ | $\vdots$ |
| $n-1$ | $f^{(n-1)}(x)$ | $g''(x)$ |
| $n$ | $f^{(n)}(x)$ | $g'(x)$ |
| $n+1$ | $f^{(n+1)}(x)$ | $g(x)$ |

Arrow labels: $+$, $-$, $+$, $\ldots$, $(-1)^{n-1}$, $(-1)^n$, and below: $(-1)^{n+1} \int dx$

As a mnemonic device, in the equality of Theorem 9.4, you may think of $k$ as an exponent in the notation $h^{(k)}(x)$, although *it is not an exponent*, and note that the sum of the "exponents" in each term is $n$, that the signs of the terms alternate starting with a $+$, and that the last term is $\pm$ the integral of $f^{(n+1)}(x)g(x)$.

## How to Use Tabular Integration

To evaluate $\int H(x)\,dx$ using tabular integration, you must write $H(x)$ as a product $f(x)h(x)$, where $f(x)$ will be differentiated $n$ times and $h(x)$ will be antidifferentiated $n$ times. Thus the choice must be made in such a way that these operations are possible. Once the choice has been made, we write $f(x)$ at the top of the second column and we differentiate $n$ times to get the other entries in that column. We also write $h(x) = g^{(n)}(x)$ in the third column and integrate $n$ times to complete the column. The value of $n$ must be such that $\int f^{(n)}(x)g(x)\,dx$ can readily be evaluated.

This generalization of integration by parts can be used most efficiently to evaluate $\int f(x)h(x)\,dx$, where $f(x)$ is a polynomial of degree $k$ and $h(x)$ can be easily antidifferentiated $k + 1$ times, because in that case, $f^{(k+1)}(x) = 0$.

## Examples

**EXAMPLE 1**  Evaluate $\displaystyle\int (x^3 + 2x^2 - 5x + 2)e^{3x}\,dx.$

**Solution**  We use tabular integration.

| $k$ | $f^{(k)}(x)$ | $g^{(4-k)}(x)$ |
|---|---|---|
| 0 | $x^3 + 2x^2 - 5x + 2$ | $e^{3x}$ |
| 1 | $3x^2 + 4x - 5$ | $\dfrac{e^{3x}}{3}$ |
| 2 | $6x + 4$ | $\dfrac{e^{3x}}{9}$ |
| 3 | $6$ | $\dfrac{e^{3x}}{27}$ |
| 4 | $0$ | $\dfrac{e^{3x}}{81}$ |

$+\displaystyle\int\,dx$

Thus,

$$\int (x^3 + 2x^2 - 5x + 2)e^{3x}\,dx = (x^3 + 2x^2 - 5x + 2)\frac{e^{3x}}{3}$$
$$- (3x^2 + 4x - 5)\frac{e^{3x}}{9} + (6x + 4)\frac{e^{3x}}{27} - 6\frac{e^{3x}}{81} + \int (0)\frac{e^{3x}}{81}\,dx$$

$$= \frac{e^{3x}}{27}(9x^3 + 9x^2 - 51x + 35) + C$$

where $C$ is an arbitrary constant.
**Do Exercise 5.**                                                              ■

The first column (giving the value of $k$) and the first row may be omitted from the table. (See Example 2.)

If at any step of a tabular integration problem $\int f(x)g^{(n+1)}(x)\,dx$, the differentiation of $f^{(k)}(x)$ becomes tedious, or the integration of $g^{(n+1-k)}(x)$ becomes difficult, or impossible, we try to simplify the product $f^{(k)}(x)g^{(n+1-k)}(x)$ and start tabular integration again with the simplified expression. We indicate this step on the table by drawing a vertical arrow pointing down. This method is illustrated in the next example.

**EXAMPLE 2**    Evaluate $\int \ln^3|x|\,dx$.

**Solution**    We start by writing $\ln^3|x| = (\ln^3|x|)(1)$ and by differentiating $\ln^3|x|$ in the first column of the table and integrating 1 in the second column. Following the table, we shall make some remarks concerning the simplifications.

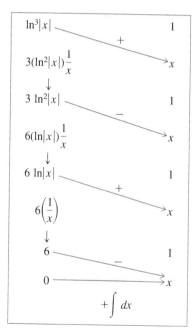

Observe that in line (2), we have $3(\ln^2|x|)\frac{1}{x}$ and $x$. Although $x$ can be integrated easily, the differentiation of $3(\ln^2|x|)\frac{1}{x}$ would have involved the Product (or Quotient) Rule and would have yielded a "messy" expression. So it was best to simplify the product $\left[3(\ln^2|x|)\frac{1}{x}\right]x$ and write the next line as "3 $\ln^2|x|$ . . . 1," before continuing the process. We indicated this simplification in the table by a vertical arrow pointing down to line (3). The reason for the other simplifications should be clear. Finally, to obtain the answer, we write only the products of those functions that are connected diagonally by arrows and ignore the functions that are at the tail of vertical arrows.

Thus, $\int \ln^3|x| \, dx = (\ln^3|x|)x - 3(\ln^2|x|)x + (6 \ln|x|)x - 6x + \int 0(x) \, dx = x(\ln^3|x| - 3 \ln^2|x| + 6 \ln|x| - 6) + C$.

**Do Exercise 11.** ∎

## Exercise Set 9.6

*Evaluate each of the following integrals using tabular integration.*

**1.** $\int x^3 e^{5x} \, dx$

**2.** $\int x^4 e^{4x} \, dx$

**3.** $\int x^5 e^{6x} \, dx$

**4.** $\int (x^2 - 5x + 6)e^{2x} \, dx$

**5.** $\int (x^3 - 7x^2 + 3x + 3)e^{-3x} \, dx$

**6.** $\int (x^4 - 5x^2 + 7x - 5)e^{-2x} \, dx$

**7.** $\int (2x^4 + 5x^3 - 7x^2 + 6x - 13)e^{2x} \, dx$

**8.** $\int (-x^5 + 3x^3 - 4x^2 + 11x - 16)2^{5x} \, dx$

**9.** $\int (x^6 + 2)3^{4x} \, dx$

**10.** $\int \ln^4|x| \, dx$

**11.** $\int \ln^5|x| \, dx$

**12.** $\int x \ln^3|x + 1| \, dx$ (*Hint:* First make the substitution $u = x + 1$.)

**13.** $\int x \ln^2|x + 3| \, dx$ (See Exercise 12.)

**14.** $\int x^2 \ln^3|x - 1| \, dx$ (See Exercise 12.)

**15.** $\int x^3 \ln^2|x + 3| \, dx$ (See Exercise 12.)

**16.** $\int x^4 \ln^3|x - 2| \, dx$ (See Exercise 12.)

**17.** $\int (x^2 - 3x + 1)\ln^2|x + 1| \, dx$ (See Exercise 12.)

**18.** $\int (4x^3 - 6x^2 + 2x - 1)\ln^3|x - 1| \, dx$ (See Exercise 12.)

**19.** Use tabular integration to evaluate the integral of Example 4, Section 9.5. (*Hint:* Let $f(x) = x^2$ and $h(x) = (25 - x^2)^{3/2}x$.)

**20.** Use tabular integration to evaluate the integral $\int x^3(16 - x^2)^{5/2} \, dx$. (*Hint:* See Exercise 19.)

**21.** In the solution of Example 1 of this section, let $\int e^{3x} dx = e^{3x}/3 + 6$, $\int (e^{3x}/3 + 6) \, dx = e^{3x}/9 + 6x + 2$, and $\int (e^{3x}/9 + 6x + 2) \, dx = e^{3x}/27 + 3x^2 + 2x + 1$, and show that we obtain the same answer.

**22.** In Exercise 21, we gave the following values to the constants of integration: $c_1 = 6$, $c_2 = 2$, $c_3 = 1$, and $c_4 = 1$. Repeat Exercise 21 with $c_1 = 3$, $c_2 = 5$, $c_3 = -4$, and $c_4 = 7$.

## 9.7 Intuitive Discussion of Limits of Sequences

Limits were discussed in Chapter 7 so that we could define derivatives. There is another type of limit that we used in Chapter 8 when we described asymptotes. You must have some understanding of this type of limit, applied to sequences, to work with definite integrals, which we will soon define. This section is aimed at giving you that understanding.

### Limits of Sequences

Almost everyone is familiar with the use of periodic decimals in writing the decimal representation of certain real numbers. When we write $\frac{4}{3}$ as 1.3333. . . or as $1.\overline{3}$, the three dots, or the bar over the 3, indicate that there are infinitely many 3s following the decimal point. We may think of an infinite sequence of real numbers defined as follows:

$$s_1 = 1, \; s_2 = 1.3, \; s_3 = 1.33, \; s_4 = 1.333, \; s_5 = 1.3333, \ldots .$$

Each $s_n$ is an approximation of the number $\frac{4}{3}$; the larger the value of the integer $n$, the better the approximation is. We say that the number $\frac{4}{3}$ is the limit of the sequence $s_n$ as $n$ tends to infinity.

As another example, consider a line segment 2 units long. Cut this segment into two equal segments, each 1 unit long. Keep the left half and divide the right half into two equal segments of $\frac{1}{2}$ unit each. Keep the segment on the left in this last division and divide the other into two equal segments of $\frac{1}{4}$ unit each. Keeping up this process, after $n$ steps we have $n + 1$ segments of lengths $1, \frac{1}{2} \; \frac{1}{4}, \frac{1}{8}, \ldots, 1/2^{n-1}$, and $1/2^{n-1}$. Let $s_n$ be the sum of the lengths of the first $n$ segments. That is,

$$s_n = 1 + \frac{1}{2} + \frac{1}{4} + \frac{1}{8} + \cdots + \frac{1}{2^{n-1}}$$

It is geometrically clear (from Figure 9.1) that $s_n$ is equal to 2, the length of the original segment, minus $1/2^{n-1}$ the length of the last segment on the right, which length was not added in defining $s_n$. (See Figure 9.1.)

**FIGURE 9.1**

Thus, $s_n = 2 - 1/2^{n-1}$.

Obviously, if $n$ is very large, $2^{n-1}$ will be very large and $1/2^{n-1}$ will be near 0. Thus, the value of $s_n$ will be close to 2 when $n$ is very large. That is, the limit of $s_n$ as $n$ tends to infinity is 2.

---

**DEFINITION 9.2:** Suppose that $s_n$ is a sequence whose range is a set of real numbers, then the real number $L$ is the *limit of $s_n$ as n tends to infinity* and we write

$$\lim_{n \to \infty} s_n = L$$

provided that we can make the value of $s_n$ arbitrarily close to $L$ by making the value of $n$ sufficiently large.*

---

*This definition is adequate for our purpose; but, in more advanced texts, the notions of "arbitrarily close" and "sufficiently large" are made precise so that theorems on limits can be proved. We shall use our intuition in working with limits. Theorem 9.5 is stated without proof.

THEOREM 9.5: Suppose that $s$ and $t$ are sequences with ranges set of real numbers and that $\lim_{n \to \infty} s_n = A$ and $\lim_{n \to \infty} t_n = B$. Then

1. $\lim_{n \to \infty} (s_n + t_n) = A + B,$

2. $\lim_{n \to \infty} (s_n - t_n) = A - B,$

3. $\lim_{n \to \infty} (s_n \cdot t_n) = A \cdot B,$

4. $\lim_{n \to \infty} \dfrac{s_n}{t_n} = \dfrac{A}{B}$, provided $B \neq 0$ and $t_n \neq 0$, and if $k$ is a positive integer,

5. $\lim_{n \to \infty} s_n{}^k = A^k.$

This theorem plus the following facts will be used to evaluate limits.

1. If $c$ is a real number and $s$ is the constant sequence defined by $s_n = c$ for all values of $n$, then

$$\lim_{n \to \infty} s_n = c$$

2. If $t$ is a sequence and $t_n$ increases without bound as $n$ tends to infinity, we say that the limit of $t_n$ as $n$ tends to infinity is infinity and we write

$$\lim_{n \to \infty} t_n = \infty$$

3. If $t$ is a sequence such that $\lim_{n \to \infty} t_n = \infty$ and $a$ is any real number, then

$$\lim_{n \to \infty} \frac{a}{t_n} = 0$$

4. If $c$ is a real number and $c > 1$, then

$$\lim_{n \to \infty} c^n = \infty$$

5. If $r$ is a real number such that $|r| < 1$, then

$$\lim_{n \to \infty} r^n = 0$$

6. If $s$, $t$, and $u$ are sequences such that $s_n \leq t_n \leq u_n$ and limit $\lim_{n \to \infty} s_n = \lim_{n \to \infty} u_n = A$, then

$$\lim_{n \to \infty} t_n = A$$

The last statement is often called the *Squeeze Theorem*.

EXAMPLE 1   Let the sequence $s$ be defined by

$$s_n = \left(1 + \frac{1}{n}\right)\left(2 + \frac{3}{2n}\right)$$

Evaluate $\lim_{n \to \infty} s_n$.

**Solution** Using the third statement, we have $\lim\limits_{n\to\infty} \dfrac{1}{n} = 0$ and $\lim\limits_{n\to\infty} \dfrac{3}{2n} = 0$. Thus,

$$\lim_{n\to\infty} s_n = \lim_{n\to\infty}\left[\left(1 + \frac{1}{n}\right)\left(2 + \frac{3}{2n}\right)\right]$$

$$= \left[\lim_{n\to\infty}\left(1 + \frac{1}{n}\right)\right] \cdot \left[\lim_{n\to\infty}\left(2 + \frac{3}{2n}\right)\right]$$

$$= \left[\lim_{n\to\infty} 1 + \lim_{n\to\infty} \frac{1}{n}\right] \cdot \left[\lim_{n\to\infty} 2 + \lim_{n\to\infty} \frac{3}{2n}\right]$$

$$= (1 + 0)(2 + 0) = 2$$

**Do Exercise 1.**

**EXAMPLE 2** Let $s$ be a sequence such that for all positive integers $n$, we have

$$\left(1 - \frac{1}{n}\right)\left(3 - \frac{5}{n}\right) < s_n < \left(1 + \frac{1}{n}\right)\left(3 + \frac{5}{n}\right)$$

Evaluate $\lim\limits_{n\to\infty} s_n$.

**Solution** Using the same steps as in Example 1, we find that

$$\lim_{n\to\infty}\left[\left(1 - \frac{1}{n}\right)\left(3 - \frac{5}{n}\right)\right] = \lim_{n\to\infty}\left[\left(1 + \frac{1}{n}\right)\left(3 + \frac{5}{n}\right)\right] = 3$$

Thus, by the Squeeze Theorem, we conclude that $\lim\limits_{n\to\infty} s_n = 3$.

**Do Exercise 11.**

## Geometric Series

The indicated sum of the terms of an unending geometric progression

$$a + ar + ar^2 + ar^3 + \cdots + ar^n + \cdots$$

is an example of an *infinite series*. It is often called a *geometric series*. The sum of the first $n$ terms of this series is called its *nth partial sum* and is denoted $S_n$. As we have seen earlier, if $r \neq 1$, then

$$S_n = \frac{a(1 - r^n)}{1 - r}$$

If $|r| < 1$, the value of $r^n$ may be made as close to 0 as we wish by making the value of $n$ sufficiently large. For example, if $r = \frac{1}{2}$, $r^2 = \frac{1}{4}$, $r^3 = \frac{1}{8}$, $r^4 = \frac{1}{16}$, $r^{10} = \frac{1}{1024}$, and so on. Thus, the value of the $n$th partial sum $S_n$ may be made as close as we please to $\frac{a}{1-r}$ by making the value of $n$ sufficiently large. By definition, the value of the geometric series is equal to $\lim\limits_{n\to\infty} S_n$, provided this limit exists. Hence, in the case $|r| < 1$, $\lim\limits_{n\to\infty} r^n = 0$, and

$$a + ar + ar^2 + ar^3 + \cdots = \lim_{n\to\infty} S_n = \frac{a}{1 - r}$$

**EXAMPLE 3**    Find the value of the geometric series

$$3 + \frac{3}{4} + \frac{3}{4^2} + \frac{3}{4^3} + \cdots + \frac{3}{4^{n-1}} + \cdots$$

**Solution**    The first term of the geometric series is 3 and the common ratio is $\frac{1}{4}$. Thus, $a = 3$ and $r = \frac{1}{4}$. Since $|r| < 1$, the value of the series is given by

$$\frac{a}{1 - r} = \frac{3}{1 - \frac{1}{4}} = \frac{3 \cdot 4}{\left(1 - \frac{1}{4}\right) \cdot 4} = \frac{12}{4 - 1} = \frac{12}{3} = 4$$

That is

$$3 + \frac{3}{4} + \frac{3}{16} + \frac{3}{4^3} + \cdots + \frac{3}{4^{n-1}} + \cdots = 4$$

Therefore the value of the geometric series is 4.
**Do Exercise 21.**

**EXAMPLE 4**    Find the second and fifth terms of a geometric series whose third term is $\frac{-3}{2}$ and whose value is $-4$.

**Solution**    We have

$$ar^{(3-1)} = ar^2 = \frac{-3}{2}, \frac{a}{1 - r} = -4, \text{ and } |r| < 1$$

Dividing the first equation by the second, we obtain

$$\frac{ar^2}{\frac{a}{1 - r}} = \frac{\frac{-3}{2}}{-4}$$

Simplifying, we get

$$8r^3 - 8r^2 + 3 = 0$$

Factoring, we have

$$(2r + 1)(4r^2 - 6r + 3) = 0$$

Thus,

$$(2r + 1) = 0 \text{ or } 4r^2 - 6r + 3 = 0$$

The first equation yields $r = \frac{-1}{2}$ and the second equation has no real solution. Since $ar^2 = \frac{-3}{2}$ and $r = \frac{-1}{2}$, $a = -6$. Therefore, the second term is $(-6)\left(\frac{-1}{2}\right)^{(2-1)} = 3$ and the fifth term is $(-6)\left(\frac{-1}{2}\right)^{(5-1)} = \frac{-3}{8}$.
**Do Exercise 31.**

**EXAMPLE 5**    Write the rational number whose decimal representation is 5.232323232323. . . . as a ratio $\frac{a}{b}$, where $a$ and $b$ are positive integers.

**Solution**   The given decimal may be considered as the number 5 plus the infinite geometric series

$$\frac{23}{100} + \frac{23}{100}\left(\frac{1}{100}\right) + \frac{23}{100}\left(\frac{1}{100}\right)^2 + \cdots + \frac{23}{100}\left(\frac{1}{100}\right)^{n-1} + \cdots$$

The first term of this series is $\frac{23}{100}$ and the common ratio is $\frac{1}{100}$. Thus the value of the infinite geometric series is given by

$$\frac{\dfrac{23}{100}}{1 - \dfrac{1}{100}} = \frac{23}{99}$$

It follows that

$$5.2323232323 \cdots = 5 + \frac{23}{99} = \frac{495 + 23}{99} = \frac{518}{99}$$

**Do Exercise 35.**                                                                  ■

**EXAMPLE 6**   Suppose that a ball rebounds one-half of the distance it falls. Ideally, it will bounce infinitely many times, but this is only an approximation of what really happens. Actually, the ball comes to rest after a certain number of bounces. If the ball is dropped from a height of 20 feet, estimate the distance it travels before coming to rest.

**Solution**   The ball drops 20 feet, then it rebounds 10 feet and drops 10 feet, then it rebounds 5 feet and drops 5 feet, then it rebounds $\frac{5}{2}$ feet and drops $\frac{5}{2}$ feet, and so on. Thus, ideally, the distance traveled by the ball is given by

$$20 + (10 + 10) + (5 + 5) + \left(\frac{5}{2} + \frac{5}{2}\right) + \cdots$$

$$= 20 + (20 + 10 + 5 + \cdots)$$

$$= 20 + \frac{20}{1 - \dfrac{1}{2}} = 20 + 40 = 60$$

since $20 + 10 + 5 + \ldots$ is an infinite geometric series with $a = 20$ and $r = \frac{1}{2}$. We conclude that before coming to rest, the ball has traveled approximately 60 feet.
**Do Exercise 45.**                                                                  ■

## More on Calculations of Limits

It is often useful to manipulate the expression defining a sequence algebraically in order to evaluate the limit of that sequence. The last two examples illustrate this method.

**EXAMPLE 7**   Evaluate $\lim\limits_{n \to \infty} \dfrac{2n + 1}{3n - 2}$.

**Solution**   Note first that

$$\frac{2n + 1}{3n - 2} = \frac{\dfrac{2n + 1}{n}}{\dfrac{3n - 2}{n}} = \frac{2 + \dfrac{1}{n}}{3 - \dfrac{2}{n}}$$

Thus,

$$\lim_{n \to \infty} \frac{2n + 1}{3n - 2} = \lim_{n \to \infty} \frac{2 + \dfrac{1}{n}}{3 - \dfrac{2}{n}}$$

$$= \frac{\lim\limits_{n \to \infty} \left(2 + \dfrac{1}{n}\right)}{\lim\limits_{n \to \infty} \left(3 - \dfrac{2}{n}\right)} = \frac{\lim\limits_{n \to \infty} 2 + \lim\limits_{n \to \infty} \dfrac{1}{n}}{\lim\limits_{n \to \infty} 3 - \lim\limits_{n \to \infty} \dfrac{2}{n}} = \frac{2 + 0}{3 - 0} = \frac{2}{3}$$

Therefore

$$\lim_{n \to \infty} \frac{2n + 1}{3n - 2} = \frac{2}{3}$$

An alternate solution may be given by first dividing $2n + 1$ by $3n - 2$ to get

$$\frac{2n + 1}{3n - 2} = \frac{2}{3} + \frac{\dfrac{7}{3}}{3n - 2}$$

**Do Exercise 3.** ■

We stated in Section 1.4 that the expression $\left(1 + \dfrac{1}{n}\right)^n$ approaches the irrational number $e$ (approximately equal to 2.71828) as $n$ increases without bound. We can now formally state that

$$\lim_{n \to \infty} \left(1 + \frac{1}{n}\right)^n = e$$

**EXAMPLE 8**   Evaluate $\lim\limits_{n \to \infty} \left(1 + \dfrac{3}{n}\right)^n$.

**Solution**   Let $n = 3k$ to obtain

$$\left(1 + \frac{3}{n}\right)^n = \left(1 + \frac{3}{3k}\right)^{3k} = \left(1 + \frac{1}{k}\right)^{3k} = \left[\left(1 + \frac{1}{k}\right)^k\right]^3$$

As $n$ increases without bound, so does $k$. Thus,

$$\lim_{n \to \infty}\left(1 + \frac{3}{n}\right)^n = \lim_{n \to \infty}\left[\left(1 + \frac{1}{k}\right)^k\right]^3$$

$$= \left[\lim_{k \to \infty}\left(1 + \frac{1}{k}\right)^k\right]^3 = e^3*$$

**Do Exercise 13.**

---

# Exercise Set 9.7

*In Exercises 1–15, evaluate* $\lim\limits_{n \to \infty} s_n$.

**1.** $s_n = \dfrac{2}{n}$

**2.** $s_n = \dfrac{-5}{n + 1}$

**3.** $s_n = \dfrac{2n + 1}{n + 3}$

**4.** $s_n = \dfrac{3n + 1}{2n - 1}$

**5.** $s_n = \dfrac{(-1)^n}{n + 1}$

**6.** $s_n = \dfrac{2 - n}{n + 3}$

**7.** $s_n = 4 - \left(\dfrac{1}{2}\right)^n$

**8.** $s_n = 2 + \left(\dfrac{1}{3}\right)^n$

**9.** $s_n = \dfrac{3}{2 - \left(\dfrac{4}{5}\right)^n}$

**10.** $s_n = \left[3 + \left(\dfrac{2}{3}\right)^n\right]\dfrac{n + 1}{3n - 2}$

**11.** $\dfrac{1}{n + 1} < s_n < \dfrac{1}{n}$

**12.** $\dfrac{2n - 1}{n + 3} < s_n < \dfrac{2n + 1}{n + 2}$

**13.** $s_n = \left(1 + \dfrac{5}{n}\right)^n$

**14.** $s_n = \left(1 + \dfrac{2}{n}\right)^{5n}$

**15.** $s_n = \left(1 + \dfrac{1}{4n}\right)^n$

**16.** Read Example 8 of this section. Then use a calculator to find the value of $\left(1 + \frac{3}{n}\right)^n$ for **a.** $n = 1000$, **b.** $n = 10,000$ and **c.** $n = 100,000$. Also, find the value of $e^3$ and compare it to your answers in parts **a**, **b**, and **c**.

**17.** Do Exercise 13. Then use a calculator to find the value of $\left(1 + \frac{5}{n}\right)^n$ for **a.** $n = 1000$, **b.** $n = 10,000$, **c.** $n = $

100,000. Compare your answers to the answer you obtained in Exercise 13.

**18.** Do Exercise 14. Then use a calculator to find the value of $\left(1 + \frac{2}{n}\right)^{5n}$ for **a.** $n = 1000$, **b.** $n = 10,000$, and **c.** $n = 100,000$. Compare your answers to the answer you obtained in Exercise 14.

**19.** Do Exercise 15. Then use a calculator to find the value of $\left(1 + \frac{1}{4n}\right)^n$ for **a.** $n = 1000$, **b.** $n = 10,000$, and **c.** $n = 100,000$. Compare your answers to the answer you obtained in Exercise 15.

*In Exercises 20–29, find the value of the geometric series.*

**20.** $18 + 6 + 2 + \cdots$

**21.** $18 - 6 + 2 - \cdots$

**22.** $6 + 3 + \dfrac{3}{2} + \cdots$

**23.** $6 - 3 + \dfrac{3}{2} - \cdots$

**24.** $2 + \dfrac{4}{3} + \dfrac{8}{9} + \cdots$

**25.** $2 - \dfrac{4}{3} + \dfrac{8}{9} - \cdots$

**26.** $2.197 + 1.69 + 1.3 + \cdots$

**27.** $2.197 - 1.69 + 1.3 - \cdots$

**28.** $7 + \sqrt{7} + 1 + \cdots$

**29.** $7 - \sqrt{7} + 1 - \cdots$

**30.** Find the second and third terms of a geometric series whose first term is 6 and whose value is 9.

**31.** Find the third and fifth terms of a geometric series whose first term is 5 and whose value is 15.

**32.** The value of a geometric series is 32 and the second term is 8 less than the first term. Find the first term.

---

*In this example, we let $n = 3k$. Thus, $n$ must be a multiple of 3. It can be shown that if $s$ is a sequence with a limit (that is, a *convergent sequence*), then the value of that limit is unique. Further if $n_i$ is an increasing sequence of positive integers, then $\lim\limits_{n \to \infty} s_n = \lim\limits_{n_i \to \infty} s_{n_i}$. Thus the argument used in the solution of this example is valid even though we took the limit of $\left(1 + \frac{3}{n}\right)^n$, as $n$ went to infinity, taking on values of multiples of 3.*

**33.** Suppose that $s_1 + s_2 + s_3 + \cdots$ is a geometric series and that $s_1 + s_3 + s_5 + \cdots = \frac{729}{4}$ and $s_2 + s_4 + s_6 + \cdots = \frac{243}{4}$. Find the first three terms of this geometric series.

**34.** Same as Exercise 33 if the sum of the odd terms is 12 and the sum of the even terms is 6.

*In Exercises 35–44, write the rational number as a ratio $\frac{a}{b}$ where a and b are positive integers.*

**35.** 3.2727272727. . .

**36.** 12.413413413413413. . .

**37.** 6.01212121212. . .

**38.** 125.47231231231231. . .

**39.** 24.625013013013013013. . .

**40.** $5.2\overline{47}$

**41.** $3.6\overline{784}$

**42.** $16.78\overline{96}$

**43.** $13.\overline{7653}$

**44.** $129.001\overline{37}$

**45.** A ball rebounds two thirds the distance it falls. If it is dropped from a height of 30 feet, estimate the distance the ball travels before it comes to rest.

**46.** A ball rebounds three fourths the distance it falls. If it is dropped from a height of 20 feet, estimate the distance the ball travels before it comes to rest.

**47.** A ball rebounds three fifths the distance it falls. If it is dropped from a height of 15 feet, estimate the distance the ball travels before it comes to rest.

**48.** Each swing of a pendulum is three fourths the length of the preceding swing. If the first swing is 30 centimeters long, how far does the pendulum travel before coming to rest?

**49.** Each swing of a pendulum is 85% of the length of the preceding swing. If the first swing is 25 inches long, how far does the pendulum travel before coming to rest?

**50.** Each swing of a pendulum is 65% the length of the preceding swing. If the first swing is 18 centimeters long, how far does the pendulum travel before coming to rest?

**51.** If $A$ square units is the area of the region bounded by the graphs of $y = x^2$, $x = 3$, and the $x$-axis, then for all positive integers $n$, the following is true:

$$\frac{9}{2}\left(1 - \frac{1}{n}\right)\left(2 - \frac{1}{n}\right) < A < \frac{9}{2}\left(1 + \frac{1}{n}\right)\left(2 + \frac{1}{n}\right)$$

Find the area of the region.

**52.** If $A$ square units is the area of the region bounded by the graphs of $y = x^3$, $x = 2$, and the $x$-axis, then for all positive integers $n$, the following is true:

$$4\left[1 - \frac{2}{n} + \left(\frac{1}{n}\right)^2\right] < A < 4\left[1 + \frac{2}{n} + \left(\frac{1}{n}\right)^2\right]$$

Find the area of the region.

*In Exercises 53 and 54, use geometric series to solve the problems. Recall that if the rate of interest of an investment is i (written as a decimal), per conversion period, then the present value P dollars of a payment of R dollars n conversion periods from now is $P = R/[(1 + i)^n]$.*

**53.** Suppose that Mrs. Tradition wishes to establish a fund, earning 8% compounded annually, to maintain a family crypt. The maintenance cost is $300 a year with the first payment due immediately. How much must she pay now to purchase an annuity that will take care of the perpetual maintenance?

**54.** Suppose that Mr. Goodfellow wishes to establish a fund, earning 9% interest compounded annually, to award a scholarship to a student majoring in business at his alma mater. The maintenance of the scholarship will cost $19,500 a year with the first payment due immediately. How much should be paid now to purchase an annuity that will take care of the perpetual maintenance of the scholarship?

# 9.8 The Definite Integral

Before reading this section, you should review Section 1.7 (The Sigma Notation) and the preceding section.

## Area Under a Curve

The derivative was introduced in Chapter 7 because we often encounter instantaneous rates of change of functions, and the derivative serves as a mathematical model for such rates of change. In the applications, we often work with another type of limit.

That is, we take the limit of a summation of a very large number of individually very small terms as the number of terms tends to infinity. An example of this type of application is finding the area in the $xy$-plane of a region bounded by several curves.

Suppose we need to find the area of a region $S$ such as the region shaded in Figure 9.2**a**. We first approximate this area as follows. We divide the interval $[a, b]$ into subintervals. Then we draw rectangles over each of these. The union of these rectangles form rectangular polygons whose areas aproximate the area of the given region. (See Figures 9.2**b** and **c**.) If we divide $[a, b]$ into more subintervals, we expect to obtain better approximations. We assume that if $R$ and $T$ are regions such that

$$R \subseteq T$$

then

$$\text{area of } R \leq \text{area of } T$$

We then find a sequence $I_n$ of inscribed rectangular polygons (see Figure 9.2**b**)) and another sequence $C_n$ of circumscribed rectangular polygons (see Figure 9.2**c**) such that the following are true:

**a.** For each $n$, the areas of $I_n$ and $C_n$ are known,
**b.** For each $n$, $I_n \subseteq S \subseteq C_n$, so that area of $I_n \leq$ area of $S \leq$ area of $C_n$, and
**c.** $\lim_{n \to \infty}(\text{area of } I_n) = \lim_{n \to \infty}(\text{area of } C_n) = A$.

Then we can conclude that the area of the region $S$ is also $A$ square units.

To carry out the details of this procedure, we first need a few definitions.

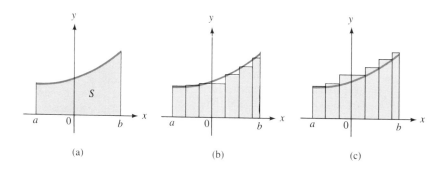

**FIGURE 9.2**

(a)          (b)          (c)

**DEFINITION 9.3:**    Let $f$ be a continuous function on the closed interval $[a, b]$ such that $f(x) > 0$ for all $x$ in $[a, b]$. Consider the region $S$ that is bounded by the graphs of $y = f(x)$, $x = a$, $x = b$, and the $x$-axis. A *partition* $p$ of the interval $[a, b]$ is an ordered set of numbers $(x_0, x_1, x_2, \ldots, x_n)$ such that $a = x_0 < x_1 < x_2 < \ldots < x_n = b$. If

$$x_1 - x_0 = x_2 - x_1 = \cdots = x_n - x_{n-1}$$

the partition is said to be *regular*. Given a partition $p$ of $[a, b]$, the interval $[x_{i-1}, x_i]$ is called the *ith subinterval* formed by the partition. By continuity of the function $f$, for each $i \in \{1, 2, 3, \ldots, n\}$, we can find $u_i$ and $v_i$ in $[x_{i-1}, x_i]$ such that

$$f(u_i) \le f(x) \le f(v_i)$$

for all $x \in [x_{i-1}, x_i]$. For each $i \in \{1, 2, 3, \ldots, n\}$, we draw a rectangle with base $[x_{i-1}, x_i]$ and height $f(u_i)$. The union of the $n$ rectangles is called the *inscribed rectangular polygon corresponding to the partition p* and is denoted $I_p$. (See Figure 9.3.) Similarly, for each $i \in \{1, 2, 3, \ldots, n\}$, we draw a rectangle with base $[x_{i-1}, x_i]$ and height $f(v_i)$. The union of these $n$ rectangles is called the *circumscribed rectangular polygon corresponding to the partition p* and is denoted $C_p$. (See Figure 9.4.)

**FIGURE 9.3**

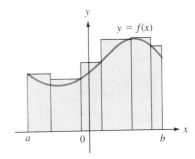

**FIGURE 9.4**

Obviously, $I_p \subseteq S \subseteq C_p$. The areas of $I_p$ and $C_p$ are easily found since they are sums of areas of the rectangles. Let $\Delta x_i = x_i - x_{i-1}$. Then for $I_p$, the dimensions of the $i$th rectangle are $\Delta x_i$ by $f(u_i)$, so that its area is $f(u_i)\Delta x_i$. Consequently, the area of $I_p$ is $\sum_{i=1}^{n} f(u_i)\Delta x_i$.

Similarly, for $C_p$, the dimensions of the $i$th rectangle are $\Delta x_i$ by $f(v_i)$, so that its area is $f(v_i)\Delta x_i$. Therefore the area of $C_p$ is $\sum_{i=1}^{n} f(v_i)\Delta x_i$. It is convenient to use the notation $\underline{A}_p$ (read: "$A$ sub $p$ lower bar") for the area of the inscribed rectangular polygon $I_p$, and $\overline{A}_p$ (read: "$A$ sub $p$ upper bar") for the area of the circumscribed rectangular polygon $C_p$. Thus we have

$$\underline{A}_p = \sum_{i=1}^{n} f(u_i)\Delta x_i$$

and

$$\overline{A}_p = \sum_{i=1}^{n} f(v_i)\Delta x_i$$

Furthermore,

$$\underline{A}_p \leq \text{area of } S \leq \overline{A}_p$$

Clearly, if the values of $\underline{A}_p$ and $\overline{A}_p$ are nearly equal, then either of these values is an approximation for the area of the region $S$. If we can find a sequence of partitions $p_n$ such that

$$\lim_{n \to \infty} \underline{A}_{p_n} = \lim_{n \to \infty} \overline{A}_{p_n} = A$$

then we conclude that the area of the region $S$ is also $A$. (This is, in fact, the way area is defined.)

In the next examples, we will use the following formulas, which can be proved by mathematical induction. (See Appendix A.)

$$\sum_{i=1}^{n} i = \frac{n(n+1)}{2}$$

$$\sum_{i=1}^{n} i^2 = \frac{n(n+1)(2n+1)}{6}$$

$$\sum_{i=1}^{n} i^3 = \left[\frac{n(n+1)}{2}\right]^2$$

Recall also that if $k$ is a constant, then

$$\sum_{i=1}^{n} k = nk$$

$$\sum_{i=1}^{n} kf(i) = k \sum_{i=1}^{n} f(i)$$

$$\sum_{i=1}^{n} [f(i) + g(i)] = \sum_{i=1}^{n} f(i) + \sum_{i=1}^{n} g(i)$$

**EXAMPLE 1**    Let $S$ be the region bounded by the graphs of $y = f(x) = 3x$, $x = 2$, and $y = 0$. Since $S$ is a right triangle, its area is easily found to be 6 square units using the formula $A = \frac{1}{2}bh$. Using this region, illustrate the techniques just described.

**a.** Let $p = \left(0, \frac{1}{3}, 1, \frac{4}{3}, 2\right)$ be a partition of the interval $[0, 2]$. Find $\underline{A}_p$ and $\overline{A}_p$.

**b.** Let $q$ be the regular partition of $[0, 2]$ which determines 6 subintervals. Find $\underline{A}_q$ and $\overline{A}_q$.

**c.** Let $r_n$ be the regular partition of $[0, 2]$ which determines $n$ subintervals. Find $\underline{A}_{r_n}$ and $\overline{A}_{r_n}$.

**d.** Show that $\lim_{n \to \infty} \underline{A}_{r_n} = \lim_{n \to \infty} \overline{A}_{r_n} = 6$.

**Solution**    Since the function $f$ is increasing on the interval $[0, 2]$, on each subinterval the minimum value of $f$ occurs at the left endpoint and the maximum value occurs at the right endpoint. (See Figure 9.5.)

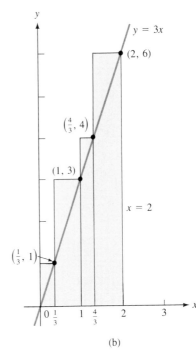

FIGURE 9.5

(a)                                    (b)

**a.** For the partition $p$, we proceed as follows: On the first subinterval $\left[0, \frac{1}{3}\right]$, $u_1 = 0$ and $v_1 = \frac{1}{3}$. Therefore $f(u_1) = 3 \cdot 0 = 0$ and $f(v_1) = 3 \cdot \frac{1}{3} = 1$. Furthermore, $\Delta x_1 = \frac{1}{3} - 0 = \frac{1}{3}$. Thus the area of the first inscribed rectangle is

$$f(u_1)\Delta x_1 = 0 \cdot \frac{1}{3} = 0$$

and the area of the first circumscribed rectangle is

$$f(v_1)\Delta x_1 = 1 \cdot \frac{1}{3} = \frac{1}{3}$$

On the second subinterval $\left[\frac{1}{3}, 1\right]$, $u_2 = \frac{1}{3}$ and $v_2 = 1$. Therefore $f(u_2) = 3 \cdot \frac{1}{3} = 1$ and $f(v_2) = 3 \cdot 1 = 3$. Also, $\Delta x_2 = 1 - \frac{1}{3} = \frac{2}{3}$. Thus the area of the second inscribed rectangle is

$$f(u_2)\Delta x_2 = 1 \cdot \frac{2}{3} = \frac{2}{3}$$

and the area of the second circumscribed rectangle is

$$f(v_2)\Delta x_2 = 3 \cdot \frac{2}{3} = 2$$

On the third subinterval $\left[1, \frac{4}{3}\right]$, $u_3 = 1$ and $v_3 = \frac{4}{3}$. Therefore $f(u_3) = 3 \cdot 1 = 3$ and $f(v_3) = 3 \cdot \frac{4}{3} = 4$. Furthermore, $\Delta x_3 = \frac{4}{3} - 1 = \frac{1}{3}$. Consequently, the area of the third inscribed rectangle is

$$f(u_3)\Delta x_3 = 3 \cdot \frac{1}{3} = 1$$

and the area of the third circumscribed rectangle is

$$f(v_3)\Delta x_3 = 4 \cdot \frac{1}{3} = \frac{4}{3}$$

On the last subinterval $\left[\frac{4}{3}, 2\right]$, $u_4 = \frac{4}{3}$ and $v_4 = 2$. Therefore $f(u_4) = 3 \cdot \frac{4}{3} = 4$ and $f(v_4) = 3 \cdot 2 = 6$. Also, $\Delta x_4 = 2 - \frac{4}{3} = \frac{2}{3}$. It follows that the area of the fourth inscribed rectangle is

$$f(u_4)\Delta x_4 = 4 \cdot \frac{2}{3} = \frac{8}{3}$$

and the area of the fourth circumscribed rectangle is

$$f(v_4)\Delta x_4 = 6 \cdot \frac{2}{3} = 4$$

Thus,

$$\underline{A}_p = 0 + \frac{2}{3} + 1 + \frac{8}{3} = \frac{13}{3}$$

and

$$\overline{A}_p = \frac{1}{3} + 2 + \frac{4}{3} + 4 = \frac{23}{3}$$

Note that

$$\frac{13}{3} < 6 < \frac{23}{3}$$

**b.** Since the partition $q$ is regular and forms six subintervals, the length of each subinterval is $\frac{2-0}{6} = \frac{1}{3}$. Thus the partition $q$ is $\left(0, \frac{1}{3}, \frac{2}{3}, 1, \frac{4}{3}, \frac{5}{3}, 2\right)$. Clearly, we have $\Delta x_1 = \Delta x_2 = \Delta x_3 = \Delta x_4 = \Delta x_5 = \Delta x_6 = \frac{1}{3}$.

Again, on each subinterval the minimum value of $f$ occurs at the left endpoint and its maximum value occurs at the right endpoint. (See Figure 9.6.)

On the first subinterval $\left[0, \frac{1}{3}\right]$, $u_1 = 0$ and $v_1 = \frac{1}{3}$. Therefore $f(u_1) = 3 \cdot 0 = 0$ and $f(v_1) = 3 \cdot \frac{1}{3} = 1$. Thus the area of the first inscribed rectangle is

$$f(u_1)\Delta x_1 = 0 \cdot \frac{1}{3} = 0$$

and the area of the first circumscribed rectangle is

$$f(v_1)\Delta x_1 = 1 \cdot \frac{1}{3} = \frac{1}{3}$$

Similarly, we find that the areas of the second, third, fourth, fifth, and sixth inscribed rectangles are $\frac{1}{3}, \frac{2}{3}, 1, \frac{4}{3}$, and $\frac{5}{3}$, respectively; and the areas of the second, third, fourth, fifth, and sixth circumscribed rectangles are $\frac{2}{3}, 1, \frac{4}{3}, \frac{5}{3}$, and 2, respectively. Therefore

$$\underline{A}_q = 0 + \frac{1}{3} + \frac{2}{3} + 1 + \frac{4}{3} + \frac{5}{3} = 5$$

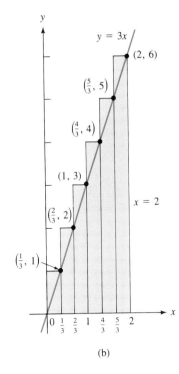

FIGURE 9.6

(a)

(b)

and

$$\bar{A}_p = \frac{1}{3} + \frac{2}{3} + 1 + \frac{4}{3} + \frac{5}{3} + 2 = 7$$

Note that

$$5 < 6 < 7$$

c. Since the partition $r_n$ is regular and forms $n$ subintervals, the length of each subinterval must be $2/n$.

Therefore the partition is $\left(0, \frac{2}{n}, 2 \cdot \frac{2}{n}, 3 \cdot \frac{2}{n}, \ldots, (n-1) \cdot \frac{2}{n}, n \cdot \frac{2}{n}\right)$. The $i$th subinterval is $\left[(i-1)\frac{2}{n}, i\frac{2}{n}\right]$. As we have seen earlier, the function $f$ is increasing on $[0, 2]$. Thus its minimum value on the $i$th subinterval occurs at the left endpoint and its maximum value occurs at the right endpoint. Hence, $u_i = (i-1)\frac{2}{n}$ and $v_i = i\frac{2}{n}$, so that $f(u_i) = 3(i-1)\frac{2}{n}$ and $f(v_i) = 3i\frac{2}{n}$. Since $\Delta x_i = \frac{2}{n}$ for each $i$, the area of the $i$th inscribed rectangle is

$$f(u_i) \cdot \Delta x_i = 3(i-1)\frac{2}{n} \cdot \frac{2}{n} = \frac{12(i-1)}{n^2}$$

and that of the $i$th circumscribed rectangle is

$$f(v_i) \cdot \Delta x_i = 3i \cdot \frac{2}{n} \cdot \frac{2}{n} = \frac{12i}{n^2}$$

Thus,

$$A_{r_n} = \sum_{i=1}^{n} f(u_i)\Delta x_i = \sum_{i=1}^{n} \frac{12(i-1)}{n^2} = \frac{12}{n^2} \sum_{i=1}^{n} (i-1)$$

$$= \frac{12}{n^2}[0 + 1 + 2 + 3 + \cdots + (n-1)]$$

$$= \frac{12}{n^2} \sum_{i=1}^{n-1} i = \frac{12}{n^2} \cdot \frac{(n-1)[(n-1)+1]}{2}$$

$$= 6\left(1 - \frac{1}{n}\right) = 6 - \frac{6}{n}$$

Also,

$$\overline{A}_{r_n} = \sum_{i=1}^{n} f(v_i)\Delta x_i = \sum_{i=1}^{n} \frac{12i}{n^2} = \frac{12}{n^2} \sum_{i=1}^{n} i$$

$$= \frac{12}{n^2} \cdot \frac{n(n+1)}{2}$$

$$= 6\left(1 + \frac{1}{n}\right) = 6 + \frac{6}{n}$$

Note that

$$6 - \frac{6}{n} < 6 < 6 + \frac{6}{n}$$

**d.** It is intuitively clear that as $n$ tends to infinity, $\frac{6}{n}$ tends to 0. Therefore

$$\lim_{n\to\infty} \underline{A}_{r_n} = \lim_{n\to\infty}\left(6 - \frac{6}{n}\right) = 6 - 0 = 6$$

and

$$\lim_{n\to\infty} \overline{A}_{r_n} = \lim_{n\to\infty}\left(6 + \frac{6}{n}\right) = 6 + 0 = 6$$

This of course was expected, since the area of the triangle $S$ was known at the outset to be 6 square units.
**Do Exercise 7.**                                                              ■

**EXAMPLE 2**    Find the area of the region bounded by the graphs of $y = f(x) = x^2 + 1$, $x = 0$, $x = 3$, and $y = 0$.

**Solution**    Let $p_n$ be a regular partition of the interval $[0, 3]$ which forms $n$ subintervals each of length $\frac{3}{n}$. The partition is $\left(0, \frac{3}{n}, 2 \cdot \frac{3}{n}, \ldots, (n-1) \cdot \frac{3}{n}, n \cdot \frac{3}{n}\right)$. Since the derivative $f'(x)$ is $2x$, it is positive on the open interval $(0, 3)$ and therefore the function $f$ is increasing. Consequently, its minimum value on the $i$th subinterval occurs at the left endpoint $(i-1)\frac{3}{n}$ and its maximum value occurs at the right endpoint $i \cdot \frac{3}{n}$. (See Figure 9.7.)

It follows that on the $i$th subinterval, the minimum and maximum values are $f\left((i-1)\frac{3}{n}\right)$ and $f\left(i\cdot\frac{3}{n}\right)$, respectively. But,

$$f\left((i-1)\frac{3}{n}\right) = \left[(i-1)\frac{3}{n}\right]^2 + 1$$

$$= (i-1)^2 \cdot \frac{9}{n^2} + 1$$

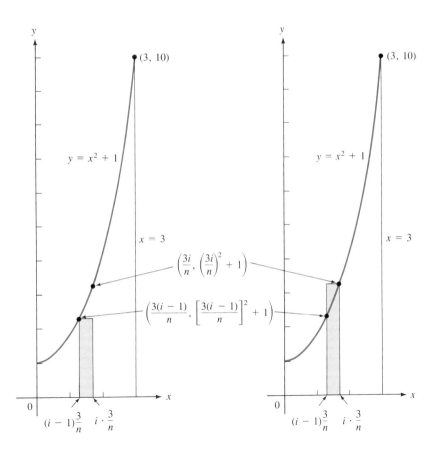

FIGURE 9.7

and

$$f\left(i\cdot\frac{3}{n}\right) = \left[i\cdot\frac{3}{n}\right]^2 + 1 = i^2 \cdot \frac{9}{n^2} + 1$$

Thus the dimensions of the $i$th inscribed rectangle are $\frac{3}{n}$ by

$$(i-1)^2 \cdot \frac{9}{n^2} + 1$$

Consequently, the area of the $i$th inscribed rectangle is

$$\left[(i-1)^2 \cdot \frac{9}{n^2} + 1\right] \cdot \frac{3}{n}$$

Similarly, the dimensions of the $i$th circumscribed rectangle are $\frac{3}{n}$ by

$$i^2 \cdot \frac{9}{n^2} + 1$$

Therefore its area is

$$\left[i^2 \cdot \frac{9}{n^2} + 1\right] \cdot \frac{3}{n}$$

Since, $\underline{A}_{p_n}$ is the sum of the areas of the $n$ inscribed rectangles, we have

$$\underline{A}_{p_n} = \sum_{i=1}^{n} \left[(i-1)^2 \cdot \frac{9}{n^2} + 1\right] \cdot \frac{3}{n}$$

$$= \sum_{i=1}^{n} \left((i-1)^2 \cdot \frac{27}{n^3} + \frac{3}{n}\right) = \frac{27}{n^3} \sum_{i=1}^{n} (i-1)^2 + \sum_{i=1}^{n} \frac{3}{n}$$

$$= \frac{27}{n^3}[0 + 1 + 2^2 + \cdots + (n-1)^2] + \sum_{i=1}^{n} \frac{3}{n}$$

$$= \frac{27}{n^3} \sum_{i=1}^{n-1} i^2 + \sum_{i=1}^{n} \frac{3}{n}$$

$$= \frac{27}{n^3} \cdot \frac{(n-1)[(n-1)+1][2(n-1)+1]}{6} + n \cdot \frac{3}{n}$$

$$= \frac{9}{2}\left(1 - \frac{1}{n}\right)\left(2 - \frac{1}{n}\right) + 3$$

Also, since $\overline{A}_{p_n}$ is the sum of the areas of the $n$ circumscribed rectangles, we have

$$\overline{A}_{p_n} = \sum_{i=1}^{n} \left[i^2 \cdot \frac{9}{n^2} + 1\right] \cdot \frac{3}{n}$$

$$= \sum_{i=1}^{n} \left[i^2 \cdot \frac{27}{n^3} + \frac{3}{n}\right] = \frac{27}{n^3} \sum_{i=1}^{n} i^2 + \sum_{i=1}^{n} \frac{3}{n}$$

$$= \frac{27}{n^3} \cdot \frac{n(n+1)(2n+1)}{6} + n \cdot \frac{3}{n}$$

$$= \frac{9}{2}\left(1 + \frac{1}{n}\right)\left(2 + \frac{1}{n}\right) + 3$$

Since $\frac{1}{n}$ is positive for each positive integer, we have

$$\frac{9}{2}\left(1 - \frac{1}{n}\right)\left(2 - \frac{1}{n}\right) + 3 < \frac{9}{2}(1)(2) + 3 = 12 < \frac{9}{2}\left(1 + \frac{1}{n}\right)\left(2 + \frac{1}{n}\right) + 3$$

Therefore

$$\underline{A}_{p_n} < 12 < \overline{A}_{p_n}$$

Furthermore, $\frac{1}{n}$ tends to 0 as $n$ tends to infinity. Hence,

$$\lim_{n \to \infty} \underline{A}_{p_n} = \lim_{n \to \infty} \left[ \frac{9}{2}\left(1 - \frac{1}{n}\right)\left(2 - \frac{1}{n}\right) + 3\right] = \frac{9}{2}(1)(2) + 3 = 12$$

and

$$\lim_{n \to \infty} \overline{A}_{p_n} = \lim_{n \to \infty} \left[ \frac{9}{2}\left(1 + \frac{1}{n}\right)\left(2 + \frac{1}{n}\right) + 3\right] = \frac{9}{2}(1)(2) + 3 = 12$$

Since

$$\lim_{n \to \infty} \underline{A}_{p_n} = \lim_{n \to \infty} \overline{A}_{p_n} = 12$$

we conclude that the area of the region bounded by the graphs of $y = x^2 + 1$, $x = 0$, $x = 3$, and $y = 0$ is 12 square units.   ∎

**EXAMPLE 3**   Let $g$ be any antiderivative of the function $f$ of the preceding example. Calculate $g(3) - g(0)$ and compare your answer to that of Example 2.

**Solution**   Let $C$ be any constant and define

$$g(x) = \frac{x^3}{3} + x + C$$

Then $g'(x) = x^2 + 1$, so $g$ is an antiderivative of $f$. Furthermore,

$$g(3) - g(0) = \left[\frac{3^3}{3} + 3 + C\right] - \left[\frac{0^3}{3} + 0 + C\right]$$
$$= 12 + C - C = 12$$

Observe that the constant of integration $C$ cancels out. Therefore we would have obtained the same result if we had chosen $C = 0$. Furthermore, the answers we obtained in Examples 2 and 3 are identical. This is not a coincidence as we shall soon see. **Do Exercise 9.**   ∎

## The Definite Integral

Examples 1 and 2 are special cases of a more general problem, which we now describe.

> **DEFINITION 9.4:**   Let $f$ be a continuous function defined on a closed bounded interval $[a, b]$, and let $p_n$ be a regular partition of $[a, b]$ which determines $n$ subintervals. By continuity of the function $f$, we know that for each $i \in \{1, 2, 3, \ldots, n\}$, we can find $u_i$ and $v_i$ in $[x_{i-1}, x_i]$ such that
>
> $$f(u_i) \le f(x) \le f(v_i)$$
>
> for all $x \in [x_{i-1}, x_i]$. We also let $w_i$ be an arbitrary point in the $i$th subinterval and we let $\Delta x$ be the length of each of the subintervals. That is,
>
> $$\Delta x = \frac{b - a}{n}$$

Then

$$\sum_{i=1}^{n} f(u_i)\Delta x$$

is called the *lower sum of f corresponding to* $p_n$ and is denoted $\underline{S}_{p_n}$ (read: "S sub p sub n lower bar"),

$$\sum_{i=1}^{n} f(v_i)\Delta x$$

is called the *upper sum of f corresponding to* $p_n$ and is denoted $\overline{S}_{p_n}$ (read: "S sub p sub n upper bar"), and

$$\sum_{i=1}^{n} f(w_i)\Delta x$$

is called a *Riemann sum of f corresponding to* $p_n$ and is denoted $R_{p_n}$.

In general, with each partition $p$ of $[a, b]$, we have infinitely many Riemann sums, since the choice of $w_i$ in each $[x_{i-1}, x_i]$ is completely arbitrary. However, we always have

$$\underline{S}_p \le R_p \le \overline{S}_p$$

since $f(u_i) \le f(w_i) \le f(v_i)$ for all $i$ in $\{1, 2, 3, \ldots, n\}$. In more advanced calculus books, it is proved that

$$\lim_{n \to \infty} \underline{S}_{p_n} = \lim_{n \to \infty} R_{p_n} = \lim_{n \to \infty} \overline{S}_{p_n}$$

**DEFINITION 9.5:**   This common limit is called *the definite integral* of $f$ from $a$ to $b$. It is denoted

$$\int_a^b f(x)\ dx$$

In this symbol, $f(x)$ is called the *integrand*, $x$ is called the *variable of integration*, and $a$ and $b$ are called the *lower limit* and *upper limit*, respectively.

The symbol $\int \ldots dx$ is the integration symbol, where $dx$ plays the same role as the subscript $x$ played in the differentiation symbol $D_x$. Note that $\int$ is an elongated "S" which reminds us that a definite integral is a limit of sums.

In more advanced texts, the definite integral is defined without requiring the partitions to be regular. Our definition is adequate for the applications we shall discuss. Also, it can be proved that if a function $f$ is continuous on a closed bounded interval $[a, b]$, then the definite inegral $\int_a^b f(x)\ dx$ exists.

Although we used the area of a region to introduce the definite integral of a function, a definite integral does not always represent an area. It occurs in other contexts as well, as we shall soon see.

EXAMPLE 4    Evaluate the definite integral $\int_0^5 (-x^3 + 1)\, dx$.

Solution    Let $p_n$ be the regular partition of the interval $[0, 5]$ which determines $n$ subintervals. Then the length of each subinterval is $\frac{5 - 0}{n}$ and we have

$$\Delta x = \frac{5}{n}$$

Since

$$\lim_{n \to \infty} \underline{S}_{p_n} = \lim_{n \to \infty} R_{p_n} = \lim_{n \to \infty} \overline{S}_{p_n}$$

we may take the limit of any Riemann sum to evaluate the definite integral. For convenience, for each $i \in \{1, 2, 3, \ldots, n\}$, we choose $w_i$ to be the right endpoint of the interval $[x_{i-1}, x_i]$. The partition $p_n$ of the interval $[0, 5]$ is $\left(0, \frac{5}{n}, 2 \cdot \frac{5}{n}, \ldots, (n-1) \cdot \frac{5}{n}, n \cdot \frac{5}{n}\right)$. Therefore the $i$th subinterval is $\left[(i-1) \cdot \frac{5}{n}, i \cdot \frac{5}{n}\right]$ and we choose $w_i = 5i/n$.

$$f(w_i) = f\left(\frac{5i}{n}\right) = -\left(\frac{5i}{n}\right)^3 + 1$$

$$= \frac{-125i^3}{n^3} + 1$$

Thus

$$R_{p_n} = \sum_{i=1}^{n} f(w_i)\Delta x$$

$$= \sum_{i=1}^{n} \left(\frac{-125i^3}{n^3} + 1\right)\frac{5}{n} = \sum_{i=1}^{n} \left(\frac{-625i^3}{n^4} + \frac{5}{n}\right)$$

$$= \frac{-625}{n^4} \sum_{i=1}^{n} i^3 + \sum_{i=1}^{n} \frac{5}{n}$$

$$= \frac{-625}{n^4} \left[\frac{n(n+1)}{2}\right]^2 + n \cdot \frac{5}{n}$$

$$= \frac{-625}{4}\left(1 + \frac{1}{n}\right)^2 + 5$$

As $n$ tends to infinity, $\frac{1}{n}$ tends to 0, so that

$$\lim_{n \to \infty} R_{p_n} = \lim_{n \to \infty}\left[\frac{-625}{4}\left(1 + \frac{1}{n}\right)^2 + 5\right]$$

$$= \frac{-625}{4}(1) + 5 = \frac{-605}{4}$$

Therefore

$$\int_0^5 (-x^3 + 1)\, dx = \frac{-605}{4}$$

**EXAMPLE 5**    Let $g$ be an antiderivative of $-x^3 + 1$ and evaluate $g(5) - g(0)$. Compare the result to the answer in Example 4.

**Solution**    Let $g(x) = -x^4/4 + x$. Then $g$ is an antiderivative of $-x^3 + 1$ since

$$D_x\left(\frac{-x^4}{4} + x\right) = -x^3 + 1$$

Thus

$$g(5) - g(0) = \left[\frac{-5^4}{4} + 5\right] - \left[\frac{-0^4}{4} + 0\right] = \frac{-625}{4} + 5 - 0$$

$$= \frac{-625 + 20}{4} = \frac{-605}{4}$$

Note that this answer is identical to the answer we obtained in Example 4. We shall soon see why we obtained the same result in both examples.
**Do Exercise 17.**    ∎

## Properties of the Definite Integral

We conclude this section by stating five properties of the definite integral.

> **Properties of the Definite Integral**
>
> 1. $\int_a^a f(x)\, dx = 0$.
>
> 2. If $f$ is a continuous function on the closed interval $[a, b]$, we define
>
> $$\int_b^a f(x)\, dx = -\int_a^b f(x)\, dx$$
>
> 3. If $f$ and $g$ are continuous functions on the closed interval $[a, b]$ and $k$ is a constant, then
>
> $$\int_a^b kf(x)\, dx = k\int_a^b f(x)\, dx$$
>
> $$\int_a^b [f(x) + g(x)]\, dx = \int_a^b f(x)\, dx + \int_a^b g(x)\, dx$$
>
> $$\int_a^b [f(x) - g(x)]\, dx = \int_a^b f(x)\, dx - \int_a^b g(x)\, dx$$
>
> 4. If $f$ is a continuous function on some interval $I$ and $a$, $b$, and $c$ are in $I$, then
>
> $$\int_a^b f(x)\, dx + \int_b^c f(x)\, dx = \int_a^c f(x)\, dx$$
>
> 5. If $f$ is a continuous function on a closed bounded interval $[a, b]$ and $f(x) \geq 0$ for all $x$ in $[a, b]$, then the area of the region bounded by the graphs of $y = f(x)$, $x = a$, $x = b$, and $y = 0$ is defined to be $\int_a^b f(x)\, dx$ square units.

# Exercise Set 9.8

*In Exercises 1–5, sketch the region bounded by the graphs of the given equations. Draw the inscribed and circumscribed rectangular polygons corresponding to the regular partition which forms n subintervals of the appropriate interval for the given n. Approximate the area of the region by calculating the areas of the inscribed and circumscribed rectangular polygons.*

**1.** $y = f(x) = x$, $x = 5$, $y = 0$, $n = 5$

**2.** $y = f(x) = \frac{1}{2}x$, $x = 8$, $y = 0$, $n = 8$

**3.** $y = f(x) = x^2$, $x = 4$, $y = 0$, $n = 8$

**4.** $y = f(x) = \frac{1}{2}x^2 + 2$, $x = 0$, $x = 5$, $y = 0$, $n = 10$

**5.** $y = f(x) = -x^2 + 10$, $x = -3$, $x = 0$, $y = 0$, $n = 6$

**6.** Let $S$ be the region bounded by the graphs of $y = f(x) = 2x$, $x = 4$, and $y = 0$. The region is a right triangle whose area is 16 square units.

   **a.** Let $p = \left(0, \frac{4}{3}, 2, \frac{10}{3}, 4\right)$ be a partition of the interval $[0, 4]$. Find $\underline{A}_p$ and $\overline{A}_p$.

   **b.** Let $q$ be the regular partition of $[0, 4]$ which determines five subintervals. Find $\underline{A}_q$ and $\overline{A}_q$.

   **c.** Let $r_n$ be the regular partition of $[0, 4]$ which determines $n$ subintervals. Find $\underline{A}_{r_n}$ and $\overline{A}_{r_n}$.

   **d.** Show that $\lim\limits_{n\to\infty} \underline{A}_{r_n} = \lim\limits_{n\to\infty} \overline{A}_{r_n} = 16$.

   **e.** Let $g$ be any antiderivative of $2x$. Verify that $g(4) - g(0) = 16$ also.

**7.** Let $S$ be the region bounded by the graphs of $y = f(x) = 4x$, $x = 2$, and $y = 0$. The region is a right triangle with area 8 square units.

   **a.** Let $p = \left(0, \frac{2}{3}, 1, \frac{5}{4}, 2\right)$ be a partition of the interval $[0, 2]$. Find $\underline{A}_p$ and $\overline{A}_p$.

   **b.** Let $q$ be the regular partition of $[0, 2]$ which determines eight subintervals. Find $\underline{A}_q$ and $\overline{A}_q$.

   **c.** Let $r_n$ be the regular partition of $[0, 2]$ which determines $n$ subintervals. Find $\underline{A}_{r_n} =$ and $\overline{A}_{r_n}$.

   **d.** Show that $\lim\limits_{n\to\infty} \underline{A}_{r_n} = \lim\limits_{n\to\infty} \overline{A}_{r_n} = 8$.

   **e.** Let $g$ be any antiderivative of $4x$. Verify that $g(2) - g(0) = 8$ also.

**8.** Let $S$ be the region bounded by the graphs of $y = f(x) = x^2$, $x = 2$, and $y = 0$.

   **a.** Show that the function $f$ is increasing on the interval $[0, 2]$.

   **b.** Let $p = \left(0, \frac{1}{3}, 1, \frac{4}{3}, 2\right)$ be a partition of the interval $[0, 2]$. Find $\underline{A}_p$ and $\overline{A}_p$.

   **c.** Let $q$ be the regular partition of $[0, 2]$ which determines six subintervals. Find $\underline{A}_q$ and $\overline{A}_q$.

   **d.** Let $r_n$ be the regular partition of $[0, 2]$ which determines $n$ subintervals. Find $\underline{A}_{r_n}$ and $\overline{A}_{r_n}$.

   **e.** Show that $\lim\limits_{n\to\infty} \underline{A}_{r_n} = \lim\limits_{n\to\infty} \overline{A}_{r_n}$.

   **f.** What is the area of the region $S$?

   **g.** Let $h$ be any antiderivative of the function $f$. Evaluate $h(2) - h(0)$ and compare your answer to that of part **e**.

**9.** Let $S$ be the region bounded by the graphs of $y = f(x) = x^3$, $x = 3$, and $y = 0$.

   **a.** Show that the function $f$ is increasing on the interval $[0, 3]$.

   **b.** Let $p = \left(0, \frac{3}{4}, 2, \frac{7}{3}, 3\right)$ be a partition of the interval $[0, 3]$. Find $\underline{A}_p$ and $\overline{A}_p$.

   **c.** Let $q$ be the regular partition of $[0, 3]$ which determines six subintervals. Find $\underline{A}_q$ and $\overline{A}_q$.

   **d.** Let $r_n$ be the regular partition of $[0, 3]$ which determines $n$ subintervals. Find $\underline{A}_{r_n}$ and $\overline{A}_{r_n}$.

   **e.** Show that $\lim\limits_{n\to\infty} \underline{A}_{r_n} = \lim\limits_{n\to\infty} \overline{A}_{r_n}$.

   **f.** What is the area of the region $S$?

   **g.** Let $h$ be any antiderivative of the function $f$. Evaluate $h(3) - h(0)$ and compare your answer to that of part **e**.

**10.** Let $S$ be the region bounded by the graphs of $x = 0$, $x = 2$, $y = 0$, and $y = f(x) = -x^2 + 5$.

   **a.** Show that the function $f$ is decreasing on the interval $[0, 2]$.

   **b.** Let $p = \left(0, \frac{1}{3}, 1, \frac{4}{3}, 2\right)$ be a partition of the interval $[0, 2]$. Find $\underline{A}_p$ and $\overline{A}_p$.

   **c.** Let $q$ be the regular partition of $[0, 2]$ which determines six subintervals. Find $\underline{A}_q$ and $\overline{A}_q$.

   **d.** Let $r_n$ be the regular partition of $[0, 2]$ which determines $n$ subintervals. Find $\underline{A}_{r_n}$ and $\overline{A}_{r_n}$.

   **e.** Show that $\lim\limits_{n\to\infty} \underline{A}_{r_n} = \lim\limits_{n\to\infty} \overline{A}_{r_n}$.

   **f.** What is the area of the region $S$?

   **g.** Let $h$ be any antiderivative of the function $f$. Evaluate $h(2) - h(0)$ and compare your answer to that of part **e**.

**11.** Let $S$ be the region bounded by the graphs of $x = -2$, $x = 0$, $y = 0$, and $y = f(x) = -x^3 + 2$.

   **a.** Show that the function $f$ is decreasing on the interval $[-2, 0]$.

   **b.** Let $p = \left(-2, -\frac{4}{3}, -1, -\frac{1}{3}, 0\right)$ be a partition of the interval $[-2, 0]$. Find $\underline{A}_p$ and $\overline{A}_p$.

   **c.** Let $q$ be the regular partition of $[-2, 0]$ which determines eight subintervals. Find $\underline{A}_q$ and $\overline{A}_q$.

**d.** Let $r_n$ be the regular partition of $[-2, 0]$ which determines $n$ subintervals. Find $\underline{A}_{r_n}$ and $\overline{A}_{r_n}$.

**e.** Show that $\lim\limits_{n \to \infty} \underline{A}_{r_n} = \lim\limits_{n \to \infty} \overline{A}_{r_n}$.

**f.** What is the area of the region $S$?

**g.** Let $h$ be any antiderivative of the function $f$. Evaluate $h(0) - h(-2)$ and compare your answer to that of part **e.**

**12.** Let $S$ be the region bounded by the graphs of $y = f(x) = x^2 + x + 1$, $x = 0$, $x = 3$, and $y = 0$.

**a.** Show that the function $f$ is increasing on the interval $[0, 3]$.

**b.** Let $p = \left(0, \frac{1}{3}, 1, \frac{4}{3}, 2, \frac{5}{2}, 3\right)$ be a partition of the interval $[0, 3]$. Find $\underline{A}_p$ and $\overline{A}_p$.

**c.** Let $q$ be the regular partition of $[0, 3]$ which determines six subintervals. Find $\underline{A}_q$ and $\overline{A}_q$.

**d.** Let $r_n$ be the regular partition of $[0, 3]$ which determines $n$ subintervals. Find $\underline{A}_{r_n}$ and $\overline{A}_{r_n}$.

**e.** Show that $\lim\limits_{n \to \infty} \underline{A}_{r_n} = \lim\limits_{n \to \infty} \overline{A}_{r_n}$.

**f.** What is the area of the region $S$?

**g.** Let $h$ be any antiderivative of the function $f$. Evaluate $h(3) - h(0)$ and compare your answer to that of part **e.**

*In Exercises 13–25 evaluate the definite integral $\int_a^b f(x)\, dx$ for the given function $f$ and limits $a$ and $b$ as follows:*

**a.** *Find the regular partition $p_n$ of $[a, b]$ which forms $n$ subintervals.*

**b.** *Find the Riemann sum $\sum\limits_{i=1}^{n} f(w_i)\Delta x$ corresponding to $p_n$ letting $w_i$ be the right endpoint of the ith subinterval.*

**c.** *Find $\lim\limits_{n \to \infty} R_{p_n}$.*

**d.** *Find any antiderivative $g$ of the function $f$ on $[a, b]$ and evaluate $g(b) - g(a)$. Then compare your answers in parts c and d.*

**13.** $\displaystyle\int_0^5 7x\, dx$

**14.** $\displaystyle\int_0^6 -4\, dx$

**15.** $\displaystyle\int_0^3 x^2\, dx$

**16.** $\displaystyle\int_0^7 x^2\, dx$

**17.** $\displaystyle\int_0^9 (x^2 + 1)\, dx$

**18.** $\displaystyle\int_0^8 (2x^2 + 3)\, dx$

**19.** $\displaystyle\int_0^6 x^3\, dx$

**20.** $\displaystyle\int_0^1 (x^3 + 3)\, dx$

**21.** $\displaystyle\int_0^8 (x^3 + 2x^2 - 5x + 3)\, dx$

**\*22.** $\displaystyle\int_1^4 5x\, dx$

**\*23.** $\displaystyle\int_1^4 x^2\, dx$

**\*24.** $\displaystyle\int_2^6 x^3\, dx$

**\*25.** $\displaystyle\int_{-1}^3 (x^2 + 3x - 1)\, dx$

# 9.9  The Definite Integral—Another Approach (Optional)

The definite integral was introduced in the preceding section using the area of a region in the $xy$-plane. We indicated at the time that a definite integral does not always represent an area. In this section, we give two different examples where solutions lead to definite integrals.

## The Definite Integral as a Distance

**EXAMPLE 1**    The instantaneous speed $v$ ft/sec of a free falling object at instant $t$ seconds after release is given by $v = 32t$. Approximate the distance the object has fallen in the first 5 seconds by dividing the time interval into

**a.** five equal time subintervals,

**b.** ten equal time subintervals, and

**c.** $n$ equal time subintervals.

Using the result of part **c**, find the exact distance the object has fallen in 5 seconds.

**Solution**    **a.** It is clear that the velocity increases with time. For example, 2 seconds after release, the object is traveling at a speed of 64 ft/sec; however, 3 seconds after release, its speed is 96 ft/sec. We divide the time interval $[0, 5]$ into five equal subintervals $[0, 1], [1, 2], [2, 3], [3, 4],$ and $[4, 5]$. On each subinterval, the minimum velocity occurs at the beginning of the interval and the maximum velocity occurs at the end. It follows that if $\Delta t$ is the length of the interval, $v_i$ and $v_t$ are the initial and terminal velocities on that interval, respectively, and $d$ is the actual distance traveled during that time interval, we must have

$$v_i \cdot \Delta t < d < v_t \cdot \Delta t$$

Applying this idea to the five equal subintervals and letting $d_1, d_2, d_3, d_4,$ and $d_5$ feet be the distances the object has fallen during the first, second, third, fourth, and fifth time intervals, respectively, and noting that $\Delta t = 1$ for each subinterval, we get

$(0)(1) < d_1 < (32)(1)$ since $v(0) = 0$ and $v(1) = 32$

$(32)(1) < d_2 < (64)(1)$ since $v(1) = 32$ and $v(2) = 64$

$(64)(1) < d_3 < (96)(1)$ since $v(2) = 64$ and $v(3) = 96$

$(96)(1) < d_4 < (128)(1)$ since $v(3) = 96$ and $v(4) = 128$

$(128)(1) < d_5 < (160)(1)$ since $v(4) = 128$ and $v(5) = 160$

Adding these inequalities, we obtain

$$32 + 64 + 96 + 128 < \sum_{i=1}^{5} d_i < 32 + 64 + 96 + 128 + 160$$

That is,

$$320 < \text{exact distance traveled} < 480$$

Thus, the distance the object has fallen is more than 320 feet but less than 480 feet.

**b.** Using the same reasoning, but with ten equal time intervals, the length of each time interval is half a second since $\frac{5}{10} = \frac{1}{2}$. Thus the subintervals will be $[0, .5],$ $[.5, 1], [1, 1.5], [1.5, 2], [2, 2.5], [2.5, 3], [3, 3.5], [3.5, 4], [4, 4.5],$ and $[4.5, 5]$. Again, the minimum velocity on a time interval occurs at the beginning of that interval and the maximum velocity at the end. As we did in part **a**, we let $d_i$ feet be the distance the object has fallen during the $i$th time interval. Since $v(t) = 32t,$ $v(0) = 0,$ $v(.5) = 16,$ $v(1) = 32,$ $v(1.5) = 48,$ $v(2) = 64,$ $v(2.5) = 80,$ $v(3) = 96,$ $v(3.5) = 112,$ $v(4) = 128,$ $v(4.5) = 144,$ and $v(5) = 160.$ Furthermore, $\Delta t = \frac{1}{2}$ and therefore

$$0 \cdot \frac{1}{2} = \quad 0 < d_1 \quad < 16 \cdot \frac{1}{2} = 8$$

$$16 \cdot \frac{1}{2} = \quad 8 < d_2 \quad < 32 \cdot \frac{1}{2} = 16$$

$$32 \cdot \frac{1}{2} = 16 < d_3 \quad < 48 \cdot \frac{1}{2} = 24$$

$$48 \cdot \frac{1}{2} = 24 < d_4 \quad < 64 \cdot \frac{1}{2} = 32$$

$$64 \cdot \frac{1}{2} = 32 < d_5 < 80 \cdot \frac{1}{2} = 40$$

$$80 \cdot \frac{1}{2} = 40 < d_6 < 96 \cdot \frac{1}{2} = 48$$

$$96 \cdot \frac{1}{2} = 48 < d_7 < 112 \cdot \frac{1}{2} = 56$$

$$112 \cdot \frac{1}{2} = 56 < d_8 < 128 \cdot \frac{1}{2} = 64$$

$$128 \cdot \frac{1}{2} = 64 < d_9 < 144 \cdot \frac{1}{2} = 72$$

$$144 \cdot \frac{1}{2} = 72 < d_{10} < 160 \cdot \frac{1}{2} = 80$$

Adding these inequalities, we obtain

$$360 < \sum_{i=1}^{10} d_i < 440$$

Thus the distance traveled by the free falling object in 5 seconds is more than 360 feet but less than 440 feet.

c. If we divide the interval [0, 5] into $n$ equal subintervals, the length of each subinterval is $\frac{5}{n}$. Thus we obtain the regular partition

$$\left( 0, \frac{5}{n}, 2 \cdot \frac{5}{n}, \ldots, (n-1) \cdot \frac{5}{n}, n \cdot \frac{5}{n} \right)$$

Note that we did not perform the multiplications $2 \cdot \frac{5}{n}$, $3 \cdot \frac{5}{n}$, and so on. The reason for this will be apparent shortly. The $i$th subinterval is $\left[ (i-1) \cdot \frac{5}{n}, i \cdot \frac{5}{n} \right]$. Thus the minimum velocity $v_m$ ft/sec on that interval occurs at $(i-1) \cdot \frac{5}{n}$, and the maximum velocity $v_M$ ft/sec occurs at $i \cdot \frac{5}{n}$. Clearly, the actual distance $d_i$ feet traveled by the free falling object during the $i$th subinterval is more than it would have been if the velocity had remained at a minimum ($v_m$ ft/sec) during that time. However, the distance is less than it would have been if the velocity had been constant and equal to the maximum velocity ($v_M$) during that time interval.

That is,

(minimum speed)(time) < distance traveled < (maximum speed)(time)

But,

$$v_m = v\left( (i-1) \cdot \frac{5}{n} \right) = 32 \cdot (i-1) \cdot \frac{5}{n} = \frac{160(i-1)}{n}$$

and

$$v_M = v\left( i \cdot \frac{5}{n} \right) = 32 \cdot i \cdot \frac{5}{n} = \frac{160i}{n}$$

Consequently,

$$\frac{160(i-1)}{n} \cdot \frac{5}{n} < d_i < \frac{160i}{n} \cdot \frac{5}{n}$$

That is,

$$\frac{800(i-1)}{n^2} < d_i < \frac{800i}{n^2}$$

The last expression represents $n$ double inequalities (one for each value of $i$), which we now add. Using the sigma notation, we get

$$\sum_{i=1}^{n} \frac{800(i-1)}{n^2} < \sum_{i=1}^{n} d_i < \sum_{i=1}^{n} \frac{800i}{n^2}$$

Since $\dfrac{800}{n^2}$ is a common factor, we can write

$$\frac{800}{n^2} \sum_{i=1}^{n} (i-1) < \sum_{i=1}^{n} d_i < \frac{800}{n^2} \sum_{i=1}^{n} i$$

Recalling the formula $\sum_{i=1}^{n} i \ \frac{n(n+1)}{2}$ and using the result of Exercise 1 of this section, we get

$$\frac{800(n-1)n}{n^2(2)} < \text{distance traveled} < \frac{800n(n+1)}{n^2(2)}$$

Simplifying, we obtain

$$400\left(1 - \frac{1}{n}\right) < \text{distance traveled} < 400\left(1 + \frac{1}{n}\right)$$

Since $\frac{1}{n}$ tends to 0 as $n$ tends to infinity, we have

$$\lim_{n \to \infty} 400\left(1 - \frac{1}{n}\right) = \lim_{n \to \infty} 400\left(1 + \frac{1}{n}\right) = 400$$

We conclude that the distance the object has fallen during the first 5 seconds is 400 feet.
**Do Exercise 5.**   ∎

Observe the following:

1. If we replace $n$ by 5 in

$$400\left(1 - \frac{1}{n}\right) < \text{distance traveled} < 400\left(1 + \frac{1}{n}\right)$$

we obtain

$$320 < \text{distance traveled} < 480$$

which is the result of part **a**. If we replace $n$ by 10, we get

$$360 < \text{distance traveled} < 440$$

which is the answer we obtained in part **b**.
2. The distance traveled in the first 5 seconds is $\int_0^5 32t \, dt$ since it is the common limit, as $n$ tends to infinity, of the upper and lower sums of the function $32t$ corresponding to the regular partition $p_n$ of the interval $[0, 5]$.

3. As a preview to the discussion of the next section, we let $g(t) = 16t^2$; then $g'(t) = 32t$. Therefore $g$ is an antiderivative of $32t$. The position function of the falling object is $h(t) = 16t^2 + C$ for some constant $C$. So $h(5)$ feet is the distance from the origin to the object at instant 5 seconds; furthermore, $h(0)$ feet is the distance from the origin to the falling object at instant 0 seconds. So $h(5) - h(0)$ is the distance traveled by the falling object. But

$$h(5) - h(0) = 16(5^2) + C - [16(0^2) + C] = 400$$

Note that 400 is the answer we obtained in part **c** of the preceding example. We have verified that

$$\int_0^5 32t \, dt = h(5) - h(0)$$

where $h(t)$ is an antiderivative of $32t$.

## The Definite Integral as the Amount Spent by Consumers

EXAMPLE 2   The Triplenight Company found that the relationship between the price $p$ dollars per racquetball racquet and the quantity $q$ racquets the consumers are willing to buy is given by

$$p = D(q) = -0.0001q^2 + 225$$

where $0 \le q \le 1400$.

a. Give a geometric interpretation of the amount the consumers have spent when 1200 racquets have been sold, assuming that the price decreased continuously from \$225 to \$81 per racquet.
b. Dividing the interval [0, 1200] into $n$ equal subintervals, approximate the amount spent by the consumers in terms of $n$.
c. Using the result of part **b**, find the amount spent by the consumers.

Solution   a. The graph of the demand equation is illustrated in Figure 9.8.

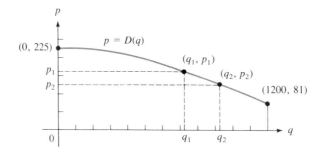

FIGURE 9.8

We can see that as the price per racquet decreases from $p_1$ to $p_2$, the number of racquets the consumers are willing to buy increases from $q_1$ to $q_2$. Thus the number of racquets sold at prices from $p_2$ to $p_1$ dollars per racquet will be $q_2 - q_1 = \Delta q$.

If that many racquets had been sold at a fixed price $p_1$ dollars per racquet, the consumers would have spent $p_1 \cdot \Delta q$ dollars. On the other hand, if the price per unit had been $p_2$ dollars, then the amount spent would have been $p_2 \cdot \Delta q$ dollars. In actuality, the amount spent is between these two quantities. That is,

$$p_2 \cdot \Delta q < \text{amount spent} < p_1 \cdot \Delta q$$

In Figure 9.8, $p_2 \cdot \Delta q$ represents the area of a rectangle which lies below the demand curve (an inscribed rectangle). On the other hand, $p_1 \cdot \Delta q$ can be interpreted as the area of a rectangle whose upper side lies above the demand curve (a circumscribed rectangle). Dividing the interval [0, 1200] into $n$ equal subintervals, and repeating the same argument for each subinterval, we see that a lower estimate for the total amount spent by the customers is given by the area of an inscribed rectangular polygon, while upper estimate is given by the area of a circumscribed rectangular polygon. (See Figures 9.9a and b.)

Let $A$ square units be the area of the region in the first quadrant bounded by the demand curve. We can verify geometrically that if the interval [0, 1200] is divided into a greater number of subintervals, the area of the inscribed rectangular polygon increases, while the area of the circumscribed rectangular polygon decreases, and that these areas seem to approach the same limit. We conclude that this limit is $A$. Therefore the total amount spent by the consumers is $A$ dollars. We should note that this is assuming that the price decreases *continuously* from $225 to $81 per racquet. In reality, this does not happen. We must remember that our example is a mathematical model, which only approximates the real situation.

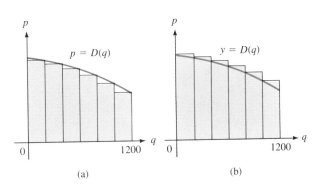

FIGURE 9.9

(a)　　　　　(b)

b. If we divide the interval [0, 1200] into $n$ equal subintervals, the length of each subinterval is $\frac{1200}{n}$. We obtain the partition $\left(0, \frac{1200}{n}, 2 \cdot \frac{1200}{n}, \ldots, (n-1) \cdot \frac{1200}{n}, n \cdot \frac{1200}{n}\right)$. The $i$th subinterval is $\left[(i-1) \cdot \frac{1200}{n}, i \cdot \frac{1200}{n}\right]$. We now find the prices corresponding to the quantities at the endpoints of this interval. We get

$$p_{i-1} = (-0.0001)\left(\frac{1200(i-1)}{n}\right)^2 + 225$$

and

$$p_i = (-0.0001)\left(\frac{1200i}{n}\right)^2 + 225$$

As the price per racquet decreases from $p_{i-1}$ to $p_i$ dollars, the number of racquets sold is $\frac{1200}{n}$. Therefore if $A_i$ dollars is the amount spent by the consumers to buy these $\frac{1200}{n}$ racquets, $A_i$ will be more than if all the racquets had been sold at $p_i$ dollars per racquet but less than if they all had been sold at $p_{i-1}$ dollars each. Thus

$$p_i \cdot \frac{1200}{n} < A_i < p_{i-1} \cdot \frac{1200}{n}$$

That is

$$\left[ -0.0001 \left( \frac{1200i}{n} \right)^2 + 225 \right] \frac{1200}{n} < A_i < \left[ -0.0001 \left( \frac{1200(i-1)}{n} \right)^2 + 225 \right] \frac{1200}{n}$$

This expression may be written

$$\frac{-172{,}800i^2}{n^3} + \frac{270{,}000}{n} < A_i < \frac{-172{,}800(i-1)^2}{n^3} + \frac{270{,}000}{n}$$

The preceding represents $n$ double inequalities, one for each value of $i$. Adding these inequalities and using the sigma notation, we may write

$$\sum_{i=1}^{n} \frac{-172{,}800i^2}{n^3} + \sum_{i=1}^{n} \frac{270{,}000}{n} < \sum_{i=1}^{n} A_i < \sum_{i=1}^{n} \frac{-172{,}800(i-1)^2}{n^3} + \sum_{i=1}^{n} \frac{270{,}000}{n}$$

Since $\frac{270{,}000}{n}$ is constant as $i$ runs from 1 to $n$, we have $\sum_{i=1}^{n} \frac{270{,}000}{n} = n \cdot \frac{270{,}000}{n} = 270{,}000$. Furthermore, $-172{,}800/n^3$ is a constant factor. Thus, we may write

$$\left[ \frac{-172{,}800}{n^3} \sum_{i=1}^{n} i^2 \right] + 270{,}000 < \sum_{i=1}^{n} A_i < \left[ \frac{-172{,}800}{n^3} \sum_{i=1}^{n} (i-1)^2 \right] + 270{,}000$$

Recalling the formula $\sum_{i=1}^{n} i^2 = \frac{n(n+1)(2n+1)}{6}$, and using the result of Exercise 2 of this section, we get

$$\frac{-172{,}800n(n+1)(2n+1)}{6n^3} + 270{,}000 < \sum_{i=1}^{n} A_i < \frac{-172{,}800(n-1)n(2n-1)}{6n^3} + 270{,}000$$

This expression may be written

$$-28{,}800 \left( 1 + \frac{1}{n} \right) \left( 2 + \frac{1}{n} \right) + 270{,}000 < A < -28{,}800 \left( 1 - \frac{1}{n} \right) \left( 2 - \frac{1}{n} \right) + 270{,}000$$

where $A = \sum_{i=1}^{n} A_i$ dollars is the total amount spent by the consumers.

c. Since $\frac{1}{n}$ tends to 0 as $n$ tends to infinity, we have

$$\lim_{n \to \infty} \left[ -28{,}800 \left( 1 + \frac{1}{n} \right) \left( 2 + \frac{1}{n} \right) + 270{,}000 \right] = \lim_{n \to \infty} \left[ -28{,}800 \left( 1 - \frac{1}{n} \right) \left( 2 - \frac{1}{n} \right) + 270{,}000 \right] = 212{,}400$$

Thus, if we assume that the price per racquet decreases continuously from \$225 to \$81, the total amount spent by the consumers is \$212,400. Note that the expressions

$$-28,800\left(1 + \frac{1}{n}\right)\left(2 + \frac{1}{n}\right) + 270,000$$

and

$$-28,800\left(1 - \frac{1}{n}\right)\left(2 - \frac{1}{n}\right) + 270,000$$

are the lower and upper sums for the function $D$, respectively, corresponding to the regular partition $p_n$ of the interval $[0, 1200]$. Therefore 212,400—which is the common limit of these expressions as $n$ tends to infinity—is the definite integral $\int_0^{1200} (-0.0001q^2 + 225)\, dq$.

**Do Exercise 13.**                                                                          ■

## Exercise Set 9.9

1. Show that $\displaystyle\sum_{i=1}^{n} (i - 1) = \frac{(n - 1)n}{2}$.

2. Show that $\displaystyle\sum_{i=1}^{n} (i - 1)^2 = \frac{(n - 1)n(2n - 1)}{6}$.

3. Show that $\displaystyle\sum_{i=1}^{n} (i - 1)^3 = \frac{(n - 1)^2 n^2}{4}$.

*In Exercises 4–10, a particle is moving along a directed line. Its velocity function on a time interval $[a, b]$ is given where distance is measured in feet and time is in seconds.*

*a. Approximate the distance traveled by the particle during the time interval by dividing that interval into six equal subintervals.*

*b. Same as part a with n equal subintervals.*

*c. Using your answer in part b and taking its limit as n tends to infinity, find the exact distance traveled by the particle in the given time interval.*

4. $v(t) = 9t$ on $[2, 5]$
5. $v(t) = t^2$ on $[0, 3]$
6. $v(t) = 4t + 1$ on $[3, 6]$

7. $v(t) = 3t^2 + 4$ on $[1, 4]$
8. $v(t) = t^3$ on $[0, 6]$
9. $v(t) = t^2 + 3t + 1$ on $[2, 5]$
10. $v(t) = t^2 - 30t + 3$ on $[2, 12]$

*In Exercises 11–15, the demand equation for a commodity is given over an interval. The price p is in dollars, and q is the number of units. If $p = D(q)$,*

*a. Show that the function D is decreasing over that interval.*

*b. Find the highest price $p_1$ dollars per unit and the lowest price $p_2$ dollars per unit on that interval.*

*c. Calculate the total amount spent by the consumers assuming that the price decreases continuously from $p_1$ to $p_2$ dollars per unit.*

11. $p = 0.01q^2 - q + 100,\ 0 \le q \le 50$
12. $p = 0.001q^2 - 3q + 2500,\ 0 \le q \le 1000$
13. $p = -0.01q^2 + 400,\ 0 \le q \le 150$
14. $p = 0.001q^3 - 0.375q^2 + 30q + 1862.5,\ 50 \le q \le 150$
15. $p = 0.0001q^3 - 0.03q^2 + 1.08q + 200,\ 20 \le q \le 180$

# 9.10 The Fundamental Theorem of Calculus

In Example 1 of the preceding section, we calculated the distance traveled by a free falling object in the time interval $[0, 5]$, knowing that the instantaneous velocity $v(t)$ ft/sec is given by the equation $v(t) = 32t$. The purpose of the example was to discuss an application whose mathematical model is a definite integral. In fact, comparing the steps to those used in defining the definite integral, we may conclude that if $d$ feet is the distance traveled by the falling object in the time interval $[0, 5]$, then

$$d = \int_0^5 v(t)\, dt$$

There is another way of finding this distance. Suppose $s(t)$ feet gives the position of the falling object on a vertical line positively oriented downward. (See Figure 9.10.)

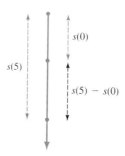

**FIGURE 9.10**

Then we observe that $v(t) = s'(t)$. Thus $s$ is an antiderivative of $v$. Furthermore, $s$ is an increasing function. Thus the distance traveled in the time interval $[0, 5]$ is $[s(5) - s(0)]$ feet, where $s(5)$ and $s(0)$ are the distances in feet from the origin to the positions of the object at instants 5 and 0 seconds, respectively. Recalling that $\int 32t\, dt = 16t^2 + C$, we have

$$s(t) = 16t^2 + C$$

where $C$ is some constant. Although we could show that the value of $C$ must be $s(0)$, we choose to keep the constant $C$ and we calculate the distance traveled as follows:

$$s(5) - s(0) = [16(5^2) + C] - [16(0^2) + C] = 400$$

Note that the constant $C$ cancels out and that our answer is the same as that obtained in Example 1 of the preceding section.

In several of the examples and exercises of Sections 9.8 and 9.9, after finding the value of a definite integral $\int_a^b f(x)\, dx$ by calculating the limit of a sequence of Riemann sums $R_{p_n}$, we showed that we would get the same answer by calculating $g(b) - g(a)$, where $g$ is any antiderivative of the function $f$. These examples illustrate a powerful method of evaluating a definite integral. The examples we discussed in the preceding two sections clearly show that even in cases where the integrands are simple, evaluating a definite integral by finding the limit of a sequence of Riemann sums can be tedious and often impossible. Therefore, the concept would not be useful without a process to perform the evaluation in a much simpler way.

## How to Evaluate a Definite Integral

At the end of this section, we state a theorem that completely justifies the method we are about to describe. We begin by outlining the process informally and by giving several examples.

## Procedure to Evaluate the Definite Integral $\int_a^b f(x)\ dx$

If $f$ is a continuous function on a closed bounded interval $[a, b]$ and we wish to evaluate the definite integral $\int_a^b f(x)\ dx$, we proceed as follows:

**STEP 1.**  Find any antiderivative $g$ of $f$.

**STEP 2.**  Evaluate $g(b)$.

**STEP 3.**  Evaluate $g(a)$.

**STEP 4.**  Calculate $g(b) - g(a)$.

The number obtained in Step 4 is the value of the definite integral.

## Examples

**EXAMPLE 1**  Evaluate $\displaystyle\int_2^5 (3x^2 + 6x + 1)\ dx$.

**Solution**  It is clear that

$$\int (3x^2 + 6x + 1)\ dx = x^3 + 3x^2 + x + C$$

where $C$ is an arbitrary constant. We may use any antiderivative of $3x^2 + 6x + 1$. Thus we let $C = 0$ so that

$$g(x) = x^3 + 3x^2 + x$$

We get

$$g(5) = 5^3 + 3(5^2) + 5 = 205 \text{ and } g(2) = 2^3 + 3(2^2) + 2 = 22$$

Thus

$$g(5) - g(2) = 205 - 22 = 183$$

We conclude that

$$\int_2^5 (3x^2 + 6x + 1)\ dx = 183$$

**Do Exercise 1.**  ∎

*Notation:* Since the expression $g(b) - g(a)$ occurs frequently when evaluating a definite integral, it is convenient to have a special symbol to denote this difference. We write

$$g(x)\big|_a^b \text{ instead of } g(b) - g(a)$$

Thus if $f$ is a continuous function on $[a, b]$ and $g$ is an antiderivative of $f$ on $[a, b]$, we have

$$\int_a^b f(x)\ dx = g(x)\big|_a^b$$

**EXAMPLE 2** Evaluate $\int_3^7 \dfrac{6x}{x^2 + 1} \, dx$.

**Solution** We must find an antiderivative of $6x/(x^2 + 1)$. Since $D_x(x^2 + 1) = 2x$, and $6x$ is a constant multiple of $2x$, we recall the formula

$$\int \frac{du}{u} = \ln|u| + C$$

and we assume that

$$\int \frac{6x}{x^2 + 1} \, dx = k \ln(x^2 + 1) + C$$

where $k$ and $C$ are constants. But,

$$D_x[k \ln (x^2 + 1) + C] = k \cdot \frac{D_x(x^2 + 1)}{x^2 + 1} + 0 = \frac{2kx}{x^2 + 1}$$

Thus we have

$$\frac{2kx}{x^2 + 1} = \frac{6x}{x^2 + 1}$$

from which we get

$$k = 3$$

Therefore,

$$\int \frac{6x}{(x^2 + 1)} \, dx = 3 \ln(x^2 + 1) + C$$

Since we are free to use any antiderivative of $6x/(x^2 + 1)$, we let $C = 0$ and obtain

$$\int_3^7 \frac{6x}{x^2 + 1} \, dx = 3 \ln(x^2 + 1)\Big|_3^7$$

$$= 3 \ln(7^2 + 1) - 3 \ln(3^2 + 1) = 3(\ln 50 - \ln 10)$$

$$= 3 \ln \frac{50}{10} = 3 \ln 5 = \ln 5^3 = \ln 125$$

**Do Exercise 17.**

**EXAMPLE 3** Evaluate $\int_0^2 e^{x^2}x \, dx$.

**Solution** $\int_0^2 e^{x^2}x \, dx = \dfrac{1}{2}e^{x^2}\Big|_0^2$

$$= \frac{1}{2}e^{2^2} - \frac{1}{2}e^{0^2}$$

$$= \frac{1}{2}e^4 - \frac{1}{2}e^0$$

$$= \frac{1}{2}(e^4 - 1)$$

**Do Exercise 13.**

EXAMPLE 4   Evaluate $\int_2^{11} \dfrac{3x}{(x^2 + 4)^{1/3}}\, dx.$

Solution   We write

$$\int_2^{11} \frac{3x}{(x^2 + 4)^{1/3}}\, dx = \int_2^{11} (x^2 + 4)^{-1/3} 3x\, dx$$

$$= \frac{9}{4}(x^2 + 4)^{2/3}\Big|_2^{11}$$

$$= \frac{9}{4}(11^2 + 4)^{2/3} - \frac{9}{4}(2^2 + 4)^{2/3}$$

$$= \frac{9}{4}(125)^{2/3} - \frac{9}{4}(8)^{2/3}$$

$$= \frac{9}{4}(25 - 4) = \frac{189}{4}$$

**Do Exercise 15.**

EXAMPLE 5   Evaluate $\int_0^2 xe^{4x}\, dx.$

Solution   In this case, we cannot easily find an antiderivative of $xe^{4x}$ by trial and error as we did in the preceding examples. We can use integration by parts and let

$$u = x \text{ and } dv = e^{4x}\, dx$$

so that

$$du = dx \text{ and } v = \frac{1}{4}e^{4x}$$

Thus

$$\int xe^{4x}\, dx = \frac{1}{4}xe^{4x} - \int \frac{1}{4}e^{4x}\, dx$$

$$= \frac{1}{4}xe^{4x} - \frac{1}{16}e^{4x} + C$$

$$= \frac{e^{4x}}{16}(4x - 1) + C$$

It follows that

$$\int_0^2 xe^{4x}\, dx = \frac{e^{4x}}{16}(4x - 1)\Big|_0^2$$

$$= \frac{e^{4(2)}}{16}[4(2) - 1] - \frac{e^{4(0)}}{16}[4(0) - 1]$$

$$= \frac{e^8}{16}(7) - \frac{e^0}{16}(-1)$$

$$= \frac{7e^8 + 1}{16}$$

**Do Exercise 19.**

Recall that it is often convenient to make a *u*-substitution to evaluate an indefinite integral. In the case of definite integrals, we must be careful when dealing with the limits of integration. We may proceed in either of two ways. We illustrate both methods in the following example.

**EXAMPLE 6**   Evaluate $\displaystyle\int_0^2 (x^3 + 3x + 1)^4(x^2 + 1)\, dx$.

**Solution**   *First method:* We first find the *indefinite* integral $\int (x^3 + 3x + 1)^4(x^2 + 1)\, dx$. We let $u = x^3 + 3x + 1$, so that $\frac{du}{dx} = 3x^2 + 3$ and $du = 3(x^2 + 1)\, dx$. Now we can write

$$\int (x^3 + 3x + 1)^4(x^2 + 1)\, dx = \frac{1}{3}\int (x^3 + 3x + 1)^4 3(x^2 + 1)\, dx$$

$$= \frac{1}{3}\int u^4\, du = \frac{1}{3}\left(\frac{u^5}{5}\right) + C$$

$$= \frac{u^5}{15} + C$$

*Before* evaluating the definite integral, we must be sure to reintroduce the variable $x$ because the limits of integration 0 and 2 were values of $x$, not of $u$. Thus,

$$\int (x^3 + 3x + 1)^4(x^2 + 1)\, dx = \frac{u^5}{15} + C$$

$$= \frac{(x^3 + 3x + 1)^5}{15} + C$$

It follows that

$$\int_0^2 (x^3 + 3x + 1)^4(x^2 + 1)\, dx = \frac{(x^3 + 3x + 1)^5}{15}\Bigg|_0^2$$

$$= \frac{1}{15}(15^5) - \frac{1}{15}(1^5)$$

$$= \frac{759{,}374}{15}$$

*Second method:* Working with the *definite* integral, we let $u = x^3 + 3x + 1$ and $du = 3(x^2 + 1)\, dx$ as we did earlier. In addition, we must change the limits of integration and note that when $x = 0$, $u = 1$ and when $x = 2$, $u = 15$. Thus,

$$\int_0^2 (x^3 + 3x + 1)^4(x^2 + 1)\, dx = \frac{1}{3}\int_0^2 (x^3 + 3x + 1)^4 3(x^2 + 1)\, dx$$

$$= \frac{1}{3}\int_1^{15} u^4\, du = \frac{1}{3}\left(\frac{u^5}{5}\right)\Bigg|_1^{15}$$

$$= \frac{15^5}{15} - \frac{1^5}{15} = \frac{759{,}374}{15}$$

**Do Exercise 11.**

Some of you will prefer the first method, others will find the second method simpler. In any application, you should choose the process that feels most comfortable. When you evaluate a definite integral, you must be careful to use as limits values of the variable of integration. For example, doing the following in Example 6 would be incorrect. Let

$$u = x^3 + 3x + 1$$

so that

$$\frac{du}{dx} = 3x^2 + 3 \text{ and } du = 3(x^2 + 1) \, dx$$

So

$$\int_0^2 (x^3 + 3x + 1)^4 (x^2 + 1) \, dx = \frac{1}{3} \int_0^2 (x^3 + 3x + 1)^4 3(x^2 + 1) \, dx$$

$$= \frac{1}{3} \int_0^2 u^4 \, du = \frac{1}{3} \left( \frac{u^5}{5} \right) \Big|_0^2 = \frac{u^5}{15} \Big|_0^2$$

$$= \frac{2^5}{15} - \frac{0^5}{15} = \frac{32}{15}$$

A wrong answer was obtained because the limits of integration 0 and 2 are values of $x$, not of $u$, and therefore should not be used as replacements for $u$.

## A "Geometric Proof" of the Fundamental Theorem of Calculus

The method used to evaluate the definite integrals in the preceding six examples is justified by a remarkable theorem. This theorem establishes a connection between differentiation and integration, and may be stated as follows:

> **THEOREM 9.6 (The Fundamental Theorem of Calculus):** Suppose $f$ is a continuous function on a closed bounded interval $[a, b]$. Then
>
> **a.** There exists at least one antiderivative of $f$ on $[a, b]$ and
> **b.** If $g$ is any antiderivative of $f$ on $[a, b]$, then
>
> $$\int_a^b f(x) \, dx = g(x)\Big|_a^b$$

**Proof:** A rigorous proof for the general case of this theorem is beyond the scope of this book. Instead, we give an outline of a "geometric proof" for a special case where the function $f$ is increasing and $f(x) \geq 0$ for all $x$ in $[a, b]$.

If $x$ is a fixed number between $a$ and $b$, it can be shown that $\int_a^x f(t) \, dt$ square units is the area of the region bounded by the graphs of $y = f(t)$, $t = a$, $t = x$, and the $t$-axis. (See Figure 9.11.) Note that we are using $t$ as the independent variable because in our proof $x$ is arbitrary but fixed. We let $A(x) = \int_a^x f(t) \, dt$. We first show that $A'(x) = f(x)$.

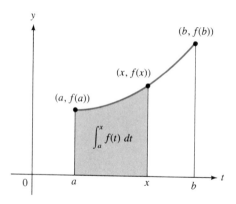

**FIGURE 9.11**

By definition,

$$A'(x) = \lim_{\Delta x \to 0} \frac{A(x + \Delta x) - A(x)}{\Delta x}$$

The area of the shaded region in Figure 9.12 represents $A(x + \Delta x) - A(x)$.

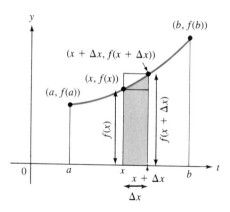

**FIGURE 9.12**

Note that the shaded region is a subset of the circumscribed rectangle whose dimensions are $\Delta x$ by $f(x + \Delta x)$ and contains the inscribed rectangle whose dimensions are $\Delta x$ by $f(x)$. Thus

$$f(x) \cdot \Delta x \leq A(x + \Delta x) - A(x) \leq f(x + \Delta x) \cdot \Delta x$$

Since $\Delta x > 0$, we may divide by $\Delta x$ to get

$$f(x) \leq \frac{A(x + \Delta x) - A(x)}{\Delta x} \leq f(x + \Delta x)$$

If $x$ is kept fixed and $\Delta x$ tends to 0, by continuity of $f$, $f(x + \Delta x)$ tends to $f(x)$. But, $\frac{A(x + \Delta x) - A(x)}{\Delta x}$ is squeezed between $f(x)$ and $f(x + \Delta x)$ for all values of $\Delta x > 0$. Thus

$$\lim_{\Delta x \to 0} \frac{A(x + \Delta x) - A(x)}{\Delta x} = f(x)$$

That is,

$$A'(x) = f(x)$$

This completes the first part of the proof; $f$ has at least one antiderivative, namely $A$.

Now let $g$ be any antiderivative of $f$. By a previous theorem (see Section 9.1), we know there is a constant $C$ such that

$$g(x) = A(x) + C$$

for all $x$ in $[a, b]$. It is clear that $A(a) = 0$. Thus

$$g(a) = 0 + C$$

so that

$$C = g(a)$$

Thus,

$$g(x) = A(x) + g(a)$$

for all $x$ in $[a, b]$.

In particular,

$$g(b) = A(b) + g(a)$$

and

$$A(b) = g(b) - g(a)$$

But,

$$A(b) = \int_a^b f(t)\, dt$$

Therefore,

$$\int_a^b f(t)\, dt = g(b) - g(a)$$

We may now replace $t$ by $x$ in the last equality and write

$$\int_a^b f(x)\, dx = g(b) - g(a) = g(x)\big|_a^b$$

The second method used to evaluate the definite integral of Example 6 is based on a change of variables. This technique is justified by the following theorem.

THEOREM 9.7:   Suppose $f$ and $g$ are functions such that $f$ is continuous, $g$ has a continuous derivative on the closed bounded interval $[a, b]$, and $f(g(x))$ is defined for all $x$ in $[a, b]$. Then, if we let $u = g(x)$, we have

$$\int_a^b f(g(x))g'(x)\, dx = \int_{g(a)}^{g(b)} f(u)\, du$$

**Proof (Optional):** Let $F$ be any antiderivative of $f$. Then

$$D_x F(g(x)) = F'(g(x))g'(x) = f(g(x))g'(x)$$

Thus $F(g(x))$ is an antiderivative of $f(g(x))g'(x)$ and

$$\int_a^b f(g(x))g'(x)\, dx = F(g(x))|_a^b$$

$$= F(g(b)) - F(g(a))$$

On the other hand, $D_u F(u) = f(u)$. Thus

$$\int_{g(a)}^{g(b)} f(u)\, du = F(u)|_{g(a)}^{g(b)}$$

$$= F(g(b)) - F(g(a))$$

Since both $\int_a^b f(g(x))g'(x)\, dx$ and $\int_{g(a)}^{g(b)} f(u)\, du$ are equal to $F(g(b)) - F(g(a))$, they are equal to each other and the proof is complete.

# Exercise Set 9.10

*In Exercises 1–24, evaluate the definite integrals.*

**1.** $\int_1^2 (4x^3 + 6x^2 + 2)\, dx$

**2.** $\int_{-1}^3 (-10x^4 + 9x^2 + 3)\, dx$

**3.** $\int_1^4 \sqrt{x}\, dx$

**4.** $\int_0^1 x^{2/3}\, dx$

**5.** $\int_0^2 e^{3x}\, dx$

**6.** $\int_3^{22} \dfrac{4}{(x + 5)^{1/3}}\, dx$

**7.** $\int_0^4 \dfrac{6}{x + 1}\, dx$

**8.** $\int_{-6}^{-3} \dfrac{4}{x + 2}\, dx$

**9.** $\int_0^1 (x^2 + 1)^5 x\, dx$

**10.** $\int_{-1}^2 (x^2 + 4x + 3)^6(x + 2)\, dx$

**11.** $\int_0^2 \sqrt{x^3 + 6x + 1}(x^2 + 2)\, dx$

**12.** $\int_0^1 \dfrac{x^3 + 2}{(x^4 + 8x + 16)^{1/4}}\, dx$

**13.** $\int_{-1}^2 x^2 e^{x^3}\, dx$

**14.** $\int_0^2 e^{x^2 + 6x + 1}(4x + 12)\, dx$

**15.** $\int_{-1}^2 \dfrac{3e^{2x}}{(e^{2x} + 1)^2}\, dx$

**16.** $\int_2^6 \dfrac{3x}{x^2 + 5}\, dx$

**17.** $\int_0^2 \dfrac{e^{5x}}{e^{5x} + 1}\, dx$

**18.** $\int_{-1}^3 x2^{x^2}\, dx$

**19.** $\int_1^3 3xe^{4x}\, dx$

**20.** $\int_{-1}^2 5xe^{2x}\, dx$

**21.** $\int_0^2 (x^2 + 1)e^{3x}\, dx$

**22.** $\int_{-1}^1 (x^3 + 2x^2 + 1)e^{3x}\, dx$

**23.** $\int_2^6 \ln(x + 1)\, dx$ (*Hint:* Use integration by parts.)

**24.** $\int_1^4 x \ln(x + 2)\, dx$

*In Exercises 25–29, a particle is moving along a directed line. Distances are measured in meters and time is in seconds. In each case, an equation defining a velocity function is given as well as a time interval. Use a definite integral to find the distance traveled by the particle during that time interval.*

**25.** $v(t) = 9.8t + 3$; $t_1 = 3$, $t_2 = 5$

**26.** $v(t) = 15t + 4$; $t_1 = 1$, $t_2 = 6$

**27.** $v(t) = 3t^2 + 1$; $t_1 = 2$, $t_2 = 7$

**28.** $v(t) = \sqrt{t} + 4$; $t_1 = 4$, $t_2 = 16$

**29.** $v(t) = e^{-3t}$; $t_1 = 0$, $t_2 = 3$

*In Exercises 30–35, a demand equation for a commodity is given, where p dollars is the price per unit and q is the number of units the consumers are willing to buy at that price. Use a definite integral to find the total amount spent by the consumers as the number of units sold increases from $q_1$ to $q_2$ and prices decrease accordingly.*

**30.** $p = -0.001q^2 - 0.25q + 750$; $q_1 = 0$, $q_2 = 500$

**31.** $p = 0.00002q^3 - 0.015q^2 + 1250$; $q_1 = 0$, $q_2 = 400$

**32.** $p = 50 - \sqrt{q + 9}$; $q_1 = 7$, $q_2 = 891$

**33.** $p = 30 - (q + 1)^{1/3}$; $q_1 = 26$, $q_2 = 7999$

**34.** $p = 50e^{-0.0007q}$; $q_1 = 0$, $q_2 = 800$

**35.** $p = \dfrac{110}{0.001q + 1}$; $q_1 = 0$, $q_2 = 1000$

**36.** Evaluate $\displaystyle\int_1^5 \left( D_x \frac{\ln(x^2 + 1)}{x^3} \right) dx$

**37.** Evaluate $\displaystyle\int_2^7 \left( D_x \frac{\ln(x + 1)}{e^{x^3}} \right) dx$

**38.** Evaluate $\displaystyle D_x \left[ \int_1^3 (x^2 + 1)e^{6x} \, dx \right]$

**39.** Evaluate $\displaystyle D_x \left[ \int_3^9 (x^3 + x^2 - 1)e^{4x} \, dx \right]$

**\*40.** A function $f$ is said to be *even* if $f(-x) = f(x)$ for all $x$ in the domain of $f$. A function $g$ is said to be *odd* if $g(-x) = -g(x)$ for all $x$ in the domain of $g$.

**a.** Show that the function $f$ defined by $f(x) = x^2$ is an even function.

**b.** Show that the function $g$ defined by $g(x) = x^3$ is an odd function.

**c.** Show that if $f$ is an odd continuous function defined on an interval $[a, b]$ and $F$ is an antiderivative of $f$, then $F$ is even. (*Hint:* Show that $F(-x)$ and $F(x)$ differ by a constant $C$ and then show that $C$ must be 0.)

**d.** Using the result of part **c**, show that if $f$ is a continuous odd function defined on an interval $[-a, a]$, then $\int_{-a}^a f(x) \, dx = 0$.

**e.** Show that if $f$ is an even continuous function defined on an interval, then one of the antiderivatives of $f$ is odd.

**f.** Using the result of part **e**, show that if $f$ is a continuous even function defined on an interval $[-a, a]$, then $\int_{-a}^a f(x) \, dx = 2 \int_0^a f(x) \, dx$. (*Hint:* If $g$ is an odd function, $g(0) = 0$.)

*In Exercises 41–46, use the result of Exercise 40d to evaluate the definite integrals. (Hint: In evaluating $\int_{-a}^a [f_1(x) + f_2(x) + \cdots + f_n(x)] \, dx$, each $f_i(x)$, where $f_i$ is an odd function, may be dropped out since its contribution to the value of the integral is zero.)*

**41.** $\displaystyle\int_{-2}^2 (x^4 + 2x^3 + 3x + 1) \, dx$

**42.** $\displaystyle\int_{-3}^3 (x^2 + 1)^{50}x \, dx$

**43.** $\displaystyle\int_{-5}^5 xe^{x^4} \, dx$

**44.** $\displaystyle\int_{-4}^4 (3x^2 + 5x\sqrt{16 - x^2}) \, dx$

**45.** $\displaystyle\int_{-3}^3 (4x^5 + 3x^3\sqrt{x^2 + 1} + 5x^2 + 6x + 1) \, dx$

**46.** $\displaystyle\int_{-1}^1 \frac{3x^5}{x^4 + x^2 + 1} \, dx$

*In the next chapter, we shall show that if we have a continuous flow of income of P dollars per year at a rate of i% compounded continuously for n years, its present value is $\int_0^n Pe^{-it} \, dt$ dollars. In Exercises 47–50, calculate the present value to the nearest dollar for the given P, i, and n.*

**47.** $P = 5000$, $i = 6$, $n = 10$.

**48.** $P = 6000$, $i = 9$, $n = 12$.

**49.** $P = 3000$, $i = 7.25$, $n = 8$.

**50.** $P = 4500$, $i = 7.25$, $n = 6$.

# 9.11 Approximate Integration (Optional)

The Fundamental Theorem of Calculus provides a convenient way to evaluate $\int_a^b f(x) \, dx$ when some antiderivative of $f$ is known. However, it is often difficult, and sometimes impossible, to find an antiderivative of a function. In such cases, we may use numerical methods to approximate the definite integral. These methods use functional values at a finite number of points in the interval $[a, b]$ and are especially effective when using calculators or computers. In Section 9.8, we introduced the definite integral $\int_a^b f(x)\ dx$ as the common limit of the sequences $\underline{S}_{p_n}$ (lower sums), $R_{p_n}$ (Riemann sums), $\overline{S}_{p_n}$ (upper sums). But in the case $f(x) \geq 0$, each of these sums is the area of a rectangular polygon corresponding to a regular partition $p_n$ of the interval $[a, b]$. For large $n$, any of these areas can be used as an approximation of $\int_a^b f(x)\ dx$.

In this section, we consider two numerical methods to approximate a definite integral, the *Trapezoidal Rule* and *Simpson's Rule*. In both methods, we assume that the function $f$ is continuous on a closed bounded interval $[a, b]$.

## The Trapezoid Rule

The idea behind this method is to sum areas of trapezoids instead of rectangles. For convenience, we assume that $f(x) \geq 0$, so that we can interpret the terms of the sums we shall get as areas. However, the rule is valid without this restriction. We first recall that the area of a trapezoid is given by the formula

$$A = h \cdot \frac{b_1 + b_2}{2}$$

where $h$ is the altitude, and $b_1$ and $b_2$ are the bases (parallel sides) of the trapezoid.

Let $p_n$ be a regular partition of $[a, b]$ which determines $n$ subintervals and let $h$ be the length of each subinterval. Then

$$h = \frac{b - a}{n}$$

The partition is $(a, a + h, a + 2h, a + 3h, \ldots, a + (n - 1)h, a + nh)$. For each $i \in \{0, 1, 2, 3, \ldots, n\}$, let $x_i = a + ih$ and $y_i = f(x_i)$, so that $x_0 = a$, $x_1 = a + h$, $x_2 = a + 2h$, and so on, and $y_0 = f(x_0) = f(a)$, $y_1 = f(x_1) = f(a + h)$, and so on.

Then draw line segments connecting the points $(x_0, y_0)$ to $(x_1, y_1)$, $(x_1, y_1)$, to $(x_2, y_2)$, $(x_2, y_2)$, to $(x_3, y_3)$, $\ldots$, $(x_{n-1}, y_{n-1})$ to $(x_n, y_n)$. (See Figure 9.13.)

We then draw $n$ trapezoids. The first trapezoid has vertices $(x_0, 0)$, $(x_1, 0)$, $(x_0, y_0)$, and $(x_1, y_1)$. Its area is given by

$$A_1 = h \cdot \frac{y_0 + y_1}{2}$$

The second trapezoid has vertices $(x_1, 0)$, $(x_2, 0)$, $(x_1, y_1)$, and $(x_2, y_2)$. Its area is given by

$$A_2 = h \cdot \frac{y_1 + y_2}{2}$$

The third trapezoid has vertices $(x_2, 0)$, $(x_3, 0)$, $(x_2, y_2)$, and $(x_3, y_3)$. Its area is given by

$$A_3 = h \cdot \frac{y_2 + y_3}{2}$$
$$\vdots$$

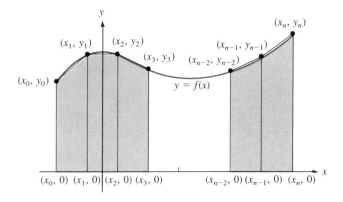

FIGURE 9.13

The $n$th trapezoid has vertices $(x_{n-1}, 0)$, $(x_n, 0)$, $(x_{n-1}, y_{n-1})$, and $(x_n, y_n)$. Its area is given by

$$A_n = h \cdot \frac{y_{n-1} + y_n}{2}$$

The sum $T_n$ of the areas of the trapezoids is given by

$$T_n = A_1 + A_2 + A_3 + \cdots + A_{n-1} + A_n$$

$$= h \cdot \frac{y_0 + y_1}{2} + h \cdot \frac{y_1 + y_2}{2} + h \cdot \frac{y_2 + y_3}{2} + \cdots$$

$$+ h \cdot \frac{y_{n-2} + y_{n-1}}{2} + h \cdot \frac{y_{n-1} + y_n}{2}$$

$$= \frac{h}{2}(y_0 + y_1 + y_1 + y_2 + y_2 + y_3 + y_3 + \cdots + y_{n-1} + y_{n-1} + y_n)$$

$$= \frac{h}{2}(y_0 + 2y_1 + 2y_2 + 2y_3 + \cdots + 2y_{n-1} + y_n)$$

It is evident that for any regular partition $p_n$ of the interval $[a, b]$, the corresponding inscribed rectangular polygon is a subset of the union of the $n$ trapezoids, and that this union is a subset of the circumscribed rectangular polygon. (See Figure 9.14.) Thus

$$\underline{S}_{p_n} \le T_{p_n} \le \overline{S}_{p_n}$$

FIGURE 9.14

Since for large values of $n$, both $\underline{S}_{p_n}$ and $\overline{S}_{p_n}$ are approximations of $\int_a^b f(x)\,dx$, so is $T_{p_n}$. We have obtained the following rule.

**The Trapezoidal Rule**

If $f$ is a continuous function on a closed bounded interval $[a, b]$ and $p_n$ is a regular partition of $[a, b]$, then

$$\int_a^b f(x)\,dx \doteq \frac{h}{2}(y_0 + 2y_1 + 2y_2 + 2y_3 + \cdots + 2y_{n-1} + y_n)$$

where $h = \frac{b-a}{n}$ and $y_i = f(a + ih)$ for $i = 0, 1, 2, 3, \ldots, n$.

**EXAMPLE 1**    Use the Trapezoidal Rule with $n = 6$ to approximate $\int_1^4 \frac{1}{x}\,dx$.

**Solution**    We first find the regular partition $p_n$. Since each subinterval has length $\frac{1}{2}$ $\left(\text{because } \frac{b-a}{n} = \frac{4-1}{6} = \frac{1}{2}\right)$, the partition is $(1, 1.5, 2, 2.5, 3, 3.5, 4)$. Thus $x_0 = 1$, $x_1 = 1.5$, $x_2 = 2$, $x_3 = 2.5$, $x_4 = 3$, $x_5 = 3.5$, and $x_6 = 4$. Next, using a calculator when necessary, we compute $f(x_0), f(x_6)$, and for each $i \in \{1, 2, 3, 4, 5\}$, we compute $2f(x_i)$ where, in this example, $f(x) = \frac{1}{x}$.

$$y_0 = f(x_0)\quad = f(1)\quad\ = \frac{1}{1}\quad\ = 1$$

$$2y_1 = 2f(x_1) = 2f(1.5) = 2 \cdot \frac{1}{1.5} \doteq 1.33333$$

$$2y_2 = 2f(x_2) = 2f(2)\quad\ = 2 \cdot \frac{1}{2}\quad = 1$$

$$2y_3 = 2f(x_3) = 2f(2.5) = 2 \cdot \frac{1}{2.5} = 0.8$$

$$2y_4 = 2f(x_4) = 2f(3)\quad\ = 2 \cdot \frac{1}{3}\quad \doteq 0.66667$$

$$2y_5 = 2f(x_5) = 2f(3.5) = 2 \cdot \frac{1}{3.5} \doteq 0.57143$$

$$y_6 = f(x_6)\quad = f(4)\quad\ = \frac{1}{4}\quad\ = \underline{0.25}$$

$$\text{Sum}\quad = 5.62143$$

Since $h = 0.5$, $\frac{h}{2} = 0.25$ and

$$\frac{h}{2}(y_0 + 2y_1 + 2y_2 + 2y_3 + 2y_4 + 2y_5 + y_6) = 0.25(5.62143) \doteq 1.40536$$

Thus

$$\int_1^4 \frac{1}{x}\,dx \doteq 1.40536$$

**Do Exercise 3.**

Observe that using the Fundamental Theorem of Calculus, we obtain

$$\int_1^4 \frac{1}{x} \, dx = \ln x \, \bigg|_1^4 = \ln 4$$

The approximate value of ln 4, found using a calculator, is 1.386294361.

Our answer is fairly accurate (approximately 1.38% error since $\frac{1.40536 - 1.38629}{1.38629} \doteq$ 0.013756). There are formulas, found in standard books on numerical analysis, for determining the accuracy of the answers. For example, it is possible to determine how large $n$ must be to get an answer correct to five decimal places.

## Simpson's Rule

We now turn to a much more powerful numerical method to approximate a definite integral. This method is known as Simpson's Rule. The basic idea behind the Trapezoidal Rule was to approximate the graph of $y = f(x)$ by a path that is a union of line segments. The idea behind Simpson's Rule is to approximate the graph of $y = f(x)$ by a path that is the union of $n$ parabolic arcs. Using the same notation but requiring **$n$ to be even**, we have, for each $i$, a parabolic arc $C_i$ starting at $(x_{i-1}, y_{i-1})$, passing through $(x_i, y_i)$, and ending at $(x_{i+1}, y_{i+1})$. The equation of each parabolic arc is of the form

$$y = g_i(x) = a_i x^2 + b_i x + c_i$$

We then use the fact that when $n$ is large,

$$\int_{x_{i-1}}^{x_{i+1}} f(x) \, dx \doteq \int_{x_{i-1}}^{x_{i+1}} g_i(x) \, dx$$

for each $i \in \{1, 2, 3, \ldots, n - 1\}$. To see this geometrically, note that the area of the region under the parabolic arc $C_i$ and above the interval $[x_{i-1}, x_{i+1}]$ is close to the area of the region under the graph of $y = f(x)$ and above $[x_{i-1}, x_{i+1}]$. (See Figure 9.15.) Using this idea, it can be shown, but we omit the details, that the following rule is true. (See Exercise *33.)

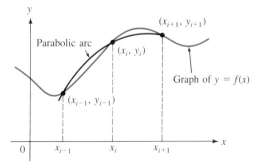

FIGURE 9.15

## Simpson's Rule

If $f$ is a continuous function on a closed bounded interval $[a, b]$, $n$ is an even positive integer, and $p_n$ is a regular partition of $[a, b]$, then

$$\int_a^b f(x)\, dx \doteq \frac{h}{3}(y_0 + 4y_1 + 2y_2 + 4y_3 + \cdots + 4y_{n-1} + y_n)$$

where $h = \frac{b-a}{n}$ and $y_i = f(a + ih)$ for $i = 0, 1, 2, 3, \ldots, n$.

**EXAMPLE 2**   Use Simpson's Rule with $n = 6$ to approximate the definite integral of Example 1.

**Solution**   The values of $x_i$, $y_i$, and $h$ are the same as in Example 1. However, the coefficients are 1, 4, 2, 4, 2, 4, and 1. Thus, we get

$$y_0 = f(x_0)\ = f(1)\quad = \frac{1}{1}\quad = 1$$

$$4y_1 = 4f(x_1) = 4f(1.5) = 4 \cdot \frac{1}{1.5} \doteq 2.66667$$

$$2y_2 = 2f(x_2) = 2f(2)\quad = 2 \cdot \frac{1}{2}\quad = 1$$

$$4y_3 = 4f(x_3) = 4f(2.5) = 4 \cdot \frac{1}{2.5} = 1.6$$

$$2y_4 = 2f(x_4) = 2f(3)\quad = 2 \cdot \frac{1}{3}\quad \doteq 0.66667$$

$$4y_5 = 4f(x_5) = 4f(3.5) = 4 \cdot \frac{1}{3.5} \doteq 1.14286$$

$$y_6 = f(x_6)\ = f(4)\quad = \frac{1}{4}\quad = \underline{0.25}$$

$$\text{Sum}\quad = 8.3262$$

Since $h = \frac{1}{2}$, $\frac{h}{3} = \frac{1}{6}$ and

$$\frac{h}{3}(y_0 + 4y_1 + 2y_2 + 4y_3 + 2y_4 + 4y_5 + y_6) \doteq \frac{1}{6}(8.3262) = 1.3877$$

Thus

$$\int_1^4 \frac{1}{x}\, dx \doteq 1.3877$$

**Do Exercise 7.**                                                                                ■

Observe that this answer is much closer to the actual value of ln 4 than the answer we obtained using the Trapezoidal Rule. The error is approximately 0.1% since $\frac{1.3877 - 1.3863}{1.3863} = 0.001$. Simpson's Rule usually leads to approximations more accurate than those obtained using the Trapezoidal Rule.

Both the Trapezoid and Simpson's Rules may be used if we know the values of a continuous function $f$ at the points $x_0, x_1, x_2, \ldots, x_n$, where $(x_0, x_1, x_2, \ldots, x_n)$ is a regular partition of $[a, b]$, even if we do not know $f$ itself.

**EXAMPLE 3**   As illustrated in Example 2, Section 9.9, if $p = D(q)$ is the demand equation for a commodity, where $p$ is the price per unit and $q$ is the number of units the consumers are willing to purchase at that price, then the total amount spent by the consumers when the quantities purchased increase from $q_1$ to $q_2$ is approximated by

$$\int_{q_1}^{q_2} D(q)\, dq$$

The following table gives the values of $p = D(q)$ for certain values of $q$, where $p$ is dollars and $q$ is in thousands of units. Assuming the function $D$ to be continuous, approximate the total amount spent by the consumers as the quantities purchased increased from 3000 units to 8000 units.

| $q$ | 3 | 3.5 | 4 | 4.5 | 5 | 5.5 | 6 | 6.5 | 7 | 7.5 | 8 |
|---|---|---|---|---|---|---|---|---|---|---|---|
| $p$ | 6 | 5.9 | 5.8 | 5.6 | 5.3 | 5 | 4.7 | 4.3 | 4 | 3.5 | 3 |

**Solution**   The total amount spent by the consumers (in thousands of dollars) is approximated by

$$\int_{3}^{8} D(q)\, dq$$

We may approximate this definite integral using Simpson's Rule since $(3, 3.5, 4, 4.5, 5, 5.5, 6, 6.5, 7, 7.5, 8)$ is a regular partition of $[3, 8]$ and the number of subintervals formed by the partition is even. Using the data from the table, we proceed as follows:

$$
\begin{aligned}
p_0 &= D(q_0) &&= D(3) && && = 6.0 \\
4p_1 &= 4D(q_1) &&= 4D(3.5) &&= 4 \cdot (5.9) &&= 23.6 \\
2p_2 &= 2D(q_2) &&= 2D(4) &&= 2 \cdot (5.8) &&= 11.6 \\
4p_3 &= 4D(q_3) &&= 4D(4.5) &&= 4 \cdot (5.6) &&= 22.4 \\
2p_4 &= 2D(q_4) &&= 2D(5) &&= 2 \cdot (5.3) &&= 10.6 \\
4p_5 &= 4D(q_5) &&= 4D(5.5) &&= 4 \cdot (5) &&= 20.0 \\
2p_6 &= 2D(q_6) &&= 2D(6) &&= 2 \cdot (4.7) &&= 9.4 \\
4p_7 &= 4D(q_7) &&= 4D(6.5) &&= 4 \cdot (4.3) &&= 17.2 \\
2p_8 &= 2D(q_8) &&= 2D(7) &&= 2 \cdot (4) &&= 8.0 \\
4p_9 &= 4D(q_9) &&= 4D(7.5) &&= 4 \cdot (3.5) &&= 14.0 \\
p_{10} &= D(q_{10}) &&= D(8) && && = \underline{\phantom{0}3.0} \\
&&&&& \text{Sum} &&= 145.8
\end{aligned}
$$

Since $h = \frac{1}{2}, \frac{h}{3} = \frac{1}{6}$ and

$$\frac{h}{3}(p_0 + 4p_1 + 2p_2 + 4p_3 + 2p_4 + 4p_5 + 2p_6 + 4p_7 + 2p_8 + 4p_9 + p_{10})$$

$$= \frac{1}{6}(145.8) = 24.3$$

Thus

$$\int_3^8 D(q) \, dq \doteq 24.3$$

Since the values of $q$ were given in thousands, we conclude that the total amount spent by the consumers is approximately \$24,300.

**Do Exercise 25.**    ∎

# Exercise Set 9.11

*In Exercises 1–20, use the Trapezoidal Rule or Simpson's Rule, as indicated, with the given value of n to estimate the definite integral. Use a calculator and round off the answers to four decimal places. When asked to approximate the error, find the exact value of the definite integral using the Fundamental Theorem of Calculus. Then calculate the percentage error by finding $\frac{A - E}{E}$, where A and E are the approximate and exact values of the integral, respectively.*

**1. a.** $\int_1^3 x^3 \, dx$; Trapezoidal Rule, $n = 4$
   **b.** Estimate the error.

**2. a.** $\int_1^3 x^3 \, dx$; Simpson's Rule, $n = 4$
   **b.** Estimate the error.

**3. a.** $\int_1^3 x^3 \, dx$; Trapezoidal Rule, $n = 8$
   **b.** Estimate the error.

**4. a.** $\int_1^3 x^3 \, dx$; Simpson's Rule, $n = 8$
   **b.** Estimate the error.

**5. a.** $\int_2^5 \frac{1}{x} \, dx$; Trapezoidal Rule, $n = 6$
   **b.** Estimate the error.

**6. a.** $\int_2^5 \frac{1}{x} \, dx$; Simpson's Rule, $n = 6$
   **b.** Estimate the error.

**7. a.** $\int_2^5 \frac{1}{x} \, dx$; Simpson's Rule, $n = 12$
   **b.** Estimate the error.

**8. a.** $\int_2^5 \frac{1}{x} \, dx$; Trapezoidal Rule, $n = 12$
   **b.** Estimate the error.

**9. a.** $\int_0^3 \frac{x}{x^2 + 1} \, dx$; Trapezoidal Rule, $n = 6$
   **b.** Estimate the error.

**10. a.** $\int_0^3 \frac{x}{x^2 + 1} \, dx$; Simpson's Rule, $n = 6$
   **b.** Estimate the error.

**11. a.** $\int_0^3 \frac{x}{x^2 + 1} \, dx$; Trapezoidal Rule, $n = 12$
   **b.** Estimate the error.

**12. a.** $\int_0^3 \frac{x}{x^2 + 1} \, dx$; Simpson's Rule, $n = 12$
   **b.** Estimate the error.

**13. a.** $\int_0^2 e^{2x} \, dx$; Trapezoidal Rule, $n = 8$
   **b.** Estimate the error.

**14. a.** $\int_0^2 e^{2x} \, dx$; Simpson's Rule, $n = 8$
   **b.** Estimate the error.

**15.** $\int_0^3 \frac{1}{x^2 + 1} \, dx$; Trapezoidal Rule, $n = 12$

**16.** $\int_0^3 \frac{1}{x^2 + 1} \, dx$; Simpson's Rule, $n = 12$

**17.** $\int_0^3 \sqrt{9 - x^2} \, dx$; Trapezoidal Rule, $n = 6$

**18.** $\int_0^3 \sqrt{9 - x^2} \, dx$; Simpson's Rule, $n = 6$

**19.** $\int_0^5 \ln(x^2 + 1) \, dx$; Trapezoidal Rule, $n = 10$

**20.** $\int_0^5 \ln(x^2 + 1) \, dx$; Simpson's Rule, $n = 10$

*In Exercises 21–24, a set of points is given. Assume that in each case the points belong to the graph of a continuous function f* on an interval $[a, b]$. Use Simpson's Rule and all the data provided to approximate $\int_a^b f(x)dx$.

**21.** $\{(1, 2), (1.5, 3), (2, 3.1), (2.5, 2.9), (3, 2.1)\}$;
$a = 1, b = 3$

**22.** $\{(0, 3), (.3, 2), (.6, 1.5), (.9, 1.9), (1.2, 2.3)\}$;
$a = 0, b = 1.2$

**23.** $\{(1, 3), (1.5, 2), (2, 2.5), (2.5, 3.1), (3, 4.2), (3.5, 5), (4, 4.8)\}$; $a = 1, b = 4$

**24.** $\{(1, 2), (1.3, 3), (1.6, 2.5), (1.9, 3), (2.2, 4), (2.5, 5.2), (2.8, 6.1)\}$; $a = 1, b = 2.8$.

In Exercises 25–28, $p = D(q)$ is the demand equation for a commodity, where $p$ is the price per unit and $q$ is the number of units the consumers are willing to purchase at that price. The total amount spent by the consumers when the quantities purchased increase from $q_1$ to $q_2$ is approximated by

$$\int_{q_1}^{q_2} D(q)\, dq$$

where we assume that the function $D$ is continuous. In each exercise, a table gives the values of $p = D(q)$ for certain values of $q$, where $p$ is in dollars and $q$ is in thousands of units. Approximate the total amount spent by the consumers as the quantities purchased increase from $q_1$ to $q_2$ units.

**25.** $q_1 = 4$, $q_2 = 9$

| q | 4 | 4.5 | 5 | 5.5 | 6 | 6.5 |
|---|---|-----|---|-----|---|-----|
| p | 9 | 8.7 | 8.5 | 8.1 | 7.9 | 7 |

| q | 7 | 7.5 | 8 | 8.5 | 9 |
|---|---|-----|---|-----|---|
| p | 6.5 | 5.9 | 5 | 3.9 | 2 |

**26.** $q_1 = 3$, $q_2 = 8$

| q | 3 | 3.5 | 4 | 4.5 | 5 | 5.5 |
|---|---|-----|---|-----|---|-----|
| p | 8 | 7.9 | 7.2 | 6.8 | 6.5 | 6 |

| q | 6 | 6.5 | 7 | 7.5 | 8 |
|---|---|-----|---|-----|---|
| p | 5.2 | 4.8 | 4 | 3.6 | 3 |

**27.** $q_1 = 5$, $q_2 = 10$

| q | 5 | 5.5 | 6 | 6.5 | 7 | 7.5 |
|---|---|-----|---|-----|---|-----|
| p | 7 | 6.5 | 6.1 | 5.9 | 5.2 | 5 |

| q | 8 | 8.5 | 9 | 9.5 | 10 |
|---|---|-----|---|-----|----|
| p | 4.5 | 4.1 | 4 | 3.2 | 3 |

**28.** $q_1 = 6$, $q_2 = 8$

| q | 6 | 6.2 | 6.4 | 6.6 | 6.8 | 7 |
|---|---|-----|-----|-----|-----|---|
| p | 9.2 | 9.1 | 8.5 | 8.1 | 7.5 | 7 |

| q | 7.2 | 7.4 | 7.6 | 7.8 | 8 |
|---|-----|-----|-----|-----|---|
| p | 6.3 | 5.7 | 4 | 3.5 | 3 |

In Exercises 29–32, $v = f(t)$ where $f$ is the velocity function of a particle moving along a directed line. It is given that $f(t) > 0$ for all $t$, and that $f$ is a continuous function. Thus, the distance traveled from instant $t_1$ to instant $t_2$ is given by

$$\int_{t_1}^{t_2} f(t)\, dt$$

In each exercise, a table gives the values of $v = f(t)$ for certain values of $t$, where $v$ is in meters per second and $t$ is in seconds. Using Simpson's Rule and all of the data provided, approximate the distance traveled by the particle from instant $t_1$ to instant $t_2$.

**29.** $t_1 = 3$, $t_2 = 8$

| t | 3 | 3.5 | 4 | 4.5 | 5 | 5.5 |
|---|---|-----|---|-----|---|-----|
| v | 6 | 7.3 | 9.2 | 8.3 | 7.6 | 7 |

| t | 6 | 6.5 | 7 | 7.5 | 8 |
|---|---|-----|---|-----|---|
| v | 6.8 | 6.1 | 6 | 5.4 | 3 |

**30.** $t_1 = 4$, $t_2 = 9$

| t | 4 | 4.5 | 5 | 5.5 | 6 | 6.5 |
|---|---|-----|---|-----|---|-----|
| v | 6 | 5.8 | 5.1 | 4.9 | 4.6 | 4 |

| t | 7 | 7.5 | 8 | 8.5 | 9 |
|---|---|-----|---|-----|---|
| v | 3.4 | 2.9 | 2 | 1.7 | 1 |

**31.** $t_1 = 5$, $t_2 = 10$

| t | 5 | 5.5 | 6 | 6.5 | 7 | 7.5 |
|---|---|-----|---|-----|---|-----|
| v | 4 | 3.4 | 3.2 | 2.8 | 2.3 | 2 |

| t | 8 | 8.5 | 9 | 9.5 | 10 |
|---|---|-----|---|-----|----|
| v | 1.6 | 1.2 | 1 | 0.6 | 1 |

**32.** $t_1 = 7$, $t_2 = 9$

| $t$ | 7 | 7.2 | 7.4 | 7.6 | 7.8 | 8 |
|-----|-----|-----|-----|-----|-----|---|
| $v$ | 6.3 | 6.2 | 5.7 | 5.2 | 4.6 | 4 |

| $t$ | 8.2 | 8.4 | 8.6 | 8.8 | 9 |
|-----|-----|-----|-----|-----|---|
| $v$ | 3.6 | 2.7 | 1 | 0.5 | 2 |

**\*33. a.** Assume $h > 0$. Evaluate $\int_{-h}^{h} (ax^2 + bx + c)\, dx$ using the Fundamental Theorem of Calculus.

   **b.** Suppose that the points $(-h, y_0)$, $(0, y_1)$, and $(h, y_2)$ are on the graph of the equation $y = ax^2 + bx + c$. Show that your answer in **a** can be written $\frac{h}{3}(y_0 + 4y_1 + y_2)$. This is used in the derivation of Simpson's Rule.

# 9.12 Chapter Review

**IMPORTANT SYMBOLS AND TERMS**

$\underline{A}_p$ [9.8]
$\overline{A}_p$ [9.8]
$g(x)\big|_a^b = g(b) - g(a)$ [9.10]
$\int_a^b f(x)\, dx$ [9.8]
$\int f(x)\, dx$ [9.1]
$\lim\limits_{n \to \infty} s_n = L$ [9.7]
$\lim\left(1 + \dfrac{1}{n}\right)^n = e$ [9.7]
$R_p$ [9.8]
$\underline{S}_{p_n}$ [9.8]
$\overline{S}_{p_n}$ [9.8]
Antiderivative [9.1]
Antidifferentiation [9.1]
Area [9.8]
Circumscribed rectangular polygon [9.8]
Definite integral [9.8]
Definite integral of $f$ from $a$ to $b$ [9.8]

Even polynomial [9.5]
Fundamental Theorem of Calculus [9.10]
Geometric series [9.7]
Indefinite integral [9.1]
Infinite series [9.7]
Inscribed rectangular polygon [9.8]
Integrand [9.2], [9.8]
Integration by parts [9.3]
Integration table [9.4]
$i$th subinterval [9.8]
Limit of $s_n$ as $n$ tends to infinity [9.7]
Lower limit [9.8]
Lower sum of $f$ corresponding to $p_n$, $\underline{S}_{p_n}$ [9.8]

$n$th partial sum [9.7]
Odd polynomial [9.5]
Partition [9.8]
Reduction formula [9.5]
Regular partition [9.8]
Riemann sum of $f$ corresponding to $p_n$, $R_{p_n}$ [9.8]
Simpson's Rule [9.11]
Tabular integration [9.6]
Trapezoid Rule [9.11]
Upper limit [9.8]
Upper sum of $f$ corresponding to $p_n$, $\overline{S}_{p_n}$ [9.8]
$u$-substitution [9.2]
Variable of integration [9.8]

**SUMMARY**

If $f$ and $g$ are two functions defined on an interval $I$ and $f'(x) = g(x)$ for every $x$ in $I$, then $f$ is said to be an antiderivative of $g$. If a function has an antiderivative, then it has infinitely many antiderivatives. However, any two antiderivatives of the same function differ at most by a constant. The symbol $\int\, dx$ is used to denote antidifferentiation with respect to $x$. That is,

$$\int f(x)\, dx = g(x) + C \qquad \text{(where } C \text{ is a constant)}$$

if, and only if, $g'(x) = f(x)$.

Often, we can obtain an antiderivative of a function by trial and error. To find $\int af(x)\, dx$ where $a$ is a constant we may be able to guess, based on our knowledge of the differentiation formulas, that $\int af(x)\, dx = g(x) + C$, where $C$ is a constant. We then differentiate $g(x) + C$ and find that $D_x(g(x) + C) = bf(x)$, where $b$ is a constant. If $a \neq b$, we divide the trial answer by the unwanted factor $b$ to eliminate that factor and we multiply by the desired factor $a$.

Antidifferentiation may be used to find the position function of a particle moving along a directed straight line when the velocity function, or acceleration function, is given.

It is often useful to make a substitution to simplify an antidifferentiation problem. If in an integral we replace $f(x)$ by $u$, then we *must* replace $f'(x)\, dx$ by $du$. The integral we obtain must be free of the variable $x$. In the following formulas, $k$, $n$, and $C$ are constants, and $f$ and $g$ are functions which have antiderivatives.

$$\int [f(u) \pm g(u)]\, du = \int f(u)\, du \pm \int g(u)\, du \tag{1}$$

$$\int kf(u)\ du = k \int f(u)\ du \tag{2}$$

$$\int u^n\ du = \frac{u^{n+1}}{n+1} + C \quad \text{(provided } (n \neq -1) \tag{3}$$

$$\int u^{-1}\ du = \ln|u| + C \tag{4}$$

$$\int e^u\ du = e^u + C \tag{5}$$

$$\int b^u\ du = \frac{b^u}{\ln b} + C \tag{6}$$

If $f$ and $g$ have continuous derivatives, we have the following integration by parts formula:

$$\int f(x)g'(x)\ dx = f(x)g(x) - \int g(x)f'(x)\ dx$$

If we make the substitutions $u = f(x)$, $du = f'(x)\ dx$, $v = g(x)$, and $dv = g'(x)\ dx$, the formula takes the following simpler form:

$$\int u\,dv = uv - \int v\,du$$

It is possible to use integration tables to evaluate an integral. We often must make a $u$-substitution to transform the integral in one of the standard forms which appears in integration tables.

In some cases where integration by parts or a formula has to be used repeatedly, it is more efficient to obtain an antiderivative by trial and error. This is particularly true in integrals of the form $\int p(x)e^{ax}\ dx$, where $a$ is a constant and $p(x)$ is a polynomial of degree $n$. Then an antiderivative is of the form $q(x)e^{ax}$, where $q(x)$ is a polynomial of degree $n$. Thus we let

$$\int p(x)e^{ax}\ dx = (a_1x^n + a_2x^{n-1} + \cdots + a_n)e^{ax} + C$$

The coefficients $a_1, a_2, \ldots, a_n$ may be found by differentiating $(a_1x^n + a_2x^{n-1} + \cdots + a_n)e^{ax}$ and setting the result equal to $p(x)e^{ax}$. A similar technique may be used to evaluate an integral of the form $\int p(x)(a^2 - x^2)^r\ dx$, where $r$ is a constant and $p(x)$ is an odd polynomial of degree $2n - 1$. An antiderivative is of the form $q(x)(a^2 - x^2)^r$ where $q(x)$ is an even polynomial of degree $2n$.

Tabular integration may often be effective in cases where integration by parts would have to be used repeatedly.

The difficulty in evaluating $\int f(x)\ dx$ often lies in recognizing which method is most likely to work. Here are a few hints.

| Type of integral | Method |
|---|---|
| $\int a[f(x)]^n f'(x)\ dx$ | Let $u = f(x)$. |
| $\int b^{f(x)} kf'(x)\ dx$ | Let $u = f(x)$. |
| $\int p(x)(ax + b)^k\ dx$, where $p(x)$ is a polynomial | Let $u = ax + b$. |
| $\int x^n b^{kx} dx$ | Use integration by parts with $u = x^n$. (Tabular integration or trial and error may also be used.) |
| $\int x^n \log_b|x|\ dx$ | Use integration by parts with $u = \log_b|x|$. |
| $\int (\log_b|x|)^n\ dx$ | Use integration by parts with $u = (\log_b|x|)^n$. (Tabular integration may also be used.) |

In this table, $a$, $b$, $c$, $k$, $n$ are constants, and $b$ is a positive constant with $b \neq 1$.

If for some sequence $s$ there is a number $L$ such that the value of $s_n$ can be made arbitrarily close to $L$ by taking $n$ sufficiently large, we say that the limit of $s_n$ as $n$ tends to infinity is $L$ and we write $\lim\limits_{n \to \infty} s_n = L$.

Suppose that $s$ and $t$ are sequences with ranges sets of real numbers and that $\lim\limits_{n \to \infty} s_n = A$ and $\lim\limits_{n \to \infty} t_n = B$. Then

**a.** $\lim\limits_{n \to \infty} (s_n \pm t_n) = A \pm B$,

**b.** $\lim\limits_{n \to \infty} (s_n \cdot t_n) = A \cdot B$,

**c.** $\lim\limits_{n \to \infty} \dfrac{s_n}{t_n} = \dfrac{A}{B}$, provided $B \neq 0$ and $t_n \neq 0$,

and if $k$ is a positive integer,

**d.** $\lim\limits_{n \to \infty} s_n^{\,k} = A^k$.

**e.** If $s$, $t$ and $u$ are sequences such that $s_n \leq t_n \leq u_n$ and $\lim\limits_{n \to \infty} s_n = \lim\limits_{n \to \infty} u_n = A$, then $\lim\limits_{n \to \infty} t_n = A$.

Part **e** is often called the "Squeeze Theorem". If $|r| < 1$, then $\lim\limits_{n \to \infty} r^n = 0$. In that case,

$$\lim_{n \to \infty} \frac{a(1 - r^n)}{1 - r} = \frac{a}{(1 - r)}$$

Thus, when $|r| < 1$, $a + ar + ar^2 + ar^3 + \cdots = \dfrac{a}{1 - r}$. Another very important limit is the following:

$$\lim_{n \to \infty} \left( 1 + \frac{1}{n} \right)^n = e$$

It is often easy to evaluate limits intuitively using facts such as $\lim\limits_{n \to \infty} \dfrac{c}{n} = 0$ whenever $c$ is a constant.

Suppose $f$ is a continuous function defined on the closed bounded interval $[a, b]$. A partition $p$ of $[a, b]$ is an ordered set of numbers $(x_0, x_1, x_2, \ldots, x_n)$ such that

$$a = x_0 < x_1 < x_2 < \cdots < x_n = b$$

If $x_1 - x_0 = x_2 - x_1 = \cdots = x_n - x_{n-1}$, the partition is said to be regular. Given a partition $p$ of $[a, b]$, the interval $[x_{i-1}, x_i]$ is called the $i$th subinterval formed by the partition. Let $p_n$ be a regular partition of $[a, b]$ with $n$ subintervals. For each $i \in \{1, 2, 3, \ldots, n\}$ we can find $u_i$ and $v_i$ in the $i$th subinterval such that

$$f(u_i) \leq f(x) \leq f(v_i)$$

for all $x \in [x_{i-1}, x_i]$. We also let $w_i$ be an arbitrary point in $[x_{i-1}, x_i]$ and $\Delta x = \dfrac{b - a}{n}$. Then

$$\sum_{i=1}^{n} f(u_i)\Delta x$$

is called the lower sum of $f$ corresponding to $p_n$ and is denoted $\underline{S}_{p_n}$ (read: "S sub $p$ sub $n$ lower bar"),

$$\sum_{i=1}^{n} f(v_i)\Delta x$$

is called the upper sum of $f$ corresponding to $p_n$ and is denoted $\overline{S}_{p_n}$ (read: "$S$ sub $p$ sub $n$ upper bar"), and

$$\sum_{i=1}^{n} f(w_i)\Delta x$$

is called a Riemann sum of $f$ corresponding to $p_n$ and is denoted $R_{p_n}$.

If $f$ is continuous, it can be proved that

$$\lim_{n\to\infty} \underline{S}_{p_n} = \lim_{n\to\infty} R_{p_n} = \lim_{n\to\infty} \overline{S}_{p_n}$$

The value of this common limit is called the definite integral of $f$ from $a$ to $b$ and is denoted $\int_a^b f(x)\,dx$.

In the special case where $f(x) \geq 0$ for all $x$ in the interval $[a, b]$, the value of $\int_a^b f(x)\,dx$ gives the area of the region below the graph of $y = f(x)$ and over the interval $[a, b]$.

If $f$ is a continuous function on the closed interval $[a, b]$, we define

$$\int_b^a f(x)\,dx = -\int_a^b f(x)\,dx$$

We also define

$$\int_a^a f(x)\,dx = 0$$

If $f$ and $g$ are continuous functions on an interval $I$, $k$ is a constant, and $a$, $b$, $c$ are in $I$, then

$$\int_a^b kf(x)\,dx = k\int_a^b f(x)\,dx$$

$$\int_a^b [f(x) \pm g(x)]\,dx = \int_a^b f(x)\,dx \pm \int_a^b g(x)\,dx$$

$$\int_a^b f(x)\,dx + \int_b^c f(x)\,dx = \int_a^c f(x)\,dx$$

A function $f$ is said to be odd if $f(-x) = -f(x)$ for all $x$ in its domain. A function $g$ is said to be even if $g(-x) = g(x)$ for all $x$ in its domain. Suppose that a function $f$ is continuous on a closed interval $[-a, a]$. Then

$$\int_{-a}^a f(x)\,dx = 0 \text{ if } f \text{ is odd}$$

and

$$\int_{-a}^a f(x)\,dx = 2\int_0^a f(x)\,dx$$

if $f$ is even.

The Fundamental Theorem of Calculus guarantees that if $f$ is a continuous function on a closed bounded interval $[a, b]$, then $f$ has an antiderivative on that interval. Furthermore, if $g$ is any antiderivative of $f$ on $[a, b]$, then

$$\int_a^b f(x)\,dx = g(x)\Big|_a^b$$

where the symbol $g(x)|_a^b$ denotes $g(b) - g(a)$.

If it is difficult to find an antiderivative of a continuous function $f$ on an interval $[a, b]$, we may use either the Trapezoid Rule or Simpson's Rule to approximate the value of the definite integral $\int_a^b f(x)\,dx$.

*Trapezoid Rule:* If $f$ is a continuous function on $[a, b]$ and $p_n = (x_0, x_1, x_2, \ldots, x_n)$ is a regular partition of $[a, b]$, we let $y_0 = f(x_0)$, $y_1 = f(x_1)$, $y_2 = f(x_2)$, $\ldots$, $y_n = f(x_n)$ and $h = \frac{b-a}{n}$. Then

$$\int_a^b f(x)\,dx \doteq \frac{h}{2}(y_0 + 2y_1 + 2y_2 + 2y_3 + \cdots + 2y_{n-1} + y_n)$$

*Simpson's Rule:* With the same hypothesis and notation as in the Trapezoid Rule but with the additional requirement that $n$ must be *even*, we have

$$\int_a^b f(x)\, dx \doteq \frac{h}{3}(y_0 + 4y_1 + 2y_2 + 4y_3 + \cdots + 4y_{n-1} + y_n)$$

**SAMPLE EXAM QUESTIONS**

1. In each of the following, verify that the function $g$ is an antiderivative of the function $f$.

   **a.** $f(x) = 12x^3 - 6x^2 + 7;\ g(x) = 3x^4 - 2x^3 + 7x - 9$
   **b.** $f(x) = 12x^3 e^{x^4};\ g(x) = 3e^{x^4} + \pi^2$
   **c.** $f(x) = (5x + 1)e^{5x};\ g(x) = xe^{5x} + \ln 5$

   **d.** $f(x) = 2x[1 + \ln(x^2 + 1)];\ g(x) = (x^2 + 1)\ln(x^2 + 1) + \dfrac{3}{4}$

   **e.** $f(x) = \dfrac{(2x - 1)e^{2x}}{x^2};\ g(x) = \dfrac{e^{2x}}{x} + \ln(e + 3)$

2. In each of the following, first guess the answer. Then test your trial antiderivative by differentiation. When necessary, adjust your trial solution by multiplying and dividing by the appropriate constants. Find

   **a.** $\displaystyle \int (3x^5 + 5)\, dx$

   **b.** $\displaystyle \int (e^{6t} + 3t + 2)\, dt$

   **c.** $\displaystyle \int \frac{z^4}{z^5 + 3}\, dz$

   **d.** $\displaystyle \int x^3 \sqrt{x^4 + 3}\, dx$

   **e.** $\displaystyle \int \frac{x^3 + 2}{x^4 + 8x + 5}\, dx$

   **f.** $\displaystyle \int \left( \frac{3}{\sqrt{t}} + e^{-5t} \right) dt$

3. In each of the following, the velocity function of a particle moving along a directed line is given. The value of the position function at a certain instant is also given. Derive the formula for the position function. Time is measured in seconds and distance is in meters.

   **a.** $v(t) = 32t + 10,\ s(0) = 15$
   **b.** $v(t) = -12t + 4,\ s(2) = 52$
   **c.** $v(t) = 9t^2 + 6t - 4,\ s(1) = 60$
   **d.** $v(t) = 3e^{.002t} + 6t^2 + 2,\ s(0) = 10$

4. In each of the following, the acceleration function of a particle moving along a directed line is given. The value of the velocity function and that of the position function at certain instants are also given. Derive the formula for the position function. Time is measured in seconds and distance is in meters.

   **a.** $a(t) = 6t + 10;\ v(0) = 2,\ s(0) = 5$
   **b.** $a(t) = 18e^{3t} + 12;\ v(0) = 4,\ s(0) = 15$
   **c.** $a(t) = 15\sqrt{t};\ v(4) = 3,\ s(9) = 17$

5. In each of the following, the marginal cost function of a commodity is given, where $q$ is the number of units of the commodity produced. In each case, the fixed cost is also given. Find the cost function.

   **a.** $M_c(q) = 0.003q^2 - 6q + 1200,\ \$50,000$

**b.** $M_c(q) = \dfrac{25}{\sqrt{q}}$, $75,000

**c.** $M_c(q) = \dfrac{10q}{q^2 + 3}$, $62,000

6. Find each of the following by first making a $u$-substitution.

   **a.** $\displaystyle \int (x^3 + 6x + 2)^{49}(x^2 + 2)\, dx$

   **b.** $\displaystyle \int (x^4 + 3x^2 + 1)^{5/2}(2x^3 + 3x)\, dx$

   **c.** $\displaystyle \int (x + 1)e^{x^2+2x+5}\, dx$

   **d.** $\displaystyle \int \dfrac{2x^3 + 1}{x^4 + 2x + 3}\, dx$

7. Use integration by parts to evaluate the following integrals.

   **a.** $\displaystyle \int 5xe^{12x}dx$

   **b.** $\displaystyle \int 4x^2e^{13x}dx$

   **c.** $\displaystyle \int \ln|x + 12|\, dx$

   **d.** $\displaystyle \int 5x \ln|x + 6|\, dx$

8. The marginal revenue function for a commodity is given by

$$M_R(q) = \frac{25}{\sqrt{q + 100}} + 0.003q^2$$

   where $q$ is the number of units sold and the revenue is in dollars. Find the revenue function.
   (*Hint:* $R(0) = 0$.)

9. The following formulas may be found in standard integration tables.

$$\int \frac{1}{a^2 - u^2}\, du = \frac{1}{2a} \cdot \ln\!\left(\frac{a + u}{a - u}\right) + C, \text{ where } u^2 < a^2$$

$$\int \frac{1}{u\sqrt{a^2 + u^2}}\, du = -\frac{1}{a} \cdot \ln\!\left(\frac{a + \sqrt{a^2 + u^2}}{u}\right) + C$$

$$\int u^n \cdot \ln|u|\, du = u^{n+1}\left[\frac{\ln|u|}{n + 1} - \frac{1}{(n + 1)^2}\right] + C, \text{ where } n \neq -1$$

$$\int u^n e^{bu}\, du = \frac{u^n e^{bu}}{b} - \frac{n}{b} \cdot \int u^{n-1} e^{bu}\, du, \text{ where } b \neq 0$$

   Evaluate each of the following with the help of these formulas.

   **a.** $\displaystyle \int \frac{3}{4x\sqrt{25 + 9x^2}}\, dx$

   **b.** $\displaystyle \int \frac{2}{25 - 9x^2}\, dx$

   **c.** $\displaystyle \int x^3 e^{7x}\, dx$

   **d.** $\displaystyle \int (x^2 + 4x + 4)\ln|x + 2|\, dx$

10. Evaluate each of the following integrals by writing an antiderivative with undetermined coefficients, differentiating the trial solution, and obtaining the values of the undetermined coefficients by comparing this derivative to the integrand in the original integral.

a. $\int (x^3 + 3x^2 - 5x + 2)e^{5x}\, dx$

b. $\int x^3(36 - x^2)^{5/2}\, dx$

c. $\int x^7 3^{x^2}\, dx$ (*Hint:* Let $u = x^2$.)

11. Let $f(x) = -x^2 + 3x + 4$, and let $p_6$ be the regular partition of $[-1, 2]$ with six subintervals. For each $i \in \{1, 2, 3, 4, 5, 6\}$ let $w_i$ be the midpoint of the $i$th subinterval. Use these $w_i$'s to find the value of the Riemann sum of $f$ corresponding to the partition $p_6$.

12. Let $y = f(x) = 2x + 1$ and let $p_6$ be the regular partition of $[1, 4]$ with six subintervals.

a. Sketch the region bounded by the graphs of $y = f(x)$, $x = 1$, $x = 4$, and the $x$-axis. Also draw the inscribed and circumscribed rectangular polygons corresponding to the partition $p_6$.

b. Evaluate $\underline{A}_{p_6}$.

c. Evaluate $\overline{A}_{p_6}$.

d. Find the area $S$ square units of the region sketched in part **a** using the formula for the area of a trapezoid.

e. Verify that $\underline{A}_{p_6} < S < \overline{A}_{p_6}$.

f. Verify that $S = \int_1^4 (2x + 1)\, dx$.

13. Let $y = g(x) = x^2 + 3$ and let $p_5$ be the regular partition of $[1, 3]$ with five subintervals.

a. Sketch the region bounded by the graphs of $y = f(x)$, $x = 1$, $x = 3$, and the $x$-axis. Also draw the inscribed and circumscribed rectangular polygons corresponding to the partition $p_5$.

b. Evaluate $\underline{A}_{p_5}$.

c. Evaluate $\overline{A}_{p_5}$.

d. Find the area $S$ square units of the region sketched in part **a** by evaluating the definite integral $\int_1^3 (x^2 + 3)\, dx$.

e. Verify that $\underline{A}_{p_5} < S < \overline{A}_{p_5}$.

14. Let $f(x) = \frac{x}{2}$ and let $p_n$ be a regular partition of $[0, 8]$ with $n$ subintervals.

a. Sketch the region bounded by the graphs of $y = f(x)$, $x = 8$, and the $x$-axis.

b. Find $\underline{S}_{p_n}$ in terms of $n$.

c. Find $\overline{S}_{p_n}$ in terms of $n$.

d. Find the area $S$ square units of the region sketched in part **a** using the formula for the area of a triangle.

e. Verify that $\underline{S}_{p_n} < S < \overline{S}_{p_n}$.

f. Verify that $S = \int_0^8 \frac{x}{2}\, dx$.

g. Verify that $\lim\limits_{n \to \infty} \underline{S}_{p_n} = \lim\limits_{n \to \infty} \overline{S}_{p_n} = S$.

*15. Let $f(x) = x^3$ and let $p_n$ be the regular partition of $[0, 5]$ with $n$ subintervals.

a. Verify that the function $f$ is increasing on the interval $[0, 5]$. (*Hint:* Show the derivative is positive.)

b. Find $\underline{S}_{p_n}$ in terms of $n$. (*Hint:* Use the formula $\sum\limits_{i=1}^{n} i^3 = \dfrac{n^2(n + 1)^2}{4}$.)

**c.** Find $\bar{S}_{P_n}$ in terms of $n$.

**d.** Find $\int_0^5 x^3 \, dx$.

**e.** Verify that $\lim_{n \to \infty} \underline{S}_{P_n} = \lim_{n \to \infty} \bar{S}_{P_n} = \int_0^5 x^3 \, dx$.

16. Use the Fundamental Theorem of Calculus, or any simpler method, to evaluate each of the following definite integrals.

  **a.** $\int_2^4 (6x^2 - 4x + 1) \, dx$

  **b.** $\int_{-1}^3 (x^2 + 1)^5 x \, dx$

  **c.** $\int_{-2}^4 3xe^{x^2} \, dx$

  **d.** $\int_{-3}^5 4x\sqrt{x^2 + 1} \, dx$

  **e.** $\int_1^3 \dfrac{x + 1}{x^2 + 2x + 5} \, dx$

  **f.** $\int_2^5 5x^2 3^{x^3} \, dx$

  **g.** $\int_{-1}^3 2xe^{5x} dx$ (*Hint:* Use integration by parts.)

  **h.** $\int_1^4 (x + 1)\ln(x + 1) \, dx$ (*Hint:* Use integration by parts.)

  **i.** $\int_{-2}^2 (x^3 + 3x)e^{x^2} \, dx$

  **j.** $\int_0^2 (2x^3 + 3x^2 - 5x + 3)e^{2x} \, dx$

  **k.** $\int_0^3 (10x^3 + 17x)\sqrt{9 - x^2} \, dx$

17. In each of the following, a demand equation for a commodity is given, where $p$ dollars is the price per unit and $q$ is the number of units the consumers are willing to buy at that price. Use a definite integral to find the total amount spent by the consumers as the number of units sold increases from $q_1$ to $q_2$ and prices decrease accordingly.

  **a.** $p = -0.001q^2 - 0.15q + 800$; $q_1 = 0$, $q_2 = 300$
  **b.** $p = 75 - \sqrt{q + 7}$; $q_1 = 2$, $q_2 = 618$

18. Evaluate each of the following:

  **a.** $\int_0^3 \left( D_x \dfrac{\ln(x + 5)}{e^{x^3}} \right) dx$

  **b.** $\int_1^5 \left( D_x \dfrac{\sqrt{x^3 + 3}}{x^4 + e^x} \right) dx$

  **c.** $D_x \left( \int_0^5 \dfrac{\ln(x^2 + 5)}{e^{x^4}} \, dx \right)$

19. A particle is moving along a directed line. The distance is measured in feet and time is in seconds. The equation defining the velocity function is

$$v(t) = 6t^2 + 4t + 2$$

Find the distance traveled by the particle during the time interval from $t_1 = 3$ to $t_2 = 10$.

20. Same as Exercise 19 if the velocity function is defined by $v(t) = \sqrt{t+3} + 4$ and the time interval is from $t_1 = 6$ to $t_2 = 22$.

21. Use the Trapezoid Rule with $n = 5$ to approximate $\int_1^6 x^2\, dx$. Then find the exact value of this integral and calculate the percentage error in using the Trapezoid Rule.

22. Use Simpson's Rule with $n = 6$ to approximate $\int_1^4 \frac{1}{x}\, dx$. Then find the exact value of this integral and calculate the percentage error in using Simpson's Rule.

23. Use the Trapezoid Rule with $n = 8$ to approximate $\displaystyle\int_1^3 \frac{1}{x^2+1}\, dx$.

24. Use Simpson's Rule with $n = 8$ to approximate $\displaystyle\int_1^3 \frac{1}{x^2+1}\, dx$.

25. Let $s_n = \left(1 + \frac{1}{2n}\right)^n$ where $n$ is a positive integer. Using a calculator, evaluate $s_n$ for **a.** $n = 10$, **b.** $n = 100$, **c.** $n = 1000$, and **d.** $n = 10{,}000$. Also evaluate $\sqrt{e}$. Based on these results, what do you conclude?

26. Let $s_n = \left(1 + \frac{1}{n}\right)^{2n}$ where $n$ is a positive integer. Using a calculator, evaluate $s_n$ for **a.** $n = 10$, **b.** $n = 100$, **c.** $n = 1000$, and **d.** $n = 10{,}000$. Also evaluate $e^2$. Based on these results, what do you conclude?

27. Using the formula

$$a + ar + ar^2 + ar^3 + \cdots = \frac{a}{1-r} \text{ whenever } |r| < 1$$

write the repeating decimal $.\overline{23}$ as the ratio of two positive integers.

28. Evaluate $\displaystyle\lim_{n\to\infty} s_n$ if

   **a.** $s_n = \dfrac{5}{2n+3}$.

   **b.** $s_n = \dfrac{3n+1}{n+2}$. $\left(Hint:\ \dfrac{3n+1}{n+2} = 3 + \dfrac{-5}{n+2}.\right)$

   **c.** $s_n = \left[5 - \left(\dfrac{6}{7}\right)^n\right]$.

29. Suppose that $s_1 + s_2 + s_3 + \ldots$ is a geometric series and that

$$s_1 + s_3 + s_5 + \cdots = \frac{16}{3}$$

and

$$s_2 + s_4 + s_6 + \cdots = \frac{4}{3}$$

Find the first three terms of this geometric series.

30. It can be show that if $A$ square units is the area of the region bounded by the graphs of $y = x^4$, $x = 5$, and the $x$-axis, then

$$\frac{625}{6}\left(1 - \frac{1}{n}\right)\left(2 - \frac{1}{n}\right)\left(3 - \frac{3}{n} - \frac{1}{n^2}\right) \le A \le \frac{625}{6}\left(1 + \frac{1}{n}\right)\left(2 + \frac{1}{n}\right)\left(3 + \frac{3}{n} + \frac{1}{n^2}\right)$$

Find $A$.

# Applications of Integration

**10**

The preceding chapter introduced the indefinite integral (antiderivative) and the definite integral. We now discuss several important applications of integration. First, we give examples from economics and business where the problems reduce to finding an unknown function using an equation that involves the function's derivatives. We will also expand on our treatment of the area between curves because many quantities in economics, such as consumers' and producers' surpluses, can be represented by areas.

# 10.1 Differential Equations (Optional)

## Interest Compounded Continuously

In many cases, the instantaneous rate of change of a function is proportional to the value of the function. Suppose for example that an amount $P_0$ dollars is invested and that the rate of interest, written as a decimal, is $i$, compounded continuously. Suppose $P(t)$ dollars is the value of that investment at instant $t$. We wish to find a formula for $P(t)$. We note that $P(0) = P_0$. As we shall see, the mathematical formulation of this problem will first give us an equation involving $P'(t)$.

If the interest is compounded $n$ times per year, the rate of interest per conversion period is $\frac{i}{n}$. The amount of time between conversion periods is $\Delta t = \frac{1}{n}$. The interest earned during a period is the value of the investment at the end of the period minus its value at the beginning. That is,

$$P(t + \Delta t) - P(t)$$

It is also the value of the investment at the start of the period multiplied by the rate of interest, or $[P(t)]\frac{i}{n}$. But $[P(t)]\frac{i}{n} = [P(t)]i\frac{1}{n} = [P(t)]i\Delta t$. Therefore,

$$P(t + \Delta t) - P(t) = [P(t)]i\Delta t$$

and, if $\Delta t \neq 0$,

$$\frac{P(t + \Delta t) - P(t)}{\Delta t} = [P(t)]i$$

Since the interest is compounded continuously, we let the number of conversion periods tend to infinity. But $\Delta t \to 0$ as $n \to \infty$. Further

$$\lim_{\Delta t \to 0} \frac{P(t + \Delta t) - P(t)}{\Delta t} = \lim_{\Delta t \to 0} [P(t)]i$$

That is,

$$P'(t) = [P(t)]i$$

Since $P(t) \neq 0$, we have

$$\frac{P'(t)}{P(t)} = i$$

Antidifferentiating both sides with respect to $t$, we get

$$\int \frac{P'(t)}{P(t)} \, dt = \int i \, dt$$

or

$$\ln |P(t)| = it + C$$

Since $P(t) > 0$, we need not write the absolute value signs. The last equation yields

$$e^{\ln P(t)} = e^{it+C} = e^C e^{it}$$

or, using the fact that $e^{\ln x} = x$,

$$P(t) = e^C e^{it}$$

Recalling that $P(0) = P_0$, we obtain

$$P_0 = P(0) = e^C e^{i(0)} = e^C e^0 = e^C$$

Thus

$$P(t) = P_0 e^{it}$$

## Differential Equations

An equation such as $P'(t) = P(t)i$ is called a differential equation. Formally, we have the following definitions.

> **DEFINITION 10.1:**   A statement of equality involving derivatives of an unknown function is called a *differential equation*. A differential equation is said to be of *order n* if an $n$th order derivative is involved in the equation, but no derivative of order higher than $n$ appears in that equation.
>
> A function $f$ is said to be a *solution* of the differential equation if the equation becomes an identity when the unknown function and all derivatives are replaced by $f$ and its corresponding derivatives.

**EXAMPLE 1**   Show that if $f_1(x) = e^x$ and $f_2(x) = e^{3x}$, then $f_1$ and $f_2$ are solutions of the differential equation

$$y'' - 4y' + 3y = 0$$

**Solution**   If $f_1(x) = e^x$, then both $f_1'(x) = e^x$ and $f_2''(x) = e^x$. Thus replacing $y$, $y'$, and $y''$ by $e^x$ in the differential equation, we get

$$e^x - 4e^x + 3e^x = 0$$

which is an identity.

Similarly, if $f_2(x) = e^{3x}$, then $f_2'(x) = 3e^{3x}$ and $f_2''(x) = 9e^{3x}$. Replacing $y$, $y'$, and $y''$ by $e^{3x}$, $3e^{3x}$, and $9e^{3x}$, respectively, we obtain

$$9e^{3x} - 4(3e^{3x}) + 3e^{3x} = 0$$

which is also an identity.

Therefore both $f_1$ and $f_2$ are solutions of the differential equation $y'' - 4y' + 3y = 0$.

**Do Exercise 1.**    ■

## Separable Differential Equations

In general, finding solutions of a differential equation is not easy. However, there are certain types of differential equations that can be easily solved. We now describe such an equation.

> **DEFINITION 10.2:**   Let $f$ and $g$ be continuous functions of $x$ and $y$, respectively. The first order differential equation
>
> $$\frac{dy}{dx} = \frac{f(x)}{g(y)}$$
>
> is said to be a *separable equation*.

Note that this equation can be written

$$g(y)\, dy = f(x)\, dx$$

with the variable $y$ appearing only on the left side of the equation and the variable $x$ appearing only on the right side. The variables have been separated.

Suppose the functions $F$ and $G$ are antiderivatives of the functions $f$ and $g$, respectively. Then

$$D_x G(y) = G'(y)\frac{dy}{dx} = g(y)\frac{dy}{dx}$$

and

$$D_x F(x) = f(x)$$

Thus the equations

$$g(y)\frac{dy}{dx} = f(x) \tag{1}$$

and

$$G(y) = F(x) + C \tag{2}$$

are equivalent.

We often say that equation (2) is the *general solution* of the differential equation (1), where in fact the solution is a function defined implicitly by equation (2). Whenever possible, we try to obtain an explicit formulation of the solution.

**Procedure for Solving a Separable Differential Equation**

The method for solving a separable differential equation consists of the following steps.

**STEP 1.** Separate the variables; that is, write the equation in the form

$$g(y) \, dy = f(x) \, dx$$

**STEP 2.** Integrate both sides.

$$\int g(y) \, dy = \int f(x) \, dx$$

**STEP 3.** The result of Step 2 is

$$G(y) = F(x) + C$$

where $C$ is a constant, $G'(y) = g(y)$, and $F'(x) = f(x)$. If possible, solve for $y$ in terms of $x$. If enough information is given to find a particular $C$, do so.

**EXAMPLE 2**   Solve the equation

$$(6y^5 + 3y^2 + 1)\frac{dy}{dx} = 2x + 1$$

**Solution**   Separating the variables, we get

$$(6y^5 + 3y^2 + 1) \, dy = (2x + 1) \, dx$$

Integrating both sides,

$$y^6 + y^3 + y = x^2 + x + C$$

This equation cannot be solved for $y$. It is the general solution and defines the solutions implicitly.
**Do Exercise 7.**                                                                                           ■

**EXAMPLE 3**   Find the general solution of the differential equation

$$\frac{dy}{dx} = -\frac{x}{y}$$

Also, find the particular solution whose graph passes through $(-3, 4)$.

**Solution**   The equation can be written

$$y \, dy = -x \, dx$$

Integrating both sides, we get

$$\int y \, dy = \int -x \, dx$$

That is,

$$\frac{y^2}{2} = -\frac{x^2}{2} + C$$

This equation can be written

$$x^2 + y^2 = K$$

where $K$ is a constant. It is the general solution.

We wish to select the particular solution whose graph passes through $(-3, 4)$. Replacing $x$ by $-3$ and $y$ by 4 in the general solution, we get

$$(-3)^2 + 4^2 = K$$

or

$$K = 25$$

Therefore $x^2 + y^2 = 25$. This equation defines two functions implicitly. Solving for $y$, we get

$$y = \sqrt{25 - x^2} \text{ or } y = -\sqrt{25 - x^2}$$

Since the value of $y$ is 4 when $x$ is $-3$, we must select $y = \sqrt{25 - x^2}$.
**Do Exercise 13.**
&#9632;

## Initial (Boundary) Conditions

As we have just seen, solving a differential equation involves finding antiderivatives, which yield arbitrary constants. Sometimes information called "initial conditions" or "*boundary conditions*" is given. With that information, it is possible to select, among all solutions, one solution which satisfies these conditions. The solution obtained is called a *particular solution* of the differential equation.

**EXAMPLE 4** The rate of change in the value of real estate in Madison at time $t$ is proportional to the value at that time. A house valued at \$94,000 three months ago is worth \$94,500 today. What will its value be one year from now?

**Solution** Let $V = V(t)$ dollars be the value of the house at time $t$. We let $t = 0$ when the value of the house was \$94,000 and $t = 3$ when the value is \$94,500. Therefore we have

$$V(0) = 94{,}000, \ V(3) = 94{,}500, \text{ and } \frac{dV}{dt} = kV$$

where $k$ is a constant. The last equation indicates that the rate of change in value $\left(\frac{dV}{dt}\right)$ and the value $(V)$ are proportional.

Separating the variables and integrating, we get

$$\frac{1}{V} \, dV = k \, dt$$

and

$$\int \frac{1}{V} \, dV = \int k \, dt$$

Therefore

$$\ln |V| = kt + C$$

To find a formulation for $V$, we use the fact that $e^{\ln x} = x$ for all $x > 0$, and since $V > 0$, we write

$$V = e^{\ln V} = e^{kt+C} = e^C e^{kt}$$

Recalling that $V = 94{,}000$ when $t = 0$, we get

$$94{,}000 = e^C e^{k(0)} = e^C e^0 = e^C$$

Thus

$$V = 94{,}000 e^{kt} \tag{1}$$

Now, using the fact that $V = 94{,}500$ when $t = 3$, we obtain

$$94{,}500 = 94{,}000 e^{3k}$$

or

$$e^{3k} = \frac{189}{188}$$

We could solve for $k$ but we find that it is simpler to evaluate $e^k$ instead by raising both sides of the equality to the one-third power. We get

$$e^k = (e^{3k})^{1/3} = \left(\frac{189}{188}\right)^{1/3}$$

Replacing $e^k$ by its value in equation (1), we obtain

$$V = 94{,}000 \left(\frac{189}{188}\right)^{t/3}$$

Since we want to find what the value of the house will be 1 year from now—that is, 15 months since it was worth \$94,000—we replace $t$ by 15 and get

$$V = 94{,}000 \left(\frac{189}{188}\right)^{15/3} = 94{,}000 \left(\frac{189}{188}\right)^5 \doteq 96{,}526.74$$

Thus a year from now the house will be valued at approximately \$96,527.
**Do Exercise 25.**                                                                        ■

**EXAMPLE 5**   When a person picks apples for the first time, the rate will be $x$ boxes of apples during the first $t$ hours. The rate of picking (number of boxes picked per hour) is proportional to $(4 + \sqrt{t})$ if $0 \le t \le 150$; after that it remains constant. If Wynne picked 16 boxes during the first 4 hours and 80 boxes in 16 hours, how many boxes can she pick in the first 144 hours?

**Solution**   We know that

$$\frac{dx}{dt} = k(4 + t^{1/2})$$

where $k$ is a constant.
   Separating the variables, we get

$$dx = k(4 + t^{1/2})\, dt$$

Integrating both sides, we obtain

$$\int dx = \int k(4 + t^{1/2})\, dt$$

That is,

$$x = k\left(4t + \frac{2}{3}t^{3/2}\right) + C$$

We know that $x = 16$ when $t = 4$ and $x = 80$ when $t = 16$. Thus

$$16 = k\left[4(4) + \frac{2}{3}(4^{3/2})\right] + C \tag{1}$$

and

$$80 = k\left[4(16) + \frac{2}{3}(16^{3/2})\right] + C \tag{2}$$

Subtracting equation (1) from equation (2), we obtain

$$64 = k\left[48 + \frac{2}{3}(64 - 8)\right]$$

or

$$64 = \frac{256k}{3}$$

Thus

$$k = \frac{3}{4}$$

Replacing $k$ by $\frac{3}{4}$ in equation (1), we get

$$16 = \frac{3}{4}\left(16 + \frac{16}{3}\right) + C$$

and

$$C = 0$$

Therefore

$$x = \frac{3}{4}\left(4t + \frac{2}{3}t^{3/2}\right) + 0 = 3t + 0.5t^{3/2}$$

To find the number of boxes picked in the first 144 hours, we replace $t$ by 144 in the last equation and obtain

$$x = 3(144) + 0.5(144^{3/2}) = 1296$$

Therefore Wynne picked 1296 boxes of apples in the first 144 hours.

Observe that Wynne picked 4 boxes per hour in the first 4 hours and averaged 9 boxes per hour over the first 144 hours. This is to be expected since performance improves with experience.

**Do Exercise 27.**

# Exercise Set 10.1

1. Let $f_1(x) = e^{2x}$ and $f_2(x) = e^{3x}$. Verify that $f_1$ and $f_2$ are solutions of the differential equation $y'' - 5y' + 6y = 0$. Show that if $C_1$ and $C_2$ are arbitrary constants and $f_3(x) = C_1 f_1(x) + C_2 f_2(x)$, then $f_3$ is also a solution.

2. Same as Exercise 1 with $f_1(x) = e^{-x}$, $f_2(x) = e^{4x}$, and $y'' - 3y' - 4y = 0$.

3. Same as Exercise 1 with $f_1(x) = e^{-2x}$, $f_2(x) = e^{3x}$, and $y'' - y' - 6y = 0$.

4. Let $f(x) = x^2 + x + 1$. Verify that $f$ is a solution of the equation $x^2y'' + 3xy' - 8y + 5x + 8 = 0$.

5. Let $g(x) = xe^{2x}$. Verify that $g$ is a solution of the equation $y'' - 3y' + 2y = e^{2x}$.

*In Exercises 6–11, find the general solution of the given differential equation.*

6. $\dfrac{dy}{dx} = \dfrac{6x^2 + 1}{4y^3 + 2y + 3}$

7. $\dfrac{dy}{dx} = \dfrac{\sqrt{x} + 4}{y^2 + 3y + 5}$

8. $\dfrac{dy}{dx} = \dfrac{4}{x(3y^2 + 1)}$

9. $\dfrac{dy}{dx} = \dfrac{e^{3x}}{2y + 6}$

10. $(y^2 + 1)\dfrac{dy}{dx} = x + e^{3x}$

11. $\dfrac{dy}{dx} = 5y$

*In Exercises 12–20, find the particular solution of the given differential equation that satisfies the initial conditions.*

12. $\dfrac{dy}{dx} = 3y;\ y = 2$ when $x = 0$

13. $\dfrac{dy}{dx} = 4xy;\ y = 3$ when $x = 0$

14. $\dfrac{dx}{dt} = 3xt^2;\ x = 5$ when $t = 0$

15. $t\dfrac{dx}{dt} + xt^2 = x;\ x = e$ when $t = 1$

16. $t^2\dfrac{dx}{dt} + x^2t^3 = x^2;\ x = -2$ when $t = 1$

17. $e^x\dfrac{dy}{dx} = xe^y;\ y = 0$ when $x = 0$

18. $e^{y-x}\dfrac{dy}{dx} = x^2;\ y = 0$ when $x = 0$

19. $3y\sqrt{x^2 + 1}\,\dfrac{dy}{dx} - xy^2 = x;\ y = 0$ when $x = \sqrt{3}$

20. $y(x^3 + 3)\dfrac{dy}{dx} - 4x^2y^2 = 4x^2;\ y = 2$ when $x = \sqrt[3]{5}$

*In Exercises 21–23, use the formula $P(t) = P_0e^{rt}$ given for interest compounded continuously at the beginning of this section.*

21. An investment of \$20,000 earns interest compounded continuously at the nominal rate of 8%. What is the value of that investment in 10 years?

22. A bank pays interest compounded continuously at a nominal annual rate of 7%. How long will it take for a deposit to double in value?

23. An initial investment of \$5,000 earns interest compounded continuously at the nominal annual rate of 9%. How long will it take for the investment to be worth \$8,000?

24. The rate of change in the value of real estate in Portland at time $t$ is proportional to the value at that time. A house valued at \$120,000 six months ago is worth \$122,000 today. What will its value be in 18 months?

25. The rate of change in the value of real estate in Vancouver at time $t$ is proportional to the square root of the value at that time. A lot valued at \$90,000 four months ago is worth \$90,601 today. How much will it be worth in two years?

26. The value of real estate in Pittsburgh is increasing at a rate proportional to the cube root of the value at time $t$. Gordon bought a lot 4 years ago for \$27,000. His lot is worth \$32,768 today. How long should he wait to sell it if he wishes to get \$59,260 for it?

27. If $x$ words are typed in $t$ minutes by a person who is learning to type, the rate of change of $x$ with respect to $t$ is proportional to $(5 - 24/\sqrt{.03t + 36})$. If Gwen has typed 342,000 words in her first 105 hours of typing, how many words will she have typed in the next 375 hours?

28. The elasticity of demand for a commodity is given by

$$\eta = \frac{-2p^2}{400 - p^2},\ \text{where } 0 < p < 20$$

Find the demand equation $q = f(p)$ if $q = 300$ when $p = 10$. (*Hint:* Use ln $C$ instead of $C$ for the constant of integration.)

**29.** The elasticity of demand for a commodity is given by

$$\eta = \frac{12p^4 - 600p^3 + 12p^2 - 600p}{3p^4 - 200p^3 + 6p^2 - 600p + 6,265,000}$$

where $0 \le p \le 50$.

Find the demand equation $x = f(p)$ if $x = 6,089,600$ when $p = 10$. (*Hint:* Use $\ln C$ instead of $C$ for the constant of integration.)

**30.** The elasticity of demand for a commodity is given by

$$\eta = \frac{-p^2}{169 - p^2}, \text{ where } 0 < p < 13$$

Find the demand equation $q = f(p)$ if $q = 1200$ when $p = 5$. (*Hint:* Use $\ln C$ instead of $C$ for the constant of integration.)

**31.** The elasticity of demand for a commodity is given by

$$\eta = \frac{-p}{4(4096 - p)}, \text{ where } 0 < p < 4000$$

Find the demand equation $q = f(p)$ if $q = 250,000$ when $p = 3471$. (*Hint:* Use $\ln C$ instead of $C$ for the constant of integration.)

**\*32.** The elasticity of demand for a commodity is given by

$$\eta = \frac{.01p^2 - 1.5p}{50 - p}, \text{ where } 0 < p < 45$$

Find the demand equation $q = f(p)$ if $q = 5000$ when $p = 0$.

**\*33.** The rate of change of the size $P = P(t)$ of a population of fruit flies on an island is proportional to $P(600,000 - P)$ where $0 < P < 600,000$. If at a certain instant it is estimated that there are 100,000 fruit flies on the island and 6 months later the estimate is 150,000, find a formula which gives the size $P(t)$ of the population $t$ months after the first estimate. $\left(Hint: \frac{1}{P(600,000 - P)} = \frac{1}{600,000}\left(\frac{1}{P} + \frac{1}{600,000 - P}\right).\right)$

**34.** The *half-life* of a radioactive material is the amount of time it takes any given amount of that material to lose half its mass by disintegration. Assume that the rate of decompo-

sition of radium at time $t$ is proportional to the amount present. It is given that 15% of the original amount is lost in 375 years. What is the half-life of radium?

**35.** Assume that the rate of decomposition of radioactive carbon at time $t$ is proportional to the amount present. It is given that 25% of the original amount is lost in 2324 years. What is the half-life of radioactive carbon? (See Exercise 34.)

**\*36.** If $a_0, a_1, a_2, \ldots, a_n$ are constants with $a_0 \ne 0$, the $n$th order differential equation $a_0 y^{(n)} + a_1 y^{(n-1)} + a_2 y^{(n-2)} + \cdots + a_n y = 0$ is called a *homogeneous linear equation with constant coefficients*. Show that if $c$ is a root of the equation $a_0 r^n + a_1 r^{n-1} + a_2 r^{n-2} + \cdots + a_n = 0$, then $y = e^{cx}$ is a solution of the differential equation.

*In Exercises 37–43, use the result of Exercise 36 to solve the equations.*

**\*37.** Find two solutions of $y'' - 3y' + 2y = 0$.

**\*38.** Find two solutions of $y'' + 5y' - 6y = 0$.

**\*39.** Find two solutions of $y'' + 3y' - 4y = 0$.

**\*40.** Find two solutions of $y'' - 5y' + 6y = 0$.

**\*41.** Find two solutions of $2y'' - 5y' - 3y = 0$.

**\*42.** Find three solutions of $y''' + 4y'' + y' - 6y = 0$.

**\*43.** Find three solutions of $y''' + 2y'' - y' - 2y = 0$.

*In Exercises 44–48, separate the variables and find the general solution of each differential equation.*

**\*44.** $\dfrac{dy}{dx} = xe^{x^2+y}$

**\*45.** $y\dfrac{dy}{dx} = x^2 e^{x^3+y^2}$

**\*46.** $4y^3 \dfrac{dy}{dx} = 3x^2 e^{x^3-y^4}$

**\*47.** $\ln|x + 1|^{(x+1)y}\dfrac{dy}{dx} = (y^2 + 1)^2$

**\*\*48.** $\dfrac{1}{y}\dfrac{dy}{dx} = e^x \ln y^x$

# 10.2 Area

Areas of regions bounded by certain curves serve as mathematical models for many types of problems in economics, business, and probability. Therefore in this section, we will describe a method for finding areas bounded by curves.

## Area Between Two Curves

Suppose $f$ and $g$ are continuous functions on a closed bounded interval $[a, b]$ over which $f(x) > g(x)$. We wish to calculate the area of the region bounded by the graphs of $y = f(x)$, $y = g(x)$, $x = a$, and $x = b$. (See Figure 10.1.) If $f$ and $g$ are linear functions, then the region is a trapezoid and the area can be calculated using a formula from geometry. In this case, the area is

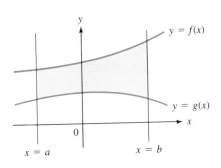

**FIGURE 10.1**

$$\frac{[f(b) - g(b)] + [f(a) - g(a)]}{2}(b - a)$$

(See Figure 10.2.)

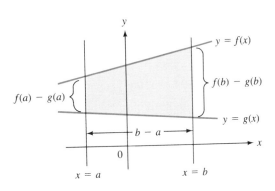

**FIGURE 10.2**

Usually, one or both of the functions will not be linear, in which case we approximate the area of the region using the area of a rectangular polygon. We let $p_n = [x_0, x_1, x_2, \ldots, x_n]$ be a regular partition of $[a, b]$. For each $i \in \{1, 2, \ldots, n\}$, we choose a point $u_i \in [x_{i-1}, x_i]$. Then $f(u_i) > g(u_i)$. Through the points $(u_i, f(u_i))$ and $(u_i, g(u_i))$, we draw lines parallel to the $x$-axis and through $(x_{i-1}, 0)$ and $(x_i, 0)$, we draw lines parallel to the $y$-axis. These four lines form a rectangle the area of which is $(f(u_i) - g(u_i))(x_i - x_{i-1})$. The union of these $n$ rectangles form a rectangular polygon whose area, if $n$ is large, closely approximates the area of our region. (See Figure 10.3.)

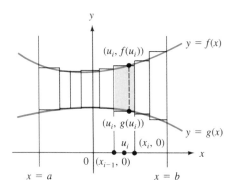

**FIGURE 10.3**

The area of the rectangular polygon is

$$[f(u_1) - g(u_1)](x_1 - x_0) + [f(u_2) - g(u_2)](x_2 - x_1) + \cdots$$
$$+ [f(u_n) - g(u_n)](x_n - x_{n-1})$$

Note that this is a Riemann sum for the function $f - g$.

Because in general the larger the value of $n$, the better the approximation, we define the area of our region to be the limit of this Riemann sum as $n$ tends to infinity. But this limit is $\int_a^b [f(x) - g(x)]\, dx$. Thus

$$A = \int_a^b [f(x) - g(x)]\, dx$$

Whenever we use this formula to find an area, we must be sure that $[f(x) - g(x)] \geq 0$ throughout the interval $[a, b]$.

**EXAMPLE 1**   Let $R$ be the region bounded by the graphs of $y = x + 2$, $y = -2x + 1$, $x = 1$, and $x = 3$.

**a.** Find the area of $R$ using the trapezoid area formula.
**b.** Find the same area using integration.

**Solution**   **a.** The region is illustrated in Figure 10.4.

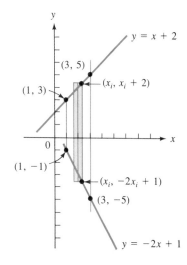

**FIGURE 10.4**

It is a trapezoid with bases of 10 and 4 units and an altitude of 2 units. Thus

$$A = \left(\frac{10 + 4}{2}\right)(2) = 14$$

The area of $R$ is 14 square units.

**b.** In Figure 10.4, we show a typical approximating rectangle whose area is $((x_i + 2) - (-2x_i + 1))(x_i - x_{i-1})$. Therefore

$$A = \int_1^3 [(x + 2) - (-2x + 1)]\, dx = \int_1^3 (3x + 1)\, dx$$

$$= \left(\frac{3x^2}{2} + x\right)\Big|_1^3 = \left(\frac{3(3^2)}{2} + 3\right) - \left(\frac{3(1^2)}{2} + 1\right) = 14$$

**Do Exercise 1.**

**EXAMPLE 2**   Find the area of the region $R$ bounded by $y = 4e^{-2x}$, $y = x^2 - 10$, $x = 0$, and $x = 3$.

**Solution**   The region is sketched in Figure 10.5.

We have shown in Figure 10.5 a typical approximating rectangle. We see that the point $(x_i, 4e^{-2x_i})$ is above the point $(x_i, x_i^2 - 10)$. The height of that rectangle is $4e^{-2x_i}$ (the ordinate of the higher point) minus $x_i^2 - 10$ (the ordinate of the lower point). The base is $x_i - x_{i-1}$. The area of the approximating rectangle is $[4e^{-2x_i} - (x_i^2 - 10)](x_i - x_{i-1})$ and therefore

$$A = \int_0^3 [4e^{-2x} - (x^2 - 10)]\, dx = \int_0^3 (4e^{-2x} - x^2 + 10)\, dx$$

$$= \left(-2e^{-2x} - \frac{x^3}{3} + 10x\right)\Big|_0^3 = (-2e^{-6} - 9 + 30) - (-2) = 23 - 2e^{-6}$$

The area of $R$ is $(23 - 2e^{-6})$ square units (approximately 22.9950425 square units).
**Do Exercise 7.**

It is good practice to check whether the answer to a problem is reasonable. The region $R$ in Figure 10.5 is approximated by a trapezoid whose vertices are $(0, 4)$, $(0, -10)$, $(3, 0)$, and $(3, -1)$. The area of this trapezoid is 22.5 square units since $\left(\frac{14 + 1}{2}\right)(3) = 22.5$. Our answer is reasonable. If you had worked this example and found that the area you obtained differed from the area of the approximating trapezoid by a large amount, then you should go back and check the details of your solution.

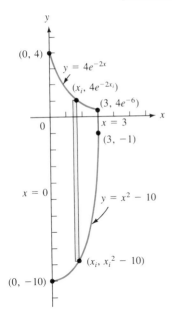

**FIGURE 10.5**

## When to Divide a Region into Subregions

In the preceding discussion of area, we required that $f(x) > g(x)$ for all $x$ in $[a, b]$. The reason for that requirement was to ensure that the dimensions of the approximating rectangle are always positive. There may be cases where we want to find the area of a region bounded by the graphs of $y = f(x)$, $y = g(x)$, $x = a$, and $x = b$ even though the condition $f(x) > g(x)$ (or $g(x) > f(x)$) for all $x$ in $[a, b]$ is not satisfied. In this case, we partition the interval $[a, b]$ into subintervals $[x_0, x_1]$, $[x_1, x_2]$, . . . , $[x_{n-1}, x_n]$ such that on each subinterval we have either $f(x) \geq g(x)$ or $g(x) \geq f(x)$.

Then for each $i$, we calculate

$$A_i = \int_{x_{i-1}}^{x_i} (f(x) - g(x))\, dx \quad (\text{if } f(x) \geq g(x) \text{ on } [x_{i-1}, x_i])$$

or

$$A_i = \int_{x_{i-1}}^{x_i} (g(x) - f(x))\, dx \quad (\text{if } g(x) \geq f(x) \text{ on } [x_{i-1}, x_i])$$

Each answer should be nonnegative. We then add $A_1 + A_2 + A_3 + \cdots + A_n$ to obtain the area of the region.

**EXAMPLE 3**    Find the area of the region $R$ bounded by the graphs of

$$y = f(x) = \frac{-x^3}{6} + \frac{x^2}{6} + x + 3$$

$$y = g(x) = \frac{x^3}{6} - \frac{x^2}{6} - x + 3$$

$$x = -3, \text{ and } x = 4$$

**Solution**    To find the points of intersection of the two graphs, we first solve the system

$$\begin{cases} y = \dfrac{-x^3}{6} + \dfrac{x^2}{6} + x + 3 \\[2mm] y = \dfrac{x^3}{6} - \dfrac{x^2}{6} - x + 3 \end{cases}$$

$\qquad\qquad\qquad\qquad\qquad\qquad\qquad\qquad\qquad\qquad\qquad\quad$ (1)

$\qquad\qquad\qquad\qquad\qquad\qquad\qquad\qquad\qquad\qquad\qquad\quad$ (2)

Subtracting equation (1) from equation (2), we obtain

$$0 = \frac{x^3}{3} - \frac{x^2}{3} - 2x$$

This equation is equivalent to

$$0 = x(x - 3)(x + 2)$$

Thus $x = -2$ or $x = 0$ or $x = 3$. Replacing $x$ by these values in equation (1), we get $y = 3$ in all three cases. Therefore the points of intersection are $(-2, 3)$, $(0, 3)$, and $(3, 3)$. The sketch of region $R$ is illustrated in Figure 10.6.

We see that $f(x) \geq g(x)$ on the intervals $[-3, -2]$ and $[0, 3]$, and $g(x) \geq f(x)$ on the intervals $[-2, 0]$ and $[3, 4]$. We find that

$$f(x) - g(x) = \left( \frac{-x^3}{6} + \frac{x^2}{6} + x + 3 \right) - \left( \frac{x^3}{6} - \frac{x^2}{6} - x + 3 \right)$$

$$= \frac{-x^3}{3} + \frac{x^2}{3} + 2x$$

and

$$g(x) - f(x) = \frac{x^3}{3} - \frac{x^2}{3} - 2x$$

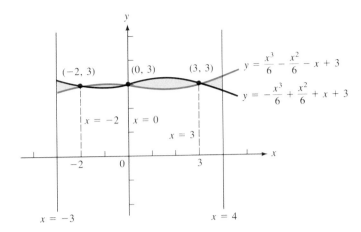

**FIGURE 10.6**

We calculate

$$A_1 = \int_{-3}^{-2} [f(x) - g(x)] \, dx = \int_{-3}^{-2} \left( \frac{-x^3}{3} + \frac{x^2}{3} + 2x \right) dx$$

$$= \left( \frac{-x^4}{12} + \frac{x^3}{9} + x^2 \right) \Bigg|_{-3}^{-2}$$

$$= \left( \frac{-16}{12} + \frac{-8}{9} + 4 \right) - \left( \frac{-81}{12} - \frac{27}{9} + 9 \right) = \frac{91}{36}$$

$$A_2 = \int_{-2}^{0} [g(x) - f(x)] \, dx = \int_{-2}^{0} \left( \frac{x^3}{3} - \frac{x^2}{3} - 2x \right) dx$$

$$= \left( \frac{x^4}{12} - \frac{x^3}{9} - x^2 \right) \Bigg|_{-2}^{0}$$

$$= 0 - \left( \frac{16}{12} - \frac{-8}{9} - 4 \right) = \frac{64}{36}$$

$$A_3 = \int_{0}^{3} [f(x) - g(x)] \, dx = \int_{0}^{3} \left( \frac{-x^3}{3} + \frac{x^2}{3} + 2x \right) dx$$

$$= \left( \frac{-x^4}{12} + \frac{x^3}{9} + x^2 \right) \Bigg|_{0}^{3}$$

$$= \left( \frac{-81}{12} + \frac{27}{9} + 9 \right) - 0 = \frac{189}{36}$$

$$A_4 = \int_{3}^{4} [g(x) - f(x)] \, dx = \int_{3}^{4} \left( \frac{x^3}{3} - \frac{x^2}{3} - 2x \right) dx$$

$$= \left( \frac{x^4}{12} - \frac{x^3}{9} - x^2 \right) \Bigg|_{3}^{4}$$

$$= \left( \frac{256}{12} - \frac{64}{9} - 16 \right) - \left( \frac{81}{12} - \frac{27}{9} - 9 \right) = \frac{125}{36}$$

Therefore,

$$A = A_1 + A_2 + A_3 + A_4 = \frac{91}{36} + \frac{64}{36} + \frac{189}{36} + \frac{125}{36} = \frac{469}{36} \doteq 13.0278$$

Thus the area of the region is approximately 13 square units.
**Do Exercise 21.**

**EXAMPLE 4**   Let the functions $f$, $g$, and $h$ be defined by

$$f(x) = -x^2 + 4x, \ 0 \le x \le 2$$

$$g(x) = \frac{16}{x^2}, \ 2 \le x \le 3$$

$$h(x) = -x, \ 0 \le x \le 3$$

Find the area of the region bounded by the graphs of $f$, $g$, and $h$ over the interval $[0, 3]$.

**Solution**   The region is sketched in Figure 10.7.

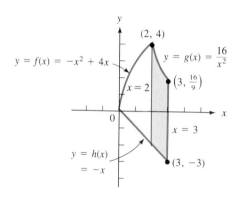

**FIGURE 10.7**

Since the region is bounded below by the graph of $h$ and above by the graphs of $f$ and $g$, we use the graph of $x = 2$ to divide the region into regions $R_1$ and $R_2$. Let $A_1$ and $A_2$ square units be the areas of $R_1$ and $R_2$, respectively. Then

$$A_1 = \int_0^2 [f(x) - h(x)] \ dx = \int_0^2 [(-x^2 + 4x) - (-x)] \ dx$$

$$= \int_0^2 (-x^2 + 5x) \ dx = \left( \frac{-x^3}{3} + \frac{5x^2}{2} \right) \Big|_0^2 = \left( \frac{-8}{3} + 10 \right) - 0 = \frac{22}{3}$$

and

$$A_2 = \int_2^3 [g(x) - h(x)] \ dx = \int_2^3 \left[ \left( \frac{16}{x^2} - (-x) \right) \right] dx$$

$$= \int_2^3 (16x^{-2} + x) \ dx = \left( -16x^{-1} + \frac{x^2}{2} \right) \Big|_2^3$$

$$= \left( \frac{-16}{3} + \frac{9}{2} \right) - \left( \frac{-16}{2} + 2 \right) = \frac{31}{6}$$

The area of the region is 12.5 square units since

$$\frac{22}{3} + \frac{31}{6} = \frac{75}{6} = 12.5$$

**Do Exercise 33.**   ▪

## Regions Bounded Above or Below by the x-Axis

If a region is bounded above or below by the x-axis, we simply use the fact that the x-axis is the graph of $y = 0$.

**EXAMPLE 5**   **a.** Find the area of the region $R_1$ bounded above by the graph of $y = 3x^2 + 1$, below by the x-axis, and by the graphs of $x = -1$ and $x = 2$.
**b.** Find the area of the region $R_2$ bounded above by the x-axis, below by the graph of $y = \frac{1}{x}$, and by the graphs of $x = -4$ and $x = -1$.

**Solution**   **a.** The region $R_1$ is sketched in Figure 10.8.
We have

$$A = \int_{-1}^{2} [(3x^2 + 1) - 0] \, dx = \int_{-1}^{2} (3x^2 + 1) \, dx$$

$$= (x^3 + x)\big|_{-1}^{2} = 10 - (-2) = 12$$

The area of $R_1$ is 12 square units.
**b.** The region $R_2$ is sketched in Figure 10.9. We have

$$A = \int_{-4}^{-1} \left(0 - \frac{1}{x}\right) dx = \int_{-4}^{-1} \frac{-1}{x} \, dx$$

$$= -\ln|x| \,\big|_{-4}^{-1} = -\ln|-1| - (-\ln|-4|) = 0 + \ln 4 = \ln 4$$

The area of $R_2$ is ln 4 square units (approximately 1.3863 square units).
**Do Exercise 5.**

**FIGURE 10.8**

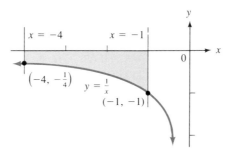

**FIGURE 10.9**   ▪

## When to Integrate with Respect to y

It is sometimes convenient to use $y$ rather than $x$ as the independent variable. We illustrate this with the next example.

**EXAMPLE 6**   Find the area of the region $R$ bounded by the graphs of $y = \sqrt{-x}$, $y = \sqrt{x-1}$, $y = 2$, and the $x$-axis.

**Solution**   The region $R$ is sketched in Figure 10.10. If we proceed as in earlier examples, we must divide the region into three subregions. But it is more convenient to treat $x$ as the dependent variable. Thus we solve for $x$ in both of the equations and obtain

$$x = f(y) = -y^2 \text{ and } x = g(y) = y^2 + 1 \text{ with } 0 \le y \le 2$$

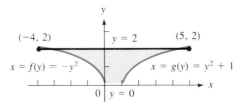

**FIGURE 10.10**

Here, the graph of $g$ is to the right of the graph of $f$ and $g(y) > f(y)$ for each $y$ in the interval $[0, 2]$. Therefore

$$A = \int_0^2 [g(y) - f(y)] \, dy = \int_0^2 [(y^2 + 1) - (-y^2)] \, dy = \int_0^2 (2y^2 + 1) \, dy$$

$$= \left( \frac{2y^3}{3} + y \right) \Bigg|_0^2 = \left( \frac{16}{3} + 2 \right) - 0 = \frac{22}{3}$$

The area of $R$ is $\frac{22}{3}$ square units.

**Do Exercise 35.**

In the preceding example, we treated $x$ as the dependent variable and found the area of the region by integrating with respect to $y$ because, if we had integrated with respect to $x$, it would have been necessary to divide the given region into three subregions. There are cases where it is useful to integrate with respect to $y$ for another reason. Suppose a region is bounded by the graphs of two one-to-one functions $f$ and $g$ where $f(x) > g(x)$ for all $x$ in the interval $(a, b)$, $f(a) = g(a)$, and $f(b) = g(b)$. (See Figure 10.11.)

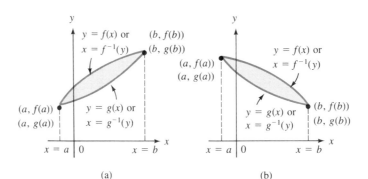

**FIGURE 10.11**

(a)

(b)

In such cases, the area of the region may be found using one of the following three equalities:

$$A = \int_a^b [f(x) - g(x)]\, dx$$

$$A = \int_{f(a)}^{f(b)} [g^{-1}(y) - f^{-1}(y)]\, dy \qquad (\text{if } f(a) = g(a) < f(b) = g(b))$$

or

$$A = \int_{f(b)}^{f(a)} [f^{-1}(y) - g^{-1}(y)]\, dy \qquad (\text{if } f(a) = g(a) > f(b) = g(b))$$

It may be that one of the integrals is easier to evaluate than the others. You should try to select the integral that will yield the simplest solution. (See Exercises 37–40.)

# Exercise Set 10.2

*In Exercises 1–40, sketch the regions bounded by the graphs of the given equations and compute the areas of these regions.*

**1.** $y = x + 3$, $y = x - 4$, $x = -1$, $x = 3$

**2.** $y = 2x + 1$, $y = -4x - 3$, $x = 1$, $x = 4$

**3.** $y = x^2 + 3$, $y = -x^2 - 1$, $x = -2$, $x = 2$

**4.** $y = 2x^2 - 4x + 3$, $y = x - 5$, $x = 0$, $x = 3$

**5.** $y = 3x^2 + 1$, $x = -1$, $x = 2$, $y = 0$

**6.** $y = x^3 + x + 1$, $x = 0$, $x = 2$, $y = 0$

**7.** $y = \sqrt{x}$, $y = -x$, $x = 1$, $x = 4$

**8.** $y = \sqrt[3]{x^2}$, $y = x + 4$, $x = 1$, $x = 8$

**9.** $y = e^x$, $y = -e^x$, $x = 0$, $x = 3$

**10.** $y = \dfrac{1}{x+2}$, $x = -1$, $x = 2$, $y = 0$

**11.** $y = \dfrac{1}{x+3}$, $x = -2$, $x = 1$, $y = -x^2 - 2$

**12.** $y = \dfrac{3}{x+4}$, $y = \dfrac{-4}{x+5}$, $x = -3$, $x = -1$

**13.** $y = x^2 + 3x - 1$, $y = 2x + 1$

**14.** $y = x^2 + 5x - 5$, $y = 3x - 2$

**15.** $y = x^2 + x - 7$, $y = -x + 1$

**16.** $y = \sqrt{x}$, $3y = x + 2$

**17.** $y = x^2 + 3x + 1$, $y = 2x^2 + 4x - 1$

**18.** $y = x^2 + 4x + 2$, $y = 2x^2 + 4x + 1$

**19.** $y = -x^2 + x + 5$, $y = x^2 - x - 7$

**20.** $y = \sqrt{x}$, $y = x^2 - \dfrac{14x}{3} + \dfrac{14}{3}$

**21.** $y = x^2 + 3x - 1$, $y = 2x + 1$, $x = -3$, $x = 2$

**22.** $y = x^2 + 5x - 5$, $y = 3x - 2$, $x = -4$, $x = 3$

**23.** $y = x^2 + x - 7$, $y = -x + 1$, $x = -7$, $x = 5$

**24.** $y = \sqrt{x}$, $3y = x + 2$, $x = 0$, $x = 6$

**25.** $y = x^2 + 3x + 1$, $y = 2x^2 + 4x - 1$, $x = -3$, $x = 3$

**26.** $y = x^2 + 4x + 2$, $y = 2x^2 + 4x + 1$, $x = -2$, $x = 2$

**27.** $y = -x^2 + x + 5$, $y = x^2 - x - 7$, $x = -4$, $x = 6$

**28.** $y = \sqrt{x}$, $y = x^2 - \dfrac{14x}{3} + \dfrac{14}{3}$, $x = 0$, $x = 5$

**29.** $y = x^3$, $y = x^2 - 2x$, $x = -1$, $x = 1$

**30.** $y = \dfrac{x^3}{3} - x^2 - \dfrac{x}{3} + 3$, $y = \dfrac{x^3}{6} - \dfrac{5x^2}{6} + 3$, $x = -3$, $x = 4$

**31.** $12y = -5x^3 - 15x^2 + 8x + 48$, $12y = -3x^3 - 13x^2 + 4x + 48$, $x = -4$, $x = 2$

**32.** $y = x^2 + 1$ for $-1 \le x \le 0$, $y = x + 1$ for $0 \le x \le 1$, $y = x^2 - 1$ for $-1 \le x \le 1$, $x = -1$, $x = 1$

**33.** $y = e^{2x}$ for $-2 \le x \le 0$, $y = e^{-x}$ for $0 \le x \le 3$, $y = -x^2$ for $-2 \le x \le 3$, $x = -2$, $x = 3$

**34.** $y = e^x$ for $-2 \leq x \leq 0$, $y = \dfrac{1}{x+1}$ for $0 \leq x \leq 2$,

$y = x^2 + 2x - \dfrac{23}{3}$ for $-2 \leq x \leq 2$, $x = -2$

**35.** $y = \sqrt[3]{x-1}$, $y = \sqrt[4]{-x-1}$, $y = 0$, $y = 2$

**36.** $y = -x^2 + 5$ for $x \leq 0$, $y = \ln x$, $y = 0$, $y = 1$

**37.** $y = \ln x$, $x = 1 + (e-1)y$

**38.** $y = \ln(2x + e)$, $2x = (e-1)(y-1)$

**39.** $y = \ln(x+1)$, $x^2 = (e-1)^2 y$

**40.** $y = \ln(x+2)$, $(e^2-1)^3 y = 2(x+1)^3$

# 10.3  Consumers' and Producers' Surplus

## Market Equilibrium

We have discussed demand and supply equations earlier. It is clear that consumers are willing to buy a greater number of units of a commodity as the price per unit decreases (the demand function is decreasing), while suppliers are willing to produce a larger number of units as the price per unit increases (the supply function is increasing). The point $(q_0, y_0)$ where the demand and supply curves intersect is called the *equilibrium point*. The numbers $q_0$ and $p_0$ are the *equilibrium quantity* and the *equilibrium price*, respectively. We shall show that when a product is sold at equilibrium price, consumers who would have been willing to pay a higher price are benefiting. The total amount they gain is called the *consumers' surplus*. Similarly, producers who were willing to sell at a lower price are benefiting also. The total amount they gain is called the *producers' surplus*. The mathematical models for both the consumers' and producers' surplus are best described using areas bounded by curves. To see how this works, consider the following examples.

## Consumers' Surplus as an Area

**EXAMPLE 1**   Suppose that among the customers in a department store, 10 would be willing to buy a certain dress if it sold for $150, 30 would buy the dress at $125, 70 at $90, and 130 at $75. Often a store will introduce a new dress at a high price, then have sales to attract customers. But this is costly to the store (inventory, cost of advertising, and so on). Suppose the store introduces the new dress at the low price of $75. Illustrate the amount saved by its customers geometrically.

**Solution**   If the new dress was priced at $150, 10 customers would have bought one and spent $1500 (1500 square units is the area of the first rectangle in Figure 10.12). Then if the dress went on sale for $125, 20 additional customers would buy one (remember that 30 were willing to pay $125, but among those were the 10 who were willing to pay $150 and they have already purchased the dress). Thus the amount spent is $2500 (since $(125)(20) = 2500$). If the price is reduced to $90, an additional 40 customers will buy the dress and spend a total of $3600. Finally, if the price per dress is reduced to $75, 60 additional customers will buy one and spend a total of $4500. The amount spent by all the customers is $12,100 (since $1500 + 2500 + 3600 + 4500 = 12,100$).

   If the dress sold at a constant price of $75, 130 customers would buy one and spend a total of $9750. The savings (consumers' surplus) the customers would enjoy would be $2350 (since $12,100 - 9750 = 2350$). This is illustrated geometrically in Figure 10.12.

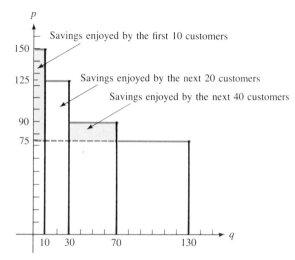

FIGURE 10.12

## Consumers' and Producers' Surplus as Integrals

In general, if $p = D(q)$ is the demand equation and $(q_0, p_0)$ is the equilibrium point, the consumers' surplus (C.S.) is approximated by the area of the region bounded by the graphs of $p = D(q)$, $p = p_0$, $q = 0$, and $q = q_0$. This approximates the savings enjoyed by the consumers when the product is sold at equilibrium price instead of starting at $D(0)$ and decreasing to $p_0$. (See Figure 10.13.)

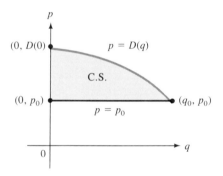

FIGURE 10.13

Using the results from Example 1, we see that the consumers' surplus is approximated by

$$C.S. = \int_0^{q_0} [D(q) - p_0]\, dq$$

A similar discussion shows that the amount gained by the suppliers when the product is sold at equilibrium price is approximated by the area of the region bounded by the graphs of $p = S(q)$, $p = p_0$, $q = 0$, and $q = q_0$. Thus the producers' surplus (P.S.) is approximately

$$P.S. = \int_0^{q_0} [p_0 - S(q)]\, dq$$

(See Figure 10.14.)

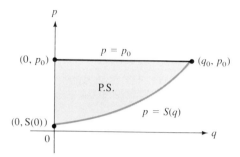

**FIGURE 10.14**

EXAMPLE 2   The demand and supply equations for a commodity are

$$p = D(q) = -q^2 - 10q + 1200, \, 0 \le q \le 25$$

and

$$p = S(q) = q^2 + 40q, \, 0 \le q \le 25$$

where the price $p$ is in dollars and the supply $q$ is in thousands of units.

**a.** Find the equilibrium price and quantity.
**b.** Find the consumers' surplus.
**c.** Find the producers' surplus.

Solution  **a.** Note that since $q$ is in thousands of units and $p$ is in dollars per unit, the consumers' and producers' surpluses will be in thousands of dollars. To find the equilibrium point, we solve the system

$$\begin{cases} p = -q^2 - 10q + 1200 \\ p = q^2 + 40q \end{cases}$$    (1)
(2)

Subtracting equation (1) from equation (2), we obtain

$$0 = 2q^2 + 50q - 1200$$

which can be written

$$0 = 2(q - 15)(q + 40)$$

Thus $q = 15$ or $q = -40$. The equilibrium quantity must be positive, so $q = 15$. We conclude that the equilibrium quantity is 15,000 units.
Set $q = 15$ in equation (1) to obtain

$$p = -15^2 - 10(15) + 1200 = 825$$

Therefore the equilibrium price is \$825 per unit.
**b.** The consumers' surplus is illustrated in Figure 10.15.

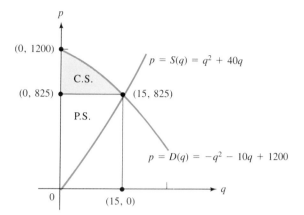

FIGURE 10.15

Clearly,

$$\text{C.S.} = \int_0^{15} [(-q^2 - 10q + 1200) - 825]\, dq$$

$$= \int_0^{15} (-q^2 - 10q + 375)\, dq = \left(\frac{-q^3}{3} - 5q^2 + 375q\right)\Bigg|_0^{15}$$

$$= \left[\frac{-15^3}{3} - 5(15^2) + 375(15)\right] - 0 = 3375$$

Since $q$ was in thousands of units, the consumers' surplus is \$3,375,000.

**c.** The producers' surplus is also illustrated in Figure 10.15. We have

$$\text{P.S.} = \int_0^{15} [825 - (q^2 + 40q)\, dq$$

$$= \int_0^{15} (825 - q^2 - 40q)\, dq = \left(825q - \frac{q^3}{3} - 20q^2\right)\Bigg|_0^{15}$$

$$= \left[825(15) - \frac{15^3}{3} - 20(15^2)\right] - 0 = 6750$$

Therefore the producers' surplus is \$6,750,000.

**Do Exercise 5.**                                                                                      ∎

## Consumers' Surplus When the Supplier has a Monopoly

In the previous discussion, we assumed that market equilibrium existed (that is, the product was sold at the equilibrium price). In general, this will not be the case if a supplier has a monopoly on a commodity. In this case, the price will be set to maximize the supplier's profit. We illustrate this situation with the following example.

**EXAMPLE 3**   A supplier has a monopoly on a certain product. The cost and demand equations are

$$C(q) = .2q^2 + 1392q + 5000, \text{ where } 0 \le q \le 40$$

and

$$p = D(q) = -q^2 - 10q + 3000, \text{ where } 0 \leq q \leq 40$$

It is given that $C(q)$ is in thousands of dollars, $q$ is in thousands of units, and $p$ is in dollars.

**a.** Find the number of units and the price per unit that will maximize the profit.

**b.** Find the consumers' surplus if the product is sold at the price that maximizes the producer's profit.

**Solution**   **a.** We first find the revenue function. Since the price is $(-q^2 - 10q + 3000)$ dollars when the number of units sold (in thousands) is $q$, the revenue is given by

$$R(q) = (-q^2 - 10q + 3000)q = -q^3 - 10q^2 + 3000q$$

where $R(q)$ is in thousands of dollars. Therefore the profit function is

$$\begin{aligned} P(q) &= R(q) - C(q) \\ &= (-q^3 - 10q^2 + 3000q) - (.2q^2 + 1392q + 5000) \\ &= -q^3 - 10.2q^2 + 1608q - 5000 \end{aligned}$$

We wish to find the value of $q$ that will maximize profit. We shall use the second derivative test. We have

$$P'(q) = -3q^2 - 20.4q + 1608 = -3(q - 20)(q + 26.8)$$

Therefore $P'(q) = 0$ whenever $q = 20$ or $q = -26.8$. We consider only the positive critical number.

Since

$$\begin{aligned} P''(q) &= -6q - 20.4 \\ P''(20) &= -6(20) - 20.4 = -140.4 \end{aligned}$$

But $P''(20) < 0$ imples that $P(20)$ is a local maximum. Note that $P''(q) < 0$ on the interval $[0, 40]$. Therefore the graph of the function $P$ is concave downward on that interval and $P(20)$ is in fact the absolute maximum on that interval. It follows that the producer's profit will be maximum when 20,000 units of the product are sold.

The price per unit is given by

$$p = -20^2 - 10(20) + 3000 = 2400$$

Thus the profit is maximum when the price per unit is $2400.

**b.** The consumers' surplus is illustrated in Figure 10.16.
The consumers' surplus is

$$\begin{aligned} \text{C.S.} &= \int_0^{20} [(-q^2 - 10q + 3000) - 2400] \, dq \\ &= \int_0^{20} (-q^2 - 10q + 600) \, dq = \left( \frac{-q^3}{3} - 5q^2 + 600q \right) \Big|_0^{20} \\ &= \left[ \frac{-20^3}{3} - 5(20^2) + 600(20) \right] - 0 \doteq 7333.\overline{3} \end{aligned}$$

Therefore the consumers' surplus is approximately $7,333,333.
**Do Exercise 21.**

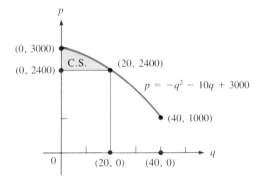

**FIGURE 10.16**

---

# Exercise Set 10.3

*In Exercises 1–20, the supply and demand equations for a commodity are given where p is in dollars and q is the number of units in thousands.*

*a. Find the equilibrium quantity and price.*
*b. Assuming that market equilibrium exists, find the consumers' surplus.*
*c. Find the producers' surplus.*

**1.** $p = S(q) = 2 + q$
   $p = D(q) = 14 - 2q, 0 \le q \le 7$

**2.** $p = S(q) = 3 + \dfrac{q}{2}$
   $p = D(q) = 15 - q, 0 \le q \le 15$

**3.** $p = S(q) = q^2 + 6q$
   $p = D(q) = -q^2 - 4q + 132, 0 \le q \le 9$

**4.** $p = S(q) = q^2 + 5q + 2$
   $p = D(q) = -q^2 - 6q + 312, 0 \le q \le 14$

**5.** $p = S(q) = q^2 + 8q + 5$
   $p = D(q) = -q^2 - 4q + 635, 0 \le q \le 23$

**6.** $p = S(q) = q^2 + 10q + 30$
   $p = D(q) = -q^2 - 3q + 5680, 0 \le q \le 70$

**7.** $p = S(q) = q^2 + 8q + 5$
   $p = D(q) = -q^2 - 7q + 1408, 0 \le q \le 32$

**8.** $p = S(q) = q^2 + 3q + 4$
   $p = D(q) = -q^2 - 9q + 786, 0 \le q \le 23$

**9.** $p = S(q) = q^3 + q^2 + 1775$
   $p = D(q) = q^3 - 1200q + 20,000, 0 \le q \le 20$

**10.** $p = S(q) = q^3 + 10q + 5895$
   $p = D(q) = q^3 - 15q^2 - 1800q + 41,000, 0 \le q \le 30$

**11.** $p = S(q) = \dfrac{q}{3} + 6$
   $p = D(q) = \sqrt{144 - 7q}, 0 \le q \le 20$

**12.** $p = S(q) = \dfrac{q}{2} + 3$
   $p = D(q) = \sqrt{169 - 3q}, 0 \le q \le 25$

**13.** $p = S(q) = \dfrac{q}{3} + 3$
   $p = D(q) = \sqrt{225 - 3q}, 0 \le q \le 50$

**14.** $p = S(q) = \sqrt{3q + 7}$
   $p = D(q) = \sqrt{121 - 3q}, 0 \le q \le 40$

**15.** $p = S(q) = \sqrt{12q + 25}$
   $p = D(q) = \sqrt{1225 - 12q}, 0 \le q \le 100$

**16.** $p = S(q) = \dfrac{q + 5}{9}$
   $p = D(q) = \sqrt[3]{1000 - 16q}, 0 \le q \le 60$

**17.** $p = S(q) = q + 20$
   $p = D(q) = \dfrac{300}{.1q + 3}, 0 \le q \le 70$

**18.** $p = S(q) = \dfrac{20q + 200}{q + 20}$
   $p = D(q) = \dfrac{400}{(.1q + 2)^2}, 0 \le q \le 80$

**19.** $p = S(q) = e^{2+.2q}$

$p = D(q) = e^{5-.1q}, 0 \le q \le 30$

**20.** $p = S(q) = \ln(q + 10)$

$p = D(q) = \ln \dfrac{750}{q + 5}, 0 \le q \le 40$

*In Exercises 21–25, a supplier has a monopoly on a certain product. The cost and demand equations are given with C(q) in thousands of dollars, q in thousands of units, and p in dollars.*

**a.** *Find the number of units and the price per unit that maximize profit.*

**b.** *Find the consumers' surplus if the product is sold at the price that maximizes profit.*

**21.** $C(q) = .5q^2 + 1000$

$p = D(q) = 150 - 2q, 0 \le q \le 70$

**22.** $C(q) = 2q^2 + 1000$

$p = D(q) = 210 - 3q, 0 \le q \le 60$

**23.** $C(q) = .1q^2 + 100$

$p = D(q) = 60 - .9q, 0 \le q \le 65$

**24.** $C(q) = q^2 + 5q + 250$

$p = D(q) = \dfrac{-q^2}{6} - q + 15, 0 \le q \le 6$

**25.** $C(q) = q^2 + 3q + 150$

$p = D(q) = -.2q^2 - 2q + 48, 0 \le q \le 10$

## 10.4  More Applications

We have seen that integration may be used effectively to estimate the consumers' surplus and the producers' surplus when a commodity is sold at a certain price. In this section, we give more integration applications for business and economics.

### Lorenz Curves

An interesting application is the study of income distribution. When a high percentage of a population receives a low percentage of the total national income, we have income inequalities. To measure the degree of inequality, we define a function $I$ as follows. We let $y = I(x)$ where $x$ is the percentage of lowest paid individuals and $y$ is the percentage of the total income. To illustrate how this function is defined, we offer the following example.

**EXAMPLE 1**  Suppose five people are working for a company. Their respective annual incomes are $10,000, $15,000, $25,000, $40,000, and $110,000. Study the income distribution of that company.

**Solution**  The total annual income for these people is $200,000. The lowest paid person (20% of the work force) earns $10,000, which is 5% of the total. Thus $I(.20) = .05$. The two lowest paid individuals earn $25,000 together. But two employees constitute 40% of the work force and $25,000 is 12.5% of $200,000. Therefore $I(.40) = .125$. Similarly, we find that the lowest paid 60% (three employees) earn $50,000, or 25% of the total income. Thus $I(.60) = .25$. We also get $I(.80) = .45$ and $I(1) = 1$.  ■

Returning to the general case, we see that $x$ and $y$ are percentages; therefore $0 \le x \le 1$ and $0 \le y \le 1$. Furthermore, we agree that 0% of the population earns 0% of the total income and that 100% of the population earns 100% of the total income. Thus we will always have $I(0) = 0$ and $I(1) = 1$. It is clear that we must have $I(x) \le x$ for all $x$ (see Exercise 35). Although the number of people in any population is finite, in our mathematical model the domain and range of the function $I$ are the closed interval $[0, 1]$. The graph of the function $I$ is called a *Lorenz curve*.*

Three different Lorenz curves are illustrated in Figure 10.17. Since $I(x) = x$ would represent an absolute equality in income distribution, Figure 10.17a represents near absolute equality, while Figure 10.17c represents great inequality. Observe in that case we have $I(.9) = 0.1$, which indicates that 90% of the population earns only 10% of the total income.

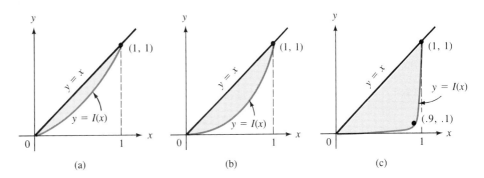

**FIGURE 10.17**                    (a)                    (b)                    (c)

## Coefficient of Inequality

Since the graph of $y = x$ represents perfect equality in income distribution, the more the graph of $y = I(x)$ differs from it, the greater the inequality. Therefore we define the *coefficient of inequality* of income distribution to be the ratio $A_1/A_2$, where $A_1$ square units is the area of the region bounded by the graphs of $y = x$ and $y = I(x)$, and $A_2$ square units is the area of the region (triangle) bounded by the graphs of $y = x$, $x = 1$, and the $x$-axis. It is clear that $0 \le A_1/A_2 \le 1$. If $A_1/A_2$ is near 0, we have near perfect equality, and if $A_1/A_2$ is near 1, we have great inequality in income distribution. Note that $A_2 = \frac{1}{2}$ square unit. Thus

$$\frac{A_1}{A_2} = \frac{A_1}{\dfrac{1}{2}} = 2A_1$$

Therefore the coefficient of inequality is given by

$$2 \int_0^1 [x - I(x)] \, dx$$

The coefficient of inequality is often called the *Gini coefficient*.**

---

*The Lorenz curve is named after its American inventor Max Otto Lorenz. However, its use was promoted by Wilford King who followed Lorenz at the University of Wisconsin.
**The Gini coefficient is named after the Italian statistician and demographer Corrodo Gini who first used the concept.

**EXAMPLE 2**   The income distribution for a country is given by

$$y = I(x) = \frac{4x^2 + x}{5}$$

**a.** Sketch the corresponding Lorenz curve.
**b.** What portion of the total income is received by the lowest paid 25% of the population?
**c.** Calculate the coefficient of inequality.

**Solution**   **a.** The Lorenz curve is sketched in Figure 10.18.

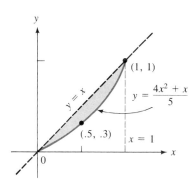

**FIGURE 10.18**

**b.** To find the percentage of income received by the lowest paid 25%, we replace $x$ by .25 and get

$$I(.25) = \frac{4(.25^2) + .25}{5} = .10$$

Therefore the lowest paid 25% of the population receive 10% of the total income.
**c.** The coefficient of inequality is given by

$$2 \int_0^1 \left( x - \frac{4x^2 + x}{5} \right) dx = 2 \int_0^1 \frac{4x - 4x^2}{5} dx$$

$$= \frac{8}{5} \int_0^1 (x - x^2) \, dx = \frac{8}{5} \left( \frac{x^2}{2} - \frac{x^3}{3} \right) \Big|_0^1$$

$$= \frac{8}{5} \left( \frac{1}{2} - \frac{1}{3} \right) - 0 = \frac{4}{15}$$

**Do Exercise 3.**   ■

## Marginal Propensity to Consume and to Save

We now discuss another concept. Suppose $C = f(I)$, where $I$ is the total national disposable income (in billions of dollars) and $C$ is the total national consumption (also in billions of dollars). Then the derivative $f'(I)$ is called the *marginal propensity to consume.* If $S$ represents national savings (in billions of dollars), then, of course, $C + S = I$. Thus

$$\frac{dC}{dI} + \frac{dS}{dI} = 1$$

The derivative $\frac{dS}{dI}$ is called the *marginal propensity to save.* This equation gives the relationship between the marginal propensity to consume and the marginal propensity to save. We illustrate these concepts with the following example.

**EXAMPLE 3** If the marginal propensity to save is given by

$$\frac{dS}{dI} = .4 - .3e^{-.3I}$$

(in billions of dollars) and consumption is $8.12 billion when disposable income is $10 billion, find the consumption function. What is consumption when disposable income is $15 billion?

**Solution** Clearly,

$$\frac{dC}{dI} = 1 - \frac{dS}{dI} = 1 - (.4 - .3e^{-.3I}) = .6 + .3e^{-.3I}$$

Thus

$$C = f(I) = \int (.6 + .3e^{-.3I}) \, dI = .6I - e^{-.3I} + K$$

where $K$ is a constant to be determined. It was given that $f(10) = 8.12$. Therefore

$$8.12 = (.6)(10) - e^{-.3(10)} + K \doteq 5.950212932 + K$$

$$K = 2.169787068$$

and

$$C = f(I) = .6I - e^{-.3I} + 2.169787068$$

To find consumption when disposable income is $15 billion, we replace $I$ by 15 and obtain

$$C = .6(15) - e^{-.3(15)} + 2.169787068 = 11.15867807$$

Thus consumption is $11.15867807 billion when disposable income is $15 billion.
**Do Exercise 9.** ∎

## Present Value of an Annuity

We now show how a definite integral can be used to estimate the present value of an annuity. Recall that if $P$ dollars are invested at a nominal rate $r$, compounded continuously, then the amount $A$ dollars in the investment $t$ years later is given by $A = Pe^{rt}$. Therefore the present value of an investment that will yield a single payment of $A$ dollars in $t$ years is given by $P = Ae^{-rt}$.

Suppose an annuity pays $B$ dollars per year distributed in $k$ equal payments each year. Then each payment will be $\frac{B}{k} = B\Delta t$ dollars where we have written $\Delta t$ for $\frac{1}{k}$.

The present value of the first payment is $(B\Delta t)e^{-r\Delta t}$ dollars, that of the second payment is $(B\Delta t)e^{-r(2\Delta t)}$ dollars, and so on. In general, the present value of the $i$th payment is given by

$$(B\Delta t)e^{-r(i\Delta t)} = Be^{-r(i\Delta t)}\Delta t = Be^{-rt_i}\Delta t$$

We have written $t_i$ instead of $i\Delta t$ since $i\Delta t$ gives the time (in years) from the present to the time of the $i$th payment. If payments are to be made for $m$ years, there will be $mk$ payments. We let $n = mk$. The present value of the annuity is

$$\sum_{i=1}^{n} Be^{-rt_i}\Delta t$$

Observe that this expression is a Riemann sum for the function $Be^{-rt}$ corresponding to the partition $[0, \Delta t, 2\Delta t, 3\Delta t, \ldots , n\Delta t]$ of the interval $[0, m]$. (Recall that $n\Delta t = n\left(\frac{1}{k}\right) = mk\left(\frac{1}{k}\right) = m$.) Thus

$$\lim_{n\to\infty} \sum_{i=1}^{n} Be^{-rt_i}\Delta t = \int_0^m Be^{-rt}\, dt$$

It follows that the value of the definite integral may be used to approximate the present value of the annuity, especially in cases where the number of payments per year is large.

**EXAMPLE 4**    Using integration, approximate the present value of an annuity that will pay $3000 per month for the next 10 years if the nominal rate of interest is 9% compounded continuously.

**Solution**    The amount paid per year is $36,000 (since $(12)(3000) = 36,000$). Therefore the present value of the annuity is approximated by $\int_0^{10} 36,000e^{-.09t}\, dt$. But,

$$\int_0^{10} 36,000e^{-.09t}\, dt = -400,000e^{-.09t}\Big|_0^{10}$$

$$= -400,000e^{-.09(10)} - (-400,000e^0) = 237,372.1361$$

Therefore the present value of the annuity is approximately $237,372.14.
**Do Exercise 15.**    ∎

## Learning Curves

Another application of definite integrals is to predict the number of work hours necessary to produce a certain number of units of a commodity. This is useful when a producer bids for a contract. Suppose $y = f(x)$ where $y$ is the number of hours required to produce the first $x$ units of a commodity. As in earlier interpretations of the derivative, $f'(x)$ approximates the amount of time (in hours) necessary to produce an additional unit when the production level is $x$ units. In general, if $g(x) = f'(x)$, then $g$ is a decreasing function because workers become more efficient as the number of units produced increases and, consequently, the time required to produce an additional unit decreases as $x$ increases. If $m$ and $n$ are positive integers, with $m < n$, $f(m)$ and $f(n)$ are the numbers of hours required to produce $m$ and $n$ units, respectively. Therefore

$f(n) - f(m)$ is the number of work hours required to increase production from a level of $m$ units to a level of $n$ units. Note that $f$ is an antiderivative of $g$. Therefore

$$f(n) - f(m) = \int_m^n g(x)\, dx$$

Thus the total number of hours needed to produce the $(m + 1)$st unit through the $n$th unit is approximated by $\int_m^n g(x)\, dx$.

**EXAMPLE 5** The Triplenight Sporting Equipment Company has found that the time $g(x)$ hours required to produce the $(x + 1)$st tennis racquet is given by

$$g(x) = 10e^{-.01x}$$

provided that $0 \le x \le 300$. Estimate the amount of time required to produce the 51st through the 200th racquet.

**Solution** The number of hours required to produce the 51st racquet through the 200th racquet is approximated by

$$\int_{50}^{200} 10e^{-.01x}\, dx$$

However,

$$\int_{50}^{200} 10e^{-.01x}\, dx = -1000e^{-.01x}\Big|_{50}^{200}$$

$$= -1000e^{-2} + 1000e^{-.5}$$

$$\doteq -135.335 + 606.530 \doteq 471.195$$

Therefore the number of hours required to produce 150 tennis racquets, once the first 50 have been produced, is approximately 471.2 hours.

**Do Exercise 23.**  ∎

The graphs of functions such as that of the preceding example are called *learning curves*. Note that these functions are decreasing because employees become more efficient as they repeat their tasks. The rate of learning is obviously not the same in all applications. By convention, the learning rate is usually described as a percentage. For example, a 90% curve indicates that each time production doubles, the newest unit of output requires 90% of the labor input of the reference unit. That is, if the first unit requires 100 hours, the second unit will require 90 hours, the fourth unit will require 81 hours, the eighth unit 72.9 hours, and so on. We illustrate this idea with the following example.

**EXAMPLE 6** If we assume that the equation of a learning curve is of the form $y = ax^b$, where $y$ is the number of hours required to produce the $x$th unit, find the values of the constants $a$ and $b$ for a 90% learning curve if 20 hours are required to produce the first unit.

**Solution** It is given that $y = 20$ when $x = 1$, and $y = 18$ when $x = 2$ (since $20(.9) = 18$). Replacing $x$ and $y$ by these values in the equation $y = ax^b$, we obtain

$$20 = a(1^b) = a \text{ and } 18 = a(2^b)$$

Thus $a = 20$. It follows that $18 = 20(2^b)$ and $.9 = 2^b$. Using the logarithmic function, we get $\ln .9 = \ln 2^b = b \ln 2$. Thus

$$b = \frac{\ln .9}{\ln 2} = -.152$$

We conclude that

$$y = 20x^{-.152}$$

**Do Exercise 25.**                                                        ■

## Profit Maximization Over Time

For our last illustration of the usefulness of the definite integral, we let $C(t)$, $R(t)$, and $P(t)$ dollars be the cost, revenue, and profit, respectively, for a firm that has operated for a period of $t$ weeks. We assume that each function is twice differentiable and that $C''(t) > 0$ and $R''(t) < 0$. This implies $C'(t)$ is increasing and $R'(t)$ is decreasing. This situation can occur in certain business operations, such as mining, where the rate of change of cost is increasing (because the extraction is done at greater depths) and the rate of increase in revenue is decreasing (because the amount of available ore decreases). In such situations, the operation will start to lose money after a certain amount of time. It is important to determine when to stop the operation in order to maximize profit. Note that

$$P(t) = R(t) - C(t)$$

Thus

$$P'(t) = R'(t) - C'(t)$$

Therefore to find the critical numbers, we set

$$P'(t) = R'(t) - C'(t) = 0$$

Suppose that $t_1$ is the solution of this equation. There is only one solution since $R'(t)$ is decreasing and $C'(t)$ is increasing as illustrated in Figure 10.19.

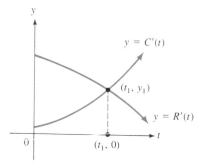

**FIGURE 10.19**

Now,

$$P''(t) = R''(t) - C''(t) < 0$$

This follows from the fact that $R''(t) < 0$ and $C''(t) > 0$. By the second derivative test, we see that $P$ has a local maximum at $t_1$. But $P''(t) < 0$ implies that the graph of the function $P$ is concave downward. Therefore, $P(t_1)$ is in fact the absolute maximum. The business should suspend its operation after $t_1$ weeks. To find the maximum profit, note that

$$R(t) = R(0) + \int_0^t R'(u) \, du$$

and

$$C(t) = C(0) + \int_0^t C'(u) \, du$$

Therefore

$$P(t) = R(t) - C(t) = R(0) - C(0) + \int_0^t [R'(u) - C'(u)] \, du$$

In general, $R(0) = 0$, but $C(0) = C_0$ is the fixed cost. Thus the maximum profit is given by

$$\text{maximum profit} = -C_0 + \int_0^{t_1} (R'(u) - C'(u)) \, du$$

We illustrate such a situation with the following example.

**EXAMPLE 7** The Childress Mining Company found that the instantaneous rates of change of revenue and cost are given by

$$R'(t) = 1500 - .00001t^3$$

and

$$C'(t) = 500 + .099t^2$$

respectively, where $R(t)$ and $C(t)$ are in thousands of dollars and $t$ is in months. The fixed cost is $250,000.

**a.** How many months should the company drill its mine to maximize total profit?
**b.** Calculate the maximum profit.

**Solution**  **a.** In order to find the number of months the company should drill its mine, we solve the equation

$$1500 - .00001t^3 = 500 + .099t^2$$

This equation is equivalent to

$$(t - 100)(.00001t^2 + .1t + 10) = 0$$

As expected, there is only one positive solution, namely 100. Therefore the company should drill its mine for 100 months.

**b.** The maximum profit is given by

$$-250 + \int_0^{100} [(1500 - .00001t^3) - (500 + .099t^2)] \, dt$$

where we wrote 250 for the fixed cost because $R(t)$ and $C(t)$ are in thousands of dollars. But,

$$-250 + \int_0^{100} [(1500 - .00001t^3) - (500 + .099t^2)] \, dt$$

$$= -250 + \int_0^{100} (-.00001t^3 - .099t^2 + 1000) \, dt$$

$$= -250 + [(-.0000025t^4 - .033t^3 + 1000t)|_0^{100}]$$

$$= -250 + [(-250 - 33,000 + 100,000) - 0] = 66,500$$

Thus the maximum profit will be $66,500,000.
**Do Exercise 29.**

---

## Exercise Set 10.4

**1.** The income distribution in a certain country is given by

$$y = I(x) = x^3$$

**a.** Sketch the corresponding Lorenz curve.
**b.** What portion of the total income is received by the lowest paid 20% of the population?
**c.** Calculate the coefficient of inequality.

**2.** The income distribution in a certain country is given by

$$y = I(x) = \frac{10x^3 + 3x}{13}$$

**a.** Sketch the corresponding Lorenz curve.
**b.** What portion of the total income is earned by the lowest paid 13% of the population?
**c.** Calculate the coefficient of inequality.

**3.** The income distribution in a country is given by

$$y = I(x) = \frac{2x^2 + 13x}{15}$$

**a.** Sketch the corresponding Lorenz curve.
**b.** What portion of the total income is received by the lowest paid 15% of the population?
**c.** Calculate the coefficient of inequality.

**4.** Suppose the income distribution in a country is given by

$$y = I(x) = \frac{2x^2 + 15x}{17}$$

**a.** Sketch the corresponding Lorenz curve.
**b.** What portion of the total income is earned by the lowest paid 17% of the population?
**c.** Calculate the coefficient of inequality.

**5.** The income distribution in a certain country is given by

$$y = I(x) = \frac{3x^2 + 32x}{35}$$

**a.** Sketch the corresponding Lorenz curve.
**b.** What portion of the total income is earned by the highest paid 65% of the population?
**c.** Calculate the coefficient of inequality.

**6.** The income distribution in some country is given by

$$y = I(x) = 1 - \sqrt{1 - x}$$

**a.** Sketch the corresponding Lorenz curve.
**b.** What portion of the total income is received by the highest paid 36% of the population?
**c.** Calculate the coefficient of inequality.

**7.** The income distribution in a certain country is given by

$$y = I(x) = \frac{x}{2 - x}$$

**a.** Sketch the corresponding Lorenz curve.
**b.** What portion of the total income is received by the highest paid 25% of the population?
**c.** Calculate the coefficient of inequality. $\left(Hint: \frac{x}{2 - x} = -1 + \frac{2}{2 - x}.\right)$

*In Exercises 8–14, the marginal propensity to consume, or to save, is given (in billions of dollars). The consumption when disposable income is $10 billion is also given (in billions of dollars). In each case **a.** find the consumption function and **b.** calculate consumption for the given disposable income.*

**8.** $\dfrac{dC}{dI} = .3 + \dfrac{.4}{\sqrt{I + 1}}$; $C = 7$ when $I = 10$, $I = 16$

9. $\frac{dC}{dI} = .4 + \frac{.2}{\sqrt{I+1}}$; $C = 8.32$ when $I = 10$, $I = 36$

10. $\frac{dC}{dI} = .4 + \frac{.2}{\sqrt[3]{I+1}}$; $C = 8.45$ when $I = 10$, $I = 27$

11. $\frac{dC}{dI} = .5 - .2e^{-0.2I}$; $C = 7.9$ when $I = 10$, $I = 20$

12. $\frac{dS}{dI} = .6 - \frac{.2}{\sqrt{I+1}}$; $C = 7.86$ when $I = 10$, $I = 64$

13. $\frac{dS}{dI} = .3 + .4e^{-0.4I}$; $C = 9$ when $I = 10$, $I = 20$

14. $\frac{dS}{dI} = .4 - \frac{\ln(I+1)}{10(I+1)}$; $C = 8.13$ when $I = 10$, $I = 25$

15. Using integration, approximate the present value of an annuity that will pay $2000 per month for the next 6 years if the nominal rate of interest is 10% compounded continuously.

16. Using integration, approximate the present value of an annuity that will pay $5000 every 6 months for the next 12 years if the nominal rate of interest is 8% compounded continuously.

17. Using integration, approximate the present value of an annuity that will pay $300 per week for the next 5 years if the nominal rate of interest is 9% compounded continuously. (Assume there are 52 weeks in a year.)

18. Mary's parents wish to establish an annuity that will provide $2500 every 3 months for the next 4 years to pay for her education. What is the present value of this annuity if the nominal rate of interest is 8% compounded continuously?

19. A business firm wishes to establish a chair at the local university that will provide $60,000 a year for the next 20 years. If the nominal rate of interest is 12% compounded continuously, approximate the present value of the annuity purchased for that purpose.

20. A successful alumnus wishes to establish a scholarship that will provide $3000 every 3 months for the next 10 years awarded to a deserving student. If the nominal rate of interest is 10% compounded continuously, approximate the present value of the annuity purchased for that purpose.

21. A successful businesswoman wishes to establish an annuity that will provide $2500 a month for the next 20 years to the local orphanage. What is the present value of this annuity if the nominal rate of interest is 11% compounded continuously?

22. An electronics firm knows that if $g(x)$ is the number of hours required by its employees to assemble the $(x + 1)$st radio, then

$$g(x) = 4x^{-.12}$$

provided that $20 \le x \le 10{,}000$. Estimate the amount of time it will take to assemble 1000 radios after the first 500 have been assembled.

23. The Udeen Electronics Company makes pocket calculators. It took 40 hours to assemble the first 20 calculators. After that, the number $g(x)$ of hours necessary to assemble the $(x + 1)$st unit (where $x$ is the number of units with 20 calculators per unit) is given by

$$g(x) = 40x^{-.15}$$

How many hours will be needed to assemble the 501st through the 1500th calculator.

24. The Gestrud Winter Sports Company found that it took its employees 100 hours to build the first 30 pairs of skis. After that, it was found that if $x$ is the number of units of 30 pairs of skis per unit, then the number $g(x)$ of hours required to build the $(x + 1)$st unit is given by

$$g(x) = 100x^{-.25}$$

Find the amount of time required to build an additional 3000 pairs of skis once the first 600 pairs have been built.

25. If we assume that the equation of a learning curve is of the form $y = ax^b$, where $y$ is the number of hours required to produce the $x$th unit, find the values of the constants $a$ and $b$ for a 90% learning curve if 10 hours are required to produce the second unit.

26. Same as Exercise 25 for an 80% learning curve knowing that 32 hours are required for the fourth unit.

27. Same as Exercise 25 for a 70% learning curve knowing that 34.3 hours are required for the eighth unit.

28. Same as Exercise 25 for a 60% learning curve knowing that 7.2 hours are required for the fourth unit.

29. The Wonderpool Lumber Company found that the rates of change of revenue and cost are given by $R'(t) = 1500 - .7t^2$ and $C'(t) = 204 + .3t^2$, respectively, where $R(t)$ and $C(t)$ are in thousands of dollars and $t$ is in months. The fixed cost is $150,000.

    a. How many months should the company continue to cut trees in order to maximize the total profit?
    b. Calculate the maximum total profit.

30. The Uhler Gold Mine Company has found that the rates of change of revenue and cost are given by $R'(t) = 2000 - .6t^2$ and $C'(t) = 400 + .4t^2$, respectively, where $R(t)$ and $C(t)$ are in thousands of dollars and $t$ is in months. The fixed cost is $120,000.

    a. How many months should the company continue to extract the gold ore in order to maximize its total profit?
    b. Calculate the maximum total profit.

31. The Udeen Oil Company has found that the rates of change of revenue and cost for its oil drilling operation are given by $R'(t) = 20{,}000 - .2t^2$ and $C'(t) = 8000 + 1t^2$, respectively, where $R(t)$ and $C(t)$ are in thousands of dollars and $t$ is in months. The fixed cost is $600,000.

   a. How many months should the company continue its drilling operation in order to maximize the total profit?
   b. Calculate the maximum total profit.

32. The Sherman Mining Company has established that the rates of change of revenue and cost for its mining operation are given by $R'(t) = 60 - 5\sqrt{t}$ and $C'(t) = 12 + 3\sqrt{t}$, respectively, where $R(t)$ and $C(t)$ are in millions of dollars and $t$ is in years. The fixed cost is $200,000.

   a. How many years should the company continue its operation in order to maximize its total profit?
   b. Calculate the maximum total profit.

33. The Rouquier Mining Company found that the rates of change of revenue and cost for its operation are given by $R'(t) = 100 - .5\sqrt[3]{t}$ and $C'(t) = 96 + .3\sqrt[3]{t}$, respectively, where $R(t)$ and $C(t)$ are in millions of dollars and $t$ is in months. The fixed cost is $5 million.

   a. How long should the company continue its operation in order to maximize its total profit?
   b. Calculate the maximum total profit.

*34. Prove that if $x_1, x_2, x_3, \ldots, x_n$ are real numbers such that $x_1 \le x_2 \le x_3 \le \ldots \le x_n$ and $a = (x_1 + x_2 + x_3 + \cdots + x_n)/n$ then $x_1 \le a \le x_n$.

*35. a. Prove that if $x_1, x_2, x_3, \ldots, x_n$ are real numbers such that $x_1 \le x_2 \le x_3 \le \ldots \le x_n$ and $k$ is an integer such that $1 \le k \le n$, then

$$\frac{x_1 + x_2 + x_3 + \cdots + x_k}{k}$$

$$\le \frac{x_1 + x_2 + x_3 + \cdots + x_n}{n}$$

(*Hint:* If $k < n$, first show that the inequality is equivalent to

$$\frac{x_1 + x_2 + x_3 + \cdots + x_k}{k}$$

$$\le \frac{x_{k+1} + x_{k+2} + \cdots + x_n}{n - k}$$

and use the result of Exercise 34.)

   b. If $y = I(x)$ describes income distribution in a country, prove that $I(x) \le x$ for all $x$. (*Hint:* Suppose there are $n$ wage earners in the country and their incomes are $I_1, I_2, I_3, \ldots, I_n$, where $I_1 \le I_2 \le I_3 \le \ldots \le I_n$. Then use the result of part **a** to show that for any integer $k$ with $1 \le k \le n$, we have

$$\frac{I_1 + I_2 + I_3 + \cdots + I_k}{I_1 + I_2 + I_3 + \cdots + I_n} \le \frac{k}{n}$$

Now note that the left and right sides of the inequality are $I(x)$ and $x$, respectively.)

# 10.5  Improper Integrals

The Fundamental Theorem of Calculus guarantees that the value of a definite integral $\int_a^b f(x)\, dx$ exists whenever $f$ is a *continuous* function defined on a *closed bounded* interval $[a, b]$. However, there are times when we wish to find values of "integrals" of functions defined over intervals although one of these conditions is not satisfied.

## Geometric Interpretation

We begin with three examples.

EXAMPLE 1    Find an upper bound for the area of the region $R$ bounded by the graph of $y = 2^{-x}$, the x-axis, and the graph of $x = 1$.

Solution    Note that the region $R$ is not bounded on the right, as illustrated in Figure 10.20.

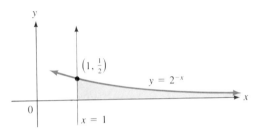

**FIGURE 10.20**

We consider the intervals $[1, 2]$, $[2, 3]$, $[3, 4]$, . . . and, using these as bases, we construct the rectangles with vertices $(i, 0)$, $(i + 1, 0)$, $(i + 1, 1/2^i)$, and $(i, 1/2^i)$ for each $i = 1, 2, 3, . . . .$ The union of these rectangles contains the region $R$. (See Figure 10.21.) Note that the area of the first rectangle is $\frac{1}{2}$ square unit, that of the second rectangle is $\frac{1}{4}$ square unit, and so on. Therefore if the area of the region $R$ is $A$ square units, we must have

$$A < \frac{1}{2} + \frac{1}{4} + \frac{1}{8} + \cdots$$

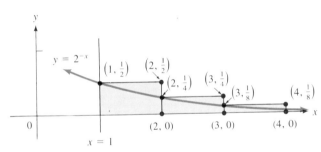

**FIGURE 10.21**

But the right side of this inequality is a geometric series with $a = \frac{1}{2}$ and $r = \frac{1}{2}$. (See Section 9.7.) So

$$\frac{1}{2} + \frac{1}{4} + \frac{1}{8} + \cdots = \frac{\dfrac{1}{2}}{1 - \dfrac{1}{2}} = 1$$

Therefore the area of the region $R$ is less than 1 square unit.  ∎

Observe that the area is finite even though the region is unbounded. This is not always the case as can be seen from the next example.

**EXAMPLE 2**  Let $R$ be the region bounded above by the graph of $y = \frac{1}{x}$, below by the $x$-axis, and to the left by the graph of $x = 1$. Show that $R$ does not have a finite area.

**Solution**  We construct rectangles over the intervals $[1, 2]$, $[2, 3]$, $[3, 4]$, . . . whose union is contained in the region $R$. That is, for each interval $[i, i + 1]$, we construct the rectangle with vertices $(i, 0)$, $(i + 1, 0)$, $\left(i, \frac{1}{i + 1}\right)$, and $\left(i + 1, \frac{1}{i + 1}\right)$. Note that the area of the first rectangle is $\frac{1}{2}$ square unit, that of the second rectangle is $\frac{1}{3}$ square unit, and so on. Thus, if $A$ square units is the area of the region $R$, we must have

$$\frac{1}{2} + \frac{1}{3} + \frac{1}{4} + \frac{1}{5} + \cdots < A$$

Again we have a sum with infinitely many terms. However, in this case, as we keep adding terms, the values of the partial sums will increase without bound. To see this, we use the following argument. We insert parentheses in the sum as follows

$$\frac{1}{2} + \left(\frac{1}{3} + \frac{1}{4}\right) + \left(\frac{1}{5} + \frac{1}{6} + \frac{1}{7} + \frac{1}{8}\right) + \left(\frac{1}{9} + \cdots + \frac{1}{16}\right) + \cdots$$

where the $n$th group of fractions inserted in parentheses has $2^n$ terms.
   Observe now that

$$\frac{1}{3} + \frac{1}{4} > \frac{1}{4} + \frac{1}{4} = \frac{1}{2},$$

$$\frac{1}{5} + \frac{1}{6} + \frac{1}{7} + \frac{1}{8} > \frac{1}{8} + \frac{1}{8} + \frac{1}{8} + \frac{1}{8} = \frac{1}{2}$$

and so on. Therefore the sum of each group of terms within a set of parentheses is larger than $\frac{1}{2}$. It follows that

$$\frac{1}{2} + \left(\frac{1}{3} + \frac{1}{4}\right) + \left(\frac{1}{5} + \frac{1}{6} + \frac{1}{7} + \frac{1}{8}\right) + \left(\frac{1}{9} + \cdots + \frac{1}{16}\right) + \cdots$$
$$> \frac{1}{2} + \frac{1}{2} + \frac{1}{2} + \cdots$$

It is clear that the values of the partial sums on the right will increase without bound as we keep adding $\frac{1}{2}$.
   Thus, the sum $\frac{1}{2} + \frac{1}{3} + \frac{1}{4} + \frac{1}{5} + \cdots$ is not finite, and we conclude that the region $R$ does not have a finite area. The region is sketched in Figure 10.22.

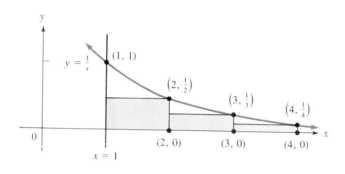

FIGURE 10.22

EXAMPLE 3    Find the area of the region $R$ bounded by the graphs of $y = 1/x^2$, $x = 1$, and the $x$-axis.

**Solution**    The region $R$ is unbounded on the right. Let $b$ be any real number larger than 1 and consider the region $R_b$ bounded by the graphs of $y = 1/x^2$, $x = 1$, $x = b$, and the $x$-axis. This region is illustrated in Figure 10.23. Then the area $A_b$ square units of that region is given by

$$A_b = \int_1^b x^{-2}\, dx = -x^{-1}\Big|_1^b = \frac{-1}{b} + 1$$

It is intuitively clear that the area of the unbounded region $R$ is approximated by $A_b$ when $b$ is large. In fact, the larger the value of $b$, the better the approximation. This leads us to consider

$$\lim_{b \to \infty} A_b = \lim_{b \to \infty} \left( \frac{-1}{b} + 1 \right) = 0 + 1 = 1$$

We conclude that the area of the unbounded region $R$ is 1 square unit.

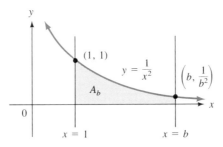

**FIGURE 10.23**

## Improper Integrals Over Unbounded Intervals

We now generalize the idea of the preceding example with the following definition.

> **DEFINITION 10.3:**    Suppose $f$ is a continuous function on the interval $[a, \infty)$. If the $\lim_{b \to \infty} \int_a^b f(x)\, dx$ exists, we say that the *improper integral* $\int_a^\infty f(x)\, dx$ is *convergent* and its value is that of the limit. Otherwise, we say that the improper integral is *divergent*.

**EXAMPLE 4**    Discuss the convergence of the following improper integrals.

**a.** $\displaystyle \int_2^\infty \frac{1}{x^3}\, dx$

**b.** $\displaystyle \int_1^\infty \frac{1}{x^{1/2}}\, dx$

**Solution**    **a.** If $b > 2$, we have

$$\int_2^b \frac{1}{x^3}\,dx = \int_2^b x^{-3}\,dx$$

$$= \frac{x^{-2}}{-2}\Big|_2^b = \frac{-1}{2b^2} + \frac{1}{2(2^2)} = \frac{1}{8} - \frac{1}{2b^2}$$

Thus

$$\lim_{b \to \infty} \int_2^b \frac{1}{x^3}\,dx = \lim_{b \to \infty}\left(\frac{1}{8} - \frac{1}{2b^2}\right)$$

$$= \frac{1}{8} - 0 = \frac{1}{8}$$

Therefore the integral $\int_2^\infty 1/x^3\,dx$ is convergent and its value is $\frac{1}{8}$.

**b.** If $b > 1$, we have

$$\int_1^b \frac{1}{x^{1/2}}\,dx = \int_1^b x^{-1/2}\,dx$$

$$= 2x^{1/2}\Big|_1^b = 2b^{1/2} - 2$$

Thus

$$\lim_{b \to \infty} \int_1^b \frac{1}{x^{1/2}}\,dx = \lim_{b \to \infty}(2b^{1/2} - 2) = \infty$$

Therefore the improper integral $\int_1^\infty 1/x^{1/2}\,dx$ diverges.
**Do Exercise 1.**

> **DEFINITION 10.4:**    If the function $f$ is continuous on the interval $[-\infty, a]$, we say that the improper integral $\int_{-\infty}^a f(x)\,dx$ is *convergent* provided that the $\lim_{b \to -\infty} \int_b^a f(x)\,dx$ exists. In that case, the value of the improper integral is the value of the limit. Otherwise, the improper integral is *divergent*.

Furthermore, if $f$ is continuous on the whole set of real numbers, then the improper integral $\int_{-\infty}^\infty f(x)\,dx$ is convergent if both improper integrals $\int_{-\infty}^a f(x)\,dx$ and $\int_a^{-\infty} f(x)\,dx$ converge for any real number $a$. In that case, the value of the first improper integral is equal to the sum of the values of the other two improper integrals. The particular choice of the number $a$ does not affect the value of the integral and often we choose 0 for the value of $a$.

**EXAMPLE 5**    Is the integral $\int_{-\infty}^\infty e^{-|x|}\,dx$ convergent or divergent? If it is convergent, find its value.

**Solution**    We consider the improper integrals $\int_{-\infty}^{0} e^{-|x|}\, dx$ and $\int_{0}^{\infty} e^{-|x|}\, dx$.
First, on the interval $[0, \infty)$, $|x| = x$. Thus

$$\int_{0}^{\infty} e^{-|x|}\, dx = \int_{0}^{\infty} e^{-x}\, dx$$

$$= \lim_{b \to \infty} \int_{0}^{b} e^{-x}\, dx = \lim_{b \to \infty} (-e^{-x}|_{0}^{b})$$

$$= \lim_{b \to \infty} \left( \frac{-1}{e^{b}} + e^{0} \right) = -0 + 1 = 1$$

On the interval $(-\infty, 0]$, $|x| = -x$. It follows that

$$\int_{-\infty}^{0} e^{-|x|}\, dx = \int_{-\infty}^{0} e^{-(-x)}\, dx = \int_{-\infty}^{0} e^{x}\, dx$$

$$= \lim_{b \to -\infty} \int_{b}^{0} e^{x}\, dx = \lim_{b \to -\infty} (e^{x}|_{b}^{0})$$

$$= \lim_{b \to -\infty} (1 - e^{b}) = 1 - 0 = 1$$

Therefore the integral $\int_{-\infty}^{\infty} e^{-|x|}\, dx$ is convergent and its value is 2 since the value of the other two improper integrals is 1.
**Do Exercise 17.**    ∎

## Improper Integrals Over Bounded Intervals

Suppose now we wish to evaluate an integral such as $\int_{0}^{1} 1/x^{1/2}\, dx$. The Fundamental Theorem of Calculus does not apply here since the integrand is not continuous at 0. However, if $0 < b < 1$, then the integrand is continuous on the interval $[b, 1]$ and

$$\int_{b}^{1} \frac{1}{x^{1/2}}\, dx = \int_{b}^{1} x^{-1/2}\, dx$$

$$= 2x^{1/2}|_{b}^{1} = (2 - 2b^{1/2})$$

Since we wish to evaluate $\int_{0}^{1} 1/x^{1/2}\, dx$, we consider values of $b$ that are near 0. In fact, we define

$$\int_{0}^{1} \frac{1}{x^{1/2}}\, dx = \lim_{b \to 0^{+}} \int_{b}^{1} \frac{1}{x^{1/2}}\, dx$$

$$= \lim_{b \to 0^{+}} (2 - 2b^{1/2}) = 2 - 0 = 2$$

Observe that it was necessary to evaluate this limit as $b$ approached 0 from the right since we assumed at the outset that $0 < b < 1$. We shall not formally define $\int_{a}^{b} f(x)\, dx$ where $f$ fails to be continuous at either $a$, $b$, or both. In each case, we proceed as we did in the preceding examples and as we illustrate in the next example.

**EXAMPLE 6**    Evaluate $\int_{0}^{1} \frac{1}{\sqrt{1 - x}}\, dx$.

**Solution**    The integrand is not continuous at 1. However, if we consider a number $a$ close to 1 such that $0 < a < 1$, then the integrand is continuous on $[0, a]$ and we can evaluate the integral over the interval $[0, a]$.

$$\int_0^a \frac{1}{\sqrt{1-x}} \, dx = \int_0^a (1-x)^{-1/2} \, dx$$

$$= -2(1-x)^{1/2}\big|_0^a = -2(1-a)^{1/2} + 2$$

Thus

$$\int_0^1 \frac{1}{\sqrt{1-x}} \, dx = \lim_{a \to 1^-} [-2(1-a)^{1/2} + 2] = 2$$

Therefore the improper integral is convergent and its value is 2.
**Do Exercise 21.**

## Present Value of a Perpetuity

In Example 4 of the preceding section we saw that the present value of an annuity that generates income over a finite amount of time can be approximated by a definite integral. Similarly, the present value of a perpetuity can be approximated by an improper integral.

**EXAMPLE 7**    An alumna wishes to make a gift to her alma mater from which it will draw $1500 per month in perpetuity to sustain a scholarship fund for business majors. Assuming that the annual rate of interest remains fixed at 10% compounded continuously, what is the present value of this endowment?

**Solution**    Since $1500 \cdot 12 = 18,000$, we first find the present value of an annuity that generates $18,000 a year for $n$ years. We use the definite integral

$$\int_0^n 18,000 e^{-.1t} \, dt$$

as was done in Example 4 of the preceding section. To find the present value of the perpetuity, we take the limit of this integral as $n \to \infty$. That is,

$$\text{present value of the perpetuity} = \lim_{n \to \infty} \int_0^n 18,000 e^{-.1t} \, dt$$

$$= \lim_{n \to \infty} (-180,000 e^{-.1t})\big|_0^n$$

$$= \lim_{n \to \infty} (-180,000 e^{-.1n} + 180,000) = 180,000$$

Therefore the present value of the annuity is $180,000.
**Do Exercise 45.**

## Exercise Set 10.5

*In Exercises 1–34, evaluate the given improper integral.*

**1.** $\int_1^\infty \dfrac{1}{\sqrt[3]{x^2}}\, dx$

**2.** $\int_1^\infty \dfrac{1}{x^3}\, dx$

**3.** $\int_1^\infty \dfrac{2}{\sqrt[4]{x^3}}\, dx$

**4.** $\int_2^\infty \dfrac{1}{x^4}\, dx$

**5.** $\int_4^\infty \dfrac{2}{x}\, dx$

**6.** $\int_2^\infty \dfrac{1}{\sqrt{x+2}}\, dx$

**7.** $\int_0^\infty e^{-3x}\, dx$

**8.** $\int_0^\infty xe^{-x^2}\, dx$

**9.** $\int_3^\infty \dfrac{1}{(x-2)^{3/2}}\, dx$

**10.** $\int_2^\infty \dfrac{1}{\sqrt[4]{x-1}}\, dx$

**11.** $\int_{-\infty}^{-2} \dfrac{1}{(x+1)^2}\, dx$

**12.** $\int_{-\infty}^{0} \dfrac{2}{(x-3)^4}\, dx$

**13.** $\int_{-\infty}^{-1} \dfrac{3}{\sqrt[3]{x^2}}\, dx$

**14.** $\int_{-\infty}^{0} \dfrac{4}{(x-1)^{8/3}}\, dx$

**15.** $\int_{-\infty}^{2} \dfrac{6}{\sqrt{3-x}}\, dx$

**16.** $\int_{-\infty}^{0} e^x\, dx$

**17.** $\int_{-\infty}^{\infty} e^{-|2x|}\, dx$

**18.** $\int_{-\infty}^{\infty} e^{-|3x|}\, dx$

**19.** $\int_{-\infty}^{\infty} e^{-|4x|}\, dx$

**20.** $\int_0^4 \dfrac{1}{\sqrt{4-x}}\, dx$

**21.** $\int_0^8 \dfrac{1}{\sqrt[3]{8-x}}\, dx$

**22.** $\int_{-1}^{2} \dfrac{2}{(x+1)^2}\, dx$

**23.** $\int_3^5 \dfrac{5}{(x-3)^3}\, dx$

**24.** $\int_{-16}^{0} \dfrac{5}{\sqrt[4]{x+16}}\, dx$

**25.** $\int_{-1}^{1} \dfrac{1}{x^3}\, dx$

**26.** $\int_{-8}^{8} \dfrac{1}{x^{2/3}}\, dx$

**27.** $\int_{-1}^{1} \dfrac{3}{x^{4/5}}\, dx$

**28.** $\int_{-4}^{1} \dfrac{2}{(x+3)^2}\, dx$

**29.** $\int_1^2 \dfrac{1}{x(\ln x)^{1/2}}\, dx$

**30.** $\int_1^3 \dfrac{1}{x(\ln x)^{1/3}}\, dx$

**31.** $\int_1^4 \dfrac{4}{x(\ln x)^2}\, dx$

**32.** $\int_{1/2}^{3/2} \dfrac{1}{x(\ln x)^{1/3}}\, dx$

**33.** $\int_2^\infty \dfrac{2}{x(\ln x)^2}\, dx$

**34.** $\int_3^\infty \dfrac{6}{x(\ln x)^3}\, dx$

*In Exercises 35–39, you may use the following fact. If $g(x)$ is a polynomial and $a$ is a positive number, then $\lim\limits_{x \to \infty} g(x)/e^{ax} = 0$. This is proved in more advanced texts.*

**35.** Sketch the region $R$ bounded by the graph of $y = xe^{-x}$ and the positive $x$-axis and find its area.

**36.** Sketch the region $R$ bounded by the graph of $y = xe^x$ and the negative $x$-axis and find its area.

**37.** Sketch the region $R$ bounded by the graph of $y = 3x^2e^{-x}$ and the positive $x$-axis and find its area.

**38.** Sketch the region $R$ bounded by the graph of $y = 5x^3e^x$ and the negative $x$-axis and find its area.

**39.** Sketch the region $R$ bounded by the graph of $y = 2xe^{-3x}$ and the positive $x$-axis and find its area.

**40.** Make a sketch of the region $R$ bounded by the graphs of $y = 1/\sqrt{x}$ and $x = 4$, in the first quadrant, and find its area.

**41.** Make a sketch of the region $R$ bounded by the graphs of $y = 8/\sqrt[3]{x}$ and $x = 8$, in the first quadrant, and find its area.

**42.** If $\int_1^\infty c/e^{3x}\, dx = 1$, what is the value of the constant $c$?

**43.** If $\int_4^\infty c/x^{3/2}\, dx = 1$, what is the value of the constant $c$?

**44.** A donor wishes to make a gift to a school from which the school will draw $2000 per month in perpetuity to sustain a scholarship fund for economics majors. Assuming that the annual rate of interest remains fixed at 9% compounded continuously, how much should the donor give the school?

**45.** An alumnus wishes to make a donation to his alma mater from which it will draw $3000 per month in perpetuity to sustain a scholarship fund for mathematics majors. Assuming that the annual rate of interest remains fixed at 12% compounded continuously, what is the present value of this endowment?

# 10.6  More on Probability (Optional)

### Review of Basic Definitions

We briefly review some definitions from probability. A *random variable* is a function whose domain is a sample space and whose range is a set of real numbers. If the range is finite, or if it can be put in a one-to-one correspondence with the set of positive integers, then $X$ is said to be a *discrete random variable*. If the range is an interval, the random variable is said to be *continuous*. If $S$ is a sample space with probability function $P$, and $X$ is a discrete random variable with domain $S$, the *probability function* (or *probability distribution*) of $X$ is the function $f$ whose domain is the range of $X$ and is defined by

$$f(x) = P(X = x)$$

for each possible value of $x$ in the range of $X$. Note that

$$P(X = x) = P(E)$$

where

$$E = \{s \mid s \in S \text{ and } X(s) = x\}$$

Suppose $S$ is a finite sample space and $X$ is a discrete random variable with probability distribution $f$ defined on the range $\{x_1, x_2, x_3, \ldots, x_n\}$ of $X$. *The mathematical expectation of $X$ is given by*

$$E(x) = x_1 f(x_1) + x_2 f(x_2) + x_3 f(x_3) + \cdots + x_n f(x_n)$$

## Probability Density Functions

When $X$ is a continuous random variable, we must use a different approach to compute probabilities. A *probability density function* (abbreviated pdf) may be used to define a probability distribution where $X$ is a continuous random variable. This function is nonnegative for all real numbers $x$, and the area between its graph and the $x$-axis is 1. (See Figure 10.24.)

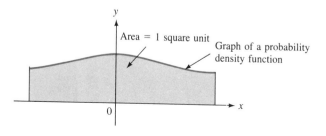

**FIGURE 10.24**

**DEFINITION 10.5:**   If $f$ is a pdf of a continuous random variable $X$ and if $[c, d]$ is a subset of the range of $X$, we define

$$P(X \text{ takes on values in } [c, d]) = \int_c^d f(x) \, dx$$

The determination of the probability density function for a continuous random variable involves some techniques beyond the scope of this book. However, in the remainder of this section, we shall illustrate the use of probability density functions without discussing how these functions are obtained.

## Uniform Density Functions

In probability, it is often the case that the outcomes in a finite sample space are equally likely. An extension of this concept for a continuous random variable is a density function which is uniformly distributed. In this case, if two intervals are subsets of the range of the random variable and have equal lengths, then the probability that the random variable is in the first interval is equal to the probability that it is in the second interval.

**EXAMPLE 1**   Passengers at the Seattle-Tacoma airport must take a train to go to certain terminals. Departures are scheduled at 3-minute intervals. If a randomly selected passenger arrives between departures, what is the probability that he/she will have to wait at least 45 seconds?

**Solution**   Let $x$ seconds be the time the passenger must wait. We wish to find $P(45 \le x \le 180)$ assuming that $x$ is uniformly distributed over the interval $[0, 180]$. In this case, the pdf is constant. Since we want the area between the graph of the pdf and the interval $[0, 180]$ to be 1, we define

$$f(x) = \begin{cases} \dfrac{1}{180} & \text{if } 0 \le x \le 180 \\ 0 & \text{elsewhere} \end{cases}$$

Then

$$P(45 \le x \le 180) = \int_{45}^{180} \frac{1}{180}\, dx = \frac{x}{180}\bigg|_{45}^{180}$$

$$= \frac{180}{180} - \frac{45}{180} = \frac{3}{4}$$

Thus the probability that the passenger will have to wait at least 45 seconds is $\frac{3}{4}$.
**Do Exercise 13.**   ■

## Expected Value of a Continuous Random Variable

We now generalize the concept of expectation defined for finite sample spaces. Suppose that a continuous random variable has a pdf which is positive only on a bounded interval $[a, b]$. Consider a regular partition of $[a, b]$, say $[x_0, x_1, x_2, \ldots, x_n]$, and define a new random variable $X^*$ as follows:

$$X^*(\alpha) = x_{i-1} \text{ whenever } X(\alpha) \text{ is in the interval } [x_{i-1}, x_i].$$

Then $X^*$ is a discrete random variable since its range has only $n$ elements. Furthermore, $X^*$ approximates $X$ when the value of $n$ is large. The probability that $X^*$ takes the value $x_{i-1}$ is equal to the probability that $X$ takes on a value in the interval $[x_{i-1}, x_i]$, which is $\int_{x_{i-1}}^{x_i} f(x)\, dx$. When $n$ is large, the length $\Delta x_i$ of each subinterval of the partition is small and the value of the definite integral is approximately $f(x_{i-1})\Delta x_i$.

Thus

$$P(X^* = x_{i-1}) \doteq f(x_{i-1})\Delta x_i$$

(See Figure 10.25.)

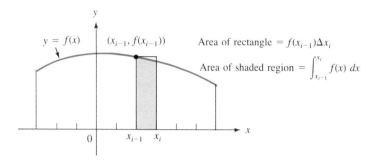

**FIGURE 10.25**

The expected value of the discrete random variable $X^*$ is given by

$$E(X^*) = \sum_{i=1}^{n} x_{i-1}P(X^* = x_{i-1}) \doteq \sum_{i=1}^{n} x_{i-1}f(x_{i-1})\Delta x_i$$

The summation on the right side of the equalities is a Riemann sum for the function $xf(x)$. Taking the limit as $n \to \infty$ and making use of the fact that $X^*$ approximates $X$ when $n$ is large, we give the following definition.

> **DEFINITION 10.6:** The *expected value* of the continuous random variable $X$ is
>
> $$E(X) = \int_a^b xf(x)\,dx$$

**EXAMPLE 2**   The Milo Electric Company is manufacturing pocket calculators. The quality control division randomly selects samples and uses these samples continuously for a maximum of 125 hours. If a calculator breaks down within that time, its lifetime $t$ hours is recorded. After testing many calculators, it is found that the probability density function for the lifetime $t$ is given by

$$f(t) = \begin{cases} \dfrac{4}{1625}(7 - \sqrt[3]{t}) & \text{if } 0 \le t \le 125 \\ 0 & \text{elsewhere} \end{cases}$$

   **a.** Verify that $f$ is indeed a probability density function.
   **b.** Find the expected lifetime of a calculator.

**Solution**   **a.** Since $f(t) = \frac{4}{1625}(7 - \sqrt[3]{t})$ when $0 < t < 125$, $f'(t) = -4t^{-2/3}/4875$, and $f'(t) < 0$ on the interval $(0, 125)$, Therefore, $f$ is decreasing on that interval. But

$f(125) = \frac{4}{1625}(7 - \sqrt[3]{125}) = \frac{8}{1625}$. Since $f(125) > 0$, and $f$ is decreasing on the interval $[0, 125]$, $f(t) > 0$ for all $t$ in the interval $[0, 125]$. Also,

$$\int_0^{125} \frac{4}{1625}(7 - t^{1/3})\, dt = \frac{4}{1625}\left(7t - \frac{3t^{4/3}}{4}\right)\Big|_0^{125}$$

$$= \frac{4}{1625}\left[7(125) - \frac{3(125^{4/3})}{4}\right] - 0 = 1$$

Therefore $f$ is a probability density function.

**b.** Let $X$ be a continuous random variable defined as follows. For any calculator $c$,

$$X(c) = \min\{\text{lifetime of } c,\ 125\}$$

Then $X$ has values in the interval $[0, 125]$ and its pdf is defined by

$$f(t) = \begin{cases} \dfrac{4}{1625}(7 - t^{1/3}) & \text{if } 0 \le t \le 125 \\ 0 & \text{elsewhere} \end{cases}$$

Thus the expected lifetime of a calculator is

$$T = \int_0^{125} t\left[\frac{4}{1625}(7 - t^{1/3})\right] dt = \int_0^{125} \frac{4}{1625}(7t - t^{4/3})\, dt$$

$$= \frac{4}{1625}\left(\frac{7t^2}{2} - \frac{3t^{7/3}}{7}\right)\Big|_0^{125} = \frac{4}{1625}\left[\frac{7(125^2)}{2} - \frac{3(125^{7/3})}{7}\right] - 0$$

$$= 52.1978$$

We conclude that the expected lifetime of a calculator is approximately 52.2 hours.
**Do Exercise 11.**   ■

## Expected Average Profit

**EXAMPLE 3**   The Milo Electronics Company offers the following warranty to its customers. It will give full, one-half, and one-third refunds if a calculator breaks down within 4, 12, and 36 weeks, respectively, from the date of purchase. Each calculator sells for $36 and its total cost to the company is $22. Assuming that 1 hour of continuous testing is equivalent to 1 week of normal use by a customer, calculate the expected average profit per calculator.

**Solution**   If $X$ is defined as in Example 2, we define a new random variable $Y$ as follows.
$Y(c) = -22$ if $0 \le X(c) < 4$, because in this case, the company gives a full refund and it loses its total cost since it does not get any revenue for this calculator.
$Y(c) = -4$ if $4 \le X(c) < 12$, because in this case, the company gives a half-refund and $36 - 22 - \frac{36}{2} = -4$.
Similarly, $Y(c) = 2$ if $12 \le X(c) < 36$, because $36 - 22 - \frac{36}{3} = 2$, and
$Y(c) = 14$ if $36 \le X(c)$, since in this case there is no refund and $36 - 22 = 14$.

The probability that a calculator breaks down within the first 4 weeks is

$$\int_0^4 \frac{4}{1625}(7 - t^{1/3})\, dt = \frac{4}{1625}\left(7t - \frac{3t^{4/3}}{4}\right)\Bigg|_0^4$$

$$= \frac{4}{1625}\left[7(4) - \frac{3(4^{4/3})}{4}\right] - 0 \doteq .057$$

The probability that a calculator breaks down between the end of the 4th week and the beginning of the 13th week after purchase is

$$\int_4^{12} \frac{4}{1625}(7 - t^{1/3})\, dt = \frac{4}{1625}\left(7t - \frac{3t^{4/3}}{4}\right)\Bigg|_4^{12}$$

$$= \frac{4}{1625}\left[7(12) - \frac{3(12^{4/3})}{4}\right] - \frac{4}{1625}\left[7(4) - \frac{3(4^{4/3})}{4}\right]$$

$$\doteq .099$$

The probability that a calculator breaks down between the end of the 12th week and the beginning of the 37th week is

$$\int_{12}^{36} \frac{4}{1625}(7 - t^{1/3})\, dt = \frac{4}{1625}\left(7t - \frac{3t^{4/3}}{4}\right)\Bigg|_{12}^{36}$$

$$= \frac{4}{1625}\left[7(36) - \frac{3(36^{4/3})}{4}\right] - \frac{4}{1625}\left[7(12) - \frac{3(12^{4/3})}{4}\right]$$

$$\doteq .245$$

Finally, the probability that a calculator does not break down in the first 36 weeks following the purchase is

$$\int_{36}^{125} \frac{4}{1625}(7 - t^{1/3})\, dt = \frac{4}{1625}\left(7t - \frac{3t^{4/3}}{4}\right)\Bigg|_{36}^{125}$$

$$= \frac{4}{1625}\left[7(125) - \frac{3(125^{4/3})}{4}\right] - \frac{4}{1625}\left[7(36) - \frac{3(36^{4/3})}{4}\right]$$

$$\doteq .599$$

Thus the probability distribution function defined on the range of $Y$ is given by $g(-22) = .057$, $g(-4) = .099$, $g(2) = .245$, and $g(14) = .599$. The average expected profit per calculator is $(-22)(.057) + (-4)(.099) + (2)(.245) + (14)(.599) = 7.226$.

We conclude that if all customers who bought calculators from the Milo Electronics Company take advantage of the warranty, the average expected profit per calculator is approximately $7.23.
**Do Exercise 31.**                                                               ■

## Another Example

Our last example involves a continuous random variable, the range of which is an unbounded interval.

**EXAMPLE 4**   Suppose $T$ is a continuous random variable that represents the lifetime, in years, of a particular brand of freezer and that $T$ has values in the interval $[0, \infty)$. Its probability density function for the lifetime $t$ is given by

$$f(t) = \begin{cases} \dfrac{1}{15} e^{-t/15} & \text{if } t \in [0, \infty) \\ 0 & \text{elsewhere} \end{cases}$$

a. Verify that $f$ is indeed a pdf.
b. What is the expected lifetime of a freezer?
c. What is the probability that a freezer will last 20 years or more?

**Solution**   a. Clearly, $\frac{1}{15} e^{-t/15} > 0$. Thus, $f(t) \geq 0$ for all $t$. Furthermore,

$$\int_0^\infty \frac{1}{15} e^{-t/15}\, dt = \lim_{u \to \infty} \int_0^u \frac{1}{15} e^{-t/15}\, dt$$

$$= \lim_{u \to \infty} \left( -e^{-t/15}\, \big|_0^u \right) = \lim_{u \to \infty} \left( -\frac{1}{e^{u/15}} + 1 \right) = 1$$

Therefore $f$ is a pdf.

b. The expected lifetime of a freezer is

$$E(T) = \int_0^\infty t \left( \frac{1}{15} e^{-t/15} \right) dt = \lim_{u \to \infty} \int_0^u \frac{t e^{-t/15}}{15}\, dt$$

To evaluate this integral, we can use integration by parts, or we can proceed as follows. We first let

$$\int \frac{t e^{-t/15}}{15}\, dt = (At + B)e^{-t/15} + C$$

where $A$ and $B$ are constants to be determined. We must have

$$D_t[(At + B)e^{-t/15} + C] = \frac{t e^{-t/15}}{15}$$

That is,

$$\left[ \frac{-At}{15} + \left( A - \frac{B}{15} \right) \right] e^{-t/15} = \frac{t e^{-t/15}}{15}$$

This yields

$$\begin{cases} \dfrac{-A}{15} = \dfrac{1}{15} \\ A - \dfrac{B}{15} = 0 \end{cases}$$

which gives $A = -1$ and $B = -15$. Therefore,

$$\int \frac{t e^{-t/15}}{15}\, dt = (-t - 15)e^{-t/15} + C$$

We conclude that

$$E(T) = \lim_{u \to \infty} \left( \frac{-t - 15}{e^{t/15}} \bigg|_0^u \right) = \lim_{u \to \infty} \left( \frac{-u - 15}{e^{u/15}} - \frac{-15}{e^0} \right) = 15 \,^*$$

The expected lifetime of a freezer is 15 years.

c. The probability that a freezer will last 20 years or more is

$$\int_{20}^{\infty} \frac{1}{15} e^{-t/15} \, dt = \lim_{u \to \infty} \int_{20}^{u} \frac{1}{15} e^{-t/15} \, dt$$

$$= \lim_{u \to \infty} (-e^{-t/15} \big|_{20}^{u}) = \lim_{u \to \infty} \left( \frac{-1}{e^{u/15}} + e^{-20/15} \right) \doteq .26$$

Therefore the probability that a freezer selected at random will last 20 years or more is approximately 0.26.

**Do Exercise 27.**

---

# Exercise Set 10.6

*In Exercises 1–10, a function is defined by a formula on a given interval and has value 0 outside that interval. **a.** Find the value of the constant c that will make the given function a probability density function of some continuous random variable X. **b.** Replace c by the value found in part **a** and calculate the expected value E(X) of the continuous random variable.*

1. $f(x) = c(4 - x)(x + 1)$ on $[1, 4]$

2. $f(x) = c(-x^2 + 7x - 10)$ on $[3, 5]$

3. $f(x) = c(-x^2 + 4x + 5)$ on $[0, 4]$

4. $f(x) = c(-x^2 + 9)$ on $[-2, 3]$

5. $f(x) = \dfrac{c}{x + 1}$ on $[0, 5]$

6. $f(x) = \dfrac{c}{x + 2}$ on $[0, 6]$

7. $f(x) = 2e^{-cx}$ on $[0, \infty)$

8. $f(x) = ce^{-5x}$ on $[0, \infty)$

9. $f(x) = ce^{-7x}$ on $[0, \infty)$

10. $f(x) = \dfrac{c}{x(x + 3)}$ on $[1, 5]$

$\left( \text{Hint: } \dfrac{3}{x(x + 3)} = \dfrac{1}{x} - \dfrac{1}{x + 3}. \right)$

11. The Triplenight Racquet Sports Company is manufacturing a new racquet ball. Its quality control division selects samples randomly and simulates continuous playing conditions with the samples for a maximum of 36 hours. If a ball breaks within that time, its lifetime $t$ is recorded. After testing many balls, it is found that the probability density function for the lifetime $t$ is given by

$$f(t) = \begin{cases} \dfrac{1}{144}(8 - \sqrt{t}) & \text{if } 0 \le t \le 36 \\ 0 & \text{elsewhere} \end{cases}$$

a. Verify that $f$ is a pdf.
b. Find the expected lifetime of a ball.

12. Same as Exercise 11 but for a different brand of ball for which the pdf for the lifetime $t$ is given by

$$f(t) = \begin{cases} \dfrac{1}{180}(13 - 2\sqrt{t}) & \text{if } 0 \le t \le 36 \\ 0 & \text{elsewhere} \end{cases}$$

13. A traffic light changes every 90 seconds. If a motorist arrives at the intersection while the light is red, what is the probability that she will have to wait at least 30 seconds for the light to turn green?

---

*We used the fact that in general, if $g(y)$ is a polynomial and $a$ is a positive constant, $\lim_{y \to \infty} g(y)/e^{ay} = 0$. This is due to the fact that $e^{ay}$ tends to infinity much faster than does $g(y)$. This can be proved formally using a technique called l'Hôpital's Rule which is discussed in more advanced calculus courses.

14. A shuttle bus leaves the hotel for downtown Reno every 30 minutes. If a visitor arrives at the bus stop and no bus is there, what is the probability that he will have to wait at least 12 minutes for the next bus?

15. A 2 hour 15 minute movie is being run continuously at a certain theater. Two spectators who have not checked the schedule arrive at the theater. What is the probability that their arrival is within 15 minutes of the start of a showing?

16. The Baldman Company is manufacturing a new hair dryer. The quality control division of the company selects hair dryers randomly and uses these samples continuously for a maximum of 225 hours. If a hair dryer breaks down within that time, its lifetime $t$ is recorded. After testing many hair dryers, it is found that the pdf for the lifetime $t$ is given by

$$f(t) = \begin{cases} \dfrac{1}{4500}(30 - t^{1/2}) & \text{on } [0, 225] \\ 0 & \text{elsewhere} \end{cases}$$

   a. Verify that $f$ is a pdf.
   b. Find the expected lifetime of a hair dryer.

17. The Sherman Electrical Company is manufacturing a new light bulb. The quality control division has found that by testing the light bulbs at high temperature, 1 minute of testing is equivalent to 1 hour of normal use by a consumer. Light bulbs are selected at random and the samples are tested for a maximum of 8000 minutes. If a light bulb burns out during that time, its lifetime $t$ is recorded. After many light bulbs are tested, it is found that the pdf for the lifetime $t$ is given by

$$f(t) = \begin{cases} \dfrac{1}{48,000}(21 - t^{1/3}) & \text{if } 0 \le t \le 8000 \\ 0 & \text{elsewhere} \end{cases}$$

   a. Verify that $f$ is a pdf.
   b. Find the expected lifetime of a light bulb.

18. Same as Exercise 17, but with pdf for the lifetime $t$ given by

$$f(t) = \begin{cases} \dfrac{.0002}{1 - e^{-1.6}}e^{-.0002t} & \text{if } 0 \le t \le 8000 \\ 0 & \text{elsewhere} \end{cases}$$

19. The quality control division of the Rouquier Electronics Company is testing a new picture tube. It has found that by testing the tubes at a certain high temperature, 1 second of testing is equivalent to 1 hour of normal use. Many samples are selected randomly and each is tested for a maximum of 21,952 seconds. If a tube burns out during that time, its lifetime $t$ is recorded. After testing many tubes, it is found that the pdf for the lifetime $t$ is given by

$$f(t) = \begin{cases} \dfrac{1}{219,520}(31 - t^{1/3}) & \text{if } 0 \le t \le 21,952 \\ 0 & \text{elsewhere} \end{cases}$$

   a. Verify that $f$ is a pdf.
   b. Find the expected lifetime of a picture tube.

20. Same as Exercise 19, but with the pdf for the lifetime $t$ given by

$$f(t) = \begin{cases} \dfrac{1}{4,390,400}(67 - t^{1/3}) & \text{if } 0 \le t \le 175,616 \\ 0 & \text{elsewhere} \end{cases}$$

21. A small-appliances manufacturer wishes to compute the expected lifetime of its new electric mixer. Its quality control division selects mixers randomly and runs each mixer continuously for a maximum of 900 hours. If a mixer breaks down during that time, its lifetime $t$ hours is recorded. After testing many mixers, it is found that the pdf for the lifetime $t$ is given by

$$f(t) = \begin{cases} \dfrac{1}{180,000}(220 - t^{1/2}) & \text{on } [0, 900] \\ 0, & \text{elsewhere} \end{cases}$$

   a. Verify that $f$ is a pdf.
   b. Find the expected lifetime of an electric mixer.

22. Same as Exercise 21, but with the pdf for the lifetime $t$ given by

$$f(t) = \begin{cases} \dfrac{1}{103,680}(104 - t^{1/2}) & \text{if } 0 \le t \le 1,296 \\ 0 & \text{elsewhere} \end{cases}$$

23. The Mason Electronics Company is manufacturing a new integrated circuit. It has found that by testing the circuits at a certain high temperature, 1 second of testing is equivalent to 1 hour of continuous use under normal conditions. Sample circuits are selected randomly and each circuit is tested for a maximum of 8100 seconds. If a circuit burns out during that time, its lifetime $t$ is recorded. After testing many circuits, the pdf for the lifetime $t$ is found to be

$$f(t) = \begin{cases} \dfrac{1}{2,430,000}(360 - t^{1/2}) & \text{if } 0 \le t \le 8100 \\ 0 & \text{elsewhere} \end{cases}$$

   a. Verify that $f$ is a pdf.
   b. Find the expected lifetime of an integrated circuit.

**24.** Same as Exercise 23, but with the pdf for the lifetime $t$ given by

$$f(t) = \begin{cases} \dfrac{1}{25,725}(19.875 - t^{1/3}) & \text{if } 0 \le t \le 2744 \\ 0 & \text{elsewhere} \end{cases}$$

**25.** A manufacturer wishes to calculate the expected lifetime of its new electric razor. Its quality control division selects razors randomly and simulates continuous use of each razor for a maximum of 343 hours. If a razor burns out during that time, its lifetime $t$ is recorded. After testing many razors, the pdf for the lifetime $t$ is found to be

$$f(t) = \begin{cases} \dfrac{4}{5145}(9 - t^{1/3}) & \text{on } [0, 343] \\ 0 & \text{elsewhere} \end{cases}$$

**a.** Verify that $f$ is a pdf.
**b.** Find the expected lifetime of a razor.

**26.** Same as Exercise 25, but with a pdf for the lifetime $t$ given by

$$f(t) = \begin{cases} \dfrac{28}{300,000}\left(\dfrac{255}{14} - t^{1/3}\right) & \text{on } [0, 1000] \\ 0 & \text{elsewhere} \end{cases}$$

**27.** Suppose $T$ is a continuous random variable that represents the lifetime, in months, of a certain brand of car battery and $T$ has values in the interval $[0, \infty)$. Its pdf for the lifetime $t$ is given by

$$f(t) = \begin{cases} \dfrac{1}{120}e^{-t/120} & \text{if } t \text{ is in } [0, \infty) \\ 0 & \text{elsewhere} \end{cases}$$

**a.** Verify that $f$ is a pdf.
**b.** What is the expected life of a battery?
**c.** If the manufacturer gives a 2-year full warranty, what is the probability that the battery will die within 6 months after the warranty expires?

**28.** Same as Exercise 27, but with a pdf for the lifetime $t$ given by

$$f(t) = \begin{cases} \dfrac{1}{96}e^{-t/96} & \text{if } t \text{ is in } [0, \infty) \\ 0 & \text{elsewhere} \end{cases}$$

**29.** Suppose $T$ is a continuous random variable that represents the lifetime, in hours, of a certain brand of integrated circuit and $T$ has values in the interval $[0, \infty)$. The pdf of the random variable $T$ is given by

$$f(t) = \begin{cases} \dfrac{1}{6000}e^{-t/6000} & \text{if } t \text{ is in } [0, \infty) \\ 0 & \text{elsewhere} \end{cases}$$

**a.** Verify that $f$ is a pdf.
**b.** What is the expected lifetime of an integrated circuit?
**c.** If, on the average, integrated circuits are used 2 hours a day and if the manufacturer gives a 2-year full warranty, what is the probability that a circuit will burn out within a year after the warranty expires?

**30.** The Triplenight Racquet Sports Company of Exercise 11 offers the following warranty. A full refund is given if the ball breaks before the company's name, which is painted on each ball, wears off. It takes 4 hours of normal use for the name to wear off. If each ball sells for $3.70 and if the total cost per ball to the company is $2.00, what is the expected profit per ball, assuming all racquetball players take advantage of the warranty?

**31.** The Baldman Company of Exercise 16 estimates that the average customer uses a hair dryer 4 minutes a day. The company offers the following warranty. A full refund is given if the dryer burns out within 60 days of the date of purchase and a half-refund if it burns out between 60 and 240 days from the date of purchase. Each hair dryer sells for $23 and the total cost to the company per hair dryer is $14. What is the expected average profit per hair dryer assuming all customers take advantage of the warranty?

**32.** The small-appliances company of Exercise 21 estimates that customers use their electric mixers an average of 6 minutes per day. The company offers the following warranty. A full refund is given if a mixer breaks down within 90 days of the date of purchase and a half-refund if it breaks down between 90 and 160 days from the date of purchase. If each mixer sells for $26 and the total cost to the company is $14 per mixer, calculate the expected average profit assuming all customers take advantage of the warranty.

**33.** The Mason Electronics Company of Exercise 23 estimates that its integrated circuits are used 90 minutes a day on the average. The company offers the following warranty. A full refund is given if a circuit burns out within the first 180 days from the date of purchase and a half-refund is given if it burns out between the 180th and the 360th day from the date of purchase. Each circuit sells for $8.50 and the total cost to the company is $4.50 per circuit. Calculate the average expected profit per circuit assuming that all customers take advantage of the warranty.

**34.** The manufacturer of Exercise 25 estimates that its razor is used 5 minutes a day on the average. It offers the following warranty. A full refund if the razor breaks down in the first 90 days from the date of purchase, a half-refund if it breaks down in the following 90 days, and a one-third refund if it breaks down in the next 90 days. The razor sells for $35 and the total cost to the company is $20 per razor. Assuming all customers take advantage of the warranty, what is the average profit the company can expect per razor?

# 10.7 Chapter Review

SUMMARY    A statement of equality involving at least one derivative of some unknown function is called a differential equation. If the highest order derivative which appears in the equation is of order $n$, the equation itself is of order $n$. A function $f$ is said to be a solution of a differential equation if the equation becomes an identity when the function and all its derivatives are replaced by $f$ and its corresponding derivatives in the equation. If $f$ and $g$ are continuous functions of $x$ and $y$, respectively, the equation $\frac{dy}{dx} = \frac{f(x)}{g(y)}$ is said to be a separable equation. To solve such an equation we proceed as follows:

**Step 1:** Write the equation in the form

$g(y)\ dy = f(x)\ dx$

**Step 2:** Integrate both sides

$$\int g(y)\ dy = \int f(x)\ dx$$

**Step 3:** The result in Step 3 is an equation of the form

$G(y) = F(x) + C$

where $C$ is an arbitrary constant.

If possible, solve for $y$ in terms of $x$. If enough information is known to solve for $C$, do so and get a particular solution.

If $f$ and $g$ are continuous functions defined on an interval $[a, b]$ and if $f(x) \geq g(x)$ for all $x$ in $[a, b]$, the area of the region bounded by the graphs of $y = f(x)$, $y = g(x)$, $x = a$, and $x = b$ is $\int_{a}^{b} [f(x) - g(x)]\ dx$ square units.

If the condition $f(x) \geq g(x)$ for all $x$ in $[a, b]$ is not satisfied, we divide the given region into a finite number of subregions so that the area of each subregion can be found. Then we add the areas of these subregions. If the demand equation and supply equation are $p = D(q)$ and $p = S(q)$, respectively, the solution $(q_0, p_0)$ of the system

$$\begin{cases} p = D(q) \\ p = S(q) \end{cases}$$

is called the equilibrium point. Also $q_0$ and $p_0$ are called the equilibrium quantity and equilibrium price, respectively. The consumers' surplus is represented by the area bounded by the graphs of $p = D(q)$, $p = p_0$, and $q = 0$. The producers' surplus is represented by the area bounded by the graphs of $p = S(q)$, $p = p_0$, and $q = 0$.

If a supplier has a monopoly on a certain product, the price per unit will be that which maximizes the profit for the supplier. If the cost and demand functions are known, we find the value of $q$, say $q_0$, that maximizes the profit. Using the demand equation, we find the price $p_0$ at which the quantity demanded is $q_0$. The consumers' surplus will then be represented by the area of the region bounded by the graphs of $p = D(q)$, $p = p_0$, and $q = 0$.

To study the income distribution of a nation, we define a function $I$ by $y = I(x)$ where $x$ is the percentage of lower paid individuals and $y$ is the percentage of the total income received by these individuals. We have $I(0) = 0$, $I(1) = 1$, and $I(x) \le x$ for all $x$. The graph of the function $I$ is called a Lorenz curve. The coefficient of inequality (Gini coefficient) is $2 \int_0^1 [x - I(x)]\, dx$. Suppose that $C = f(I)$ where $I$ is the total disposable national income (in billions of dollars) and $C$ is the national consumption (also in billions of dollars). The derivative $f'(I)$ is called the marginal propensity to consume. If $S$ represents the national savings (in billions of dollars), we have $C + S = I$ so that

$$\frac{dC}{dI} + \frac{dS}{dI} = 1$$

The derivative $\frac{dS}{dI}$ is called the marginal propensity to save. If an annuity pays $B$ dollars per year distributed in $k$ equal payments each year, and if the rate of interest is $r$ (written as a decimal) compounded continuously and the payments are to be made for $m$ years, the present value of the annuity is approximately $\int_0^m Be^{-rt}\, dt$ dollars. Under the same conditions, the present value of a perpetuity is $\int_0^\infty Be^{-rt}\, dt$ dollars.

Suppose $y = f(x)$ where $y$ is the number of hours required to produce $x$ units of a certain commodity, and let $g(x) = f'(x)$. Then $f(n) - f(m)$ is the number of hours required to increase production from a level of $m$ units to a level of $n$ units. Since $f$ is an antiderivative of $g$, by the Fundamental Theorem of Calculus, we have

$$\int_m^n g(x)\, dx = f(n) - f(m)$$

Therefore the total number of hours required to produce the $(m + 1)$st unit through the $n$th unit is approximated by

$$\int_m^n g(x)\, dx$$

The graph of the equation $y = g(x)$ is called a learning curve. In general, $g$ is a decreasing function because employees become more efficient as they repeat their tasks. The learning rate is often described as a percentage. For example, a 90% curve indicates that each time production doubles, the newest unit of output requires 90% of the labor input of the reference unit. That is, if the first unit requires 10 hours, the second unit will require 9 hours, the fourth unit will require 8.1 hours, and so on. Often the equation of a learning curve is of the form $y = ax^b$. The constants $a$ and $b$ may be found by replacing $x$ and $y$ by given data to obtain a system of two equations in $a$ and $b$, and by solving the system.

Let $C(t)$, $R(t)$, and $P(t)$ dollars be the cost, revenue, and profit, respectively, for a firm which has operated its business for a period of $t$ weeks. Assume that $C$, $R$, and $P$ are twice differentiable, that $C''(t) > 0$, and that $R''(t) < 0$, which implies that $C'(t)$ is increasing and $R'(t)$ is decreasing. We wish to determine when to stop the operation in order to maximize profit. Since

$$P(t) = R(t) - C(t)$$

we have

$$P'(t) = R'(t) - C'(t)$$

To find the critical numbers, we set

$$P'(t) = R'(t) - C'(t) = 0$$

This equation has a unique solution $t_1$. Furthermore, $P''(t_1) < 0$. Thus $P(t_1)$ is in fact an absolute maximum of the function $P$ and the business should suspend its operation after $t_1$ weeks. To find the maximum profit, we note that

$$R(t) = R(0) + \int_0^t R'(u)\, du$$

and

$$C(t) = C(0) + \int_0^t C'(u)\, du$$

Thus

$$P(t) = R(t) - C(t) = R(0) - C(0) + \int_0^t [R'(u) - C'(u)]\, du$$

In general, $R(0) = 0$, but $C(0) = C_0$ is the fixed cost. Therefore the maximum profit is given by

$$P(t_1) = -C_0 + \int_0^{t_1} [R'(u) - C'(u)]\, du$$

We may define improper integrals as follows:

**Case 1:** The function $f$ is continuous on $[a, \infty)$. If $\lim_{b\to\infty} \int_a^b f(x)\, dx$ exists, we say that the improper integral $\int_a^\infty f(x)\, dx$ is convergent and its value is the value of the limit. Otherwise, we say that the improper integral is divergent.
**Case 2:** The function $f$ is continuous on $(-\infty, b]$. If $\lim_{a\to-\infty} \int_a^b f(x)\, dx$ exists, we say that the improper integral $\int_{-\infty}^b f(x)\, dx$ is convergent and its value is the value of the limit. Otherwise, we say that the improper integral is divergent.
**Case 3:** If the function $f$ is continuous on the set of real numbers, and if for any real number $a$, both integrals $\int_{-\infty}^a f(x)\, dx$ and $\int_a^\infty f(x)\, dx$ are convergent, then we say that the integral $\int_{-\infty}^\infty f(x)\, dx$ is convergent and its value is the sum of $\int_{-\infty}^a f(x)\, dx$ and $\int_a^\infty f(x)\, dx$.
**Case 4:** The function is continuous on the half-open interval $[a, b)$ but is not continuous at $b$. If $\lim_{t\to b^-} \int_a^t f(x)\, dx$ exists, we say that the improper integral $\int_a^b f(x)\, dx$ is convergent and its value is that of the limit.
**Case 5:** The function is continuous on the half-open interval $(a, b]$ but is not continuous at $a$. If $\lim_{t\to a^+} \int_t^b f(x)\, dx$ exists, we say that the improper integral $\int_a^b f(x)\, dx$ is convergent and its value is that of the limit.

A probability density function (abbreviated pdf) is used to define a probability distribution in the case that $X$ is a continuous random variable. This function is nonnegative for all real numbers $x$, and the area between its graph and the $x$-axis is 1. If $f$ is a pdf of a continuous random variable $X$ and if $[c, d]$ is a subset of the range of $X$, we define

$$P(X \text{ takes on values in } [c, d]) = \int_c^d f(x)\, dx$$

If the continuous random variable $X$ has a uniform distribution over $[a, b]$, then we define the pdf by

$$f(x) = \begin{cases} \dfrac{1}{b - a} & \text{if } a \leq x \leq b \\ 0 & \text{elsewhere} \end{cases}$$

because we wish the area between the graph of the pdf and the $x$-axis to be 1. Suppose now we have a pdf that is positive only on a bounded interval $[a, b]$. The expected value of the random variable $X$ is defined by

$$E(X) = \int_a^b xf(x)\, dx$$

This concept is used in situations such as the following. Suppose that by testing randomly selected samples of a commodity continuously for $a$ units of time and by recording the lifetime $t$ hours if a sample breaks down within $a$ units of time, we find a pdf $f$ where $f(t) \geq 0$ if $0 \leq t \leq a$, and $f(t) = 0$ otherwise. We define a continuous random variable as follows. For any unit $c$ of the commodity, we let $X(c) = \min\{\text{lifetime of } c, a\}$. Then $X$ has values in the interval $[0, a]$ and its pdf is the function found by testing the samples. Thus the expected lifetime of the commodity is given by

$$T = \int_0^a tf(t)\, dt$$

The average expected profit per unit of a commodity when a company offers a warranty may be calculated using these concepts.

Suppose $T$ is a continuous random variable which represents the lifetime, in years, of a certain commodity, $T$ has values in the interval $[0, \infty)$, and the pdf for the lifetime $t$ is $f(t)$, where $f(t) \geq 0$ if $t \in [0, \infty)$ and $f(t) = 0$ elsewhere. Then, the expected lifetime of that commodity is given by

$$E(T) = \int_0^\infty tf(t)\, dt$$

and the probability that it will last $n$ years or more is given by $\int_n^\infty f(t)\, dt$.

**SAMPLE EXAM QUESTIONS**

1. Let $f_1(x) = e^{-4x}$ and $f_2(x) = e^{5x}$. Verify that $f_1$ and $f_2$ are solutions of the equation $y'' - y' - 20y = 0$.

2. Let $f_1(x) = e^{-2x}$, $f_2(x) = e^x$ and $f_3(x) = e^{3x}$. Verify that $f_1$, $f_2$, and $f_3$ are solutions of the equation $y''' - 2y'' - 5y' + 6y = 0$.

3. Let $f(x) = x^2 + 3x + 1$. Verify that $f$ is a solution of the equation $x^2y'' + 2xy' - 6y + 12x + 6 = 0$.

4. Find the general solution of each differential equation.

   a. $\dfrac{dy}{dx} = \dfrac{5x^2 + 3}{2y^3 - 5y + 6}$

   b. $\dfrac{dy}{dx} = \dfrac{\sqrt{x^3}}{e^y + 1}$

   c. $(y^3 + 5)\dfrac{dy}{dx} = e^{5x} + x^2 + 1$

   d. $\dfrac{dy}{dx} = x^3 e^{x^4 + 2y}$

**e.** $(x + 3)y^3 \dfrac{dy}{dx} = \ln(x + 3)^{y^4+1}$

5. Find the particular solution of the differential equation that satisfies the given conditions.

**a.** $\dfrac{dx}{dt} = 4xt^3$; $x = e^4$ when $t = -1$

**b.** $e^{2y-x} \dfrac{dy}{dx} = 3x$; $y = 2$ when $x = 0$

**c.** $\ln(y + e)^{x^2+1} \dfrac{dy}{dx} = x(y + e)$; $y = 0$ when $x = 0$

6. An investment of \$15,000 earns interest compounded continuously at the nominal rate of 6%. What is the value of that investment in 12 years?

7. A bank pays interest compounded continuously at a nominal rate of 8%. **a.** How long will it take for a deposit to triple in value? **b.** How long will it take a \$6000 deposit to be worth \$9000?

8. The rate of change in the value of real estate in Quebec at instant $t$ is proportional to the cube root of the value at that time. A lot valued at \$64,000 nine months ago is worth \$68,921 today. How much will the lot be worth 18 months from now?

9. The elasticity of demand for a certain commodity is given by $\eta = -3p^2/(2500 - p^2)$, where $0 < p < 50$. Find the demand equation $q = f(p)$ if $q = 192,000$ when $p = 30$.

10. **a.** Suppose $a$, $b$, and $c$ are constants with $a \neq 0$ and $r$ is a solution of the quadratic equation $ax^2 + bx + c = 0$. Show that $e^{rx}$ is a solution of the differential equation $ay'' + by' + cy = 0$.

   **b.** Find two solutions of the equation $y'' + 3y' - 10y = 0$.

   **c.** Find two solutions of the equation $2y'' - y' - 21y = 0$.

11. **a.** Sketch the region $R$ that is bounded by the graphs of $y = x^2 - 6x + 10$, $y = -x^2 + 4x - 5$, $x = 1$, and $x = 4$.

   **b.** Find the area of the region $R$.

12. **a.** Sketch the region $R$ that is bounded by the graphs of $y = \frac{1}{x}$, $y = \frac{-2}{x}$, $x = 1$, and $x = e$.

   **b.** Find the area of the region $R$.

13. **a.** Sketch the region $R$ that is bounded by the graphs of $y = \sqrt{1 - x}$, $y = \sqrt{x - 3}$, $y = 0$, and $y = 3$.

   **b.** Find the area of the region $R$.

14. Let the functions $f$, $g$, and $h$ be defined as follows:

$$y = f(x) = \sqrt{x} \text{ if } 0 \le x \le 4$$
$$y = g(x) = x^2 - 16x + 50 \text{ if } 4 \le x \le 6$$
$$y = h(x) = -2x, \text{ if } 0 \le x \le 6$$

   **a.** Sketch the region $R$ that is bounded by the graphs of $y = f(x)$, $y = g(x)$, $y = h(x)$, and $x = 6$.

   **b.** Find the area of the region $R$.

15. **a.** Sketch the region $R$ that is bounded by the graphs of $y = .25x^2 - 1.5x + 2$, $y = -.25x^2 + 2x - 3$, $x = 1$, and $x = 6$.

   **b.** Find the area of the region $R$.

16. **a.** Sketch the region $R$ that is bounded by the graphs of $y = \ln x$ and $e(e^2 - 1)y = 2x + e(e^2 - 3)$.

   **b.** Find the area of the region $R$. (*Hint:* The two points of intersection are $(e, 1)$ and $(e^3, 3)$.)

**17.** The supply and demand equations for a commodity are

$$p = S(q) = \frac{q}{3}$$

$$p = D(q) = \frac{-q}{4} + 14, \text{ where } 0 \le q \le 56$$

In the equations, $p$ is the price in dollars per unit of the commodity and $q$ is the quantity in thousands.

**a.** Find the equilibrium quantity and the equilibrium price.
**b.** Assuming that market equilibrium exists, find the consumers' surplus.
**c.** Find the producers' surplus.

**18.** Same as Exercise 17 if

$$p = S(q) = .25q^2 + 10q$$
$$p = D(q) = -.5q^2 - 6q + 448, \text{ where } 0 \le q \le 24$$

**19.** Same as Exercise 17 if

$$p = S(q) = 3\sqrt{q}$$

and the demand equation is linear and its graph passes through the points (40, 45) and (80, 35).

**20.** A supplier has a monopoly on a certain product. The cost and demand functions are defined where $C(q)$ is in thousands of dollars, $q$ is in thousands of units, and $p$ is in dollars.

$$C(q) = .35q^2 + 200$$
$$p = D(q) = 30 - .15q, \text{ where } 0 \le q \le 200$$

**a.** Find the number of units and the price per unit that will maximize profit.
**b.** Find the consumers' surplus if the product is sold at the price that maximizes profit.

**21.** The income distribution in a certain country is given by

$$y = I(x) = \frac{3x^2 + 8x}{11}$$

**a.** Sketch the corresponding Lorenz curve.
**b.** What portion of the total income is received by the lowest paid 25% of the population?
**c.** Calculate the coefficient of inequality.

**22.** Same as Exercise 21 if $y = I(x) = \dfrac{x}{3 - 2x}$. $\left(Hint: \dfrac{x}{3 - 2x} = \dfrac{-1}{2} + \dfrac{\frac{3}{2}}{3 - 2x}.\right)$

**23.** Same as Exercise 21 if $y = I(x) = \dfrac{(1 - e^{-x})x}{1 - e^{-1}}$.

**24.** Suppose the marginal propensity to save is given by

$$\frac{dS}{dI} = .35 + .02e^{-.02I}$$

(in billions of dollars) and the consumption is 5.7 billion when the national disposable income is 6 billion.

**a.** Find the consumption function.
**b.** What is the consumption when disposable income is 50 billion dollars?

**25.** Using integration, approximate the present value of an annuity that will pay $4500 per month for the next 15 years if the nominal rate of interest is 8% compounded continuously.

**26.** A company has found that the time $g(x)$ hours required to manufacture the $(x + 1)$st gadget is given by

$$g(x) = 4e^{-.03x}$$

provided that $0 \leq x \leq 250$. Estimate the amount of time required to manufacture the 21st through the 100th gadget.

**27.** If we assume that the equation of a learning curve is of the form $y = ax^b$ where $y$ is the number of hours required to produce the $x$th unit, find the values of the constants $a$ and $b$ for a 90% learning curve if 30 hours of work are required to produce the eighth unit.

**28.** The Montagne Mining Company has found that the instantaneous rates of change of revenue and cost are given by

$$R'(t) = 52 + 150e^{-.002t}$$

and

$$C'(t) = 52 + te^{-.002t}$$

respectively, where $0 \leq t \leq 500$, $R(t)$ and $C(t)$ are in thousands of dollars, and $t$ is in months. The fixed cost is $37,500.

    **a.** Show that $R'$ is a decreasing function and $C'$ is an increasing function.
    **b.** How many months should the company drill its mine in order to maximize the total profit?
    **c.** Calculate the maximum total profit.

**29.** Evaluate each of the following improper integrals. If some integral is divergent, give the reason.

    **a.** $\displaystyle\int_1^\infty \frac{4}{x^2}\, dx$

    **b.** $\displaystyle\int_{-\infty}^2 e^{3x}\, dx$

    **c.** $\displaystyle\int_2^\infty \frac{3x}{x^2 + 1}\, dx$

    **d.** $\displaystyle\int_{-\infty}^0 \frac{e^x}{e^x + 5}\, dx$

    **e.** $\displaystyle\int_{-\infty}^\infty \frac{3x^2}{(x^3 + 1)^2}\, dx$

    **f.** $\displaystyle\int_0^3 \frac{3}{\sqrt{x}}\, dx$

**30.** A traffic light changes every 150 seconds. If a motorist arrives at the intersection while the light is red, what is the probability that she will have to wait at least 40 seconds?

**31.** An electronics company is manufacturing VCRs. The quality control division randomly selects samples and uses these samples continuously for a maximum of 121 hours. If a VCR is found to be defective during that time, its lifetime $t$ hours is recorded. Following the testing of many VCRs, the pdf for the lifetime $t$ is found to be

$$f(t) = \begin{cases} \dfrac{3}{2057}(13 - \sqrt{t}) & \text{if } 0 \leq t \leq 121 \\ 0, & \text{elsewhere} \end{cases}$$

    **a.** Verify that $f$ is indeed a pdf.

    **b.** Assuming that 1 hour of continuous testing by the company is equivalent to 1 week of normal use by a customer, find the expected lifetime of a VCR.

**32.** The company of Exercise 31 offers the following warranty to its customers. It will give full, one-half, and one-fourth refunds if a VCR breaks down within 12, 24, and 48 weeks, respectively, from the date of purchase. Each VCR sells for \$500 and its total cost to the company is \$275. Again assuming that 1 hour of continuous testing by the company is equivalent to 1 week of normal use by a customer, calculate the expected average profit per VCR.

**33.** Suppose that $T$ is a continuous random variable which represents the lifetime, in years, of a certain brand of refrigerator and that $T$ has values in the interval $[0, \infty)$. Its pdf for the lifetime $t$ is given by

$$f(t) = \begin{cases} \dfrac{1}{23}e^{-t/23} & \text{if } t \in [0, \infty) \\ 0 & \text{elsewhere} \end{cases}$$

    **a.** Verify that $f$ is indeed a pdf.

    **b.** What is the expected lifetime of a refrigerator?

    **c.** What is the probability that a refrigerator will last 15 years or more?

# Functions of Several Variables

**11**

The functions studied until now have had domains and ranges that are sets of real numbers. They are called functions of one variable because the values of the dependent variable are determined by the values of a single independent variable. However, in applications, it often happens that the value of one variable depends on the values of two or more variables. For example, the number of units of a commodity sold by a company may depend on the price per unit and on the amount spent for advertising. In this chapter, we focus on functions of two variables.

# 11.1  Functions of Several Variables

## Functions of Two or More Independent Variables

DEFINITION 11.1:    Suppose that $D$ is a set of ordered pairs of real numbers and that a rule is given which assigns to each member $(x, y)$ of $D$ a unique real number. This rule is called a *function of two variables*. The set $D$ is called the *domain* of the function. If $f$ is the name of the function, then for each ordered pair $(x, y)$ in $D$, the number corresponding to that $(x, y)$ is denoted $f(x, y)$ and is called the *image* of $(x, y)$. The set of all images is called the *range* of the function.

## Domain of a Function of Several Variables

We adopt the following convention. If the domain of a function of two variables is not specified, then it is understood to be the set of all ordered pairs of real numbers that are permissible in the expression defining the function. That is, $(a, b)$ is permissible in the expression $z = f(x, y)$ if, when $x$ and $y$ are replaced by $a$ and $b$, respectively, $f(a, b)$ is defined and is a real number. We illustrate with examples.

EXAMPLE 1    Let $f(x, y) = \dfrac{4}{(x - 2)(y - 3)}$. Find the domain of the function $f$.

Solution    Since the expression defining $f$ involves division, we recall that division by 0 is not defined. Thus we may not replace $x$ and $y$ by 2 and 3, respectively. Therefore, the ordered pairs $(2, c)$ and $(d, 3)$ for any real numbers $c$ and $d$ are not permissible. It follows that the domain of the function $f$ is the set of all ordered pairs of real numbers except those listed above. Geometrically, it is the set of all points in the $xy$-plane except the graphs of the equations $x = 2$ and $y = 3$. (See Figure 11.1.)

y

These lines are not
in the domain

$y = 3$

0

x

$x = 2$

**FIGURE 11.1**

**EXAMPLE 2**   Let $g(x, y) = \sqrt{9 - x^2 - y^2}$. Find the domain of the function $g$.

**Solution**   Since $\sqrt{9 - x^2 - y^2}$ is a real number only if $9 - x^2 - y^2$ is nonnegative, we must have

$$9 - x^2 - y^2 \geq 0$$

That is,

$$x^2 + y^2 \leq 9$$

Recalling that $x^2 + y^2$ is the square of the distance from the origin to the point $(x, y)$, we conclude that this distance must be less than or equal to 3. Therefore the domain of $g$ is the set of all points in the $xy$-plane either on or inside the circle of radius 3 and center the origin. (See Figure 11.2.)

**FIGURE 11.2**

**EXAMPLE 3**   Evaluate $f(-1, 2), f(2, 3)$, and $f(-3, 4)$ if $f(x, y) = 2x^2 + \sqrt{3 - y}$.

**Solution**   The value of $f(-1, 2)$ is obtained by replacing $x$ and $y$ by $-1$ and 2, respectively, in the expression defining $f(x, y)$. We obtain

$$f(-1, 2) = 2(-1)^2 + \sqrt{3 - 2} = 2 + 1 = 3$$

Similarly,

$$f(2, 3) = 2(2^2) + \sqrt{3 - 3} = 8 + 0 = 8$$

and

$$f(-3, 4) = 2(-3)^2 + \sqrt{3 - 4}$$
$$= 18 + \sqrt{-1}$$

But, $\sqrt{-1}$ is not a real number. Thus $(-3, 4)$ is not in the domain of $f$ and $f(-3, 4)$ is not defined.

**Do Exercises 3 and 9.**

If the domain of a function $f$ is a set $D$ of ordered triples $(x, y, z)$ of real numbers, we say that $f$ is a function of three variables. More generally, if the domain is a set of ordered $n$-tuples $(x_1, x_2, \ldots, x_n)$ of real numbers, the function is said to be a function of $n$ variables. The method used to determine the domain is the same as in the case of a function of two variables.

**EXAMPLE 4**   Let $w = f(x, y, z) = \dfrac{-6}{x^2 + y^2 + z^2}$. Determine the domain of $f$ and calculate $f(-1, 1, 2)$.

**Solution**   The quotient $\dfrac{-6}{x^2 + y^2 + z^2}$ is not defined only when $x^2 + y^2 + z^2 = 0$. But this occurs only if $x = y = z = 0$. Thus the domain of $f$ is the set of all ordered triples of real numbers except $(0, 0, 0)$. To calculate $f(-1, 1, 2)$, we replace $x$, $y$, and $z$ by $-1$, 1, and 2, respectively, in the expression defining $f$. We get

$$f(-1, 1, 2) = \frac{-6}{(-1)^2 + 1^2 + 2^2} = -1$$

**Do Exercise 13.**   ■

## Application

**EXAMPLE 5**   A store in Dallas carries two brands of perfume. The demand for each increases not only when its own price decreases but also when the price of the other increases. Suppose that when the prices of brand A and brand B are $p_A$ and $p_B$ dollars per ounce, respectively, the demand for brand A is $(800 - 30p_A + p_B^2)$ ounces per week and that for brand B is $(600 - 40p_B + 25p_A)$ ounces per week. Find the revenue per week from the sale of these two brands of perfume in terms of $p_A$ and $p_B$. Then calculate the revenue when

**a.** $P_A = 15$ and $P_B = 10$
**b.** $P_A = 10$ and $P_B = 15$
**c.** $P_A = 30$ and $P_B = 10$

**Solution**   In general, the revenue is found by multiplying the number of units sold by the price per unit. Thus, the revenue from brand A is $(800 - 30p_A + p_B^2)p_A$ dollars and the revenue from brand B is $(600 - 40p_B + 25p_A)p_B$ dollars. The total revenue $R(p_A, p_B)$ is the sum of the two revenues. That is,

$$R(p_A, p_B) = (800 - 30p_A + p_B^2)p_A + (600 - 40p_B + 25p_A)p_B$$
$$= p_A p_B^2 - 30p_A^2 + 25p_A p_B - 40p_B^2 + 800p_A + 600p_B$$

Now we have

**a.** $R(15, 10) = 15(10^2) - 30(15^2) + 25(15)(10) - 40(10^2) + 800(15) + 600(10)$
       $= 12{,}500.$

Similarly, we get

**b.** $R(10, 15) = 11,000$

and

**c.** $R(30, 10) = 9500$.

Note how the revenue changes as the price of each brand of perfume changes. It is useful to know how to price each brand to maximize the revenue. You will learn how to do this soon.

**Do Exercise 41.** ■

## Graph of a Function of Two Variables

Many of the properties of a function of one variable are best understood by looking at the graph of that function. In general, the graph of a function of one variable is a curve on a coordinate plane. The graph of a function of two variables is a surface in a three-dimensional coordinate system. At this point, you should read Section 4.2, Three-Dimensional Coordinate System.

**DEFINITION 11.2:** If $f$ is a function of two variables, the *graph* of $f$ is the set of all points in the *xyz*-space whose coordinates $(a, b, c)$ satisfy the following two conditions:

1. $(a, b)$ is in the domain of the function $f$.
2. $c = f(a, b)$.

**EXAMPLE 6** Find the coordinates of three points on the graph of the function $g$ of Example 2 and plot them.

**Solution** Recall that $g(x, y) = \sqrt{9 - x^2 - y^2}$ and that the domain of $g$ is the set of points in the *xy*-plane on and inside the circle with center at the origin and radius 3. The points $(0, 0)$, $(1, 2)$, and $(1, -2)$ clearly are within that circle. We find that

$$g(0, 0) = \sqrt{9 - 0^2 - 0^2} = \sqrt{9} = 3$$
$$g(1, 2) = \sqrt{9 - 1^2 - 2^2} = \sqrt{4} = 2$$
$$g(1, -2) = \sqrt{9 - 1^2 - (-2)^2} = \sqrt{4} = 2$$

Thus the points $(0, 0, 3)$, $(1, 2, 2)$, and $(1, -2, 2)$ are on the graph of $g$. These points are plotted in Figure 11.3.

**Do Exercise 27.** ■

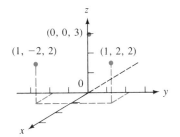

**FIGURE 11.3**

In general, the graph of a function of two variables is a surface. For most functions of two variables, it is difficult to sketch their graphs by hand. However, many computers are capable of drawing the graphs of many of these functions. The graph of a function drawn by a computer is shown in Figure 11.4.

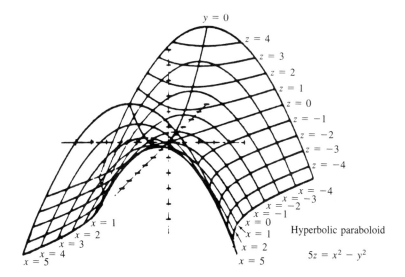

**FIGURE 11.4**

$y = 0$

$z = 4$
$z = 3$
$z = 2$
$z = 1$
$z = 0$
$z = -1$
$z = -2$
$z = -3$
$z = -4$

$x = -4$
$x = -3$
$x = -2$
$x = -1$
$x = 0$
$x = 1$
$x = 2$
$x = 5$

$x = 1$
$x = 2$
$x = 3$
$x = 4$
$x = 5$

Hyperbolic paraboloid

$5z = x^2 - y^2$

In applications for business and economics, it is seldom necessary to draw the graph of a function of two variables by hand; therefore we shall not discuss techniques for sketching these graphs. Those interested may find these techniques described in more advanced calculus textbooks. However, it is useful to look at graphs to visualize the basic ideas. We have sketched the graphs of three functions in Figures 11.5, 11.6, and 11.7.

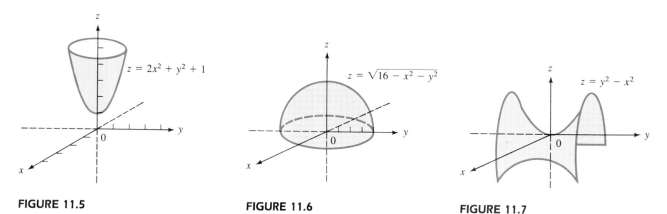

**FIGURE 11.5**

$z = 2x^2 + y^2 + 1$

**FIGURE 11.6**

$z = \sqrt{16 - x^2 - y^2}$

**FIGURE 11.7**

$z = y^2 - x^2$

## Level Curves

A method used by map makers for many years is useful for studying three-dimensional surfaces and that is by drawing two-dimensional graphs. All points on a map representing points on the earth that are at the same elevation are connected by curves, often of different colors, where the color depends on the elevation. We do a similar thing when studying the graph of a function of two variables. Given a number $c$ in the range of a function $f$, the set of all points in the domain of $f$ whose coordinates satisfy the equation $f(x, y) = c$ is called a *level curve of f*. Essentially, such a level curve represents all the points in the domain of $f$ which have the property that the surface (the graph of $f$) over these points is at a distance $c$ from the $xy$-plane, above the plane if $c > 0$ and below it if $c < 0$.

**EXAMPLE 7** Sketch the level curves of $f$ corresponding to $c = 1$ and $c = 6$ if $f(x, y) = 10 - x^2 - y^2$.

**Solution** For $c = 1$, the level curve is the set of points $(x, y)$ which satisfy

$$1 = 10 - x^2 - y^2$$

This equation reduces to $x^2 + y^2 = 9$, which is the equation of the circle with center the origin and radius 3.

For $c = 6$, we get

$$6 = 10 - x^2 - y^2$$

which may be written $x^2 + y^2 = 4$. The corresponding level curve is the circle whose center is the origin and whose radius is 2. These two level curves are shown in Figure 11.8.

**Do Exercise 39.**

**FIGURE 11.8**

# Exercise Set 11.1

*In Exercises 1–10, find the domain of the function f and find the value of f(x, y) at the indicated points.*

**1.** $f(x, y) = x^2 + 3xy + y^2$; (1, 2), (−2, 3), (3, −4)

**2.** $f(x, y) = \dfrac{4}{x - y}$; (1, 2), (−3, 5), (2, −6)

**3.** $f(x, y) = \dfrac{3x}{(x - 5)(y - 7)}$; (0, −1), (−2, −5), (3, 2)

**4.** $f(x, y) = \dfrac{2y}{(x + 3)(y + 2)}$; (−3, −5), (−2, −5), (0, 1)

**5.** $f(x, y) = \dfrac{x + 2y}{(x + 2)(x - y)}$; (3, 4), (−1, 6), (3, −2)

**6.** $f(x, y) = 5x - \sqrt{6 - y}$; (2, 1), (3, −4), (−12, −3)

**7.** $f(x, y) = 2xy + \sqrt{4 - x^2}$; (2, 4), (1, 5), (0, 3)

**8.** $f(x, y) = \sqrt{9 - x^2} + \sqrt{16 - y^2}$; (0, 0), (1, −2), (2, −3)

**9.** $f(x, y) = \sqrt{25 - x^2 - y^2}$; (1, 2), (3, 4), (−1, 2)

**10.** $f(x, y) = \dfrac{4xy}{\sqrt{36 - x^2 - y^2}}$; (−1, 2), (2, 3), (−3, −2)

*In Exercises 11–15, find the domain of the function f and find the value of f(x, y, z) at the indicated points.*

**11.** $f(x, y, z) = x^2 + xy + 2z^2$; (1, 1, 2), (−1, 2, −3), (2, 4, 0)

**12.** $f(x, y, z) = 2x + 3y - \sqrt{4 - z}$; (3, 5, 1), (0, 3, −2), (2, 4, 3)

**13.** $f(x, y, z) = 3z + \sqrt{9 - x^2 - y^2}$; (1, 1, 3), (0, 2, 5), (−1, 2, −3)

**14.** $f(x, y, z) = \dfrac{x + y}{(x - 3)(y - 4)(z + 5)}$; (2, 3, 4), (−2, −4, 2), (0, 0, 1)

**15.** $f(x, y, z) = x^2 + 3y^2 - \sqrt{16 - z^2}$; (1, 3, 2), (0, 3, −4), (−2, 4, −3)

*In Exercises 16–25, plot the points with the given coordinates.*

**16.** (1, 2, 4)  **17.** (−2, 3, 5)

**18.** (3, −2, 4)  **19.** (0, 2, 4)

**20.** (−1, 2, 0)  **21.** (−3, 0, 4)

**22.** $(4, 2, 0)$

**23.** $(0, 0, 5)$

**24.** $(-2, -2, 3)$

**25.** $(-2, -2, -2)$

**26.** Find three points on the graph of the function $f$ of Exercise 1 and plot these three points.

**27.** Same as Exercise 26 for the function of Exercise 2.

**28.** Same as Exercise 26 for the function of Exercise 3.

**29.** Same as Exercise 26 for the function of Exercise 4.

**30.** Same as Exercise 26 for the function of Exercise 5.

**31.** Same as Exercise 26 for the function of Exercise 6.

*In Exercises 32–40, sketch the level curve of $f$ corresponding to the given value of $c$.*

**32.** $f(x, y) = 5x - 2y, c = 10$

**33.** $f(x, y) = y - 2x^2, c = 5$

**34.** $f(x, y) = y - x^2 - 3x, c = 4$

**35.** $f(x, y) = x^2 - y, c = -3$

**36.** $f(x, y) = xy, c = 4$

**37.** $f(x, y) = x^2 + y^2 + 2, c = 6$

**38.** $f(x, y) = y - \dfrac{x(y - 2)}{x + 3}, c = \dfrac{-1}{2}$

**39.** $f(x, y) = x^2 + y^2, c = 16$

**40.** $f(x, y) = 4x^2 + 9y^2, c = 36$

**41.** A store carries two brands of pocket calculators. When the price of brand A is $p_A$ dollars per calculator and that of brand B is $p_B$ dollars per calculator, the demands per month are given by

$$D_A(p_A, p_B) = 300 - 10p_A + 20p_B \text{ for brand A}$$

and

$$D_B(p_A, p_B) = 450 - 15p_B + 25p_A \text{ for brand B}$$

Find the revenue per month from the sale of these two brands of calculators in terms of $p_A$ and $p_B$.

**42.** A company manufactures two types of television sets at a cost of \$200 and \$300 per set, respectively. If $p_1$ and $p_2$ dollars are the prices of these television sets, the demand functions are given by

$$D_1(p_1, p_2) = 3000 - 2p_1 + 3p_2$$

and

$$D_2(p_1, p_2) = 4500 - p_2 + 3p_1$$

where $D_1(p_1, p_2)$ and $D_2(p_1, p_2)$ are the numbers of television sets of type 1 and type 2, respectively, sold each month. Express the company's monthly profit in terms of $p_1$ and $p_2$.

**43.** A rectangular tank must hold 3000 cubic feet. The cost to build the base is \$20/sq ft and the cost to build the sides is \$12/sq ft. The tank is to be covered with a canvas costing \$3/sq ft. If the length and width of the tank are $x$ and $y$ feet, respectively, express the total cost to build the tank in terms of $x$ and $y$.

# 11.2 Partial Differentiation

## Directional Derivatives

We have seen that the derivative of a function of one variable gives the rate of change of the dependent variable with respect to the independent variable. A geometric interpretation of this fact was given by considering the graph of a differentiable function $f$ and defining the slope of the tangent line to the graph of $f$ at the point $(a, f(a))$ to be the value of the derivative at $a$. We used this definition extensively when we studied optimization because techniques such as the First Derivative Test are best understood geometrically.

In this section, we discuss differentiation of functions of two variables. In general, the graph of a function of two variables is a surface in the $xyz$-space. Leaving the realm of mathematics for a moment, imagine that you are somewhere on a mountain where you have gone for a hike. The ground is a surface. It is clear that at any instant, your rate of change of elevation depends on the direction you are walking in. In fact, it is conceivable that if you are going north, you are walking uphill; while walking east may take you downhill. We shall see examples of similar surfaces later on.

Going back to mathematics, let $f$ be a function of two variables, and let $(x_0, y_0)$ be a point in the domain of $f$. Let $L$ be a directed line segment passing through $(x_0, y_0)$ and whose points are in the domain of $f$. The directional derivative of $f$ at $(x_0, y_0)$ in the direction of $L$ is the instantaneous rate of change of $f$ in the direction of that line segment. More specifically, we have the following definition.

> **DEFINITION 11.3:**   If $\Delta s$ is the directed distance from $(x_0, y_0)$ to $(x, y)$ where $(x, y)$ is on the segment $L$, then
>
> $$\lim_{\Delta s \to 0} \frac{f(x, y) - f(x_0, y_0)}{\Delta s}$$
>
> if it exists, is the *directional derivative* of $f$ at $(x_0, y_0)$ in the direction of $L$. (See Figure 11.9.)

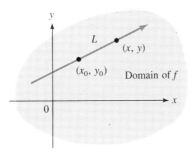

**FIGURE 11.9**

## Partial Derivatives

In this text, we are interested only in the derivatives in the directions of the $x$-axis and $y$-axis. In these cases, the directional derivatives are called *partial derivatives*. Specifically, let $(x_0, y_0)$ be a point in the domain of $f$ and let $L$ be a line segment parallel to the $x$-axis passing through the point $(x_0, y_0)$ and having each of its points in the domain of $f$. Any point on that segment will have coordinates $(x, y_0)$. Let $x - x_0 = h$ so that $x = x_0 + h$ and points on the segment $L$ have coordinates $(x_0 + h, y_0)$. The directed distance from the point $(x_0, y_0)$ to the point $(x_0 + h, y_0)$ is $h$. (See Figure 11.10.)

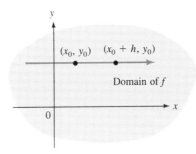

**FIGURE 11.10**

We give the following definition.

> **DEFINITION 11.4:**   The directional derivative of the function $f$ at the point $(x_0, y_0)$ in the direction of $L$ is
>
> $$\lim_{h \to 0} \frac{f(x_0 + h, y_0) - f(x_0, y_0)}{h}$$
>
> This limit, if it exists, is called the *partial derivative of f with respect to x*.

It is denoted by any one of the following symbols:

$$\frac{\partial f}{\partial x}(x_0, y_0), f_x(x_0, y_0), f_1(x_0, y_0), \text{ or } \frac{\partial z}{\partial x}\bigg|_{(x_0, y_0)} \text{ in the case } z = f(x, y)$$

Of course, if $z = f(x, y)$, the partial derivative of $f$ with respect to $x$ at an arbitrary point $(x, y)$ is denoted by any one of the following symbols:

$$\frac{\partial f}{\partial x}(x, y), f_x(x, y), f_1(x, y), \text{ or } \frac{\partial z}{\partial x}$$

It should be noted that we cannot use the same Leibniz notation as we used for ordinary derivatives. That is, $\frac{dz}{dx}$ is not the same as $\frac{\partial z}{\partial x}$.

## Computation of Partial Derivatives

When considering the partial derivative with respect to $x$, the value of $y$ is constant. Thus if $f$ is a function of two variables and we wish to find its partial derivative with respect to $x$, we treat $y$ as a constant and we think of $f$ as a function whose only independent variable is $x$. We may then use any of the differentiation formulas for functions of one variable.

**EXAMPLE 1**   Let $f(x, y) = 3x^2 y^3$. Find $f_x(x, y)$.

**Solution**   Since we treat $y$ as a constant, $y^3$ is also considered a constant. We now use the formula

$$D_x(ku) = kD_x u$$

where $k$ is a constant and $u$ is a differentiable function of $x$. We obtain

$$f_x(x, y) = 3y^3 D_x(x^2) = 3y^3(2x) = 6xy^3 \qquad \blacksquare$$

**EXAMPLE 2**   Use the definition of partial derivative to give a different solution for Example 1.

**Solution**   The definition states that

$$f_x(x, y) = \lim_{h \to 0} \frac{f(x + h, y) - f(x, y)}{h}$$

provided this limit exists. In the case where $f(x, y) = 3x^2y^3$, we have

$$f(x + h, y) = 3(x + h)^2y^3 = 3x^2y^3 + 6xhy^3 + 3h^2y^3$$

Therefore,

$$f(x + h, y) - f(x, y) = (3x^2y^3 + 6xhy^3 + 3h^2y^3) - 3x^2y^3$$
$$= h(6xy^3 + 3hy^3)$$

and

$$\frac{f(x + h, y) - f(x, y)}{h} = \frac{h(6xy^3 + 3hy^3)}{h} = 6xy^3 + 3hy^3$$

It follows that

$$f_x(x, y) = \lim_{h \to 0} \frac{f(x + h, y) - f(x, y)}{h}$$
$$= \lim_{h \to 0}(6xy^3 + 3hy^3) = 6xy^3$$

since $3hy^3$ tends to 0 as $h$ tends to 0.
**Do Exercise 1.** ∎

The partial derivative of a function $f$ with respect to $y$ is simply the directional derivative in the direction of the $y$-axis.

---

**DEFINITION 11.5:** If $f$ is a function of two variables and $(x_0, y_0)$ is a point in the domain of $f$, then the *partial derivative of $f$ with respect to $y$* at $(x_0, y_0)$ is the value of

$$\lim_{k \to 0} \frac{f(x_0, y_0 + k) - f(x_0, y_0)}{k}$$

provided this limit exists.

---

If $z = f(x, y)$, to find the partial derivative with respect to $y$, we treat $x$ as a constant and we consider $f$ as a function of the independent variable $y$. We may then use any of the formulas of differentiation. Any one of the symbols $\frac{\partial f}{\partial y}(x, y)$, $f_y(x, y)$, $f_2(x, y)$, or $\frac{\partial z}{\partial y}$ denotes the partial derivative of $f$ with respect to $y$.

For convenience, we recall the basic formulas of differentiation that were established earlier. If $f$ and $g$ are differentiable functions, $u = f(x)$, $v = g(x)$, and $k$ and $r$ are real constants, we have the following formulas.

## Basic Formulas of Differentiation

1. $D_x(u + v) = D_x u + D_x v$
2. $D_x(u - v) = D_x u - D_x v$
3. $D_x(uv) = u D_x v + v D_x u$
4. $D_x\left(\dfrac{u}{v}\right) = \dfrac{v D_x u - u D_x v}{v^2}$
5. $D_x(ku) = k D_x u$
6. $D_x u^r = r u^{r-1} D_x u$
7. $D_x e^u = e^u D_x u$
8. $D_x b^u = b^u (\ln b) D_x u$, where $b$ is a positive constant.
9. $D_x \ln|u| = \dfrac{D_x u}{u}$
10. $D_x \log_b|u| = \dfrac{D_x u}{(\ln b)u}$

**EXAMPLE 3**   Find $\dfrac{\partial z}{\partial y}$ if $z = f(x, y) = x^3 \ln(x^2 + y^2)$.

**Solution**   Since we must treat $x$ as a constant, $x^2$ and $x^3$ are also constants. Thus,

$$\frac{\partial z}{\partial y} = x^3 \frac{\partial}{\partial y} \ln(x^2 + y^2) \qquad \text{(since } x^3 \text{ is a constant)}$$

$$= x^3 \, \frac{\dfrac{\partial}{\partial y}(x^2 + y^2)}{x^2 + y^2} \qquad \text{(by Formula 9)}$$

$$= x^3 \frac{0 + 2y}{x^2 + y^2} = \frac{2x^3 y}{x^2 + y^2}$$

**Do Exercise 19.**

## Geometric Interpretation of Partial Derivatives

There is a geometric interpretation of the partial derivatives of a function of two variables. Let $f$ be a function defined in a neighborhood of the point $(x_0, y_0)$ of the $xy$-plane and let $z_0 = f(x_0, y_0)$. Then the point $(x_0, y_0, z_0)$ is on the graph of $f$, which is a surface in the $xyz$-space. Consider planes perpendicular to the $xy$-plane, passing through the point $(x_0, y_0)$ and parallel to the $x$ and $y$ axes, respectively. The first plane intersects the surface along a curve $C_1$ and the second plane intersects it along a curve $C_2$. Both $C_1$ and $C_2$ pass through the point $(x_0, y_0, z_0)$. The curve $C_1$ has a tangent line at $(x_0, y_0, z_0)$ and the slope of that tangent line is $f_x(x_0, y_0)$. The curve $C_2$ has a tangent line at $(x_0, y_0, z_0)$ and the slope of that tangent line is $f_y(x_0, y_0)$. (See Figure 11.11.)

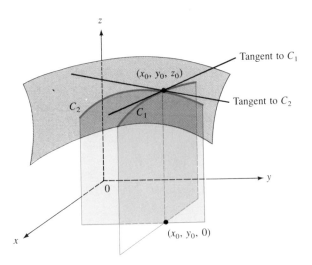

**FIGURE 11.11**

## Higher Order Partial Derivatives

Suppose $f$ is a function of two variables. Then, generally, both partial derivatives are also functions of two variables. Their partial derivatives, if they exist, are called the *second partial derivatives*. There are four second partial derivatives, the different notations for which are the following:

$$\frac{\partial}{\partial x}\left(\frac{\partial z}{\partial x}\right) \text{ and } (f_x)_x \text{ are denoted } \frac{\partial^2 z}{\partial x^2} \text{ and } f_{xx}, \text{ respectively}$$

$$\frac{\partial}{\partial y}\left(\frac{\partial z}{\partial y}\right) \text{ and } (f_y)_y \text{ are denoted } \frac{\partial^2 z}{\partial y^2} \text{ and } f_{yy}, \text{ respectively}$$

$$\frac{\partial}{\partial x}\left(\frac{\partial z}{\partial y}\right) \text{ and } (f_y)_x \text{ are denoted } \frac{\partial^2 z}{\partial x \partial y} \text{ and } f_{yx}, \text{ respectively}$$

$$\frac{\partial}{\partial y}\left(\frac{\partial z}{\partial x}\right) \text{ and } (f_x)_y \text{ are denoted } \frac{\partial^2 z}{\partial y \partial x} \text{ and } f_{xy}, \text{ respectively}$$

Note the order in which the differentiation is performed. The notation $\frac{\partial^2 z}{\partial x \partial y}$ indicates that we found the partial derivative with respect to $y$ first, then with respect to $x$. This is natural since $\frac{\partial^2 z}{\partial x \partial y}$ means $\frac{\partial}{\partial x}\left(\frac{\partial z}{\partial y}\right)$. On the other hand, when we use the subscript notation, $f_{xy}$ indicates that differentiation with respect to $x$ is performed first, then with respect to $y$ since $f_{xy} = (f_x)_y$.

**EXAMPLE 4**   Find the four second partial derivatives of $f(x, y) = e^{x^2 y^3}$.

**Solution**   The partial derivative with respect to $x$ is

$$f_x(x, y) = e^{x^2 y^3} \frac{\partial}{\partial x}(x^2 y^3) \qquad \text{(by Formula 7)}$$

$$= e^{x^2 y^3}(y^3)\left(\frac{\partial}{\partial x}x^2\right) = e^{x^2 y^3}(y^3)(2x) = 2xy^3 e^{x^2 y^3}$$

Similarly,

$$f_y(x, y) = e^{x^2y^3}\frac{\partial}{\partial y}(x^2y^3) \qquad \text{(by Formula 7)}$$

$$= e^{x^2y^3}(x^2)\left(\frac{\partial}{\partial y}y^3\right) = e^{x^2y^3}(x^2)(3y^2) = 3x^2y^2e^{x^2y^3}$$

We find $f_{xx}$ by differentiating $f_x$ with respect to $x$.

$$f_{xx}(x, y) = \frac{\partial}{\partial x}(2xy^3e^{x^2y^3})$$

$$= (2xy^3)\frac{\partial}{\partial x}e^{x^2y^3} + e^{x^2y^3}\left(\frac{\partial}{\partial x}2xy^3\right) \qquad \text{(by Formula 3)}$$

$$= (2xy^3)e^{x^2y^3}\frac{\partial}{\partial x}(x^2y^3) + e^{x^2y^3}\left(2y^3\frac{\partial}{\partial x}x\right) \qquad \text{(by Formula 7)}$$

$$= (2xy^3)e^{x^2y^3}(y^3)(2x) + e^{x^2y^3}(2y^3)(1) \qquad \text{(by Formula 5)}$$

$$= 2y^3e^{x^2y^3}(2x^2y^3 + 1)$$

Also,

$$f_{yy}(x, y) = \frac{\partial}{\partial y}(3x^2y^2e^{x^2y^3})$$

$$= (3x^2y^2)\frac{\partial}{\partial y}e^{x^2y^3} + e^{x^2y^3}\left(\frac{\partial}{\partial y}3x^2y^2\right) \qquad \text{(by Formula 3)}$$

$$= (3x^2y^2)e^{x^2y^3}\frac{\partial}{\partial y}(x^2y^3) + e^{x^2y^3}\left(3x^2\frac{\partial}{\partial y}y^2\right) \qquad \text{(by Formula 7)}$$

$$= (3x^2y^2)e^{x^2y^3}(x^2)(3y^2) + e^{x^2y^3}(3x^2)(2y) \qquad \text{(by Formula 5)}$$

$$= 3x^2ye^{x^2y^3}(3x^2y^3 + 2)$$

We find $f_{xy}$ by differentiating $f_x$ with respect to $y$.

$$f_{xy}(x, y) = \frac{\partial}{\partial y}(2xy^3e^{x^2y^3})$$

$$= (2xy^3)\frac{\partial}{\partial y}e^{x^2y^3} + e^{x^2y^3}\left(\frac{\partial}{\partial y}2xy^3\right) \qquad \text{(by Formula 3)}$$

$$= (2xy^3)e^{x^2y^3}\frac{\partial}{\partial y}(x^2y^3) + e^{x^2y^3}\left(2x\frac{\partial}{\partial y}y^3\right) \qquad \text{(by Formula 7)}$$

$$= (2xy^3)e^{x^2y^3}(x^2)(3y^2) + e^{x^2y^3}(2x)(3y^2)$$

$$= 6xy^2e^{x^2y^3}(x^2y^3 + 1)$$

Similarly, we find $f_{yx}$ by differentiating $f_y$ with respect to $x$.

$$f_{yx}(x, y) = \frac{\partial}{\partial x}(3x^2y^2e^{x^2y^3})$$

$$= (3x^2y^2)\frac{\partial}{\partial x}e^{x^2y^3} + e^{x^2y^3}\left(\frac{\partial}{\partial x}3x^2y^2\right) \qquad \text{(by Formula 3)}$$

$$= (3x^2y^2)e^{x^2y^3}\frac{\partial}{\partial x}(x^2y^3) + e^{x^2y^3}\left(3y^2\frac{\partial}{\partial x}x^2\right) \quad \text{(by Formula 7)}$$

$$= (3x^2y^2)e^{x^2y^3}(y^3)(2x) + e^{x^2y^3}(3y^2)(2x) \quad \text{(by Formula 5)}$$

$$= 6xy^2e^{x^2y^3}(x^2y^3 + 1)$$

**Do Exercise 47.**

Observe that in this example, we have $f_{xy}(x, y) = f_{yx}(x, y)$. In more advanced texts, it is shown that if the mixed partial derivatives of a function $f$ of two variables are continuous, then they are equal. More specifically, we have the following theorem.

> **THEOREM 11.1:**   If a function $f$ of two variables has continuous mixed second partial derivatives $f_{xy}$ and $f_{yx}$ in an open rectangle (that is, a rectangle without its boundary), then $f_{xy}(x, y) = f_{yx}(x, y)$ for every point $(x, y)$ in that rectangle.

### Usefulness of Partial Derivatives

When a variable quantity $z$ depends on more than one other variable quantity, it is difficult to study the effect that each of these quantities has on $z$ if we let all of them vary at the same time. For example, the pressure $P$ of a gas depends on both the volume $V$ and the temperature $T$ of that gas. It is customary to hold volume constant and let the temperature vary to study the effect temperature has on pressure, or to keep the temperature constant and vary the volume to study the effect that volume has on pressure. This is in fact why partial differentiation is useful in studying functions of several variables. If we have a function of $n$ independent variables, we keep all but one of the independent variables constant and we differentiate with respect to the remaining independent variable. The result gives us the instantaneous rate of change of the dependent variable with respect to that independent variable, while all other independent variables are held constant.

# Exercise Set 11.2

*In Exercises 1–5, use the definition of partial derivatives to find the values of $f_x$ and $f_y$ at the indicated points. (See Example 2.)*

**1.** $f(x, y) = x^2y^2$ at $(1, 3)$

**2.** $f(x, y) = 4x^3y^2$ at $(-2, 1)$

**3.** $f(x, y) = -3x^2y^4$ at $(-1, -3)$

**4.** $f(x, y) = 6x^4y$ at $(3, -3)$

**5.** $f(x, y) = 2xy^3$ at $(5, 2)$

*In Exercises 6–30, find all first partial derivatives of the given function using the differentiation formulas.*

**6.** $f(x, y) = x^4y^3$

**7.** $f(s, t) = -3s^3t^7$

**8.** $h(u, v) = 9u^{1/2}v^{1/4}$

**9.** $f(x, y) = -3x^2y^5 + 4x^3y$

**10.** $g(s, t) = \dfrac{4st}{3s^2 + 5s^3t}$

**11.** $h(x, y) = \dfrac{3x^2 + 5x^4y^3}{2x^5y^2}$

**12.** $f(x, y) = \dfrac{5xy^3 - 2x^2y^5}{3x^2y^{1/2}}$

**13.** $f(x, y) = 3e^{x^2y^5}$

**14.** $g(s, t) = -2e^{s^3t^4}$

**15.** $h(u, v) = 5e^{-u^2v^6}$

**16.** $f(s, t) = 4s^2e^{3st}$

**17.** $g(x, y) = -4x^3y^2e^{x^3y}$

**18.** $h(s, t) = 3s^4t^{1/4}e^{-3s^2t^5}$

**19.** $f(x, y) = \dfrac{5e^{3xy}}{2x^2y^3}$

**20.** $g(u, v) = 5u^2 3^{2u^2v^4}$

**21.** $h(x, y) = 7xy5^{x^4y^3}$

**22.** $f(s, t) = \ln(s^4 + t^6)$

**23.** $g(x, y) = \ln(3x^2 + 5y^6)$

**24.** $h(u, v) = (u^2 + 4v^6)\ln(u^2 + 4v^6)$

**25.** $f(x, y) = \dfrac{\ln(3x^{10} + 5x^2y^4)}{e^{x^2y^4}}$

**26.** $g(s, t) = \log_2(3s^2 + 5t^2)$

**27.** $h(x, y) = \log_3(x^2y^6 + 6)$

**28.** $f(s, t) = -3t^4 \log_5(s^6 + 10t^8)$

**29.** $g(x, y) = \dfrac{-3\log_7(4x^2 + y^2)}{5x^3y^2 + 1}$

**30.** $h(u, v) = 2^{uv}\log_2(u^4 + 5u^2v^6)$

*In Exercises 31–40, find $f_{xx}(x, y)$ and $f_{yy}(x, y)$.*

**31.** $f(x, y) = 5x^2y^5$

**32.** $f(x, y) = 3e^{x^4y^5}$

**33.** $f(x, y) = 3^{x^5y^3}$

**34.** $f(x, y) = \ln(x^4 + 6y^8)$

**35.** $f(x, y) = \log_5(3x^2 + 5y^4)$

**36.** $f(x, y) = (x^2 + y^2)\ln(x^2 + y^2)$

**37.** $f(x, y) = \dfrac{e^{-3xy}}{x^2y}$

**38.** $f(x, y) = \dfrac{2e^{3xy}}{x^{1/4}}$

**39.** $f(x, y) = e^5 \log_3(e^3x^2 + y^e)$

**40.** $f(x, y) = (x^2 + y^3)5^{-3xy}$

*In Exercises 41–50, verify that $f_{xy}(x, y) = f_{yx}(x, y)$.*

**41.** $f(x, y) = 5x^5y^7$

**42.** $f(x, y) = 3e^{x^6y^2}$

**43.** $f(x, y) = e^{(3x^2 + 5y^3)}$

**44.** $f(x, y) = 3^{x^3y^4}$

**45.** $f(x, y) = x^2e^{5y}$

**46.** $f(x, y) = x^3\ln(2x^4 + 3y^2)$

**47.** $f(x, y) = (x^2 + y^3)e^{xy}$

**48.** $f(x, y) = (3x + 5x^2y^3)\log_2(x^2 + y^2)$

**49.** $f(x, y) = \dfrac{\ln(x^2 + 6y^8)}{e^{5xy}}$

**50.** $f(x, y) = e^{xy} + (xy)^e$

---

# 11.3  Chain Rule (Optional)

## Composition of Functions of Several Variables

In many applications, a quantity may be expressed as a function of two or more variables, each of which is a function of yet another variable. For example, the demand for a commodity A may be a function of the price of that commodity and of the price of another commodity B. As we will see in Section 11.4, this happens if the commodities are complement or substitute. The price of each of these commodities may be increasing with time and consequently each is a function of time. It follows that the demand for commodity A is also a function of time.

**EXAMPLE 1**   The demand for commodity X is given by

$$D_x = 50,000 - 30p_x^2 - 150p_y^2$$

where $p_X$ and $p_Y$ are the prices (in dollars) per unit of commodities X and Y, respectively. Market conditions indicate that $t$ months from now, these prices will be

$$p_X = 6 + .04t$$

and

$$p_Y = 5 + .2\sqrt[3]{t}$$

**a.** Express the demand for commodity X as a function of time.
**b.** What will the demand for commodity X be 8 months from now?

**Solution**   **a.** We substitute $6 + .04t$ for $p_X$ and $5 + .2\sqrt[3]{t}$ for $p_Y$ in the demand equation and obtain

$$D_X = 50{,}000 - 30(6 + .04t)^2 - 150(5 + .2\sqrt[3]{t})^2$$

**b.** Replacing $t$ by 8 in this equation, we obtain

$$D_X = 50{,}000 - 30[6 + .04(8)]^2 - 150(5 + .2\sqrt[3]{8})^2 = 44{,}427.728$$

Therefore 8 months from now the demand for commodity X will be approximately 44,428 units.
**Do Exercise 3.**                                                                              ■

## The Chain Rule

Suppose our goal in Example 1 had been to find the rate of change of demand for commodity X with respect to time. Having expressed $D_X$ as a function of $t$, we would need only to differentiate that expression with respect to $t$. This approach, however, is not always practical. For example, the different functions may be defined by complicated expressions. It is often simpler to use the Chain Rule. Recall that in the case of functions of one variable, we saw that if $y = f(u)$, $u = g(x)$, and $h(x) = f(g(x))$, then under the proper conditions

$$h'(x) = f'(g(x))g'(x)$$

or using Leibniz notation

$$\frac{dy}{dx} = \frac{dy}{du} \cdot \frac{du}{dx}$$

We shall state without proof the multivariable analogue of this result. For simplicity, we begin with the case of a function of two variables $x$ and $y$, each of which is a function of a single independent variable $t$.

THEOREM 11.2:    Let $z = f(x, y)$ be a function with continuous partial derivatives and suppose $x = g(t)$ and $y = h(t)$ where both $g$ and $h$ have continuous derivatives. Then the composite function defined by $z = f(g(t), h(t))$ is differentiable and

$$\frac{dz}{dt} = \frac{\partial z}{\partial x} \cdot \frac{dx}{dt} + \frac{\partial z}{\partial y} \cdot \frac{dy}{dt}$$

**EXAMPLE 2**   Let $z = 3x^2y^4$, $x = e^{3t}$, and $y = e^{2t}$. Find $\frac{dz}{dt}$.

**a.** First writing $z$ as a function of $t$.
**b.** Using the Chain Rule.

**Solution** **a.** Substituting $e^{3t}$ for $x$ and $e^{2t}$ for $y$ in $z = 3x^2y^4$, we obtain

$$z = 3(e^{3t})^2(e^{2t})^4 = 3e^{14t}$$

Thus

$$\frac{dz}{dt} = D_t 3e^{14t} = 3e^{14t}(14) = 42e^{14t}$$

**b.** To use the Chain Rule, we first calculate

$$\frac{\partial z}{\partial x} = \frac{\partial}{\partial x}(3x^2y^4) = 6xy^4$$

$$\frac{\partial z}{\partial y} = \frac{\partial}{\partial y}(3x^2y^4) = 12x^2y^3$$

$$\frac{dx}{dt} = D_t e^{3t} = 3e^{3t}$$

$$\frac{dy}{dt} = D_t e^{2t} = 2e^{2t}$$

Therefore

$$\frac{dz}{dt} = \frac{\partial z}{\partial x} \cdot \frac{dx}{dt} + \frac{\partial z}{\partial y} \cdot \frac{dy}{dt}$$

$$= 6xy^4 \cdot 3e^{3t} + 12x^2y^3 \cdot 2e^{2t}$$

To check that the answers are the same, we substitute $e^{3t}$ for $x$ and $e^{2t}$ for $y$ to obtain

$$\frac{dz}{dt} = 6e^{3t}(e^{2t})^4 \cdot 3e^{3t} + 12(e^{3t})^2(e^{2t})^3 \cdot 2e^{2t} = 42e^{14t}$$

as expected.
**Do Exercise 5.**

We now state the Chain Rule for the case of a function of two variables $u$ and $v$, each of which is a function of two variables $x$ and $y$.

---

**THEOREM 11.3:** Let $z = f(u, v)$, $u = g(x, y)$, $v = h(x, y)$, and $z = F(x, y) = f(g(x, y), h(x, y))$. If $f$, $g$, and $h$ have continuous partial derivatives, then

$$\frac{\partial z}{\partial x} = \frac{\partial z}{\partial u} \cdot \frac{\partial u}{\partial x} + \frac{\partial z}{\partial v} \cdot \frac{\partial v}{\partial x}$$

and

$$\frac{\partial z}{\partial y} = \frac{\partial z}{\partial u} \cdot \frac{\partial u}{\partial y} + \frac{\partial z}{\partial v} \cdot \frac{\partial v}{\partial y}$$

These expressions are generalizations of the simpler case that we stated first. To see this, observe that to find $\frac{\partial z}{\partial x}$, $y$ is held constant and we differentiate with respect to $x$. But, when $y$ is held constant, $u$ and $v$ reduce to functions of the single variable $x$, and we can apply the method of the simpler case at once. Similarly for $\frac{\partial z}{\partial y}$.

**EXAMPLE 3**    Let $z = \ln|u + 3v|$, $u = x^2 + e^y$, and $v = x^2y^3$. Find $\frac{\partial z}{\partial x}$ in terms of $x$ and $y$.

**Solution**    We first find

$$\frac{\partial z}{\partial u} = \frac{\partial}{\partial u}\ln|u + 3v| = \frac{1}{u + 3v} \cdot \frac{\partial}{\partial u}(u + 3v) = \frac{1}{u + 3v} \cdot 1 = \frac{1}{u + 3v}$$

$$\frac{\partial z}{\partial v} = \frac{\partial}{\partial v}\ln|u + 3v| = \frac{1}{u + 3v} \cdot \frac{\partial}{\partial v}(u + 3v) = \frac{1}{u + 3v} \cdot 3 = \frac{3}{u + 3v}$$

$$\frac{\partial u}{\partial x} = \frac{\partial}{\partial x}(x^2 + e^y) = 2x$$

$$\frac{\partial v}{\partial x} = \frac{\partial}{\partial x}(x^2y^3) = 2xy^3$$

Thus

$$\frac{\partial z}{\partial x} = \frac{\partial z}{\partial u} \cdot \frac{\partial u}{\partial x} + \frac{\partial z}{\partial v} \cdot \frac{\partial v}{\partial x}$$

$$= \frac{1}{u + 3v} \cdot 2x + \frac{3}{u + 3v} \cdot 2xy^3 = \frac{2x + 6xy^3}{u + 3v}$$

But $u = x^2 + e^y$ and $v = x^2y^3$. Therefore

$$\frac{\partial z}{\partial x} = \frac{2x + 6xy^3}{(x^2 + e^y) + 3x^2y^3} = \frac{2x(1 + 3y^3)}{x^2 + e^y + 3x^2y^3}$$

**Do Exercise 9.**    ■

## A Further Generalization of the Chain Rule

For a further generalization of the Chain Rule, let $z = F(u, v, w)$, $u = f(x, y)$, $v = g(x, y)$, $w = h(x, y)$, and $z = G(x, y) = F(f(x, y), g(x, y), h(x, y))$. If $F, f, g$, and $h$ have continuous partial derivatives, then

$$\frac{\partial z}{\partial x} = \frac{\partial z}{\partial u} \cdot \frac{\partial u}{\partial x} + \frac{\partial z}{\partial v} \cdot \frac{\partial v}{\partial x} + \frac{\partial z}{\partial w} \cdot \frac{\partial w}{\partial x}$$

and

$$\frac{\partial z}{\partial y} = \frac{\partial z}{\partial u} \cdot \frac{\partial u}{\partial y} + \frac{\partial z}{\partial v} \cdot \frac{\partial v}{\partial y} + \frac{\partial z}{\partial w} \cdot \frac{\partial w}{\partial y}$$

The generalizations of the Chain Rule for functions of many variables are easy to remember if we observe that in the last generalization, $z = F(u, v, w)$ is a function of *three* variables and each of these is a function of the *two* variables $x$ and $y$. So the composition is a function of $x$ and $y$, $z = G(x, y)$. Consequently, we expect *two* partial derivatives $\frac{\partial z}{\partial x}$ and $\frac{\partial z}{\partial y}$.

Observe the following:

1. Each of these partial derivatives is expressed as a sum of *three* terms, one for each of the variables $u$, $v$, and $w$.
2. Each of these terms is a product of two partial derivatives, the first factor has one of the variables $u$, $v$, or $w$ in its "denominator" and the second factor has the same variable in its "numerator."
3. In each of the three terms on the right side, the "numerator" of the first factor and the "denominator" of the second factor are the same as the "numerator" and "denominator" of the partial derivative on the left side.

As a mnemonic device, you may recall that the dot product of two $n$-dimensional vectors is defined as follows:

$$(x_1, x_2, \ldots, x_n) \cdot (y_1, y_2, \ldots, y_n) = x_1 y_1 + x_2 y_2 + \cdots + x_n y_n$$

Then

$$\frac{\partial z}{\partial x} = \frac{\partial z}{\partial u} \cdot \frac{\partial u}{\partial x} + \frac{\partial z}{\partial v} \cdot \frac{\partial v}{\partial x} + \frac{\partial z}{\partial w} \cdot \frac{\partial w}{\partial x}$$

may be written

$$\frac{\partial z}{\partial x} = \left( \frac{\partial z}{\partial u}, \frac{\partial z}{\partial v}, \frac{\partial z}{\partial w} \right) \cdot \left( \frac{\partial u}{\partial x}, \frac{\partial v}{\partial x}, \frac{\partial w}{\partial x} \right)$$

Similarly,

$$\frac{\partial z}{\partial y} = \left( \frac{\partial z}{\partial u}, \frac{\partial z}{\partial v}, \frac{\partial z}{\partial w} \right) \cdot \left( \frac{\partial u}{\partial y}, \frac{\partial v}{\partial y}, \frac{\partial w}{\partial y} \right)$$

Using this pattern, we get a natural generalization of the Chain Rule for a function $f(u_1, u_2, \ldots, u_n)$ of $n$ variables where each of these variables is a function of $m$ variables $x_1, x_2, \ldots, x_m$. (See Exercise 30.)

**EXAMPLE 4**  Let $z = e^{3u^2 vw}$, $u = x^2 + 3y$, $v = 2x^3 - 4y^2$, and $w = \ln(x^2 + 3y^4)$. Find $\dfrac{\partial z}{\partial y}$.

**Solution**  We first find the partial derivatives of $z$ with respect to $u$, $v$, and $w$.

$$\frac{\partial z}{\partial u} = \frac{\partial}{\partial u} e^{3u^2 vw} = e^{3u^2 vw} \cdot \frac{\partial}{\partial u}(3u^2 vw) = 6uvw e^{3u^2 vw}$$

$$\frac{\partial z}{\partial v} = \frac{\partial}{\partial v} e^{3u^2 vw} = e^{3u^2 vw} \cdot \frac{\partial}{\partial v}(3u^2 vw) = 3u^2 w e^{3u^2 vw}$$

$$\frac{\partial z}{\partial w} = \frac{\partial}{\partial w} e^{3u^2 vw} = e^{3u^2 vw} \cdot \frac{\partial}{\partial w}(3u^2 vw) = 3u^2 v e^{3u^2 vw}$$

Since we are seeking the partial derivative of $z$ with respect to $y$, we must also find the partials of $u$, $v$, and $w$ with respect to $y$.

$$\frac{\partial u}{\partial y} = \frac{\partial}{\partial y}(x^2 + 3y) = 3$$

$$\frac{\partial v}{\partial y} = \frac{\partial}{\partial y}(2x^3 - 4y^2) = -8y$$

$$\frac{\partial w}{\partial y} = \frac{\partial}{\partial y}\ln(x^2 + 3y^4) = \frac{1}{x^2 + 3y^4}\frac{\partial}{\partial y}(x^2 + 3y^4) = \frac{12y^3}{x^2 + 3y^4}$$

Using the Chain Rule, we obtain

$$\frac{\partial z}{\partial y} = \frac{\partial z}{\partial u} \cdot \frac{\partial u}{\partial y} + \frac{\partial z}{\partial v} \cdot \frac{\partial v}{\partial y} + \frac{\partial z}{\partial w} \cdot \frac{\partial w}{\partial y}$$

$$= 6uvwe^{3u^2vw}(3) + 3u^2we^{3u^2vw}(-8y) + 3u^2ve^{3u^2vw}\left(\frac{12y^3}{x^2 + 3y^4}\right)$$

$$= 6ue^{3u^2vw}\left(3vw - 4yuw + \frac{6y^3uv}{x^2 + 3y^4}\right)$$

For practice, replace $u$, $v$, and $w$ by their values in terms of $x$ and $y$ to obtain $\frac{\partial z}{\partial y}$ in terms of $x$ and $y$.
**Do Exercise 13.**                                                                 ■

## Application

**EXAMPLE 5**   A manufacturer found that the total cost of producing $q_X$ units of commodity X and $q_Y$ units of commodity Y is

$$C = 5000 + 25q_X + .000002q_Xq_Y^2 + 12q_Y$$

If the price of commodities X and Y are $p_X$ and $p_Y$ in dollars per unit, respectively, the demand for X is

$$q_X = 4000 - 20p_X - 35p_Y$$

and the demand for Y is

$$q_Y = 3500 - .05p_Y^2 - 540\sqrt{p_X}$$

Find the rate of change of the total cost with respect to the price of commodity X when $p_X = 9$ and $p_Y = 40$.

**Solution**   We first find

$$\frac{\partial C}{\partial q_X} = \frac{\partial}{\partial q_X}(5000 + 25q_X + .000002q_Xq_Y^2 + 12q_Y) = 25 + .000002q_Y^2$$

$$\frac{\partial C}{\partial q_Y} = \frac{\partial}{\partial q_Y}(5000 + 25q_X + .000002q_Xq_Y^2 + 12q_Y) = .000004q_Xq_Y + 12$$

$$\frac{\partial q_X}{\partial p_X} = \frac{\partial}{\partial p_X}(4000 - 20p_X - 35p_Y) = -20$$

$$\frac{\partial q_Y}{\partial p_X} = \frac{\partial}{\partial p_X}(3500 - .05p_Y^2 - 540\sqrt{p_X}) = \frac{-270}{\sqrt{p_X}}$$

Using the Chain Rule, we obtain

$$\frac{\partial C}{\partial p_X} = \frac{\partial C}{\partial q_X} \cdot \frac{\partial q_X}{\partial p_X} + \frac{\partial C}{\partial q_Y} \cdot \frac{\partial q_Y}{\partial p_X}$$

$$= (25 + .000002q_Y^2)(-20) + (.000004q_Xq_Y + 12)\left(\frac{-270}{\sqrt{p_X}}\right)$$

When $p_X = 9$ and $p_Y = 40$,

$$q_X = 4000 - 20(9) - 35(40) = 2420$$

and

$$q_Y = 3500 - .05(40)^2 - 540\sqrt{9} = 1800$$

Substituting these values in the expression for $\partial C/\partial p_X$, we obtain

$$\frac{\partial C}{\partial p_X} = [25 + .000002(1800)^2](-20)$$

$$+ [.000004(2420)(1800) + 12]\left(\frac{-270}{\sqrt{9}}\right) = -3277.76$$

We conclude that when the price of commodity Y is held at $40 per unit and the price of commodity X is increased from $9 to $10, the total cost of production decreases by approximately $3278.
**Do Exercise 17.**                                                                 ∎

## Implicit Differentiation

Using the Chain Rule, we may calculate derivatives of functions defined implicitly. (See also Section 7.8.) Suppose for example that $F(x, y) = 0$ defines $y$ implicitly as a function of $x$. That is, there is a function $f$ such that $F(x, f(x)) = 0$ is an identity. Then, assuming the functions involved have continuous derivatives, we use the Chain Rule to obtain

$$\frac{dF(x, f(x))}{dx} = \frac{\partial F(x, y)}{\partial x} \cdot \frac{dx}{dx} + \frac{\partial F(x, y)}{\partial y} \cdot \frac{dy}{dx} = 0$$

Since $\dfrac{dx}{dx} = 1$, we get

$$\frac{dy}{dx} = \frac{-\dfrac{\partial F(x, y)}{\partial x}}{\dfrac{\partial F(x, y)}{\partial y}}$$

provided that $\dfrac{\partial F(x, y)}{\partial y} \neq 0$.

**EXAMPLE 6**   The equation $x^2y^3 + x^2 = y^4 + 3x^3 + 2$ defines a function $f$ where $y = f(x)$. Find $D_xy$.

**Solution**    We first write

$$F(x, y) = x^2y^3 + x^2 - y^4 - 3x^3 - 2 = 0$$

Thus,

$$\frac{\partial F(x, y)}{\partial x} = 2xy^3 + 2x - 9x^2$$

and

$$\frac{\partial F(x, y)}{\partial y} = 3x^2y^2 - 4y^3$$

It follows that

$$\frac{dy}{dx} = \frac{-(2xy^3 + 2x - 9x^2)}{3x^2y^2 - 4y^3} = \frac{9x^2 - 2xy^3 - 2x}{3x^2y^2 - 4y^3}$$

provided $3x^2y^2 - 4y^3 \neq 0$.

Note that we obtained the same result in Example 2, Section 7.8.
**Do Exercise 23.**                                                                      ■

This technique can be generalized to calculate partial derivatives. Suppose that $F(x, y, z) = 0$ defines $z = f(x, y)$ implicitly. Then, if the functions involved have continuous partial derivatives, it can be shown that

$$\frac{\partial z}{\partial x} = -\frac{\dfrac{\partial F(x, y, z)}{\partial x}}{\dfrac{\partial F(x, y, z)}{\partial z}} \quad \text{and} \quad \frac{\partial z}{\partial y} = -\frac{\dfrac{\partial F(x, y, z)}{\partial y}}{\dfrac{\partial F(x, y, z)}{\partial z}}$$

provided that $\dfrac{\partial F(x, y, z)}{\partial z} \neq 0$. (See Exercise 29.)

**EXAMPLE 7**    Suppose that $x^2y^4 + 3xz = 2 - e^{2z}$ defines $z = f(x, y)$ implicitly, and assume that all functions involved have continuous partial derivatives. Find $\frac{\partial z}{\partial x}$ and $\frac{\partial z}{\partial y}$.

**Solution**    Write

$$F(x, y, z) = x^2y^4 + 3xz - 2 + e^{2z} = 0$$

Then,

$$\frac{\partial F(x, y, z)}{\partial x} = 2xy^4 + 3z$$

$$\frac{\partial F(x, y, z)}{\partial y} = 4x^2y^3$$

$$\frac{\partial F(x, y, z)}{\partial z} = 3x + 2e^{2z}$$

It follows that

$$\frac{\partial z}{\partial x} = \frac{-\dfrac{\partial F(x, y, z)}{\partial x}}{\dfrac{\partial F(x, y, z)}{\partial z}} = \frac{-(2xy^4 + 3z)}{3x + 2e^{2z}}$$

and

$$\frac{\partial z}{\partial y} = \frac{-\dfrac{\partial F(x, y, z)}{\partial y}}{\dfrac{\partial F(x, y, z)}{\partial z}} = \frac{-4x^2y^3}{3x + 2e^{2z}}$$

provided that $3x + 2e^{2z} \neq 0$.
**Do Exercise 27.** ∎

# Exercise Set 11.3

**1.** The demand for commodity X is given by

$$q_X = 30,000 - 430\sqrt{p_X} - 350\sqrt[3]{p_Y}$$

where $p_X$ and $p_Y$ are the prices (in dollars) per unit of commodities X and Y, respectively. Market conditions show that $t$ months from now, these prices will be $p_X = 76.2 + .3t$ and $p_Y = 22.2 + 1.2\sqrt{t}$.

**a.** Express the demand for commodity X as a function of time.
**b.** What will the demand for commodity X be 16 months from now?

**2.** Same as Exercise 1 with $q_X = 25,000 - 320\sqrt[3]{p_X} - 250\sqrt{p_Y}$, $p_X = 54.4 + .6t$, and $p_Y = 20.2 + 1.2\sqrt{t}$.

**3.** A department store carries two brands of cologne. Past sales figures show that if brand A is sold for $p_A$ dollars per bottle and brand B for $p_B$ dollars per bottle, the demand for brand A will be $q_A$ bottles per month, where

$$q_A = 500 - 12\sqrt{p_A} - 123\sqrt[4]{p_B}$$

It is estimated that $t$ months from now the prices of brands A and B will be $p_A = 20 + .2t$ and $p_B = 14.75 + .002t^2$ dollars per bottle, respectively.

**a.** Express the demand for brand A as a function of time.
**b.** What will the demand for brand A be 25 months from now?

**4.** A company manufactures products X and Y. The total cost of producing and selling $q_X$ units of X and $q_Y$ units of Y per month is given by

$$C = 10,000 + 350\sqrt{q_X} + .02q_Y^2$$

where $C$ is in dollars. Furthermore, the demand equations for these products are $q_X = 2970 - 10p_X - 35p_Y$ and $q_Y = 1950 - 100\sqrt{p_X} - .25p_Y^2$, where $p_X$ and $p_Y$ are the prices (in dollars) per unit of X and Y, respectively, and $q_X$ and $q_Y$ units are the quantities demanded per month of X and Y, respectively.

**a.** Express the total cost $C$ as a function of $p_X$ and $p_Y$.
**b.** What is the total cost when $p_X = 25$ and $p_Y = 32$?

*In Exercises 5–8, use the Chain Rule to find $\frac{dz}{dt}$. a. First writing z as a function of t. b. Using the Chain Rule.*

**5.** $z = 4x^3y^5$; $x = e^{-2t}$, $y = e^{3t}$
**6.** $z = \sqrt{x^2 + 3y^3}$; $x = e^{4t}$, $y = e^{-5t}$
**7.** $z = \ln(x^2 + y^4)$; $x = 2^t$, $y = t^3$
**8.** $z = e^{x^2y^3}$; $x = t^3$, $y = 3t^2$

*In Exercises 9–16, use the Chain Rule to find $\frac{\partial z}{\partial x}$ and $\frac{\partial z}{\partial y}$ in terms of x and y.*

**9.** $z = \ln(u^2 + 5v^2)$, $u = x^2 + 3y^6$, $v = e^{x^2y}$
**10.** $z = 5e^{u^2v^3}$, $u = x^3 + 3y$, $v = x^4 + \sqrt{y}$
**11.** $z = 5u^2e^{uv}$, $u = \ln(x^2 + y^2)$, $v = 6x^2y^5$
**12.** $z = \sqrt{u^2 + v^2}$, $u = 3x + 2y$, $v = e^{xy}$
**13.** $z = e^{5u^3v^2w^4}$, $u = x^3 + 6y$, $v = 5x^2 + 6y^2$, $w = 5x^6 + 3y^4$
**14.** $z = \ln(u^2 + v^2 + 4w^6)$, $u = 2x^2 - 3y$, $v = 2x^3 + 4y^4$, $w = 3x^2 + 3y$
**15.** $z = \sqrt{r^2 + s^2 + t^4}$, $r = e^{xy}$, $s = x^2y^3$, $t = x + 3y$
**16.** $z = \dfrac{5u^2}{v^2 + w^3}$, $u = 4x + 3y$, $v = 3x^2y^5$, $w = 4e^{xy}$

17. A manufacturer found that the total cost of producing $q_X$ units of commodity X and $q_Y$ units of commodity Y is

$$C = 3000 + 16q_X + .00003q_X q_Y{}^2 + 17q_Y$$

If the price of commodities X and Y are $p_X$ and $p_Y$ in dollars per unit, respectively, the demand for X is

$$q_X = 5000 - 15p_X - 26p_Y$$

and the demand for Y is

$$q_Y = 4500 - .03p_X{}^2 - 540\sqrt{p_Y}$$

Find the rate of change of the total cost with respect to the price of commodity X when $p_X = 15$ and $p_Y = 36$.

18. In Exercise 17, find the rate of change of the total cost with respect to the price of commodity Y when $p_X = 17$ and $p_Y = 25$.

19. For the company of Exercise 4, use the Chain Rule to find the rate of change of total cost with respect to $p_X$ when $p_X = 25$ and $p_Y = 32$.

20. The radius and height of a right circular cone are increasing at the rates of 1.5 in./min and 2 in./min, respectively. How fast is the volume increasing when the radius is 30 inches and the height is 100 inches?

21. Motorist A is traveling south toward an intersection at a speed of 88 ft/sec while motorist B is traveling east, away from the same intersection at a speed of 73 ft/sec. How fast is the distance between the two motorists changing when A and B are 4000 feet and 5000 feet from the intersection, respectively?

*In Exercises 22–24, an equation defines y implicitly as a differentiable function of x. Use the method of Example 6 to find $\frac{dy}{dx}$.*

22. $x^2 y^3 + 5x^4 = 3xy^5 - 7y^4 + 10$

23. $4x^3 + e^{xy} = 7x^2 y^3 + e^x + 3$

24. $\ln(x^2 + y^2) + x^5 = 8y^2 + e^y - 5e^x$

*In Exercises 25–28, an equation defines z as a function of x and y. Use the method of Example 7 to find $\frac{\partial z}{\partial x}$ and $\frac{\partial z}{\partial y}$.*

25. $x^2 + y^2 + z^2 = 36$

26. $x^2 - 5y^3 + z^2 = 15 - x^2 y z^3$

27. $e^{xy} + e^{xz} + e^{yz} = 50$

28. $\ln(x^2 + z^2) + z - x^2 y = 10$

29. Suppose that $F(x, y, z) = 0$ defines $z = f(x, y)$ implicitly. Assuming that the functions involved have continuous partial derivatives, show that

$$\frac{\partial z}{\partial x} = \frac{-\dfrac{\partial F(x, y, z)}{\partial x}}{\dfrac{\partial F(x, y, z)}{\partial z}} \quad \text{and} \quad \frac{\partial z}{\partial y} = \frac{-\dfrac{\partial F(x, y, z)}{\partial y}}{\dfrac{\partial F(x, y, z)}{\partial z}}$$

provided that $\frac{\partial F(x, y, z)}{\partial z} \neq 0$. (*Hint:* Start with $u = F(x, y, z) = 0$. Use the Chain Rule to find $\frac{\partial u}{\partial x}$, then use the fact that $\frac{\partial x}{\partial x} = 1$, $\frac{\partial y}{\partial x} = 0$ (since y is treated as a constant), and $\frac{\partial u}{\partial x} = 0$. Then solve for $\frac{\partial z}{\partial x}$. (Proceed similarly to find $\frac{\partial z}{\partial y}$.)

30. Let $w = f(u, x, y, z)$, where $u$, $x$, $y$, and $z$ are functions of the three variables $r$, $s$, and $t$. Write the expressions for $\frac{\partial u}{\partial r}$, $\frac{\partial u}{\partial s}$, $\frac{\partial u}{\partial t}$.

# 11.4   Applications of Partial Differentiation

## Approximations Using Partial Derivatives

Differentials were used in Section 8.10 to approximate the value of a function $f$ near $x_0$, knowing the value of $f(x_0)$. We generalize this method in order to approximate the value of $f$, a function of two variables, near $(x_0, y_0)$, knowing the value $f(x_0, y_0)$.

Just as the equation of the tangent line to the graph of $y = f(x)$ at $(x_0, y_0)$ is

$$y - y_0 = f'(x_0)(x - x_0)$$

provided $f'(x_0)$ exists, the equation of the tangent plane to the graph of $z = f(x, y)$ at the point $(x_0, y_0, z_0)$ is

$$z - z_0 = f_x(x_0, y_0)(x - x_0) + f_y(x_0, y_0)(y - y_0)$$

provided that $f_x(x_0, y_0)$ and $f_y(x_0, y_0)$ exist. At the point $(x_0+\Delta x, y_0+\Delta y)$, we have

$$z - z_0 = f_x(x_0, y_0)(x_0 + \Delta x - x_0) + f_y(x_0, y_0)(y_0 + \Delta y - y_0)$$
$$= f_x(x_0, y_0)\Delta x + f_y(x_0, y_0)\Delta y$$

Thus

$$z = f_x(x_0, y_0)\Delta x + f_y(x_0, y_0)\Delta y + z_0 \qquad\qquad (*)$$

Over a small neighborhood of $(x_0, y_0)$, the plane tangent to the graph of $z = f(x, y)$ at the point $(x_0, y_0, z_0)$ approximates the graph. Observe that the point $(x_0+\Delta x, y_0+\Delta y, z)$ for the value of $z$ obtained in equation (*) is on the tangent plane. Thus it is near the point $(x_0+\Delta x, y_0+\Delta y, f(x_0+\Delta x, y_0+\Delta y))$, which is on the graph of $f$. Consequently, the value of $z$ obtained in (*) is a good approximation of $f(x_0+\Delta x, y_0+\Delta y)$ provided that $\Delta x$ and $\Delta y$ are small in absolute value. (See Figure 11.12.)

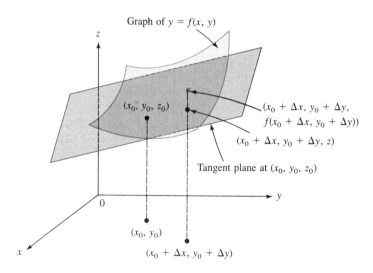

**FIGURE 11.12**

EXAMPLE 1   Let $f(x, y) = x^2y^3$.

a. Calculate the exact value of $f(3.01, 2.03)$.
b. Approximate the value of $f(3.01, 2.03)$ by first finding the value of $f(3, 2)$ and using partial derivatives.
c. Compare the two results.

**Solution**   a. The exact value is

$$f(3.01, 2.03) = (3.01)^2(2.03)^3 = 75.7916051627$$

b. We also have

$$f(3, 2) = (3^2)(2^3) = 72$$

Furthermore,

$$f_x(x, y) = 2xy^3 \text{ and } f_y(x, y) = 3x^2y^2$$

Thus,

$$f_x(3, 2) = 2(3)(2^3) = 48 \text{ and } f_y(3, 2) = 3(3^2)(2^2) = 108$$

Here, $\Delta x = 3.01 - 3 = .01$ and $\Delta y = 2.03 - 2 = .03$. Therefore

$$f(3.01, 2.03) \doteq f_x(3, 2)\Delta x + f_y(3, 2)\Delta y + f(3, 2)$$
$$= (48)(.01) + (108)(.03) + 72 = 75.72$$

c. Using the approximate value gives an error of .09% since

$$\frac{75.7916051627 - 75.72}{75.7916051627} \doteq .000944$$

**Do Exercise 1.**

■

EXAMPLE 2    A store knows that if the price of choice beef is $p_B$ dollars per pound and the price of frozen salmon is $p_S$ dollars per pound, the number of pounds of choice beef sold each week is

$$f(p_B, p_S) = 15,000 - 1000p_B^{1/2} + 600p_S^{1/2}$$

a. Calculate the number of pounds of choice beef sold in one week if the price of choice beef and frozen salmon are $4 per pound and $2.25 per pound, respectively.
b. Using the result of part **a**, estimate the number of pounds of choice beef sold during a week during which the price of choice beef is $4.10 per pound and the price of frozen salmon is $2.05 per pound.

Solution    a. We have

$$f(4, 2.25) = 15,000 - 1000(4^{1/2}) + 600(2.25^{1/2}) = 13,900$$

Thus 13,900 pounds of choice beef will be sold when the price of beef is $4.00 per pound and the price of salmon is $2.25 per pound.
b. Here, $\Delta p_B = 4.10 - 4 = .10$ and $\Delta p_S = 2.05 - 2.25 = -.20$. Also,

$$f_{p_B}(p_B, p_S) = -500p_B^{-1/2} \text{ and } f_{p_S}(p_B, p_S) = 300p_S^{-1/2}$$

Therefore

$$f_{p_B}(4, 2.25) = -500(4^{-1/2}) = -250 \text{ and } f_{p_S}(4, 2.25)$$
$$= 300(2.25^{-1/2}) = 200$$

It follows that

$$f(4.10, 2.05) \doteq f_{p_B}(4, 2.25)\Delta p_B + f_{p_S}(4, 2.25)\Delta p_S + f(4, 2.25)$$
$$= (-250)(.10) + (200)(-.20) + 13,900 = 13,835$$

Therefore, approximately 13,835 pounds of choice beef will be sold during a week when the prices of choice beef and frozen salmon are $4.10 and $2.05 per pound, respectively.
**Do Exercise 7.**

■

We now discuss other applications of partial derivatives. Consider two commodities X and Y and suppose that the price per unit of commodity X is $p_X$ dollars and the price per unit of commodity Y is $p_Y$ dollars. Suppose further that the demand functions for these commodities are

$$D_X = f(p_X, p_Y) \text{ and } D_Y = g(p_X, p_Y)$$

We shall study the significance of the partial derivatives of these two functions. We shall also expand on the concept of elasticity discussed in Section 8.2.

### Marginal Analysis

The partial derivative $\partial D_X/\partial p_X$ is the *marginal demand for commodity X with respect to $p_X$*. It approximates the amount by which the demand for commodity X changes per unit increase in the price of commodity X when the price of commodity Y remains constant.

Similarly, $\partial D_X/\partial p_Y$ is the *marginal demand for commodity X with respect to $p_Y$* and approximates the change in the demand for commodity X per unit increase in the price of commodity Y when the price of commodity X remains constant.

The partial derivatives $\partial D_Y/\partial p_X$ and $\partial D_Y/\partial p_Y$ have similar interpretations.

**EXAMPLE 3**   The demand functions for commodities X and Y are given by

$$D_X = 4000 - 5p_X^2 - 12p_Y^3 \text{ and } D_Y = 4500 - 4p_X^3 - 10p_Y^2$$

Find the values of the four marginal demand functions and interpret the results when $p_X = 6$ and $p_Y = 5$.

**Solution**   By definition, the marginal demand for commodity X with respect to $p_X$ is given by

$$\frac{\partial D_X}{\partial p_X} = -10p_X$$

When $p_X = 6$, $\partial D_X/\partial p_X = -60$. Thus if the price of commodity Y remains fixed, an increase in the price of commodity X from \$6 to \$7 will cause a decrease in the demand for commodity X of 60 units.

Similarly,

$$\frac{\partial D_X}{\partial p_Y} = -36p_Y^2$$

When $p_Y = 5$, $\partial D_X/\partial p_Y = -900$. Therefore if the price of commodity X is held constant, an increase in the price of commodity Y from \$5 to \$6 per unit will result in a decrease in the demand for commodity X of 900 units.

Also,

$$\frac{\partial D_Y}{\partial p_X} = -12p_X^2 \quad \text{and} \quad \frac{\partial D_Y}{\partial p_Y} = -20p_Y$$

When $p_X = 6$ and $p_Y = 5$, we have

$$\frac{\partial D_Y}{\partial p_X} = -432 \quad \text{and} \quad \frac{\partial D_Y}{\partial p_Y} = -100$$

We conclude that if the price of commodity Y is held constant, an increase in the price of commodity X from $6 to $7 per unit will cause a decrease in demand for commodity Y of 432 units. Furthermore, if the price of commodity X remains constant, an increase in the price of commodity Y from $5 to $6 per unit will result in a decrease in the demand for commodity Y of 100 units.

**Do Exercise 11.**

As we have seen earlier, demand functions are decreasing functions. Suppose that we have two commodities X and Y, whose prices per unit are $p_X$ and $p_Y$, respectively. Suppose further that the demand equations are

$$D_X = f(p_X, p_Y) \text{ and } D_Y = g(p_X, p_Y)$$

Then, if the price of commodity Y is held constant, the demand for commodity X will decrease as its price $p_X$ increases. Therefore $\partial D_X/\partial p_X \leq 0$. Similarly, $\partial D_Y/\partial p_Y \leq 0$.

Note that in the preceding example, the partial derivatives $\partial D_X/\partial p_Y$ and $\partial D_Y/\partial p_X$ were both negative. In general, what can be said about the signs of these partial derivatives? Pairs of commodities are classified as *substitutes (competitive)* or *complements* by observing the interaction between the demands and prices of these commodities. For example, an increase in the cost of racquetball court time may result in a decrease in the demand for racquetballs even if the price of balls remains constant. On the other hand, an increase in the price of meat may cause an increase in the demand for fish although the price of fish remains constant because some people will change from meat to fish owing to the price increase of meat. Therefore meat and fish are classified as substitutes (competitive).

> **DEFINITION 11.6:** The commodities X and Y are *substitutes* provided that $\partial D_X/\partial p_Y > 0$ and $\partial D_Y/\partial p_X > 0$. On the other hand, the two commodities are *complements* if, and only if, $\partial D_X/\partial p_Y < 0$ and $\partial D_Y/\partial p_X < 0$.

**EXAMPLE 4**   The demand functions for commodities X and Y are given by

$$D_X = 5000 - 3p_X^2 - 15p_Y^2 \text{ and } D_Y = 4500 - 5p_X^2 - 12p_Y^2$$

respectively, where $p_X$ dollars is the price per unit of commodity X and $p_Y$ dollars is the price per unit of commodity Y.

**a.** Find the four marginal demand functions.

**b.** Determine whether the commodities are substitutes or complements.

**Solution**   **a.** By definition, the four marginal demand functions are the four partial derivatives.

$$\frac{\partial D_X}{\partial p_X} = -6p_X, \frac{\partial D_X}{\partial p_Y} = -30p_Y, \frac{\partial D_Y}{\partial p_X} = -10p_X, \text{ and } \frac{\partial D_Y}{\partial p_Y} = -24p_Y$$

**b.** Since $p_X$ and $p_Y$ are positive, we see that $\partial D_X/\partial p_Y$ and $\partial D_Y/\partial p_X$ are both negative. Thus, the two commodities are complements.

**Do Exercise 17.**

### Cross Elasticities

We now expand on the concept of elasticity that was introduced in Section 8.2. Again, suppose we have two commodities X and Y, whose prices per unit are $p_X$ and $p_Y$, respectively. Suppose further that the demand equations are

$$D_X = f(p_X, p_Y) \text{ and } D_Y = g(p_X, p_Y)$$

*The elasticity of demand for commodity X is defined by*

$$\eta_{XX} = \frac{\partial D_X}{\partial p_X} \cdot \frac{p_X}{D_X}$$

As was done in Section 8.2, it can be shown that $\eta_{XX}$ approximates the ratio of the percentage change in demand for commodity X to the percentage change in the price of commodity X as the price of commodity Y remains constant.

Now let $\Delta p_Y$ be the change in the price of commodity Y, $p_Y$ be the original price of commodity Y, $\Delta D_X$ the resulting change in quantity demanded of commodity X, and $D_X$ the original quantity demanded of commodity X. Then the ratio of the percentage change in demand for commodity X to the percentage change in the price of commodity Y is

$$\frac{\dfrac{\Delta D_X}{D_X}}{\dfrac{\Delta p_Y}{p_Y}}$$

This ratio can be written

$$\frac{\Delta D_X}{\Delta p_Y} \cdot \frac{p_Y}{D_X}$$

In the limit, as $\Delta p_Y$ approaches 0, the foregoing expression yields

$$\frac{\partial D_X}{\partial p_Y} \cdot \frac{p_Y}{D_X}$$

This leads to the following definition.

**DEFINITION 11.7:** The *cross elasticity of demand of commodity X with respect to the price of commodity Y* is defined by

$$\eta_{XY} = \frac{\partial D_X}{\partial p_Y} \cdot \frac{p_Y}{D_X}$$

Thus $\eta_{XY}$ can be interpreted as an approximation of the ratio of the percentage change in the demand for commodity X to the percentage change in the price of commodity Y as the price of commodity X remains fixed. Of course, $\eta_{YY}$ and $\eta_{YX}$ are defined similarly. We observe that if commodities X and Y are substitutes (competitive), then $\eta_{XY}$ and $\eta_{YX}$ are positive. On the other hand, when commodities X and Y are complements, $\eta_{XY}$ and $\eta_{YX}$ are negative.

**EXAMPLE 5**  The demand functions for commodities X and Y are given by

$$D_X = 4500 - 4p_X^2 + 12p_Y \text{ and } D_Y = 6000 + 5p_X - 3p_Y^2$$

respectively.

**a.** Find the elasticity of demand for X when $p_X = 25$ and $p_Y = 10$, and interpret the result.
**b.** Find the cross elasticity of demand for X with respect to $p_Y$ when $p_X = 25$ and $p_Y = 10$, and interpret the result.
**c.** Find the cross elasticity of demand for Y with respect to $p_X$ when $p_X = 25$ and $p_Y = 10$, and interpret the result.

**Solution**  **a.** By definition, the elasticity of demand for X is

$$\eta_{XX} = \frac{\partial D_X}{\partial p_X} \cdot \frac{p_X}{D_X}$$

In this case, $\partial D_X / \partial p_X = -8p_X$. When $p_X = 25$ and $p_Y = 10$, we have

$$\frac{\partial D_X}{\partial p_X} = (-8)(25) = -200$$

and

$$D_X = 4500 - 4(25^2) + 12(10) = 2120$$

Thus

$$\eta_{XX} = -200 \cdot \frac{25}{2120} \doteq -2.36$$

Therefore if the price of commodity Y remains constant, an increase of 1% in the price of commodity X will result in an approximate decrease of 2.36% in the demand for that commodity.

**b.** By definition, the cross elasticity of demand for X with respect to $p_Y$ is

$$\eta_{XY} = \frac{\partial D_X}{\partial p_Y} \cdot \frac{p_Y}{D_X}$$

We have

$$\frac{\partial D_X}{\partial p_Y} = 12$$

Thus

$$\eta_{XY} = 12 \cdot \frac{10}{2120} \doteq .057$$

Therefore if the price of commodity X remains constant, an increase of 1% in the price of commodity Y will result in an approximate increase of 0.057% in the demand for commodity X.

**c.** By definition

$$\eta_{YX} = \frac{\partial D_Y}{\partial p_X} \cdot \frac{p_X}{D_Y}$$

We have

$$\frac{\partial D_Y}{\partial p_X} = 5$$

and when $p_X = 25$ and $p_Y = 10$, we get

$$D_Y = 6000 + 5(25) - 3(10^2) = 5825$$

Thus

$$\eta_{YX} = 5 \cdot \frac{25}{5825} \doteq .021$$

We conclude that if the price of commodity Y remains constant, an increase of 1% in the price of commodity X will cause an approximate increase of .021% in the demand for commodity Y.

**Do Exercise 29.**    ■

# Exercise Set 11.4

**1.** Let $f(x, y) = \sqrt{x^2 + y^2}$.

  **a.** Find $f(4, 3)$.

  **b.** Using partial derivatives and the result of part **a**, approximate $f(4.1, 2.8)$.

**2.** Let $g(x, y) = \sqrt{x} \cdot \sqrt[3]{y^2}$.

  **a.** Find $g(25, 8)$.

  **b.** Using partial derivatives and the result of part **a**, approximate $g(24.8, 8.1)$.

**3.** Let $h(x, y) = \dfrac{\sqrt[3]{x^2}}{\sqrt{y}}$.

  **a.** Find $h(27, 4)$.

  **b.** Using the result of part **a** and partial derivatives, approximate $h(26, 4.5)$.

**4.** Let $f(x, y) = \sqrt{x^2 + y^3}$.

  **a.** Calculate $f(66, 12)$.

  **b.** Using the result of part **a** and partial derivatives, approximate $f(65, 12.1)$.

**5.** Let $g(x, y) = \sqrt{x^2 - y^2}$.

  **a.** Calculate $g(13, 5)$.

  **b.** Using the result of part **a** and partial derivatives, approximate $g(12.9, 4.2)$.

**6.** Let $h(x, y) = \dfrac{\sqrt{x + y}}{x - y}$.

  **a.** Find $h(85, 84)$.

  **b.** Using partial derivatives and the result of part **a**, approximate $h(85.2, 83.9)$.

**7.** A store has found that if the price of lamb is $p_L$ dollars per pound and the price of pork is $p_P$ dollars per pound, the number of pounds of lamb sold each week is

$$L(p_L, p_P) = 6000 - 400p_L^{1/2} + 800p_P^{1/2}$$

  **a.** Calculate the number of pounds of lamb sold in one week if the price of lamb is $4 per pound and the price of pork is $2.25 per pound.

  **b.** Using the result of part **a**, estimate the number of pounds of lamb sold in one week if the prices of lamb and pork are $3.95 and $2.30 per pound, respectively.

**8.** The store in Exercise 7 also found that the number of pounds of pork sold per week is

$$P(p_L, p_P) = 4000 + 300p_L^{1/3} - 500p_P^{1/2}$$

  **a.** Calculate the number of pounds of pork sold during one week if the prices of lamb and pork are $8 and $4 per pound, respectively.

  **b.** Using the result of part **a** and partial derivatives, estimate how many pounds of pork will be sold in one week if the price of lamb is $7.50 per pound and that of pork is $4.10 per pound.

9. At a certain factory, the production $Q$ units per week of a certain commodity is given by the equation

$$Q = 250K^{1/2}L^{1/2}$$

where $K$ denotes the capital investment measured in thousands of dollars and $L$ is the number of worker hours per week. The current capital investment is \$900,000 and the present number of worker hours per week is 8100.

a. Calculate the number of units produced per week at the present level.
b. Use partial derivatives to estimate the level of production per week if the capital investment is decreased to \$898,000 and the number of hours per week is increased to 8140.

10. An editor found that if $x$ thousand dollars is spent on advertising and $y$ thousand dollars is spent on production, the number $Q$ of copies of a new mathematics book that will be sold is

$$Q = .25x^{3/2}y^{4/3}$$

The editor budgeted \$36,000 for advertising and \$512,000 for production.

a. If that level of expenditure is kept, how many books will be sold?
b. Using partial derivatives and the result of part a, estimate the number of copies that will be sold if the amount spent on advertising is increased by \$1000 and the amount spent on production is decreased by \$2000.

11. From experience, the manager of a store knows that if the price of brand A chocolate is $p_A$ dollars per pound and that of brand B chocolate is $p_B$ dollars per pound, then the daily profit derived from these two brands of chocolate will be given by

$$P(p_A, p_B) = -4p_A^2 + 8p_Ap_B - 6p_B^2 + 57p_A$$
$$+ 88p_B - 970$$

a. Find the marginal profit with respect to $p_A$ when $p_A = 10$ and $p_B = 8$, and interpret the result.
b. Find the marginal profit with respect to $p_B$ when $p_A = 10$ and $p_B = 8$, and interpret your result.

12. The manager of a gas station has found that when the price of regular gasoline is $p_R$ dollars per gallon and the price of premium gasoline is $p_P$ dollars per gallon, the daily profit will be given by

$$P(p_R, p_P) = -200p_R^2 + 450p_Rp_P - 200p_P^2$$
$$+ 3030p_R + 1930p_P - 4700$$

a. Find the marginal profit with respect to $p_R$ when $p_R = 1.05$ and $p_P = 1.25$, and interpret your result.
b. Find the marginal profit with respect to $p_P$ when $p_R = 1.05$ and $p_P = 1.25$, and interpret your result.

13. A store manager knows that if the prices of commodities A and B are $p_A$ and $p_B$ dollars per unit, respectively, the demand for commodity A is given by some function $D = f(p_A, p_B)$ but has no knowledge of the formulation of that function. However, the following information is available. When $p_A = 5$ and $p_B = 7$, $D = 8000$. Furthermore, at these prices, the marginal demand with respect to $p_A$ is 120 and the marginal demand with respect to $p_B$ is 180. Approximate the demand for commodity A when $p_A = 5.20$ and $p_B = 6.90$.

14. The store manager of Exercise 13 also knows that the demand for commodity B is given by some function $D = g(p_A, p_B)$ but has no specific formulation of that function. The following information is available. When $p_A = 6$ and $p_B = 8$, $D = 7500$. Also, at these prices, the marginal demand with respect to $p_A$ is 250 and the marginal demand with respect to $p_B$ is 300. Approximate the demand for commodity B when $p_A = 5.90$ and $p_B = 8.25$.

15. A hardware store carries two brands of paint. Studies have shown that if brand A sells for $p_A$ dollars per gallon and brand B sells for $p_B$ dollars per gallon, the demand for brand A will be given by

$$D_A = 950 - 5p_A^2 + 16p_B^{1/2}$$

and that for brand B will be given by

$$D_B = 120 + 12p_A - 15p_B^{1/2}$$

where $D_A$ and $D_B$ are the numbers of gallons sold each week. Find the values of the four marginal demand functions when $p_A = 13$ and $p_B = 16$, and interpret your results.

16. A store carries two kinds of detergents. If $D_A$ and $D_B$ are the respective demands for brand A and brand B in thousands of boxes per week, then when the price of brand A is $p_A$ dollars per box and the price of brand B is $p_B$ dollars per box, we have

$$D_A = 650 - 12p_A^2 + 10p_B^3$$

and

$$D_B = 800 + 15p_A^3 - 13p_B^2$$

Find the values of the four marginal demand functions when $p_A = 4.50$ and $p_B = 5.10$, and interpret your results.

*In Exercises 17–26, two demand functions are given for commodities X and Y where $p_X$ and $p_Y$ dollars are the prices per unit of commodities X and Y, respectively. In each case, find the four* *marginal demand functions and conclude whether the commodities are substitutes (competitive) or complements.*

**17.** $D_X = 500 - 5p_X + 4p_Y$, $D_Y = 700 + 6p_X - 5p_Y$

**18.** $D_X = 4000 - 2p_X^2 + 3p_Y^2$, $D_Y = 5000 + 3p_X^2 - 5p_Y^2$

**19.** $D_X = 3750 - 4p_X^{3/2} - 5p_Y^2$, $D_Y = 4250 - 4p_X^2 - 7p_Y^{5/2}$

**20.** $D_X = 7250 - 5p_X^2 - 7p_Y^{4/3}$, $D_Y = 8000 - 4p_X^{3/2} - 9p_Y^3$

**21.** $D_X = 8500 - 6p_X^{7/3} + 8p_Y^{5/2}$, $D_Y = 9000 + 3p_X^{5/3} - 4p_Y^{3/2}$

**22.** $D_X = 4700 - 3p_X^{5/2} - 5p_Y^{6/5}$, $D_Y = 5250 - 4p_X^2 - 6p_Y^{4/3}$

**23.** $D_X = 500 - \ln p_X + 3p_Y^2$, $D_Y = 600 + \ln(p_X + 1) - 2p_Y^3$

**24.** $D_X = 1250 - .12p_X^2 - \ln(p_Y + 1)$,
$D_Y = 2000 - 2p_X^{3/2} - \ln(p_Y^3 + 1)$

**25.** $D_X = \dfrac{150p_Y}{p_X^{1/2}}$, $D_Y = \dfrac{175p_X}{p_Y^{1/4}}$

**26.** $D_X = \dfrac{180p_Y}{p_X^{1/3}}$, $D_Y = \dfrac{230p_X^2}{p_Y^{4/3}}$

*In Exercises 27–34, two demand functions are given for commodities X and Y where $p_X$ and $p_Y$ dollars are the prices per unit of commodities X and Y, respectively. In each case, do the following for the given functions and the given prices and interpret your results.*

**a.** *Find the elasticity of demand for commodity X.*
**b.** *Find the elasticity of demand for commodity Y.*
**c.** *Find the cross elasticity of demand for X with respect to $p_Y$.*
**d.** *Find the cross elasticity of demand for Y with respect to $p_X$.*

**27.** $D_X = 450 - 3p_X + 5p_Y$, $D_Y = 600 + 5p_X - 4p_Y$; $p_X = 6$, $p_Y = 4$

**28.** $D_X = 4500 - 3p_X^2 + 4p_Y^2$, $D_Y = 6000 + 4p_X^2 - 3p_Y^2$; $p_X = 10$, $p_Y = 12$

**29.** $D_X = 3500 - 4p_X^{3/2} - 6p_Y^2$, $D_Y = 4000 - 4p_X^{5/2} - 6p_Y^2$; $p_X = 9$, $p_Y = 6$

**30.** $D_X = 7250 - 5p_X^2 - 7p_Y^{4/3}$, $D_Y = 8000 - 4p_X^{3/2} - 9p_Y^3$; $p_X = 4$, $p_Y = 8$

**31.** $D_X = 8500 - 6p_X^{7/3} + 8p_Y^{5/2}$, $D_Y = 9000 + 3p_X^{5/3} - 4p_Y^{3/2}$; $p_X = 8$, $p_Y = 16$

**32.** $D_X = 4700 - 3p_X^{5/2} - 5p_Y^{6/5}$, $D_Y = 5250 - 4p_X^2 - 6p_Y^{7/5}$; $p_X = 16$, $p_Y = 32$

**33.** $D_X = \dfrac{100p_Y}{p_X^{1/2}}$, $D_Y = \dfrac{120p_X}{p_Y^{1/3}}$; $p_X = 9$, $p_Y = 27$

**34.** $D_X = \dfrac{150p_Y}{p_X^{1/4}}$, $D_Y = \dfrac{230p_X^{3/2}}{p_Y^{3/4}}$; $p_X = 16$, $p_Y = 16$

# 11.5 Optimization

Maxima and minima of a function of one variable were discussed in Sections 8.3–8.7. We now discuss optimization of functions of two variables.

## Critical Points

In geometric terms, a relative (local) maximum of a function of two variables is indicated by a peak—a point on the graph of the function that is higher than any nearby point on that graph. On the other hand, a relative (local) minimum is indicated on the surface as the bottom of a valley—a point on the graph that is lower than any other nearby point on that graph. (See Figure 11.13.)

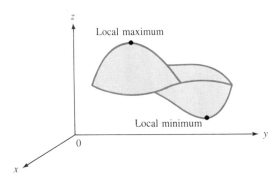

**FIGURE 11.13**

Formally, we have the following definitions.

---

**DEFINITION 11.8:**  Let $f$ be a function of two variables. The function is said to have a *relative (local) maximum at* $(a, b)$ provided there is a circle centered at $(a, b)$ in the $xy$-plane such that $f(x, y) \leq f(a, b)$ for all $(x, y)$ within that circle. The number $f(a, b)$ is called a *relative maximum of the function*. If $f(x, y) < f(a, b)$ for all $(x, y)$ distinct from $(a, b)$ within the circle, then the number $f(a, b)$ is said to be a *strict relative maximum*. Similarly, if $g$ is a function of two variables, then $g$ is said to have a *relative (local) minimum at* $(c, d)$ provided there is a circle centered at $(c, d)$ in the $xy$-plane such that $g(c, d) \leq g(x, y)$ for all $(x, y)$ within that circle. The number $g(c, d)$ is called a *relative (local) minimum of the function*. If $g(c, d) < g(x, y)$ for all $(x, y)$ distinct from $(c, d)$ within the circle, then $g(c, d)$ is said to be a *strict relative minimum*.

---

Recall that if a function $f$ of one variable has a local maximum or minimum at an interior point $c$ of its domain, then $f'(c) = 0$ or $f'(c)$ does not exist, and we called $c$ a critical number of the function $f$. We also showed that a function of one variable can have an extremum only at a critical number or at a boundary point of its domain.

Suppose $g$ is a function of two variables that has a local maximum, or minimum, at an interior point of its domain, say $(a, b)$. Let $f(x) = g(x, b)$. Then $f$ is a function of one variable that has a local maximum, or minimum, at $a$. Thus $f'(a) = 0$ or $f'(a)$ does not exist. Therefore $g_x(a, b) = 0$ or $g_x(a, b)$ does not exist since $f'(x) = g_x(x, b)$. Geometrically, this means that if $C$ is the intersection of the graph of $g$ and a plane parallel to the $xz$-plane through the point $(a, b)$, then the point $(a, b, g(a, b))$ is a high or low point of $C$. (See Figure 11.14.)

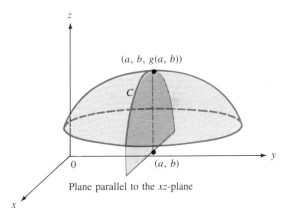

$(a, b, g(a, b))$

$C$

$(a, b)$

Plane parallel to the $xz$-plane

**FIGURE 11.14**

Similarly, if we let $h(y) = g(a, y)$, then $h$ is a function of one variable that has a local maximum, or minimum, at $b$. Consequently, $h'(b) = 0$ or $h'(b)$ is not defined. Therefore $g_y(a, b) = 0$ or $g_y(a, b)$ is not defined since $h'(y) = g_y(a, y)$. Geometrically, this means that if $C$ is the intersection of the graph of $g$ and a plane parallel to the $yz$-plane and passing through $(a, b)$, then $(a, b, g(a, b))$ is a high or low point of $C$. (See Figure 11.15.)

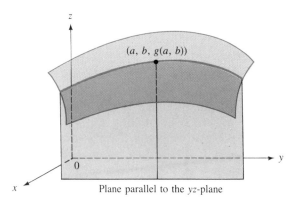

**FIGURE 11.15**

$x$            Plane parallel to the $yz$-plane

In view of this discussion, we give the following definition.

> **DEFINITION 11.9:** Suppose $f$ is a function of two variables. An interior point $(a, b)$ of the domain of $f$ is said to be a *critical point of f* if, and only if, two of the following four statements are true: $f_x(a, b) = 0$, $f_y(a, b) = 0$, $f_x(a, b)$ is not defined, and $f_y(a, b)$ is not defined.

It can be shown that a function of two variables can have a local maximum or minimum in the interior of its domain only at critical points. In this book, we shall consider only critical points for which $f_x(a, b) = f_y(a, b) = 0$.

As in the case of functions of one variable, a function of two variables may have a critical point $(a, b)$ and yet have neither a maximum nor a minimum at $(a, b)$, as illustrated in the following example.

**EXAMPLE 1**  Let $f(x, y) = x^2 - y^2$. Show that $(0, 0)$ is a critical point of $f$ but that $f$ has neither a maximum nor a minimum at the origin.

**Solution**  We have $f_x(x, y) = 2x$ and $f_y(x, y) = -2y$. Therefore

$$f_x(0, 0) = f_y(0, 0) = 0$$

and $(0, 0)$ is a critical point of $f$. Clearly,

$$f(0, 0) = 0^2 - 0^2 = 0$$

However, for any circle centered at the origin in the $xy$-plane, if we choose a point $(x, 0)$ on the $x$-axis and within the circle, we have

$$f(x, 0) = x^2 - 0^2 = x^2$$

Therefore if $x \neq 0$, $f(x, 0)$ is positive and greater than $f(0, 0)$. On the other hand, if we choose a point $(0, y)$ on the $y$-axis and within the circle, we have

$$f(0, y) = 0^2 - y^2 = -y^2$$

It follows that if $y \neq 0$, $f(0, y)$ is negative and smaller than $f(0, 0)$. We conclude that $f(0, 0)$ is neither a local maximum nor a local minimum of the function $f$. (See Figure 11.16.)

**FIGURE 11.16**

The surface of the graph shown in Figure 11.16 resembles a saddle. Therefore the point where the tangent plane to the surface is horizontal, but near the point the surface is partly below and partly above the tangent plane, is called a *saddle point*. As in the case of functions of one variable, geometric interpretations are also helpful in locating relative maxima and minima of a function of two variables. However, geometric interpretation alone can be misleading. It is possible, for example, that *every* section of a graph of a function $f$ of two variables by a vertical plane through a point $(a, b, c)$ is a curve with a low point at $(a, b, c)$ and yet the function $f$ does not have a local minimum at $(a, b)$. (See Exercise 46.)

## The Second Partials Test

As in the case of a function of one variable, we have a test for extrema involving second derivatives. We call it the *Second Partials Test*. It is stated as a theorem the proof of which is beyond the scope of this book.

> **THEOREM 11.4:** Let $f$ be a function of two variables. Suppose that the partial derivatives of $f$ are continuous in a disk centered at $(a, b)$ and that $f_x(a, b) = f_y(a, b) = 0$.
>   Let $A = f_{xx}(a, b)$, $B = f_{yy}(a, b)$, $C = f_{xy}(a, b) = f_{yx}(a, b)$; and $D = AB - C^2$. Then, if
>
> **a.** $D > 0$, $f$ has a local maximum at $(a, b)$ in the case $A < 0$ and a local minimum at $(a, b)$ in the case $A > 0$, and
> **b.** $D < 0$, $f$ has neither a maximum nor a minimum at $(a, b)$; the surface has a saddle point.

We note the following:

> The Second Partials Test cannot be used if $D = 0$.
>   Since $D = AB - C^2$, $D > 0$ implies that $AB > 0$. Thus $A$ and $B$ must have the same sign. Therefore $f_{xx}(a, b)$ and $f_{yy}(a, b)$ are both positive, or both negative, in the case $D > 0$.

You should recall the Second Derivative Test for a function of one variable. If $f'(a) = 0$ and $f''(a) < 0$, the graph of $f$ is concave down (sad) at $a$ and therefore $f$ has a local maximum at $a$. On the other hand, if $f'(a) = 0$ and $f''(a) > 0$, then the graph of $f$ is concave up (smiling) at $a$ and $f$ has a local minimum at $a$.

Note that the result is similar in the case of a function of two variables. If $D > 0$ and $A < 0$ (thus $B < 0$ also), both sections of the graph of $f$ by planes parallel to the $xz$-plane and $yz$-plane and through the point $(a, b)$ are concave down. This suggests (but does not prove) that $f$ has a local maximum at $(a, b)$.

If $D > 0$ and $A > 0$ (thus $B > 0$ also), both sections of the graph of $f$ by planes parallel to the $xz$-plane and $yz$-plane and through the point $(a, b)$ are concave up. This suggests that $f$ has a local minimum at $(a, b)$.

**EXAMPLE 2** Locate the relative extrema of

$$f(x, y) = x^3 + y^3 + 3y^2 - 3x - 9y + 2$$

**Solution** Since relative extrema can occur only at critical points, we first find all critical points of $f$. We find that

$$f_x(x, y) = 3x^2 - 3 \text{ and } f_y(x, y) = 3y^2 + 6y - 9$$

We set

$$\begin{cases} 3x^2 - 3 \quad\;\; = 0 \\ 3y^2 + 6y - 9 = 0 \end{cases}$$

This system can be written

$$\begin{cases} 3(x - 1)(x + 1) = 0 \\ 3(y + 3)(y - 1) = 0 \end{cases}$$

Therefore the critical points of $f$ are $(1, -3)$, $(1, 1)$, $(-1, -3)$, and $(-1, 1)$. We now find the second partial derivatives.

$$f_{xx}(x, y) = 6x, f_{yy}(x, y) = 6y + 6, f_{xy}(x, y) = f_{yx}(x, y) = 0$$

At $(1, -3)$, we have

$$A = f_{xx}(1, -3) = 6, B = f_{yy}(1, -3) = 6(-3) + 6 = -12$$
$$C = f_{xy}(1, -3) = 0$$

It follows that

$$D = AB - C^2 = 6(-12) - 0^2 = -72$$

Thus $f$ has neither a maximum nor a minimum at $(1, -3)$. Its graph has a saddle point since $D < 0$.

At $(1, 1)$, we have

$$A = f_{xx}(1, 1) = 6, B = f_{yy}(1, 1) = 6(1) + 6 = 12, C = f_{xy}(1, 1) = 0$$

It follows that

$$D = AB - C^2 = 6(12) - 0^2 = 72$$

Since $D > 0$ and $A > 0$, $f$ has a local minimum at $(1, 1)$.

At $(-1, -3)$, we get

$$A = f_{xx}(-1, -3) = 6(-1) = -6, B = f_{yy}(-1, -3) = 6(-3) + 6 = -12$$
$$C = f_{xy}(-1, -3) = 0$$

It follows that

$$D = AB - C^2 = -6(-12) - 0^2 = 72$$

Since $D > 0$ and $A < 0$, $f$ has a local maximum at $(-1, -3)$.

Finally, at $(-1, 1)$, we have

$$A = f_{xx}(-1, 1) = 6(-1) = -6, B = f_{yy}(-1, 1) = 6(1) + 6 = 12$$
$$C = f_{xy}(-1, 1) = 0$$

It follows that

$$D = AB - C^2 = -6(12) - 0^2 = -72$$

Since $D < 0$, $f$ has neither a maximum nor a minimum at $(-1, 1)$. Its graph has a saddle point.

**Do Exercise 15.**                                                                    ∎

## Absolute Extrema

Thus far we have discussed only relative (local) extrema of a function of two variables. Suppose that a function $f$ of two variables is defined on a region $D$. If

$$f(a, b) \le f(x, y) \le f(c, d)$$

for all $(x, y)$ in $D$, then $f(a, b)$ is the *absolute minimum* of $f$ and $f(c, d)$ is the *absolute maximum* of $f$ on $D$. In more advanced texts, it is proved that if $f$ is continuous and $D$ is closed and bounded, then $f$ has both an absolute minimum and an absolute maximum on $D$. As in the case of a function of one variable, an absolute extremum must occur either at a boundary point of $D$ or at a critical point of $f$. However, the case of a function of two variables is much more difficult to handle since, in general, the boundary of a region $D$ in the $xy$-plane has infinitely many points. Therefore in the next example, and in the applied exercises at the end of this section, we shall assume that the local extrema we obtain are also the absolute extrema.

## An Application

**EXAMPLE 3**    The total cost of grade A beef is \$3 per pound and that of grade B beef is \$2 per pound. If the retail prices of grade A and grade B beef are $p_A$ and $p_B$ dollars per pound, respectively, the demand for grade A beef is given by

$$D_A = 2750 - 700p_A + 200p_B$$

and that for grade B beef is given by

$$D_B = 2400 + 150p_A - 800p_B$$

where $D_A$ and $D_B$ are the quantities (in pounds) of grade A and grade B beef sold per day.

What should the retail price of each grade of beef be to maximize the daily profit? What is the maximum daily profit?

**Solution**    It is useful to first enter the available information in a table.

|         | Price per pound | Cost per pound | Profit per pound | Quantity sold | Total profit |
|---------|-----------------|----------------|------------------|---------------|--------------|
| Grade A | $p_A$           | 3              | $p_A - 3$        | $D_A$         | $(p_A - 3)D_A$ |
| Grade B | $p_B$           | 2              | $p_B - 2$        | $D_B$         | $(p_B - 2)D_B$ |

It is now easy to see that the profit per pound is $(p_A - 3)$ dollars for grade A beef and $(p_B - 2)$ dollars for grade B beef. Thus the daily profit realized from grade A beef is

$$(p_A - 3)D_A = (p_A - 3)(2750 - 700p_A + 200p_B) \text{ dollars}$$

and that realized from grade B beef is

$$(p_B - 2)D_B = (p_B - 2)(2400 + 150p_A - 800p_B) \text{ dollars}$$

Therefore the total daily profit is given by

$$P(p_A, p_B) = (p_A - 3)(2750 - 700p_a + 200p_B) + (p_B - 2)(2400 + 150p_A - 800p_B)$$
$$= -700p_A^2 + 350p_A p_B - 800p_B^2 + 4550p_A + 3400p_B - 13050$$

Thus

$$P_{p_A}(p_A, p_B) = -1400p_A + 350p_B + 4550$$

and

$$P_{p_B}(p_A, p_B) = 350p_A - 1600p_B + 3400$$

To find the critical points of $P$, we set both partial derivatives equal to 0.

$$\begin{cases} -1400p_A + 350p_B + 4550 = 0 & \quad(1) \\ 350p_A - 1600p_B + 3400 = 0 & \quad(2) \end{cases}$$

Multiplying equation (2) by 4 and adding the result to equation (1), we obtain

$$-6050p_B + 18{,}150 = 0$$

This yields $p_B = 3$. Replacing $p_B$ by 3 in equation (1), we get

$$-1400p_A + 350(3) + 4550 = 0$$

which gives $p_A = 4$.

Therefore the critical point of $P$ is $(4, 3)$. For the Second Partials Test, we find the second partial derivatives.

$$P_{p_A p_A}(p_A, p_B) = \frac{\partial}{\partial p_A}(-1400p_A + 350p_B + 4550) = -1400$$

$$P_{p_B p_B}(p_A, p_B) = \frac{\partial}{\partial p_B}(350p_A - 1600p_B + 3400) = -1600$$

$$P_{p_A p_B}(p_A, p_B) = \frac{\partial}{\partial p_A}(-1400p_A + 350p_B + 4550) = 350$$

Observe that $P_{p_B p_A}(p_A, p_B) = 350$ also, as expected.

Now,

$$A = P_{p_A p_A}(4, 3) = -1400, \ B = P_{p_B p_B}(4, 3) = -1600$$
$$C = P_{p_A p_B}(4, 3) = 350$$

It follows that

$$D = AB - C^2 = (-1400)(-1600) - 350^2 = 2{,}117{,}500$$

Since $D > 0$ and $A < 0$, the function $P$ has a maximum at $(4, 3)$. Therefore the maximum daily profit will be realized if the retail price of grade A beef is \$4 per pound and that of grade B beef is \$3 per pound. Since

$$P(4, 3) = -700(4^2) + 350(4)(3) - 800(3^2) + 4550(4) + 3400(3) - 13{,}050$$
$$= 1150$$

the maximum daily profit is \$1150.
**Do Exercise 39.**    ■

## How to Find Critical Points

To find the critical points of a function $f$ of two variables, we must solve the system

$$\begin{cases} f_x(x, y) = 0 & (1) \\ f_y(x, y) = 0 & (2) \end{cases}$$

In Example 2, equations (1) and (2) were in one unknown only. Thus the system was easy to solve. In Example 3, the system was a system of linear equations and therefore could also be readily solved. In general, the situation is not that simple. If the system of equations is not linear, solving it may be tedious. A technique which often works is substitution. We illustrate this method in the next two examples.

**EXAMPLE 4**    Locate the relative extrema of

$$f(x, y) = x^3 + y^3 - 8xy$$

**Solution**    We first find the partial derivatives.

$$f_x(x, y) = 3x^2 - 8y \text{ and } f_y(x, y) = 3y^2 - 8x$$

We must solve the system

$$\begin{cases} 3x^2 - 8y = 0 & (1) \\ 3y^2 - 8x = 0 & (2) \end{cases}$$

Equation (1) yields $y = 3x^2/8$.
   Substituting $3x^2/8$ for $y$ in equation (2), we obtain

$$3\left(\frac{3x^2}{8}\right)^2 - 8x = 0$$

That is,

$$\frac{27x^4}{64} - 8x = 0$$

or

$$x\left(\frac{27x^3}{64} - 8\right) = 0$$

Thus

$$x = 0 \text{ or } \frac{27x^3}{64} - 8 = 0$$

It follows that

$$x = 0 \text{ or } x = \frac{8}{3}$$

Since $y = 3x^2/8$, $x = 0$ yields $y = 0$, and $x = \frac{8}{3}$ yields $y = \frac{8}{3}$. Therefore the critical points are $(0, 0)$ and $\left(\frac{8}{3}, \frac{8}{3}\right)$. Furthermore, $f_{xx}(x, y) = 6x$, $f_{yy}(x, y) = 6y$, and $f_{xy}(x, y) = -8$.

At $(0, 0)$,

$$A = f_{xx}(0, 0) = 0, B = f_{yy}(0, 0) = 0, C = f_{xy}(0, 0) = f_{yx}(0, 0) = -8$$

Thus

$$D = AB - C^2 = (0)(0) - (-8)^2 = -64$$

Since $D < 0$, the function has neither a maximum nor a minimum at $(0, 0)$. Its graph has a saddle point.

At $\left(\frac{8}{3}, \frac{8}{3}\right)$, we have

$$A = f_{xx}\left(\frac{8}{3}, \frac{8}{3}\right) = 6\left(\frac{8}{3}\right) = 16, B = f_{yy}\left(\frac{8}{3}, \frac{8}{3}\right) = 6\left(\frac{8}{3}\right) = 16$$

$$C = f_{xy}\left(\frac{8}{3}, \frac{8}{3}\right) = -8$$

Hence,

$$D = AB - C^2 = (16)(16) - (-8)^2 = 192$$

Since $D > 0$ and $A > 0$, the function has a relative minimum at $\left(\frac{8}{3}, \frac{8}{3}\right)$.
**Do Exercise 21.**                                                                    ■

**EXAMPLE 5**    Find the relative extrema of

$$F(x, y) = x^3 - 2x^2y + xy^2 + y^2 - 4y + 10$$

**Solution**    We first find the partial derivatives.

$$F_x(x, y) = 3x^2 - 4xy + y^2 \text{ and } F_y(x, y) = -2x^2 + 2xy + 2y - 4$$

The system of equations is

$$\begin{cases} 3x^2 - 4xy + y^2 = 0 & (1) \\ -2x^2 + 2xy + 2y - 4 = 0 & (2) \end{cases}$$

Equation (1) may be written

$$(3x - y)(x - y) = 0$$

Therefore

$$3x - y = 0 \text{ or } x - y = 0$$

It follows that

$$y = 3x \text{ or } x = y$$

We substitute these expressions in equation (2). For $y = 3x$, we get

$$-2x^2 + 2x(3x) + 2(3x) - 4 = 0$$

which yields

$$4x^2 + 6x - 4 = 0$$

whose solutions are $-2$ and $\frac{1}{2}$. Since $y = 3x$, we get the two critical points $(-2, -6)$ and $\left(\frac{1}{2}, \frac{3}{2}\right)$. Furthermore, $y = x$ gives

$$-2x^2 + 2x(x) + 2x - 4 = 0$$

whose only solution is 2. Since in this case $y = x$, the third critical point is $(2, 2)$.

To use the Second Partials Test, we calculate

$$F_{xx}(x, y) = 6x - 4y, \ F_{yy}(x, y) = 2x + 2, \text{ and } F_{xy}(x, y) = -4x + 2y$$

(Again, note that $F_{yx}(x, y) = -4x + 2y$ also.)

We now test each critical point.

**Case 1:** The critical point is $(-2, -6)$.

$$A = F_{xx}(-2, -6) = 6(-2) - 4(-6) = 12$$
$$B = F_{yy}(-2, -6) = 2(-2) + 2 = -2$$
$$C = F_{xy}(-2, -6) = -4(-2) + 2(-6) = -4$$
$$D = AB - C^2 = (12)(-2) - (-4)^2 = -40$$

Since $D < 0$, the function $F$ has neither a maximum nor a minimum at $(-2, -6)$. Its graph has a saddle point.

**Case 2:** The critical point is $\left(\frac{1}{2}, \frac{3}{2}\right)$.

$$A = F_{xx}\left(\frac{1}{2}, \frac{3}{2}\right) = 6\left(\frac{1}{2}\right) - 4\left(\frac{3}{2}\right) = -3$$

$$B = F_{yy}\left(\frac{1}{2}, \frac{3}{2}\right) = 2\left(\frac{1}{2}\right) + 2 = 3$$

$$C = F_{xy}\left(\frac{1}{2}, \frac{3}{2}\right) = -4\left(\frac{1}{2}\right) + 2\left(\frac{3}{2}\right) = 1$$

$$D = AB - C^2 = (-3)(3) - 1^2 = -10$$

Since $D < 0$, the function has neither a maximum nor a minimum at $\left(\frac{1}{2}, \frac{3}{2}\right)$. Its graph has a saddle point.
Its graph has a saddle point.

**Case 3:** The critical point is $(2, 2)$.

$$A = F_{xx}(2, 2) = 6(2) - 4(2) = 4$$
$$B = F_{yy}(2, 2) = 2(2) + 2 = 6$$
$$C = F_{xy}(2, 2) = -4(2) + 2(2) = -4$$
$$D = AB - C^2 = (4)(6) - (-4)^2 = 8$$

Since $D > 0$ and $A > 0$, the function $F$ has a local minimum at $(2, 2)$.
**Do Exercise 25.**

## Procedure to Find Extrema

A function $f$ of two variables can have relative extrema only at critical points.

**STEP 1.**    In this text, we consider only those functions whose partial derivatives are defined at the critical points. To locate the critical points, we find the first partial derivatives and solve the system

$$\begin{cases} f_x(x, y) = 0 \\ f_y(x, y) = 0 \end{cases}$$

**STEP 2.**    Find $f_{xx}(x, y), f_{yy}(x, y)$, and $f_{xy}(x, y)$. Note that $f_{xy}(x, y)$ and $f_{yx}(x, y)$ should be equal since the functions we use have continuous partial derivatives.

**STEP 3.**    If $(a, b)$ is a critical point found in Step 1, evaluate $A = f_{xx}(a, b)$, $B = f_{yy}(a, b)$, $C = f_{xy}(a, b)$, and $D = AB - C^2$. If $D = 0$, the test fails. If $D < 0$, the function has neither a maximum nor a minimum—its graph has a saddle point. If $D > 0$, continue to Step 4.

**STEP 4.**    If the value of $A$ is positive (in that case the value of $B$ must also be positive since $AB - C^2 > 0$), then the function has a local minimum at $(a, b)$. If the value of $A$ is negative (in this case the value of $B$ is also negative), the function has a local maximum at $(a, b)$. Repeat Steps 3 and 4 for each critical point obtained in Step 1.

## Exercise Set 11.5

*In Exercises 1–33, locate the critical points of the given functions and determine whether the functions have relative maxima, relative minima, or saddle points at these critical points.*

**1.** $f(x, y) = x^2 + y^2 - 25$

**2.** $f(x, y) = 36 - 4x^2 - 9y^2$

**3.** $f(x, y) = x^2 + y^2 - 2x - 6y + 12$

**4.** $f(x, y) = x^2 + y^2 + 4x - 2y + 6$

**5.** $f(x, y) = x^2 - y^2 - 8x + 6y - 15$

**6.** $f(x, y) = 2x^2 + 5xy + 3y^2 - 6x + 5y - 13$

**7.** $f(x, y) = x^2 - 5xy + 6y^2 + 3x - 6y + 10$

**8.** $f(x, y) = 4x^2 + 9xy + 5y^2 - 6x + 3y - 7$

**9.** $f(x, y) = 2x^2 - 9xy + 10y^2 + 7x - 5y + 16$

**10.** $f(x, y) = 3x^2 - 7xy + 4y^2 - 5x + 3y - 13$

**11.** $f(x, y) = x^2 + 7xy + 12y^2 - 7x + 5y + 15$

**12.** $f(x, y) = x^3 + y^2 - 3x + 10y - 15$

**13.** $f(x, y) = x^2 + y^3 - 4x + 9y + 7$

**14.** $f(x, y) = x^3 + y^2 - 12x - 4y + 13$

**15.** $f(x, y) = 2x^3 + 3x^2 + y^2 - 12x - 6y + 15$

**16.** $f(x, y) = x^3 + 3x^2 + 2y^2 - 24x + 12y - 13$

**17.** $f(x, y) = y^3 + x^2 + 6y^2 + 6x - 15y + 17$

**18.** $f(x, y) = y^3 + x^2 + 9y^2 - 6x - 21y + 21$

**19.** $f(x, y) = x^3 + y^3 + 6x^2 - 3y^2 - 15x - 9y + 10$

**20.** $f(x, y) = x^3 + y^3 - 6x^2 + 3y^2 + 9x - 24y - 13$

**21.** $f(x, y) = y^3 + x^2 - 4xy + 4y + 12$

22. $f(x, y) = x^3 + y^3 - 3xy + 14$

23. $f(x, y) = y^3 + x^2 - 6xy + 24y + 15$

24. $f(x, y) = x^3 - 6x^2y + 9xy^2 + 3y^2 - 18y + 15$

25. $f(x, y) = y^3 + 9x^2y - 6xy^2 + 6x^2 - 144x + 6$

26. $f(x, y) = 9x^3 - 12x^2y + 5xy^2 - 450y + 18$

27. $f(x, y) = xy + \dfrac{8}{x} + \dfrac{1}{y}$

28. $f(x, y) = 3xy + \dfrac{27}{x} + \dfrac{1}{3y}$

29. $f(x, y) = e^{2xy}$

30. $f(x, y) = e^{(x^2 + xy + y^2 + 3y)}$

31. $f(x, y) = (x - 2)\ln xy$

32. $f(x, y) = xy + 2y^2 + \ln x$

33. $f(x, y) = xy - 2x^2 - \ln y$

34. A company manufactures two brands of a certain commodity. If $q_A$ units of brand A and $q_B$ units of brand B are produced each day, the daily profit $P(q_A, q_B)$ dollars is given by

$$P(q_A, q_B) = -30q_A^2 + 90q_Aq_B - 70q_B^2 + 300q_A$$
$$+ 150q_B - 6750$$

How many units of each brand should be produced daily to maximize profit? What is the maximum daily profit?

35. Same as Exercise 34 with the daily profit given by

$$P(q_A, q_B) = -7q_A^2 + 5q_Aq_B - q_B^2 + 120q_A$$
$$+ 180q_B - 16,400$$

36. The Sherman Company has found that if it manufactures $q_D$ deluxe and $q_S$ standard models of its bicycles each day, the cost $C$ dollars and revenue $R$ dollars per day are given by

$$C = 5q_D^2 - 4q_Dq_S + q_S^2 + 60q_D + 50q_S + 2600$$

and

$$R = 180q_D + 250q_S$$

How many of each model should the company manufacture to maximize profit? What is the maximum daily profit?

37. The Triplenight Company has found that if it manufactures $q_S$ thousand standard and $q_D$ thousand deluxe models of its racquetball racquets each month, the monthly revenue $R$ and monthly cost $C$ are

$$R = 80q_S + 120q_D$$

and

$$C = 3q_S^2 - 3q_Sq_D + q_D^2 + 50q_S + 75q_D + 1125$$

where $R$ and $C$ are in thousands of dollars. How many of each model should the company produce monthly to maximize profit? What is the maximum monthly profit?

38. The manager of a department store can buy brand X and brand Y cologne at a cost of $8 and $10 per bottle, respectively. It is estimated that if brand X is sold for $p_X$ dollars per bottle and brand Y is sold for $p_Y$ dollars per bottle, the number of brand X sold weekly will be $D_X$ and the number of bottles of brand Y sold weekly will be $D_Y$ where

$$D_X = 150 - 50p_X + 40p_Y$$

and

$$D_Y = 190 + 60p_X - 70p_Y$$

How should each brand be priced to maximize the weekly profit? What is the maximum weekly profit?

39. The proshop manager of an athletic club can buy brand A and brand B racquetballs at a cost of $2 and $3 per ball, respectively. It is estimated that if brand A balls sell for $p_A$ dollars each and brand B balls sell for $p_B$ dollars each, the numbers $D_A$ and $D_B$ of each brand of balls sold each week are

$$D_A = 820 - 1800p_A + 1000p_B$$

and

$$D_B = 1100 + 1000p_A - 1000p_B$$

How should each brand be priced to maximize the weekly profit? What is the maximum weekly profit?

40. A topless rectangular tank is to be built to have a volume of 135 cubic feet. The base will be made of material costing $5/sq ft and material costing $4/sq ft will be used for the sides. What should the dimensions be in order to minimize the cost of material?

41. Same as Exercise 40 with volume 1152 cubic feet, and the cost of material for the base and sides is $8/sq ft and $6/sq ft, respectively.

42. The total costs of brand A and brand B hairdryers are $28 and $30 each, respectively. If the retail prices (in dollars) of brand A and brand B are $p_A$ and $p_B$ each, respectively, the demand for brand A is

$$D_A = 7736 - 320p_A + 150p_B$$

and that for brand B is

$$D_B = 8260 + 200p_A - 400p_B$$

where $D_A$ and $D_B$ are the numbers of hairdryers of brand A and brand B sold per year, respectively. What should the price of each brand be to maximize the yearly profit? What is the maximum yearly profit?

**43.** The Baldhead Company makes two types of hairdryers. If $D_A$ and $D_B$ are the respective demands when the prices are $p_A$ and $p_B$ dollars each, then

$$D_A = 456 - 30p_A + 5p_B$$

and

$$D_B = 666 + 10p_A - 40p_B$$

It costs the company $12 to produce a type A hairdryer and $15 to produce a type B hairdryer. At what prices should the company sell the hairdryers to maximize profit? What is the maximum profit?

**44.** A manufacturer can produce a commodity at two locations. If $q_A$ units are produced at location A and $q_B$ units are produced at location B, the costs $C_A$ and $C_B$ dollars, at locations A and B are

$$C_A = .01q_A^2 + 100q_A + 40,000$$

and

$$C_B = .04q_B^2 + 28q_B + 50,000$$

When $q_A$ and $q_B$ units are produced at these locations the manufacturer can sell each unit for $(220 - .04(q_A + q_B))$

dollars. How many units should be produced at each location to maximize profit?

**45.** Same as Exercise 44 with

$$C_A = .02q_A^2 + 80q_A + 10,000$$
$$C_B = .03q_B^2 + 60q_B + 14,000$$

and the selling price per unit $(196 - .05(q_A + q_B))$ dollars.

**\*46.** Let $f(x, y) = (y - x^2)(y - 2x^2)$.

**a.** Show that the intersection of any plane passing through the $z$-axis with the graph of the function $f$ is a curve $C$ which has a low point at the origin. (*Hint:* Let $y = mx$ where $m$ is a nonzero constant. Then the function becomes a function of one variable

$$g(x) = (mx - x^2)(mx - 2x^2)$$
$$= 2x^4 - 3mx^3 + m^2x^2$$

Show that for each value of $m$, $g$ has a local minimum at $x = 0$.)

**b.** Because of the result of part **a**, it would appear that the function $f$ has a minimum at the origin. Yet this is not the case. To show this, calculate $f(0, 0)$ and argue that in any neighborhood of the origin we can choose a point $(x_1, y_1)$ such that $x_1^2 < y_1 < 2x_1^2$. Then show that $f(x_1, y_1) < f(0, 0)$. Therefore $f(0, 0)$ cannot be a local minimum.

# 11.6  Lagrange Multipliers (Optional)

## Optimization Problems with Constraints

We have seen that an important application of differentiation is optimization. In the examples, we were asked to locate the extrema of functions of one variable (Sections 8.3–8.7) and of two variables (Section 11.5). There are many situations when it is desirable to locate the extrema of a function over a proper subset of its domain. These situations occur when the independent variables are subject to some restrictions (constraints). For example, an orchardist might want to decide how many acres of apple trees and pear trees she should plant in order to maximize profit. It might be known from experience that the profit per year is $f(x, y)$ dollars when $x$ acres of apple trees and $y$ acres of pear trees are planted. A natural requirement, of course, is $x \geq 0$ and $y \geq 0$. If there are no other restrictions, the graph of the function $f$ is a surface whose highest point would indicate the highest profit. (See Figure 11.17.)

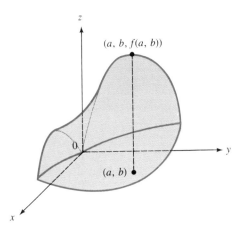

**FIGURE 11.17**

In Figure 11.17, we see that the function $f$ assumes its maximum value at $(a, b)$. Thus the orchardist should plant $a$ acres of apple trees and $b$ acres of pear trees to maximize profit. However, we usually have additional restrictions. For example, there are only $k$ acres of land to be planted. Thus the question should be rephrased as follows. What is the maximum value of $f$ if $x + y = k$? Therefore, we must consider only the part of the graph of $f$ that is above the line $x + y = k$. (See Figure 11.18.)

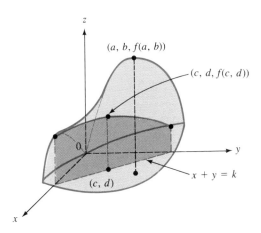

**FIGURE 11.18**

In Figure 11.18, we see that the point $(c, d)$ on the line $x + y = k$ is such that the function $f$ is maximum at that point. In this case, we could solve the problem simply by finding $y$ in terms of $x$ to obtain $y = k - x$, and then replace $y$ by $k - x$ in the profit function. We thus obtain a function of one variable only

$$P(x) = f(x, k - x)$$

It would then be sufficient to locate where the function $P$ attains its maximum value using the First or Second Derivative Test of Sections 8.5 and 8.6. We illustrate this idea with the following simple example.

**EXAMPLE 1** Mrs. Greenthumb wishes to enclose a rectangular region against an existing wall for her vegetable garden. She has 200 feet of fence available. Determine the dimensions of the rectangle that will maximize the area. Assume that the existing wall is at least 200 feet long.

**Solution** Let $x$ feet be the length of the side parallel to the existing wall and $y$ feet be the length of each of the sides perpendicular to the wall. The area is

$$A = xy$$

Since the amount of fence available is 200 feet, we must also have

$$x + y + y = x + 2y = 200$$

(See Figure 11.19.)
We must find the maximum value of

$$A = f(x, y) = xy$$

subject to the constraint

$$x + 2y = 200$$

The second equation yields

$$x = 200 - 2y$$

Thus

$$A = (200 - 2y)y = -2y^2 + 200y$$

We now have the area as a function of $y$.

$$A = g(y) = -2y^2 + 200y$$

We use the Second Derivative Test to locate the maximum value of $g$. We find

$$g'(y) = -4y + 200$$

and

$$g''(y) = -4$$

We let

$$g'(y) = -4y + 200 = 0$$

Thus

$$4y = 200 \text{ and } y = 50$$

Therefore, 50 is the only critical point of $g$. Since $g''(50) = -4$, we conclude that the function $g$ has a maximum at $y = 50$ and $x = 200 - 2(50) = 100$. ∎

Existing wall

$y$ ft.                    $y$ ft.

$x$ ft.

**FIGURE 11.19**

## The Method of Lagrange Multipliers

There are situations where it is difficult, even impossible, to solve for $y$ in terms of $x$. In these situations, a method derived by the French mathematician Joseph-Louis Lagrange (1736–1813) may be used. The method of Lagrange multipliers gives a necessary condition for an extremum and is described as follows.

Suppose that $z = f(x, y)$ and that we wish to locate its extrema where $x$ and $y$ are subject to the constraint $g(x, y) = 0$. We define an auxiliary function $F$ by

$$F(x, y, \lambda) = f(x, y) - \lambda g(x, y)$$

The new variable $\lambda$ is the Greek letter read "lambda" and is called the *Lagrange multiplier*. It can be shown that if $(x_0, y_0, \lambda_0)$ is a critical point of $F$, then $(x_0, y_0)$ is a critical point of $f$, subject to the constraint $g(x, y) = 0$, and conversely, if $(x_0, y_0)$ is a critical point of $f$ subject to the constraint $g(x, y) = 0$, then there is a number $\lambda_0$ such that $(x_0, y_0, \lambda_0)$ is a critical point of $F$.

Thus to find the critical points of $f$ subject to the constraint $g(x, y) = 0$, we find instead the critical points of the auxiliary function $F$. These are found by solving the system

$$\begin{cases} F_x(x, y, \lambda) = 0 \\ F_y(x, y, \lambda) = 0 \\ F_\lambda(x, y, \lambda) = 0 \end{cases}$$

Since $F(x, y, \lambda) = f(x, y) - \lambda g(x, y)$, we have

$$F_x(x, y, \lambda) = f_x(x, y) - \lambda g_x(x, y)$$

$$F_y(x, y, \lambda) = f_y(x, y) - \lambda g_y(x, y)$$

$$F_\lambda(x, y, \lambda) = -g(x, y)$$

Thus the system can be written

$$\begin{cases} f_x(x, y) - \lambda g_x(x, y) = 0 \\ f_y(x, y) - \lambda g_y(x, y) = 0 \\ -g(x, y) \qquad\qquad = 0 \end{cases}$$

Using the method of Lagrange multipliers we solve this system of three equations in the three unknowns $x$, $y$, and $\lambda$. Here $\lambda$ seems to be an extraneous variable. An interpretation of its value will be given at the end of this section. Note that if $x = a$, $y = b$, and $\lambda = \lambda_0$ is a solution of the system, the point $(a, b)$ must be tested to determine whether it yields a maximum, a minimum, or neither.

There is a complicated criterion which can be used to distinguish between maxima and minima; however, in practice, it is easier to rely on other considerations, such as geometric or physical characteristics. In more advanced calculus books, it is proved that a function which is continuous on a closed, bounded set will always assume a maximum and a minimum value on that set. We need not define in this elementary treatment of the subject the meaning of the words "closed" and "bounded," but we will point out that in the applications the constraints are usually of the form $g(x, y) = k$, $x \geq 0$, and $y \geq 0$, and they usually define a set of points that is closed and bounded. We therefore expect a continuous function $f$, subject to constraints, to have a maximum and a minimum value. However, we must keep in mind that these extrema could occur on the boundary of the region defined by the constraints and in that case would not necessarily be obtained by solving the system of Lagrange's equations. (See Example 4.)

We now solve Example 1 using the method of Lagrange multipliers. Previously, we found that

$$A = f(x, y) = xy$$

subject to the constraints

$$x + 2y = 200, x \geq 0, \quad \text{and} \quad y \geq 0$$

We let $g(x, y) = x + 2y - 200$ so that the constraint is of the form $g(x, y) = 0$. Then we define the auxiliary function $F$ by

$$F(x, y, \lambda) = f(x, y) - \lambda g(x, y) = xy - \lambda(x + 2y - 200)$$

Then,

$$F_x(x, y, \lambda) = y - \lambda$$

$$F_y(x, y, \lambda) = x - \lambda(2) = x - 2\lambda$$

and

$$F_\lambda(x, y, \lambda) = -(x + 2y - 200) = -x - 2y + 200$$

Thus we must solve the system

$$\begin{cases} y - \lambda & = 0 & (1) \\ x - 2\lambda & = 0 & (2) \\ -x - 2y + 200 = 0 & & (3) \end{cases}$$

Multiplying equation (1) by $-2$ and adding the result to equation (2), we obtain $x - 2y = 0$. Thus $x = 2y$. Replacing $x$ by $2y$ in equation (3) yields

$$-2y - 2y + 200 = 0$$

and

$$y = 50$$

Therefore

$$x = 2(50) = 100$$

The constraints $x + 2y = 200$, $x \geq 0$, and $y \geq 0$ describe a line segment $L$ whose endpoints are $(0, 100)$ and $(200, 0)$. We evaluate the function $f$ at these endpoints and at the point $(100, 50)$ found by the method of Lagrange multipliers. We obtain

$$f(0, 100) = (0)(100) \quad = 0$$
$$f(200, 0) = (200)(0) \quad = 0$$
$$f(100, 50) = (100)(50) = 5000$$

The largest of these three functional values is 5000. Thus, $f$ has a maximum value at $(100, 50)$, as before.
**Do Exercise 1.**

## Lagrange Multipliers for Functions of Three Variables

The method of Lagrange multipliers may be extended to functions of three variables. We first discuss how to find partial derivatives of a function of three variables. If $w = f(x, y, z)$, we may find the partial derivative of $f$ with respect to $x$, denoted $f_x(x, y, z)$, by treating $y$ and $z$ as constants and differentiating with respect to $x$. The partial derivatives $f_y(x, y, z)$ and $f_z(x, y, z)$ are found similarly. For example, if

$$w = f(x, y, z) = 2x^2y^3z^5$$

then

$$f_x(x, y, z) = 4xy^3z^5, \; f_y(x, y, z) = 6x^2y^2z^5, \text{ and } f_z(x, y, z) = 10x^2y^3z^4$$

For a function of three variables, the method of Lagrange multipliers can be described as follows.

> Suppose $w = f(x, y, z)$ and we wish to find the extrema of this function where $x$, $y$, and $z$ are subject to the constraint $g(x, y, z) = 0$. Assume that $f$ and $g$ have continuous partial derivatives on a domain $D$ and define an auxiliary function $F$ by
>
> $$F(x, y, z, \lambda) = f(x, y, z) - \lambda g(x, y, z)$$
>
> To find the critical points of $f$ subject to $g(x, y, z) = 0$, we find instead the critical points of $F$. As in the preceding case, if $(x_0, y_0, z_0, \lambda_0)$ is a critical point of $F$, then $(x_0, y_0, z_0)$ is a critical point of $f$ subject to the constraint $g(x, y, z) = 0$.

We still must decide whether $f$ has a maximum, a minimum, or neither at that critical point. In most applications, the nature of the problem will indicate whether a maximum or minimum exists at that point without further testing. Of course, the critical points of $F$ are found by solving the system of Lagrange equations

$$\begin{cases} F_x(x, y, z, \lambda) = 0 \\ F_y(x, y, z, \lambda) = 0 \\ F_z(x, y, z, \lambda) = 0 \\ F_\lambda(x, y, z, \lambda) = 0 \end{cases}$$

We illustrate the method with the following example.

**EXAMPLE 2**   A rectangular container, without a top, is to have a volume of 2700 cubic feet. The material for the bottom costs $8/sq ft, while the material for the sides costs $5/sq ft. Find the dimensions that will minimize the cost of material for that box.

**Solution**   Suppose the bottom is $x$ feet by $y$ feet and the height is $z$ feet. Then the volume is $xyz$ cubic feet. We therefore have the constraint

$$xyz = 2700$$

We first write this constraint as

$$g(x, y, z) = xyz - 2700 = 0$$

The cost of material for the bottom is $8xy$ dollars, and the cost of the material for the sides is $5(2xz + 2yz)$ dollars. Thus, the total cost is given by

$$C(x, y, z) = 8xy + 5(2xz + 2yz) = 8xy + 10xz + 10yz$$

We define the auxiliary function $F$ by

$$\begin{aligned} F(x, y, z, \lambda) &= C(x, y, z) - \lambda g(x, y, z) \\ &= 8xy + 10xz + 10yz - \lambda(xyz - 2700) \end{aligned}$$

Therefore

$$\begin{aligned} F_x(x, y, z, \lambda) &= 8y + 10z - \lambda yz \\ F_y(x, y, z, \lambda) &= 8x + 10z - \lambda xz \\ F_z(x, y, z, \lambda) &= 10x + 10y - \lambda xy \\ F_\lambda(x, y, z, \lambda) &= -(xyz - 2700) = 2700 - xyz \end{aligned}$$

We must solve the system

$$\begin{cases} 8y + 10z - \lambda yz = 0 & (1) \\ 8x + 10z - \lambda xz = 0 & (2) \\ 10x + 10y - \lambda xy = 0 & (3) \\ 2700 - xyz = 0 & (4) \end{cases}$$

Because of the nature of the problem, we may assume that $x > 0$ and $y > 0$. Thus $xy \neq 0$ and by equation (3) we have

$$\lambda = \frac{10x + 10y}{xy}$$

Substituting $\dfrac{10x + 10y}{xy}$ for $\lambda$ in equations (1) and (2) and simplifying, we get

$$\begin{cases} 8y + 10z - \dfrac{10(x + y)z}{x} = 0 & (5) \\ 8x + 10z - \dfrac{10(x + y)z}{y} = 0 & (6) \\ \qquad\qquad xyz = 2700 & (7) \end{cases}$$

This system can be written

$$\begin{cases} 8xy - 10yz = 0 & (8) \\ 8xy - 10xz = 0 & (9) \\ \qquad xyz = 2700 & (10) \end{cases}$$

Subtracting equation (9) from equation (8), we obtain

$$10xz - 10yz = 0$$

which can be written

$$10z(x - y) = 0$$

Thus $z = 0$ or $x - y = 0$. Since $z = 0$ is not a feasible solution, we must have $x - y = 0$; that is, $x = y$. Replacing $y$ by $x$ in equations (9) and (10), we obtain

$$\begin{cases} 8x^2 - 10xz = 0 & (11) \\ \qquad x^2z = 2700 & (12) \end{cases}$$

Equation (11) yields

$$8x^2 = 10xz$$

or

$$z = \frac{4x}{5} \quad \text{(since } x \neq 0)$$

Substituting $\frac{4x}{5}$ for $z$ in equation (12) gives

$$x^2\left(\frac{4x}{5}\right) = 2700$$

or

$$x^3 = 3375$$

Thus,

$$x = 15$$

It follows that

$$y = 15$$

and replacing $x$ and $y$ by 15 in equation (10), we obtain

$$(15)(15)z = 2700$$

which yields

$$z = 12$$

Thus, (15, 15, 12) is the only feasible critical point. By the nature of the problem, we may assume that the function $C$ has a minimum value, subject to the constraint $xyz = 2700$. This minimum value must occur at the point (15, 15, 12) and is

$$C(15, 15, 12) = 8(15^2) + 10(15)(12) + 10(15)(12) = 5400$$

Thus the container should be 15 feet by 15 feet at the bottom and 12 feet high. The minimum cost of the material will be $5400.
**Do Exercise 15.**    ■

EXAMPLE 3    Find the smallest distance between the point (5, 12, 25) and a point which is on the plane whose equation is $2x + 7y + 26z = 15$.

**Solution** Let $(x, y, z)$ be any point on the plane. Then we have

$$2x + 7y + 26z = 15$$

and we write

$$g(x, y, z) = 2x + 7y + 26z - 15 = 0 \qquad (*)$$

The distance between the point $(5, 12, 25)$ and $(x, y, z)$ is

$$d = \sqrt{(x - 5)^2 + (y - 12)^2 + (z - 25)^2}$$

We wish to find the smallest value of $d$ subject to the constraint given by equation (*). The work is simplified considerably by finding the minimum value of $d^2$ since $d$ will assume its minimum value at the same point where $d^2$ assumes its minimum. We thus let

$$f(x, y, z) = d^2 = (x - 5)^2 + (y - 12)^2 + (z - 25)^2$$

and define the auxiliary function $F$ by

$$\begin{aligned} F(x, y, z, \lambda) &= f(x, y, z) - \lambda g(x, y, z) \\ &= (x - 5)^2 + (y - 12)^2 + (z - 25)^2 - \lambda(2x + 7y + 26z - 15) \end{aligned}$$

Thus,

$$\begin{aligned} F_x(x, y, z, \lambda) &= 2(x - 5) - 2\lambda = 2x - 2\lambda - 10 \\ F_y(x, y, z, \lambda) &= 2(y - 12) - 7\lambda = 2y - 7\lambda - 24 \\ F_z(x, y, z, \lambda) &= 2(z - 25) - 26\lambda = 2z - 26\lambda - 50 \\ F_\lambda(x, y, z, \lambda) &= -(2x + 7y + 26z - 15) = -2x - 7y - 26z + 15 \end{aligned}$$

The system of Lagrange equations is

$$\begin{cases} 2x - 2\lambda - 10 = 0 & (1) \\ 2y - 7\lambda - 24 = 0 & (2) \\ 2z - 26\lambda - 50 = 0 & (3) \\ -2x - 7y - 26z + 15 = 0 & (4) \end{cases}$$

We use the techniques introduced in Section 2.2 to find $x = 3$, $y = 5$, $z = -1$, and $\lambda = -2$.

Thus the only possible extremum value of $f$ occurs at $(3, 5, -1)$. We know that a minimum distance must exist, so we conclude that the minimum value of $d^2$ is 729, since

$$f(3, 5, -1) = (3 - 5)^2 + (5 - 12)^2 + (-1 - 25)^2 = 729$$

The minimum value of $d$ is 27, since $\sqrt{729} = 27$. Therefore the minimum distance between the point $(5, 12, 25)$ and the points on the given plane is 27 units.
**Do Exercise 7.** ∎

## A Word of Caution

In the preceding examples, we had only one solution of the system of Lagrange equations, and the nature of the problem assured us of the existence of an extremum.

We therefore assumed that the extremum occurred at the point given by the solution of the system. It is wise to proceed with care in making such assumptions, because sometimes erroneous conclusions can result. We illustrate with the following example.

**EXAMPLE 4**   Find the extremum of

$$f(x, y) = 8y^4 + 8xy^3 - 12y^3 - 12xy^2 + 6y^2 + 6xy - x - y$$

subject to the constraints $x + y = 1$, $x \geq 0$, and $y \geq 0$.

**Solution**   Let $g(x, y) = x + y - 1$, and

$$F(x, y, \lambda) = f(x, y) - \lambda g(x, y)$$
$$= 8y^4 + 8xy^3 - 12y^3 - 12xy^2 + 6y^2 + 6xy - x - y - \lambda(x + y - 1)$$

We find

$$F_x(x, y, \lambda) = 8y^3 - 12y^2 + 6y - 1 - \lambda,$$
$$F_y(x, y, \lambda) = 32y^3 + 24xy^2 - 36y^2 - 24xy + 12y + 6x - 1 - \lambda$$
$$F_\lambda(x, y, \lambda) = -(x + y - 1) = -x - y + 1$$

The Lagrange system is

$$\begin{cases} 8y^3 - 12y^2 + 6y - 1 - \lambda = 0 & (1) \\ 32y^3 + 24xy^2 - 36y^2 - 24xy + 12y + 6x - 1 - \lambda = 0 & (2) \\ -x - y + 1 = 0 & (3) \end{cases}$$

Subtracting equation (1) from equation (2), we obtain

$$24y^3 + 24xy^2 - 24y^2 - 24xy + 6y + 6x = 0$$

which can be written

$$24y^2(x + y) - 24y(x + y) + 6(x + y) = 0 \qquad (4)$$

Since by equation (3) we have $x + y = 1$, we may replace each of the factors $(x + y)$ by 1 in equation (4) and obtain

$$24y^2 - 24y + 6 = 0$$

or

$$6(2y - 1)^2 = 0$$

The only solution is $y = \frac{1}{2}$. Replacing $y$ by $\frac{1}{2}$ in equation (3), we obtain $x = \frac{1}{2}$. Thus we assume that $f$ has an extremum at $\left(\frac{1}{2}, \frac{1}{2}\right)$. However, we now show that this is a wrong assumption. First note that

$$f\left(\frac{1}{2}, \frac{1}{2}\right) = 8\left(\frac{1}{2}\right)^4 + 8\left(\frac{1}{2}\right)\left(\frac{1}{2}\right)^3 - 12\left(\frac{1}{2}\right)^3 - 12\left(\frac{1}{2}\right)\left(\frac{1}{2}\right)^2$$
$$+ 6\left(\frac{1}{2}\right)^2 + 6\left(\frac{1}{2}\right)\left(\frac{1}{2}\right) - \frac{1}{2} - \frac{1}{2} = 0$$

Also, $f(1, 0) = -1$ and $f(0, 1) = 8 - 12 + 6 - 1 = 1$. Therefore the maximum value of $f$, subject to the given constraints, is 1 and occurs at $(0, 1)$ and its minimum value is $-1$ and occurs at $(1, 0)$. At the point $\left(\frac{1}{2}, \frac{1}{2}\right)$, $f$ has neither a maximum, nor a minimum. (See Exercise 40.)    ∎

## Interpretation of the Lagrange Multiplier $\lambda$

**A CLOSER LOOK**    We now give an interpretation of the value of the Lagrange multiplier.

Suppose that $z = f(x, y)$ is subject to the constraint $g(x, y) = C$, where $C$ is a constant. It is clear that if $E$ is an extremum value of $f$ subject to that constraint, then the value of $E$ depends on that of $C$. Suppose now for each value of $C$, there is only one corresponding value of $E$, so that $E$ is a function of $C$. In more advanced calculus texts, the following is proved.

If $x = a$, $y = b$, and $\lambda = \lambda_0$ is a solution of a system of Lagrange equations, then $\lambda_0 = \frac{dE}{dC}$. As we have seen earlier, a derivative gives a rate of change. Consequently, $\lambda_0$ approximates the change in the extremum value of $f$ when the value of $C$ is increased by 1 in the constraint equation. Although the proof of this fact is beyond the scope of this book, we give the following example to illustrate the basic idea.

**EXAMPLE 5**    An aluminum cylindrical can with top and bottom has a surface area of $C$ square inches.

**a.** Using the method of Lagrange multipliers find the maximum volume $V_m$ in terms of $C$.
**b.** Find the value of Lagrange's multiplier in terms of $C$.
**c.** Find the derivative $dV_m/dC$ and compare the answer with that of part **b**.
**d.** Find the maximum volume of the can if the surface area is 64 square inches.
**e.** Using the result of part **b**, approximate the increase in maximum volume if the surface area is increased to 65 square inches.
**f.** Using the result of part **a**, calculate the maximum volume when the surface area is 65 square inches and compare the answer to that of part **e**.

**Solution**    **a.** Let the radius of the can be $r$ inches and its height be $h$ inches. Then the volume is

$$V = f(r, h) = \pi r^2 h \text{ cubic inches}$$

The surface area is

$$S = g(r, h) = 2\pi r^2 + 2\pi r h \text{ square inches}$$

We must maximize $f(r, h)$ subject to the constraints $g(r, h) = C$, $r > 0$, and $h > 0$. We first write the constraint equation as

$$G(r, h) = 2\pi r^2 + 2\pi r h - C$$

and define the auxiliary function $F$ by

$$F(r, h, \lambda) = f(r, h) - \lambda G(r, h)$$
$$= \pi r^2 h - \lambda(2\pi r^2 + 2\pi rh - C)$$

We find

$$F_r(r, h, \lambda) = 2\pi rh - \lambda(4\pi r + 2\pi h)$$
$$F_h(r, h, \lambda) = \pi r^2 - \lambda(2\pi r)$$
$$F_\lambda(r, h, \lambda) = -(2\pi r^2 + 2\pi rh - C)$$

The system of Lagrange equations is

$$\begin{cases} 2\pi rh - \lambda(4\pi r + 2\pi h) = 0 & \text{(1)} \\ \pi r^2 - \lambda(2\pi r) = 0 & \text{(2)} \\ -(2\pi r^2 + 2\pi rh - C) = 0 & \text{(3)} \end{cases}$$

Equation (2) yields $\lambda = \frac{r}{2}$. Substituting $\frac{r}{2}$ for $\lambda$ in equation (1), we obtain

$$2\pi rh - \frac{r}{2}(4\pi r + 2\pi h) = 0$$

Simplifying, we get

$$\pi rh - 2\pi r^2 = 0$$

and

$$\pi r(h - 2r) = 0$$

which yields

$$h = 2r \qquad \text{(since } r \neq 0\text{)}$$

Replacing $h$ by $2r$ in equation (3), we obtain

$$-[2\pi r^2 + 2\pi r(2r) - C] = 0$$

and

$$6\pi r^2 = C$$

Thus

$$r = \sqrt{\frac{C}{6\pi}}$$

It follows that

$$h = 2\sqrt{\frac{C}{6\pi}}$$

and

$$\lambda = \frac{1}{2}\sqrt{\frac{C}{6\pi}}$$

Assuming that there is a maximum volume, it must be attained when the radius is $\sqrt{C/6\pi}$ inches and the height is $2\sqrt{C/6\pi}$ inches. Therefore, the maximum volume is

$$V_m = \pi\left(\sqrt{\frac{C}{6\pi}}\right)^2\left(2\sqrt{\frac{C}{6\pi}}\right) = \frac{1}{3\sqrt{6\pi}}C^{3/2}$$

**b.** The value of the Lagrange multiplier is the value of $\lambda$ found in part **a**. It is

$$\frac{1}{2\sqrt{6\pi}} \cdot C^{1/2}$$

**c.** Differentiating $V_m$ with respect to $C$, we obtain

$$\frac{dV_m}{dC} = D_C\left(\frac{1}{3\sqrt{6\pi}}C^{3/2}\right) = \frac{1}{3\sqrt{6\pi}}\left(\frac{3}{2}\right)C^{1/2} = \frac{1}{2\sqrt{6\pi}}C^{1/2}$$

Observe that $dV_m/dC = \lambda$.

**d.** If $C = 64$, we get

$$V_m = \frac{1}{3\sqrt{6\pi}}(64^{3/2}) \doteq 39.309557$$

Thus, when the surface area is 64 square inches, the maximum volume is approximately 39.31 cubic inches.

**e.** When the value of $C$ is increased by 1 (to 65), the increase in maximum volume is approximated by the value of the derivative $dV_m/dC$. When $C = 64$, the value of $\lambda$ and therefore of $dV_m/dC$, is $\frac{1}{2\sqrt{6\pi}}(64^{1/2}) \doteq .921318$. Thus when $C = 65$, the maximum volume is approximately 40.230875 cubic inches since $39.309557 + .921318 = 40.230875$.

**f.** Using the result of part **a** to calculate the maximum volume when the surface area is 65 square inches, we obtain

$$V_m = \frac{1}{3\sqrt{6\pi}}(65^{3/2}) \doteq 40.234464 \doteq 40.23$$

Observe that the approximation obtained in part **e** was correct to two decimal places.

**Do Exercise 41.** ∎

## Lagrange Multiplier Procedure

Suppose that $z = f(x, y)$ and that the independent variables $x$ and $y$ are subject to the constraint $g(x, y) = 0$. Suppose also that $f$ and $g$ have continuous partial derivatives on a domain $D$. To find the local extrema of $f$, subject to $g(x, y) = 0$, we first define

$$F(x, y, \lambda) = f(x, y) - \lambda g(x, y)$$

We then find the critical points of $F$ by solving the system

$$\begin{cases} F_x(x, y, \lambda) = 0 \\ F_y(x, y, \lambda) = 0 \\ F_\lambda(x, y, \lambda) = 0 \end{cases}$$

For each solution $x = a$, $x = b$, and $\lambda = \lambda_0$ of the system, the point $(a, b)$ is also a critical point of $f$ and must be tested to determine whether it yields a maximum, a minimum, or neither. In most applications in this book, there will be only one such point and the existence of an extremum at that point will be obvious.

If the constraint equation is of the form $h(x, y) = C$ and for each value of $C$ in the constraint equation, there is only one local extremum $E = f(a, b)$, then $E$ is a function of $C$.

In this case,

$$\frac{dE}{dC} = \lambda_0$$

Thus, the value of the Lagrange multiplier found in solving the system of Lagrange equations provides an estimate of the change in the value of the extremum of $f$ as the value of $C$ is increased by one unit.

The case where both $f$ and $g$ are functions of three variables is entirely analogous.

# Exercise Set 11.6

1. Find the smallest distance between the origin and a point on the line $3x + 4y = 25$.

2. Find the smallest distance between the origin and a point on the line $5x - 12y + 169 = 0$.

3. Find the smallest distance between $(-1, 3)$ and a point on the line $x + y = 8$.

4. Find the smallest distance between $(2, -5)$ and a point on the line $x + 2y = 2$.

5. Find the smallest distance between the origin and a point on the plane $2x - y + 3z = 14$.

6. Find the smallest distance between the origin and a point on the plane $x + 5y - 2z = 30$.

7. Find the smallest distance between $(-1, 2, 3)$ and a point on the plane $x + 2y + 2z = 27$.

8. Find the smallest distance between $(2, 1, -3)$ and a point on the plane $3x + 6y + 2z = 55$.

9. Let $x$, $y$, $z$ be positive numbers and $x + y + z = K$ where $K$ is a positive constant. What is the maximum value of $xyz$?

How does the geometric mean $(xyz)^{1/3}$ compare to the arithmetic mean $\frac{x + y + z}{3}$?

10. Find the least amount of plywood needed to build a topless rectangular box of volume 125/cu in.

11. Same as Exercise 10 if the box has a top.

12. A capsule is to contain 10/cu mm of medicine. The capsule is a right circular cylinder with half a sphere at each end. Find the dimensions that will minimize its surface area.

13. Find the dimensions of a rectangular box, open at the top, with maximum volume if the surface area is 108 square inches.

14. Same as Exercise 13 if the surface area is 192 square inches.

15. The material for the bottom of a rectangular topless box costs $1.50/sq ft, while the material for the sides costs $1/sq ft. Find the greatest volume such a box can have if the total amount of money available is $288.

16. Same as Exercise 15 with the cost of material for the bottom and sides at $5/sq ft and $3/sq ft, respectively, and the total amount of money available is $135.

**17.** Same as Exercise 15 with the cost of material for the bottom and sides at $3/sq ft and $2/sq ft, respectively, and the total amount of money available is $144.

**\*18.** A tent is shaped as shown in the figure. The floor of the tent is made up of canvas, as are the sides and the conical top. If the volume of the tent must be 120 cubic feet, what should the values of $r$, $h_1$, and $h_2$ be in order to minimize the area of the canvas used to build the tent?

**19.** A topless rectangular tackle box is to be made as illustrated. Its volume must be 144/cu in. What should the dimensions be in order to minimize the total area of the material used?

**20.** A farmer wishes to build a fence to enclose a rectangular pasture on a straight riverbank along which no fencing is necessary. Find the dimensions of the pasture that will require the least amount of fencing if the total area is to be 28,800/sq ft.

**21.** Suppose the farmer of the preceding exercise has 1600 feet of fencing available. Find the dimensions that will maximize the area of the rectangular pasture.

**22.** Find the dimensions of a soft drink can if the volume of the can is 21/cu in. and the least amount of aluminum is used to make the can.

**23.** A cylindrical can is to hold 25 cubic inches of frozen juice. The sides are made of cardboard and the cost per square inch is one-third of the cost per square inch of the metal top and bottom of the can. What dimensions of the can will minimize the cost of material?

**24.** Find the dimensions of the largest rectangular package in volume that can be mailed given that regulations require that the sum of the length and girth cannot exceed 90 inches. (*Hint:* The girth is the length of the perimeter of a cross section of the package that is perpendicular to its longest side.)

*In Exercises 25–28, the Cobb-Douglas production function\* for a certain manufacturer is given. The cost per unit of labor, the cost per unit of capital, and the total amount of money available for labor and capital are also given. In each case, find the maximum production level.*

**25.** $f(L, K) = 320L^{.4}K^{.6}$; labor: $30 per unit
   capital: $70 per unit
   amount available: $210,000.

**26.** $f(L, K) = 500L^{.3}K^{.7}$; labor: $35 per unit
   capital: $90 per unit
   amount available: $189,000.

**27.** $f(L, K) = 750L^{.6}K^{.4}$; labor: $50 per unit
   capital: $120 per unit
   amount available: $600,000.

**28.** $f(L, K) = 630L^{.75}K^{.25}$; labor: $35 per unit
   capital: $52 per unit
   amount available: $182,000.

*In Exercises 29–32, the Cobb-Douglas production function for a certain manufacturer is given. The cost per unit of labor, the cost per unit of capital, and the number of units of production are also given. In each case, find the minimum total cost of labor and capital.*

**29.** $f(L, K) = 400L^{.25}K^{.75}$; labor: $20 per unit
   capital: $960 per unit
   production needed: 100,000 units.

**30.** $f(L, K) = 350L^{1/3}K^{2/3}$; labor: $25 per unit
   capital: $1350 per unit
   production needed: 700,000 units.

---

\*In order to study the relationship between labor, capital, and production, economists use the Cobb-Douglas production function which is described as follows.

$$Q = f(L, K) = AL^cK^d$$

where $c + d = 1$, $0 < c < 1$, $L$ is the number of units of labor, $K$ is the number of units of capital, and $f(L, K)$ is the number of units produced. The coefficient $A$ is a constant.

**31.** $f(L, K) = 420L^{2/3}K^{1/3}$; labor: $30 per unit
capital: $405 per unit
production needed: 336,000 units.

**32.** $f(L, K) = 620L^{.2}K^{.8}$; labor: $25 per unit
capital: $3200 per unit
production needed: 527,000 units.

**33.** Use the Lagrange multiplier method to find the marginal productivity of money* in Exercise 25. Interpret your result.

**34.** Same as Exercise 33 for Exercise 26.

**35.** Same as Exercise 33 for Exercise 27.

**36.** Same as Exercise 33 for Exercise 28.

**37.** An editor estimates that if $x$ thousand dollars is spent on development and $y$ thousand dollars is spent on promotion, approximately $N$ new books will be sold, where

$$N = 3x^{5/2}y^{3/2}$$

If the editor has $120,000 to spend on development and promotion, how much should be allocated to each in order to maximize the number of books sold? Using the Lagrange multiplier, estimate the increase in the maximum number of books sold if the editor has $1000 more to spend on development and production.

**38.** A manufacturer estimates that if $x$ thousand dollars is spent on development and $y$ thousand dollars is spent on promotion, approximately $N$ units of a new product will be sold, where

$$N = 90x^{1.4}y^{.6}$$

If the manufacturer wishes to spend $150,000 on development and promotion, how much should be allocated to each in order to maximize sales? Using the Lagrange multiplier method, estimate the increase in the maximum number of units sold if the manufacturer increases the budget for development and promotion by $1000.

**39.** A manufacturer can produce a commodity at two different locations. If $q_A$ units are produced at site $A$ and $q_B$ units are produced at site $B$, the total cost $C$ dollars is

$$C = .6q_A{}^2 + .5q_Aq_B + .4q_B{}^2 + 150q_A$$
$$+ 100q_B + 20,000$$

If the manufacturer must fill an order for 600 units of the commodity, how many units should be produced at each site in order to minimize the cost?

**40.** In Example 4 of this section, substitute $1 - x$ for $y$ in the expression defining the function $f(x, y)$, thus getting a function $g$ of $x$ alone. Then show that $\frac{1}{2}$ is a critical point of the function $g$ and that $g$ has neither a maximum nor a minimum at $\frac{1}{2}$.

**\*41.** A rectangular region is to be enclosed against an existing wall by a fence. The side against the wall does not need fencing, one side perpendicular to the wall will cost $30/ft, while the other two sides will cost $20/ft. Suppose $C$ dollars is spent on fencing.

 a. Using Lagrange's method, find the maximum area $A_m$ in terms of $C$.
 b. Find the value of the Lagrange multiplier in terms of $C$.
 c. Find the derivative $dA_m/dC$ and compare the answer with that of part **b**.
 d. Find the maximum area of the region if the value of $C$ is 800.
 e. Using the result of part **b**, approximate the increase in maximum area if the value of $C$ is increased to 801.
 f. Using the result of part **a**, calculate the maximum area when the value of $C$ is 801 and compare the answer to that of part **e**.

**\*42.** A topless box with square bottom is to be built with material costing $60/sq ft for the bottom and $15/sq ft for the four sides. Suppose $C$ dollars is to be spent on material.

 a. Using Lagrange's method, find the maximum volume $V_m$ in terms of $C$.
 b. Find the value of the Lagrange multiplier in terms of $C$.
 c. Find the derivative $dV_m/dC$ and compare the answer with that of part **b**.
 d. Find the maximum volume of the box if the value of $C$ is 10,800.
 e. Using the result of part **b**, approximate the increase in maximum volume if the value of $C$ is increased to 10,801.
 f. Using the result of part **a**, calculate the maximum volume when the value of $C$ is 10,801 and compare the answer to that of part **e**.

---

*If $Q = f(L, K)$ is a production function and $g(L, K) = C$ is the constraint representing the total cost of labor and capital, then the maximum production $M$ is a function of $C$ and $\frac{dM}{dC} = \lambda_0$. Economists call this derivative the *marginal productivity of money*. Its value approximates the change in maximum production for an additional dollar spent on labor and capital.

# 11.7 The Method of Least Squares (Optional)

In the applications, it is frequently necessary to construct a mathematical model to represent a situation. We have presented many such mathematical models. In writing a textbook, an author has the luxury of making up examples with functions and data to provide reasonable, simple solutions. However, in practice, a manufacturer does not have the freedom of making up a cost or demand function. This must be done on the basis of observed data and reasonable assumptions. This observed data may be collected during normal operations or by experimentation.

## The Least Square Criterion

In this section, we discuss a method for obtaining a mathematical model from observed data. In general, a good model must be accurate, yet simple. The first step in making up a mathematical model is to guess which type of function will best fit the observed data. In Figure 11.20a and 11.20b, points have been plotted to display the data from two basic experiments. Looking at the scattered points on each figure, it is reasonable to assume that the points in Figure 11.20a could belong to the graph of a function of the form $y = Ae^{kx}$. In Figure 11.20b, the points appear to be nearly on a straight line; therefore these points could belong to the graph of an equation of the form $y = mx + b$.

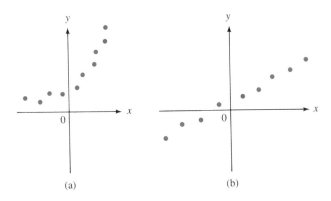

FIGURE 11.20                    (a)                              (b)

The second step in creating a mathematical model is to choose, among all the functions selected in the first step, the function which best fits the collection of points plotted from the observed data. A method introduced by the French mathematician Adrien Marie Legendre (1752–1833) is frequently used. We now describe this method.

Suppose that $n$ points $(x_1, y_1), (x_2, y_2), \ldots, (x_n, y_n)$ have been plotted to represent observed data from an experiment. Suppose further that we have selected a linear function $f(x) = mx + b$ to be the model for that experiment. To test how well the model fits the data, for each $i$, we consider the difference between the ordinate of the point $(x_i, mx_i + b)$, which belongs to the graph of the model, and the ordinate of the point $(x_i, y_i)$, which belongs to the data. Each of the differences $mx_i + b - y_i$ is called a *deviation*. It is positive if the point on the line is above the corresponding data point, negative if it is below. (See Figure 11.21.)

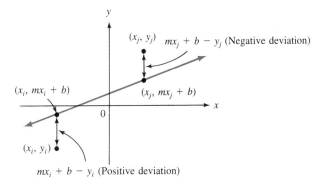

FIGURE 11.21

If the model is accurate, we expect each of these deviations to be small in absolute value. To measure the overall closeness of the fit, we might want to add all the deviations. However, some of these may be positive, and others negative. Consequently, their sum could be small even though the deviations could be large. To avoid this difficulty, we calculate the sum of the squares of the deviations and we try to find the linear function that will minimize that sum. For this reason, this technique is called the *Method of Least Squares*. The line with equation $y = mx + b$ so obtained is often called the *line of best fit* for the $n$ given points. It is also called the *line of regression*, especially when it is used in statistics.

**EXAMPLE 1**    Determine the line that best fits the following set of points in the sense of least squares: $(2, 3)$, $(3, 4)$, $(4, 4)$, $(5, 5)$, $(6, 5)$.

**Solution**    Let the line be $y = mx + b$. We tabulate the coordinates of the data points, of the corresponding points on the line, and the deviations.

| $i$ | $x_i$ | $y_i$ | $mx_i + b$ | Deviations $(mx_i + b - y_i)$ |
|-----|-------|-------|------------|-------------------------------|
| 1 | 2 | 3 | $2m + b$ | $2m + b - 3$ |
| 2 | 3 | 4 | $3m + b$ | $3m + b - 4$ |
| 3 | 4 | 4 | $4m + b$ | $4m + b - 4$ |
| 4 | 5 | 5 | $5m + b$ | $5m + b - 5$ |
| 5 | 6 | 5 | $6m + b$ | $6m + b - 5$ |

We must minimize the sum of the squares of the deviations. Let

$$f(m, b) = (2m + b - 3)^2 + (3m + b - 4)^2 + (4m + b - 4)^2$$
$$+ (5m + b - 5)^2 + (6m + b - 5)^2$$

To use the Second Partial Derivative Test we first find

$$f_m(m, b) = 2(2m + b - 3)2 + 2(3m + b - 4)3 + 2(4m + b - 4)4$$
$$+ 2(5m + b - 5)5 + 2(6m + b - 5)6 = 180m + 40b - 178$$

$$f_b(m, b) = 2(2m + b - 3) + 2(3m + b - 4) + 2(4m + b - 4)$$
$$+ 2(5m + b - 5) + 2(6m + b - 5) = 40m + 10b - 42$$

$$f_{mm}(m, b) = 180, f_{bb}(m, b) = 10, \text{ and } f_{mb}(m, b) = 40$$

We set $f_m(m, b)$ and $f_b(m, b)$ equal to 0 and obtain the system

$$\begin{cases} 180m + 40b - 178 = 0 \\ 40m + 10b - 42 = 0 \end{cases}$$

The solution is $m = .5$, $b = 2.2$. Also, $A = f_{mm}(.5, 2.2) = 180$, $B = f_{bb}(.5, 2.2) = 10$, $C = f_{mb}(.5, 2.2) = 40$, and $D = AB - C^2 = (180)(10) - 40^2 = 200$. Since $D > 0$ and $A > 0$, the function $f$ has a minimum at $(.5, 2.2)$. Therefore the equation of the line that best fits the data points in the sense of least squares is $y = .5x + 2.2$.
**Do Exercise 3.**                                                                                   ■

## Formulas for Slope and y-Intercept of the Line of Best Fit

**A CLOSER LOOK**    It is possible to obtain formulas that give the slope and the y-intercept of the line of best fit in terms of the coordinates of the data points. We begin with the following theorem.

> **THEOREM 11.5:** If $x_1$, $x_2$, $x_3$, . . . , $x_n$ are $n$ real numbers, at least two of which are distinct, then
>
> $$n\left(\sum x_i^2\right) > \left(\sum x_i\right)^2 {}^*$$

**Proof:** Let $a$ be the average (arithmetic mean) of the $n$ numbers. That is,

$$a = \frac{1}{n}\sum x_i$$

Since at least two of the $x_i$'s are distinct, $a - x_i \neq 0$ for at least one $i$. Thus, $(a - x_i)^2 > 0$ for at least one $i$. Since $(a - x_i)^2 \geq 0$ for each $i$, we must have

$$\sum (a - x_i)^2 > 0$$

Thus

$$\sum (a^2 - 2ax_i + x_i^2) > 0$$

and

$$\sum a^2 - \sum 2ax_i + \sum x_i^2 > 0$$

It follows that

$$na^2 - 2a \sum x_i + \sum x_i^2 > 0$$

Since $\sum x_i = na$, this can be written

$$na^2 - 2a(na) + \sum x_i^2 > 0$$

---

*For convenience, we are writing $\sum$ instead of $\sum\limits_{i=1}^{n}$.

That is,

$$-na^2 + \sum x_i^2 > 0$$

Hence,

$$\sum x_i^2 > na^2 = n\left(\frac{\sum x_i}{n}\right)^2 = \frac{\left(\sum x_i\right)^2}{n}$$

It follows that

$$n \sum x_i^2 > \left(\sum x_i\right)^2$$

We are now ready to state and prove the Least Squares Theorem.

> **THEOREM 11.6:**   Let $(x_1, y_1)$, $(x_2, y_2)$, $\ldots$ , $(x_n, y_n)$ be a set of data points where at least two of the $x_i$'s are distinct. If $y = mx + b$ is the equation of the line of best fit, then
>
> $$m = \frac{nP - XY}{nS - X^2} \text{ and } b = \frac{SY - PX}{nS - X^2}$$
>
> where $P = \sum x_i y_i$, $S = \sum x_i^2$, $X = \sum x_i$, and $Y = \sum y_i$.

**Proof:** The line $y = mx + b$ will best fit the data points if the sum $\sum(mx_i + b - y_i)^2$ is minimum. Observe that all $x_i$'s and $y_i$'s are constants and $m$ and $b$ are unknowns (variables).

Thus we let

$$f(m, b) = \sum (mx_i + b - y_i)^2$$

Our goal is to find values of $m$ and $b$ for which the value of the function $f$ is minimum. We shall use the Second Partials Test introduced in Section 11.5. Since the derivative of a sum of functions is equal to the sum of their derivatives, we find that

$$f_m(m, b) = \sum 2(mx_i + b - y_i)\frac{\partial}{\partial m}(mx_i + b + y_i)$$

$$= \sum 2(mx_i + b - y_i)x_i = \sum 2(mx_i^2 + bx_i - x_i y_i)$$

and

$$f_b(m, b) = \sum 2(mx_i + b - y_i)\frac{\partial}{\partial b}(mx_i + b + y_i)$$

$$= \sum 2(mx_i + b - y_i)(1) = \sum 2(mx_i + b - y_i)$$

In order to find the critical points, we set each partial derivative equal to 0 and obtain the following system

$$\begin{cases} \sum 2(mx_i^2 + bx_i - x_i y_i) = 0 \\ \sum 2(mx_i + b - y_i) = 0 \end{cases}$$

which can be written

$$\begin{cases} 2m \sum x_i^2 + 2b \sum x_i - 2 \sum x_i y_i = 0 \\ 2m \sum x_i + 2b \sum 1 - 2 \sum y_i = 0 \end{cases}$$

If we let $P = \sum x_i y_i$, $S = \sum x_i^2$, $X = \sum x_i$, and $Y = \sum y_i$, we get

$$\begin{cases} 2Sm + 2Xb - 2P = 0 & \text{(1)} \\ 2Xm + 2nb - 2Y = 0 & \text{(2)} \end{cases}$$

Multiplying equation (1) by $\frac{X}{2}$ and equation (2) by $\frac{S}{2}$ and subtracting the results, we obtain

$$(nS - X^2)b = SY - PX$$

Since $nS - X^2 = n \sum x_i^2 - (\sum x_i)^2$, we know by Theorem 11.5 that $nS - X^2 > 0$. Therefore,

$$b = \frac{SY - PX}{nS - X^2}$$

Substituting this for $b$ in equation (1), we get

$$2Sm + 2X\left(\frac{SY - PX}{nS - X^2}\right) - 2P = 0$$

from which we obtain

$$m = \frac{nP - XY}{nS - X^2}$$

Thus the critical point of the function $f$ is

$$\left(\frac{nP - XY}{nS - X^2}, \frac{SY - PX}{nS - X^2}\right)$$

It remains to show that $f$ actually does assume a minimum value at that point. We shall use the Second Partials Test. We find that

$$f_{mm}(m, b) = \sum 2x_i^2$$

$$f_{bb}(m, b) = \sum 2 = 2n$$

$$f_{mb}(m, b) = \sum 2x_i$$

Since each of these partial derivatives is a constant, their values at the critical point is that constant value, and we have

$$D = AB - C^2 = \left(2 \sum x_i^2\right)(2n) - \left(\sum 2x_i\right)^2 = 4\left[n \sum x_i^2 - \left(\sum x_i\right)^2\right]$$

Thus $D > 0$, since $n \sum x_i^2 - (\sum x_i)^2 > 0$ by Theorem 11.5. Furthermore, the value of $f_{mm}$ at the critical point is positive since $f_{mm}(m, b) = \sum 2x_i^2$. Thus the function $f$ has a minimum at the critical point. ∎

**EXAMPLE 2**   Find the line of best fit for the data of Example 1, using Theorem 11.6.

**Solution**    Let the line be $y = mx + b$. We tabulate the quantities which, according to Theorem 11.6, are needed to determine $m$ and $b$.

| $i$ | $x_i$ | $y_i$ | $x_i y_i$ | $x_i^2$ |
|---|---|---|---|---|
| 1 | 2 | 3 | 6 | 4 |
| 2 | 3 | 4 | 12 | 9 |
| 3 | 4 | 4 | 16 | 16 |
| 4 | 5 | 5 | 25 | 25 |
| 5 | 6 | 5 | 30 | 36 |
| $n = 5$ | $X = \sum x_i = 20$ | $Y = \sum y_i = 21$ | $P = \sum x_i y_i = 89$ | $S = \sum x_i^2 = 90$ |

Thus

$$m = \frac{nP - XY}{nS - X^2} = \frac{5(89) - (20)(21)}{5(90) - 20^2} = \frac{1}{2}$$

and

$$b = \frac{SY - PX}{nS - X^2} = \frac{(90)(21) - (89)(20)}{5(90) - 20^2} = 2.2$$

Therefore the equation of the line which best fits the set of points in the sense of least squares is

$$y = \frac{1}{2}x + 2.2$$

**Do Exercise 5.**    ■

## Simplified Formulas for Slope and y-Intercept of the Line of Best fit

The expressions giving the values of $m$ and $b$ are rather involved and can be made simpler by a translation of the $y$-axis.

Observe that if $X = 0$ in the following formulas

$$m = \frac{nP - XY}{nS - X^2} \text{ and } b = \frac{SY - PX}{nS - X^2}$$

both expressions take the simpler form

$$m = \frac{P}{S} \text{ and } b = \frac{Y}{n}$$

Recall that if the coordinates of a point are $(x, y)$ and the $y$-axis is translated so that the new origin is at the point $(h, 0)$, then the new coordinates are $(x', y')$ where

$$x' = x - h \text{ and } y' = y$$

(See Figure 11.22.)

**FIGURE 11.22**

If we choose $h$ to be the average (arithmetic mean) of the $x_i$'s, then each $x_i'$ is equal to $x_i - h$ and

$$\sum x_i' = \sum (x_i - h) = 0$$

(See Exercise 23.)

We illustrate this idea with the following example.

**EXAMPLE 3** Determine the line of regression for the points: (3, 4), (5, 5), (7, 5), (9, 8).

**Solution** The points (3, 4), (5, 5), (7, 5), and (9, 8) have been plotted in Figure 11.23.

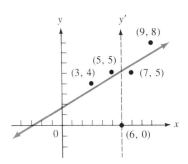

**FIGURE 11.23**

Note that the average of the abscissas is 6 since

$$\frac{3 + 5 + 7 + 9}{4} = 6$$

Thus we translate the $y$-axis so that the new origin is at (6, 0). With respect to the new system of coordinates, the data points become $(-3, 4)$, $(-1, 5)$, $(1, 5)$, and $(3, 8)$. We now proceed as we did in Example 1.

| $i$ | $x_i$ | $y_i$ | $x_iy_i$ | $x_i^2$ |
|---|---|---|---|---|
| 1 | $-3$ | 4 | $-12$ | 9 |
| 2 | $-1$ | 5 | $-5$ | 1 |
| 3 | 1 | 5 | 5 | 1 |
| 4 | 3 | 8 | 24 | 9 |
| $n = 4$ | $X = \sum x_i = 0$ | $Y = \sum y_i = 22$ | $P = \sum x_iy_i = 12$ | $S = \sum x_i^2 = 20$ |

Thus

$$m = \frac{P}{S} = \frac{12}{20} = .6 \quad \text{and} \quad b = \frac{Y}{n} = \frac{22}{4} = 5.5$$

The regression line is $y = .6x' + 5.5$, where we use the prime because the equation is with respect to the new coordinate system. If we wish to write the equation with respect to the original system, we simply note that $x' = x - 6$ (this is easy to remember since $x' = 0$ when $x = 6$) and we obtain

$$y = .6(x - 6) + 5.5 = .6x + 1.9$$

Therefore with respect to the original coordinate system, the regression line is

$$y = .6x + 1.9$$

**Do Exercise 7.**

## Application

**EXAMPLE 4** A manufacturer of a certain product has received orders during the years 1985–1991, as described in this table.

| Year | 1985 | 1986 | 1987 | 1988 | 1989 | 1990 | 1991 |
|------|------|------|------|------|------|------|------|
| **Number of orders (in thousands)** | 20 | 18 | 19 | 22 | 17 | 23 | 21 |

**a.** Find the least squares regression line for these data.
**b.** Use the result of part **a** to estimate the number of orders the manufacturer should receive in 1992.

**Solution** **a.** To simplify our work, we let the origin on the $x$-axis represent the year 1988. Then the years 1985, 1986, 1987, 1989, 1990, and 1991 will be represented by $-3$, $-2$, $-1$, $1$, $2$, and $3$, respectively. It follows that the set of data points can be written

| $x_i$ | $-3$ | $-2$ | $-1$ | 0 | 1 | 2 | 3 |
|-------|------|------|------|---|---|---|---|
| $y_i$ | 20 | 18 | 19 | 22 | 17 | 23 | 21 |

Again, using the method described in Example 1, we get

| $i$ | $x_i$ | $y_i$ | $x_i y_i$ | $x_i^2$ |
|-----|-------|-------|-----------|---------|
| 1 | $-3$ | 20 | $-60$ | 9 |
| 2 | $-2$ | 18 | $-36$ | 4 |
| 3 | $-1$ | 19 | $-19$ | 1 |
| 4 | 0 | 22 | 0 | 0 |
| 5 | 1 | 17 | 17 | 1 |
| 6 | 2 | 23 | 46 | 4 |
| 7 | 3 | 21 | 63 | 9 |
| $n = 7$ | $X = \sum x_i = 0$ | $Y = \sum y_i = 140$ | $P = \sum x_i y_i = 11$ | $S = \sum x_i^2 = 28$ |

Therefore,

$$m = \frac{P}{S} = \frac{11}{28} \quad \text{and} \quad b = \frac{Y}{n} = \frac{140}{7} = 20$$

An equation of the regression line is $y = \frac{11}{28}x + 20$.

**b.** The year 1992 is represented by $x = 4$. When $x = 4$,

$$y = \frac{11}{28}(4) + 20 \doteq 21.5714$$

The manufacturer should receive approximately 21,571 orders in 1992.
**Do Exercise 13.**                                                                            ∎

## Alternate Formulas for Slope and y-Intercept of the Line of Best Fit

The expressions giving the values of $m$ and $b$ for the equation of the regression line are not simple, although they are easy to use, as illustrated in the preceding three examples. There are alternate formulas for $m$ and $b$. To facilitate the statement of these formulas, we introduce the following notation. If $a_1, a_2, a_3, \ldots, a_n$ is a set of $n$ real numbers, their average (arithmetic mean) is denoted $\bar{a}$. That is,

$$\bar{a} = \frac{1}{n} \sum a_i$$

Using this convention, it can be shown that in the statement of Theorem 11.5, we can write

$$m = \frac{\sum (x_i - \bar{x})(y_i - \bar{y})}{\sum (x_i - \bar{x})^2}$$

and

$$b = \bar{y} - m\bar{x}$$

(See Exercises 22–25.) This expression for $m$ has a certain symmetry; therefore it is easier to remember than the first expression. It can be shown (see Exercise 21) that the point $(\bar{x}, \bar{y})$ is on the regression line. Using this fact, the formula giving the value of $b$ can be obtained by replacing $x$ by $\bar{x}$ and $y$ by $\bar{y}$ in the equation $y = mx + b$. (See Exercise 22.)

**EXAMPLE 5**   Find an equation of the regression line for the data of Example 1 using the alternate form of the formulas for $m$ and $b$.

**Solution**   We first calculate the averages $\bar{x}$ and $\bar{y}$ as follows.

$$\bar{x} = \frac{2 + 3 + 4 + 5 + 6}{5} = 4$$

and

$$\bar{y} = \frac{3 + 4 + 4 + 5 + 5}{5} = 4.2$$

Then we make up the following table.

| $i$ | $x_i$ | $y_i$ | $x_i - \bar{x}$ | $y_i - \bar{y}$ | $(x_i - \bar{x})(y_i - \bar{y})$ | $(x_i - \bar{x})^2$ |
|---|---|---|---|---|---|---|
| 1 | 2 | 3 | $-2$ | $-1.2$ | 2.4 | 4 |
| 2 | 3 | 4 | $-1$ | $-.2$ | .2 | 1 |
| 3 | 4 | 4 | 0 | $-.2$ | 0 | 0 |
| 4 | 5 | 5 | 1 | .8 | .8 | 1 |
| 5 | 6 | 5 | 2 | .8 | 1.6 | 4 |

Adding the entries in the sixth column, we get

$$\sum (x_i - \bar{x})(y_i - \bar{y}) = 5$$

and adding the entries in the seventh column, we obtain

$$\sum (x_i - \bar{x})^2 = 10$$

Thus

$$m = \frac{\sum (x_i - \bar{x})(y_i - \bar{y})}{\sum (x_i - \bar{x})^2} = \frac{5}{10} = \frac{1}{2}$$

and

$$b = \bar{y} - m\bar{x} = 4.2 - \frac{1}{2}(4) = 2.2$$

Therefore the regression line is

$$y = \frac{1}{2}x + 2.2$$

Observe that the same equation was obtained in Example 1.
**Do Exercise 9.**

# Exercise Set 11.7

*In Exercises 1–10, find the line that best fits the given data points in the sense of least squares. Draw a figure showing the data points and the regression line.*

**1.** (1, 5), (2, 7), (3, 8), (4, 9)

**2.** (−3, −12), (−1, −7), (1, −3), (3, 2)

**3.** (−2, 8), (−1, 7), (0, 6), (1, 5), (2, 4)

**4.** (1, 5), (2, 6), (3, 8), (4, 10), (5, 11)

**5.** (−4, 12), (−1, 6), (2, 0), (5, −6), (8, −13)

**6.** (−5, −6), (−3, −2), (−1, 3), (1, 7), (3, 12)

**7.** (−2, 7), (−1, 5), (0, 3), (1, 1), (2, −1), (3, −3)

**8.** (1, −1), (4, 9), (5, 12), (7, 18), (10, 28)

**9.** (−4, −17), (−3, −14), (−1, −7), (3, 6), (8, 21)

**10.** (−5, 17) (−4, 14), (−1, 5), (3, −7), (3, −8), (4, −10)

**11.** A manufacturer of a certain product has received orders during the years 1983–1991 as shown in the table.

| Year | 1983 | 1984 | 1985 | 1986 | 1987 |
|---|---|---|---|---|---|
| Number of orders (in thousands) | 42 | 40 | 43 | 45 | 44 |

| Year | 1988 | 1989 | 1990 | 1991 |
|---|---|---|---|---|
| Number of orders (in thousands) | 48 | 48 | 51 | 57 |

**a.** Find the least squares regression line for this set of data.

**b.** Use the result of part **a** to predict the number of orders the manufacturer should receive in 1992.

**12.** Same as Exercise 11 with the following set of data.

| Year | 1983 | 1984 | 1985 | 1986 | 1987 |
|---|---|---|---|---|---|
| Number of orders (in thousands) | 27 | 30 | 35 | 31 | 36 |

| Year | 1988 | 1989 | 1990 | 1991 |
|---|---|---|---|---|
| Number of orders (in thousands) | 40 | 38 | 45 | 44 |

**13.** The admissions office at a college has compiled the following data over the last 5 years relating the number of freshmen admitted by July 1 to the number enrolling the following fall.

| Number of freshmen admitted (in hundreds) | 15 | 16 | 16.5 | 17 | 19 |
|---|---|---|---|---|---|
| Number of freshmen enrolled (in hundreds) | 9 | 9.5 | 10 | 10.7 | 12 |

**a.** Find the least squares regression line for this set of data.
**b.** If 2300 freshmen have been admitted by July 1 of the following year, approximately how many freshmen will enroll the following fall?

**14.** A company sells its new product in five comparable cities in order to study the relationship between the price $p$ dollars per unit and the volume of sales $q$ thousand units. The following data was compiled over a 6-month period.

| $q$ | 21 | 19.5 | 23 | 25.5 | 20.5 |
|---|---|---|---|---|---|
| $p$ | 12 | 13 | 11.5 | 10 | 12.25 |

**a.** Find the least squares regression line for this set of data.
**b.** Using your answer of part **a**, estimate the volume of sales if the price is $14 per unit.
**c.** The equation of the regression line is the demand equation for the product. What price per unit will yield the maximum revenue?

**15.** A university conducted a study of the relationship between students' performance on the entrance exam and their performance during their freshman year. The university chose eight students randomly and compiled the following data.

The $x$ is the score on the entrance exam and $y$ is the grade point average at the end of the freshman year.

| $x$ | 85 | 75 | 60 | 72 |
|---|---|---|---|---|
| $y$ | 3.1 | 2.7 | 2.2 | 3.0 |

| $x$ | 90 | 70 | 75 | 88 |
|---|---|---|---|---|
| $y$ | 3.5 | 2.8 | 3.0 | 3.0 |

**a.** Find the equation of the regression line for the given data.
**b.** Using the result of part **a**, estimate a student grade point average at the end of the freshman year if that student scored 79 on the entrance exam.

**16.** A sociology student wanted to study the relationship between a family's annual income and the amount spent on entertainment by that family yearly. She chose six families randomly and obtained the following data. Each family's annual income is represenid by $x$, and the corresponding amount spent yearly on entertainment is represented by $y$. The entries in the table are in thousands of dollars.

| $x$ | 50 | 12 | 30 | 40 | 15 | 25 |
|---|---|---|---|---|---|---|
| $y$ | 2.5 | .7 | 1.5 | 2.1 | .75 | 1.5 |

**a.** Find the equation of the regression line for the data gathered by the student.
**b.** Using the result of part **a**, estimate the amount spent on entertainment in one year by a family whose annual income is $45,000.

**17.** To study the effectiveness of a certain poison, various dosages of that poison were fed to groups of 100 rats, and the following data were obtained. The dosage in milligrams is represented by $x$ and the number of deaths in the corresponding group is represented by $y$.

| $x$ | 15 | 23 | 31 | 39 | 47 | 55 | 63 | 71 | 79 | 87 |
|---|---|---|---|---|---|---|---|---|---|---|
| $y$ | 5 | 12 | 25 | 30 | 51 | 60 | 75 | 82 | 89 | 93 |

**a.** Find an equation of the regression line for the data.
**b.** Using the answer of part **a**, estimate the number of deaths in a group of 100 rats if that group is fed 35 milligrams of the poison.

**18.** Ten specimens of a certain alloy were produced at different temperatures and the strength of each specimen was tested. The table gives the observed temperatures as $x$ (in degrees Fahrenheit) and the tested strength as $y$ (in coded units).

| $x$ | 1200 | 1300 | 1400 | 1500 | 1600 |
|---|---|---|---|---|---|
| $y$ | 145 | 162 | 165 | 183 | 187 |

| $x$ | 1700 | 1800 | 1900 | 2000 | 2100 |
|---|---|---|---|---|---|
| $y$ | 202 | 205 | 221 | 225 | 238 |

**a.** Find an equation of the regression line for the data.
**b.** Using the answer of part **a**, estimate the strength (in coded units) of the alloy if it is produced at 1750° F.

**19.** To study the relationship between the age of a certain model car and the yearly expenses on repairs, the following information was collected on eight randomly selected cars of different ages. In the table, $x$ represents the age of the car (in years) and $y$ represents the yearly expenses on repairs (in dollars).

| $x$ | 1 | 2 | 3 | 4 | 5 | 6 | 7 | 8 |
|---|---|---|---|---|---|---|---|---|
| $y$ | 100 | 120 | 210 | 350 | 425 | 550 | 635 | 750 |

**a.** Find an equation of the regression line for the data.
**b.** Using the result of part **a**, estimate the amount spent yearly on a car that is 10 years old.

**20.** A grocery store chain wished to study the relationship between the amount of shelf space used to display a certain brand of coffee and the amount of that brand of coffee sold each week. It selected eight of its stores in comparable areas and gathered the following data. In the table, $x$ represents the amount of shelf space used (in square feet), and $y$ represents the number of pounds of coffee sold each week.

| $x$ | 15 | 16 | 17 | 18 | 19 | 20 | 21 | 22 |
|---|---|---|---|---|---|---|---|---|
| $y$ | 200 | 250 | 270 | 250 | 290 | 295 | 310 | 340 |

**a.** Find the equation of the regression line for the data.

**b.** On the basis of the answer in part **a**, estimate the number of pounds of that brand of coffee sold per week if the amount of shelf space used for display is 24 square feet.

**\*21.** Show that if $(x_1, y_1), (x_2, y_2), \ldots, (x_n, y_n)$ is a set of data points and $\bar{x}, \bar{y}$ are the arithmetic means of the $x_i$'s and $y_i$'s, respectively, then the point $(\bar{x}, \bar{y})$ is on the line of regression. (*Hint:* Replace $y$ by $\bar{y}$ and $x$ by $\bar{x}$ in the equation

$$y = \frac{nP - XY}{nS - X^2}x + \frac{SY - PX}{nS - X^2}$$

and show that a true statement is obtained. Recall that $\bar{x} = \frac{X}{n}$ and $\bar{y} = \frac{Y}{n}$.)

**\*22.** Using the result of Exercise 21, show that the $y$-intercept of the line of regression is given by

$$b = \bar{y} - m\bar{x}$$

**\*23.** Show that if $a_1, a_2, \ldots, a_n$ are $n$ real numbers, then $\Sigma (a_i - \bar{a}) = 0$.

**\*24.** Suppose that $(x_1, y_1), (x_2, y_2), \ldots, (x_n, y_n)$ are $n$ points, and define $P, S, X$, and $Y$ as in Theorem 11.6. Show that

**a.** $\Sigma (x_i - \bar{x})(y_i - \bar{y}) = \dfrac{nP - XY}{n}$

and

**b.** $\Sigma (x_i - \bar{x})^2 = \dfrac{nS - X^2}{n}$

(*Hint:* Write $\Sigma (x_i - \bar{x})(y_i - \bar{y}) = \Sigma (x_i - \bar{x})y_i - \Sigma (x_i - \bar{x})\bar{y} = \Sigma (x_i - \bar{x})y_i - \bar{y} \Sigma (x_i - \bar{x}) = \Sigma x_i y_i - \bar{x} \Sigma y_i - \bar{y}(0)$ (by Exercise 23), and so on.)

**\*25.** Using the result of Exercise 24, show that the slope of the regression line may be expressed as

$$m = \frac{\Sigma (x_i - \bar{x})(y_i - \bar{y})}{\Sigma (x_i - \bar{x})^2}$$

# 11.8 Double Integrals

We have generalized the concept of differentiation to functions of several variables. In this section we give a brief introduction to the generalization of integration to functions of two variables. We first evaluate *indefinite* integrals such as $\int f(x,y) \, dx$ and $\int f(x,y) \, dy$. We also evaluate *definite* integrals such as $\int_a^b f(x,y) \, dx$, $\int_a^b f(x, y) \, dy$, $\int_c^d \int_a^b f(x, y) \, dx \, dy$, and others. We then define, as in the case of functions of one variable, the definite integral as a limit of Riemann sums. Finally, we give a sampling of applications, most of which are generalizations of applications presented in Chapter 10.

## Iterated Integrals

Since antidifferentiation is the reverse of differentiation, we may antidifferentiate a function of two or more variables with respect to one of the variables by treating all other variables as constants and proceeding using the techniques introduced in Chapter 9. Thus $\int f(x, y)\, dx$ will represent a set of functions such that $g(x, y)$ is a member of that set if, and only if, $\frac{\partial}{\partial x} g(x, y) = f(x, y)$. Similarly, $\int f(x, y)\, dy$ will represent a set of functions such that $h(x, y)$ is a member of that set if, and only if, $\frac{\partial}{\partial y} h(x, y) = f(x, y)$.

**EXAMPLE 1**  Evaluate the following integrals.

**a.** $\displaystyle\int (20x^3y^4 + 28xy^6)\, dx$

**b.** $\displaystyle\int (20x^3y^4 + 28xy^6)\, dy$

**Solution**  **a.** The $dx$ tells us that we antidifferentiate with respect to $x$, holding $y$ constant. Thus

$$\int (20x^3y^4 + 28xy^6)\, dx = \int 20x^3y^4\, dx + \int 28xy^6\, dx$$

$$= 20y^4 \int x^3\, dx + 28y^6 \int x\, dx$$

$$= 20y^4 \cdot \frac{x^4}{4} + 28y^6 \cdot \frac{x^2}{2} + C(y)$$

$$= 5x^4y^4 + 14x^2y^6 + C(y)$$

where $C(y)$ is any function of $y$.

Observe that we added $C(y)$ as a constant of integration, since when $y$ is treated as a constant, so is $C(y)$. We may verify that our answer is correct as follows.

$$\frac{\partial}{\partial x}(5x^4y^4 + 14x^2y^6 + C(y)) = 20x^3y^4 + 28xy^6 + 0$$

$$= 20x^3y^4 + 28xy^6$$

**b.** The $dy$ tells us that we antidifferentiate with respect to $y$, holding $x$ constant. Thus

$$\int (20x^3y^4 + 28xy^6)\, dy = \int 20x^3y^4\, dy + \int 28xy^6\, dy$$

$$= 20x^3 \int y^4\, dy + 28x \int y^6\, dy$$

$$= 20x^3 \cdot \frac{y^5}{5} + 28x \cdot \frac{y^7}{7} + K(x)$$

$$= 4x^3y^5 + 4xy^7 + K(x)$$

where $K(x)$ is any function of $x$.

Note again that we added $K(x)$ as a constant of integration. Since

$$\frac{\partial}{\partial y}(4x^3y^5 + 4xy^7 + K(x)) = 20x^3y^4 + 28xy^6 + 0$$

$$= 20x^3y^4 + 28xy^6$$

our answer is correct.

**Do Exercise 1.**

We may also evaluate definite integrals such as $\int_a^b f(x, y)\, dx$ and $\int_c^d g(x, y)\, dy$. In the first one, we integrate with respect to $x$ and therefore the limits $a$ and $b$ are values of $x$, while in the second integral we integrate with respect to $y$, so the limits $c$ and $d$ are values of $y$.

**EXAMPLE 2** Evaluate the following definite integrals.

a. $\displaystyle\int_1^3 (20x^3y^4 + 28xy^6)\, dx$

b. $\displaystyle\int_1^3 (20x^3y^4 + 28xy^6)\, dy$

**Solution** a. In Example 1, we found that

$$\int (20x^3y^4 + 28xy^6)\, dx = 5x^4y^4 + 14x^2y^6 + C(y)$$

Thus

$$\int_1^3 (20x^3y^4 + 28xy^6)\, dx = (5x^4y^4 + 14x^2y^6 + C(y))\big|_1^3$$

$$= [5(3^4)y^4 + 14(3^2)y^6 + C(y)] - [5(1)^4y^4 + 14(1^2)y^6 + C(y)]$$

$$= 112y^6 + 400y^4$$

b. Notice that $C(y)$ canceled out. Therefore in the evaluation of such integrals, we usually let the constant of integration be 0 at the outset.

b. We found in Example 1 that

$$\int (20x^3y^4 + 28xy^6)\, dy = 4x^3y^5 + 4xy^7 + K(x)$$

Thus

$$\int_1^3 (20x^3y^4 + 28xy^6)\, dy = (4x^3y^5 + 4xy^7)\big|_1^3$$

$$= [(4x^3(3^5) + 4x(3^7)] - [4x^3(1^5) + 4x(1^7)]$$

$$= 968x^3 + 8744x$$

**Do Exercise 7.**

Of course expressions such as $\int_{g(y)}^{h(y)} f(x, y)\, dx$ will make sense since the limits $g(y)$ and $h(y)$ are treated as constants when we integrate with respect to $x$. We illustrate this in the next example.

**EXAMPLE 3** Evaluate

$$\int_{2y}^{y^3} (20x^3y^4 + 28xy^6)\, dx$$

**Solution** In Example 1 we found that

$$\int (20x^3y^4 + 28xy^6)\, dx = 5x^4y^4 + 14x^2y^6 + C(y)$$

Therefore,

$$\int_{2y}^{y^3} (20x^3y^4 + 28xy^6)\, dx = (5x^4y^4 + 14x^2y^6)\big|_{2y}^{y^3}$$

$$= [5(y^3)^4y^4 + 14(y^3)^2y^6] - [5(2y)^4y^4 + 14(2y)^2y^6]$$
$$= 5y^{16} + 14y^{12} - 80y^8 - 56y^8$$
$$= 5y^{16} + 14y^{12} - 136y^8$$

**Do Exercise 11.**

We see that in general $\int_{g(y)}^{h(y)} f(x, y)\, dx$ is a function of $y$, and $\int_{g(x)}^{h(x)} f(x, y)\, dy$ is a function of $x$. Thus we may evaluate expressions such as

$$\int_a^b \left[ \int_{g(y)}^{h(y)} f(x, y)\, dx \right] dy$$

and

$$\int_a^b \left[ \int_{g(x)}^{h(x)} f(x, y)\, dy \right] dx$$

Such expressions are called *iterated integrals* and are usually written

$$\int_a^b \int_{g(y)}^{h(y)} f(x, y)\, dx\, dy$$

and

$$\int_a^b \int_{g(x)}^{h(x)} f(x, y)\, dy\, dx$$

respectively.

**EXAMPLE 4**    Evaluate

$$\int_1^3 \int_{2y}^{y^3} (20x^3y^4 + 28xy^6)\, dx\, dy$$

**Solution**    In Example 3 we found that

$$\int_{2y}^{y^3} (20x^3y^4 + 28xy^6)\, dx = 5y^{16} + 14y^{12} - 136y^8$$

Thus,

$$\int_1^3 \int_{2y}^{y^3} (20x^3y^4 + 28xy^6)\, dx\, dy = \int_1^3 \left[ \int_{2y}^{y^3} (20x^3y^4 + 28xy^6)\, dx \right] dy$$

$$= \int_1^3 (5y^{16} + 14y^{12} - 136y^8)\, dy$$

$$= \left( 5 \cdot \frac{y^{17}}{17} + 14 \cdot \frac{y^{13}}{13} - 136 \cdot \frac{y^9}{9} \right) \Big|_1^3$$

$$= \frac{5(3^{17} - 1)}{17} + \frac{14(3^{13} - 1)}{13} - \frac{136(3^9 - 1)}{9}$$

**Do Exercise 15.**

 Observe that when evaluating $\int_a^b \int_{g(y)}^{h(y)} f(x, y) \, dx \, dy$, we integrate with respect to $x$ first, and when evaluating $\int_a^b \int_{g(x)}^{h(x)} f(x, y) \, dy \, dx$, we integrate with respect to $y$ first.

**EXAMPLE 5**   Evaluate

**a.** $\displaystyle\int_1^3 \int_2^4 16xy^3 \, dx \, dy$ and **b.** $\displaystyle\int_2^4 \int_1^3 16xy^3 \, dy \, dx$

**Solution**   **a.** $\displaystyle\int_1^3 \int_2^4 16xy^3 \, dx \, dy = \int_1^3 \left[ \int_2^4 16xy^3 \, dx \right] dy$

$$= \int_1^3 [8x^2y^3|_2^4] \, dy = \int_1^3 [8(4^2)y^3 - 8(2^2)y^3] \, dy$$

$$= \int_1^3 96y^3 \, dy = 24y^4|_1^3$$

$$= 24(3^4) - 24(1^4) = 1920$$

**b.** $\displaystyle\int_2^4 \int_1^3 16xy^3 \, dy \, dx = \int_2^4 \left[ \int_1^3 16xy^3 \, dy \right] dx$

$$= \int_2^4 [4xy^4|_1^3] \, dx = \int_2^4 [4x(3^4) - 4x(1^4)] \, dx$$

$$= \int_2^4 320x \, dx = 160x^2|_2^4$$

$$= 160(4^2) - 160(2^2) = 1920$$

Note that the two answers are identical.
**Do Exercise 19.**

## Double Integrals

The definite integral of a continuous function of one variable was defined on a closed, bounded interval $[a, b]$. (Section 9.8.) We now define the double integral of a function $f$ of two variables over a region $R$. We require $f$ to be continuous. Furthermore, $R$ must be closed and bounded, and its intersection with any rectangle whose sides are parallel to the axes must have a well-defined area.

Let $f$ and $R$ be as described and

$$Q = \{(x, y) \mid a \le x \le b, c \le y \le d\}$$

where $a$, $b$, $c$, and $d$ have been chosen so that $R \subseteq Q$. Let $n$ be a positive integer. Consider the regular partitions $p_n$ and $q_n$ of $[a, b]$ and $[c, d]$, respectively. Through each point of $p_n$ (on the $x$-axis), draw a line parallel to the $y$-axis and through each point of $q_n$ (on the $y$-axis) draw a line parallel to the $x$-axis. These lines divide the rectangle $Q$ into $n^2$ small rectangles. (See Figure 11.24.)

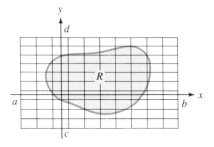

**FIGURE 11.24**

Ignore those rectangles that have empty intersection with $R$ and label the nonempty intersections of the small rectangles with $R$, say $R_1, R_2, R_3, \ldots, R_m$. Necessarily, we must have $m \leq n^2$. In each $R_i$ choose a point $(x_i, y_i)$ and let $\Delta R_i$ denote the area of $R_i$. Then the sum

$$S_m = \sum_{i=1}^{m} f(x_i, y_i)\Delta R_i$$

is called a *Riemann sum*. We should note that $S_m$ depends on the choice of $(x_i, y_i)$ in each $R_i$. However, in more advanced calculus texts it is shown that when the function $f$ is continuous and the region $R$ has the required properties, $\lim_{m \to \infty} S_m$ exists, is finite, and does not depend on the choice of $(x_i, y_i)$ in each $R_i$.

**DEFINITION 11.10:** The common limit just described is called the *double integral of f over R* and is denoted

$$\iint\limits_{R} f(x, y)\, dA$$

The function $f$ is called the *integrand* and $R$ is the *region of integration*.

In dividing the rectangle $Q$ into subrectangles, the partitions of $[a, b]$ and $[c, d]$ need not be regular. We used regular partitions because we wished only to illustrate that a double integral is a limit of Riemann sums while keeping the details as simple as possible. We also did not consider upper and lower sums, although this could have been done.

## Evaluation of Double Integrals

The evauation of a double integral using the definition is usually difficult. Therefore we must have another way of calculating double integrals. It can be shown that a double integral is a number that can be found by evaluating an iterated integral, as we shall illustrate in the next examples. First, we must describe certain types of regions.

The set $\{(x, y) \mid a \le x \le b, g_1(x) \le y \le g_2(x)\}$, where $g_1$ and $g_2$ are continuous functions defined on $[a, b]$, is called a *type 1 region*, while the set $\{(x, y) \mid h_1(y) \le x \le h_2(y), c \le y \le d\}$, where $h_1$ and $h_2$ are continuous functions defined on $[c, d]$, is called a *type 2 region*. An *elementary region* is a region that can be decomposed into a finite number of regions, each of which is type 1, type 2, or both. (See Figure 11.25.)

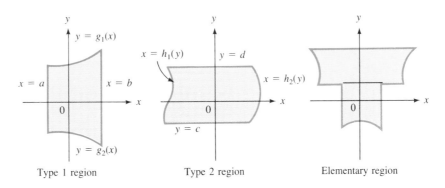

**FIGURE 11.25**

Type 1 region        Type 2 region        Elementary region

The following theorem is stated without proof. The proof may be found in more advanced calculus books.

**THEOREM 11.7:** Suppose $f$ is a continuous function defined on a type 1 region $R$, where

$$R = \{(x, y) \mid a \le x \le b, g_1(x) \le y \le g_2(x)\}$$

then

$$\iint\limits_{R} f(x, y)\, dA = \int_a^b \int_{g_1(x)}^{g_2(x)} f(x, y)\, dy\, dx$$

Also, if $f$ is a continuous function defined on a type 2 region $Q$, where

$$Q = \{(x, y) \mid h_1(y) \le x \le h_2(y), c \le y \le d\}$$

then

$$\iint\limits_{Q} f(x, y)\, dA = \int_c^d \int_{h_1(y)}^{h_2(y)} f(x, y)\, dx\, dy$$

It is important to note the order of integration in each case. When we evaluate a double integral over a type 1 region, the parts of the boundary of the region which are line segments are vertical and therefore they are pieces of the lines $x = a$ and $x = b$. Thus the constant limits are values of $x$. Since *the constant limits of integration should be used last*, because the value of the integral is a number, we must integrate with respect to $x$ last, and with respect to $y$ first. On the other hand, when we evaluate a double integral over a type 2 region, we must integrate with respect to $y$ last, and with respect to $x$ first, for similar reasons.

**EXAMPLE 6**   Let $R = \{(x, y) \mid 1 \le x \le 3, -x - 1 \le y \le x^2\}$ and $f(x, y) = 4x^2y$. Evaluate $\iint\limits_R f(x, y)\, dA$.

**Solution**   The region $R$ is sketched in Figure 11.26. Clearly, $R$ is a type 1 region. Thus in evaluating the appropriate iterated integral, we must integrate with respect to $x$ last, and with respect to $y$ first. The lower and upper limits of integration for $y$ will be $y = -x - 1$ and $y = x^2$, respectively. We write

$$\iint\limits_R 4x^2y\, dA = \int_1^3 \int_{-x-1}^{x^2} 4x^2y\, dy\, dx$$

$$= \int_1^3 [2x^2y^2 \, |_{-x-1}^{x^2}]\, dx$$

$$= \int_1^3 [2x^2(x^2)^2 - 2x^2(-x - 1)^2]\, dx$$

$$= \int_1^3 (2x^6 - 2x^4 - 4x^3 - 2x^2)\, dx$$

$$= \frac{2x^7}{7} - \frac{2x^5}{5} - x^4 - \frac{2x^3}{3} \Big|_1^3$$

$$= \left[\frac{2(3^7)}{7} - \frac{2(3^5)}{5} - 3^4 - \frac{2(3^3)}{3}\right] - \left[\frac{2(1^7)}{7} - \frac{2(1^5)}{5} - 1^4 - \frac{2(1^3)}{3}\right]$$

$$= \frac{45,196}{105}$$

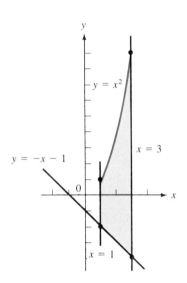

**FIGURE 11.26**

Do Exercise 23.   ■

A region may be both a type 1 and type 2 region. In this case, we may integrate with respect to either variable first. However, even in such cases, it is often necessary to choose one order of integration over the other. We illustrate this with the next example.

**EXAMPLE 7**   Let $R$ be the region bounded by the $y$-axis and the graphs of $y = x$ and $y = 1$. Evaluate the double integral $\iint\limits_R x^2e^{y^4}\, dA$.

**Solution**   The region $R$ is sketched in Figure 11.27. We can describe $R$ as the set $\{(x, y) \mid 0 \le x \le 1, x \le y \le 1\}$, which is a type 1 region, or $\{(x, y) \mid 0 \le x \le y, 0 \le y \le 1\}$, which is a type 2 region. Thus we may write

$$\iint\limits_R x^2e^{y^4}\, dA = \int_0^1 \int_x^1 x^2e^{y^4}\, dy\, dx$$

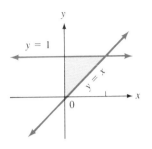

**FIGURE 11.27**

and

$$\iint\limits_R x^2 e^{y^4} \, dA = \int_0^1 \int_0^y x^2 e^{y^4} \, dx \, dy$$

If we try to evaluate $\int_0^1 \int_x^1 x^2 e^{y^4} \, dy \, dx$, we run into difficulty immediately because $e^{y^4}$ does not have an elementary antiderivative. Thus we proceed as follows.

$$\iint\limits_R x^2 e^{y^4} \, dA = \int_0^1 \int_0^y x^2 e^{y^4} \, dx \, dy = \int_0^1 \left[ \frac{x^3 e^{y^4}}{3} \Big|_0^y \right] dy$$

In this first step, we treated $e^{y^4}$ as a constant and antidifferentiated $x^2$ with respect to $x$. Furthermore,

$$\int_0^1 \left[ \frac{x^3 e^{y^4}}{3} \Big|_0^y \right] dy = \int_0^1 \left[ \frac{y^3 e^{y^4}}{3} - \frac{0^3 e^{y^4}}{3} \right] dy = \int_0^1 \frac{y^3 e^{y^4}}{3} \, dy$$

$$= \frac{e^{y^4}}{12} \Big|_0^1 = \frac{e^{1^4}}{12} - \frac{e^{0^4}}{12} = \frac{e-1}{12}$$

**Do Exercise 31.**

Suppose now $f$ is a continuous function defined on a region $R$ which is neither type 1 nor type 2, but is elementary. We decompose $R$ into regions $R_1, R_2, \ldots, R_n$, each of which is either type 1 or type 2, and we use the fact that

$$\iint\limits_R f(x, y) \, dA = \iint\limits_{R_1} f(x, y) \, dA + \iint\limits_{R_2} f(x, y) \, dA + \cdots + \iint\limits_{R_n} f(x, y) \, dA$$

We illustrate this method in the following example.

**EXAMPLE 8**   Let $A$, $B$, $C$, $D$, and $E$ have coordinates $(0, 2)$, $(4, 2)$, $(4, -4)$, $(0, -4)$, and $(0, 0)$, respectively, and let $R$ be the region inside the rectangle $ABCD$ and outside the triangle $BCE$. Also let $f(x, y) = ye^x$. Evaluate $\iint\limits_R f(x, y) \, dA$.

**Solution**   Let $R_1$ and $R_2$ be the regions bounded by the triangles $ABE$ and $CDE$, respectively. Then $R = R_1 \cup R_2$. (See Figure 11.28.) Thus

$$\iint\limits_R ye^x \, dA = \iint\limits_{R_1} ye^x \, dA + \iint\limits_{R_2} ye^x \, dA$$

The equation of the line through the points $(0, 0)$ and $(4, 2)$ is $x = 2y$, so $R_1$ can be described as

$$\{(x, y) \mid 0 \le x \le 2y, \ 0 \le y \le 2\}$$

which is a type 2 region. Thus

$$\iint\limits_{R_1} ye^x \, dA = \int_0^2 \int_0^{2y} ye^x \, dx \, dy$$

**FIGURE 11.28**

First integrating with respect to $x$ and treating $y$ as a constant, we get

$$\int_0^2 \int_0^{2y} ye^x \, dx \, dy = \int_0^2 [ye^x \, |_0^{2y}] \, dy$$

$$= \int_0^2 [ye^{2y} - ye^0] \, dy = \int_0^2 (ye^{2y} - y) \, dy$$

Using integration by parts (or tabular integration), we find that

$$\int ye^{2y} \, dy = \frac{2y - 1}{4} e^{2y} + C$$

Therefore,

$$\int_0^2 (ye^{2y} - y) \, dy = \left[\frac{2y - 1}{4} e^{2y} - \frac{y^2}{2}\right]\bigg|_0^2$$

$$= \left[\frac{2(2) - 1}{4} e^{2(2)} - \frac{2^2}{2}\right] - \left[\frac{2(0) - 1}{4} e^{2(0)} - \frac{0^2}{2}\right]$$

$$= \frac{3}{4} e^4 - 2 + \frac{1}{4} = \frac{3e^4 - 7}{4}$$

Similarly, the equation of the line through the points $(0, 0)$ and $(4, -4)$ is $x = -y$, so $R_2$ can be described as

$$\{(x, y) \mid 0 \le x \le -y, -4 \le y \le 0\}$$

which is also a type 2 region. Thus

$$\iint_{R_2} ye^x \, dA = \int_{-4}^0 \int_0^{-y} ye^x \, dx \, dy$$

First integrating with respect to $x$ and treating $y$ as a constant, we get

$$\int_{-4}^0 \int_0^{-y} ye^x \, dx \, dy = \int_{-4}^0 [ye^x|_0^{-y}] \, dy$$

$$= \int_{-4}^0 [ye^{-y} - ye^0] \, dy = \int_{-4}^0 (ye^{-y} - y) \, dy$$

Again using integration by parts (or tabular integration), we find that

$$\int ye^{-y} \, dy = -e^{-y}(y + 1) + C$$

Therefore

$$\int_{-4}^0 (ye^{-y} - y) \, dy = \left[-e^{-y}(y + 1) - \frac{y^2}{2}\right]\bigg|_{-4}^0$$

$$= \left[-e^{-0}(0 + 1) - \frac{0^2}{2}\right] - \left[-e^{-(-4)}(-4 + 1) - \frac{(-4)^2}{2}\right]$$

$$= 7 - 3e^4$$

It follows that

$$\iint_R ye^x \, dA = \frac{3e^4 - 7}{4} + (7 - 3e^4) = \frac{21 - 9e^4}{4}$$

**Do Exercise 33.**                                              ■

## Average Value of a Function

It may be helpful to first look at the case of a function of one variable. Suppose $f$ is continuous and nonnegative on $[a, b]$. Let $R$ be the region bounded by the $x$-axis and the graphs of $y = f(x)$, $x = a$, and $x = b$. We know that

$$\text{area of } R = \int_a^b f(x) \, dx \text{ square units}$$

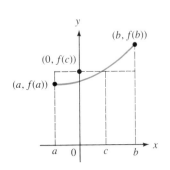

**FIGURE 11.29**

It is intuitively clear (see Figure 11.29), that there is a point $c$, between $a$ and $b$, such that the area of the rectangle of length $(b - a)$ units and of height $f(c)$ units is equal to the area of the region $R$. Thus

$$\int_a^b f(x) \, dx = (b - a)f(c)$$

When $f$ is continuous on a closed bounded interval $[a, b]$, the existence of such a point $c$ is guaranteed by the *Mean Value Theorem for Integrals*, even if the continuous function $f$ is not nonnegative. We write

$$f(c) = \frac{1}{(b - a)} \int_a^b f(x) \, dx$$

and we call $f(c)$ the *average value of $f$ on* $[a, b]$. In other words, we obtain the average value of $f$ on $[a, b]$ by dividing $\int_a^b f(x) \, dx$ by the length of the interval $[a, b]$. Similarly, suppose $f$ is a continuous function defined on the rectangle $R$, where $R = \{(x, y) \mid a \le x \le b, c \le y \le d\}$. Then we obtain the average value of $f$ on the rectangle $R$ by dividing the double integral $\iint_R f(x, y) \, dA$ by the area of the rectangle $R$. Formally, we give the following definition.

---

**DEFINITION 11.11:**   Let $f$ be a continuous function defined on the rectangle $R$ where

$$R = \{(x, y) \mid a \le x \le b, c \le y \le d\}$$

Then the *average value of $f$ over $R$* is

$$\frac{1}{(b - a)(d - c)} \iint_R f(x, y) \, dA$$

**EXAMPLE 9** Let $f(x, y) = 8x^2y^3$ and

$$R = \{(x, y) \mid -1 \le x \le 2, -1 \le y \le 3\}$$

Find the average value of $f$ over the rectangle $R$.

**Solution** The average value is obtained as follows:

$$\frac{1}{(b-a)(d-c)} \iint_R f(x, y)\, dA = \frac{1}{[2-(-1)][3-(-1)]} \iint_R 8x^2y^3\, dA$$

$$= \frac{1}{12} \int_{-1}^{3} \int_{-1}^{2} 8x^2y^3\, dx\, dy$$

$$= \frac{1}{12} \int_{-1}^{3} \left[ \frac{8x^3y^3}{3} \Big|_{-1}^{2} \right] dy$$

$$= \frac{1}{12} \int_{-1}^{3} \left[ \frac{8(2^3)y^3}{3} - \frac{8(-1)^3y^3}{3} \right] dy = \frac{1}{12} \int_{-1}^{3} 24y^3\, dy$$

$$= \frac{1}{12} [6y^4 \big|_{-1}^{3}] = \frac{1}{12}[6(3^4) - 6(-1)^4] = \frac{480}{12} = 40$$

Thus the average value of $f$ over the rectangle $R$ is 40.
**Do Exercise 35.**

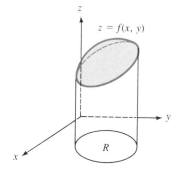

**FIGURE 11.30**

## Applications

We now discuss a sampling of applications, most of which are generalizations of applications discussed in Chapter 10. We have seen that if $f$ is continuous and non-negative on an interval $[a, b]$, then the area of the region bounded by the graphs of $y = f(x)$, $x = a$, $x = b$, and the $x$-axis is $\int_a^b f(x)\, dx$ square units. Similarly, if $f$ is a continuous and nonnegative function defined on a region $R$ and $R$ has the required properties for the double integral $\iint_R f(x, y)\, dA$ to be defined, then the volume of the solid below the graph of $z = f(x, y)$ and directly above the region $R$ is $\iint_R f(x, y)\, dA$ cubic units. (See Figure 11.30.)

**EXAMPLE 10** Find the volume of the solid in the first octant bounded by the three coordinate planes and below the plane $2x + 3y + 5z = 60$.

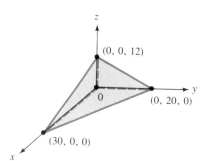

**FIGURE 11.31**

**Solution**   The solid is sketched in Figure 11.31. It is easy to see that the $x$, $y$, and $z$ intercepts are $(30, 0, 0)$, $(0, 20, 0)$, and $(0, 0, 12)$, respectively. Thus the base of the solid is the triangle with vertices $(30, 0, 0)$, $(0, 20, 0)$, and $(0, 0, 0)$. The region $R$ is that triangle. In the $xy$-plane, it can be described as the region bounded by the graphs of $2x + 3y = 60$, $y = 0$, and $x = 0$. (See Figure 11.32.) This region is of type 1 because when we draw a line through one of its points $(x, y)$ and parallel to the $y$-axis, that line intersects the lower part of $R$ at $(x, 0)$ and the upper part at $\left(x, 20 - \frac{2x}{3}\right)$. Observe that the ordinate of the second point is not constant. Also, the line segment through $(x, 0)$ and $\left(x, 20 - \frac{2x}{3}\right)$ will sweep over $R$ as $x$ increases from 0 to 30. Therefore

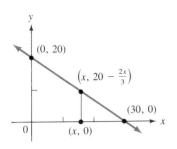

**FIGURE 11.32**

$$R = \left\{(x, y) \mid 0 \le x \le 30, \ 0 \le y \le 20 - \frac{2x}{3}\right\}$$

Using the equation of the plane to solve for $z$ in terms of $x$ and $y$, we get

$$z = f(x, y) = 12 - \frac{2x}{5} - \frac{3y}{5}$$

Thus the volume of the solid is

$$\iint\limits_{R} \left(12 - \frac{2x}{5} - \frac{3y}{5}\right) dA \text{ cubic units}$$

But,

$$\iint\limits_{R} \left(12 - \frac{2x}{5} - \frac{3y}{5}\right) dA = \int_0^{30} \int_0^{20 - 2x/3} \left(12 - \frac{2x}{5} - \frac{3y}{5}\right) dy \, dx$$

Treating $12 - \dfrac{2x}{5}$ as a constant and integrating with respect to $y$, we get

$$\int_0^{30} \int_0^{20 - 2x/3} \left(12 - \frac{2x}{5} - \frac{3y}{5}\right) dy \, dx = \int_0^{30} \left[\left(12 - \frac{2x}{5}\right)y - \frac{3y^2}{10}\right)\Bigg|_0^{20 - 2x/3}\right] dx$$

$$= \int_0^{30} \left[\left(12 - \frac{2x}{5}\right)\left(20 - \frac{2x}{3}\right) - \frac{3}{10}\left(20 - \frac{2x}{3}\right)^2 - 0\right] dx$$

$$= \int_0^{30} \left(\frac{2x^2}{15} - 8x + 120\right) dx = \left(\frac{2x^3}{45} - 4x^2 + 120x\right)\Bigg|_0^{30}$$

$$= \left[\frac{2(30^3)}{45} - 4(30^2) + 120(30)\right] - 0 = 1200$$

Thus the volume of the solid is 1200 cubic units.
**Do Exercise 39.**                                                                      ■

**EXAMPLE 11**   If the price of choice beef is $p_B$ dollars per pound and the price of frozen salmon is $P_S$ dollars per pound, the number of pounds of choice beef sold each week at a certain store is given by

$$f(p_B, p_S) = 15{,}000 - 900p_B^{1/2} + 600p_S^{1/2}$$

During a certain year, the price of choice beef increased steadily from \$4 to \$5.29 per pound and that of frozen salmon increased steadily from \$2.25 to \$2.89 per pound. What is the weekly average number of pounds of choice beef sold during the year?

**Solution** Let $R = \{(p_B, p_S) \mid 4 \le p_B \le 5.29 \text{ and } 2.25 \le p_S \le 2.89\}$. We must find the average value of the function $f$ over the rectangle $R$. Let this value be $W$. The area of $R$ is .8256 since $(5.29 - 4)(2.89 - 2.25) = .8256$.

Then

$$W = \frac{1}{.8256} \iint_R (15{,}000 - 900 p_B^{1/2} + 600 p_S^{1/2})\, dA$$

$$= \frac{1}{.8256} \int_{2.25}^{2.89} \int_{4}^{5.29} (15{,}000 - 900 p_B^{1/2} + 600 p_S^{1/2})\, dp_B\, dp_S$$

Treating $p_S$ as a constant and integrating with respect to $p_B$, we obtain

$$W = \frac{1}{.8256} \int_{2.25}^{2.89} (15{,}000\, p_B - 600 p_B^{3/2} + 600 p_S^{1/2} p_B)\Big|_4^{5.29}\, dp_S$$

$$= \frac{1}{.8256} \int_{2.25}^{2.89} [15{,}000(5.29 - 4) - 600(5.29^{3/2} - 4^{3/2}) + 600(5.29 - 4)p_S^{1/2}]\, dp_S$$

$$= \frac{1}{.8256} \int_{2.25}^{2.89} (16{,}849.8 + 774 p_S^{1/2})\, dp_S$$

$$= \frac{1}{.8256} (16{,}849.8 p_S + 516 p_S^{3/2})\Big|_{2.25}^{2.89}$$

$$= \frac{1}{.8256} [16{,}849.8(2.89 - 2.25) + 516(2.89^{3/2} - 2.25^{3/2})]$$

$$\doteq 14{,}023.11047$$

Thus during that year, the store will sell an average of approximately 14,023 pounds of choice beef per week.
**Do Exercise 41.** ■

The next optional example should be read only if you are familiar with the content of Section 10.6. We first give the following definition.

---

**DEFINITION 11.12:** A nonnegative function $f$ defined on the $xy$-plane is called a *joint probability density function* of the random variables $x$ and $y$ if, and only if,

$$P[(x, y) \in R] = \iint_R f(x, y)\, dA$$

for any region $R$ in the $xy$-plane. Of course, $P[(x, y) \in R]$ denotes the probability that $(x, y)$ belongs to the region $R$.

EXAMPLE 12    Let $x$ be the number of minutes that a person stands in line to purchase an airline ticket and let $y$ minutes be the time that a person waits at the gate before boarding. Suppose that $f(x, y)$, the joint probability density function, is defined by

$$f(x, y) = \begin{cases} \dfrac{1}{61,509,375} 2x^2y, & \text{if } 0 \le x \le 45 \text{ and } 0 \le y \le 45 \\ 0, & \text{elsewhere} \end{cases}$$

If you stand in line to buy a ticket, then wait at the gate to board, what is the probability that your total waiting time does not exceed 25 minutes?

Solution    Let $x$ minutes be the time that you stand in line to buy a ticket and $y$ minutes be the time that you wait at the gate. We must find the probability that $x + y \le 25$. Clearly, $x \ge 0$ and $y \ge 0$. The set of points which satisfy these three inequalities is the triangle $R$ whose vertices are $(0, 0)$, $(0, 25)$, and $(25, 0)$. (See Figure 11.33.) Thus

$$P[(x, y) \in R] = \iint\limits_{R} \frac{1}{61,509,375} 2x^2y \, dA$$

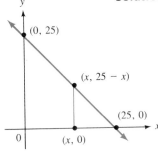

FIGURE 11.33

The intersection with $R$ of a line through a point $(x, y)$ of $R$ and parallel to the $y$ axis is a line segment whose lowest point is $(x, 0)$ and highest point is $(x, 25 - x)$. This line segment will sweep $R$ as $x$ increases from 0 to 25. (See Figure 11.33.) Thus $R$ is a type 1 region and in evaluating the double integral we integrate with respect to $y$ first. We write

$$P[(x, y) \in R] = \int_0^{25} \int_0^{25-x} \frac{1}{61,509,375} 2x^2y \, dy \, dx$$

Treating $x$ as a constant and integrating with respect to $y$, we obtain

$$\int_0^{25} \int_0^{25-x} \frac{1}{61,509,375} 2x^2y \, dy \, dx = \int_0^{25} \frac{1}{61,509,375} x^2y^2 \Big|_0^{25-x} \, dx$$

$$= \int_0^{25} \frac{1}{61,509,375} (625x^2 - 50x^3 + x^4) \, dx$$

$$= \frac{1}{61,509,375} \left( \frac{625x^3}{3} - \frac{50x^4}{4} + \frac{x^5}{5} \right) \Big|_0^{25}$$

$$= \frac{1}{61,509,375} \left( \frac{625(25^3)}{3} - \frac{50(25^4)}{4} + \frac{(25^5)}{5} \right) - 0$$

$$\doteq .005292214$$

The probability that your total waiting time does not exceed 25 minutes is .0053.
**Do Exercise 43.**

# Exercise Set 11.8

*In Exercises 1–6, find the antiderivatives.*

**1.** $\int (24x^2y^3 + 20xy^4)\, dy$    **2.** $\int 5x^2 \sqrt[3]{y}\, dx$

**3.** $\int 6x^2(x^3 + 1)^4 y\, dx$    **4.** $\int \dfrac{8xy}{x^2 + 1}\, dx$

**5.** $\int 6x^2 e^{3y}\, dx$    **6.** $\int 5e^{xy}\, dy$

*In Exercises 7–13, evaluate the given integral.*

**7.** $\int_1^2 (24x^2y^3 + 20xy^4)\, dy$    **8.** $\int_1^2 5x^2 \sqrt[3]{y}\, dx$

**9.** $\int_0^1 6x^2(x^3 + 1)^4 y\, dx$    **10.** $\int_0^2 \dfrac{8xy}{x^2 + 1}\, dx$

**11.** $\int_{3y}^{y^2} (24x^2y^3 + 20xy^4)\, dx$

**12.** $\int_{2y}^{3y^2} 6x^2(x^3 + 1)^4 y\, dx$

**13.** $\int_{\sqrt{y}}^{3y} \dfrac{8xy}{x^2 + 1}\, dx$

*In Exercises 14–18, evaluate the iterated integrals.*

**14.** $\int_1^2 \int_{3y}^{y^2} (24x^2y^3 + 20xy^4)\, dx\, dy$

**15.** $\int_1^2 \int_{x^3}^{8x^6} 5x^2 \sqrt[3]{y}\, dy\, dx$

**16.** $\int_1^4 \int_0^y 6x^2(x^3 + 1)^4 y^2\, dx\, dy$

**17.** $\int_1^2 \int_0^{\sqrt{y}} \dfrac{8xy}{x^2 + 1}\, dx\, dy$

**18.** $\int_0^1 \int_1^{\sqrt[3]{y}} 6x^2 e^{3y}\, dx\, dy$

*In Exercises 19–22, evaluate the iterated integrals of parts a and b and compare the answers.*

**19. a.** $\int_1^3 \int_2^5 (24x^2y^3 + 20xy^4)\, dx\, dy$

**b.** $\int_2^5 \int_1^3 (24x^2y^3 + 20xy^4)\, dy\, dx$

**20. a.** $\int_{-1}^2 \int_0^1 6x^2(x^3 + 1)^4 y\, dx\, dy$

**b.** $\int_0^1 \int_{-1}^2 6x^2(x^3 + 1)^4 y\, dy\, dx$

**21. a.** $\int_2^4 \int_0^2 \dfrac{8xy}{x^2 + 1}\, dx\, dy$

**b.** $\int_0^2 \int_2^4 \dfrac{8xy}{x^2 + 1}\, dy\, dx$

**22. a.** $\int_2^4 \int_1^2 5e^x y\, dx\, dy$

**b.** $\int_1^2 \int_2^4 5e^x y\, dy\, dx$

*In Exercises 23–30, a function f and a region R are defined. For each, evaluate the double integral $\iint\limits_R f(x, y)\, dA$.*

**23.** $f(x, y) = 3x^2 + 4xy$,
 $R = \{(x, y) \mid -1 \le x \le 3,\ 1 \le y \le 2\}$

**24.** $f(x, y) = xe^y$, $R = \{(x, y) \mid -2 \le x \le 3,\ 0 \le y \le 2\}$

**25.** $f(x, y) = \dfrac{x^2 y^3}{y^4 + 4}$,
 $R = \{(x, y) \mid -2 \le x \le 3,\ -1 \le y \le 2\}$

**26.** $f(x, y) = x^3 + 5xy + y^2$,
 $R = \{(x, y) \mid 0 \le x \le 3,\ -x - 1 \le y \le 2x + 2\}$

**27.** $f(x, y) = 3x^2 - 5xy + 3y^3$,
 $R = \{(x, y) \mid 1 - y \le x \le 2y + 5,\ -1 \le y \le 1\}$

**28.** $f(x, y) = \dfrac{2y}{x}$, $R = \{(x, y) \mid 1 \le x \le e,\ \ln x \le y \le x^2\}$

**29.** $f(x, y) = y + 2$,
 $R = \{(x, y) \mid -e^y \le x \le e^y,\ \ln 2 \le y \le \ln 5\}$

**30.** $f(x, y) = x^2 e^{2y}$,
 $R = \{(x, y) \mid -\sqrt[3]{y^2} \le x \le 3\sqrt[3]{y^2},\ -8 \le y \le -1\}$

*In Exercises 31 and 32, a function f and a region R are described. In each case do the following: **a.** Describe the region R as a type 1 region. **b.** Describe the region R as a type 2 region. **c.** Evaluate the double integral $\iint\limits_R f(x, y)\, dA$.*

**31.** The region R is bounded by the graphs of $x = 0$, $y = 0$, and $x + y = 2$. Also $f(x, y) = xe^{(2-y)^3}$.

**32.** The region R is bounded by the graphs of $y = e^x$, $x = 0$, and $y = 3$. Also $f(x, y) = \dfrac{4x}{y}$.

**33.** Let $A$, $B$, $C$, $D$, and $E$ have coordinates $(1, 4)$, $(5, 4)$, $(5, 1)$, $(1, 1)$, and $(5, 2)$, respectively, and let R be the region inside the rectangle $ABCD$ and outside the triangle $ADE$. Also, let $f(x, y) = 5x^2 y + 2x - 3y$. Evaluate $\iint\limits_R f(x, y)\, dA$.

**34.** The region $R$ is bounded by the graphs of $y = e^{-2}$, $y = e^x$, $y = e^{-x}$, and $y = e^3$. Also, $f(x, y) = 2xe^{2x}y$. Evaluate the double integral $\iint_R f(x, y)\, dA$.

*In Exercises 35–37, find the average value of the function f over the rectangle R.*

**35.** $f(x, y) = 3x^2 + 4xy$,
$R = \{(x, y) \mid -1 \le x \le 3, 1 \le y \le 2\}$

**36.** $f(x, y) = xe^y$, $R = \{(x, y) \mid -2 \le x \le 3, 0 \le y \le 2\}$

**37.** $f(x, y) = \dfrac{x^2 y^3}{y^4 + 4}$,
$R = \{(x, y) \mid -2 \le x \le 3, -1 \le y \le 2\}$

**38.** Find the volume in the first octant bounded by the three coordinate planes and below the plane $2x + 5y + 3z = 90$.

**39.** Let $R = \{(x, y) \mid 0 \le x \le 1, 0 \le y \le 1 - x^2\}$. Find the volume of the solid in the first octant above $R$ and below the plane $x + y + z = 2$.

**40.** If the Baldhead Company spends $K$ million dollars on capital and $L$ million dollars on labor in a year, it will produce $P(L, K)$ million hairdryers, where

$$P(L, K) = .16L^{1/3}K^{2/3}$$

Over a certain number of years, the company has increased its expenditures from $64 million to $125 million per year on labor and from $27 million to $42.875 million per year

on capital. What is the average yearly production of hairdryers over these years?

**41.** If the price of coffee is $p_C$ dollars per pound and the price of tea is $P_T$ dollars per pound, the number of pounds of coffee sold each week at a certain store is given by

$$f(p_C, p_T) = 2000 - 350p_C^{1/2} + 75p_T^{1/3}$$

During a certain year, the prices of coffee and tea increased steadily from $4 to $4.41 per pound and from $8 to $8.74 per pound, respectively. What is the weekly average number of pounds of coffee sold during the year?

**42.** A county is represented by the region $R$ bounded by the rectangle with vertices $(0, 0)$, $(9, 0)$, $(9, 16)$, and $(0, 16)$. Suppose the value of land at the point $(x, y)$ is $V(x, y)$ dollars per acre where

$$V(x, y) = 160x^{3/2}y^{3/4}$$

Find the average value of land in the county.

**43.** Let $x$ and $y$ represent the number of yards gained by team A, running and passing, respectively, during a football game. If the joint probability function for $x$ and $y$ is $f(x, y) = .00002e^{-.005x}e^{-.004y}$, find the probability that team A is held to not more than 250 yards in total offense.

**44.** Same as Exercise 43 if the joint probability function for $x$ and $y$ is $f(x, y) = .000012e^{-.004x}e^{-.003y}$.

# 11.9 Chapter Review

**IMPORTANT SYMBOLS AND TERMS**

$\partial f/\partial x(x_0, y_0)$ [11.2]
$\partial f/\partial y(x_0, y_0)$ [11.2]
$z = f(x, y)$ [11.1]
$\partial z/\partial x$ [11.2]
$\partial z/\partial y$ [11.2]
$\partial^2 z/\partial x^2$ [11.2]
$\partial^2 z/\partial y^2$ [11.2]
$\partial^2 z/\partial x\partial y$ [11.2]
$\partial^2 z/\partial y\partial x$ [11.2]
$f_x(x_0, y_0)$ [11.2]
$f_y(x_0, y_0)$ [11.2]
$f_{xx}$ [11.2]
$f_{yy}$ [11.2]
$f_{yx}$ [11.2]
$f_{xy}$ [11.2]
$\int f(x, y)\, dy$ [11.8]
$\iint f(x, y)\, dx\, dy$ [11.8]

$\iint f(x, y)\, dy\, dx$ [11.8]
$\int_a^b \int_{h_1(y)}^{h_2(y)} f(x, y)\, dx\, dy$ [11.8]
$\int f(x, y)\, dx$ [11.8]
$\eta_{xx}$ [11.4]
$\eta_{xy}$ [11.4]
Absolute extremum [11.5]
Auxiliary function [11.6]
Average value of a function [11.8]
Chain Rule [11.3]
Cobb-Douglas production function [11.6]
Complements [11.4]
Composition of functions [11.3]
Coordinates [11.1]
Critical point [11.5]

Cross elasticity of demand [11.4]
Directional derivative [11.2]
Domain [11.1]
Double integrals [11.8]
Elasticity of demand [11.4]
Elementary region [11.8]
Function of two variables [11.1]
Graph [11.1]
Image [11.1]
Implicit differentiation [11.3]
Iterated integrals [11.8]
Joint probability density function [11.8]
Lagrange multiplier [11.6]

SUMMARY   The function concept may be extended to the case where the domain is a set of ordered pairs (triples) of real numbers. We say that the function is a function of two (three) variables and we write $z = f(x, y)$ ($w = f(x, y, z)$). In general, the graph of a function of two variables is a surface in a three-dimensional coordinate system. If $f$ is a function of two variables and $c$ is a number in the range of $f$, the set of all $(x, y)$ such that $f(x, y) = c$ is called a level curve. If $f$ is a function of two variables, $(a, b)$ is in the domain of $f$; the instantaneous rate of change of $f$ at $(a, b)$ in a certain direction is called a directional derivative. In the special case where the direction is that of the $x$-axis, the directional derivative is called the partial derivative with respect to $x$ and is denoted by any of the following symbols:

$$f_x(a, b), \left.\frac{\partial z}{\partial x}\right|_{(a,b)}, \frac{\partial f(a, b)}{\partial x}$$

The partial derivative with respect to $y$ is defined and denoted similarly. To find $f_x(x, y)$, treat $y$ as a constant and differentiate $f$ with respect to $x$. Similarly, to find $f_y(x, y)$, treat $x$ as a constant and differentiate $f$ with respect to $y$. Higher order derivatives may be obtained. We use the following notation:

$f_{xx}$ denotes the partial derivative of $f_x$ with respect to $x$,

$f_{xy}$ denotes the partial derivative of $f_x$ with respect to $y$, and

$f_{yy}$ and $f_{yx}$ have similar interpretations.

Using the $\partial$-notation:

$$\frac{\partial^2 z}{\partial x^2} \text{ means } f_{xx}, \frac{\partial^2 z}{\partial y^2} \text{ means } f_{yy}, \frac{\partial^2 z}{\partial x \partial y} \text{ means } f_{yx}, \text{ and } \frac{\partial^2 z}{\partial y \partial x} \text{ means } f_{xy}.$$

The Chain Rule states that if $z = f(x, y)$, $x = g(t)$, and $y = h(t)$, then under certain conditions $z = f(g(t), h(t))$ is differentiable and

$$\frac{dz}{dt} = \frac{\partial z}{\partial x} \cdot \frac{dx}{dt} + \frac{\partial z}{\partial y} \cdot \frac{dy}{dt}$$

More generally, if $z = f(u, v)$, $u = g(x, y)$, $v = h(x, y)$, and $z = F(x, y) = f(g(x, y), h(x, y))$, then if $f$, $g$, and $h$ have continuous partial derivatives

$$\frac{\partial z}{\partial x} = \frac{\partial z}{\partial u} \cdot \frac{\partial u}{\partial x} + \frac{\partial z}{\partial v} \cdot \frac{\partial v}{\partial x}$$

and

$$\frac{\partial z}{\partial y} = \frac{\partial z}{\partial u} \cdot \frac{\partial u}{\partial y} + \frac{\partial z}{\partial v} \cdot \frac{\partial v}{\partial y}$$

Suppose $F(x, y) = 0$ defines $y$ implicitly as a function of $x$. Then assuming the functions involved have continuous derivatives, we have

$$\frac{dy}{dx} = \frac{-F_x(x, y)}{F_y(x, y)}$$

provided that $F_y(x, y) \neq 0$.

Furthermore, suppose $F(x, y, z) = 0$ defines $z = f(x, y)$ implicitly. Then if the functions involved have continuous partial derivatives,

$$\frac{\partial z}{\partial x} = \frac{-F_x(x, y, z)}{F_z(x, y, z)} \text{ and } \frac{\partial z}{\partial y} = \frac{-F_y(x, y, z)}{F_z(x, y, z)}$$

provided that $F_z(x, y, z) \neq 0$.

Partial derivatives are used in many applications. If $z = f(x, y)$, the value of the expression

$$f_x(x_0, y_0)\Delta x + f_y(x_0, y_0)\Delta y + z_0$$

is an approximation of the value of the function $f$ at $(x_0 + \Delta x, y_0 + \Delta y)$.

If $D_x = f(p_x, p_y)$ and $D_y = g(p_x, p_y)$ are the demand equations of two commodities X and Y, then the partial derivative $\partial D_x/\partial p_x$ is the marginal demand of commodity X with respect to $p_x$ and approximates the amount by which the demand for commodity X changes per unit increase in the price of commodity X when the price of commodity Y remains constant. Similarly for $\partial D_x/\partial p_y$. The partial derivatives $\partial D_y/\partial p_x$ and $\partial D_y/\partial p_y$ have similar interpretations.

Pairs of commodities are classified as substitutes (competitive) or complements. The elasticity of demand for commodity X is defined by

$$\eta_{XX} = \frac{\partial D_x}{\partial p_x} \cdot \frac{p_x}{D_x}$$

and approximates the ratio of the percentage change in demand for commodity X to the percentage change in the price of commodity X as the price of commodity Y remains constant. The cross elasticity of demand of commodity X with respect to the price of commodity Y is defined by

$$\eta_{XY} = \frac{\partial D_x}{\partial p_y} \cdot \frac{p_y}{D_x}$$

and approximates the ratio of the percentage change in demand for commodity X to the percentage change in the price of commodity Y as the price of commodity X remains constant. We note that $\eta_{YX}$ is defined and interpreted similarly. If commodities X and Y are substitutes, $\eta_{XY}$ and $\eta_{YX}$ are positive. On the other hand, if commodities X and Y are complements, $\eta_{XY}$ and $\eta_{YX}$ are negative.

An interior point $(a, b)$ of the domain of a function $f$ of two variables is said to be a critical point of $f$ if two of the following four statements are true: $f_x(a, b) = 0$, $f_y(a, b) = 0$, $f_x(a, b)$ is not defined, and $f_y(a, b)$ is not defined. A function of two variables can have a local maximum or minimum in the interior of its domain only at critical points. We consider only critical points for which $f_x(a, b) = f_y(a, b) = 0$. We use the Second Partials Test to locate where a function of two variables attains local extrema. The absolute maximum or absolute minimum of a function $f$ can occur only at a critical point of $f$ or on the boundary of the domain of $f$. We often rely on the nature of the problem to decide whether a local extremum is in fact an absolute extremum.

To find critical points of a function of two variables subject to some constraint, we use the method of the Lagrange multiplier. Suppose $z = f(x, y)$ and the independent variables $x$ and $y$ are subject to the constraint $g(x, y) = 0$. Suppose also that $f$ and $g$ have continuous partial derivatives on a domain $D$. To locate the local extrema of $f$, subject to the constraint $g(x, y) = 0$, we proceed as follows. We first define an auxiliary function $F$ by

$$F(x, y, \lambda) = f(x, y) - \lambda g(x, y)$$

where $\lambda$ is called the Lagrange multiplier.

We then find the critical points of $F$ by solving the system

$$\begin{cases} F_x(x, y, \lambda) = 0 \\ F_y(x, y, \lambda) = 0 \\ F_\lambda(x, y, \lambda) = 0 \end{cases}$$

This system is equivalent to

$$\begin{cases} f_x(x, y) = \lambda g_x(x, y) \\ f_y(x, y) = \lambda g_y(x, y) \\ g(x, y) = 0 \end{cases}$$

For each solution $x = a$, $y = b$, and $\lambda = \lambda_0$, the point $(a, b)$ is also a critical point of $f$ subject to the constraint $g(x, y) = 0$. The point $(a, b)$ must be tested to determine whether it yields a maximum, a minimum, or neither.

In an experiment, we may obtain a set of data points $(x_i, y_i)$, where $i \in \{1, 2, \ldots, n\}$ and the relationship between the $x_i$'s and $y_i$'s is almost linear. We can find a straight line $L$ which best fits the data. This line is often called a line of regression. We find the equation $y = mx + b$ of the line of regression by requiring that the sum of the squares of the vertical distances from each data point to the line be minimum. The result is

$$m = \frac{nP - XY}{nS - X^2} \text{ and } b = \frac{SY - PX}{nS - X^2}$$

where $P = \sum_{i=1}^{n} x_i y_i$, $S = \sum_{i=1}^{n} x_i^2$, $X = \sum_{i=1}^{n} x_i$, and $Y = \sum_{i=1}^{n} y_i$.

If we translate the $y$-axis so that the new origin is at the point $(h, 0)$, where $h$ is the average (arithmetic mean) of the $x_i$'s, then these expressions take the much simpler form

$$m = \frac{P}{S} \text{ and } b = \frac{Y}{n}$$

Of course, the equation that is obtained is in terms of the new coordinates.

We also generalize the concept of integration working with functions of two variables. A double integral is defined as a limit of Riemann sums. However, iterated integrals are used to calculate the values of double integrals. For example, if $R = \{(x, y) \mid a \le x \le b, h_1(x) \le y \le h_2(x)\}$ and $f$ is continuous on $R$, then

$$\iint_R f(x, y) \, dR = \int_a^b \int_{h_1(x)}^{h_2(x)} f(x, y) \, dy \, dx$$

To evaluate this last integral, we first hold $y$ constant and antidifferentiate $f(x, y)$ with respect to $x$. We then evalute the result between $h_1(x)$ and $h_2(x)$, obtaining a function $g(x)$. We then evaluate $\int_a^b g(x) \, dx$.

Double integrals may be used to find the volume of a solid, the average value of a continuous function over a rectangle, and in calculating probabilities involving two variables. The probability that $x$ is between $a$ and $b$, and $y$ is between $c$ and $d$ is given by

$$P[a \le x \le b \text{ and } c \le y \le d] = \int_c^d \int_a^b f(x, y) \, dx \, dy$$

where $f$ is a joint probability density function (pdf).

**SAMPLE EXAM QUESTIONS**

1. Find the domain of the function $f$ and find the value of the function at the indicated points.

   **a.** $f(x, y) = x^2 + 6xy - y^2$; $(1, 3)$, $(-2, -1)$

   **b.** $f(x, y) = \dfrac{x + y}{2x - y}$; $(1, 3)$, $(0, 1)$

   **c.** $f(x, y) = \dfrac{3x}{(x - 2)(y + 3)}$; $(-1, 4)$, $(3, -2)$

   **d.** $f(x, y) = \sqrt{100 - x^2 - y^2}$; $(2, 5)$, $(8, 5)$

   **e.** $f(x, y) = \dfrac{x + 2y}{1 - e^{2x-y}}$; $(2, -1)$, $(1, 1)$

   **f.** $f(x, y, z) = 3x^2 + 6xyz - z^3$; $(-1, 1, 3)$, $(0, 2, -1)$

   **g.** $f(x, y, z) = \dfrac{x + 2y - 3z}{(x - 3)(y + 2)(z - 5)}$; $(2, -1, 4)$, $(5, 2, 3)$

   **h.** $f(x, y, z) = \sqrt{(x + 3)(y - 5)(z^2 + 9)}$; $(-2, 6, 0)$, $(2, 7, -1)$

2. Plot the points with coordinates:

   **a.** $(2, 4, 3)$      **b.** $(-1, 3, 2)$      **c.** $(0, 3, 5)$      **d.** $(1, 2, 4)$

3. Find the coordinates of three points that are on the following graphs.

   **a.** $f(x, y) = x^3 + 2xy - y^3$

   **b.** $f(x, y) = \dfrac{x + y}{3x - y}$

   **c.** $f(x, y) = \dfrac{2x}{(x - 5)(y + 2)}$

   **d.** $f(x, y) = \sqrt{36 - x^2 - y^2}$

   **e.** $f(x, y) = \dfrac{x - 3y}{1 - e^{3x-2y}}$

4. Sketch the level curve of the function $f$ corresponding to the given value of $c$.

   **a.** $f(x, y) = 3x - 4y$, $c = 24$
   **b.** $f(x, y) = y - 3x^2$, $c = 5$
   **c.** $f(x, y) = y - x^2 - 6x$, $c = 4$
   **d.** $f(x, y) = 2x^2 - y$, $c = 2$
   **e.** $f(x, y) = x^2 + y^2 + 3$, $c = 19$

5. A store carries two brands of tennis balls. When the prices of brand A and brand B are $p_A$ and $p_B$ dollars per can, respectively, the demand per week is

   $$D_A(p_A, p_B) = 500 - 8p_A + 12p_B \quad \text{for brand A}$$

   and

   $$D_B(p_A, p_B) = 400 + 15p_A - 9p_B \quad \text{for brand B}$$

Find the revenue per week, in terms of $p_A$ and $p_B$, from the sale of these two brands of tennis balls.

**6.** If in Problem 5 the cost to the store per can is \$2.50 for brand A and \$2.80 for brand B, find the profit per week, in terms of $p_A$ and $p_B$, generated by the sale of these two brands of tennis balls.

**7.** Find the partial derivatives $f_x(x, y)$ and $f_y(x, y)$.

   **a.** $f(x, y) = 3x^4y^6$

   **b.** $f(x, y) = \dfrac{x^3y^2}{x^2 + y^4}$

   **c.** $f(x, y) = (x^2 + y^5)e^{xy}$

   **d.** $f(x, y) = (3x^2y + 5y)\ln(x^2 + y^4)$

   **e.** $f(x, y) = 5e^{2x^2y + 3xy^2}$

   **f.** $f(x, y) = \dfrac{\ln(x + y) + 5}{x + y}$

   **g.** $f(x, y) = \sqrt{5x^2 + 6y^4}$

**8.** Find the second partial derivatives $f_{xx}$, $f_{yy}$, $f_{xy}$, and $f_{yx}$.

   **a.** $f(x, y) = 3x^2y^7$

   **b.** $f(x, y) = 5e^{5xy}$

   **c.** $f(x, y) = \ln(x^2 + 5y^6)$

   **d.** $f(x, y) = (xy)^e + e^{xy}$

**9.** Find the following second order partial derivatives $\partial^2 z/\partial x^2$, $\partial^2 z/\partial y^2$, $\partial^2 z/\partial x \partial y$, and $\partial^2 z/\partial y \partial x$ if

   **a.** $z = 5x^7y^3$

   **b.** $z = e^{5x^2y^3}$

   **c.** $z = \ln(5x^2y + 6xy^2)$

**10.** Verify that $f_{xy}(x, y) = f_{yx}(x, y)$ if

   **a.** $f(x, y) = 6x^2y^3 + 5e^{2xy}$

   **b.** $f(x, y) = \ln(e^x + 3e^{2y})$

**11.** Let $f(x, y) = 3x^2y^2$.

   **a.** Calculate the exact value of $f(2.98, 3.01)$.

   **b.** Approximate the value of $f(2.98, 3.01)$ by first finding the value of $f(3, 3)$ and using the partial derivatives.

   **c.** Compare the two results.

**12.** The demand equations for commodities X and Y are

$$D_x = 26{,}500 - 6p_x{}^2 - 7p_y{}^3 \text{ and } D_y = 35{,}800 - 2p_x{}^3 - 5p_y{}^2$$

where $p_x$ and $p_y$ dollars are the prices per unit of commodities X and Y, respectively.

   **a.** Find the values of the four marginal demand functions and interpret the results when $p_x = 12$ and $p_y = 10$.

   **b.** Are these commodities substitutes or complements?

**13.** The demand equations for commodities X and Y are

$$D_x = 14{,}300 - 5p_x{}^2 + 3p_y{}^3 \text{ and } D_y = 25{,}800 + 6p_x{}^2 - 5p_y{}^2$$

where $p_x$ and $p_y$ dollars are the prices per unit of commodities X and Y, respectively.

a. Find the values of the four marginal demand functions and interpret the results when $p_x = 8$ and $p_y = 7$.

b. Are these commodities substitutes or complements?

14. The demand equations for commodities X and Y are

$$D_x = 13{,}500 - 3p_x{}^2 + 15p_y \text{ and } D_y = 16{,}700 + 7p_x - 4p_y{}^{3/2}$$

respectively.

a. Find the elasticity of demand for commodity X when $p_x = 10$ and $p_y = 9$, and interpret the result.

b. Find the cross elasticity of demand for commodity X with respect to $p_y$ when $p_x = 12$ and $p_y = 16$, and interpret the result.

c. Find the cross elasticity of demand for commodity Y with respect to $p_x$ when $p_x = 8$ and $p_y = 9$, and interpret the result.

15. Find all critical points of each of the functions defined below. Then determine whether the function has a relative maximum, a relative minimum, or neither at each of these points.

a. $f(x, y) = 2x^2 + 6xy + 5y^2 + 4x + 8y + 3$
b. $f(x, y) = x^3 + y^3 + 2y^2 - 6x - 7y + 13$
c. $f(x, y) = 3xy - 9x^2 - \ln y^8$
d. $f(x, y) = e^{(x^2 + xy + y^2 - 6x)}$

16. Let

$$f(x, y) = 5y^3 + 3x^2y - 4x^2 - 7xy^2 + 13x + 5$$

Verify that the points $(3, 1)$, $(5, 3)$, $\left(\frac{39}{11}, \frac{13}{11}\right)$, and $\left(\frac{65}{27}, \frac{13}{9}\right)$ are critical points of $f$. Then determine whether the function has a relative maximum, a relative minimum, or neither at each critical point.

17. Same as Exercise 16 for

$$G(x, y) = 4x^3 - 13x^2y + 4xy^2 - 8y^2 + 52y + 7$$

and the points $(2, 1)$, $(2, 12)$, $\left(\frac{26}{35}, \frac{156}{35}\right)$, and $\left(\frac{-26}{9}, \frac{-13}{9}\right)$.

18. A company found that if it manufactures and sells $q_A$ units of commodity A and $q_B$ units of commodity B per week, the profit per week derived from these is given by

$$P(q_A, q_B) = -60q_A{}^2 + 70q_Aq_B - 80q_B{}^2 + 9600q_A + 8700q_B - 42{,}000$$

How many units of each commodity should be produced and sold each week in order to maximize profit?

19. A certain company found that if it manufactures $q_D$ deluxe and $q_S$ standard models of its cameras each day, the cost $C$ dollars and revenue $R$ dollars are

$$C = 3.5q_D{}^2 - 5q_Dq_S + 2.8q_S{}^2 + 30q_D + 34q_S + 2000$$

and

$$R = 80q_D + 120q_S$$

How many of each model should the company manufacture each day to maximize the daily profit? What is the maximum daily profit?

**20.** A store found that if the price of beef is $p_B$ dollars per pound and the price of salmon is $p_S$ dollars per pound, then it will sell $D_B$ pounds of beef and $D_S$ pounds of salmon each day, where

$$D_B = 1700 - 850p_B + 400p_S$$

and

$$D_S = 2000 + 200p_B - 350p_S$$

If the store pays $3 per pound for beef and $4 per pound for salmon, what should the prices of beef and salmon be in order to maximize the daily profit? What is the maximum daily profit?

**21.** A topless rectangular container is to have a volume of 5120 cubic meters. The material for the bottom costs $12/sq m, while the material for the sides costs $4.80/sq m. Using the Lagrange multiplier, find the dimensions of the container that will minimize the cost of material.

**22.** Using the Lagrange multiplier, find the least distance between $(5, 7)$ and the line $3x + 4y = 18$.

**23.** Using the Lagrange multiplier, find the least distance between $(3, 6, 2)$ and the plane $2x + 4y + 5z + 5 = 0$

**24.** The Cobb-Douglas production function for a certain manufacturer is $f(L, K) = 650L^{.2}K^{.8}$, where $L$ represents the number of units of labor, $K$ the number of units of capital, and $f(L, K)$ the number of units produced. If the cost of labor is $45 per unit, the cost of capital is $95 per unit, and the amount available is $855,000, use the Lagrange multiplier method to determine the maximum production level.

**25.** The Cobb-Douglas production function for a certain manufacturer is $f(L, K) = 375L^{2/3}K^{1/3}$, where $L$ represents the number of units of labor, $K$ the number of units of capital, and $f(L, K)$ the number of units produced. If the cost of labor is $55 per unit, the cost of capital is $220 per unit, and the number of units produced must be 112,500, use the Lagrange multiplier method to determine the minimum total cost of labor and capital.

**26.** An editor estimates that if $x$ thousand dollars is spent on development and $y$ thousand dollars is spent on promotion, approximately $N$ new books will be sold where

$$N = 4x^{2.5}y^{1.5}$$

If the editor has $250,000 to spend on development and promotion, how much should be allocated to each in order to maximize the number of new books sold? Using the Lagrange multiplier method, estimate the increase in the maximum number of books sold if the editor has $1000 more to spend on development and promotion.

**27. a.** Find the equation of the line which best fits the following set of points in the sense of least squares: $(2, 5)$, $(4, 4)$, $(6, 2)$, $(8, 1)$.
   **b.** Plot the four points and sketch the graph of the equation obtained in part **a** on the same coordinate plane.

**28.** If $z = x^2y^3 + e^{xy}$, $x = t^2 + 1$, and $y = 3t + 2$, find $\frac{dz}{dt}$.

**29.** If $z = 3e^{u^3v^2}$, $u = x^2 + 4y^3$, $v = \sqrt{x^3} + y^2$, find $\frac{\partial z}{\partial x}$ and $\frac{\partial z}{\partial y}$ in terms of $x$ and $y$.

**30.** Find $\frac{\partial z}{\partial x}$ and $\frac{\partial z}{\partial y}$ if $x^2 - 6y^3 + 3z^2 = 24 - x^2y^3z^4$.

**31.** Find

   **a.** $\int (12x^3y^2 + 13xy^3) \, dy$
   **b.** $\int 3x^3 \sqrt[5]{y} \, dx$

c. $\int_1^2 (4x^3y^2 + 10xy^3)\, dy$

d. $\int_1^2 \int_{3y}^{y^2} 15x^3y^4\, dx\, dy$

32. Evaluate each integral and compare the answers.

a. $\int_1^3 \int_2^5 4x^3y^2\, dx\, dy$

b. $\int_2^5 \int_1^3 4x^3y^2\, dy\, dx$

33. Evaluate $\iint\limits_R f(x, y)\, dA$ if $f(x, y) = 2x^2 + 5xy$ and $R = \{(x, y) \mid -1 \le x \le 3,$

$1 \le y \le 2\}$.

34. Evaluate $\iint\limits_R f(x, y)\, dA$ if $f(x, y) = 2xy$ and $R = \{(x, y) \mid -e^y \le x \le e^y,$

$\ln 3 \le y \le \ln 7\}$.

35. Let $A, B, C, D,$ and $E$ have coordinates $(0, 6), (10, 6), (10, 0), (0, 0),$ and $(0, 3)$, respectively and let $R$ be the region inside the rectangle $ABCD$ and outside the triangle $BCE$. Also let $f(x, y) = 3x^2 + xy$. Evaluate $\iint\limits_R f(x, y)\, dA$.

36. Find the average value of the function $f$ over the rectangle $R$ if $R = \{(x, y) \mid -2 \le x \le 4, 1 \le y \le 5\}$ and $f(x, y) = 2x^2 - 4xy^2$.

37. Find the volume in the first octant bounded by the three coordinate planes and below the plane $x + 3y + 2z = 18$.

38. If the price of butter is $p_B$ dollars per pound and the price of margarine is $p_M$ dollars per pound, the number of pounds of butter sold each week at a certain store is given by

$$f(p_B, p_M) = 1500 - 250p_B^{1/2} + 25p_M^{1/3}$$

During a certain year, the prices of butter and margarine increased steadily from $1.95 to $2.15 per pound and from $0.85 to $0.95 per pound, respectively. What is the weekly average number of pounds of butter sold during the year?

39. Suppose that $x$ and $y$ represent the numbers of yards gained by team A running and passing, respectively, during a football game. If the joint probability function for $x$ and $y$ is $f(x, y) = .000008e^{-.004x}e^{-.002y}$, find the probability that team A is held to not more than 300 yards in total offense.

# Mathematical Induction

## The Principle of Mathematical Induction

In Chapter 5, Section 5.1, you learned that if $a_1, a_2, a_3, \ldots$ is an arithmetic progression, then the sum of the first $n$ terms is given by

$$a_1 + a_2 + \cdots + a_n = \frac{(a_1 + a_n)n}{2}$$

and, if the first term and common ratio of a geometric progression are $a$ and $r$, respectively, then

$$a + ar + ar^2 + \cdots + ar^{n-1} = \frac{a(1 - r^n)}{1 - r}$$

provided that $r \neq 1$. These two formulas are statements involving an integer $n$. It was proved that both are true for all positive integers. Often, however, there is a case where we wish to show that a statement $S(n)$ involving an integer is true for all positive integers. The principle of mathematical induction is the tool that we can use in such a case.

In the experimental sciences we might test a formula with a variable $n$ for $n = 1$, $2, 3, \ldots, k$ and, if it is true for these tests, we would conclude that the formula is probably true for all positive integers $n$, especially if the value of $k$ is large. This procedure is *not* valid in mathematics. Consider the statement "$n^2 - n + 41$ is a prime number." If we test this statement for $n = 1, 2, 3, \ldots, 40$, it is true in all 40 cases. We might be tempted to conclude it is true for all positive integers! However, let us try one more value and replace $n$ by 41. We get "$41^2 - 41 + 41$ is a prime number" and that statement is false since $41^2 - 41 + 41 = 41^2$ which is not prime because it is divisible by 41.* An example that might be even more convincing is the equality

$$1 + 2 + \cdots + n = \frac{n(n + 1)}{2} + (n - 1)(n - 2)(n - 3) \ldots (n - 10{,}000)$$

---

*Numbers in the form $2^n - 1$, where $n$ is a positive integer, are called *Mersenne numbers* after the French monk, Marin Mersenne (1588–1648). Mersenne primes are Mersenne numbers that are prime. It has been verified that $2^p - 1$ is prime for $p = 2, 3, 5, 7, 13, 17, 19, 31, 61, 89, 107$, and 127 and is composite for all other $p < 257$. (See Exercises 28 and 29.)

which becomes a true statement if we replace $n$ successively by the first 10,000 positive integers, but it is false thereafter.

The following example might help make the principle of mathematical induction intuitively plausible. Suppose a person wishes to climb a ladder with infinitely many rungs. If that person can step on the first rung of the ladder and after having stepped on any rung he or she can step on the next rung, then it will be possible for that person to climb the ladder. This idea is similar to the idea now formally stated as an axiom.

---

**PRINCIPLE OF MATHEMATICAL INDUCTION:**    Let $S(n)$ be a proposition involving an integer. If

1. *The proposition is true for $n = 1$ and*
2. *The truth of the proposition for $n = k$ implies the truth of the proposition for the next value $n = k + 1$,*

*then the proposition is true for all positive integers.*\*

---

## Proofs by Induction

**EXAMPLE 1**    In Chapter 5, Section 5.1, we derived a formula for the sum of the first $n$ terms of a geometric progression. Use the principle of mathematical induction to prove that formula.

**Solution**    The first term of the G.P. is $a$ and the common ratio is $r$, where $r \neq 1$. We must prove that

$$a + ar + ar^2 + \cdots + ar^{n-1} = \frac{a(1 - r^n)}{1 - r}$$

is true for all positive integral values of $n$.

1. First, check the formula for $n = 1$. When we have only one term on the left side the result is

$$a = \frac{a(1 - r^1)}{1 - r}$$

which is true. The first step of the induction proof is complete.

2. Suppose now that the formula is true for $n = k$. That is, suppose that

$$a + ar + ar^2 + \cdots + ar^{k-1} = \frac{a(1 - r^k)}{1 - r} \tag{1}$$

is true.

---

\*In some instances, a statement $S(n)$ may be true for all integers $n \geq n_1$ where $n_1$ is some integer. In such cases, induction starts at $n_1$. In other words, the first step is to verify that the statement is true for $n = n_1$, the smallest integral value of $n$ for which the statement is to be proven true. The second step consists of verifying that if we assume that the statement is true for $n = k$ where $k \geq n_1$, then the statement is true for $n = k + 1$. (See, for example, Exercises 21 and 22.)

We must therefore show that the formula is true for $n = k + 1$. That is, we must prove that

$$a + ar + ar^2 + \cdots + ar^{(k+1)-1} = \frac{a(1 - r^{k+1})}{1 - r} \tag{2}$$

is also true.

Adding $ar^k$ to both sides of the true equality (1), we obtain

$$a + ar + ar^2 + \cdots + ar^{k-1} + ar^k = \frac{a(1 - r^k)}{1 - r} + ar^k$$

$$= \frac{a(1 - r^k) + ar^k(1 - r)}{1 - r}$$

$$= \frac{a - ar^k + ar^k - ar^{k+1}}{1 - r}$$

$$= \frac{a(1 - r^{k+1})}{1 - r}$$

Therefore,

$$a + ar + ar^2 + \cdots + ar^{(k+1)-1} = \frac{a(1 - r^{k+1})}{1 - r}$$

is true. This completes the second step of the induction proof and we conclude that if $r \neq 1$, the formula

$$a + ar + ar^2 + \cdots + ar^{n-1} = \frac{a(1 - r^n)}{1 - r}$$

is true for all positive integral values of $n$.
**Do Exercise 1.**                                                                        ■

**EXAMPLE 2**     Prove that the sum of the squares of the first $n$ positive integers is equal to $\frac{n(n + 1)(2n + 1)}{6}$.

**Solution**     1. First, verify that the statement is true for $n = 1$. We have

$$1^2 = \frac{1(1 + 1)(2 + 1)}{6}$$

which yields the true statement $1 = 1$. The first step of the induction proof is complete.

*2. Next, suppose that the statement is true for $n = k$. That is, suppose that

$$\sum_{i=1}^{k} i^2 = \frac{k(k + 1)(2k + 1)}{6} \tag{1}$$

is true.

---

*See Section 1.7 for the meaning and use of the symbol $\Sigma$.

We must prove that the statement is also true for $n = k + 1$. That is, we must show that

$$\sum_{i=1}^{k+1} i^2 = \frac{(k + 1)[(k + 1) + 1][2(k + 1) + 1]}{6} \tag{2}$$

is also true.

Add $(k + 1)^2$ to both sides of the true equality (1) to get

$$\left(\sum_{i=1}^{k} i^2\right) + (k + 1)^2 = \frac{k(k + 1)(2k + 1)}{6} + (k + 1)^2$$

This can be written

$$\begin{aligned}
\sum_{i=1}^{k+1} i^2 &= \frac{k(k + 1)(2k + 1) + 6(k + 1)^2}{6} \\
&= \frac{(k + 1)[k(2k + 1) + 6(k + 1)]}{6} \\
&= \frac{(k + 1)(2k^2 + 7k + 6)}{6} \\
&= \frac{(k + 1)(k + 2)(2k + 3)}{6} \\
&= \frac{(k + 1)[(k + 1) + 1][2(k + 1) + 1]}{6}
\end{aligned}$$

This completes the second step of the induction proof. We conclude that the formula

$$\sum_{i=1}^{n} i^2 = \frac{n(n + 1)(2n + 1)}{6}$$

is true for all positive integral values of $n$.
**Do Exercise 3.**

EXAMPLE 3  Prove that for each positive integer $n$, the following is true:

$$1 + \frac{1}{2} + \frac{1}{4} + \frac{1}{8} + \cdots + \frac{1}{2^n} < 2$$

Solution  1. Let $n = 1$ to obtain the inequality

$$1 + \frac{1}{2} < 2$$

which is true since $1.5 < 2$. The first step of the induction proof is complete.
2. Assume that the statement is true for $n = k$. That is, assume that

$$1 + \frac{1}{2} + \frac{1}{4} + \frac{1}{8} + \cdots + \frac{1}{2^k} < 2 \tag{1}$$

is true.

We must therefore show that the statement is true for $n = k + 1$ also. That is, that

$$1 + \frac{1}{2} + \frac{1}{4} + \frac{1}{8} + \cdots + \frac{1}{2^{k+1}} < 2 \qquad (2)$$

is also true.

Multiplying both sides of the true inequality (1) by $\frac{1}{2}$, we get

$$\frac{1}{2} + \frac{1}{4} + \frac{1}{8} + \frac{1}{16} + \cdots + \frac{1}{2^{k+1}} < 1$$

Adding 1 to both sides of the inequality we obtain

$$1 + \frac{1}{2} + \frac{1}{4} + \frac{1}{8} + \frac{1}{16} + \cdots + \frac{1}{2^{k+1}} < 2$$

This completes the second step of the induction proof. Therefore the inequality

$$1 + \frac{1}{2} + \frac{1}{4} + \frac{1}{8} + \cdots + \frac{1}{2^{n}} < 2$$

is true for all positive integral values of $n$.
**Do Exercise 11.**

The following two examples illustrate that both parts of the principle of mathematical induction must be verified to confirm that both are true.*

**EXAMPLE 4**   Consider the formula

$$5 + 10 + 15 + \cdots + 5n = \frac{5n(n + 1)}{2} + (n - 1)$$

Show that the formula is true for $n = 1$, but not true for another positive integer.

**Solution**   Replace $n$ by 1 to have only one term on the left side of the formula. The equality

$$5 = \frac{5 \cdot 1 \cdot (1 + 1)}{2} + (1 - 1)$$

is true. However, if we replace $n$ by 2, we have two terms on the left side.

$$5 + 10 = \frac{5 \cdot 2 \cdot (2 + 1)}{2} + (2 - 1)$$

This can be simplified to the **false** statement $15 = 16$.

---

*The analogy of the infinite ladder should make it clear that the truth of *both* parts of the principle of mathematical induction must be verified if this principle is to be used. Knowing that a person can step on the first rung of a ladder does not guarantee that the whole ladder can be climbed unless we are sure that after a certain rung is stepped on, the next one can also be stepped on. However, knowing that a person can indeed step on any rung of a ladder, whenever he or she has stepped on the previous rung, also requires that the first rung be stepped on before we can actually be sure that the person can climb the whole ladder.

**EXAMPLE 5**    Show that the statement "$1 + 2 + 3 + \cdots + n = \frac{n(n + 1)}{2} + 1$ holds for all positive integers" is false although it satisfies the second condition of the principle of mathematical induction.

**Solution**    If we replace $n$ by some positive integer, say 2, we get

$$1 + 2 = \frac{2(2 + 1)}{2} + 1$$

which yields the false statement $3 = 4$.

We now show that the second condition of the principle of mathematical induction is satisfied. Assume that the formula is true for $n = k$. Then the equality

$$1 + 2 + 3 + \cdots + k = \frac{k(k + 1)}{2} + 1$$

is true. Adding $(k + 1)$ to both sides of this equality, we obtain

$$\begin{aligned}
1 + 2 + 3 + \cdots + k + (k + 1) &= \frac{k(k + 1)}{2} + 1 + (k + 1) \\
&= \frac{k(k + 1) + 2(k + 1)}{2} + 1 \\
&= \frac{(k + 1)(k + 2)}{2} + 1 \\
&= \frac{(k + 1)[(k + 1) + 1]}{2} + 1
\end{aligned}$$

Thus, we have shown that if we assume the truth of the formula for $n = k$, the truth of the formula for $n = k + 1$ will follow. Thus, the second condition of the principle of mathematical induction is satisfied, but the statement "$1 + 2 + 3 + \cdots + n = \frac{n(n + 1)}{2} + 1$ holds for all positive integers" is false.

**Do Exercise 27.**    ■

# Exercise Set A

*In Exercises 1–22, use mathematical induction to prove that the given equation, or statement, is true for all positive integral values of n.*

**1.** $1 + 2 + 3 + \cdots + n = \dfrac{n(n + 1)}{2}$

**2.** $1^3 + 2^3 + 3^3 + \cdots + n^3 = \dfrac{n^2(n + 1)^2}{4}$

**3.** $1^4 + 2^4 + 3^4 + \cdots + n^4 =$
$\dfrac{n(n + 1)(2n + 1)(3n^2 + 3n - 1)}{30}$

**4.** $1 \cdot 2 + 2 \cdot 3 + 3 \cdot 4 + \cdots + n(n + 1) =$
$\dfrac{n(n + 1)(n + 2)}{3}$

**5.** $2 + 2^2 + 2^3 + \cdots + 2^n = 2^{n+1} - 2$

**6.** $1 + 3 + 5 + \cdots + (2n - 1) = n^2$

**7.** $\dfrac{1}{1 \cdot 2} + \dfrac{1}{2 \cdot 3} + \dfrac{1}{3 \cdot 4} + \cdots + \dfrac{1}{n(n + 1)} = \dfrac{n}{n + 1}$

**8.** $\dfrac{1}{1 \cdot 3} + \dfrac{1}{2 \cdot 4} + \dfrac{1}{3 \cdot 5} + \cdots + \dfrac{1}{n(n + 2)} =$
$\dfrac{n(3n + 5)}{4(n + 1)(n + 2)}$

**9.** $\dfrac{5}{1 \cdot 2} \cdot \dfrac{1}{3} + \dfrac{7}{2 \cdot 3} \cdot \dfrac{1}{3^2} + \cdots + \dfrac{2n + 3}{n(n + 1)} \cdot \dfrac{1}{3^n} =$
$1 - \dfrac{1}{n + 1} \cdot \dfrac{1}{3^n}$

**10.** $1 + \dfrac{1}{3} + \dfrac{1}{9} + \dfrac{1}{27} + \cdots + \dfrac{1}{3^n} < \dfrac{3}{2}$

**11.** $1 + \dfrac{1}{4} + \dfrac{1}{16} + \dfrac{1}{64} + \cdots + \dfrac{1}{4^n} < \dfrac{4}{3}$

**12.** $2^{2n} - 1$ is a multiple of 3.

**13.** $2n^3 + 3n^2 + n$ is a multiple of 6.

**14.** $\displaystyle\sum_{i=1}^{n} (4i - 3) = n(2n - 1)$

**15.** $\displaystyle\sum_{i=1}^{n} (3i + 2) = \dfrac{n(3n + 7)}{2}$

**16.** $\displaystyle\sum_{i=1}^{n} i(i + 2) = \dfrac{n(n + 1)(2n + 7)}{6}$

**17.** $\displaystyle\sum_{i=1}^{n} (7i - 5) = \dfrac{n(7n - 3)}{2}$

**18.** $\displaystyle\sum_{i=1}^{n} (2i - 3)^2 = \dfrac{n(4n^2 - 12n + 11)}{3}$

**19.** $(a - b)$ is a factor of $a^n - b^n$. (*Hint:* $a^{k+1} - b^{k+1} = (a^{k+1} - ab^k) + (ab^k - b^{k+1})$.)

**20.** $(a + b)$ is a factor of $a^{2n-1} + b^{2n-1}$. (*Hint:* See Exercise 19.)

**21.** $2^n < n!$ for all integers $n > 3$. (*Hint:* Recall that $n! = n(n - 1)(n - 2) \cdot \ldots \cdot 2 \cdot 1$. Show that the statement is true for $n = 4$ and that the truth of the statement for $n = k$, where $k > 3$, implies the truth of the statement for $n = k + 1$.)

**22.** Prove that if $a \neq 0$ and $a > -1$, then

$$(1 + a)^n > 1 + na$$

is true for each integer $n > 1$.

**23.** Note that

$$1 = 1$$
$$1 - 4 = -(1 + 2)$$
$$1 - 4 + 9 = 1 + 2 + 3$$
$$1 - 4 + 9 - 16 = -(1 + 2 + 3 + 4)$$
$$1 - 4 + 9 - 16 + 25 = 1 + 2 + 3 + 4 + 5$$

Guess a general statement involving the variable $n$ that the foregoing five equalities suggest. Then prove that your statement is true for all positive integral values of $n$.

**24.** Repeat Exercise 23 using the following six equalities:

$$1 = 1$$
$$1 - 3 = -2$$
$$1 - 3 + 5 = 3$$
$$1 - 3 + 5 - 7 = -4$$
$$1 - 3 + 5 - 7 + 9 = 5$$
$$1 - 3 + 5 - 7 + 9 - 11 = -6$$

**25.** Repeat Exercise 23 using the following five equalities:

$$1 - 1/2 = 1/2$$
$$(1 - 1/2)(1 - 1/3) = 1/3$$
$$(1 - 1/2)(1 - 1/3)(1 - 1/4) = 1/4$$
$$(1 - 1/2)(1 - 1/3)(1 - 1/4)(1 - 1/5) = 1/5$$
$$(1 - 1/2)(1 - 1/3)(1 - 1/4)(1 - 1/5)(1 - 1/6) = 1/6$$

**26.** Consider the statement

$$2 + 4 + 6 + \cdots + 2n = n^2 + n + 1$$

Prove that the truth of this equality for $n = k$ implies its truth for $n = k + 1$. There are, however, positive integers that are not solutions of the sentence. Does this contradict the induction principle? Are there any positive integers that are solutions of the given equality?

**27.** Repeat Exercise 26 with the statement

$$1 + 2 + 3 + \cdots + n = \dfrac{(2n + 1)^2}{8}$$

**28.** Show that $2^{11} - 1$ is not prime.

**29.** Show that $2^n - 1$ is not prime whenever $n = n_1 \cdot n_2$, where $n_1$ and $n_2$ are integers greater than 1. (*Hint:* Use the fact that $a - b$ is a factor of $a^k - b^k$, for each positive integer $k$, and $c^{pq} = (c^p)^q$.)

# Table of Selected Integrals

# B

We have selected only formulas that are necessary to solve integration problems in this book. More complete tables may be found in most calculus texts and in books of standard mathematical tables. The constant of integration $C$ is omitted from the formulas.

*Forms Containing $(au^n + b)$*

1. $\displaystyle \int u^n \, du = \frac{u^{n+1}}{n+1}, \; n \neq -1$

2. $\displaystyle \int \frac{du}{u} = \ln |u|$

3. $\displaystyle \int \frac{du}{u(au^n + b)} = \frac{1}{bn} \ln \left| \frac{u^n}{au^n + b} \right|$

4. $\displaystyle \int \frac{du}{u^2(au + b)} = -\frac{1}{bu} + \frac{a}{b^2} \ln \left| \frac{au + b}{u} \right|$

5. $\displaystyle \int u(au + b)^n \, du = \begin{cases} \dfrac{1}{a^2(n+2)}(au+b)^{n+2} - \dfrac{b}{a^2(n+1)}(au+b)^{n+1}, \\ \qquad n \neq -1, -2 \\[2mm] \dfrac{1}{a^2}[au + b - b \ln |au + b|], \; n = -1 \\[2mm] \dfrac{1}{a^2}\left[ \ln |au + b| + \dfrac{b}{au + b} \right] n = -2 \end{cases}$

6. $\displaystyle \int u^2(au + b)^n \, du = \frac{1}{a^3}\left[ \frac{(au+b)^{n+3}}{n+3} - 2b\frac{(au+b)^{n+2}}{n+2} + b^2\frac{(au+b)^{n+1}}{n+1} \right],$
$\qquad n \neq -1, -2, -3$

*Forms Containing $\sqrt{au + b}$*

7. $\displaystyle \int u\sqrt{au + b} \, du = \frac{2(3au - 2b)}{15a^2}(au + b)^{3/2}$

8. $\displaystyle \int u^2\sqrt{au + b} \, du = \frac{2(8b^2 - 12abu + 15a^2u^2)}{105a^3}(au + b)^{3/2}$

937

9. $\displaystyle\int \frac{du}{u\sqrt{au + b}} = \frac{1}{\sqrt{b}} \ln \left| \frac{\sqrt{au + b} - \sqrt{b}}{\sqrt{au + b} + \sqrt{b}} \right|$

*Forms Containing $a^2 - u^2$*

10. $\displaystyle\int \frac{du}{a^2 - u^2} = \frac{1}{2a} \ln \left| \frac{a + u}{a - u} \right|$

11. $\displaystyle\int \frac{du}{u^2 - a^2} = \frac{1}{2a} \ln \left| \frac{u - a}{u + a} \right|$

*Forms Containing $\sqrt{a^2 - u^2}$*

12. $\displaystyle\int \frac{du}{u\sqrt{a^2 - u^2}} = -\frac{1}{a} \ln \left| \frac{a + \sqrt{a^2 - u^2}}{u} \right|$

13. $\displaystyle\int \frac{\sqrt{a^2 - u^2}}{u}\, du = \sqrt{a^2 - u^2} - a \ln \left| \frac{a + \sqrt{a^2 - u^2}}{u} \right|$

*Forms Containing $\sqrt{u^2 \pm a^2}$*

14. $\displaystyle\int \sqrt{u^2 \pm a^2}\, du = \frac{u}{2}\sqrt{u^2 \pm a^2} \pm \frac{a^2}{2} \ln \left| u + \sqrt{u^2 \pm a^2} \right|$

15. $\displaystyle\int u^2\sqrt{u^2 - a^2}\, du = \frac{u}{8}(2u^2 - a^2)\sqrt{u^2 - a^2} - \frac{a^4}{8} \ln \left| u + \sqrt{u^2 - a^2} \right|$

16. $\displaystyle\int u^3\sqrt{u^2 - a^2}\, du = \frac{1}{5}(u^2 - a^2)^{5/2} + \frac{a^2}{3}(u^2 - a^2)^{3/2}$

17. $\displaystyle\int u^3\sqrt{u^2 + a^2}\, du = \left( \frac{1}{5}u^2 - \frac{2}{15}a^2 \right)(u^2 + a^2)^{3/2}$

18. $\displaystyle\int \frac{du}{\sqrt{u^2 \pm a^2}} = \ln \left| u + \sqrt{u^2 \pm a^2} \right|$

19. $\displaystyle\int \frac{\sqrt{u^2 + a^2}}{u}\, du = \sqrt{u^2 + a^2} - a \ln \left| \frac{a + \sqrt{u^2 + a^2}}{u} \right|$

20. $\displaystyle\int \frac{\sqrt{u^2 \pm a^2}}{u^2}\, du = -\frac{\sqrt{u^2 \pm a^2}}{u} + \ln \left| u + \sqrt{u^2 \pm a^2} \right|$

21. $\displaystyle\int \frac{du}{(u^2 - a^2)^{3/2}} = -\frac{u}{a^2\sqrt{u^2 - a^2}}$

22. $\displaystyle\int (u^2 \pm a^2)^{3/2}\, du = \frac{1}{4}\left[ u(u^2 \pm a^2)^{3/2} \pm \frac{3a^2u}{2}(u^2 \pm a^2)^{1/2} \right.$
$$\left. + \frac{3a^4}{2} \ln \left| u + (u^2 \pm a^2)^{1/2} \right| \right]$$

23. $\displaystyle\int u^3(u^2 + a^2)^{3/2}\, du = \frac{1}{7}(u^2 + a^2)^{7/2} - \frac{a^2}{5}(u^2 + a^2)^{5/2}$

24. $\displaystyle\int u^3(u^2 - a^2)^{3/2}\, du = \frac{1}{7}(u^2 - a^2)^{7/2} + \frac{a^2}{5}(u^2 - a^2)^{5/2}$

25. $\displaystyle \int \frac{u^2}{(u^2 \pm a^2)^{3/2}}\, du = \frac{-u}{(u^2 \pm a^2)^{1/2}} + \ln \left| u + \sqrt{u^2 \pm a^2} \right|$

*Exponential Forms*

26. $\displaystyle \int ue^{au}\, du = \frac{e^{au}}{a^2}(au - 1)$

27. $\displaystyle \int u^n e^{au}\, du = \frac{1}{a}u^n e^{au} - \frac{n}{a} \int u^{n-1}\, e^{au}\, du$

28. $\displaystyle \int \frac{e^{au}}{u^n}\, du = -\frac{e^{au}}{(n-1)u^{n-1}} + \frac{a}{n-1} \int \frac{e^{au}}{u^{n-1}}\, du,\ n \neq 1$

29. $\displaystyle \int \frac{du}{1 + e^u} = u - \ln(1 + e^u)$

30. $\displaystyle \int \frac{du}{a + be^{mu}} = \frac{u}{a} - \frac{1}{am} \ln \left| a + be^{mu} \right|,\ m \neq 0$

*Logarithmic Forms*

31. $\displaystyle \int \ln |u|\, du = u \ln |u| - u$

32. $\displaystyle \int u^n \ln |u|\, du = \begin{cases} \dfrac{u^{n+1}}{n+1} \ln |u| - \dfrac{u^{n+1}}{(n+1)^2},\ n \neq -1 \\[2ex] \dfrac{1}{2} \ln^2 |u|,\ n = -1 \end{cases}$

33. $\displaystyle \int \ln^n |u|\, du = u \ln^n |u| - n \int \ln^{n-1} |u|\, du$

# Answers to Odd-Numbered Exercises

C

If you wish to see detailed solutions of the odd-numbered exercises, you may want to obtain a copy of the Student Solutions Manual. For further help, a Study Guide is available where you will find suggested strategies to study the chapters, additional solved examples, detailed solutions to the even-numbered sample exam questions from the chapter summaries, and a complete set of $3 \times 5$ study cards. Your instructor may be able to order these supplements from your college bookstore.

## Getting Started (Pages xxix–xxxii)

**1.** $(x - y)(x^6 + x^5y + x^4y^2 + x^3y^3 + x^2y^4 + xy^5 + y^6)$ **3.** $(2x - 5y)(4x^2 + 10xy + 25y^2)$

**5.** $(2a - 3c)(16a^4 + 24a^3c + 36a^2c^2 + 54ac^3 + 81c^4)$ **7.** $(x + y)(x^2 - xy + y^2)$ **9.** $\left(\dfrac{x}{2} + 3y\right)\left(\dfrac{x^2}{4} - \dfrac{3}{2}xy + 9y^2\right)$

**11.** 1 6 15 20 15 6 1 **13.** $x^6 + 6x^5y + 15x^4y^2 + 20x^3y^3 + 15x^2y^4 + 6xy^5 + y^6$

**15.** $x^7 + 7x^6y + 21x^5y^2 + 35x^4y^3 + 35x^3y^4 + 21x^2y^5 + 7xy^6 + y^7$ **17.** $8x^3 + 36x^2y + 54xy^2 + 27y^3$

**19.** $81a^4 - 216a^3b + 216a^2b^2 - 96ab^3 + 16b^4$

**21.** $256x^8 + 512x^7y + 448x^6y^2 + 224x^5y^3 + 70x^4y^4 + 14x^3y^5 + \dfrac{7}{4}x^2y^6 + \dfrac{xy^7}{8} + \dfrac{y^8}{256}$

**23. a.** 12 **b.** $792a^7b^5$ **c.** $220a^9b^3$ **25. a.** 14 **b.** $364a^3b^{11}$ **c.** $2002a^5b^9$

**27. a.** 17 **b.** $2380a^4b^{13}$ **c.** $12{,}376a^6b^{11}$ **29. a.** 20 **b.** $15{,}504a^{15}b^5$ **c.** $1140a^{17}b^3$

**31. a.** 15 **b.** $5005a^6b^9$ **c.** $6435a^8b^7$ **33.** 2600 **35.** 1275

**37. a.** 9, 3 **b.** 36, 6 **c.** 100, 10 **d.** 225, 15 **e.** 441, 21 **f.** 784, 28, $\dfrac{n^2(n + 1)^2}{4}$ **39.** 5083

**41.** 7269.29 **43.** 25.832571 **45.** 30.36845238 **47.** 947 **49.** 3833 **51.** $-4059$ **53.** 803.3827

**55.** 989,052.3819 **57.** 41,260,314,020 **59.** $-21{,}000.93812$ **61.** 62.25143833 **63.** 74.96647626

**65.** $-1{,}592{,}648.342$ **67.** $-5.23401763$ **69. a.** $\$11{,}102.59$ **b.** $\$11{,}351.89$ **c.** $\$11{,}483.61$ **d.** $\$11{,}574.23$

**71.** $\$871.11$ **73.** $\$37{,}460.61$ **75.** $\$300.00$

## Exercise Set 0.1 (Pages 10–11) The Real Numbers

**1.** $-10$ **3.** $-7$ **5.** $\dfrac{2}{9}$ **7.** $\dfrac{-22}{437}$ **9.** $\dfrac{-51}{845}$ **11.** $\dfrac{-15}{11}$ **13.** $\dfrac{42}{5}$ **15.** $1.8\overline{3}$ **17.** $1.625$ **19.** $1.\overline{54}$

**21.** $16.\overline{538461}$ **23.** $12.\overline{8823529411764705}$ **25.** $-9x^4$ **27.** $\dfrac{128a^3}{b^2}$ **29.** $\dfrac{-x^3}{8y}$ **31.** $\dfrac{-x^{11}y^3}{7}$

**33.** $\dfrac{1 \pm i}{2}$, imaginary **35.** $3, \dfrac{-4}{3}$, real **45.** no

## Exercise Set 0.2 (Pages 16–17) Elementary Introduction to Sets

**1.** $\{1, 2, 3, 4, 5, 6\}$ **3.** $\{1\}$ **5.** {Wisconsin, Washington, West Virginia, Wyoming} **7.** $\{c, d, e\}$

**9.** {Johnson, Nixon, Ford, Carter, Reagan, Bush} **11.** $\{7\}$ **13.** $\{6\}$ **15.** $\{2\}$ **17.** $\varnothing$ **19.** $\{1\}$

**21.** Lincoln, Jefferson, Truman    **23.** 3 and 4 are the only members of the set.    **25.** 5, 9, 233    **27.** 1, 2, 4
**29.** Harry Reasoner, Barbara Walters, Phil Donahue    **31.** false    **33.** true    **35.** false
**37.** 4, $\varnothing$, {1, 2}, {1}, {2}
**39.** 32, $\varnothing$, {1}, {2}, {3}, {4}, {5}, {1, 2}, {1, 3}, {1, 4}, {1, 5}, {2, 3}, {2, 4}, {2, 5}, {3, 4}, {3, 5}, {4, 5}, {3, 4, 5}, {2, 4, 5},
{2, 3, 5}, {2, 3, 4}, {1, 4, 5}, {1, 3, 5}, {1, 3, 4}, {1, 2, 5}, {1, 2, 4}, {1, 2, 3}, {1, 2, 3, 4}, {1, 2, 3, 5}, {1, 2, 4, 5}, {1, 3, 4, 5},
{2, 3, 4, 5}, {1, 2, 3, 4, 5}    **41.** 3    **43.** 15    **45.** $2^n - 1$
**47. a.** 1    **b.** 6    **c.** 15    **d.** 20    **e.** 15    **f.** 6    **g.** 1    **49.** {0, 7, 8, 9}    **51.** {0, 1, 2, 5, 6, 7, 9}
**53.** {1, 2, 3, 4, 5, 6}, {2, 5}    **55.** {2, 3, 4, 5, 8}, $\varnothing$    **57.** {0, 1, 3, 4, 6, 7, 8, 9}, {0, 1, 3, 4, 6, 7, 8, 9}
**59.** {0, 1, 2, 5, 6, 7, 8, 9}, {0, 1, 2, 5, 6, 7, 8, 9}

### Exercise Set 0.3 (Pages 21–22) Venn Diagrams

**1.**
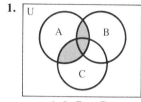
$A \cap (B \cup C)$

**3.**
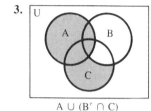
$A \cup (B' \cap C)$

**5.**
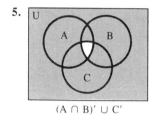
$(A \cap B)' \cup C'$

**7.**

$A \cap B$

$(A \cap B)'$

$A'$

**9.**
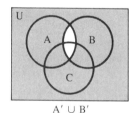
$A \subseteq (A \cup B)$

$B'$

$A' \cup B'$

**11.**

$B \cap C$

$A \cup (B \cap C)$

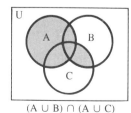
$A \cup B$

$A \cup C$

$(A \cup B) \cap (A \cup C)$

**13. a.** 6   **b.** 8   **c.** 281   **15. a.** 15   **b.** 55   **c.** 115   **d.** 100   **e.** 60   **17. a.** 13   **b.** 20   **c.** 134

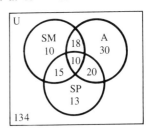

**19. a.** 50   **b.** 60   **c.** 50   **d.** 20

## Exercise Set 0.4 (Pages 29–30) Polynomials

**1.** $-3x^4 + 6x^3 + 5x^2 + 10x - 17$, $3x^4 - 2x^3 + 5x^2 - 10x - 3$   **3.** $7x^6 + 5x^3 + 10x - 21$, $7x^6 - 5x^3 + 5$

**5.** $4x^5 - 3x^4 + x^3 + 11x^2 + 3x$, $4x^5 + 3x^4 - x^3 + x^2 + x - 10$

**7.** $2x^6 + 2x^4 + 7x^2 - 3x + 1$, $-2x^6 - 7x^2 + 9x - 13$   **9.** $\frac{17}{12}x^3 + \frac{5}{42}x^2 + \frac{15}{56}x - \frac{17}{52}$, $\frac{-1}{12}x^3 + \frac{65}{42}x^2 - \frac{113}{56}x + \frac{103}{52}$

**11.** $12x^2 - x - 20$   **13.** $x^2 - 4$   **15.** $x^3 + 3x^2 - 22x + 12$   **17.** $6x^3 - x^2 - x + 21$

**19.** $-6x^7 - 7x^6 + 20x^5 + 50x^4 - 4x^3 - 35x^2 - 100x + 70$

**21.** $35x^9 + 35x^7 - 91x^6 + 25x^4 - 40x^3 + 25x^2 - 105x + 104$

**23.** $-12x^9 + 4x^8 + 20x^7 - 14x^6 + 20x^5 + 47x^4 + 17x^3 + 7x^2 + 5x - 25$

**25.** $2x^{10} + x^8 + 6x^7 - 5x^6 - 3x^5 + x^4 + 21x^3 - 60x^2 + 57x - 42$   **27.** $q(x) = 3x + 14$, $r(x) = 36$

**29.** $q(x) = x + 2$, $r(x) = 0$   **31.** $q(x) = 5x^3 + 22x^2 + 112x + 555$, $r(x) = 2786$

**33.** $q(x) = x^2 - 3x + 8$, $r(x) = -30x + 22$   **35.** $q(x) = 3x^4 + 12x^3 + 46x^2 + 160x + 543$, $r(x) = 1858x - 1093$

**37.** 3678   **39.** $-2128$

## Exercise Set 0.5 (Pages 32–33) Synthetic Division

**1.** $q(x) = 5x + 20$, $r(x) = 53$   **3.** $q(x) = 4x^2 + 13x + 20$, $r(x) = 47$   **5.** $q(x) = 3x^2 + x - 6$, $r(x) = 20$

**7.** $q(x) = 4x^4 + 24x^3 + 149x^2 + 889x + 5337$, $r(x) = 32{,}020$

**9.** $q(x) = 6x^6 + 29x^5 + 113x^4 + 458x^3 + 1824x^2 + 7301x + 29{,}198$, $r(x) = 116{,}793$

**11.** $q(x) = 6x^8 - 30x^7 + 142x^6 - 710x^5 + 3554x^4 - 17{,}770x^3 + 88{,}845x^2 - 444{,}225x + 2{,}221{,}131$,

$r(x) = -11{,}105{,}660$   **13.** $r(x) = -29{,}245{,}885$, $q(x) = 7x^{10} - 28x^9 + 112x^8 - 448x^7 + 1785x^6 - 7140x^5 + 28{,}560x^4 -$

$114{,}240x^3 + 456{,}966x^2 - 1{,}827{,}869x + 7{,}311{,}472$   **15.** 65   **17.** 4   **19.** 15   **21.** 1167   **23.** 1,065,018

## Exercise Set 0.6 (Page 38) Factoring

**1. a.** yes   **b.** yes   **c.** yes   **d.** yes   **e.** no, $2^2 \cdot 3^3 \cdot 5$

**3. a.** yes   **b.** yes   **c.** yes   **d.** yes   **e.** yes, $2^3 \cdot 3^4 \cdot 5^2 \cdot 11$

**5. a.** yes   **b.** yes   **c.** yes   **d.** yes   **e.** no, $2 \cdot 3^5 \cdot 5^4$   **7.** $(x - 7)^2$   **9.** $(x - 3)(x + 2)$

**11.** $(x - 7)(x + 5)$   **13.** $(x - 2)(x + 2)$   **15.** $(2x + 3)(x - 4)$   **17.** $(3x + 5)(x + 3)$   **19.** $(x + 1)(x + 5)(x - 2)$

**21.** $(x + 2)(x - 7)(x + 3)$   **23.** $(x + 1)(3x + 1)(x + 3)$   **25.** $(x + 1)(x + 2)(x - 3)(x - 5)$

**27.** $(x - \sqrt{3})(x + \sqrt{3})$   **29.** $(x - 2 - \sqrt{3})(x - 2 + \sqrt{3})$   **31.** $(x - \sqrt{3})(x + \sqrt{3})(x^2 + 3)$

**33.** $(x + 1)(x - 2 - \sqrt{5})(x - 2 + \sqrt{5})$   **35.** $(x - 5)(x - 1 - \sqrt{5})(x - 1 + \sqrt{5})$   **37.** not factorable

**39.** $(z + 3w)(x + 2y)$   **41.** $(2x + y)(z - 2w)$   **43.** $(2y + 3w)(2x - 3z)$   **45.** $(x - y)(x^2 + xy + y^2 + x + y)$

## Exercise Set 0.7 (Pages 48–50) Equations in One Variable

**1.** all real numbers except $-2$, 3, and $-5$   **3.** all real numbers except $-1$, 1, and $-6$

**5.** all real numbers greater than or equal to 4   **7.** $\{-3\}$   **9.** $\{2\}$   **11.** $\{-2\}$   **13.** $\{2\}$   **15.** $\{1, 5\}$   **17.** $\{2, -9\}$

**19.** $\{5, -15\}$   **21.** $\{6\}$   **23.** $\{-5, 2\}$   **25.** $\left\{\frac{-28}{5}, 7\right\}$   **27.** $\left\{\frac{-20}{3}, 3\right\}$   **29.** $\left\{2, \frac{-7}{3}\right\}$   **31.** $\{8\}$   **33.** $\{-2\}$

**35.** $\{2\}$   **37.** \$10,000 at 12% and ⌐30,000 at 8%   **39.** \$44   **41.** 41   **43.** 62 hours, \$27

**45.** 80 pounds of the first kind and 40 pounds of the second   **47.** $3\frac{1}{3}$ quarts   **49.** 300   **51.** 12 gallons

**53.** 70 cm by 70 cm   **55.** 20 ft (along existing wall) by 25 ft

**Exercise Set 0.8 (Pages 60–61) Inequalities**

**1.** $(-5, +\infty)$         **3.** $(-\infty, 7]$         **5.** $\left(\frac{130}{21}, +\infty\right)$         **7.** $(-1, 3)$         **9.** no solutions

**11.** $[-10, 2]$         **13.** $[-2, 5]$         **15.** $(-\infty, 1.5)$         **17.** $[-1.5, 1.5]$

**19.** $(-\infty, -\sqrt{19} - 4) \cup (\sqrt{19} - 4, +\infty)$         **21.** $(-5, -4) \cup (1, +\infty)$         **23.** $(-\infty, 2) \cup (3, 5)$

**25.** $(-4, 2) \cup (4, +\infty)$         **27.** $(-\infty, -5)$         **29.** $(-\infty, -\sqrt{3} - 1) \cup (-1, \sqrt{3} - 1)$

**31.** $\left(\frac{-3 - \sqrt{33}}{2}, -3\right) \cup \left(\frac{-3 + \sqrt{33}}{2}, +\infty\right)$         **33.** $\left(-\sqrt{5}, \frac{3 - \sqrt{3}}{3}\right] \cup \left[\frac{3 + \sqrt{3}}{3}, \sqrt{5}\right)$

**35.** $(-\infty, -1 - 3\sqrt{2}) \cup (-5, -1 + 3\sqrt{2})$         **37.** $(-2/3, 2) \cup (3, 4)$         **39.** $(-\infty, -2) \cup (-1, +\infty)$

**45.** 25         **47.** $10 < p < 30$         **49.** $25 < p < 45$         **51.** at least 40,000 units but no more than 60,000 units
**53.** between 2 and 4 quarts         **55.** $(-\infty, -5) \cup (-3, 5)$

**Exercise Set 0.9 (Pages 65–66) Absolute Value**

**1.** $(-1, 7)$         **3.** $(-\infty, -10) \cup (-4, +\infty)$         **5.** $\left(-\frac{2}{3}, 2\right)$         **7.** $(-\infty, -4] \cup [3, +\infty)$

**9.** $\varnothing$         **11.** $[0, 4]$         **13.** $(2, +\infty)$         **15.** $(-\infty, 0) \cup (2, +\infty)$

**17.** $(-\infty, \infty)$         **19.** $\{-5, 11\}$         **21.** $\{-3, 5\}$         **23.** $\left\{-\frac{7}{4}, 3\right\}$         **25. a.** 13         **b.** $-13$         **c.** 13
**27. a.** $\frac{5}{6}$         **b.** $-\frac{5}{6}$         **c.** $\frac{5}{6}$

**Chapter Review 0.10 (Pages 68–70) Sample Exam Questions**

**1. a.** $-32$         **b.** $-9$         **c.** $\frac{34}{5}$         **d.** $\frac{47}{21}$         **e.** $\frac{-297}{43}$         **f.** $\frac{11}{8}$         **g.** $-\frac{1}{20}$         **h.** $\frac{-4}{3}$
**3. a.** $\{3\}$         **b.** $\{9\}$         **c.** $\left\{\frac{1}{3}\right\}$         **5. a.** $\{5\}$         **b.** $\{2, -3\}$         **7. b.** $-9971$
**9.** The remainder is 7.

**11. a.** $(x - 1)(x - \sqrt{3})(x + \sqrt{3})$   **b.** $(x^2 + 5)(x - 3)$   **c.** $(x - 2)\left(x + \frac{1}{2} - \frac{\sqrt{5}}{2}\right)\left(x + \frac{1}{2} + \frac{\sqrt{5}}{2}\right)$

**13. a.** $(4, \infty)$      **b.** $[5, \infty)$         **c.** $(-3, 7)$               **d.** $[-5, -1]$

**e.** $(-\infty, -3] \cup [7, \infty)$       **f.** $\varnothing$    **g.** $\left(\frac{7}{8}, \frac{17}{8}\right)$               **h.** $(-14, 1)$

**i.** $(-3, -2] \cup [2, +\infty)$        **j.** $(-\infty, -3) \cup (-2, 1) \cup (5, +\infty)$     **k.** $(-\infty, 2) \cup (3, +\infty)$       **l.** $(2, +\infty)$

**15. a.** false   **b.** true   **c.** false   **d.** false   **e.** true   **f.** true   **g.** true   **h.** false   **i.** false   **17.** 31

**19.** not be paid

## Exercise Set 1.1 (Pages 80–81) Function

**1. a.** $\{2, 4, 6, 8\}$   **b.** $y, z$   **3. a.** $\{0, 1, 2, 3, 4\}$   **b.** 2, 6, 8

**5. a.** $\{0, 1, 2, 3, 4\}$   **b.** $f(0) = 0, f(2) = 8, f(4) = 64$   **7. a.** $S$   **b.** 5, 5, 37   **9. a.** $S$   **b.** $-3, 4, 3, 9$

**11.** No   **13.** $\{x \mid x \neq 4\}$   **15.** $\{x \mid x \neq 4 \text{ or } -4\}$   **17.** $[2, +\infty)$   **19.** $2, -.5, -\frac{2}{7}$   **21.** $\frac{1}{3}, \frac{1}{15}, \frac{1}{3}$

**23.** $0, \sqrt{19}, 3\sqrt{13}$

**25.** $S(x) = x^2 + 5x - 1$. The domain is $(-\infty, +\infty)$. $D(x) = -x^2 - x + 7$. The domain is $(-\infty, +\infty)$.
$P(x) = 2x^3 + 9x^2 + x - 12$. The domain is $(-\infty, +\infty)$. $Q(x) = \frac{2x + 3}{(x + 4)(x - 1)}$. The domain is $(-\infty, -4) \cup (-4, 1) \cup (1, +\infty)$.

**27.** $S(x) = \dfrac{5x - 5}{(x - 3)(x + 2)}$. The domain of $S(x)$, $D(x)$, $P(x)$, and $Q(x)$ is $(-\infty, -2) \cup (-2, 3) \cup (3, +\infty)$.
$D(x) = \dfrac{-x + 13}{(x - 3)(x + 2)}$. $P(x) = \dfrac{6}{x^2 - x - 6}$, $Q(x) = \dfrac{2x + 4}{3x - 9}$.

**29.** $S(x) = \dfrac{x^3 - x^2 + 6x + 33}{(x + 5)(x^2 - 4)}$. The domain of $S(x)$, $D(x)$, $P(x)$, and $Q(x)$ is $(-\infty, -5) \cup (-5, -2) \cup (-2, 2) \cup (2, +\infty)$.
$D(x) = \dfrac{x^3 - 3x^2 - 14x - 17}{(x + 5)(x^2 - 4)}$. $P(x) = \dfrac{1}{x + 2}$. $Q(x) = \dfrac{(x - 2)^2(x + 2)}{(x + 5)^2}$.

**31.** $S(x) = \dfrac{x^3 - 4x^2 - 21x + 104}{(x + 7)(x - 5)}$. The domains of $S(x)$, $D(x)$, and $P(x)$ exclude $-7$ and 5.
The domain of $Q(x)$ excludes $-7$, 5, and 3. $D(x) = \dfrac{x^3 - 6x^2 - 29x + 146}{(x + 7)(x - 5)}$. $P(x) = \dfrac{x^2 + 2x - 15}{x + 7}$. $Q(x) = \dfrac{(x - 5)^2(x + 5)}{(x + 7)(x - 3)}$.

**33.**

| $x$ | 1 | 6 | 8 |
|---|---|---|---|
| $S(x)$ | 5 | $-1$ | 9 |
| $D(x)$ | $-1$ | $-1$ | $-1$ |
| $P(x)$ | 6 | 0 | 20 |
| $Q(x)$ | $\frac{2}{3}$ | | $\frac{4}{5}$ |

The domains of $S(x)$, $D(x)$ and $P(x)$ are $\{1, 6, 8\}$.
The domain of $Q(x)$ is $\{1, 8\}$.

**35.**

| $x$ | 1 | 3 | 5 |
|---|---|---|---|
| $S(x)$ | 2 | 6 | 2 |
| $D(x)$ | 6 | $-4$ | 2 |
| $P(x)$ | $-8$ | 5 | 0 |
| $Q(x)$ | $-2$ | .2 | |

The domains of $S(x)$, $D(x)$, and $P(x)$ are $\{1, 3, 5\}$.
The domain of $Q(x)$ is $\{1, 3\}$.

**37.**

**39.**

**41.**

**43.** $2x^2 - 1$, $4x^2 - 12x + 10$, $4x - 9$, $x^4 + 2x^2 + 2$
**45.** $x^4 - 6x^3 + 11x^2 - 6x - 3$, $-x^4 - 4x^3 + 5x^2 + 18x - 20$, $x^4 + 4x^3 - 8x$, $-x^4 + 6x^3 - 16x^2 + 21x - 12$
**47.** $\big||x + 2| - 3\big|$, $\big||x - 3| + 2\big|$, $\big||x - 3| - 3\big|$, $\big||x + 2| + 2\big|$     **49.** $h(x) = 5x + 2$ and $g(x) = \sqrt{x}, \ x \geq -.4$
**51.** $h(x) = 3x + 2$ and $g(x) = x^9$     **53.** Yes     **55.** Yes

## Exercise Set 1.2 (Pages 90–91) Inverse of a Function

**7. b.**

| $x$ | $-7$ | 1 | 2 | 3 | 5 |
|---|---|---|---|---|---|
| $f^{-1}(x)$ | $e$ | $d$ | $a$ | $c$ | $b$ |

**9. b.**

| $x$ | $s$ | $t$ | $u$ | $v$ | $w$ | $y$ | $z$ |
|---|---|---|---|---|---|---|---|
| $f^{-1}(x)$ | 1 | 2 | 3 | 4 | 5 | 6 | 7 |

**11. b.**

| $x$ | 0 | 1 | 4 | 9 | 16 | 25 |
|---|---|---|---|---|---|---|
| $f^{-1}(x)$ | 0 | $-1$ | 2 | $-3$ | 4 | 5 |

**13. b.** $f^{-1}(x) = 3 + x^2$ for $x$ in the interval $[0, +\infty)$

**15. b.** $f^{-1}(x) = \frac{7 - x}{2}$     **17.** $f^{-1}(x) = \frac{5 - 3x}{x - 1}$     **19.** $h^{-1}(x) = \frac{x - 2}{3 + x}$     **21.** $g^{-1}(x) = \frac{x}{x - 1}$

**23. a.** $B = \{-1, 3, 7, 8, 15\}$     **b.**

| $x$ | $a$ | $b$ | $c$ | $d$ | $e$ |
|---|---|---|---|---|---|
| $f^{-1}(x)$ | 3 | 7 | $-1$ | 8 | 15 |

**25. a.** $[-2, +\infty)$     **b.** $y = 2 + \sqrt{x - 1}, \ x \geq 1$     **27. a.** $[5, +\infty)$     **b.** $h^{-1}(x) = \sqrt{x - 2} + 5, \ x \geq 2$
**29. a.** $[2, \infty)$     **b.** $g^{-1}(x) = 2 + \sqrt{1 - x}$     **31. a.** $[3, +\infty)$     **b.** $y = 3 + \sqrt[4]{x - 2}$
**33.**                          **35.**                                         **37.**                          **39.**

**41.**

**43.**

**45.**

**47.**

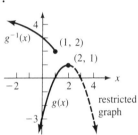

## Exercise Set 1.3 (Pages 100–102) Some Special Functions

**1.** $y = 2x - 1$   **3.** $y = 4x + 18$   **5.** $y = 2$   **7.** $y = -x + 11$   **9.** $y = 9$   **11.** $x + 2y + 7 = 0$

**13.**

**15.**

**17.**

**19. a.** 2
**b.**

**21. a.** 2
**b.**

**23. a.** 2
**b.**

**25. a.** 3
**b.**

**27. a.** The domain excludes 1 and $-3$.   **b.** $\frac{2}{5}, \frac{8}{21}, \frac{22}{45}$   **29. a.** The domain excludes 1 and $-1$.   **b.** $\frac{-17}{9}, 3, -1$

**31. a.**   $(f \circ g)(x)$   **b.**   $(g \circ f)(x)$   **c.**   $(f \circ h)(x)$   **d.** $(h \circ f)(x)$

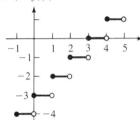

**33.** 1, 1, 1, $-1$, $-1$, $-1$, 1, 1, 1, $-1$, $-1$, $-1$, 1

**35.**

**37.**

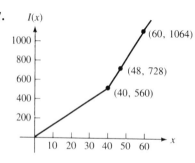

**39. a.** $p = -\frac{1}{3}x + 75$     **b.** $r = -\frac{1}{3}x^2 + 75x$
**c.** 81, $1,987

**41.**

26.25 feet

20 feet  20 feet

26.25 feet

**43.** $10 per credit hour     **45. a.** $p = -\frac{1}{2}x + 20$     **b.** $r = -\frac{1}{2}x^2 + 20x$     **c.** 16

## Exercise Set 1.4 (Pages 113–114) The Exponential Function

**1.** 540     **3.** $\frac{81}{49}$     **5.** 972     **7.** 2304     **9.** $3^{-88/15}$     **11.**

**13.**

**15.**

**17. a.**

**b.**

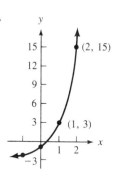

**19.** $10,199.44, $5,199.44     **21.** $9,861.69, $1,861.69     **23.** Bank C     **25.** 8.24%     **27.** 7.25%     **29.** $15,429.70
**31.** $96,006     **33.** $464,372     **35.** 4,661,528     **37.** .625 gm     **39.** .6931 gm
**41. a.** 49.63     **b.** 76.87     **c.** 100.02     **43. a.** $21,675     **b.** $15,660     **c.** $8,175
**45. a.** $6,782     **b.** $4,560     **c.** $3,066

## Exercise Set 1.5 (Pages 120–121) The Logarithmic Function

**1. a.** 1.8060     **b.** 1.5562     **c.** 1.7323     **3. a.** .1204     **b.** .357825     **c.** −.0178428
**5. a.** .6309     **b.** 2.3776     **c.** 1.8927     **7. a.** .4055     **b.** −.2876     **c.** .5234
**9. a.** $3 + \ln 5$     **b.** $\frac{2}{3}$     **c.** 2.2958     **11.** {39}     **13.** $\left\{-\frac{20}{3}\right\}$     **15.** ∅     **17.** {6}     **19.** {−3}     **21.** {−6, 1}
**23.** {4}     **25.** ∅     **27.** −8.4603     **29.** 56 quarters     **31.** 221 months     **33.** 13.73 years     **35.** March of 1991
**37.** $6\frac{1}{2}$ weeks     **39.** In the year 1993     **41.** 701.4 years     **47.** 13.1 years

## Exercise Set 1.6 (Pages 126–127) More on Graphs

**1.**

**3. a.** Option 1   **b.** Option 2

**5.**

**7.**

**9.**

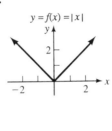

Shifting 2 units to the right

Shifting 5 units down

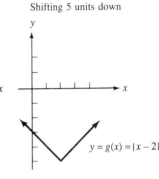

$y = g(x) = |x - 2| - 5$

**11.**

Shifting 5 units to the right

Shifting 3 units up

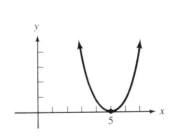

$y = (x - 5)^2 + 3$

**13.**

Shifting 2 units left

Shifting 3 units down

$y = g(x) = (x + 2)^3 - 3$

**15.**

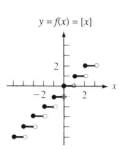

$y = f(x) = [x]$

Shifting 5 units to the right

$y$

Shifting 2 units up

$y = g(x) = [x - 5] + 2$

**17.** $g(x) = |x + 1| - 3$    **19.** $g(x) = (x + 2)^2 + 5$

**21.**

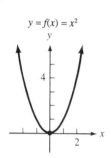

$y = f(x) = x^2$

$y$

Shifting the graph of $f$ two units to the right and then up one unit we obtain

$y = g(x)$

**23.**

$y = f(x) = x^3$

$y$

Shifting the graph of $f$ one unit to the left and up 5 units we obtain

$y = g(x) = (x + 1)^3 + 5$

$y$

$(-1, 5)$    $(0, 6)$

**25.**

$y = f(x) = x^4$

$y$

Shifting the graph two units to the left and up three units we obtain

$y = g(x) = (x + 2)^4 + 3$

$(0, 19)$    20

10

$(-2, 3)$

**27.** Unemployment was increasing in February through May, November and December, and decreasing in other months.

**29.**

$y$

$y = x^2$

$x$

$y = -x^2$

**31.**

$$y = [x]$$

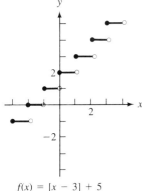

$$f(x) = [x - 3] + 5$$

$$-f(x) = -[x - 3] - 5$$

## Exercise Set 1.7 (Page 130) The Sigma Notation

**1.** 67   **3.** 55   **5.** 862   **7.** 0   **9.** 85   **11.** 244   **13.** 15   **15.** 1,136,275   **17.** 12,060

**19.** 207,917,325   **21.** 102,541,820   **23.** $-1674$   **25.** $\displaystyle\sum_{i=1}^{9} (2i - 1)$   **27.** $\displaystyle\sum_{i=1}^{7} (2 + 3i)$   **29.** $\displaystyle\sum_{i=1}^{8} i^2$   **31.** $\displaystyle\sum_{i=2}^{6} i^3$

**33.** $\displaystyle\sum_{i=2}^{11} (-1)^i$   **35.** $\displaystyle\sum_{i=1}^{9} i^2(-1)^{i+1}$

## Chapter Review 1.8 (Pages 133–135) Sample Exam Questions

**1. a.** $\{a, b, c, d\}$   **b.** $\{3, 9, 1\}$   **c.** Not one-to-one

**3. a.** The set of all real numbers   **b.** Set of all real numbers   **c.** One-to-one, $f^{-1}(x) = \sqrt[3]{x + 2}$

**5. a.** The set of all humans   **b.** The set of all human females who have had a child   **c.** Not one-to-one   **7.** $[2, +\infty)$

**9.** $(-\infty, -9) \cup (2, +\infty)$   **11.** $(-\infty, 3) \cup (3, +\infty)$

**13.** $S(x) = \dfrac{x^3 - 2x + 5}{x - 1}$ The domains of $S(x)$, $D(x)$ and $P(x)$ are $(-\infty, 1) \cup (1, +\infty)$. $D(x) = \dfrac{x^3 - 4x - 1}{x - 1}$, $P(x) = x^2 + 5x + 6$,

$Q(x) = \dfrac{(x - 1)^2(x + 2)}{(x + 3)}$ The domain of $Q(x)$ is $(-\infty, -3) \cup (-3, 1) \cup (1, +\infty)$.   **15.** $4x^2 + 16x + 12, \ 2x^2 + 4x - 3$

**17.**

$f(x)$

**19.**

$h(x)$

**21.**

$g(x)$

**23.** $h(x) = 2x + 3$ and $g(x) = x^4$

**25. b.** $g^{-1}(x) = \frac{1}{3}x - \frac{2}{3}$   **c.**

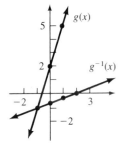

**27. a.** $y = \frac{1}{2}x + \frac{5}{2}$  **b.** $y - 5 = -\frac{2}{3}(x - 2)$  **c.** $y = 2x + 5$  **d.** $x = 1$  **e.** $y = 4$

**29.**

**31.** $837.68, $137.68

**33. a.** .7781  **b.** .9030  **c.** .2385  **d.** $-.8293$  **e.** 2.398  **f.** 3.3222

**35.** 23.2 years  **37. a.** 2870  **b.** 150  **c.** $\frac{100}{101}$  **39.** 20 feet

**41. a.** $p = -\frac{1}{2}q + 110$  **b.** $R(q) = -\frac{1}{2}q^2 + 110q$  **c.** 50 racquets, $85, $1,000

**43.**

Shifting the graph 3 units to the right and up 5
units we obtain the graph of $y = |x - 3| + 5$

**45.** $f(x) = |x - 2| - 8$

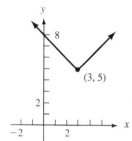

(3, 5)

## Exercise Set 2.1 (Pages 147–148) Systems of Two Linear Equations in Two Variables

**1.** $\{(1, -2)\}$  **3.** $\{(3, -1)\}$  **5.** $\varnothing$  **7.** $\{(4, -3)\}$  **9.** $\{(-2, -3)\}$  **11.** $\left\{\left(c, \frac{5-c}{3}\right) \mid c \text{ is a real number}\right\}$

**13.** $\{(7, -1)\}$  **15.** $\left\{\left(\frac{-66}{41}, \frac{277}{41}\right)\right\}$  **17.** $\left\{\left(\frac{1}{3}, \frac{1}{2}\right)\right\}$  **19.** $\{(-1, 2)\}$  **21.** $\{(-3, 5)\}$  **23.** $\{(5, -2)\}$

**25.** $\left\{\left(\frac{3c-2}{4}, c\right) \mid c \text{ is an arbitrary real number}\right\}$  **27.** $\left\{\left(\frac{1}{3}, \frac{1}{2}\right)\right\}$  **29.** $\{(2, 5)\}$  **31.** $\varnothing$  **33.** $\{(-1, 5)\}$  **35.** $\{(1, -2)\}$

**37.** $\{(3, -1)\}$  **39.** $\varnothing$  **41.** $\{(4, -3)\}$  **43.** $\{(-2, -3)\}$  **45.** $\left\{\left(c, \frac{5-c}{3}\right) \mid c \text{ is a real number}\right\}$  **47.** $\{(7, -1)\}$

**49.** $\left\{\left(\frac{-66}{41}, \frac{277}{41}\right)\right\}$  **51.** $\left\{\left(\frac{1}{3}, \frac{1}{2}\right)\right\}$  **53.** $(2, 5), (2, -5), (-2, 5), (-2, -5)$  **55.** $\left\{\left(-\frac{1}{2}, -\frac{1}{3}\right)\right\}$  **57.** $\left\{\left(\frac{1}{3}, -1\right)\right\}$

**59.** $110 with the Dallas fan and $130 with the L.A. fan  **61.** Mix 30 pounds of Brand A with 70 pounds of Brand B.

**63.** $32,000 must be invested at 6.5% and $18,000 at 9.5%.

## Exercise Set 2.2 (Pages 157–159) Systems of Three Linear Equations in Three Variables

**1.** $\{(-1, 2, 1)\}$  **3.** $\{(-2, 2, 3)\}$  **5.** $\varnothing$  **7.** $\{(2, 1, 5)\}$  **9.** $\{(-3, 1, -1)\}$  **11.** $\{(2, 2, 2)\}$

**13.** $\{(-1, -3, 2)\}$  **15.** $\{(-1, 2, 1)\}$  **17.** $\{(-2, 2, 3)\}$  **19.** $\{(-3, 2, 0)\}$

**21.** $\left\{\left(\frac{5}{7} + \frac{9}{7}c, \frac{13}{7}c + \frac{8}{7}, c\right) \mid c \text{ is a real number}\right\}$  **23.** $\{(2, 2, 2)\}$  **25.** $\{(-1, -3, 2)\}$  **27.** $\{(-1, 2, 1)\}$

**29.** 300 pounds of kind A, 80 pounds of B, and 20 pounds of C

**31.** $25,000 in bond A, $15,000 in bond B and $10,000 in bond C  **33.** X costs $9.25 per unit, Y $8.50 and Z $13.75.

**35.** 5 houses of type A, 3 of type B, and 6 of type C  **37.** The wife's age is 40, the professor's 44, and the son's 16.

**39.** 0 lbs of A, $\frac{2600}{17}$ lbs of B, and $\frac{800}{17}$ lbs of C

## Exercise Set 2.3 (Page 164) Graphical Solutions of Systems of Equations

**1.** $\{(1, 4), (-5, -8)\}$  **3.** $\{(2, 5), (-2, -3)\}$  **5.** $\{(-2, -1), (1, 5)\}$  **7.** $\{(1, 9), (-2, -6)\}$

**9.** $\{(-1, 5), (2, 2.11)\}$  **11.** $\{(1, 2), (3, 8)\}$  **13.** $\{(.5, .5)\}$  **15.** $\{(-.98, 1.40)\}$

**Exercise Set 2.4 (Pages 167–169) Systems of Linear Inequalities in Two Variables**

**1.**

**3.**

**5.**

**7.**

**9.**

**11.**

**13.**

**15.**

**17.**

**19.**

**21.**

**23.**

**25.**

**27.**

**29.**

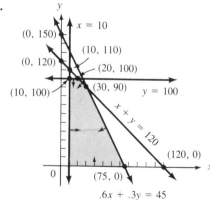

**Exercise Set 2.5 (Pages 180–184) Applications to Business and Economics**

**1. a.** $25,000   **b.** 12,500   **3. a.** 300   **b.** $300 < q < 700$   **5. a.** 50   **b.** $50 < q < 80$
**7.** $350, 33,000 units

**9. a.** $30, 1200 thousand units
  **b.** $36, 1452 thousand units
  **c.**

**11. a.** $45, 2775 thousand units
  **b.** $48, 3138 thousand units
  **c.**

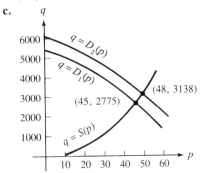

**13. a.** $21, 54 thousand units    **b.** $24, 39 thousand units
**15. a.** $D(p) = -32p + 1862$    **b.** $S(p) = 24p - 98$    **c.** $35, 742 units    **d.** $38, 646 units
**17. a.** $D(p) = -10p + 560$    **b.** $S(p) = 15p - 40$    **c.** $24, 320 units    **d.** $21, 350 units
**19. a.** $S(p) = 12.5p - 16$    **b.** 8 francs/liter, 84 million liters    **c.** 4.4 francs/liter
**21. a.** $S(p) = 30p - 69$    **b.** $14/unit, 351 thousand units    **c.** $5.70/unit
**23. a.** $3.20, $2.50 are the equilibrium prices    **b.** The equilibrium quantities are 7.1 thousand units and 16.1 thousand units.
**25.** The equilibrium prices are $4.50, $3.20, and $5.10 per unit. The equilibrium quantities are 551.9, 209.4, and 452.1 thousand units.
**27.** The equilibrium prices are $9.80, $10.20, and $8.75 per unit. The equilibrium quantities are 1386.1, 993.85, and 1465.4 thousand units.

## Chapter Review 2.6 (Pages 187–189) Sample Exam Questions

**1.** $\{(2, -3)\}$    **3.** $\{(2, -5)\}$    **5.** $\{(2, -1)\}$    **7.** $\{(-3, 2)\}$    **9.** $\{(2, 3), (2, -3), (-2, 3), (-2, -3)\}$
**11.** $\left\{\left(-\frac{1}{3}, \frac{1}{2}\right)\right\}$    **13.** $\{(2, 21), (-1, 3)\}$    **15.** $\{(1, 12), (-2, 0)\}$    **17.** $\{(9, 10, -12)\}$    **19.** $\{(14, -15, -13)\}$
**21.** $\varnothing$    **23.** $\{(2, 3, 1)\}$    **25.** $\left\{\left(\frac{1}{2}, \frac{-3}{2}, 0\right)\right\}$    **27.** $\left\{\left(c, \frac{-17}{3} + \frac{7}{3}c, \frac{-5c}{3} + \frac{16}{3}\right) \mid c \text{ is a real number}\right\}$    **29.** $\{(3, 2, -1)\}$

**31.**

**33.**

**35.** The value of the first store is $750,000 and that of the second store is $550,000.
**37.** The adults' wage is $72 and the teens' is $32.
**39.** Kris can do the job in 16 days and Pete can do it in 20 days.

**41.**

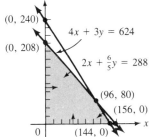

**43. a.** $4.95 per pound and $2.50 per pound    **b.** 202,500 pounds and 499,000 pounds

Exercise Set 3.1 (Pages 196–199) Matrix Notation and Matrix Arithmetic

**1.** $2 \times 3, 3 \times 2, 3 \times 4$

**3.** $\begin{bmatrix} 5 & 7 & 9 \\ 8 & 10 & 12 \end{bmatrix}$

**5.** $\begin{bmatrix} 3 & 5 & 7 \\ 6 & 8 & 10 \\ 11 & 13 & 15 \end{bmatrix}$

**7.** $\begin{bmatrix} 0 & 4 & 5 & 6 \\ 5 & 0 & 7 & 8 \\ 7 & 8 & 0 & 10 \\ 9 & 10 & 11 & 0 \end{bmatrix}$

**9.** $\begin{bmatrix} 1 & 5 & 6 & 7 \\ 7 & 4 & 9 & 10 \\ 10 & 11 & 9 & 13 \end{bmatrix}$

**11.** $\begin{bmatrix} -1 & 2 & 2 & 2 \\ 4 & 1 & 4 & 4 \\ 6 & 6 & -1 & 6 \\ 8 & 8 & 8 & 1 \end{bmatrix}$

**13.** $\begin{bmatrix} -9 & -12 & 6 \\ -6 & 12 & 0 \\ -21 & -18 & -36 \end{bmatrix}$

**15.** Not possible

**17.** $\begin{bmatrix} 30 & 19 \\ 15 & 9 \\ 17 & -2 \\ 3 & 10 \end{bmatrix}$

**19.** $\begin{bmatrix} 11 & 58 & -3 \\ 25 & -1 & -24 \\ 35 & 10 & 11 \end{bmatrix}$

**21.** $y = 3, x = 5, z = 7$

**23.** $x = 1, y = 2, z = 4, w = 3$

**25.** $x = -2, y = 3$

**27.** $y = 2, x = 1, z = 1$

**29.** $\begin{bmatrix} 1 & 2 & 9 & 9 \\ 8 & 2 & 5 & 6 \\ 6 & 5 & 2 & 11 \\ 0 & 6 & 6 & 7 \\ 7 & 7 & 2 & 0 \end{bmatrix}$

**31. c.** $\begin{bmatrix} 0 & 5 & 3 & 3 & 4 \\ 1 & 0 & 5 & 2 & 5 \\ 3 & 1 & 2 & 5 & 0 \\ 4 & 4 & 1 & 0 & 3 \\ 2 & 1 & 6 & 3 & 0 \end{bmatrix}$

**d.** Player 1

**33. a.** $\begin{bmatrix} 60 & 39 & 57 \\ 71 & 46 & 40 \end{bmatrix}$

**b.** $\begin{bmatrix} 74 & 48 & 70 \\ 87 & 56 & 49 \end{bmatrix}$

Exercise Set 3.2 (Pages 211–214) Matrix Multiplication

**1.** $\begin{bmatrix} -3 & 18 \\ 41 & 20 \\ 16 & 34 \end{bmatrix}$, undefined, undefined. $\begin{bmatrix} 1 & 5 & 12 \\ 15 & 39 & -18 \end{bmatrix}$, $\begin{bmatrix} 61 & 4 & 23 \\ 42 & 6 & 18 \\ 11 & 14 & 13 \end{bmatrix}$, $\begin{bmatrix} 11 & 5 \\ 39 & 69 \end{bmatrix}$

**3.** Undefined, undefined, undefined, undefined, undefined, $\begin{bmatrix} 51 & 24 & -1 \end{bmatrix}$

**5.** Undefined, $\begin{bmatrix} 70 & 1 & 62 & 5 \\ 15 & 7 & 10 & 20 \end{bmatrix}$, undefined, undefined, $\begin{bmatrix} 48 \\ 21 \end{bmatrix}$, undefined

**7.** $\begin{bmatrix} 1356 & 960 & 1360 \\ -72 & -192 & -152 \end{bmatrix}$

**9.** $\begin{bmatrix} 46 & 43 & 62 \\ 72 & 76 & 109 \\ 6 & 23 & 32 \end{bmatrix}$

**11.** $\begin{bmatrix} 1 & 15 \\ 47 & -1 \end{bmatrix}$, $\begin{bmatrix} -2 & 15 & -17 \\ 16 & 0 & 28 \\ 14 & 25 & 2 \end{bmatrix}$

**13.** $\begin{bmatrix} -24 & 52 \\ 24 & 5 \end{bmatrix}$, $\begin{bmatrix} 9 & 23 & 31 & -12 \\ 0 & -12 & -6 & -9 \\ 9 & 31 & 35 & -6 \\ 18 & 10 & 44 & -51 \end{bmatrix}$

**15.** $\begin{bmatrix} 1 & 0 \\ 0 & 1 \end{bmatrix}$, $\begin{bmatrix} 1 & 0 \\ 0 & 1 \end{bmatrix}$

**17.** $\begin{bmatrix} 30 & 24 & 18 \\ 84 & 69 & 54 \\ 138 & 114 & 90 \end{bmatrix}, \begin{bmatrix} 90 & 114 & 138 \\ 54 & 69 & 84 \\ 18 & 24 & 30 \end{bmatrix}$    **19.** $\begin{bmatrix} 7 & 14 & -11 \\ 17 & 40 & 19 \\ 9 & 15 & -17 \end{bmatrix}, \begin{bmatrix} 0 & 14 & -4 \\ 12 & 46 & -3 \\ -12 & 3 & -16 \end{bmatrix}$

**21.** $A = \begin{bmatrix} 1 & 2 \\ -1 & 3 \end{bmatrix}, B = \begin{bmatrix} -1 & 8 \\ -4 & 7 \end{bmatrix}$    **23.** $\begin{bmatrix} 1 & -2 \\ 3 & -6 \end{bmatrix}$    **27.** $\begin{bmatrix} 1 & 0 \\ 0 & 1 \end{bmatrix}, \begin{bmatrix} 1 & 0 \\ 0 & 1 \end{bmatrix}$

**29.** $\begin{bmatrix} 1 & 0 \\ 0 & 1 \end{bmatrix}$ and $\begin{bmatrix} 1 & 0 & 121 \\ 0 & 1 & 71 \\ 0 & 0 & 0 \end{bmatrix}$    **31.** $\begin{bmatrix} 5 & 2 \\ 7 & 3 \end{bmatrix}\begin{bmatrix} x \\ y \end{bmatrix} = \begin{bmatrix} -3 \\ 5 \end{bmatrix}$    **33.** $\begin{bmatrix} 11 & 6 \\ 9 & 5 \end{bmatrix}\begin{bmatrix} x \\ y \end{bmatrix} = \begin{bmatrix} 7 \\ -5 \end{bmatrix}$

**35.** $\begin{bmatrix} 1 & -4 & 0 & 1 \\ 5 & 0 & -5 & 4 \\ 0 & 1 & 3 & 1 \\ 2 & 2 & -1 & 2 \end{bmatrix}\begin{bmatrix} x \\ y \\ z \\ w \end{bmatrix} = \begin{bmatrix} -2 \\ 3 \\ 5 \\ -1 \end{bmatrix}$

**37. a.** $\begin{bmatrix} 33,700 & 36,700 & 36,300 & 34,575 \\ 30,400 & 33,300 & 32,900 & 31,560 \\ 29,500 & 33,450 & 32,525 & 31,575 \end{bmatrix}$ The number in the $i$th row and $j$th column is the cost of project $j$ using supplier $i$.

**b.** Use supplier C for project I, supplier B for project II, supplier C for project III, and supplier B for project IV.

**39.**

| From   To → ↓ | Vancouver | Victoria |
|---|---|---|
| Quebec | 37 | 19 |
| Montreal | 37 | 16 |
| Toronto | 34 | 19 |

**41.** $\begin{bmatrix} 83.65 & 96.25 & 91.25 & 90.35 & 95.95 & 42.4 \\ 87.4 & 94.70 & 90.2 & 87.6 & 92.2 & 35.7 \\ 85.45 & 95.40 & 90.85 & 88.7 & 94.55 & 38.55 \end{bmatrix}$ The number in the $i$th row and $j$th column is the final grade given to student $j$ by professor $i$.

**43. a.** $AB = BA = \begin{bmatrix} 12 & 6 \\ 36 & 18 \end{bmatrix}$

**b.** det $A = 0$ and det $B = -24$. So, there does not exist a $2 \times 2$ matrix $C$ such that $A$ and $B$ are powers of $C$, otherwise 0 and $-24$ would be powers of det$C$, and this is impossible.

## Exercise Set 3.3 (Pages 231–235) Solution of Linear Systems by Row Reduction

**1.** $\begin{bmatrix} 2 & 3 & | & -5 \\ 3 & -5 & | & 17 \end{bmatrix}$    **3.** $\begin{bmatrix} -7 & 2 & 5 & | & 4 \\ 3 & 0 & 7 & | & 0 \\ 0 & -5 & 2 & | & 7 \end{bmatrix}$    **5.** $\begin{cases} 2x & = 3 \\ 5x + 3y = -2 \end{cases}$    **7.** $\begin{cases} 2x & + 3z + 4w = 7 \\ -3x + 2y + 5z & = 2 \\ 2x - y & + 4w = -3 \end{cases}$

**9.** $\varnothing$    **11.** $\left\{ \left( \frac{1}{2}(4 - 3c), c \right) \mid c \text{ is a real number} \right\}$    **13.** $\varnothing$    **15.** $\{(1, 1, 2)\}$    **17.** $\varnothing$    **19.** $\{(2, 2, 1)\}$

**21.** $\{(3 + 10a, -4 - 17a, a) \mid a \text{ is a real number}\}$    **23.** $\{(3, 4, -2)\}$    **25.** $\{(5, 2, 3)\}$

**27.** $w = 1, x = -1, y = 0,$ and $z = 2$    **29.** $\{(3, 0, 1, -2)\}$    **31.** $\varnothing$    **33.** $\varnothing$

**35.** $\left\{ \left( 2 - \frac{a}{2}, 1, a, 3 \right) \mid a \text{ is a real number} \right\}$    **37.** $\{(2 - a, -2 + 4a, a, 3) \mid a \text{ is a real number}\}$

**39.** $\{(7 - 3a, a - 3, a, -3) \mid a \text{ is a real number}\}$    **41.** $\left\{ \left( -\frac{a}{2}, 0, a, 0 \right) \mid a \text{ is a real number} \right\}$

**43.** $\{(-a, 4a, a, 0) \mid a \text{ is a real number}\}$    **45.** $\{(-3a, a, a, 0) \mid a \text{ is a real number}\}$

**47. a.** $\{(45, 19)\}$    **b.** $\{(29, 13)\}$    **c.** $\{(57, 24)\}$    **49. a.** $\{(38, 7, 18)\}$    **b.** $\{(-10, -1, -5)\}$    **c.** $\{(71, 8, 32)\}$

**51. a.** $\{(76, 23, -64)\}$    **b.** $\{(-71, -21, 64)\}$    **c.** $\{(177, 55, -152)\}$

**53. b.** $\{(5, 3)\}, \{(6, 2)\}, \{(4, 5)\}, \{(6, 6)\}, \{(7, 4)\}$    **55. b.** $\{(60, 70, 90)\}, \{(50, 75, 100)\}$, no feasible solution

**57.** 0, 27, 37.5, $3,120     **59.** 56, 0, 28, $4,200     **61.** 50, 60, 40, 30, $10,700
**63.** The equilibrium price for commodity X is $4.50, for Y is $3.20, and for Z is $5.10. The equilibrium quantity for X is 551,900, for Y is 209,400, and for Z is 452,100.
**65.** The equilibrium prices for commodities X, Y, and Z are $9.80, $10.20, and $8.75, respectively. The corresponding equilibrium quantities are 1,386,100,   993,850, and 1,465,400.

## Exercise Set 3.4 (Pages 246–247) Multiplicative Inverse

**5.** $\begin{bmatrix} 3 & -7 \\ -2 & 5 \end{bmatrix}$   **7.** $\begin{bmatrix} 3 & -13 \\ -2 & 9 \end{bmatrix}$   **9.** $\begin{bmatrix} -3 & 2 \\ \frac{5}{4} & -\frac{3}{4} \end{bmatrix}$   **11.** $\begin{bmatrix} 7 & 6 & -17 \\ 2 & 2 & -5 \\ -6 & -5 & 15 \end{bmatrix}$

**13.** $\begin{bmatrix} 5 & -5 & 4 \\ 0 & 3 & 1 \\ 2 & -1 & 2 \end{bmatrix}$   **15.** No inverse   **17.** $\frac{1}{5}\begin{bmatrix} -1 & 1 & -1 \\ -9 & 19 & -24 \\ 7 & -12 & 17 \end{bmatrix}$   **19.** $\frac{1}{273}\begin{bmatrix} -18 & -2 & 63 \\ 6 & 31 & -21 \\ 45 & 5 & -21 \end{bmatrix}$

**21.** $\begin{bmatrix} 35 & -54 & 34 & -73 \\ 1 & -2 & 1 & -2 \\ 10 & -16 & 10 & -21 \\ -31 & 47 & -30 & 65 \end{bmatrix}$   **23.** $\begin{bmatrix} 128 & 35 & -28 & -73 \\ 4 & 1 & -1 & -2 \\ 37 & 10 & -8 & -21 \\ -114 & -31 & 25 & 65 \end{bmatrix}$   **25.** $\{(-29, 21)\}$

**27.** $\{(-74, 51)\}$   **29.** $\{(37, -14)\}$   **31.** $\{(-58, -17, 52)\}$   **33.** $\{(13, 16, 11)\}$   **35.** No solutions
**37.** $\left\{\left(\frac{-7}{5}, \frac{-128}{5}, \frac{94}{5}\right)\right\}$   **39.** $\left\{\left(\frac{-95}{273}, \frac{244}{273}, \frac{374}{273}\right)\right\}$   **41.** $\{(-604, -18, -175, 535)\}$   **43.** $\{(478, 16, 139, -425)\}$
**45.** At the central store 160 pounds of kind A are mixed with 40 pounds of kind B. At the northern store 180 pounds of kind A are mixed with 120 pounds of kind B. At the southern store 100 pounds of kind A are mixed with 150 pounds of kind B.
**47.** On weekdays sell 4000 seats in section A and 2000 in B. On Saturday sell 1000 seats in section A and 5000 in B. On Sunday sell 3000 seats in section A and 3000 in B.     **49.** $\{(10, 6)\}, \{(9, 7)\}, \{(11, 5)\}, \{(10, 7)\}, \{(9, 8)\}$
**51.** $\{(20, 15)\}, \{(22, 16)\}, \{(21, 20)\}, \{(25, 18)\}$     **53.** $\{(3, 5, 4)\}, \{(4, 6, 3)\}, \{(6, 3, 5)\}$

## Exercise Set 3.5 (Pages 261–262) Determinants

**1.** $-7$   **3.** 2   **5.** $-21$   **7.** $-35$   **9.** 18   **11.** 10   **13.** $-1$   **15.** $-3$   **17.** $x = 3$
**19.** $x = 6$ or $x = -1$   **21.** $\{(-16, 23)\}$   **23.** $\{(2, 1)\}$   **25.** $\{(-5, 3)\}$   **27.** $\{(-3, 1)\}$   **29.** $\{(-3, 7)\}$
**31.** $\{(-1, 2, 1)\}$   **33.** $\{(-2, 2, 3)\}$   **35.** Cramer's rule cannot be used to solve the system.   **37.** $\{(2, 1, 5)\}$
**39.** $\{(-3, 1, -1)\}$   **41.** $\{(2, 2, 2)\}$   **43.** $\{(-1, -3, 2)\}$   **45.** $\{(-1, 2, 1)\}$
**47.** 300 pounds of A, 80 pounds of B, and 20 pounds of C
**49.** $25,000 in Bond A, $15,000 in Bond B, and $10,000 in Bond C
**51.** X costs $9.25 per unit, Y costs $8.50 per unit, and Z costs $13.75 per unit.
**53.** 5 type A houses, 3 type B houses, and 6 type C houses
**55.** The wife's age is 40, the husband's is 44, and the son's is 16.     **57.** $-x + y = 1$     **59.** $13x + 2y = 27$
**61.** $x - 9y = 22$   **63.** 75   **65.** $-162$   **67.** $-6$   **69.** 12 sq. units   **71.** 28 sq. units

## Exercise Set 3.6 (Page 268) Inverse of a Square Matrix Using Adjoints

**5.** $\begin{bmatrix} 3 & -7 \\ -2 & 5 \end{bmatrix}$   **7.** $\begin{bmatrix} 3 & -13 \\ -2 & 9 \end{bmatrix}$   **9.** $\begin{bmatrix} -3 & 2 \\ \frac{5}{4} & \frac{-3}{4} \end{bmatrix}$   **11.** $\begin{bmatrix} 7 & 6 & -17 \\ 2 & 2 & -5 \\ -6 & -5 & 15 \end{bmatrix}$   **13.** $\begin{bmatrix} 5 & -5 & 4 \\ 0 & 3 & 1 \\ 2 & -1 & 2 \end{bmatrix}$

**15.** $A^{-1}$ does not exist.   **17.** $.2\begin{bmatrix} -1 & 1 & -1 \\ -9 & 19 & -24 \\ 7 & -12 & 17 \end{bmatrix}$   **19.** $\frac{1}{273}\begin{bmatrix} -18 & -2 & 63 \\ 6 & 31 & -21 \\ 45 & 5 & -21 \end{bmatrix}$

**21.** $\begin{bmatrix} 35 & -54 & 34 & -73 \\ 1 & -2 & 1 & -2 \\ 10 & -16 & 10 & -21 \\ -31 & 47 & -30 & 65 \end{bmatrix}$     **23.** $\begin{bmatrix} 128 & 35 & -28 & -73 \\ 4 & 1 & -1 & -2 \\ 37 & 10 & -8 & -21 \\ -114 & -31 & 25 & 65 \end{bmatrix}$

## Exercise Set 3.7 (Pages 280–282) Applications in Economics and Cryptography: Input-Output Analysis and Coded Messages

**1. a.** Industry I should produce 1296 units and Industry II should produce 1464 units.
   **b.** Industry I should produce 1668 units and Industry II should produce 1862 units.
   **c.** Industry I should produce 1492 units and Industry II should produce 1478 units.
**3. a.** Industry I should produce 5440 units, Industry II should produce 3990 units.
   **b.** Industry I should produce 6120 units, Industry II should produce 4570 units.
   **c.** Industry I should produce 7660 units, Industry II should produce 5885 units.
**5. a.** Industry I should produce 3380.8 units, Industry II should produce 4184 units.
   **b.** Industry I should produce 3220 units, Industry II should produce 4100 units.
   **c.** Industry I should produce 3512.8 units, Industry II should produce 4044 units.
**7. a.** Industries I, II, and III must produce approximately 3966, 4743, and 4076 units, respectively.
   **b.** Industries I, II, and III must produce approximately 4155, 4019, and 4178 units, respectively.
   **c.** Industries I, II, and III must produce approximately 4223, 4242, and 4752 units, respectively.
**9. a.** The units required of Industries I, II, and III are 4392, 5034, and 5198, respectively.
   **b.** 5660, 6560, 6000 units are required of the respective industries.
   **c.** 5006, 5795, and 5291 units are required of the respective industries.
**11. a.** Industries I and II must produce 2360 and 2640 millions of dollars per year, respectively.
   **b.** The industries must produce 3120 and 3380 millions of dollars per year, respectively.
   **c.** The industries must produce 3544 and 3796 millions of dollars per year, respectively.
**13. a.** Industries I and II must produce 2860 and 2235 millions of dollars per year, respectively.
   **b.** The industries must produce 4600 and 3650 millions of dollars per year, respectively.
   **c.** The industries must produce 4440 and 3290 millions of dollars per year, respectively.
**15. a.** Industries I and II must produce 3401.6 and 4068 millions of dollars per year, respectively.
   **b.** The industries must produce 3514.4 and 4112 millions of dollars per year, respectively.
   **c.** The industries must produce 3153.6 and 4028 millions of dollars per year, respectively.
**17. a.** Industries I, II, and III must produce approximately 5286, 4231, and 4981 million dollars per year, respectively.
   **b.** The industries must produce approximately 5633, 5181, and 5659 million dollars per year, respectively.
**19. a.** In millions of dollars per year, Industries I, II, and III must produce approximately 4614, 4675, and 4235, respectively.
   **b.** The industries must produce approximately 6811, 8107, and 7124 millions of dollars per year, respectively.
**21.** DRUGS KILL     **23.** STUDY EVERY DAY
**25.** 5, 61, 31, 3, 50, 26, 9, 70, 32, 27, 128, 73, 27, 136, 81, 16, 84, 41, 22, 99, 56, 5, 32, 15

## Exercise Set 3.8 (Pages 289–291) Markov Chains

**1.** Not a state vector     **3.** Not a state vector     **5.** A state vector     **7.** Not a transition matrix
**9.** Not a transition matrix     **11.** A transition matrix     **13.** T is regular.     **15.** T is not regular.     **17.** T is regular.
**19.** [ .3   .7 ]   **21.** [ .45   .55 ]   **23.** [ .25   .35   .4 ]   **25.** [ .25   .75 ]   **27.** [ .17   .83 ]
**29.** [ .2   .6   .2 ]
**31. a.** The stock has approximately a 36.8% chance of increasing, a 29.6% chance of staying the same, and a 33.7% chance of decreasing on January 9.
   **b.** The stock has approximately a 36.6% chance of increasing, 33.8% chance of decreasing, and 29.6% chance of staying the same on March 28.
   **c.** The stock has approximately a 36.6% chance of increasing, a 29.6% chance of staying the same, and a 33.8% chance of decreasing on September 29.
**33. a.** 28.75%     **b.** 47.875%     **c.** 19.9%     **35. a.** 31.95%     **b.** 35.0925%     **c.** 38.41925%

## Chapter Review 3.9 (Pages 294–299) Sample Exam Questions

**1.** $\begin{bmatrix} -1 & -4 \\ 1 & -2 \end{bmatrix}$
**3.** $\begin{bmatrix} 0 & 5 & 8 & 11 \\ -1 & 0 & 5 & 8 \\ -6 & -3 & 0 & 3 \\ -13 & -10 & -7 & 0 \end{bmatrix}$

**5. a.** $x = 3, y = 4, z = 7$ **b.** $x = 6, y = 2, z = 3$ **c.** $x = 3, y = -4, z = 1$

**7. a.** $\mathbf{A} = \begin{bmatrix} 1 & 2 \\ 3 & 4 \end{bmatrix}$ and $\mathbf{B} = \begin{bmatrix} 5 & 6 \\ 7 & 8 \end{bmatrix}$ **b.** $\mathbf{A} = \begin{bmatrix} 2 & 0 \\ 0 & 2 \end{bmatrix}$ and $\mathbf{B} = \begin{bmatrix} \frac{1}{2} & 0 \\ 0 & \frac{1}{2} \end{bmatrix}$

**9. a.** $\{(a - 3.2, 3.4 + .5a, a) \mid a$ is a real number$\}$ **b.** $\{(2, -1, 3)\}$
**11. a.** $\{(72, -29, 11)\}$ **b.** $\{(54, -23, 10)\}$ **c.** $\{(-25, 8, -1)\}$
**13. a.** $\{(37, -30)\}$ **b.** $\{(-24, -18, 39)\}$ **c.** $\{(43, -28, 23)\}$ **15. a.** $\{(2, -3)\}$ **b.** $\{(-1, 2, 4)\}$
**17. a.** Industries I, II, and III must produce approximately 4759, 4817, and 4891 million dollars per year, respectively.
 **b.** The industries must produce approximately 5817, 5627, and 5850 million dollars per year, respectively.
 **c.** The industries must produce approximately 8446, 8483, and 9504 million dollars per year, respectively.
**19.** GOOD LUCK
**21. a.** Not a transition matrix **b.** A transition matrix **c.** Not a transition matrix **d.** A transition matrix
**23.** [ .48 .52 ] **25. a.** 10.25% **b.** 37.875% **c.** 47.83625%

**27. a.** $\begin{cases} 8x + 5y = 70 \\ 3x + 2y = 20 \end{cases}$ $\begin{cases} 8x + 5y = 75 \\ 3x + 2y = 30 \end{cases}$ $\begin{cases} 8x + 5y = 64 \\ 3x + 2y = 28 \end{cases}$ $\begin{cases} 8x + 5y = 75 \\ 3x + 2y = 33 \end{cases}$ $\begin{cases} 8x + 5y = 62 \\ 3x + 2y = 25 \end{cases}$

**b.** $\{(40, -50)\}, \{(0, 15)\}, \{(-12, 32)\}, \{(-15, 39)\}$, and $\{(-1, 14\}$. However, only on Tuesday is the solution feasible.

## Exercise Set 4.1 (Pages 308–310) The Geometric Approach

**1.** 59, 3 **3.** 68, −7 **5.** 22.65, 7.5 **7.** 250, 0 **9.** 47, −239 **11.** 427, 187
**13.** No maximum value, minimum value of 38 **15.** No maximum or minimum **17.** No maximum, no minimum
**19.** $z = 370x + 290y$, 600 acres of wheat and 200 acres of barley
**21.** $z = 350x + 275y$, 2500 units of the deluxe model and 15,250 units of the standard model
**23.** $z = 80x + 70y$, 50 shares of stock A and 25 shares of stock B **25.** 23 **27.** 6 baskets and 4 vases
**29.** 30 first-class rooms and 90 regular rooms

## Exercise Set 4.2 (Page 315) The Three-Dimensional Coordinate System

**1. a.**

**b.**

**c.**

**3. a.**

**b.**

**c.**

**5.**

**7.**

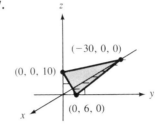

**9. a.** $(2, 3, 4)$
**b.**

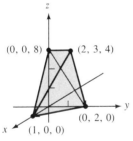

**11. a.** $(7, 5, 6)$
**b.**

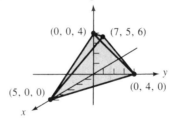

**13. a.** $(8, 7, 4)$
**b.**

**15.** 15      **17.** 28      **19.** 10      **21.** 84

**Exercise Set 4.3 (Pages 324–325) Introduction to the Simplex Method**

**1.** $3x + y + s = 15$          **3.** $2x + 3y + s = 12$          **5.** $2x + y + s = 14$

**7.** $4x + 5y + 4z + s = 40$

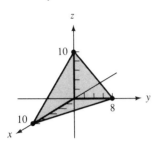

**9.** $4x + y + 4z + s = 16$

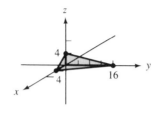

**11.** $2x + 2y + 3z + s = 12$

**13. a.**

**b.** $x + y + r = 7$
$3x + y + s = 15$

**15. a.**

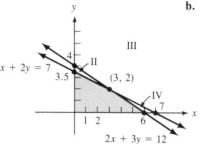

**b.** $x + 2y + r = 7$
$2x + 3y + s = 12$

**17. a.**

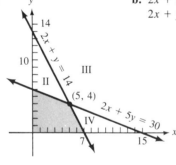

**b.** $2x + 5y + r = 30$
$2x + y + s = 14$

**19. a.**

**b.** $x + y + r = 10$
$x + 2y + s = 17$
$3x + 2y + t = 27$

**21. a.**

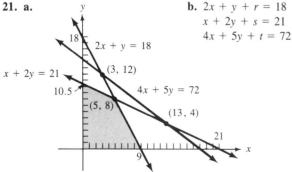

**b.** $2x + y + r = 18$
$x + 2y + s = 21$
$4x + 5y + t = 72$

**23. a.**

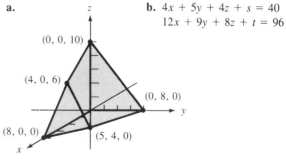

**b.** $4x + 5y + 4z + s = 40$
$12x + 9y + 8z + t = 96$

**25. a.** 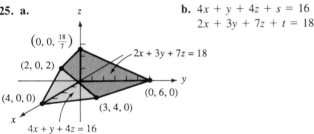   **b.** $4x + y + 4z + s = 16$
$2x + 3y + 7z + t = 18$

**27. a.** 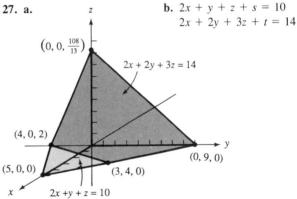   **b.** $2x + y + z + s = 10$
$2x + 2y + 3z + t = 14$

## Exercise Set 4.4 (Page 338) Moving from Corner to Corner

**1.** $(0, -3, 0, 5, 0)$   **3.** $(13, 0, -2, 0, 17, 0)$   **5.** $(-2, 3, 0, 5, 0, 13)$   **7.** $(0, 0), (0, 7), (4, 3)$
**9.** $(0, 0), \left(0, \frac{7}{2}\right), (3, 2)$   **11.** $(0, 0), (7, 0), (5, 4)$   **13.** $(0, 0), (9, 0), (7, 3), (3, 7)$
**15.** $(0, 0), (9, 0), (5, 8), (0, 10.5)$   **17.** $(0, 0, 0), (8, 0, 0), (5, 4, 0), (4, 0, 6)$
**19.** $(0, 0, 0), (4, 0, 0), (2, 0, 2), (3, 4, 0)$   **21.** $(0, 0, 0), (5, 0, 0), (3, 4, 0), (4, 0, 2)$
**23.** $(0, 0, 0), (0, 0, 15), (3, 0, 12), (3, 0, 0), (6, 4, 5), (6, 4, 0)$

## Exercise Set 4.5 (Pages 354–357) The Simplex Method

**1.** 82   **3.** No maximum   **5.** 400   **7.** 672   **9.** 684   **11.** 52   **13.** 158   **15.** 1925   **17.** 80   **19.** 195
**21.** 116   **23.** 125
**25.** The maximum profit is $15,000, which occurs when the complete 100 acres is planted in tomatoes.
**27.** The maximum profit is $50,000, which occurs when all 100 acres are planted in oranges and none in lemons.
**29.** The maximum profit is $7,000, which occurs when 100 units of product C is produced, 30 units of product B, and none of product A.
**31.** The maximum profit is $14,000, which occurs when 20 acres are planted in wheat, 20 in barley, and none in peanuts.
**33.** The maximum revenue is $11,600 and occurs for 800 pounds of variety I, 1800 pounds of variety II, and 400 pounds of variety III.
**35.** The maximum profit is $550, which occurs when 35 units of A, 0 units of B, and 5 units of C are produced.
**37.** The maximum profit is $9,000,000, which occurs when $(1 - t)40$ type A houses, $(1 - t)172 + 180t$ type B houses, and $(1 - t)22 + 30t$ type C houses are constructed for $0 \le t \le 1$.

## Exercise Set 4.6 (Pages 373–375) Minimization and Duality

**1.** 1539   **3.** 5700   **5.** 1260   **7.** 98   **9.** 684   **11.** 270   **13.** 3850   **15.** 600   **17.** 1960   **19.** 80
**21.** 58
**23.** The minimum cost is $47,500, which occurs when $25,000 is spent on radio advertising and $22,500 is spent on TV advertising.
**25.** The minimum cost is $1,818,000. This occurs when plants A, B, and C are operated for 30, 12, and 18 days, respectively.
**27.** The minimum cholesterol is 88 units, which occurs for 0, 6, and 10 ounces of foods A, B, and C respectively.
**29.** The minimum cost is $774,000, which occurs when the plants at Arlington, Bellingham, and Mount Vernon operate for 0, 9, and 27 days respectively.
**31.** The minimum cost is $9,900, and this occurs when 450 units are sent from Site A to Distributor I, 150 units from Site A to Distributor II, no units from Site B to Distributor I, and 600 units from Site B to Distributor II.
**33.** The minimum cost is $9,625, which occurs when 20, 5, 0, and 25 units are shipped from Factory A to Distributors I and II and from Factory B to Distributors I and II respectively.

## Exercise Set 4.7 (Pages 393–395) The Big *M* Method

**1.** 34 **3.** No maximum **5.** 90 **7.** 52 **9.** 39 **11.** 54 **13.** 45 **15.** 42
**17.** $P = 400x + 300y$, 5 acres of peanuts and 13 acres of corn
**19.** $P = 120x + 95y$, 27,000 units of model A and 2000 units of model B
**21.** $R = 1.05x + 1.15y$, 500 gallons of regular and 1100 gallons of premium
**23.** $C = 20x + 35y + 25z$, 175 shares of stock A, 75 shares of stock B, and 250 shares of stock C
**25.** $C = 3x + 3.75y$, 10 pounds of Type I and 10 pounds of Type II
**27.** $C = 8x_1 + 6x_2 + 7x_3 + 5x_4$, 100 motors are shipped from site A to plant I, 700 motors are shipped from site A to plant II, 500 motors are shipped from site B to plant I, and 0 motors are shipped from site B to plant II.

## Chapter Review 4.8 (Pages 396–399) Sample Exam Questions

**1.** The maximum value of $z$ is 25. The minimum value is 8. **3.** The maximum value of $w$ is 52. The minimum value is 10.
**5.** The maximum value of $z$ is 50.62$\overline{6}$. The minimum value is 17.
**7. a.**

**b.**

**c.**

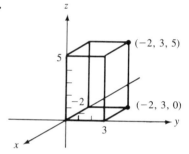

**9.** 10
**11. a.** (3, 4, 6)
  **b.**

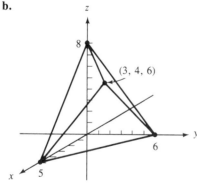

  **c.** 44

**13. a.** (0, 0), (12, 0), (10, 4) **b.** (0, 0), (0, 15), (5, 10), (8, 5), (10, 0)
  **c.** (0, 0, 0), (5, 0, 0), (0, 6, 0), (0, 10, 15)
**15.** No maximum
**17.** 44
**19.** 2430
**21.** 90
**23.** The maximum profit is $15,600, which occurs for 40 gadgets A and 100 gadgets B.
**25.** The Arlington plant operates for 18 hours and the Marysville plant operates for 3 hours.
**27.** 39
**29.** 51

## Exercise Set 5.1 (Pages 408–409) Sequences, Arithmetic and Geometric Progressions

**1.** $2, \frac{2}{3}, \frac{2}{5}, \frac{2}{7}, \frac{2}{9}$ **3.** $-1, -\frac{2}{5}, -\frac{2}{10}, -\frac{2}{17}, -\frac{2}{26}$ **5.** $\frac{1}{2}, -\frac{2}{3}, \frac{3}{4}, -\frac{4}{5}, \frac{5}{6}$ **7.** $2, 1, -\frac{3}{4}, -\frac{7}{11}, -\frac{8}{14}$
**9.** $-4, -1, 2, 5, 8, 11, 14, 17, 20, 23$ **11.** $-5, -7, -9, -11, -13, -15, -17, -19$ **13.** $-3, 1, 5, 9, 13$
**15.** 10, 8, 6, 4, 2 **17.** $-8, -5, -2, 1, 4, 7, 10, 13$ **19.** 12,915 **21.** 16,020 **23.** 15,050 **25.** 19,665
**27.** $22,500, $20,000, $17,500, $15,000, $12,500, $10,000, $7500, $5000
**29.** $285,000, $270,750, $257,212.50, $244,351.875, $232,134.2813
**31.** $502.50, $505.01, $507.54, $510.08, $512.62, $515.19 **33.** 805.33, 810.70, 816.11, 821.55, 827.02, 832.54

**35.** $603.50, $607.02, $610.56, $614.12, $617.71, $621.31     **37.** 10, 50, 250, 1250     **39.** 256, 128, 64, 32
**41.** If $r = 3$, $a = 4$, $a_2 = 12$, $a_3 = 36$. If $r = -3$, $a = 4$, $a_2 = -12$, $a_3 = 36$     **43.** $-2, -6, -18$     **45.** 729, 243, 81
**47.** $\dfrac{5^{20} - 1}{2}$     **49.** 1023.9999     **51.** $2(3^{25} - 1)$ or $1 + 3^{25}$     **53.** $1 - 3^{27}$     **55.** $\dfrac{3^7 - 3^{-21}}{2}$     **57.** $3884.14
**59.** $3691.21     **61.** $10,489.68     **63.** 191.75 inches     **65.** $506,098.24     **67.** Second option
**69.** If the player loses 9 straight hands, he could not double his last bet of $256 because $572 is over the maximum limit.

## Exercise Set 5.2 (Pages 422–424) Annuities

**1.** 1.576899264     **3.** 2.772469785     **5.** .591457366     **7.** 10.113249     **9.** 15.813679     **11.** 22.562977
**13.** 11.618933     **15.** 15.562251     **17.** $13,247.65, $1247.65     **19.** $115.25     **21.** $8504.07
**23.** $21,602.21, $3602.21     **25.** $8674.16, $1674.16     **27.** $17,968.23     **29.** $20,243.08     **31.** $8116.88
**33.** $42,620.68     **35.** $838.42     **37.** $2179.36     **39.** $260,383.36     **41.** $128,250     **45.** $18,550.45
**47.** $10,365.34     **49.** $146,182.02

## Exercise Set 5.3 (Pages 433–434) Mortgages

**1. a.** $211.06     **b.** $126.06     **c.** $85.00     **d.** $7,916.73     **e.** $34,118
**3. a.** $207.58     **b.** $48.96     **c.** $158.62     **d.** $5456.25     **e.** $2,454.80
**5. a.** $990.98     **b.** $678.52     **c.** $312.45     **d.** $33,862.52     **e.** $147,835.20
**7. a.** $1214.89     **b.** $1267.28     **c.** $1321.73     **d.** $1417.07     **e.** $1597.47     **f.** $1995.14     **9.** $143,745.18
**11.** $128,947.89     **13. a.** $265.71     **b.** $43.57     **c.** $222.14     **d.** $2,754.78     **15.** 47 months     **17.** 66 months

## Chapter Review 5.4 (Pages 436–438) Sample Exam Questions

**1. a.** $1, \frac{4}{7}, \frac{4}{10}, \frac{4}{13}, \frac{4}{16}$     **b.** $-\frac{2}{4}, \frac{4}{5}, -\frac{6}{6}, \frac{8}{7}, -\frac{10}{8}$     **c.** $6, \frac{11}{8}, \frac{18}{19}, \frac{27}{34}, \frac{38}{53}$     **3. a.** $-3, 0, 3, 6$     **b.** 59,100
**5. a.** $-8, 16, -32$     **b.** $-1,398,100$     **7.** $20,090.42     **9.** $3103.21     **11.** $9743.96     **13.** $8983.35
**15.** $8,504.08     **17.** $149,251.72     **19.** $12,706.66     **21.** $337.28     **23.** $157,231.11     **25.** 70 months

## Exercise Set 6.1 (Pages 446–447) Introduction

**1. a.**

| Relative frequency $\dfrac{s}{n}$ |
| --- |
| .03 |
| .025 |
| .02$\overline{6}$ |
| .0275 |
| .03 |
| .028$\overline{3}$ |
| .031428571 |
| .03125 |
| .03$\overline{2}$ |
| .032 |
| .031$\overline{8}$ |
| .031$\overline{6}$ |
| .03076923 |
| .03142857 |
| .031$\overline{3}$ |

**b.** .031$\overline{3}$     **3. a.**

| Relative frequency $\dfrac{s}{n}$ |
| --- |
| .02 |
| .0225 |
| .018$\overline{3}$ |
| .02125 |
| .021 |
| .021$\overline{6}$ |
| .021428571 |
| .02125 |
| .019$\overline{4}$ |
| .0195 |

**b.** .0195

**5.** $\frac{5}{36}$     **7.** $\frac{5}{36}$     **9.** $\frac{1}{36}$
**11. a.** $\{(2, 6), (3, 5), (4, 4), (5, 3), 6, 2)\}, \frac{5}{36}$     **b.** $\{(3, 6), (4, 5), (5, 4), (6, 3)\}, \frac{1}{9}$
 **c.** $\{(2, 6), (3, 5), (4, 4), (5, 3), (6, 2), (3, 6), (4, 5), (5, 4), (6, 3)\}, \frac{1}{4}$     **d.** $\emptyset, 0$     **e.** $\frac{31}{36}$

**13. a.** {(T, T, T), (T, T, H), (T, H, T), (H, T, T), (H, H, T), (H, T, H), (T, H, H), (H, H, H)}
   **b. i.** {(T, T, H), (T, H, T), (H, T, T)}, $\frac{3}{8}$
   **ii.** {(T, T, H), (T, H, T), (H, T, T), (T, H, H), (H, T, H), (H, H, T), (H, H, H)}, $\frac{7}{8}$    **iii.** {(H, H, H)}, $\frac{1}{8}$
   **iv.** {(H, H, T), (H, T, H), (T, H, H), (H, H, H)}, $\frac{1}{2}$    **15.** $\frac{5}{6}$    **17.** 2 to 23    **19.** 7 to 13    **21.** .19

## Exercise Set 6.2 (Pages 455–456) Counting Techniques

**1.** 5,040    **3.** 100,100    **5.** 720    **7.** 10    **9.** 1,716    **11.** 120    **13.** 210
**15.** (T, T, T, T), (T, T, T, H), (T, T, H, T), (T, T, H, H), (T, H, T, T),
(T, H, T, H), (T, H, H, T), (T, H, H, H), (H, T, T, T), (H, T, T, H),
(H, T, H, T), (H, T, H, H), (H, H, T, T), (H, H, T, H), (H, H, H, T),
(H, H, H, H)
**17.** (T, T, T, R), (T, T, T, G), (T, T, T, B), (T, T, H, R), (T, T, H, G),
(T, T, H, B), (T, H, T, R), (T, H, T, G), (T, H, T, B), (T, H, H, R),
(T, H, H, G), (T, H, H, B), (H, H, H, R), (H, H, H, G), (H, H, H, B),
(H, H, T, R), (H, H, T, G), (H, H, T, B), (H, T, H, R), (H, T, H, G),
(H, T, H, B), (H, T, T, R), (H, T, T, G), (H, T, T, B)

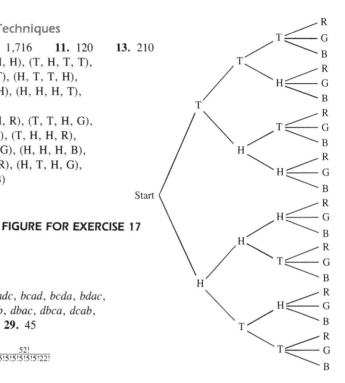

**FIGURE FOR EXERCISE 17**

**21.** 24, *abcd, abdc, acbd, acdb, adcb, adbc, bacd, badc, bcad, bcda, bdac,*
*bdca, cabd, cadb, cbad, cbda, cdab, cdba, dabc, dacb, dbac, dbca, dcab,*
*dcba*    **23.** 4,989,600    **25.** 35    **27.** 1,716    **29.** 45
**31.** 220    **33.** 479,001,600, 30.4 years    **35.** 28
**37. a.** 174,420    **b.** 277,134    **39.** 630    **41.** $\frac{52!}{5!5!5!5!5!5!22!}$

## Exercise Set 6.3 (Pages 469–470) Conditional Probability

**1.** $\frac{5}{24}$    **3. a.** $\frac{2}{3}$    **b.** $\frac{18}{73}$    **c.** $\frac{9}{41}$    **5.** $\frac{25}{49}$    **7.** $\frac{64}{169}$    **9.** $\frac{2}{13}$    **11.** $\frac{4}{13}$    **13.** $\frac{2}{3}$    **15.** .116948178
**17.** .31500766    **19.** .706317891    **21. a.** $\frac{38}{59}$    **b.** $\frac{35}{73}$    **c.** $\frac{21}{59}$    **23.** $\frac{1}{7}$    **25.** .1738    **27.** .1876525
**29.** .8375275    **31.** .453122696    **33.** $\frac{4}{7}$    **35.** $\frac{6}{11}$    **37.** $\frac{9}{19}$    **39.** $\frac{4}{19}$

## Exercise Set 6.4 (Pages 475–477) Bayes' Formula

**1.** .3151    **3.** .4611    **5. a.** .6528    **b.** .3472    **7. a.** .3478    **b.** .6522    **9.** .7303
**11. a.** .1763    **b.** .4232    **c.** .2494    **d.** .1511    **13. a.** .2349    **b.** .2761    **c.** .2698    **15.** .8978
**17.** .3245    **19. a.** .6    **b.** .4286    **c.** one

## Exercise Set 6.5 (Pages 486–488) Expected Value

**1.** $53.80    **3.** Play game 1.    **5.**

| $x$ | $f(x)$ |
|---|---|
| 0 | $\frac{1}{32}$ |
| 1 | $\frac{5}{32}$ |
| 2 | $\frac{10}{32}$ |
| 3 | $\frac{10}{32}$ |
| 4 | $\frac{5}{32}$ |
| 5 | $\frac{1}{32}$ |

**7.** $f(k) = \left(\frac{5}{6}\right)^{k-1}\left(\frac{1}{6}\right)$, $k = 1, 2, 3, \ldots$
**9.** 2
**11.** 7
**13.** $278.75
**15.** $11.375
**17.** 80¢
**19.** 1.79

**21. a.** $E(X) = E(Y) = 6$   **b.** 6.6, 2.569   **c.** 10.6, 3.256   **23.** $5.8\overline{3}$, 2.415

**d.**

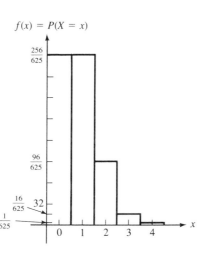
$f(x) = P_1(X = x)$          $g(y) = P_2(Y = y)$

## Exercise Set 6.6 (Pages 495–497) The Binomial Distribution

**1.** $a^6 + 6a^5b + 15a^4b^2 + 20a^3b^3 + 15a^2b^4 + 6ab^5 + b^6$   **3.** $\frac{200}{16,807}$   **5.** $\frac{27}{1024}$   **7.** .119   **9.** .281   **11.** .0577
**13.** .144   **15. a.** .2936   **b.** .2668   **c.** .0243

**17.**

| $x$ | $f(x)$ |
|---|---|
| 0 | .4096 |
| 1 | .4096 |
| 2 | .1536 |
| 3 | .0256 |
| 4 | .0016 |

$f(x) = P(X = x)$

**19.**

| $x$ | $f(x)$ |
|---|---|
| 0 | .004096 |
| 1 | .036864 |
| 2 | .13824 |
| 3 | .27648 |
| 4 | .31104 |
| 5 | .186624 |
| 6 | .046656 |

**21.** .671   **23. a.** .2013   **b.** .8791   **25.** .890625   **27.** .97398
**29. a.** $(.2)^{20}$   **b.** $(.2)^{18}(124.84)$   **c.** .00000018   **31.** 4   **33.** 4.8, 2.88, 1.697   **35.** 5.4, 3.78, 1.944   **37.** .9270

## Exercise Set 6.7 (Pages 504–506) More on Markov Chains

**1.** The steady-state vector is $\begin{bmatrix} \frac{10}{17} & \frac{7}{17} \end{bmatrix}$.   **3.** The steady-state vector is $[\ .8 \quad .2\ ]$.
**5.** The steady-state vector is $\begin{bmatrix} \frac{32}{95} & \frac{7}{95} & \frac{56}{95} \end{bmatrix}$.   **9.** $x = .1$, $y = .2$, and $z = .4$   **11.** $x = .1$, $y = .2$, and $z = .3$
**13. a.** $[\ .26 \quad .52 \quad .22\ ]$   **b.** $[\ .26 \quad .41 \quad .33\ ]$   **c.** $[\ .251 \quad .504 \quad .245\ ]$
**d.** $[\ .2237 \quad .5255 \quad .2508\ ]$

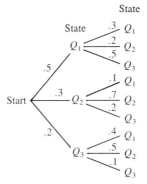

**15. a.** 20%   **b.** 14.3%   **17. a.** 55%, 29.5%, and 15.5%   **b.** 60%, 24.4%, and 15.6%   **19.** .6125

**21. a.** .836325   **b.** .1088122   **c.** .055   **23.** $\begin{bmatrix} \frac{10}{17} & \frac{7}{17} \\ \frac{10}{17} & \frac{7}{17} \end{bmatrix}$   **25.** $\begin{bmatrix} .8 & .2 \\ .8 & .2 \end{bmatrix}$   **27.** $\begin{bmatrix} \frac{32}{95} & \frac{7}{95} & \frac{56}{95} \\ \frac{32}{95} & \frac{7}{95} & \frac{56}{95} \\ \frac{32}{95} & \frac{7}{95} & \frac{56}{95} \end{bmatrix}$   **29.** Yes

## Chapter Review 6.8 (Pages 508–511) Sample Exam Questions

**1. a.** 120   **b.** 132   **c.** $5.\overline{6}$   **d.** 210   **e.** 40,320   **f.** 220   **g.** 360,360   **h.** 9,189,180   **3.** 64,864,800
**5.** 1,058   **7.** 2,520
**9. a.** The sample space consists of all ordered 4-tuples $(x, y, z, w)$ where $x, y, z$, and $w$ are members of the set {H, T}.
**b.** The probabilities are $\frac{1}{4}, \frac{15}{16}, \frac{11}{16}$, and $\frac{1}{16}$   **11.** 8 to 17   **13.** $\frac{2}{9}$   **15.** $\frac{9}{64}$   **17.** .8136   **19.** $\frac{25}{91}$   **21.** $\frac{4}{9}$
**23.** lose approximately 39¢   **25.** $241.20   **27.** .3456

**29. a.**

| $x$ | $f(x)$ |
|-----|--------|
| 0 | .343 |
| 1 | .441 |
| 2 | .189 |
| 3 | .027 |

**b.** .9, .63, .794   **31.** .0341   **33.** $x = .6, y = .4$, and $z = .2$   **35. a.** 12.2%   **b.** $8\frac{1}{3}$%

## Exercise Set 7.1 (Pages 517–518) Rate of Change

**1.** $2,400   **3.** 6, $-30$   **5.** $-6, -24$   **7.** 5, 1   **9.** $e^3 - e$, 10   **11.** $-3, \frac{26}{3}$   **13.** 3 seconds, 432 feet
**15.** .01 seconds, 1.2816 feet   **17.** 96 feet/second   **19.** 161.6 feet/second   **21.** $16(6 + h)$ feet/second   **23.** 5, 5, 1
**25.** $a, 7a + a^2, 7 + a$   **27.** $2,400, $3,600, $1,200, $12/doll, $18/doll, $6/doll   **29.** 7.835 words/week
**31.** 2503 feet/second   **33. a.** 9   **b.** $7 + h$   **35. a.** $\frac{1}{8}$   **b.** $\dfrac{1}{\sqrt{9 + h} + 3}$   **37. a.** 1.75   **b.** $\dfrac{h + 2^{h-1} + 1/2}{h}$

## Exercise Set 7.2 (Pages 527–528) Intuitive Description of Limit

**1.** 8   **3.** 6   **5.** 32   **7.** 23   **9.** 3   **11.** Undefined   **13.** 0   **15.** $-6$   **17.** Does not exist   **19.** $\frac{1}{4}$
**21.** $-3$   **23.** $-1.5$   **25.** $\frac{5}{64}$   **27.** $\frac{32}{135}$   **29.** 8   **31.** 64 ft/sec
**33. a.** $f(2) = 9$   **b.** $5 + h$   **c.** 5   **d.** $y = 5x - 1$   **e.**

**35.** $-3$
**37.** 19
**39.** $-\frac{5}{3}$

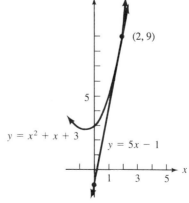

## Exercise Set 7.3 (Pages 534–535) Continuity and One-Sided Limits

**1.** not continuous   **3.** not continuous   **5.** continuous   **7.** not continuous   **9.** not continuous

**11.** $\lim\limits_{x \to -3^+} f(x) = -1$

$\lim\limits_{x \to -3^-} f(x) = -2$

**13.** $\lim\limits_{t \to -2^-} h(t) = 4$

$\lim\limits_{t \to -2^+} h(t) = 3$

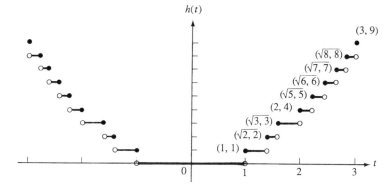

**15.** $\lim\limits_{x \to 2^+} g(x) = 3$

$\lim\limits_{x \to 2^-} g(x) = 3$

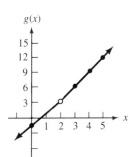

**17.** $\lim\limits_{x \to 9^+} f(x) = 7$

$\lim\limits_{x \to 9^-} f(x) = \frac{1}{6}$

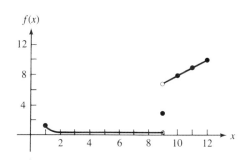

**19. a.** $-2$   **b.** $-2$   **c.** $-4$
**21. a.** $1$   **b.** $3$   **c.** $3$
**25.**

$$y = \frac{2}{x + 3}$$

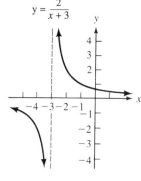

**27.** $4.4$

### Exercise Set 7.4 (Pages 542–543) The Derivative

**1.** $3$   **3.** $4$   **5.** $\frac{1}{4}$   **7.** $-\frac{1}{18}$   **9.** $5$   **11.** $2x + 5$   **13.** $\dfrac{3}{2\sqrt{3x + 1}}$   **15.** $-1(x + 5)^{-3/2}$

**17. a.** $f(2) = 11$   **b.**

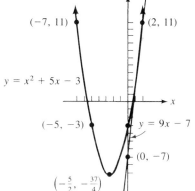

**c.** $9$   **d.** $y = 9x - 7$

**19. a.** $g(2) = 2$
**b.**

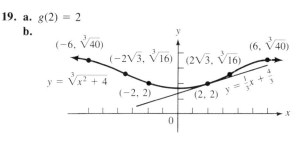

**c.** $\frac{1}{3}$   **d.** $y = \frac{1}{3}x + \frac{4}{3}$

**21.** 11 ft/sec    **23.** 128 ft/sec    **25.** $\frac{-20}{9}$ ft/sec    **27. a.** 0    **b.** 0    **c.** 0    **d.** Yes    **e.** 1    **f.** $-1$    **g.** No
**29. a.** 11    **b.** 11    **c.** 11    **d.** Yes    **e.** 13    **f.** 7    **g.** No

## Exercise Set 7.5 (Pages 548–549) Basic Differentiation Formulas

**1.** 0    **3.** 0    **5.** $4t + 3$    **7.** $12x^3 + 20x - 6$    **9.** $\frac{3}{2}x^{1/2}$    **11.** $6s^5 + 12s^3 + 6s$    **13.** $15t^4 + 9t^2 + 4t$
**15.** $\frac{7}{3}x^{-1/15} + \frac{2}{5}x^{-11/15} - x^{-7/5}$    **17.** $-2.4u^{-1.6}$    **19.** $y = 7x + 8$    **21.** $39x - 8y - 52 = 0$    **23.** 26 ft/sec
**25.** $\frac{5}{3}$ ft/sec    **27.** 4.875 ft/sec    **29. a.** $150 - 4q + .03q^2$    **b.** \$50    **c.** \$49.01
**31. a.** 12 seconds    **b.** 2304 feet    **c.** 9 seconds    **d.** 24 seconds
**33. a.** 52 seconds    **b.** 43,264 feet    **c.** 49 seconds    **d.** 104 seconds    **35.** $(2, -34)$ and $(-3, 91)$    **37.** $(2, -33)$
**39.** $(2.2, -4.16)$    **41. a.** 43    **b.** $h(x) = 3(x + 5)$    **c.** $6x + 6$    **d.** 24, 24

## Exercise Set 7.6 (Pages 553–554) Basic Differentiation Formulas (continued)

**9.** $6x^5 + 55x^4 + 12x^3 + 10x + 55$    **11.** $70x^6 + 30x^4 + 64x^3 + 75x^2 + 15$    **13.** $\frac{5}{2}x^{3/2} + \frac{9}{2}x^{1/2} - \dfrac{1}{2\sqrt{x}} + 6x + 9$

**15.** $\dfrac{-2x^5 - 4x^3 + 6x}{(x^4 + 3)^2}$    **17.** $\dfrac{18x^6 + 100x^4 + 24x}{(3x^2 + 10)^2}$    **19.** $\left(x - 3\sqrt{x} - \dfrac{3}{2\sqrt{x}}\right)(\sqrt{x} - 2)^{-2}$

**21. a.** $20 - \dfrac{30{,}000}{\sqrt{q}(\sqrt{q} + 10)^2}$    **b.** \$12.50

**23. a.** $100q(q^2 + 3q + 3)/(q + 2)(q + 3)$    **b.** $\dfrac{(q + 3)(200q + 600) - (100q^2 + 600q)}{(q + 3)^2}$

   **c.** $\dfrac{(q + 2)(400q + 500) - (200q^2 + 500q)}{(q + 2)^2}$    **d.** $P'(q) = R'(q) - C'(q)$

**25. a.** $\frac{1}{1250}(-q^4 + 100q^3 - q^2 + 100q) - 1000 - 50q - \dfrac{30{,}000q}{q^2 + 500}, \ 0 \le q \le 100$    **b.** $50 + \dfrac{30{,}000(-q^2 + 500)}{(q^2 + 500)^2}$

   **c.** $\frac{1}{1250}(-4q^3 + 300q^2 - 2q + 100)$    **d.** $P'(q) = R'(q) - C'(q)$
**27. a.** $D_x \dfrac{1}{x^{10}} = \dfrac{x^{10}D_x(1) - 1 \cdot D_x x^{10}}{(x^{10})^2} = \dfrac{-10x^9}{x^{20}} = \dfrac{-10}{x^{11}} = -10x^{-11}$

## Exercise Set 7.7 (Pages 559–560) The Chain Rule

**1.** $40(2x + 6)(x^2 + 6x + 3)^{39}$    **3.** $-65(2x + 3)(x^2 + 3x + 1)^{-14}$    **5.** $\frac{5}{4}(2u + 5)\sqrt[4]{u^2 + 5u + 1}$
**7.** $3(x^3 + 1)^4(x^4 + 3x^2 + 2)^2(4x^3 + 6x) + (x^4 + 3x^2 + 2)^3(12x^2)(x^3 + 1)^3$
**9.** $\dfrac{(u^3 + 2u^2 + 3)^5(4)(u^2 + 1)^3(2u) - (u^2 + 1)^4(5)(u^3 + 2u^2 + 3)^4(3u^2 + 4u)}{(u^3 + 2u^2 + 3)^{10}}$
**11.** $\dfrac{(t^3 + 4)^2 \cdot \frac{1}{3}(t^2 + 1)^{-2/3}(2t) - (t^2 + 1)^{1/3}(2)(t^3 + 4)(3t^2)}{(t^3 + 4)^4}$    **13.** $\frac{1}{3}(1 + t)^{-2/3}$    **15.** $\dfrac{12(t^2 + 3)^{11}(-2t^5 + 2t - 12t^3)}{(t^4 + 1)^{13}}$
**17.** $20\left(\dfrac{3s + 1}{\sqrt{s + 2}}\right)^{19} \cdot \dfrac{3\sqrt{s + 2} - (3s + 1) \cdot \frac{1}{2}(s + 2)^{-1/2}}{s + 2}$    **19.** $17\left(y^2 + \dfrac{1}{\sqrt{y}}\right)^{16}\left(2y - \frac{1}{2}y^{-3/2}\right)$    **21.** $y = 160x - 128$
**23.** $y = -1{,}376{,}000(x + 3) - 10{,}000$    **25.** $y = 0$    **27. a.** $\frac{12}{5}$    **b.** $-\frac{5}{12}$    **29.** 474 ft/sec
**31. a.** $30 - \frac{1}{5}(q^2 + 625)^{-4/5}(2q)$    **b.** \$29.968
**33. a.** $\left(10 + \dfrac{q^2}{q^3 + 50}\right)^3 + 3q\left(10 + \dfrac{q^2}{q^3 + 50}\right)^2\left[\dfrac{2q(q^3 + 50) - q^2(3q^2)}{(q^3 + 50)^2}\right]$    **b.** \$1003.88    **35.** .60375 items/week
**37.** $\dfrac{4t^3}{t^4 + 4}$

## Exercise Set 7.8 (Pages 565–566) Implicit Differentiation

**1.** $f_1(x) = \dfrac{-5 + \sqrt{25 + 4x^2}}{2}$ and $f_2(x) = \dfrac{-5 - \sqrt{25 + 4x^2}}{2}$    **3.** $f_1(x) = 2x, f_2(x) = -2x$    **5.** $\dfrac{6xy^2 + 6 - 3x^2 - 2xy}{x^2 - 6x^2y + 7}$

**7.** $\dfrac{24x^2 - 10xy^4 - 6}{20x^2y^3 - 14y}$   **9.** $\dfrac{2xy^4e^6 - 4x^3\log_5 3}{42y^6 - 4x^2y^3e^6}$   **11.** $\dfrac{2xy - \frac{3}{7}x^{-4/7}y^2 - y^{3/4}}{\frac{3}{4}xy^{-1/4} + 2yx^{3/7} - x^2}$   **13. b.** $y = \frac{35}{57}x - \frac{184}{57}$

**15. b.** $2x - 9y = -5$   **17.** $-\frac{3}{2} \pm \dfrac{9x}{2\sqrt{9x^2 + 16}}$   **19.** $\dfrac{2y^2 + 10xy - 18x^2}{3y^2 - 4xy - 5x^2}$

## Exercise Set 7.9 (Pages 572–573) Derivative of Exponential and Logarithmic Functions

**1.** $5x^4e^{x^5}$   **3.** $(4x + 5)e^{2x^2+5x+1}$   **5.** $2(x^2 + 5x + 1)(2x + 5)e^{(x^2+5x+1)^2}$   **7.** $e^{x^2+3}(2x^3 + 6x^2 + 16x + 3)$

**9.** $e^{x^2+7}\left(2x\sqrt{x^2 + 7} + \dfrac{x}{\sqrt{x^2 + 7}}\right)$   **11.** $\dfrac{e^{5x}(5x - 2)}{x^3}$   **13.** $\dfrac{4x^3 - 5x^4 - 10}{e^{5x}}$   **15.** $\dfrac{x/\sqrt{x^2 + 1} - 3\sqrt{x^2 + 1}}{e^{3x+2}}$

**17.** $\dfrac{6x^5 + 2x}{x^6 + x^2 + 1}$   **19.** $\dfrac{5x^4 + 6x}{x^5 + 3x^2 - 6}$   **21.** $\dfrac{(x^4 + 6x + 1)3x^2}{x^3 - 1} + (4x^3 + 6)\ln|x^3 - 1|$

**23.** $(x^2 + 6)^{1/3} \cdot \dfrac{2x}{x^2 + 6} + \ln(x^2 + 6) \cdot \frac{1}{3}(x^2 + 6)^{-2/3}(2x)$   **25.** $\dfrac{(x^5 + 2x + 1) \cdot \left(\dfrac{20x^3}{5x^4 + 2}\right) + (5x^4 + 2)\ln(5x^4 + 2)}{(x^5 + 2x + 1)^2}$

**27.** $\dfrac{\ln(x^2 + 2)(4x^3/(x^4 + 1)) - \ln(x^4 + 1)(2x/(x^2 + 2))}{\ln^2(x^2 + 2)}$

**29.** $\dfrac{2x}{x^2 + 2} + \dfrac{4x^3}{x^4 - 3} - 2(3 + \ln|5x|)e^{x^2} \cdot \frac{1}{x} - 2xe^{x^2}(3 + \ln|5x|)^2$   **31.** $\dfrac{5}{(5x + 1)\ln 3}$   **33.** $\dfrac{2x + 5}{(x^2 + 5x + 2)\ln 10}$

**35.** $\dfrac{1}{\ln 6}\left[\dfrac{(x^2 + 6)(3x^2 + 3)}{x^3 + 3x + 7} + 2x \ln|x^3 + 3x + 7|\right]$   **37.** $\dfrac{20xy^4 - 2xe^{x^2} - 3x^2y^2e^{x^3y^2}}{2x^3ye^{x^3y^2} - 40x^2y^3}$   **39.** $\dfrac{6x^2y^4 - xy^2e^{xy} - 6y}{2x + x^2ye^{xy} - 9x^3y^3}$

**41.** $\dfrac{10y + 14x^2y - 70x^2y^4}{105x^3y^3 - 5\pi x}$   **43.** $y - \sqrt{e} = \dfrac{-1}{2\sqrt{e}}(x - e)$   **47.** 2,955.6 bacteria/hour

**49. a.** $75e^{-.00002q^2}e^{.002}[1 - .00004q^2]$   **b.** \$36.92
**51. a.** $.002e^{.002q}(-85,000q + 60,000,000) - 85,000e^{.002q}$   **b.** \$12,823.66   **53.** .24 m/min   **55.** $x < -2$ or $x > 1$

## Exercise Set 7.10 (Pages 578–579) Logarithmic Differentiation

**1.** $(x^2 + 1)(x^3 + 3x + 2)(x^4 + x^2 + 1)\left(\dfrac{2x}{x^2 + 1} + \dfrac{3x^2 + 3}{x^3 + 3x + 2} + \dfrac{4x^3 + 2x}{x^4 + x^2 + 1}\right)$

**3.** $(x^3 + 5x^2 + 6x + 2)^2\sqrt{x^3 + 1}\left(\dfrac{2(3x^2 + 10x + 6)}{x^3 + 5x^2 + 6x + 2} + \dfrac{3x^2}{2(x^3 + 1)}\right)$

**5.** $\dfrac{(3x^5 - 4x^2 + 6)^4(x^2 - 6x + 7)^3}{(x^2 + 1)^{1/3}(x^3 - 4x^2 + 6)^2}\left[\dfrac{4(15x^4 - 8x)}{3x^5 - 4x^2 + 6} + \dfrac{3(2x - 6)}{x^2 - 6x + 7} - \dfrac{2x}{3(x^2 + 1)} - \dfrac{2(3x^2 - 8x)}{x^3 - 4x^2 + 6}\right]$

**7.** $\dfrac{e^{x^2}}{(x^4 + 3)^2}\left(2x - \dfrac{8x^3}{x^4 + 3}\right)$   **9.** $\dfrac{-5}{(x^4 + 1)(x^2 + 6)^2\sqrt[3]{x^3 + x^2 + x + 1}}\left[\dfrac{4x^3}{x^4 + 1} + \dfrac{4x}{x^2 + 6} + \dfrac{3x^2 + 2x + 1}{3(x^3 + x^2 + x + 1)}\right]$

**11.** $5^{5x^2-6x+10}(10x - 6)\ln 5$   **13.** $\dfrac{4\sqrt{x^2+3}x \ln 4}{\sqrt{x^2 + 3}}$   **15.** $(x^2)(2^x)\left(\frac{2}{x} + \ln 2\right)$   **17.** $(e^x + 4)^{x^2}\left(2x \ln(e^x + 4) + \dfrac{x^2e^x}{e^x + 4}\right)$

**19.** $(x^6 + 3x^2 + 5)e^{3x}\left[3e^{3x}\ln(x^6 + 3x^2 + 5) + \dfrac{6e^{3x}(x^5 + x)}{x^6 + 3x^2 + 5}\right]$

**21.** $[(x^2 + 1)^{5x+2}]^{(x^2+6)}\left[(15x^2 + 4x + 30)\ln(x^2 + 1) + \dfrac{(2x)(5x^3 + 2x^2 + 30x + 12)}{x^2 + 1}\right]$   **23.** $y = \frac{4385}{204}(x - 2) + 34$

**29.** $x^2(x^4 + 3)^{x^2-1}(4x^3) + (x^4 + 3)^{x^2}\ln(x^4 + 3)(2x)$

## Exercise Set 7.11 (Pages 584–585) Higher Order Derivatives

**1.** $4x + 5, 4, 0$   **3.** $5x^4 + 6x - 6, 20x^3 + 6, 60x^2$   **5.** $4t^3e^{t^4}, e^{t^4}(12t^2 + 16t^6), e^{t^4}(144t^5 + 64t^9 + 24t)$

**7.** $e^{5u}(5u + 1), 5e^{5u}(5u + 2), 25e^{5u}(5u + 3)$   **9.** $2x(1 + \ln(x^2 + 1)), 2 + 2\ln(x^2 + 1) + \dfrac{4x^2}{x^2 + 1}, \dfrac{4x^3 + 12x}{(x^2 + 1)^2}$

**11.** $\frac{4}{3}t^{-1/3}, -\frac{4}{9}t^{-4/3}, \frac{16}{27}t^{-7/3}$

**13.** $\dfrac{-2v^5 - 4v^3 + 2v}{(v^4 + 1)^2}$, $(6v^8 + 20v^6 - 24v^4 - 12v^2 + 2)(v^4 + 1)^{-3}$, $(48v^7 + 120v^5 - 96v^3 - 24v)(v^4 + 1)^{-3} +$
$(6v^8 + 20v^6 - 24v^4 - 12v^2 + 2)(-3)(v^4 + 1)^{-4}(4v^3)$ **15. a.** 194 cm/sec **b.** 70 cm/sec$^2$
**17. a.** 29.75 cm/sec **b.** 10.25 cm/sec$^2$ **19.** $-\frac{1}{4}, \frac{7}{16}, \frac{-21}{128}$ **21.** 3, $-\frac{43}{3}$, 204 **23.** 0, $-1$, 0
**25. a.** $100e^{-.001q} - .1qe^{-.001q}$ **b.** \$30.33 **c.** $(-.2 + .0001q)e^{-.001q}$ **d.** 9¢/radio/radio
**27. a.** $10 \ln(q^2 + 10,000) + 20 - \dfrac{200,000}{q^2 + 10,000} - 10 \ln(10,000)$ **b.** \$10.69 **c.** $\dfrac{20q}{q^2 + 10,000} + \dfrac{400,000q}{(q^2 + 10,000)^2}$
**d.** .2 **29.** $e^{5x}(125x^5 + 375x^4 + 300x^3 + 60x^2)$ **31.** $e^{x^2}(32x^7 + 320x^5 + 760x^3 + 360x)$

## Chapter Review 7.12 (Pages 588–590) Sample Exam Questions

**1. a.** 3, 42, 14 **b.** $-4, -12, 3$ **3.** $15\pi$ cm$^2$/cm **5. a.** 9 **b.** 8 **c.** 7.1 **d.** 7 **e.** $y = 7x - 6$
**7. a.** 13 **b.** 5 **c.** $2x + 7$ **d.** 13, 5 **9. a.** 3 **b.** 2, is not continuous **13.** $\frac{1}{6}$
**15.** $\dfrac{-2xy^3 - 5y^2 - 2x}{10xy - 5e^y + 3x^2y^2}$ **17. a.** 42 **b.** $-\frac{1}{108}$ **c.** $\frac{110}{343}$ **d.** 0 **e.** $-\dfrac{1}{e^2}$ **f.** 1024 **g.** 0
**19.** $-\frac{11}{20}, \frac{91}{400}$, .08865 **21.** $10x(x^2 + 1)^{-2/3}$

## Exercise Set 8.1 (Pages 596–597) Marginal Analysis

**1. a.** $-4q^2 + 2400q - 14,000 - 13,428\sqrt{q}$ **b.** $\dfrac{6714}{\sqrt{q}}$ **c.** $-8q + 2400$ **d.** $2400 - 8q - \dfrac{6714}{\sqrt{q}}$ **e.** \$559.50
**f.** \$1248 **g.** \$688.50
**3. a.** $100qe^{-.02q} - (-.25q^2 + 25q + 600)$ **b.** $-.5q + 25$ **c.** $100e^{-.02q} - 2qe^{-.02q}$
**d.** $100e^{-.02q} - 2qe^{-.02q} - (-.5q + 25)$ **e.** \$5 **f.** $20e^{-.8}$ **g.** \$3.99
**5. a.** $-.0125q^2 + 62.5q - 5000 - 1000\sqrt{q}$ **b.** $\dfrac{500}{\sqrt{q}}$ **c.** $-.025q + 62.5$ **d.** $-.025q + 62.5 - \dfrac{500}{\sqrt{q}}$ **e.** \$12.50
**f.** \$22.50 **g.** \$10.00
**7. a.** $-.05q^2 + 120q - 60,000e^{-.001q}$ **b.** $30 - 60e^{-.001q}$ **c.** $-.1q + 150$ **d.** $-.1q + 120 + 60e^{-.001q}$
**e.** $-\$14.45$ **f.** \$120 **g.** \$134.45 **9. a.** $\dfrac{1}{q}\left[\dfrac{-6714}{\sqrt{q}} - \dfrac{14,000}{q}\right]$ **b.** $-2.72$
**11. a.** $-.25 - \dfrac{600}{q^2}$ **b.** $-.7398$ **13. a.** $-\dfrac{500}{q\sqrt{q}} - \dfrac{5000}{q^2}$ **b.** $-.00684$
**15. a.** $\dfrac{e^{-.001q}}{q}\left(-60 - \dfrac{60,000}{q}\right)$ **b.** $-.0882$
**17. a.** $-.001q^3 + 640q$ **b.** $-.003q^2 + 640$ **c.** \$120
**19. a.** $p = -\dfrac{q}{8} + 100$ **b.** $-\dfrac{q^2}{8} + 100q$ **c.** $-\dfrac{q}{4} + 100$ **d.** \$12.50 **21.** $y - 4 = -3(x + 2)$
**23.** $y - 5 = \frac{15}{8}(x - 4)$ **25.** $y = 2$ **27.** 27 ft/s, 10 ft/s$^2$ **29.** $54e^2$ ft/s, $90e^2$ ft/s$^2$

## Exercise Set 8.2 (Pages 601–602) Elasticity

**1.** $-2.571$ **3.** $-1.976$ **5.** $-.55\overline{5}$ **7.** $-.1736$ **9.** 5.142% **11.** $-5.93\%$ **13.** 3.6% **15.** 3.31%
**17. a.** inelastic **b.** inelastic **c.** elastic **19. a.** inelastic **b.** inelastic **c.** elastic
**21. a.** inelastic **b.** elastic **c.** elastic **23. a.** inelastic **b.** inelastic **c.** inelastic
**25.** The demand is elastic if $6 < p \le 9$, inelastic if $3 \le p \le 6$, and has unit elasticity if $p = 6$. **29.** Not necessarily

## Exercise Set 8.3 (Pages 610–612) Optimization

**1. a.** $(-2, 2)$ and $(5, 7)$ **b.** $(-4, -2)$ and $(2, 5)$ **c.** $(-2, 1), (2, 5)$, and $(5, -1)$
**d.** Local maxima occur when $x = -4, 2$, and 7. **e.** Local minima occur when $x = -2, 5$. **f.** 5 **g.** $-1$
**3. a.** $(-4, -2)$ and $(1, 3)$ **b.** $(-5, -4), (-2, 1)$, and $(3, 5)$ **c.** $(-4, -3), (-2, 2), (1, -1)$, and $(3, 2)$
**d.** Local maxima occur when $x = -5, x = -2$, and $x = 3$. **e.** Local minima occur when $x = -4, x = 1$, and $x = 5$.
**f.** 2 **g.** $-3$

**5. a.** $(3, +\infty)$   **b.** $(-\infty, 3)$   **7. a.** $(0, +\infty)$   **b.** $(-\infty, 0)$   **9. a.** $(-\infty, -1)$ and $(2, +\infty)$   **b.** $(-1, 2)$
**c., d.**

**c., d.**

**c., d.**

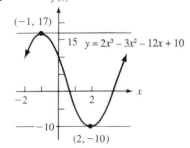

**11. a.** $(-\infty, -4)$ and $(3, +\infty)$   **b.** $(-4, 3)$     **13. a.** $(2, +\infty)$   **b.** $(-\infty, 2)$
**c., d.**

**c., d.**

**15. a.** $(0, +\infty)$   **b.** $(-\infty, 0)$   **c.** $0$   **d.** none   **17. a.** $(-4, +\infty)$   **b.** $(-\infty, -4)$   **c.** $-4$   **d.** none
**19. a.** $\left(-\infty, -\frac{1}{e} - 1\right)$ and $\left(\frac{1}{e} - 1, +\infty\right)$   **b.** $\left(-\frac{1}{e} - 1, -1\right)$ and $\left(-1, \frac{1}{e} - 1\right)$   **c.** $-\frac{1}{e} - 1, \frac{1}{e} - 1$   **d.** none
**21. a.** $(-\infty, 0)$ and $(0, +\infty)$   **b.** none   **c.** none   **d.** $0$
**23. a.** none   **b.** $\left(-\infty, \frac{4}{3}\right)$ and $\left(\frac{4}{3}, +\infty\right)$   **c.** none   **d.** none   **25.** $(-3, 2)$ and $(5, 6)$   **27.** $-3, -2,$ or $5$
**29.** $(-5, -4)$ and $(-1, 4)$   **31. a.** $(5, 6)$ and $(0, 1)$   **b.** $(1, 5)$   **c.** $(0, 1)$
**33. a.** $-q^3 + 48q$   **b.** $0 < q < 4$   **c.** $4 < q < 6$
**35. b.** $(10q - q^2)e^{-.1q}$   **c.** $R$ is decreasing on $(15 - 5\sqrt{5}, 10)$ and is increasing on $(0, 15 - 5\sqrt{5})$.
**37. a.** $f'(x) = 7x^6 + 10x^4 + 9x^2 + 5 \geq 0$ since each of the first 3 terms is nonnegative and 5 is positive.   **b.** Since $f'(x) > 0$
for all $x$, $f$ is strictly increasing. Thus it is one-to-one.

## Exercise Set 8.4 (Page 615) More on Optimization

**1. a.** $5$
**b.**

**3. a.** $\frac{25}{4}$
**b.**

**5. a.** $6$
**b.**

**7. a.** $\sqrt{\frac{343}{27}}$

**b.**

**9. a.** $-1, -3$

**b.**

**13. a.** 13 ft/s

## Exercise Set 8.5 (Page 625) The First Derivative Test

**1.** $-\frac{5}{2}$.    **3.** 0    **5.** none    **7.** $-6$ and 3    **9.** 0 and $-\frac{2}{3}$    **11.** $-1$    **13.** $\frac{1}{e} - 1$ and $-\frac{1}{e} - 1$

**15.** local maximum at $x = 3$    **17.** local minimum at $x = 2$, local maxima at $x = -4$ and $x = 4$

**19.** local maximum at $x = -4$, local minima at $x = -5$ and $x = 1$    **21.** local minimum at $x = -\frac{1}{3}$

**23.** local minimum at $x = 0$    **25.** local minimum when $x = 3$, local maximum at $x = -1$

**27.** local maximum occurs at $x = -1$, local minimum occurs at $x = 1$

**29.** absolute minimum of 1, absolute maximum of 16

**31.** absolute maximum is $g(-4) = 478$, absolute minimum is $g(3) = -159$

**33.** absolute minimum of $h(0) = 0$, no absolute maximum

**35.** The absolute minimum is $f(-1) = -e^{-1/4}$ and the absolute maximum is $f(4) \doteq 1.47$.

**37.** The absolute maximum is $f(3) = 29$ and the absolute minimum is $f(1) = -15$.

## Exercise Set 8.6 (Pages 630–631) The Second Derivative Test

**1.** local minimum at $x = \frac{-5}{2}$    **3.** local minimum when $x = 0$    **5.** none

**7.** local minimum at $x = 3$ and a local maximum at $x = -6$

**9.** local maximum when $x = \frac{-2}{3}$ and local minimum when $x = 0$    **11.** local minimum when $x = -1$

**13.** local minimum when $x = -1 + \frac{1}{e}$ and a local maximum when $x = -1 - \frac{1}{e}$    **15.** local maximum when $x = 3$

**17.** local minimum at $x = 2$    **19.** local maximum when $x = -4$    **21.** local minimum when $t = \frac{-1}{3}$

**23.** local minimum when $x = 0$    **25.** local maximum when $t = -1$ and a local minimum when $t = 3$

**27.** local minimum when $x = 1$, local maximum when $x = -1$    **29.** The Second Derivative Test cannot be used.

**31.** local minimum when $x = 3$    **33.** local minimum when $x = 0$ and a local maximum when $x = -\frac{2}{3}$

**35.** local maximum when $x = 4$    **37.** local minimum at $x = 1$ and local maximum at $x = -1$

## Exercise Set 8.7 (Pages 640–644) Applications

**1.** 100 feet by 100 feet    **3.** $\sqrt[3]{32}$ feet by $\sqrt[3]{32}$ feet by $4^{4/3}$ feet

**5.** The expensive side is 30 feet, the opposite side is 30 feet, and the other 2 sides are 40 feet.

**7.** The bottom is 21 ft by 12 ft, the front and back are 21 ft by 7 ft, and the sides are 12 ft by 7 ft.

**9.** The ends are 10 feet and the front is 32 feet.    **11.** 30 ft by 42 ft with the fence on the 42 ft side

**13.** in 3 weeks, $87.48    **15.** $9    **17.** 12 miles down the shore, $15,400

**19.** 200 acres in Grain A and 300 acres in Grain B    **21.** 50    **23.** 55 mph    **25.** 50 mph

**29. a.** $\left(\frac{3}{2}, 5\right)$    **b.** 1    **c.** 4, 2    **d.** 4    **33.** 5348    **35.** 72    **37.** 10,491    **39.** 20    **41.** 7

**43.** 18 miles west of the point to which he is going

**45. a.**          $d \doteq 3.59$ feet and $x \doteq 3.59$ feet with $x$ and $d$ as shown    **b.** $d \doteq 3.35$ feet and $x \doteq 4.0$ feet

$x$ ft.

$d$ ft.

## Exercise Set 8.8 (Page 653) Curve Sketching

**1.** $f(x) = x^2 + 6x - 2$

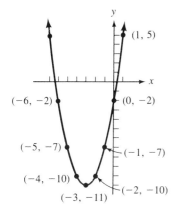

**3.** $y = h(x) = 2x^2 - 12x + 5$

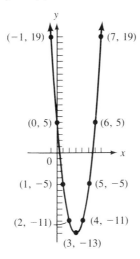

**5.** $g(x) = x^3 + 3x^2 - 24x + 5$

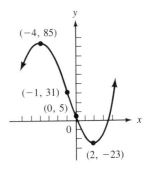

**7.** $f(x) = -x^3 - 3x^2 + 9x + 6$

**9.** $h(x) = x^4 - 4x + 3$

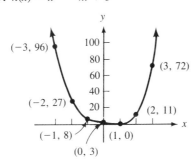

**11. a.** $\frac{1}{3}$   **b.** $\frac{1}{3}$
**13. a.** $-1$   **b.** $-1$
**15. a.** $2$   **b.** $2$
**17. a.** $0$   **b.** $0$
**19. a.** $+\infty$   **b.** $-\infty$
**21. a.** $-\infty$   **b.** $+\infty$
**23.** $+\infty$
**25. a.** $+\infty, -\infty$   **b.** $-\infty, +\infty$
**27. a.** $+\infty, -\infty$   **b.** $+\infty, -\infty$

**29.** $g(x) = x^4 - 8x^3 + 24x^2 - 32x + 16 = (x - 2)^4$

**31.** none
**33.** $(-2, f(-2))$ and $(4, f(4))$
**35.** $(0, 0)$
**37. d.**

$f(x) = x^4 + x$

**Exercise Set 8.9 (Pages 663–664) More on Curve Sketching**

**1.** $y = f(x) = \frac{x-3}{x+2}$

**3.**

**5.**

**7.**

**9.**

**11.**

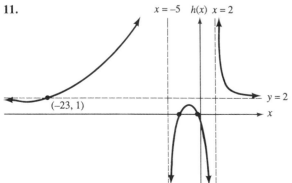

**13.** $y = f(x) = xe^{-3x}$

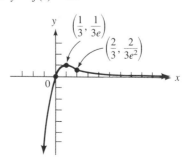

**15.** $y = (x + 1)\ln|x + 1|$

**19. a.** $-.000074996$     **b.** $-224.99$

**21.**

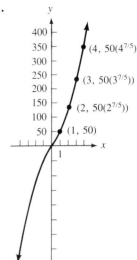

## Exercise Set 8.10 (Pages 668–669) Differentials

**1. a.** 2.84     **b.** 2.8     **3. a.** 2.901     **b.** 2.8     **5. a.** .01     **b.** .010003̄
**7. a.** $-.027028672$     **b.** $-.026̄$     **9.** 8.01562     **11.** 3.004938272
**13.** 940,000     **15. a.** .03 and 3%     **b.** .06 and 6%     **17.** .03, 3%
**19.** .28 in$^3$     **21.** .0143, 1.43%     **23.** .0064, .64%

## Exercise Set 8.11 (Pages 674–675) Relative Rates

**1.** 75 mph     **3.** 6.5 ft/s     **5.** $-.0597$ in./s     **7.** 64 mph     **9.** $\frac{5}{3}$ mph     **11.** 1.54 mph     **13.** $-.0005968$ ft/min
**15.** .172 ft/min     **17.** 264 units/second     **19.** \$450/month
**21. a.** $-\$4476$/month     **b.** $-\$6000$/month     **c.** $-\$1524$/month
**23. a.** \$30/day     **b.** $-\$3.718$/day     **c.** $-\$33.718$/day     **25.** $-\$100,000$/year

## Chapter Review 8.12 (Pages 677–680) Sample Exam Questions

**1. a.** $30/\sqrt{q + 50}$     **b.** \$2
**3. a.** $-1.2q + 1200$     **b.** $-1.2q^2 + 1200q$     **c.** $-.2q^2 + 220q - 20,000$     **d.** 100 units     **e.** $-.4q + 220$     **f.** \$100
  **g.** 550 units     **h.** $-q + 980 + \dfrac{20,000}{q}$     **i.** $-1 - \dfrac{20,000}{q^2}$     **5. a.** inelastic     **b.** elastic
**7. a.** $(-1.5, \infty)$, $(-\infty, -1.5)$, $(-1.5, 116.5)$
  **b.** The graph is concave upward on $(-\infty, -2)$ and $(1, +\infty)$. It is concave downward on $(-2, 1)$. The points of inflection are
  $(-2, 0)$ and $\left(1, \frac{12}{e}\right)$     **c.** $(-\infty, 0)$, $(0, \infty)$, $(0, 0)$     **9.** local minimum when $x = 5$ and a local maximum when $x = -3$
**11.** 50 and $-6$
**13. a.** $(-\infty, -3)$ and $(1, 5)$     **b.** $(-3, 1)$ and $(5, +\infty)$     **c.** $(-\infty, -1)$ and $(3, +\infty)$     **d.** $(-1, 3)$     **e.** $-3, 1,$ and 5
  **f.** $-1$ and 3     **15. a.** $(2x + 5)\, dx$     **b.** $e^{3t}t(2 + 3t)\, dt$     **c.** $\dfrac{4x^3(1 - \ln(x^4 + 3))}{(x^4 + 3)^2}\, dx$     **17.** 24.96     **19.** 70.9 mph
**21. a.** $\left(\frac{1}{2}, 2\right)$, $\left(-\infty, \frac{1}{2}\right)$ and $(2, \infty)$     **b.** $(-\infty, 1)$ and $(3.5, \infty)$, $(1, 3.5)$     **c.** $\left(1, \frac{4}{e}\right)$ and $(3.5, 44e^{-3.5})$
  **d.** local minimum when $x = \frac{1}{2}$, local maximum when $x = 2$     **e.**

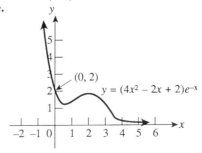

**23.** The line $y = 1$ is a horizontal asymptote to the left and right. The lines $x = 5$ and $x = \frac{13}{3}$ are vertical asymptotes. The $x$-intercepts are 4 and 7. The $y$-intercept is $\frac{84}{65}$.

**25.** The line $y = 4$ is a horizontal asymptote to the left and right. The lines $x = 2$ and $x = -2$ are vertical asymptotes. $-\frac{1}{2}$ is the $y$-intercept and there are no $x$-intercepts.

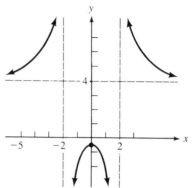

**27.**

**29.** $q^3 - 36q^2 + 432q,\ 0 \le q \le 18,\ 18{,}000$

**31.** $7

## Exercise Set 9.1 (Pages 687–688) Antidifferentiation

**13.** $\frac{1}{2}x^4 + 5x + c$    **15.** $\frac{1}{4}\ln|t^4 + 1| + c$    **17.** $\frac{3}{204}(x^4 + 1)^{51} + c$    **19.** $\frac{7}{4}e^{s^4} + c$    **21.** $\frac{1}{3}\ln|x^3 + 3x + 1| + c$

**23.** $\frac{4}{7}t^{7/2} + 3t^2 + t + c$    **25.** $\frac{3}{20}(x^2 + 1)^{10/3} + c$    **27.** $\frac{5}{4}\ln|x^4 + 1| + c$    **29.** $.5e^{t^2 + 2t + 5} + c$    **31.** $16t^2 + 15t + 20$

**33.** $2t^3 + 2t^2 + t - 10$    **35.** $\frac{1}{3}\ln|t^3 + 3t + 1| + 50$    **37.** $-6t^2 + 3t + 60,\ -2t^3 + 1.5t^2 + 60t - 114.5$

**39.** $\frac{100}{3}e^{.03t} - \frac{85}{3},\ \frac{10{,}000}{9}e^{.03t} - \frac{85}{3}t - \frac{9100}{9}$    **41.** $\frac{.01}{3}q^3 - 3q^2 + 15q + 45{,}000$    **43.** $-1500e^{-.002q} + 36{,}500$

**45.** $g(t) = 5t^{3/2} + 6$

## Exercise Set 9.2 (Pages 694–695) The U-Substitution

**1.** $\frac{1}{33}(x^3 + 5)^{11} + c$    **3.** $\dfrac{-(x^4 + 3)^{-4}}{16} + c$    **5.** $\dfrac{(x^2 + 2x + 5)^{11}}{22} + c$    **7.** $\frac{15}{8}(x^2 + 6x + 1)^{4/3} + c$

**9.** $\frac{7}{5}\ln|5x + 4| + c$    **11.** $\frac{1}{2}\ln|e^{2x} + 1| + c$    **13.** $\frac{2}{65}(e^{5x} + 1)^{13} + c$    **15.** $\frac{1}{6}e^{6x} + c$    **17.** $\frac{1}{3}e^{x^3 + 6x + 10} + c$

**19.** $2e^{\sqrt{x}} + c$    **21.** $\frac{1}{2\ln 2}2^{x^2} + c$    **23.** $\left(\frac{5}{3}x - \frac{5}{9}\right)e^{3x} + c$    **25.** $(t^2 + 1)^{3/2} + 9$    **27.** $\frac{-3}{2}e^{-t^2} + \frac{11}{2}$

**29.** $\frac{2}{\ln 5}5^{t^2 + 1} + 25 - \frac{10}{\ln 5}$    **31.** $\frac{2}{3}\sqrt[3]{q^2 + 6q + 1} + 29998.5$    **33.** $\ln|q^2 + q + 5| + 40{,}000 - \ln 5$    **41.** $\frac{1}{16}e^{4t} + \frac{47}{4}t + \frac{399}{16}$

**43.** $\frac{1}{9}te^{3t} - \frac{2}{27}e^{3t} + \frac{73t}{9} + \frac{1082}{27}$

## Exercise Set 9.3 (Pages 700–701) Integration by Parts

**1.** $\frac{3}{8}xe^{8x} - \frac{3}{64}e^{8x} + c$    **3.** $\frac{1}{2}(5x + 6)e^{2x} - \frac{5}{4}e^{2x} + c$    **5.** $\frac{1}{4}(3x + 1)e^{4x} - \frac{3}{16}e^{4x} + c$    **7.** $\frac{1}{3}x^3e^{x^3} - \frac{1}{3}e^{x^3} + c$

**9.** $x\ln|x^3| - 3x + c$    **11.** $x^2\ln|x + 3| - \dfrac{x^2}{2} + 3x - 9\ln|x + 3| + c$    **13.** $\dfrac{-1}{2x^2}\ln|x| - \dfrac{1}{4x^2} + c$

**15.** $2\sqrt{x}\ln|x| - 4\sqrt{x} + c$    **17.** $\dfrac{x}{\ln 3}(3^x) - \dfrac{1}{(\ln 3)^2} \cdot 3^x + c$

**19.** $\frac{1}{4}x^4\ln|x + 4| - \frac{1}{16}x^4 + \frac{1}{3}x^3 + 2x^2 - 16x + 64\ln|x + 4| + c$    **21.** $\frac{3x}{4}(x + 2)^{4/3} - \frac{9}{28}(x + 2)^{7/3} + c$

**23.** $2x\sqrt{x+4} - \frac{4}{3}(x+4)^{3/2} + c$   **25.** $\frac{x^2}{21}(x+6)^{21} - \frac{x}{21(11)}(x+6)^{22} + \frac{(x+6)^{23}}{21(11)(23)} + c$

**27.** $\frac{1}{10}x^2e^{10x} - \frac{1}{50}xe^{10x} + \frac{1}{500}e^{10x} + c$   **29.** $\frac{-1}{2}x^4e^{-2x} - x^3e^{-2x} - \frac{3}{2}x^2e^{-2x} - \frac{3}{2}xe^{-2x} - \frac{3}{4}e^{-2x} + c$

**31.** $10e^{-.1q}(q^2 + 20q - 9800) + 113{,}000$

**33. a.** $-\frac{1}{2}te^{-2t} - \frac{1}{4}e^{-2t} + 25.25$   **b.** $\frac{1}{4}te^{-2t} + \frac{1}{4}e^{-2t} + 25.25t - 101.25 - 1.5e^{-10}$

## Exercise Set 9.4 (Pages 703–704) Integration Tables

**1.** $\ln|x + \sqrt{x^2 - 25}| + C$   **3.** $\frac{1}{2}[x\sqrt{x^2 + 16} + 16\ln|x + \sqrt{x^2 + 16}|] + C$

**5.** $\frac{1}{2}[e^x\sqrt{e^{2x} + 16} + 16\ln|e^x + \sqrt{e^{2x} + 16}|] + C$   **7.** $\frac{2}{375}(15x - 16)(5x + 8)^{3/2} + C$   **9.** $\frac{1}{14}\ln\left|\frac{7+x}{7-x}\right| + C$

**11.** $\frac{x}{5} - \frac{1}{5}\ln(5 + 3e^x) + C$   **13.** $\frac{1}{10}x^2e^{10x} - \frac{1}{500}(10x - 1)e^{10x} + C$   **15.** $\frac{\ln^6|x|}{6} + C$   **17.** $\frac{3}{20}\ln\left|\frac{2x+5}{2x-5}\right| + C$

**19.** $\frac{1}{2}x^4e^{x^2} - (x^2 - 1)e^{x^2} + C$   **21.** $\frac{x}{32}(8x^2 - 25)\sqrt{4x^2 - 25} - \frac{625}{64}\ln|2x + \sqrt{4x^2 - 25}| + C$   **23.** $\frac{-e^x}{25\sqrt{e^{2x} - 25}} + C$

**25.** $\frac{1}{567}(9x^2 + 64)^{7/2} - \frac{64}{405}(9x^2 + 64)^{5/2} + C$   **27.** $\frac{1}{3125}(25x^2 - 121)^{5/2} + \frac{121}{1875}(25x^2 - 121)^{3/2} + C$

**29.** $\left(\frac{4}{5}x^4 - \frac{128}{15}\right)(x^4 + 16)^{3/2} + C$

**31.** $\frac{x^2 + 2}{8}(x^4 + 4x^2 + 13)^{3/2} + \frac{27}{16}(x^2 + 2)\sqrt{x^4 + 4x^2 + 13} + \frac{243}{16}\ln|x^2 + 2 + \sqrt{x^4 + 4x^2 + 13}| + C$

**33.** $\frac{1}{13}\ln\left|\frac{x+4}{9-x}\right| + C$   **35.** $\frac{-\sqrt{x^4 + 6x^2 - 16}}{2(x^2 + 3)} + \frac{1}{2}\ln|x^2 + 3 + \sqrt{x^4 + 6x^2 - 16}| + C$

## Exercise Set 9.5 (Page 709) Guessing Again

**1.** $\frac{3}{52}(x^4 + 3)^{26} + C$   **3.** $\frac{9}{20}(x^5 + 4)^{4/3} + C$   **5.** $2(x^3 + 6x + 1)^{1/3} + C$   **7.** $\frac{2}{51}(e^{2x} + 1)^{51} + C$

**9.** $\frac{x}{3}e^{3x} - \frac{x}{9}e^{3x} + C$   **11.** $\left(-\frac{1}{3}x^3 - \frac{1}{3}x^2 - \frac{2}{9}x - \frac{2}{27}\right)e^{-3x} + K$   **13.** $\left(\frac{x^2}{\ln 5} - \frac{2}{(\ln 5)^2}x + \frac{2}{(\ln 5)^3}\right)5^x + K$

**15.** $(x^2 + 5x + 1)e^{4x} + K$   **17.** $(x^3 + 2x^2 + 5x + 1)e^{3x} + K$   **19.** $\left(\frac{1}{9}x^4 - \frac{125}{63}x^2 - \frac{1250}{63}\right)(25 - x^2)^{5/2} + K$

**21.** $\left(\frac{2}{15}x^6 - \frac{32}{55}x^4 - \frac{4096}{385}x^2 - \frac{262{,}144}{1155}\right)(16 - x^2)^{3/4} + K$   **23.** $\left(\frac{1}{9}x^4 - \frac{107}{63}x^2 - \frac{80}{63}\right)(16 - x^2)^{5/2} + K$

**25.** $(4x^4 - 67x^2 + 48)(16 - x^2)^{5/2} + K$   **27.** $(2x^4 + 3x^2 - 44)(4 - x^2)^{1/2} + K$

## Exercise Set 9.6 (Page 713) Tabular Integration

**1.** $\frac{e^{5x}}{625}(125x^3 - 75x^2 + 30x - 6) + C$   **3.** $\frac{e^{6x}}{46{,}656}(7776x^5 - 6480x^4 + 4320x^3 - 2160x^2 + 720x - 120) + C$

**5.** $\frac{e^{-3x}}{81}(-27x^3 + 162x^2 + 27x - 72) + C$   **7.** $\frac{e^{2x}}{32}(32x^4 + 16x^3 - 136x^2 + 232x - 324) + C$

**9.** $\frac{3^{4x}}{16{,}384\ln^73}[(4096\ln^63)x^6 - (6144\ln^53)x^5 + (7680\ln^43)x^4 - (7680\ln^33)x^3 + (5760\ln^23)x^2 - (2880\ln 3)x + 720 +$
$8192\ln^63] + C$   **11.** $x\ln^5|x| - 5x\ln^4|x| + 20x\ln^3|x| - 60x\ln^2|x| + 120x\ln|x| - 120x + C$

**13.** $\frac{(x - 3)(x + 3)}{2}\ln^2|x + 3| - \frac{1}{2}\ln|x + 3|(x^2 - 6x - 27) + \frac{x^2}{4} - \frac{9x}{2} + C$

**15.** $\frac{1}{4}(x^4 - 81)\ln^2|x + 3| - \frac{1}{8}(x + 3)(x^3 - 7x^2 + 39x - 225)\ln|x + 3| + \frac{1}{8}\left(\frac{x^4}{4} - \frac{7}{3}x^3 + \frac{39}{2}x^2 - 225x\right) + C$

**17.** $\frac{1}{6}(x + 1)(2x^2 - 11x + 17)\ln^2|x + 1| - \frac{1}{18}(x + 1)(4x^2 - 37x + 139)\ln|x + 1| + \frac{1}{18}\left(\frac{4}{3}x^3 - \frac{37}{2}x^2 + 139x\right) + C$

**19.** $-\frac{x^2}{5}(25 - x^2)^{5/2} - \frac{2}{35}(25 - x^2)^{7/2} + C$   **21.** $\frac{e^{3x}}{27}(9x^3 + 9x^2 - 51x + 35) + K$

## Exercise Set 9.7 (Pages 720–721) Intuitive Discussion of Limits

**1.** 0   **3.** 2   **5.** 0   **7.** 4   **9.** 1.5   **11.** 0   **13.** $e^5$   **15.** $\sqrt[4]{e}$

**17. a.** 146.5756256   **b.** 148.2278203   **c.** 148.3946092, 148.4131591

**19. a.** 1.283985298   **b.** 1.284021404   **c.** 1.284025015, 1.284025417   **21.** 13.5   **23.** 4   **25.** 1.2   **27.** $\frac{28{,}561}{23{,}000}$

**29.** $\dfrac{7\sqrt{7}}{\sqrt{7}+1}$  **31.** $\dfrac{20}{9}, \dfrac{80}{81}$  **33.** 162, 54, 18  **35.** $\dfrac{324}{99}$  **37.** $\dfrac{992}{165}$  **39.** $\dfrac{24,600,388}{999,000}$  **41.** $\dfrac{183,740}{49,950}$  **43.** $\dfrac{45,880}{3333}$

**45.** 150 feet  **47.** 60 feet  **49.** $166\frac{2}{3}$ inches  **51.** 9 square units  **53.** $4050

## Exercise Set 9.8 (Pages 735–736) The Definite Integral

**1.** $10 \leq A \leq 15$  **3.** $17.5 \leq A \leq 25.5$  **5.** $18.625 \leq A \leq 23.125$

Inscribed rectangles

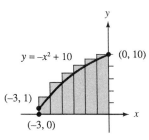

Circumscribed rectangles

**7. a.** $5.63\overline{8}$, 10.36  **b.** 7, 9  **c.** $\dfrac{8(n-1)}{n}$, $\dfrac{8(n+1)}{n}$

**9. b.** 11.666, 32.551  **c.** 14.0625, 27.5625  **d.** $\dfrac{81(n-1)^2}{4n^2}$, $\dfrac{81(n+1)^2}{4n^2}$  **f.** $\dfrac{81}{4}$  **g.** $\dfrac{81}{4}$

**11. b.** 5.938, 10.802  **c.** 7.0625, 9.0625

**d.** $4 + 16\left(1 - \dfrac{3(n+1)}{2n} + \dfrac{(n+1)(2n+1)}{2n^2} - \dfrac{(n+1)^2}{4n^2}\right)$, $4 + 16\left(1 - \dfrac{3}{2}\left(1 - \dfrac{1}{n}\right) + \dfrac{1}{2}\left(1 - \dfrac{1}{n}\right)\left(2 - \dfrac{1}{n}\right) - \dfrac{\left(1 - \frac{1}{n}\right)^2}{4}\right)$  **f.** 8

**g.** 8  **13. a.** $\left(0, \dfrac{5}{n}, 2 \cdot \dfrac{5}{n}, 3 \cdot \dfrac{5}{n}, \ldots, n \cdot \dfrac{5}{n} = 5\right)$  **b.** $\dfrac{175(n+1)}{2n}$  **c.** 87.5  **d.** 87.5

**15. a.** $\left(0, \dfrac{3}{n}, 2\left(\dfrac{3}{n}\right), 3\left(\dfrac{3}{n}\right), \ldots, n\left(\dfrac{3}{n}\right) = 3\right)$  **b.** $\dfrac{9(n+1)(2n+1)}{2n^2}$  **c.** 9  **d.** 9

**17. a.** $\left(0, \dfrac{9}{n}, 2\left(\dfrac{9}{n}\right), 3\left(\dfrac{9}{n}\right), \ldots, n\left(\dfrac{9}{n}\right) = 9\right)$  **b.** $\dfrac{243(n+1)(2n+1)}{2n^2} + 9$  **c.** 252  **d.** 252

**19. a.** $\left(0, \dfrac{6}{n}, 2\left(\dfrac{6}{n}\right), 3\left(\dfrac{6}{n}\right), \ldots, n\left(\dfrac{6}{n}\right) = 6\right)$  **b.** $\dfrac{324(n+1)^2}{n^2}$  **c.** 324  **d.** 324

**21. a.** $\left(0, \dfrac{8}{n}, 2\left(\dfrac{8}{n}\right), 3\left(\dfrac{8}{n}\right), \ldots, n\left(\dfrac{8}{n}\right) = 8\right)$  **b.** $8\left[128\left(1 + \dfrac{1}{n}\right)^2 + \dfrac{64}{3}\left(1 + \dfrac{1}{n}\right)\left(2 + \dfrac{1}{n}\right) - 20\left(1 + \dfrac{1}{n}\right) + 3\right]$  **c.** $\dfrac{3688}{3}$  **d.** $\dfrac{3688}{3}$

**23. a.** $\left(1, 1 + \dfrac{3}{n}, 1 + 2\left(\dfrac{3}{n}\right), \ldots, 1 + n\left(\dfrac{3}{n}\right) = 4\right)$  **b.** $\dfrac{3}{n}\left(n + 3(n+1) + \dfrac{3}{2}\dfrac{(n+1)(2n+1)}{n}\right)$  **c.** 21  **d.** 21

**25. a.** $\left(-1, -1, + \dfrac{4}{n}, -1 + 2\left(\dfrac{4}{n}\right), \ldots, -1 + n\left(\dfrac{4}{n}\right) = 3\right)$  **b.** $\dfrac{4}{n}\left(-n + 2 + \dfrac{8}{3}\dfrac{(n+1)(2n+1)}{n}\right)$  **c.** $\dfrac{52}{3}$  **d.** $\dfrac{52}{3}$

## Exercise Set 9.9 (Page 743) The Definite Integral: Another Approach

**5. a.** $\dfrac{55}{8} \leq d \leq \dfrac{91}{8}$  **b.** $\dfrac{9(n-1)(2n-1)}{2n^2} \leq d \leq \dfrac{9(n+1)(2n+1)}{2n^2}$  **c.** 9 ft

**7. a.** $64.125 \le d \le 86.625$    **b.** $21 + 27\left(1 - \frac{1}{n}\right) + \frac{27}{2}\left(1 - \frac{1}{n}\right)\left(2 - \frac{1}{n}\right) \le d \le 21 + 27\left(1 + \frac{1}{n}\right) + \dfrac{27\left(2 + \frac{1}{n}\right)\left(2 + \frac{1}{n}\right)}{2}$

**c.** 75 feet

**9. a.** $66.125 \le d \le 81.125$    **b.** $33 + \frac{63}{2}\left(1 - \frac{1}{n}\right) + \frac{9}{2}\left(1 - \frac{1}{n}\right)\left(2 - \frac{1}{n}\right) \le d \le 33 + \frac{63}{2}\left(1 + \frac{1}{n}\right) + \frac{9}{2}\left(1 + \frac{1}{n}\right)\left(2 + \frac{1}{n}\right)$

**11. b.** \$100 per unit, \$75 per unit    **c.** \$4166.67    **13. b.** \$400, \$175    **c.** \$48,750

**15. b.** \$210.40, \$5.60    **c.** \$17,280

## Exercise Set 9.10 (Pages 752–753) The Fundamental Theorem of Calculus

**1.** 31    **3.** $\frac{14}{3}$    **5.** $\dfrac{e^6 - 1}{3}$    **7.** $6 \ln 5$    **9.** $\frac{21}{4}$    **11.** $\frac{2}{9}(21^{3/2} - 1)$    **13.** $\frac{1}{3}(e^8 - e^{-1})$

**15.** $\frac{-3}{2}(e^4 + 1)^{-1} + \frac{3}{2}(e^{-2} + 1)^{-1}$    **17.** $\frac{1}{5}\ln(e^{10} + 1) - \frac{1}{5}\ln 2$    **19.** $\dfrac{33e^{12} - 9e^4}{16}$    **21.** $e^6\left(\frac{35}{27}\right) - \frac{11}{27}$

**23.** $7 \ln 7 - 3 \ln 3 - 4$    **25.** 84.4 meters    **27.** 340 meters    **29.** $\dfrac{1 - e^{-9}}{3}$ meters    **31.** \$308,000

**33.** \$119,250.79    **35.** \$76,246.19    **37.** $\dfrac{\ln 8}{e^{343}} - \dfrac{\ln 3}{e^8}$    **39.** 0    **41.** $\frac{84}{5}$    **43.** 0    **45.** 96    **47.** \$37,599

**49.** \$18,211

## Exercise Set 9.11 (Pages 760–762) Approximate Integration

**1. a.** 20.5    **b.** 2.5%    **3. a.** 20.125    **b.** .625%    **5. a.** .92063492    **b.** .47%

**7. a.** .916298378    **b.** .00083%    **9. a.** 1.128183024    **b.** −2.15%    **11. a.** 1.145634045    **b.** −.0063

**13. a.** 27.35507653    **b.** 2.07%    **15.** 1.2487    **17.** 6.8894618    **19.** 9.04529    **21.** 5.65    **23.** $10.2\overline{6}$

**25.** \$33,500    **27.** \$24,750    **29.** 34.1 meters    **31.** 10.2 meters

## Chapter Review 9.12 (Pages 766–770) Sample Exam Questions

**3. a.** $16t^2 + 10t + 15$    **b.** $-6t^2 + 4t + 68$    **c.** $3t^3 + 3t^2 - 4t + 58$    **d.** $1500e^{.002t} + 2t^3 + 2t - 1490$

**5. a.** $.001q^3 - 3q^2 + 1200q + 50,000$    **b.** $C(q) = 50\sqrt{q} + 75,000$    **c.** $5 \ln(q^2 + 3) + 62,000 - 5 \ln 3$

**7. a.** $\frac{5x}{12}e^{12x} - \frac{5}{144}e^{12x} + C$    **b.** $\frac{4}{13}x^2e^{13x} - \frac{8}{169}xe^{13x} + \frac{8}{169(13)}e^{13x} + C$    **c.** $x \ln|x + 12| - x + 12 \ln|x + 12| + C$

**d.** $\frac{5}{2}x^2 \ln|x + 6| - \frac{5}{2}\left(\dfrac{x^2}{2} - 6x + 36 \ln|x + 6|\right) + C$

**9. a.** $-\frac{3}{20}\ln\left(\dfrac{5 + \sqrt{25 + 9x^2}}{3x}\right) + C$    **b.** $\frac{1}{15}\ln\left|\dfrac{5 + 3x}{5 - 3x}\right| + C$    **c.** $\dfrac{x^3e^{7x}}{7} - \dfrac{3x^2e^{7x}}{49} + \dfrac{6xe^{7x}}{343} - \dfrac{6}{343(7)}e^{7x} + C$

**d.** $(x + 2)^3\left(\frac{\ln|x + 2|}{3} - \frac{1}{9}\right) + C$    **11.** $\frac{217}{16}$

**13. a.**

$y = x^2 + 3$

**b.** 13.12    **c.** 16.32    **d.** $\frac{46}{3}$    **15. b.** $\frac{625}{4}\left(1 - \frac{1}{n}\right)^2$    **c.** $\dfrac{625(n + 1)^2}{4n^2}$    **d.** $\frac{625}{4}$

**17. a.** \$224,250    **b.** \$35,801.33

**19.** 2142 feet

**21.** $\displaystyle\int_1^6 x^2\, dx \doteq 72.5$, $\displaystyle\int_1^6 x^2\, dx = 72.\overline{6}$, The percentage error is 1.16%.

**23.** .4659

**25. a.** 1.6288946    **b.** 1.6466685

**c.** 1.6485153 $\sqrt{e} \doteq 1.6487213$ $\displaystyle\lim_{n \to \infty}\left(1 + \frac{1}{2n}\right)^n = \sqrt{e}$

**27.** $\frac{23}{99}$

**29.** $5, \frac{5}{4}, \frac{5}{16}$

**Exercise Set 10.1 (Pages 779–780) Differential Equations**

**7.** $\frac{2}{3}x^{3/2} + 4x + C$     **9.** $\frac{1}{3}e^{3x} + C$     **11.** $y = Ce^{5x}$     **13.** $y = 3e^{2x^2}$     **15.** $x = te^{(-t^2+3)/2}$     **17.** $y = x - \ln(x + 1)$

**19.** $3\ln(1 + y^2) = 2\sqrt{x^2 + 1} - 4$     **21.** \$44,510.82     **23.** 5.22 years     **25.** $V = \$94,249$     **27.** 1,770,000

**29.** $f(p) = 3p^4 - 200p^3 + 6p^2 - 600p + 6,265,000$     **31.** $f(p) = 50,000\sqrt[4]{4096 - p}$     **33.** $P(t) = \dfrac{600,000}{5\left(\frac{5}{3}\right)^{-t/6} + 1}$

**35.** 6000 years     **37.** $y = e^{2x}, y = e^x$     **39.** $y = e^{-4x}, y = e^x$     **41.** $y = e^{-x/2}, y = e^{3x}$

**43.** $y = e^{-2x}, y = e^x$ and $y = e^{-x}$     **45.** $y^2 = -\ln\left(C_1 - \frac{2}{3}e^{x^3}\right)$     **47.** $y^2 + 1 = \dfrac{1}{-2\ln\left|\ln|x + 1|\right| + C_1}$

**Exercise Set 10.2 (Pages 789–790) Area**

**1.** 28

**3.** $\frac{80}{3}$

**5.** 12

**7.** $\frac{73}{6}$

**9.** $2e^3 - 2$

**11.** $9 + \ln 4$

**13.** 4.5

**15.** 36

**17.** 4.5

**19.** $\frac{125}{3}$

**21.** 15

**23.** 108

**25.** 15

**27.** 120

**29.** 2.5

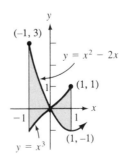

**31.** $\frac{55}{12}$

**33.** $\frac{79}{6} - \frac{1}{2}e^{-4} - e^{-3}$

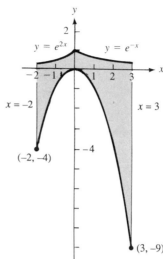

**35.** $\frac{72}{5}$

**37.** $1.5 - \frac{e}{2}$

**39.** $\frac{4}{3} - \frac{e}{3}$

### Exercise Set 10.3 (Pages 795–796) Consumers' and Producers' Surplus

**1. a.** 4,000 units, $6/unit    **b.** $16,000    **c.** $8,000    **3. a.** 6,000 units, $72/unit    **b.** $216,000    **c.** $252,000
**5. a.** 15,000 units, $350/unit    **b.** $2,700,000    **c.** $3,150,000
**7. a.** 23,000 units, $718/unit    **b.** $9,962,833    **c.** $10,227,333
**9. a.** 15,000 units, $5,375/unit    **b.** $97,031,250    **c.** $40,218,750
**11. a.** 9,000 units, $9/unit    **b.** $14,143    **c.** $13,500.00
**13. a.** 27,000 units, $12/unit    **b.** $42,000    **c.** $121,500
**15. a.** 50,000 units, $25/unit    **b.** $263,889    **c.** $388,889
**17. a.** 30,000 units, $50/unit    **b.** $579,442    **c.** $450,000
**19. a.** 10,000 units, $54,598.15/unit    **b.** $392,169    **c.** $309,936    **21. a.** 30,000, $90    **b.** $900,000
**23. a.** 30,000, $33    **b.** $405,000    **25. a.** 5,000, $33    **b.** $41,666.67

### Exercise Set 10.4 (Pages 804–806) More Applications

**1. a.**    **b.** .008    **c.** $\frac{1}{2}$    **3. a.**    **b.** .133    **c.** $.0\overline{4}$

**5. a.**  **b.** .6695 **c.** .02857 **7. a.**  **b.** .4 **c.** .2274

**9. a.** $.4I + .4(I + 1)^{1/2} + 4.32 - .4(11)^{1/2}$ **b.** \$19.82 billion **11. a.** $.5I - e^{-.2I} + 2.9 - e^{-2}$ **b.** \$12.78 billion
**13. a.** $.7I + e^{-.4I} + 2 - e^{-4}$ **b.** \$16.02 billion **15.** \$108,285.21 **17.** \$62,811.12 **19.** \$454,641.02
**21.** \$242, 508.23 **23.** 1120.98 hours **25.** $11.\overline{1}, -.152$ **27.** $100, -.515$ **29. a.** 36 **b.** \$30,954,000
**31. a.** 200 months **b.** \$1,599,400,000 **33. a.** 125 months **b.** \$120,000,000

## Exercise Set 10.5 (Page 813) Improper Integrals

**1.** divergent **3.** divergent **5.** divergent **7.** $\frac{1}{3}$ **9.** 2 **11.** 1 **13.** divergent **15.** divergent **17.** 1
**19.** $\frac{1}{2}$ **21.** 6 **23.** divergent **25.** divergent **27.** 30 **29.** $2\sqrt{\ln 2}$ **31.** divergent **33.** $\frac{2}{\ln 2}$
**35.** 1 **37.** 6 **39.** $\frac{2}{9}$ **41.** 48

**43.** 1 **45.** \$300,000

## Exercise Set 10.6 (Pages 820–822) More on Probability

**1. a.** $\frac{2}{27}$ **b.** $\frac{117}{54}$ **3. a.** $\frac{3}{92}$ **b.** 2 **5. a.** $\frac{1}{\ln 6}$ **b.** $\frac{1}{\ln 6}(5 - \ln 6)$ **7. a.** 2 **b.** $\frac{1}{2}$ **9. a.** 7 **b.** $\frac{1}{7}$
**11. b.** 14.4 **13.** $\frac{2}{3}$ **15.** $\frac{1}{9}$ **17. b.** 2571.43 minutes **19. b.** 7683.2 hours **21. b.** 441 hours
**23. b.** 3888 hours **25. b.** 137.2 hours **27. b.** 120 months **c.** .0399 **29. b.** 6000 hours **c.** .09
**31.** \$7.59 **33.** \$3.51

## Chapter Review 10.7 (Pages 826–830) Sample Exam Questions

**5. a.** $x = e^{t^4+3}$ **b.** $e^{2y} = 6xe^x - 6e^x + 6 + e^4$ **c.** $(\ln(y + e))^2 = 1 + \ln(x^2 + 1)$
**7. a.** 13.7327 years **b.** 5.068 years **9.** $q = 3(2500 - p^2)^{3/2}, 0 < p < 50$
**11. a.**  **b.** 12 **13. a.**  **b.** 24

**15. a.**

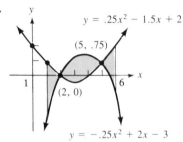

$y = .25x^2 - 1.5x + 2$

$(5, .75)$

$(2, 0)$

$y = -.25x^2 + 2x - 3$

**b.** $\frac{49}{12}$

**17. a.** 24,000 units, \$8/unit   **b.** \$72,000   **c.** \$96,000
**19. a.** 100,000 units, \$30/unit   **b.** \$1,250,000
**c.** \$1,000,000

**21. a.**

$y = x$

$y = \dfrac{3x^2 + 8x}{11}$

**b.** .1989   **c.** $.\overline{09}$

**23. a.**

$y = x$

$y = \dfrac{(1 - e^{-x})x}{1 - e^{-1}}$

**b.** .087   **c.** .254

**25.** \$471,694   **27.** 41.15, $-.152$   **29. a.** 4   **b.** $\dfrac{e^6}{3}$   **c.** divergent   **d.** $\ln(1.2)$   **e.** divergent   **f.** $6\sqrt{3}$

**31. b.** 44.84 weeks   **33. b.** 23 years   **c.** .52

## Exercise Set 11.1 (Pages 837–838) Functions of Several Variables

**1.** the set of all ordered pairs of real numbers, 11, $-5$, $-11$
**3.** the set of all ordered pairs of real numbers except for those of the form $(5, c)$ or $(d, 7)$, 0, $-\frac{1}{14}$, $\frac{9}{10}$
**5.** the set of all ordered pairs of real numbers except those of the form $(-2, c)$ or $(d, d)$, $-2.2$, $-\frac{11}{7}$, $-\frac{1}{25}$
**7.** the set of all ordered pairs of real numbers except those of the form $(x, y)$ where $4 - x^2 < 0$, 16, $10 + \sqrt{3}$, 2
**9.** The domain consists of points inside or on the circle with center $(0, 0)$ and radius 5. $2\sqrt{5}$, 0, $2\sqrt{5}$
**11.** the set of all ordered triples of real numbers, 10, 17, 12
**13.** the set of all ordered triples or real numbers except for those of the form $(x, y, z)$ where $x^2 + y^2 > 9$, $9 + \sqrt{7}$, $15 + \sqrt{5}$, $-7$
**15.** the set of all ordered triples of real numbers except for those of the form $(x, y, z)$ where $z > 4$ or $z < -4$, $28 - 2\sqrt{3}$, 27, $52 - \sqrt{7}$

**17.,19.,21.**

$(-3, 0, 4)$   $(-2, 3, 5)$
$(0, 2, 4)$

**23.,25.**

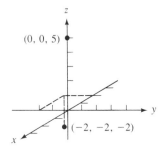

$(0, 0, 5)$

$(-2, -2, -2)$

**27.**

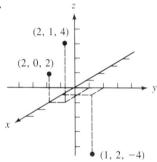

$(2, 1, 4)$
$(2, 0, 2)$

$(1, 2, -4)$

**29.**

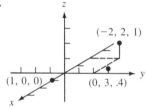

$(-2, 2, 1)$
$(1, 0, 0)$
$(0, 3, .4)$

**31.**

$(1, 2, 3)$
$(1, -3, 2)$

$(0, 6, 0)$

**33.**

$y = 2x^2 + 5$

**35.**

$x^2 - y = -3$

**37.** $4 = x^2 + y^2$
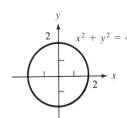
$x^2 + y^2 = 4$

**39.** $16 = x^2 + y^2$
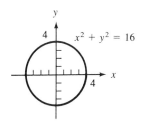
$x^2 + y^2 = 16$

**41.** $R(p_A, p_B) = 300p_A - 10p_A{}^2 + 450p_B - 15p_B{}^2 + 45p_Ap_B$

**43.** $C(x, y) = $ cost of top $+$ cost of base $+$ cost of sides

$xyh = 3000$

$h = \frac{3000}{xy}$

$C(x, y) = 23xy + \frac{72{,}000}{y} + \frac{72{,}000}{x}$

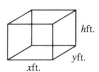

hft.

yft.

xft.

## Exercise Set 11.2 (Pages 845–846) Partial Differentiation

**1.** 18, 6   **3.** 486, 324   **5.** 16, 120   **7.** $-9t^7s^2, -21s^3t^6$   **9.** $-6xy^5 + 12x^2y, -15x^2y^4 + 4x^3$

**11.** $\dfrac{-4.5}{x^4y^2} - \dfrac{5y}{2x^2}, \dfrac{-3}{x^3y^3} + \dfrac{5}{2x}$   **13.** $6xy^5e^{x^2y^5}, 15x^2y^4e^{x^2y^5}$   **15.** $-10uv^6e^{-u^2v^6}, -30u^2v^5e^{-u^2v^6}$

**17.** $-12x^2y^2e^{x^3y} - 12x^5y^3e^{x^3y}, -8x^3ye^{x^3y} - 4x^6y^2e^{x^3y}$   **19.** $\dfrac{(15xy - 10)e^{3xy}}{2x^3y^3}, \dfrac{15e^{3xy}(xy - 1)}{2x^2y^4}$

**21.** $(28x^4y^4\ln 5 + 7y)5^{x^4y^3}, 5^{x^4y^3}(21x^5y^3\ln 5 + 7x)$   **23.** $\dfrac{6x}{3x^2 + 5y^6}, \dfrac{30y^5}{3x^2 + 5y^6}$

**25.** $-2xy^4e^{-x^2y^4}\ln(3x^2 + 5x^2y^4) + \dfrac{e^{-x^2y^4}(30x^9 + 10xy^4)}{3x^{10} + 5x^2y^4}, -4x^2y^3e^{-x^2y^4}\ln(3x^{10} + 5x^2y^4) + \dfrac{e^{-x^2y^4}(20x^2y^3)}{3x^{10} + 5x^2y^4}$

**27.** $\dfrac{2xy^6}{(\ln 3)(x^2y^6 + 6)}, \dfrac{6x^2y^5}{(\ln 3)(x^2y^6 + 6)}$

**29.** $\dfrac{-24x}{(\ln 7)(4x^2 + y^2)(5x^3y^2 + 1)} + 3(5x^3y^2 + 1)^{-2}(15x^2y^2)\log_7(4x^2 + y^2), \dfrac{-6y}{(\ln 7)(4x^2 + y^2)(5x^3y^2 + 1)} +$
$3(5x^3y^2 + 1)^{-2}(10x^3y)\log_7(4x^2 + y^2)$   **31.** $10y^5, 100x^2y^3$

**33.** $(20x^3y^3 + 25x^8y^6\ln 3)(\ln 3)3^{x^5y^3}, (\ln 3)3^{x^5y^3}(6x^5y + 9x^{10}y^4\ln 3)$   **35.** $\dfrac{6(5y^4 - 3x^2)}{(\ln 5)(3x^2 + 5y^4)^2}, \dfrac{20(9x^2y^2 - 5y^6)}{(\ln 5)(3x^2 + 5y^4)^2}$

**37.** $e^{-3xy}(6x^{-3} + 9x^{-2}y + 6x^{-4}y^{-1} + 6x^{-3}), 3x^{-1}y^{-2}e^{-3xy} + 9y^{-1}e^{-3xy} + 2x^{-2}y^{-3}e^{-3xy} + 3x^{-1}y^{-2}e^{-3xy}$

**39.** $\dfrac{2e^8(y^e - e^3x^2)}{(e^3x^2 + y^e)^2\ln 3}, \dfrac{e^6}{\ln 3}((e - 1)y^{e-2}(e^3x^2 + y^e) - y^{e-1}ey^{e-1})\dfrac{1}{(e^3x^2 + y^e)^2}$

**41.** $f_{xy}(x, y) = f_{yx}(x, y) = 175x^4y^6$   **43.** $f_{xy}(x, y) = f_{yx}(x, y) = 90xy^2e^{(3x^2+5y^3)}$   **45.** $f_{xy}(x, y) = f_{yx}(x, y) = 10xe^{5y}$

**47.** $f_{xy}(x, y) = f_{yx}(x, y) = (xy^4 + x^3y + 3x^2 + 4y^3)e^{xy}$

**49.** $f_{xy}(x, y) = f_{yx}(x, y) = \dfrac{-96xy^7 - 300x^2y^8 - 1440y^{16} - 10x^4}{(x^2 + 6y^8)^2e^{5xy}} + \dfrac{(25xy - 5)\ln(x^2 + 6y^8)}{e^{5xy}}$

## Exercise Set 11.3 (Pages 854–855) Chain Rule

**1. a.** $30{,}000 - 430\sqrt{76.2 + .3t} - 350\sqrt[3]{22.2 + 1.2\sqrt{t}}$   **b.** 25,080 units

**3. a.** $500 - 12\sqrt{20 + .2t} - 123\sqrt[4]{14.75 + .002t^2}$   **b.** 194 bottles   **5. a.** $36e^{9t}$   **b.** $36e^{9t}$

**7. a.** $\dfrac{2(\ln 2)2^{2t} + 12t^{11}}{2^{2t} + t^{12}}$   **b.** $\dfrac{2(2^{2t})\ln 2 + 12t^{11}}{2^{2t} + t^{12}}$   **9.** $\dfrac{4x(x^2 + 3y^6) + 20xye^{2x^2y}}{(x^2 + 3y^6)^2 + 5e^{2x^2y}}, 36\dfrac{(x^2 + 3y^6)y^5 + 10x^2e^{2x^2y}}{(x^2 + 3y^6)^2 + 5e^{2x^2y}}$

**11.** $\frac{\partial z}{\partial x} = \frac{10e^{uv}u(2 + uv)x}{(x^2 + y^2)} + 60u^3xy^5e^{uv}$, $\frac{\partial z}{\partial y} = \frac{5ue^{uv}(2 + uv)2y}{x^2 + y^2} + 5u^3e^{uv}(30x^2y^4)$, finally, replace $u$ by $\ln(x^2 + y^2)$ and $v$ by $6x^2y^5$

**13.** $\frac{\partial z}{\partial x} = 5u^2vw^3e^{5u^3v^2w^4}(9x^2vw + 20uwx + 120uvx^5)$, $\frac{\partial z}{\partial y} = u^2vw^3e^{5u^3v^2w^4}(90vw + 120uwy + 240uvy^3)$, now replace $u$ by $x^3 + 6y$, $v$ by $5x^2 + 6y$ and $w$ by $5x^6 + 3y^4$

**15.** $\frac{\partial z}{\partial x} = \frac{rye^{xy} + 3xy^3s + 2t^3}{\sqrt{r^2 + s^2 + t^4}}$; now replace $r$ by $e^{xy}$, $s$ by $x^2y^3$ and $t$ by $x + 3y$     **17.** $-\$1221.89$ per dollar

**19.** $-\$103.45$ per dollar     **21.** 2.03 ft/sec     **23.** $\frac{12x^2 + ye^{xy} - 14xy^3 - e^x}{21x^2y^2 - xe^{xy}}$     **25.** $\frac{-x}{z}, \frac{-y}{z}$

**27.** $\frac{-(ye^{xy} + ze^{xz})}{xe^{xz} + ye^{yz}}, \frac{-xe^{xy} - ze^{yz}}{xe^{xz} + ye^{yz}}$

## Exercise Set 11.4 (Pages 862–864) Applications of Partial Differentiation

**1. a.** 5     **b.** 4.96     **3. a.** $\frac{9}{2}$     **b.** 4.1     **5. a.** 12     **b.** 12.225     **7. a.** 6400     **b.** 6418.33
**9. a.** 675,000     **b.** 675,917     **11. a.** 41     **b.** 72     **13.** 8006 units
**15.** $\frac{\partial D_A}{\partial P_A} = -130, \frac{\partial D_A}{\partial P_B} = 2, \frac{\partial D_B}{\partial P_A} = 12, \frac{\partial D_B}{\partial P_B} = -\frac{15}{8}$     **17.** substitutes     **19.** complements     **21.** substitutes
**23.** substitutes     **25.** substitutes     **27. a.** .039823     **b.** −.026     **c.** .044     **d.** .0497
**29. a.** −.051     **b.** −.154     **c.** −.136     **d.** −.864     **31. a.** −.113     **b.** −.043     **c.** 1.286     **d.** .018
**33. a.** −.5     **b.** −.333     **c.** 1     **d.** 1

## Exercise Set 11.5 (Pages 874–876) Optimization

**1.** local minimum at (0, 0)     **3.** local minimum at (1, 3)     **5.** saddle point when $x = 4$ and $y = 3$
**7.** saddle point when $x = 6$ and $y = 3$     **9.** saddle point when $x = 95$ and $y = 43$
**11.** saddle point when $x = -203$ and $y = 59$     **13.** no critical points
**15.** local minimum at (1, 3), saddle point at (−2, 3)     **17.** local minimum at (−3, 1), saddle point at (−3, −5)
**19.** local minimum at (1, 3), saddle point at (1, −1), saddle point at (−5, 3), local maximum at (−5, −1)
**21.** local minimum at (4, 2), saddle point at $\left(\frac{4}{3}, \frac{2}{3}\right)$     **23.** saddle point at (6, 2), local minimum at (12, 4)
**25.** saddle point at (3, 3), saddle point at (−4, −4), local minimum at (12, 36)     **27.** local minimum at $\left(4, \frac{1}{2}\right)$
**29.** saddle point at (0, 0)     **31.** saddle point at $\left(2, \frac{1}{2}\right)$     **33.** saddle point at $\left(\frac{1}{2}, 2\right)$
**35.** 380 units of A, 1040 units of B, $100,000     **37.** 65,000 standard models, 120,000 deluxe models, $2,550,000
**39.** brand A sells for $2.20, brand B sells for $3.25, $34.50     **41.** 12 feet by 12 feet by 8 feet in height
**43.** A sells for $15.60, B sells for $18, $586.80     **45.** 400 units at location A and 600 at location B

## Exercise Set 11.6 (Pages 889–891) Lagrange Multipliers

**1.** 5     **3.** $3\sqrt{2}$     **5.** $\sqrt{14}$     **7.** 6     **9.** $\frac{K^3}{27}$. The geometric mean is not larger than the arithmetic mean.

**11.** 150 in.$^2$     **13.** 6 inches by 6 inches by 3 inches in height     **15.** 384 ft$^3$     **17.** 48 ft$^3$
**19.** $w = 2\sqrt[3]{12}$ inches, $l = 6\sqrt[3]{12}$ inches and $h = \sqrt[3]{12}$ inches
**21.** The ends of the pasture are 400 feet and the side adjacent the river is 800 feet.
**23.** The radius is $\sqrt[3]{\frac{25}{6\pi}}$ in. and the height is $\sqrt[3]{\frac{900}{\pi}}$ in.     **25.** 687,348 units     **27.** 3,234,989 units     **29.** $160,000
**31.** $108,000     **33.** 3.27     **35.** 5.39
**37.** $75,000 is spent on development and $45,000 is spent on promotion, 1,470,000.
**39.** 130 units at site A and 470 units at site B
**41. a.** $C^2/4000$     **b.** $\frac{C}{2000}$     **c.** $\frac{C}{2000}$     **d.** 160 ft$^2$     **e.** .4 ft$^2$     **f.** 160.40025; the increase in area is .40025 ft$^2$

## Exercise Set 11.7 (Pages 901–903) The Method of Least Squares

**1.** $y = 1.3x + 4$

**3.** $y = -x + 6$

**5.** $y = \frac{-31}{15}x + \frac{59}{15}$

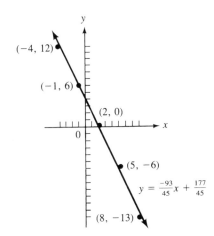

**7.** $y = -2x + 3$

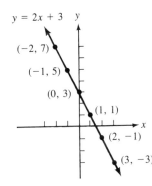

**9.** $y = 3.\overline{185}x - 4.\overline{1}$

**11. a.** $y = \frac{53}{30}x + \frac{418}{9}$
where 1987 corresponds
to $x = 0$   **b.** 55.3

**13. a.** $y = \frac{343}{440}x - \frac{12,225}{4400}$
   **b.** 1515
**15. a.** $y = \frac{1825}{57,990}x + \frac{28,599}{57,990}$
   **b.** 2.98
**17. a.** $y = \frac{4}{3}x - 15.8$
   **b.** 31
**19. a.** $y = \frac{685}{7}x - \frac{335}{7}$
   **b.** \$930.71

## Exercise Set 11.8 (Pages 918–919) Double Integrals

**1.** $6x^2y^4 + 4xy^5 + C(x)$   **3.** $\frac{2}{5}y(x^3 + 1)^5 + C(y)$   **5.** $2x^3e^{3y} + C(y)$   **7.** $90x^2 + 124x$   **9.** $\frac{62}{5}y$
**11.** $18y^9 - 90y^7 - 216y^6$   **13.** $4y \ln(9y^2 + 1) - 4y \ln(y + 1)$   **15.** 11,097.4   **17.** $6 \ln 3 - 1$
**19. a.** 28,884   **b.** 28.884   **21. a.** $8 \ln(3.4)$   **b.** $8 \ln(3.4)$
**23.** 52   **25.** $\frac{35 \ln 4}{12}$   **27.** 250.53

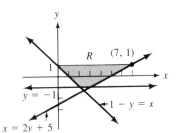

**29.** $10 \ln 5 - 4 \ln 2 + 6$

$x = -e^y$    $y = \ln 2$    $x = e^y$

**31. a.** $\{(x, y) \mid 0 \le x \le 2, 0 \le y \le 2 - x\}$
   **b.** $\{(x, y) \mid 0 \le y \le 2, 0 \le x \le 2 - y\}$
   **c.** $\frac{1}{6}e^8 - \frac{1}{6}$

**33.** $\int_1^2 \int_{4y-3}^5 (5x^2y + 2x - 3y) \, dx \, dy + \int_2^4 \int_{9-2y}^5 (5x^2y + 2x - 3y) \, dx \, dy = 1120$

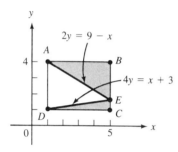

**35.** $13$    **37.** $\frac{7}{36} \ln 4$    **39.** $\frac{49}{60}$    **41.** $1435$ lb    **43.** $.3066$

## Chapter Review 11.9 (Pages 923–927) Sample Exam Questions

**1. a.** the set of all ordered pairs of real numbers; $10, 15$
   **b.** the set of all ordered pairs of real numbers except those of the form $(x, y)$ where $y = 2x$; $-4, -1$
   **c.** the set of all ordered pairs of real numbers except for those of the form $(x, y)$ where $x = 2$ or $y = -3$; $\frac{1}{7}, 9$
   **d.** all ordered pairs of real numbers except for those of the form $(x, y)$ where $x^2 + y^2 > 100$; $\sqrt{71}, \sqrt{11}$
   **e.** all ordered pairs of real numbers except for those of the form $(x, y)$ where $2x - y = 0$; $0, \frac{3}{1-e}$
   **f.** all ordered triples of real numbers; $-42, 1$
   **g.** the set of all ordered triples of real numbers except for those of the form $(x, y, z)$ where $x = 3, y = -2$ or $z = 5$; $-12, 0$
   **h.** the set of all ordered triples of real numbers except for those of the form $(x, y, z)$ where $x > -3$ and $y < 5$, or $x < -3$ and $y > 5$; $3, 10$

**3. a.** $(0, -1, 1), (2, 0, 8), (1, 2, -3)$    **b.** $(0, 1, -1), \left(2, 0, \frac{1}{3}\right)$, and $(1, -1, 0)$    **c.** $(0, 3, 0), (3, 0, -1.5)$, and $(0, 0, 0)$
   **d.** $(0, 0, 6), (0, 6, 0)$, and $(6, 0, 0)$    **e.** $\left(1, 0, \frac{1}{1 - e^3}\right), \left(0, 1, \frac{-3}{1 - e^{-2}}\right), \left(2, 1, \frac{-1}{1 - e^4}\right)$

**5.** $500p_A - 8p_A{}^2 + 400p_B - 9p_B{}^2 + 27p_Ap_B$

**7. a.** $12x^3y^6, 18x^4y^5$    **b.** $\dfrac{x^4y^2 + 3x^2y^6}{(x^2 + y^4)^2}, \dfrac{2x^5y - 2y^5x^3}{(x^2 + y^4)^2}$    **c.** $2xe^{xy} + (x^2 + y^5)ye^{xy}, 5y^4e^{xy} + (x^2 + y^5)xe^{xy}$
   **d.** $(6xy)\ln(x^2 + y^4) + (3x^2 + 5y)\left(\dfrac{2x}{x^2 + y^4}\right), (3x^2 + 5)\ln(x^2 + y^4) + (3x^2y + 5y)\left(\dfrac{4y^3}{x^2 + y^4}\right)$
   **e.** $5(4xy + 3y^2)e^{2x^2y + 3xy^2}, 5(2x^2 + 6xy)e^{2x^2y + 3xy^2}$    **f.** $\dfrac{-4 - \ln(x + y)}{(x + y)^2}, \dfrac{-4 - \ln(x + y)}{(x + y)^2}$
   **g.** $5x(5x^2 + 6y^4)^{-1/2}, 12y^3(5x^2 + 6y^4)^{-1/2}$
**9. a.** $210x^5y^3, 30x^7y, 105x^6y^2, 105x^6y^2$
   **b.** $10y^3e^{5x^2y^3} + 100x^2y^6e^{5x^2y^3}, 30x^2ye^{5x^2y^3} + 225x^4y^4e^{5x^2y^3}, 30xy^2e^{5x^2y^3} + 150x^3y^5e^{5x^2y^3}, 30xy^2e^{5x^2y^3} + 150x^3y^5e^{5x^2y^3}$
   **c.** $\dfrac{-1}{x^2} - \dfrac{25}{(5x + 6y)^2}, \dfrac{-1}{y^2} - \dfrac{36}{(5x + 6y)^2}, \dfrac{-30}{(5x + 6y)^2}, \dfrac{-30}{(5x + 6y)^2}$

**11. a.** 241.3719361    **b.** 241.38    **c.** The percentage error is $-.0033408\%$

**13. a.** $-80, 441, 1152, -70$. When the price of $y$ is held fixed at \$7, the demand for $x$ decreases by approximately 80 units and the demand for $y$ increases by 1152 units as the price of $x$ increases from \$8 to \$9. When the price of $x$ is held fixed at \$8, the demand for $x$ increases by approximately 441 units and the demand for $y$ decreases by 70 units, as the price of $y$ increases from \$7 to \$8.    **b.** substitutes

**15. a.** critical point is $(2, -2)$, local minimum at $(2, -2)$    **b.** critical points are $(\sqrt{2}, 1)$, $\left(\sqrt{2}, -\frac{7}{3}\right)$, $(-\sqrt{2}, 1)$, $\left(-\sqrt{2}, -\frac{7}{3}\right)$, local minimum at $(\sqrt{2}, 1)$, saddle point at $\left(\sqrt{2}, -\frac{7}{3}\right)$, saddle point at $(-\sqrt{2}, 1)$, local maximum at $\left(-\sqrt{2}, -\frac{7}{3}\right)$

**c.** critical points are $\left(\frac{2}{3}, 4\right)$, $\left(-\frac{2}{3}, -4\right)$, saddle points at $\left(\frac{2}{3}, 4\right)$ and $\left(-\frac{2}{3}, -4\right)$

**d.** critical point is $(4, -2)$, local minimum at $(4, -2)$

**17.** saddle point at $(2, 1)$, saddle point at $(2, 12)$, local maximum at $\left(\frac{26}{35}, \frac{156}{35}\right)$, saddle point at $\left(-\frac{26}{9}, -\frac{13}{9}\right)$

**19.** 50 deluxe models and 60 standard models, \$1,830    **21.** 16 m by 16 m by 20 m in height    **23.** $3\sqrt{5}$

**25.** \$49,500    **27. a.** $y = -.7x + 6.5$    **b.**

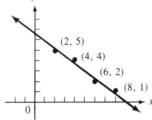

**29.** $18(x^2 + 4y^3)^2 x(x^{3/2} + y^2)^2 e^{(x^2+4y^3)^3(\sqrt{x^3}+y^2)^2} + 9(x^2 + 4y^3)^3(\sqrt{x^3} + y^2)\sqrt{x}e^{(x^2+4y^3)^3(\sqrt{x^3}+y)^2}$,

$e^{(x^2+4y^3)^3(\sqrt{x^3}+y^2)^2}(108(x^2 + 4y^3)^2(\sqrt{x^3} + y^2)^2 y^2 + 12(x^2 + 4y^3)^3(\sqrt{x^3} + y^2)y$

**31. a.** $4x^3y^3 + \frac{13}{4}xy^4 + C(x)$    **b.** $\frac{3}{4}x^4\sqrt[5]{y} + K(y)$    **c.** $\frac{28}{3}x^3 + \frac{75x}{2}$    **d.** $-14,883$    **33.** $\frac{146}{3}$

**35.** 1800    **37.** 162    **39.** .2035

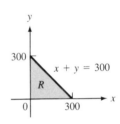

# Index